Final Exam R May 15, 2003
10:15 – 12:15
Office Hrs Fri May 9th 10-noon
Mon May 12 10-noon



## CONVERSION FACTORS

### Length

1 m = 39.37 in. = 3.281 ft
1 in. = 2.54 cm
1 km = 0.621 mi
1 mi = 5280 ft = 1.609 km
1 lightyear (ly) = $9.461 \times 10^{15}$ m
1 angstrom (Å) = $10^{-10}$ m

### Mass

1 kg = $10^3$ g = $6.85 \times 10^{-2}$ slug
1 slug = 14.59 kg
1 u = $1.66 \times 10^{-27}$ kg

### Time

1 min = 60 s
1 h = 3600 s
1 day = $8.64 \times 10^4$ s
1 yr = 365.242 days = $3.156 \times 10^7$ s

### Volume

1 L = 1000 $cm^3$ = $3.531 \times 10^{-2}$ $ft^3$
1 $ft^3$ = $2.832 \times 10^{-2}$ $m^3$
1 gal = 3.786 L = 231 $in.^3$

### Angle

180° = $\pi$ rad
1 rad = 57.30°
1° = 60 min = $1.745 \times 10^{-2}$ rad

### Speed

1 km/h = 0.278 m/s = 0.621 mi/h
1 m/s = 2.237 mi/h = 3.281 ft/s
1 mi/h = 1.61 km/h = 0.447 m/s = 1.47 ft/s

### Force

1 N = 0.2248 lb = $10^5$ dynes
1 lb = 4.448 N
1 dyne = $10^{-5}$ N = $2.248 \times 10^{-6}$ lb

### Work and energy

1 J = $10^7$ erg = 0.738 ft · lb = 0.239 cal
1 cal = 4.186 J
1 ft · lb = 1.356 J
1 Btu = $1.054 \times 10^3$ J = 252 cal
1 J = $6.24 \times 10^{18}$ eV
1 eV = $1.602 \times 10^{-19}$ J
1 kWh = $3.60 \times 10^6$ J

### Pressure

1 atm = $1.013 \times 10^5$ N/$m^2$ (or Pa) = 14.70 lb/$in.^2$
1 Pa = 1 N/$m^2$ = $1.45 \times 10^{-4}$ lb/$in.^2$
1 lb/$in.^2$ = $6.895 \times 10^3$ N/$m^2$

### Power

1 hp = 550 ft · lb/s = 0.746 kW
1 W = 1 J/s = 0.738 ft · lb/s
1 Btu/h = 0.293 W

# College Physics

## SIXTH EDITION

### *Volume 2*

# COLLEGE PHYSICS

## SIXTH EDITION

### *Volume 2*


RAYMOND A. SERWAY

JAMES MADISON UNIVERSITY

JERRY S. FAUGHN

EASTERN KENTUCKY UNIVERSITY

CONTRIBUTING AUTHOR:

**CLEMENT J. MOSES**

Australia · Canada · Mexico · Singapore · Spain

United Kingdom · United States

Editor in Chief: Michelle Julet
Senior Development Editor: Susan Dust Pashos
Assistant Editor: Alyssa White
Editorial Assistant: Jessica Howard
Technology Project Manager: Samuel Subity
Marketing Manager: Kelley McAllister
Marketing Assistant: Sandra Perin
Advertising Project Manager: Stacey Purviance
Senior Project Manager, Editorial Production: Charlene Catlett Squibb, Teri Hyde
Print/Media Buyer: Karen Hunt
Production Service: Progressive Publishing Alternatives



Printed in Canada
1 2 3 4 5 6 7 06 05 04 03 02

For more information about our products, contact us at:
**Thomson Learning Academic Resource Center**
**1-800-423-0563**
For permission to use material from this text, contact us by:
**Phone:** 1-800-730-2214
**Fax:** 1-800-730-2215
**Web:** http://www.thomsonrights.com



Brooks/Cole—Thomson Learning
511 Forest Lodge Road
Pacific Grove, CA 93950
USA

**Asia**
Thomson Learning
60 Albert Street, #15-01
Albert Complex
Singapore 189969

Text Designer: Lisa Adamitis
Art Director: Carol Bleistine
Photo Researcher: Dena Betz
Copy Editor: Progressive Publishing Alternatives
Illustrator: Rolin Graphics
Cover Designer: Larry Didona
Cover Image: Aurora Borealis over Trees, © Yutaka Iijima/Photonica
Cover Printer: Transcontinental Printing
Compositor: Progressive Information Technologies
Printer: Transcontinental Printing

**Australia**
Nelson Thomson Learning
102 Dodds Street
South Melbourne, Victoria 3205
Australia

**Canada**
Nelson Thomson Learning
1120 Birchmount Road
Toronto, Ontario M1K 5G4
Canada

**Europe/Middle East/Africa**
Thomson Learning
Berkshire House
168-173 High Holborn
London WC1V 7AA
United Kingdom

**Latin America**
Thomson Learning
Seneca, 53
Colonia Polanco
11560 Mexico D.F.
Mexico

**Spain**
Paraninfo Thomson Learning
Calle/Magallanes, 25
28015 Madrid, Spain

Library of Congress Control Number: 2002113279

*We dedicate this book*

To our wives, Elizabeth Ann Serway and Mary Ann Faughn, to our children, Mark Serway, Michele Austin, David Serway, Jennifer Serway, David Faughn, and Laura Faughn

*and*

To our grandchildren, Nicole Austin, Christopher Austin, Brian Serway, Kaitlyn Serway, Nathan Serway, and Stone Evan Faughn.

We have been truly blessed by receiving so much love and support from these warm and thoughtful family members through all the years.

# Contents Overview

## VOLUME 1



## VOLUME 2

T1.1

1.9 $q_1 = 12\times10^{-9}C$ $q_2 = -18\times10^{-9}C$ $F_{12}$ = from 1 on 2

.3m

a) $F_{12} = \frac{k_e \cdot |q_1| \cdot |q_2|}{r^2}$ plug in #s $= 2.1576\times10^{-5}$ $F_{12} = 2.2\times10^{-5}$ attraction

b) but when connected $-12\times10^{-9}c$ of negative charge neutralizes the $+12\times10^{-9}C$ of (+) charge $F_{12} = -F_{21}$

Remaining $-6\times10^{-9}C$ of negative charge equally distributes itself btwn the 2 conducting spheres, so $q_1 = q_2 = -3\times10^{-9}C$

$F_{12} = \frac{k_e(3\times10^{-9})^2}{.3^2} = 8.99\times10^{-7}N$ $F_{12} = 9\times10^{-7}N$ Repulsion

---

1.12 Note that $nC = \times10^{-9}C$ & Positive charges produce repulsive forces

$q_1 = 3\times10^{-9}C$

.5m

.5m .5m $q_3 = 6nC$

$q_2 = 2nC$

Calculate the Coulomb forces on $q_3$:

$F_{13} = \frac{k_e \cdot 3\times10^{-9} \cdot 6\times10^{-9}}{(\sqrt{.5^2+.5^2})^2} = 3.2364\times10^{-7}N$ @ 45° below +x axis

$F_{23} = \frac{k_e \cdot 2nC \cdot 6nC}{(\sqrt{.5^2+.5^2})^2} = 2.1576\times10^{-7}N$ @ 45° above +x axis

Break each force into components:

| | x | y |
|---|---|---|
| $F_{13}$: | $F_{13}\cos45° = 2.288\times10^{-7}N$ | $-F_{13}\sin45° = -2.288\cdot10^{-7}$ |
| $F_{23}$: | $F_{23}\cos45 = 1.526\times10^{-7}N$ | $F_{23}\sin45° = 1.526\times10^{-7}N$ |
| | $\Sigma F_x = 3.814\times10^{-7}$ | $\Sigma F_y = -.763\times10^{-7}N$ |

Individual forces on $q_3$

$F_{23}$ $F_{23y}$ $F_{13y}$ $F_{13}$

$F_R = \sqrt{(3.814\times10^{-7})^2 + (-.763\times10^{-7})^2}$

$F_R = 3.89\times10^{-7}N$

Resultant force on $q_3$

$F_R$

$\theta = \tan^{-1}\left(\frac{.763\times10^{-7}}{3.814\times10^{-7}}\right) = 11.3°$ below +x axis

---

1.23 $E = 640\ N/C$ $v_f = 1.2\times10^6\ m/s$ $q_p = 1.6\times10^{-19}C$ $m_p = 1.67\times10^{-27}kg$

a $F = ma$ $F = q_0E$ $a = \frac{qE}{m}$

b $v_f = v_0 + at$ ($v_0 = 0$) $t = \frac{v_f}{a}$

c. $\Delta x = v_0t + \frac{1}{2}at^2$ ($v_0 = 0$)

$\Delta x = \frac{1}{2}at^2$

d. $KE = \frac{1}{2}mv^2$

# CONTENTS

## VOLUME 2

## PART 4: Electricity and Magnetism *xxxii*

## CHAPTER 20 Induced Voltages and Inductance 620



## CHAPTER 21 Alternating Current Circuits and Electromagnetic Waves 652



# PART 5: Light and Optics *684*

## CHAPTER 22 Reflection and Refraction of Light 686



## CHAPTER 23 Mirrors and Lenses 714



## CHAPTER 24 Wave Optics 747



## CHAPTER 25 Optical Instruments 781



# PART 6: Modern Physics *804*

## CHAPTER 26 Relativity 806

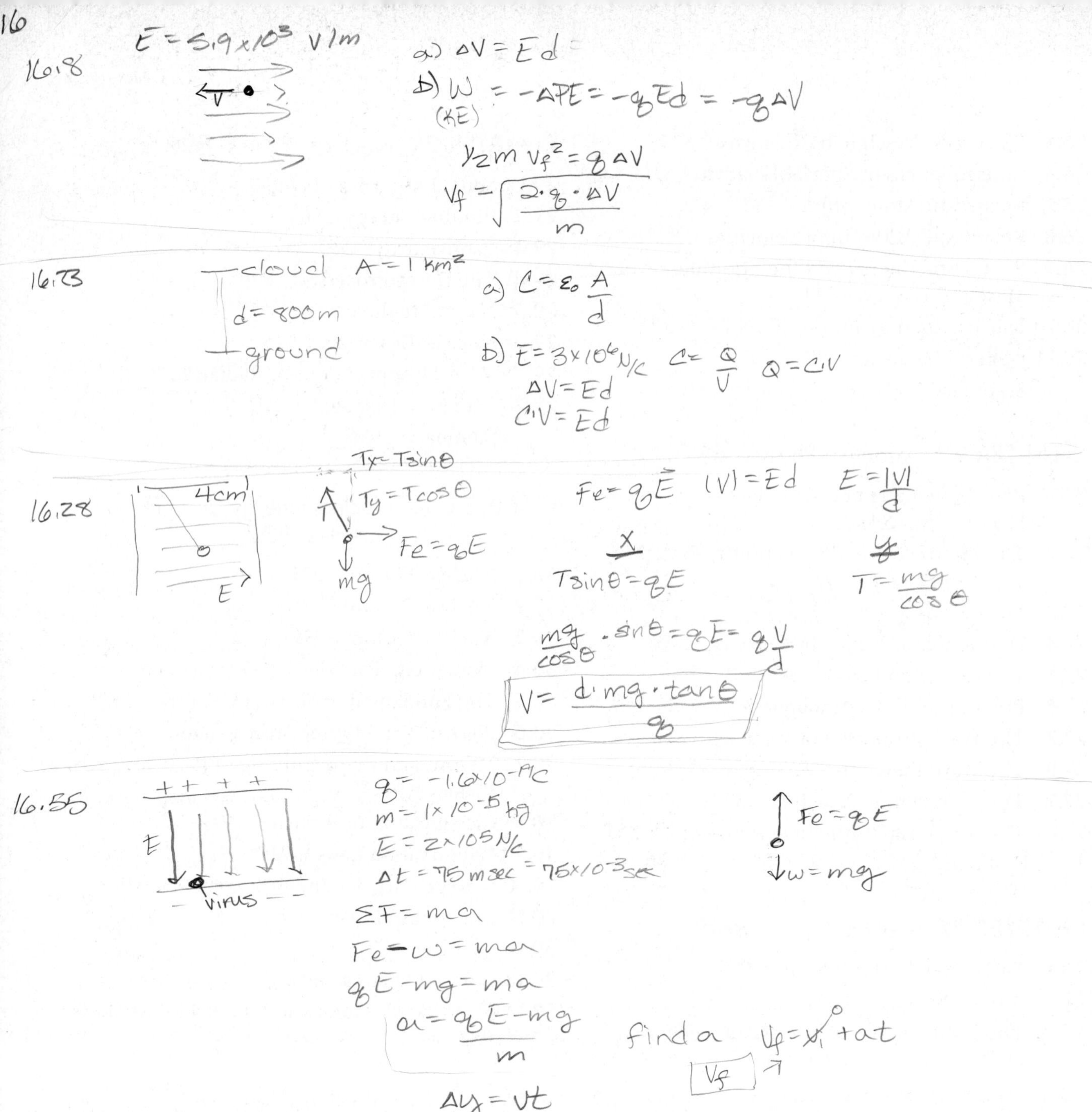

16

16.18 $E = 5.9 \times 10^3$ V/m

a) $\Delta V = Ed$

b) $W_{(KE)} = -\Delta PE = -qEd = -q\Delta V$

$\frac{1}{2} m v_f^2 = q\Delta V$

$v_f = \sqrt{\frac{2 \cdot q \cdot \Delta V}{m}}$

16.23 cloud $A = 1\ km^2$

$d = 800$ m

ground

a) $C = \varepsilon_0 \frac{A}{d}$

b) $E = 3 \times 10^6$ N/C $\quad C = \frac{Q}{V} \quad Q = CV$

$\Delta V = Ed$

$CV = Ed$

16.28 4cm

E

$T_x = T\sin\theta$

$T_y = T\cos\theta$

T

$F_e = qE$

mg

$F_e = q\vec{E} \quad |V| = Ed \quad E = \frac{|V|}{d}$

x: $T\sin\theta = qE$

y: $T = \frac{mg}{\cos\theta}$

$\frac{mg}{\cos\theta} \cdot \sin\theta = qE = q\frac{V}{d}$

$$V = \frac{d \cdot mg \cdot \tan\theta}{q}$$

16.55 + + + +

E

virus

− −

$q = -1.6 \times 10^{-19}$ C

$m = 1 \times 10^{-15}$ kg

$E = 2 \times 10^5$ N/C

$\Delta t = 75$ msec $= 75 \times 10^{-3}$ sec

$F_e = qE$

$w = mg$

$\Sigma F = ma$

$F_e - w = ma$

$qE - mg = ma$

$a = \frac{qE - mg}{m}$

find a $\quad v_f = v_i^{0} + at$

$v_f$

$\Delta y = vt$

# About the Authors

**Raymond A. Serway** received his doctorate at the Illinois Institute of Technology and is Professor Emeritus at James Madison University. Dr. Serway began his teaching career at Clarkson University, where he conducted research and taught from 1967 to 1980. His second academic appointment was at James Madison University as Professor of Physics and Head of the Physics Department from 1980 to 1986. He remained at James Madison University until his retirement in 1997. He was the recipient of the Madison Scholar Award at James Madison University in 1990, the Distinguished Teaching Award at Clarkson University in 1977, and the Alumni Achievement Award from Utica College in 1985. As Guest Scientist at the IBM Research Laboratory in Zurich, Switzerland, he worked with K. Alex Müller, 1987 Nobel Prize recipient. Dr. Serway also held research appointments at Rome Air Development center from 1961 to 1963, at IIT Research Institute from 1963 to 1967, and as a visiting scientist at Argonne National Laboratory, where he collaborated with his mentor and friend, Sam Marshall. Dr. Serway is also the senior author of *Principles of Physics,* 3rd edition; *Physics for Scientists and Engineers,* 5th edition; *Modern Physics,* 2nd edition with Dr. Moses; and the high-school textbook *Physics* with Dr. Faughn (the latter published by Holt, Rinehart, & Winston). In addition, Dr. Serway has published more than 40 research papers in the field of condensed matter physics and has given more than 70 presentations at professional meetings. Dr. Serway and his wife Elizabeth enjoy traveling, golfing, and spending quality time with their four children and five grandchildren.

**Jerry S. Faughn** earned his doctorate at the University of Mississippi. He is Professor Emeritus and former Chair of the Department of Physics and Astronomy at Eastern Kentucky University. Dr. Faughn has also written a microprocessor interfacing text for upper-division physics students. He is co-author of a nonmathematical physics text and a physical science text for general education students, and (with Dr. Serway) the high-school textbook *Physics,* published by Holt, Rinehart, & Winston. He has taught courses ranging from the lower division to the graduate level, but his primary interest is in students just beginning to learn physics. He has been director of a number of NSF and state grants, many of which were devoted to the improvement of physics education. He believes that there is no greater calling than to be a teacher and an interpreter of physics for others. Dr. Faughn has a wide variety of hobbies, among which are reading, travel, genealogy, and old-time radio. His wife Mary Ann is an avid gardener, and he contributes to her efforts by staying out of the way. His daughter Laura is in family practice and his son David is an attorney.

**Clement J. Moses** is Emeritus Professor of Physics at Utica College of Syracuse University. He was born and brought up in Utica, New York, and holds an A. B. from Hamilton College, an M. S. from Cornell, and a Ph.D. from State University of New York at Binghamton. He has over 30 years of science writing and teaching experience at the college level, and is a co-author of the text *Modern Physics,* 2nd edition with Dr. Serway. His research work, in both industrial and university settings, has dealt with defects in solids, solar cells, and the dynamics of atoms at surfaces. In addition to science writing, Professor Moses likes cooking, fishing, singing, and going to operas.

# PREFACE

*College Physics* is written for a one-year course in introductory physics usually taken by students majoring in biology, the health professions, and other disciplines including environmental, earth, and social sciences, and technical fields such as architecture. The mathematical techniques used in the book include algebra, geometry, and trigonometry, but not calculus.

The main objectives of this introductory textbook are twofold: to provide the student with a clear and logical presentation of the basic concepts and principles of physics, and to strengthen an understanding of the concepts and principles through a broad range of interesting applications to the real world. To meet these objectives, we have emphasized sound physical arguments and problem-solving methodology. At the same time, we have attempted to motivate the student through practical examples that demonstrate the role of physics in other disciplines.

This textbook, which covers the standard topics in classical physics and 20th century physics, is divided into six parts. Part I (Chapters 1–9) deals with Newtonian mechanics and the physics of fluids; Part II (Chapters 10–12) is concerned with heat and thermodynamics; Part III (Chapters 13–14) covers wave motion and sound; Part IV (Chapters 15–21) is concerned with electricity and magnetism; Part V (Chapters 22–25) treats the properties of light and the field of geometric and wave optics; Part VI (Chapters 26–30) represents an introduction to special relativity, quantum physics, atomic, and nuclear physics.

## CHANGES TO THE SIXTH EDITION

A number of new features, changes, and improvements have been added to this edition. Based on comments from users of the fifth edition and reviewers' suggestions, a major effort was made to improve clarity of presentation, precision of language, and accuracy throughout. The new pedagogical features added to this edition are based on current trends in science education. The following represent the major changes in the sixth edition.

### PEDAGOGICAL CHANGES

- **Quick Quizzes** Several questions labeled Quick Quizzes are now included in each chapter to provide students with opportunities to test their understanding of the physical concepts presented. The questions require students to make decisions on the basis of sound reasoning, and some are intended to help students overcome common misconceptions. Most questions are presented in multiple choice format, and can be adapted for assessing student performance in the classroom. Answers to all questions are found at the end of the textbook, while answers with detailed explanations are provided in the Instructor's Manual.
- **Tips** These new features, placed in the margins of the text, address common student misconceptions that often cause students to follow unproductive paths. The "Tips" should help students avoid common mistakes and misunderstandings.
- **Problems and Conceptual Questions** A substantial revision of the end-of-chapter problems and conceptual questions was made in this edition. Most of the new problems that have been added are intermediate in level, and all have been carefully edited and reworded where necessary. Many new problems require

students to make order-of-magnitude calculations. Solutions to approximately 12 problems per chapter are included in the *Student Solutions Manual and Study Guide.* Boxed numbers identify these problems. A smaller subset of solutions will be posted on the World Wide Web **(http://info.brookscole.com/serway)** and will be accessible to students and instructors using *College Physics.* The **web** icon identifies these problems.

- **Group Activities** This new feature at the end of each chapter is included to encourage students to engage in activities outside of the classroom. Some are simple experiments or demonstrations the student can perform individually or with the assistance of a classmate. Others are problems, frequently consisting of both conceptual and numerical parts, that will stimulate group discussion.
- **Webnotes** Useful World Wide Web addresses are provided as marginal notes to encourage students to explore extensions of material beyond what is covered in the text.

## CONTENT CHANGES

Although the overall content and organization of the textbook is similar to that of the fifth edition, several changes were implemented.

- Chapter 2 now includes several new examples on kinematics and the revised presentation places an increased emphasis on qualitative explanations of equations and graphs.
- Chapter 5 was extensively rewritten, and now includes a discussion of the various ways that energy can be transferred into or out of a system. We also added a new discussion of energy and power considerations when someone jumps vertically into the air, and a new section on work done by a variable force.
- Chapter 6 includes a new section on rocket propulsion emphasizing that the operation of a rocket can be understood from the law of conservation of momentum as applied to the rocket plus its ejected fuel. The chapter also presents new applications on injury in automobile collisions and finding the impact force on landing in a jump.
- We have added a discussion of arches and the ultimate strength of materials, including biological substances, in Section 9.2.
- A number of changes were made to Chapter 12. The first law of thermodynamics is now expressed as $\Delta U = Q + W$, where positive $W$ is defined as the work done *on* the system. This form of the first law is used in most modern chemistry textbooks, and has been recommended by a committee appointed by AAPT (the American Association of Physics Teachers). Chapter 12 places an increased emphasis on the application of the first law to internal combustion engines and cyclic processes. The chapter also discusses how the human body constantly transforms internal energy into other forms of energy.
- Chapter 16 now offers more explanation of how to relate circuit schematics to actual circuits. This chapter also discusses the operation of the defibrillator as a practical example of capacitance.
- In Chapter 17, we added a section entitled "Electrical Activity in the Heart," which is of special interest to pre-medical and life science students.
- Chapter 18 includes a section entitled "Conduction of Electrical Signals by Neurons." This discussion is a recast of an earlier essay by Paul Davidovits on the nervous system and will be of special interest to pre-medical and life science students.
- Chapter 19 now includes a new application on electric motors.
- A new application that discusses the refraction of a laser beam in a DVD player is presented in Chapter 22.
- Chapter 23 places an increased emphasis on ray diagrams.

- New to this edition is Section 24.5 entitled "Using Interference to Read CD's and DVD's." This discussion represents an interesting application of interference to modern technology.
- Chapter 26 was extensively rewritten and includes a new example on the conversion of mass to energy.
- Chapter 27 now includes a discussion of how X-ray diffraction is used to determine the structure of biologically important molecules.
- Chapter 28 was extensively revised and now includes two new sections entitled "Energy Bands in Solids" and "Semiconductor Devices." These topics are important for understanding the electronic properties of solids, and are basic to understanding the behavior of semiconductor devices such as diodes and transistors.
- Section 29.8, new to this edition, discusses various types of radiation detectors such as Geiger counters, bubble chambers, and scintillation counters. These practical applications will be of special interest to students preparing for health related professions.
- Some sections in the fifth edition were either deleted or moved to more appropriate locations. Section 1.7 on mathematical notation was moved to an appendix.

Most of the section entitled "The Diffraction Grating" was moved from Chapter 25 to Chapter 24 because it logically follows the discussion of diffraction. The section entitled "Pair Production and Annihilation" was moved from Chapter 27 to Chapter 26 because the topic is a verification of the equivalence of mass and other forms of energy as predicted by the theory of relativity.

The QuickLabs that were included in the fifth edition have been deleted, but the most popular QuickLabs are now incorporated into the new **Group Activities** feature described earlier.

Multiple-choice questions that appeared in the fifth edition have been removed, but some have been rewritten as Quick Quiz questions.

## TEXTBOOK FEATURES

Most instructors would agree that the textbook assigned in a course should be the student's primary guide for understanding and learning the subject matter. Furthermore, the textbook should be easily accessible and written in a style that facilitates instruction and learning. With this in mind, we have included many pedagogical features that are intended to enhance the textbook's usefulness to both students and instructors. These features are as follows:

**STYLE** We have attempted to write the book in a style that is clear, logical, relaxed, and pleasing to the reader. At the same time, we have attempted to keep the presentation accurate and precise. New terms are carefully defined, and we have avoided the use of jargon.

**PREVIEWS** All chapters begin with a preview that includes a brief discussion of the chapter's objectives and content.

**ORGANIZATION** The book is divided into the following six parts: mechanics, thermodynamics, vibrations and wave motion, electricity and magnetism, light and optics, and modern physics. Each part includes an overview of the subject matter to be covered in that part and some historical perspectives.

**UNITS** The international system of units (SI) is used throughout the book. The U.S. customary system of units is used only to a limited extent in the problem sets of the early chapters on mechanics.

**MARGINAL NOTES** Comments and notes appearing in the margin can be used to locate important statements, equations, definitions, and concepts in the text.

**PROBLEM-SOLVING STRATEGIES** General strategies and suggestions are included for solving the types of problems featured in both the worked examples and end-of-chapter problems. This feature, highlighted by a surrounding box, is intended to help students identify the essential steps in solving problems and increase their skills as problem solvers.

**PHYSICS IN ACTION** This boxed material focuses on photographs of interesting demonstrations and phenomena in physics, accompanied by detailed explanations. The material can also serve as a source of information for initiating classroom discussions.

**LIFE SCIENCE TOPICS** Many chapters include text, worked examples and problems dealing with applications of physics to the life sciences. These are identified by the DNA icon ( ).

**WORKED EXAMPLES** A large number of worked examples, including many new ones, are presented as an aid in understanding and/or reinforcing physical concepts. In many cases, these examples serve as models for solving end-of-chapter problems. The examples are set off from the text for ease of location, and all examples are given titles to describe their content. Many examples include a **Reasoning** section to illustrate the underlying concepts and methodology used in arriving at a correct solution. This will help students understand the logic behind the solution and the advantage of using a particular approach to solve the problem. The solution answer is highlighted with a light blue screen. Many worked examples are followed immediately by exercises with answers. These exercises represent extensions of the worked examples and are intended to sharpen student's problem-solving skills and test their understanding of concepts. Students who work through these exercises on a regular basis should find the end-of-chapter problems less intimidating.

**IMPORTANT STATEMENTS AND EQUATIONS** Most important statements and definitions are set in **boldface type** or are highlighted with a background screen for added emphasis and ease of review. Similarly, important equations are highlighted with a gold screen to facilitate location.

**ILLUSTRATIONS AND PHOTOGRAPHS** The text material, worked examples, and end-of-chapter questions and problems are accompanied by numerous figures, photographs, and tables. Full color is used to add clarity to the figures and to make the visual presentation as realistic and pleasing as possible. Three-dimensional effects are rendered with the use of shaded and lightened areas, where appropriate. Vectors are color coded, and curves in *xy* plots are drawn in color. Color photographs have been carefully selected, and their accompanying captions have been written to serve as an added instructional tool. A complete description of the pedagogical use of color appears on the inside front cover.

**SUMMARIES** Each chapter contains a summary which reviews the important concepts and equations discussed in that chapter.

**CONCEPTUAL QUESTIONS** A set of conceptual questions is provided at the end of each chapter. The **Applying Physics** examples presented in the text should serve as models for students when conceptual questions are assigned or used in tests. The questions provide the student with a means of self-testing the concepts presented in the chapter. Some conceptual questions are appropriate for initiating classroom discussions. Answers to all odd-numbered conceptual questions are located in the answer section at the end of the book.

**END-OF-CHAPTER PROBLEMS** An extensive set of problems is included at the end of each chapter. Answers to odd-numbered problems are given at the end of the book. For the convenience of both the student and instructor, about two thirds of the problems are keyed to specific sections of the chapter. The remaining problems, labeled "Additional Problems," are not keyed to specific sections. There are three levels of problems according to their level of difficulty. Straightforward problems are numbered in black, intermediate level problems are numbered in blue, and the most challenging problems are numbered in magenta. Those problems with a focus on the life sciences are identified by the DNA icon ( ). The set of problems under the heading of **Group Activities,** new to this edition, have been included to encourage students to engage in scientific activities outside the classroom.

**APPENDICES** Several appendices are provided at the end of the book. Most of the appendix material represents a review of mathematical techniques used in the book, such as scientific notation, algebra, geometry, and trigonometry. References to these appendices is made as needed throughout the book. Most of the mathematical review sections include worked examples and exercises with answers. Some appendices contain useful tables that supplement textual information. For easy reference, the front endpapers contain a chart explaining the use of color throughout the book and a list of often-used conversion factors.

## TEACHING OPTIONS

This book contains more than enough material for a one-year course in introductory physics. This serves two purposes. First, it gives the instructor more flexibility in choosing topics for a specific course. Second, the book becomes more useful as a resource for students. On the average, it should be possible to cover about one chapter each week for a class that meets three hours per week. Those sections, examples, and end-of-chapter problems dealing with applications of physics to the life sciences are identified with the DNA icon ( ). We offer the following suggestions for shorter courses for those instructors who choose to move at a slower pace through the year:

***Option A:*** If you choose to place more emphasis on contemporary topics in physics, you should consider omitting all or parts of Chapter 8 (Rotational Equilibrium and Rotational Dynamics), Chapter 21 (Alternating Current Circuits and Electromagnetic Waves), and Chapter 25 (Optical Instruments).

***Option B:*** If you choose to place more emphasis on classical physics, you could omit all or parts of Part VI of the textbook, which deals with special relativity and other topics in 20th century physics.

The Instructor's Manual offers additional suggestions for specific sections and topics that may be omitted without loss of continuity if time presses.

## STUDENT ANCILLARIES

Thomson·Brooks/Cole offers several items to supplement and enhance the classroom experience. These ancillaries will allow instructors to customize the textbook to their students' needs and to their own style of instruction. One or more of these ancillaries may be shrink-wrapped with the text at a reduced price:

***STUDENT SOLUTIONS MANUAL AND STUDY GUIDE*** by John R. Gordon, Charles Teague, and Raymond A. Serway. Now offered in two volumes, this manual features detailed solutions to approximately 12 problems per chapter. These problems are indicated in the text with boxed numbers. The manual also features a skills section, important notes from key sections of the text, and a list of important equations and concepts. Volume 1 contains Chapters 1–14 and Volume 2 contains Chapters 15–30.

***CORE CONCEPTS IN COLLEGE PHYSICS* CD-ROM, VERSION 2.0** The *Core Concepts in College Physics* CD-ROM applies the power of multimedia to the introductory physics course, offering full-motion animation and video, engaging interactive graphics, clear and concise text, and guiding narration. Drawing from topics in mechanics, thermodynamics, electromagnetism, and optics, *Core Concepts in College Physics* focuses on those concepts students typically find most difficult in the course. The CD-ROM also presents step-by-step explorations of essential mathematics, problem-solving strategies, and animations of problems to promote conceptual understanding and sharpen problem-solving skills. The accompanying *Workbook* contains practical physics problems coordinating with the CD, along with worked solutions.

***PHYSICS LABORATORY MANUAL,*** 2nd edition by David Loyd. This manual supplements the learning of basic physical principles while introducing laboratory procedures and equipment. Each chapter of the manual includes a pre-laboratory assignment, objectives, an equipment list, the theory behind the experiment, experimental procedures, graphs, and questions. A laboratory report is provided for each experiment so the student can record data, calculations, and experimental results. Students are encouraged to apply statistical analysis to their data in order to develop their ability to judge the validity of their results.

***ADDITIONAL ONLINE RESOURCES*** In addition to the companion Web site for this textbook **(http://info.brookscole.com/serway),** students using *College Physics* by Serway and Faughn are encouraged to visit the Brooks/Cole Physics Resource Center at the address below for features such as online quizzing and additional Weblinks. See **http://physics.brookscole.com**

## INSTRUCTOR ANCILLARIES

Ancillaries offered in two volumes are split as follows: Volume 1 contains Chapters 1–14 and Volume 2 contains Chapters 15–30.

***INSTRUCTOR'S MANUAL*** by Jerry Faughn and Charles Teague. Available in two volumes, this manual consists of complete solutions to all the problems in the text, answers to the even-numbered problems and conceptual questions, full answers with explanations to the Quick Quizzes, and a list of suggested readings from journals and other resources.

***PRINTED TEST BANK*** by Ed Oberhofer. Available in two volumes, the comprehensive test bank contains approximately 1 750 problems and questions in multiple choice format. Answers are provided in a separate key. Instructors may duplicate pages for distribution to students.

***COMPUTERIZED TEST BANK*** Available for Windows and Macintosh, the computerized test bank allows instructors to create, deliver and customize tests and quizzes using questions from the printed Test Bank. Instructors may rearrange and edit existing questions, or add their own.

***Overhead Transparency Acetates*** The collection of transparencies in two volumes consists of approximately 200 full-color figures and photographs from the text to enhance lectures. These transparencies feature large print for easy viewing in the classroom.

***Instructor's Manual for Physics Laboratory Manual,*** 2nd edition by David Loyd. Each chapter contains a discussion of the experiment, teaching hints, answers to selected questions from the student laboratory manual, and a post-laboratory quiz with short answers and essay questions. The author has also included a list of the suppliers of scientific equipment and a summary of the equipment needed for all the experiments in the manual.

***Multimedia Presentation Manager for Introductory Physics 2003: A Microsoft® PowerPoint® Link Tool*** This one-stop lecture tool and instructional resource makes it easy to assemble, edit, publish and present custom media-enhanced lectures for your course, using Microsoft® PowerPoint®. The two-volume cross-platform (Win/Mac) CD set includes electronic files of the ***Instructor's Manual*** and ***Test Bank,*** plus digital files of textbook art and additional video clips and animations.

Also found on the ***Multimedia Presentation Manager for Introductory Physics 2003*** are simulations for ***Interactive Physics***™ ***2000,*** the highly acclaimed software from MSC Software. Many simulations are keyed to specific worked examples and end-of-chapter problems in *College Physics,* while others stand alone as laboratory exercises. The instructor or department must own a multi-user license for ***Interactive Physics***™ in order to assign the simulations for student use.

***WebTutor™ Advantage on WebCT and Blackboard*** *WebTutor Advantage* enables the instructor to create and manage a course Web site. WebTutor's course management tool gives the instructor the ability to provide virtual office hours, post syllabi, set up threaded discussions, track student progress with the quizzing material, and much more. WebTutor also provides robust communication tools, such as a course calendar, asynchronous discussion, real time chat, a whiteboard, and an integrated e-mail system.

For students, WebTutor offers real-time access to a full array of study tools, including chapter outlines, summaries, learning objectives, glossary flashcards (with audio), practice quizzes, and Web links.

WebTutor is available for Semesters I (Chapters 1–14) and II (Chapters 15–30) in both WebCT and Blackboard.

***MyCourse 2.0*** *MyCourse 2.0* offers instructors a simple solution for a custom course Web site that allows for assignments, tracking and reporting student progress, syllabus loading, and more. Contact your Thomson·Brooks/Cole representative for details. To see a demo of *MyCourse 2.0,* visit **http://mycourse.thomsonlearning.com**

***Instructor Options for Online Homework*** For detailed explanations and demonstrations, contact your Thomson·Brooks/Cole representative or visit the following Web sites:

- ***WebAssign: A Web-Based Homework System***
  **http://www.webassign.net** or contact WebAssign at **webassign@ncsu.edu**
- ***Homework Service***
  **http://hw.ph.utexas.edu/hw.html** or contact **moore@physics.utexas.edu**

***InfoTrac® College Edition*** InfoTrac® is an online university library offering the full text of articles from almost 4 000 scholarly and popular publications, updated daily and going back as much as 22 years. This is available bundled free with new copies of *College Physics,* 6th edition by Serway and Faughn, and both adopters and their students receive unlimited access for four months.

Available to qualified adopters. Please consult your local sales representative for details.

## ACKNOWLEDGMENTS

In preparing the 6th edition of this textbook, we have been guided by the expertise of many people who have reviewed manuscript and/or provided prerevision suggestions. We wish to acknowledge the following reviewers and express our sincere appreciation for their helpful suggestions, criticisms, and encouragement.

**Sixth edition reviewers:**

Marilyn Akins, *Broome Community College*
Lawrence Anderson-Huang, *University of Toledo*
Lattie F. Collins, *East Tennessee State University*
Michael Dennin, *University of California, Irvine*
N. John DiNardo, *Drexel University*
Leonard X. Finegold, *Drexel University*
Eric Ganz, *University of Minnesota*
Grant W. Hart, *Brigham Young University*
Christopher Herbert, *New Jersey City University*
George W. Kattawar, *Texas A & M University*
Ivan Kramer, *University of Maryland, Baltimore County*
Michael LoPresto, *Henry Ford Community College*
Ed Oberhofer, *Lake Sumter Community College*
J. Scott Payson, *Wayne State University*
Patrick Polley, *Beloit College*
W. Steve Quon, *Ventura College*
Barry Robertson, *Queen's University*
Larry Rowan, *University of North Carolina, Chapel Hill*
Perry A. Tompkins, *Samford University*
Bernard Whiting, *University of Florida*

We thank the following people for their suggestions and assistance during the preparation of earlier editions of this textbook:

Albert Altman, *University of Lowell*
John Anderson, *University of Pittsburgh*
Subhash Antani, *Edgewood College*
Neil W. Ashcroft, *Cornell University*
Charles R. Bacon, *Ferris State University*
Dilip Balamore, *Nassau Community College*
Ralph Barnett, *Florissant Valley Community College*
Lois Barrett, *Western Washington University*
Paul D. Beale, *University of Colorado at Boulder*
Paul Bender, *Washington State University*
David H. Bennum, *University of Nevada at Reno*
Jeffery Braun, *University of Evansville*
John Brennan, *University of Central Florida*
Michael Bretz, *University of Michigan, Ann Arbor*
Michael E. Browne, *University of Idaho*
Joseph Cantanzarite, *Cypress College*
Ronald W. Canterna, *University of Wyoming*
Clinton M. Case, *Western Nevada Community College*
Neal M. Cason, *University of Notre Dame*
Roger W. Clapp, *University of South Florida*
Giuseppe Colaccico, *University of South Florida*
Lawrence B. Coleman, *University of California, Davis*
Jorge Cossio, *Miami Dade Community College*
Terry T. Crow, *Mississippi State College*
Stephen D. Davis, *University of Arkansas at Little Rock*
John DeFord, *University of Utah*
Chris J. DeMarco, *Jackson Community College*
Robert J. Endorf, *University of Cincinnati*
Paul Feldker, *Florissant Valley Community College*
Tom French, *Montgomery County Community College*
Albert Thomas Fromhold, Jr., *Auburn University*
Lothar Frommhold, *University of Texas at Austin*
Teymoor Gedayloo, *California Polytechnic State University*

Simon George, *California State University, Long Beach*
John R. Gordon, *James Madison University*
George W. Greenlees, *University of Minnesota*
Wlodzimierz Guryn, *Brookhaven National Laboratory*
James Harmon, *Oklahoma State University*
Grant W. Hart, *Brigham Young University*
John Ho, *State University of New York at Buffalo*
Murshed Hossain, *Rowan University*
Robert C. Hudson, *Roanoke College*
Fred Inman, *Mankato State University*
Ronald E. Jodoin, *Rochester Institute of Technology*
Drasko Jovanovic, *Fermilab*
Frank Kolp, *Trenton State University*
Joan P. S. Kowalski, *George Mason University*
Ivan Kramer, *University of Maryland, Baltimore County*
Sol Krasner, *University of Chicago*
Karl F. Kuhn, *Eastern Kentucky University*
David Lamp, *Texas Tech University*
Harvey S. Leff, *California State Polytechnic University*
Joel Levine, *Orange Coast College*
Michael Lieber, *University of Arkansas*
James Linbald, *Saddleback Community College*
Bill Lochslet, *Pennslyvania State University*
Bo Lou, *Ferris State University*
Jeffery V. Mallow, *Loyola University of Chicago*
David Markowitz, *University of Connecticut*
Steven McCauley, *California State Polytechnic University, Pomona*
Joe McCauley, Jr., *University of Houston*
Ralph V. McGrew, *Broome Community College*
Bill F. Melton, *University of North Carolina at Charlotte*
H. Kent Moore, *James Madison University*
John Morack, *University of Alaska, Fairbanks*
Steven Morris, *Los Angeles Harbor College*
Carl R. Nave, *Georgia State University*
Martin Nikolo, *Saint Louis University*
Blaine Norum, *University of Virginia*
M. E. Oakes, *University of Texas at Austin*
Lewis J. Oakland, *University of Minnesota*
Ed Oberhofer, *University of North Carolina at Charlotte*
Lewis O'Kelly, *Memphis State University*
David G. Onn, *University of Delaware*
T. A. K. Pillai, *University of Wisconsin, La Crosse*
Lawrence S. Pinsky, *University of Houston*
William D. Ploughe, *Ohio State University*
Brooke M. Pridmore, *Clayton State University*
Joseph Priest, *Miami University*
James Purcell, *Georgia State University*
Michael Ram, *State University of New York at Buffalo*
Kurt Reibel, *Ohio State University*
Virginia Roundy, *California State University, Fullerton*
William R. Savage, *The University of Iowa*
Reinhard A. Schumacher, *Carnegie Mellon University*
John Simon, *University of Toledo*
Donald D. Snyder, *Indiana University at Southbend*
Carey E. Stronach, *Virginia State University*
Thomas W. Taylor, *Cleveland State University*
L. L. Van Zandt, *Purdue University*
Howard G. Voss, *Arizona State University*
Larry Weaver, *Kansas State University*
Donald H. White, *Western Oregon State College*
George A. Williams, *The University of Utah*
Jerry H. Wilson, *Metropolitan State College*
Robert M. Wood, *University of Georgia*
Clyde A. Zaidins, *University of Colorado at Denver*

We are delighted to acknowledge the creative and original new worked examples, applied content, questions, problems and illustrations contributed by Clement J. Moses. Edward McCliment of the University of Iowa and Doug Davis of Eastern Illinois University generously contributed many of the new end-of-chapter problems and questions, especially those of interest to the life sciences. Edward F. Redish of the University of Maryland graciously allowed us to list some of his problems from the Activity Based Physics Project as Group Activities. Some of the remaining Group Activities were written by Robert J. Beichner of North Carolina State University and appeared as QuickLabs in the fifth edition. Steve Quon of Ventura College wrote most of the Webnotes.

This book was carefully checked and re-checked for accuracy by John W. Jewett of the California State University at Pomona, Robert Ehrlich of George Mason University, and Ralph McGrew of Broome Community College. Though responsibility for any remaining errors rests with us, we thank them for their untiring efforts.

We are extremely grateful to the publishing team at the Brooks/Cole Publishing Company for their expertise and outstanding work in all aspects of this project. Susan Pashos carefully analyzed the reviews of the sixth edition, provided important guidance in all aspects of the project, and made many important improvements in the final manuscript. Donna King of Progressive Information Technologies managed the book through all production stages. We appreciate the fine work of Alyssa White in managing the development of the ancillary materials. Samuel Subity developed the Web site and other electronic media. Carol Bleistine managed the overall art and design program. Dena Betz acquired many new and excellent photos. We also recognize important contributions by Charlene Squibb (Senior Production Manager), Kelley McAllister (Marketing Manager), Lisa Adamitis (book designer), and Stacey Purviance (Manager of Creative Development). We thank Angus McDonald, Executive Editor, for his support and problem-solving spirit.

Finally, we dedicate this book to our wives and children, for their love, support, and long-term sacrifices.

**Raymond A. Serway**
Leesburg, Virginia

**Jerry S. Faughn**
Richmond, Kentucky

# APPLICATIONS

Although physics is relevant to so much in our modern lives, this may not be obvious to students in an introductory course. In this sixth edition of *College Physics,* we continue a design feature begun in the previous edition. This feature makes the relevance of physics to everyday life more obvious by pointing out specific applications in the form of a marginal note. Some of these applications pertain to the life sciences and are marked with the DNA icon ( ). The list below is not intended to be a complete listing of all the applications of the principles of physics found in this textbook. Many other applications are to be found within the text and especially in the worked examples, Conceptual Questions, and end-of-chapter problems.

**APPLICATION**

DIRECTS YOU TO SECTIONS DISCUSSING APPLIED PRINCIPLES OF PHYSICS

---

# TO THE STUDENT

It is appropriate to offer some words of advice that should be of benefit to you, the student. Before doing so, we assume that you have read the Preface, which describes the various features of the text that will help you through the course.

## HOW TO STUDY

Very often instructors are asked, "How should I study physics and prepare for examinations?" There is no simple answer to this question, but we would like to offer some suggestions that are based on our own experiences in learning and teaching over the years.

First and foremost, maintain a positive attitude toward the subject matter, keeping in mind that physics is the most fundamental of all natural sciences. Other science courses that follow will use the same physical principles, so it is important that you understand and are able to apply the various concepts and theories discussed in the text.

## CONCEPTS AND PRINCIPLES

It is essential that you understand the basic concepts and principles before attempting to solve assigned problems. You can best accomplish this goal by carefully reading the textbook before you attend your lecture on the covered material. When reading the text, you should jot down those points that are not clear to you. We've purposely left wide margins in the text to give you space for doing this. Also be sure to make a diligent attempt at answering the questions in the Quick Quizzes as you come to them in your reading. We have worked hard to prepare questions that help you judge for yourself how well you understand the material. Pay careful attention to the many Tips in the margins of the text. These will help you to avoid misconceptions, mistakes, and misunderstandings and will help you to maximize the efficiency of your time by minimizing adventures along fruitless paths. During class, take careful notes and ask questions about those ideas that are unclear to you. Keep in mind that few people are able to absorb the full meaning of scientific material after only one reading. Several readings of the text and your notes may be necessary. Your lectures and laboratory work supplement reading of the textbook and should clarify some of the more difficult material. You should minimize your memorization of material. Successful memorization of passages from the text, equations, and derivations does not necessarily indicate that you understand the material. Your understanding of the material will be enhanced through a combination of efficient study habits, discussions with other students and with instructors, and your ability to solve the problems presented in the textbook. Ask questions whenever you feel clarification of a concept is necessary.

## STUDY SCHEDULE

It is important that you set up a regular study schedule, preferably one that calls for daily involvement with course work. Make sure that you to read the syllabus for the course and adhere to the schedule set by your instructor. The lectures will be much more illuminating if you read the corresponding textbook material *before* attending them. As a general rule, you should devote about two hours of study time for every hour you are in class. If you are having trouble with the course, seek the advice of the instructor or other students who have taken the course. You may find it necessary to seek further instruction from experienced students. Very often, instructors offer review sessions in addition to regular class periods. It is important that you avoid the practice of delaying study until a day or two before an exam. More often than not, this approach has disastrous results. Rather than undertake an all-night study session, briefly review the basic concepts and equations and get a good night's rest. If you feel you need additional help in understanding the concepts, in preparing for exams, or in problem-solving, we suggest that you acquire a copy of the *Student Solutions Manual and Study Guide* that accompanies this textbook; this manual should be available at your college bookstore.

## USE THE FEATURES

You should make full use of the various features of the text discussed in the Preface. For example, marginal notes are useful for locating and describing important equations and concepts, and **boldfaced** type indicates important statements and definitions. Many useful tables are contained in the Appendices, but most are incorporated in the text where they are most often referenced. Appendix A is a convenient review of mathematical techniques.

Answers to odd numbered questions and problems and all Quick Quizzes are given at the end of the textbook. The exercises (with answers) that follow some worked examples represent extensions of those examples; in most of these exercises, you are expected to perform a simple calculation. Their purpose is to test your problem-solving skills as you read through the text. Problem-Solving Strategies and Hints are included in selected chapters throughout the text and give you additional information about how you should solve problems. The Table of Contents provides an overview of the entire text, while the Index enables you to locate specific material quickly. Footnotes sometimes are used to supplement the text or to cite other references on the subject discussed.

After reading a chapter, you should be able to define any new quantities introduced in that chapter and to discuss the principles and assumptions that were used to arrive at certain key relations. The chapter summaries and the review sections of the *Student Solutions Manual and Study Guide* should help you in this regard. In some cases, it may be necessary for you to refer to the index of the text to locate certain topics. You should be able to correctly associate with each physical quantity the symbol used to represent that quantity and the unit in which the quantity is specified. Furthermore, you should be able to express each important relation in a concise and accurate prose statement.

## PROBLEM-SOLVING

R. P. Feynman, Nobel laureate in physics, once said, "You do not know anything until you have practiced." In keeping with this statement, we strongly advise that you develop the skills necessary to solve a wide range of problems. Your ability to solve problems will be one of the main tests of your knowledge of physics, and therefore you should try to solve as many problems as possible. It is essential that you understand basic concepts and principles before attempting to solve problems. It is good practice to try to find alternate solutions to the same problem. For example, you can solve problems in mechanics using Newton's laws, but very often an alternative method that draws on energy considerations is more direct. You should not deceive yourself into thinking that you understand a problem merely because you have seen it solved in class. You must be able to solve the problem and similar problems on your own.

The approach to solving problems should be carefully planned. A systematic plan is especially important when a problem involves several concepts. First, read the problem several times until you are confident you understand what is being asked. Look for any key words that will help you interpret the problem and perhaps allow you to make certain assumptions. Your ability to interpret a question properly is an integral part of problem-solving. Second, you should acquire the habit of writing down the information given in a problem and those quantities that need to be found; for example, you might construct a table listing both the quantities given and the quantities to be found. This procedure is sometimes used in the worked examples of the textbook. Finally, after you have decided on the method you feel is appropriate for a given problem, proceed with your solution. General problem-solving strategies of this type are included in the text and are highlighted with a surrounding box. If you follow the steps of this procedure, you will not only find it easier to come up with a solution, but you will also gain more from your efforts.

Often, students fail to recognize the limitations of certain equations or physical laws in a particular situation. It is very important that you understand and remember the assumptions that underlie a particular theory or formalism. For example, certain equations in kinematics apply only to a particle moving with constant acceleration. These equations are not valid for describing motion whose acceleration is not constant, such as the motion of an object connected to a spring or the motion of an object through a fluid.

## EXPERIMENTS

Physics is a science based on experimental observations. In view of this fact, we recommend that you try to supplement the text by performing various types of "hands-on" experiments, either at home or in the laboratory. For example, the common Slinky™ toy is excellent for studying traveling waves; a ball swinging on the end of a long string can be used to investigate pendulum motion; various objects attached to the end of a vertical spring or rubber band can be used to investigate elasticity; an old pair of Polaroid sunglasses and some discarded lenses and a magnifying glass are the components of various experiments in optics; and an approximate measure of the free-fall acceleration can be determined simply by measuring with a stopwatch the time it takes for a ball to drop from a known height. The list of such experiments is endless. When physical models are not available, be imaginative and try to develop models of your own.

Some of the Group Activities at the end of each chapter can be performed on your own or with the help of a friend and will give you specific guidelines for acquiring "hands-on" experience.

## CLOSING COMMENTS

Someone once said that there are only two professions in which people truly enjoy what they are doing, professional sports and physics. It is our sincere hope that you too will find physics an exciting and enjoyable experience and that you will profit from this experience, regardless of your chosen profession.

Welcome to the exciting world of physics!

*To see the World in a Grain of Sand*
*And a Heaven in a Wild Flower,*
*Hold infinity in the palm of your hand*
*And Eternity in an hour.*

W. Blake, "Auguries of Innocence"

# College Physics

## SIXTH EDITION

### *Volume 2*

# PART Four

## Electricity and Magnetism

We now begin a study of the branch of physics concerned with electric and magnetic phenomena. The laws of electricity and magnetism play central roles in the operation of many devices such as DVDs, VCRs, electric motors, computers, high-energy accelerators, and a host of electronic devices used in medicine. More fundamentally, we know that the interatomic and intermolecular forces that are responsible for the formation of solids and liquids and most basic life processes, such as the replication of chromosomes, are electric in origin. Furthermore, such forces as the pushes and pulls between objects and the elastic force in a spring arise from electric forces at the atomic level.

The ancient Greeks observed electric and magnetic phenomena as early as 700 B.C. They found that a piece of amber, when rubbed, became electrified and attracted pieces of straw or feathers. The existence of magnetic forces was known as a result of observations that a naturally occurring stone called *magnetite* ($Fe_2O_3$) was attracted to iron. (The word *electric* comes from the Greek word for amber, *elecktron.* The word *magnetic* comes from the name of the country where magnetite was found, *Magnesia,* now Turkey.)

In 1600 William Gilbert discovered that electrification was not limited to amber but is a general phenomenon. In 1785 Charles Coulomb described experiments that confirmed the inverse-square force law for electricity.

It was not until the early part of the 19th century that scientists established that electricity and magnetism are, in fact, related phenomena. In 1820 Hans Oersted discovered that a compass needle is deflected when placed near a wire carrying an electric current. A few years later, Michael Faraday showed that when a wire is moved near a magnet (or, equivalently, when a magnet is moved near a wire), an electric current is observed in the wire. From 1865 to 1873 James Clerk Maxwell used these observations and other experimental facts as bases for formulating the laws of electromagnetism as we now know them. (*Electromagnetism* is a name given to the combined subjects of electricity and magnetism.) Shortly thereafter, Heinrich Hertz verified Maxwell's predictions by producing electromagnetic waves in the laboratory. Guglielmo Marconi, inspired by the clever experiments of Hertz and others, invented the radio and filed the first patent for wireless telegraphy in 1896. The rest is history as today we sit bathed in electromagnetic waves from hundreds of practical devices ranging from hand-held computers to pagers.

Maxwell's contributions to the science of electromagnetism were especially significant because the laws he formulated are basic to *all* forms of electromagnetic phenomena including light. His work is comparable in importance to Newton's discovery of the laws of motion and the theory of gravitation.

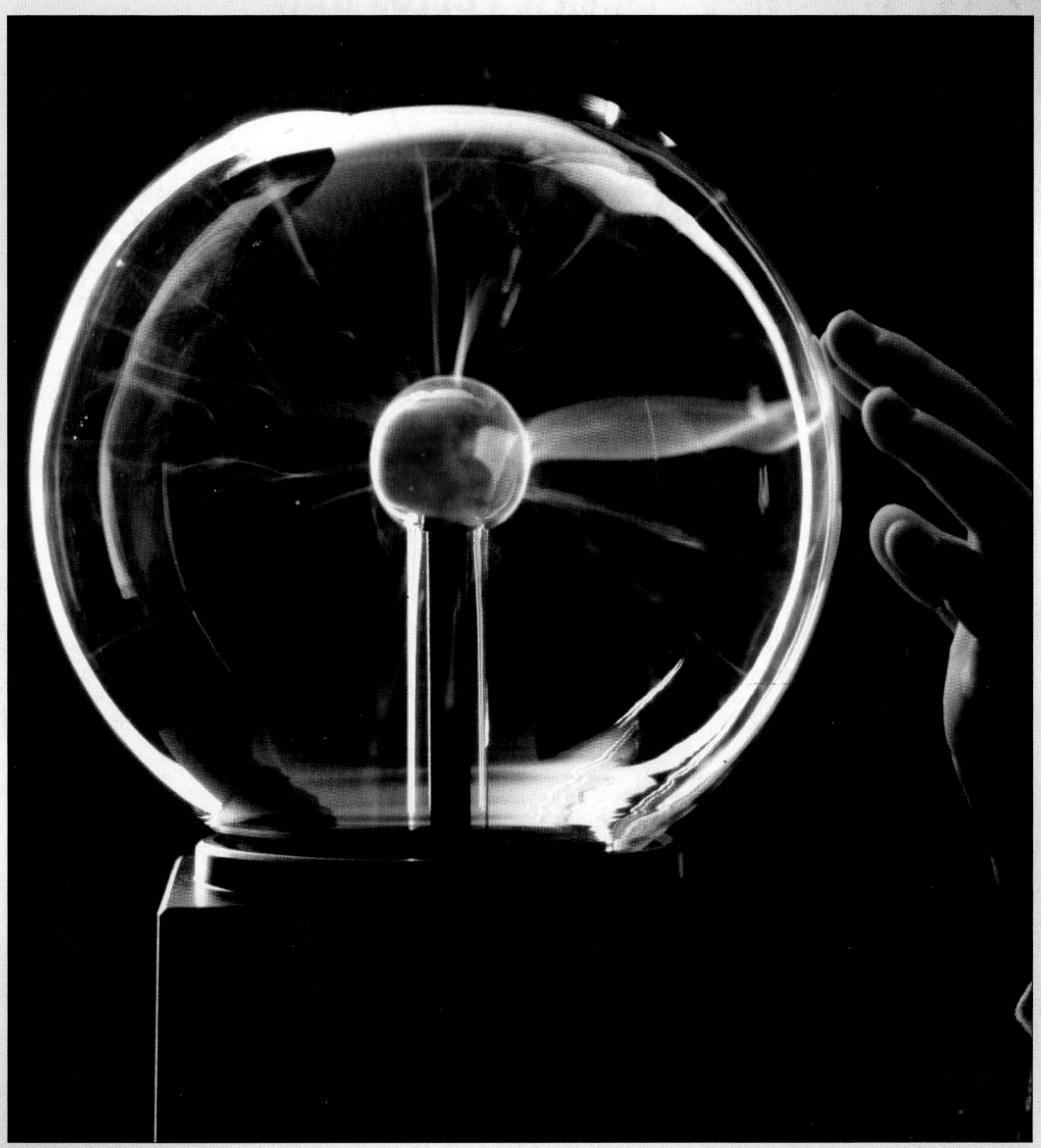

In plasma globes such as this, a high voltage (typically 5 kV) is applied between the center of the globe and the outer grounded glass surface. The interior of the globe contains a mixture of inert gases, and the applied voltage partially ionizes the gases. The ionized gases create filaments that conduct current as shown by the colorful paths in the photograph. When the outer surface is touched, the electrical discharge is drawn to your hand because your body is the path of least resistance to Earth compared to that of the grounded glass globe. ***(Richard Megna/Fundamental Photographs)***

15

# Electric Forces and Electric Fields

## Chapter Outline

**This nighttime view of multiple bolts of lightning was photographed in Tucson, Arizona. During a thunderstorm, a high concentration of electrical charge in a thundercloud creates a higher than normal electric field between the thundercloud and the negatively charged Earth's surface. This strong electric field creates an electric discharge between the charged cloud and ground—an enormous spark. Other discharges that are observed in the sky include cloud-to-cloud discharges and the more frequent intracloud discharges.** *(© Keith Kent/Science Photo Library/Photo Researchers, Inc.)*

The earliest known study of electricity was conducted by the Greeks about 700 B.C. By modern standards, their contributions to the field were modest. However, from those roots have sprung the enormous electrical distribution systems and sophisticated electronic instruments that are so much a part of our world today. It all began, apparently, when someone noticed that a fossil material called amber attracted small objects after being rubbed with wool. Since then we have learned that this phenomenon is not restricted to amber and wool but occurs (to some degree) when almost any two nonconducting substances are rubbed together.

In this chapter we use this effect—charging by friction—to begin an investigation of electric forces. We then discuss Coulomb's law, which is the fundamental law of force between any two stationary charged particles. The concept of an electric field associated with charges is then introduced and its effects on other charged particles described. We include a brief discussion of the Van de Graaff generator, and end by describing a way to calculate electric fields in certain situations.

**BENJAMIN FRANKLIN (1706–1790)**

Franklin was a printer, author, physical scientist, inventor, diplomat, and a founding father of the United States. This original and innovative thinker invented the Franklin stove, bifocals, the lightning rod, and the glass harmonica. He never patented any of his inventions, preferring that they be used for the benefit of all people. Franklin's clear thoughts, economy of expression, and quintessential optimism are evident in both his diplomatic and scientific works.

Franklin's work on electricity in the late 1740s changed a jumbled, unrelated set of observations into a connected science of electricity. Briefly, Franklin believed that electricity consisted of a single fluid and that an excess of fluid charged a body "positively" and a deficit "negatively." He conducted careful experiments to show that the natural state of an object was one of electrical neutrality. However, if an object was charged positively by rubbing it with a donor material, the donor was found to have an equal negative charge. He interpreted this to mean that the electric fluid added to the first object was just equal to that lost by the donor. Thus, electric fluid could be collected, transferred, and circulated, but it was *always* conserved. Franklin proposed several experiments involving lightning in order to "ascertain its sameness with the electric fluid." During a lightning storm, Franklin performed the dangerous kite experiment, and later used its positive outcome to practical use with the invention of the lightning rod. *(Science VU/Visuals Unlimited)*

## 15.1 PROPERTIES OF ELECTRIC CHARGES

A number of simple experiments demonstrate the existence of electrostatic forces. For example, after running a plastic comb through your hair, you will find that the comb attracts bits of paper. The attractive force is often strong enough to suspend the paper from the comb. The same effect occurs with other rubbed materials, such as glass and hard rubber.

Another simple experiment is to rub an inflated balloon with wool (or across your hair). On a dry day, the rubbed balloon will then stick to the wall of a room, often for hours. When materials behave in this way, they are said to have become **electrically charged.** You can give your body an electric charge by vigorously rubbing your shoes on a wool rug or by sliding across a car seat. You can then feel—and remove—the charge on your body by lightly touching another person. Under the right conditions, a visible spark can be seen when you touch, and a slight tingle is felt by both parties. (Experiments such as these work best on a dry day because excessive moisture can provide a pathway for charge to leak off a charged object.)

Experiments also demonstrate that there are two kinds of electric charge, which Benjamin Franklin (1706–1790) named **positive** and **negative.** Figure 15.1

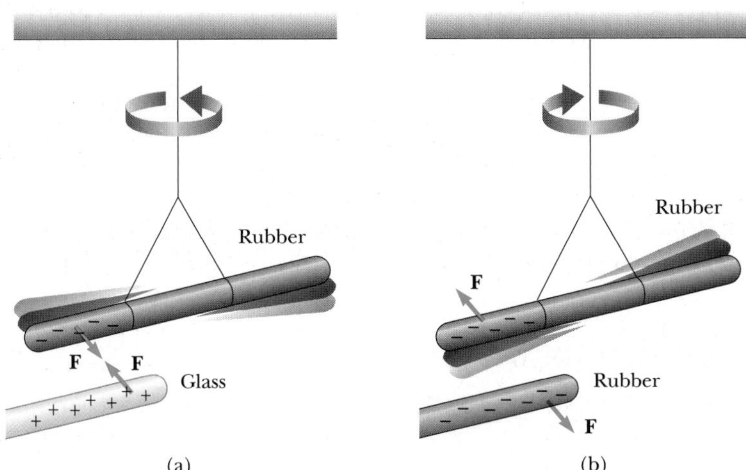


**FIGURE 15.1** (a) A negatively charged rubber rod suspended by a thread is attracted to a positively charged glass rod. (b) A negatively charged rubber rod is repelled by another negatively charged rubber rod.

illustrates the interaction of the two charges. A hard rubber (or plastic) rod that has been rubbed with fur is suspended by a piece of string. When a glass rod that has been rubbed with silk is brought near the rubber rod, the rubber rod is attracted toward the glass rod (Fig. 15.1a). If two charged rubber rods (or two charged glass rods) are brought near each other, as in Figure 15.1b, the force between them is repulsive. This observation demonstrates that the rubber and glass have different kinds of charge. We use the convention suggested by Franklin, where the electric charge on the glass rod is called positive and that on the rubber rod is called negative. On the basis of observations such as these, we conclude that **like charges repel one another and unlike charges attract one another.** Note that every object usually contains equal amounts of positive and negative charge—electrical forces occur when excess charge is transferred from one object to another.

Like charges repel; unlike charges attract

Nature's basic carrier of positive charge is the proton, a particle which, along with neutrons, is located in the nucleus of an atom. Because the nucleus of an atom is held firmly in place inside a solid, protons are never moved from one material to another. Thus, when an object becomes charged, it does so because it has either gained or lost nature's basic carrier of negative charge, the electron. It is convenient to picture the electron circling about the nucleus of an atom like a planet revolving about the Sun. The usual state of an atom is neutral; that is, for every positive charge in the nucleus there is an electron orbiting the nucleus to cancel the charge. Any uncharged object in our large-scale world contains an enormous number of electrons and protons—about $10^{23}$ of each per cubic centimeter. However, for every negative electron a positively charged proton is present.

### Webnote 15.1

For a more all-inclusive view of the electron and its function in the atom, read the article at *http://encarta.msn.com/find/concise.asp?z=1&pg=2&ti=01A8A000#s3*

Charge has a natural tendency to be transferred between unlike materials. Rubbing the two materials together serves to increase the area of contact and thus to enhance the charge transfer process.

An important characteristic of charge is that **electric charge is always conserved.** That is, when two initially neutral objects are charged by being rubbed together, charge is not created in the process. The objects become charged because **negative charge is transferred from one object to the other.** One object gains some amount of negative charge while the other loses an equal amount of negative charge and hence is left with a positive charge. For example, when a glass rod is rubbed with silk, as in Figure 15.2, the silk obtains a negative charge that is equal in magnitude to the positive charge on the glass rod as negatively charged electrons are transferred from the glass to the silk in the rubbing process. Likewise, when rubber is rubbed with fur, electrons are transferred from the fur to the rubber.

Charge is conserved; charge is quantized

In 1909 Robert Millikan (1886–1953) discovered that if an object is charged, its charge is always a multiple of a fundamental unit of charge, which we designate with the symbol $e$. In modern terms, the charge is said to be **quantized.** This means that charge occurs as discrete bundles in nature. Thus, an object may have a charge of $\pm e$, or $\pm 2e$, or $\pm 3e$, and so on, but never,[1] say, a fractional charge of $\pm 1.5e$. Other experiments in Millikan's time showed that the electron has a charge of $-e$ and the proton has an equal and opposite charge, $+e$. Some particles, such as a neutron, have no net charge. A neutral atom (one with no net charge) contains as many protons as electrons. The value of $e$ is now known to be $1.602\ 19 \times 10^{-19}$ C. (The SI unit of electric charge, the **coulomb** [C], will be defined more precisely in a later section.)

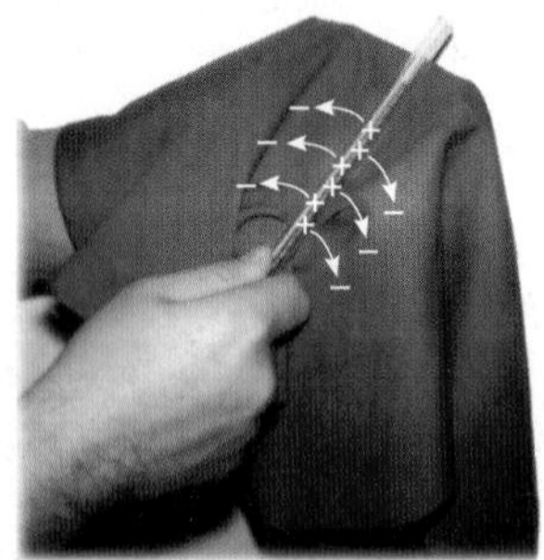

**FIGURE 15.2** When a glass rod is rubbed with silk, electrons are transferred from the glass to the silk. Because of conservation of charge, each electron adds negative charge to the silk, and an equal positive charge is left behind on the rod. Also, because the charges are transferred in discrete bundles, the charges on the two objects are $\pm e$ or $\pm 2e$ or $\pm 3e$, and so on.

## 15.2 INSULATORS AND CONDUCTORS

It is convenient to classify substances in terms of their ability to conduct electric charge.

[1] Recent developments have shown the existence of fundamental particles called **quarks** that have charges of $\pm e/3$ or $\pm 2e/3$. Note that charge is *still* quantized, but in units of $\pm e/3$. A more complete discussion of quarks and their properties is presented in Chapter 30.

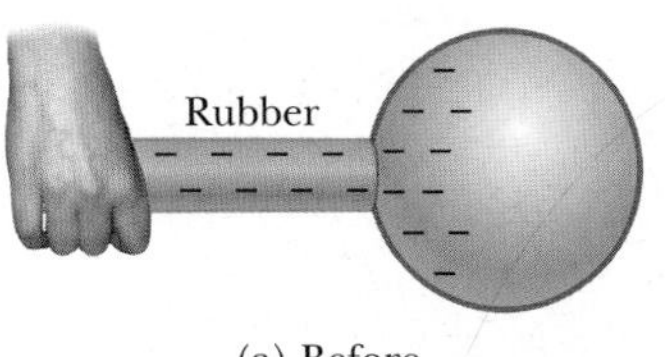


(a) Before

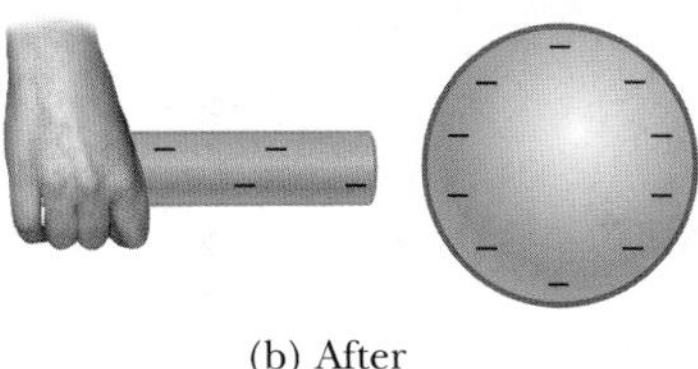

(b) After

**FIGURE 15.3** Charging a metal object by conduction. (a) The charged rubber rod is placed in contact with the insulated metal sphere. Some electrons move from the rod onto the sphere. (b) When the rod is removed, the sphere is left with a negative charge.

**Conductors are materials in which electric charges move freely, and insulators are materials in which electric charges do not move freely.**

Glass and rubber are insulators. When such materials are charged by rubbing, only the rubbed area becomes charged, and there is no tendency for the charge to move into other regions of the material. In contrast, materials such as copper, aluminum, and silver are good conductors. When such materials are charged in some small region, the charge readily distributes itself over the entire surface of the material. If you hold a copper rod in your hand and rub the rod with wool or fur, it will not attract a piece of paper. This might suggest that a metal cannot be charged. However, if you hold the copper rod with an insulator and then rub it with wool or fur, the rod remains charged and attracts the paper. In the first case, the electric charges produced by rubbing readily move from the copper through your body and finally to ground. In the second case, the insulating handle prevents the flow of charge to ground.

**Semiconductors** are a third class of materials, and their electrical properties are somewhere between those of insulators and those of conductors. Silicon and germanium are well-known semiconductors that are widely used in the fabrication of a variety of electronic devices.

## CHARGING BY CONDUCTION

Consider a negatively charged rubber rod brought into contact with a neutral conducting sphere that is insulated so that there is no conducting path for charges to leave the sphere. Some electrons on the rubber rod are now able to move onto the sphere, as in Figure 15.3. When the rubber rod is removed, the sphere is left with a negative charge. This process is referred to as charging by **conduction.** The object being charged in such a process (the sphere) is always left with a charge having the same sign as the object doing the charging (the rubber rod).

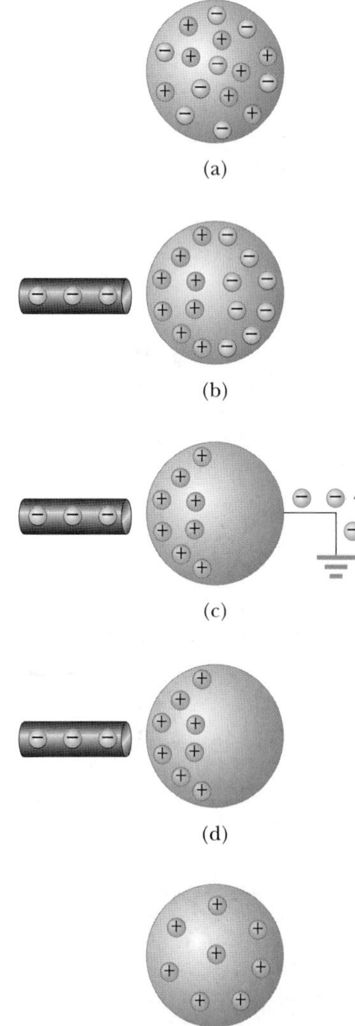


**FIGURE 15.4** Charging a metal object by induction. (a) A neutral metallic sphere, with equal numbers of positive and negative charge. (b) The charge on the neutral sphere is redistributed when a charged rubber rod is placed near the sphere. (c) When the sphere is grounded, some of its electrons leave through the ground wire. (d) When the ground connection is removed, the sphere has excess positive charge that is nonuniformly distributed. (e) When the rubber rod is removed, the excess positive charge becomes uniformly distributed over the surface of the sphere.

## CHARGING BY INDUCTION

When an object is connected to a conducting wire or copper pipe buried in Earth, it is said to be **grounded.** Earth can be considered an infinite reservoir for electrons; this means that it can accept or supply an unlimited number of electrons. With this in mind, we can understand the charging of a conductor by a process known as **induction.**

Consider a negatively charged rubber rod brought near a neutral (uncharged) conducting sphere that is insulated so that there is no conducting path to ground (Fig. 15.4a). The repulsive force between the electrons in the rod and those in the sphere causes a redistribution of charge on the sphere so that some

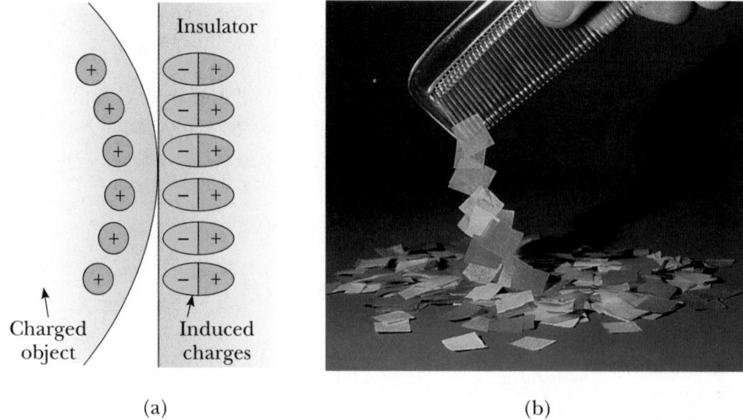


**FIGURE 15.5** (a) The charged object on the left induces charges on the surface of an insulator. (b) A charged comb attracts bits of paper because charges are displaced in the paper. *(© 1968 Fundamental Photographs)*

electrons move to the side of the sphere farthest away from the rod (Fig. 15.4b). The region of the sphere nearest the negatively charged rod has an excess of positive charge because of the migration of electrons away from this location. If a grounded conducting wire is then connected to the sphere, as in Figure 15.4c, some of the electrons leave the sphere and travel to ground. If the wire to ground is then removed (Fig. 15.4d), the conducting sphere is left with an excess of induced positive charge. Finally, when the rubber rod is removed from the vicinity of the sphere (Fig. 15.4e), the induced positive charge remains on the ungrounded sphere. Even though the positively charged atomic nuclei remain fixed, this excess positive charge becomes uniformly distributed over the surface of the ungrounded sphere because of the repulsive forces among the like charges and the high mobility of electrons in a metal.

In the process of inducing a charge on the sphere, the charged rubber rod loses none of its negative charge because it never came in contact with the sphere. Furthermore, the sphere is left with a charge opposite to that of the rubber rod. Note that **charging an object by induction requires no contact with the object inducing the charge.**

A process very similar to charging by induction in conductors also takes place in insulators. In most neutral atoms or molecules, the center of positive charge coincides with the center of negative charge. However, in the presence of a charged object, these centers may separate slightly, resulting in more positive charge on one side of the molecule than on the other. This effect is known as **polarization.** The realignment of charge within individual molecules produces an induced charge on the surface of the insulator, as shown in Figure 15.5a. Knowing about induction in insulators, we can explain why a comb that has been rubbed through hair attracts bits of electrically neutral paper (Fig. 15.5b) and why a balloon that has been rubbed against our clothing is able to stick to an electrically neutral wall.

## APPLYING PHYSICS 15.1

A positively charged ball hanging from a nonconducting string is brought near a nonconducting object. Based on the behavior of the ball-string combination, the ball is seen to be attracted to the object. From this experiment, it is not possible to determine whether the object is negatively charged or neutral. Why not? What additional experiment would help you decide between these two possibilities?

**Explanation** The attraction between the ball and the object could be an attraction of unlike charges, or it could be an attraction between a charged object and a neutral object as a result of polarization of the molecules of the neutral object. Two additional experiments help us determine whether the object is charged. First, a known neutral ball is brought near the object—if there is an attraction, the object is negatively charged. Another possibility is to bring a known negatively charged ball near the object—if there is a repulsion, then the object is negatively charged. If there is no attraction, then the object is neutral.

| Quick Quiz 15.1 | If a suspended object A is attracted to object B, which is charged, we can conclude that (a) object A is uncharged, (b) object A is charged, (c) object B is positively charged, or (d) object A may be either charged or uncharged. |
|---|---|

## 15.3 COULOMB'S LAW

CHARLES COULOMB (1736–1806)

Coulomb's major contribution to science was in the field of electrostatics and magnetism. During his lifetime, he also investigated the strengths of materials and determined the forces that affect objects on beams, thereby contributing to the field of structural mechanics. *(Photo courtesy of AIP Niels Bohr Library, E. Scott Barr Collection)*

In 1785 Charles Coulomb (1736–1806) established the fundamental law of electric force between two stationary charged particles. Experiments show that

**An electric force has the following properties:**

1. **It is inversely proportional to the square of the separation $r$ between the two particles and is along the line joining them.**
2. **It is proportional to the product of the magnitudes of the charges $|q_1|$ and $|q_2|$ on the two particles.**
3. **It is attractive if the charges are of opposite sign and repulsive if the charges have the same sign.**

From these observations, we can express the magnitude of the electric force between two charges separated by a distance $r$ as

$$F = k_e \frac{|q_1||q_2|}{r^2} \quad [15.1]$$

Coulomb's law

where $k_e$ is a constant called the *Coulomb constant.* Note that Equation 15.1, known as **Coulomb's law,** applies only to point charges and to spherical distributions of charges.

The value of the Coulomb constant in Equation 15.1 depends on the choice of units. The SI unit of charge is the **coulomb** (C). From experiment, we know that the **Coulomb constant** in SI units has the value

$$k_e = 8.987\,5 \times 10^9 \text{ N}\cdot\text{m}^2/\text{C}^2 \quad [15.2]$$

To simplify our calculations, we shall use the approximate value

$$k_e \approx 8.99 \times 10^9 \text{ N}\cdot\text{m}^2/\text{C}^2 \quad [15.3]$$

The charge on the proton has a magnitude of $e = 1.6 \times 10^{-19}$ C. Therefore, it would take $1/e = 6.3 \times 10^{18}$ protons to create a total charge of $+1$ C. Likewise, $6.3 \times 10^{18}$ electrons would have a total charge of $-1$ C. Compare this with the number of free electrons in 1 cm$^3$ of copper, which is on the order of $10^{23}$. Even so, 1 C is a large amount of charge. In typical electrostatic experiments, where a rubber or glass rod is charged by friction, a net charge on the order of $10^{-6}$ C ($= 1\ \mu$C) is obtained. In other words, only a very small fraction of the total available charge is transferred between the rod and rubbing material. Table 15.1 lists the charges and masses of the electron, proton, and neutron.

When dealing with Coulomb's force law, remember that force is a vector quantity and must be treated accordingly. Figure 15.6a shows the electric force of repulsion between two positively charged particles. Electric forces obey Newton's third law, and hence the forces $\mathbf{F}_{12}$ and $\mathbf{F}_{21}$ are equal in magnitude but opposite in direction. (The notation $\mathbf{F}_{12}$ denotes the force exerted by particle 1 on particle 2. Likewise, $\mathbf{F}_{21}$ is the force exerted by particle 2 on particle 1.) It bears repeating

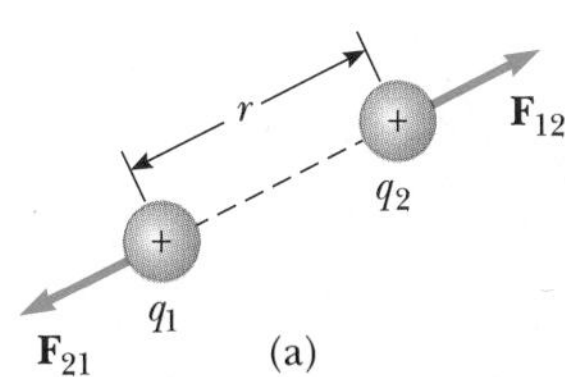


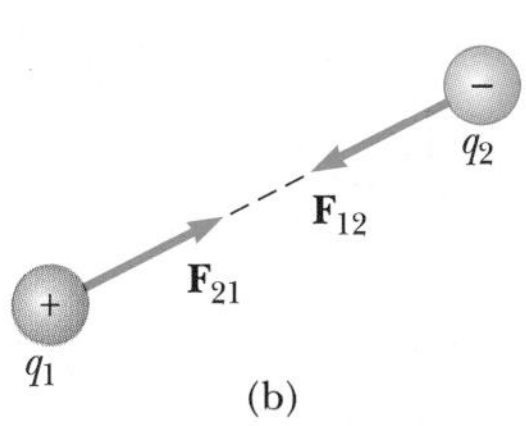


**FIGURE 15.6** Two point charges separated by a distance $r$ exert a force on each other given by Coulomb's law. The force on $q_1$ is equal in magnitude and opposite in direction to the force on $q_2$. (a) When the charges are of the same sign, the force is repulsive. (b) When the charges are of opposite sign, the force is attractive.

**TABLE 15.1 Charge and Mass of the Electron, Proton, and Neutron**

| Particle | Charge (C) | Mass (kg) |
|---|---|---|
| Electron | $-1.60 \times 10^{-19}$ | $9.11 \times 10^{-31}$ |
| Proton | $+1.60 \times 10^{-19}$ | $1.67 \times 10^{-27}$ |
| Neutron | 0 | $1.67 \times 10^{-27}$ |

that $F_{12}$ and $F_{21}$ are always equal regardless of whether $q_1$ and $q_2$ have the same magnitude.

The Coulomb force is the second example we have seen of a *field force*—a force exerted by one object on another even though there is no physical contact between them. Recall that another example of a field force is gravitational attraction. The mathematical form of the Coulomb force is the same as that of the gravitational force. That is, they are both inversely proportional to the square of the distance of separation. However, there are some important differences between electric and gravitational forces. Electric forces can be either attractive or repulsive, but gravitational forces are always attractive.

**Quick Quiz 15.2**

Object A has a charge of $+2\ \mu$C, and object B has a charge of $+6\ \mu$C. Which statement is true: (a) $\mathbf{F}_{AB} = -3\mathbf{F}_{BA}$, (b) $\mathbf{F}_{AB} = -\mathbf{F}_{BA}$, or (c) $3\mathbf{F}_{AB} = -\mathbf{F}_{BA}$

## Example 15.1 The Electric Force and the Gravitational Force

The electron and proton of a hydrogen atom are separated (on the average) by a distance of about $5.3 \times 10^{-11}$ m. Find the magnitudes of the electric force and the gravitational force that each particle exerts on the other.

**Solution** From Coulomb's law, we find that the attractive electric force has the magnitude

$$F_e = k_e \frac{|e|^2}{r^2} = \left(8.99 \times 10^9 \frac{\text{N}\cdot\text{m}^2}{\text{C}^2}\right) \frac{(1.6 \times 10^{-19}\text{ C})^2}{(5.3 \times 10^{-11}\text{ m})^2} = 8.2 \times 10^{-8}\text{ N}$$

From Newton's law of universal gravitation and Table 15.1, we find that the gravitational force has the magnitude

$$F_g = G\frac{m_e m_p}{r^2} = \left(6.67 \times 10^{-11} \frac{\text{N}\cdot\text{m}^2}{\text{kg}^2}\right) \frac{(9.11 \times 10^{-31}\text{ kg})(1.67 \times 10^{-27}\text{ kg})}{(5.3 \times 10^{-11}\text{ m})^2}$$

$$= 3.6 \times 10^{-47}\text{ N}$$

Because $F_e/F_g \approx 2 \times 10^{39}$, the gravitational force between the charged atomic particles is negligible compared with the electric force.

### THE SUPERPOSITION PRINCIPLE

Frequently, more than two charges are present and it is necessary to find the resultant electric force on one of them. This can be accomplished by noting that the electric force between any pair of charges is given by Equation 15.1. Therefore, the resultant force on any one charge equals the vector sum of the forces exerted by the other individual charges that are present. This is another example of the **superposition principle.** For example, if you have three charges and you want to find the force exerted by charges 2 and 3 on charge 1, you first find the force exerted by charge 2 on charge 1 and the force exerted by charge 3 on charge 1. You then add these two forces together *vectorially* to get the resultant force on charge 1. The following example illustrates this procedure.

**Webnote 15.2**

You can check your answer to a Coulomb's law calculation at *http://hyperphysics.phy-astr.gsu.edu/hbase/electric.elefor.html#c1*

## Example 15.2 Using the Superposition Principle

Consider three point charges at the corners of a triangle, as in Figure 15.7, where $q_1 = 6.00 \times 10^{-9}$ C, $q_3 = 5.00 \times 10^{-9}$ C, $q_2 = -2.00 \times 10^{-9}$ C, and the distances of separation are as shown in the figure. Find the resultant force on $q_3$.

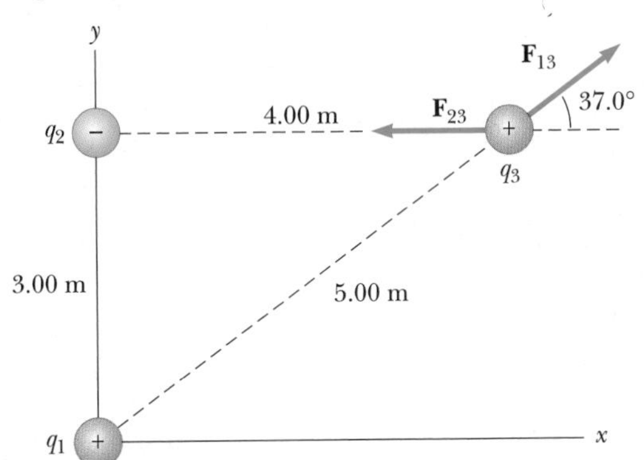


**FIGURE 15.7** (Example 15.2) The force exerted by $q_1$ on $q_3$ is $\mathbf{F}_{13}$. The force exerted by $q_2$ on $q_3$ is $\mathbf{F}_{23}$. The *resultant force* exerted on $q_3$ is the *vector* sum $\mathbf{F}_{13} + \mathbf{F}_{23}$.

**Reasoning** It is first necessary to find the direction of the forces exerted by $q_1$ and $q_2$ on $q_3$. The force $\mathbf{F}_{23}$ exerted by $q_2$ on $q_3$ is attractive because $q_2$ and $q_3$ have opposite signs. The force $\mathbf{F}_{13}$ exerted by $q_1$ on $q_3$ is repulsive because both $q_1$ and $q_3$ are positive. To find the resultant force on $q_3$, it is necessary to find the magnitudes of $\mathbf{F}_{23}$ and $\mathbf{F}_{13}$ by use of Coulomb's law and then add the two forces vectorially.

**Solution** The magnitude of the force exerted by $q_2$ on $q_3$ is

$$F_{23} = k_e \frac{|q_2||q_3|}{r^2} = (8.99 \times 10^9 \text{ N}\cdot\text{m}^2/\text{C}^2)\frac{(2.00 \times 10^{-9}\text{ C})(5.00 \times 10^{-9}\text{ C})}{(4.00\text{ m})^2}$$

$$= 5.62 \times 10^{-9}\text{ N}$$

The magnitude of the force exerted by $q_1$ on $q_3$ is

$$F_{13} = k_e \frac{|q_1||q_3|}{r^3} = (8.99 \times 10^9 \text{ N}\cdot\text{m}^2/\text{C}^2)\frac{(6.00 \times 10^{-9}\text{ C})(5.00 \times 10^{-9}\text{ C})}{(5.00\text{ m})^2}$$

$$= 1.08 \times 10^{-8}\text{ N}$$

The force $\mathbf{F}_{13}$ makes an angle of 37.0° with the *x* axis and is directed along the line connecting $q_1$ and $q_3$. Therefore, the *x* component of this force has the magnitude $F_{13}\cos 37.0° = 8.63 \times 10^{-9}$ N, and the *y* component of $\mathbf{F}_{13}$ has the magnitude $F_{13}\sin 37.0° = 6.50 \times 10^{-9}$ N. The force $\mathbf{F}_{23}$ is in the negative *x* direction. Hence, the *x* and *y* components of the resultant force on $q_3$ are

$$F_x = 8.63 \times 10^{-9}\text{ N} - 5.62 \times 10^{-9}\text{ N} = 3.01 \times 10^{-9}\text{ N}$$

$$F_y = 6.50 \times 10^{-9}\text{ N}$$

The magnitude of the resultant force on the charge $q_3$ is therefore

$$\sqrt{(3.01 \times 10^{-9}\text{ N})^2 + (6.50 \times 10^{-9}\text{ N})^2} = 7.16 \times 10^{-9}\text{ N}$$

and the force vector makes an angle with the *x* axis of

$$\theta = \tan^{-1}\left(\frac{F_y}{F_x}\right) = \tan^{-1}\left(\frac{6.50 \times 10^{-9}\text{ N}}{3.01 \times 10^{-9}\text{ N}}\right) = 65.2°$$

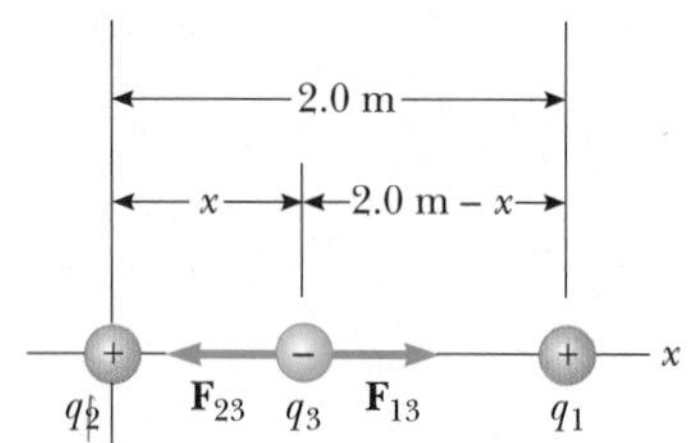


**FIGURE 15.8** (Example 15.3) Three point charges are placed along the $x$ axis. The charge $q_3$ is negative, whereas $q_1$ and $q_2$ are positive. If the resultant force on $q_3$ is zero, then the force $\mathbf{F}_{13}$ exerted by $q_1$ on $q_3$ must be equal in magnitude and opposite the force $\mathbf{F}_{23}$ exerted by $q_2$ on $q_3$.

**Example 15.3 Where Is the Resultant Force Zero?**

Three charges lie along the $x$ axis as in Figure 15.8. The positive charge $q_1 = 15\ \mu\text{C}$ is at $x = 2.0$ m, and the positive charge $q_2 = 6.0\ \mu\text{C}$ is at the origin. Where must a *negative* charge $q_3$ be placed on the $x$ axis so that the resultant force on it is zero?

**Reasoning** Because $q_3$ is negative and $q_1$ and $q_2$ are positive, the forces $\mathbf{F}_{13}$ and $\mathbf{F}_{23}$ are both attractive, as indicated in Figure 15.8. The only location where the force exerted by $q_2$ on $q_3$ is opposite the force exerted by $q_1$ on $q_3$ lies on the $x$ axis between $q_1$ and $q_2$. Because we require that the resultant force on $q_3$ be zero, then $F_{13}$ must equal $F_{23}$.

**Solution** If we let $x$ be the coordinate of $q_3$, then the forces $\mathbf{F}_{13}$ and $\mathbf{F}_{23}$ have the magnitudes

$$F_{13} = k_e \frac{(15 \times 10^{-6}\ \text{C})|q_3|}{(2.0 - x)^2} \quad \text{and} \quad F_{23} = k_e \frac{(6.0 \times 10^{-6}\ \text{C})|q_3|}{x^2}$$

If the resultant force on $q_3$ is zero, then $F_{13} = F_{23}$, or

$$k_e \frac{(15 \times 10^{-6}\ \text{C})|q_3|}{(2.0 - x)^2} = k_e \frac{(6.0 \times 10^{-6}\ \text{C})|q_3|}{x^2}$$

Because $k_e$, $10^{-6}$, and $q_3$ are common to both sides, they can be cancelled from the equation, and we have (after some reduction)

$$(2.0 - x)^2(6.0) = x^2(15)$$

This can be expanded to a quadratic equation, which can then be solved for $x$. An easier approach is first to take the positive square root of both sides:

$$(2.0 - x)\sqrt{6.0} = x\sqrt{15}$$

$$2.0 - x = x(1.58)$$

$$x = 0.78\ \text{m}$$

## 15.4 THE ELECTRIC FIELD

Two different field forces have been introduced into our discussions so far: the gravitational force and the electrostatic force. As pointed out earlier, these forces are capable of acting through space, producing an effect even when there is no physical contact between the objects involved. Field forces can be discussed in a variety of ways, but an approach developed by Michael Faraday (1791–1867) is of such practical value that we shall devote much attention to it in the next few chapters. In this approach, an **electric field** is said to exist in the region of space around a charged object. When another charged object enters this electric field, the field is what exerts a force on the second charged object. As an example, consider Figure 15.9, which shows an object with a small positive charge $q_0$ placed near a second object with a larger positive charge $Q$.

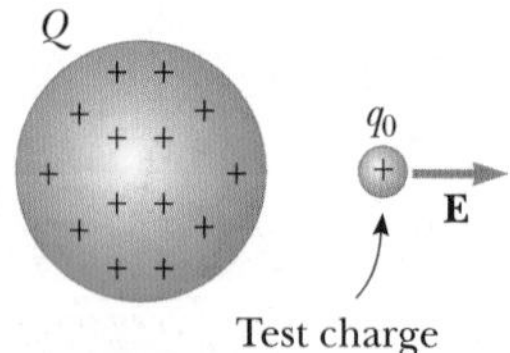


**FIGURE 15.9** A small object with a positive charge $q_0$ placed near an object with a larger positive charge $Q$ experiences an electric field $\mathbf{E}$ directed as shown. The magnitude of the electric field at the location of $q_0$ is defined as the electric force on $q_0$ divided by the charge $q_0$.

We define the electric field at the location of the small "test" charge to be the electric force acting on it divided by the charge $q_0$ of the test charge:

$$\mathbf{E} \equiv \frac{\mathbf{F}}{q_0} \qquad [15.4]$$

Note that this is the electric field at the location of $q_0$ produced by the charge $Q$, not the field produced by $q_0$. The electric field is a vector quantity having the SI units newtons per coulomb (N/C). **The direction of E at a point is defined to be the direction of the electric force that would be exerted on a small posi-**

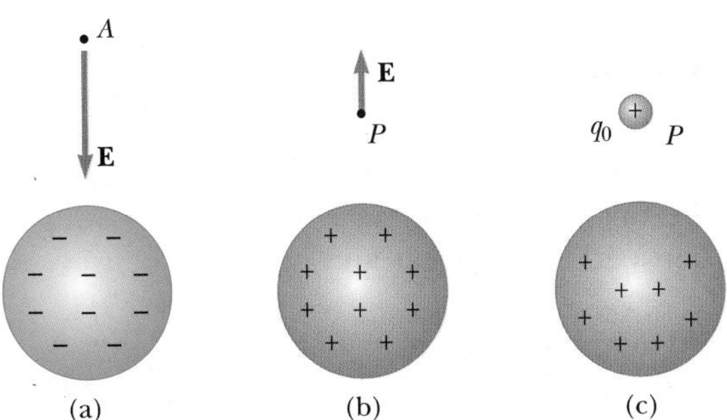


**FIGURE 15.10** (a) The electric field at $A$ due to the negatively charged sphere is downward, toward the negative charge. (b) The electric field at $P$ due to the positively charged conducting sphere is upward, away from the positive charge. (c) A test charge $q_0$ placed at $P$ will cause a rearrangement of charge on the sphere unless $q_0$ is very small compared with the charge on the sphere.

**tive test charge placed at that point.** Thus, in Figure 15.9, the direction of the electric field is horizontal and to the right. The electric field at point $A$ in Figure 15.10a is vertical and downward because at this point a positive charge would experience a force of attraction toward the negatively charged sphere. Note that the test charge $q_0$ is required to be a small charge, in order to cause no significant rearrangement of the charge on the large sphere as shown in Figure 15.10c.

Once the electric field is known at some point, the force on *any* particle with charge $q$ placed at that point can be calculated from a rearrangement of Equation 15.4:

$$\mathbf{F} = q\mathbf{E} \quad [15.5]$$

Consider a point charge $q$ located a distance $r$ from a test charge $q_0$. According to Coulomb's law, the *magnitude* of the force on the test charge is

$$F = k_e \frac{|q||q_0|}{r^2}$$

Because the magnitude of the electric field at the position of the test charge is defined as $E = F/q_0$, we see that the *magnitude* of the electric field due to the charge $q$ at the position of $q_0$ is

$$E = k_e \frac{|q|}{r^2} \quad [15.6]$$

Electric field due to a charge $q$

If $q$ is *positive,* as in Figure 15.11a, the field at $P$ due to this charge is *radially outward* from $q$. If $q$ is *negative,* as in Figure 15.11b, the field at $P$ is directed *toward* $q$. Equation 15.6 points out an important property of electric fields that makes them useful quantities for describing electrical phenomena. As the equation indicates, an electric field at a given point depends only on the charge $q$ on the object setting up the field and the distance $r$ from that object to a specific point in space. As a result, we can say that an electric field exists at point $P$ in Figure 15.11 whether or not there is a test charge at $P$.

The principle of superposition holds when the electric field due to a group of point charges is calculated. We first use Equation 15.6 to calculate the electric field produced by each charge individually at a point, and then add these electric fields together as vectors.

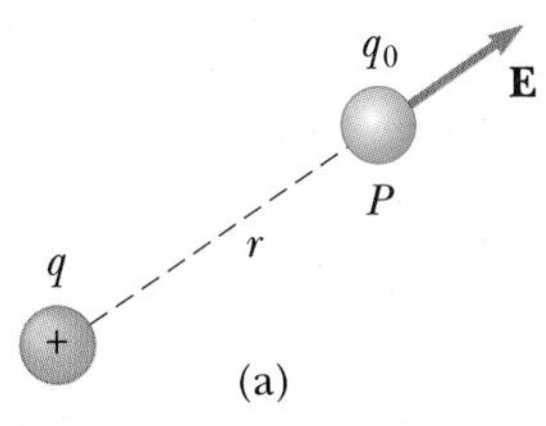


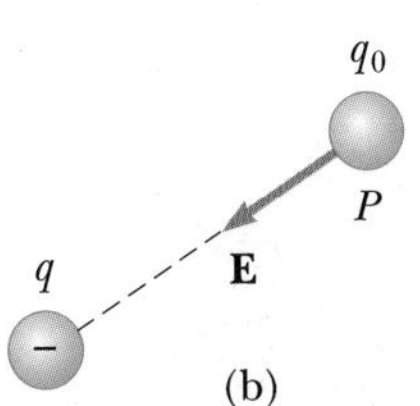


**FIGURE 15.11** A test charge $q_0$ at $P$ is a distance $r$ from a point charge $q$. (a) If $q$ is positive, the electric field at $P$ points radially *outward* from $q$. (b) If $q$ is negative, the electric field at $P$ points radially *inward* toward $q$.

**PROBLEM-SOLVING STRATEGY** **Electric Forces and Fields**

1. **Units.** When performing calculations that use the Coulomb constant $k_e$, charges must be in coulombs and distances in meters. If they are given in other units, you must convert them.
2. **Applying Coulomb's law to point charges.** It is important to use the superposition principle properly when dealing with a collection of

interacting point charges. If several charges are present, the resultant force on any one of them is found by determining the individual force exerted on it by every other charge and then determining the vector sum of all these forces. The force that any charged object exerts on another is given by Coulomb's law, and the direction of the force is found by noting that the forces are repulsive between like charges and attractive between unlike charges.

3. **Calculating the electric field of point charges.** Because the electric field is a vector quantity, the superposition principle can be applied. To find the resultant electric field at a given point due to several charges, first calculate the electric field at the point due to each individual charge. The vector sum of the fields due to all the individual charges is the resultant field at the point.

## APPLYING PHYSICS 15.2

An electron moving horizontally passes between two horizontal plates, the upper charged negatively, the lower positively. A uniform, upward-directed electric field exists in this region, and this field exerts an electric force downward on the electron. Describe the movement of the electron in this region.

**Explanation** The magnitude of the electric force on the electron of charge $e$ due to a uniform electric field $\mathbf{E}$ is $F = eE$. Thus, the force is constant. Compare this to the force on a projectile of mass $m$ moving in the gravitational field of Earth. The magnitude of the gravitational force is $mg$. In both cases, the particle is subject to a constant force in the vertical direction and has an initial velocity in the horizontal direction. Thus, the path will be the same in each case—the electron will move as a projectile with an acceleration in the vertical direction and constant velocity in the horizontal direction. Once the electron leaves the region between the plates, the electric field disappears, and the electron continues moving in a straight line according to Newton's first law.

## Example 15.4 Electric Force on a Proton

Find the electric force on a proton placed in an upward electric field of magnitude $2.0 \times 10^4$ N/C.

**Solution** Because the charge on a proton is $+q = +1.6 \times 10^{-19}$ C, the electric force acting on the proton is

$$F = qE = (1.6 \times 10^{-19}\ \text{C})(2.0 \times 10^4\ \text{N/C}) = \boxed{3.2 \times 10^{-15}\ \text{N}}$$

where the force is upward, in the positive $y$ direction. The magnitude of the gravitational force acting downward on the proton has the value $m_pg = (1.67 \times 10^{-27}\ \text{kg})(9.80\ \text{m/s}^2) = 1.64 \times 10^{-26}$ N. Hence, the magnitude of the gravitational force is negligible compared with that of the electric force.

## Example 15.5 Electric Field Due to Two Point Charges

Charge $q_1 = 7.00\ \mu\text{C}$ is at the origin, and charge $q_2 = -5.00\ \mu\text{C}$ is on the $x$ axis, 0.300 m from the origin (Fig. 15.12). Find the electric field at point $P$, which has coordinates (0, 0.400) m.

**Reasoning** It is first necessary to find the direction of the field at $P$ set up by each charge. The field $\mathbf{E}_1$ at $P$ due to $q_1$ is vertically upward as in Figure 15.12 because $\mathbf{E}_1$ is in the direction that a positive test charge would move if it were placed at $P$. Likewise, the field $\mathbf{E}_2$ at $P$ due to $q_2$ is directed toward $q_2$ as in Figure 15.12. The magnitudes of the fields can be found from $E = k_e|q|/r^2$ and then added together vectorially.

**Solution** The magnitudes of $\mathbf{E}_1$ and $\mathbf{E}_2$ are

$$E_1 = k_e\frac{|q_1|}{{r_1}^2} = (8.99 \times 10^9\ \text{N}\cdot\text{m}^2/\text{C}^2)\frac{(7.00 \times 10^{-6}\ \text{C})}{(0.400\ \text{m})^2} = 3.93 \times 10^5\ \text{N/C}$$

$$E_2 = k_e\frac{|q_2|}{{r_2}^2} = (8.99 \times 10^9\ \text{N}\cdot\text{m}^2/\text{C}^2)\frac{(5.00 \times 10^{-6}\ \text{C})}{(0.500\ \text{m})^2} = 1.80 \times 10^5\ \text{N/C}$$

The vector $\mathbf{E}_1$ has an $x$ component of zero. The vector $\mathbf{E}_2$ has an $x$ component given by $E_2 \cos\theta = \frac{3}{5}E_2 = 1.08 \times 10^5$ N/C and a negative $y$ component given by $-E_2 \sin\theta = -\frac{4}{5}E_2 = -1.44 \times 10^5$ N/C. Hence, the resultant component in the $x$ direction is

$$E_x = 1.08 \times 10^5\ \text{N/C}$$

and the resultant component in the $y$ direction is

$$E_y = E_{y1} + E_{y2} = 3.93 \times 10^5\ \text{N/C} - 1.44 \times 10^5\ \text{N/C} = 2.49 \times 10^5\ \text{N/C}$$

From the Pythagorean theorem ($E = \sqrt{{E_x}^2 + {E_y}^2}$), we find that $\mathbf{E}$ has a magnitude $2.72 \times 10^5$ N/C and makes an angle $\phi$ with the positive $x$ axis given by

$$\phi = \tan^{-1}\left(\frac{E_y}{E_x}\right) = \tan^{-1}\left(\frac{2.49 \times 10^5\ \text{N/C}}{1.08 \times 10^5\ \text{N/C}}\right) = 66.5°$$

**EXERCISE** Find the force on a positive test charge of $2.00 \times 10^{-8}$ C placed at $P$.

**ANSWER** $5.44 \times 10^{-3}$ N in the same direction as $\mathbf{E}$.

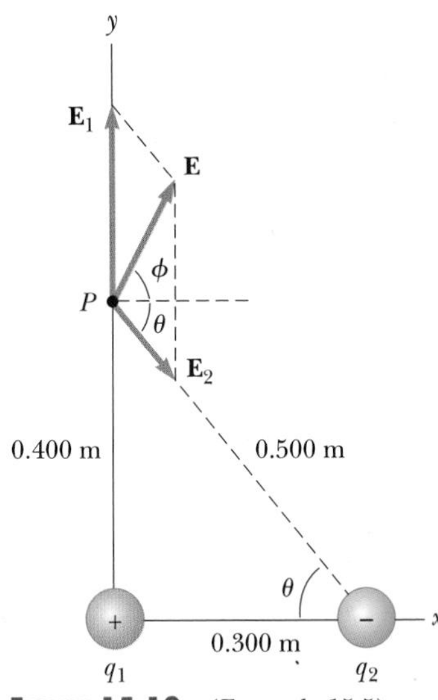


**FIGURE 15.12** (Example 15.5) The resultant electric field $\mathbf{E}$ at $P$ equals the vector sum $\mathbf{E}_1 + \mathbf{E}_2$, where $\mathbf{E}_1$ is the field due to the positive charge $q_1$ and $\mathbf{E}_2$ is the field due to the negative charge $q_2$.

**Quick Quiz 15.3** A test charge of $+3\ \mu$C is at a point $P$ where the electric field due to other charges is directed to the right and has a magnitude of $4 \times 10^6$ N/C. If the test charge is replaced with a charge of $-3\ \mu$C, the electric field at $P$ (a) has the same magnitude but changes direction, (b) increases in magnitude and changes direction, (c) remains the same, or (d) decreases in magnitude and changes direction.

**Quick Quiz 15.4** A Styrofoam ball covered with a conducting paint has a mass of $5.0 \times 10^{-3}$ kg and a charge of 4.0 $\mu$C. What electric field directed upward will produce an electric force on the ball that will balance the weight of the ball? (a) $8.2 \times 10^2$ N/C (b) $1.2 \times 10^4$ N/C (c) $2.0 \times 10^{-2}$ N/C (d) $5.1 \times 10^6$ N/C

**Quick Quiz 15.5** A circular ring of radius $b$ has a total charge $q$ uniformly distributed around it. The magnitude of the electric field at the center of the ring is (a) 0, (b) $k_eq/b^2$, (c) $k_eq^2/b^2$, (d) $k_eq^2/b$, or (e) none of these.

**Quick Quiz 15.6** A "free" electron and "free" proton are placed in an identical electric field. Which of the following statements are true? (a) Each particle experiences the same electric force and the same acceleration. (b) The electric force on the proton is greater in magnitude than the force on the electron but in the opposite direction. (c) The electric force on the proton is equal in magnitude to the force on the electron, but in the opposite direction. (d) The magnitude of the acceleration of the electron is greater than that of the proton. (e) Both particles experience the same acceleration.

## 15.5 ELECTRIC FIELD LINES

A convenient aid for visualizing electric field patterns is to draw lines pointing in the direction of the electric field vector at any point. These lines introduced by Michael Faraday—called **electric field lines**—are related to the electric field in any region of space in the following manner:

1. The electric field vector **E** is tangent to the electric field lines at each point.
2. The number of lines per unit area through a surface perpendicular to the lines is proportional to the strength of the electric field in a given region.

Thus, **E** is large when the field lines are close together and small when they are far apart.

Figure 15.13a shows some representative electric field lines for a single positive point charge. Note that this two-dimensional drawing contains only the field lines that lie in the plane containing the point charge. The lines are actually directed radially outward from the charge in *all* directions, somewhat like the quills of an angry porcupine. Because a positive test charge placed in this field would be repelled by the charge $q$, the lines are directed radially away from the positive charge. In a similar way, the electric field lines for a single negative point charge are directed toward the charge (Fig. 15.13b). In either case, the lines are radial and extend all the way to infinity. Note that the lines are closer together as they get near the charge, indicating that the strength of the field is increasing. Equation 15.6 verifies that this should indeed be the case.

The rules for drawing electric field lines for any charge distribution are as follows:

1. The lines for a group of point charges must begin on positive charges and end on negative charges. In the case of an excess of charge, some lines will begin or end infinitely far away.

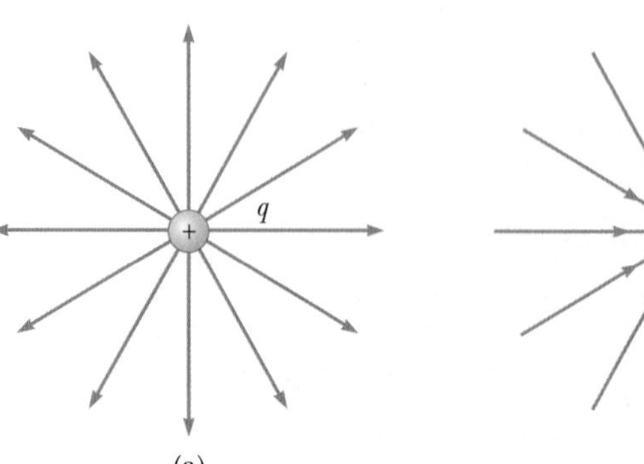


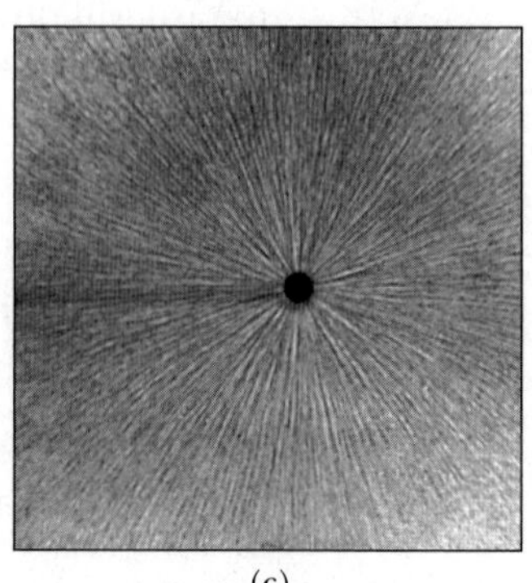


**FIGURE 15.13** The electric field lines for a point charge. (a) For a positive point charge, the lines radiate outward. (b) For a negative point charge, the lines converge inward. Note that the figures show only those field lines that lie in the plane containing the charge. (c) The dark lines are small pieces of thread suspended in oil, which align with the electric field produced by a small charged conductor at the center. *(Photo courtesy of Harold M. Waage, Princeton University)*

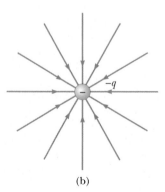

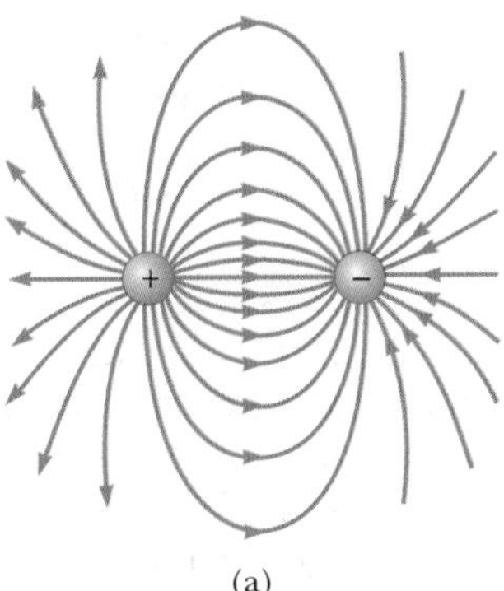


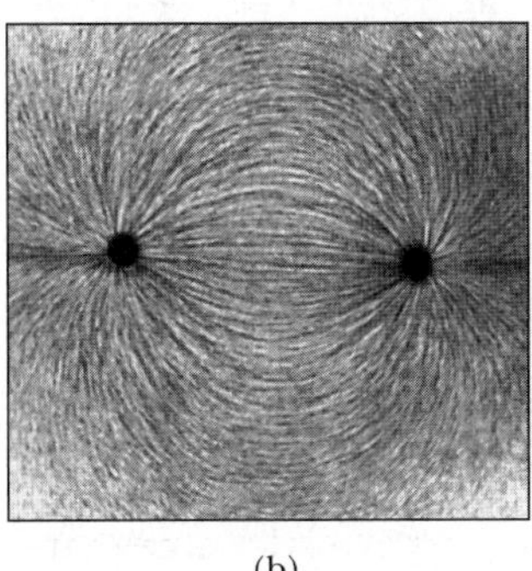

(a) (b)

**FIGURE 15.14** (a) The electric field lines for two equal and opposite point charges (an electric dipole). Note that the number of lines leaving the positive charge equals the number terminating at the negative charge. (b) The dark lines are small pieces of thread suspended in oil, which align with the electric field produced by two charged conductors. *(Photo courtesy of Harold M. Waage, Princeton University)*

2. The number of lines drawn leaving a positive charge or ending on a negative charge is proportional to the magnitude of the charge.
3. No two field lines can cross each other.

Figure 15.14 shows the beautifully symmetric electric field lines for two point charges of equal magnitude but opposite sign. This charge configuration is called an **electric dipole.** In this case, the number of lines that begin at the positive charge must equal the number that terminate at the negative charge. At points very near either charge, the lines are nearly radial. The high density of lines between the charges indicates a strong electric field in this region.

Figure 15.15 shows the electric field lines in the vicinity of two equal positive point charges. Again, close to either charge the lines are nearly radial. The same number of lines emerges from each charge because the charges are equal in magnitude. At great distances from the charges, the field is approximately equal to that of a single point charge of magnitude $2q$. The bulging out of the electric field lines between the charges indicates the repulsive nature of the electric force between like charges. Also, the low density of field lines between the charges indicates a weak field in this region, unlike the dipole.

Finally, Figure 15.16 is a sketch of the electric field lines associated with the positive charge $+2q$ and the negative charge $-q$. In this case, the number of lines leaving charge $+2q$ is twice the number terminating on charge $-q$. Hence, only

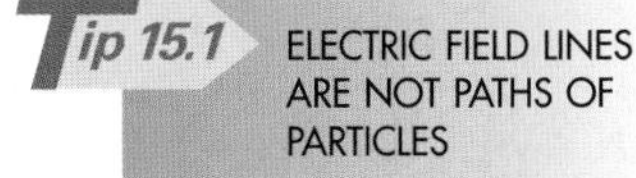

Tip 15.1
ELECTRIC FIELD LINES ARE NOT PATHS OF PARTICLES


Electric field lines are *not* material objects. They are used only as a pictorial representation of the electric field at various locations. Except in special cases, they *do not* represent the path of a charged particle released in an electric field.

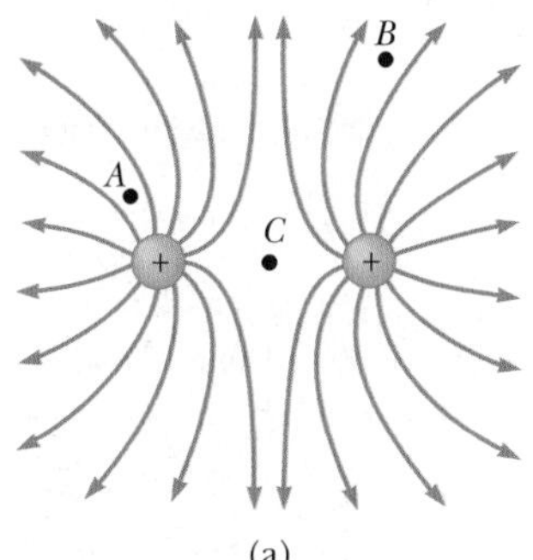


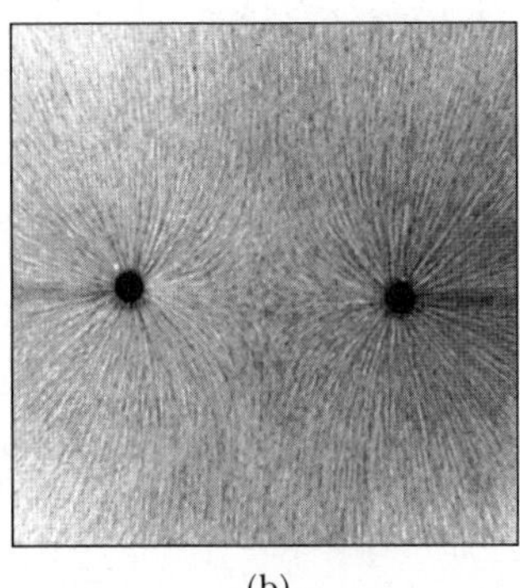

(a) (b)

**FIGURE 15.15** (a) The electric field lines for two positive point charges. The points *A*, *B*, and *C* will be discussed in Quick Quiz 15.7. (b) The dark lines are small pieces of thread suspended in oil, which align with the electric field produced by two charged conductors. *(Photo courtesy of Harold M. Waage, Princeton University)*

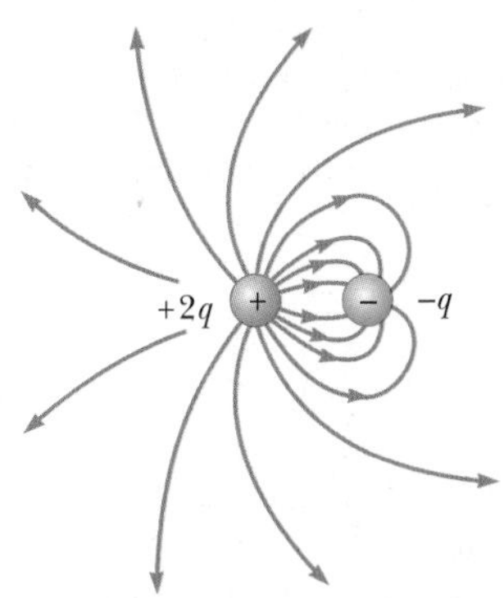


**FIGURE 15.16** The electric field lines for a point charge $+2q$ and a second point charge $-q$. Note that two lines leave the charge $+2q$ for every line that terminates on $-q$.

half of the lines that leave the positive charge end at the negative charge. The remaining half terminate on negative charges that we assume to be located at infinity. At great distances from the charges (great compared with the charge separation), the electric field lines are equivalent to those of a single charge $+q$.

**Quick Quiz 15.7** Rank the magnitudes of the electric field at points $A$, $B$, and $C$ in Figure 15.15, largest magnitude first.

## APPLYING PHYSICS 15.3

**APPLICATION**
MEASURING ATMOSPHERIC ELECTRIC FIELDS

The electric field near Earth's surface in fair weather is about 100 N/C downward. Under a thundercloud, the electric field can be very large, on the order of 20 000 N/C. How are these electric fields measured?

**Explanation** A device for measuring these fields is called the *field mill.* Figure 15.17 shows the fundamental components of a field mill—two metal plates parallel to the ground. Each plate is connected to ground with a wire, with an ammeter (a low-resistance device for measuring flow of charge, to be discussed in Section 19.6) in one path. Consider first just the lower plate. Because it is connected to ground, and the ground happens to carry a negative charge, the plate is negatively charged. Thus, electric field lines, directed downward, end on the plate, as in Figure 15.17a. Now, imagine that the upper plate is suddenly moved over the lower plate, as in Figure 15.17b. This plate is also connected to ground, and is also negatively charged, so the field lines now end on the upper plate. The negative charges in the lower plate are repelled by those on the upper plate and must pass through the ammeter, registering a flow of charge. The amount of charge that was on the lower plate is related to the strength of the electric field. Thus, the flow of charge through the ammeter can be calibrated to measure the electric field. The plates are normally designed like the blades of a fan, with the upper plate rotating so that the lower plate is alternately covered and uncovered. As a result, charges flow back and forth continually through the ammeter, and the reading can be related to the electric field strength.

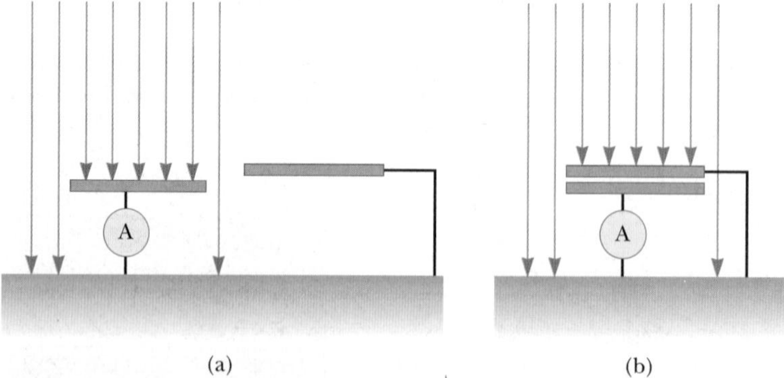


**FIGURE 15.17** (Applying Physics 15.3) In (a), electric field lines end on negative charges on the lower plate. In (b), the second plate is moved above the lower plate. Electric field lines now end on the upper plate, and the negative charges in the lower plate are repelled through the ammeter.

## 15.6 CONDUCTORS IN ELECTROSTATIC EQUILIBRIUM

A good electric conductor, such as copper, contains charges (electrons) that are not bound to any atom and are free to move about within the material. When no net motion of charge occurs within a conductor, the conductor is said to be in

**electrostatic equilibrium.** An isolated conductor (one that is insulated from ground) has the following properties:

Properties of an isolated conductor

1. **The electric field is zero everywhere inside the conducting material.**
2. **Any excess charge on an isolated conductor resides entirely on its surface.**
3. **The electric field just outside a charged conductor is perpendicular to the conductor's surface.**
4. **On an irregularly shaped conductor, the charge accumulates at locations where the radius of curvature of the surface is smallest—that is, at sharp points.**

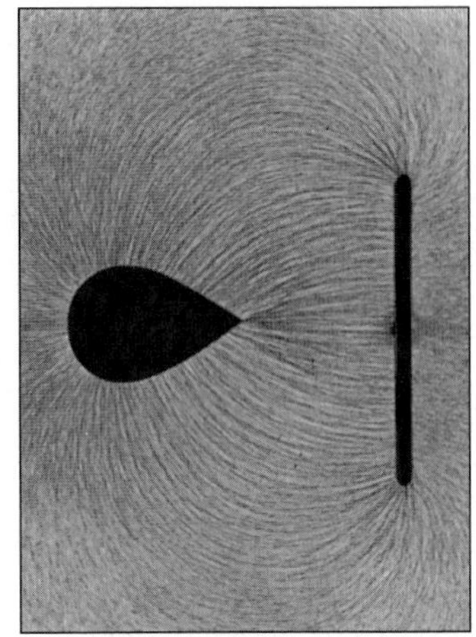

Electric field pattern of a charged conducting plate near an oppositely charged pointed conductor. Small pieces of thread suspended in oil align with the electric field lines. Note that the electric field is most intense near the pointed part of the conductor where the radius of curvature is the smallest. Also, the lines are perpendicular to the conductors. *(Courtesy of Harold M. Waage, Princeton University)*

The first property can be understood by examining what would happen if it were *not* true. If there were an electric field inside a conductor, the free charge there would move and a flow of charge, or current, would be created. However, if there were a net movement of charge, the conductor would no longer be in electrostatic equilibrium.

Property 2 is a direct result of the $1/r^2$ repulsion between like charges described by Coulomb's law. If by some means an excess of charge is placed inside a conductor, the repulsive forces between the like charges push them as far apart as possible, causing them to quickly migrate to the surface. (We do not prove it here, but it is of interest to note that the excess charge resides on the surface due to the fact that Coulomb's law is an inverse-square law. With any other power law, an excess of charge would exist on the surface, but there would be a distribution of charge, of either the same or opposite sign, inside the conductor.)

Property 3 can be understood by again considering what would happen if it were not true. If the electric field in Figure 15.18a were not perpendicular to the surface, the electric field would have a component along the surface, which would cause the free charges of the conductor to move (to the left in the figure). If the charges moved, however, a current would be created and the conductor would no longer be in electrostatic equilibrium. Hence, **E** must be perpendicular to the surface.

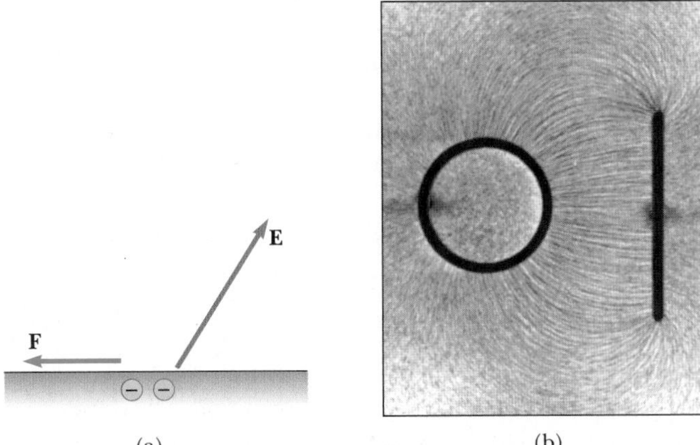


**FIGURE 15.18** (a) Negative charges at the surface of a conductor. If the electric field were at an angle to the surface as shown, an electric force would be exerted on the charges along the surface and they would move to the left. Because the conductor is assumed to be in electrostatic equilibrium, **E** cannot have a component along the surface and hence must be perpendicular to it. (b) The electric field pattern of a charged conducting plate near an oppositely charged conducting cylinder. Small pieces of thread suspended in oil align with the electric field lines. Note that (1) the electric field lines are perpendicular to the conductors and (2) there are no lines inside the cylinder ($\mathbf{E} = 0$). *(Courtesy of Harold M. Waage, Princeton University)*

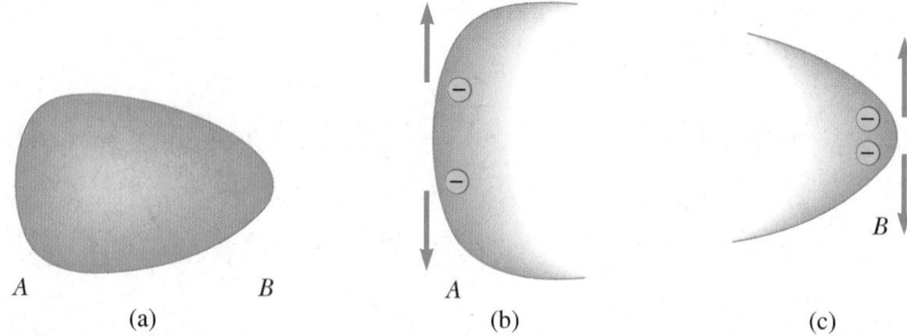


**FIGURE 15.19** (a) A conductor with a flatter end A and a relatively sharp end B. Excess charge placed on a conductor resides entirely at its surface and is distributed so that (b) there is less charge per unit area on the flatter end and (c) there is a large charge per unit area on the sharper end.

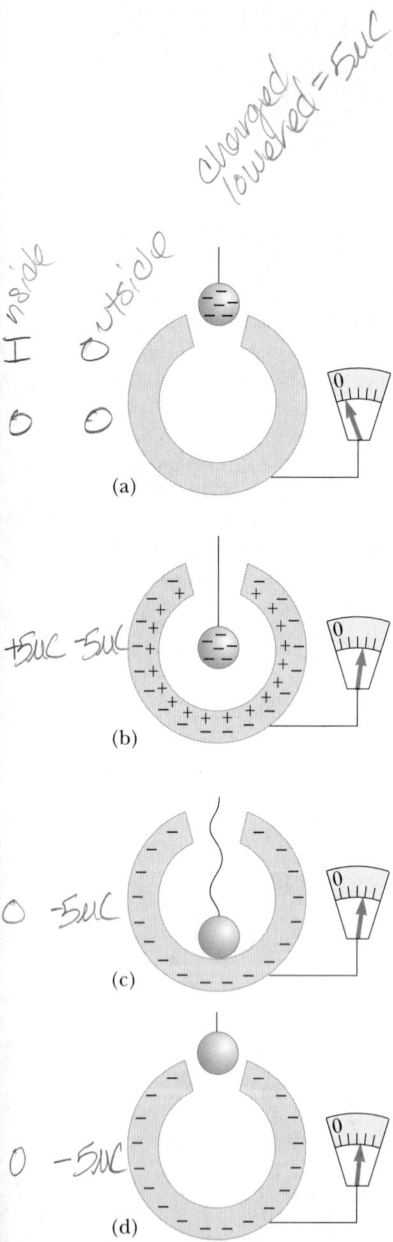


**FIGURE 15.20** An experiment showing that any charge transferred to a conductor resides on its surface in electrostatic equilibrium. The hollow conductor is insulated from ground, and the small metal ball is supported by an insulating thread.

To see why property 4 must be true, consider Figure 15.19a, which shows a conductor that is fairly flat at one end and relatively pointed at the other. Any excess charge placed on the object moves to its surface. Figure 15.19b shows the forces between two such charges at the flatter end of the object. These forces are predominantly directed parallel to the surface. Thus, the charges move apart until repulsive forces from other nearby charges create an equilibrium situation. At the sharp end, however, the forces of repulsion between two charges are directed predominantly away from the surface, as in Figure 15.19c. As a result, there is less tendency for the charges to move apart along the surface here, and the amount of charge per unit area is greater than at the flat end. The cumulative effect of many such outward forces from nearby charges at the sharp end produces a large resultant force directed away from the surface that can be great enough to cause charges to leap from the surface into the surrounding air.

Many experiments have shown that the net charge on a conductor resides on its surface. The experiment described here was first performed by Michael Faraday. A metal ball having a negative charge was lowered at the end of a silk thread (an insulator) into an uncharged hollow conductor insulated from ground, as in Figure 15.20a. (This experiment is referred to as **Faraday's ice-pail experiment** because he used a metal ice pail as the hollow conductor.) As the ball was lowered into the pail, the needle on an electrometer attached to the outer surface of the pail was observed to deflect. (An electrometer is a device used to measure charge.) The needle deflected because the charged ball induced a positive charge on the inner wall of the pail, which left an equal negative charge on the outer wall (Fig. 15.20b).

Faraday next touched the inner surface of the pail with the ball and noted that the needle deflection did not change, either when the ball touched the inner surface of the pail (Fig. 15.20c) or when it was removed (Fig. 15.20d). Furthermore, he found that the ball was now completely uncharged. Apparently, when the ball touched the inside of the pail, the excess negative charge on the ball was neutralized by the induced positive charge on the inner surface of the pail. Thus, Faraday discovered the useful result that *all* the excess charge on an object can be transferred to an already charged metal shell if the object is touched to the *inside* of the shell. As we shall shortly see, this is the principle of operation of the Van de Graff generator.

Faraday concluded that because the electrometer deflection did not change when the charged ball touched the inside of the pail, the positive charge induced on the inside surface of the pail was just enough to neutralize the negative charge on the ball. As a result of his investigations, he concluded that a charged object suspended inside a metal container causes a rearrangement of charge on the container in such a manner that the sign of the charge on the inside surface of the container is *opposite* the sign of the charge on the suspended object. This produces a charge on the outside surface of the container of the same sign as that on the suspended object.

Faraday also found that if the electrometer was connected to the inside surface of the pail after the experiment was run, the needle showed no deflection. Thus, the *excess* charge acquired by the pail when contact was made between ball and pail appeared on the outer surface of the pail.

If a metal rod having sharp points is attached to a house, most of any charge on the house passes through these points, thus eliminating the induced charge on the house produced by storm clouds. In addition, a lightning discharge striking the house passes through the metal rod and is safely carried to the ground through wires leading from the rod to Earth. Lightning rods using this principle were first developed by Benjamin Franklin. It is an interesting sidelight to American history to note that some European countries could not accept the fact that such a worthwhile idea could have originated in the New World. As a result, they "improved" the design by eliminating the sharp points.

**APPLICATION**

LIGHTNING RODS

## APPLYING PHYSICS 15.4

Suppose a point charge $+Q$ is in empty space. Wearing rubber gloves, you proceed to surround the charge with a concentric spherical conducting shell. What effect does this have on the field lines from the charge?

**Explanation** When the spherical shell is placed around the charge, the charges in the shell rearrange to satisfy the rules for a conductor in equilibrium. A net charge of $-Q$ moves to the interior surface of the conductor, so that the electric field inside the conductor becomes zero. That is, the field lines originating on the $+Q$ charge terminate on the $-Q$ charges. The movement of the $-Q$ charges to the inner surface of the sphere leaves a net charge of $+Q$ on the outer surface of the sphere. Thus, the only change in the field lines from the initial situation will be the absence of field lines within the conductor.

## APPLYING PHYSICS 15.5

Why is it safe to stay inside an automobile during a lightning storm?

**Explanation** Although many people believe that this is safe because of the insulating characteristics of the rubber tires, this is not true. Lightning is able to travel through several kilometers of air, so it can certainly penetrate a centimeter of rubber. The interior of the car is safe because charges on the car's metal shell reside on its outer surface, as noted in property 2 discussed earlier. Thus, an occupant in the automobile touching the inner surfaces is not in danger.

**APPLICATION**

DRIVER SAFETY DURING ELECTRICAL STORMS

## 15.7 THE MILLIKAN OIL-DROP EXPERIMENT

From 1909 to 1913, Robert Andrews Millikan (1868–1953) performed a brilliant set of experiments at the University of Chicago in which he measured the elementary charge $e$ of the electron and demonstrated the quantized nature of the electronic charge. The apparatus he used, diagrammed in Figure 15.21, contains two parallel metal plates. Oil droplets that have been charged by friction in an atomizer are allowed to pass through a small hole in the upper plate. A horizontal light beam illuminates the oil droplets, which are viewed by a telescope whose axis is at right angles to the beam. When the droplets are viewed in this manner, they appear as shining stars against a dark background, and the rate of fall of individual drops can be determined.

Let us assume that a single drop having a mass $m$ and carrying a charge $q$ is being viewed, and that its charge is negative. If no electric field is present between the plates, the two forces acting on the charge are the force of gravity $m\mathbf{g}$ acting

### Webnote 15.3

You can read the 1923 Nobel Prize presentation speech and download Millikan's Nobel lecture at *http://nobel.sdsc.edu/physics/laureates/1923/millikan-bio.html*

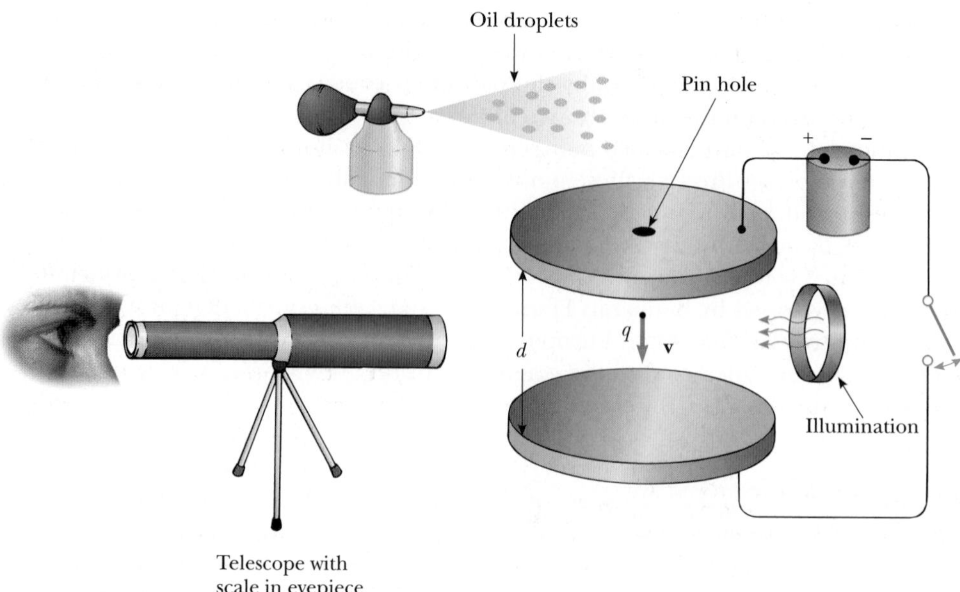


**FIGURE 15.21** A schematic view of the Millikan oil-drop apparatus.

downward and an upward viscous drag force **D** (Fig. 15.22a). The drag force is proportional to the speed of the drop. When the drop reaches its terminal speed $v$, the two forces balance each other ($mg = D$).

Now suppose that an electric field is set up between the plates by a battery connected so that the upper plate is positively charged. In this case, a third force $q\mathbf{E}$ acts on the charged drop. Because $q$ is negative and **E** is downward, the electric force is *upward* as in Figure 15.22b. If this force is great enough, the drop moves upward and the drag force $\mathbf{D}'$ acts downward. When the upward electric force $q\mathbf{E}$ balances the sum of the force of gravity and the drag force, both acting downward, the drop reaches a new terminal speed $v'$.

With the field turned on, a drop moves slowly upward, typically at rates of *hundredths* of a centimeter per second. The rate of fall in the absence of a field is comparable. Hence, a single droplet with constant mass and radius can be followed for hours as it alternately rises and falls, by simply turning the electric field on and off.

After making measurements on thousands of droplets, Millikan and his coworkers found that every drop, to within about 1% precision, had a charge equal to some positive or negative integer multiple of the elementary charge $e$. That is,

$$q = ne \qquad n = 0, \pm 1, \pm 2, \pm 3, \ldots \qquad [15.7]$$

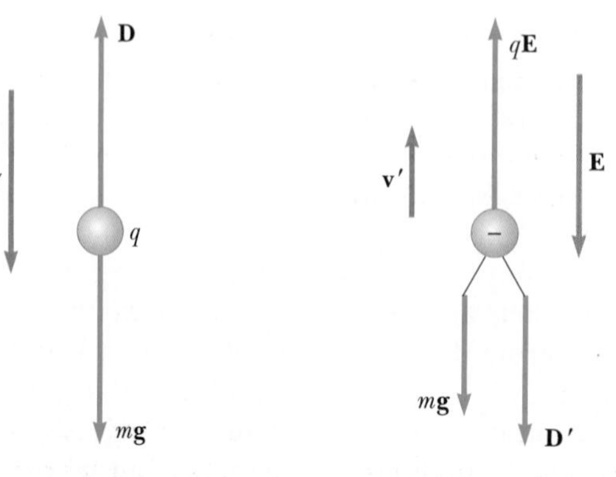


**FIGURE 15.22** The forces on a charged oil droplet in the Millikan experiment.

where $e = 1.60 \times 10^{-19}$ C. It was later established that positive integer multiples of $e$ arose when an oil droplet lost one or more electrons. Likewise, negative integer multiples of $e$ occurred when a drop gained one or more electrons. Gains or losses in integral numbers provide conclusive evidence that charge is quantized. In 1923, Millikan was awarded the Nobel Prize in physics for this work.

**Webnote 15.4**

Interesting historical photos of Van de Graaff generators built for research (including a giant one on railroad tracks) can be found at *http://www.mos.org/sln/toe/history.html*

## 15.8 The Van De Graaff Generator

In 1929 Robert J. Van de Graaff designed and built an electrostatic generator that is used extensively in nuclear physics research. The principles of its operation can be understood with the help of the properties of electric fields and charges already presented in this chapter. Figure 15.23 shows the basic construction details of this device. A motor-driven pulley $P$ moves a belt past positively charged comb-like metallic needles positioned at $A$. Negative charges are attracted to these needles from the belt, leaving the left side of the belt with a net positive charge. The positive charges attract electrons onto the belt as it moves past a second comb of needles at $B$, increasing the excess positive charge on the dome. Because the electric field inside the metal dome is negligible, the positive charge on it can easily be increased regardless of how much charge is already present. The result is that the dome is left with a large amount of positive charge.

This accumulation of charge on the dome cannot continue indefinitely, because eventually an electric discharge through the air takes place. To understand why, consider that, as more and more charge appears on the surface of the dome, the magnitude of the electric field at the surface of the dome is also increasing. Finally, the strength of the field becomes great enough to partially ionize the air near the surface, thus making the air partially conducting. Charges on the dome now have a pathway to leak off into the air, which can produce some spectacular "lightning bolts" as the discharge occurs. As noted earlier, charges find it easier to leap off a surface at points where the curvature is great. As a result, one way to inhibit the electric discharge and to increase the amount of charge that can be stored on the dome is to increase its radius. Another method for inhibiting discharge is to place the entire system in a container filled with a high-pressure gas, which is significantly more difficult to ionize than air at atmospheric pressure.

If protons (or other charged particles) are introduced into a tube attached to the dome, the large electric field outside the dome exerts a repulsive force on the protons, causing them to accelerate to energies high enough to initiate nuclear reactions between the protons and various target nuclei.

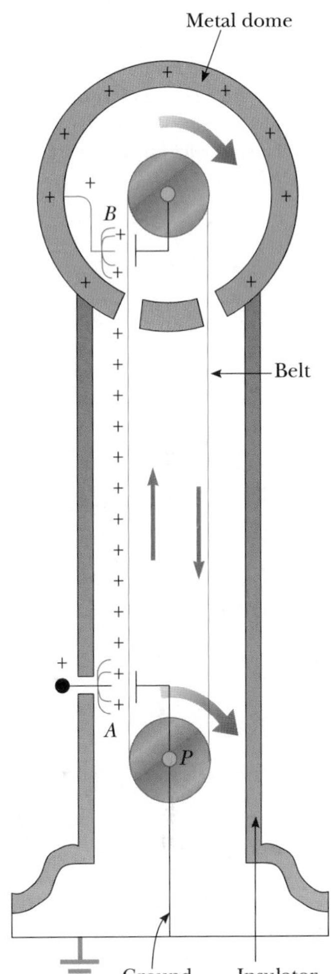


**Figure 15.23** A diagram of a Van de Graaff generator. Charge is transferred to the dome by means of a rotating belt. The charge is deposited on the belt at point $A$ and transferred to the dome at point $B$.

## 15.9 Electric Flux and Gauss's Law

In this section, we introduce the concept of electric flux and then show that it can be used to calculate the electric field of a charged object. This technique for calculating electric fields was developed by Karl Friedrich Gauss (1777–1855). Gauss's law provides a direct connection between the electric flux through a closed surface and the total charge inside that surface producing the electric field. It is basically a statement that the number of electric field lines coming out of a closed surface is proportional to the positive charge inside the surface, a consequence of field lines starting on positive charges.

### Electric Flux

Consider an electric field that is uniform in both magnitude and direction as in Figure 15.24. The electric field lines penetrate a surface of area $A$, which is perpendicular to the field. The technique used for drawing a figure such as Figure 15.24 is that the number of lines per unit area $N/A$ is proportional to the

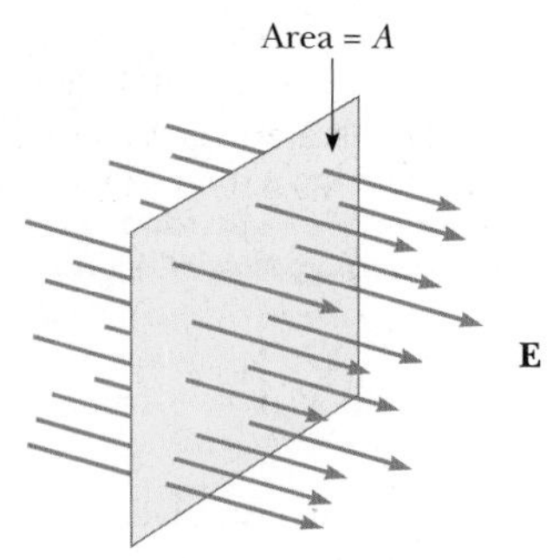


**FIGURE 15.24** Field lines of a uniform electric field penetrating a plane of area $A$ perpendicular to the field. The electric flux $\Phi_E$ through this area is equal to $EA$.

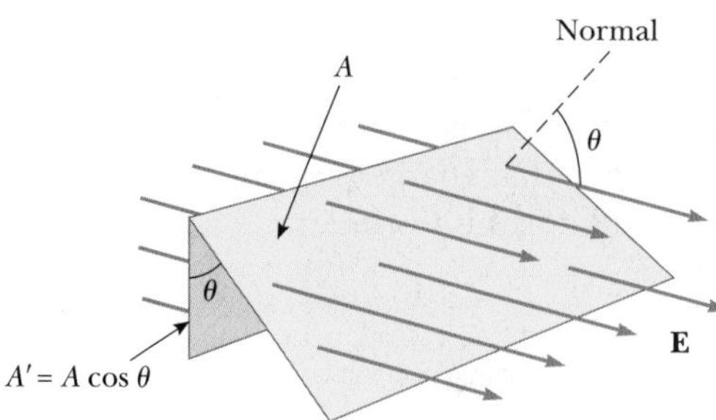


**FIGURE 15.25** Field lines for a uniform electric field through an area $A$ that is at an angle to the field. Because the number of lines that go through the shaded area $A'$ is the same as the number that go through $A$, we conclude that the flux through $A'$ is equal to the flux through $A$ and is given by $\Phi_E = EA \cos \theta$.

magnitude of the electric field, or $E \propto N/A$. We can rewrite this as $N \propto EA$, which means that the number of field lines is proportional to the *product* of $E$ and $A$. This product of $E$ and $A$ is called the **electric flux** and is represented by the symbol $\Phi_E$, where

$$\Phi_E = EA \qquad [15.8]$$

Note that $\Phi_E$ has SI units of $N \cdot m^2/C$ and is essentially the number of field lines that pass through some area $A$ oriented perpendicular to the field. (It is called flux by analogy with the term flux in fluid flow, which is just the volume of liquid flowing through a perpendicular area per second.). If the surface under consideration is not perpendicular to the field as in Figure 15.25, the expression for the electric flux is

Electric flux 

$$\Phi_E = EA \cos \theta \qquad [15.9]$$

where the perpendicular ("normal") to the area $A$ is at an angle $\theta$ with respect to the field. The number of lines that cross this area is equal to the number that cross the projected area $A'$, which is perpendicular to the field. We see that the two areas are related by $A' = A \cos \theta$. From Equation 15.9 we see that the flux through a surface of fixed area has the maximum value $EA$ when the surface is perpendicular to the field (when $\theta = 0°$) and that the flux is zero when the surface is parallel to the field (when $\theta = 90°$). **When the area is constructed such that a closed surface is formed, we shall adopt the convention that flux lines passing into the interior of the volume are negative and those passing out of the interior of the volume are positive.**

## Example 15.6 Flux Through a Cube

Consider a uniform electric field oriented in the $x$ direction. Find the net electric flux through the surface of a cube of edges $L$ oriented as shown in Figure 15.26.

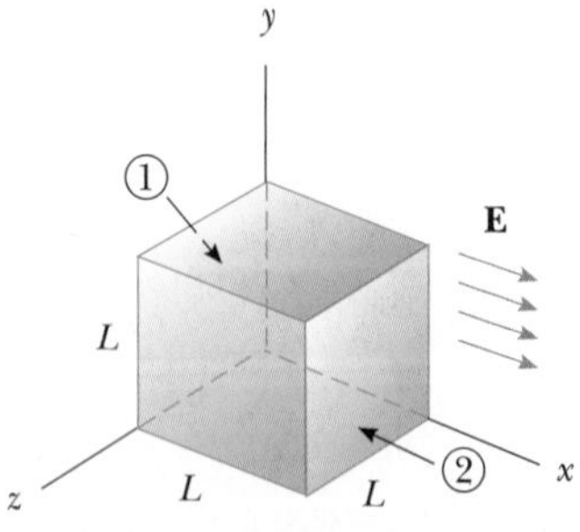


**FIGURE 15.26** (Example 15.6) A hypothetical surface in the shape of a cube in a uniform electric field parallel to the $x$ axis. The net flux through the surface is zero when the net charge inside the cube is zero.

**Solution** The net flux can be evaluated by summing up the fluxes through each face of the cube. First, note that the flux through four of the faces is zero, because **E** is parallel to the surface in these areas. These surfaces are those that are parallel to the $xy$ and $xz$ planes. For these surfaces, $\theta = 90°$, so $\Phi_E = EA \cos 90° = 0$. For surface 1, which lies in the $yz$ plane in Figure 15.26, the flux lines pass into the interior of the cube, and the flux is taken to be negative. We have

$$\Phi_{E1} = -EA = -EL^2$$

For surface 2 at the right face of the cube, the flux is positive and given by

$$\Phi_{E2} = EA = EL^2$$

The net flux through the surface of the cube is

$$\Phi_{E,\text{ net}} = \sum EA = \Phi_{E1} + \Phi_{E2} = -EL^2 + EL^2 = 0$$

## Gauss's Law

Consider a point charge $q$ surrounded by a spherical surface of radius $r$ centered on the charge, as in Figure 15.27. The magnitude of the electric field everywhere on the surface of the sphere is

$$E = k_e \frac{q}{r^2}$$

Note that the electric field is perpendicular to the spherical surface at all points on the surface. Therefore, the electric flux through the surface is $EA$, where $A = 4\pi r^2$ is the surface area of the sphere:

$$\Phi_E = EA = k_e \frac{q}{r^2}(4\pi r^2) = 4\pi k_e q$$

It is convenient to express $k_e$ in terms of another constant $\epsilon_0$, where $k_e = 1/(4\pi\epsilon_0)$. The constant $\epsilon_0$ is called the **permittivity of free space** and has the value

$$\epsilon_0 = \frac{1}{4\pi k_e} = 8.85 \times 10^{-12} \text{ C}^2/\text{N} \cdot \text{m}^2 \qquad [15.10]$$

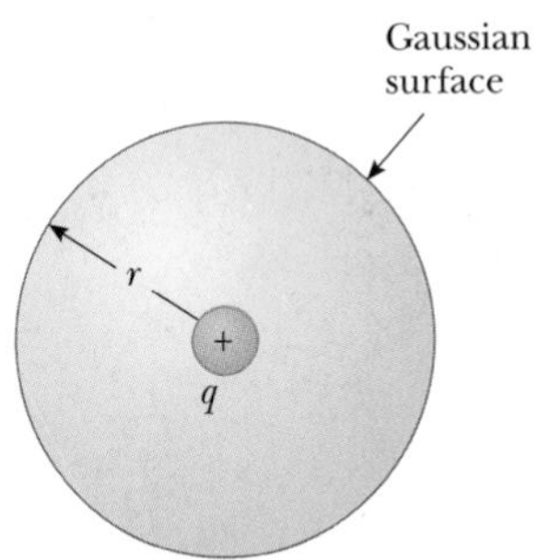


**Figure 15.27** The flux through the spherical surface of radius $r$ surrounding a point charge $q$ is $\Phi_E = q/\epsilon_0$. The flux through any arbitrary surface surrounding the charge is also equal to $q/\epsilon_0$.

The electric flux through the closed spherical surface that surrounds the charge $q$ can now be expressed as

$$\Phi_E = 4\pi k_e q = \frac{q}{\epsilon_0}$$

This result says that the electric flux through a sphere that surrounds a charge $q$ is equal to the charge divided by the constant $\epsilon_0$. Using calculus, we can show that this simple result is true for *any* closed surface that surrounds the charge $q$. For example, if the surface surrounding $q$ is irregular, the flux through that surface is also $q/\epsilon_0$. This result is a special case of Gauss's law and the closed surface is referred to as a *gaussian surface.*

**Tip 15.2 Gaussian Surfaces Are Not Real**

Note that a gaussian surface is an imaginary surface. It does not have to coincide with the surface of a physical object.

In general, **Gauss's law** states that **the electric flux through any closed surface is equal to the net charge $Q$ inside the surface divided by $\epsilon_0$:**

$$\Phi_E = \frac{Q}{\epsilon_0} \qquad [15.11]$$

Gauss's law

It is the fundamental law describing how charges create electric fields. In principle, Gauss's law can always be used to calculate the electric field of a system of charges or a continuous distribution of charge. In practice, the technique is useful only in a limited number of situations in which there is a high degree of symmetry.

### Applying Physics 15.6

A spherical surface surrounds a point charge $q$. Describe what happens to the total flux through the surface if (a) the charge is tripled, (b) the volume of the sphere is doubled, (c) the surface is changed to a cube, (d) the charge is moved to another location inside the surface, and (e) the charge is moved outside the surface.

**Explanation** (a) If the charge is tripled, the flux through the surface is also tripled, because the net flux is proportional to the charge inside the surface. (b) The flux remains constant when the volume changes because the surface surrounds the same amount of charge, regardless of its volume. (c) The flux does not change when the shape of the closed surface changes. (d) The flux through the closed surface remains unchanged as the charge inside the surface is moved to another location inside that surface. (e) The flux is zero because the charge inside the surface is zero. All of these conclusions are arrived at through an understanding of Gauss's law.

**Quick Quiz 15.8** For a surface through which the net flux is zero, the following four statements *could* be true. Which of the statements *must* be true? (a) There are no charges inside the surface. (b) The net charge inside the surface is zero. (c) The electric field is zero everywhere on the surface. (d) The number of electric field lines entering the surface equals the number leaving the surface.

## Example 15.7 The Electric Field of a Charged Thin Spherical Shell

A thin spherical shell of radius $R$ has a total charge $Q$ distributed uniformly over its surface (Fig. 15.28). Find the electric field at points (a) outside and (b) inside the shell.

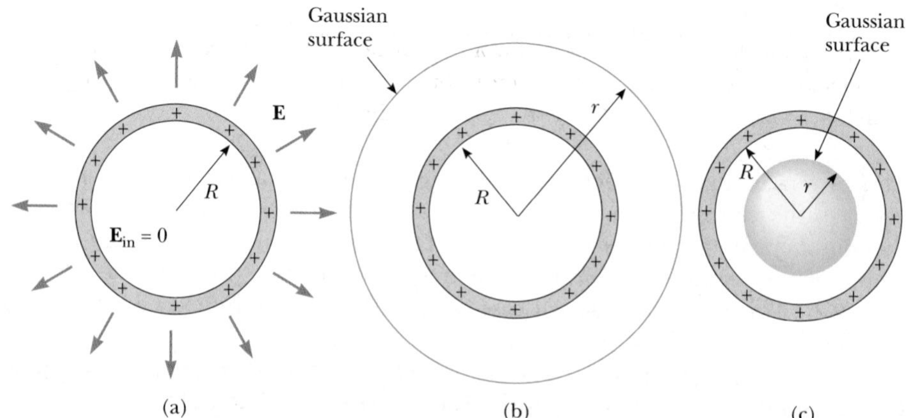


**FIGURE 15.28** (Example 15.7) (a) The electric field inside a uniformly charged spherical shell is *zero*. The field outside is the same as that of a point charge having a total charge $Q$ located at the center of the shell. (b) The construction of a gaussian surface for calculating the electric field *outside* a spherical shell. (c) The construction of a gaussian surface for calculating the electric field *inside* a spherical shell.

### Solution

**A** The calculation of the field outside the shell is identical to that carried out in the body of the text for a point charge. Because the charge distribution is spherically symmetric, we imagine a spherical gaussian surface of radius $r > R$, concentric with the shell, as in Figure 15.28b. Following the same line of reasoning as that for a point charge, the flux through this surface is $\Phi_E = E(4\pi r^2) = Q/\epsilon_0$. Thus, the magnitude of the electric field everywhere on this surface is

$$E = \frac{Q}{4\pi r^2 \epsilon_0} = k_e \frac{Q}{r^2}$$

Note that the field at a point outside the shell is equivalent to that of a point charge located at the center of the shell.

**B** The electric field inside the spherical shell is *zero*. This follows from Gauss's law applied to a spherical gaussian surface of radius $r < R$, placed inside the shell, as shown in Figure 15.28c. Because the net charge inside this surface is zero, $Q = 0$, and we see that $\mathbf{E}_{in} = 0$ in the region inside the shell.

## Example 15.8 A Nonconducting Plane Sheet of Charge

Find the electric field due to a nonconducting infinite plane sheet of charge with uniform charge per unit area $\sigma$.

**Solution** The symmetry of the situation shows that the electric field must be perpendicular to the plane and that the direction of the field on one side of the plane must be opposite its direction on the other side, as shown in Figure 15.29. It is convenient to choose for our gaussian surface a small cylinder whose axis is perpendicular to the plane and whose ends each have an area $A$. Note that no electric field lines pass through the curved surface of the cylinder. Field lines pass only through the two ends of the cylinder. Hence, the total flux through the cylinder is $\Phi_E = E(2A)$. Because the total charge *inside* the gaussian surface is $Q = \sigma A$, Gauss's law gives

$$\Phi_E = 2EA = \frac{\sigma A}{\epsilon_0}$$

or

$$E = \frac{\sigma}{2\epsilon_0}$$

Because the distance of the surfaces from the plane does not appear in our result, we conclude that $E = \sigma/2\epsilon_0$ at *any* distance from the infinite plane. That is, the electric field is uniform everywhere, depending only on the charge per unit area $\sigma$.

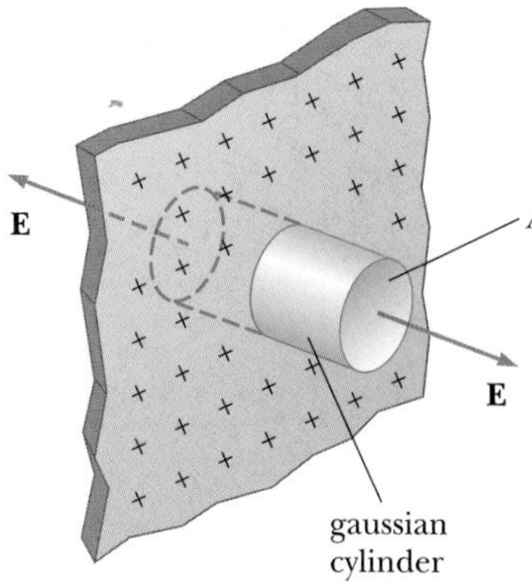


**FIGURE 15.29** (Example 15.8) A cylindrical gaussian surface penetrating an infinite sheet of charge. The flux through each end of the gaussian surface is $EA$. There is no flux through the cylinder's curved surface.

## SUMMARY

**Electric charges** have the following important properties:

1. Unlike charges attract one another and like charges repel one another.
2. Electric charge is always conserved.
3. Charge is quantized—that is, it exists in discrete packets that are integral multiples of the electronic charge.
4. The force between charged particles varies as the inverse square of their separation.

**Conductors** are materials in which charges move freely. **Insulators** are materials that do not readily transport charge.

**Coulomb's law** states that the electric force between two stationary charged particles separated by a distance $r$ has the magnitude

$$F = k_e \frac{|q_1||q_2|}{r^2} \qquad [15.1]$$

where $|q_1|$ and $|q_2|$ are the magnitudes of the charges on the particles in coulombs and $k_e$ is the **Coulomb constant,** which has the approximate value

$$k_e \approx 8.99 \times 10^9 \text{ N} \cdot \text{m}^2/\text{C}^2 \qquad [15.3]$$

An electric field **E** exists at some point in space if a small positive test charge $q_0$ placed at that point experiences an electric force **F**. The electric field is defined as

$$\mathbf{E} \equiv \frac{\mathbf{F}}{q_0} \qquad [15.4]$$

The **direction** of the electric field at a point in space is defined to be the direction of the electric force that would be exerted on a small positive charge placed at that point.

The magnitude of the electric field due to a point charge $q$ at distance $r$ from the point charge is

$$E = k_e \frac{|q|}{r^2} \qquad [15.6]$$

**Electric field lines** are useful for describing the electric field in any region of space. The electric field vector **E** is tangent to the electric field lines at every point. Furthermore, the number of electric field lines per unit area through a surface perpendicular to the lines is proportional to the strength of the electric field in that region.

A **conductor in electrostatic equilibrium** has the following properties:

1. The electric field is zero everywhere inside the conducting material.
2. Any excess charge on an isolated conductor must reside entirely on its surface.
3. The electric field just outside a charged conductor is perpendicular to the conductor's surface.
4. On an irregularly shaped conductor, charge accumulates where the radius of curvature of the surface is smallest—that is, at sharp points.

**Gauss's law** states that the electric flux through any closed surface is equal to the net charge $Q$ inside the surface divided by the permittivity of free space, $\epsilon_0$:

$$\Phi_E = \frac{Q}{\epsilon_0} \qquad [15.11]$$

## CONCEPTUAL QUESTIONS

1. Because of a higher moisture content, air is a better conductor of electricity in the summer than in the winter. Would you expect the shocks from static electricity to be more severe in summer or winter? Explain.
2. Hospital personnel must wear special conducting shoes when they work around oxygen in an operating room. Why? Contrast with what might happen when personnel wear rubber-soled shoes.
3. Two insulated rods are oppositely charged on the ends. They are mounted at the centers so that they are free to rotate, and then held in the position shown in a view from above (Fig. Q15.3). The rods rotate in the plane of the paper. Will the rods stay in the positions when released? If not, into what position(s) will they move? Will the final configuration(s) be stable?

**FIGURE Q15.3**

4. Explain from an atomic viewpoint why charge is usually transferred by electrons.
5. Explain how a positively charged object can leave another metallic object with a net negative charge. Discuss the motion of charges during this process.
6. Why are electrostatic effects often seen on clothes just out of a dryer?
7. If a metal object receives a positive charge, does its mass appreciably increase, decrease, or stay the same? What happens to its mass if the object receives a negative charge?
8. When defining the electric field, why is it necessary to specify that the magnitude of the test charge be very small?
9. In fair weather, there is an electric field at Earth's surface, pointing down into the ground. What is the electric charge on the ground in this situation?
10. A student stands on a thick piece of insulating material, places her hand on top of a Van de Graaff generator, and then turns on the generator. Does she receive a shock?
11. An uncharged, metal-coated Styrofoam ball is suspended in the region between two vertical metal plates. If the two plates are charged, one positive and one negative, describe the motion of the ball after it is brought into contact with one of the plates.
12. Is it possible for an electric field to exist in empty space? Explain.
13. There are great similarities between electric and gravitational fields. A room can be electrically shielded so that there are no electric fields in the room by surrounding it with a conductor. Can a room be gravitationally shielded? Why or why not? (*Hint:* There are two kinds of charge in nature, but only one kind of mass.)
14. Would life be different if the electron were positively charged and the proton were negatively charged? Does the choice of signs have any bearing on physical and chemical interactions? Explain.
15. Explain why Gauss's law cannot be used to calculate the electric field of (a) a polar molecule consisting of a positive and a negative charge separated by a very small distance, (b) a charged disk, and (c) three point charges at the corner of a triangle.
16. Why should a ground wire be connected to the metal support rod for a television antenna?
17. A balloon is negatively charged by rubbing and then clings to a wall. Does this mean that the wall is positively charged? Why does the balloon eventually fall?
18. Why is it not a good idea to seek shelter under a tree during a lightning storm?
19. A charged comb often attracts small bits of dry paper that then fly away when they touch the comb. Explain.

## PROBLEMS

1, 2, 3 = straightforward, intermediate, challenging ☐ = full solution available in Student Solutions Manual/Study Guide

**web** = solution posted at **http://info.brookscole.com/serway** = biomedical application

### Section 15.3 Coulomb's Law

**1.** A $4.5 \times 10^{-9}$ C charge is located 3.2 m from a $-2.8 \times 10^{-9}$ C charge. Find the electrostatic force exerted by one charge on the other.

**2.** Two neighboring cells are ionized (charged) by x-radiation. (a) If the charge on each cell is equal to $1.60 \times 10^{-14}$ C and the cells are 2.5 $\mu$m apart, what is the force exerted by each cell? (b) If the distance between the cells is doubled, what happens to the force between them?

**3.** An alpha particle (charge $= +2.0e$) is sent at high speed toward a gold nucleus (charge $= +79e$). What is the electrical force acting on the alpha particle when it is $2.0 \times 10^{-14}$ m from the gold nucleus?

**4.** Determine what the mass of a proton would be if the gravitational force between two of them were equal to the electrical force between them.

**5.** The nucleus of $^8$Be, which consists of 4 protons and 4 neutrons, is very unstable and spontaneously breaks into two alpha particles (helium nuclei, each consisting of 2 protons and 2 neutrons). (a) What is the force between the two alpha particles when they are $5.00 \times 10^{-15}$ m apart, and (b) what will be the magnitude of the acceleration of the alpha particles due to this force? Note that the mass of an alpha particle is 4.0026 u.

**6.** A molecule of DNA (deoxyribonucleic acid) is 2.17 $\mu$m long. The ends of the molecule become singly ionized—negative on one end, positive on the other. The helical molecule acts like a spring and compresses 1.00% upon becoming charged. Determine the effective spring constant of the molecule.

**7.** Suppose that 1.00 g of hydrogen is separated into electrons and protons. Suppose also that the protons are placed at Earth's North Pole and the electrons are placed at the South Pole. What is the resulting compressional force on Earth?

**8.** An electron is released a short distance above Earth's surface. A second electron directly below it exerts an electrostatic force on the first electron just great enough to cancel the gravitational force on it. How far below the first electron is the second?

**9.** Two identical conducting spheres are placed with their centers 0.30 m apart. One is given a charge of $12 \times 10^{-9}$ C and the other a charge of $-18 \times 10^{-9}$ C. (a) Find the electrostatic force exerted by one sphere on the other. (b) The spheres are connected by a conducting wire. After equilibrium has occurred, find the electrostatic force between the two.

**10.** Calculate the magnitude and direction of the Coulomb force on each of the three charges in Figure P15.10.

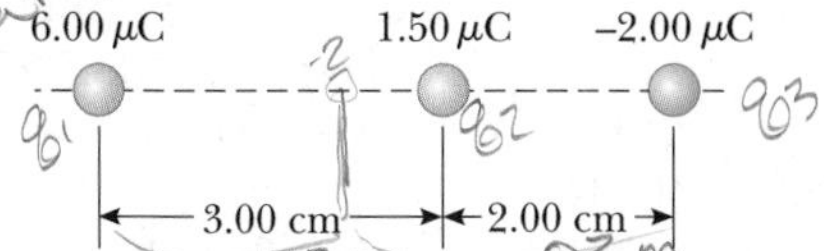

6.00 $\mu$C 1.50 $\mu$C −2.00 $\mu$C

3.00 cm 2.00 cm

**FIGURE P15.10** (Problems 10 and 18)

**11.** Three charges are arranged as shown in Figure P15.11. Find the magnitude and direction of the electrostatic force on the charge at the origin.

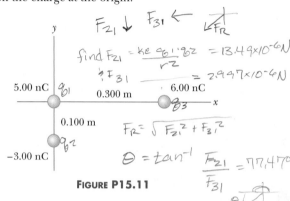

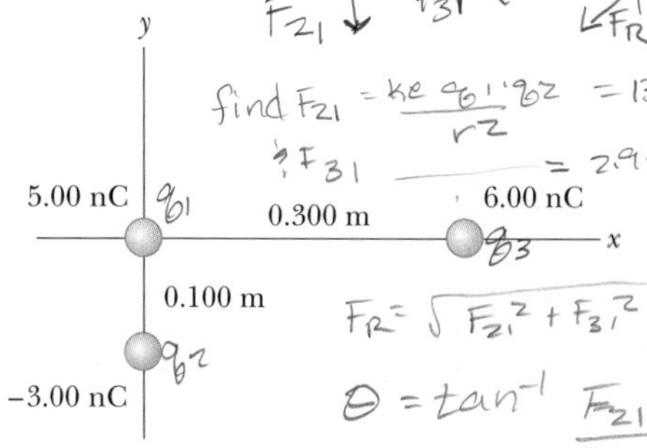

y; 5.00 nC; 0.300 m; 6.00 nC; x; 0.100 m; −3.00 nC

**FIGURE P15.11**

**12.** Three charges are arranged as shown in Figure P15.12. Find the magnitude and direction of the electrostatic force on the 6.00-nC charge.

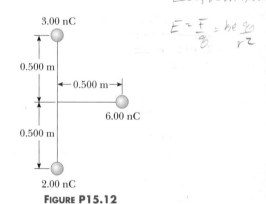

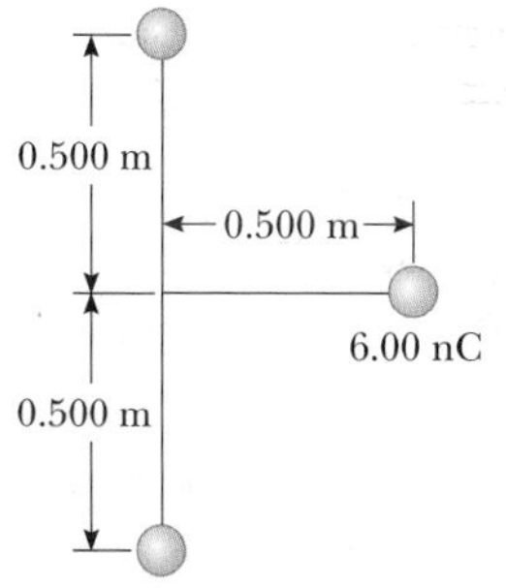

3.00 nC; 0.500 m; 0.500 m; 6.00 nC; 0.500 m; 2.00 nC

**FIGURE P15.12**

**13.** Three point charges are located at the corners of an equilateral triangle as in Figure P15.13. Calculate the net electric force on the 7.00-$\mu$C charge.

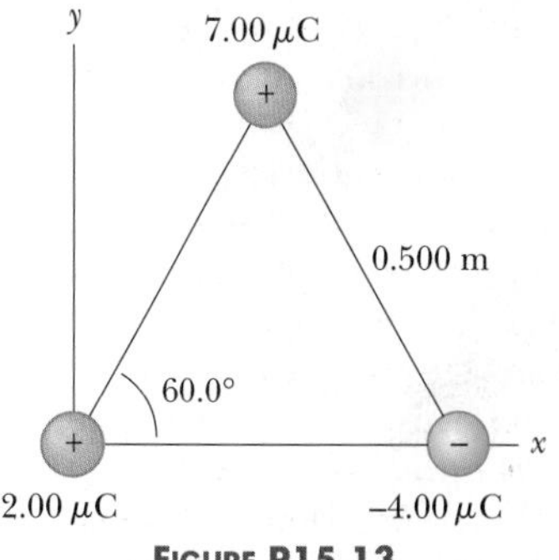


**FIGURE P15.13**

**14.** Two small beads having positive charges $3q$ and $q$ are fixed at the opposite ends of a horizontal, insulating rod, extending from the origin to the point $x = d$. As in Figure

P15.14, a third small charged bead is free to slide on the rod. At what position is the third bead in equilibrium? Can it be in stable equilibrium?

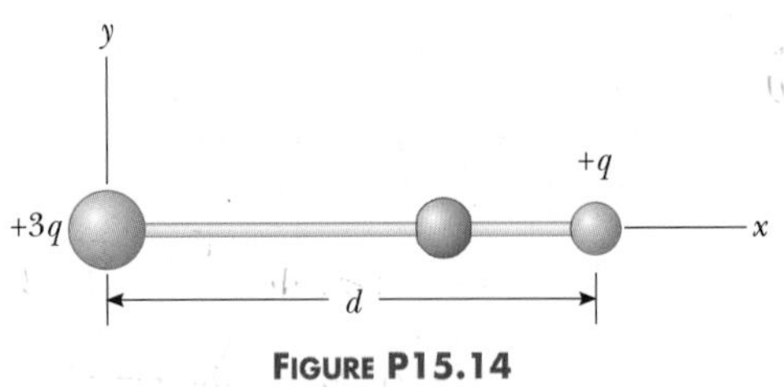


**FIGURE P15.14**

15. Two small metallic spheres, each of mass 0.20 g, are suspended as pendulums by light strings from a common point as shown in Figure P15.15. The spheres are given the same electric charge, and it is found that the two come to equilibrium when each string is at an angle of 5.0° with the vertical. If each string is 30.0 cm long, what is the magnitude of the charge on each sphere?

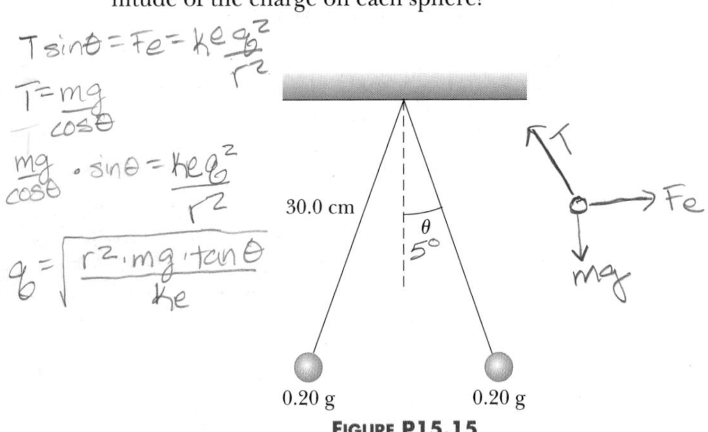


**FIGURE P15.15**

16. A charge of $6.00 \times 10^{-9}$ C and a charge of $-3.00 \times 10^{-9}$ C are separated by a distance of 60.0 cm. Find the position at which a third charge of $12.0 \times 10^{-9}$ C can be placed so that the net electrostatic force on it is zero.

## Section 15.4 The Electric Field

17. In a hydrogen atom, what are the magnitude and direction of the electric field set up by the proton at the location of the electron ($0.51 \times 10^{-10}$ m away from the proton)?

18. (a) Determine the electric field strength at a point 1.00 cm to the left of the middle charge shown in Figure P15.10. (b) If a charge of $-2.00$ $\mu$C is placed at this point, what are the magnitude and direction of the force on it?

19. An airplane is flying through a thundercloud at a height of 2 000 m. (This is a very dangerous thing to do because of updrafts, turbulence, and the possibility of electric discharge.) If there are charge concentrations of $+40.0$ C at height 3 000 m within the cloud and $-40.0$ C at height 1 000 m, what is the electric field **E** at the aircraft?

20. An electron is accelerated by a constant electric field of magnitude 300 N/C. (a) Find the acceleration of the electron. (b) Use the equations of motion with constant acceleration to find the electron's speed after $1.00 \times 10^{-8}$ s, assuming it starts from rest.

21. A piece of aluminum foil of mass $5.00 \times 10^{-2}$ kg is suspended by a string in an electric field directed vertically upward. If the charge on the foil is 3.00 $\mu$C, find the strength of the field that will reduce the tension in the string to zero.

22. An electron with a speed of $3.00 \times 10^{6}$ m/s moves into a uniform electric field of 1 000 N/C. The field is parallel to the electron's motion. How far does the electron travel before it is brought to rest?

23. A proton accelerates from rest in a uniform electric field of 640 N/C. At some later time, its speed is $1.20 \times 10^{6}$ m/s. (a) Find the magnitude of the acceleration of the proton. (b) How long does it take the proton to reach this speed? (c) How far has it moved in this interval? (d) What is its kinetic energy at the later time?

24. Positive charges are situated at three corners of a rectangle, as shown in Figure P15.24. Find the electric field at the fourth corner.

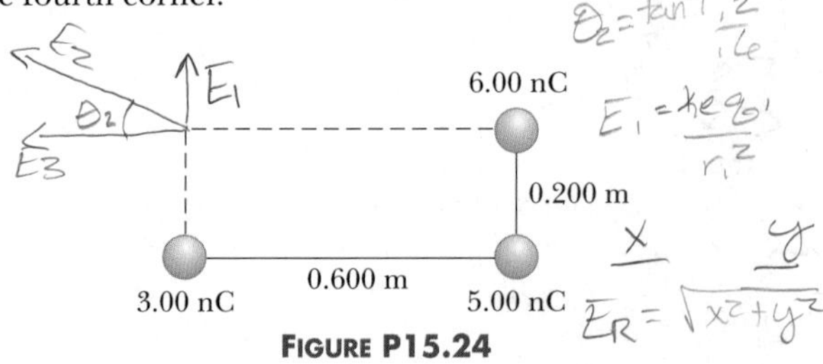


**FIGURE P15.24**

25. Three identical charges ($q = -5.0$ $\mu$C) are along a circle of 2.0-m radius at angles of 30°, 150°, and 270°, as shown in Figure P15.25. What is the resultant electric field at the center of the circle?

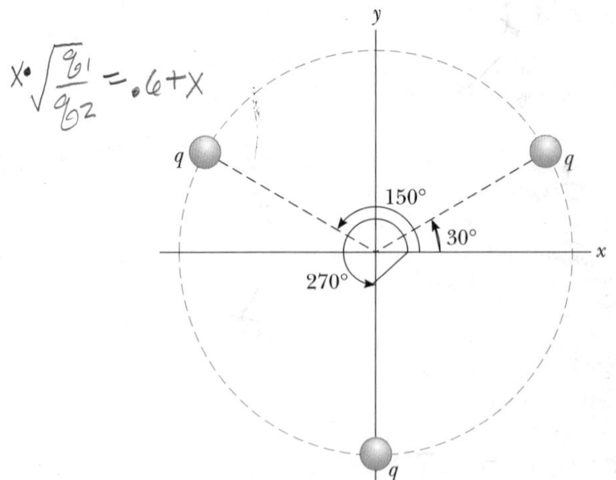


**FIGURE P15.25**

26. Two point charges lie along the $y$ axis. A charge of $q_1 = -9.0$ $\mu$C is at $y = 6.0$ m, and a charge of $q_2 = -8.0$ $\mu$C is at $y = -4.0$ m. Locate the point (other than infinity) at which the total electric field is zero.

27. web In Figure P15.27, determine the point (other than infinity) at which the total electric field is zero.

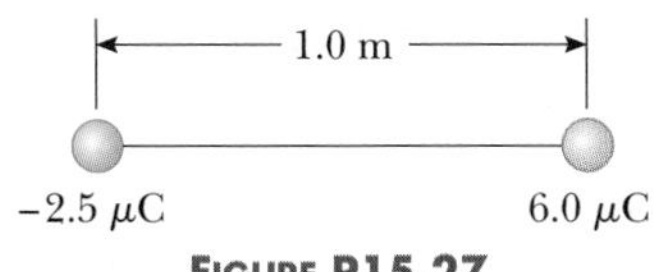


**FIGURE P15.27**

## Section 15.5 Electric Field Lines

## Section 15.6 Conductors in Electrostatic Equilibrium

**28.** Figure P15.28 shows the electric field lines for two point charges separated by a small distance. (a) Determine the ratio $q_1/q_2$. (b) What are the signs of $q_1$ and $q_2$?

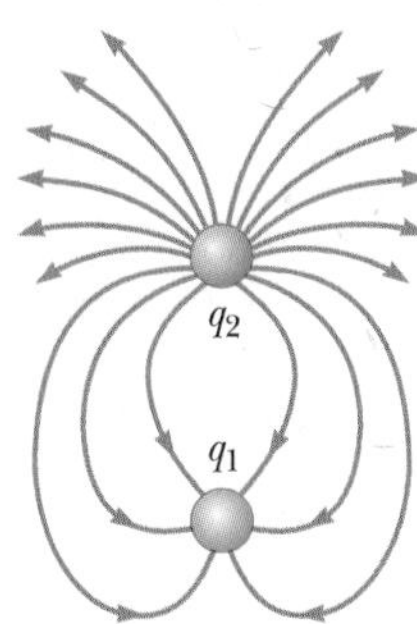


**FIGURE P15.28**

**29.** (a) Sketch the electric field lines around an isolated point charge, $q > 0$. (b) Sketch the electric field pattern around an isolated negative point charge of magnitude $-2q$.

**30.** (a) Sketch the electric field pattern around two positive point charges of magnitude 1 $\mu$C placed close together. (b) Sketch the electric field pattern around two negative point charges of $-2$ $\mu$C placed close together. (c) Sketch the pattern around two point charges of $+1$ $\mu$C and $-2$ $\mu$C placed close together.

**31.** Two point charges are a small distance apart. (a) Sketch the electric field lines for the two if one has a charge four times that of the other and both charges are positive. (b) Repeat for the case in which both charges are negative.

**32.** (a) Sketch the electric field pattern set up by a positively charged hollow sphere. Include the regions both inside and outside the sphere. (b) A conducting cube is given a positive charge. Sketch the electric field pattern both inside and outside the cube.

**33.** Refer to Figure 15.20. The charge lowered into the center of the hollow conductor has a magnitude of 5 $\mu$C. Find the magnitude and sign of the charge on the inside and outside of the hollow conductor when the charge is as shown in (a) Figure 15.20a, (b) Figure 15.20b, (c) Figure 15.20c, and (d) Figure 15.20d.

## Section 15.8 The Van de Graaff Generator

**34.** The dome of a Van de Graaff generator receives a charge of $2.0 \times 10^{-4}$ C. Find the strength of the electric field (a) inside the dome; (b) at the surface of the dome, assuming it has a radius of 1.0 m; and (c) 4.0 m from the center of the dome. (*Hint:* See Section 15.6 to review properties of conductors in electrostatic equilibrium. Also use the fact that the points on the surface are outside a spherically symmetric charge distribution; the total charge may be considered as located at the center of the sphere.)

**35.** If the electric field strength in air exceeds $3.0 \times 10^6$ N/C, the air becomes a conductor. Using this fact, determine the maximum amount of charge that can be carried by a metal sphere 2.0 m in radius. (See the hint in Problem 34.)

**36.** Air breaks down (loses its insulating quality) and sparking results if the field strength is increased to about $3.0 \times 10^6$ N/C. (a) What acceleration does an electron experience in such a field? (b) If the electron starts from rest, in what distance does it acquire a speed equal to 10% of the speed of light?

**37.** A Van de Graaff generator is charged so that the electric field at its surface is $3.0 \times 10^4$ N/C. Find (a) the electric force exerted on a proton released at its surface and (b) the acceleration of the proton at this instant of time.

## Section 15.9 Electric Flux and Gauss's Law

**38.** A flat surface having an area of 3.2 m$^2$ is rotated in a uniform electric field of magnitude $E = 6.2 \times 10^5$ N/C. Determine the electric flux through this area (a) when the electric field is perpendicular to the surface and (b) when the electric field is parallel to the surface.

**39.** An electric field of intensity 3.50 kN/C is applied along the $x$ axis. Calculate the electric flux through a rectangular plane 0.350 m wide and 0.700 m long if (a) the plane is parallel to the $yz$ plane; (b) the plane is parallel to the $xy$ plane; and (c) the plane contains the $y$ axis and its normal makes an angle of 40.0° with the $x$ axis.

**40.** The electric field everywhere on the surface of a thin spherical shell of radius 0.750 m is measured to be equal to 890 N/C and points radially toward the center of the sphere. (a) What is the net charge within the sphere's surface? (b) What can you conclude about the nature and distribution of the charge inside the spherical shell?

**41.** A 40-cm-diameter loop is rotated in a uniform electric field until the position of maximum electric flux is found. The flux in this position is measured to be $5.2 \times 10^5$ N·m$^2$/C. Calculate the electric field strength in this region.

**42.** A point charge of $+5.00$ $\mu$C is located at the center of a sphere with a radius of 12.0 cm. Determine the electric flux through the surface of the sphere.

**43.** A point charge $q$ is located at the center of a spherical shell of radius $a$, which has a charge $-q$ uniformly distributed on its surface. Find the electric field (a) for all points outside the spherical shell and (b) for a point inside the shell a distance $r$ from the center.

**44.** Use Gauss's law and the fact that the electric field inside any closed conductor in electrostatic equilibrium is zero to show that any excess charge placed on the conductor must reside on its surface.

**45.** **web** An infinite plane conductor has charge spread out on its surface as shown in Figure P15.45. Use Gauss's law to show that the electric field at any point outside the conductor is given by $E = \sigma/\epsilon_0$, where $\sigma$ is the charge per unit area on the conductor. (*Hint:* Choose a gaussian surface in the shape of a cylinder with one end inside the conductor and one end outside the conductor.)

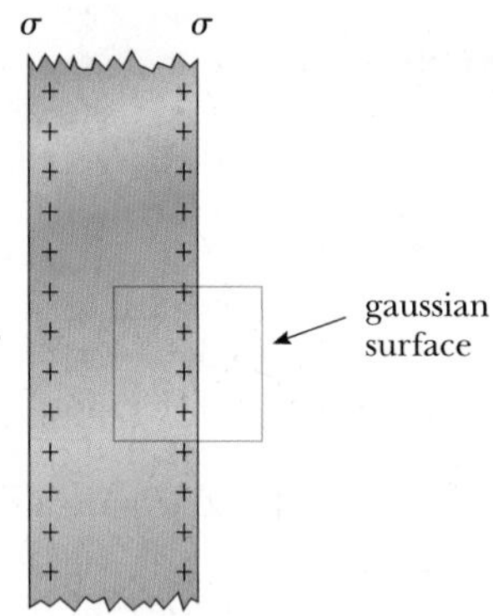


**FIGURE P15.45**

46. Show that the electric field just outside the surface of a good conductor of any shape is given by $E = \sigma/\epsilon_0$, where $\sigma$ is the charge per unit area on the conductor. (*Hint:* The electric field just outside the surface of a charged conductor is perpendicular to its surface.)

## ADDITIONAL PROBLEMS

**47.** Two protons in an atomic nucleus are typically separated by a distance of $2 \times 10^{-15}$ m. The electric repulsion force between the protons is huge, but the attractive nuclear force is even stronger, and keeps the nucleus from bursting apart. What is the magnitude of the electrical force between two protons separated by $2.00 \times 10^{-15}$ m?

48. In the Bohr theory of the hydrogen atom, an electron moves in a circular orbit about a proton, where the radius of the orbit is $0.53 \times 10^{-10}$ m. (a) Find the electrostatic force acting on each particle. (b) If this force causes the centripetal acceleration of the electron, what is the speed of the electron?

**49.** Three point charges are aligned along the $x$ axis as shown in Figure P15.49. Find the electric field at the position $x = +2.0$ m, $y = 0$.

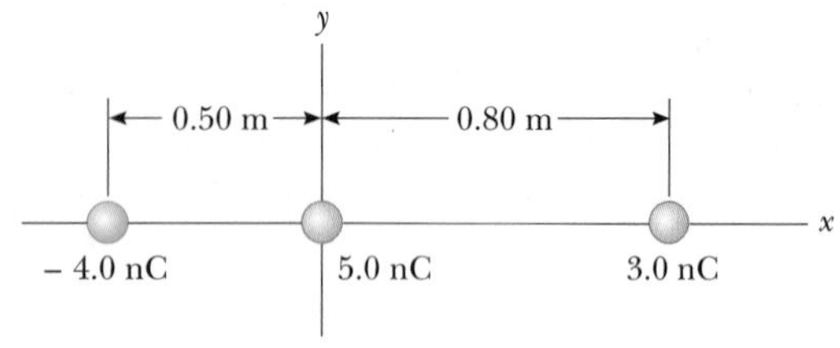


**FIGURE P15.49**

50. A small 2.00-g plastic ball is suspended by a 20.0-cm-long string in a uniform electric field, as shown in Figure P15.50. If the ball is in equilibrium when the string makes a 15.0° angle with the vertical as indicated, what is the net charge on the ball?

51. (a) Two identical point charges $+q$ are located on the $y$ axis at $y = +a$ and $y = -a$. What is the electric field along the $x$ axis at $x = b$? (b) A circular ring of charge of radius $a$ has a total positive charge $Q$ distributed uniformly around it. The ring is in the $x = 0$ plane with its center at the origin. What is the electric field along the $x$ axis at $x = b$ due to the ring of charge? (*Hint:* Consider the charge $Q$ to consist of very many pairs of identical point charges positioned at ends of diameters of the ring.)

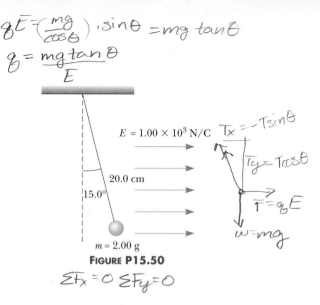

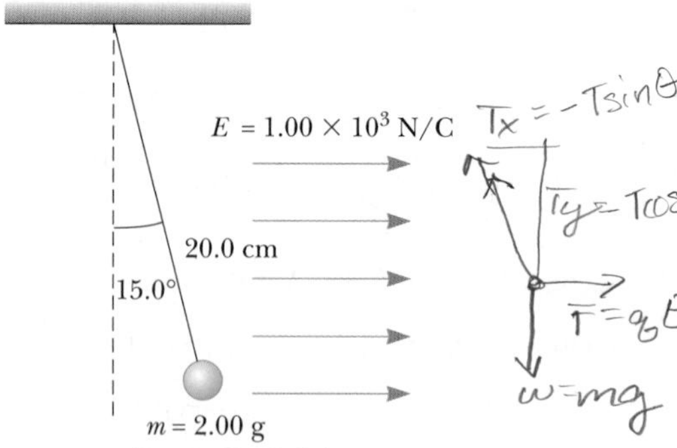

**FIGURE P15.50**

52. A positively charged bead having a mass of 1.00 g falls from rest in a vacuum from a height of 5.00 m in a uniform vertical electric field with a magnitude of $1.00 \times 10^4$ N/C. The bead hits the ground at a speed of 21.0 m/s. Determine (a) the direction of the electric field (up or down), and (b) the charge on the bead.

53. A solid conducting sphere of radius 2.00 cm has a charge of 8.00 $\mu$C. A conducting spherical shell of inner radius 4.00 cm and outer radius 5.00 cm is concentric with the solid sphere and has a charge of $-4.00$ $\mu$C. Find the electric field at (a) $r = 1.00$ cm, (b) $r = 3.00$ cm, (c) $r = 4.50$ cm, and (d) $r = 7.00$ cm from the center of this charge configuration.

54. Two small silver spheres, each with a mass of 100 g, are separated by 1.00 m. Calculate the fraction of the electrons in one sphere that must be transferred to the other in order to produce an attractive force of $1.00 \times 10^4$ N (about a ton) between the spheres. (The number of electrons per atom of silver is 47, and the number of atoms per gram is Avogadro's number divided by the molar mass of silver, 107.87.)

55. A vertical electric field of magnitude $2.00 \times 10^4$ N/C exists above Earth's surface on a day when a thunderstorm is brewing. A car with a rectangular size of 6.00 m by 3.00 m is traveling along a roadway sloping downward at 10.0°. Determine the electric flux through the bottom of the car.

56. A 2.00-$\mu$C charged 1.00-g cork ball is suspended vertically on a 0.500-m-long light string in the presence of a uniform downward-directed electric field of magnitude $E = 1.00 \times 10^5$ N/C. If the ball is displaced slightly from the vertical, it oscillates like a simple pendulum. (a) Determine the period of this oscillation. (b) Should gravity be included in the calculation for part (a)? Explain.

57. Two 2.0-g spheres are suspended by 10.0-cm-long light **web** strings (Fig. P15.57). A uniform electric field is applied in the $x$ direction. If the spheres have charges of $-5.0 \times 10^{-8}$ C and $+5.0 \times 10^{-8}$ C, determine the electric field intensity that enables the spheres to be in equilibrium at $\theta = 10°$.

58. Two point charges like those in Figure P15.58 are called an electric dipole. Show that the electric field at a distant point along the $x$ axis is given by the expression $E_x = 4k_e qa/x^3$.

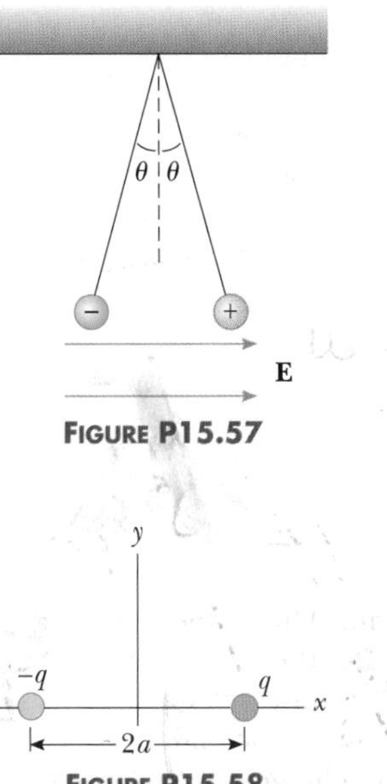


FIGURE P15.57

FIGURE P15.58

59. A charged cork ball of mass 1.00 g is suspended on a light string in the presence of a uniform electric field as in Figure P15.59. When the electric field has an $x$ component of $3.00 \times 10^5$ N/C and a $y$ component of $5.00 \times 10^5$ N/C, the ball is in equilibrium at $\theta = 37.0°$. Find (a) the charge on the ball and (b) the tension in the string.

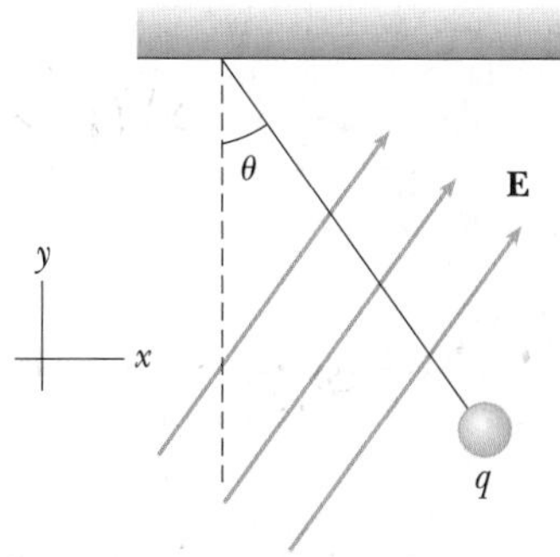


FIGURE P15.59

## GROUP ACTIVITIES

**G.1** Listed below are a number of experiments that you can perform to investigate static electricity.

(a) Attach two inflated balloons to the ends of a light string having a length of about 2 m. Tape the center of the string to the top of an open doorway as in Fig. GA15.1. Note that the balloons touch each other as they hang freely. Now rub each balloon several times with a wool cloth and let them hang freely once again. Why are the balloons no longer touching each other but are now separated?

(b) Rub an inflated balloon with a piece of wool and press it against a wall. Note that the balloon adheres to the wall. Why?

(c) Rub nylon hose with a plastic dry cleaner bag and observe how the hose expands like a balloon. Why?

(d) Tear some paper into very small pieces. Comb your hair and then bring the comb close to the paper pieces. Notice that they are accelerated toward the comb. How does the magnitude of the electric force compare with the magnitude of the gravitational force exerted on the paper? Keep watching and you might see a few pieces jump away from the comb. They do not fall away; they are repelled. What causes this?

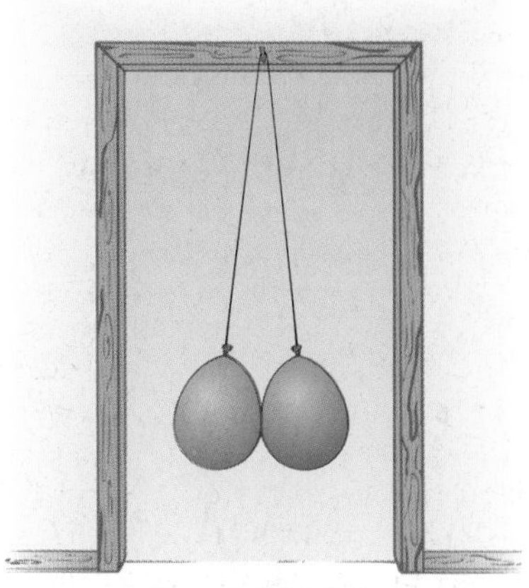

FIGURE GA15.1

**G.2** For this experiment, you will need two 20-cm strips of transparent tape (mass of each is about 65 mg). Fold about 1 cm of tape over at one end of each strip to create a handle. Press both pieces of tape side by side onto a table top, rubbing your finger back and forth across the strips. Quickly pull the strips off the surface so that they become charged. Hold the tape handles together and the strips will repel each other, forming an inverted V shape. Measure the angle between the pieces and estimate the excess charge on each strip. Assume that the charges act as if they were located at the center of mass of each strip.

**G.3** Rub an inflated balloon with a piece of wool cloth and place the balloon near a fine stream of water falling from a faucet. The stream of water deflects toward the balloon. Why? Vary the distance between the balloon and the stream and observe the displacement of the water stream for different distances. What is the relationship between the displacement of the stream and the distance of separation?

**G.4** If you have access to a Van de Graaff generator, here are a few interesting things to try. (a) Stack several aluminum pie plates on top of the generator, then turn the generator on. What do you think is going to happen? Why? Try it. (b) Tape one pie plate on top of the generator and pour in some puffed rice or paper pieces. What do you think will happen when you turn the generator on? Why? Try it and see. (c) Tape a glass beaker to the top of the generator and pour in some puffed rice. What do you think will happen when you turn the generator on? Try it. (d) Bring a fluorescent bulb close to a charged generator and observe what happens. Why?

**G.5** A point charge of magnitude 5.00 $\mu$C is at the origin of a coordinate system while a charge of $-4.00$ $\mu$C is at the point $x = 1.00$ m. There is a point on the $x$ axis at $x$ less than infinity where the electric field goes to zero. (a) Show by conceptual arguments that this point cannot be located between the charges. (b) Show by conceptual arguments that the point cannot be at any location between $x = 0$ and negative infinity. (c) Show by conceptual arguments that the point must be between $x = 1.00$ m and $x =$ positive infinity. (d) Use the values given to find the point and show that it is consistent with your conceptual argument.

**G.6** Fill in the blanks using $+Q$, 0, or $-Q$ and defend your answer. A spherical conducting object A with a charge of $+Q$ is lowered through a hole into a metal container B, which is initially uncharged.

(a) When A is at the center of B but not touching it, the charge on the inner surface of B is ____________.

(b) The charge on the outer surface of B is ____________.

(c) Object A now is allowed to touch the inner surface of B. The charge on A is now ____________.

(d) The charge on the inner surface of B is now ____________.

(e) The charge on the outer surface of B is now ____________.

**G.7** In the Millikan oil-drop experiment, an atomizer (a sprayer with a fine nozzle) is used to introduce many tiny droplets of oil between two oppositely charged parallel metal plates. Some of the droplets pick up one or more excess electrons. The charge on the plates is adjusted so that the electric force on the excess electrons exactly balances the weight of the droplet. The idea is to look for a droplet that has the smallest electric force and assume that it has only one excess electron. This lets the observer measure the charge on the electron. Suppose we are using an electric field of $3 \times 10^4$ N/C. The charge on one electron is about $1.6 \times 10^{-19}$ C. Estimate the radius of an oil drop whose weight could be balanced by the electric force of this field on one electron.

**G.8** Two hard rubber spheres of mass 10 g are rubbed vigorously with fur on a dry day. They are then suspended from a rod with two insulating strings of length 5.0 cm. They are observed to hang at equilibrium as shown in Fig. GA15.8, each at an angle of 10° with the vertical. Estimate the amount of charge that is found on each sphere.

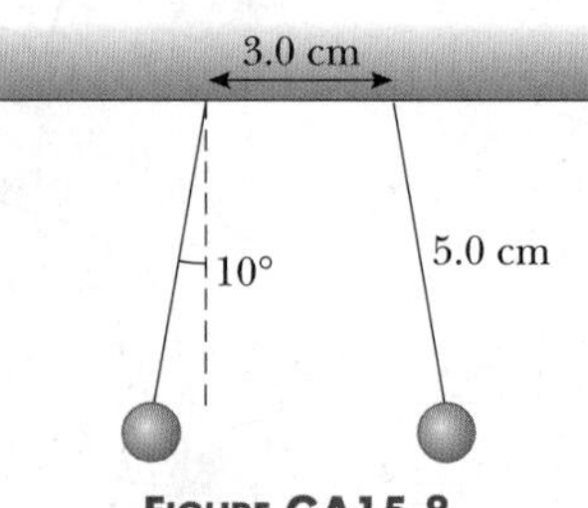


**FIGURE GA15.8**

(Problems 7 and 8 are courtesy of E. F. Redish. For more problems of this type, visit **http://www.physics.umd.edu/perg/**)

# 16

# Electrical Energy and Capacitance

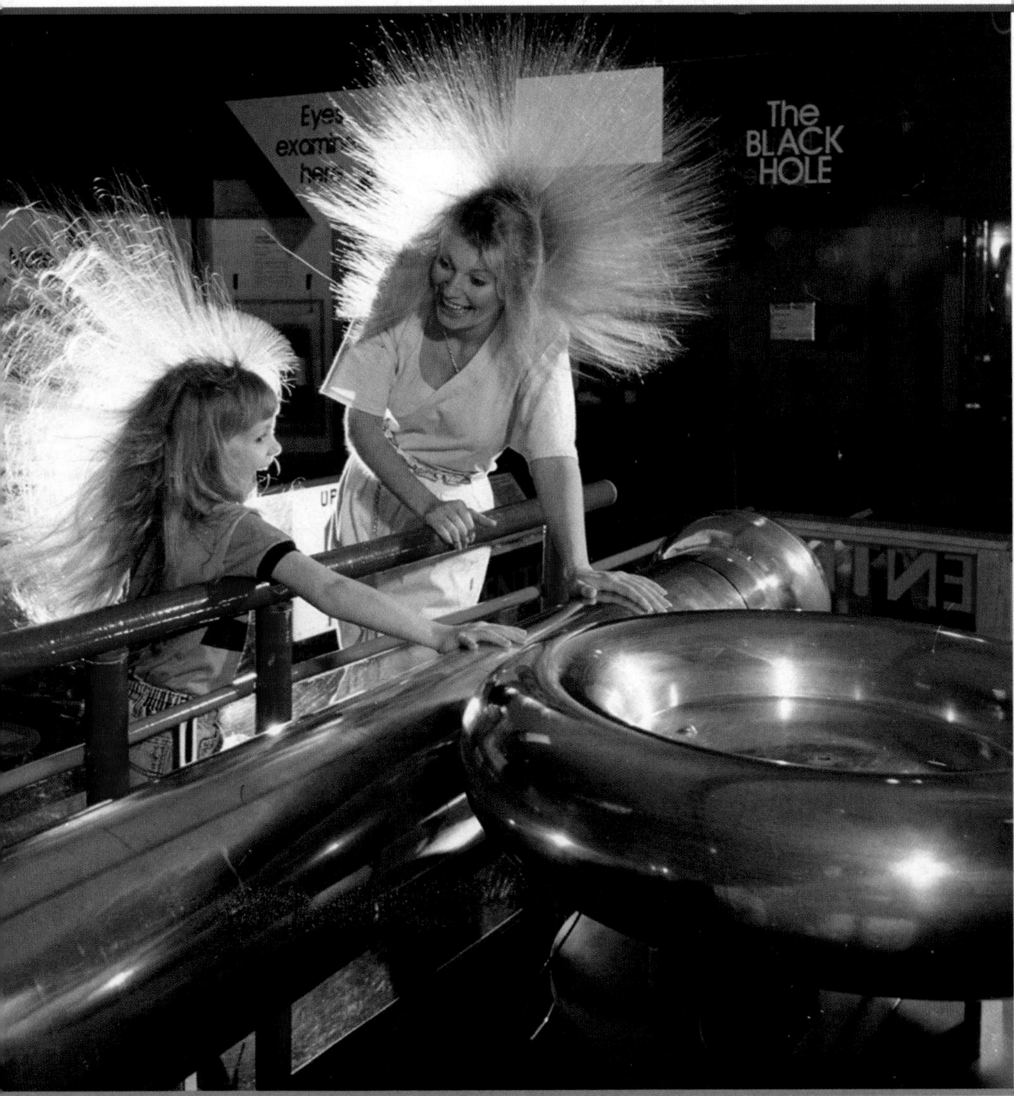


**High Voltage! Everything in the foreground of this picture is at the same electrical potential of many kilovolts. With no differences in potential, no charge is moving and no one gets a shock.** *(Courtesy of Resonance Research Corporation)*

## Chapter Outline

The concept of potential energy was first introduced in Chapter 5. A potential energy function can be defined for any conservative force, such as the force of gravity. By using the principle of conservation of energy, we were often able to avoid working directly with forces when solving problems. In this chapter we discover that the potential energy concept is also useful in the study of electricity. Because the Coulomb force is conservative, we can define an electrical potential energy corresponding to the Coulomb force. This concept of potential energy is of value, but perhaps even more valuable is a quantity called electric potential, defined as potential energy per unit charge.

Electric potential is of great practical value for dealing with electric circuits. For example, when we speak of a voltage applied between two points, we are actually referring to an electric potential difference between those points. We take our first steps toward understanding circuits with a discussion of electric potential, followed by an investigation of a common circuit element called a capacitor.

## 16.1 POTENTIAL DIFFERENCE AND ELECTRIC POTENTIAL

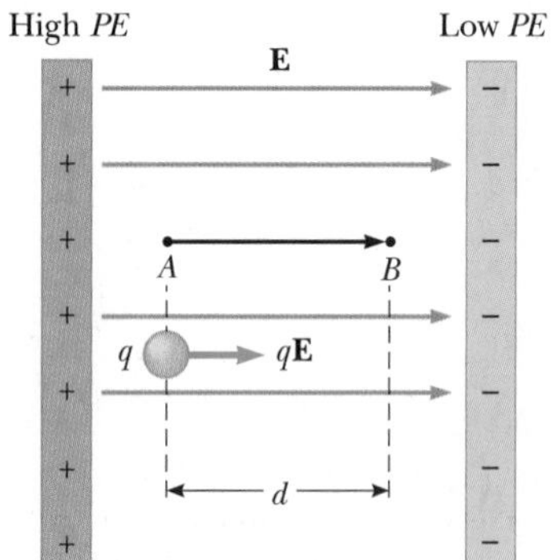


**FIGURE 16.1** When a charge $q$ moves in a uniform electric field **E** from point $A$ to point $B$, the work done by the electric force on the charge is $qEd$.

In Chapter 5 we showed that the gravitational force is conservative. This means that the work done by this force on an object depends only on the initial and final positions of the object and not on the path connecting the two positions. Furthermore, because the gravitational force is conservative, it is possible to define a potential energy function, which we call gravitational potential energy. Because the Coulomb force law is of the same form as the universal law of gravity, it follows that **the electrostatic force is also conservative.** Therefore, it is possible to define an electrical potential energy function associated with this force.

Let us consider electrical potential energy from the point of view of the particular situation shown in Figure 16.1. Imagine a small positive charge placed at point $A$ in a *uniform* electric field of magnitude $E$. (The electric field between equally and oppositely charged parallel metal plates is uniform.) As the charge moves from point $A$ to point $B$, under the influence of the electric force of magnitude $qE$ exerted on it, the work done by the electric field on the charge is positive, and given by $W_{AB} = Fd = qEd$. Therefore, the charge accelerates to the right, gaining kinetic energy. **As the charged particle gains kinetic energy, it loses an equal amount of potential energy.**

By definition, **the work done by a conservative force equals the negative of the change in potential energy, $\Delta PE$.** The change in electrical potential energy is therefore

Change in electrical potential energy

$$\Delta PE = -W_{AB} = -qEd \quad [16.1]$$

Note that although potential energy can be defined for any electric field, **Equation 16.1 is valid only for the case of a uniform electric field.** In subsequent sections we shall examine situations in which the electric field is not uniform.

In the coming pages we often have occasion to use electrical potential energy, but of even more practical importance in the study of electricity is the concept of potential difference.

**The potential difference between points $A$ and $B$, $V_B - V_A$, is defined as the change in potential energy (final value minus initial value) of a charge $q$ moved from $A$ to $B$ divided by the size of the charge.**

Potential difference between two points

$$\Delta V \equiv V_B - V_A = \frac{\Delta PE}{q} \quad [16.2]$$

Potential difference should not be confused with potential energy. The change in electric potential between two points is proportional to the change in electrical potential energy of a charge as it moves between the points, and we see from Equation 16.2 that the two are related as $\Delta PE = q\,\Delta V$. Because potential energy is a scalar quantity, **electric potential is also a scalar quantity.** From Equation 16.2 we see that electric potential difference is a measure of energy per unit charge. Alternatively, electrical potential difference is the work done to move a charge from a point *A* to a point *B* divided by the magnitude of the charge. Thus, the SI units of electric potential are[1] joules per coulomb, called volts (V):

$$1\ \text{V} \equiv 1\ \text{J/C} \quad [16.3]$$

This says that 1 J of work must be done to move a 1-C charge between two points that are at a potential difference of 1 V. In the process of moving through a potential difference of 1 V, the 1-C charge gains (or loses) 1 J of energy.

For the special case of a uniform electric field such as that between charged parallel plates, dividing Equation 16.1 by $q$ gives

$$\frac{\Delta PE}{q} = V_B - V_A = -Ed \quad [16.4]$$

This equation shows that potential difference also has units of electric field times distance. From this, it follows that the SI units of electric field, newtons per coulomb, can also be expressed as volts per meter:

$$1\ \text{N/C} = 1\ \text{V/m}$$

Because Equation 16.4 is directly related to Equation 16.1, remember that it is valid only for the case of a uniform electric field.

Let us examine the changes in energy associated with movements of charge in the electric field pictured in Figure 16.2a. Because the positive charge $q$ tends to move in the direction of the electric field, we must apply an upward external force on the charge to move it from *B* to *A*. Work is done by the external force on the charge, and this means that **a positive charge gains electrical potential energy when it is moved in a direction opposite the electric field.** This is analogous to an object gaining gravitational potential energy when it rises to higher elevations in the presence of gravity, as in Figure 16.2b. If a positive charge is released from rest at point *A*, it experiences a force $q\mathbf{E}$ in the direction of the field (downward in Figure 16.2a). Therefore, it accelerates downward, gaining kinetic energy. **As it gains kinetic energy, it loses an equal amount of electrical potential energy.** Also, as Equation 16.4 shows, if a positive charge moves from *A* to *B*, its electric potential decreases.

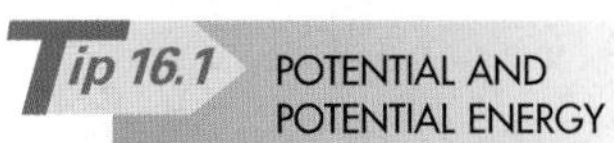


Note that electric potential is characteristic of the field only, independent of a test charge that may be placed in that field. On the other hand, potential energy is a characteristic of the charge-field system due to an interaction between the field and a charge placed in the field.

By contrast, if the test charge $q$ is negative, the situation is reversed. **A negative charge loses electrical potential energy when it moves in the direction**

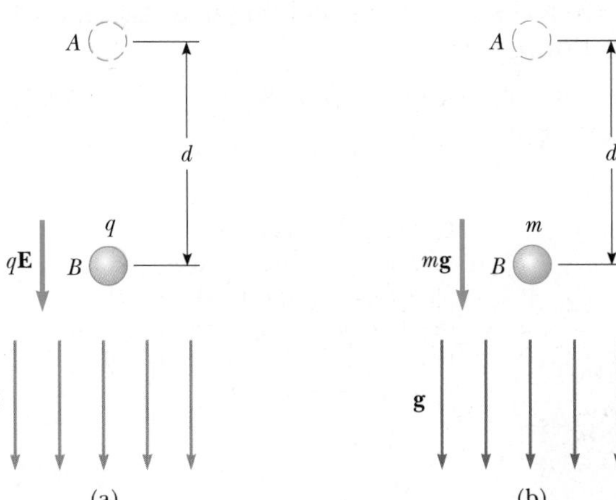


**FIGURE 16.2** (a) When the electric field **E** is directed downward, point *B* is at a lower electric potential than point *A*. A positive test charge that moves from *A* to *B* loses electric potential energy. (b) An object of mass *m* moving downward in the direction of the gravitational field **g** loses gravitational potential energy.

[1] Note that the symbol *V* (italic) represents potential, whereas V (roman) is the symbol for the unit of this quantity—volts. Do not confuse these two symbols.

**opposite the electric field.** That is, a negative charge released from rest in the field **E** accelerates in a direction *opposite* the field.

Thus, when a positive charge is placed in an electric field, it moves in the direction of the field, from a point of high potential to a point of lower potential. In the process, its electrical potential energy decreases and its kinetic energy increases.

When a negative charge is placed in an electric field, it moves opposite to the direction of the field, from a point of low potential to a point of higher potential. In the process, it also undergoes a decrease in electrical potential energy and an increase in kinetic energy.

**APPLICATION**

AUTOMOBILE BATTERIES

Let us pause briefly to discuss a situation that illustrates the concept of electric potential difference. Consider the common 12-V automobile battery. Such a battery maintains a potential difference across its terminals, where the positive terminal is 12 V higher in potential than the negative terminal. In practice, the negative terminal is usually connected to the metal body of the car, which can be considered at a potential of zero volts. The battery becomes a useful device when it is connected by conducting wires to such things as headlights, a radio, power windows, motors, and so forth. Now consider a charge of + 1 C, to be moved around a circuit that contains the battery connected to some of these external devices. As the charge is moved inside the battery from the negative terminal (at 0 V) to the positive terminal (at 12 V), the work done by the battery on the charge is 12 J. Thus, every coulomb of positive charge that leaves the positive terminal of the battery carries an energy of 12 J. As this charge moves through the external circuit toward the negative terminal, it gives up its 12 J of electrical energy to the external devices. When the charge reaches the negative terminal, its electrical energy is zero. At this point, the battery takes over and restores 12 J of energy to the charge as it is moved from the negative to the positive terminal, enabling it to make another transit of the circuit. The actual amount of charge that leaves the battery each second and traverses the circuit depends on the properties of the external devices, as we shall see in the next chapter.

**Quick Quiz 16.1**

If an electron is released from rest in a uniform electric field, the electric potential energy of the charge-field system (a) increases, (b) decreases, or (c) remains the same.

## Example 16.1 The Field Between Two Parallel Plates of Equal and Opposite Charge

Figure 16.3 illustrates a situation in which a constant electric field can be set up. A 12-V battery is connected between two parallel metal plates separated by 0.30 cm. Find the magnitude of the electric field.

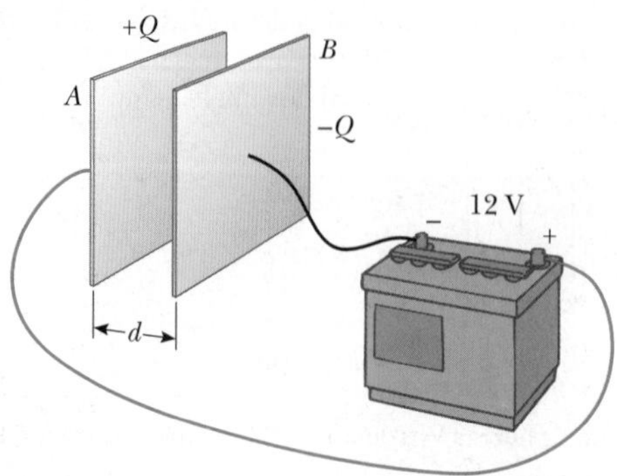


**FIGURE 16.3** (Example 16.1) A 12-V battery connected to two parallel plates. The electric field between the plates has a magnitude given by the potential difference divided by the plate separation *d*.

**Reasoning** The electric field is uniform (except near the edges of the metal plates), and thus the relationship between potential difference and the magnitude of the field is given by Equation 16.4:

$$V_B - V_A = -Ed$$

Chemical forces inside a battery maintain one electrode, called the positive terminal, at a higher potential than a second electrode, the negative terminal. Thus, in Figure 16.3, plate *B*, which is connected to the negative terminal, must be at a lower potential than plate *A*, which is connected to the positive terminal.

**Solution** We have

$$V_B - V_A = -12 \text{ V}$$

This gives a value for *E* of

$$E = -\frac{(V_B - V_A)}{d} = -\frac{(-12 \text{ V})}{0.30 \times 10^{-2} \text{ m}} = \boxed{4.0 \times 10^3 \text{ V/m}}$$

The direction of this field is from the positive plate to the negative plate. A device consisting of two plates separated by a small distance is called a parallel-plate capacitor (to be discussed later in this chapter).

## Example 16.2 Motion of a Proton in a Uniform Electric Field

A proton is released from rest in a uniform electric field of magnitude $8.0 \times 10^4$ V/m, directed along the positive *x* axis (Fig. 16.4). The proton undergoes a displacement of 0.50 m in the direction of the field.

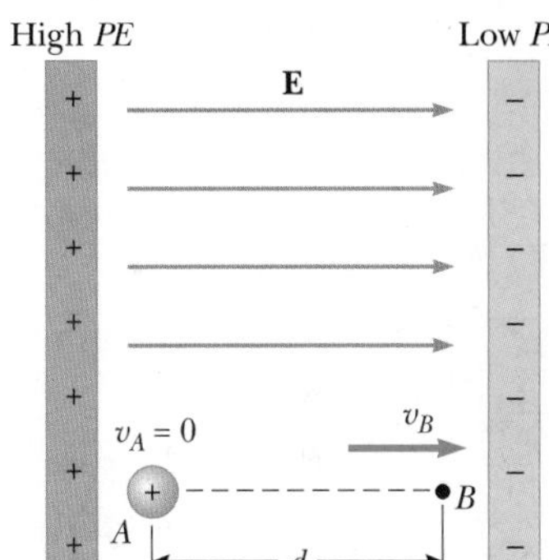


**FIGURE 16.4** (Example 16.2) A proton accelerates from *A* to *B* in the direction of the uniform electric field.

**A** Find the change in electric potential of the proton as a result of this displacement.

**Solution** From Equation 16.4, we have

$$\Delta V = V_B - V_A = -Ed = -(8.0 \times 10^4 \text{ V/m})(0.50 \text{ m}) = \boxed{-4.0 \times 10^4 \text{ V}}$$

Thus, the electric potential of the proton decreases as it moves from *A* to *B*.

**B** Find the change in electrical potential energy of the proton for this displacement and explain the physical meaning of the sign of the change.

**Solution**

$$\Delta PE = q\Delta V = e\Delta V = (1.6 \times 10^{-19} \text{ C})(-4.0 \times 10^4 \text{ V}) = \boxed{-6.4 \times 10^{-15} \text{ J}}$$

The negative sign here means that the electrical potential energy of the proton decreases as it moves in the direction of the electric field. This makes sense because, as the proton *accelerates* in the direction of the field, it gains kinetic energy and at the same time loses electrical potential energy (mechanical energy is conserved).

**C** Find the speed of the proton after it has moved 0.50 m, starting from rest.

**Solution** If no forces other than the conservative electrical force are acting on the proton, we can apply the principle of conservation of mechanical energy in the form

$$KE_i + PE_i = KE_f + PE_f$$

In our case, $KE_i = 0$; hence, the preceding expression gives

$$KE_f = PE_i - PE_f = -\Delta PE$$

With this equation and the results of part B, we find that

$$\tfrac{1}{2}m_p v_f^2 = 6.4 \times 10^{-15} \text{ J}$$

and

$$v_f^2 = \frac{2(6.4 \times 10^{-15}\ \text{J})}{1.67 \times 10^{-27}\ \text{kg}} = 7.66 \times 10^{12}\ \text{m}^2/\text{s}^2$$

$$v_f = 2.8 \times 10^6\ \text{m/s}$$

## 16.2 ELECTRIC POTENTIAL AND POTENTIAL ENERGY DUE TO POINT CHARGES

In electric circuits, a point of zero electric potential is often defined by grounding (connecting to Earth) some point in the circuit. For example, if the negative plate in Example 16.1 were grounded, it would be considered to have a potential of zero and the positive plate to have a potential of 12 V. In this section, we wish to define the *electric potential due to a point charge* at a point in space. In this case, the point of zero electric potential is taken to be at an infinite distance from the charge, far from its influence and the influence of any other charges. With this choice, the methods of calculus can be used to show that the electric potential created by a point charge $q$ at any distance $r$ from the charge is given by

Electric potential created by a point charge

$$V = k_e \frac{q}{r} \qquad [16.5]$$

Equation 16.5 shows that the potential, or work per unit charge to move a test charge in from infinity to a distance $r$ from a positive point charge $q$ increases the closer the positive test charge is moved to $q$. The plot of Equation 16.5 shown in Figure 16.5 shows that the potential associated with a point charge decreases as $1/r$ as we move away from the charge in contrast to the magnitude of the electric field of a point charge, which decreases as $1/r^2$.

Equation 16.5 points out a significant property of electric potential that makes it an important quantity in the study of electricity: The potential at a given point depends only on the charge $q$ of the object setting up the potential and the distance $r$ from that object to a specific point in space. As a result, we can say that a potential exists at some point in space whether or not there is a test charge at that point.

Superposition principle

The electric potential of two or more charges is obtained by applying the **superposition principle.** That is, **the total electric potential at some point *P* due to several point charges is the algebraic sum of the electric potentials**

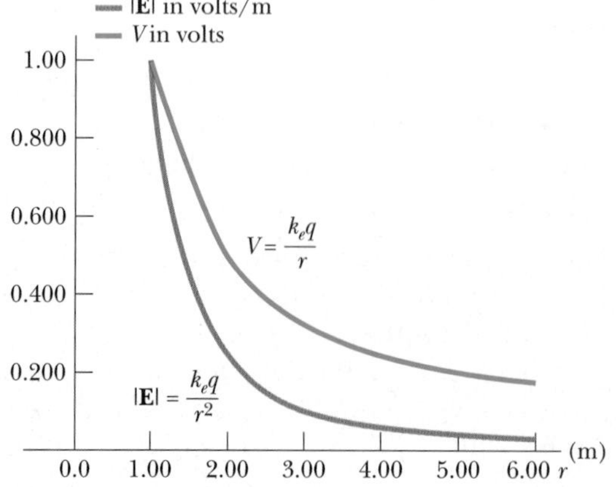


**FIGURE 16.5** Electric field and potential versus distance from a point charge of $1.11 \times 10^{-10}$ C. $V$ decreases as $1/r$, whereas $E$ decreases as $1/r^2$.

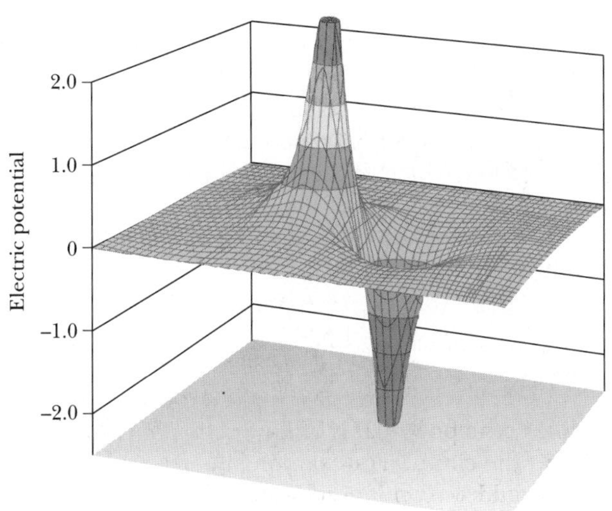


**FIGURE 16.6** The electric potential (in arbitrary units) in the plane containing an electric dipole. Potential is plotted in the vertical dimension.

**due to the individual charges.** This is similar to the method used in Chapter 15 to find the resultant electric field at a point in space. However, note that in the case of potentials, we must evaluate an *algebraic sum* of individual potentials to obtain the total, because **potentials are scalar quantities.** Thus, it is much easier to evaluate the electric potential at some point due to several charges than to evaluate the electric field, which is a vector quantity.

Figure 16.6 is a computer-generated plot of the electric potential associated with an electric dipole. The charges (equal magnitude, opposite sign) lie in a horizontal plane at the center of the potential spikes. The value of the potential is plotted in the vertical dimension. The computer program has added the potential due to each charge to arrive at total values of the potential.

We now consider the electrical potential energy of the interaction of a system of two charged particles. If $V_1$ is the electric potential due to charge $q_1$ at a point $P$, then the work required to bring charge $q_2$ from infinity to $P$ without acceleration is $q_2V_1$. By definition, this work equals the potential energy $PE$ of the two-particle system when the particles are separated by a distance $r$ (Fig. 16.7).

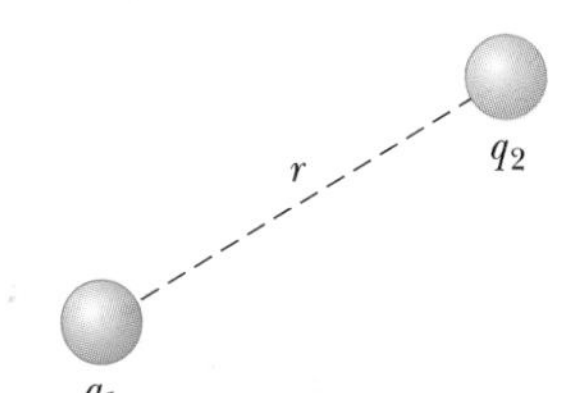

**FIGURE 16.7** If two point charges are separated by the distance $r$, the potential energy of the pair is $k_eq_1q_2/r$.

Therefore, we can express the electrical potential energy of the *pair* of charges as

$$PE = q_2V_1 = k_e\frac{q_1q_2}{r} \qquad [16.6]$$

Potential energy of a pair of charges

Note that if the charges are of the *same* sign, $PE$ is positive. This is consistent with the fact that like charges repel, and so positive work must be done on the system by an external agent to force the two charges near one another. Conversely, if the charges are of *opposite* sign, the force is attractive and $PE$ is negative. This means that negative work must be done to hold back unlike charges from accelerating as they are brought close together.

**Quick Quiz 16.2**

If the electric potential at some point is zero, you can conclude that (a) no charges exist in the vicinity of that point, (b) some charges are positive and some are negative, or (c) all charges in the vicinity have the same sign. Choose each correct answer.

**Quick Quiz 16.3** A spherical balloon contains a positively charged particle at its center. As the balloon is inflated to a larger volume while the charged particle remains at the center, which of the following changes? (a) the electric potential at the surface of the balloon, (b) the magnitude of the electric field at the surface of the balloon, (c) the electric flux through the balloon.

**PROBLEM-SOLVING STRATEGY** **Electric Potential**

When you solve problems involving electric potential, remember that potential is a *scalar quantity* (rather than a vector quantity, like the electric field), so there are no components to worry about. Therefore, when using the superposition principle to evaluate the electric potential at a point due to a system of point charges, simply take the algebraic sum of the potentials due to all charges. You must keep track of signs, however. The potential due to each positive charge is positive, and the potential due to each negative charge is negative. Use the basic equation $V = k_e q/r$.

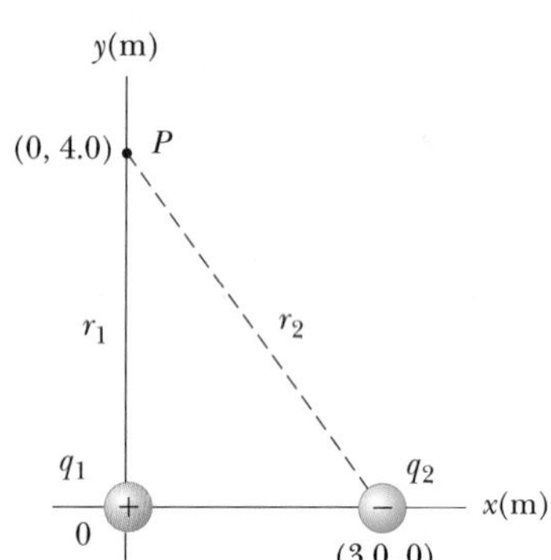


**FIGURE 16.8** (Example 16.3) The electric potential at point $P$ due to the point charges $q_1$ and $q_2$ is the algebraic sum of the potentials due to the individual charges.

## Example 16.3 Finding the Electric Potential

A 5.0-$\mu$C point charge is at the origin, and a point charge $q_2 = -2.0\ \mu$C is on the $x$ axis at (3.0, 0) m, as in Figure 16.8.

**A** If the electric potential is taken to be zero at infinity, find the total electric potential due to these charges at point $P$, with coordinates (0, 4.0) m.

**Reasoning** The electric potential at $P$ due to each charge can be calculated from $V = k_e q/r$. The total electric potential is the scalar sum of these two potentials.

**Solution** The electric potential at $P$ due to the 5.0-$\mu$C charge at the origin is

$$V_1 = k_e \frac{q_1}{r_1} = \left(8.99 \times 10^9 \frac{\text{N}\cdot\text{m}^2}{\text{C}^2}\right)\left(\frac{5.0 \times 10^{-6}\text{ C}}{4.0\text{ m}}\right) = 1.12 \times 10^4\text{ V}$$

and the electric potential due at $P$ to the charge of $-2.0\ \mu$C is

$$V_2 = k_e \frac{q_2}{r_2} = \left(8.99 \times 10^9 \frac{\text{N}\cdot\text{m}^2}{\text{C}^2}\right)\left(\frac{-2.0 \times 10^{-6}\text{ C}}{5.0\text{ m}}\right) = -0.360 \times 10^4\text{ V}$$

and

$$V_P = V_1 + V_2 = 7.6 \times 10^3\text{ V}$$

**B** How much work is required to bring a third point charge of 4.0 $\mu$C from infinity to $P$?

**Solution**

$$W = q_3 V_P = (4.0 \times 10^{-6}\text{ C})(7.6 \times 10^3\text{ V})$$

Because 1 V = 1 J/C, $W$ reduces to

$$W = 3.1 \times 10^{-2}\text{ J}$$

**EXERCISE** Find the magnitude and direction of the electric field at point $P$.

**ANSWER** $2.3 \times 10^3$ N/C at an angle of 79° with the $x$ axis.

## 16.3 POTENTIALS AND CHARGED CONDUCTORS

In order to determine the electric potential at all points on a charged conductor, let us combine Equations 16.1 and 16.2. From Equation 16.1 we see that the work done by electric forces on a charge is related to the change in electrical potential energy of the charge by

$$W = -\Delta PE$$

Furthermore, from Equation 16.2 we see that the change in electrical potential energy between two points $A$ and $B$ is related to the potential difference between these points by

$$\Delta PE = q(V_B - V_A)$$

Combining these two equations, we find that

$$W = -q(V_B - V_A) \qquad [16.7]$$

As we see from this result, in general, **no work is required to move a charge between two points that are at the same electric potential. That is, $W = 0$ when $V_B = V_A$.**

In Chapter 15 we found that when a conductor is in electrostatic equilibrium, a net charge placed on it resides entirely on its surface. Furthermore, we showed that the electric field just outside the surface of a charged conductor in electrostatic equilibrium is perpendicular to the surface and that the field inside the conductor is zero. We shall now show that **all points on the surface of a charged conductor in electrostatic equilibrium are at the same potential.**

Properties of a charged conductor in equilibrium

Consider a surface path connecting any points $A$ and $B$ on a charged conductor, as in Figure 16.9. The electric field **E** is always perpendicular to the displacement along this path; therefore, no work is done by the electric field if a charge is moved between these points. From Equation 16.7 we see that if the work done is zero, the difference in electric potential $V_B - V_A$ is also zero. Therefore, **the electric potential is a constant everywhere on the surface of a charged conductor in equilibrium.** Furthermore, because the electric field inside a conductor is zero, no work is required to move a charge between two points inside the conductor. Again, Equation 16.7 shows that if the work done is zero, the difference in electric potential between any two points inside a conductor must also be zero. Thus, we conclude that the electric potential is constant everywhere inside a conductor.

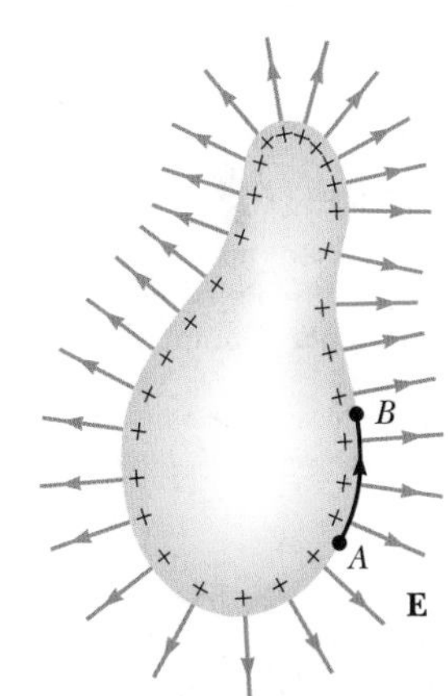


**FIGURE 16.9** An arbitrarily shaped conductor with an excess positive charge. When the conductor is in electrostatic equilibrium, all of the charge resides at the surface, $\mathbf{E} = 0$ inside the conductor, and the electric field just outside the conductor is perpendicular to the surface. The potential is constant inside the conductor and is equal to the potential at the surface.

Finally, because one of the points inside the conductor could be arbitrarily close to the surface of the conductor, we conclude that **the electric potential is constant everywhere inside a conductor and equal to its value at the surface.** As a consequence, no work is required to move a charge from the interior of a charged conductor to its surface. (Note that the potential inside a conductor is not necessarily zero even though the interior electric field is zero.)

### THE ELECTRON VOLT

A unit of energy commonly used in atomic and nuclear physics because of its convenient size is the electron volt (eV). Electrons in normal atoms typically have energies of tens of electron volts, excited electrons in atoms emitting x-rays have energies of thousands of electron volts, and high-energy gamma rays (electromagnetic waves) emitted by the nucleus have energies of millions of electron volts.

Definition of the electron volt

**The electron volt is defined as the energy that an electron (or proton) gains when accelerated through a potential difference of 1 V.**

Because $1\ V = 1\ J/C$ and because the magnitude of charge on the electron or proton is $1.60 \times 10^{-19}$ C, we see that the electron volt is related to the joule by

$$1\ eV = 1.60 \times 10^{-19}\ C \cdot V = 1.60 \times 10^{-19}\ J \qquad [16.8]$$

## APPLYING PHYSICS 16.1

Suppose scientists had chosen to measure small energies in proton volts rather than electron volts. What difference would this make?

**Explanation** There would be no change at all. An electron volt is the kinetic energy gained by an electron in being accelerated through a potential difference of 1 V. A proton accelerated through 1 V would have the same kinetic energy, because it carries the same charge as the electron (except for sign). The proton would be moving in the opposite direction and more slowly after accelerating through 1 V, due to its opposite charge and its larger mass, but it would still gain 1 electron volt, or 1 proton volt, of kinetic energy.

## 16.4 EQUIPOTENTIAL SURFACES

A surface on which all points are at the same potential is called an **equipotential surface.** The potential difference between any two points on an equipotential surface is zero. Hence, **no work is required to move a charge at constant speed on an equipotential surface.** Equipotential surfaces have a simple relationship to the electric field. **The electric field at every point on an equipotential surface is perpendicular to the surface.** If the electric field **E** had a component parallel to the surface, this component would produce an electric force on a charge placed on the surface. This force would do work on the charge as it moved from one point to another, in contradiction to the definition of an equipotential surface.

**Webnote 16.1**

For an excellent applet demonstrating electric field lines and equipotentials of point charges, visit *http://www.cco.caltech.edu/~phys1/java/phys1/EField/EField.html*

It is convenient to represent equipotential surfaces on a diagram by drawing equipotential contours, which are two-dimensional views of the intersections of the equipotential surfaces with the plane of the drawing. We shall refer to these equipotential contours as simply **equipotentials.** Figure 16.10a shows the equipotentials (in blue) associated with a positive point charge. Note that the equipotentials are perpendicular to the electric field lines (in red) at all points. Recall that the electric potential created by a point charge $q$ is given by $V = k_e q/r$. This rela-

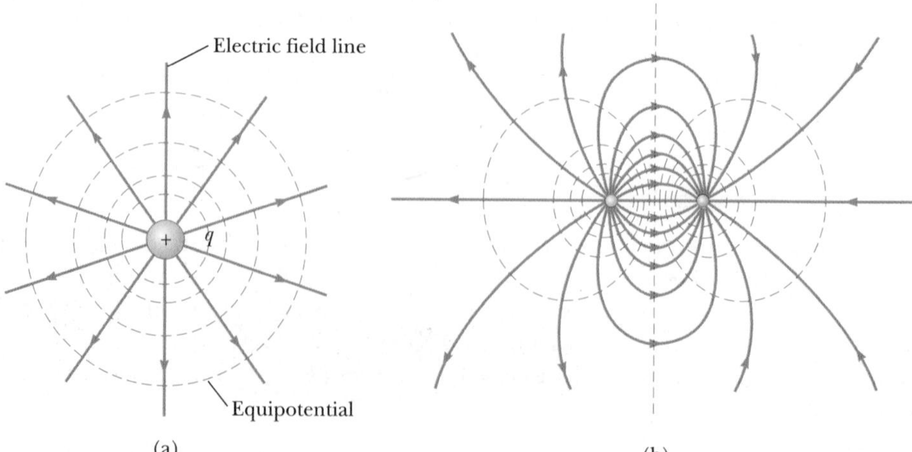


**FIGURE 16.10** Equipotentials (dashed blue lines) and electric field lines (red lines) for (a) a positive point charge and (b) two point charges of equal magnitude and opposite sign. In all cases, the equipotentials are *perpendicular* to the electric field lines at every point.

tion shows that, for a single point charge, the potential is constant on any surface for which $r$ is constant. Therefore, the equipotentials of a point charge are a family of spheres centered on the point charge. Figure 16.10b shows the equipotentials associated with two charges of equal magnitude but opposite sign (an electric dipole).

## 16.5 APPLICATIONS

### THE ELECTROSTATIC PRECIPITATOR

One important application of electric discharge in gases is a device called an *electrostatic precipitator.* It is used to remove particulate matter from combustion gases, thereby reducing air pollution. It is especially useful in coal-burning power plants and in industrial operations that generate large quantities of smoke. Systems currently in use can eliminate approximately 90% by mass of the ash and dust from the smoke. Unfortunately, a very high percentage of the lighter particles still escape, and these contribute significantly to smog and haze.

**APPLICATION**

THE ELECTROSTATIC PRECIPITATOR

Figure 16.11 illustrates the basic idea of the electrostatic precipitator. A high voltage (typically 4–100 kV) is maintained between a wire running down the center of a duct and the outer wall, which is grounded. The wire is maintained at a negative electric potential with respect to the wall, and so the electric field is directed toward the wire. The electric field near the wire reaches a high enough value to cause a discharge around the wire and the formation of positive ions, electrons, and negative ions, such as $O_2^-$. As the electrons and negative ions are accelerated toward the outer wall by the nonuniform electric field, the dirt particles in the streaming gas become charged by collisions and ion capture. Because most of the charged dirt particles are negative, they are also drawn to the outer wall by the electric field. When the duct is shaken, the particles fall loose and are collected at the bottom.

In addition to reducing the amounts of harmful gases and particulate matter in the atmosphere, the electrostatic precipitator recovers valuable metal oxides from the stack.

**APPLICATION**

THE ELECTROSTATIC AIR CLEANER

An *electrostatic air cleaner,* used in homes to relieve the discomfort of allergy sufferers, uses many of the same principles as the precipitator. Air laden with dust and pollen is drawn into the device across a positively charged mesh screen. The

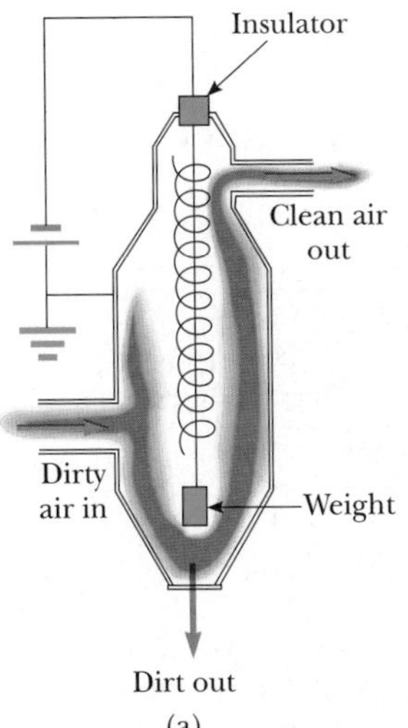


(a)

(b)

(c)

**FIGURE 16.11** (a) Diagram of an electrostatic precipitator. The high voltage maintained on the central wires creates an electric discharge in the vicinity of the wire. Compare air pollution when the electrostatic precipitator is (b) operating and (c) turned off. *(b, Rei O'Hara/Black Star/PNI; c, Greig Cranna/Stock, Boston/PNI)* airborne particles become positively charged when they make intimate contact with the screen. The particles then pass through a second, negatively charged mesh screen. The electrostatic force of attraction between the positively charged particles in the air and the negatively charged screen causes the particles to precipitate out onto the surface of the screen. In this fashion, a very high percentage of contaminants are removed from the air stream.

## XEROGRAPHY AND LASER PRINTERS

**APPLICATION**

XEROGRAPHIC COPIERS

The process of xerography is widely used for making photocopies of printed materials. The basic idea behind the process was developed by Chester Carlson, who was granted a patent for his invention in 1940. In 1947 the Xerox Corporation launched a full-scale program to develop automated duplicating machines using Carlson's process. The huge success of that development is quite evident; today, practically all offices and libraries have one or more duplicating machines, and the capabilities of these machines continue to evolve.

Some features of the xerographic process involve simple concepts from electrostatics and optics. However, the one idea that makes the process unique is the use of photoconductive material to form an image. (A photoconductor is a material that is a poor conductor of electricity in the dark but becomes a reasonably good electric conductor when exposed to light.)

Figure 16.12 illustrates the steps in the xerographic process. First, the surface of a plate or drum is coated with a thin film of the photoconductive material (usually selenium or some compound of selenium), and the photoconductive surface is given a positive electrostatic charge in the dark (Fig. 16.12a). The page to be copied is then projected onto the charged surface (Fig. 16.12b). The photocon-

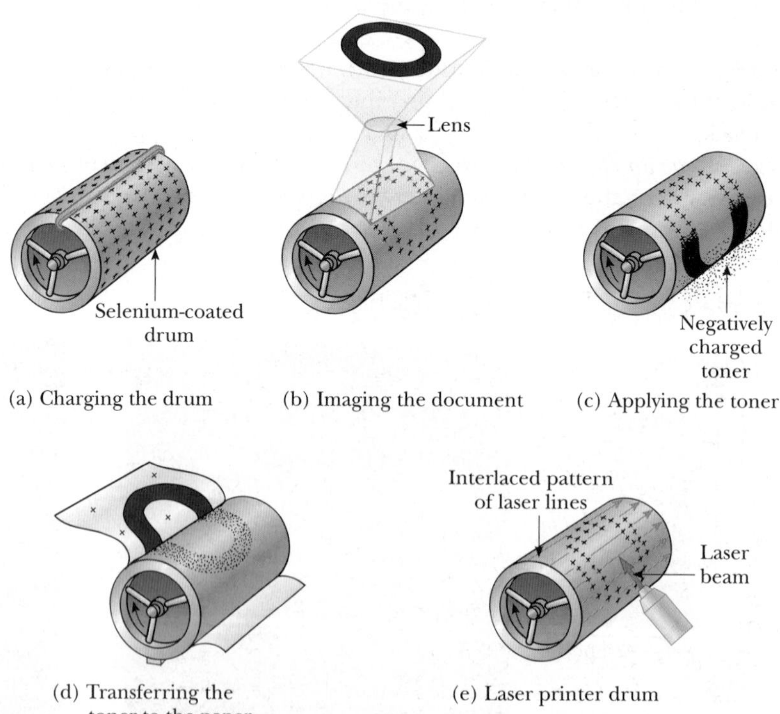


**FIGURE 16.12** The xerographic process. (a) The photoconductive surface is positively charged. (b) Through the use of a light source and lens, a hidden image is formed on the charged surface in the form of positive charges. (c) The surface containing the image is covered with a negatively charged powder, which adheres only to the image area. (d) A piece of paper is placed over the surface and given a charge. This transfers the image to the paper, which is then heated to "fix" the powder to the paper. (e) The image on the drum of a laser printer is produced by turning a laser beam on and off as it sweeps across the selenium-coated drum.

ducting surface becomes conducting only in areas where light strikes; there the light produces charge carriers in the photoconductor, which neutralize the positively charged surface. The charges remain on those areas of the photoconductor not exposed to light, however, leaving a hidden image of the object in the form of a positive surface charge distribution.

Next, a powder called a *toner* is negatively charged and dusted onto the photoconducting surface (Fig. 16.12c). The charged powder adheres only to the areas that contain the positively charged image. At this point, the image becomes visible. It is then transferred to the surface of a sheet of positively charged paper. Finally, the toner is "fixed" to the surface of the paper by heat (Fig. 16.12d). This results in a permanent copy of the original.

**APPLICATION**

LASER PRINTERS

The steps for producing a document on a laser printer are similar to those used in a photocopy machine, in that parts (a), (c), and (d) of Figure 16.12 remain essentially the same. The difference between the two techniques lies in the way the image is formed on the selenium-coated drum. In a laser printer, the command to print the letter O, for instance, is sent to a laser from the memory of a computer. A rotating mirror inside the printer causes the beam of the laser to sweep across the selenium-coated drum in an interlaced pattern (Fig. 16.12e). Electrical signals generated by the printer turn the laser beam on and off in a pattern that traces out the letter O in the form of positive charges on the selenium. Toner is then applied to the drum, and the transfer to paper is accomplished as in a photocopy machine.

## 16.6 CAPACITANCE

A **capacitor** is a device used in a variety of electric circuits—for example, to tune the frequency of radio receivers, eliminate sparking in automobile ignition systems, or to store short-term energy for rapid release in electronic flash units. Figure 16.13 shows a typical design for a capacitor. It consists of two parallel metal plates separated by a distance $d$. When used in an electric circuit, the plates are connected to the positive and negative terminals of a battery or some other voltage source. When this connection is made, electrons are pulled off one of the plates, leaving it with a charge of $+Q$, and transferred through the battery to the other plate, leaving it with a charge of $-Q$, as shown in the figure. This charge transfer stops when the potential difference across the plates equals the potential difference of the battery. Thus, a charged capacitor is a device that stores energy that can be reclaimed when needed for a specific application.

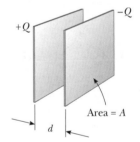

**FIGURE 16.13** A parallel-plate capacitor consists of two parallel plates, each of area $A$, separated by a distance $d$. The plates carry charges of equal magnitude and opposite sign.

**The capacitance $C$ of a capacitor is defined as the ratio of the magnitude of the charge on either conductor (plate) to the magnitude of the potential difference between the conductors (plates):**

$$C \equiv \frac{Q}{\Delta V} \qquad [16.9]$$

Capacitance of a pair of conductors

Capacitance has the SI units coulombs per volt, called **farads** (F) in honor of Michael Faraday. That is,

$$1\text{ F} \equiv 1\text{ C/V}$$

The farad is a very large unit of capacitance. In practice, most typical capacitors have capacitances ranging from microfarads ($1\ \mu\text{F} = 1 \times 10^{-6}$ F) to picofarads ($1\text{ pF} = 1 \times 10^{-12}$ F).

For example, if a 3.0-$\mu$F capacitor is connected to a 12-V battery, the magnitude of the charge on each plate of the capacitor is

$$Q = C\Delta V = (3.0 \times 10^{-6}\text{ F})(12\text{ V}) = 36\ \mu\text{C}$$

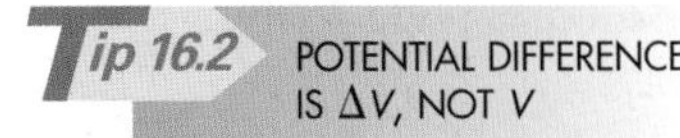

Tip 16.2
POTENTIAL DIFFERENCE IS ΔV, NOT V


In this book, we use the symbol $\Delta V$ for the potential difference across a circuit element or a device. Note that almost every other book uses the symbol $V$ for potential difference, without the delta sign, which can lead to confusion.

From Equation 16.9, we see that a large capacitance is needed to store a large amount of charge for a given applied voltage.

## 16.7 THE PARALLEL-PLATE CAPACITOR

The capacitance of a device depends on the geometric arrangement of the conductors. For example, the capacitance of a parallel-plate capacitor whose plates are separated by air (see Fig. 16.13) is

Capacitance of a parallel-plate capacitor

$$C = \epsilon_0 \frac{A}{d} \qquad [16.10]$$

where $A$ is the area of one of the plates, $d$ is the distance between the plates, and $\epsilon_0$ is the permittivity of free space.

Although we do not derive Equation 16.10, let us attempt to make it seem plausible. As you can see from the definition of capacitance, $C \equiv Q/\Delta V$, the amount of charge a given capacitor can store for a given potential difference across its plates increases as the capacitance increases. Therefore, it seems reasonable that a capacitor constructed from plates with large areas should be able to store a large charge. Furthermore, if the oppositely charged plates are close together, the attractive force between them will be large. In fact, for a given potential difference, the charge on the plates increases with decreasing plate separation.

**APPLICATION**

CAMERA FLASH ATTACHMENTS

One practical device that uses a capacitor is the flash attachment on a camera. A battery is used to charge the capacitor, and this stored charge is then released when the shutter-release button is pressed to take a picture. The stored charge is delivered to a flash tube very quickly, illuminating the subject at the instant more light is needed.

**APPLICATION**

COMPUTER KEYBOARDS

Computers make use of capacitors in many ways. For example, one type of computer keyboard has capacitors at the bases of its keys, as in Figure 16.14. Each key is connected to a movable plate, which represents one side of the capacitor; the fixed plate on the bottom of the keyboard represents the other side of the capacitor. When a key is pressed, the capacitor spacing decreases, causing an increase in capacitance. External electronic circuits recognize each key by the *change* in its capacitance when it is pressed.

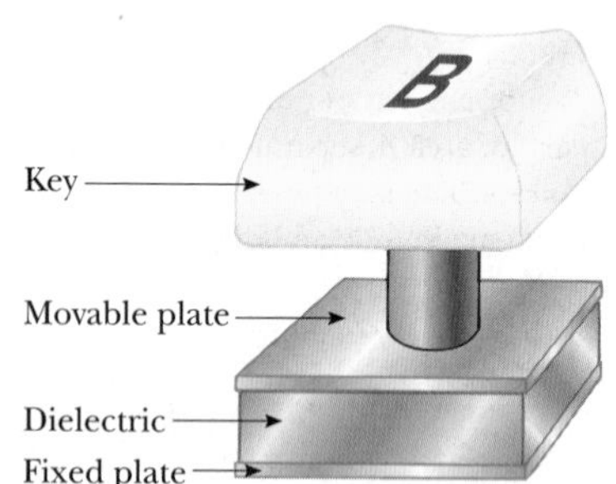


**FIGURE 16.14** When the key of one type of keyboard is pressed, the capacitance of a parallel-plate capacitor increases as the plate spacing decreases. The substance labeled "dielectric" is an insulating material as described in Section 16.10.

### Example 16.4 Calculating *C* for a Parallel-Plate Capacitor

A parallel-plate capacitor has an area of $A = 2.00 \text{ cm}^2 = 2.00 \times 10^{-4} \text{ m}^2$ and a plate separation of $d = 1.00 \text{ mm} = 1.00 \times 10^{-3} \text{ m}$. Find its capacitance.

**Solution** From $C = \epsilon_0 A/d$ we find that

$$C = \epsilon_0 \frac{A}{d} = (8.85 \times 10^{-12} \text{ C}^2/\text{N} \cdot \text{m}^2)\left(\frac{2.00 \times 10^{-4} \text{ m}^2}{1.00 \times 10^{-3} \text{ m}}\right)$$

$$= 1.77 \times 10^{-12} \text{ F} = 1.77 \text{ pF}$$

**EXERCISE** Show that $1 \text{ C}^2/\text{N} \cdot \text{m}$ equals 1 F.

### SYMBOLS FOR CIRCUIT ELEMENTS AND CIRCUITS

The symbol that is commonly used to represent a capacitor in a circuit is —||—, or sometimes —|(—. Do not confuse this with the circuit symbol —$\overset{+}{|}\overset{-}{|}$— used to designate a battery (or any other direct current source). The positive terminal of the battery is at the higher potential and is represented by the longer vertical line in the battery symbol. In the next chapter we

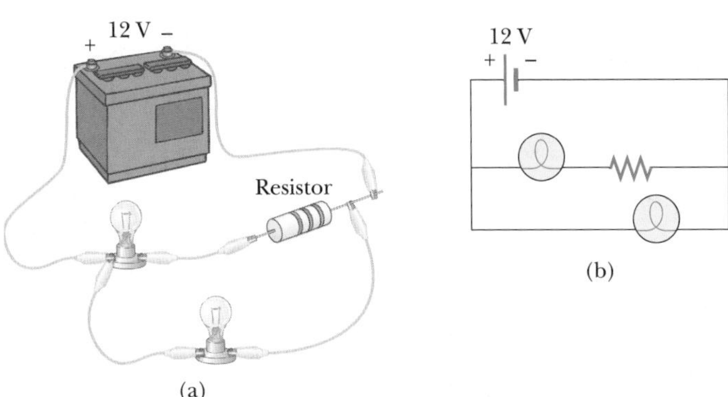


**FIGURE 16.15** (a) A real circuit and (b) its equivalent circuit diagram.

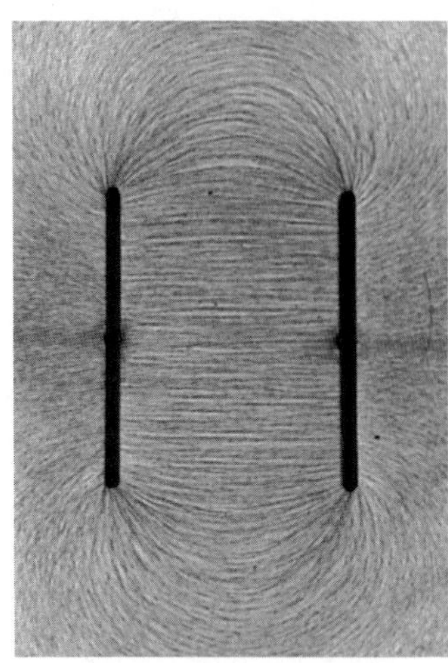

The electric field pattern of two oppositely charged conducting parallel plates. Note the nonuniform nature of the electric field at the ends of the plates. *(Courtesy of Harold M. Waage, Princeton University)*

shall discuss another circuit element called a resistor, represented by the symbol . The wires in a circuit that do not have appreciable resistance compared to other elements in the circuit will be represented by straight lines.

Although we will discuss circuits in more detail in the next chapter, it is important to realize that a circuit is a collection of real objects usually containing a source of electrical energy (such as a battery) connected to elements that convert electrical energy to other forms (light, heat, sound), or store the energy in electric or magnetic fields for later retrieval. A real circuit and its diagram are sketched in Figure 16.15. The circuit symbol for a lightbulb shown in Figure 16.15b is . If you are not familiar with circuit diagrams, trace the path of the real circuit with your finger to see that it is equivalent to the geometrically regular schematic diagram.

## 16.8 COMBINATIONS OF CAPACITORS

Two or more capacitors can be combined in circuits in several ways. The equivalent capacitances of certain combinations can be calculated with methods described in this section. The results are of practical use, for example, if you wish to build a circuit having a specific capacitance and you have only specific values of capacitance at your disposal.

### PARALLEL COMBINATION

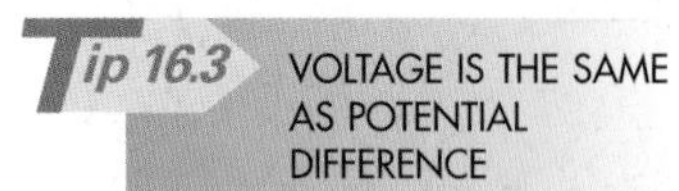

Tip 16.3
VOLTAGE IS THE SAME AS POTENTIAL DIFFERENCE


A voltage *across* a device, such as a capacitor, has the same meaning as the potential difference across the device. For example, if we say that the voltage across a capacitor is 12 V, we mean that the potential difference between its plates is 12 V.

Two capacitors connected as shown in Figure 16.16a are known as a *parallel combination* of capacitors. The left plate of each capacitor is connected by a conducting wire to the positive terminal of the battery, and the left plates are therefore at the same potential. Likewise, the right plates are connected to the negative terminal of the battery. When the capacitors are first connected in the circuit, electrons are transferred from the left plates through the battery to the right plates, leaving the left plates positively charged and the right plates negatively charged. The energy source for this charge transfer is the internal chemical energy stored in the battery, which is converted to electrical energy. The flow of charge ceases when the voltage across the capacitors equals that of the battery. The capacitors reach their maximum charge when the flow of charge ceases. Let us call the maximum charges on the two capacitors $Q_1$ and $Q_2$. Then the *total charge* $Q$ stored by the two capacitors is

$$Q = Q_1 + Q_2 \qquad [16.11]$$

We can replace these two capacitors with one equivalent capacitor having a capacitance of $C_{eq}$. This equivalent capacitor must have exactly the same external effect on the circuit as the original two. That is, it must store $Q$ units of charge. We also see from Figure 16.16b that **the potential differences across the capacitors**

$C_1$
+
–
$C_2$
+
–
+
–
$\Delta V$
$\Delta V_1 = \Delta V_2 = \Delta V$
$C_1$
$Q_1$
$C_2$
$Q_2$
+
–
$\Delta V$
$C_{eq} = C_1 + C_2$
+
–
$\Delta V$
(a)
(b)
(c)


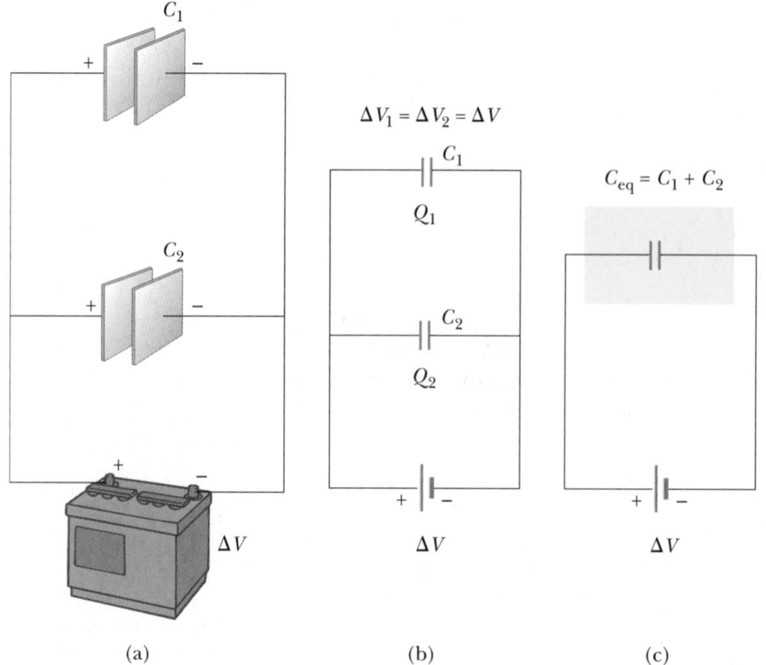

**FIGURE 16.16** (a) A parallel connection of two capacitors. (b) The circuit diagram for the parallel combination. (c) The potential differences across the capacitors are the same, and the equivalent capacitance is $C_{eq} = C_1 + C_2$.

**in a parallel circuit are the same; each is equal to the voltage $\Delta V$ of the battery.** From Figure 16.16c, we see that the voltage across the equivalent capacitor is also $\Delta V$. Thus, the charge on each capacitor is

$$Q_1 = C_1 \Delta V \quad \text{and} \quad Q_2 = C_2 \Delta V$$

The charge on the equivalent capacitor is

$$Q = C_{eq} \Delta V$$

Substituting these relations into Equation 16.11 gives

$$C_{eq} \Delta V = C_1 \Delta V + C_2 \Delta V$$

or

$$C_{eq} = C_1 + C_2 \quad \left(\begin{array}{c}\text{parallel} \\ \text{combination}\end{array}\right) \tag{16.12}$$

If we extend this treatment to three or more capacitors connected in parallel, the equivalent capacitance is found to be

$$C_{eq} = C_1 + C_2 + C_3 + \cdots \quad \left(\begin{array}{c}\text{parallel} \\ \text{combination}\end{array}\right) \tag{16.13}$$

Thus, we see that **the equivalent capacitance of a parallel combination of capacitors is greater than any of the individual capacitances.**

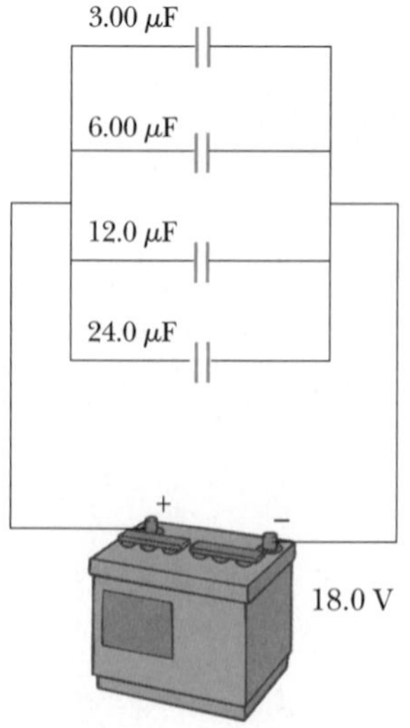

3.00 μF
6.00 μF
12.0 μF
24.0 μF
+
–
18.0 V


**FIGURE 16.17** (Example 16.5) Four capacitors connected in parallel.

## Example 16.5 Four Capacitors Connected in Parallel

Determine the capacitance of the single capacitor that is equivalent to the parallel combination of capacitors shown in Figure 16.17, and find the charge on the 12.0-$\mu$F capacitor.

**Solution** The equivalent capacitance is found by use of Equation 16.13:

$$\begin{aligned} C_{eq} &= C_1 + C_2 + C_3 + C_4 \\ &= 3.00\ \mu\text{F} + 6.00\ \mu\text{F} + 12.0\ \mu\text{F} + 24.0\ \mu\text{F} = 45.0\ \mu\text{F} \end{aligned}$$

The potential difference across the 12.0-$\mu$F capacitor (and all other capacitors in this case) is equal to the voltage of the battery, and so

$$Q = C\Delta V = (12.0 \times 10^{-6}\text{ F})(18.0\text{ V}) = 216 \times 10^{-6}\text{ C} = \boxed{216\ \mu\text{C}}$$

## SERIES COMBINATION

Now consider two capacitors connected in *series,* as illustrated in Figure 16.18a. **For a series combination of capacitors, the magnitude of the charge must be the same on all the plates.** To see why this must be true, let us consider the charge transfer process in some detail. We start with uncharged capacitors. When a battery is connected to the circuit, electrons are transferred from the left plate of $C_1$ to the right plate of $C_2$ through the battery. As this negative charge accumulates on the right plate of $C_2$, an equivalent amount of negative charge is removed from the left plate of $C_2$, leaving it with an excess positive charge. The negative charge leaving the left plate of $C_2$ accumulates on the right plate of $C_1$, where again an equivalent amount of negative charge is removed from the left plate. The result of this is that **all of the right plates gain charges of $-Q$ and all the left plates have charges of $+Q$.** (This is a consequence of the conservation of charge.)

Q is the same for all capacitors connected in series

We can find an equivalent capacitor that performs the same function as the series combination. After it is fully charged, **the equivalent capacitor must end up with a charge of $-Q$ on its right plate and a charge of $+Q$ on its left plate.** By applying the definition of capacitance to the circuit in Figure 16.18b, we have

$$\Delta V = \frac{Q}{C_{eq}}$$

where $\Delta V$ is the potential difference between the terminals of the battery and $C_{eq}$ is the equivalent capacitance. From Figure 16.18a we see that

$$\Delta V = \Delta V_1 + \Delta V_2 \qquad [16.14]$$

where $\Delta V_1$ and $\Delta V_2$ are the potential differences across capacitors $C_1$ and $C_2$. (This is a consequence of the conservation of energy.) The potential difference across any number of capacitors (or other circuit elements) in series equals the sum of the potential differences across the individual capacitors. Because $Q = C\,\Delta V$ can be applied to each capacitor, the potential differences across them

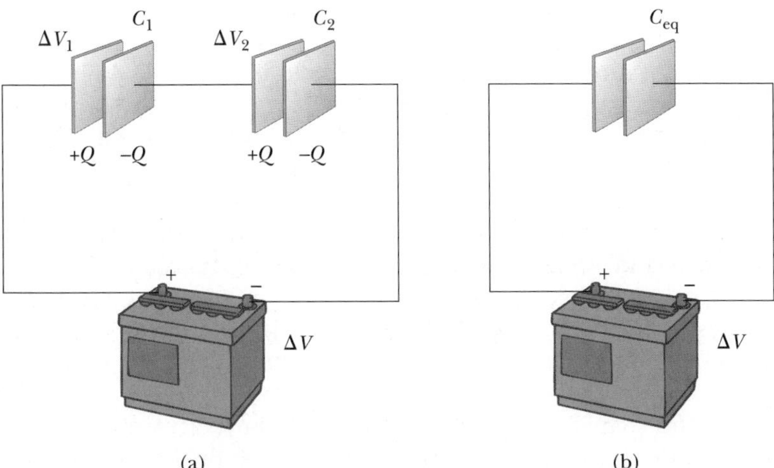


**FIGURE 16.18** A series combination of two capacitors. The charges on the capacitors are the same, and the equivalent capacitance can be calculated from the reciprocal relationship $1/C_{eq} = (1/C_1) + (1/C_2)$.

are given by

$$\Delta V_1 = \frac{Q}{C_1} \qquad \Delta V_2 = \frac{Q}{C_2}$$

Substituting these expressions into Equation 16.14, and noting that $\Delta V = Q/C_{eq}$, we have

$$\frac{Q}{C_{eq}} = \frac{Q}{C_1} + \frac{Q}{C_2}$$

Canceling $Q$, we arrive at the relationship

$$\frac{1}{C_{eq}} = \frac{1}{C_1} + \frac{1}{C_2} \qquad \left(\begin{array}{c}\text{series} \\ \text{combination}\end{array}\right) \qquad [16.15]$$

If this analysis is applied to three or more capacitors connected in series, the equivalent capacitance is found to be

$$\frac{1}{C_{eq}} = \frac{1}{C_1} + \frac{1}{C_2} + \frac{1}{C_3} + \cdots \qquad \left(\begin{array}{c}\text{series} \\ \text{combination}\end{array}\right) \qquad [16.16]$$

As we shall demonstrate in Example 16.6, this implies that **the equivalent capacitance of a series combination is always less than any individual capacitance in the combination.**

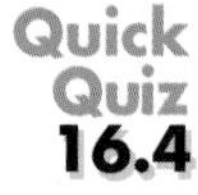


A capacitor is designed so that one plate is large and the other is small. If the plates are connected to a battery, (a) the large plate has a greater charge than the small plate, (b) the large plate has less charge than the small plate, or (c) the plates have charges equal in magnitude but opposite in sign.

## PROBLEM-SOLVING STRATEGY Capacitors

1. Be careful with your choice of units. To calculate the capacitance of a device in farads, make sure that distances are in meters and use the SI value of $\epsilon_0$.
2. When two or more unequal capacitors are connected in *series*, they carry the same charge, but the potential differences across them are not the same. Their capacitances add as reciprocals, and the equivalent capacitance of the combination is always *less* than the smallest individual capacitor.
3. When two or more capacitors are connected in *parallel*, the potential differences across them are the same. The charge on each capacitor is proportional to its capacitance; hence, the capacitances add directly to give the equivalent capacitance of the parallel combination.
4. A complicated circuit consisting of capacitors can often be reduced to a simple circuit containing only one capacitor. To do this, examine your initial circuit and replace any capacitors in series or any in parallel with equivalent capacitors, using the rules in 2 and 3. After making these changes, sketch your new circuit. Examine it and replace any series or parallel combinations again. Continue this process until a single equivalent capacitor is found.
5. To find the charge on, or the potential difference across, one of the capacitors in the complicated circuit, start with the final circuit found in 4, and gradually work your way back through the circuits using $C = Q/\Delta V$ and the information in 2 and 3 above.

## Example 16.6 Four Capacitors Connected in Series

Four capacitors are connected in series with a battery, as in Figure 16.19.

**A** Find the capacitance of the equivalent capacitor.

**Solution** The equivalent capacitance is found from Equation 16.16:

$$\frac{1}{C_{eq}} = \frac{1}{3.0\ \mu F} + \frac{1}{6.0\ \mu F} + \frac{1}{12\ \mu F} + \frac{1}{24\ \mu F}$$

$$C_{eq} = 1.6\ \mu F$$

Note that the equivalent capacitance is less than the capacitance of any of the individual capacitors in the combination.

**B** Find the charge on the 12-$\mu$F capacitor.

**Solution** We find the charge on the equivalent capacitor:

$$Q = C_{eq}\,\Delta V = (1.6 \times 10^{-6}\ F)(18\ V) = 29\ \mu C$$

This is also the charge on each of the capacitors it replaced. Thus, the charge on the 12-$\mu$F capacitor in the original circuit is 29 $\mu$C.

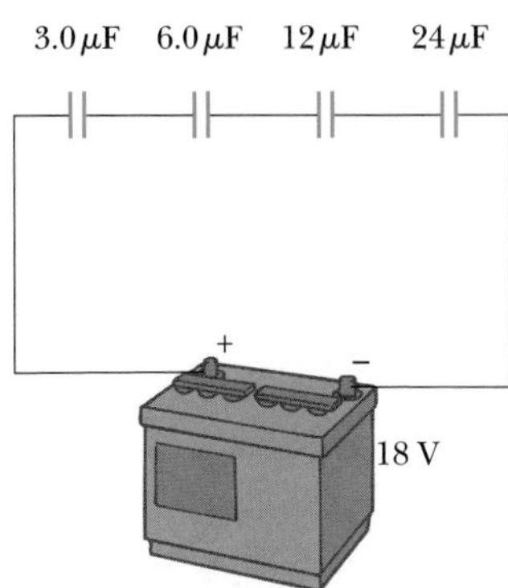


**FIGURE 16.19** (Example 16.6) Four capacitors connected in series.

## Example 16.7 Equivalent Capacitance

Find the equivalent capacitance between $a$ and $b$ for the combination of capacitors shown in Figure 16.20a. All capacitances are in microfarads.

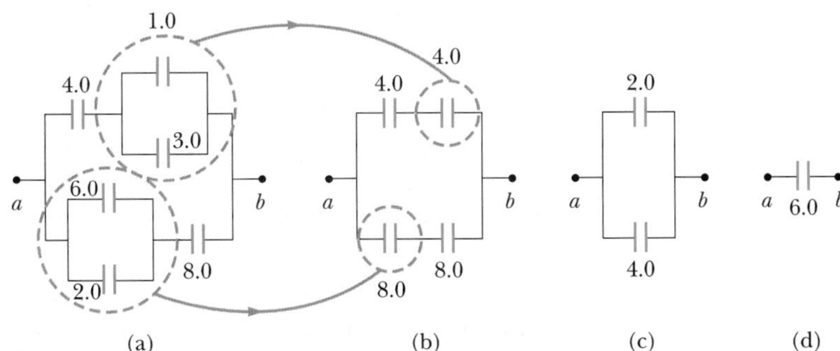


**FIGURE 16.20** (Example 16.7) To find the equivalent capacitance of the circuit in (a), the circuit is reduced in steps—as indicated in (b), (c), and (d)—using the series and parallel rules described in the text.

**Solution** Using Equations 16.13 and 16.16, we reduce the combination step by step as indicated in the figure. The 1.0-$\mu$F and 3.0-$\mu$F capacitors are in *parallel* and combine according to $C_{eq} = C_1 + C_2$. Their equivalent capacitance is 4.0 $\mu$F. Likewise, the 2.0-$\mu$F and 6.0-$\mu$F capacitors are also in *parallel* and have an equivalent capacitance of 8.0 $\mu$F. The upper branch in Figure 16.20b now consists of two 4.0-$\mu$F capacitors in *series,* which combine according to

$$\frac{1}{C_{eq}} = \frac{1}{C_1} + \frac{1}{C_2} = \frac{1}{4.0\ \mu F} + \frac{1}{4.0\ \mu F} = \frac{1}{2.0\ \mu F}$$

$$C_{eq} = 2.0\ \mu F$$

Likewise, the lower branch in Figure 16.20b consists of two 8.0-$\mu$F capacitors in *series* and has an equivalent capacitance of 4.0 $\mu$F. Finally, the 2.0-$\mu$F and 4.0-$\mu$F capacitors in Figure 16.20c are in *parallel* and have an equivalent capacitance of 6.0 $\mu$F. Hence, the equivalent capacitance of the circuit is 6.0 $\mu$F .

## 16.9 ENERGY STORED IN A CHARGED CAPACITOR

Almost everyone who works with electronic equipment has at some time verified that a capacitor can store energy. If the plates of a charged capacitor are connected by a conductor such as a wire, charge transfers from one plate to the other until the two are uncharged. The discharge can often be observed as a visible spark. If you accidentally touched the opposite plates of a charged capacitor, your fingers would act as a pathway by which the capacitor could discharge, inflicting an electric shock. The degree of shock would depend on the capacitance and voltage applied to the capacitor. Where high voltages and large quantities of charge are present, as in the power supply of a television set, such a shock can be fatal.

If a capacitor is initially uncharged (both plates neutral), so that the plates are at the same potential, almost no work is required to transfer a small amount of charge $\Delta Q$ from one plate to the other. However, once this charge has been transferred, a small potential difference $\Delta V = \Delta Q/C$ appears between the plates. Therefore, work must be done to transfer additional charge against this potential difference. We must look at the way the voltage changes at each moment. As more charge is transferred from one plate to the other, the potential difference increases in proportion to the charge. If the potential difference at any instant during the charging process is $\Delta V$, the work required to move more charge $\Delta Q$ through this potential difference is $\Delta V \Delta Q$; that is,

$$\Delta W = \Delta V \Delta Q$$

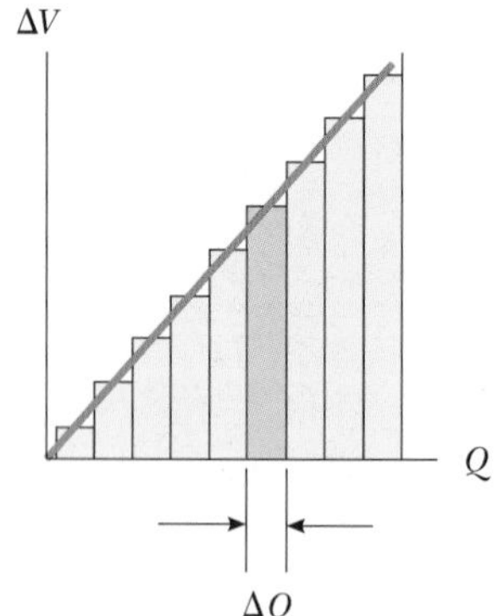


**FIGURE 16.21** A plot of voltage versus charge for a capacitor is a straight line with the slope $1/C$. The work required to move a charge of $\Delta Q$ through a potential difference of $\Delta V$ across the capacitor plates is $\Delta W = \Delta V \Delta Q$, which equals the area of the blue rectangle. The *total work* required to charge the capacitor to a final charge of $Q$ is the area under the straight line, which equals $Q\Delta V/2$.

We know that $\Delta V = Q/C$ for a capacitor that has a total charge of $Q$. Therefore, a plot of voltage versus total charge gives a straight line with a slope of $1/C$, as shown in Figure 16.21. Because the work $\Delta W$ is the area of the shaded rectangle, the total work done in charging the capacitor to a final voltage $\Delta V$ is the area under the voltage-charge curve, which in this case equals the area under the straight line. Because the area under this line is the area of a triangle (which is one half of the product of the base and height), the total work done is

$$W = \tfrac{1}{2} Q \Delta V \qquad [16.17]$$

Note that this is also the energy stored in the capacitor, because the work required to charge the capacitor equals the energy stored in the capacitor after it is charged. From the definition of capacitance, we find $Q = C\Delta V$; hence, we can express the energy stored as

$$\text{Energy stored} = \tfrac{1}{2} Q \Delta V = \tfrac{1}{2} C(\Delta V)^2 = \frac{Q^2}{2C} \qquad [16.18]$$

For example, the amount of energy stored in a 5.0-$\mu$F capacitor when it is connected across a 120-V battery is

$$\text{Energy stored} = \tfrac{1}{2} C(\Delta V)^2 = \tfrac{1}{2}(5.0 \times 10^{-6}\ \text{F})(120\ \text{V})^2 = 3.6 \times 10^{-2}\ \text{J}$$

In practice, there is a limit to the maximum energy (or charge) that can be stored in a capacitor, because electrical breakdown ultimately occurs between the plates of the capacitor at a sufficiently large value of $\Delta V$. For this reason, capacitors are usually labeled with a maximum operating voltage.

DEFIBRILLATORS

Large capacitors can store enough electrical energy to cause severe burns or even death if they are discharged so that the flow of charge can pass through the heart. Under the proper conditions, however, they can be used to sustain life by stopping cardiac fibrillation in heart attack victims. When fibrillation occurs, the heart produces a rapid, irregular pattern of beats. A fast discharge of electrical energy through the heart can return the organ to its normal beat pattern. Emergency medical teams use portable defibrillators that contain batteries capable of charging a capacitor to a high voltage. (The circuitry actually permits the capacitor to be charged to a much higher voltage than the battery.) In this case and others (camera flash units and lasers used for fusion experiments), capacitors serve as energy reservoirs that can be slowly charged and then discharged quickly to provide large amounts of energy in a short pulse. The stored electrical energy is released through the heart by conducting electrodes, called paddles, that are placed on both sides of the victim's chest. The paramedics must wait between applications of the electrical energy due to the time necessary for the capacitors to become fully charged. The high voltage on the capacitor can be obtained from a low-voltage battery in a portable machine with the phenomenon of *electromagnetic induction,* to be studied in Chapter 20.

**Webnote 16.2**

Researchers report evidence suggesting that the heart's erratic shivering during cardiac fibrillation is a form of chaos. Read more in *http://www.sciencenews.org/sn_arc97/1_25_97/fob1.htm*

### Example 16.8 Typical Voltage, Energy, and Discharge Time for a Defibrillator

A fully charged defibrillator contains 1.2 kJ of energy stored in a 110-$\mu$F capacitor. In a discharge through a patient, 600 J of electrical energy are delivered in 2.5 ms. (This is impossible for a battery, which cannot supply such a large amount of energy in such a short time.)

**A** Find the voltage needed to store 1.2 kJ in the unit.

**Solution** Because we know the energy stored and the capacitance, we can use Equation 16.18 to find the required voltage:

$$\text{Energy stored} = \tfrac{1}{2}C\,\Delta V^2$$

$$\Delta V = \sqrt{\frac{2 \times (\text{energy stored})}{C}} = \sqrt{\frac{2(1.2 \times 10^3\text{ J})}{1.1 \times 10^{-4}\text{ F}}} = 4.7 \times 10^3\text{ V}$$

**B** What power is delivered to the patient in watts?

**Solution** Because 600 J of energy are delivered in 2.5 ms, we can find the power delivered to the patient as follows:

$$\mathcal{P} = \frac{\text{Energy delivered}}{\Delta t} = \frac{600\text{ J}}{2.5 \times 10^{-3}\text{ s}} = 2.4 \times 10^5\text{ W}$$

### APPLYING PHYSICS 16.2

You have three capacitors and two batteries. How should you connect the batteries to all three capacitors so that the capacitors will store the maximum possible energy?

**Explanation** The energy stored in the capacitor is proportional to the capacitance and the square of the potential difference. Thus, we would like to maximize each of these quantities. We can do this by connecting the three capacitors in parallel (so that the capacitances add) across the two batteries in series (so that the potential differences add).

**Quick Quiz 16.5**

You charge a parallel-plate capacitor, remove it from the battery, and prevent the wires connected to the plates from touching each other. When you pull the plates farther apart, do the following quantities increase, decrease, or stay the same? (a) $C$, (b) $Q$, (c) $E$ between the plates, (d) $\Delta V$, (e) energy stored in the capacitor.

## 16.10 CAPACITORS WITH DIELECTRICS

A **dielectric** is an insulating material, such as rubber, plastic, or waxed paper. When a dielectric is inserted between the plates of a capacitor, the capacitance increases. If the dielectric completely fills the space between the plates, the capacitance is multiplied by the factor $\kappa$, called the **dielectric constant.**

The following experiment illustrates the effect of a dielectric in a capacitor. Consider a parallel-plate capacitor of charge $Q_0$ and capacitance $C_0$ in the absence of a dielectric. The potential difference across the capacitor plates can be measured, and it is given by $\Delta V_0 = Q_0/C_0$ (Fig. 16.22a). Because the capacitor is not connected to an external circuit, there is no pathway for charge to leave or be added to the plates. If a dielectric is now inserted between the plates as in Figure 16.22b, it is found that the voltage across the plates is *reduced* by the factor $\kappa$ to the value $\Delta V$, where

$$\Delta V = \frac{\Delta V_0}{\kappa}$$

Because $\kappa > 1$, $\Delta V$ is less than $\Delta V_0$. Because the charge $Q_0$ on the capacitor does not change, we conclude that the capacitance $C$ in the presence of the dielectric must change to the value

$$C = \frac{Q_0}{\Delta V} = \frac{Q_0}{\Delta V_0/\kappa} = \frac{\kappa Q_0}{\Delta V_0}$$

or

$$C = \kappa C_0 \qquad [16.19]$$

According to this result, the capacitance is *multiplied* by the factor $\kappa$ when the dielectric completely fills the region between the plates. For a parallel-plate capacitor, where the capacitance in the absence of a dielectric is $C_0 = \epsilon_0 A/d$, we can express the capacitance in the presence of a dielectric as

$$C = \kappa \epsilon_0 \frac{A}{d} \qquad [16.20]$$

From this result it appears that the capacitance could be made very large by decreasing $d$, the plate separation. In practice, the lowest value of $d$ is limited by the electric discharge that can occur through the dielectric material separating the plates. For any given plate separation, there is a maximum electric field that can be produced in the dielectric before it breaks down and begins to conduct. This maximum electric field is called the **dielectric strength,** and for air its value is

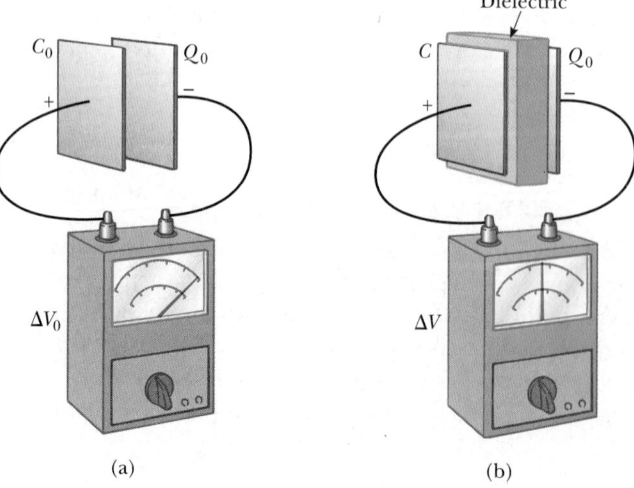


**FIGURE 16.22** (a) With air between the plates, the voltage across the capacitor is $\Delta V_0$, the capacitance is $C_0$, and the charge is $Q_0$. (b) With a dielectric between the plates, the charge remains at $Q_0$, but the voltage and capacitance both change.

**TABLE 16.1 Dielectric Constants and Dielectric Strengths of Various Materials at Room Temperature**

| Material | Dielectric Constant, $\kappa$ | Dielectric Strength (V/m) |
|---|---|---|
| Vacuum | 1.000 00 | — |
| Air | 1.000 59 | $3 \times 10^6$ |
| Bakelite | 4.9 | $24 \times 10^6$ |
| Fused quartz | 3.78 | $8 \times 10^6$ |
| Pyrex glass | 5.6 | $14 \times 10^6$ |
| Polystyrene | 2.56 | $24 \times 10^6$ |
| Teflon | 2.1 | $60 \times 10^6$ |
| Neoprene rubber | 6.7 | $12 \times 10^6$ |
| Nylon | 3.4 | $14 \times 10^6$ |
| Paper | 3.7 | $16 \times 10^6$ |
| Strontium titanate | 233 | $8 \times 10^6$ |
| Water | 80 | — |
| Silicone oil | 2.5 | $15 \times 10^6$ |

Dielectric breakdown in air. Sparks are produced when a large alternating voltage is applied across the wires using a high-voltage induction coil power supply. *(© Loren Winters/Visuals Unlimited)*

about $3 \times 10^6$ V/m. Most insulating materials have dielectric strengths greater than that of air, as indicated by the values in Table 16.1.

Commercial capacitors are often made using metal foil interlaced with thin sheets of paraffin-impregnated paper or mylar, which serves as the dielectric material. These alternating layers of metal foil and dielectric are then rolled into a small cylinder (Fig. 16.23a). A high-voltage capacitor commonly consists of a number of interwoven metal plates immersed in silicone oil (Fig. 16.23b). Small capacitors are often constructed from ceramic materials. Variable capacitors (typically 10–500 pF) usually consist of two interwoven sets of metal plates, one fixed and the other movable, with air as the dielectric.

An electrolytic capacitor (Fig. 16.23c) is often used to store large amounts of charge at relatively low voltages. It consists of a metal foil in contact with an electrolyte—a solution that conducts charge by virtue of the motion of the ions contained in it. When a voltage is applied between the foil and the electrolyte, a thin layer of metal oxide (an insulator) is formed on the foil, and this layer serves as the dielectric. Enormous capacitances can be attained because the dielectric layer is very thin.

When electrolytic capacitors are used in circuits, the polarity (the plus and minus signs on the device) must be observed. If the polarity of the applied voltage is opposite that intended, the oxide layer will be removed and the capacitor will conduct rather than store charge. Furthermore, reversing the polarity can result in such a large current that the capacitor may either burn or produce steam and explode.

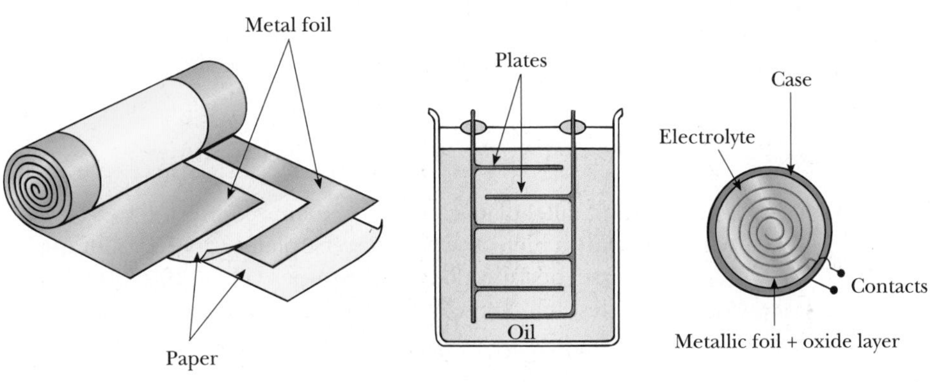


**FIGURE 16.23** Three commercial capacitor designs. (a) A tubular capacitor whose plates are separated by paper and then rolled into a cylinder, (b) a high-voltage capacitor consisting of many parallel plates separated by oil, and (c) an electrolytic capacitor.

A collection of capacitors used in a variety of applications. *(Paul Silverman/Fundamental Photographs)*

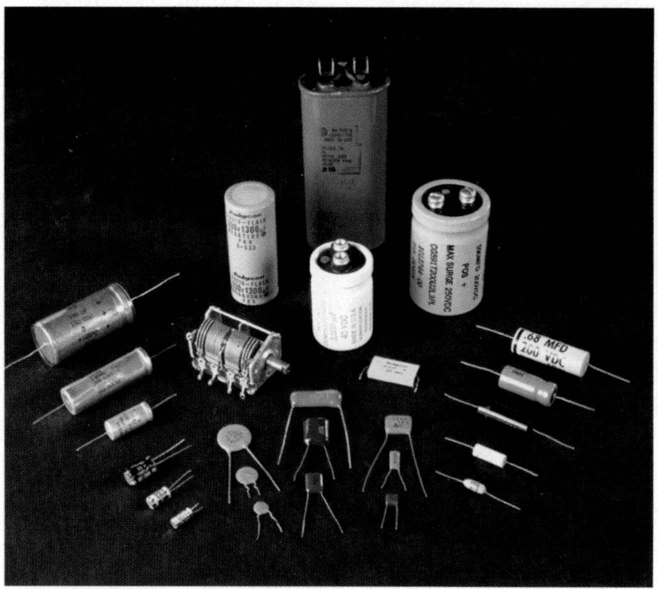

**APPLICATION**

STUD FINDERS

## APPLYING PHYSICS 16.3

If you have ever tried to hang a picture on a wall securely, you know that it can be difficult to locate a wooden stud in which to anchor your nail or screw. The principles discussed in this section can be used to detect a stud electronically. The primary element of an electronic stud finder is a capacitor with its plates arranged side by side instead of facing one another, as in Figure 16.24. How does this device work?

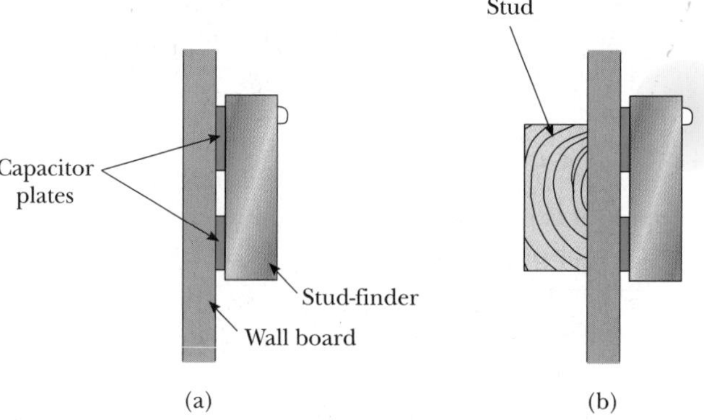


**FIGURE 16.24** (Applying Physics 16.3) A stud finder. (a) The materials between the plates of the capacitor are the drywall and the air behind it. (b) The materials become drywall and wood when the detector moves across a stud in the wall. The change in the dielectric constant causes a signal light to illuminate.

**Explanation** As the detector is moved along a wall, its capacitance changes when it passes across a stud because the dielectric constant of the material "between" the plates changes. The change in capacitance can be used to cause a light to come on, signaling the presence of the stud.

### Webnote 16.3

At home you might have a touch-sensitive lamp (turns on when you touch it). This device elegantly makes use of capacitance. Visit *http://www.howstuffworks.com/question42.htm*

**Quick Quiz 16.6**

A fully charged parallel-plate capacitor remains connected to a battery while you slide a dielectric between the plates. Do the following quantities increase, decrease, or stay the same? (a) $C$, (b) $Q$, (c) $E$ between the plates, (d) $\Delta V$, (e) energy stored in the capacitor.

## Example 16.9 A Paper-Filled Capacitor

A parallel-plate capacitor has plates 2.0 cm by 3.0 cm. The plates are separated by a 1.0-mm thickness of paper.

**A** Find the capacitance of this device.

**Solution** Because $\kappa = 3.7$ for paper (Table 16.1), we get

$$C = \kappa\epsilon_0 \frac{A}{d} = 3.7\left(8.85 \times 10^{-12} \frac{\text{C}^2}{\text{N} \cdot \text{m}^2}\right)\left(\frac{6.0 \times 10^{-4}\ \text{m}^2}{1.0 \times 10^{-3}\ \text{m}}\right)$$

$$= 20 \times 10^{-12}\ \text{F} = 20\ \text{pF}$$

**B** Find the maximum charge that can be placed on the capacitor.

**Solution** From Table 16.1 we see that the dielectric strength of paper is equal to $16 \times 10^6$ V/m. Because the paper thickness is 1.0 mm, the maximum voltage that can be applied before electrical breakdown occurs can be calculated using Equation 16.4:

$$\Delta V_{\text{max}} = E_{\text{max}} d = (16 \times 10^6\ \text{V/m})(1.0 \times 10^{-3}\ \text{m}) = 16 \times 10^3\ \text{V}$$

Hence, the maximum charge that can be placed on the capacitor is

$$Q_{\text{max}} = C\Delta V_{\text{max}} = (20 \times 10^{-12}\ \text{F})(16 \times 10^3\ \text{V}) = 0.32\ \mu\text{C}$$

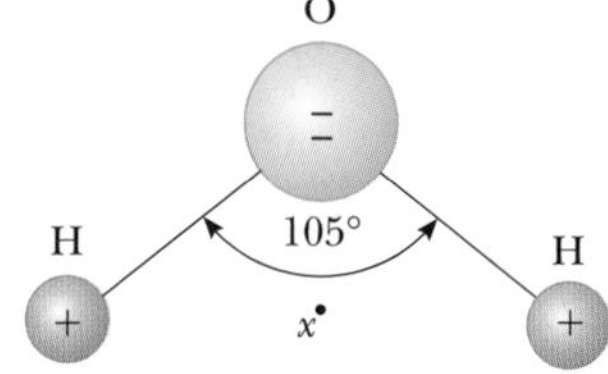


**FIGURE 16.25** The water molecule, $H_2O$, has a permanent polarization resulting from its bent geometry. The point labeled *x* is the center of positive charge.

### AN ATOMIC DESCRIPTION OF DIELECTRICS

The explanation of why a dielectric increases the capacitance of a capacitor is based on an atomic description of the material, which in turn involves a property of some molecules called **polarization.** A molecule is said to be polarized when there is a separation between the "centers of gravity" of its negative charge and its positive charge. In some molecules, such as water, this condition is always present. To see why, consider the geometry of a water molecule (Fig. 16.25).

The molecule is arranged so that the negative oxygen atom is bonded to the positively charged hydrogen atoms with a 105° angle between the two bonds. The center of negative charge is at the oxygen atom, and the center of positive charge lies at a point midway along the line joining the hydrogen atoms (point *x* in the diagram). Materials composed of molecules that are permanently polarized in this fashion have large dielectric constants, and indeed Table 16.1 shows that the dielectric constant of water is quite large ($\kappa = 80$).

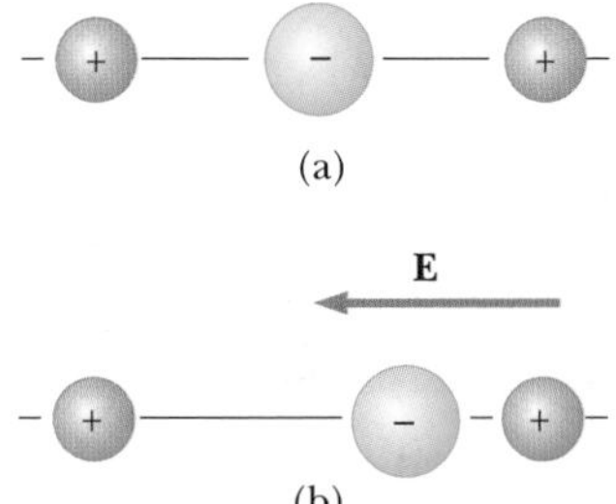


**FIGURE 16.26** (a) A symmetric molecule has no permanent polarization. (b) An external electric field induces a polarization in the molecule.

A symmetric molecule (Fig. 16.26a) can have no permanent polarization, but a polarization can be induced by an external electric field. A field directed to the left, as in Figure 16.26b, would cause the center of positive charge to shift to the left from its initial position, and the center of negative charge to shift to the right. This *induced polarization* is the effect that predominates in most materials used as dielectrics in capacitors.

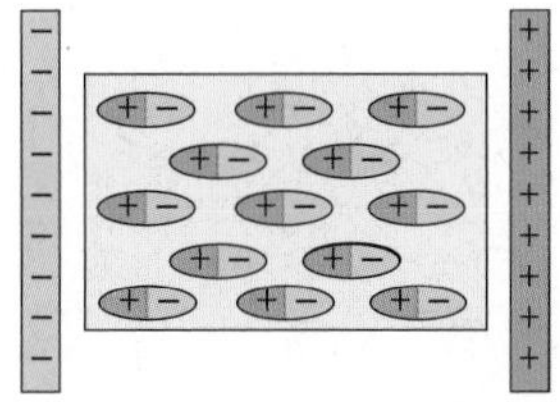

**FIGURE 16.27** When a dielectric is placed between the plates of a charged parallel-plate capacitor, the dielectric becomes polarized. This creates a net positive induced charge on the left side of the dielectric and a net negative induced charge on the right side. As a result, the capacitance of the device is multiplied by the factor $\kappa$.

To understand why the polarization of a dielectric can affect capacitance, consider Figure 16.27, which shows a slab of dielectric placed between the plates of a parallel-plate capacitor. The dielectric becomes polarized as shown because it is in the electric field that exists between the metal plates. Notice that a net positive charge appears on the dielectric surface adjacent to the negatively charged metal plate. The presence of this positive charge on the dielectric effectively reduces some of the negative charge on the metal, allowing more negative charge to be stored on the capacitor plates for a given applied voltage. From the definition of capacitance, $C = Q/\Delta V$, we see that, because the plates can store more charge for a given voltage, the capacitance must increase.

## APPLYING PHYSICS 16.4

Consider a parallel-plate capacitor with a dielectric material between the plates. Is the capacitance higher on a cold day or a hot day?

**Explanation** The polarization of the molecules in the dielectric increases the capacitance when the dielectric is added. As the temperature increases, there is more vibrational motion of the polarized molecules. This disturbs the orderly arrangement of the polarized molecules, and the net polarization decreases. Thus, the capacitance must decrease as the temperature increases.

## SUMMARY

The **difference in electric potential $\Delta V$** between two points $A$ and $B$ is

$$\Delta V \equiv V_B - V_A = \frac{\Delta PE}{q} \quad [16.2]$$

where $\Delta PE$ is the *change* in electrical potential energy experienced by a charge $q$ as it moves between $A$ and $B$. The units of potential difference are joules per coulomb, or **volts;** $1\ \mathrm{J/C} = 1\ \mathrm{V}$.

The **electric potential difference** between two points $A$ and $B$ in a uniform electric field **E** is

$$V_B - V_A = -Ed \quad [16.4]$$

where $d$ is the distance between $A$ and $B$ and $E$ is the strength of the electric field in that region.

The **electric potential** due to a point charge $q$ at distance $r$ from the point charge is

$$V = k_e \frac{q}{r} \quad [16.5]$$

The **electrical potential energy** of a pair of point charges separated by distance $r$ is

$$PE = k_e \frac{q_1 q_2}{r} \quad [16.6]$$

Every point on the surface of a charged conductor in electrostatic equilibrium is at the same potential. Furthermore, the potential is constant everywhere inside the conductor and equals its value on the surface.

The **electron volt** is defined as the energy that an electron (or proton) gains when accelerated through a potential difference of 1 V. The conversion between electron volts and joules is

$$1\ \mathrm{eV} = 1.60 \times 10^{-19}\ \mathrm{C \cdot V} = 1.60 \times 10^{-19}\ \mathrm{J} \quad [16.8]$$

A **capacitor** consists of two metal plates with charges that are equal in magnitude but opposite in sign. The capacitance ($C$) of any capacitor is the ratio of the magnitude of the charge $Q$ on either plate to the potential difference $\Delta V$ between them:

$$C \equiv \frac{Q}{\Delta V} \quad [16.9]$$

Capacitance has the units coulombs per volt, or farads; $1\ \mathrm{C/V} = 1\ \mathrm{F}$.

The capacitance of two parallel metal plates of area $A$ separated by distance $d$ is

$$C = \epsilon_0 \frac{A}{d} \quad [16.10]$$

where $\epsilon_0$ is a constant called the **permittivity of free space,** with the value $\epsilon_0 = 8.85 \times 10^{-12}\ \mathrm{C^2/N \cdot m^2}$.

The **equivalent capacitance of a parallel combination** of capacitors is

$$C_{\text{eq}} = C_1 + C_2 + C_3 + \cdots \qquad [16.13]$$

If two or more capacitors are connected in series, the **equivalent capacitance of the series combination** is

$$\frac{1}{C_{\text{eq}}} = \frac{1}{C_1} + \frac{1}{C_2} + \frac{1}{C_3} + \cdots \qquad [16.16]$$

Three equivalent expressions for calculating the **energy stored** in a charged capacitor are

$$\text{Energy stored} = \tfrac{1}{2}Q\,\Delta V = \tfrac{1}{2}C(\Delta V)^2 = \frac{Q^2}{2C} \qquad [16.18]$$

When a nonconducting material, called a **dielectric,** is placed between the plates of a capacitor, the capacitance is multiplied by the factor $\kappa$, which is called the **dielectric constant** and is a property of the dielectric material. The capacitance of a parallel-plate capacitor filled with a dielectric is

$$C = \kappa\epsilon_0 \frac{A}{d} \qquad [16.20]$$

## CONCEPTUAL QUESTIONS

1. Criticize this statement: "There is no potential in a region where the electric field is uniform."
2. "The equipotentials of a uniformly charged conducting sphere are spheres." Is this statement true? Discuss.
3. A parallel-plate capacitor is charged by a battery, and the battery is then disconnected from the capacitor. Because the charges on the capacitor plates are opposite in sign, they attract each other. Hence, it takes positive work to increase the plate separation. Show that the external work done when the plate separation is increased leads to an increase in the energy stored in the capacitor.
4. Distinguish between electric potential and electrical potential energy.
5. Suppose you are sitting in a car and a 20-kV power line drops across the car. Should you stay in the car or get out? The power line potential is 20 kV compared to the potential of the ground.
6. Why is it important to avoid sharp edges or points on conductors used in high-voltage equipment?
7. Explain why, under static conditions, all points in a conductor must be at the same electric potential.
8. If you are given three different capacitors $C_1$, $C_2$, and $C_3$, how many different combinations of capacitance can you produce, using all capacitors in your circuits?
9. Why is it dangerous to touch the terminals of a high-voltage capacitor even after the voltage source that charged the battery is disconnected from the capacitor? What can be done to make the capacitor safe to handle after the voltage source has been removed?
10. The plates of a capacitor are connected to a battery. What happens to the charge on the plates if the connecting wires are removed from the battery? What happens to the charge if the wires are removed from the battery and connected to each other?
11. Can electric field lines ever cross? Why or why not? Can equipotentials ever cross? Why or why not?
12. What happens to the energy stored in a capacitor (a) if the potential difference across the plates is doubled? (b) if the plate separation is doubled while the capacitor is connected to a battery?
13. (a) What happens to the *charge* stored in a capacitor if the potential difference across the plates is doubled? (b) What happens to the *energy* stored in a capacitor if the plate separation is doubled while the charge on the capacitor remains constant (as it would after being disconnected from a battery)?
14. Is it always possible to reduce a combination of capacitors to one equivalent capacitor with the rules developed in this chapter? Explain.
15. If you were asked to design a capacitor where small size and large capacitance were required, what factors would be important in your design?
16. Explain why a dielectric increases the maximum operating voltage of a capacitor although the physical size of the capacitor does not change.
17. What happens to the energy stored in a capacitor (a) if a dielectric is inserted into the capacitor while it is connected to a battery? (b) if a dielectric is inserted into the capacitor while the charge on the capacitor remains constant (as it would after being disconnected from a battery)?

## PROBLEMS

1, 2, 3 = straightforward, intermediate, challenging ☐ = full solution available in Student Solutions Manual/Study Guide

web = solution posted at **http://info.brookscole.com/serway** = biomedical application

### Section 16.1 Potential Difference and Electric Potential

1. A proton moves 2.00 cm parallel to a uniform electric field with $E = 200$ N/C. (a) How much work is done by the field on the proton? (b) What change occurs in the potential energy of the proton? (c) Through what potential difference did the proton move?
2. A uniform electric field of magnitude 250 V/m is directed in the positive $x$ direction. A positive 12-$\mu C$ charge moves from the origin to the point $(x, y) = (20 \text{ cm}, 50 \text{ cm})$. (a) What is the change in the potential energy of this charge? (b) Through what potential difference does the charge move?
3. A potential difference of 90 mV exists between the inner and outer surfaces of a cell membrane. The inner surface is negative relative to the outer surface. How much work is required to eject a positive sodium ion ($Na^+$) from the interior of the cell?
4. An ion accelerated through a potential difference of 115 V experiences an increase in kinetic energy of $7.37 \times 10^{-17}$ J. Calculate the charge on the ion.
5. The difference in potential between the accelerating plates of a TV set is about 25 kV. If the distance between these plates is 1.5 cm, find the magnitude of the uniform electric field in this region.
6. To recharge a 12-V battery, a battery charger must move $3.6 \times 10^5$ C of charge from the negative terminal to the positive terminal. How much work is done by the battery charger? Express your answer in joules.
7. web A pair of oppositely charged parallel plates are separated by 5.33 mm. A potential difference of 600 V exists between the plates. (a) What is the magnitude of the electric field between the plates? (b) What is the magnitude of the force on an electron between the plates? (c) How much work must be done on the electron to move it to the negative plate if it is initially positioned 2.90 mm from the positive plate?
8. Suppose an electron is released from rest in a uniform electric field whose strength is $5.90 \times 10^3$ V/m. (a) Through what potential difference will it have passed after moving 1.00 cm? (b) How fast will the electron be moving after it has traveled 1.00 cm?
9. A 4.00-kg block carrying a charge $Q = 50.0$ $\mu$C is connected to a spring for which $k = 100$ N/m. The block lies on a frictionless horizontal track, and the system is immersed in a uniform electric field of magnitude $E = 5.00 \times 10^5$ V/m directed as in Figure P16.9. (a) If the block is released at rest when the spring is unstretched (at $x = 0$), by what maximum amount does the spring expand? (b) What is the equilibrium position of the block?

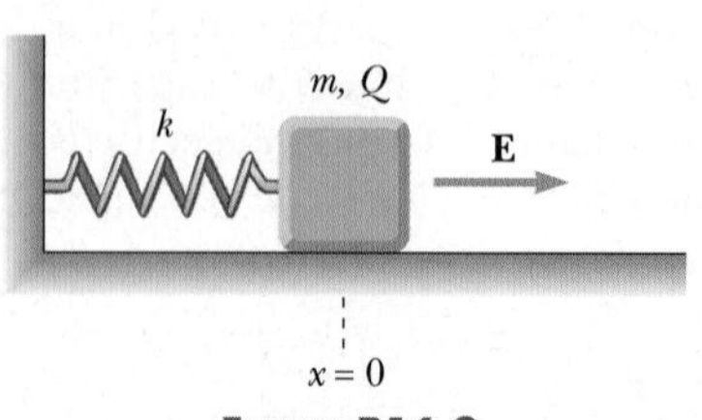

**FIGURE P16.9**

10. On planet Tehar, the free-fall acceleration is the same as that on Earth, but there is also a strong downward electric field that is uniform close to the planet's surface. A 2.00-kg ball having a charge of 5.00 $\mu$C is thrown upward at a speed of 20.1 m/s, and it hits the ground after an interval of 4.10 s. What is the potential difference between the starting point and the top point of the trajectory?

### Section 16.2 Electric Potential and Potential Energy Due to Point Charges

### Section 16.3 Potentials and Charged Conductors

### Section 16.4 Equipotential Surfaces

11. (a) Find the potential 1.00 cm from a proton. (b) What is the potential difference between two points that are 1.00 cm and 2.00 cm from a proton?
12. Find the potential at point $P$ for the rectangular grouping of charges shown in Figure P16.12.

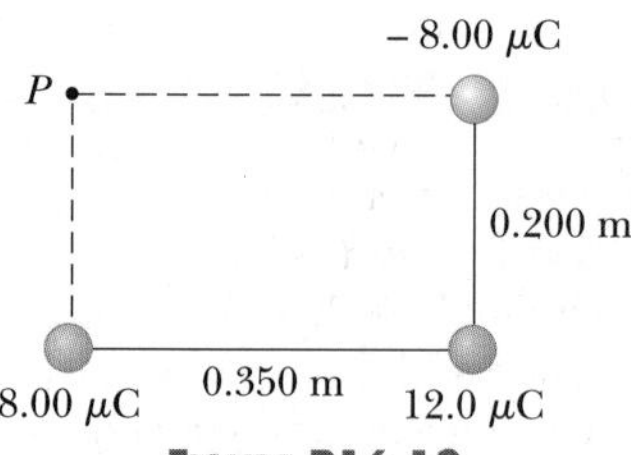

**FIGURE P16.12**

13. (a) Find the electric potential, taking zero at infinity, at the upper-right corner (the corner without a charge) of the rectangle in Figure P16.13. (b) Repeat if the 2.00-$\mu$C charge is replaced with a charge of $-2.00$ $\mu$C.

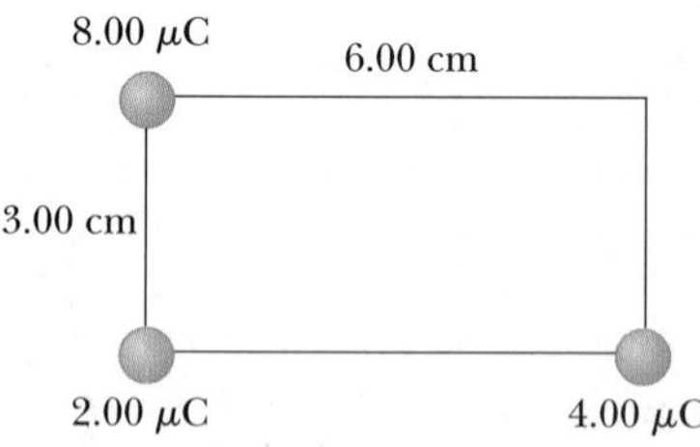

**FIGURE P16.13** (Problems 13 and 14)

14. Three charges are situated at corners of a rectangle as in Figure P16.13. How much energy would be expended in moving the 8.00-$\mu$C charge to infinity?
15. Two point charges $Q_1 = +5.00$ nC and $Q_2 = -3.00$ nC are separated by 35.0 cm. (a) What is the electric potential at a point midway between the charges? (b) What is the potential energy of the pair of charges? What is the significance of the algebraic sign of your answer?
16. Calculate the speed of (a) an electron that has a kinetic energy of 1.00 eV and (b) a proton that has a kinetic energy of 1.00 eV.

17. The three charges in Figure P16.17 are at the vertices of an isosceles triangle. Calculate the electric potential at the midpoint of the base, taking $q = 7.00$ nC.

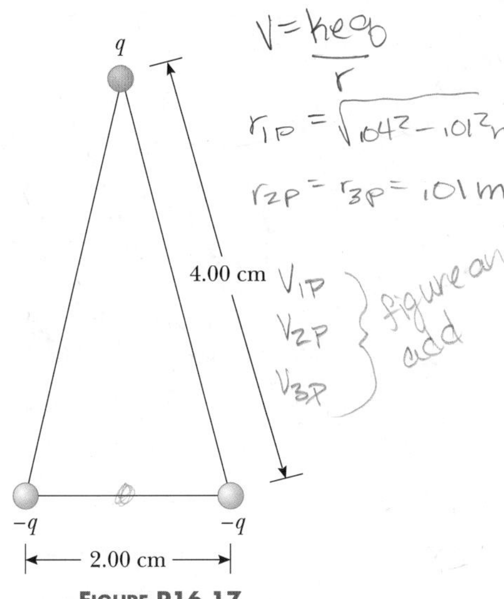


**FIGURE P16.17**

18. An electron starts from rest 3.00 cm from the center of a uniformly charged sphere of radius 2.00 cm. If the sphere carries a total charge of $1.00 \times 10^{-9}$ C, how fast will the electron be moving when it reaches the surface of the sphere?

19. web In Rutherford's famous scattering experiments (which led to the planetary model of the atom), alpha particles (having charges of $+2e$ and masses of $6.64 \times 10^{-27}$ kg) were fired toward a gold nucleus with charge $+79e$. An alpha particle, initially very far from the gold nucleus, is fired at $2.00 \times 10^{7}$ m/s directly toward the gold nucleus as in Figure P16.19. How close does the alpha particle get to the gold nucleus before turning around? Assume the gold nucleus remains stationary.

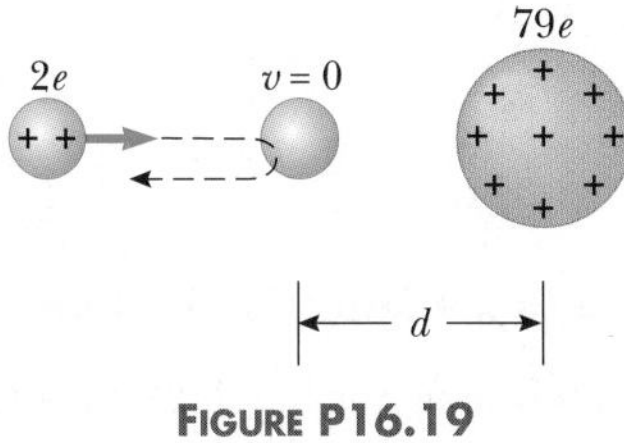


**FIGURE P16.19**

20. Starting with the definition of work, prove that at every point on an equipotential surface the surface must be perpendicular to the local electric field.

21. A small spherical object carries a charge of 8.00 nC. At what distance from the center of the object is the potential equal to 100 V? 50.0 V? 25.0 V? Is the spacing of the equipotentials proportional to the change in potential?

### Section 16.6 Capacitance

### Section 16.7 The Parallel-Plate Capacitor

22. (a) How much charge is on each plate of a 4.00-$\mu$F capacitor when it is connected to a 12.0-V battery? (b) If this same capacitor is connected to a 1.50-V battery, what charge is stored?

23. Consider Earth and a cloud layer 800 m above Earth to be the plates of a parallel-plate capacitor. (a) If the cloud layer has an area of 1.0 km$^2$ = $1.0 \times 10^{6}$ m$^2$, what is the capacitance? (b) If an electric field strength greater than $3.0 \times 10^{6}$ N/C causes the air to break down and conduct charge (lightning), what is the maximum charge the cloud can hold?

24. The potential difference between a pair of oppositely charged parallel plates is 400 V. (a) If the spacing between the plates is doubled without altering the charge on the plates, what is the new potential difference between the plates? (b) If the plate spacing is doubled while the potential difference between the plates is kept constant, what is the ratio of the final charge on one of the plates to the original charge?

25. An air-filled capacitor consists of two parallel plates, each with an area of 7.60 cm$^2$, separated by a distance of 1.80 mm. If a 20.0-V potential difference is applied to these plates, calculate (a) the electric field between the plates, (b) the capacitance, and (c) the charge on each plate.

26. A 1-megabit computer memory chip contains many $60.0 \times 10^{-15}$ F capacitors. Each capacitor has a plate area of $21.0 \times 10^{-12}$ m$^2$. Determine the plate separation of such a capacitor (assume a parallel-plate configuration). The diameter of an atom is on the order of $10^{-10}$ m = 1 Å. Express the plate separation in angstroms.

27. The plates of a parallel-plate capacitor are separated by 0.100 mm. If the material between the plates is air, what plate area is required to provide a capacitance of 2.00 pF?

28. A small object with a mass of 350 mg carries a charge of 30.0 nC and is suspended by a thread between the vertical plates of a parallel-plate capacitor. The plates are separated by 4.00 cm. If the thread makes an angle of 15.0° with the vertical, what is the potential difference between the plates?

### Section 16.8 Combinations of Capacitors

29. A series circuit consists of a 0.050-$\mu$F capacitor, a 0.100-$\mu$F capacitor, and a 400-V battery. Find the charge (a) on each of the capacitors and (b) on each of the capacitors if they are reconnected in parallel across the battery.

30. Three capacitors $C_1 = 5.00$ $\mu$F, $C_2 = 4.00$ $\mu$F, and $C_3 = 9.00$ $\mu$F are connected together. Find the effective capacitance of the group (a) if they are all in parallel, and (b) if they are all in series.

31. (a) Find the equivalent capacitance of the group of capacitors in Figure P16.31. (b) Find the charge on and the potential difference across each.

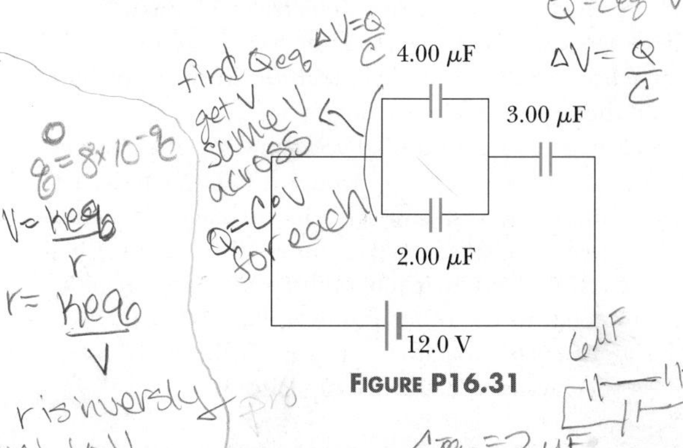


**FIGURE P16.31**

32. Two capacitors when connected in parallel give an equivalent capacitance of 9.00 pF and give an equivalent capacitance of 2.00 pF when connected in series. What is the capacitance of each capacitor?

33. Four capacitors are connected as shown in Figure P16.33. (a) Find the equivalent capacitance between points *a* and *b*. (b) Calculate the charge on each capacitor if a 15.0-V battery is connected across points *a* and *b*.

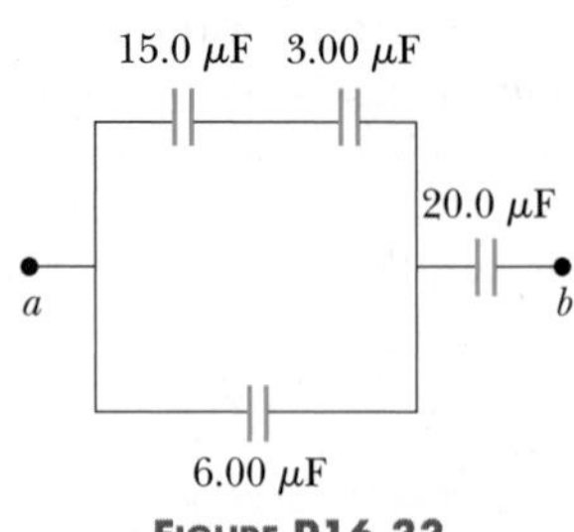


**FIGURE P16.33**

34. Consider the combination of capacitors in Figure P16.34. (a) What is the equivalent capacitance of the group? (b) Determine the charge on each capacitor.

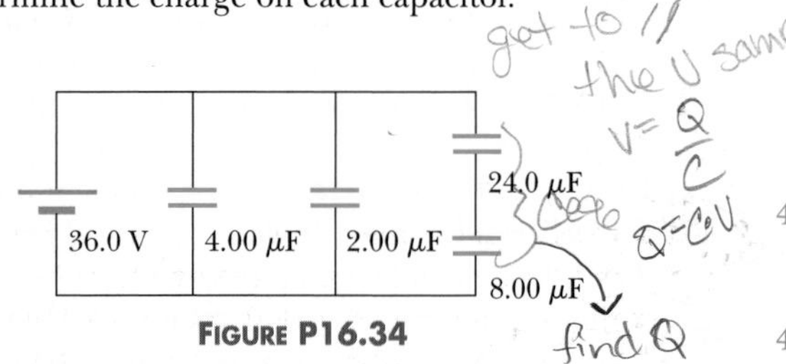


**FIGURE P16.34**

35. Find the charge on each of the capacitors in Figure P16.35.

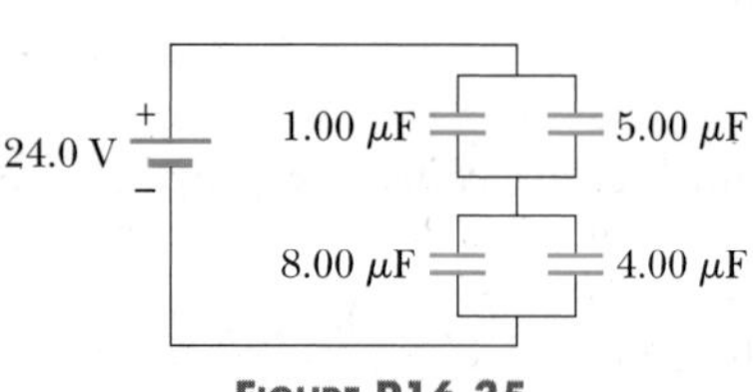


**FIGURE P16.35**

36. To repair a power supply for a stereo amplifier, an electronics technician needs a 100-$\mu$F capacitor capable of withstanding a potential difference of 90 V between the plates. The only available supply is a box of five 100-$\mu$F capacitors, each having a maximum voltage capability of 50 V. Can the technician substitute a combination of these capacitors that has the proper electrical characteristics, and if so, what will be the maximum voltage across any of the capacitors used? (*Hint:* The technician may not have to use all the capacitors in the box.)

37. web A 25.0-$\mu$F capacitor and a 40.0-$\mu$F capacitor are charged by being connected across separate 50.0-V batteries. (a) Determine the resulting charge on each capacitor. (b) The capacitors are then disconnected from their batteries and connected to each other, with each negative plate connected to the other positive plate. What is the final charge of each capacitor, and what is the final potential difference across the 40.0-$\mu$F capacitor?

38. A 10.0-$\mu$F capacitor is fully charged across a 12.0-V battery. The capacitor is then disconnected from the battery and connected across an initially uncharged capacitor, *C*. The resulting voltage across each capacitor is 3.00 V. What is the capacitance *C*?

39. A 1.00-$\mu$F capacitor is first charged by being connected across a 10.0-V battery. It is then disconnected from the battery and connected across an uncharged 2.00-$\mu$F capacitor. Determine the resulting charge on each capacitor.

40. Find the equivalent capacitance between points *a* and *b* for the group of capacitors connected as shown in Figure P16.40 if $C_1 = 5.00\ \mu F$, $C_2 = 10.0\ \mu F$, and $C_3 = 2.00\ \mu F$.

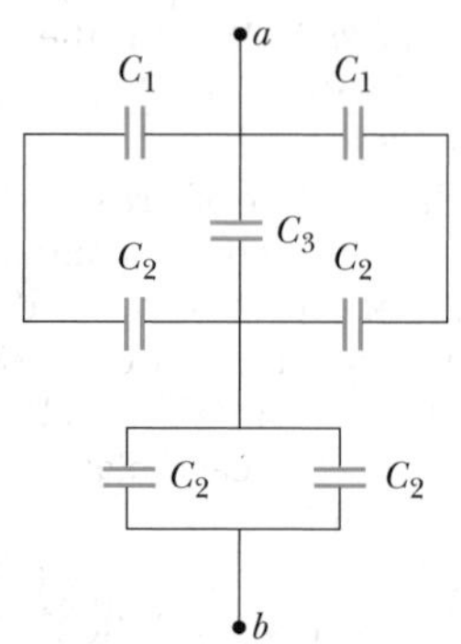


**FIGURE P16.40** (Problems 40 and 41)

41. For the network described in the previous problem if the potential between points *a* and *b* is 60.0 V, what charge is stored on $C_3$?

42. Find the equivalent capacitance between points *a* and *b* in the combination of capacitors shown in Figure P16.42.

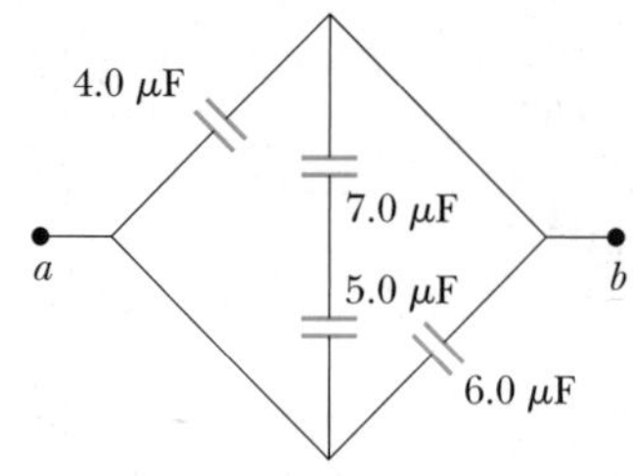


**FIGURE P16.42**

## Section 16.9 Energy Stored in a Charged Capacitor

43. A parallel-plate capacitor has 2.00-cm$^2$ plates that are separated by 5.00 mm with air between them. If a 12.0-V battery is connected to this capacitor, how much energy does it store?

44. Two capacitors $C_1 = 25.0\ \mu F$ and $C_2 = 5.00\ \mu F$ are connected in parallel and charged with a 100-V power supply. (a) Calculate the total energy stored in the two capacitors. (b) What potential difference would be required across the same two capacitors connected in *series* in order that the combination store the same energy as in (a)?

45. Consider the parallel-plate capacitor formed by Earth and a cloud layer as described in Problem 16.23. Assume this capacitor will discharge (that is, lightning occurs) when the electric field strength between the plates reaches $3.0 \times 10^6$ N/C. What is the energy released if the capacitor discharges completely during a lightning strike?

46. A certain storm cloud has a potential difference of $1.00 \times 10^8$ V relative to a tree. If, during a lightning storm, 50.0 C of charge is transferred through this potential difference and 1.00% of the energy is absorbed by the tree, how much water (sap in the tree) initially at 30.0°C can be boiled away? Water has a specific heat of 4 186 J/kg·°C, a boiling point of 100°C, and a heat of vaporization of $2.26 \times 10^6$ J/kg.

## Section 16.10 Capacitors with Dielectrics

47. A capacitor with air between its plates is charged to 100 V and then disconnected from the battery. When a piece of glass is placed between the plates, the voltage across the capacitor drops to 25 V. What is the dielectric constant of this glass? (Assume the glass completely fills the space between the plates.)

48. Two parallel plates, each of area 2.00 cm$^2$, are separated by 2.00 mm with purified nonconducting water between them. A voltage of 6.00 V is applied between the plates. Calculate (a) the magnitude of the electric field between the plates, (b) the charge stored on each plate, and (c) the charge stored on each plate if the water is removed and replaced with air.

49. Determine (a) the capacitance and (b) the maximum voltage that can be applied to a Teflon-filled parallel-plate capacitor having a plate area of 175 cm$^2$ and insulation thickness of 0.040 0 mm.

50. A commercial capacitor is constructed as in Figure 16.23a. This particular capacitor is made from two strips of aluminum separated by a strip of paraffin-coated paper. Each strip of foil and paper is 7.00 cm wide. The foil is 0.004 00 mm thick, and the paper is 0.025 0 mm thick and has a dielectric constant of 3.70. What length should the strips be if a capacitance of $9.50 \times 10^{-8}$ F is desired before the capacitor is rolled up? (Use the parallel-plate formula. Adding a second strip of paper and rolling up the capacitor doubles its capacitance by allowing both surfaces of each strip of foil to store charge.)

51. A model of a red blood cell portrays the cell as a spherical capacitor—a positively charged liquid sphere of surface area $A$, separated by a membrane of thickness $t$ from the surrounding negatively charged fluid. Tiny electrodes introduced into the interior of the cell show a potential difference of 100 mV across the membrane. The membrane's thickness is estimated to be 100 nm and its dielectric constant to be 5.00. (a) If an average red blood cell has a mass of $1.00 \times 10^{-12}$ kg, estimate the volume of the cell and thus find its surface area. The density of blood is 1 100 kg/m$^3$. (b) Estimate the capacitance of the cell. (c) Calculate the charge on the surface of the membrane. How many electronic charges does this represent?

## ADDITIONAL PROBLEMS

52. Three parallel-plate capacitors are constructed, each having the same plate spacing $d$, and with $C_1$ having plate area $A_1$, $C_2$ having area $A_2$, and $C_3$ having area $A_3$. Show that the total capacitance $C$ of these three capacitors connected in parallel is the same as a capacitor having plate spacing $d$ and plate area $A = A_1 + A_2 + A_3$.

53. Three parallel-plate capacitors are constructed, each having the same plate area $A$, and with $C_1$ having plate spacing $d_1$, $C_2$ having plate spacing $d_2$, and $C_3$ having plate spacing $d_3$. Show that the total capacitance $C$ of these three capacitors connected in series is the same as a capacitor of plate area $A$ and with plate spacing $d = d_1 + d_2 + d_3$.

web

54. Charges of equal magnitude $1.00 \times 10^{-15}$ C and opposite sign are distributed over the inner and outer surfaces of the cell wall in Figure P16.54. Find the force on the potassium ion ($K^+$) if the ion is (a) 2.70 $\mu$m from the center of the cell, (b) 2.92 $\mu$m from the center, and (c) 4.00 $\mu$m from the center.

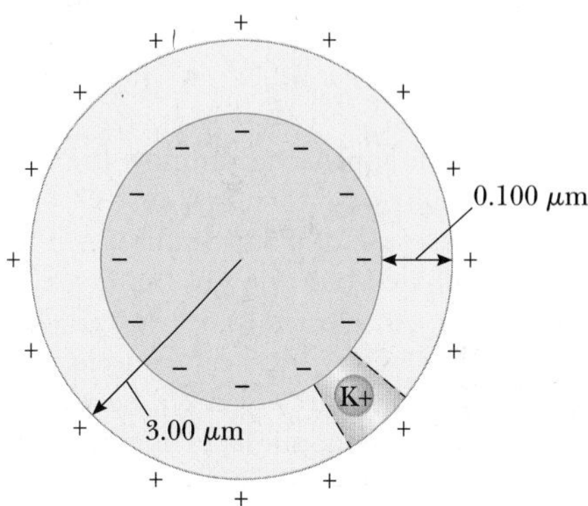


FIGURE P16.54

55. A virus rests on the bottom plate of oppositely charged parallel plates in the vacuum chamber of an electron microscope. The electric field strength between the plates is $2.00 \times 10^5$ N/C, and the bottom plate is negative. If the virus has a mass of $1.00 \times 10^{-15}$ kg and suddenly acquires a charge of $-1.60 \times 10^{-19}$ C, what are its velocity and position 75.0 ms later? Do not disregard gravity.

56. A plastic pellet has a mass of $7.50 \times 10^{-10}$ kg and carries a charge of $2.00 \times 10^{-9}$ C. The pellet is fired horizontally between a pair of oppositely charged parallel plates, as shown in Figure P16.56. If the electric field strength between the plates is $9.20 \times 10^4$ N/C and the pellet enters the plates with an initial speed of 150 m/s, determine the deflection angle $\delta$. (*Hint:* This exercise is similar to those involving the motion of a projectile under gravity.)

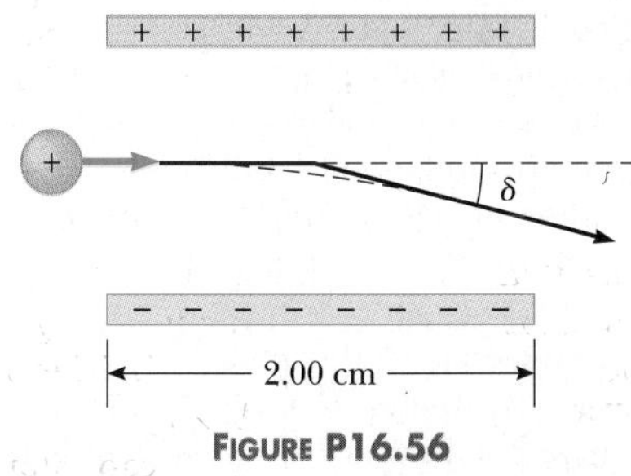


FIGURE P16.56

57. Find the equivalent capacitance of the group of capacitors in Figure P16.57.

58. A spherical capacitor consists of a spherical conducting shell of radius $b$ and charge $-Q$ concentric with a smaller conducting sphere of radius $a$ and charge $Q$. (a) Find the capacitance of this device. (b) Show that as the radius $b$ of

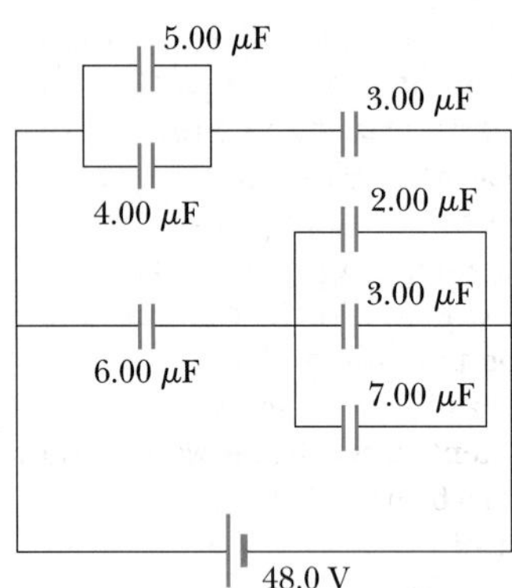


**FIGURE P16.57**

the outer sphere approaches infinity, the capacitance approaches the value of $a/k_e = 4\pi\epsilon_0 a$.

**59.** The immediate cause of many deaths is ventricular fibrillation, uncoordinated quivering of the heart as opposed to proper beating. An electric shock to the chest can cause momentary paralysis of the heart muscle, after which the heart will sometimes start organized beating again. A *defibrillator* is a device that applies a strong electric shock to the chest over a time of a few milliseconds. The device contains a capacitor of several microfarads, charged to several thousand volts. Electrodes called paddles, about 8 cm across and coated with conducting paste, are held against the chest on both sides of the heart. Their handles are insulated to prevent injury to the operator, who calls, "Clear!" and pushes a button on one paddle to discharge the capacitor through the patient's chest. Assume that an energy of 300 W · s is to be delivered from a 30.0 μF capacitor. To what potential difference must it be charged?

**60.** When a certain air-filled parallel-plate capacitor is connected across a battery, it acquires a charge (on each plate) of 150 μC. While the battery connection is maintained, a dielectric slab is inserted into and fills the region between the plates. This results in the accumulation of an additional charge of 200 μC on each plate. What is the dielectric constant of the dielectric slab?

61. Capacitors $C_1 = 6.0\ \mu F$ and $C_2 = 2.0\ \mu F$ are charged as a parallel combination across a 250-V battery. The capacitors are disconnected from the battery and from each other. They are then connected positive plate to negative plate and negative plate to positive plate. Calculate the resulting charge on each capacitor.

62. Capacitors $C_1 = 4.0\ \mu F$ and $C_2 = 2.0\ \mu F$ are charged as a series combination across a 100-V battery. The two capacitors are disconnected from the battery and from each other. They are then connected positive plate to positive plate and negative plate to negative plate. Calculate the resulting charge on each capacitor.

63. The charge distribution shown in Figure P16.63 is referred to as a linear quadrupole. (a) Show that the electric potential at a point on the $x$ axis where $x > d$ is

$$V = \frac{2k_eQd^2}{x^3 - xd^2}$$

(b) Show that the expression obtained in (a) when $x \gg d$ reduces to

$$V = \frac{2k_eQd^2}{x^3}$$

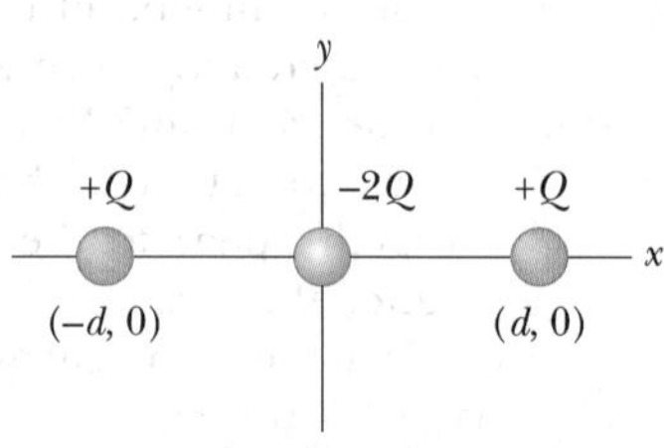


**FIGURE P16.63**

64. The energy stored in a 52.0-μF capacitor is used to melt a 6.00-mg sample of lead. To what voltage must the capacitor be initially charged, assuming that the initial temperature of the lead is 20.0°C? (Lead has a specific heat of 128 J/kg · °C, a melting point of 327.3°C, and a latent heat of fusion of 24.5 kJ/kg.)

65. Consider a parallel-plate capacitor with charge $Q$ and area $A$, filled with dielectric material having dielectric constant $\kappa$. It can be shown that the magnitude of the attractive force exerted by each plate on the other is given by $F = Q^2/(2\kappa\epsilon_0 A)$. When a potential difference of 100 V exists between the plates of an air-filled 20-μF parallel-plate capacitor, what force does each plate exert on the other if they are separated by 2.0 mm?

66. An electron is fired at a speed of $v_0 = 5.6 \times 10^6$ m/s and at an angle of $\theta_0 = -45°$ between two parallel conducting plates that are $D = 2.0$ mm apart, as in Figure P16.66. If the potential difference between the plates is $\Delta V = 100$ V, determine (a) how close $d$ the electron will get to the bottom plate and (b) where the electron will strike the top plate.

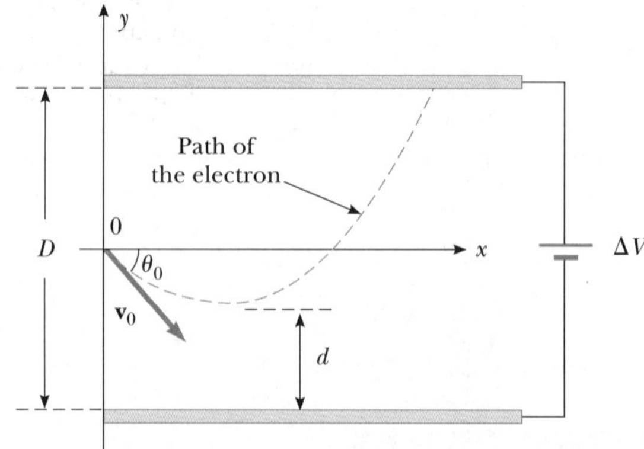


**FIGURE P16.66**

## GROUP ACTIVITIES

**G.1** It takes an electric field of about 30 kV/cm to cause a spark in dry air. Shuffle across a rug and reach toward a doorknob. By estimating the length of the spark, determine the electric potential difference between your finger and the doorknob that existed just before you touched the knob. Try this experiment again on a very humid day and you will find that the spark is much shorter or is imperceptible. Why?

**G.2** Suppose you are given a battery, a capacitor, two switches, a lightbulb and several pieces of connecting wire. On a sheet of paper, design a circuit that will do the following. (1) When switch 1 is closed and switch 2 is open, the capacitor charges but no current moves through the lightbulb. (2) Then when switch 1 is opened and switch 2 closed, the lightbulb is connected to the capacitor but not to the battery. Describe the motion of charge in the circuit when switch 1 is closed and switch 2 is open. Is energy being stored in the capacitor? What measurements would you have to make to determine how much energy is stored if any? What happens to the lightbulb when switch 1 is opened after the capacitor has charged and switch 2 then closed? Will the bulb light and stay lit? What happens to the charge on the capacitor when switch 2 is closed in this way?

**G.3** (a) Under what conditions can the equation $V_B - V_A = -Ed$ be used? (b) Can this equation be used to find the difference in potential between two points in an electric field set up by a point charge? Defend your answer. (c) Can this equation be used to find the difference in potential between two points in the electric field between the plates of a charged, parallel-plate capacitor? Defend your answer. (d) What is the strength of the electric field between two parallel metal plates attached to a battery of 12 V when the plates are separated by 5 mm?

**G.4** Two charges of 1.0 $\mu$C and $-2.0$ $\mu$C are 0.5 m apart at two vertices of an equilateral triangle. (a) What is the direction of the potential set up by the 1.0-$\mu$C charge at the third vertex of the triangle? Be careful, and defend your answer. (b) What is the direction of the potential set up by the charge of $-2.0$ $\mu$C at this third vertex? Again be very careful before you defend your answer too strongly. (c) Find the electric potential at this point.

**G.5** (a) A group of capacitors connected in parallel have the same (i) charge on them, (ii) potential difference across them, or (iii) neither of the above.
(b) A group of capacitors connected in series have the same (i) charge on them, (ii) potential difference across them, or (iii) neither of the above.
(c) The equivalent capacitance for a group of capacitors connected in parallel is (i) greater than any of the capacitors in the group, (ii) less than any of the capacitors in the group, or (iii) neither of the above.
(d) The equivalent capacitance for a group of capacitors connected in series is (i) greater than any of the capacitors in the group, (ii) less than any of the capacitors in the group, or (iii) neither of the above.
(e) Defend your answers by examining a circuit consisting of a 3.0-$\mu$F and a 5.0-$\mu$F capacitor used with a 12-V battery. Be sure that the answers you find are consistent with your answers to parts (a) through (d).

**G.6** Figure GA16.6 shows a contour map of a hilly island. The outer part of the figure is at sea level (marked 0). Each contour line from the region marked zero shows a level 10 m higher than the previous. The maximum height is 70 m and is shown by the number 70. Answer the following questions by giving the pair of grid markers (a letter and a number) closest to the point being requested.
(a) Where is there a steep cliff?
(b) Where is there a pass between two hills?
(c) Where is the easiest slope to climb?

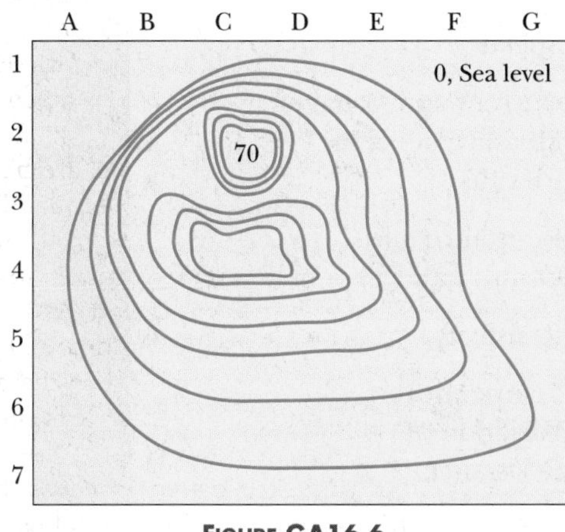


**FIGURE GA16.6**

Group Activity 6 is courtesy of Edward F. Redish. For more problems of this type, see
**http://www.physics.umd.edu/perg/**

# 17 Current and Resistance

## Chapter Outline



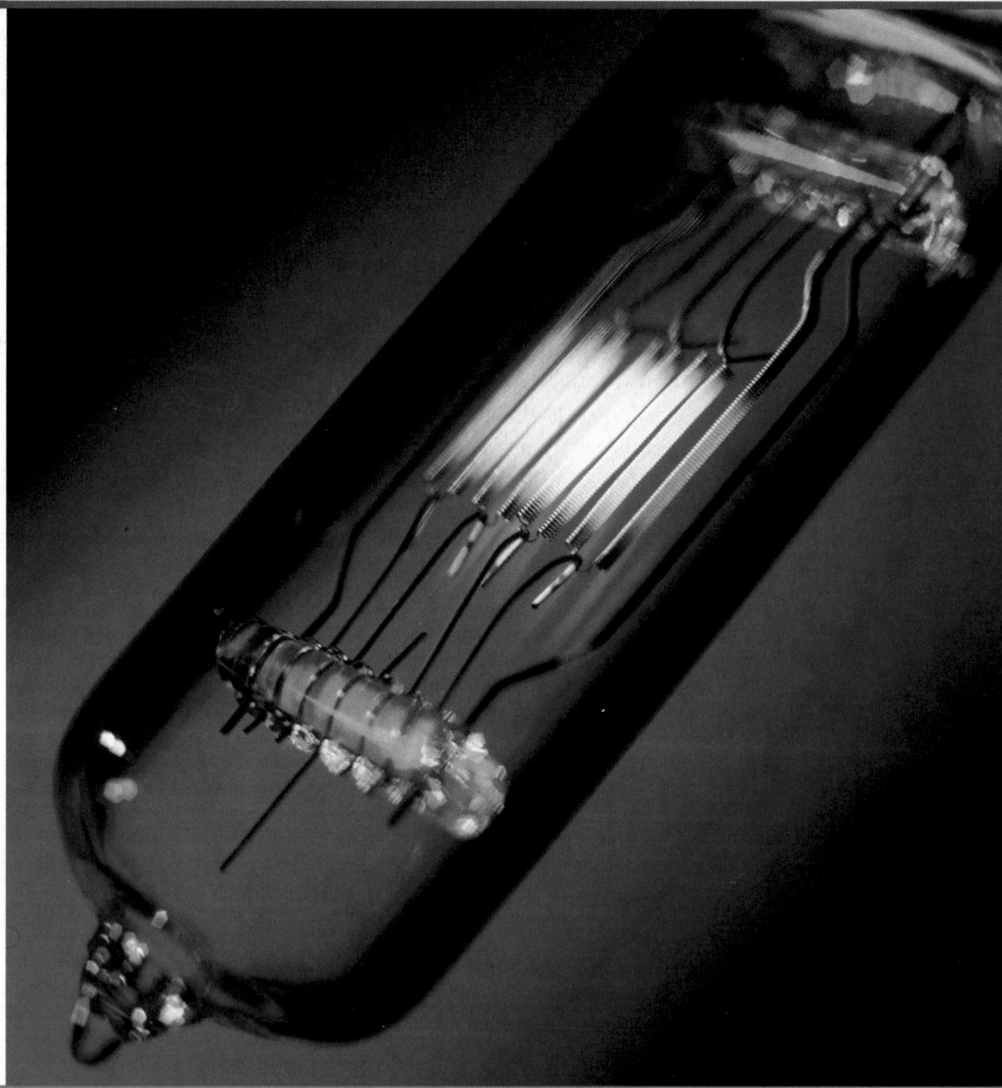

**This lightbulb emits light when its tungsten filaments carry sufficient current. Tungsten is an ideal filament material because it is strong and has a high melting point (about 3 410°C). The interior of the bulb contains an inert gas, such as a mixture of argon and nitrogen. Air must be avoided because the oxygen it contains would corrode the tungsten filaments.** *(Tek Image/Science Photo Library/Photo Researchers, Inc.)*

Many practical applications and devices are based on the principles of static electricity, but electricity was destined to become an inseparable part of our daily lives when scientists learned how to produce a continuous flow of charge for relatively long periods of time by using the battery. The battery or voltaic cell was invented in 1800 by the Italian physicist Alessandro Volta. The battery is a device that supplies a continuous flow of charge at low potential, as opposed to the previously invented electrostatic machines, which produced a tiny flow of charge at high potential for brief periods. This steady source of electric current allowed scientists to perform experiments to learn how to control the flow of electric charges in circuits. Today, electric currents power our lights, radios, television sets, air conditioners, and refrigerators; they ignite the gasoline in automobile engines, travel through miniature components making up the chips of microcomputers, and perform countless other invaluable tasks.

In this chapter we define current and discuss some of the factors that contribute to the resistance to flow of charge in conductors. We also discuss energy transformations in electric circuits. These topics will be the foundation for additional work with circuits in later chapters.

## 17.1 ELECTRIC CURRENT

Whenever electric charges of like signs move, a *current* is said to exist. To define current more precisely, suppose the charges are moving perpendicularly to a surface of area $A$, as in Figure 17.1. (This area could be the cross-sectional area of a wire, for example.) **The current is the rate at which charge flows through this surface.** If $\Delta Q$ is the amount of charge that passes through this area in a time interval of $\Delta t$, the current $I$ is equal to the ratio of the charge to the time interval:

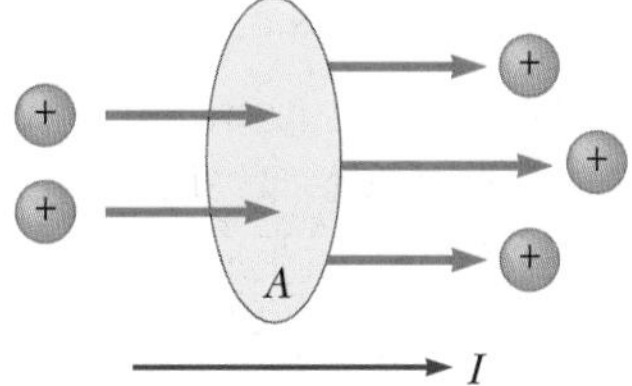


**FIGURE 17.1** Charges in motion through an area $A$. The time rate of flow of charge through the area is defined as the current $I$. The direction of the current is the direction of flow of positive charges.

$$I \equiv \frac{\Delta Q}{\Delta t} \quad [17.1]$$

The SI unit of current is the **ampere (A):**

$$1 \text{ A} = 1 \text{ C/s} \quad [17.2]$$

Thus, 1 A of current is equivalent to 1 C of charge passing through the cross-sectional area in a time interval of 1 s.

When charges flow through a surface as in Figure 17.1, they can be positive, negative, or both. **The direction of conventional current used in this book is the direction positive charge would flow.** (This historical convention originated about 200 years ago when the ideas of positive and negative charge were introduced.) In a common conductor, such as copper, the current is due to the motion of negatively charged electrons. Therefore, when we speak of current in such a conductor, the direction of the current is opposite the direction of motion of electrons. On the other hand, if one considers a beam of positively charged protons in an accelerator, the current is in the direction of motion of the protons. In some cases—gases and electrolytes, for example—the current is the result of the flows of both positive and negative charges. It is common to refer to a moving charge (whether it is positive or negative) as a *charge carrier.* In a metal, for example, the charge carriers are electrons.

The phrases *flow of current* or *current flow* are commonly used but are incorrect because current *is* a flow (of charge). We shall avoid these redundant phrases and speak of *flow of charge* or *charge flow.*

### Example 17.1 The Current in a Lightbulb

The amount of charge that passes through the filament of a certain lightbulb in 2.00 s is 1.67 C. Find (a) the current in the lightbulb and (b) the number of electrons that pass through the filament in 1 s.

**Solution**

**A** From Equation 17.1 we have

$$I = \frac{\Delta Q}{\Delta t} = \frac{1.67 \text{ C}}{2.00 \text{ s}} = 0.835 \text{ A}$$

**B** In 1 s, 0.835 C of charge must pass the cross-sectional area of the filament. This total charge per second is equal to the number of electrons, $N$, times the charge on a single electron.

$$Nq = N(1.60 \times 10^{-19} \text{ C/electron}) = 0.835 \text{ C}$$

$$N = 5.22 \times 10^{18} \text{ electrons}$$

**Quick Quiz 17.1** Consider positive and negative charges moving horizontally through the four regions in Figure 17.2. Rank the currents in these four regions, from lowest to highest.

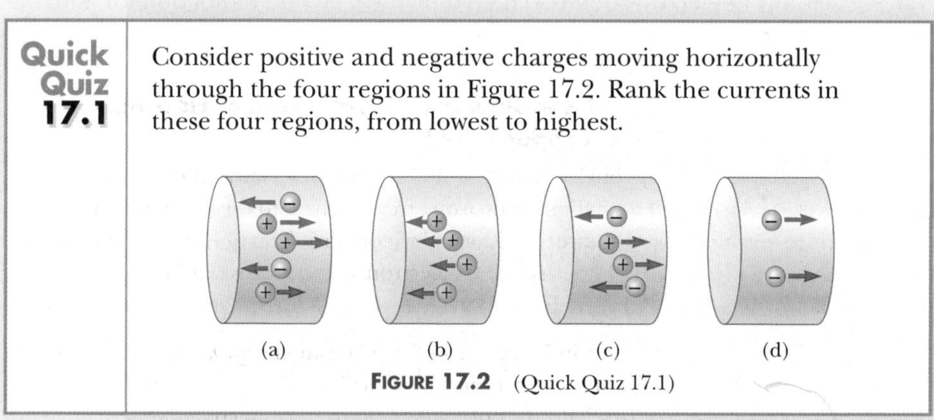


**FIGURE 17.2** (Quick Quiz 17.1)

## 17.2 A MICROSCOPIC VIEW: CURRENT AND DRIFT SPEED

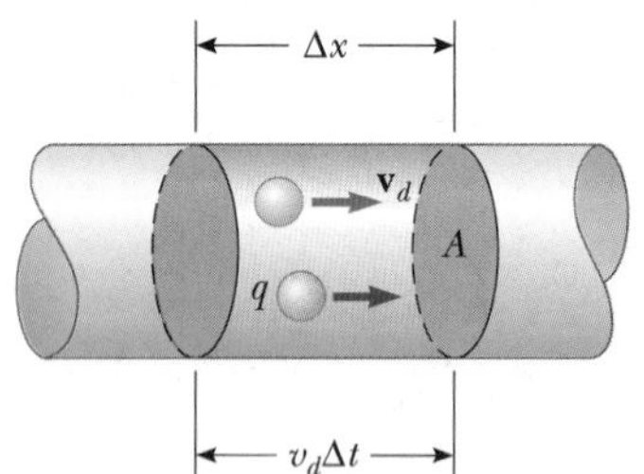


**FIGURE 17.3** A section of a uniform conductor of cross-sectional area $A$. The charge carriers move with a speed $v_d$, and the distance they travel in the time $\Delta t$ is given by $\Delta x = v_d \Delta t$. The number of mobile charge carriers in the section of length $\Delta x$ is given by $nAv_d \Delta t$, where $n$ is the number of mobile carriers per unit volume.

It is instructive to develop a model that will allow us to relate the macroscopic current to the motion of the microscopic charged particles that make up the current. We will show that current depends on the average speed of the charge carriers in the direction of the current, the number of charge carriers per unit volume, and the size of the charge carried by each.

Consider identically charged particles moving in a conductor of cross-sectional area $A$ (Fig. 17.3). The volume of an element of length $\Delta x$ of the conductor is $A\,\Delta x$. If $n$ represents the number of mobile charge carriers per unit volume, then the number of carriers in the volume element is $nA\,\Delta x$. The mobile charge $\Delta Q$ in this element is therefore

$$\Delta Q = \text{number of carriers} \times \text{charge per carrier} = (nA\,\Delta x)q$$

where $q$ is the charge on each carrier. If the carriers move with a constant average speed $v_d$, called the **drift speed,** the distance they move in time interval $\Delta t$ is $\Delta x = v_d\,\Delta t$. Therefore, we can write $\Delta Q$ as

$$\Delta Q = (nAv_d\,\Delta t)q$$

If we divide both sides of this equation by $\Delta t$, we see that the current in the conductor is

Current is proportional to the drift speed

$$I = \frac{\Delta Q}{\Delta t} = nqv_dA \qquad [17.3]$$

To understand the meaning of drift speed, consider a conductor in which the charge carriers are free electrons. If the conductor is isolated, these electrons undergo random motion similar to that of gas molecules. Note that the drift speed is normally much smaller than the free electrons' average speed between collisions with the fixed atoms of the conductor. When a potential difference is applied between the ends of the conductor (say, by means of a battery), an electric field is set up in the conductor, creating an electric force on the electrons and hence a current. In reality, the electrons do not simply move in straight lines along the conductor. Instead, they undergo repeated collisions with the metal atoms, and the result is a complicated zigzag motion with only a small average drift speed along the wire (Fig. 17.4). The energy transferred from the electrons to the metal atoms during collision increases the vibrational energy of the atoms and causes a corresponding increase in the temperature of the conductor. However, despite the collisions, the electrons move slowly along the conductor in a direction opposite **E** with the drift velocity $\mathbf{v}_d$.

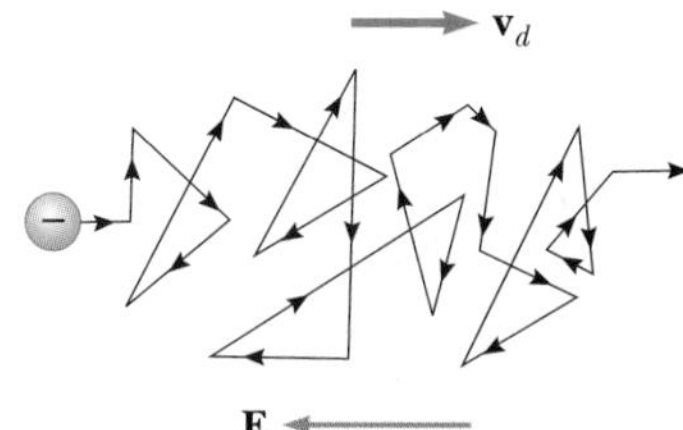


**FIGURE 17.4** A schematic representation of the zigzag motion of a charge carrier in a conductor. The sharp changes in direction are due to collisions with atoms in the conductor. Note that the net motion of electrons is opposite the direction of the electric field.

## Example 17.2 Just How Slow Is the Drift Speed of Electrons in a Copper Wire?

A copper wire of cross-sectional area $3.00 \times 10^{-6}$ m$^2$ carries a current of 10.0 A. (a) Assuming that each copper atom contributes one free electron to the metal, find the drift speed of the electrons in this wire. The density of copper is 8.95 g/cm$^3$. (b) Compare the drift speed to the random rms speed an electron would have at 20°C using the ideal gas model.

**Reasoning** All the variables in Equation 17.3 are known except $n$, the number of free charge carriers per unit volume. We can find $n$ by recalling that one mole of copper contains Avogadro's number ($6.02 \times 10^{23}$) of atoms, and each atom contributes one charge carrier to the metal. The volume of one mole can be found from copper's known density and its atomic mass.

### Solution

**A** From the periodic table of the elements (inside back cover), we find that the mass of one mole of copper is 63.5 g. Knowing the density of copper enables us to calculate the volume occupied by 63.5 g of copper:

$$V = \frac{m}{\rho} = \frac{63.5 \text{ g}}{8.95 \text{ g/cm}^3} = 7.09 \text{ cm}^3$$

If we now assume that each copper atom contributes one free electron to the body of the material, we have

$$n = \frac{6.02 \times 10^{23} \text{ electrons}}{7.09 \text{ cm}^3} = 8.48 \times 10^{22} \text{ electrons/cm}^3$$

$$= \left(8.48 \times 10^{22} \frac{\text{electrons}}{\text{cm}^3}\right)\left(10^6 \frac{\text{cm}^3}{\text{m}^3}\right) = 8.48 \times 10^{28} \text{ electrons/m}^3$$

From Equation 17.3, we find that the drift speed is

$$v_d = \frac{I}{nqA} = \frac{10.0 \text{ C/s}}{(8.48 \times 10^{28} \text{ electrons/m}^3)(1.60 \times 10^{-19} \text{ C})(3.00 \times 10^{-6} \text{ m}^2)}$$

$$= 2.46 \times 10^{-4} \text{ m/s}$$

**B** Equation 10.18 states that the randomly directed rms speed of an ideal gas particle of mass $m$ at absolute temperature $T$ is given by

$$v_{\text{rms}} = \sqrt{\frac{3k_B T}{m}}$$

Because the free electrons in a metal behave approximately as an ideal gas, we have for the rms speed of an electron

$$v_{\text{rms}} = \sqrt{\frac{3(1.38 \times 10^{-23}\text{ J/K})(293\text{ K})}{9.11 \times 10^{-31}\text{ kg}}} = 1.15 \times 10^{5}\text{ m/s}$$

Thus, the drift speed of an electron in a wire is only about one billionth of its random thermal speed.

**Tip 17.2 ELECTRONS ARE EVERYWHERE IN THE CIRCUIT**

Electrons do not have to travel from the light switch to the light for the light to operate. Electrons already in the filament of the lightbulb move in response to the electric field set up by a battery (or other source of potential difference). Also, the battery does *not* provide electrons to the circuit.

Example 17.2 shows that drift speeds are typically very small. In fact, the drift speed is much smaller than the average speed between collisions; for instance, electrons traveling at $2.46 \times 10^{-4}$ m/s would take about 68 min to travel 1 m! In view of this low speed, you might wonder why a light turns on almost instantaneously when a switch is thrown. Think of the flow of water through a pipe. If a drop of water is forced into one end of a pipe that is already filled with water, a drop must be pushed out the other end of the pipe. Although it may take an individual drop a long time to make it through the pipe, a flow initiated at one end produces a similar flow at the other end very quickly. Another familiar analogy is the motion of a bicycle chain. When the sprocket moves one link, the other links all move more or less immediately, even though it takes a given link some time to make a complete rotation. In a conductor, the electric field that drives the free electrons travels with a speed close to that of light. Thus, when you flip a light switch, the message for the electrons to start moving through the wire (the electric field) reaches them at a speed on the order of $10^{8}$ m/s.

## APPLYING PHYSICS 17.1

Suppose a current-carrying wire has a cross-sectional area that gradually becomes smaller along the wire, so that the wire has the shape of a very long cone. How does the drift velocity vary along the wire?

**Explanation** Every portion of the wire is carrying the same current because charge cannot build up at any point on the wire. Thus, as the cross-sectional area decreases, the drift velocity must increase to maintain the constant value of the current. This increased drift velocity is a result of the electric field lines in the wire being compressed into a smaller area, thus increasing the strength of the field.

## APPLYING PHYSICS 17.2

We have seen that an electric field must exist inside a conductor that carries a current. How is this possible in view of the fact that in electrostatics we concluded that the electric field is zero inside a conductor?

**Explanation** In the electrostatic case in which charges are stationary, the internal electric field must be zero because a nonzero field would produce a current (by interacting with the free electrons in the conductor), which would violate the condition of static equilibrium. In this chapter we deal with conductors that carry current, a nonelectrostatic situation. The current arises because of a potential difference applied between the ends of the conductor, which produces an internal electric field. So there is no paradox.

## 17.3 CURRENT AND VOLTAGE MEASUREMENTS IN CIRCUITS

So far we have tried to keep things simple and concentrate on concepts like current without mentioning how one measures current or even sets up a circuit to make charge flow. In order to understand better the concepts covered later in this chapter, let us digress briefly to discuss what a circuit is and how you measure current in and voltage across some circuit element.

Look at the circuit shown in Figure 17.5a. This is the actual circuit that would have to be set up to measure the current in Example 17.1. Figure 17.5b shows a stylized figure called a **circuit diagram** (as first introduced in Chapter 16) that represents the actual circuit of Figure 17.5a. This circuit consists of only a battery and a lightbulb. The word *circuit* means that current circulates around a closed loop of some sort. The battery pumps charge through the bulb and around the loop. No charge would flow if you did not have a complete conducting path from the positive terminal of the battery into one side of the bulb, out the other, and through the copper conducting wires back to the negative terminal of the battery. The most important quantities that characterize how the bulb will work in different situations are the current $I$ in the bulb and the potential difference $\Delta V$ *across* the bulb. To measure the current in the bulb, we place an ammeter, the device for measuring current, in line with the bulb so that there is no path for the current to sneak around the meter; all of the charge passing through the bulb must also pass through the ammeter. The voltmeter measures the potential difference or voltage between the two ends of the bulb's filament. If we use two meters simultaneously as in Figure 17.5a, we can remove the voltmeter and see if its presence effects the current reading. Figure 17.5c shows a digital **multimeter,** a convenient device with a digital readout that can be used to measure voltage, current, or resistance. An advantage of using a digital multimeter as a voltmeter is that it usually will not affect the current, because a digital meter has enormous resistance to the flow of charge in the voltmeter mode.

At this point, you can measure the current as a function of voltage (the $I$-$\Delta V$ curve) of various devices in the lab. All you need is a variable voltage supply (an adjustable battery) capable of supplying potential differences from about $-5$ V to $+5$ V, a bulb, a resistor, some wires and alligator clips, and a couple of multimeters. Always start your measurements using the highest multimeter scales, that is, 10 A and 1 000 V, and increase the sensitivity one scale at a time to obtain the highest accuracy without overloading the meters. (To increase the sensitivity means that we lower the maximum current or voltage that the scale reads.) Also note that the meters must be connected with the proper polarity with respect to the voltage supply, as shown in Figure 17.5b. Finally, follow your instructor's directions carefully to avoid damaged meters and a soaring lab fee.

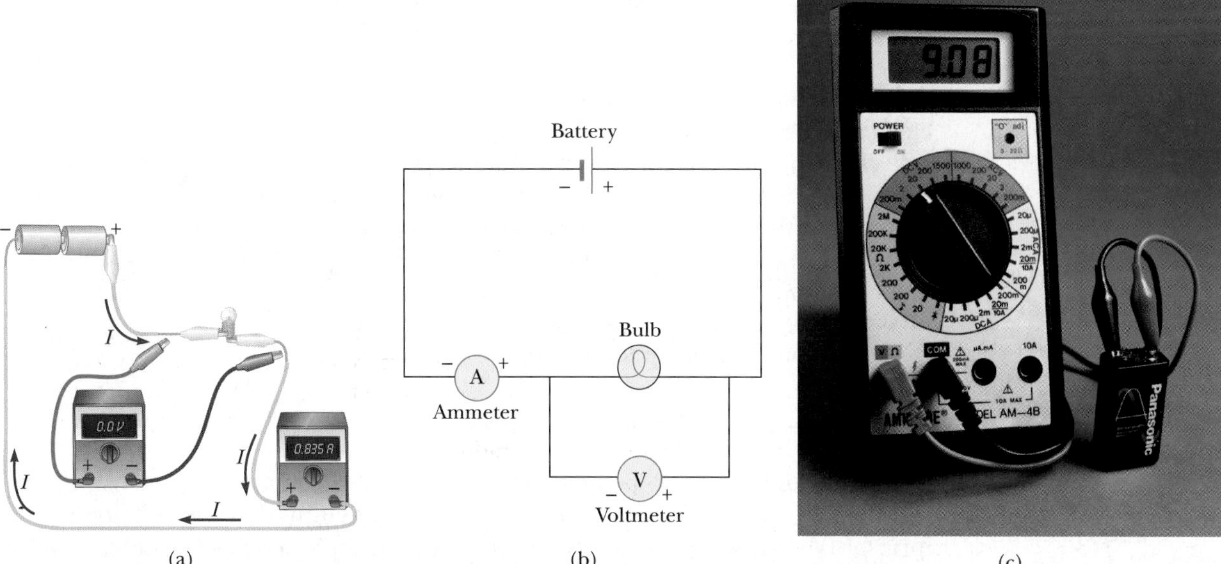


**FIGURE 17.5** (a) A sketch of an actual circuit used to measure the current in a flashlight bulb and the potential difference across it. (b) A schematic diagram of the circuit shown in part (a). (c) A digital multimeter that can be used to measure both currents and potential differences. Here, the meter is measuring the potential difference across the terminals of a battery. *(c, Michael Dalton, Fundamental Photographs)*

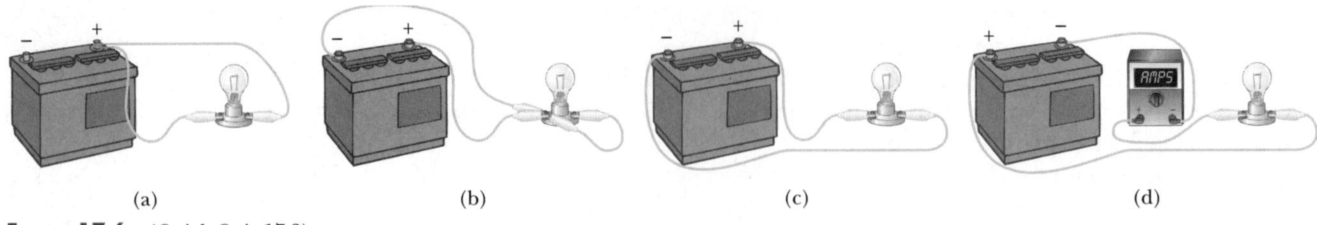


**FIGURE 17.6** (Quick Quiz 17.2)

**Quick Quiz 17.2** Look at the four "circuits" shown in Figure 17.6 and select those that will light the bulb.

## 17.4 RESISTANCE AND OHM'S LAW

When a voltage (potential difference) $\Delta V$ is applied across the ends of a metallic conductor as in Figure 17.7, the current in the conductor is found to be proportional to the applied voltage; that is, $I \propto \Delta V$. If the proportionality holds, we can write $\Delta V = IR$, where the proportionality constant $R$ is called the resistance of the conductor. In fact, we define the **resistance** as the ratio of the voltage across the conductor to the current it carries:

Resistance 

$$R \equiv \frac{\Delta V}{I} \qquad [17.4]$$

Resistance has the SI units volts per ampere, called **ohms** ($\Omega$). Thus, if a potential difference of 1 V across a conductor produces a current of 1 A, the resistance of the conductor is 1 $\Omega$. For example, if an electrical appliance connected to a 120-V source carries a current of 6 A, its resistance is 20 $\Omega$.

It is useful to compare the concepts of electric current, voltage, and resistance with the flow of water in a river. As water flows downhill in a river of constant width and depth, the flow rate (water current) depends on the steepness of descent of the river and the effects of rocks, the river bank, and other obstructions. Based on this analogy, it seems reasonable that increasing the voltage applied to a circuit should increase the current in the circuit just as increasing the steepness of descent increases the water current. Also, increasing the obstructions in the river's path will reduce the water current just as increasing the resistance in a circuit will lower the electric current. Resistance in a circuit arises due to collisions between the electrons carrying the current with fixed atoms inside the conductor. These collisions inhibit the movement of charges in much the same way as would a force of friction. For many materials, including most metals, experiments show that **the resistance remains constant over a wide range of applied voltages or currents.** This statement is known as **Ohm's law** after Georg Simon Ohm (1789–1854), who was the first to conduct a systematic study of electrical resistance.

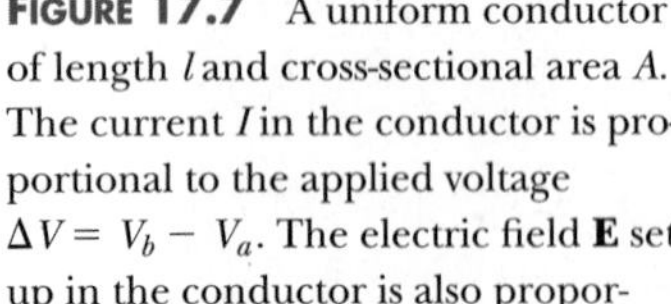


**FIGURE 17.7** A uniform conductor of length $l$ and cross-sectional area $A$. The current $I$ in the conductor is proportional to the applied voltage $\Delta V = V_b - V_a$. The electric field **E** set up in the conductor is also proportional to the current.

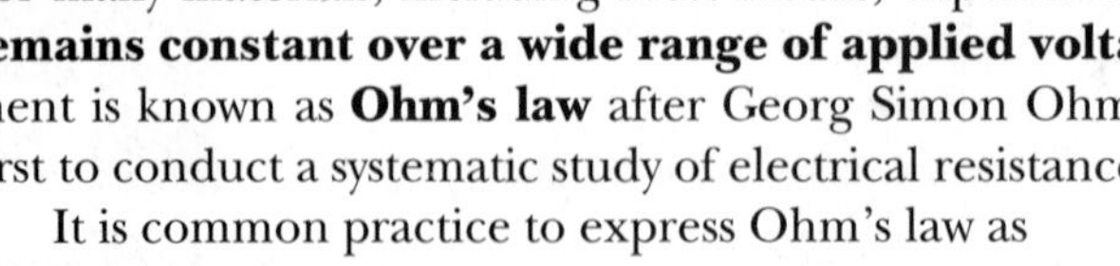

It is common practice to express Ohm's law as

Ohm's law 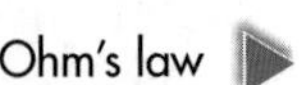

$$\Delta V = IR \qquad [17.5]$$

where $R$ is understood to be independent of $\Delta V$, the potential drop across the resistor, and $I$, the current in the resistor. We shall continue to use this traditional form of Ohm's law when discussing electrical circuits. A **resistor** is a conductor that provides a specified resistance in an electric circuit. The symbol for a resistor in circuit diagrams is a zigzag line, —/\/\/\—.

Ohm's law is an empirical relationship that is valid only for certain materials. Materials that obey Ohm's law, and hence have a constant resistance over a wide

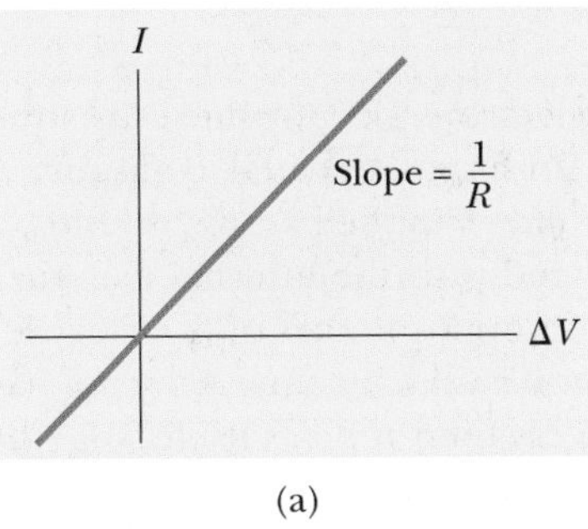


(a)

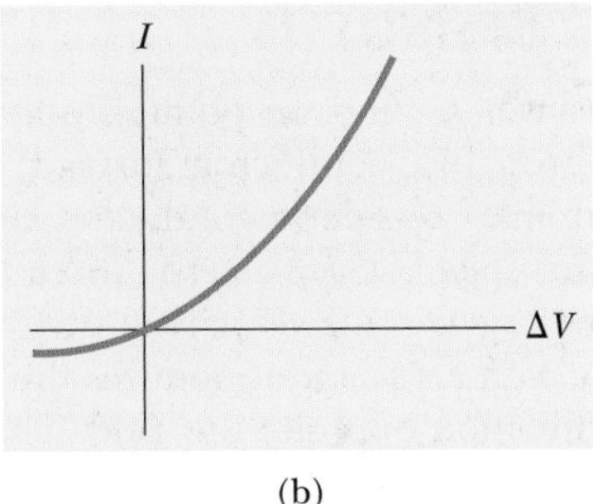


(b)

**FIGURE 17.8** (a) The current-voltage curve for an ohmic material. The curve is linear, and the slope gives the resistance of the conductor. (b) A nonlinear current-voltage curve for a semiconducting diode. This device does not obey Ohm's law.

**GEORG SIMON OHM (1787–1854)**

Ohm, a high school teacher in Cologne and later a professor at Munich, formulated the concept of resistance and discovered the proportionalities expressed in Equation 17.6. But Henry Cavendish had earlier determined the proportionality of current to voltage that we call Ohm's law. *(© Bettmann/CORBIS)*

range of voltages, are said to be **ohmic.** Materials whose resistance changes with voltage or current are **nonohmic.** Ohmic materials have a linear current-voltage relationship over a large range of applied voltages (Fig. 17.8a). Nonohmic materials have a nonlinear current-voltage relationship (Fig. 17.8b). One common semiconducting device that is nonohmic is the **diode,** a circuit element that acts like a one-way valve for current (see Problem 62). Its resistance is small for currents in one direction (positive $\Delta V$) and large for currents in the reverse direction (negative $\Delta V$). Most modern electronic devices, such as transistors, have nonlinear current-voltage relationships; their operation depends on the particular ways in which they violate Ohm's law.

**Quick Quiz 17.3** In Figure 17.8b, does the resistance of the diode (a) increase or (b) decrease as the positive voltage $\Delta V$ increases?

## Example 17.3 The Resistance of a Steam Iron

All electric devices are required to have identifying plates that specify their electrical characteristics. The plate on a certain steam iron states that the iron carries a current of 6.4 A when connected to a 120-V source. What is the resistance of the steam iron?

**Solution** From Ohm's law, we find the resistance to be

$$R = \frac{\Delta V}{I} = \frac{120 \text{ V}}{6.4 \text{ A}} = 19 \text{ }\Omega$$

**EXERCISE** The resistance of a hot plate is 48 Ω. How much current does the plate carry when connected to a 120-V source?

**ANSWER** 2.5 A

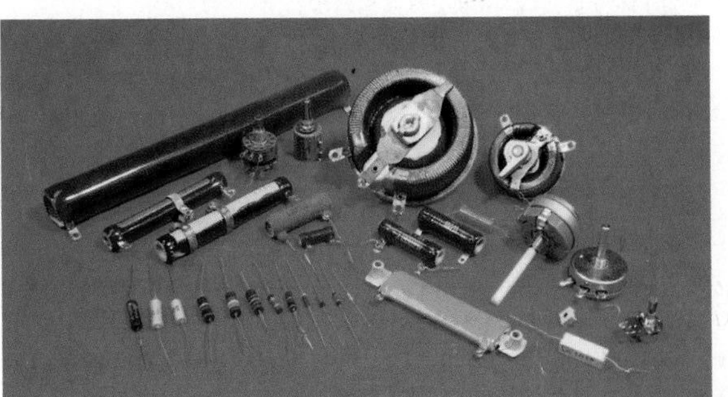

An assortment of resistors used for a variety of applications in electronic circuits. *(Courtesy of Henry Leap and Jim Lehman)*

## 17.5 RESISTIVITY

In an earlier section we pointed out that electrons do not move in straight-line paths through a conductor. Instead, they undergo repeated collisions with the metal atoms. Consider a conductor with a voltage applied across its ends. An electron gains speed as the electric force associated with the internal electric field accelerates it, giving it a velocity in the direction opposite that of the electric field. A collision with an atom randomizes the electron's velocity, thus reducing its velocity in the direction opposite the field. The process then repeats itself. Together these collisions affect the electron somewhat as a force of internal friction would. This is the origin of a material's resistance. The resistance of an ohmic conductor is proportional to its length $l$ and inversely proportional to its cross-sectional area $A$. That is,

$$R = \rho \frac{l}{A} \qquad [17.6]$$

where the constant of proportionality $\rho$ is called the **resistivity** of the material.[1] Every material has a characteristic resistivity that depends on its electronic structure and on temperature. Good electric conductors have very low resistivities, and good insulators have very high resistivities. Table 17.1 lists the resistivities of a variety of materials at 20°C. Because resistance values are in ohms, resistivity values must be in ohm-meters.

Equation 17.6 states that the resistance of a cylindrical conductor is proportional to its length and inversely proportional to its cross-sectional area. This may be understood by analogy to the flow of liquid through a pipe. As the length of the pipe is increased, the resistance to liquid flow increases because of a gain in friction between the fluid and the walls of the pipe. As its cross-sectional area is in-

**TABLE 17.1 Resistivities and Temperature Coefficients of Resistivity for Various Materials (at 20°C)**

| Material | Resistivity $\rho$ ($\Omega \cdot$m) | Temperature Coefficient of Resistivity $\alpha$ [(°C)$^{-1}$] |
|---|---|---|
| Silver | $1.59 \times 10^{-8}$ | $3.8 \times 10^{-3}$ |
| Copper | $1.7 \times 10^{-8}$ | $3.9 \times 10^{-3}$ |
| Gold | $2.44 \times 10^{-8}$ | $3.4 \times 10^{-3}$ |
| Aluminum | $2.82 \times 10^{-8}$ | $3.9 \times 10^{-3}$ |
| Tungsten | $5.6 \times 10^{-8}$ | $4.5 \times 10^{-3}$ |
| Iron | $10.0 \times 10^{-8}$ | $5.0 \times 10^{-3}$ |
| Platinum | $11 \times 10^{-8}$ | $3.92 \times 10^{-3}$ |
| Lead | $22 \times 10^{-8}$ | $3.9 \times 10^{-3}$ |
| Nichrome* | $150 \times 10^{-8}$ | $0.4 \times 10^{-3}$ |
| Carbon | $3.5 \times 10^{5}$ | $-0.5 \times 10^{-3}$ |
| Germanium | 0.46 | $-48 \times 10^{-3}$ |
| Silicon | 640 | $-75 \times 10^{-3}$ |
| Glass | $10^{10}$–$10^{14}$ | |
| Hard rubber | $\approx 10^{13}$ | |
| Sulfur | $10^{15}$ | |
| Quartz (fused) | $75 \times 10^{16}$ | |

* A nickel-chromium alloy commonly used in heating elements.

All values are at 20°C

[1] The symbol $\rho$ used for resistivity should not be confused with the same symbol used earlier in the book for density. Very often, a single symbol is used to represent different quantities.

creased, the pipe can transport more fluid in a given time interval, so its resistance drops.

## APPLYING PHYSICS 17.3

**APPLICATION**

DIMMING OF AGING LIGHTBULBS

It is a common observation that as a lightbulb ages, it gives off less light than when new. Why?

**Explanation** There are two reasons for this, one electrical and one optical, but both are related to the same phenomenon occurring within the bulb. The filament of a lightbulb is made of a tungsten wire that, in an old lightbulb, has been kept at a high temperature for many hours. These high temperatures cause tungsten to be evaporated from the filament, decreasing its radius. From $R = \rho\, l/A$, we see that a decreased cross-sectional area leads to an increase in resistance of the filament. This increasing resistance with age means that the filament will carry less current for the same applied voltage. With less current in the filament, there is less light output, and the filament glows more dimly.

At the high operating temperature of the filament, tungsten atoms leave the surface of the filament, much as water molecules evaporate from a puddle of water. These atoms are carried away by convection currents in the gas in the bulb and are deposited on the inner surface of the glass. In time, the glass becomes less transparent because of this tungsten coating, which decreases the amount of light that passes through the glass.

## Example 17.4 The Resistance of Nichrome Wire

**A** Calculate the resistance per unit length of a 22-gauge nichrome wire of radius 0.321 mm.

**Solution** The cross-sectional area of this wire is

$$A = \pi r^2 = \pi(0.321 \times 10^{-3}\ \text{m})^2 = 3.24 \times 10^{-7}\ \text{m}^2$$

The resistivity of nichrome is $1.5 \times 10^{-6}\ \Omega \cdot \text{m}$ (Table 17.1). Thus, we can use Equation 17.6 to find the resistance per unit length:

$$\frac{R}{l} = \frac{\rho}{A} = \frac{1.5 \times 10^{-6}\ \Omega \cdot \text{m}}{3.24 \times 10^{-7}\ \text{m}^2} = \boxed{4.6\ \Omega/\text{m}}$$

**B** If a potential difference of 10.0 V is maintained across a 1.0-m length of the nichrome wire, what is the current in the wire?

**Solution** Because a 1.0-m length of this wire has a resistance of 4.6 Ω, Ohm's law gives

$$I = \frac{\Delta V}{R} = \frac{10.0\ \text{V}}{4.6\ \Omega} = \boxed{2.2\ \text{A}}$$

Note from Table 17.1 that the resistivity of nichrome is about 100 times that of copper, a typical good conductor. Therefore, a copper wire of the same radius would have a resistance per unit length of only 0.052 Ω/m, and a 1.0-m length of copper wire of the same radius would carry the same current (2.2 A) with an applied voltage of only 0.11 V.

Because of its resistance to oxidation, nichrome is often used for heating elements in toasters, irons, and electric heaters.

**EXERCISE** What is the resistance of a 6.0-m length of 22-gauge nichrome wire? How much current does it carry when connected to a 120-V source?

**ANSWER** 28 Ω; 4.3 A

**Quick Quiz 17.4** Aliens with strange powers visit Earth and double every linear dimension of every object on the surface of the Earth. Does the electrical cord from the wall socket to your floor lamp now have (a) more resistance than before, (b) less resistance, or (c) the same resistance? Does the lightbulb filament glow (d) more brightly than before, (e) less brightly, or (f) the same? (Assume the resistivities of materials remain the same before and after the doubling.)

**Quick Quiz 17.5** A voltage $\Delta V$ is applied across the ends of a nichrome heater wire having a cross-sectional area $A$ and length $L$. The same voltage is applied across the ends of a second heater wire having a cross-sectional area $A$ and length $2L$. Which wire gets hotter? (a) The shorter wire, (b) the longer wire, or (c) not enough information to say.

## 17.6 TEMPERATURE VARIATION OF RESISTANCE

An old-fashioned carbon-filament incandescent lamp. The resistance of such a lamp is typically 10 Ω, but it changes with temperature. *(Courtesy of Central Scientific Company)*

The resistivity, and hence the resistance, of a conductor depends on a number of factors. One of the most important is the temperature of the metal. For most metals, resistivity increases with increasing temperature. This correlation can be understood as follows. As the temperature of the material increases, its constituent atoms vibrate with increasingly greater amplitudes. Just as it is more difficult to weave one's way through a crowded room when the people are in motion than when they are standing still, so do the electrons find it more difficult to pass atoms vibrating with large amplitudes. The increased electron scattering with increasing temperature results in increased resistivity.

For most metals, resistivity increases approximately linearly with temperature over a limited temperature range, according to the expression

$$\rho = \rho_0[1 + \alpha(T - T_0)] \tag{17.7}$$

where $\rho$ is the resistivity at some temperature $T$ (in Celsius degrees), $\rho_0$ is the resistivity at some reference temperature $T_0$ (usually taken to be 20°C), and $\alpha$ is a parameter called the **temperature coefficient of resistivity.** The temperature coefficients for various materials are provided in Table 17.1. The interesting negative values of $\alpha$ for semiconductors arise because these materials possess weakly bound charge carriers that become free to move and contribute to the current as the temperature is raised.

Because the resistance of a conductor with uniform cross section is proportional to the resistivity, according to Equation 17.6 ($R = \rho l/A$), the temperature variation of resistance can be written

$$R = R_0[1 + \alpha(T - T_0)] \tag{17.8}$$

Precise temperature measurements are often made using this property, as shown by the following example.

### Example 17.5 A Platinum Resistance Thermometer

A resistance thermometer, which measures temperature by measuring the change in resistance of a conductor, is made of platinum and has a resistance of 50.0 Ω at 20.0°C. When the device is immersed in a vessel containing melting indium, its resistance increases to 76.8 Ω. From this information, find the melting point of indium.

**Solution** If we solve Equation 17.8 for $T - T_0$ and get $\alpha$ for platinum from Table 17.1, we obtain

$$T - T_0 = \frac{R - R_0}{\alpha R_0} = \frac{76.8\ \Omega - 50.0\ \Omega}{[3.92 \times 10^{-3}\ (°\text{C})^{-1}][50.0\ \Omega]} = 137°\text{C}$$

Because $T_0 = 20.0°\text{C}$, we find that the melting point of indium is

$$T = 157°\text{C}$$

## 17.7 SUPERCONDUCTORS

There is a class of metals and compounds whose resistances fall virtually to *zero* below a certain temperature $T_c$ called the *critical temperature.* These materials are known as **superconductors.** The resistance-temperature graph for a superconductor follows that of a normal metal at temperatures above $T_c$ (Fig. 17.9). When the temperature is at or below $T_c$, the resistance suddenly drops to zero. This phenomenon was discovered in 1911 by the Dutch physicist H. Kamerlingh Onnes as he and a graduate student worked with mercury, which is a superconductor below 4.1 K. Recent measurements have shown that the resistivities of superconductors below $T_c$ are less than $4 \times 10^{-25}\ \Omega \cdot \text{m}$—around $10^{17}$ times smaller than the resistivity of copper and in practice considered to be zero.

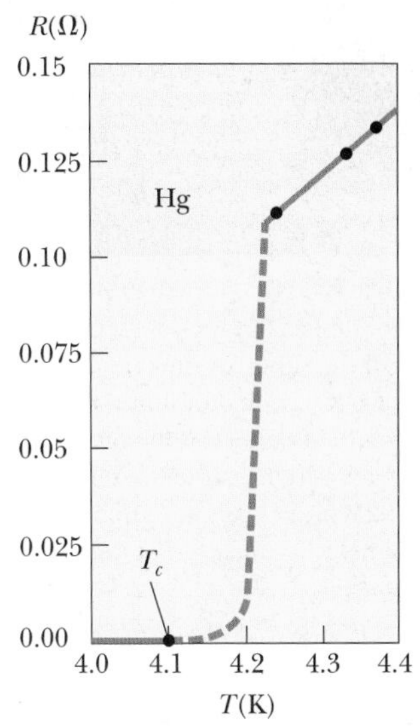


**FIGURE 17.9** Resistance versus temperature for a sample of mercury. The graph follows that of a normal metal above the critical temperature $T_c$. The resistance drops to zero at the critical temperature, which is 4.1 K for mercury, and remains at zero for lower temperatures.

Today, thousands of superconductors are known, including such common metals as aluminum, tin, lead, zinc, and indium. Table 17.2 lists the critical temperatures of several superconductors. The value of $T_c$ is sensitive to chemical composition, pressure, and crystalline structure. Interestingly, copper, silver, and gold, which are excellent conductors, do not exhibit superconductivity.

One of the truly remarkable features of superconductors is the fact that, once a current is set up in them, it persists *without any applied voltage* (because $R = 0$). In fact, steady currents in superconducting loops have been observed to persist for many years with no apparent decay!

**TABLE 17.2**

**Critical Temperatures for Various Superconductors**

| Material | $T_c$ (K) |
|---|---|
| Zn | 0.88 |
| Al | 1.19 |
| Sn | 3.72 |
| Hg | 4.15 |
| Pb | 7.18 |
| Nb | 9.46 |
| $Nb_3Sn$ | 18.05 |
| $Nb_3Ge$ | 23.2 |
| $YBa_2Cu_3O_7$ | 90 |
| Bi-Sr-Ca-Cu-O | 105 |
| Tl-Ba-Ca-Cu-O | 125 |
| $HgBa_2Ca_2Cu_3O_8$ | 134 |

An important development in physics that created much excitement in the scientific community was the discovery of high-temperature copper-oxide-based superconductors. The excitement began with a 1986 publication by J. Georg Bednorz and K. Alex Müller, scientists at the IBM Zurich Research Laboratory in Switzerland, in which they reported evidence for superconductivity at a temperature near 30 K in an oxide of barium, lanthanum, and copper. Bednorz and Müller were awarded the Nobel prize for physics in 1987 for their discovery. This discovery was remarkable in view of the fact that the critical temperature was significantly higher than that of any previously known superconductor. Shortly thereafter, a new family of compounds was investigated, and research activity in the field of superconductivity proceeded vigorously. In early 1987, groups at the University of Alabama at Huntsville and the University of Houston announced the discovery of superconductivity at about 92 K in an oxide of yttrium, barium, and copper ($YBa_2Cu_3O_7$), shown as the gray disk in Figure 17.10. Late in 1987, teams of scientists from Japan and the United States reported superconductivity at 105 K in an oxide of bismuth, strontium, calcium, and copper. More recently, scientists have reported superconductivity at temperatures as high as 150 K in an oxide containing mercury. At this point one cannot rule out the possibility of room-temperature superconductivity, and the search for novel superconducting materials continues. It is an important search both for scientific reasons and because practical applications become more probable and widespread as the critical temperature is raised and the expense of cooling materials decreases.

### Webnote 17.1

The Department of Energy has a good Web site on superconductivity at *http://www.eren.doe.gov/superconductivity/*

An important and useful application is superconducting magnets in which the magnetic field intensities are about ten times greater than those of the best normal electromagnets. Such magnets are being considered as a means of storing

**FIGURE 17.10** A small permanent magnet floats freely above a ceramic disk of the superconductor $YBa_2Cu_3O_7$ cooled by liquid nitrogen at 77 K. The superconductor has zero electric resistance at temperatures below 92 K and expels any applied magnetic field. *(Courtesy of IBM Research Laboratory)*

energy. The idea of using superconducting power lines for transmitting power efficiently is also receiving serious consideration. Modern superconducting electronic devices consisting of two thin-film superconductors separated by a thin insulator have been constructed. They include magnetometers (magnetic-field measuring devices) and various microwave devices.

## 17.8 ELECTRICAL ENERGY AND POWER

If a battery is used to establish an electric current in a conductor, chemical energy stored in the battery is continuously transformed into kinetic energy of the charge carriers. This kinetic energy is quickly lost as a result of collisions between the charge carriers and fixed atoms in the conductor, causing an increase in the temperature of the conductor. Thus, the chemical energy stored in the battery is continuously transformed into internal energy.

In order to understand the process of energy transfer in a simple circuit, consider a battery whose terminals are connected to a resistor (Fig. 17.11). (Remember that the positive terminal of the battery is always at the higher potential.) Now imagine following a quantity of positive charge $\Delta Q$ around the circuit from point $A$ through the battery and resistor and back to $A$. Point $A$ is a reference point that is grounded (the ground symbol is ⏚), and its potential is taken to be zero. As the charge moves from $A$ to $B$ through the battery, whose potential difference is $\Delta V$, the electrical potential energy of the system increases by the amount $\Delta Q\,\Delta V$, and the chemical potential energy in the battery decreases by the same amount. (Recall from Chapter 16 that $\Delta PE = q\,\Delta V$.) However, as the charge moves from $C$ to $D$ through the resistor, it loses this electrical potential energy during collisions with atoms in the resistor. In this process, the energy is transformed to internal energy corresponding to increased vibrational motion of the atoms in the resistor. Because we have ignored the resistance of the interconnecting wires, no energy transformation occurs for paths $BC$ and $DA$. When the charge returns to point $A$, the net result is that some of the chemical energy in the battery has been delivered to the resistor and causes its temperature to rise. Feeling the resistor, we would often notice that it was warmer than its surroundings.

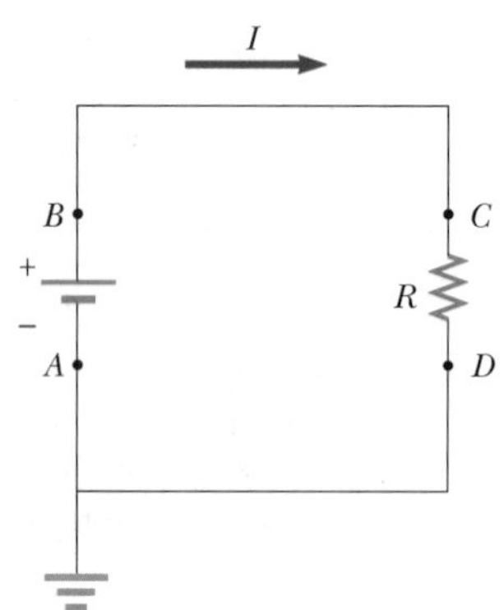


**FIGURE 17.11** A circuit consisting of a battery and resistance $R$. Positive charge flows clockwise from the positive to the negative terminal of the battery. Point A is grounded.

The charge $\Delta Q$ loses energy $\Delta Q\,\Delta V$ as it passes through the resistor. If $\Delta t$ is the time it takes the charge to pass through the resistor, then the rate at which it loses electrical potential energy is

$$\frac{\Delta Q}{\Delta t}\,\Delta V = I\,\Delta V$$

where $I$ is the current in the resistor and $\Delta V$ is the potential difference across it. Of course, the charge regains this energy when it passes through the battery, at the expense of chemical energy in the battery. The rate at which the system loses potential energy as the charge passes through the resistor is equal to the rate at which the system gains internal energy in the resistor. Thus, the power $\mathcal{P}$ representing the rate at which energy is delivered to the resistor is

$$\mathcal{P} = I\,\Delta V \qquad [17.9]$$

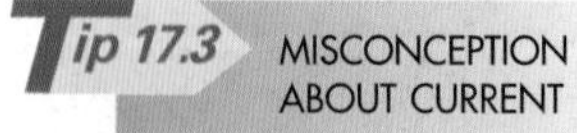


Note that current is *not* "used up" in the resistor. The current is the same *everywhere* in the circuit. Charges flow in the same rotational sense at *all* points in the circuit.

We have developed this result by considering a battery delivering energy to a resistor. However, Equation 17.9 can be used to determine the power transferred from a voltage source to *any* device carrying a current $I$ and having a potential difference $\Delta V$ between its terminals.

Using Equation 17.9 and the fact that $\Delta V = IR$ for a resistor, we can express the power delivered to the resistor in the alternative forms

Power delivered to a resistor

$$\mathcal{P} = I^2R = \frac{(\Delta V)^2}{R} \qquad [17.10]$$

When $I$ is in amperes, $\Delta V$ in volts, and $R$ in ohms, the SI unit of power is the watt (introduced in Chapter 5). The power delivered to a conductor of resistance $R$ is often referred to as an *$I^2R$ loss.* Note that Equation 17.10 applies only to resistors and not to nonohmic devices like lightbulbs and diodes.

Regardless of the ways in which you use electrical energy in your home, you ultimately must pay for it or risk having your power turned off. The unit of energy used by electric companies to calculate consumption, the **kilowatt-hour,** is defined in terms of the unit of power and the amount of time it is supplied. One kilowatt-hour (kWh) is the energy converted or consumed in 1 h at the constant rate of 1 kW. It has the numerical value

$$1 \text{ kWh} = (10^3 \text{ W})(3\,600 \text{ s}) = 3.60 \times 10^6 \text{ J} \quad [17.11]$$

The kilowatt-hour is a unit of energy

On an electric bill, the amount of electricity used in a given period is usually stated in multiples of kilowatt-hours.

## APPLYING PHYSICS 17.4

When is more power delivered to a lightbulb—just after it is turned on and the glow of the filament is increasing or after it has been on for a few seconds and the glow is steady?

**Explanation** Once the switch is closed, the line voltage is applied across the lightbulb. As the voltage is applied across the cold filament when first turned on, the resistance of the filament is low, the current is high, and a relatively large amount of power is delivered to the bulb. As the filament warms, its resistance rises and the current decreases. As a result, the power delivered to the bulb decreases. The large current spike at the beginning of operation is the reason that lightbulbs often fail just after they are turned on.

**Quick Quiz 17.6** For the two resistors shown in Figure 17.12, rank the currents at points $a$ through $f$, from largest to smallest.

**Quick Quiz 17.7** Two resistors, A and B, are connected across the same potential difference. The resistance of A is twice that of B. (a) Which resistor dissipates more power? (b) Which carries the greater current?

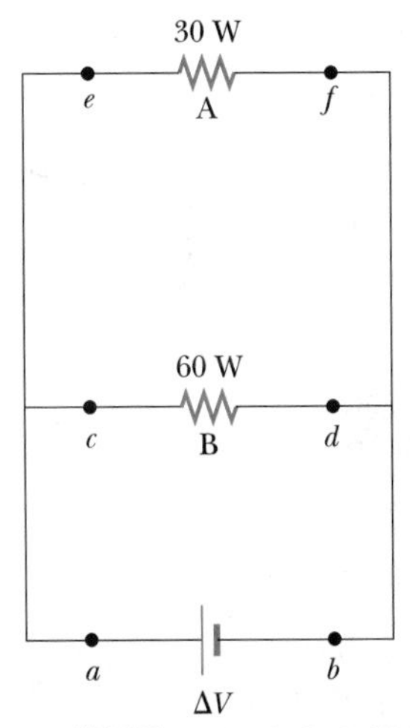


**FIGURE 17.12** (Quick Quiz 17.6)

### Example 17.6 The Power Converted by an Electric Heater

An electric heater is operated by applying a potential difference of 50.0 V to a nichrome wire of total resistance 8.00 Ω. Find the current carried by the wire and the power rating of the heater.

**Solution** Because $\Delta V = IR$, we have

$$I = \frac{\Delta V}{R} = \frac{50.0 \text{ V}}{8.00\ \Omega} = 6.25 \text{ A}$$

We can find the power rating using $\mathcal{P} = I^2R$:

$$\mathcal{P} = I^2R = (6.25 \text{ A})^2(8.00\ \Omega) = 313 \text{ W}$$

**EXERCISE** If we doubled the applied voltage to the heater, what would happen to the current and power?

**ANSWER** The current would double, and the power would quadruple.

## PHYSICS *IN ACTION*

The photo shows electrical workers at dusk, trying to restore power to the eastern Ontario town of St. Isadore that was without power for several days in January 1998 because of a severe ice storm. When power transmission lines fall because of lightning, ice storms, or earthquakes, it is very dangerous to touch them due to the high electric potential, possibly hundreds of thousands of volts relative to the ground. When transporting electrical energy through power lines, utility companies seek to minimize the power transformed into internal energy in the lines and maximize the energy delivered to the consumer. Because $\mathcal{P} = I\,\Delta V$, the same amount of power can be transported either at high currents and low potential differences or at low currents and high potential differences. Utility companies choose to transport electrical energy at low currents and high potential differences primarily for economic reasons. Copper wire is very expensive, and so it is cheaper to use high-resistance wire (that is, wire having a small cross-sectional area; see Eq. 17.6). In the expression for power delivered to a resistor, $\mathcal{P} = I^2R$, the resistance of the transmission wire is fixed at a relatively high value for economic considerations. The $I^2R$ loss can be reduced by keeping the current $I$ as low as possible.

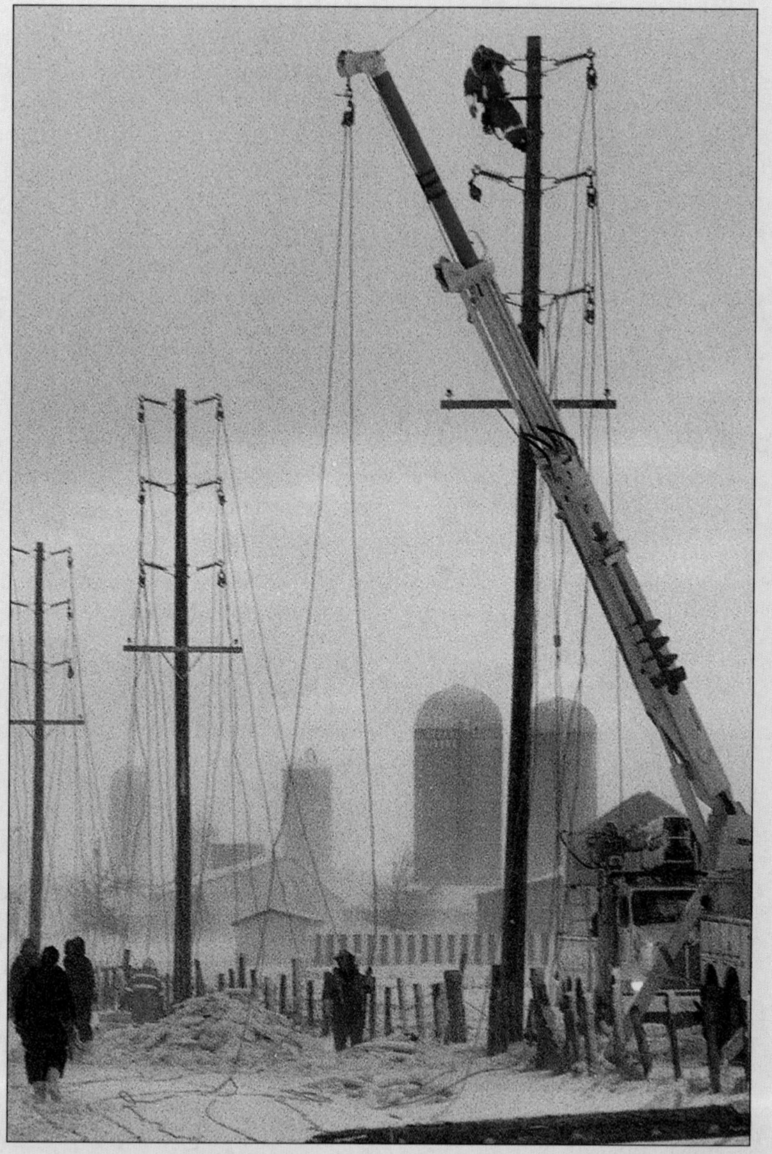

*(AP/Wide World Photos/Fred Chartrand)*

### Example 17.7 Lighting Up Your Life

A circuit provides a current of 20.0 A at an operating voltage of 120 V. How many 75-W bulbs can operate with this voltage source?

**Solution** The total power that can be delivered by the voltage source is given by Equation 17.9:

$$\mathcal{P}_{\text{total}} = I\Delta V = (20.0\text{ A})(120\text{ V}) = 2.40 \times 10^3\text{ W}$$

Because each bulb requires a power of 75 W, the number of bulbs that can operate from this voltage source is

$$\text{Number of bulbs} = \frac{\mathcal{P}_{\text{total}}}{\mathcal{P}_{\text{bulb}}} = \frac{2.40 \times 10^3\text{ W}}{75\text{ W}} = 32$$

### Example 17.8 The Cost of Operating a Lightbulb

How much does it cost to burn a 100-W lightbulb for 24 h if electric energy costs 12 cents or \$0.12 per kilowatt-hour?

**Solution** A 100-W lightbulb is equivalent to a 0.10-kW bulb. Because energy consumed equals power × time, the amount of energy you must pay for, expressed in kilowatt-hours, is

$$\text{Energy} = (0.10\text{ kW})(24\text{ h}) = 2.4\text{ kWh}$$

If energy is purchased at 12 cents or \$0.12 per kilowatt-hour, the 24-h cost is

$$\text{Cost} = (2.4\text{ kWh})(\$0.12/\text{kWh}) = \$0.29$$

This is a small amount of money, but when larger and more complex electric devices are used, the cost goes up rapidly.

**Exercise** If electric energy costs \$0.12/kWh, what does it cost to operate an electric oven, which operates at 20.0 A and 220 V, for 5.0 h?

**Answer** \$2.70

## 17.9 Electrical Activity in the Heart

### Electrocardiograms

Every action involving the body's muscles is initiated by electrical activity. The voltages produced by muscular action in the heart are particularly important to physicians. Voltage pulses cause the heart to beat, and the waves of electrical excitation that sweep across the heart associated with the heartbeat are conducted through the body via the body fluids. These voltage pulses are large enough to be detected by suitable monitoring equipment attached to the skin. A sensitive voltmeter making good electrical contact with the skin by means of contacts attached with conducting paste can be used to measure heart pulses that are typically of the order of 1 mV at the surface of the body. The voltage pulses can be recorded on an instrument called an **electrocardiograph,** and the pattern recorded by this instrument is called an **electrocardiogram** (EKG). In order to understand the information contained in an EKG pattern, it is necessary first to describe the underlying principles concerning electrical activity in the heart.

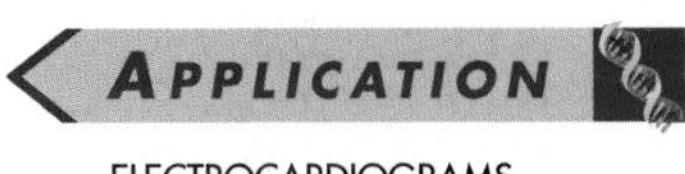


ELECTROCARDIOGRAMS

The right atrium of the heart contains a specialized set of muscle fibers called the SA (sinoatrial) node, which initiate the heartbeat (Fig. 17.13). Electric impulses that originate in these fibers gradually spread from cell to cell throughout the right and left atrial muscles, causing them to contract. The pulse that passes through the muscle cells is often called a *depolarization wave* because of its effect on individual cells. If an individual muscle cell were examined in its resting state, a double-layered electric charge distribution would be found on its surface, as shown in Figure 17.14a. The impulse generated by the SA node momentarily and locally allows positive charge on the outside of the cell to flow in and neutralize the negative charge on the inside layer. This changes the cell's charge distribution to that shown in Figure 17.14b. Once the depolarization wave has passed through a cell, an individual heart muscle cell recovers the resting state charge distribution (positive out, negative in) shown in Figure 17.14a in about 250 ms. When the impulse reaches the AV (atrioventricular) node (Fig. 17.13), the muscles of the atria begin to relax, and the pulse is directed by the AV node to the ventricular muscles. The muscles of the ventricles contract as the depolarization wave spreads through the ventricles along a group of fibers called the *Purkinje fibers.* The ventricles then

Sinoatrial (SA) node
Purkinje fibers
LA
RA
LV
RV
Atrioventricular (AV) node

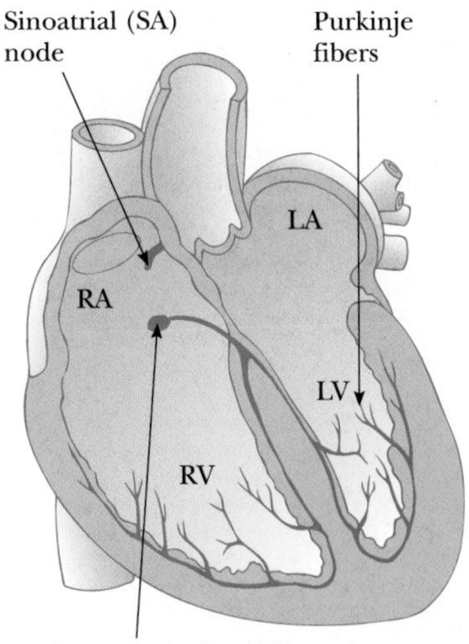

**Figure 17.13** The electrical conduction system of the human heart. (RA: right atrium; LA: left atrium; RV: right ventricle; LV: left ventricle.)

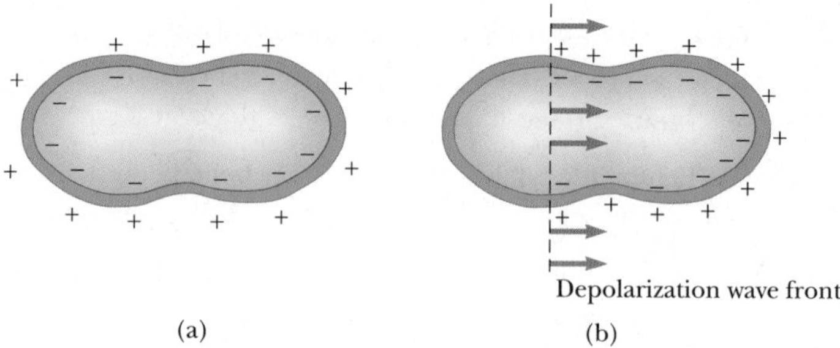


**FIGURE 17.14** (a) Charge distribution of a muscle cell in the atrium before a depolarization wave has passed through the cell. (b) Charge distribution as the wave passes.

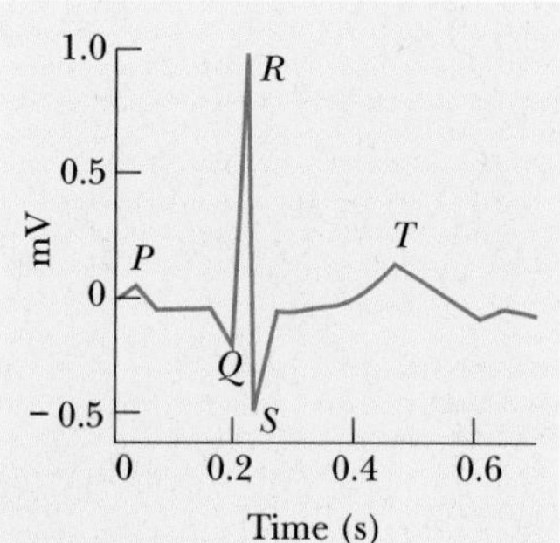


**FIGURE 17.15** An EKG response for a normal heart.

## Webnote 17.2

Electrocardiograms have unique signatures associated with malfunctions. To see and learn about these signatures, visit the home page of the following EKG library: *http://www.ecglibrary.com/ecghome.html*

relax after the pulse has passed through. At this point, the SA node is again triggered and the cycle is repeated.

A sketch of the electrical activity registered on an EKG for one beat of a normal heart is shown in Figure 17.15. The pulse indicated by *P* occurs just before the atria begin to contract. The *QRS* pulse occurs in the ventricles just before they contract, and the *T* pulse occurs when the cells in the ventricles begin to recover. EKGs for an abnormal heart are shown in Figure 17.16. The *QRS* portion of the pattern shown in Figure 17.16a is wider than normal. This indicates that the patient may have an enlarged heart. (Why?) Figure 17.16b indicates that there is no constant relationship between the *P* pulse and the *QRS* pulse. This suggests a blockage in the electrical conduction path between the SA and AV nodes, which results in the atria and ventricles beating independently, leading to inefficient heart pumping. Finally, Figure 17.16c shows a situation in which there is no *P* pulse and an irregular spacing between the *QRS* pulses. This is symptomatic of irregular atrial contraction, which is called *fibrillation.* In this situation, the atrial and ventricular contractions are irregular.

As noted previously, the sinoatrial node directs the heart to beat at the appropriate rate, usually about 72 beats per minute. However, disease or the aging process can damage the heart and slow its beating, and a medical assist may be

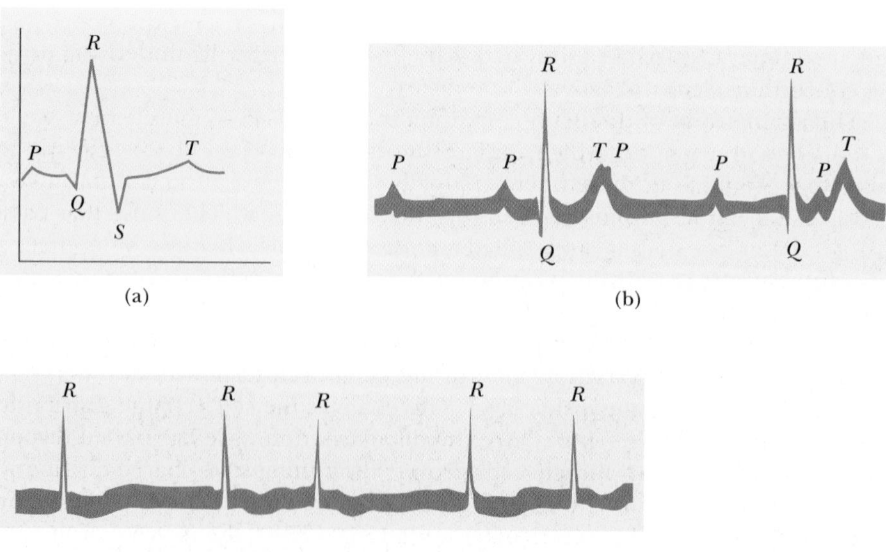


**FIGURE 17.16** Abnormal EKGs.

necessary in the form of a *cardiac pacemaker* attached to the heart. This matchbox-sized electrical device implanted under the skin has a lead that is connected to the wall of the right ventricle. Pulses from this lead stimulate the heart to maintain its proper rhythm. In general, a pacemaker is designed to produce pulses at a rate of about 60 per minute, slightly slower than the normal beats per minute but sufficient to maintain life. The circuitry basically consists of a capacitor charging up from a lithium battery to a certain voltage and then discharging. The design of the circuit is such that if the heart is beating normally the capacitor is never allowed to charge completely and send pulses to the heart.

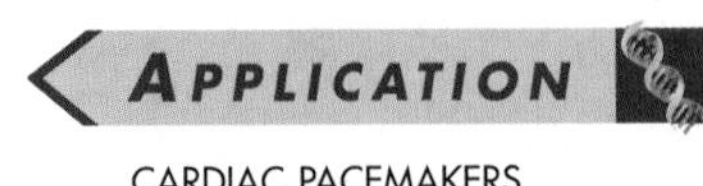


CARDIAC PACEMAKERS

## An Emergency Room in Your Chest

The operation in June 2001 on Vice President Dick Cheney focused attention on the progress in treating heart problems with tiny implanted electrical devices. Aptly termed "an emergency room in your chest" by Cheney's attending physician, current devices called Implanted Cardioverter Defibrillators **(ICDs)** can monitor, record, and logically process heart signals and then supply different corrective signals to hearts beating too slowly, too rapidly, or irregularly, even monitoring and sending independent signals to atria and ventricles! Figure 17.17a shows a sketch of an ICD with conducting leads that are implanted in the heart. Figure 17.17b shows an actual titanium-encapsulated dual chamber ICD.

IMPLANTED CARDIOVERTER DEFIBRILLATORS

The latest ICDs are quite sophisticated devices capable of

1. monitoring both atrial and ventricular chambers to differentiate between atrial and potentially fatal ventricular arrhythmias, which require prompt regulation;
2. storing about $\frac{1}{2}$ hour of heart signals, which can easily be read out by a physician;
3. being easily reprogrammed with an external magnetic wand;
4. performing complicated signal analysis and comparison;

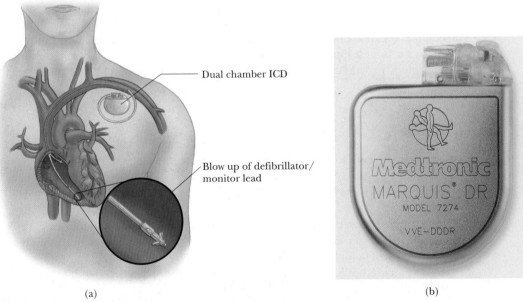

Dual chamber ICD

Blow up of defibrillator/ monitor lead

(a)

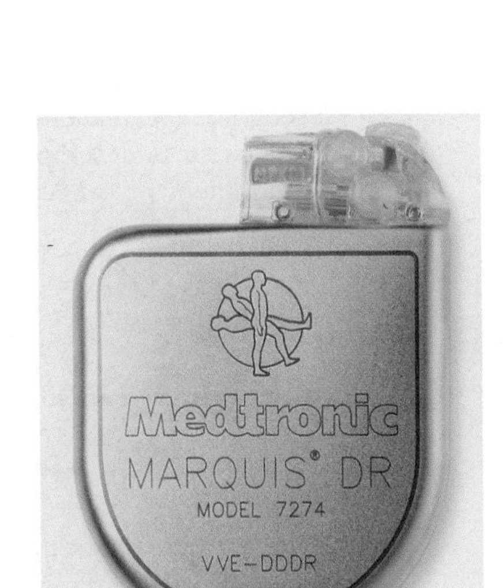


(b)

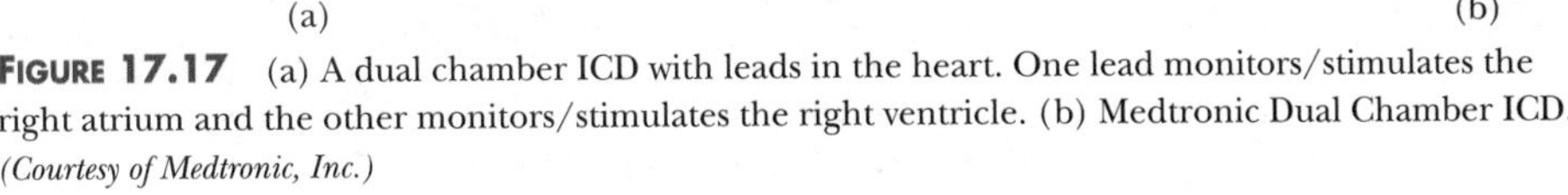

**Figure 17.17** (a) A dual chamber ICD with leads in the heart. One lead monitors/stimulates the right atrium and the other monitors/stimulates the right ventricle. (b) Medtronic Dual Chamber ICD. *(Courtesy of Medtronic, Inc.)*

**TABLE 17.3 Properties of Implanted Cardioverter Defibrillators[a]**

| | |
|---|---|
| **Physical Specifications** | |
| Mass (g) | 85 |
| Size (cm) | 7.3 × 6.2 × 1.3 (about 5 stacked silver dollars) |
| **Antitachycardia Pacing** | ICD delivers a burst of critically timed, low-energy pulses |
| Number of bursts | 1–15 |
| Burst cycle length (ms) | 200–552 |
| Number of pulses per burst | 2–20 |
| Pulse amplitude (V) | 7.5 or 10 |
| Pulsewidth (ms) | 1.0 or 1.9 |
| **High-Voltage Defibrillation** | |
| Pulse energy (J) | 37 stored/33 delivered |
| Pulse amplitude (V) | 801 |
| **Bradycardia Pacing** | A dual chamber ICD can steadily deliver repetitive pulses to both the atrium and ventricle |
| Base frequency (beats/min) | 40–100 |
| Pulse amplitude (V) | 0.25–7.5 |
| Pulsewidth (ms) | 0.05, 0.1–1.5, 1.9 |

[a] For more information, see www.photonicd.com/specs.html

5. supplying 0.25- to 10-V repetitive pacing signals to speed up or slow down a malfunctioning heart, or a high-voltage pulse of about 800 V to halt the potentially fatal condition of ventricular fibrillation, in which the heart rapidly quivers rather than beats (people who have experienced such a high-voltage jolt say that it feels like a kick or a bomb going off in the chest);
6. automatically adjusting the number of pacing pulses per minute to match the patient's activity.

ICDs are powered by lithium batteries and have implanted lifetimes of 4–6 years. Some basic properties of these adjustable ICDs are given in Table 17.3. In this table, *tachycardia* means rapid heartbeat and *bradycardia* means slow heartbeat. A key factor in developing tiny electrical implants that serve as defibrillators is the development of capacitors with relatively large capacitance (125 $\mu$F) and small physical size.

## SUMMARY

The **electric current** $I$ in a conductor is defined as

$$I \equiv \frac{\Delta Q}{\Delta t} \qquad [17.1]$$

where $\Delta Q$ is the charge that passes through a cross section of the conductor in time $\Delta t$. The SI unit of current is the **ampere** (A); 1 A = 1 C/s. By convention, the direction of current is in the direction of flow of positive charge.

The current in a conductor is related to the motion of the charge carriers by

$$I = nqv_dA \qquad [17.3]$$

where $n$ is the number of mobile charge carriers per unit volume, $q$ is the charge on each carrier, $v_d$ is the drift speed of the charges, and $A$ is the cross-sectional area of the conductor.

The **resistance** $R$ of a conductor is defined as the ratio of the potential difference across the conductor to the current:

$$R \equiv \frac{\Delta V}{I} \qquad [17.4]$$

The SI units of resistance are volts per ampere, or **ohms** ($\Omega$); $1\ \Omega = 1\ \text{V/A}$.

**Ohm's law** describes many conductors for which the applied voltage is directly proportional to the current it causes. The proportionality constant is the resistance:

$$\Delta V = IR \qquad [17.5]$$

If a conductor has length $l$ and cross-sectional area $A$, its **resistance** is

$$R = \rho \frac{l}{A} \qquad [17.6]$$

where $\rho$ is an intrinsic property of the conductor called the **resistivity.** The SI unit of resistivity is the **ohm-meter** ($\Omega \cdot \text{m}$).

The resistivity of a conductor varies with temperature over a limited temperature range, according to the expression

$$\rho = \rho_0[1 + \alpha(T - T_0)] \qquad [17.7]$$

where $\alpha$ is the **temperature coefficient of resistivity** and $\rho_0$ is the resistivity at some reference temperature $T_0$ (usually taken to be 20°C).

The resistance of a conductor varies with temperature according to the expression

$$R = R_0[1 + \alpha(T - T_0)] \qquad [17.8]$$

If a potential difference $\Delta V$ is maintained across an electrical device, the **power,** or rate at which energy is supplied to the device, is

$$\mathcal{P} = I\Delta V \qquad [17.9]$$

Because the potential difference across a resistor is $\Delta V = IR$, the **power delivered to a resistor** can be expressed as

$$\mathcal{P} = I^2R = \frac{(\Delta V)^2}{R} \qquad [17.10]$$

A **kilowatt-hour** is the amount of energy converted or consumed in 1 h by a device supplied with power at the rate of 1 kW. This is equivalent to

$$1\ \text{kWh} = 3.60 \times 10^6\ \text{J} \qquad [17.11]$$

## CONCEPTUAL QUESTIONS

1. A coulomb is a very large unit of charge, and yet currents of several amperes are quite common. How is this possible?
2. Edison's original lightbulb contained a carbon filament. How should carbon be described electrically?
3. Why don't the free electrons in a metal fall to the bottom of the metal due to gravity? And charges in a conductor are supposed to reside on the surface—why don't the free electrons all go to the surface?
4. In an analogy between traffic flow and electrical current, what would correspond to the charge $Q$? What would correspond to the current $I$?
5. Newspaper articles often have statements such as, "10 000 volts of electricity surged *through* the victim's body." What is wrong with this statement?
6. "A microvoltmeter measured the resistance of the sciatic nerve." Criticize this statement. How might it be corrected?
7. When the voltage across a certain conductor is doubled, the current is observed to triple. What can you conclude about the conductor?
8. There is an old admonition given to experimenters to "keep one hand in the pocket" when working around high voltages. Why might this be a good idea?
9. What factors affect the resistance of a conductor?
10. Some homes have light dimmers that are operated by rotation of a knob. What is being changed in the electric circuit when the knob is rotated?
11. Two wires A and B with circular cross section are made of the same metal and have equal lengths, but the resistance of wire A is three times greater than that of wire B. What is the ratio of their cross-sectional areas? How do the radii compare?
12. What single experimental requirement makes superconducting devices expensive to operate? In principle, can this limitation be overcome?

13. What could happen to the drift velocity of the electrons in a wire and to the current in the wire if the electrons could move freely without resistance through this metallic conductor?

14. Use the atomic theory of matter to explain why the resistance of a material should increase as its temperature increases.

## PROBLEMS

1, 2, 3 = straightforward, intermediate, challenging □ = full solution available in Student Solutions Manual/Study Guide

web = solution posted at **http://info.brookscole.com/serway** = biomedical application

### Section 17.1 Electric Current

### Section 17.2 A Microscopic View: Current and Drift Speed

1. If a current of 80.0 mA exists in a metal wire, how many electrons flow past a given cross section of the wire in 10.0 min? Sketch the directions of the current and the electrons' motion.
2. A small sphere that carries a charge $q$ is whirled in a circle at the end of an insulating string. The angular frequency of rotation is $\omega$. What average current does this rotating charge represent?
3. A total charge of 6.0 mC passes through a cross-sectional area of a wire in 2.0 s. What is the current in the wire?
4. In a particular television picture tube, the measured beam current is 60.0 $\mu$A. How many electrons strike the screen every second?
5. In the Bohr model of the hydrogen atom, an electron in the lowest energy state moves at a speed of $2.19 \times 10^6$ m/s in a circular path having a radius of $5.29 \times 10^{-11}$ m. What is the effective current associated with this orbiting electron?
6. If $3.25 \times 10^{-3}$ kg of gold is deposited on the negative electrode of an electrolytic cell in a period of 2.78 h, what is the current through the cell in this period? Assume that the gold ions carry one elementary unit of positive charge.
7. A 200-km-long high-voltage transmission line 2.0 cm in diameter carries a steady current of 1 000 A. If the conductor is copper with a free charge density of $8.5 \times 10^{28}$ electrons per cubic meter, how long (in years) does it take one electron to travel the full length of the cable?
8. web An aluminum wire with a cross-sectional area of $4.0 \times 10^{-6}$ m$^2$ carries a current of 5.0 A. Find the drift speed of the electrons in the wire. The density of aluminum is 2.7 g/cm$^3$. (Assume that one electron is supplied by each atom.)
9. If the current carried by a conductor is doubled, what happens to the (a) charge carrier density and (b) electron drift velocity?

### Section 17.4 Resistance and Ohm's Law

### Section 17.5 Resistivity

10. A lightbulb has a resistance of 240 Ω when operating at a voltage of 120 V. What is the current through the lightbulb?
11. A person notices a mild shock if the current along a path through the thumb and index finger exceeds 80 $\mu$A. Compare the maximum allowable voltage without shock across the thumb and index finger with a dry-skin resistance of $4.0 \times 10^5$ Ω and a wet-skin resistance of 2 000 Ω.
12. Suppose that you wish to fabricate a uniform wire out of 1.00 g of copper. If the wire is to have a resistance of $R = 0.500$ Ω, and if all of the copper is to be used, what will be (a) the length and (b) the diameter of this wire?
13. Calculate the diameter of a 2.0-cm length of tungsten filament in a small lightbulb if its resistance is 0.050 Ω.
14. Eighteen-gauge wire has a diameter of 1.024 mm. Calculate the resistance of 15 m of 18-gauge copper wire at 20°C.
15. A potential difference of 12 V is found to produce a current of 0.40 A in a 3.2-m length of wire with a uniform radius of 0.40 cm. What is (a) the resistance of the wire and (b) the resistivity of the wire?
16. A length $L_0$ of copper wire has a resistance $R_0$. The wire is cut into three pieces of equal length. The pieces are then connected as parallel lengths between points $A$ and $B$. What resistance will this new "wire" of length $L_0/3$ have between points $A$ and $B$?
17. A wire 50.0 m long and 2.00 mm in diameter is connected to a source with a potential difference of 9.11 V, and the current is found to be 36.0 A. Assume a temperature of 20°C and, using Table 17.1, identify the metal of the wire.
18. A rectangular block of copper has sides of length 10 cm, 20 cm, and 40 cm. If the block is connected to a 6.0-V source across opposite faces of the rectangular block, what are (a) the maximum current and (b) minimum current that can be carried?
19. The breathing monitor shown in Figure P17.19 girds the patient with a mercury-filled rubber tube and measures the variation of the tube resistance. The tube has an unstretched length of 1.25 m and an inside diameter of 2.51 mm. The monitor is connected to a 100-mV power supply, and the total resistance of the circuit is that due to

**FIGURE P17.19**

the mercury *plus* 1.00 Ω (an internal resistance of the power supply). Determine the change of current through the monitor as a patient draws in a breath and stretches the hose by 10.0 cm. Take $\rho_{Hg} = 9.40 \times 10^{-7}\ \Omega \cdot m$.

## Section 17.6 Temperature Variation of Resistance

**20.** A certain lightbulb has a tungsten filament with a resistance of 19 Ω when cold and 140 Ω when hot. Assume that Equation 17.8 can be used over the large temperature range involved here, and find the temperature of the filament when it is hot. Assume an initial temperature of 20°C.

**21.** While taking photographs in Death Valley on a day when the temperature is 58.0°C, Bill Hiker finds that a certain voltage applied to a copper wire produces a current of 1.000 A. Bill then travels to Antarctica and applies the same voltage to the same wire. What current does he register there if the temperature is −88.0°C? Assume that no change occurs in the wire's shape and size.

**22.** If a silver wire has a resistance of 10.0 Ω at 20.0°C, what resistance does it have at 40.0°C? Neglect any change in length or cross-sectional area resulting from the change in temperature.

**23.** At 20°C the carbon resistor in an electric circuit, connected to a 5.0-V battery, has a resistance of 200 Ω. What is the current in the circuit when the temperature of the carbon rises to 80°C?

**24.** At 40.0°C, the resistance of a segment of gold wire is 100.0 Ω. When the wire is placed in a liquid bath, the resistance decreases to 97.0 Ω. What is the temperature of the bath? (*Hint:* First determine the resistance of the gold wire at room temperature, 20°C.)

**25.** The copper wire used in a house has a cross-sectional area of 3.00 mm². If 10.0 m of this wire is used to wire a circuit in the house at 20.0°C, find the resistance of the wire at temperatures of (a) 30.0°C and (b) 10.0°C.

26. An aluminum rod has a resistance of 1.234 Ω at 20.0°C. Calculate the resistance of the rod at 120°C by accounting for the changes in both the resistivity and the dimensions of the rod.

27. (a) A 34.5-m length of copper wire at 20.0°C has a radius of 0.25 mm. If a potential difference of 9.0 V is applied across the length of the wire, determine the current in the wire. (b) If the wire is heated to 30.0°C while the 9.0-V potential difference is maintained, what is the resulting current in the wire?

**web**

28. A toaster rated at 1 050 W operates on a 120-V household circuit and has a 4.00-m length of nichrome wire as its heating element. The operating temperature of this element is 320°C. What is the cross-sectional area of the wire?

29. In one form of plethysmograph (a device for measuring volume), a rubber capillary tube with an inside diameter of 1.00 mm is filled with mercury at 20°C. The resistance of the mercury is measured with the aid of electrodes sealed into the ends of the tube. If 100.00 cm of the tube is wound in a spiral around a patient's upper arm, the blood flow during a heartbeat causes the arm to expand, stretching the tube to a length of 100.04 cm. From this observation (assuming cylindrical symmetry) you can find the change in volume of the arm, which gives an indication of blood flow. (a) Calculate the resistance of the mercury. (b) Calculate the fractional change in resistance during the heartbeat. (*Hint:* The fraction by which the cross-sectional area of the mercury thread decreases is the fraction by which the length increases, since the volume of mercury is constant.) Take $\rho_{Hg} = 9.4 \times 10^{-7}\ \Omega \cdot m$.

30. A platinum resistance thermometer has resistances of 200.0 Ω when placed in a 0°C ice bath and 253.8 Ω when immersed in a crucible containing melting potassium. What is the melting point of potassium? (*Hint:* First determine the resistance of the platinum resistance thermometer at room temperature, 20°C.)

## Section 17.8 Electrical Energy and Power

**31.** A toaster is rated at 600 W when connected to a 120-V source. What current does the toaster carry, and what is its resistance?

**32.** The output power of the Sun is $4.0 \times 10^{26}$ W. Calculate at eight cents per kilowatt-hour the cost of running the Sun for one second.

**33.** How many 100-W lightbulbs can you use in a 120-V circuit without tripping a 15-A circuit breaker? (The bulbs are connected in parallel, which means that the potential difference across each lightbulb is 120 V.)

**34.** A high-voltage transmission line with a resistance of 0.31 Ω/km carries a current of 1 000 A. The line is at a potential of 700 kV at the power station and carries the current to a city located 160 km from the power station. (a) What is the power loss due to resistance in the line? (b) What fraction of the transmitted power does this loss represent?

35. The heating element of a coffee maker operates at 120 V and carries a current of 2.00 A. Assuming that the water absorbs all of the energy converted by the resistor, calculate how long it takes to heat 0.500 kg of water from room temperature (23.0°C) to the boiling point.

**36.** The power supplied to a typical black-and-white television set is 90 W when the set is connected to 120 V. (a) How much electric energy does this set consume in one hour? (b) A color television set draws about 2.5 A when connected to 120 V. How much time is required for it to consume the same energy as the black-and-white model consumes in one hour?

**37.** What is the required resistance of an immersion heater that will increase the temperature of 1.50 kg of water from 10.0°C to 50.0°C in 10.0 min while operating at 120 V?

38. A certain toaster has a heating element made of nichrome resistance wire. When the toaster is first connected to a 120-V source of potential difference (and the wire is at a temperature of 20.0°C) the initial current is 1.80 A. However, the current begins to decrease as the resistive element warms up. When the toaster has reached its final operating temperature, the current has dropped to 1.53 A. (a) Find the power the toaster converts when it is at its operating temperature. (b) What is the final temperature of the heating element?

39. A copper cable is designed to carry a current of 300 A with a power loss of 2.00 W/m. What is the required radius of this cable?

**40.** A small motor draws a current of 1.75 A from a 120-V line. The output power of the motor is 0.20 hp. (a) At a rate of \$0.060/kWh, what is the cost of operating the motor for 4.0 h? (b) What is the efficiency of the motor?

41. We estimate that there are 270 million plug-in electric clocks in the United States, approximately one clock for each person. The clocks convert energy at the average rate of 2.50 W. To supply this energy, how many metric tons of coal are burned per hour in coal-fired electric-generating plants that are, on average, 25.0% efficient? The heat of combustion for coal is 33.0 MJ/kg.

42. The cost of electricity varies widely throughout the United States; \$0.120/kWh is one typical value. At this unit price, calculate the cost of (a) leaving a 40.0-W porch light on for two weeks while you are on vacation, (b) making a piece of dark toast in 3.00 min with a 970-W toaster, and (c) drying a load of clothes in 40.0 min in a 5 200-W dryer.

43. How much does it cost to watch a complete 21-hour-long World Series on a 180-W television set? Assume that electricity costs \$0.070/kWh.

44. A house is heated by a 24-kW electric furnace using resistance heating. The rate for electrical energy is \$0.080/kWh. If the heating bill for January is \$200, how long must the furnace have been running on an average January day?

45. An 11-W energy-efficient fluorescent lamp is designed to produce the same illumination as a conventional 40-W lamp. How much does the energy-efficient lamp save during 100 hours of use? Assume a cost of \$0.080/kWh for electrical energy.

46. An electric resistance heater is to deliver 1 500 kcal/h to a room using 110-V electricity. If fuses come in 10-A, 20-A, and 30-A sizes, what is the smallest fuse that can safely be used in the heater circuit?

47. web The heating coil of a hot water heater has a resistance of 20 Ω and operates at 210 V. If electrical energy costs \$0.080/kWh, what does it cost to raise the 200 kg of water in the tank from 15°C to 80°C? (See Chapter 11.)

## ADDITIONAL PROBLEMS

48. One lightbulb is marked "25 W 120 V" and another "100 W 120 V"; this means that each converts its respective power when plugged into a constant 120-V potential difference. (a) Find the resistance of each bulb. (b) How long does it take for 1.00 C to pass through the dim bulb? How is this charge different upon its exit versus its entry into the bulb? (c) How long does it take for 1.00 J to pass through the dim bulb? How is this energy different upon its exit versus its entry into the bulb? (d) Find the cost of running the dim bulb continuously for 30.0 days if the electric company sells its product at \$0.070 0 per kWh. What physical quantity *does* the electric company sell? What is its price for one SI unit of this quantity?

49. A particular wire has a resistivity of $3.0 \times 10^{-8}\ \Omega \cdot \text{m}$ and a cross-sectional area of $4.0 \times 10^{-6}\ \text{m}^2$. A length of this wire is to be used as a resistor that will develop 48 W of power when connected across a 20-V battery. What length of wire is required?

50. A steam iron draws 6.0 A from a 120-V line. (a) How many joules of internal energy are produced in 20 min? (b) How much does it cost, at \$0.080/kWh, to run the steam iron for 20 min?

51. An experiment is conducted to measure the electrical resistivity of nichrome in the form of wires with different lengths and cross-sectional areas. For one set of measurements, a student uses 30-gauge wire, which has a cross-sectional area of $7.30 \times 10^{-8}\ \text{m}^2$. The student measures the potential difference across the wire and the current in the wire with a voltmeter and an ammeter, respectively. For each of the measurements given in the table taken on wires of three different lengths, calculate the resistance of the wires and the corresponding values of the resistivity. What is the average value of the resistivity, and how does this value compare with the value given in Table 17.1?

| $L$ (m) | $\Delta V$ (V) | $I$ (A) | $R$ (Ω) | $\rho$(Ω·m) |
|---|---|---|---|---|
| 0.540 | 5.22 | 0.500 | | |
| 1.028 | 5.82 | 0.276 | | |
| 1.543 | 5.94 | 0.187 | | |

52. Birds resting on high-voltage power lines are a common sight. The copper wire on which a bird stands is 2.2 cm in diameter and carries a current of 50 A. If the bird's feet are 4.0 cm apart, calculate the potential difference across its body.

53. A small sphere that carries a charge of 8.00 nC is whirled in a circle at the end of an insulating string. The angular speed is $100\pi$ rad/s. What average current does this rotating charge represent?

54. The current in a conductor varies in time as shown in Figure P17.54. (a) How many coulombs of charge pass through a cross section of the conductor in the interval $t = 0$ to $t = 5.0$ s? (b) What constant current would transport the same total charge during the 5.0-s interval as does the actual current?

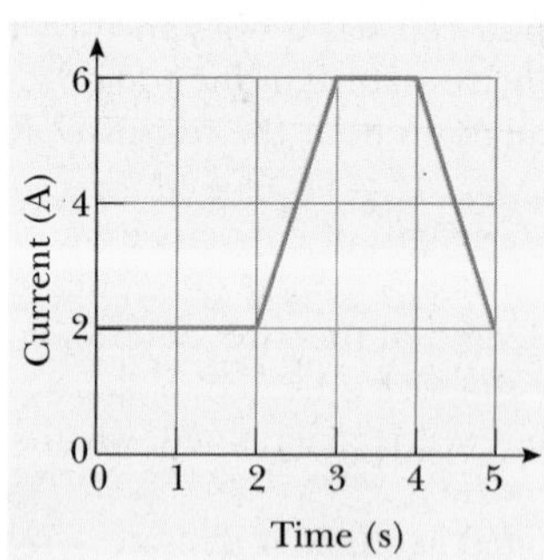


**FIGURE P17.54**

55. An electric car is designed to run off a bank of 12.0-V batteries with a total energy storage of $2.00 \times 10^7$ J. (a) If the electric motor draws 8.00 kW, what is the current delivered to the motor? (b) If the electric motor draws 8.00 kW as the car moves at a steady speed of 20.0 m/s, how far will the car travel before it is "out of juice"?

56. (a) A 115-g mass of aluminum is formed into a right circular cylinder, shaped so that its diameter equals its height. Calculate the resistance between the top and bottom faces of the cylinder at 20°C. (b) Calculate the resistance between opposite faces if the same mass of aluminum is formed into a cube.

57. A length of metal wire has a radius of $5.00 \times 10^{-3}$ m and a resistance of 0.100 Ω. When the potential difference across the wire is 15.0 V, the electron drift speed is found to be $3.17 \times 10^{-4}$ m/s. Based on these data, calculate the density of free electrons in the wire.

58. A carbon wire and a nichrome wire are connected one after the other. If the combination has total resistance of 10.0 kΩ at 20°C, what is the resistance of each wire at 20°C so that the resistance of the combination does not change with temperature?

59. (a) Determine the resistance of a lightbulb marked 100 W @ 120 V. (b) Assuming that the filament is tungsten and has a cross-sectional area of 0.010 mm$^2$, determine the length of the wire inside the bulb when the bulb is operating. (c) Why do you think the wire inside the bulb is tightly coiled? (d) If the temperature of the tungsten wire is 2 600°C when the bulb is operating, what is the length of the wire after the bulb is turned off and has cooled to 20°C? (See Chapter 10, and use $4.5 \times 10^{-6}$/°C as the coefficient of linear expansion for tungsten.)

60. In a certain stereo system, each speaker has a resistance of 4.00 Ω. The system is rated at 60.0 W in each channel. Each speaker circuit includes a fuse rated at a maximum current of 4.00 A. Is this system adequately protected against overload?

61. A resistor is constructed by forming a material of resistivity $3.5 \times 10^5$ Ω·m into the shape of a hollow cylinder of length 4.0 cm and inner and outer radii of 0.50 cm and 1.2 cm, respectively. In use, a potential difference is applied between the ends of the cylinder, producing a current parallel to the length of the cylinder. Find the resistance of the cylinder.

62. The graph in Figure P17.62a shows the current $I$ in a diode as a function of potential difference $\Delta V$ across the diode. Figure P17.62b shows the circuit used to make the measurements. The symbol —▶|— represents the diode. (a) Using Equation 17.4, make a table of the resistance of the diode for different values of $\Delta V$ in the range from −1.5 V to +1.0 V. (b) Based on your results, what amazing electrical property does a diode possess?

63. An x-ray tube used for cancer therapy operates at 4.0 MV, with a beam current of 25 mA striking the metal target. Nearly all the power in this beam is transferred to a stream of water flowing through holes drilled in the target. What rate of flow, in kilograms per second, is needed if the temperature rise ($\Delta T$) of the water is not to exceed 50°C?

64. A 50.0-g sample of a conducting material is all that is available. The resistivity of the material is measured to be $11 \times 10^{-8}$ Ω·m, and the density is 7.86 g/cm$^3$. The material is to be shaped into a solid cylindrical wire that has a total resistance of 1.5 Ω. (a) What length is required? (b) What must be the diameter of the wire?

65. **web** (a) A sheet of copper ($\rho = 1.7 \times 10^{-8}$ Ω·m) is 2.0 mm thick and has surface dimensions of 8.0 cm × 24 cm. If the long edges are joined to form a tube 24 cm in length, what is the resistance between the ends? (b) What mass of copper is required to manufacture a 1 500-m-long spool of copper cable with a total resistance of 4.5 Ω?

66. When a straight wire is heated, its resistance changes according to the equation

$$R = R_0[1 + \alpha(T - T_0)]$$

where $\alpha$ is the temperature coefficient of resistivity. (a) Show that a more precise result, which includes the fact that the length and area of a wire change when it is heated, is

$$R = \frac{R_0[1 + \alpha(T - T_0)][1 + \alpha'(T - T_0)]}{[1 + 2\alpha'(T - T_0)]}$$

where $\alpha'$ is the coefficient of linear expansion (see Chapter 10). (b) Compare these two results for a 2.00-m-long copper wire of radius 0.100 mm, starting at 20.0°C and heated to 100.0°C.

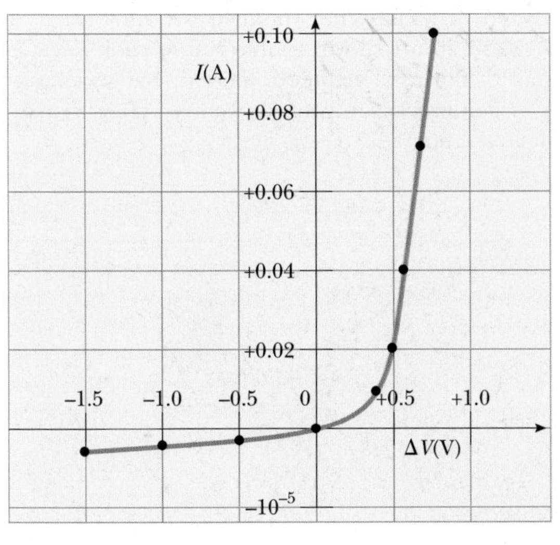


(a)

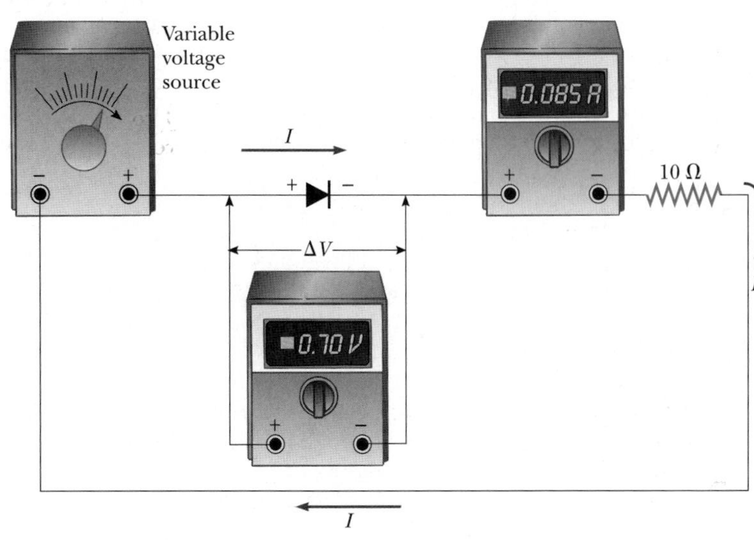


(b)

**FIGURE P17.62**

## GROUP ACTIVITIES

**G.1** Connect one terminal of a D-cell battery to the base of a flashlight bulb using insulated wire, tape a second wire to the other battery terminal, and tape a third wire to the center conductor of the bulb as in Figure GA17.1. Make sure to remove about 1 cm of insulation from the ends of all wires before making connections. Now bridge the gap between the open wires with different objects, such as a plastic pen, an aluminum can, a penny, a rubber band, and a spoon. Which objects make the bulb light up? Explain your observations.

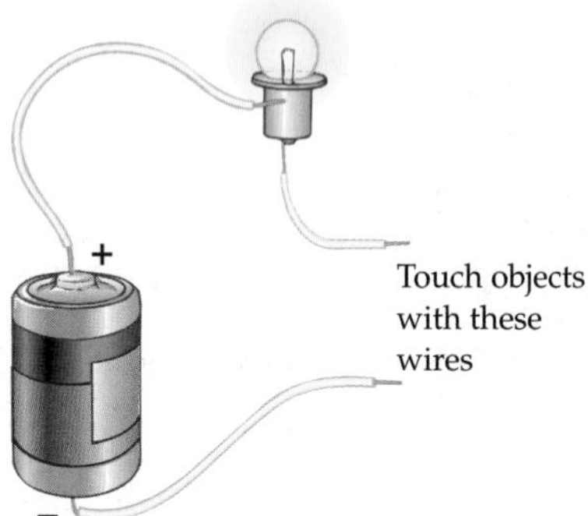


**FIGURE GA17.1**

**G.2** When the lightbulbs in your home are used, they are always connected across the same potential difference. Which do you believe has a filament with the highest resistance when cool, a 60-W bulb or a 100-W bulb? To check your prediction, ask your instructor to lend you a device called an ohmmeter and to instruct you in its use. A resistor must always be disconnected from a circuit when its resistance is measured with an ohmmeter.

**G.3** Examine the label on several household appliances such as a toaster, television, lamp, stereo, air conditioner, and clock. From the label, determine the power rating of the device in watts. Check the billing statement from your electric utility company to find the cost of electrical energy per kilowatt-hour. (Prices usually range from about a nickel to twenty cents.) Calculate the cost of running each appliance for 1 h. Estimate how many hours per day each appliance is used. Then calculate the monthly cost of using each appliance based on your daily estimate.

**G.4** A piece of material is connected to a battery. All of the following factors affect the current in the wire. Explain why these influence the current and discuss whether the current is directly proportional or inversely proportional to each factor, or has another pattern of dependence. (a) The cross-sectional area of the material, the length of the material, the temperature of the material, the number of free charge carriers in the material. (b) "When I turn on a wall switch, the current in a room seems to come on instantly. Thus, I know that the drift velocity of the electrons in a wire is extremely fast." Argue for or against this statement. (c) Assume a certain material has $9.00 \times 10^{28}$ free electrons per cubic meter, a cross-sectional area of $3.00 \times 10^{-6}$ m$^2$, and a current of 0.300 A. Find the drift speed of the electrons in the material.

**G.5** (a) Explain why temperature affects the resistance of a material. (b) Explain why length affects the resistance of a material. (c) Explain why cross-sectional area affects the resistance of a material. (d) A tungsten-filament lightbulb has a resistance of 240 Ω when connected to a 120-V outlet, and it is found that the temperature of the filament is about 2 000°C under these conditions. What is the current in the bulb when it is connected to a 2.00-V battery?

**G.6** Two identical, parallel copper wires used as cable TV lines are placed underground between two points 1.00 mile apart. These wires are usually not connected, but a construction accident shorts the wires together at some point. Problem: Where is that point? To try to isolate the spot so that repair can be initiated, a technician goes to end A of the lines and finds that a 12.0-V battery connected across the wires at that end produces a current of 1.00 A. Doing the same at the other end, end B, produces a current of 0.20 A. (a) Is the break closer to end A or to end B? (b) How far from the 0.20-A end should digging begin?

**G.7** You are cooking breakfast for yourself and a friend using a 1 200-W waffle iron and a 500-W coffee pot. Usually, you operate these appliances from a 110-V outlet for 0.50 h each day. (a) At eight cents per kWh, how much do you spend to cook breakfast each month? (b) You find yourself addicted to waffles and would like to upgrade to a 2 400-W waffle iron that will enable you to cook twice as many waffles during a half-hour period, but you know that the circuit breaker in your kitchen is a 20-A breaker. Can you do the upgrade?

# Direct Current Circuits

# 18

## Chapter Outline



**This versatile circuit enables the experimenter to examine the properties of circuit elements such as capacitors and resistors and their effects on circuit behavior.** *(Courtesy of Central Scientific Company)*

This chapter analyzes some simple circuits whose elements include batteries, resistors, and capacitors in varied combinations. Such analysis is simplified by the use of two rules known as Kirchhoff's rules, which follow from the principle of conservation of energy and the law of conservation of charge. Most of the circuits are assumed to be in *steady state,* which means that the currents are constant in magnitude and direction. We close the chapter with a discussion of circuits containing resistors and capacitors, in which current varies with time.

## 18.1 SOURCES OF emf

The source that maintains the current in a closed circuit is called a source of emf.[1] Any devices (such as batteries and generators) that increase the potential energy of charges circulating in circuits are sources of emf. One can think of such a source as a "charge pump" that forces electrons to move in a direction opposite the electrostatic field inside the source. The emf $\mathcal{E}$ of a source is the work done per unit charge, and hence the SI unit of emf is the volt.

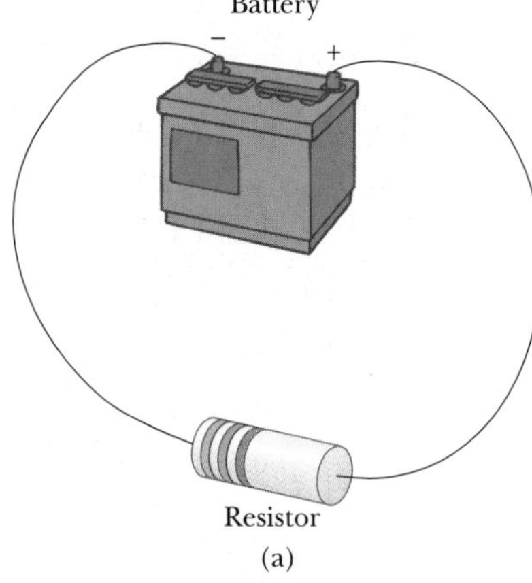

(a)

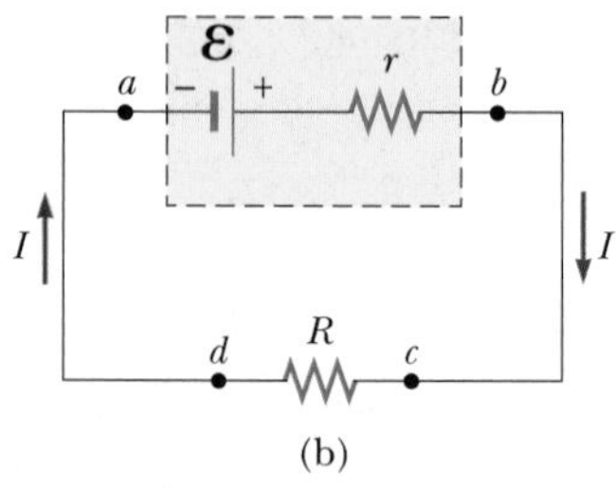

(b)

**FIGURE 18.1** (a) A circuit consisting of a resistor connected to the terminals of a battery. (b) A circuit diagram of a source of emf $\mathcal{E}$ having internal resistance $r$ connected to an external resistor $R$.

Consider the circuit in Figure 18.1a, consisting of a battery connected to a resistor. We assume that the connecting wires have no resistance. If we neglect the internal resistance of the battery, the potential difference across the battery (the terminal voltage) equals the emf of the battery. However, because a real battery always has some internal resistance $r$, the terminal voltage is not equal to the emf. The circuit of Figure 18.1a can be described schematically by the diagram in Figure 18.1b. The battery, represented by the dashed rectangle, consists of a source of emf $\mathcal{E}$ in series with an internal resistance $r$. Now imagine a positive charge moving from $a$ to $b$ in Figure 18.1b. As the charge passes from the negative to the positive terminal of the battery, the potential of the charge increases by $\mathcal{E}$. However, as the charge moves through the resistance $r$ its potential decreases by the amount $Ir$, where $I$ is the current in the circuit. Thus, the terminal voltage of the battery, $\Delta V = V_b - V_a$, is

$$\Delta V = \mathcal{E} - Ir \qquad [18.1]$$

Note from this expression that **$\mathcal{E}$ is equal to the terminal voltage when the current is zero,** called the **open-circuit voltage.** By inspecting Figure 18.1b, we see that the terminal voltage $\Delta V$ must also equal the potential difference across the external resistance $R$, often called the **load resistance;** that is, $\Delta V = IR$. Combining this with Equation 18.1, we see that

$$\mathcal{E} = IR + Ir \qquad [18.2]$$

An assortment of batteries. *(George Semple)*

[1] The term was originally an abbreviation for electromotive force, but emf is not really a force, so the long form is discouraged.

Solving for the current gives

$$I = \frac{\mathcal{E}}{R + r}$$

This shows that the current in this simple circuit depends on both the resistance external to the battery and the internal resistance. If $R$ is much greater than $r$, we can neglect $r$ in our analysis and we do, for many circuits. If we multiply Equation 18.2 by the current $I$, we get

$$I\mathcal{E} = I^2R + I^2r$$

This equation tells us that the total power output $I\mathcal{E}$ of the source of emf is converted to the rate $I^2R$ at which energy is delivered to the load resistance *plus* the rate $I^2r$ at which energy is delivered to the internal resistance. Again, if $r \ll R$, most of the power delivered by the battery is transferred to the load resistance.

Unless otherwise stated, we will assume in our examples and end-of-chapter problems that the internal resistance of a battery in a circuit is negligible.

**Tip 18.1 WHAT IS CONSTANT IN A BATTERY?**

Equation 18.2 shows that the current in a circuit depends on the resistance of the battery. Note that the battery is not a source of constant current. Furthermore, the terminal voltage of a battery given by Equation 18.1 is also not constant. A battery is a source of constant emf.

## 18.2 RESISTORS IN SERIES

When two or more resistors are connected end-to-end as in Figure 18.2, they are said to be in *series*. The resistors could be simple devices such as lightbulbs or heating elements. When the two resistors $R_1$ and $R_2$ are connected to a battery as in Figure 18.2, **the current is the same in the two resistors because any charge that flows through $R_1$ must also flow through $R_2$.** This is analogous to water flowing through a pipe with two constrictions, corresponding to $R_1$ and $R_2$. Whatever volume of water flows in one end in a given time interval must exit the opposite end in the same time interval.

For a series connection of resistors, the current is the same in all the resistors.

Because the potential difference between $a$ and $b$ in Figure 18.2b equals $IR_1$ and the potential difference between $b$ and $c$ equals $IR_2$, the potential difference between $a$ and $c$ is

$$\Delta V = IR_1 + IR_2 = I(R_1 + R_2)$$

Regardless of how many resistors we have in series, the sum of the potential differences across the resistors is equal to the total potential difference across the combination. As we will show later, this is a consequence of the conservation of energy. Figure 18.2c shows an equivalent resistor $R_{eq}$ that can replace the two resistors of the original circuit. The equivalent resistance has the same effect on the circuit as

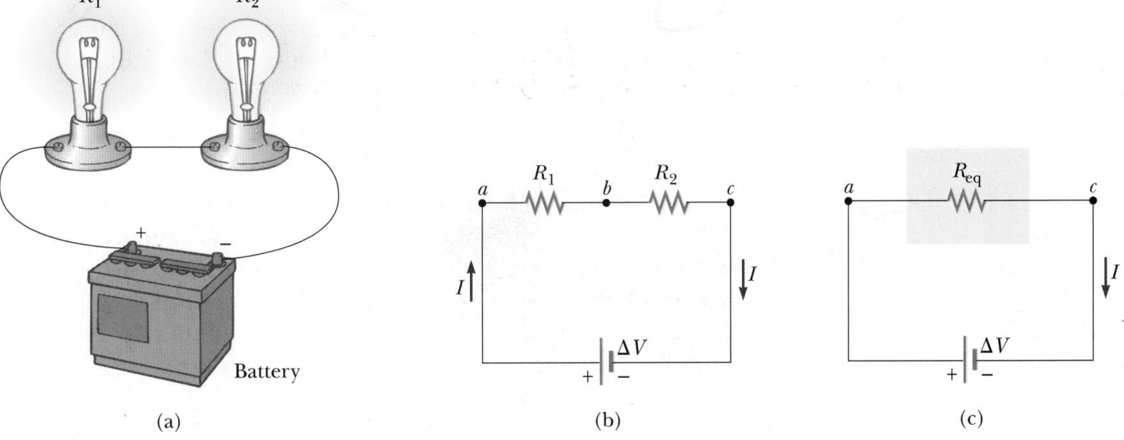


**FIGURE 18.2** A series connection of two resistors $R_1$ and $R_2$. The currents in the resistors are the same, and the equivalent resistance of the combination is given by $R_{eq} = R_1 + R_2$.

the combination of resistors because it results in the same current in the circuit. Applying Ohm's law to this resistor, we have

$$\Delta V = IR_{\text{eq}}$$

Equating the preceding two expressions, we have

$$IR_{\text{eq}} = I(R_1 + R_2)$$

or

$$R_{\text{eq}} = R_1 + R_2 \qquad [18.3]$$

An extension of this analysis shows that the equivalent resistance of three or more resistors connected in series is

Equivalent resistance of a series combination of resistors.

$$R_{\text{eq}} = R_1 + R_2 + R_3 + \cdots \qquad [18.4]$$

Therefore, **the equivalent resistance of a series combination of resistors is the algebraic sum of the individual resistances and is always greater than any individual resistance.**

Note that if the filament of one lightbulb in Figure 18.2 were to fail, the circuit would no longer be complete (an open-circuit condition would exist) and the second bulb would also go out.

## APPLYING PHYSICS 18.1

**APPLICATION**

CHRISTMAS LIGHTS IN SERIES

A new design for Christmas tree lights allows them to be connected in series. One might expect that a failed bulb in such a string would result in an open circuit and all of the bulbs would go out. How can the bulbs be designed to prevent this from happening?

**Explanation** If the string of lights contained normal bulbs, a failed bulb would be hard to locate. Each bulb would have to be replaced with a good bulb, one by one, until the failed bulb is found. If there happened to be two or more failed bulbs in the string of lights, finding them would be a formidable task.

Christmas lights use specially designed bulbs that have an insulated loop of wire (jumper) across the conducting supports to the bulb filaments (Fig. 18.3). If the filament breaks and the bulb fails, the resistance of this bulb increases dramatically. As a result, most of the applied voltage appears across the loop of wire. This voltage causes the insulation around the loop of wire to melt, causing the metal wire to make electrical contact with the supports. This produces a conducting path through the bulb, enabling the other bulbs to remain lit.

### Webnote 18.1

Holiday lights are a tradition at the end of the year. Learn how these colorful and festive decorations are constructed at *http://www.howstuffworks.com/christmas-lights.htm*

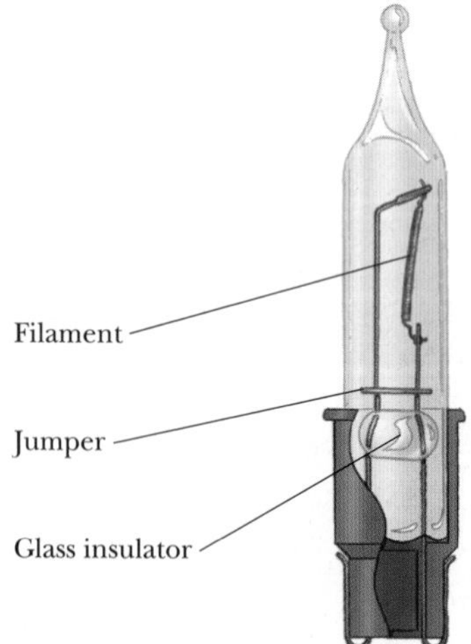


**FIGURE 18.3** (Applying Physics 18.1) Diagram of a modern miniature holiday lightbulb, with a jumper connection to provide a current if the filament breaks.

**Quick Quiz 18.1** When a piece of wire is used to connect points *b* and *c* in Figure 18.2b, the brightness of bulb $R_1$ (a) increases, (b) decreases, or (c) stays the same. The brightness of bulb $R_2$ (a) increases, (b) decreases, or (c) stays the same.

**Quick Quiz 18.2** With the switch in the circuit of Figure 18.4a closed, no current exists in $R_2$ because the current has an alternate zero-resistance path through the switch. Current does exist in $R_1$ and this current is measured with the ammeter at the right side of the circuit. If the switch is opened (Fig. 18.4b), current exists in $R_2$. After the switch is opened, the reading on the ammeter (a) increases, (b) decreases, (c) does not change.

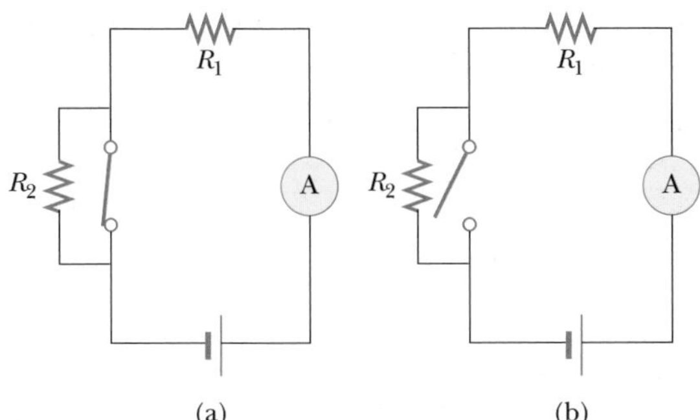


**FIGURE 18.4** (Quick Quiz 18.2)

## Example 18.1 Four Resistors in Series

Four resistors are arranged as shown in Figure 18.5a. Find (a) the equivalent resistance and (b) the current in the circuit if the emf of the battery is 6.0 V.

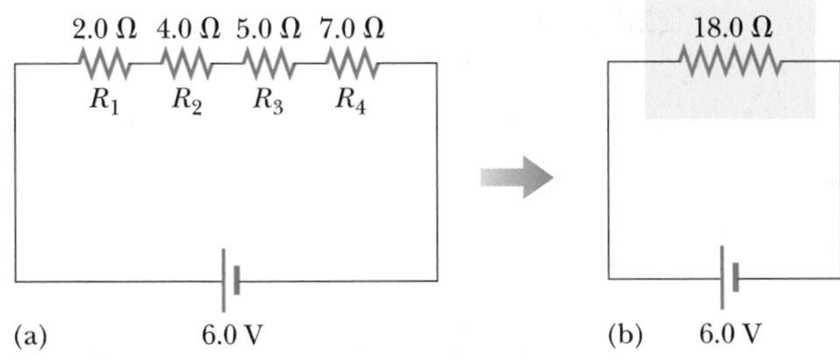


**FIGURE 18.5** (Example 18.1) (a) Four resistors connected in series. (b) The equivalent resistance of the circuit in (a).

### Solution

**A** The equivalent resistance is found from Equation 18.4:

$$R_{eq} = R_1 + R_2 + R_3 + R_4 = 2.0\ \Omega + 4.0\ \Omega + 5.0\ \Omega + 7.0\ \Omega = 18.0\ \Omega$$

**B** If we apply Ohm's law to the equivalent resistor in Figure 18.5b, we find the current in the circuit to be

$$I = \frac{\Delta V}{R_{eq}} = \frac{6.0\ \text{V}}{18.0\ \Omega} = \tfrac{1}{3}\ \text{A}$$

**EXERCISE** Because the current in the equivalent resistor is $\frac{1}{3}$ A, this must also be the current in each resistor of the original circuit. Find the voltage drop across each resistor.

**ANSWER** $\Delta V_{2\Omega} = \frac{2}{3}$ V; $\Delta V_{4\Omega} = \frac{4}{3}$ V; $\Delta V_{5\Omega} = \frac{5}{3}$ V; $\Delta V_{7\Omega} = \frac{7}{3}$ V

## 18.3 RESISTORS IN PARALLEL

Now consider two resistors connected in parallel, as in Figure 18.6. In this case, **the potential differences across the resistors are the same because each is connected directly across the battery terminals.** The currents are generally not the same. When the charge reaches point $a$ (called a junction) in Figure 18.6b, the current splits into two parts, $I_1$ going through $R_1$ and $I_2$ going through $R_2$. If $R_1$ is greater than $R_2$, then $I_1$ is less than $I_2$. That is, the charge tends to follow the path of least resistance. **Because charge is conserved, the current *I* that enters point a must equal the total current leaving that point, $I_1 + I_2$.** That is,

$$I = I_1 + I_2$$

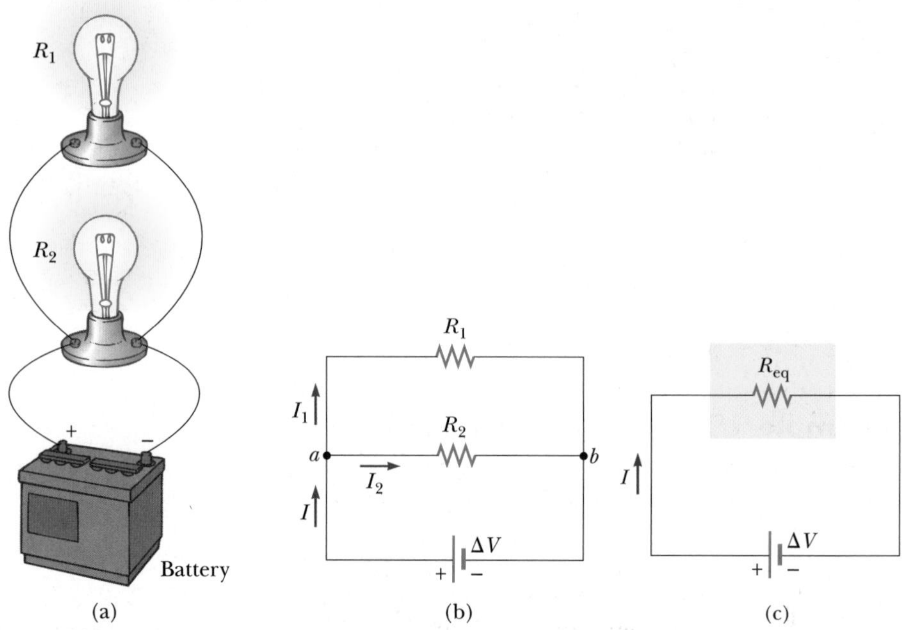


**FIGURE 18.6** (a) A parallel connection of two resistors. (b) A circuit diagram for the parallel combination. (c) The voltages across the resistors are the same, and the equivalent resistance of the combination is given by the reciprocal relationship $1/R_{eq} = 1/R_1 + 1/R_2$.

The potential drop must be the same for the two resistors and must also equal the potential drop across the battery. Ohm's law applied to each resistor gives

$$I_1 = \frac{\Delta V}{R_1} \qquad I_2 = \frac{\Delta V}{R_2}$$

Ohm's law applied to the equivalent resistor in Figure 18.6c gives

$$I = \frac{\Delta V}{R_{eq}}$$

When these expressions for the current are substituted into the equation $I = I_1 + I_2$, and $\Delta V$ is cancelled, we obtain

$$\frac{1}{R_{eq}} = \frac{1}{R_1} + \frac{1}{R_2} \qquad \text{(parallel combination)} \qquad [18.5]$$

An extension of this analysis to three or more resistors in parallel produces the following general expression for the equivalent resistance:

Equivalent resistance of a parallel combination of resistors.

$$\frac{1}{R_{eq}} = \frac{1}{R_1} + \frac{1}{R_2} + \frac{1}{R_3} + \cdots \qquad [18.6]$$

From this expression, it can be seen that **the inverse of the equivalent resistance of two or more resistors connected in parallel is the algebraic sum of the inverses of the individual resistances. The equivalent resistance is always less than the smallest resistance in the group.**

Household circuits are always wired so that the electrical devices are connected in parallel, as in Figure 18.6a. In this manner, each device operates independently of the others, so that if one is switched off, the others remain on. For example, if one of the lightbulbs in Figure 18.6 were removed from its socket, the other would continue to operate. Equally important, each device operates on the same voltage. If the devices were connected in series, the voltage across any one device would depend on how many devices were in the combination and on their individual resistances.

**APPLICATION**

CIRCUIT BREAKERS

In many household circuits, circuit breakers are used in series with other circuit elements for safety purposes. A circuit breaker is designed to switch off and open the circuit at some maximum current (typically 15 A or 20 A) whose value depends on the nature of the circuit. If a circuit breaker were not used, excessive currents caused by operating several devices simultaneously could result in excessive wire temperatures and perhaps cause a fire. In older home construction, fuses were used in place of circuit breakers. When the current in a circuit exceeds some value, the conductor in a fuse melts and opens the circuit. The disadvantage of fuses is that they are destroyed in the process of opening the circuit, whereas circuit breakers can be reset.

### PROBLEM-SOLVING STRATEGY Resistors

1. When two or more unequal resistors are connected in *series,* they carry the same current, but the potential differences across them are not the same. The resistors add directly to give the equivalent resistance of the series combination.
2. When two or more unequal resistors are connected in *parallel,* the potential differences across them are the same. Because the current is inversely proportional to the resistance, the currents through them are not the same. The equivalent resistance of a parallel combination of resistors is found through reciprocal addition, and the equivalent resistance is always *less* than the smallest individual resistor in the combination.
3. A complicated circuit consisting of several resistors and batteries can often be reduced to a simple circuit with only one resistor. To do so, examine the initial circuit and replace any resistors in series or any in parallel using the procedures outlined in Steps 1 and 2. Sketch the new circuit after these changes have been made. Examine the new circuit and replace any series or parallel combinations. Continue this process until a single equivalent resistance is found.
4. If the current in or the potential difference across a resistor in the complicated circuit is to be identified, start with the final circuit found in Step 3 and gradually work back through the circuits, using $\Delta V = IR$ and the procedures of Steps 1 and 2.

### APPLYING PHYSICS 18.2

Compare the brightness of the four identical bulbs shown in Figure 18.7. What happens if bulb A fails, so that it cannot conduct current? What if C fails? What if D fails?

**Explanation** Bulbs A and B are connected in series across the emf of the battery, whereas bulb C is connected by itself across the battery. Thus, the emf is split between bulbs A and B. As a result, bulb C will be brighter than bulbs A and B, which should be equally as bright as each other. Bulb D has a wire connected across it. Thus, there is no potential difference across D, and it does not glow at all. If bulb A fails, B goes out, but C stays lit. If C fails, there is no effect on the other bulbs. If D fails, the event is undetectable, because D was not glowing initially.

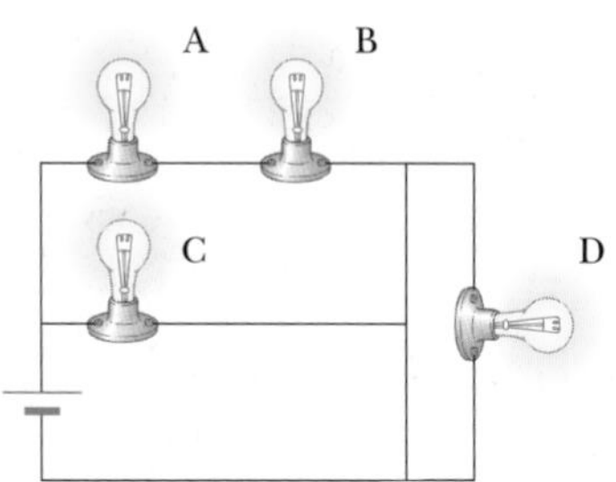


**FIGURE 18.7** (Applying Physics 18.2)

## APPLYING PHYSICS 18.3

**APPLICATION**

THREE-WAY LIGHTBULBS

Figure 18.8 illustrates how a three-way lightbulb is constructed to provide three levels of light intensity. The socket of the lamp is equipped with a three-way switch for selecting different light intensities. The bulb contains two filaments. Why are the filaments connected in parallel? Explain how the two filaments are used to provide three different light intensities.

**Explanation** If the filaments were connected in series and one of them were to fail, there would be no current in the bulb and the bulb would give no illumination, regardless of the switch position. However, when the filaments are connected in parallel and one of them (say the 75-W filament) fails, the bulb will still operate in one of the switch positions because there is current in the other (100-W) filament. The three light intensities are made possible by selecting one of three values of filament resistance using a single value of 120 V for the applied voltage. The 75-W filament offers one value of resistance, the 100-W filament offers a second value, and the third resistance is obtained by combining the two filaments in parallel. When switch $S_1$ is closed and switch $S_2$ is opened, only the 75-W filament carries current. When switch $S_1$ is open and switch $S_2$ is closed, only the 100-W filament carries current. When both switches are closed, both filaments carry current and a total illumination corresponding to 175 W is obtained.

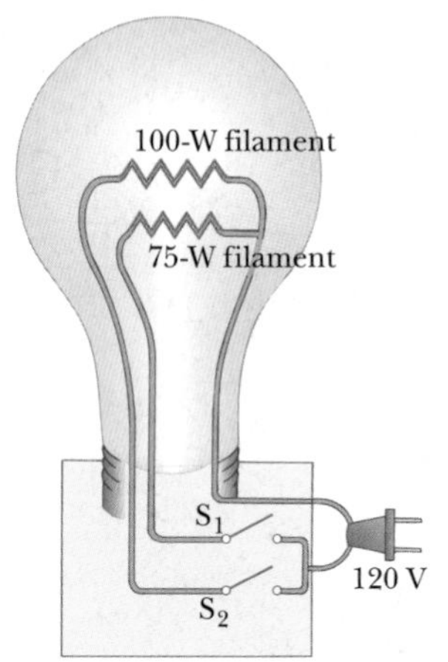


**FIGURE 18.8** (Applying Physics 18.3)

**Quick Quiz 18.3**

With the switch in the circuit of Figure 18.9a open, there is no current in $R_2$. There is current in $R_1$ and this current is measured with the ammeter at the right side of the circuit. If the switch is closed (Fig. 18.9b), there is current in $R_2$. When the switch is closed, the reading on the ammeter (a) increases, (b) decreases, or (c) remains the same.

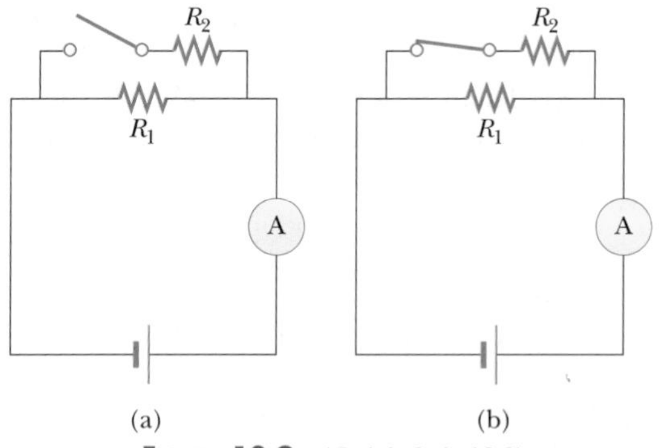


**FIGURE 18.9** (Quick Quiz 18.3)

**Quick Quiz 18.4**

You have a large supply of lightbulbs and a battery. You start with one lightbulb connected to the battery and notice its brightness. You then add one lightbulb at a time, each new bulb being added in parallel to the previous bulbs. As the lightbulbs are added, what happens (a) to the brightness of the bulbs, (b) to the current in the bulbs, (c) to the power delivered by the battery, (d) to the lifetime of the battery, (e) to the terminal voltage of the battery? *Hint:* Do not ignore the internal resistance of the battery.

## Example 18.2 Three Resistors in Parallel

Three resistors are connected in parallel as in Figure 18.10. A potential difference of 18 V is maintained between points *a* and *b*.

**A** Find the current in each resistor.

**Solution** Because the resistors are in parallel, the potential difference across each is 18 V. Let us apply $\Delta V = IR$ to find the current in each resistor:

$$I_1 = \frac{\Delta V}{R_1} = \frac{18\text{ V}}{3.0\ \Omega} = 6.0\text{ A}$$

$$I_2 = \frac{\Delta V}{R_2} = \frac{18\text{ V}}{6.0\ \Omega} = 3.0\text{ A}$$

$$I_3 = \frac{\Delta V}{R_3} = \frac{18\text{ V}}{9.0\ \Omega} = 2.0\text{ A}$$

There is a saying that holds that current, like politicians, follows the path of least resistance. This is only partially correct, as our example demonstrates. The smallest (3.0-Ω) resistor carries the largest current, while the other larger resistors of 6.0 Ω and 9.0 Ω carry smaller currents.

**B** Calculate the power delivered to each resistor and the total power delivered to the three resistors.

**Solution** Applying $\mathcal{P} = I^2R$ to each resistor gives

$$3\ \Omega: \quad \mathcal{P}_1 = I_1{}^2R_1 = (6.0\text{ A})^2(3.0\ \Omega) = 110\text{ W}$$

$$6\ \Omega: \quad \mathcal{P}_2 = I_2{}^2R_2 = (3.0\text{ A})^2(6.0\ \Omega) = 54\text{ W}$$

$$9\ \Omega: \quad \mathcal{P}_3 = I_3{}^2R_3 = (2.0\text{ A})^2(9.0\ \Omega) = 36\text{ W}$$

(Note that you can also use $\mathcal{P} = (\Delta V)^2/R$ to find the power dissipated by each resistor.) Summing the three quantities gives a total power of 200 W.

**EXERCISE** Calculate the equivalent resistance of the three resistors, and from this result find the total power dissipated.

**ANSWER** $\frac{18}{11}\ \Omega$; 200 W

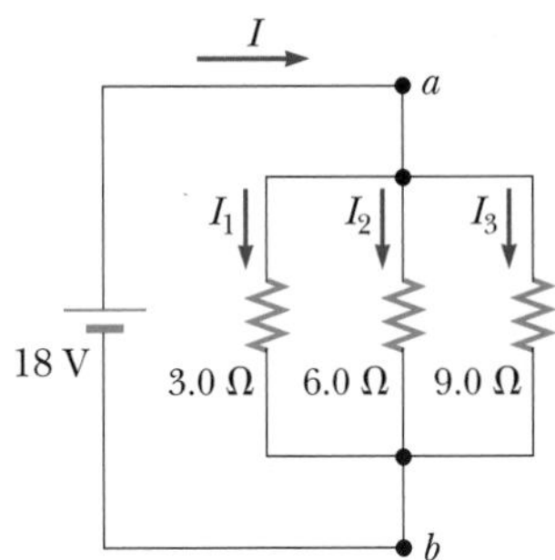


**FIGURE 18.10** (Example 18.2) Three resistors connected in parallel. The voltage across each resistor is 18 V.

## Example 18.3 Equivalent Resistance

Four resistors are connected as shown in Figure 18.11a.

**A** Find the equivalent resistance between points *a* and *c*.

**Solution** The circuit can be reduced in steps, as shown in Figures 18.11b and 18.11c. The 8.0-Ω and 4.0-Ω resistors are in series, and so the equivalent resistance between *a* and *b* is 12 Ω (Eq. 18.4). The 6.0-Ω and 3.0-Ω resistors are in parallel, and so from Equation 18.6 we find that the equivalent resistance from *b* to *c* is 2.0 Ω. Hence, the equivalent resistance from *a* to *c* is 14 Ω.

**B** What is the current in each resistor if the terminals of a 42-V battery are connected between *a* and *c*?

**Solution** The current *I* is the same in the 8.0-Ω and 4.0-Ω resistors because they are in series. Using Ohm's law and the results of (a), we get

$$I = \frac{\Delta V_{ac}}{R_{\text{eq}}} = \frac{42\text{ V}}{14\ \Omega} = 3.0\text{ A}$$

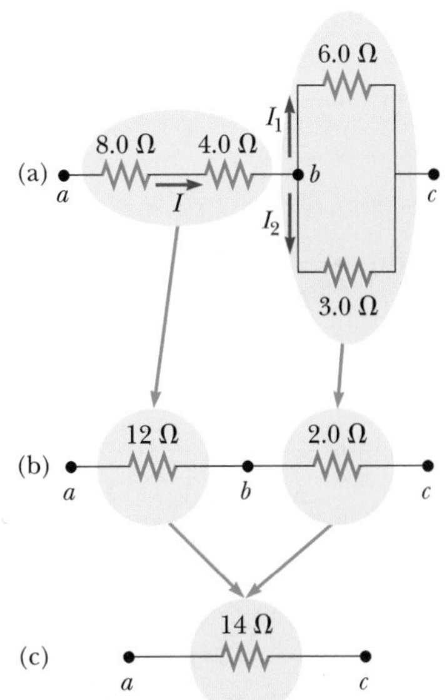


**FIGURE 18.11** (Example 18.3) The four resistors shown in (a) can be reduced in steps to an equivalent 14-Ω resistor.

When this current enters the junction at $b$, it splits; part of it passes through the 6.0-Ω resistor ($I_1$), and part passes through the 3.0-Ω resistor ($I_2$). Because the potential difference $\Delta V_{bc}$ across these resistors is the *same* (they are in parallel), $(6\ \Omega)I_1 = (3\ \Omega)I_2$, or $I_2 = 2I_1$. Using this result and the fact that $I_1 + I_2 = 3.0$ A, we find that $I_1 =$ 1.0 A and $I_2 =$ 2.0 A. We could have guessed this from the start by noting that the current in the 3.0-Ω resistor has to be twice the current in the 6.0-Ω resistor in view of their relative resistances and the fact that the same voltage is applied to both.

As a final check, note that $\Delta V_{bc} = (6\ \Omega)I_1 = (3\ \Omega)I_2 = 6.0$ V and $\Delta V_{ab} = (12\ \Omega)I_1 = 36$ V; therefore, $\Delta V_{ac} = \Delta V_{ab} + \Delta V_{bc} = 42$ V, as expected.

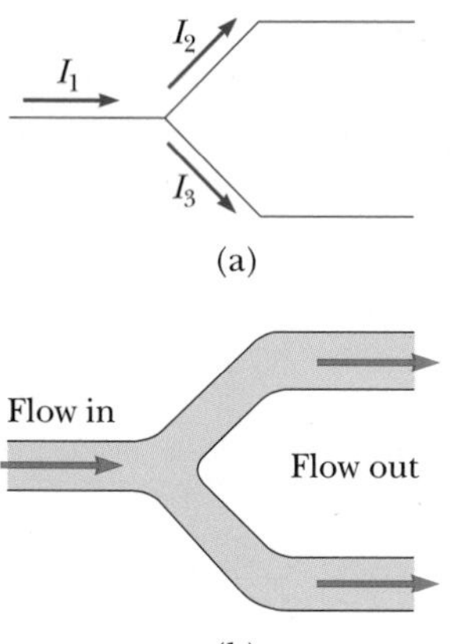


**FIGURE 18.12** (a) A schematic diagram illustrating Kirchhoff's junction rule. Conservation of charge requires that whatever current enters a junction must leave that junction. Therefore, in this case, $I_1 = I_2 + I_3$. (b) A mechanical analog of the junction rule: The net flow in must equal the net flow out.

## 18.4 KIRCHHOFF'S RULES AND COMPLEX DC CIRCUITS

As demonstrated in the preceding section, we can analyze simple circuits using Ohm's law and the rules for series and parallel combinations of resistors. However, there are many ways in which resistors can be connected so that the circuits formed cannot be reduced to a single equivalent resistor. The procedure for analyzing more complex circuits is greatly simplified by the use of two simple rules called **Kirchhoff's rules:**

1. **The sum of the currents entering any junction must equal the sum of the currents leaving that junction. (This rule is often referred to as the junction rule.)**
2. **The sum of the potential differences across all the elements around any closed-circuit loop must be zero. (This rule is usually called the loop rule.)**

The junction rule is a statement of *conservation of charge.* Because charge must be conserved, whatever current enters a point in a circuit must equal the total current leaving that point. If we apply this rule to the junction in Figure 18.12a, we get

$$I_1 = I_2 + I_3$$

Figure 18.12b represents a mechanical analog to this situation, in which water flows through a branched pipe with no leaks. The flow rate into the pipe equals the total flow rate out of the two branches.

The loop rule is equivalent to the principle of *conservation of energy.* Any charge that moves around any closed loop in a circuit (starting and ending at the same point) must gain as much energy as it loses. It gains energy as it is pumped through a source of emf. Its energy may decrease in the form of a potential drop, $-IR$, across a resistor or as a result of flowing backward through a source of emf—that is, from the positive to the negative terminal inside the battery. In the latter case, electrical energy is converted to chemical energy as the battery is charged.

When applying Kirchhoff's rules, you must make two decisions at the beginning of a problem.

1. Assign symbols and directions to the currents in all branches of the circuit. Do not be alarmed that you might guess the direction of a current incorrectly; the resulting answer will be negative, but *its magnitude will be correct.*
2. When applying the loop rule, you must choose a direction for traversing the loop and be consistent in going either clockwise or counterclockwise. As you traverse the loop, record voltage drops and rises according to the rules stated next. They are summarized in Figure 18.13, where it is assumed that the traversal is from point $a$ toward point $b$:

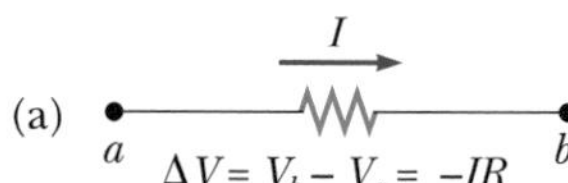


**FIGURE 18.13** Rules for determining the potential changes across a resistor and a battery, assuming the battery has no internal resistance.

(a) If a resistor is traversed in the direction of the current, the change in electric potential across the resistor is $-IR$ (Fig. 18.13a).
(b) If a resistor is traversed in the direction opposite the current, the change in electric potential across the resistor is $+IR$ (Fig. 18.13b).
(c) If a source of emf is traversed in the direction of the emf (from $-$ to $+$ on the terminals), the change in electric potential is $+\mathcal{E}$ (Fig. 18.13c).
(d) If a source of emf is traversed in the direction opposite the emf (from $+$ to $-$ on the terminals), the change in electric potential is $-\mathcal{E}$ (Fig. 18.13d).

There are limits to the numbers of times the junction rule and the loop rule can be used. You can use the junction rule as often as needed so long as, each time you write an equation, you include in it a current that has not been used in a previous junction-rule equation. (If this procedure is not followed, a new equation will not be produced.) In general, the number of times the junction rule can be used is one fewer than the number of junction points in the circuit. The loop rule can be used as often as needed so long as a new circuit element (resistor or battery) or a new current appears in each new equation. In general, **to solve a particular circuit problem, you need as many independent equations as you have unknowns.**

**GUSTAV KIRCHHOFF, GERMAN PHYSICIST (1824–1887)**

Kirchhoff, a professor at Heidelberg, together with Robert Bunsen invented the spectroscopy that we study in Chapter 28. He also formulated another rule that states "a cool substance will absorb light of the same wavelengths that it emits when hot." *(AIP ESVA, W. F. Meggers Collection)*

**PROBLEM-SOLVING STRATEGY** **Kirchhoff's Rules**

1. First, draw the circuit diagram and assign labels and symbols to all the known and unknown quantities. Assign *directions* to the currents in each part of the circuit. Although the assignment of current directions is arbitrary, you must stick with them throughout as you apply Kirchhoff's rules.
2. Apply the junction rule to any junction in the circuit. The junction rule may be applied as many times as a new current (one not used in a previous application) appears in the resulting equation.
3. Now apply Kirchhoff's loop rule to as many loops in the circuit as are needed to solve for the unknowns. In order to apply this rule, you must correctly identify the change in electric potential as you cross each element in traversing the closed loop. Watch out for signs! (Use the conventions in Fig. 18.13.)
4. Solve the equations simultaneously for the unknown quantities. Be careful in your algebraic steps, and check your numerical answers for consistency.

## Example 18.4 Applying Kirchhoff's Rules

Find the currents in the circuit shown in Figure 18.14.

**Reasoning** There are three unknown currents in this circuit, and so we must obtain three independent equations. We can find the equations with one application of the junction rule and two applications of the loop rule.

**Solution** The first step is to assign a current to each branch of the circuit; these are our unknowns and are labeled $I_1$, $I_2$, and $I_3$ in Figure 18.14. It is also necessary to guess directions for the currents. Your experience with circuits such as this should tell you that the directions of all three have been chosen correctly. (However, recall that if a current direction is chosen incorrectly, the numerical answer will turn out negative, but the magnitude will be correct. This point will be demonstrated in the next example.)

We now apply Kirchhoff's rules. First we can apply the junction rule using either *c* or *d*, the only two junctions in the circuit. Let us choose junction *c*. The net current into

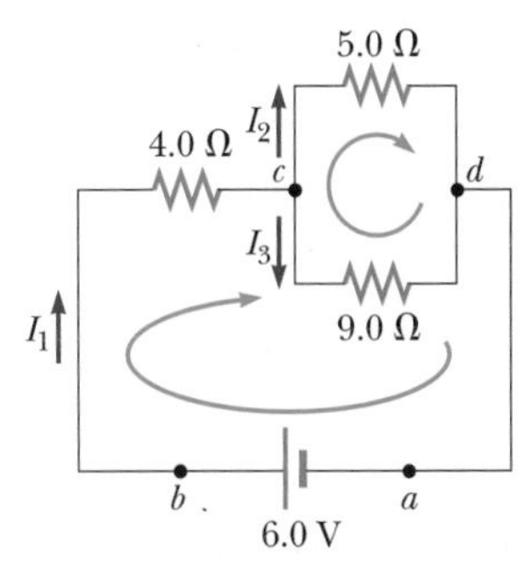


**FIGURE 18.14** (Example 18.4) A multiloop circuit.

this junction is $I_1$, and the net current leaving it is $I_2 + I_3$. Thus, the junction rule applied to $c$ gives

$$I_1 = I_2 + I_3$$

Recall that you may apply the junction rule over and over until you reach a situation in which no new currents appear in an equation. In this example we have reached that point with one application. If we apply the junction rule at $d$, we find that $I_1 = I_2 + I_3$, exactly the same equation.

We have three unknowns in our problem, $I_1$, $I_2$, and $I_3$; thus, we need two more independent equations before we can find a solution. We obtain these equations by applying the loop rule to the two loops indicated in the figure. Note that there are actually three loops in the circuit, but these two are sufficient to complete the problem. (Where is the loop that we do not use?)

When applying the loop rule, we must first choose the loops to be traversed, and then the directions in which to traverse them. We have selected the two loops indicated in the figure and have decided to traverse both of them clockwise. Other choices could be made, but the final result would be the same.

Starting at point $a$ and moving clockwise around the large loop, we encounter the following voltage changes (see Fig. 18.13 for the basic sign conventions):

- From $a$ to $b$, we encounter a voltage change of 6.0 V.
- From $b$ to $c$ through the 4.0-Ω resistor, we encounter a voltage change of $-(4.0\ \Omega)I_1$.
- From $c$ to $d$ through the 9.0-Ω resistor, we encounter a voltage change of $-(9.0\ \Omega)I_3$.

**Tip 18.2 CURRENT DOES NOT TAKE THE PATH OF LEAST RESISTANCE**

The statement "current takes the path of least resistance" is incorrect in reference to a parallel combination of current paths. In this context, the current takes all paths. Those paths with lower resistance have larger currents, whereas those with higher resistance have lower currents.

No voltage change occurs from $d$ back to $a$. Now that we have made a complete traversal of the loop, we can equate the sum of the voltage changes to zero:

$$6.0\ \text{V} - (4.0\ \Omega)I_1 - (9.0\ \Omega)I_3 = 0$$

Moving clockwise around the small loop from point $c$, we encounter the following:

- From $c$ to $d$ through the 5.0-Ω resistor, a voltage change of $-(5.0\ \Omega)I_2$
- From $d$ to $c$ through the 9.0-Ω resistor, a voltage change of $+(9.0\ \Omega)I_3$

We find that

$$-(5.0\ \Omega)I_2 + (9.0\ \Omega)I_3 = 0$$

Thus, we have the following three equations to be solved for the three unknowns:

$$I_1 - I_2 - I_3 = 0$$
$$6.0 - 4.0I_1 - 9.0I_3 = 0$$
$$-5.0I_2 + 9.0I_3 = 0$$

where we have dropped units of volts and ohms and realize that the currents will have units of amperes.

If you need help in solving three equations with three unknowns, see Example 18.5. You should be able to obtain the following answers:

$$I_1 = 0.83\ \text{A} \qquad I_2 = 0.53\ \text{A} \qquad I_3 = 0.30\ \text{A}$$

**Exercise** Solve this same problem by using the methods learned earlier for series and parallel combinations of resistors. First find the equivalent resistance of the circuit, which you can then use to obtain $I_1$.

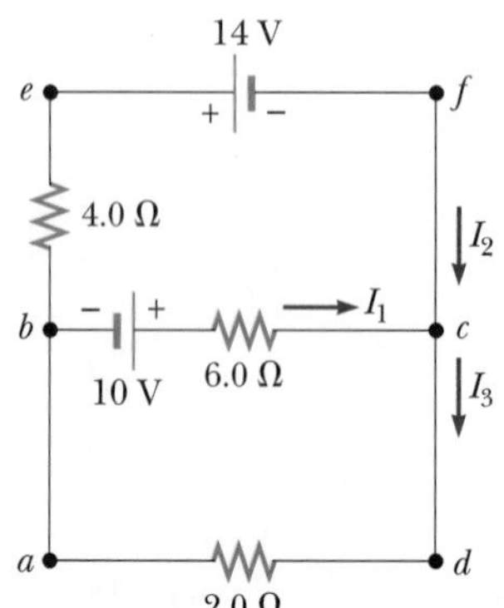

**Figure 18.15** (Example 18.5) A circuit containing three loops.

## Example 18.5 Another Application of Kirchhoff's Rules

Find $I_1$, $I_2$, and $I_3$ in Figure 18.15.

**Reasoning** To find the three unknown currents, we apply the junction rule once and the loop rule twice.

**Solution** We choose the directions of the currents as shown in the figure. Applying Kirchhoff's first rule to junction $c$ gives

$$I_1 + I_2 = I_3 \quad (1)$$

The circuit has three loops: *abcda, befcb,* and *aefda.* We need only two loop equations to determine the unknown currents. The third loop equation would give no new information. Applying Kirchhoff's second rule to loops *abcda* and *befcb* and traversing these loops clockwise, we obtain the following expressions:

$$\text{Loop } abcda: \quad 10\text{ V} - (6.0\ \Omega)I_1 - (2.0\ \Omega)I_3 = 0 \quad (2)$$

$$\text{Loop } befcb: \quad -14\text{ V} + (6.0\ \Omega)I_1 - 10\text{ V} - (4.0\ \Omega)I_2 = 0 \quad (3)$$

Note that in loop *befcb,* a positive sign is obtained when the 6.0-$\Omega$ resistor is traversed, because the direction of the path is opposite the direction of the current $I_1$. A third loop equation for *aefda* gives $-14\text{ V} - (2.0\ \Omega)I_3 - (4.0\ \Omega)I_2 = 0$, which is just the sum of (2) and (3). Expressions (1), (2), and (3) represent three linear, independent equations with three unknowns.

We can solve the problem as follows: substitution of (1) into (2) gives, with units ignored for the moment,

$$10 - 6.0I_1 - 2.0(I_1 + I_2) = 0$$

$$10 = 8.0I_1 + 2.0I_2 \quad (4)$$

Dividing each term in (3) by 2 and rearranging the equation gives

$$-12 = -3.0I_1 + 2.0I_2 \quad (5)$$

Subtracting (5) from (4) eliminates $I_2$, giving

$$22 = 11I_1$$

$$I_1 = 2.0\text{ A}$$

Using this value of $I_1$ in (5) yields a value for $I_2$:

$$2.0I_2 = 3.0I_1 - 12 = 3.0(2.0) - 12 = -6.0$$

$$I_2 = -3.0\text{ A}$$

Finally, $I_3 = I_1 + I_2 = -1.0$ A. Hence, the currents have the values

$$I_1 = 2.0\text{ A} \qquad I_2 = -3.0\text{ A} \qquad I_3 = -1.0\text{ A}$$

The fact that $I_2$ and $I_3$ are both negative merely indicates that we chose the wrong directions for these currents. The numerical values are correct.

**EXERCISE** Find the potential difference between junctions $b$ and $c$.

**ANSWER** $V_b - V_c = 2.0$ V

## 18.5 *RC* CIRCUITS

So far, we have been concerned with circuits with constant currents. We now consider direct-current circuits containing capacitors, in which the currents vary with time. Consider the series circuit in Figure 18.16a. Let us assume that the capacitor is initially uncharged with the switch opened. After the switch is closed, the battery begins to charge the plates of the capacitor and a current passes through the resistor. The charging process continues until the capacitor is charged to its maximum equilibrium value $Q = C\mathcal{E}$, where $\mathcal{E}$ is the maximum voltage across the capacitor. Once the capacitor is fully charged, the current in the circuit is zero. If we assume that the capacitor is uncharged before the switch is closed, and the switch is closed

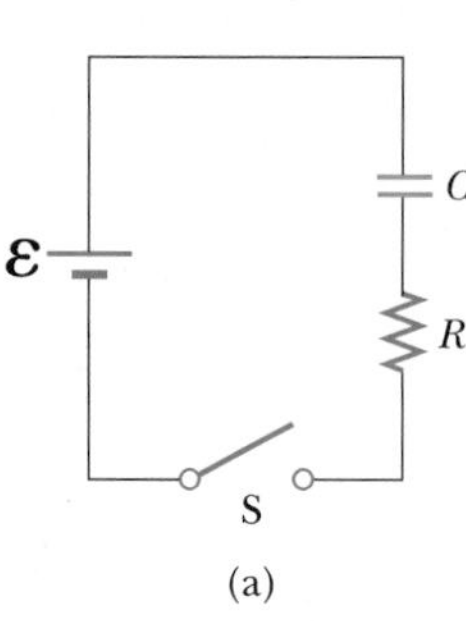


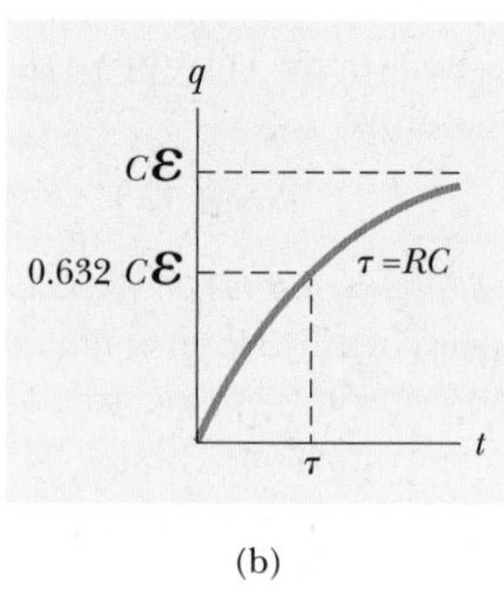


**FIGURE 18.16** (a) A capacitor in series with a resistor, a battery, and a switch. (b) A plot of the charge on the capacitor versus time after the switch for the circuit is closed. After one time constant $\tau$, the charge is 63% of the maximum value $C\mathcal{E}$. The charge approaches its maximum value as $t$ approaches infinity.

at $t = 0$, we find that the charge on the capacitor varies with time according to the expression

$$q = Q(1 - e^{-t/RC}) \qquad [18.7]$$

where $e = 2.718\ldots$ is Euler's constant, the base of the natural logarithms. Figure 18.16b is a graph of this expression. Note that the charge is zero at $t = 0$ and approaches its maximum value $Q$ as $t$ approaches infinity. The voltage $\Delta V$ across the capacitor at any time is obtained by dividing the charge by the capacitance. That is, $\Delta V = q/C$.

As you can see from Equation 18.7, it takes an infinite amount of time for the capacitor to become fully charged. The term $RC$ that appears in Equation 18.7, is called the **time constant,** $\tau$ (Greek letter tau), so

$$\tau = RC \qquad [18.8]$$

The time constant represents the time required for the charge to increase from zero to 63.2% of its maximum equilibrium value. That is, in one time constant, the charge on the capacitor increases from zero to $0.632Q$. This can be seen by substituting $t = \tau = RC$ in Equation 18.7 and solving for $q$. (Note that $1/e = 0.632$.) It is important to note that a capacitor charges very slowly in a circuit with a long time constant, whereas it charges very rapidly in a circuit with a short time constant.

Now consider the circuit in Figure 18.17a, consisting of a capacitor with an initial charge $Q$, a resistor, and a switch. Before the switch is closed, the potential difference across the charged capacitor is $Q/C$. Once the switch is closed, the charge begins to flow through the resistor from one capacitor plate to the other until the capacitor is fully discharged. If the switch is closed at $t = 0$, it can be shown that the charge $q$ on the capacitor varies with time according to the expression

$$q = Qe^{-t/RC} \qquad [18.9]$$

That is, the charge decreases exponentially with time as shown in Figure 18.17b. In the interval $t = \tau = RC$, the charge decreases from its initial value $Q$ to $0.368Q$. In other words, in one time constant, the capacitor loses 63.2% of its initial charge. Because $\Delta V = q/C$, we see that the voltage across the capacitor also decreases ex-

### Webnote 18.2

You can see the effects of charging and discharging the capacitor in an *RC* circuit by studying the following java applet: *http://www.phy.ntnu.edu.tw/java/rc/rc.html*

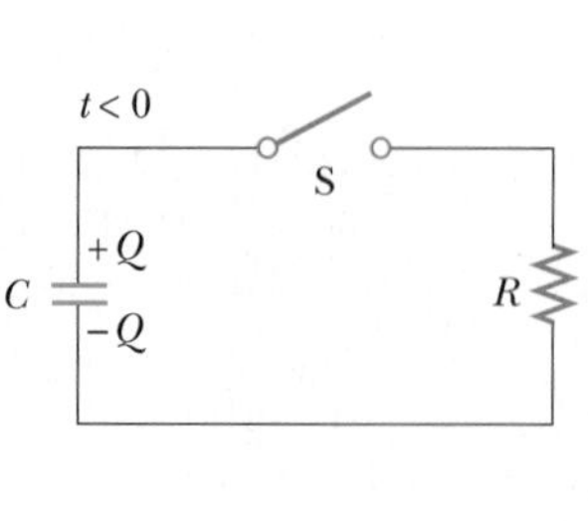


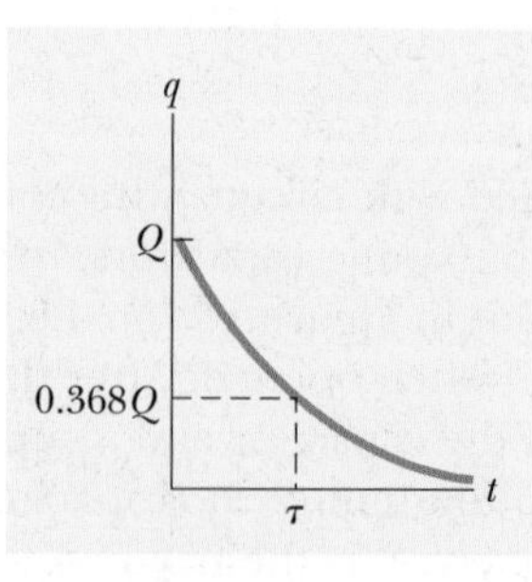


**FIGURE 18.17** (a) A charged capacitor connected to a resistor and a switch. (b) A graph of the charge on the capacitor versus time after the switch is closed.

ponentially with time according to the expression $\Delta V = \mathcal{E}e^{-t/RC}$, where $\mathcal{E}$ (which equals $Q/C$) is the initial voltage across the fully charged capacitor.

## APPLYING PHYSICS 18.4

**APPLICATION**

TIMED WINDSHIELD WIPERS

Many automobiles are equipped with windshield wipers that can be used intermittently during a light rainfall. How does the operation of this feature depend on the charging and discharging of a capacitor?

**Explanation** The wipers are part of an *RC* circuit whose time constant can be varied by selecting different values of *R* through a multiposition switch. The brief time that the wipers remain on and the time they are off are determined by the value of the time constant of the circuit.

## APPLYING PHYSICS 18.5

**APPLICATION**

BACTERIAL GROWTH

In biological applications concerned with population growth, an equation is used that is similar to the exponential equations encountered in the analysis of *RC* circuits. It is

$$N_f = N_i 2^n$$

where $N_f$ is the number of bacteria present after $n$ doubling times, $N_i$ is the number present initially, and $n$ is the number of growth cycles or doubling times. Doubling times vary according to the organism. The doubling time for the bacteria responsible for leprosy is about 30 days, and that for the salmonella bacteria responsible for food poisoning is about 20 minutes. For those of you with a biology inclination, consider that only 10 salmonella bacteria find their way onto a turkey leg after your Thanksgiving meal. Four hours later you come back for a midnight snack. How many bacteria are present now?

**Explanation** The number of doubling times is 240 min/20 min = 12. Thus, we have

$$N_f = N_i 2^n = (10 \text{ bacteria})(2^{12}) = 40\,960 \text{ bacteria}$$

So, your system will have to deal with an invading host of about 41 000 bacteria, which are going to continue to double in a very promising environment.

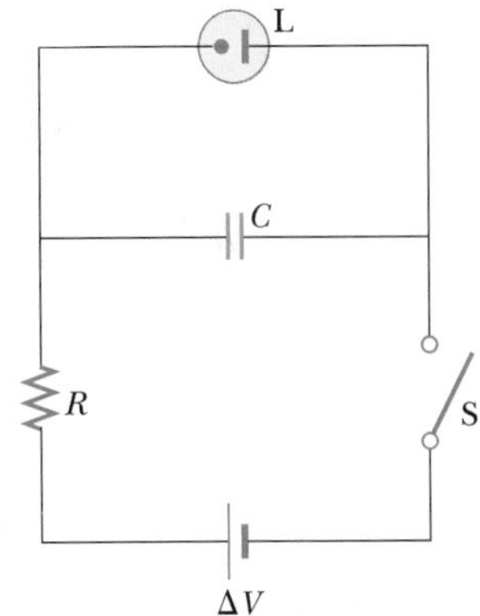


**FIGURE 18.18** (Applying Physics 18.6)

## APPLYING PHYSICS 18.6

**APPLICATION**

ROADWAY FLASHERS

Many roadway construction sites have flashing yellow lights to warn motorists of possible dangers. What causes the lights to flash?

**Explanation** A typical circuit for such a flasher is shown in Figure 18.18. The lamp L is a gas-filled lamp that acts as an open circuit until a large potential difference causes a discharge, which gives off a bright light. During this discharge, charge flows through the gas between the electrodes of the lamp. When the switch is closed, the battery charges up the capacitor. At the beginning, the current is high and the charge on the capacitor is low, so most of the potential difference appears across the resistance *R*. As the capacitor charges, more potential difference appears across it, reflecting the lower current and, thus, lower potential difference across the resistor. Eventually, the potential difference across the capacitor reaches a value at which the lamp will conduct, causing a flash. This discharges the capacitor through the lamp and the process of charging begins again. The period between flashes can be adjusted by changing the time constant of the *RC* circuit.

**Example 18.6 Charging a Capacitor in an *RC* Circuit**

An uncharged capacitor and a resistor are connected in series to a battery, as in Figure 18.16a. If $\mathcal{E} = 12$ V, $C = 5.0\ \mu$F, and $R = 8.0 \times 10^5\ \Omega$, find the time constant of the circuit, the maximum charge on the capacitor, and the charge on the capacitor after one time constant—that is, when $t = RC$.

**Solution** The time constant of the circuit is

$$\tau = RC = (8.0 \times 10^5\ \Omega)(5.0 \times 10^{-6}\ \text{F}) = \boxed{4.0\ \text{s}}$$

The maximum charge on the capacitor is

$$Q = C\mathcal{E} = (5.0 \times 10^{-6}\ \text{F})(12\ \text{V}) = \boxed{60\ \mu\text{C}}$$

After one time constant, the charge on the capacitor is 63.2% of its maximum value:

$$q = 0.632Q = 0.632(60 \times 10^{-6}\ \text{C}) = \boxed{38\ \mu\text{C}}$$

**EXERCISE** Find the charge on the capacitor and the voltage across the capacitor after time $t$ has elapsed.

**ANSWER** $q = (60\ \mu\text{C})(1 - e^{-t/4})$; $\Delta V = (12\ \text{V})(1 - e^{-t/4})$

**Example 18.7 Discharging a Capacitor in an *RC* Circuit**

Consider a capacitor $C$ being discharged through a resistor $R$ as in Figure 18.17a. After how many time constants does the charge on the capacitor drop to one fourth of its initial value?

**Solution** The charge on the capacitor varies with time according to Equation 18.9:

$$q = Qe^{-t/RC}$$

where $Q$ is the initial charge on the capacitor. To find the time it takes the charge $q$ to drop to one fourth of its initial value, we substitute $q = Q/4$ into this expression and solve for $t$:

$$\tfrac{1}{4}Q = Qe^{-t/RC}$$

$$\tfrac{1}{4} = e^{-t/RC}$$

Taking logarithms of both sides, we find that

$$\ln(\tfrac{1}{4}) = -\ln 4 = -\frac{t}{RC}$$

$$t = RC \ln 4 = \boxed{1.39RC}$$

**EXERCISE** If $R = 8.0 \times 10^5\ \Omega$, $C = 5.0\ \mu$F, and the initial voltage across the capacitor is 6.0 V, what is the voltage across the capacitor after time $t$ has elapsed?

**ANSWER** $\Delta V = (6.0\ \text{V})e^{-t/4}$

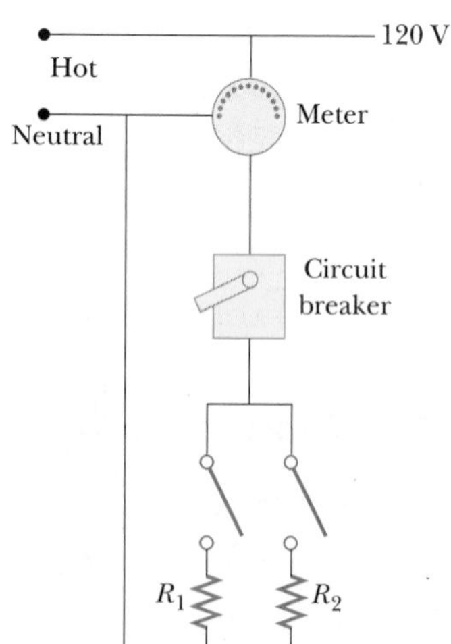


**FIGURE 18.19** A wiring diagram for a household circuit. The resistances $R_1$ and $R_2$ represent appliances or other electrical devices that operate at an applied voltage of 120 V.

## 18.6 HOUSEHOLD CIRCUITS

Household circuits are a very practical application of some of the ideas presented in this chapter. In a typical installation, the utility company distributes electric power to individual houses with a pair of wires, or power lines. Electrical devices in a house are then connected in parallel to these lines, as shown in Figure 18.19.

The potential drop between the two wires is about 120 V. (These are actually alternating currents and voltages, but for the present discussion we shall assume that they are direct, constant currents and voltages.) One of the wires is connected to ground, and the other wire, sometimes called the "hot" wire, is at a potential of 120 V. A meter and a circuit breaker or fuse are connected in series with the wire entering the house, as indicated in Figure 18.19.

**APPLICATION**

FUSES AND CIRCUIT BREAKERS

In modern homes, circuit breakers are used in place of fuses. When the current in a circuit exceeds some value (typically 20 A), the circuit breaker acts as a switch and opens the circuit. Figure 18.20 shows one design for a circuit breaker. Current passes through a bimetallic strip, the top of which bends to the left when excessive current heats it. If the strip bends far enough to the left, it settles into a groove in the spring-loaded metal bar. When this occurs, the bar drops enough to open the circuit at the contact point. The bar also flips a switch that indicates that the circuit breaker has interrupted the current. (After the overload is removed, the switch can be flipped back on.) Circuit breakers based on this design have the disadvantage that some time is required for the heating of the strip, and thus the circuit is not opened rapidly enough when it is overloaded. As a consequence, many circuit breakers are now designed to use electromagnets, which we shall discuss in Chapter 19.

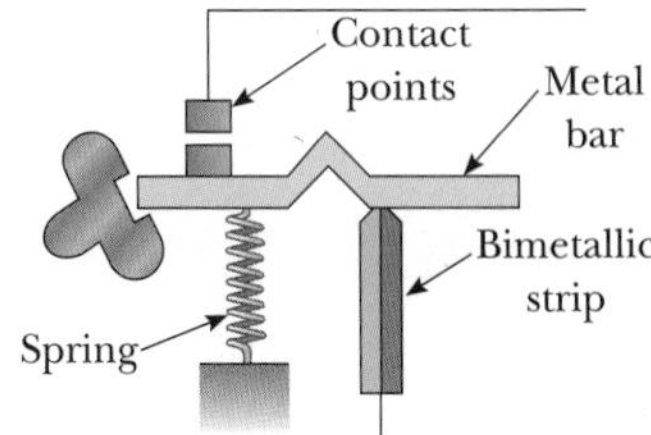


**FIGURE 18.20** A circuit breaker that uses a bimetallic strip for its operation.

The wire and circuit breaker are carefully selected to meet the current demands of a circuit. If the circuit is to carry currents as large as 30 A, a heavy-duty wire and appropriate circuit breaker must be used. Household circuits that are normally used to power lamps and small appliances often require only 20 A. Each circuit has its own circuit breaker to accommodate its maximum safe current.

As an example, consider a circuit that powers a toaster, a microwave oven, and a heater (represented by $R_1$, $R_2$, . . . in Fig. 18.19). We can calculate the current carried by each appliance using the equation $\mathcal{P} = I\,\Delta V$. The toaster, rated at 1 000 W, draws a current of 1 000/120 = 8.33 A. The microwave oven, rated at 800 W, draws a current of 6.67 A, and the heater, rated at 1 300 W, draws a current of 10.8 A. If the three appliances are operated simultaneously, they draw a total current of 25.8 A. Therefore, the breaker should be able to handle at least this much current, or else it will be tripped. As an alternative, one could operate the toaster and microwave oven on one 20-A circuit and the heater on a separate 20-A circuit.

### Webnote 18.3

At home you probably have at least one light that is controlled by two switches. Flipping either switch changes the light from off to on, or from on to off. Discover how this works at *http://www.howstuffworks.com/three-way.htm*

Many heavy-duty appliances, such as electric ranges and clothes dryers, require 240 V to operate. The power company supplies this voltage by providing, in addition to a live wire that is 120 V above ground potential, a wire, also considered live, that is 120 V below ground potential (Fig. 18.21). Therefore, the potential drop across the two live wires is 240 V. An appliance operating from a 240-V line requires half the current of one operating from a 120-V line; therefore, smaller wires can be used in the higher-voltage circuit without becoming overheated.

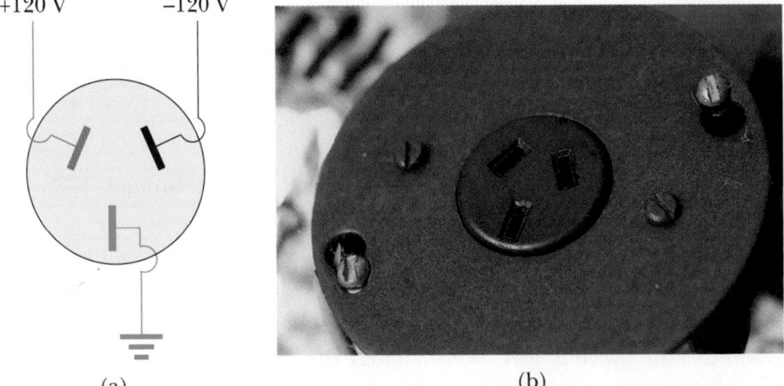


**FIGURE 18.21** Power connections for a 240-V appliance. *(b, George Semple)*

## 18.7 ELECTRICAL SAFETY

A person can be electrocuted by touching a live wire (which commonly is live because of a frayed cord and exposed conductors) while the person is in contact with ground. The ground contact might be made by touching a water pipe (which is normally at ground potential) or by standing on the ground with wet feet, because impure water is a good conductor. Obviously, such situations should be avoided at all costs.

Electric shock can result in fatal burns, or it can cause the muscles of vital organs, such as the heart, to malfunction. The degree of damage to the body depends on the magnitude of the current, the length of time it acts, and the part of the body through which it passes. Currents of 5 mA or less can cause a sensation of shock but ordinarily do little or no damage. If the current is larger than about 10 mA, the hand muscles contract and the person may be unable to let go of the live wire. If a current of about 100 mA passes through the body for just a few seconds, it can be fatal. Such large currents paralyze the respiratory muscles. In some cases, currents of about 1 A through the body produce serious burns.

**APPLICATION**

THIRD WIRE ON CONSUMER APPLIANCES

As an additional safety feature for consumers, electrical equipment manufacturers now use electrical cords that have a third wire, called a ground. To understand how this works, consider the drill being used in Figure 18.22. A two-wire device has one wire, called the "hot" wire, connected to the high-potential (120-V)

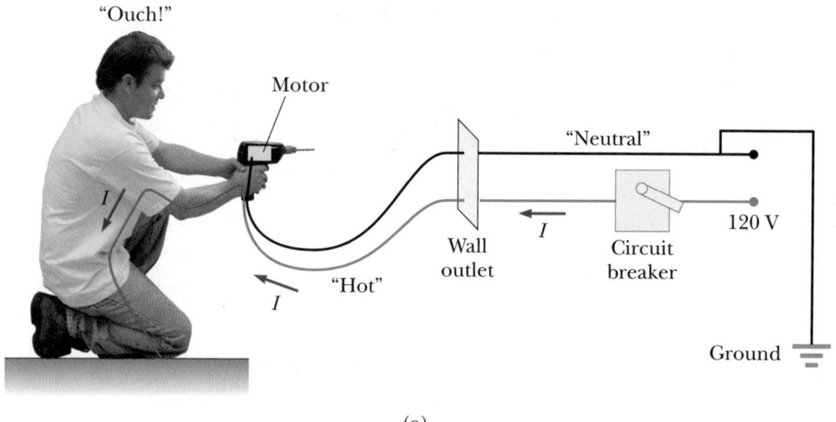


(a)

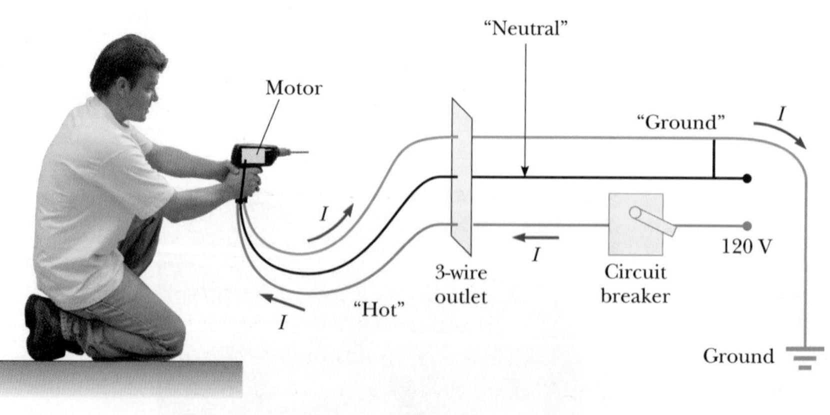


(b)

**FIGURE 18.22** The "hot" wire, at 120 V, always includes a circuit breaker for safety. (a) When the drill is operated with two wires, the normal current path is from the "hot" wire through the motor connections and back to ground through the "neutral" wire. However, here the high-voltage wire has come in contact with the drill case so that the person holding the drill receives an electrical shock. (b) Shock can be prevented by a third wire running from the drill case to the ground.

side of the input power line, and the second wire is connected to ground (0 V). If the high-voltage wire comes in contact with the case of the drill (Fig. 18.22a), a "short circuit" occurs. In this undesirable circumstance, the pathway for the current is from the high-voltage wire through the person holding the drill and to Earth—a pathway that can kill. Protection is provided by a third wire, connected to the case of the drill (Fig. 18.22b). In this situation, if a short occurs, the path of least resistance for the current is from the high-voltage wire through the case and back to ground through the third wire. The resulting high current produced will trip a circuit breaker before the person is injured.

Special power outlets called ground-fault interrupters (GFIs) are now being used in kitchens, bathrooms, basements, and other hazardous areas of new homes. They are designed to protect people from electrical shock by sensing small currents—approximately 5 mA and greater—leaking to ground. When current above this level is detected, the device shuts off (interrupts) the current in less than a millisecond. Ground-fault interrupters will be discussed further in Chapter 19.

**Webnote 18.4**

Read more about the electrical safety features of three-prong cords at *http://www.howstuffworks.com/question110.htm*

## 18.8 Conduction of Electrical Signals by Neurons[2]

The most remarkable use of electrical phenomena in living organisms is found in the nervous system of animals. Specialized cells in the body called **neurons** form a complex network that receives, processes, and transmits information from one part of the body to another. The center of this network is located in the brain, which has the ability to store and analyze information. Based on this information, the nervous system controls parts of the body.

The nervous system is very complex and consists of about $10^{10}$ interconnected neurons. Neurons are the basic units of the nervous system. Some aspects of the nervous system are well known. Over the past 45 years, the method of signal propagation through the nervous system has been established. The messages are voltage pulses called *action potentials* transmitted by neurons. When a neuron receives a strong enough stimulus, it produces repetitive voltage pulses that are actively propagated along its structure. The strength of the stimulus is conveyed by the number of pulses produced. When the pulses reach the end of the neuron, they activate either muscle cells or other neurons. It is interesting that there is a "firing threshold" for neurons. That is, action potentials will propagate along a neuron only if the stimulus is sufficiently strong.

Neurons can be divided into three classes: sensory neurons, motor neurons, and interneurons. The sensory neurons receive stimuli from sensory organs that monitor the external and internal environment of the body. Depending on their specialized functions, the sensory neurons convey messages about factors such as light, temperature, pressure, muscle tension, and odor to higher centers in the nervous system. The motor neurons carry messages that control the muscle cells. These messages are based on the information provided by the sensory neurons and by the brain. The interneurons transmit information from one neuron to another.

Each neuron consists of a cell body to which are attached input ends called **dendrites** and a long tail called the **axon,** which transmits the signal away from the cell (Fig. 18.23). The far end of the axon branches into nerve endings that transmit the signal across small gaps to other neurons or to muscle cells. A simple sensory-motor neuron circuit is shown in Figure 18.24. A stimulus from a muscle produces nerve impulses that travel to the spine. Here the signal is transmitted to a motor neuron, which in turn sends impulses to control the muscle. Figure 18.25 shows an electron microscope image of neurons in the brain.

[2] This section is based upon an essay by Paul Davidovits of Boston College.

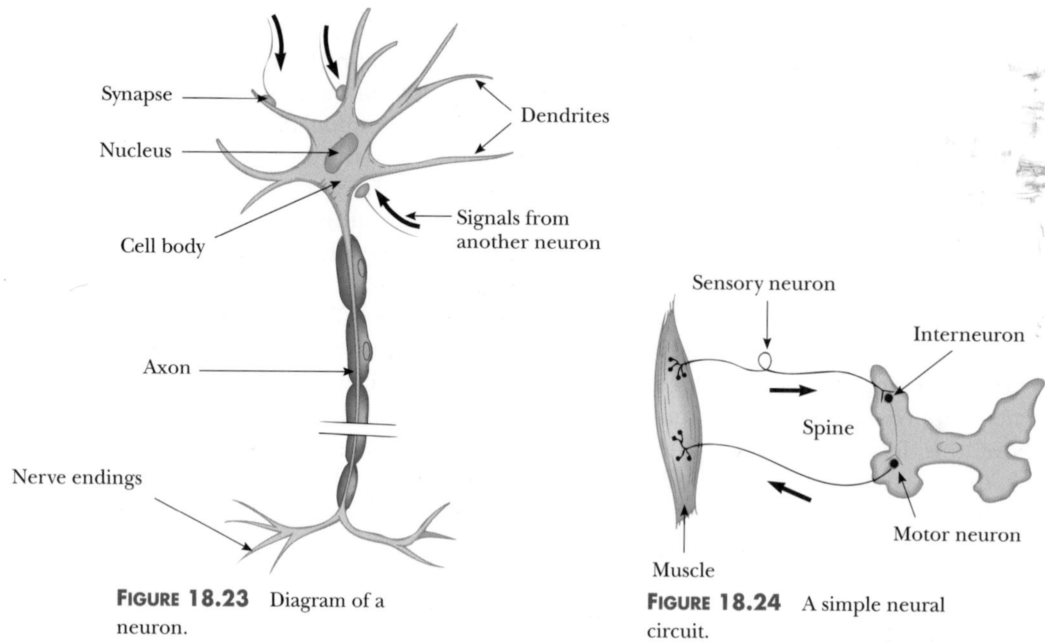


**FIGURE 18.23** Diagram of a neuron.

**FIGURE 18.24** A simple neural circuit.

The axon, which is an extension of the neuron cell, conducts the electric impulses away from the cell body. Some axons are extremely long. In humans, for example, the axons connecting the spine with the fingers and toes are more than 1 m long. The neuron can transmit messages because of the special active electrical characteristics of the axon. (The axon acts as an *active* source of energy like a battery rather than like a *passive* stretch of resistive wire.) Much of the information about the electrical and chemical properties of the axon is obtained by inserting small needle-like probes into the axon. Figure 18.26 shows an experimental setup. Note that the outside of the axon is grounded so that all measured voltages are with respect to a zero potential on the outside of the axon. With such probes it is possible to inject current into the axon, measure the resulting action potential as a function of time at a fixed point, and sample the cell's chemical composition. Such experiments are usually difficult to run because the diameter of most axons is very small. Even the largest axons in the human nervous system have a diameter of only about $20 \times 10^{-4}$ cm. The giant squid, however, has an axon with a diameter of about 0.5 mm, which is large enough for the convenient insertion of probes. Much of the information about signal transmission in the nervous system has come from experiments with the squid axon.

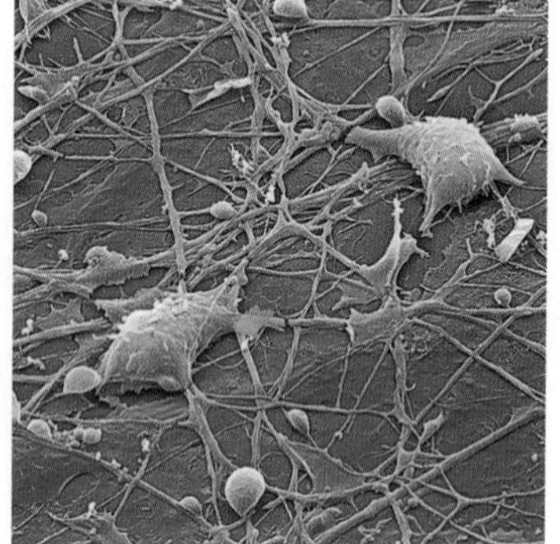

**FIGURE 18.25** Stellate neuron from human cortex. *(Juergen Berger, Max-Planck Institute/Science Photo Library/Photo Researchers, Inc.)*

In the aqueous environment of the body, salts and other molecules dissociate into positive and negative ions. As a result, body fluids are relatively good conductors of electricity. The inside of the axon is filled with an ionic fluid that is sepa-

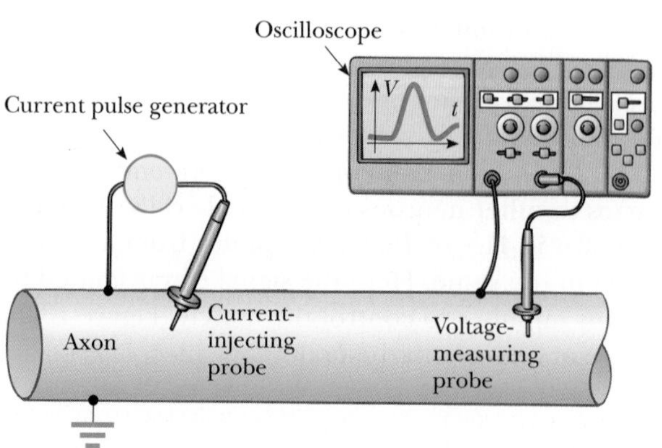


**FIGURE 18.26** An axon stimulated electrically. The left probe injects a short current pulse and the right probe measures the resulting action potential as a function of time.

rated from the surrounding body fluid by a thin membrane that is only about 5 to 10 nm thick. The resistivities of the internal and external fluids are about the same, but their chemical compositions are substantially different. The external fluid is similar to sea water. Its ionic solutes are mostly positive sodium ions and negative chloride ions. Inside the axon, the positive ions are mostly potassium ions and the negative ions are mostly large organic ions.

Because there is a large concentration of sodium ions outside the axon and a large concentration of potassium ions inside, one might wonder why the concentrations are not equalized by diffusion. In other words, why don't the sodium ions leak into the axon and the potassium ions leak out of it? The answer lies in the properties of the axon membrane, which, as part of a living cell with an energy supply, can change its permeability on the time scale of milliseconds!

When the axon is not conducting an electric pulse, the axon membrane is highly permeable to potassium ions, slightly permeable to sodium ions, and impermeable to large organic ions. Thus, although sodium ions cannot easily enter the axon, potassium ions can leave it. As the positive potassium ions leave the axon, however, they leave behind large negative organic ions, which cannot follow them through the membrane. As a result, a negative potential builds up inside the axon with respect to the outside. The final negative potential reached, which has been measured at about $-70$ mV, holds back the outflow of potassium ions so that at equilibrium, the concentration of ions is as we have stated.

The mechanism for the production of an electric signal by the neuron is conceptually simple but was experimentally difficult to sort out. When a neuron changes its resting potential because of an appropriate stimulus, the properties of its membrane change locally. As a result, there is a sudden flow of sodium ions into the cell that lasts for about 2 ms. This produces the $+30$-mV peak in the action potential shown in Figure 18.27a. Immediately after, there is an increase in potassium ion flow out of the cell, which restores the resting action potential of $-70$ mV in an additional 3 ms. Both the $Na^+$ and $K^+$ ion flows have been

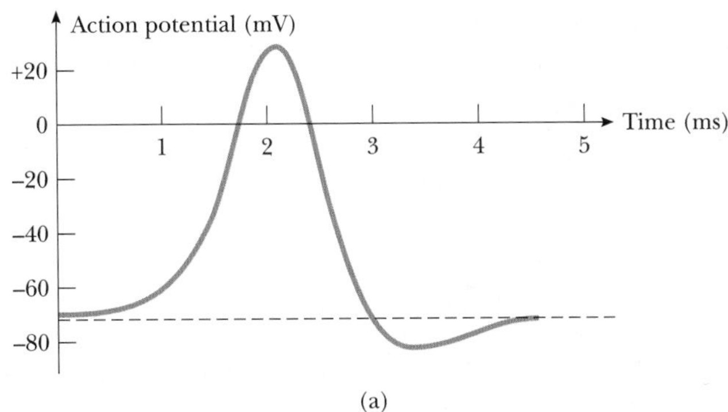


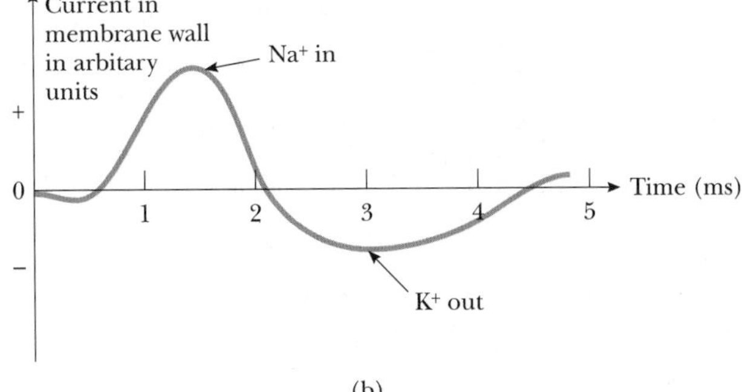


**FIGURE 18.27** (a) Typical action potential as a function of time. (b) Current in the axon membrane wall as a function of time.

measured by using radioactive Na and K tracers. The nerve signal has been measured to travel along the axon at speeds of 50 m/s to about 150 m/s. This flow of charged particles (or signal transmission) in a nerve axon is *unlike* signal transmission in a metal wire. In an axon, charges move in a direction perpendicular to the direction of travel of the nerve signal, and the nerve signal moves much slower than a voltage pulse traveling along a metallic wire.

Although the axon is a complex structure, and much of how ion channels selectively pass $Na^+$ and $K^+$ is not understood,[3] standard electric circuit concepts of current and capacitance can be used to analyze axons. It is left as a problem to show that the axon, having equal and opposite charges separated by a thin dielectric membrane, acts like a capacitor (see Problem 18.41).

## SUMMARY

A **source of emf** is any device that transforms nonelectrical energy into electrical energy.

The **equivalent resistance** of a set of resistors connected in **series** is

$$R_{eq} = R_1 + R_2 + R_3 + \cdots \qquad [18.4]$$

The **equivalent resistance** of a set of resistors connected in **parallel** is

$$\frac{1}{R_{eq}} = \frac{1}{R_1} + \frac{1}{R_2} + \frac{1}{R_3} + \cdots \qquad [18.6]$$

Complex circuits are conveniently analyzed by using **Kirchhoff's rules:**

1. The sum of the currents entering any junction must equal the sum of the currents leaving that junction.
2. The sum of the potential differences across all the elements around any closed circuit loop must be zero.

The first rule is a statement of **conservation of charge.** The second is a statement of **conservation of energy.**

As a capacitor is charged by a battery through a resistor, the charge increases from zero to some maximum value. The **time constant** $\tau = RC$ represents the time it takes the charge on the capacitor to increase from zero to 63% of its maximum value.

## CONCEPTUAL QUESTIONS

1. Is the direction of current through a battery always from the negative terminal to the positive one? Explain.
2. Given three lightbulbs and a battery, sketch as many different circuits as you can.
3. If the energy transferred to a dead battery during charging is $W$, is the total energy transferred out of the battery to an external circuit during use in which it completely discharges also $W$?
4. How would you connect resistors so that the equivalent resistance is larger than the individual resistances? Give an example involving two or three resistors.
5. If you have your headlights on while you start your car, why do they dim while the car is starting?
6. How would you connect resistors so that the equivalent resistance is smaller than the individual resistances? Give an example involving two or three resistors.
7. Electrical devices are often rated with a voltage and a current—for example, 120 V, 5 A. Batteries, however, are only rated with a voltage—for example, 1.5 V. Why?
8. A "short circuit" is a circuit containing a path of very low resistance in parallel with some other part of the circuit. Discuss the effect of a short circuit on the portion of the circuit it parallels. Use a lamp with a frayed cord as an example.
9. Connecting batteries in series increases the emf applied to a circuit. What advantage might there be to connecting them in parallel?
10. If electrical power is transmitted over long distances, the resistance of the wires becomes significant. Why? Which mode of transmission would result in less energy loss—high current and low voltage or low current and high voltage? Discuss.

[3] For recent developments on this topic, see the November 1, 2001 issue of *Nature.*

11. In Figure Q18.11, describe what happens to the lightbulb after the switch is closed. Assume the capacitor has a large capacitance and is initially uncharged, and assume that the light illuminates when connected directly across the battery terminals.

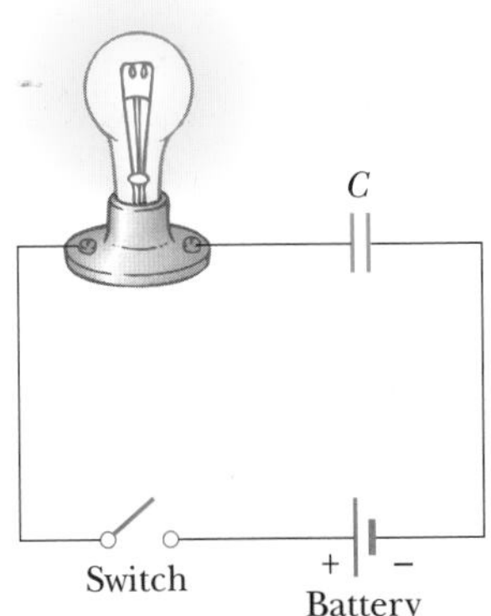


**FIGURE Q18.11**

12. Two sets of Christmas tree lights are available. For set A, when one bulb is removed, the remaining bulbs remain illuminated. For set B, when one bulb is removed, the remaining bulbs do not operate. Explain the difference in wiring for the two sets.
13. Why is it possible for a bird to sit on a high-voltage wire without being electrocuted?

**FIGURE Q18.13** Birds on a high-voltage wire. *(Superstock)*

14. Are the two headlights on a car wired in series or in parallel? How can you tell?
15. Embodied in Kirchhoff's rules are two conservation laws. What are they?
16. A ski resort consists of a few chairlifts and several interconnected downhill runs on the side of a mountain, with a lodge at the bottom. The lifts are analogous to batteries and the runs are analogous to resistors. Describe how two runs can be in series. Describe how three runs can be in parallel. Sketch a junction of one lift and two runs. One of the skiers is carrying an altimeter. State Kirchhoff's junction rule and Kirchhoff's loop rule for ski resorts.
17. Suppose you are flying a kite when it strikes a high-voltage wire (a very dangerous situation). What factors determine how great a shock you will receive?
18. Why is it dangerous to turn on a light when you are in a bathtub?
19. Suppose a parachutist lands on a high-voltage wire and grabs the wire as she prepares to be rescued. Will she be electrocuted? If the wire then breaks, should she continue to hold onto the wire as she falls to the ground?
20. Would a fuse or circuit breaker work successfully if it were placed in parallel with the device it was supposed to protect?
21. A series circuit consists of three identical lamps connected to a battery as in Figure Q18.21. When the switch S is closed, what happens (a) to the intensities of lamps A and B, (b) to the intensity of lamp C, (c) to the current in the circuit, and (d) to the voltage drop across the three lamps? (e) Does the power dissipated in the circuit increase, decrease, or remain the same?

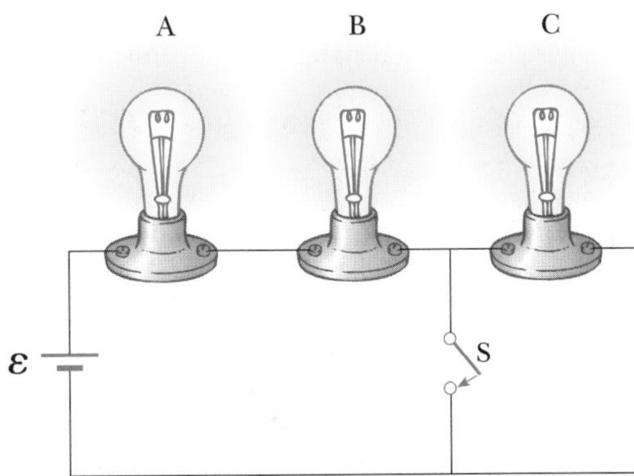


**FIGURE Q18.21**

22. Figure Q18.22 shows a series connection of three lamps, all rated at 120 V, with power ratings of 60 W, 75 W, and 200 W. Why do the intensities of the lamps differ? Which lamp has the greatest resistance? How would their intensities differ if they were connected in parallel?

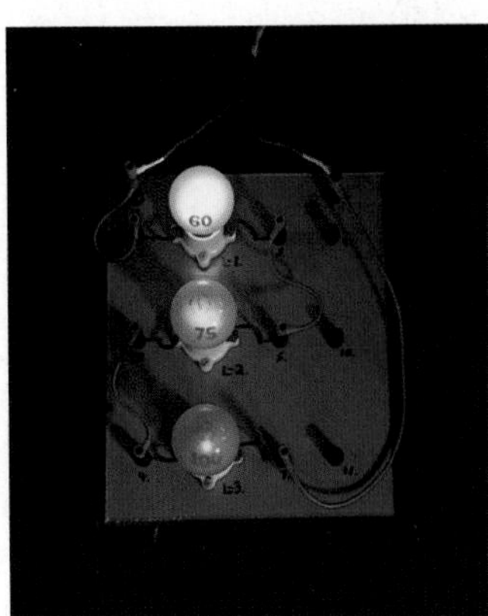


**FIGURE Q18.22** *(Courtesy of Henry Leap and Jim Lehman)*

## PROBLEMS

1, 2, 3 = straightforward, intermediate, challenging □ = full solution available in Student Solutions Manual/Study Guide

web = solution posted at **http://info.brookscole.com/serway** = biomedical application

### Section 18.1 Sources of emf

### Section 18.2 Resistors in Series

### Section 18.3 Resistors in Parallel

1. A battery having an emf of 9.00 V delivers 117 mA when connected to a 72.0-Ω load. Determine the internal resistance of the battery.
2. A 4.0-Ω resistor, an 8.0-Ω resistor, and a 12-Ω resistor are connected in series with a 24-V battery. What are (a) the equivalent resistance and (b) the current in each resistor? (c) Repeat for the case in which all three resistors are connected in parallel across the battery.
3. A lightbulb marked "75 W [at] 120 V" is screwed into a socket at one end of a long extension cord in which each of the two conductors has a resistance of 0.800 Ω. The other end of the extension cord is plugged into a 120-V outlet. Draw a circuit diagram, and find the actual power of the bulb in this circuit.
4. A 9.0-Ω resistor and a 6.0-Ω resistor are connected in series with a power supply. (a) The voltage drop across the 6.0-Ω resistor is measured to be 12 V. Find the voltage output of the power supply. (b) The two resistors are connected in parallel across a power supply, and the current through the 9.0-Ω resistor is found to be 0.25 A. Find the voltage setting of the power supply.
5. (a) Find the equivalent resistance between points $a$ and $b$ in Figure P18.5. (b) Calculate the current in each resistor if a potential difference of 34.0 V is applied between points $a$ and $b$.

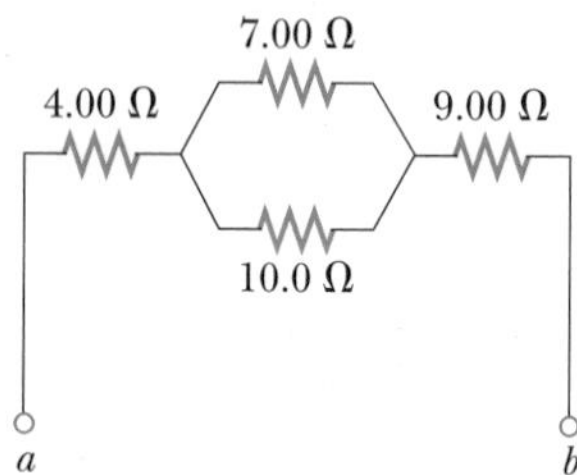


**FIGURE P18.5**

6. Find the equivalent resistance of the circuit in Figure P18.6.

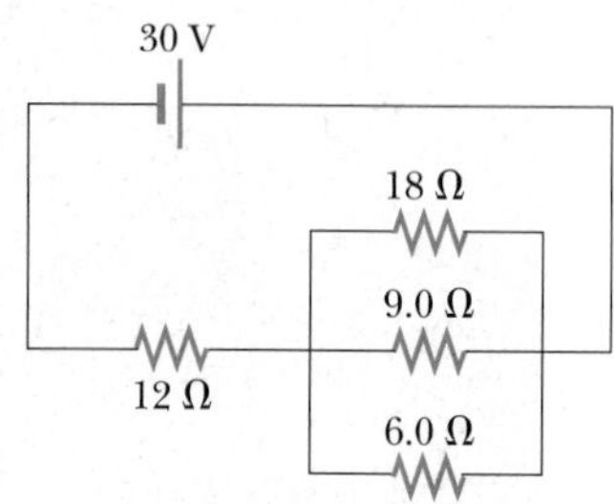


**FIGURE P18.6**

7. Find the equivalent resistance of the circuit in Figure P18.7.

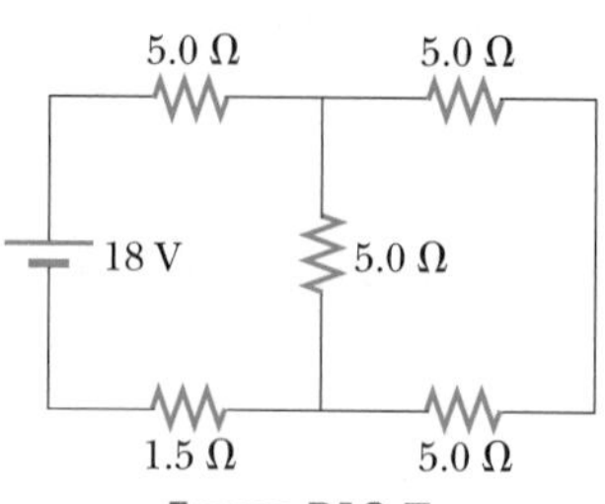


**FIGURE P18.7**

8. (a) Find the equivalent resistance of the circuit in Figure P18.8. (b) If the total power supplied to the circuit is 4.00 W, find the emf of the battery.

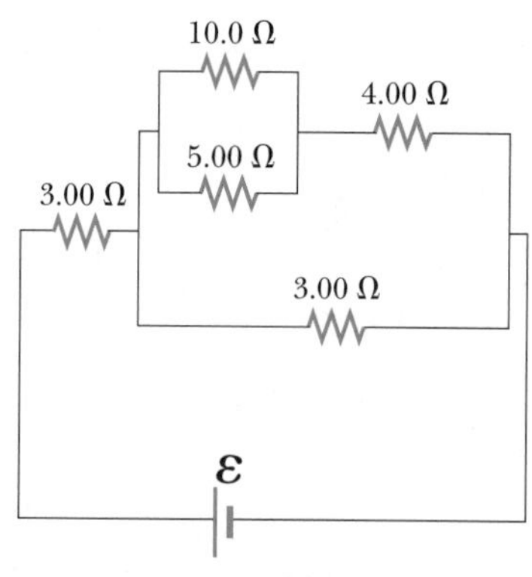


**FIGURE P18.8**

9. Consider the circuit shown in Figure P18.9. Find (a) the current in the 20.0-Ω resistor and (b) the potential difference between points $a$ and $b$.

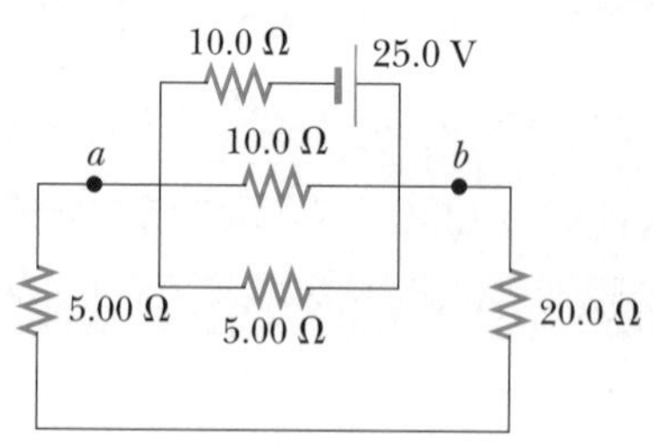


**FIGURE P18.9**

10. Two resistors, $A$ and $B$, are connected in parallel across a 6.0-V battery. The current through $B$ is found to be 2.0 A. When the two resistors are connected in series to the 6.0-V battery, a voltmeter connected across resistor $A$ measures a voltage of 4.0 V. Find the resistances of $A$ and $B$.
11. The resistance between terminals $a$ and $b$ in Figure P18.11 is 75 Ω. If the resistors labeled $R$ have the same value, determine $R$.
12. Three 100-Ω resistors are connected as shown in Figure P18.12. The maximum power that can safely be delivered to any one resistor is 25.0 W. (a) What is the maximum voltage that can be applied to the terminals $a$ and $b$?

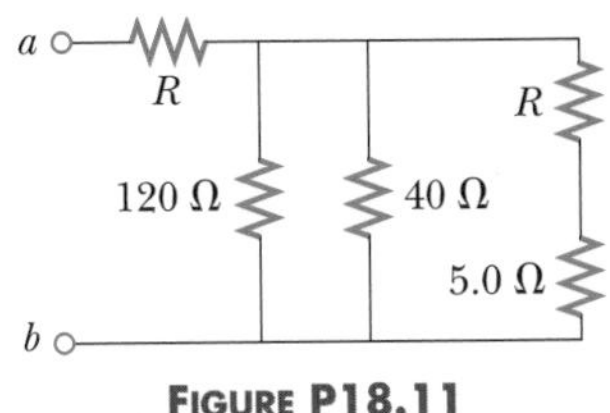


FIGURE P18.11

(b) For the voltage determined in part (a), what is the power delivered to each resistor? What is the total power delivered?

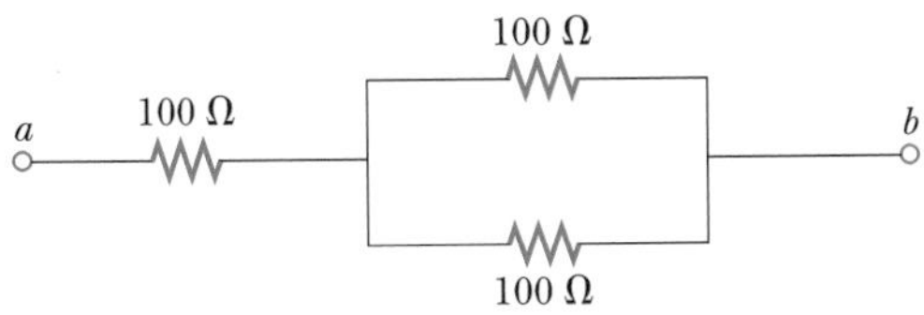


FIGURE P18.12

13. Find the current in the 12-Ω resistor in Figure P18.13. web

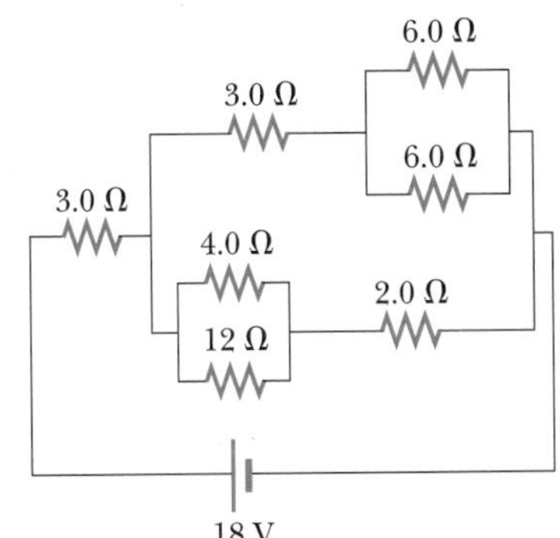


FIGURE P18.13

14. Calculate the power delivered to each resistor in the circuit shown in Figure P18.14.

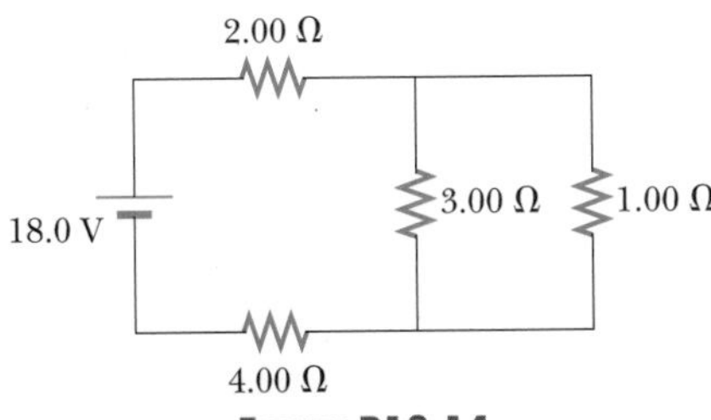


FIGURE P18.14

15. (a) You need a 45-Ω resistor, but the stockroom has only 20-Ω and 50-Ω resistors. How can the desired resistance be achieved under these circumstances? (b) What can you do if you need a 35-Ω resistor?

## Section 18.4 Kirchhoff's Rules and Complex DC Circuits

*Note:* The currents are not necessarily in the direction shown for some circuits.

16. The ammeter shown in Figure P18.16 reads 2.00 A. Find $I_1$, $I_2$, and $\mathcal{E}$.

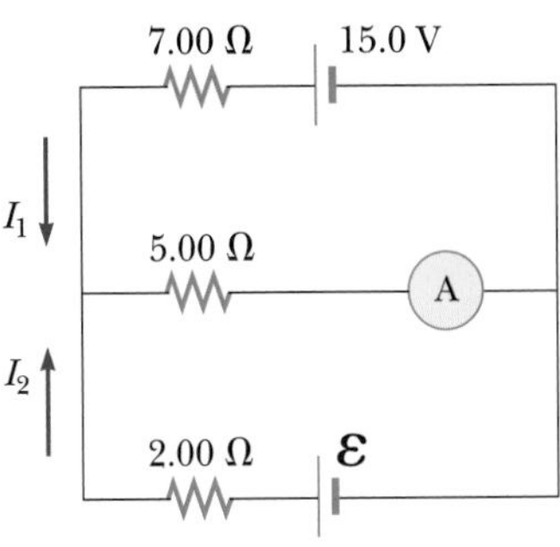


FIGURE P18.16

17. Determine the current in each branch of the circuit shown in Figure P18.17.

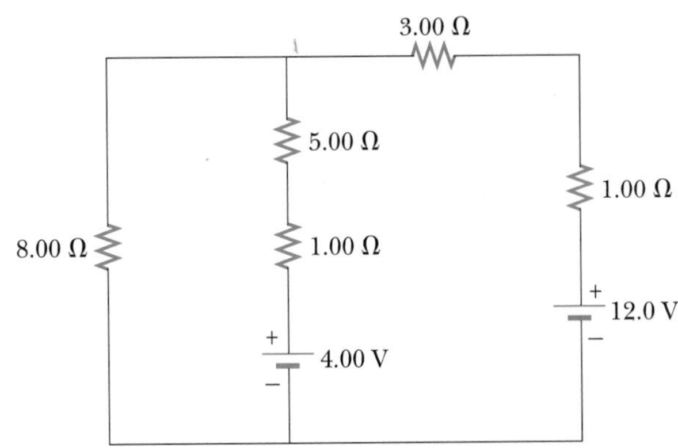


FIGURE P18.17 (Problems 17 and 18)

18. In Figure P18.17, show how to add just enough ammeters to measure every different current in the circuit. Show how to add just enough voltmeters to measure the potential difference across each resistor and across each battery.

19. Figure P18.19 shows a circuit diagram. Determine (a) the current, (b) the potential of wire $A$ relative to ground, and (c) the voltage drop across the 1 500-Ω resistor.

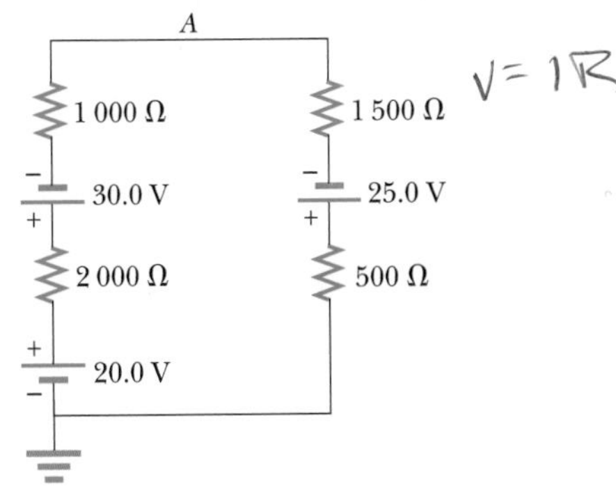


FIGURE P18.19

20. In the circuit of Figure P18.20, the current $I_1$ is 3.0 A and the values of $\mathcal{E}$ and $R$ are unknown. What are the currents $I_2$ and $I_3$?

21. What is the emf $\mathcal{E}$ of the battery in the circuit of Figure P18.21?

22. Find the current in each of the three resistors of Figure P18.22 (a) by the rules for resistors in series and parallel and (b) by the use of Kirchhoff's rules.

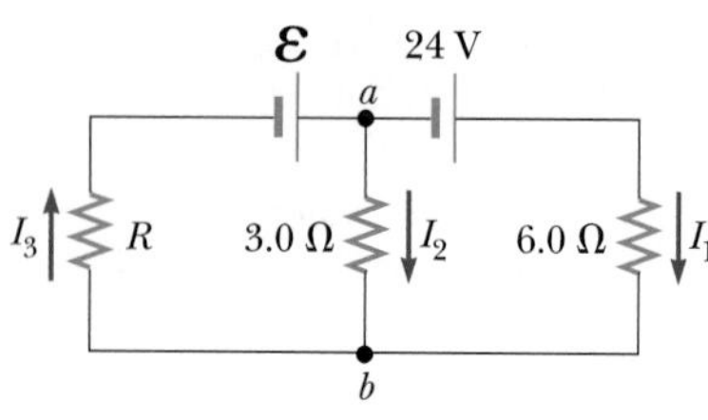

FIGURE P18.20

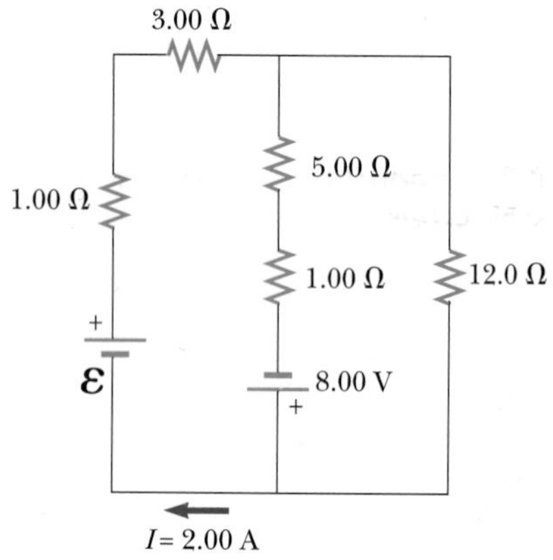

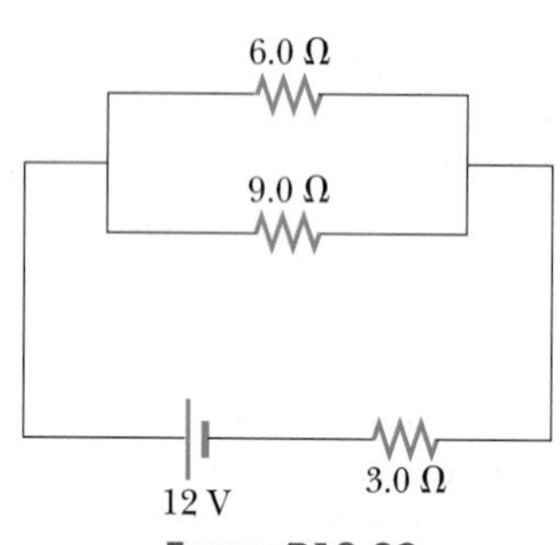

FIGURE P18.21

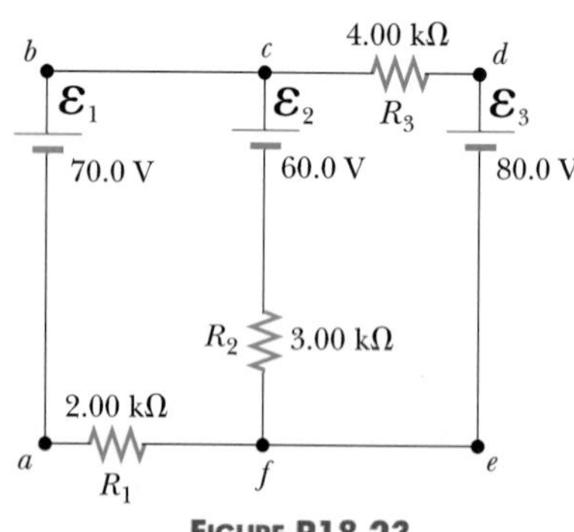

FIGURE P18.22

**23.** (a) Using Kirchhoff's rules, find the current in each resistor shown in Figure P18.23 and (b) find the potential difference between points *c* and *f*.

FIGURE P18.23

24. Two 1.50-V batteries—with their positive terminals in the same direction—are inserted in series into the barrel of a flashlight. One battery has an internal resistance of 0.255 Ω, the other an internal resistance of 0.153 Ω. When the switch is closed, a current of 0.600 A passes through the lamp. (a) What is the lamp's resistance? (b) What fraction of the power dissipated is dissipated in the batteries?

25. Calculate each of the unknown currents $I_1$, $I_2$, and $I_3$ for the circuit of Figure P18.25.

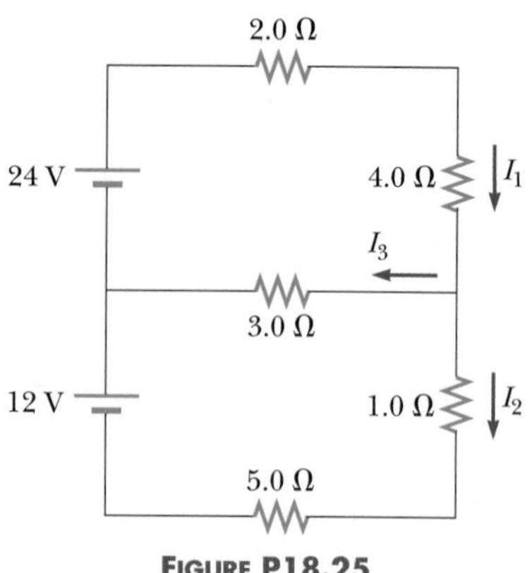

FIGURE P18.25

**26.** A dead battery is charged by connecting it to the live battery of another car with jumper cables (Fig. P18.26). Determine the current in the starter and in the dead battery.

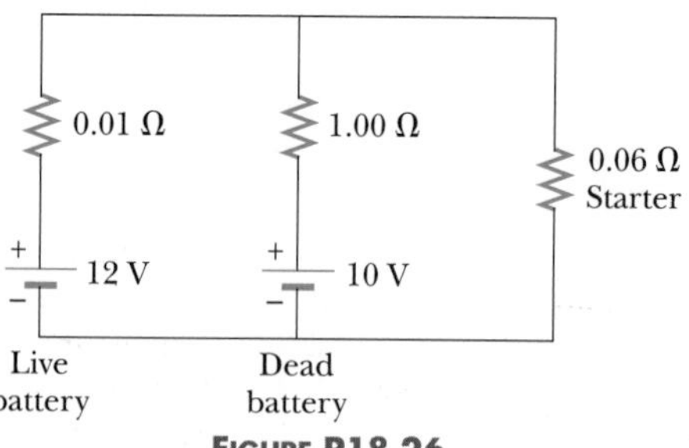

FIGURE P18.26

27. Find the current in each resistor in Figure P18.27.

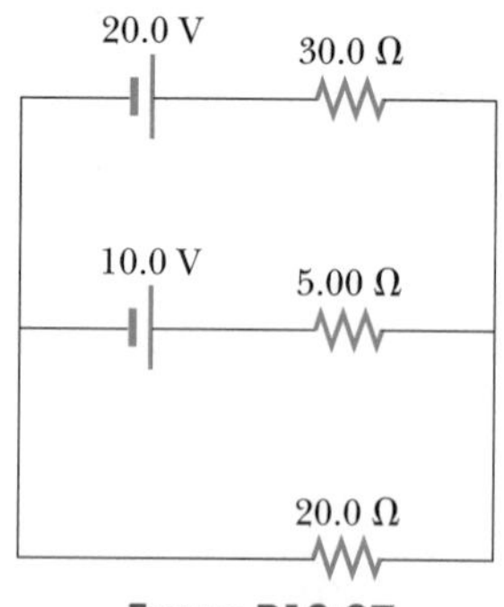

FIGURE P18.27

28. (a) Determine the potential difference $\Delta V_{ab}$ for the circuit in Figure P18.28. Note that each battery has an internal resistance as indicated in the figure. (b) If points *a* and *b* are connected by a 7.0-Ω resistor, what is the current through this resistor?

29. Find the potential difference across each resistor in Figure P18.29.

## Section 18.5 *RC* Circuits

**30.** Show that $\tau = RC$ has units of time.

**31.** Consider a series *RC* circuit for which $C = 6.0\ \mu\text{F}$, $R = 2.0 \times 10^6\ \Omega$, and $\mathcal{E} = 20$ V. Find (a) the time con-

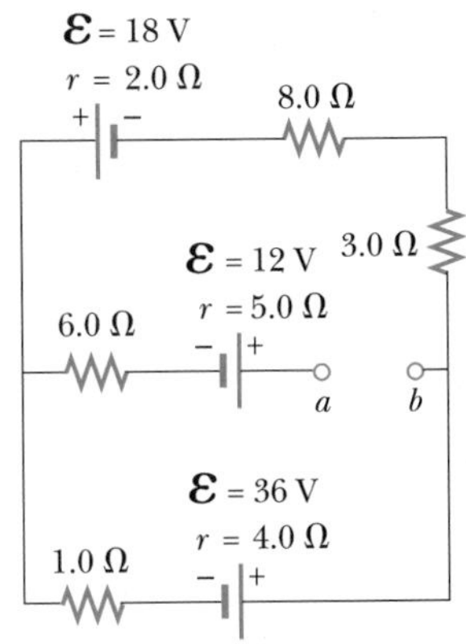


**FIGURE P18.28**

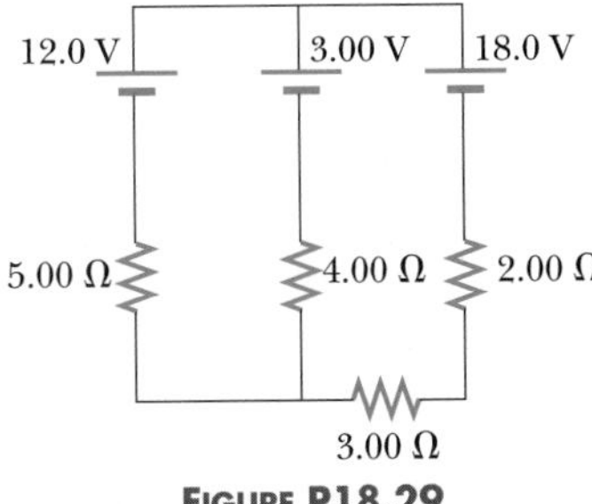


**FIGURE P18.29**

stant of the circuit and (b) the maximum charge on the capacitor after a switch in the circuit is closed.

32. An uncharged capacitor and a resistor are connected in series to a source of emf. If $\mathcal{E} = 9.00$ V, $C = 20.0\ \mu$F, and $R = 100\ \Omega$, find (a) the time constant of the circuit, (b) the maximum charge on the capacitor, and (c) the charge on the capacitor after one time constant.

33. Consider a series *RC* circuit for which $R = 1.0$ MΩ, $C = 5.0\ \mu$F, and $\mathcal{E} = 30$ V. The capacitor is initially uncharged when the switch is open. Find the charge on the capacitor 10 s after the switch is closed.
web

34. A series combination of a 12-kΩ resistor and an unknown capacitor is connected to a 12-V battery. One second after the circuit is completed, the voltage across the capacitor is 10 V. Determine the capacitance.

35. A capacitor in an *RC* circuit is charged to 60.0% of its maximum value in 0.900 s. What is the time constant of the circuit?

36. A series *RC* circuit has a time constant of 0.960 s. The battery has an emf of 48.0 V, and the maximum current in the circuit is 0.500 mA. What are (a) the value of the capacitance and (b) the charge stored in the capacitor 1.92 s after the switch is closed?

## Section 18.6 Household Circuits

37. An electric heater is rated at 1 300 W, a toaster is rated at 1 000 W, and an electric grill is rated at 1 500 W. The three appliances are connected in parallel to a common 120-V circuit. (a) How much current does each appliance draw? (b) Is a 30.0-A circuit breaker sufficient in this situation? Explain.

38. A lamp ($R = 150\ \Omega$), an electric heater ($R = 25\ \Omega$), and a fan ($R = 50\ \Omega$) are connected in parallel across a 120-V line. (a) What total current is supplied to the circuit? (b) What is the voltage across the fan? (c) What is the current in the lamp? (d) What power is expended in the heater?

39. A heating element in a stove is designed to dissipate 3 000 W when connected to 240 V. (a) Assuming that the resistance is constant, calculate the current in this element if it is connected to 120 V. (b) Calculate the power it dissipates at this voltage.

40. Your toaster oven and coffeemaker each dissipate 1 200 W of power. Can you operate them together if the 120-V line that feeds them has a circuit breaker rated at 15 A? Explain.

## Section 18.8 Conduction of Electrical Signals by Neurons

41. Assume that a length of axon membrane of about 10 cm is excited by an action potential. (Length excited = nerve speed × pulse duration = 50 m/s × 2.0 ms = 10 cm.) In the resting state, the outer surface of the axon wall is charged positively with $K^+$ ions and the inner wall has an equal and opposite charge of negative organic ions as shown in Figure P18.41. Model the axon as a parallel plate capacitor and use $C = \kappa\epsilon_0 A/d$ and $Q = C\Delta V$ to investigate the charge as follows. Use typical values for a cylindrical axon of cell wall thickness $d = 1.0 \times 10^{-8}$ m, axon radius $r = 10\ \mu$m, and cell wall dielectric constant $\kappa = 3.0$. (a) Calculate the positive charge on the outside of a 10-cm piece of axon when it is not conducting an electric pulse. How many $K^+$ ions are on the outside of the axon? Is this a large charge per unit area? [*Hint:* Calculate the charge per unit area in terms of the number of square angstroms (Å$^2$) per electronic charge. An atom has a cross section of about 1 Å$^2$ (1 Å = $10^{-10}$ m).] (b) How much positive charge must flow through the cell membrane to reach the excited state of +30 mV from the resting state of −70 mV? How many sodium ions is this? (c) If it takes 2.0 ms for the $Na^+$ ions to enter the axon, what is the average current in the axon wall in this process? (d) How much energy does it take to raise the potential of the inner axon wall to +30 mV starting from the resting potential of −70 mV?

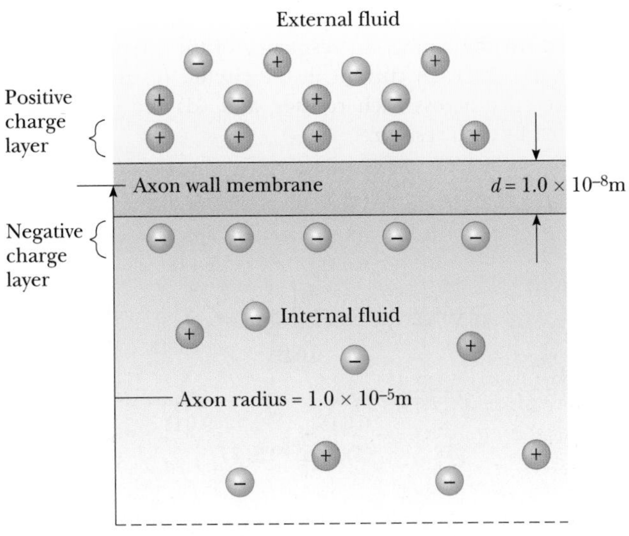


**FIGURE P18.41** (Problems 41 and 42)

42. Continuing with the model of the axon as a capacitor from Problem 41 and Figure P18.41, (a) how much energy does it take to restore the inner wall of the axon to − 70 mV starting from +30 mV? (b) Find the average current in the axon wall during this process, if it takes 3.0 ms.

43. Using Figure 18.27b and the results of Problem 18.41d and Problem 18.42a, find the average power supplied by the axon during firing and recovery.

## ADDITIONAL PROBLEMS

44. Consider an $RC$ circuit in which the capacitor is being charged by a battery connected in the circuit. After a time equal to two time constants, what percent of the *final* charge is present on the capacitor?

45. Find the equivalent resistance between points $a$ and $b$ in Figure P18.45.

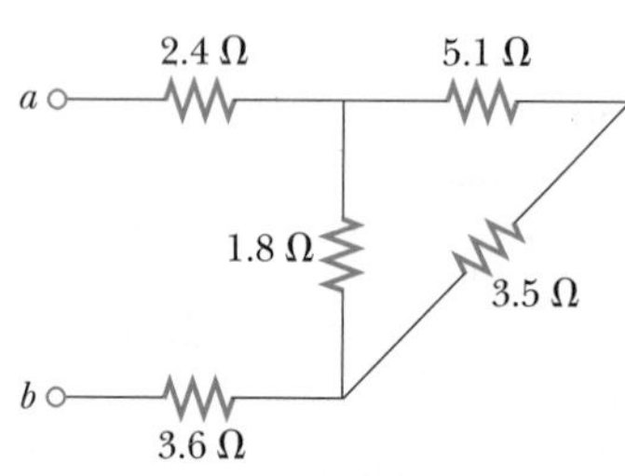


**FIGURE P18.45**

46. For the circuit in Figure P18.46, calculate (a) the equivalent resistance of the circuit and (b) the power dissipated by the entire circuit. (c) Find the current in the 5.0-Ω resistor.

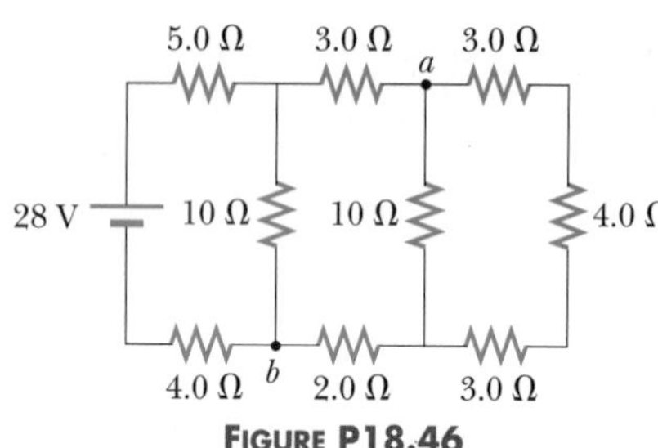


**FIGURE P18.46**

47. Find (a) the equivalent resistance of the circuit in Figure P18.47, (b) each current in the circuit, (c) the potential difference across each resistor, and (d) the power dissipated by each resistor.

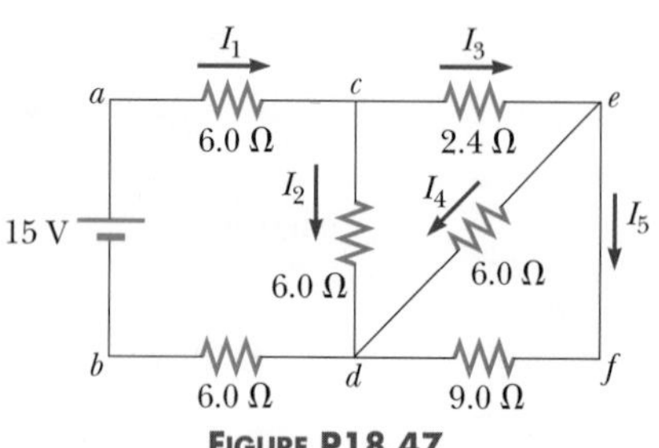


**FIGURE P18.47**

48. Three 60.0-W, 120-V lightbulbs are connected across a 120-V power source, as shown in Figure P18.48. Find (a) the total power delivered to the three bulbs and (b) the potential difference across each. Assume that the resistance of each bulb is constant (even though in reality the resistance increases markedly with current).

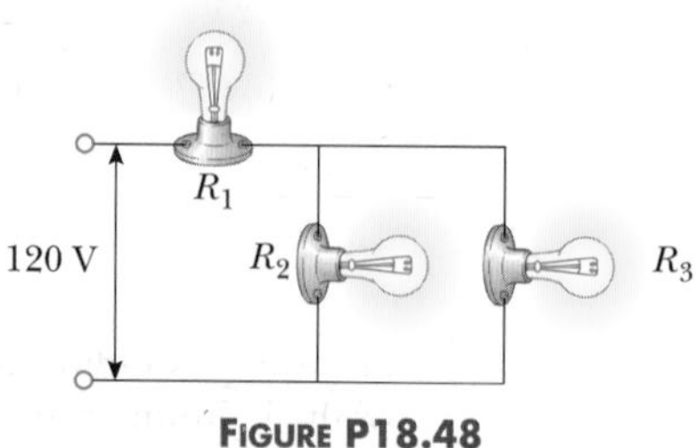


**FIGURE P18.48**

49. web An automobile battery has an emf of 12.6 V and an internal resistance of 0.080 Ω. The headlights have total resistance of 5.00 Ω (assumed constant). What is the potential difference across the headlight bulbs (a) when they are the only load on the battery and (b) when the starter motor is operated, taking an additional 35.0 A from the battery?

50. In Figure P18.50, suppose that the switch has been closed for a length of time sufficiently long for the capacitor to become fully charged. (a) Find the steady-state current in each resistor and (b) find the charge on the capacitor.

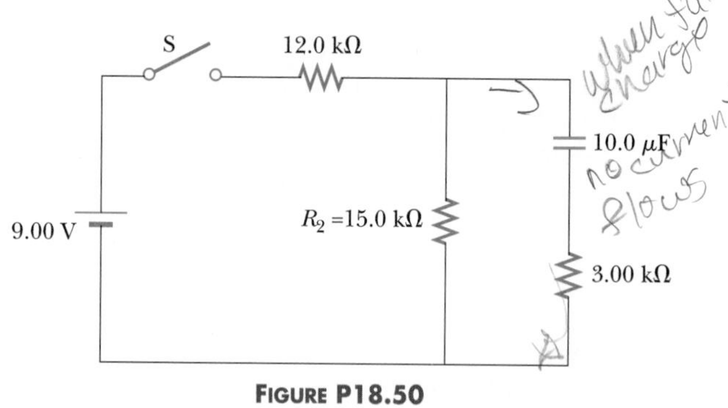


**FIGURE P18.50**

51. Find the values of $I_1$, $I_2$, and $I_3$ for the circuit in Figure P18.51.

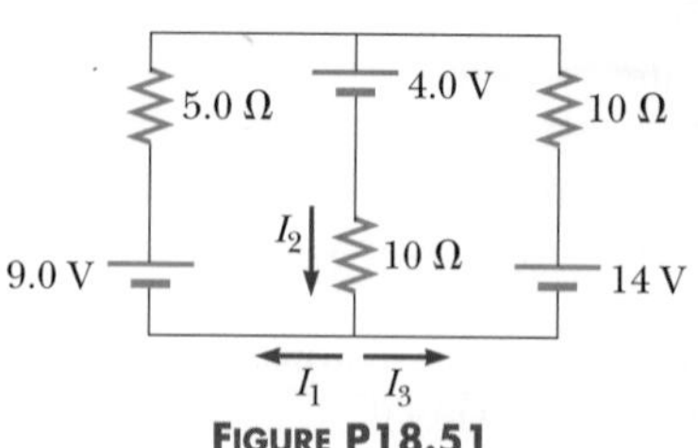


**FIGURE P18.51**

52. The resistance between points $a$ and $b$ in Figure P18.52 drops to one half its original value when switch S is closed. Determine the value of $R$.

53. A generator has a terminal voltage of 110 V when it delivers 10.0 A, and 106 V when it delivers 30.0 A. Calculate the emf and the internal resistance of the generator.

54. An emf of 10 V is connected to a series $RC$ circuit consisting of a resistor of $2.0 \times 10^6$ Ω and a capacitor of 3.0 μF. Find the time required for the charge on the capacitor to reach 90% of its final value.

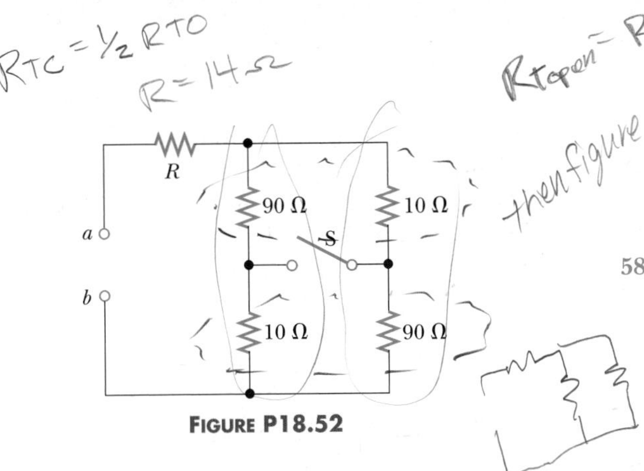


FIGURE P18.52

55. The student engineer of a campus radio station wishes to verify the effectiveness of the lightning rod on the antenna mast (Fig. P18.55). The unknown resistance $R_x$ is between points $C$ and $E$. Point $E$ is a "true ground" but is inaccessible for direct measurement since this stratum is several meters below the Earth's surface. Two identical rods are driven into the ground at $A$ and $B$, introducing an unknown resistance $R_y$. The procedure is as follows: measure resistance $R_1$ between points $A$ and $B$, then connect $A$ and $B$ with a heavy conducting wire and measure resistance $R_2$ between points $A$ and $C$. (a) Derive a formula for $R_x$ in terms of the observable resistances $R_1$ and $R_2$. (b) A satisfactory ground resistance would be $R_x < 2.0\ \Omega$. Is the grounding of the station adequate if measurements give $R_1 = 13\ \Omega$ and $R_2 = 6.0\ \Omega$?

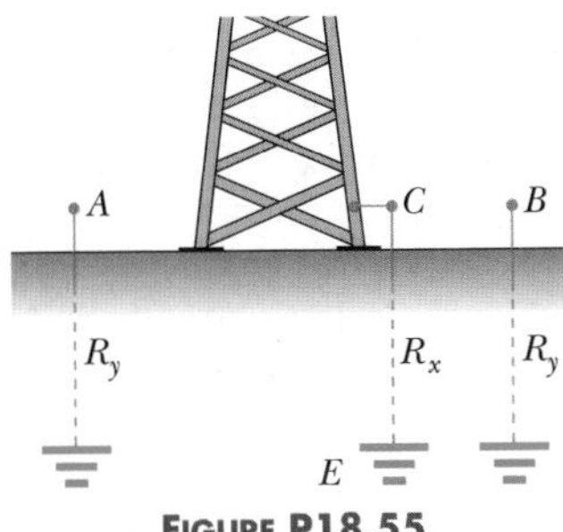


FIGURE P18.55

56. The resistor $R$ in Figure P18.56 dissipates 20 W of power. Determine the value of $R$.

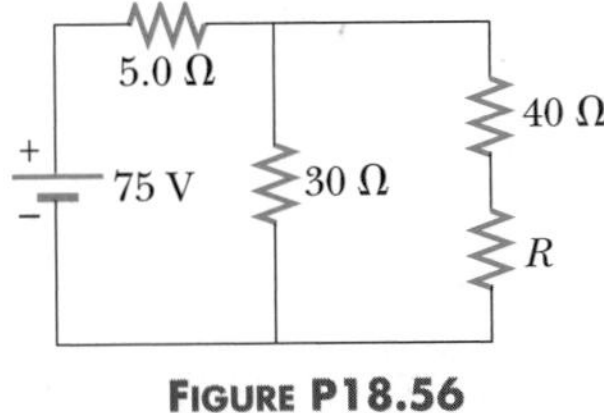


FIGURE P18.56

57. A voltage $\Delta V$ is applied to a series configuration of $n$ resistors, each of value $R$. The circuit components are reconnected in a parallel configuration, and voltage $\Delta V$ is again applied. Show that the power consumed by the series configuration is $1/n^2$ times the power consumed by the parallel configuration.

web

58. For the network in Figure P18.58, show that the resistance between points $a$ and $b$ is $R_{ab} = \frac{27}{17}\ \Omega$. (*Hint:* Connect a battery with emf $\mathcal{E}$ across points $a$ and $b$ and determine $\mathcal{E}/I$, where $I$ is the current in the battery.)

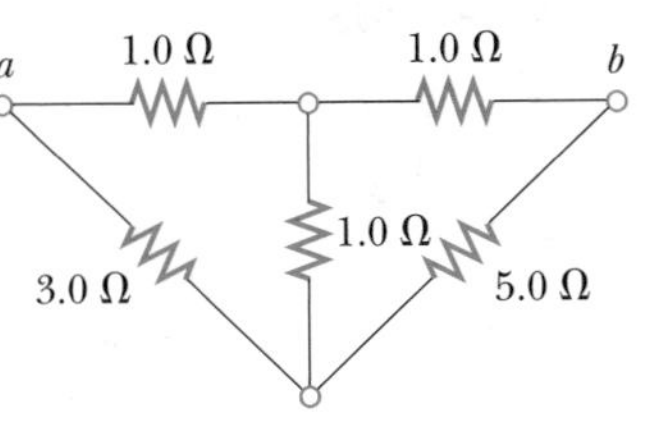


FIGURE P18.58

59. A battery with an internal resistance of 10.0 Ω produces an open-circuit voltage of 12.0 V. A variable load resistance with a range of 0 to 30.0 Ω is connected across the battery. (*Note:* A battery has a resistance that depends on the condition of its chemicals and increases as the battery ages. This internal resistance can be represented in a simple circuit diagram as a resistor in series with the battery.) (a) Graph the power dissipated in the load resistor as a function of the load resistance. (b) With your graph, demonstrate the following important theorem: *the power delivered to a load is a maximum if the load resistance equals the internal resistance of the source.*

60. The circuit in Figure P18.60 contains two resistors, $R_1 = 2.0\ \text{k}\Omega$ and $R_2 = 3.0\ \text{k}\Omega$, and two capacitors, $C_1 = 2.0\ \mu\text{F}$ and $C_2 = 3.0\ \mu\text{F}$, connected to a battery with emf $\mathcal{E} = 120$ V. If there are no charges on the capacitors before switch S is closed, determine as functions of time the charges $q_1$ and $q_2$ on capacitors $C_1$ and $C_2$, respectively, after the switch is closed. (*Hint:* First reconstruct the circuit so that it becomes a simple *RC* circuit containing a single resistor and single capacitor in series, connected to the battery, and then determine the total charge $q$ stored in the circuit.)

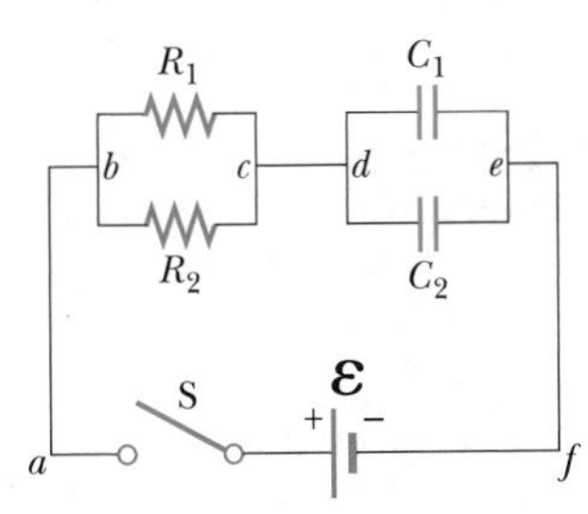


FIGURE P18.60

## GROUP ACTIVITIES

G.1 Connect one terminal of a D-cell battery to the base of a flashlight bulb using insulated wire, tape a second wire to the other battery terminal, and tape a third wire to the center conductor of the bulb as shown in the Figure GA18.1a. Make sure to remove about 1 cm of insulation from the ends of all wires before making connections. Connect the two open wires together to complete the circuit, and note the illumination of the bulb. Now add a second D-cell battery to the circuit as in Figure GA18.1b to give a total voltage of 3.0 V, connect the two open wires together to complete the circuit, and note the illumination of the bulb. Why does the bulb grow brighter in this case?

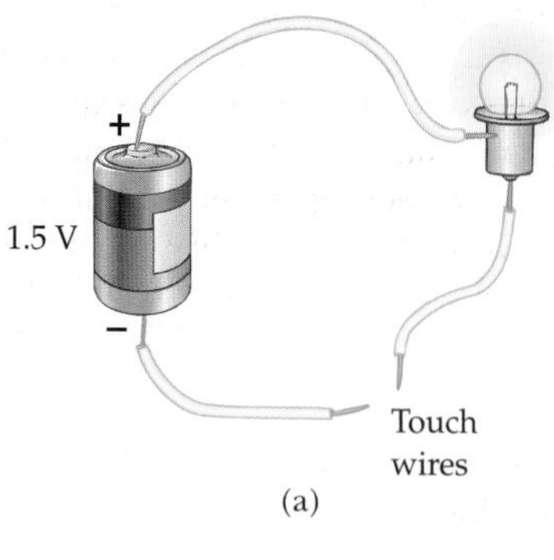

(a)

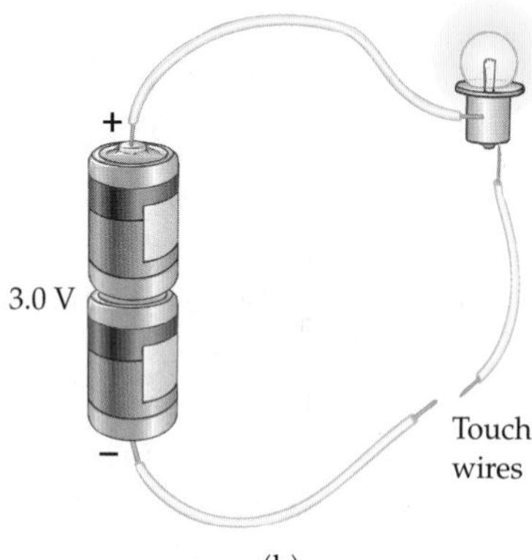

(b)

**FIGURE GA18.1**

**G.2** Use the basic equipment of activity 1 plus a few more items to test some additional features of circuits. First, note the brightness of a single bulb connected to the battery. Now connect two bulbs in series with each other and the battery. Predict whether the bulbs will be dimmer or brighter than when operated separately. Try it and see. Continue for three bulbs in series.

**G.3** Repeat activity 2, but in this activity you must connect the bulbs in parallel with each other. Predict how the brightness will change as you add more bulbs in parallel. Why?

**G.4** Continue your experimentation with circuits by connecting the battery to one bulb followed in the same circuit by two bulbs in parallel then back to the battery. Predict the brightness of each bulb in this situation before you connect the circuit. Explain the results you obtain.

**G.5** (a) A group of resistors connected in parallel have the same (i) current in them, (ii) potential difference across them, or (iii) neither of the above. (b) A group of resistors connected in series have the same (i) current in them, (ii) potential difference across them, or (iii) neither of the above. (c) The equivalent resistance for a group of resistors connected in parallel is (i) greater than any of the resistors in the group, (ii) less than any of the resistors in the group, or (iii) neither of the above. (d) The equivalent resistance of a group of resistors connected in series is (i) greater than any of the resistors in the group, (ii) less than any of the resistors in the group, or (iii) neither of the above. (e) Defend your answers by examining a circuit consisting of a 3.0-Ω resistor and a 5.0-Ω resistor connected across a 12-V battery. Be sure that the answers you find are consistent with your answers above.

**G.6** (a) Two resistors are connected in series. The power converted by each resistor is (i) the same, or (ii) not necessarily the same. (b) Two resistors are connected in parallel. The power converted by each resistor is (i) the same, or (ii) not necessarily the same. (c) Suppose two resistors of 100 Ω and 200 Ω are connected to a 10.0-V power supply. Find the power converted by the two resistors in the configurations of parts (a) and (b). Are your answers consistent with your selections in parts (a) and (b)?

**G.7** In this activity, select the correct answer from within the parentheses. (a) You traverse a 20-Ω resistor in a circuit such that you are moving in the same direction as a current of 2.0 A. In Kirchhoff's loop rule, the potential difference is (− 40 V, 40 V). (b) You traverse the same resistor moving against the current. The potential difference encountered is (− 40 V, 40 V). (c) You traverse a 12-V battery such that you are moving from the negative terminal to the positive terminal. In Kirchhoff's loop rule the potential difference is considered to be (12 V, − 12 V). (d) You traverse the same battery moving from the positive terminal to the negative terminal. The potential difference encountered is (12 V, − 12 V). (e) When working a Kirchhoff's rules problem, if you make an incorrect guess for the direction of the current in a branch, the final answer found for that current will be (positive, negative, meaningless).

**G.8** Consider the two arrangements of batteries and bulbs shown in Figure GA18.8. All four bulbs are identical and have resistance $R$, and the two batteries are identical with output voltage $\Delta V$. Answer the following questions and explain your answer. (a) In case 1, which bulb is brighter? (b) In case 1, what is the current in each bulb and the power supplied to each bulb? (c) In case 2, which bulb is brighter? (d) In case 2, what is the current in each bulb, the voltage drop across each bulb, and the power delivered to each bulb? (e) Which bulbs are brighter, those in case 1 or those in case 2? (f) If one bulb in each case fails, will the other go out as well? If the other bulb doesn't fail, will it get brighter or stay the same?

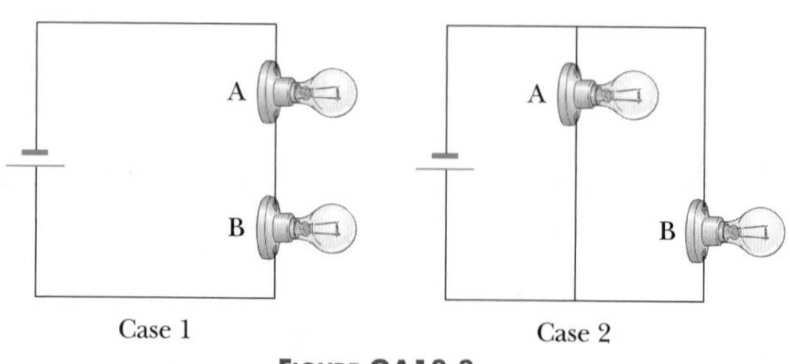

**FIGURE GA18.8**

**G.9** In Figure GA18.9, some bulbs and batteries are connected together in different ways. The bulbs are all identical. For each case, give the letter that tells which of the bulbs will be brighter. If two or more will be equal in brightness, give the letters of all the bulbs that will be of the greatest brightness.

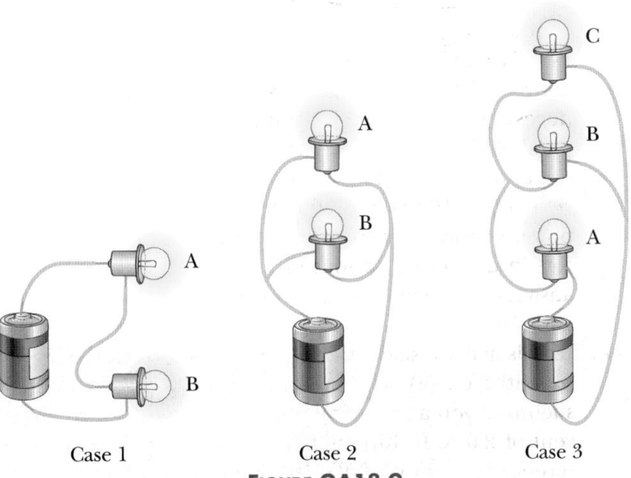


**FIGURE GA18.9**

**G.10** (a) All of the bulbs in Figure GA18.10a have the same resistance $R$. If bulb B is removed from the circuit, what happens to the current in bulb A, bulb D, and the battery? Indicate whether it increases, decreases or remains the same. For each case explain your reasoning. (b) A wire is added to the circuit as in Figure GA18.10b. What happens to the current in bulb A, bulb D, and the battery? Indicate whether it increases, decreases, or remains the same. For each case, explain your reasoning. (c) What is the equivalent resistance of the network in Figure GA18.10c? Again, all the bulbs have the same resistance $R$. If the current in the bulb on the right is $I$, what is the current in each bulb and in the battery?

(Problems 8, 9, and 10 are courtesy of E. F. Redish. For other problems of this type, visit http://www.physics.umd.edu/perg/)

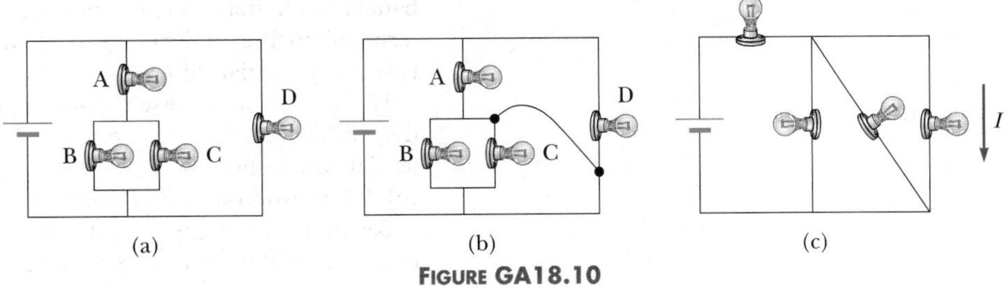


**FIGURE GA18.10**

# 19 Magnetism

## Chapter Outline



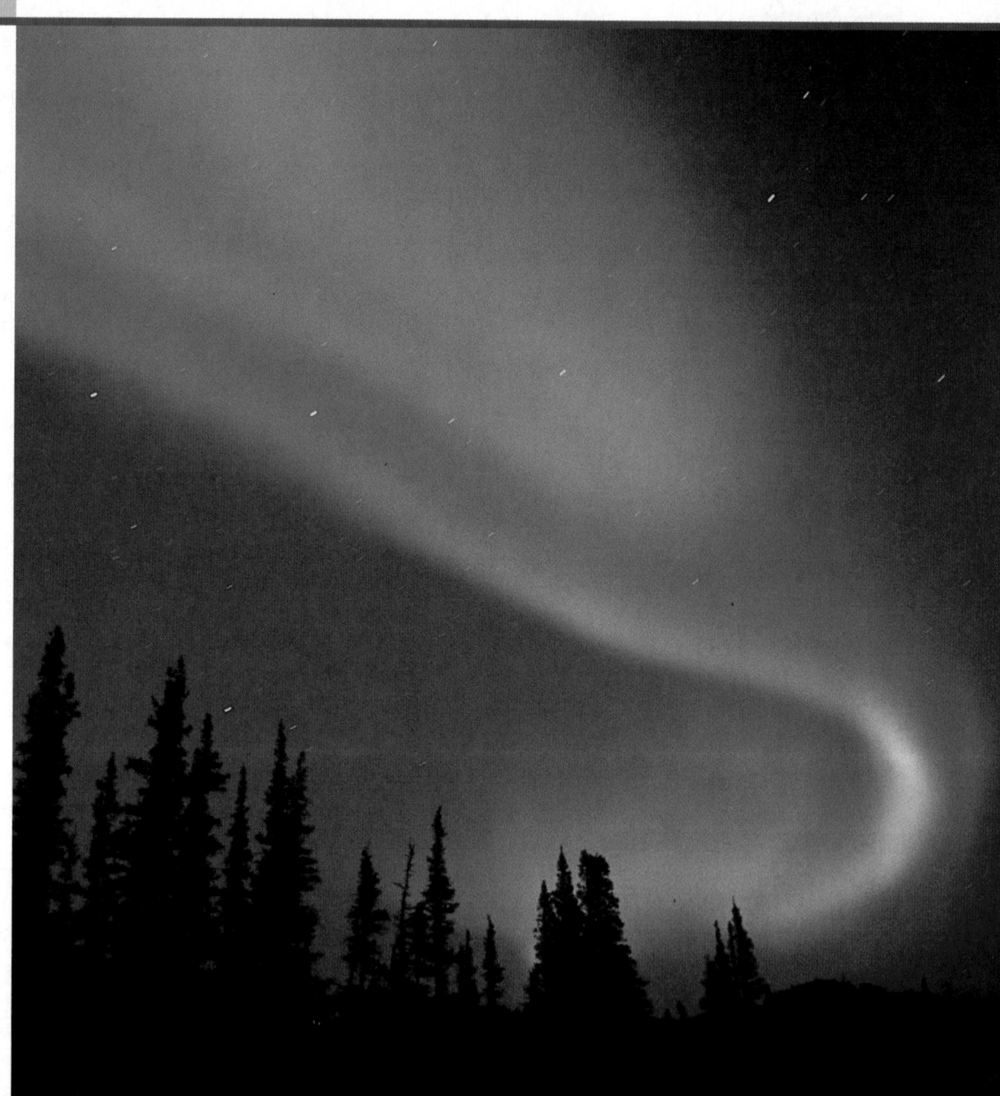

**Aurora borealis, the Northern Lights. These displays are caused by cosmic ray particles trapped in the magnetic field of Earth. When the particles collide with atoms in the atmosphere, they cause the atoms to emit visible light.** *(Johnny Johnson/Stone/Getty)*

The list of important technological applications of magnetism is very long. For instance, large electromagnets are used to pick up heavy loads. Magnets are used in such devices as meters, motors, and loudspeakers. Magnetic tapes are routinely used in sound and video recording equipment and for computer data storage. Intense magnetic fields are currently being used in magnetic resonance imaging devices (MRI) to explore the human body with better resolution and greater safety than x-rays.

As we investigate magnetism in this chapter, we shall find that this subject cannot be divorced from electricity. For example, magnetic fields affect moving charges and moving charges produce magnetic fields. The ultimate source of all magnetic fields is electric current.

## 19.1 MAGNETS

Most people have had experience with some form of magnet. You are most likely familiar with the common iron horseshoe magnet that can pick up iron-containing objects such as paper clips and nails. Several commercially available magnets are shown in Figure 19.1. In the discussion that follows, we shall assume that the magnet has the shape of a bar. Iron objects are most strongly attracted to the ends of such a bar magnet, called its **poles.** One end is called the **north pole** and the other the **south pole.** The names come from the behavior of a magnet in the presence of Earth's magnetic field. If a bar magnet is suspended from its midpoint by a piece of string so that it can swing freely in a horizontal plane, it will rotate until its north pole points to the north and its south pole points to the south. The same idea is used to construct a simple compass. Magnetic poles also exert attractive or repulsive forces on each other similar to the electrical forces between charged objects. In fact, simple experiments with two bar magnets show that **like poles repel each other and unlike poles attract each other.** Although the force between two magnetic poles is similar to the force between two electric charges, there is an important difference between electric and magnetic phenomena. Electric charges can be isolated (witness the proton and the electron), but magnetic poles cannot be isolated. In fact, no matter how many times a permanent magnet is cut, each piece always has a north pole and a south pole. Thus, magnetic poles always occur in north-south pairs. There is some theoretical basis for the speculation that magnetic monopoles (isolated north or south poles) may exist in nature, and attempts to detect them are currently an active experimental field of investigation. However, none of these attempts has yet proven successful.

**FIGURE 19.1** An assortment of commercially available magnets. The four red magnets and the large black magnet on the left are made of an alloy of iron, aluminum, and cobalt. The six horseshoe magnets on the right are made of different nickel alloy steels. The rectangular magnets on the lower right are ceramics made of iron, nickel, and beryllium oxides. *(Courtesy of Central Scientific Company)*

Another similarity between electric and magnetic effects concerns methods for making magnets. In Chapter 15 we learned that when two materials such as rubber and wool are rubbed together, each becomes charged, one positively and the other negatively. In a somewhat analogous fashion, an unmagnetized piece of iron can be magnetized by stroking it with a magnet. Magnetism can also be induced in iron (and other materials) by other means. For example, if a piece of unmagnetized iron is placed near a strong permanent magnet, the piece of iron eventually becomes magnetized. The process can be accelerated by heating and cooling the iron. Naturally occurring magnetic materials, such as magnetite, achieve their magnetism in this manner, because they have been subjected to Earth's magnetic field over very long periods of time. The extent to which a piece of material retains its magnetism depends on whether it is classified as being magnetically hard or soft. **Soft** magnetic materials, such as iron, are easily magnetized but also tend to lose their magnetism easily. In contrast, **hard** magnetic materials such as cobalt and nickel are difficult to magnetize but tend to retain their magnetism.

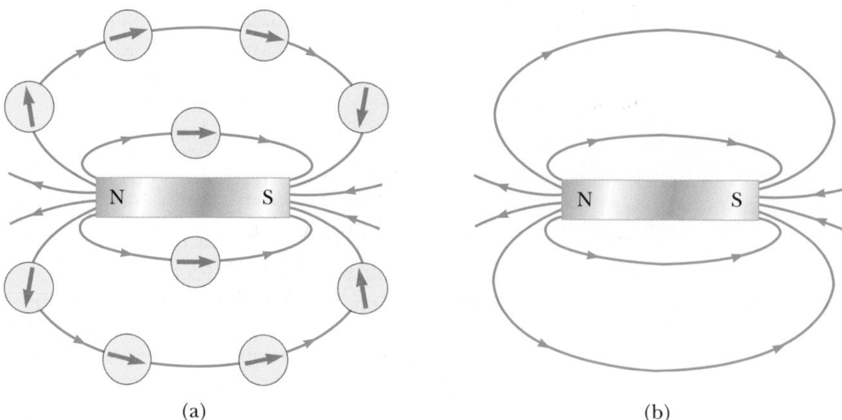


**FIGURE 19.2** (a) Tracing the magnetic field of a bar magnet. (b) Several magnetic field lines of a bar magnet.

In earlier chapters we described the interaction between charged objects in terms of electric fields. Recall that an electric field surrounds any stationary electric charge. The region of space surrounding a *moving* charge also includes a magnetic field. A magnetic field also surrounds any magnetic material.

To describe any type of vector field, we must define its magnitude, or strength, and its direction. The direction of the magnetic field **B** at any location is the direction in which the north pole of a compass needle points at that location. Figure 19.2a shows how the magnetic field of a bar magnet can be traced with the aid of a compass, defining a **magnetic field line.** Several magnetic field lines of a bar magnet traced out in this manner appear in the two-dimensional representation in Figure 19.2b. Magnetic field patterns can be displayed by small iron filings placed in the vicinity of a magnet, as in Figure 19.3.

**APPLICATION**

DUSTING FOR FINGERPRINTS

Forensic scientists use a technique similar to that shown in Figure 19.3 to find fingerprints at a crime scene. One way to find latent, or invisible, prints is by sprinkling a powder of iron dust on a surface. The iron adheres to perspiration or body oils present and can be spread around on the surface with a magnetic brush that never comes into contact with the powder or the surface.

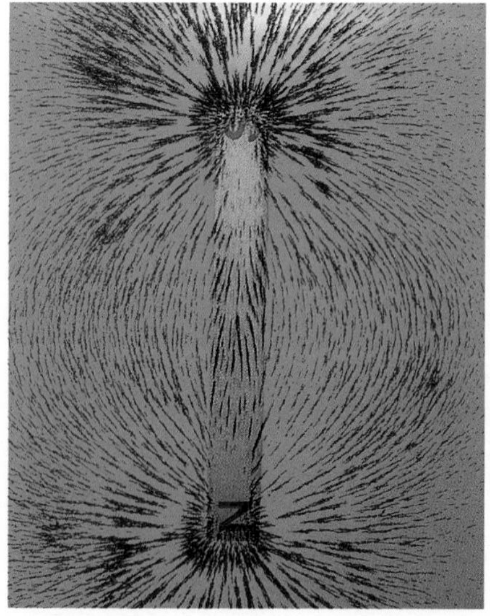

(a)

(b)

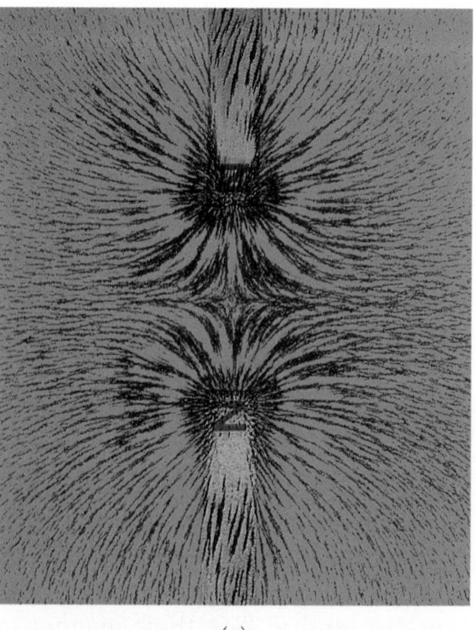

(c)

**FIGURE 19.3** (a) The magnetic field pattern of a bar magnet, displayed by iron filings on a sheet of paper. (b) The magnetic field pattern between *unlike* poles of two bar magnets, displayed by iron filings. (c) The magnetic field pattern between two *like* poles. *(Courtesy of Henry Leap and Jim Lehman)*

## 19.2 EARTH'S MAGNETIC FIELD

When we speak of a small bar magnet as having north and south poles, we should more properly say that it has a "north-seeking" pole and a "south-seeking" pole. By this we mean that if such a magnet is used as a compass, one end will seek, or point to, the geographic North Pole of the Earth. Thus, we conclude that **Earth's geographic North Pole corresponds to a magnetic south pole, and its geographic South Pole corresponds to a magnetic north pole.** In fact, the configuration of Earth's magnetic field, pictured in Figure 19.4, very much resembles what would be achieved by burying a huge bar magnet deep in Earth's interior.

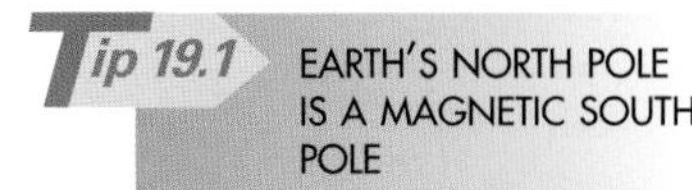


Don't be confused because the north pole of a magnet points to the geographic North Pole of Earth. Earth's geographic North Pole is really a magnetic south pole.

If a compass needle is suspended in bearings that allow it to rotate in the vertical plane as well as in the horizontal plane, the needle is horizontal with respect to Earth's surface only near the Equator. As the device is moved northward, the needle rotates so that it points more and more toward Earth's surface. The angle between the direction of the magnetic field and the horizontal is called the **dip angle.** Finally, at a point just north of Hudson Bay in Canada, the north pole of the needle points directly downward, and the dip angle is 90°. This site, first found in 1832, is considered to be the location of Earth's south magnetic pole. It is approximately 1 300 mi from Earth's geographic North Pole and its position varies with time. Similarly, Earth's magnetic north pole is about 1 200 miles from its geographic South Pole. Thus, it is only approximately correct to say that a compass needle points north. The difference between true north, defined as the geographic North Pole, and north indicated by a compass varies from point to point on Earth, and the difference is referred to as *magnetic declination.* For example, along a line through South Carolina and the Great Lakes, a compass indicates true north, whereas in Washington state it aligns 25° east of true north (Fig. 19.5).

Although Earth's magnetic field pattern is similar to the pattern that would be set up by a bar magnet placed at its center, we can easily explain why the source of Earth's field cannot be large masses of permanently magnetized material. Earth does have large deposits of iron ore deep beneath its surface, but the high temperatures in its core prevent the iron from retaining any permanent magnetization. The most likely source of Earth's magnetic field is believed to be electric current in the liquid part of its core. This current is not well understood but is thought to be driven by an interaction between Earth's rotation and convective motion in the hot liquid core. There is some evidence that the strength of a planet's magnetic field is related to the planet's rate of rotation. For example, Jupiter rotates faster

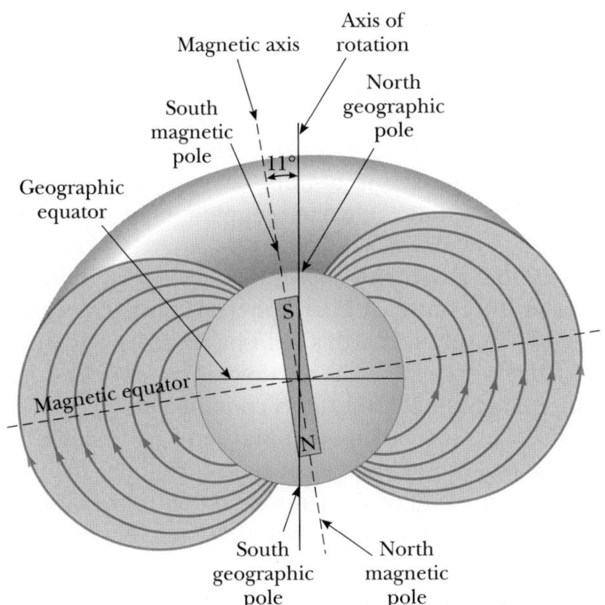


**FIGURE 19.4** Earth's magnetic field lines. Note that magnetic south is at the north geographic pole, and magnetic north is at the south geographic pole.

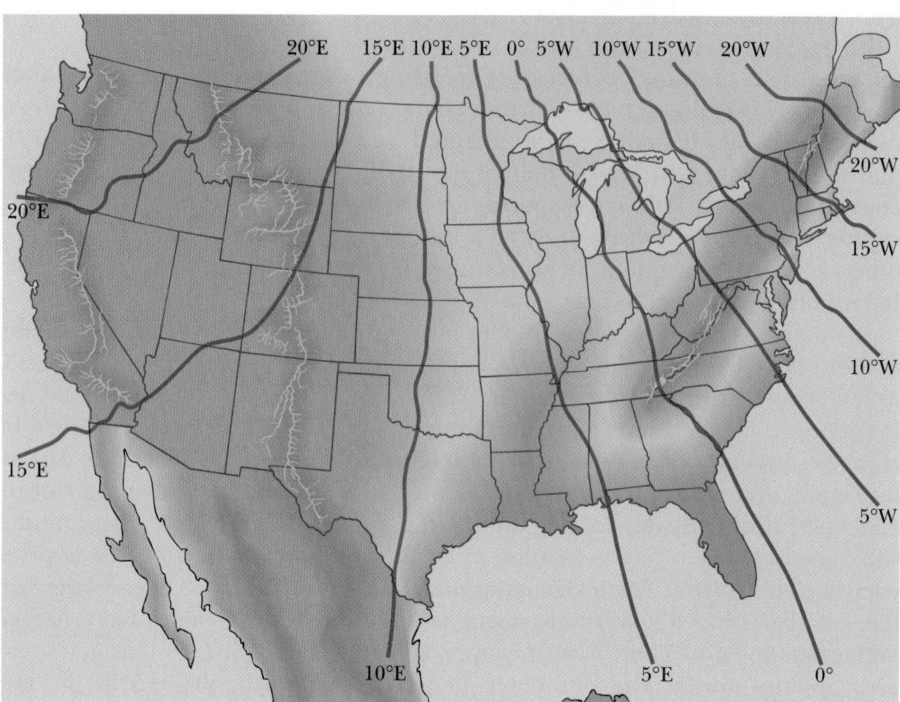


**FIGURE 19.5** A map of the lower forty-eight United States showing the declination of a compass from true north.

## Webnote 19.1

Bacteria that have their own built-in directional magnetic sensors are described as magnetotactic. Learn more about these creatures that existed long before humans navigated Earth with compasses.

*http://www.wpi.edu/~ppham/bacteria.html*

*http://www.calpoly.edu/~rfrankel/mtbcalpoly.html*

Tips for collecting your own sample of magnetotactic bacteria can be found at

*http://www.home.duq.edu/~stolz/magbachunt.html*

than Earth, and recent space probes indicate that Jupiter's magnetic field is stronger than ours even though Jupiter lacks an iron core. Venus, on the other hand, rotates more slowly than Earth, and its magnetic field is found to be weaker. Investigation into the cause of Earth's magnetism continues.

An interesting sidelight concerning Earth's magnetic field is that the field direction reverses every few million years. Evidence for this is provided by basalt (an iron-containing rock) that is sometimes spewed forth by volcanic activity on the ocean floor. As the lava cools, it solidifies and retains a picture of Earth's magnetic field direction. When the basalt deposits are dated, they provide evidence for periodic reversals of the magnetic field. The origin of these field reversals is still not understood.

MAGNETIC BACTERIA

It has long been speculated that some animals, such as birds, use Earth's magnetic field to guide their migrations. Studies have shown that a type of anaerobic bacterium that lives in swamps has a magnetized chain of magnetite as part of its internal structure. (The term *anaerobic* means that these bacteria live and grow without oxygen; in fact, oxygen is toxic to them.) The magnetized chain acts as a compass needle that enables the bacteria to align with Earth's magnetic field. When they find themselves out of the mud on the bottom of the swamp, they return to their oxygen-free environment by following the magnetic field lines. Further evidence for their magnetic sensing ability is the fact that bacteria found in the Northern Hemisphere have internal magnetite chains that are opposite in polarity to those of similar bacteria in the Southern Hemisphere. This is consistent with the fact that in the Northern Hemisphere Earth's field has a downward component, whereas in the Southern Hemisphere it has an upward component. Recently, a meteorite originating on Mars has been found to contain a chain of magnetite. NASA scientists believe it may be a fossil of ancient Martian bacterial life.

APPLICATION

LABELING AIRPORT RUNWAYS

Earth's magnetic field is used to label runways at airports according to their direction. A large number is painted on the end of the runway so that it can be read by the pilot of an incoming airplane. This number describes the direction in which the airplane is traveling, expressed as the magnetic heading, in degrees measured clockwise from magnetic north divided by 10. Thus, a runway marked 9 would be directed toward the east (90° divided by 10), and one marked 18 would be directed toward the magnetic south.

**APPLYING PHYSICS 19.1**

On a business trip to Australia, you take along your American-made compass that you use on camping trips. Does this compass work correctly in Australia?

**Explanation** You will have no problem using the compass in Australia. The north pole of the magnet in the compass is attracted to the south magnetic pole near the geographic North Pole, just as it is in the United States. The only difference in the magnetic field lines is that they have an upward component in Australia, whereas they have a downward component in the United States. Your compass held in a horizontal plane cannot detect this, however—it only displays the direction of the horizontal component of the magnetic field.

## 19.3 MAGNETIC FIELDS

Experiments show that a stationary charged particle does not interact with a static magnetic field. However, **when moving through a magnetic field, a charged particle experiences a magnetic force.** This force has its maximum value when the charge moves perpendicularly to the magnetic field lines, decreases in value at other angles, and becomes zero when the particle moves along the field lines. We can use these observations to describe the magnetic field.

In our discussion of electricity, we defined the electric field at some point in space as the electric force per unit charge acting on some test charge placed at that point. In a similar manner, we can describe the properties of the magnetic field **B** at some point in terms of the magnetic force exerted on a test charge at that point. Our test object is a charge $q$ moving with velocity **v**. It is found experimentally that the strength of the magnetic force on the particle is proportional to the magnitude of the charge $q$, the magnitude of the velocity **v**, the strength of the external magnetic field **B**, and the sine of the angle $\theta$ between the direction of **v** and the direction of **B**.

These observations can be summarized by writing the magnitude of the magnetic force as

$$F = qvB \sin \theta \qquad [19.1]$$

This expression is used to define the magnitude of the magnetic field as

$$B \equiv \frac{F}{qv \sin \theta} \qquad [19.2]$$

If $F$ is in newtons, $q$ in coulombs, and $v$ in meters per second, the SI unit of magnetic field is the **tesla** (T), also called the **weber** (Wb) **per square meter** (that is, $1 \text{ T} = 1 \text{ Wb/m}^2$). Thus, if a 1-C charge moves through a magnetic field of magnitude 1 T with a speed of 1 m/s perpendicularly to the field (so that $\sin \theta = 1$), the magnetic force exerted on the charge is 1 N. We can express the units of **B** as

$$[\text{B}] = \text{T} = \frac{\text{Wb}}{\text{m}^2} = \frac{\text{N}}{\text{C} \cdot \text{m/s}} = \frac{\text{N}}{\text{A} \cdot \text{m}} \qquad [19.3]$$

In practice, the cgs unit for magnetic field, the **gauss** (G), is often used. The gauss is related to the tesla through the conversion

$$1 \text{ T} = 10^4 \text{ G}$$

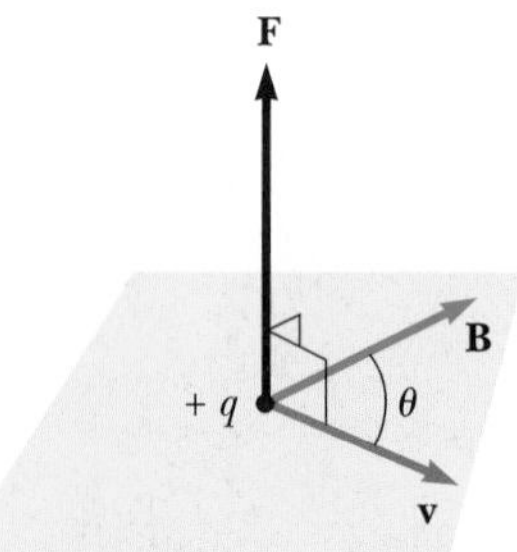


**FIGURE 19.6** The direction of the magnetic force on a positively charged particle moving with a velocity **v** in the presence of a magnetic field. When **v** is at an angle $\theta$ with **B**, the magnetic force is perpendicular to both **v** and **B**.

Conventional laboratory magnets can produce magnetic fields as large as about 25 000 G, or 2.5 T. Superconducting magnets that can generate magnetic fields as great as $3 \times 10^5$ G, or 30 T, have been constructed. These values can be compared with the value of Earth's magnetic field near its surface, which is about 0.5 G, or $0.5 \times 10^{-4}$ T.

From Equation 19.1 we see that the force on a charged particle moving in a magnetic field has its maximum value when the particle moves *perpendicularly* to the magnetic field, corresponding to $\theta = 90°$, so that $\sin\theta = 1$. The magnitude of this maximum force has the value

$$F_{\text{max}} = qvB \quad [19.4]$$

Also, note from Equation 19.1 that $F$ is zero when **v** is parallel to **B** (corresponding to $\theta = 0°$ or $180°$). Thus, no magnetic force is exerted on a charged particle when it moves in the direction of the magnetic field or opposite the field.

Experiments show that the direction of the magnetic force is always perpendicular to both **v** and **B**, as shown in Figure 19.6 for a positively charged particle. To determine the direction of the force, we employ **right-hand rule #1:**

**Hold your right hand open as illustrated in Figure 19.7, and then place your fingers in the direction of B with your thumb pointing in the direction of v. The force F on a positive charge is directed *out* of the palm of your hand.**

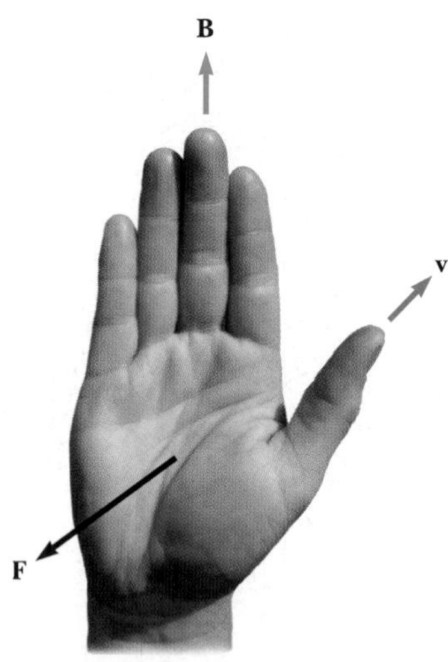


**FIGURE 19.7** Right-hand rule #1 for determining the direction of the magnetic force on a positive charge moving with a velocity **v** in a magnetic field **B**. With your thumb in the direction of **v** and your four fingers in the direction of **B**, the force is directed out of the palm of your hand.

If the charge is negative rather than positive, the force is directed *opposite* that shown in Figures 19.6 and 19.7. That is, if $q$ is negative, simply use the right-hand rule to find the direction of **F** for positive $q$, and then reverse this direction for the negative charge.

**Quick Quiz 19.1** A charged particle moves in a straight line through a certain region of space. The magnetic field in that region (a) has a magnitude of zero, (b) has a zero component perpendicular to the particle's velocity, and (c) has a zero component parallel to the particle's velocity.

**Quick Quiz 19.2** The north-pole end of a bar magnet is held near a stationary positively charged piece of plastic. Is the plastic (a) attracted, (b) repelled, or (c) unaffected by the magnet?

## Example 19.1 A Proton Traveling in Earth's Magnetic Field

A proton moves with a speed of $1.0 \times 10^5$ m/s through Earth's magnetic field, which has a value of 55 $\mu$T at a particular location. When the proton moves eastward, the magnetic force acting on it is a maximum, and when it moves northward, no magnetic force acts on it. What is the strength of the magnetic force, and what is the direction of the magnetic field?

**Solution** The magnitude of the force can be found from Equation 19.4:

$$F_{\text{max}} = qvB = (1.6 \times 10^{-19}\ \text{C})(1.0 \times 10^5\ \text{m/s})(55 \times 10^{-6}\ \text{T})$$
$$= 8.8 \times 10^{-19}\ \text{N}$$

The direction of the magnetic field cannot be determined precisely from the information given in the problem. Because no magnetic force acts on a charged particle when it is moving parallel to the field, all that we can say for sure is that the magnetic field is directed either northward or southward.

**EXERCISE** Calculate the gravitational force on the proton and compare it with the magnetic force. Note that the mass of the proton is $1.67 \times 10^{-27}$ kg.

**ANSWER** $1.6 \times 10^{-26}$ N; $F_{grav}/F_{max} = 1.9 \times 10^{-8}$

## Example 19.2 A Proton Moving in a Magnetic Field

A proton moves at $8.0 \times 10^6$ m/s along the $x$ axis. It enters a region in which there is a magnetic field of magnitude 2.5 T, directed at an angle of 60° with the $x$ axis and lying in the $xy$ plane (Fig. 19.8). Calculate the initial force on and acceleration of the proton.

**Solution** From Equation 19.1 we get

$$F = qvB \sin\theta = (1.6 \times 10^{-19}\ \text{C})(8.0 \times 10^6\ \text{m/s})(2.5\ \text{T})(\sin 60°)$$
$$= 2.8 \times 10^{-12}\ \text{N}$$

Use right-hand rule #1, noting that the charge is positive, to see that the force is in the positive $z$ direction. Verify that the units of $F$ in the calculation reduce to newtons.

Because the mass of the proton is $1.67 \times 10^{-27}$ kg, its initial acceleration is

$$a = \frac{F}{m} = \frac{2.8 \times 10^{-12}\ \text{N}}{1.67 \times 10^{-27}\ \text{kg}} = 1.7 \times 10^{15}\ \text{m/s}^2$$

in the positive $z$ direction.

**EXERCISE** Calculate the acceleration of an electron that moves through the same magnetic field at the same speed as the proton. The mass of an electron is $9.11 \times 10^{-31}$ kg.

**ANSWER** $3.0 \times 10^{18}$ m/s$^2$ in the negative $z$ direction

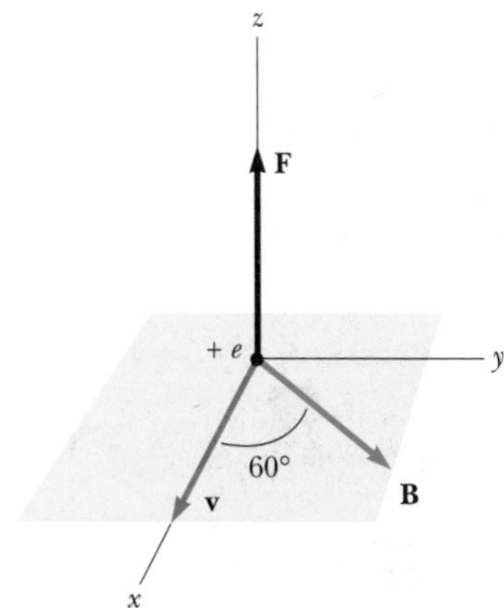


**FIGURE 19.8** (Example 19.2) The magnetic force **F** on a proton is in the positive $z$ direction when **v** and **B** lie in the $xy$ plane.

## 19.4 MAGNETIC FORCE ON A CURRENT-CARRYING CONDUCTOR

If a force is exerted on a single charged particle when it moves through a magnetic field, it should be no surprise that a current-carrying wire also experiences a force when placed in a magnetic field (see Figure 19.9). This follows from the fact that the current is a collection of many charged particles in motion; hence, the resultant force on the wire is due to the sum of the individual forces on the charged particles. The force on the particles is transmitted to the "bulk" of the wire through collisions with the atoms making up the wire.

Before we continue, some explanation is in order concerning notation in many of the figures. To indicate the direction of **B**, we use the following convention.

**If B is directed *into* the page, as in Figure 19.10, we use a series of blue crosses, representing the tails of arrows. If B is directed *out* of the page, we use a series of blue dots, representing the tips of arrows. If B lies in the plane of the page, we use a series of blue field lines with arrowheads.**

The force on a current-carrying conductor can be demonstrated by hanging a wire between the poles of a magnet, as in Figure 19.10. In this figure, the magnetic

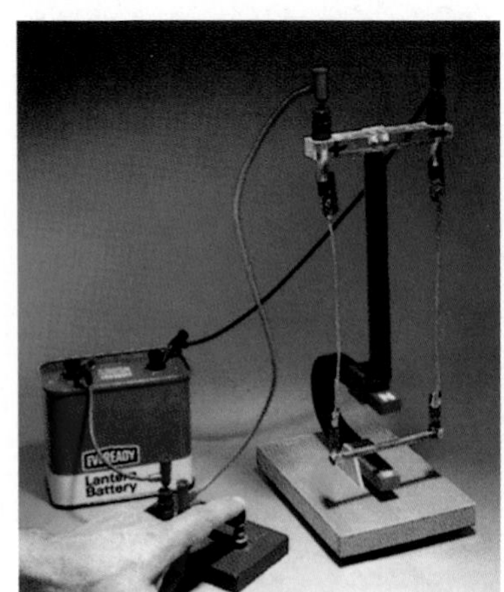

**FIGURE 19.9** This apparatus demonstrates the force on a current-carrying conductor in an external magnetic field. Why does the bar swing *away* from the magnet after the switch is closed? *(Courtesy of Henry Leap and Jim Lehman)*

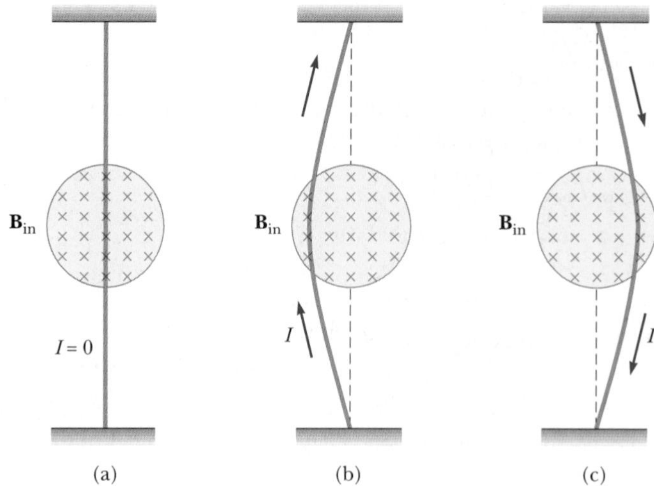


**FIGURE 19.10** A segment of a flexible vertical wire partially stretched between the poles of a magnet, with the field (blue crosses) directed into the page. (a) When there is no current in the wire, it remains vertical. (b) When the current is upward, the wire deflects to the left. (c) When the current is downward, the wire deflects to the right.

field is directed into the page and covers the region within the shaded circle. The wire deflects to the right or left when a current is passed through it.

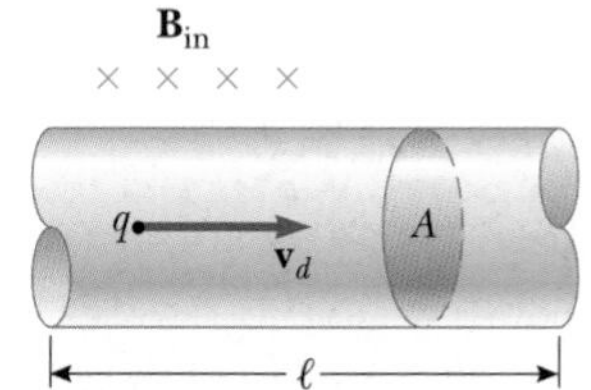


**FIGURE 19.11** A section of a wire containing moving charges in an external magnetic field **B**.

Let us quantify this discussion by considering a straight segment of wire of length $\ell$ and cross-sectional area $A$ carrying current $I$ in a uniform external magnetic field **B**, as in Figure 19.11. We assume that the magnetic field is perpendicular to the wire and is directed into the page. Each charge carrier in the wire experiences a force of magnitude $F_{max} = qv_dB$, where $v_d$ is the drift velocity of the charge. To find the total force on the wire, we multiply the force on one charge carrier by the number of carriers in the segment. Because the volume of the segment is $A\ell$, the number of carriers is $nA\ell$, where $n$ is the number of carriers per unit volume. Hence, the magnitude of the total magnetic force on the wire of length $\ell$ is

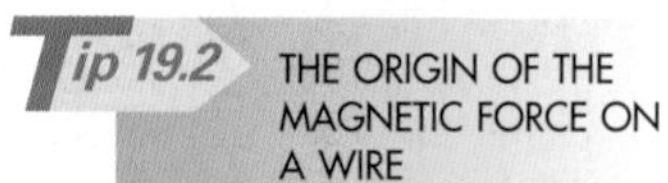

**Tip 19.2 THE ORIGIN OF THE MAGNETIC FORCE ON A WIRE**

When a magnetic field is applied at some angle to a wire carrying a current, a magnetic force is exerted on each moving charge in the wire. The total magnetic force on the wire is the sum of all the magnetic forces on the individual charges producing the current.

Total force = force on each charge carrier × total number of carriers

$$F_{max} = (qv_dB)(nA\ell)$$

From Chapter 17, however, we know that the current in the wire is given by $I = nqv_dA$. Therefore, $F_{max}$ can be expressed as

$$F_{max} = BI\ell \quad [19.5]$$

**This equation can be used *only* when the current and the magnetic field are at right angles to each other.**

If the wire is not perpendicular to the field but is at some arbitrary angle, as in Figure 19.12, the magnitude of the magnetic force on the wire is

$$F = BI\ell \sin\theta \quad [19.6]$$

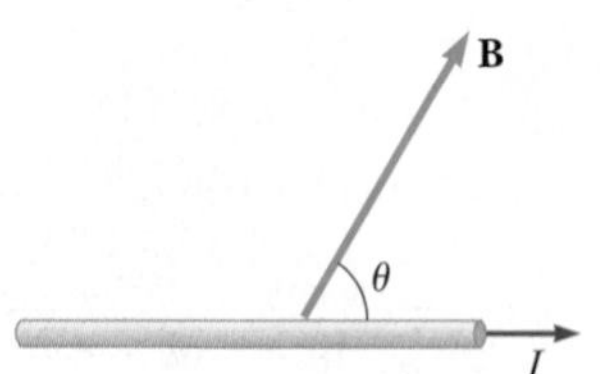


**FIGURE 19.12** A wire carrying a current $I$ in the presence of an external magnetic field **B** that makes an angle $\theta$ with the wire.

where $\theta$ is the angle between **B** and the direction of the current. The direction of this force can be obtained by using right-hand rule #1. However, in this case you must place your thumb in the direction of the current rather than in the direction of **v**. In Figure 19.12, the direction of the magnetic force on the wire is out of the page.

Finally, when the current is either in the direction of the field or opposite the direction of the field, the magnetic force on the wire is zero.

**APPLICATION**

LOUDSPEAKER OPERATION

The fact that a magnetic force acts on a current-carrying wire in a magnetic field is the operating principle of most speakers in sound systems. One speaker design, shown in Figure 19.13, consists of a coil of wire (called the voice coil), a flexible paper cone that acts as the speaker, and a permanent magnet. The coil of wire surrounding the north pole of the magnet is shaped so that the magnetic field lines are directed radially outward from the coil's axis. When an electrical signal is sent to the coil, producing a current in the coil as in Figure 19.13, a magnetic force to the left acts on the coil. (This can be seen by applying right-hand rule #1 to each turn of wire.) When the current reverses direction, as it would for a sinusoidally varying current, the magnetic force on the coil also reverses direction, and the cone, which is attached to the coil, accelerates to the right. An alternating current through the coil causes an alternating force on the coil, which results in vibrations of the cone. The vibrating cone creates sound waves as it pushes and pulls on the air in front of it. In this way a 1-kHz electrical signal is converted to a 1-kHz sound wave.

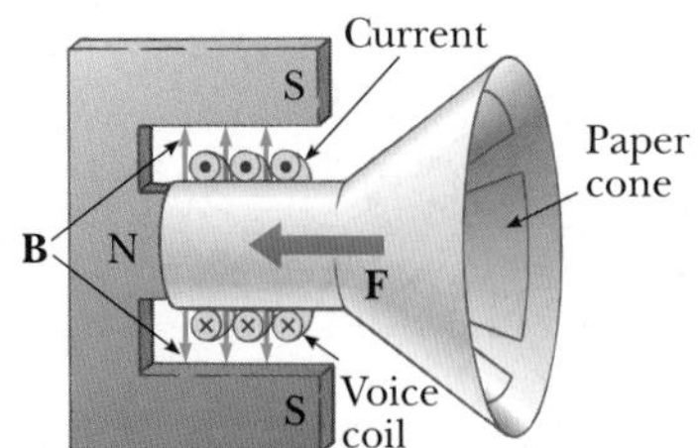


**FIGURE 19.13** Diagram of a loudspeaker.

An unusual application of the force on a current-carrying conductor is illustrated by the electromagnetic pump shown in Figure 19.14. Artificial hearts require a pump to keep the blood flowing, and kidney dialysis machines also require a pump to assist the heart in pumping blood that is to be cleansed. Ordinary mechanical pumps create difficulties because they damage the blood cells as cells move through the pump. The mechanism shown in Figure 19.14 has demonstrated some promise in such applications. A magnetic field is established across a segment of the tube containing the blood, flowing in the direction shown by the velocity **v**. An electric current passing through the fluid perpendicular to the blood velocity as shown has a magnetic force acting on it in the direction of **v**, as application of right-hand rule #1 shows. (Place your right thumb in the direction of the *current*!) This force of magnetic origin helps to keep the blood in motion.

ELECTROMAGNETIC PUMPS FOR ARTIFICIAL HEARTS AND KIDNEYS

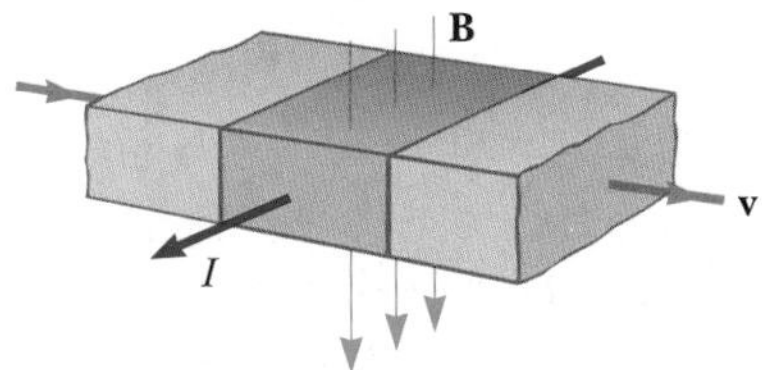


**FIGURE 19.14** A simple electromagnetic pump has no moving parts to damage a conducting fluid, such as blood, passing through. Application of right-hand rule #1 (fingers along **B**, thumb along $I$ in this case) shows that the force on the current-carrying segment of the fluid is in the direction of the velocity.

## APPLYING PHYSICS 19.2

In a lightning strike, there is a rapid movement of negative charge from a cloud to the ground. In what direction is a lightning strike deflected by Earth's magnetic field?

**Explanation** The downward flow of negative charge in a lightning stroke is equivalent to an upward-moving current. Thus, we have an upward-moving current in a northward-directed magnetic field. According to right-hand rule #1, the lightning strike deflects toward the west.

### Example 19.3 A Current-Carrying Wire in Earth's Magnetic Field

A wire carries a current of 22 A from east to west. Assume that at this location Earth's magnetic field is horizontal and directed from south to north, and that it has a magnitude of $0.50 \times 10^{-4}$ T. Find the magnetic force on a 36-m length of wire. How does the force change if the current runs west to east?

**Solution** Because the directions of the current and magnetic field are at right angles, we can use Equation 19.5. The magnitude of the magnetic force is

$$F_{max} = BI\ell = (0.50 \times 10^{-4}\ \mathrm{T})(22\ \mathrm{A})(36\ \mathrm{m}) = \boxed{4.0 \times 10^{-2}\ \mathrm{N}}$$

Right-hand rule #1 shows that the force on the wire is directed toward Earth.

If the current is directed from west to east, the force has the same magnitude but its direction is upward, away from Earth.

**EXERCISE** If the current is directed north to south, what is the magnetic force on the wire?

**ANSWER** Zero

## 19.5 TORQUE ON A CURRENT LOOP AND ELECTRIC MOTORS

In the preceding section we showed how a force is exerted on a current-carrying conductor when the conductor is placed in an external magnetic field. With this as a starting point, we now show that a torque is exerted on a current loop placed in

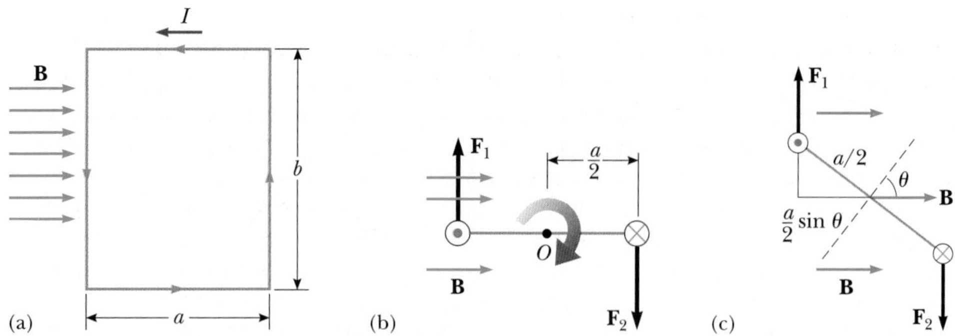


**FIGURE 19.15** (a) Top view of a rectangular loop in a uniform magnetic field **B**. No magnetic forces act on the sides of length *a* parallel to **B**, but forces do act on the sides of length *b*. (b) Side view of the rectangular loop shows that the forces $\mathbf{F}_1$ and $\mathbf{F}_2$ on the sides of length *b* create a torque that tends to twist the loop clockwise. (c) If **B** is at an angle $\theta$ with a line perpendicular to the plane of the loop, the torque is given by $BIA \sin \theta$.

a magnetic field. The results of this analysis will be of great practical value when we discuss generators and motors, here and in Chapter 20.

Consider a rectangular loop carrying current $I$ in the presence of an external uniform magnetic field in the plane of the loop, as shown in Figure 19.15a. The forces on the sides of length $a$ are zero because these wires are parallel to the field. The magnitude of the magnetic forces on the sides of length $b$, however, is

$$F_1 = F_2 = BIb$$

The direction of $\mathbf{F}_1$, the force on the left side of the loop, is out of the page, and that of $\mathbf{F}_2$, the force on the right side of the loop, is into the page. If we view the loop from the side, as in Figure 19.15b, the forces are directed as shown. If we assume that the loop is pivoted so that it can rotate about point $O$, we see that these two forces produce a torque about $O$ that rotates the loop clockwise. The magnitude $\tau_{max}$ of this torque is

$$\tau_{max} = F_1 \frac{a}{2} + F_2 \frac{a}{2} = (BIb) \frac{a}{2} + (BIb) \frac{a}{2} = BIab$$

where the moment arm about $O$ is $a/2$ for both forces. Because the area of the loop is $A = ab$, the torque can be expressed as

$$\tau_{max} = BIA \qquad [19.7]$$

Note that this result is valid only when the magnetic field is parallel to the plane of the loop, as in the view in Figure 19.15b. If the field makes an angle $\theta$ with a line perpendicular to the plane of the loop, as in Figure 19.15c, the moment arm for each force is given by $(a/2) \sin \theta$. An analysis such as the one just used produces, for the magnitude of the torque,

$$\tau = BIA \sin \theta \qquad [19.8]$$

This result shows that the torque has the *maximum* value $BIA$ when the field is parallel to the plane of the loop ($\theta = 90°$) and is *zero* when the field is perpendicular to the plane of the loop ($\theta = 0$). As seen in Figure 19.15c, the loop tends to rotate to smaller values of $\theta$ (so that the normal to the plane of the loop rotates toward the direction of the magnetic field).

Although this analysis is for a rectangular loop, a more general derivation would indicate that Equation 19.8 applies regardless of the shape of the loop. Furthermore, the torque on a coil with $N$ turns is

$$\tau = NBIA \sin \theta \qquad [19.9]$$

## Example 19.4 The Torque on a Circular Loop in a Magnetic Field

A circular wire loop of radius 50.0 cm is placed in a magnetic field of magnitude 0.50 T. The normal to the plane of the loop makes an angle of 30.0° with the magnetic field as shown in an edge view in Figure 19.16. The current in the loop is 2.0 A in the direction shown. Find the magnitude of the torque at this instant.

**Solution** Regardless of the shape of the loop, Equation 19.8 is valid:

$$\tau = BIA \sin \theta = (0.50 \text{ T})(2.0 \text{ A})[\pi(0.50 \text{ m})^2](\sin 30.0°) = 0.39 \text{ N} \cdot \text{m}$$

**Exercise** Find the torque on the loop if it has three turns rather than one.

**Answer** The torque is 3 times that on the one-turn loop, or 1.2 N · m.

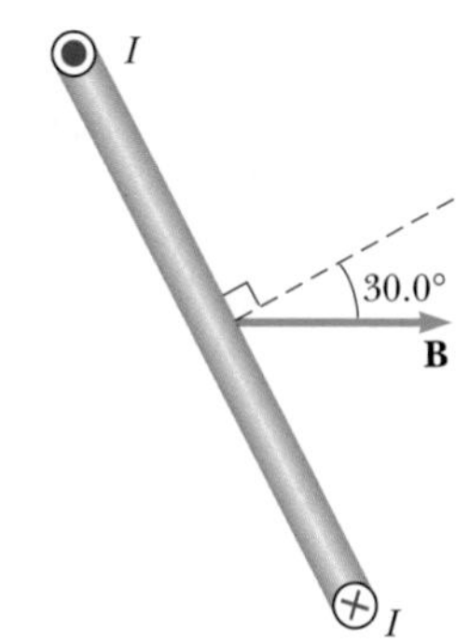


**Figure 19.16** (Example 19.4) An edge view of a circular current loop in an external magnetic field **B**.

## Electric Motors

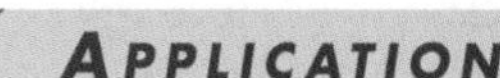


Electric Motors

It is difficult to imagine life in the 21st century without electric motors. Among the numerous appliances that contain motors are computer disk drives, CD players, VCR and DVD players, food processors and blenders, car starters, furnaces, and air conditioners. The motors convert electrical energy to kinetic energy of rotation, and consist basically of a rigid current-carrying loop that rotates when placed in the field of a magnet.

As we have just seen (Figure 19.15), the torque on such a loop rotates the loop to smaller values of $\theta$ until the torque becomes zero when the magnetic field is perpendicular to the plane of the loop and $\theta = 0$. If the loop turns past this angle, and the current remains in the direction shown in Figure 19.15, the torque reverses direction and turns the loop in the opposite direction, that is, counterclockwise. To overcome this difficulty and provide continuous rotation in one direction, the current in the loop must periodically reverse direction. In alternating current (ac) motors, this occurs naturally as the current reverses direction 120 times each second. In direct current (dc) motors, the current reversal is accomplished mechanically with split-ring contacts (commutators) and brushes as shown in Figure 19.17.

Although actual motors contain many current loops and commutators, for simplicity Figure 19.17 shows only a single loop and a single set of split-ring

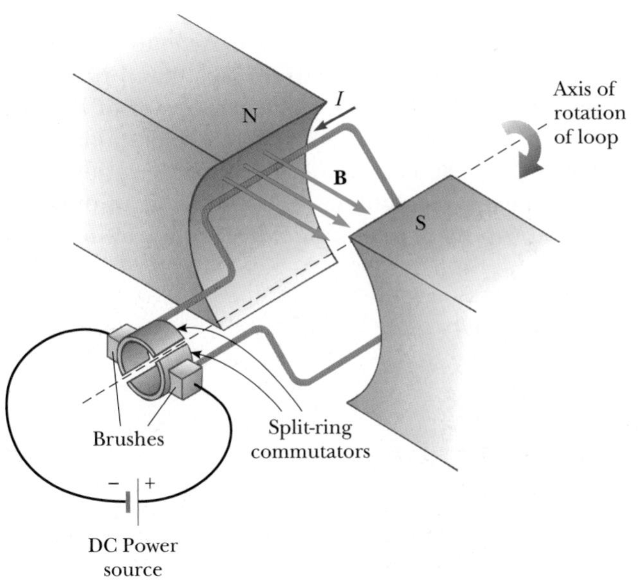


**Figure 19.17** Simplified sketch of a dc electric motor.

**FIGURE 19.18** The Honda Insight combines a three-cylinder gasoline automobile engine with a thin electric motor for improved efficiency and added power when needed. The electric motor (circled) also acts as a generator during braking or coasting downhill to recharge the batteries, with the result that they never need to be recharged by the owner. *(Courtesy of America Honda Motor Co., Inc.)*

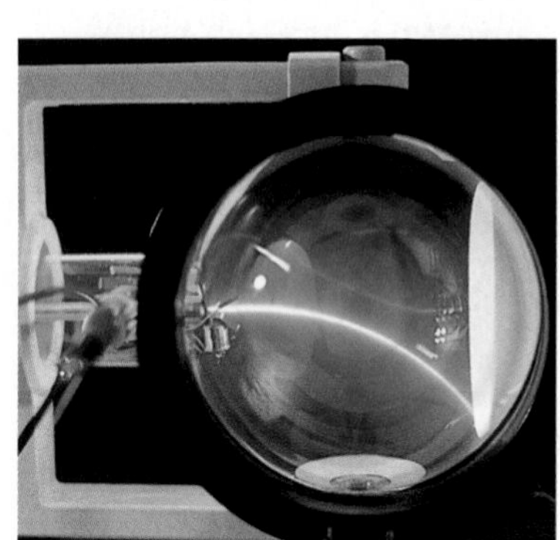

The bending of an electron beam in an external magnetic field. The tube contains gas at very low pressure, and the beam is made visible as the electrons collide with the gas atoms, which in turn emit visible light. The beam is projected to the right and deflects downward in the presence of a magnetic field produced by a pair of current-carrying coils. What is the direction of the magnetic field? *(Courtesy of Central Scientific Company)*

contacts rigidly attached to and rotating with the loop. A set of electrical stationary contacts called *brushes* are maintained in electrical contact with the rotating split ring. These brushes are usually made of graphite because graphite is a good electrical conductor as well as a good lubricant. Just as the loop becomes perpendicular to the magnetic field and the torque becomes zero, inertia carries the loop forward in the clockwise direction and the brushes cross the gaps in the ring, causing the loop current to reverse its direction. This provides another pulse of torque in the clockwise direction for another 180°, the current reverses, and the process repeats itself. Figure 19.18 shows a modern motor used to provide added power to a hybrid gasoline-electric car.

## 19.6 MOTION OF A CHARGED PARTICLE IN A MAGNETIC FIELD

Consider the case of a positively charged particle moving in a uniform magnetic field so that the direction of the particle's velocity is *perpendicular to the field,* as in Figure 19.19. The label $\mathbf{B}_{\text{in}}$ and the blue crosses indicate that $\mathbf{B}$ is directed into the page. Application of right-hand rule #1 at point $P$ shows that the direction of the magnetic force $\mathbf{F}$ at this location is upward. This causes the particle to alter its direction of travel and to follow a curved path. Application of right-hand rule #1 at any point shows that **the magnetic force is always *toward* the center of the circular path;** therefore, the magnetic force causes a centripetal acceleration, which changes only the direction of $\mathbf{v}$ and not its magnitude. Because $\mathbf{F}$ produces the centripetal acceleration, we can equate its magnitude, $qvB$ in this case, to the mass of the particle multiplied by the centripetal acceleration $v^2/r$. From Newton's second law, we find that

$$F = qvB = \frac{mv^2}{r}$$

which gives

$$r = \frac{mv}{qB} \qquad [19.10]$$

This says that the radius of the path is proportional to the momentum $mv$ of the particle and is inversely proportional to the magnetic field.

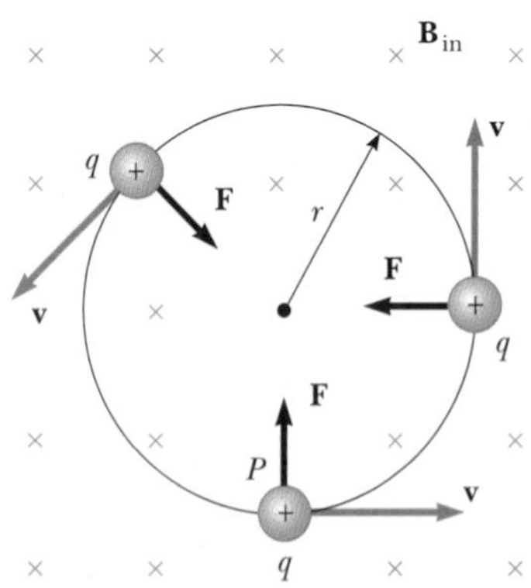


**FIGURE 19.19** When the velocity of a charged particle is perpendicular to a uniform magnetic field, the particle moves in a circle whose plane is perpendicular to **B**, which is directed into the page. (The crosses represent the tails of the magnetic field vectors.) The magnetic force **F** on the charge is always directed toward the center of the circle.

If the initial direction of the velocity of the charged particle is not perpendicular to the magnetic field but instead is directed at an angle to the field, as shown in Figure 19.20, the path followed by the particle is a spiral (called a helix) along the magnetic field lines.

### Webnote 19.2

See an animation of positive and negative charges being deflected by a magnetic field at *http://www.lightlink.com/sergey/java/java/partmagn/index.html*

## APPLYING PHYSICS 19.3

Suppose a uniform magnetic field exists in a finite region of space. Can you inject a charged particle into this region from the outside and have it stay trapped in the region by the magnetic force?

**Explanation** Let us consider separately the components of the particle velocity parallel and perpendicular to the field lines in the region. For the component parallel to the field lines, there is no force on the particle—it continues to move with the parallel component of the velocity. Now consider the component of velocity perpendicular to the field lines. This component results in a magnetic force that is perpendicular to both the field lines and the velocity component. The path of a particle for which the force is always perpendicular to the velocity is a circle. Thus, the particle follows a circular arc and exits the field on the other side of the circle, as shown in Figure 19.21 for a particle with constant kinetic energy. On the other hand, a particle can become trapped if it loses some kinetic energy in a collision after entering the field, as in Figure 19.20.

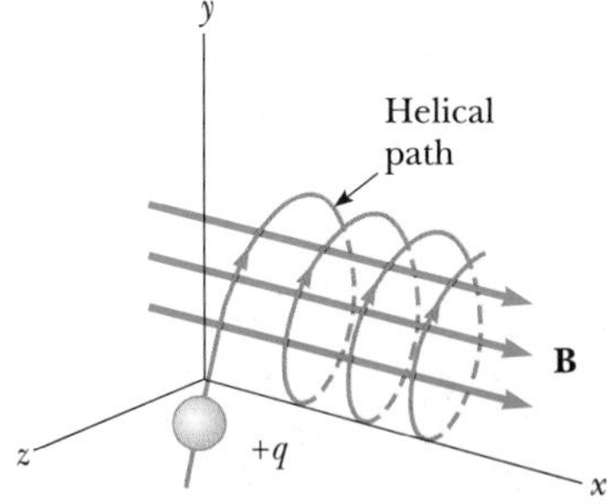


**FIGURE 19.20** A charged particle having a velocity directed at an angle with a uniform magnetic field moves in a helical path.

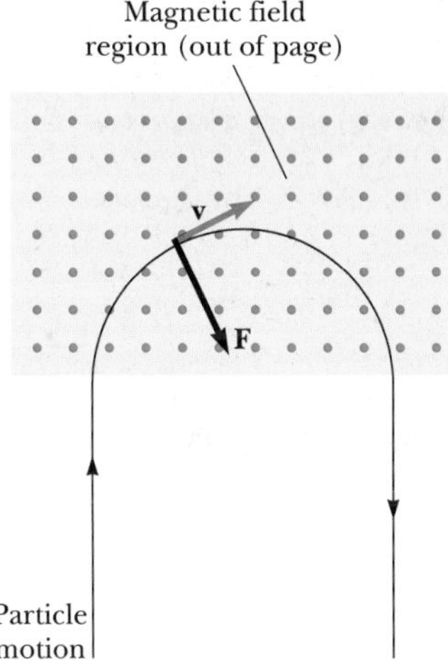


**FIGURE 19.21** (Applying Physics 19.3)

**Quick Quiz 19.3** As a charged particle moves freely in a circular path in the presence of a constant magnetic field applied perpendicular to the particle's velocity, its kinetic energy (a) remains constant, (b) increases, or (c) decreases.

**Quick Quiz 19.4** Two charged particles are projected into a region in which a magnetic field is perpendicular to their velocities. After they enter the magnetic field, you can conclude that (a) the charges are deflected in opposite directions, (b) the charges continue to move in a straight line, (c) the charges move in circular paths, or (d) the charges move in circular paths but in opposite directions.

## Example 19.5 A Proton Moving Perpendicularly to a Uniform Magnetic Field

A proton is moving in a circular orbit of radius 14 cm in a uniform magnetic field of magnitude 0.35 T, directed perpendicularly to the velocity of the proton. Find the orbital speed of the proton.

**Solution** From Equation 19.10, we get

$$v = \frac{qBr}{m} = \frac{(1.6 \times 10^{-19}\ \text{C})(0.35\ \text{T})(14 \times 10^{-2}\ \text{m})}{1.67 \times 10^{-27}\ \text{kg}} = 4.7 \times 10^{6}\ \text{m/s}$$

**EXERCISE** If an electron moves perpendicularly to the same magnetic field with this speed, what is the radius of its circular orbit?

**ANSWER** $7.6 \times 10^{-5}$ m

**Webnote 19.3**

To see how mass spectrometry is used to screen newborns for genetic disorders, visit *http://www.duke.edu/~mdfeezor/NSHome/index.html*

**APPLICATION**

MASS SPECTROMETERS

### Example 19.6 The Mass Spectrometer

Two singly ionized atoms move out of a slit at point *S* in Figure 19.22 and into a magnetic field of 0.10 T. Each has a speed of $1.0 \times 10^6$ m/s. The nucleus of the first atom contains one proton and has a mass of $1.67 \times 10^{-27}$ kg, and the nucleus of the second atom contains a proton and a neutron and has a mass of $3.34 \times 10^{-27}$ kg. Atoms with the same number of protons in the nucleus but different masses are called isotopes. The two isotopes here are hydrogen and deuterium. Find their distance of separation when they strike a photographic plate at *P*.

**Solution** The radius of the circular path followed by the lighter isotope, hydrogen, is

$$r_1 = \frac{m_1 v}{qB} = \frac{(1.67 \times 10^{-27}\text{ kg})(1.0 \times 10^6\text{ m/s})}{(1.6 \times 10^{-19}\text{ C})(0.10\text{ T})} = 0.10\text{ m}$$

The radius of the path of the heavier isotope, deuterium, is

$$r_2 = \frac{m_2 v}{qB} = \frac{(3.34 \times 10^{-27}\text{ kg})(1.0 \times 10^6\text{ m/s})}{(1.6 \times 10^{-19}\text{ C})(0.10\text{ T})} = 0.21\text{ m}$$

The distance of separation is

$$x = 2r_2 - 2r_1 = 0.21\text{ m}$$

The concepts used in this example underlie the operation of a device called a **mass spectrometer,** which is sometimes used to separate isotopes according to their mass-to-charge ratios, but more often is used to measure masses.

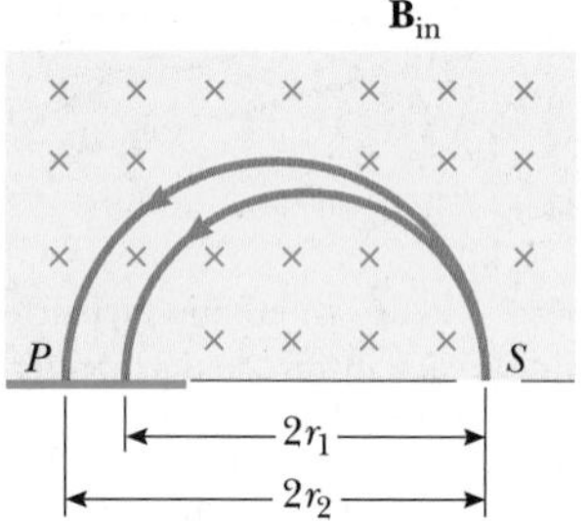


**FIGURE 19.22** (Example 19.6) Two isotopes leave the slit at point *S* and travel in different circular paths before striking a photographic plate at *P*.

## 19.7 MAGNETIC FIELD OF A LONG, STRAIGHT WIRE AND AMPÈRE'S LAW

During a lecture demonstration in 1819, the Danish scientist Hans Oersted (1777–1851) found that an electric current in a wire deflected a nearby compass needle. This momentous discovery, linking a magnetic field with an electric current for the first time, was the beginning of our understanding of the origin of magnetism.

A simple experiment first carried out by Oersted in 1820 clearly demonstrates that a current-carrying conductor produces a magnetic field. In this experiment, several compass needles are placed in a horizontal plane near a long, vertical wire, as in Figure 19.23a. When there is no current in the wire, all needles point in the same direction (that of Earth's field), as we would expect. However, when the wire

**HANS CHRISTIAN OERSTED (1777–1851), DANISH PHYSICIST AND CHEMIST.**

Oersted is best known for observing that a compass needle deflects when placed near a wire carrying a current. This important discovery was the first evidence of the connection between electric and magnetic phenomena. Oersted was also the first to prepare pure aluminum. *(North Wind Picture Archives)*

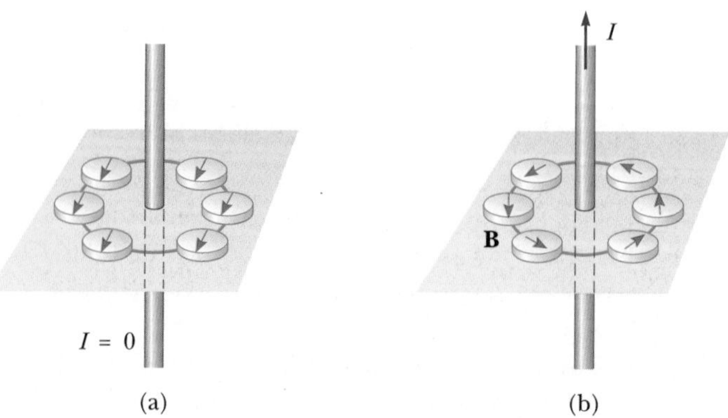


**FIGURE 19.23** (a) When there is no current in the vertical wire, all compass needles point in the same direction. (b) When the wire carries a strong current, the compass needles deflect in directions tangent to the circle, pointing in the direction of **B** due to the current.

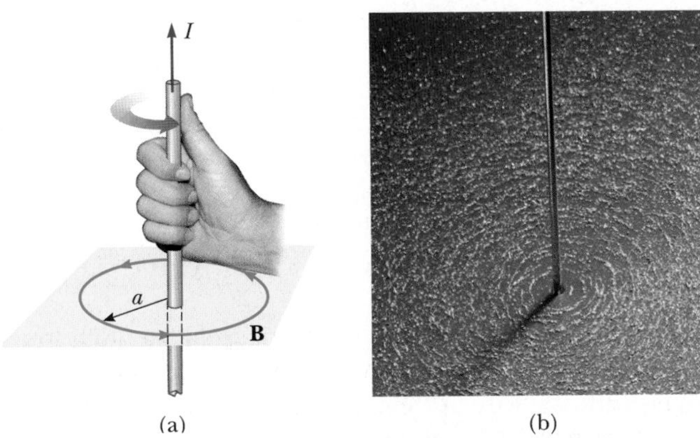


**FIGURE 19.24** (a) Right-hand rule #2 for determining the direction of the magnetic field due to a long, straight wire carrying a current. Note that the magnetic field lines form circles around the wire. (b) Circular magnetic field lines surrounding a current-carrying wire, displayed by iron filings. *(© Richard Megna, Fundamental Photographs)*

**ANDRÉ-MARIE AMPÈRE (1775–1836)**

Ampère, a Frenchman, is credited with the discovery of electromagnetism—the relationship between electric currents and magnetic fields. Ampère's genius, particularly in mathematics, became evident by the age of 12, but his personal life was filled with tragedy. His father, a wealthy city official, was guillotined during the French Revolution, and his wife died young, in 1803. Ampère died at the age of 61 of pneumonia. His judgment of his life is clear from the epitaph he chose for his gravestone: *Tandem felix* (Happy at last). *(Leonard de Selva/CORBIS)*

carries a strong, steady current, the needles all deflect in directions tangent to the circle, as in Figure 19.23b. These observations show that the direction of **B** is consistent with the following convenient rule, **right-hand rule #2:**

**If the wire is grasped in the right hand with the thumb in the direction of the current, as in Figure 19.24a, the fingers will curl in the direction of B.**

When the current is reversed, the filings in Figure 19.24b also reverse.

Because the filings point in the direction of **B**, we conclude that the lines of **B** form circles about the wire. By symmetry, the magnitude of **B** is the same everywhere on a circular path centered on the wire and lying in a plane perpendicular to the wire. By varying the current and distance from the wire, we find that **B** is proportional to the current and inversely proportional to the distance from the wire.

Shortly after Oersted's discovery, scientists arrived at an expression for the strength of the magnetic field due to the current in a long, straight wire. The magnetic field strength at distance $r$ from a wire carrying current $I$ is

$$B = \frac{\mu_0 I}{2\pi r} \quad [19.11]$$

Magnetic field due to a long, straight wire

This result shows that the magnitude of the magnetic field is proportional to the current and decreases as the distance from the wire increases, as we might intuitively expect. The proportionality constant $\mu_0$, called the **permeability of free space,** has the value

$$\mu_0 \equiv 4\pi \times 10^{-7}\ \mathrm{T\cdot m/A} \quad [19.12]$$

**Tip 19.3 USE YOUR RIGHT HAND, NOT THE LEFT**

Note that we have introduced two right-hand rules in this chapter. Be sure to use *only* your right hand when applying these rules.

## AMPÈRE'S LAW AND A LONG, STRAIGHT WIRE

Equation 19.11 enables us to calculate the magnetic field due to a long, straight wire carrying a current. A general procedure for deriving such equations was proposed by the French scientist André-Marie Ampère (1775–1836); it provides a relation between the current in an arbitrarily shaped wire and the magnetic field produced by the wire.

Consider an arbitrary closed path surrounding a current as in Figure 19.25. The path consists of many short segments, each of length $\Delta\ell$. Let us now multiply one of these lengths by the component of the magnetic field parallel to that segment, where the product is labeled $B_{\parallel}\,\Delta\ell$. According to Ampère, the sum of all such products over the closed path is equal to $\mu_0$ times the net current $I$ that passes through the surface bounded by the closed path. This statement, known as **Ampère's circuital law,** can be written

Ampère's circuital law

$$\sum B_{\parallel}\,\Delta\ell = \mu_0 I \qquad [19.13]$$

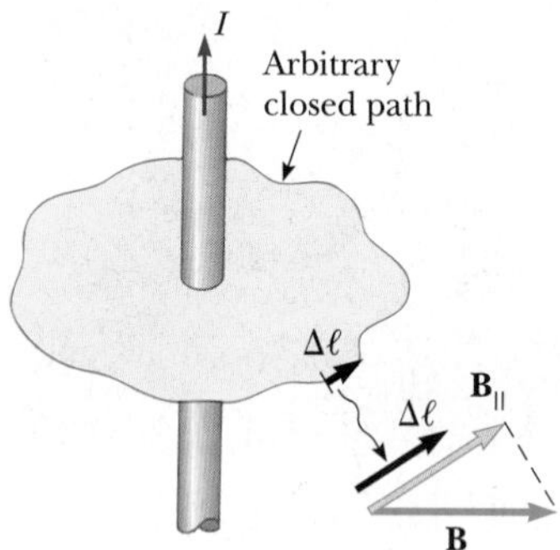


**FIGURE 19.25** An arbitrary closed path around a current is used to calculate the magnetic field due to the current by use of Ampère's rule.

where $B_{\parallel}$ is the component of **B** parallel to the segment of length $\Delta\ell$ and $\Sigma\, B_{\parallel}\,\Delta\ell$ means that we take the sum over all the products $B_{\parallel}\,\Delta\ell$ around the closed path. Ampère's law is the fundamental law describing how electric currents create magnetic fields in the surrounding empty space.

We can use Ampère's circuital law to derive the magnetic field due to a long, straight wire carrying a current $I$. As discussed earlier, the magnetic field lines of this configuration form circles with the wire at their centers. The magnetic field is tangent to this circle at every point and has the same value $B$ over the entire circumference of a circle of radius $r$, so that $B_{\parallel} = B$, as shown in Figure 19.26. We now calculate the sum $\Sigma\, B_{\parallel}\,\Delta\ell$ over the circular path and note that $B_{\parallel}$ can be removed from the sum (because it has the same value $B$ for each element on the circle). Equation 19.13 then gives

$$\sum B_{\parallel}\,\Delta\ell = B_{\parallel}\sum \Delta\ell = B(2\pi r) = \mu_0 I$$

Dividing both sides by the circumference $2\pi r$, we obtain

$$B = \frac{\mu_0 I}{2\pi r}$$

This is identical to Equation 19.11, which is the magnetic field of a long, straight current.

Ampère's circuital law provides an elegant and simple method for calculating the magnetic fields of highly symmetric current configurations. However, it cannot be easily used to calculate magnetic fields for complex current configurations that lack symmetry. Furthermore, Ampère's circuital law is valid only when the currents and fields do not change with time.

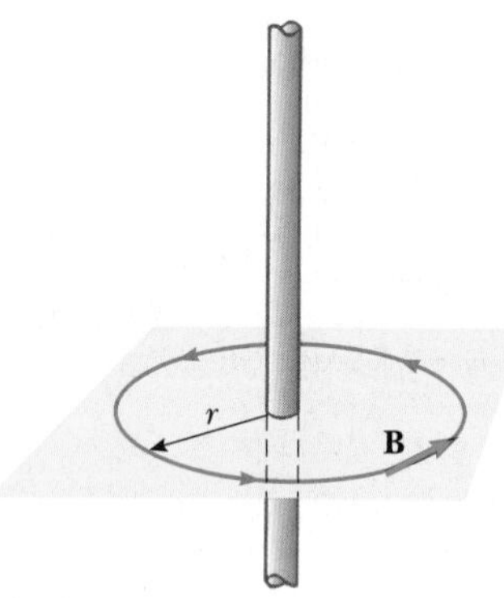


**FIGURE 19.26** A closed circular path of radius $r$ around a long, straight current-carrying wire is used to calculate the magnetic field set up by the wire.

## APPLYING PHYSICS 19.4

Consider a plastic ring encircling a long, straight wire, which is coming out of the page in Figure 19.27. On the plastic ring are fastened two bar magnets as shown. If a current flows in the wire in a direction out of the page, is there a net torque on the ring-magnet combination? If so, in which direction?

**Explanation** Each of the poles on the bar magnets experiences a magnetic force due to the circular magnetic field of the wire. The magnetic field falls off inversely with distance from the wire. Thus, the force on the north poles is smaller than the force on the south poles. The torque, however, increases linearly with distance from the wire. Thus, the effects of the decrease in field strength and the increase in torque with distance cancel, and there is no net torque on the ring-magnet combination.

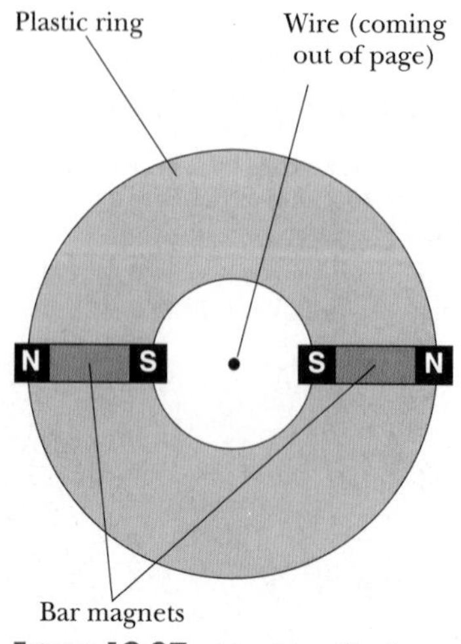


**FIGURE 19.27** (Applying Physics 19.4)

### Example 19.7 The Magnetic Field of a Long Wire

A long, straight wire carries a current of 5.00 A. At one instant, a proton, 4.00 mm from the wire, travels at $1.50 \times 10^3$ m/s parallel to the wire and in the same direction as the current (Fig. 19.28). Find the magnitude and direction of the magnetic force that is acting on the proton because of the magnetic field produced by the wire.

**Solution** From Equation 19.11, the magnitude of the magnetic field produced by the current at a point 4.00 mm from the wire is

$$B = \frac{\mu_0 I}{2\pi r} = \frac{(4\pi \times 10^{-7}\ \text{T}\cdot\text{m/A})(5.00\ \text{A})}{2\pi(4.00 \times 10^{-3}\ \text{m})} = 2.50 \times 10^{-4}\ \text{T}$$

This field is directed into the page at the location of the proton, as shown by right-hand rule #2 for a long, straight wire (see Fig. 19.24a).

The magnitude of the magnetic force on the proton is

$$F = qvB = (1.60 \times 10^{-19}\ \text{C})(1.50 \times 10^{3}\ \text{m/s})(2.50 \times 10^{-4}\ \text{T})$$
$$= 6.00 \times 10^{-20}\ \text{N}$$

The force is directed toward the wire, as shown by right-hand rule #1 for the force on a moving charge (see Fig. 19.7).

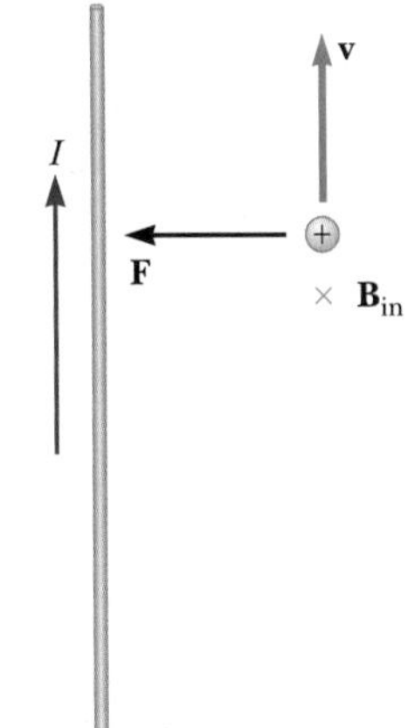


**FIGURE 19.28** (Example 19.7) The magnetic field due to the current is into the page at the location of the proton, and the magnetic force on the proton is to the left.

## 19.8 MAGNETIC FORCE BETWEEN TWO PARALLEL CONDUCTORS

As we have seen, a magnetic force acts on a current-carrying conductor when the conductor is placed in an external magnetic field. Because a conductor carrying a current creates a magnetic field around itself, it is easy to understand that two current-carrying wires placed close together exert magnetic forces on each other. Consider two long, straight, parallel wires separated by the distance $d$ and carrying currents $I_1$ and $I_2$ in the same direction, as shown in Figure 19.29. Let us determine the magnetic force on one wire due to a magnetic field set up by the other wire.

Wire 2, which carries current $I_2$, sets up magnetic field $\mathbf{B}_2$ at wire 1. The direction of $\mathbf{B}_2$ is perpendicular to the wire, as shown in the figure. Using Equation 19.11, we see that the magnitude of this magnetic field is

$$B_2 = \frac{\mu_0 I_2}{2\pi d}$$

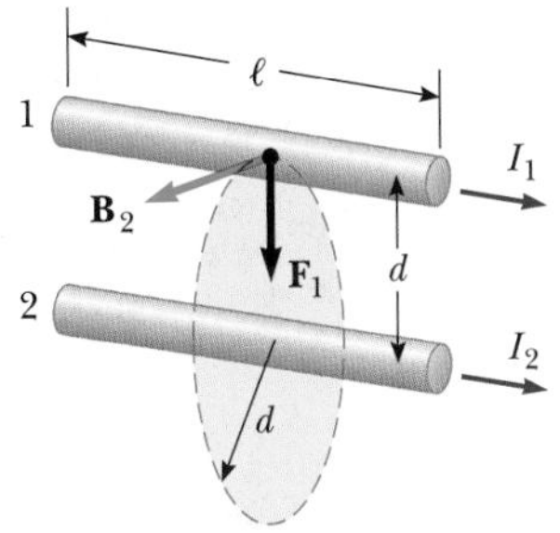

**FIGURE 19.29** Two parallel wires, each carrying a steady current, exert forces on each other. The field $\mathbf{B}_2$ at wire 1 due to wire 2 produces a force on wire 1 given by $F_1 = B_2 I_1 \ell$. The force is attractive if the currents have the same direction, as shown, and repulsive if the two currents have opposite directions.

According to Equation 19.5, the magnitude of the magnetic force on wire 1 in the presence of field $\mathbf{B}_2$ due to $I_2$ is

$$F_1 = B_2 I_1 \ell = \left(\frac{\mu_0 I_2}{2\pi d}\right) I_1 \ell = \frac{\mu_0 I_1 I_2 \ell}{2\pi d}$$

We can rewrite this in terms of the force per unit length:

$$\frac{F_1}{\ell} = \frac{\mu_0 I_1 I_2}{2\pi d} \quad [19.14]$$

The direction of $\mathbf{F}_1$ is downward, toward wire 2, as indicated by right-hand rule #1. If we consider the field set up at wire 2 due to wire 1, the force $\mathbf{F}_2$ on wire 2 is found to be equal in magnitude and opposite in direction to $\mathbf{F}_1$. This is what we would expect from Newton's third law of action-reaction.

We have shown that parallel conductors carrying currents in the same direction *attract* each other. You should use the approach indicated by Figure 19.29 and the steps leading to Equation 19.14 to show that parallel conductors carrying currents in opposite directions *repel* each other.

The force between two parallel wires carrying a current is used to define the SI unit of current, the **ampere** (A), as follows:

Definition of the ampere

**If two long, parallel wires 1 m apart carry the same current, and the magnetic force per unit length on each wire is $2 \times 10^{-7}$ N/m, then the current is defined to be 1 A.**

Definition of the coulomb

The SI unit of charge, the **coulomb** (C), can now be defined in terms of the ampere as follows:

**If a conductor carries a steady current of 1 A, then the quantity of charge that flows through any cross section in 1 s is 1 C.**

**Quick Quiz 19.5** If $I_1 = 2$ A and $I_2 = 6$ A in Figure 19.29, which of the following is true: (a) $F_1 = 3F_2$, (b) $F_1 = F_2$, or (c) $F_1 = F_2/3$?

### Example 19.8 Levitating a Wire

Two wires, each having a weight per unit length of $1.0 \times 10^{-4}$ N/m, are strung parallel to one another above Earth's surface, one directly above the other. The wires are aligned in a north-south direction so that Earth's magnetic field will not affect them. When their distance of separation is 0.10 m, what must be the current in each in order for the lower wire to levitate the upper wire? Assume that the wires carry the same currents, traveling in opposite directions.

**Solution** If the upper wire is to float, it must be in equilibrium under the action of two forces: the force of gravity and magnetic repulsion. The weight per unit length—here $1.0 \times 10^{-4}$ N/m—must be equal and opposite the magnetic force per unit length given in Equation 19.14. Because the currents are the same, we have

$$\frac{F_1}{\ell} = \frac{mg}{\ell} = \frac{\mu_0 I^2}{2\pi d}$$

$$1.0 \times 10^{-4}\ \mathrm{N/m} = \frac{(4\pi \times 10^{-7}\ \mathrm{T \cdot m/A})(I^2)}{(2\pi)(0.10\ \mathrm{m})}$$

We solve for the current to find

$$I = 7.1\ \mathrm{A}$$

**EXERCISE** If the current in each wire is doubled, what is the equilibrium separation of the two wires?

**ANSWER** 0.40 m

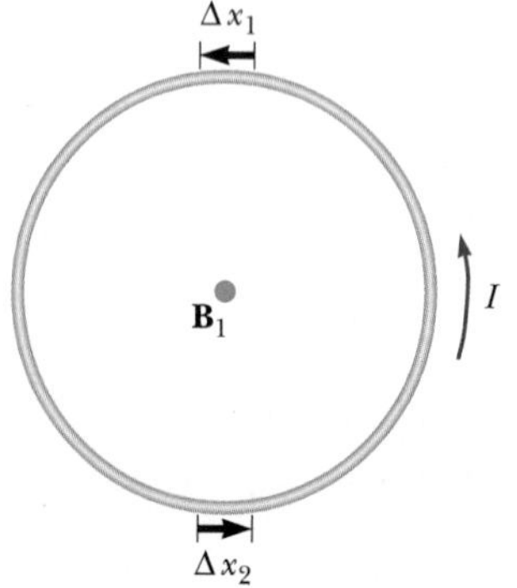

**FIGURE 19.30** All segments of the current loop produce a magnetic field at the center of the loop, directed *out of the page.*

## 19.9 MAGNETIC FIELD OF A CURRENT LOOP

The strength of the magnetic field set up by a piece of wire carrying a current can be enhanced at a specific location if the wire is formed into a loop. You can understand this by considering the effect of several small segments of the current loop, as in Figure 19.30. The small segment at the top of the loop, labeled $\Delta x_1$, produces at the loop's center a magnetic field of magnitude $B_1$, directed out of the page. The direction of **B** can be verified using right-hand rule #2 for a long, straight wire. Imagine holding the wire with your right hand, with your thumb pointing in the direction of the current. Your fingers curl around in the direction of **B**.

A segment at the bottom of the loop of length $\Delta x_2$ also contributes to the field at the center, thus increasing its strength. The field produced at the center of the

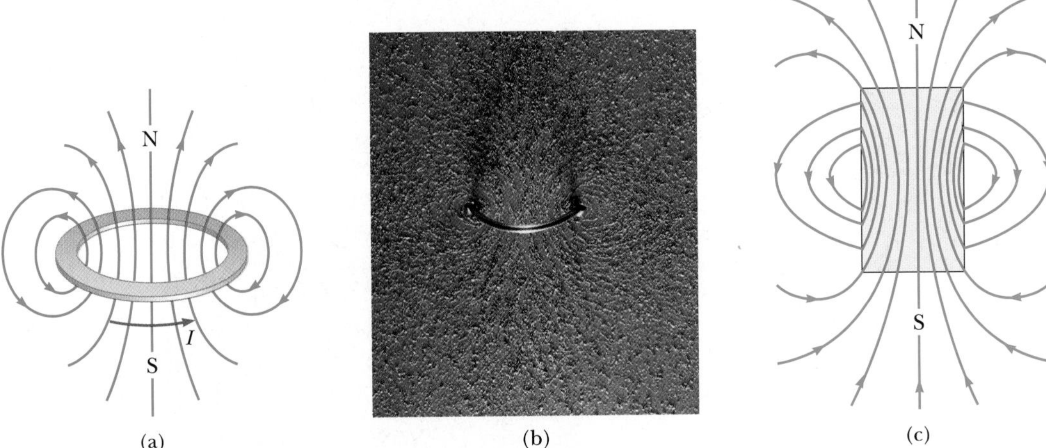


**FIGURE 19.31** (a) Magnetic field lines for a current loop. Note that the magnetic field lines of the current loop resemble those of a bar magnet. (b) Field lines of a current loop, displayed by iron filings. *(© Richard Megna, Fundamental Photographs)* (c) The magnetic field of a bar magnet is similar to that of a current loop.

current loop by the segment $\Delta x_2$ has the same magnitude as $B_1$ and is also directed out of the page. Similarly, all other such segments of the current loop contribute to the field. The net effect is a magnetic field for the current loop as pictured in Figure 19.31a.

Notice in Figure 19.31a that the magnetic field lines enter at the bottom of the current loop and exit at the top. Thus, one side of the loop acts as though it were the north pole of a magnet, and the other acts as a south pole. The fact that the field set up by such a current loop bears a striking resemblance to the field of a bar magnet (Fig. 19.31c) will be of interest to us in a future section.

## APPLYING PHYSICS 19.5

In electrical circuits, it is often the case that wires carrying currents in opposite directions are twisted together. What is the advantage of doing this?

**Explanation** If the wires are not twisted together, the combination of the two wires forms a current loop, which produces a relatively strong magnetic field. In fact, the magnetic field generated by the loop could be strong enough to affect adjacent circuits or components.

## 19.10 MAGNETIC FIELD OF A SOLENOID

If a long, straight wire is bent into a coil of several closely spaced loops, the resulting device is a **solenoid,** often called an **electromagnet.** This device is important in many applications because it acts as a magnet only when it carries a current. As we shall see, the magnetic field inside a solenoid increases with the current and is proportional to the number of coils per unit length.

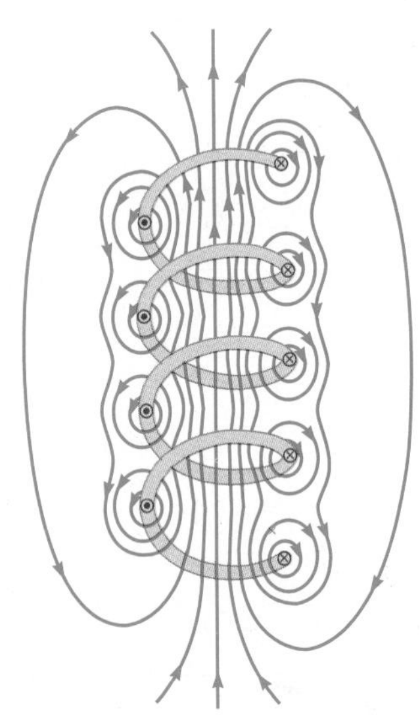

**FIGURE 19.32** The magnetic field lines for a loosely wound solenoid.

Figure 19.32 shows the magnetic field lines of a loosely wound solenoid of length $\ell$ and total number of turns $N$. Note that the field lines inside the solenoid are nearly parallel, uniformly spaced, and close together. This indicates that the field inside the solenoid is nearly uniform and strong. The exterior field at the sides of the solenoid is nonuniform, is much weaker than the interior field, and is in the *opposite direction* to the field inside the solenoid.

If the turns are closely spaced, the field lines are as shown in Figure 19.33a, entering at one end of the solenoid and emerging at the other. This means that

**FIGURE 19.33** (a) Magnetic field lines for a tightly wound solenoid of finite length carrying a steady current. The field inside the solenoid is nearly uniform and strong. Note that the field lines resemble those of a bar magnet, so the solenoid effectively has north and south poles. (b) The magnetic field pattern of a bar magnet, displayed by small iron filings on a sheet of paper. Compare with Figure 19.31c. *(Courtesy of Henry Leap and Jim Lehman)*

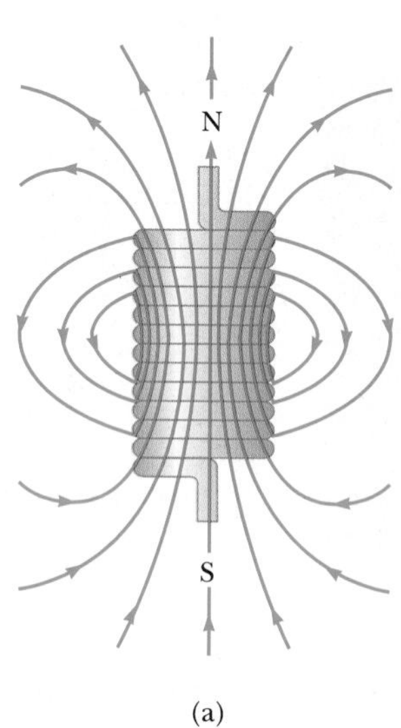


(a)

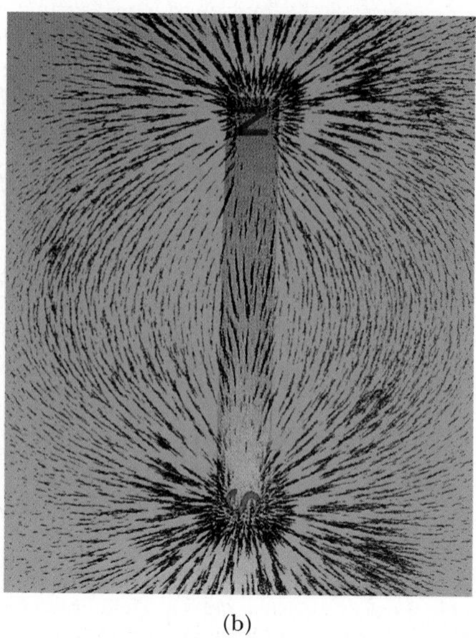


(b)

one end of the solenoid acts as a north pole and the other end acts as a south pole. If the length of the solenoid is much greater than its radius, the lines that leave the north end of the solenoid spread out over a wide region before returning to enter the south end. Hence, as you can see in Figure 19.33a, the magnetic field lines outside are widely separated, indicative of a weak field. This is in contrast to a much stronger field inside the solenoid, where the lines are close together. Also, the field inside the solenoid has a constant magnitude at all points far from its ends. The expression for the magnetic field inside the solenoid is

The magnetic field inside a solenoid 

$$B = \mu_0 n I \qquad [19.15]$$

where $n = N/\ell$ is the number of turns per unit length of the solenoid.

## Example 19.9 The Magnetic Field Inside a Solenoid

A certain solenoid consists of 100 turns of wire and has a length of 10.0 cm.

**A** Find the magnetic field inside the solenoid when it carries a current of 0.500 A.

**Solution** The number of turns per unit length is

$$n = \frac{N}{\ell} = \frac{100 \text{ turns}}{0.10 \text{ m}} = 1\,000 \text{ turns/m}$$

so

$$B = \mu_0 n I = (4\pi \times 10^{-7} \text{ T}\cdot\text{m/A})(1\,000 \text{ turns/m})(0.500 \text{ A})$$
$$= 6.28 \times 10^{-4} \text{ T}$$

By comparison, this is about 10 times stronger than Earth's magnetic field.

**B** Assume that the field has one half of this value just outside the solenoid, at the point labeled N in Figure 19.33a. Find the magnitude and direction of the magnetic force acting on an electron that is moving from right to left in the figure, through point N, at 375 m/s.

**Solution** The magnitude of the magnetic force on the electron is

$$F = qvB = (1.60 \times 10^{-19} \text{ C})(375 \text{ m/s})(3.14 \times 10^{-4} \text{ T}) = 1.88 \times 10^{-20} \text{ N}$$

By use of right-hand rule #1 as in Figure 19.7, the direction of this force is found to be out of the page. (Remember to change the direction of the magnetic force for the negatively charged electron.) This force will deflect the electron from its original direction of motion.

**EXERCISE** How many turns should the solenoid have (assuming it carries the same current) if the field inside is to be 5 times as great?

**ANSWER** 500 turns

So-called steering magnets placed along the neck of the picture tube in a television set, as in Figure 19.34, are used to make the electron beam move to the desired locations on the screen, thus tracing out the images of your favorite program.

**APPLICATION**

CONTROLLING THE ELECTRON BEAM IN A TELEVISION SET

## AMPÈRE'S LAW APPLIED TO A SOLENOID

We can use Ampère's law to obtain the expression for the magnetic field inside a solenoid carrying a current $I$. A cross section taken along the length of part of our solenoid is shown in Fig. 19.35. **B** inside the solenoid is uniform and parallel to the axis, and **B** outside is approximately zero. Consider a rectangular path of length $L$ and width $w$ as shown in Figure 19.35. We can apply Ampère's law to this path by evaluating the sum of $B_{\|}\,\Delta\ell$ over each side of the rectangle. The contribution along side 3 is clearly zero, because $\mathbf{B} = 0$ in this region. The contributions from sides 2 and 4 are both zero, because **B** is perpendicular to $\Delta\ell$ along these paths. Side 1 of length $L$ gives a contribution of $BL$ to the sum, because **B** along this path is uniform and parallel to $\Delta\ell$. Therefore, the sum over the closed rectangular path has the value

$$\sum B_{\|}\,\Delta\ell = BL$$

The right side of Ampère's law involves the total current that passes through the area bounded by the path chosen. In our case, the total current through the rectangular path equals the current through each turn of the solenoid multiplied by the number of turns. If $N$ is the number of turns in the length $L$, then the total current through the rectangular path equals $NI$. Therefore, Ampère's law applied to this path gives

$$\sum B_{\|}\,\Delta\ell = BL = \mu_0 NI$$

$$B = \mu_0 \frac{N}{L} I = \mu_0 n I$$

where $n = N/L$ is the number of turns per unit length.

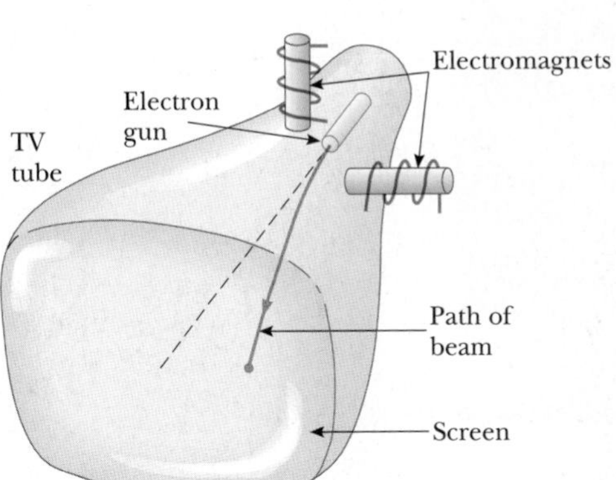


**FIGURE 19.34** Electromagnets are used to deflect electrons to desired positions on the screen of a television tube.

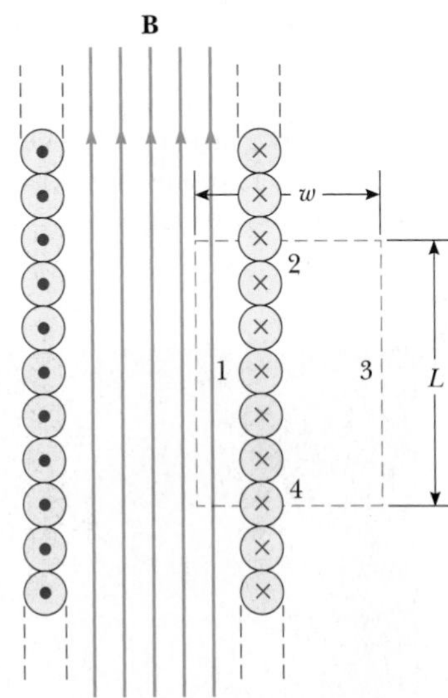


**FIGURE 19.35** A cross-sectional view of a tightly wound solenoid. If the solenoid is long relative to its radius, we can assume that the magnetic field inside is uniform and the field outside is zero. Ampère's law applied to the red dashed rectangular path can then be used to calculate the field inside the solenoid.

## 19.11 MAGNETIC DOMAINS

The magnetic field produced by a current in a coil of wire gives us a hint as to what might cause certain materials to exhibit strong magnetic properties. A single coil like that in Figure 19.31a has a north pole and a south pole, but if this is true for a coil of wire, it should also be true for any current confined to a circular path. *In particular, an individual atom should act as a magnet because of the motion of the electrons about the nucleus.* Each electron, with its charge of $1.6 \times 10^{-19}$ C, circles the atom once in about $10^{-16}$ s. If we divide the electronic charge by this time interval, we see that the orbiting electron is equivalent to a current of $1.6 \times 10^{-3}$ A. Such a current produces a magnetic field on the order of 20 T at the center of the circular path. From this we see that a very strong magnetic field would be produced if several of these atomic magnets could be aligned inside a material. This does not occur, however, because the simple model we have described is not the complete story. A thorough analysis of atomic structure shows that the magnetic field produced by one electron in an atom is often canceled by an oppositely revolving electron in the same atom. The net result is that **the magnetic effect produced by the electrons orbiting the nucleus is either zero or very small for most materials.**

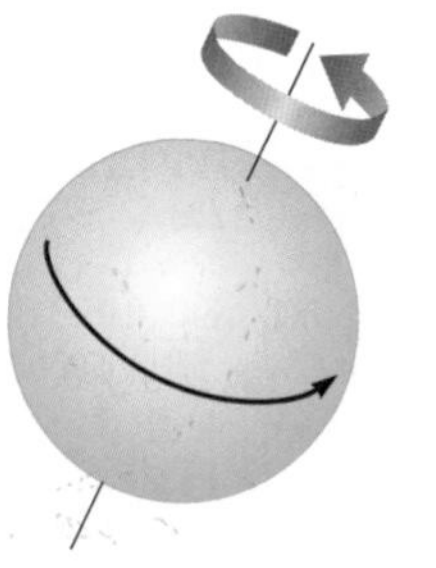

**FIGURE 19.36** Classical model of a spinning electron.

The magnetic properties of many materials are explained by the fact that an electron not only circles in an orbit but also spins on its axis like a top (Fig. 19.36). (This classical description should not be taken too literally. The property of *spin*

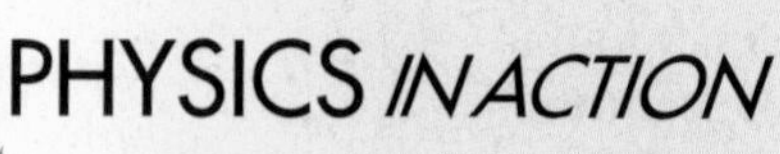

PHYSICS *IN ACTION*

### The Motion of Charged Particles in Magnetic Fields

The white arc in the photograph on the left indicates the circular path followed by an electron beam in a magnetic field. The vessel contains gas at very low pressure, and the beam is made visible as the electrons collide with the gas atoms, which emit visible light. The magnetic field is produced by two coils (not shown). The apparatus can be used to measure the ratio of $e/m$ for the electron.

In the photograph on the right, oxygen, a paramagnetic substance, is attracted to a magnetic field. The liquid oxygen in this photograph is suspended between the poles of a permanent magnet. Paramagnetic substances contain atoms (or ions) that behave as tiny bar magnets. These magnetic atoms interact weakly with each other and are randomly oriented in the absence of an external magnetic field. When the substance is placed in an external magnetic field, the tiny bar magnets associated with the atoms of the substance tend to line up with the field.

*(Courtesy of Central Scientific Company)*

*(Courtesy of Leon Lewandowski)*

can be understood only with the methods of quantum mechanics, which we shall not discuss here.) The spinning electron represents a charge in motion that produces a magnetic field. The field due to the spinning is generally stronger than the field due to the orbital motion. In atoms containing many electrons, the electrons usually pair up with their spins opposite each other, so that their fields cancel each other. That is why most substances are not magnets. However, in certain strongly magnetic materials such as iron, cobalt, and nickel, the magnetic fields produced by the electron spins do not cancel completely. Such materials are said to be **ferromagnetic.** In ferromagnetic materials, strong coupling occurs between neighboring atoms, forming large groups of atoms called **domains** in which the spins are aligned. Typically, the sizes of these domains range from about $10^{-4}$ cm to 0.1 cm. In an unmagnetized substance the domains are randomly oriented, as shown in Figure 19.37a. When an external field is applied, domains that are already aligned with the field tend to grow at the expense of the others (Fig. 19.37b and c). This causes the material to become magnetized.

In what are called hard magnetic materials, domain alignment persists after the external field is removed; the result is a **permanent magnet.** In soft magnetic materials, such as iron, once the external field is removed, thermal agitation produces domain motion and the material quickly returns to an unmagnetized state.

The alignment of domains explains why the strength of an electromagnet is increased dramatically by the insertion of an iron core into the magnet's center. The magnetic field produced by the current in the loops causes alignment of the domains, thus producing a large net external field. The use of iron as a core is also advantageous because it is a soft magnetic material and loses its magnetism almost instantaneously after the current in the coils is turned off.

**Tip 19.4 THE ELECTRON DOES NOT SPIN**

Even though we use the word *spin,* the electron is not physically spinning. The electron has an intrinsic angular momentum *as if it were spinning,* but the concept of spin angular momentum is actually a relativistic effect.

### Webnote 19.4

To see actual photographs of magnetic domains in ferromagnetic materials and the equipment used to record them, visit *http://www.physics.brown.edu/Studies/Demo/em/demo/5g2020.htm*

### Webnote 19.5

Try this experiment! Grapes are repelled by a magnet. Find out why. *http://www.exploratorium.edu/snacks/diamagnetism_www/index.html*

## APPLYING PHYSICS 19.6

Suppose a cardboard tube is filled with very fine iron filings. Each filing is in the form of a sliver—long and thin. If the tube of filings is shaken, what is the likelihood that all of them will wind up aligned in the same direction? Suppose the shaking is repeated while the tube is in a strong magnetic field. What is the likelihood now that the filings will be aligned? Is this a macroscopic analog to anything discussed in this section?

**Explanation** When the tube is shaken outside of a magnetic field, the orientation of the filings will be random—it is extremely unlikely that they will all be aligned in the same direction. If the shaking occurs in a magnetic field, the filings, which become magnetized, can align with the field while they are momentarily airborne and free to rotate. Thus, it is likely that a large number of filings will be aligned with the field after the shaking. The filings are an analog to magnetic domains. When the material is placed in a magnetic field, more domains end up aligned with the field than misaligned, similar to the iron filings.

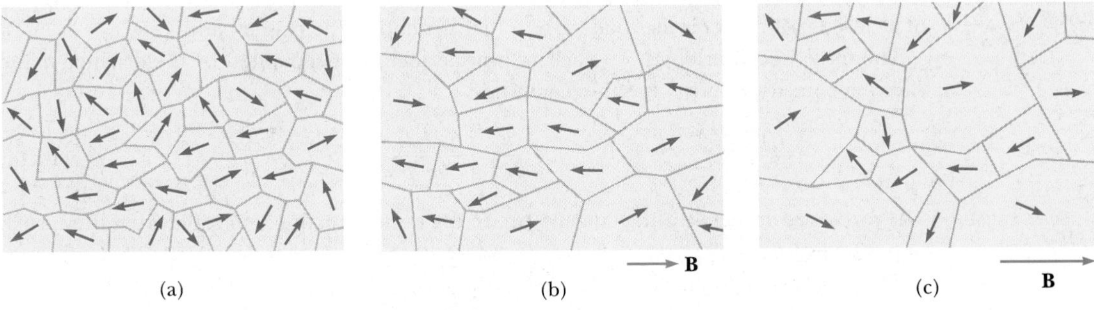


**FIGURE 19.37** (a) Random orientation of domains in an unmagnetized substance. (b) When an external magnetic field **B** is applied, the domains aligned with **B** grow larger. (c) As the field is made even stronger, the domains not aligned with the external field become very small.

## SUMMARY

The **magnetic force** that acts on a charge $q$ moving with velocity $\mathbf{v}$ in a magnetic field $\mathbf{B}$ has the magnitude

$$F = qvB \sin \theta \qquad [19.1]$$

where $\theta$ is the angle between $\mathbf{v}$ and $\mathbf{B}$.

To find the direction of this force, you can use **right-hand rule #1:** Place the fingers of your open right hand in the direction of $\mathbf{B}$ and point your thumb in the direction of the velocity $\mathbf{v}$. The force $\mathbf{F}$ on a positive charge is directed out of the palm of your hand. If the charge is *negative* rather than positive, the force is directed opposite the force given by the right-hand rule.

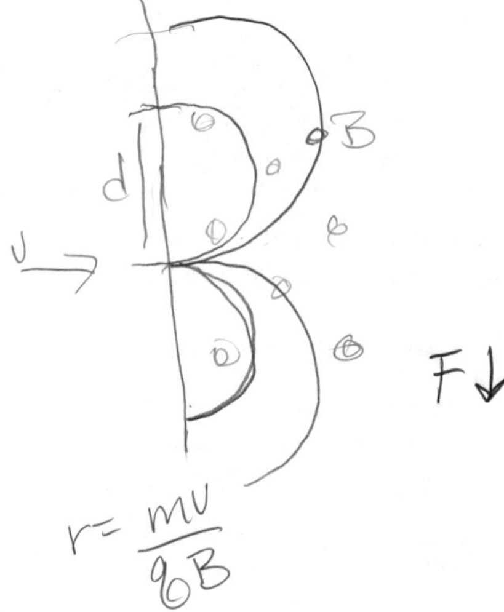

The SI unit of magnetic field is the **tesla** (T), or weber per square meter ($\text{Wb/m}^2$). An additional commonly used unit for magnetic field is the **gauss** (G); $1\text{ T} = 10^4\text{ G}$.

If a straight conductor of length $\ell$ carries current $I$, the magnetic force on that conductor when it is placed in a uniform external magnetic field $\mathbf{B}$ is

$$F = BI\ell \sin \theta \qquad [19.6]$$

Right-hand rule #1 also gives the direction of the magnetic force on the conductor. In this case, however, you must place your thumb in the direction of the current rather than in the direction of $\mathbf{v}$.

The torque $\tau$ on a current-carrying loop of wire in a magnetic field, $\mathbf{B}$, has the magnitude

$$\tau = BIA \sin \theta \qquad [19.8]$$

where $I$ is the current in the loop and $A$ is its cross-sectional area. The angle between $\mathbf{B}$ and a line drawn perpendicularly to the plane of the loop is $\theta$.

If a charged particle moves in a uniform magnetic field so that its initial velocity is perpendicular to the field, it will move in a circular path whose plane is perpendicular to the magnetic field. The radius $r$ of the circular path is

$$r = \frac{mv}{qB} \qquad [19.10]$$

where $m$ is the mass of the particle and $q$ is its charge.

The magnetic field at distance $r$ from a **long, straight wire** carrying current $I$ has the magnitude

$$B = \frac{\mu_0 I}{2\pi r} \qquad [19.11]$$

where $\mu_0 = 4\pi \times 10^{-7}\text{ T}\cdot\text{m/A}$ is the **permeability of free space.** The magnetic field lines around a long, straight wire are circles concentric with the wire.

**Ampère's law** can be used to find the magnetic field around certain simple current-carrying conductors. It can be written

$$\sum B_{\parallel}\, \Delta\ell = \mu_0 I \qquad [19.13]$$

where $B_{\parallel}$ is the component of $\mathbf{B}$ tangent to a small current element of length $\Delta\ell$ that is part of a closed path, and $I$ is the total current that penetrates the closed path.

The force per unit length on each of two parallel wires separated by the distance $d$ and carrying currents $I_1$ and $I_2$ has the magnitude

$$\frac{F}{\ell} = \frac{\mu_0 I_1 I_2}{2\pi d} \qquad [19.14]$$

The forces are attractive if the currents are in the same direction and repulsive if they are in opposite directions.

The magnetic field inside a solenoid has the magnitude

$$B = \mu_0 n I \qquad [19.15]$$

where $n$ is the number of turns of wire per unit length, $n = N/\ell$.

## CONCEPTUAL QUESTIONS

1. Is a net force exerted on a bar magnet in a uniform magnetic field?
2. A coil with a current alternating in direction demagnetizes a watch or tape head. Explain why an alternating current is necessary to achieve this effect.
3. A proton moving horizontally enters a region where a uniform magnetic field is directed perpendicular to the proton's velocity, as shown in Figure Q19.3. Describe the subsequent motion of the proton. How would an electron behave under the same circumstances?

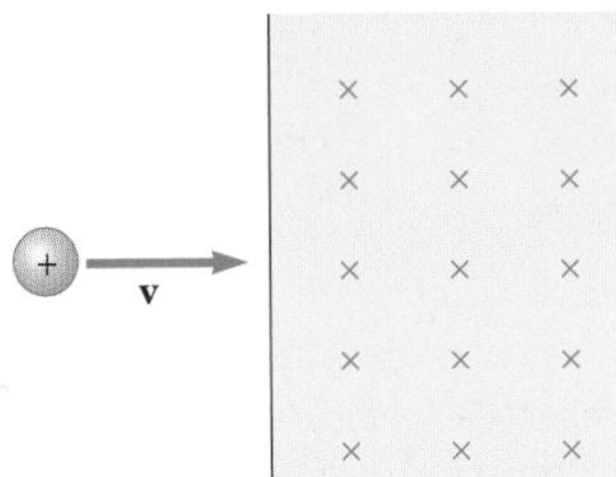


**FIGURE Q19.3**

4. A current-carrying conductor experiences no magnetic force when placed in a certain manner in a uniform magnetic field. Explain.
5. How can the motion of a charged particle be used to distinguish between a magnetic field and an electric field in a certain region?
6. Which way would a compass point if you were at Earth's north magnetic pole?
7. Why does the picture on a television screen become distorted when a magnet is brought near the screen as in Figure Q19.7? (*Caution:* You should not do this at home on a color television set, because it may permanently affect the television picture quality.)
8. A magnet attracts a piece of iron. The iron can then attract another piece of iron. On the basis of domain alignment, explain what happens in each piece of iron.
9. A Hindu ruler once suggested that he be entombed in a magnetic coffin with the polarity arranged so that he could be forever suspended between heaven and Earth. Is such magnetic levitation possible? Discuss.
10. Will a nail be attracted to either pole of a magnet? Explain what is happening inside the nail when it is placed near the magnet.

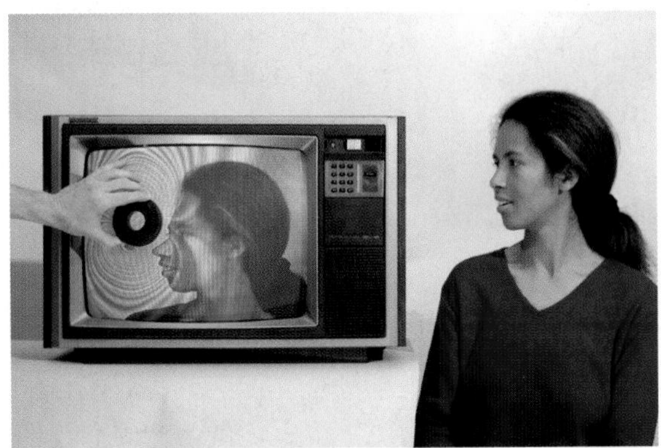

**FIGURE Q19.7** (*© Loren Winters/Visuals Unlimited*)

11. Suppose you move along a wire at the same speed as the drift speed of the electrons in the wire. Do you now measure a magnetic field of zero?
12. Why does hitting a magnet with a hammer cause the magnetism to be reduced?
13. Can you use a compass to detect the currents in wires in the walls near light switches in your home?
14. It is found that charged particles from outer space, called cosmic rays, strike Earth more frequently at the poles than at the Equator. Why?
15. Two wires carry currents in opposite directions and are oriented parallel, with one above the other. The wires repel each other. Is the upper wire in a stable levitation over the lower wire? Suppose the current in one wire is reversed, so that the wires now attract. Is the lower wire hanging in a stable attraction to the upper wire?
16. How can a current loop be used to determine the presence of a magnetic field in a given region of space?
17. A hanging Slinky toy is attached to a powerful battery and a switch. When the switch is closed so that current suddenly flows through the Slinky, does the Slinky compress or expand?
18. Is it possible to orient a current loop in a uniform magnetic field such that the loop will not tend to rotate?
19. Parallel wires exert magnetic forces on each other. What about perpendicular wires? Imagine two wires oriented perpendicular to each other and almost touching. Each wire carries a current. Is there a force between the wires?
20. Can a constant magnetic field set into motion an electron at rest? Explain your answer.

## PROBLEMS

1, 2, 3 = straightforward, intermediate, challenging □ = full solution available in Student Solutions Manual/Study Guide

**web** = solution posted at **http://info.brookscole.com/serway** [icon] = biomedical application

### Section 19.3 Magnetic Fields

1. An electron gun fires electrons into a magnetic field that is directed straight downward. Find the direction of the force exerted by the field on an electron for each of the following directions of the electron's velocity: (a) horizontal and due north; (b) horizontal and 30° west of north; (c) due north, but at 30° below the horizontal; (d) straight upward. (Remember that an electron has a negative charge.)
2. (a) Find the direction of the force on a proton (a positively charged particle) moving through the magnetic fields in Figure P19.2, as shown. (b) Repeat part (a), assuming the moving particle is an electron.

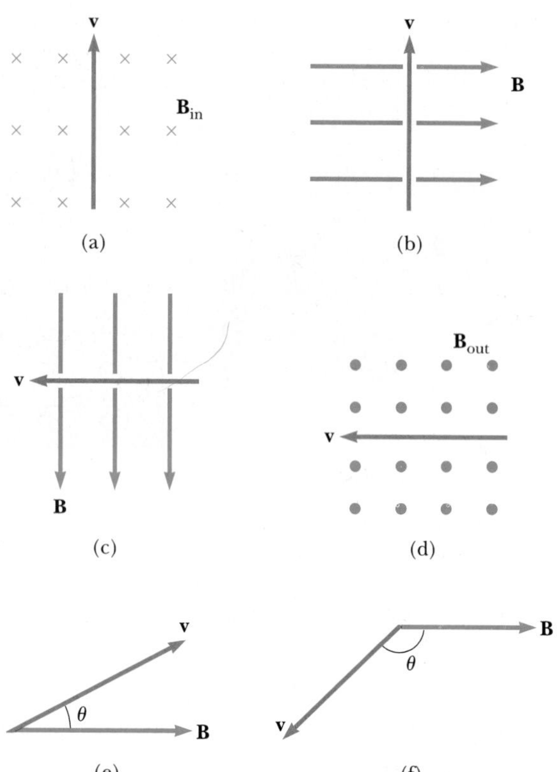


**FIGURE P19.2** (Problems 2 and 13) For Problem 13, replace the velocity vector with a current in that direction.

3. Find the direction of the magnetic field acting on the positively charged particle moving in the various situations shown in Figure P19.3, if the direction of the magnetic force acting on it is as indicated.

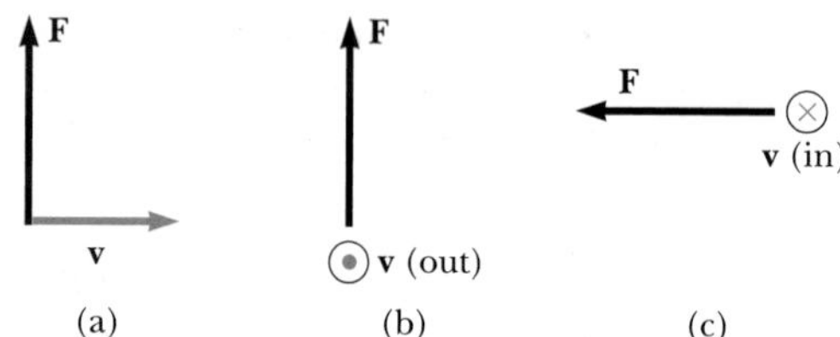


**FIGURE P19.3** (Problems 3 and 12) For Problem 12, replace the velocity vector with a current in that direction.

4. Determine the initial direction of the deflection of charged particles as they enter the magnetic fields shown in Figure P19.4.
5. At the Equator near Earth's surface, the magnetic field is approximately 50.0 $\mu$T northward and the electric field is about 100 N/C downward in fair weather. Find the gravitational, electric, and magnetic forces on an electron with an instantaneous velocity of $6.00 \times 10^6$ m/s directed to the east in this environment.
6. A proton travels with a speed of $3.0 \times 10^6$ m/s at an angle of 37° with the direction of a magnetic field of 0.30 T in the $+y$ direction. What are (a) the magnitude of the magnetic force on the proton and (b) the proton's acceleration?

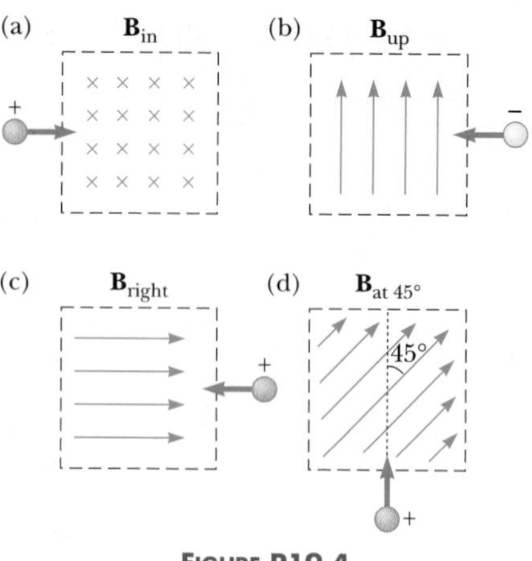


**FIGURE P19.4**

7. web What velocity would a proton need to circle Earth 1 000 km above the magnetic equator, where Earth's magnetic field is directed horizontally north and has a magnitude of $4.00 \times 10^{-8}$ T?
8. An electron is accelerated through 2 400 V from rest and then enters a region where there is a uniform 1.70-T magnetic field. What are the (a) maximum and (b) minimum magnitudes of the magnetic force this charge can experience?
9. A proton moves perpendicularly to a uniform magnetic field **B** at $1.0 \times 10^7$ m/s and experiences an acceleration of $2.0 \times 10^{13}$ m/s² in the $+x$ direction when its velocity is in the $+z$ direction. Determine the magnitude and direction of the field.
10. Sodium ions ($Na^+$) move at 0.851 m/s through a bloodstream in the arm of a person standing near a large magnet. The magnetic field has a strength of 0.254 T and makes an angle of 51.0° with the motion of the sodium ions. The arm contains 100 cm³ of blood with $3.00 \times 10^{20}$ $Na^+$ ions per cubic centimeter. If no other ions were present in the arm, what would be the magnetic force on the arm?

## Section 19.4 Magnetic Force on a Current-Carrying Conductor

11. A current $I = 15$ A is directed along the positive $x$ axis and perpendicularly to a magnetic field. The conductor experiences a magnetic force per unit length of 0.12 N/m in the negative $y$ direction. Calculate the magnitude and direction of the magnetic field in the region through which the current passes.
12. In Figure P19.3, assume that in each case the velocity vector shown is replaced with a wire carrying a current in the direction of the velocity vector. For each case, find the direction of the magnetic field that will produce the magnetic force shown.
13. In Figure P19.2, assume that in each case the velocity vector shown is replaced with a wire carrying a current in the direction of the velocity vector. For each case, find the direction of the magnetic force acting on the wire.
14. A wire carries a steady current of 2.40 A. A straight section of the wire is 0.750 m long and lies along the $x$ axis within a uniform magnetic field of magnitude 1.60 T in the posi-

tive $z$ direction. If the current is in the $+x$ direction, what is the magnetic force on the section of wire?

**15.** A wire carries a current of 10.0 A in a direction that makes an angle of 30.0° with the direction of a magnetic field of strength 0.300 T. Find the magnetic force on a 5.00-m length of the wire.

16. At a certain location, Earth has a magnetic field of $0.60 \times 10^{-4}$ T pointing 75° below the horizontal in a north-south plane. A 10.0-m-long straight wire carries a 15-A current. (a) If the current is directed horizontally toward the east, what are the magnitude and direction of the magnetic force on the wire? (b) What are the magnitude and direction of the force if the current is directed vertically upward?

17. A wire with a mass per unit length of 1.00 g/cm is placed on a horizontal surface with a coefficient of friction of 0.200. The wire carries a current of 1.50 A eastward and moves horizontally to the north. What are the magnitude and the direction of the *smallest* vertical magnetic field that enables the wire to move in this fashion?

18. A conductor suspended by two flexible wires as shown in Figure P19.18 has a mass per unit length of 0.040 0 kg/m. What current must exist in the conductor for the tension in the supporting wires to be zero when the magnetic field is 3.60 T into the page? What is the required direction for the current?

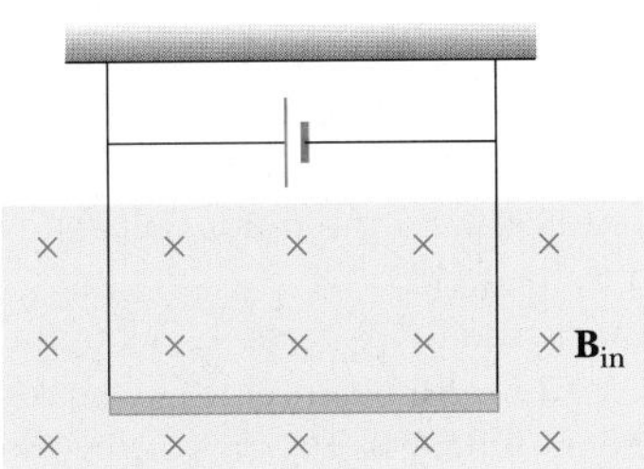


**FIGURE P19.18**

19. **web** An unusual message delivery system is pictured in Figure P19.19. A 15-cm length of conductor that is free to move is held in place between two thin conductors. When a 5.0-A current is directed as shown in the figure, the wire segment moves upward at a constant velocity. If the mass of the wire is 15 g, find the magnitude and direction of the minimum magnetic field that is required to move the wire. (The wire slides without friction on the two vertical conductors.)

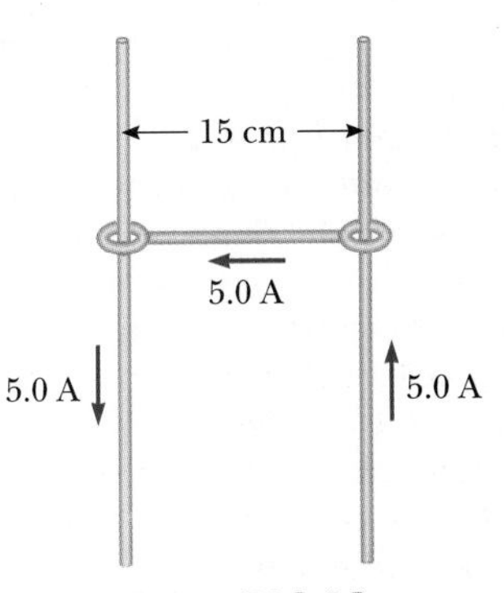


**FIGURE P19.19**

20. A thin, horizontal copper rod is 1.00 m long and has a mass of 50.0 g. What is the minimum current in the rod that can cause it to float in a horizontal magnetic field of 2.00 T?

21. In Figure P19.21, the cube is 40.0 cm on each edge. Four straight segments of wire—*ab, bc, cd,* and *da*—form a closed loop that carries a current $I = 5.00$ A, in the direction shown. A uniform magnetic field of magnitude $B = 0.020\,0$ T is in the positive $y$ direction. Determine the magnitude and direction of the magnetic force on each segment.

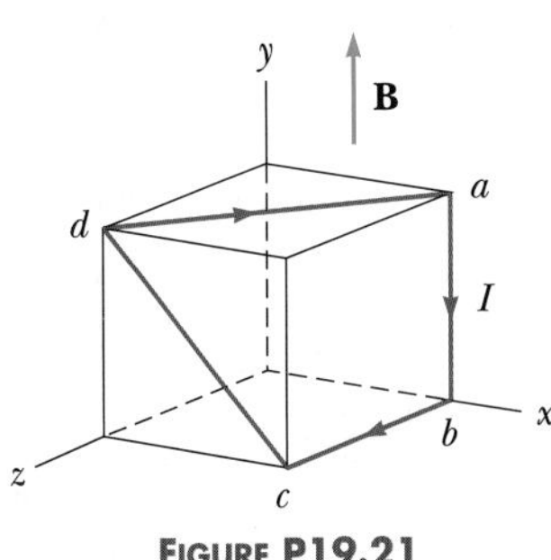


**FIGURE P19.21**

## Section 19.5 Torque on a Current Loop and Electric Motors

**22.** A current of 17.0 mA is maintained in a single circular loop with a circumference of 2.00 m. A magnetic field of 0.800 T is directed parallel to the plane of the loop. What is the magnitude of the torque exerted by the magnetic field on the loop?

**23.** An 8-turn coil encloses an elliptical area having a major axis of 40.0 cm and a minor axis of 30.0 cm (Fig. P19.23). The coil lies in the plane of the page and has a 6.00-A current flowing clockwise around it. If the coil is in a uniform magnetic field of $2.00 \times 10^{-4}$ T, directed toward the left of the page, what is the magnitude of the torque on the coil? (*Hint:* The area of an ellipse is $A = \pi ab$, where $a$ and $b$ are the semi-major and semi-minor axes of the ellipse.)

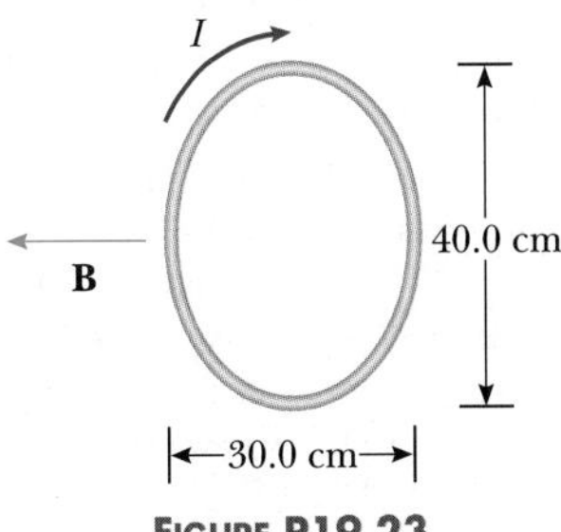


**FIGURE P19.23**

24. A rectangular loop consists of 100 closely wrapped turns and has dimensions 0.40 m by 0.30 m. The loop is hinged along the $y$ axis, and the plane of the coil makes an angle of 30.0° with the $x$ axis (Fig. P19.24). What is the magnitude of the torque exerted on the loop by a uniform magnetic field of 0.80 T directed along the $x$ axis, when the current in the windings has a value of 1.2 A in the direction shown? What is the expected direction of rotation of the loop?

25. A long piece of wire with a mass of 0.100 kg and a total length of 4.00 m is used to make a square coil with a side

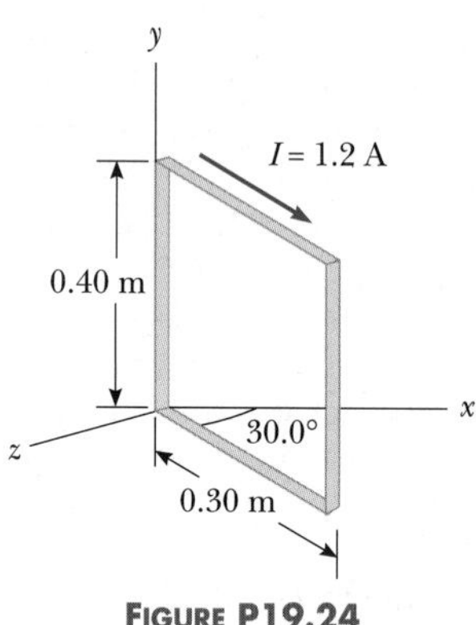


FIGURE P19.24

of 0.100 m. The coil is hinged along a horizontal side, carries a 3.40-A current, and is placed in a vertical magnetic field with a magnitude of 0.010 0 T. (a) Determine the angle that the plane of the coil makes with the vertical when the coil is in equilibrium. (b) Find the torque acting on the coil due to the magnetic force at equilibrium.

26. A copper wire is 8.00 m long, and has a cross-sectional area of $1.00 \times 10^{-4}$ m$^2$. This wire forms a 1-turn loop in the shape of a square and is then connected to a battery that applies a potential difference of 0.100 V. If the loop is placed in a uniform magnetic field of magnitude 0.400 T, what is the maximum torque that can act on it? The resistivity of copper is $1.70 \times 10^{-8}\ \Omega \cdot$m.

## Section 19.6 Motion of a Charged Particle in a Magnetic Field

27. A particle with a +2.0 $\mu$C charge and a kinetic energy of 0.090 J is fired into a uniform magnetic field of magnitude 0.10 T. If the particle moves in a circular path of radius 3.0 m, determine its mass.

28. A cosmic-ray proton in interstellar space has an energy of 10.0 MeV and executes a circular orbit having a radius equal to that of Mercury's orbit around the Sun ($5.80 \times 10^{10}$ m). What is the magnetic field in that region of space?

**29.** Figure P19.29a is a diagram of a device called a velocity selector, in which particles of a specific velocity pass through undeflected but those with greater or lesser velocities are deflected either upward or downward. An electric field is directed perpendicularly to a magnetic field. This produces on the charged particle an electric force and a magnetic force that can be equal in magnitude and opposite in direction (Fig. P19.29b), and hence cancel. Show that particles with a speed of $v = E/B$ will pass through undeflected.

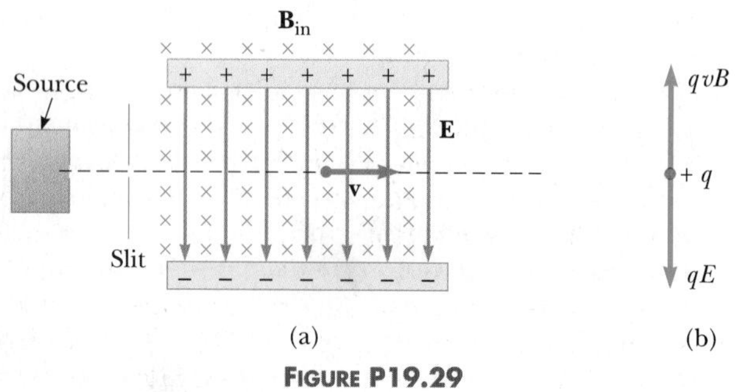


FIGURE P19.29

30. Consider the mass spectrometer shown schematically in Figure P19.30. The electric field between the plates of the velocity selector is 950 V/m, and the magnetic fields in both the velocity selector and the deflection chamber have magnitudes of 0.930 T. Calculate the radius of the path in the system for a singly charged ion with mass $m = 2.18 \times 10^{-26}$ kg. (*Hint:* See Problem 29.)

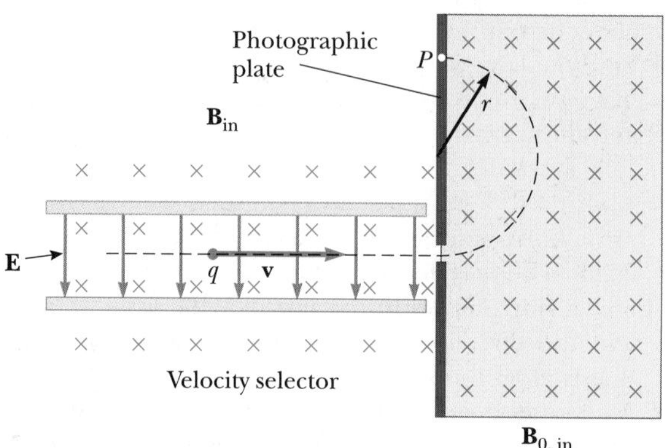


FIGURE P19.30 A mass spectrometer. Charged particles are first sent through a velocity selector. They then enter a region where a magnetic field $\mathbf{B}_0$ (inward) causes positive ions to move in a semicircular path and strike a photographic film at *P*.

31. A singly charged positive ion has a mass of $2.50 \times 10^{-26}$ kg. After being accelerated through a potential difference of 250 V, the ion enters a magnetic field of 0.500 T, in a direction perpendicular to the field. Calculate the radius of the ion's path in the field.

32. A mass spectrometer is used to examine the isotopes of uranium. Ions in the beam emerge from the velocity selector at a speed of $3.00 \times 10^5$ m/s and enter a uniform magnetic field of 0.600 T directed perpendicularly to the velocity of the ions. What is the distance between the impact points formed on the photographic plate by singly charged ions of $^{235}$U and $^{238}$U?

33. An electron moves in a circular path perpendicular to a constant magnetic field with a magnitude of 1.00 mT. If the angular momentum of the electron about the center of the circle is $4.00 \times 10^{-25}$ J·s, determine (a) the radius of the circular path and (b) the speed of the electron.

## Section 19.7 Magnetic Field of a Long, Straight Wire and Ampère's Law

**34.** Find the direction of the current in the wire in Figure P19.34 that would produce a magnetic field directed as shown, in each case.

**35.** A lightning bolt may carry a current of $1.00 \times 10^4$ A for a short period of time. What is the resulting magnetic field 100 m from the bolt? Suppose that the bolt extends far above and below the point of observation.

**36.** In 1962, measurements of the magnetic field of a large tornado were made at the Geophysical Observatory in Tulsa, Oklahoma. If the tornado's field was $B = 1.50 \times 10^{-8}$ T pointing north when the tornado was 9.00 km east of the observatory, what current was carried up or down the funnel of the tornado? Model the vortex as a long straight wire carrying a current.

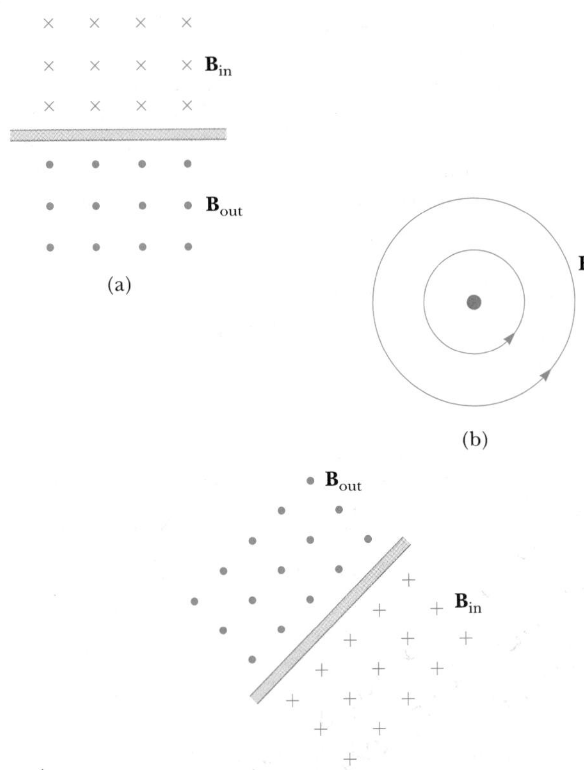

FIGURE P19.34

**37.** At what distance from a long, straight wire carrying a current of 5.0 A is the magnetic field due to the wire equal to the strength of Earth's field, approximately $5.0 \times 10^{-5}$ T?

**38.** The two wires shown in Figure P19.38 carry currents of 5.00 A in opposite directions and are separated by 10.0 cm. Find the direction and magnitude of the net magnetic field (a) at a point midway between the wires, (b) at point $P_1$ (10.0 cm to the right of the wire on the right), and (c) at point $P_2$ (20.0 cm to the left of the wire on the left).

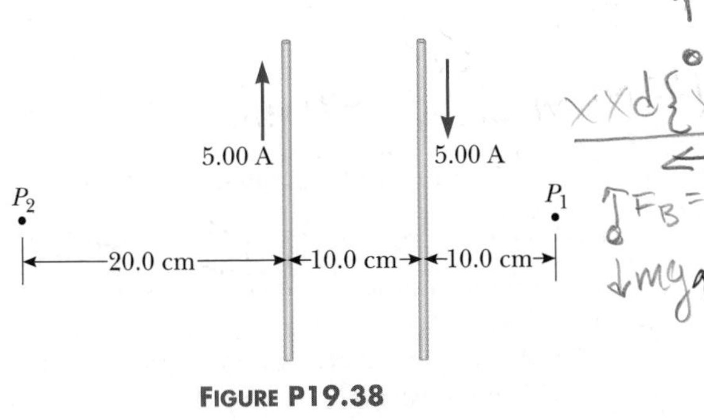

FIGURE P19.38

**39.** Four long, parallel conductors carry equal currents of $I = 5.00$ A. Figure P19.39 is an end view of the conductors. The current direction is into the page at points $A$ and $B$ (indicated by the crosses) and out of the page at $C$ and $D$ (indicated by the dots). Calculate the magnitude and direction of the magnetic field at point $P$, located at the center of the square of edge length 0.200 m.

**40.** The two wires in Figure P19.40 carry currents of 3.00 A and 5.00 A in the direction indicated. (a) Find the direction and magnitude of the magnetic field at a point midway between the wires. (b) Find the magnitude and direction of the magnetic field at point $P$, located 20.0 cm above the wire carrying the 5.00-A current.

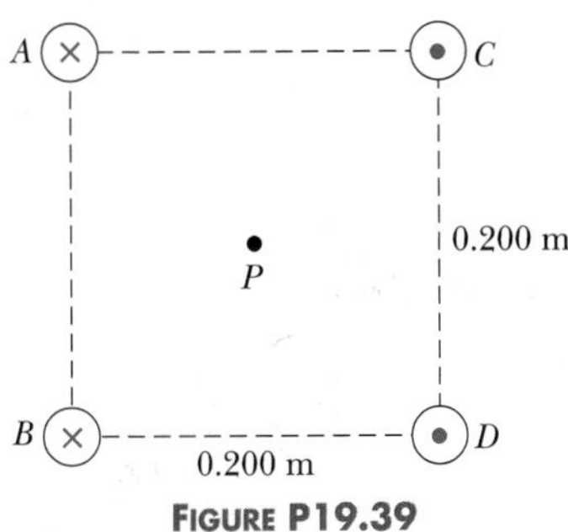

FIGURE P19.39

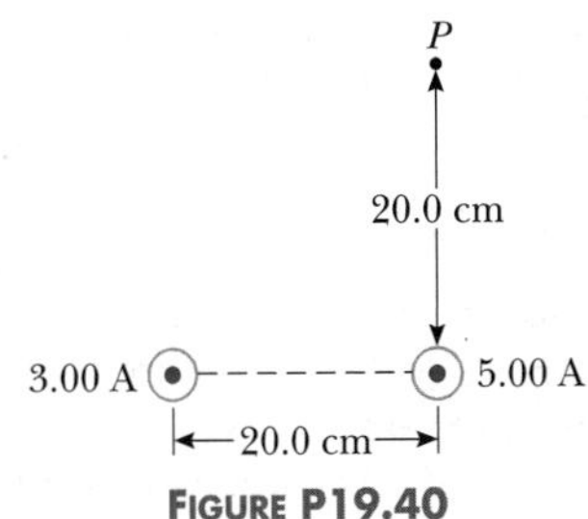

FIGURE P19.40

**41.** A wire carries a 7.00-A current along the $x$ axis and another wire carries a 6.00-A current along the $y$ axis as shown in Figure P19.41. What is the magnetic field at point $P$ located at $x = 4.00$ m, $y = 3.00$ m?

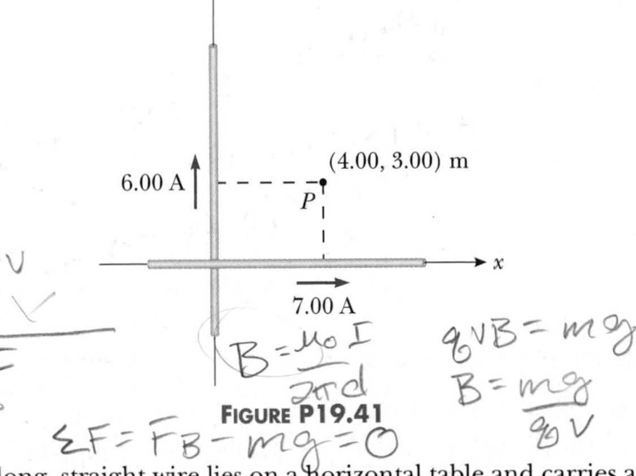

FIGURE P19.41

**42.** A long, straight wire lies on a horizontal table and carries a current of 1.20 μA. In a vacuum, a proton moves parallel to the wire (opposite the current) with a constant velocity of $2.30 \times 10^4$ m/s at a constant distance $d$ above the wire. Determine the value of $d$. You may ignore the magnetic field due to Earth.

**43.** The magnetic field 40.0 cm away from a long, straight wire carrying current 2.00 A is 1.00 μT. (a) At what distance is it 0.100 μT? (b) At one instant, the two conductors in a long household extension cord carry equal 2.00-A currents in opposite directions. The two wires are 3.00 mm apart. Find the magnetic field 40.0 cm away from the middle of the straight cord, in the plane of the two wires. (c) At what distance is it one tenth as large? (d) The center wire in a coaxial cable carries current 2.00 A in one direction, and the sheath around it carries current 2.00 A in the opposite direction. What magnetic field does the cable create at points outside?

## Section 19.8 Magnetic Force Between Two Parallel Conductors

**44.** Two parallel wires are 10.0 cm apart, and each carries a current of 10.0 A. (a) If the currents are in the same direction, find the force per unit length exerted by one of the wires on the other. Are the wires attracted or repelled? (b) Repeat the problem with the currents in opposite directions.

45. **web** A wire with a weight per unit length of 0.080 N/m is suspended directly above a second wire. The top wire carries a current of 30.0 A, and the bottom wire carries a current of 60.0 A. Find the distance of separation between the wires so that the top wire will be held in place by magnetic repulsion.

46. In Figure P19.46, the current in the long, straight wire is $I_1 = 5.00$ A, and the wire lies in the plane of the rectangular loop, which carries 10.0 A. The dimensions are $c = 0.100$ m, $a = 0.150$ m, and $\ell = 0.450$ m. Find the magnitude and direction of the net force exerted by the magnetic field due to the straight wire on the loop.

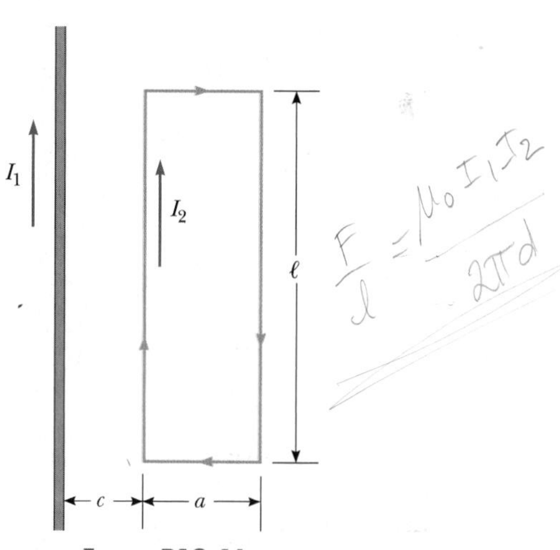


**FIGURE P19.46**

## Section 19.10 Magnetic Field of a Solenoid

**47.** What current is required in the windings of a long solenoid that has 1 000 turns uniformly distributed over a length of 0.400 m in order to produce a magnetic field of magnitude $1.00 \times 10^{-4}$ T at the center of the solenoid?

48. It is desired to construct a solenoid that has a resistance of 5.00 Ω (at 20°C) and that produces a magnetic field at its center of $4.00 \times 10^{-2}$ T when it carries a current of 4.00 A. The solenoid is to be constructed from copper wire having a diameter of 0.500 mm. If the radius of the solenoid is to be 1.00 cm, determine (a) the number of turns of wire needed and (b) the length the solenoid should have.

49. A single-turn square loop of wire, 2.00 cm on a side, carries a counterclockwise current of 0.200 A. The loop is inside a solenoid, with the plane of the loop perpendicular to the magnetic field of the solenoid. The solenoid has 30 turns per centimeter and carries a counterclockwise current of 15.0 A. Find the force on each side of the loop and the torque acting on it.

50. An electron moves at a speed of $1.0 \times 10^4$ m/s in a circular path of radius of 2.0 cm inside a solenoid. The magnetic field of the solenoid is perpendicular to the plane of the electron's path. Find (a) the strength of the magnetic field inside the solenoid and (b) the current in the solenoid if it has 25 turns per centimeter.

## *ADDITIONAL PROBLEMS*

**51.** A circular coil consisting of a single loop of wire has a radius of 30.0 cm and carries a current of 25 A. It is placed in an external magnetic field of 0.30 T. Find the torque on the wire when the plane of the coil makes an angle of 35° with the direction of the field.

52. An electron enters a region of magnetic field of magnitude 0.010 0 T, traveling perpendicular to the linear boundary of the region. The direction of the field is perpendicular to the velocity of the electron. (a) Determine the time it takes for the electron to leave the "field-filled" region, noting that its path is a semicircle. (b) Find the kinetic energy of the electron if the radius of its semicircular path is 2.00 cm.

53. Two long, straight wires cross each other at right angles, as shown in Figure P19.53. (a) Find the direction and magnitude of the magnetic field at point *P*, which is in the same plane as the two wires. (b) Find the magnetic field at a point 30.0 cm above the point of intersection (30.0 cm out of the page, toward you).

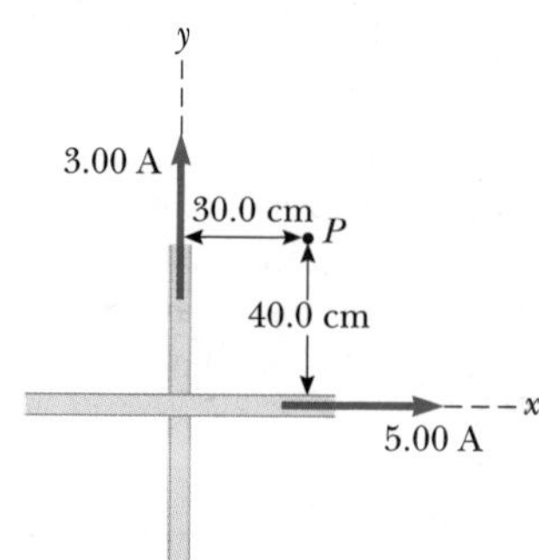


**FIGURE P19.53**

54. A 0.200-kg metal rod carrying a current of 10.0 A glides on two horizontal rails 0.500 m apart. What vertical magnetic field is required to keep the rod moving at a constant speed if the coefficient of kinetic friction between the rod and rails is 0.100?

55. **web** Two species of singly charged positive ions of masses $20.0 \times 10^{-27}$ kg and $23.4 \times 10^{-27}$ kg enter a magnetic field at the same location with a speed of $1.00 \times 10^5$ m/s. If the strength of the field is 0.200 T, and the ions move perpendicularly to the field, find their distance of separation after they complete one half of their circular path.

56. Two parallel conductors carry currents in opposite directions, as shown in Figure P19.56. One conductor carries a current of 10.0 A. Point *A* is the midpoint between the wires, and point *C* is 5.00 cm to the right of the 10.0-A current. *I* is adjusted so that the magnetic field at *C* is zero. Find (a) the value of the current *I* and (b) the value of the magnetic field at *A*.

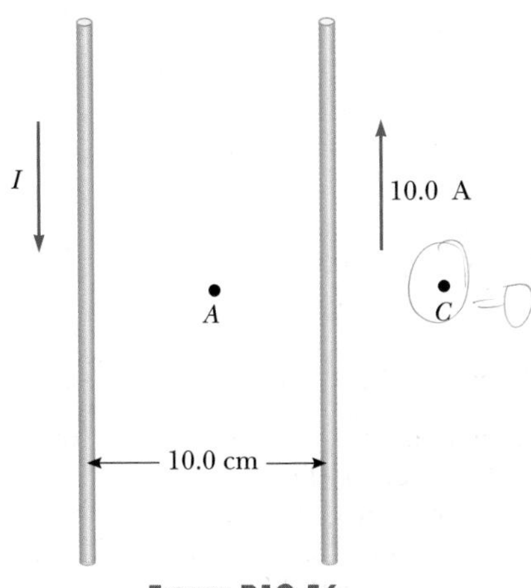


FIGURE P19.56

57. A heart surgeon monitors the flow rate of blood through an artery using an electromagnetic flowmeter (shown schematically in Fig. P19.57). Electrodes $A$ and $B$ make contact with the outer surface of the blood vessel, which has interior diameter 3.00 mm. (a) For a magnetic field magnitude of 0.040 0 T, a potential difference of 160 $\mu$V appears between the electrodes. Calculate the speed of the blood. (b) Verify that electrode $A$ is positive, as shown. Does the sign of the emf depend on whether the mobile ions in the blood are predominantly positively or negatively charged? Explain.

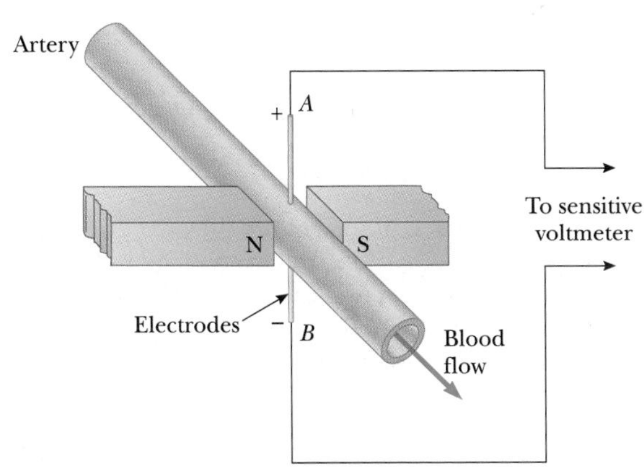


FIGURE P19.57

58. Two circular loops are parallel, coaxial, and almost in contact, 1.00 mm apart (Fig. P19.58). Each loop is 10.0 cm in radius. The top loop carries a clockwise current of 140 A. The bottom loop carries a counterclockwise current of 140 A. (a) Calculate the magnetic force that the bottom loop exerts on the top loop. (b) The upper loop has a mass of 0.021 0 kg. Calculate its acceleration, assuming that the only forces acting on it are the force in part (a) and its weight. (*Hint:* The distance between the loops is small in comparison to the radius of curvature, so the loops may be treated as long, straight, parallel wires.)

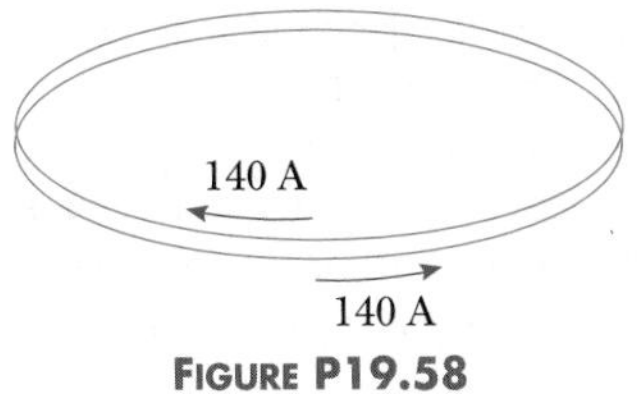


FIGURE P19.58

59. A 1.00-kg ball having net charge $Q = 5.00\ \mu$C is thrown out of a window horizontally at a speed $v = 20.0$ m/s. The window is at a height $h = 20.0$ m above the ground. A uniform horizontal magnetic field of magnitude $B = 0.010\ 0$ T is perpendicular to the plane of the ball's trajectory. Find the magnitude of the magnetic force acting on the ball just before it hits the ground. (*Hint:* Ignore magnetic forces in finding the ball's final velocity.)

60. At the Fermilab accelerator in Batavia, Illinois, protons having momentum $4.80 \times 10^{-16}$ kg·m/s are held in a circular orbit of radius 1.00 km by an upward magnetic field. What is the magnitude of this field?

61. Two long, parallel conductors carry currents $I_1 = 3.00$ A and $I_2 = 3.00$ A, both directed into the page in Figure P19.61. Determine the magnitude and direction of the resultant magnetic field at $P$.

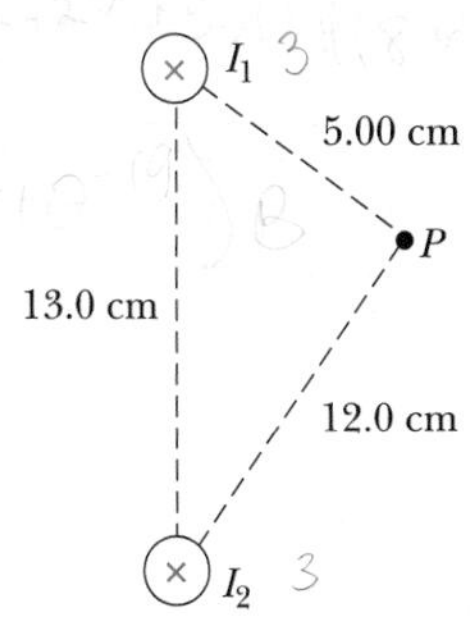


FIGURE P19.61

62. A uniform horizontal wire with a linear mass density of 0.50 g/m carries a 2.0-A current. It is placed in a constant magnetic field, with a strength of $4.0 \times 10^{-3}$ T, that is horizontal and perpendicular to the wire. As the wire moves upward starting from rest, (a) what is its acceleration and (b) how long does it take to rise 50 cm? Neglect the magnetic field of Earth.

63. Three long, parallel conductors carry currents of $I = 2.0$ A. Figure P19.63 is an end view of the conductors, with each current coming out of the page. Given that $a = 1.0$ cm, determine the magnitude and direction of the magnetic field at points (a) $A$, (b) $B$, and (c) $C$.

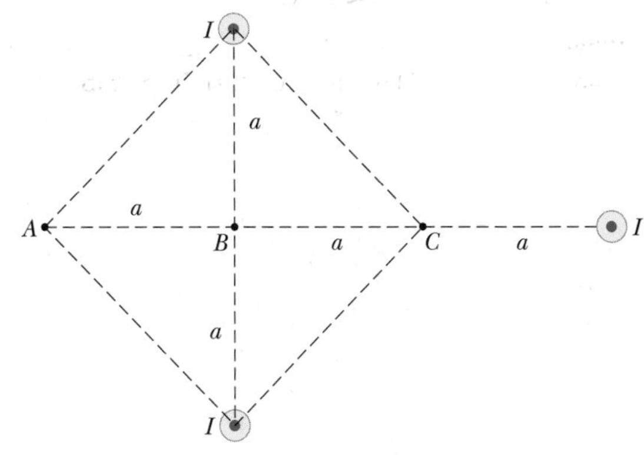


FIGURE P19.63

64. Two long, parallel wires, each with a mass per unit length of 40 g/m, are supported in a horizontal plane by 6.0-cm-long strings, as shown in Figure P19.64. Each wire carries

the same current $I$, causing the wires to repel each other so that the angle $\theta$ between the supporting strings is 16°. (a) Are the currents in the same or opposite directions? (b) Determine the magnitude of each current.

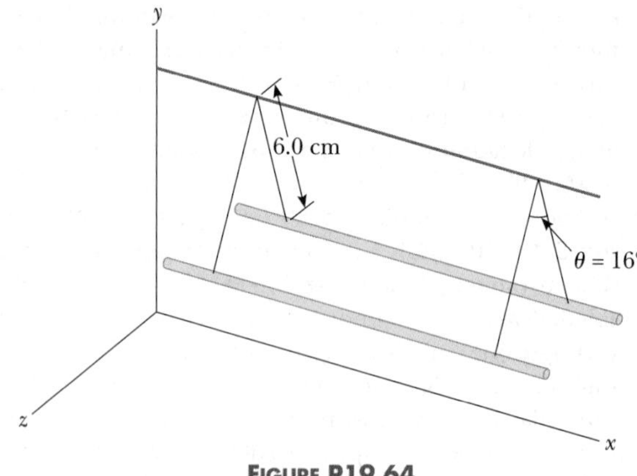

FIGURE P19.64

65. Protons having a kinetic energy of 5.00 MeV are moving in the positive $x$ direction and enter a magnetic field of 0.050 0 T in the $z$ direction, out of the plane of the page, and extending from $x = 0$ to $x = 1.00$ m as in Figure P19.65. (a) Calculate the $y$ component of the protons' momentum as they leave the magnetic field. (b) Find the angle $\alpha$ between the initial velocity vector of the proton beam and the velocity vector after the beam emerges from the field. (*Hint:* Neglect relativistic effects and note that 1 eV = $1.60 \times 10^{-19}$ J.)

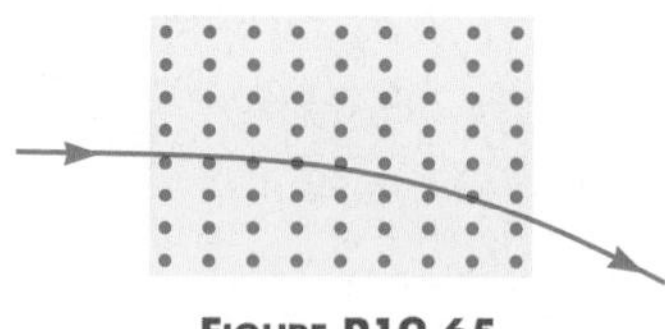
FIGURE P19.65

66. A straight wire of mass 10.0 g and length 5.0 cm is suspended from two identical springs that, in turn, form a closed circuit (Fig. P19.66). The springs stretch a distance of 0.50 cm under the weight of the wire. The circuit has a total resistance of 12 Ω. When a magnetic field is turned on, directed out of the page (indicated by the dots in Fig. P19.66), the springs are observed to stretch an additional 0.30 cm. What is the strength of the magnetic field? (The upper portion of the circuit is fixed.)

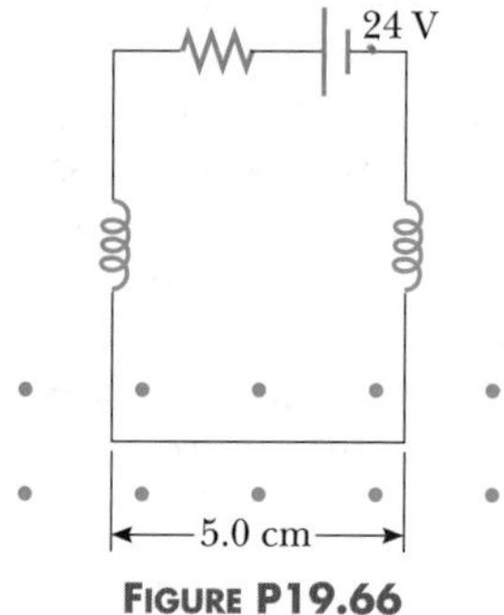

FIGURE P19.66

## GROUP ACTIVITIES

G.1 For this experiment, you will need a small bar magnet, a small plastic container, and a bowl of water. Tape the magnet to the bottom of the small container, and float the container and magnet on the surface of the bowl as in Figure GA19.1. The magnet and small container should rotate and come to equilibrium with the magnet pointing along a north-south line. The compass you have constructed is similar to the type used in early sailing vessels. How can you determine which direction is north and which is south?

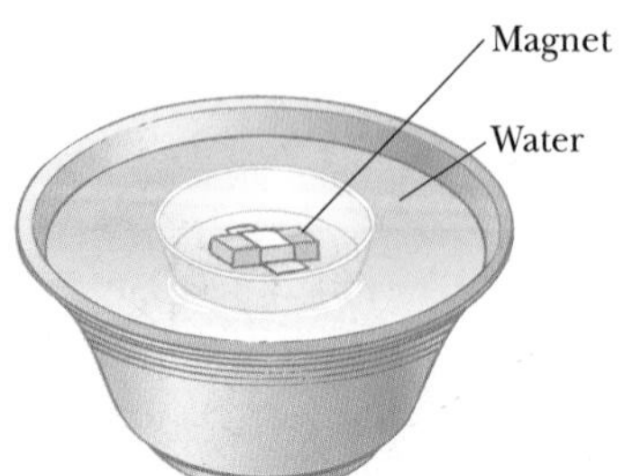

FIGURE GA19.1

G.2 In the northern hemisphere the direction of Earth's magnetic field becomes more and more nearly vertical the farther north one goes. To find the variation from the horizontal of the magnetic field in your locale, try the following. Press an unmagnetized needle through a Ping-Pong ball and balance the combination between two drinking glasses that are lined up along an east-west line. Now press a magnetized needle through the ball at right angles to the unmagnetized needle so that the needle points north. The magnetized needle can now rotate in the vertical direction and will point in the direction of Earth's magnetic field, which is at some angle below the horizontal. Take several measurements of this dip angle and obtain an average value.

G.3 Construct an electromagnet by wrapping about 1 m of small-diameter insulated wire around a steel nail. Tape the ends of the wires to a D-cell battery as in Figure GA19.3. How many staples or paper clips can you pick up with your electromagnet? How would you increase the magnetic field set up by the nail? Disconnect the wires from the battery and test to see how much magnetism is retained by the nail by seeing how many staples it can pick up. A convenient way to test the strength of a magnet is to attach a paper clip to a rubber band. Note how far the rubber band is stretched before the clip comes free of the magnet. Test your electromagnet in this way. Where is the magnetic field of the electromagnet strongest, at the ends of the nail or near its center? When you have your nail magnetized, bang it against a table or the floor and then check its magnetism. Why does the nail lose its magnetism by this procedure?

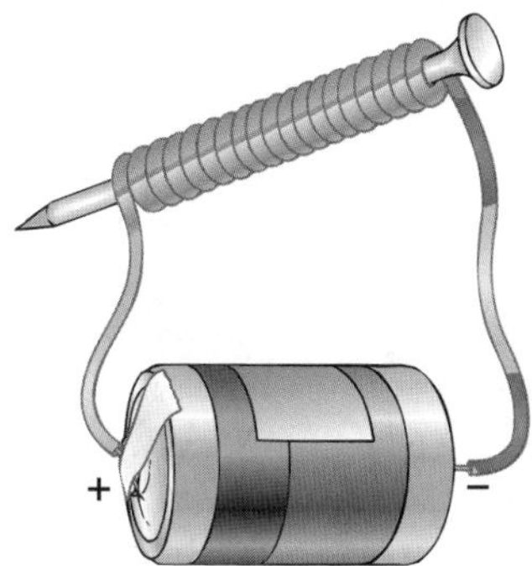


**FIGURE GA19.3**

**G.4** You can trace out the field pattern of a magnet with iron filings. Any machine shop will supply the filings, which should be soaked in a soap solution to remove grit and oil, and then dried. Scatter them lightly over the surface of a paper covering the magnet, and then lightly tap the paper to jar the filings into alignment. Explain why the filings form their pattern. Examine the field pattern set up in the following situations. (a) Arrange two bar magnets about 4 cm apart, aligned with opposite poles facing each other. (b) Use two bar magnets about 4 cm apart, aligned with like poles facing each other. (c) Examine the pattern set up by a horseshoe magnet.

**G.5** In your home television set, a beam of electrons moves from the back of the picture tube to the screen where they strike a fluorescent dot that glows with a particular color when hit. The electrons are accelerated through a voltage of about 20 000 V at the back of the tube. (a) Does the orientation of the television set in Earth's magnetic field affect the beam? (b) In which direction(s) should the set be oriented so that the beam is affected the most? (c) What is the force on an individual electron in direction and magnitude when the beam is moving from east to west? (Assume the magnetic field is horizontal at the location of the television set and has a value of $0.50 \times 10^{-4}$ T.)

**G.6** A beam of alpha particles ($m = 6.64 \times 10^{-27}$ kg, $q = 3.2 \times 10^{-19}$ C) is to be used to produce some radioactive nuclei in a hospital setting where the nuclei are needed as cancer-fighting agents. The beam must enter through one wall of a room directly at the center of the wall, turn in a 90° angle and exit directly at the center of the adjacent wall. The room is 2.5 m by 2.5 m, and it is known that the alpha particles travel at a speed of $2.00 \times 10^5$ m/s. (a) Sketch a design of the room and the direction of the magnetic field in the room that will allow this movement to take place. (b) If a beam of electrons of the same speed is directed along the same initial path into the same magnetic field used for the alpha particles, how will they move? Show this trajectory on your sketch. Is the radius of curvature larger or smaller? (c) What is the magnitude of the magnetic field you need to cause the alpha particles to follow the desired path?

$l = .15$ m
$I = 5$ A
$m = .018$ kg

$\Sigma F = IBL \sin\theta - mg = 0$

$B = .196$ T — the magnitude & direction of the min. B that is required to move the wire

+2 μC charge  KE .09 J  B = .1 T  circular path r = 3 m

find mass

$KE = \frac{1}{2}mv^2 = .09$

$v = \sqrt{\frac{.18}{m}}$

$F = ma_c = \frac{mv^2}{r}$

$F = qvB \sin\theta$

$qvB = \frac{mv^2}{r}$  $m = \frac{rqB}{v}$

# 20

# Induced Voltages and Inductance

## Chapter Outline

**The vibrating strings induce a voltage in pickup coils that detect and amplify the musical sounds being produced. The details of how this works are discussed in this chapter.** *(Photo Disc/Getty Images)*

In 1819 Hans Christian Oersted discovered that a magnetic compass experiences a force in the vicinity of an electric current. Although there had long been speculation that such a relationship existed, this was the first evidence of a link between electricity and magnetism. Because nature is often symmetric, the discovery that electric currents produce magnetic fields led scientists to suspect that magnetic fields could produce electric currents. Indeed, experiments conducted by Michael Faraday in England and independently by Joseph Henry in the United States in 1831 showed that a changing magnetic field could induce an electric current in a circuit. The results of these experiments led to a basic and important law known as Faraday's law. In this chapter we discuss Faraday's law and several practical applications, one of which is the production of electrical energy in power generation plants throughout the world.

**MICHAEL FARADAY, BRITISH PHYSICIST AND CHEMIST (1791–1867)**

Faraday is often regarded as the greatest experimental scientist of the 1800s. His many contributions to the study of electricity include the invention of the electric motor, electric generator, and transformer, as well as the discovery of electromagnetic induction and the laws of electrolysis. Greatly influenced by religion, he refused to work on military poison gas for the British government. *(By kind permission of the President and Council of the Royal Society)*

## 20.1 INDUCED emf AND MAGNETIC FLUX

We begin this chapter by describing an experiment, first conducted by Faraday, demonstrating that a current can be produced by a changing magnetic field. The apparatus shown in Figure 20.1 consists of a coil connected to a switch and a battery. We refer to this coil as the *primary coil* and to the corresponding circuit as the primary circuit. The coil is wrapped around an iron ring to intensify the magnetic field produced by the current in the coil. A second coil, at the right, is wrapped around the iron ring and is connected to an ammeter. We refer to this as the *secondary coil* and to the corresponding circuit as the secondary circuit. Note that there is no battery in the secondary circuit. The only purpose of this circuit is to detect any current that might be produced by a magnetic field.

At first glance, you might guess that no current would ever be detected in the secondary circuit. However, when the switch in the primary circuit is suddenly closed or opened, something quite amazing happens. Just after the switch is closed, the ammeter in the secondary circuit deflects in one direction and then returns to zero. When the switch is opened, the ammeter deflects in the opposite direction and again returns to zero. Finally, when there is a steady current in the primary circuit, the ammeter reads zero.

From observations such as these, Faraday concluded that an electric current can be produced by a *changing* magnetic field. (A steady magnetic field cannot produce a current.) The current produced in the secondary circuit occurs only for an instant while the magnetic field through the secondary coil is changing. In effect, the secondary circuit behaves as though a source of emf were connected to it

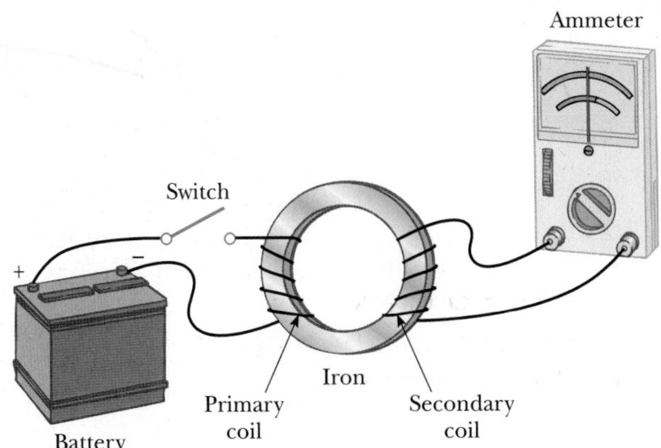


**FIGURE 20.1** Faraday's experiment. When the switch in the primary circuit at the left is closed, the ammeter in the secondary circuit at the right deflects momentarily. The emf in the secondary circuit is induced by the changing magnetic field through the coil in this circuit.

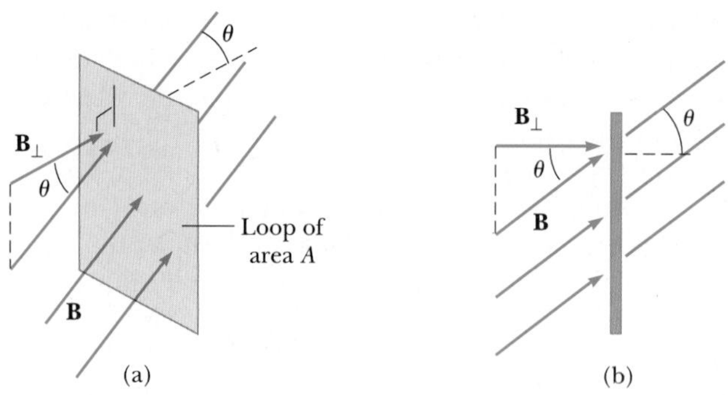


**FIGURE 20.2** (a) A uniform magnetic field **B** making an angle $\theta$ with the normal to the plane of a wire loop of area $A$. (b) An edge view of the loop.

for a short time. It is customary to say that **an induced emf is produced in the secondary circuit by the changing magnetic field.**

## MAGNETIC FLUX

In order to evaluate induced emfs quantitatively, it is first necessary to understand what factors affect the phenomenon. The emf is actually induced by a change in a quantity called the magnetic flux rather than simply by a change in the magnetic field. Magnetic flux is defined in a similar manner to the electric flux (Section 15.10) and is proportional to both the strength of the magnetic field passing through the plane of a loop of wire and the area of the loop.

Consider a loop of wire in the presence of a uniform magnetic field **B**. If the loop has an area $A$, the **magnetic flux $\Phi_B$** through the loop is defined as

Magnetic flux 

$$\Phi_B = B_\perp A = BA \cos \theta \qquad [20.1]$$

where $B_\perp$ is the component of **B** perpendicular to the plane of the loop as in Figure 20.2a, and $\theta$ is the angle between **B** and the normal (perpendicular) to the plane of the loop. Figure 20.2b is an edge view of the loop and the penetrating magnetic field lines. When the field is perpendicular to the plane of the loop as in Figure 20.3a, $\theta = 0$ and $\Phi_B$ has a maximum value, $\Phi_{B,\max} = BA$. When the plane of the loop is parallel to **B** as in Figure 20.3b, $\theta = 90°$ and $\Phi_B = 0$. The flux can also be negative. For example, when $\theta = 180°$, the flux is equal to $-BA$. Because $B$ is in SI units of teslas, or webers per square meter, the units of flux are $\text{T} \cdot \text{m}^2$, or webers (Wb).

We can emphasize the qualitative meaning of Equation 20.1 by first drawing magnetic field lines, as in Figure 20.3. The number of lines per unit area increases as the field strength increases. **The value of the magnetic flux is proportional to the total number of lines passing through the loop.** Thus, we see that the most lines pass through the loop when its plane is perpendicular to the field as in Figure 20.3a, and so the flux has its maximum value. As Figure 20.3b shows, no lines pass through the loop when its plane is parallel to the field, and so in this case $\Phi_B = 0$.

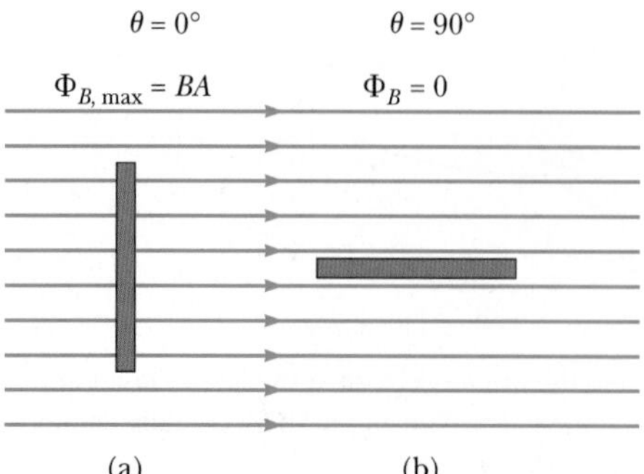


**FIGURE 20.3** An edge view of a loop in a uniform magnetic field. (a) When the field lines are perpendicular to the plane of the loop, the magnetic flux through the loop is a maximum and equal to $\Phi_B = BA$. (b) When the field lines are parallel to the plane of the loop, the magnetic flux through the loop is zero.

## APPLYING PHYSICS 20.1

Argentina has more land area ($2.8 \times 10^6$ km$^2$) than Greenland ($2.2 \times 10^6$ km$^2$). Yet the magnetic flux of Earth's magnetic field is larger through Greenland than through Argentina. Why?

**Explanation** There are two reasons for the larger magnetic flux through Greenland. Greenland (latitude 60°N to 80°N) is closer to a magnetic pole than is Argentina (latitude 20°S to 50°S). As a result, the angle that the magnetic field lines make with the vertical is smaller in Greenland than in Argentina. Thus, more field lines penetrate the surface in Greenland. The second reason is also a result of Greenland's proximity to a magnetic pole. The field strength is larger in the vicinity of a magnetic pole than it is farther from the pole. Therefore, the magnitude of **B** in the definition of flux is larger for Greenland than for Argentina. These two influences dominate over the slightly larger area of Argentina.

## 20.2 FARADAY'S LAW OF INDUCTION

The usefulness of the concept of magnetic flux is made obvious by another simple experiment that demonstrates the basic idea of electromagnetic induction. Consider a wire loop connected to an ammeter as in Figure 20.4. If a magnet is moved toward the loop, the ammeter needle deflects in one direction as in Figure 20.4a. When the magnet is held stationary as in Figure 20.4b, the needle is not deflected. If the magnet is moved away from the loop, the needle deflects in the opposite direction as in Figure 20.4c. If the magnet is held stationary and the loop is moved

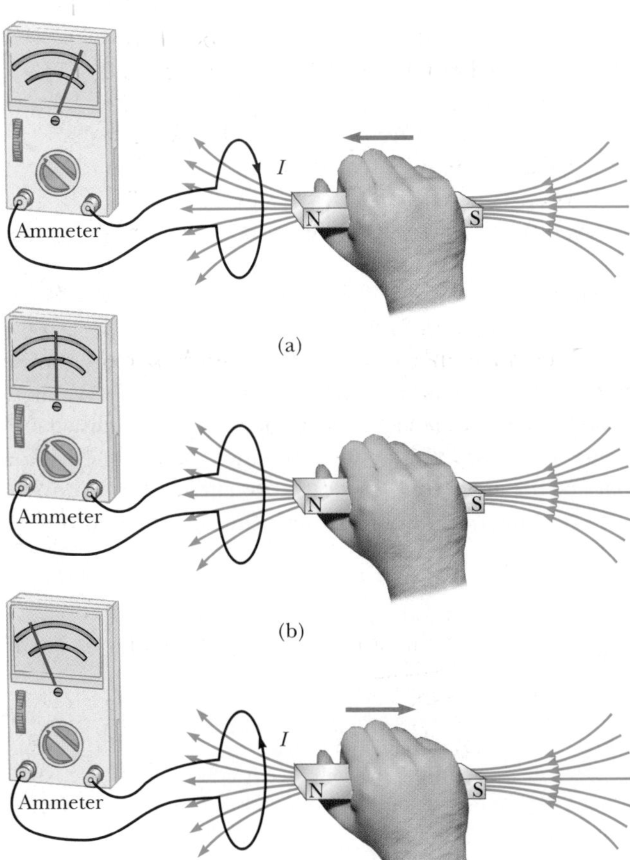


**FIGURE 20.4** (a) When a magnet is moved toward a wire loop connected to an ammeter, the ammeter deflects as shown, indicating that a current $I$ is induced in the loop. (b) When the magnet is held stationary, no current is induced in the loop, even when the magnet is inside the loop. (c) When the magnet is moved away from the loop, the ammeter deflects in the opposite direction, indicating that the induced current is opposite that shown in part (a).

either toward or away from the magnet, the needle also deflects. From these observations, it can be concluded that **a current is set up in the circuit as long as there is relative motion between the magnet and the loop.** Note that the same experimental results are found whether the loop moves or the magnet moves. We call such a current an **induced current** because it is produced by an **induced emf.**

This experiment has something in common with the Faraday experiment discussed in Section 20.1. In each case, an emf is induced in a circuit when the magnetic flux through the circuit changes with time. In fact, we can make the following general summary of such experiments involving an induced emf and a changing magnetic flux:

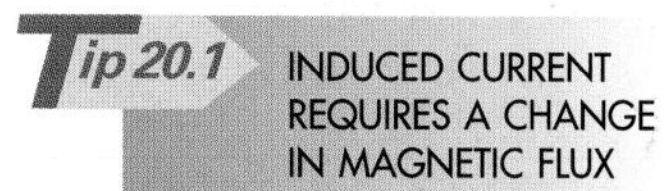

Tip 20.1
INDUCED CURRENT REQUIRES A CHANGE IN MAGNETIC FLUX


Note that the existence of magnetic flux through an area is not sufficient to create an induced emf. A *change* in the magnetic flux over some time interval $\Delta t$ must occur for an emf to be induced.

**The instantaneous emf induced in a circuit equals the time rate of change of magnetic flux through the circuit.**

If a circuit contains $N$ tightly wound loops and the flux through each loop changes by the amount $\Delta\Phi_B$ during the interval $\Delta t$, the average emf induced in the circuit during time $\Delta t$ is

Faraday's law 

$$\mathcal{E} = -N\frac{\Delta\Phi_B}{\Delta t} \qquad [20.2]$$

This is a statement of **Faraday's law of magnetic induction.** Because $\Phi_B = BA\cos\theta$, note that a change with time of any of the factors $B$, $A$, or $\theta$ produces an emf. We explore the effect of a change of each of these factors in the following sections. The minus sign in Equation 20.2 is included to indicate the polarity of the induced emf, which can be found by use of **Lenz's law:**

Lenz's law

**The polarity of the induced emf is such that it produces a current whose magnetic field opposes the change in magnetic flux through the loop. That is, the induced current tends to maintain the original flux through the circuit.**

We consider several applications of Lenz's law in Section 20.4.

GROUND FAULT INTERRUPTERS

The ground fault interrupter (GFI) is an interesting safety device that protects people against electric shock when they touch appliances and power tools. Its operation makes use of Faraday's law. Figure 20.5 shows the essential parts of a ground fault interrupter. Wire 1 leads from the wall outlet to the appliance to be protected, and wire 2 leads from the appliance back to the wall outlet. An iron ring surrounds the two wires to confine the magnetic field set up by each wire. A sensing coil, which can activate a circuit breaker when changes in magnetic flux occur, is wrapped around part of the iron ring. Because the currents in the wires are in opposite directions, the net magnetic field through the sensing coil due to

Alternating current
Circuit breaker
Sensing coil
Iron ring
1
2

**FIGURE 20.5** Essential components of a ground fault interrupter (contents of the gray box in Fig. 20.6a). In newer homes, such devices are built directly into wall outlets. The purpose of the sensing coil and circuit breaker is to cut off the current before damage is done.

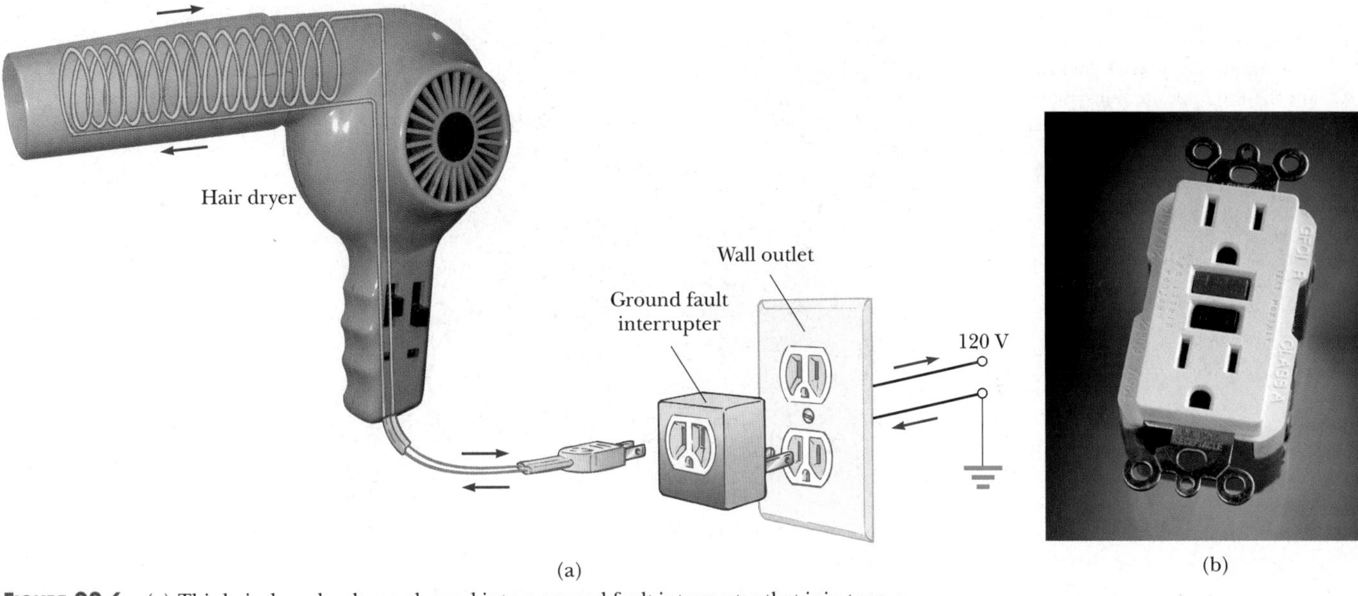


**FIGURE 20.6** (a) This hair dryer has been plugged into a ground fault interrupter that is in turn plugged into an unprotected wall outlet. (b) You have likely seen such a ground fault interrupter in a hotel bathroom, where hair dryers and electric shavers will be used by people just out of the shower or who might touch a water pipe, providing a ready path to ground in the event of a short circuit. *(b, Richard Megna, Fundamental Photographs, NYC)*

the currents is zero. However, if a short circuit occurs in the appliance so that there is no returning current, the net magnetic field through the sensing coil is no longer zero. This can happen, for example, if one of the wires loses its insulation, providing a path through you to ground if you happen to be touching the appliance and are grounded as in Figure 18.22a. Because the current is alternating, the magnetic flux through the sensing coil changes with time, producing an induced voltage in the coil. This induced voltage is used to trigger a circuit breaker, stopping the current quickly in about a millisecond before it reaches a level that might be harmful to the person using the appliance. A ground fault interrupter provides faster and more complete protection than even the case-ground-and-circuit-breaker combination shown in Figure 18.22b. For this reason, ground fault interrupters are commonly found in bathrooms, where electricity poses a hazard to people (see Figure 20.6).

Another interesting application of Faraday's law is the production of sound in an electric guitar. A vibrating string induces an emf in a coil (Fig. 20.7). The

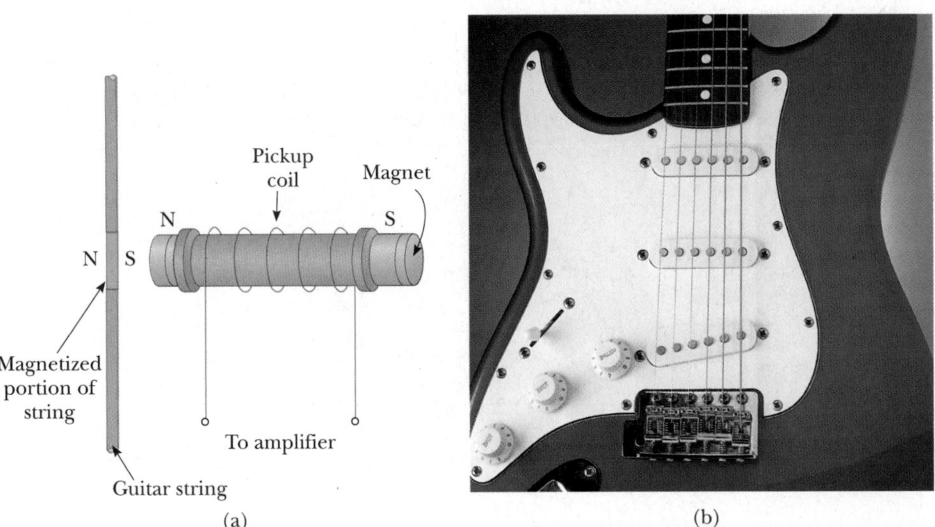


**FIGURE 20.7** (a) In an electric guitar, a vibrating string induces a voltage in the pickup coil. (b) Several pickups allow the vibration to be detected from different portions of the string. *(b, Charles D. Winters)*

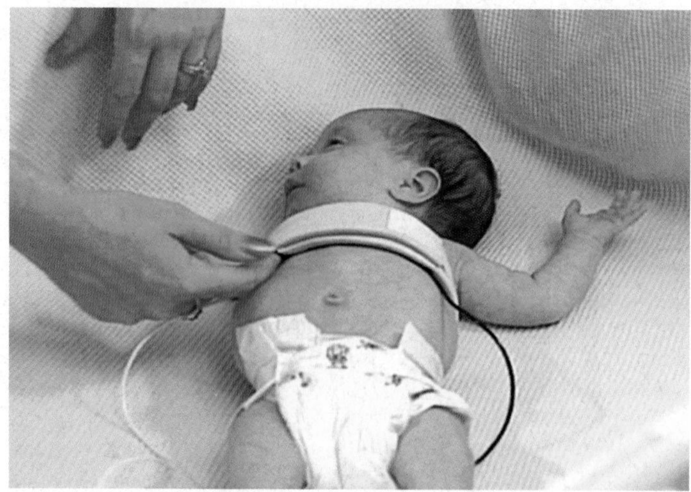

**FIGURE 20.8** This infant is wearing a monitor designed to alert caregivers if breathing stops. Note the two wires attached to opposite sides of the chest. *(Courtesy of PedsLink Pediatric Healthcare Resources, Newport Beach, CA)*

APPLICATION

ELECTRIC GUITAR PICKUPS

pickup coil is placed near the vibrating guitar string, which is made of a metal that can be magnetized. The permanent magnet inside the coil magnetizes the portion of the string nearest the coil. When the guitar string vibrates at some frequency, its magnetized segment produces a changing magnetic flux through the pickup coil. The changing flux induces a voltage in the coil; the voltage is fed to an amplifier. The output of the amplifier is sent to the loudspeakers, producing the sound waves that we hear.

APPLICATION

APNEA MONITORS

Sudden Infant Death Syndrome (SIDS) is a devastating affliction in which a baby suddenly stops breathing during sleep without an apparent cause. One type of monitoring device sometimes used to alert caregivers of the cessation of breathing uses induced currents, as shown in Figure 20.8. A coil of wire attached to one side of the chest carries an alternating current. The varying magnetic flux produced by this current passes through a pickup coil attached to the opposite side of the chest. Expansion and contraction of the chest caused by breathing or movement changes the strength of the voltage induced in the pickup coil. However, if breathing stops, the pattern of the induced voltage stabilizes, and external circuits monitoring the voltage sound an alarm to the caregivers after a momentary pause to ensure that a problem actually does exist.

**Quick Quiz 20.1**

Figure 20.9 is a graph of the magnitude $B$ versus time $t$ for a magnetic field that passes through a fixed loop and is oriented perpendicular to the plane of the loop. Rank the magnitudes of the emf generated in the loop at the three instants indicated ($a$, $b$, $c$), from largest to smallest.

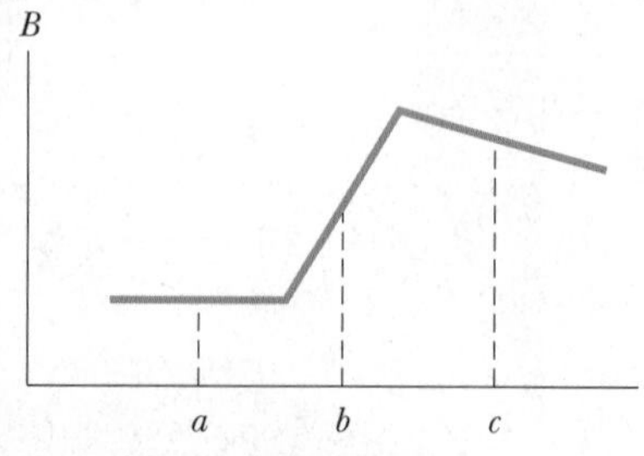


**FIGURE 20.9** (Quick Quiz 20.1)

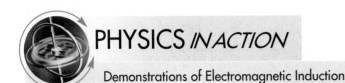

## Demonstrations of Electromagnetic Induction

On the left is a simple demonstration involving a permanent magnet, a conducting coil, and a *galvanometer,* which is a device used in the construction of analog ammeters and voltmeters. The galvanometer needle can deflect to either side of the zero mark, indicating the direction of a current. When a strong magnet is moved toward or away from the coil attached to the galvanometer, a current is induced in the coil, indicated by the momentary deflection of the galvanometer during the movement of the magnet. What is the cause of this induced current? Would the galvanometer needle deflect if the coil were moved toward a stationary magnet?

On the right, to demonstrate electromagnetic induction, an ac voltage is applied to the lower coil in the apparatus. A voltage is induced in the upper coil, as indicated by the illuminated flashlight bulb connected to this coil. What do you think happens to the bulb's intensity as the upper coil is moved over the vertical tube? To answer this question, note that the magnetic field associated with the lower coil varies along the axis of the tube.

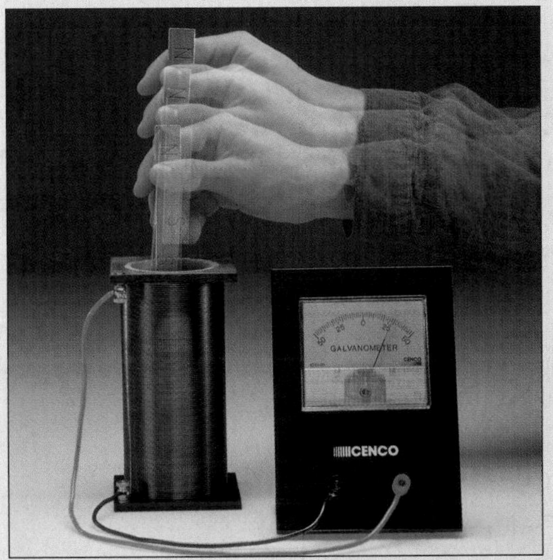


*(Richard Megna, Fundamental Photographs)*

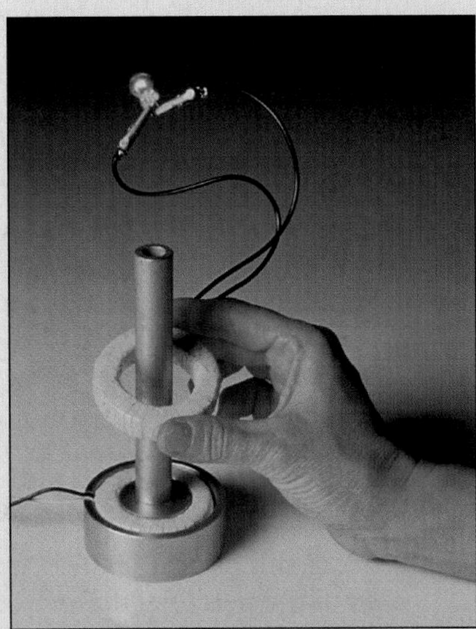

*(Courtesy of Central Scientific Company)*

## Example 20.1 Application of Faraday's Law (Change of *B* with time)

A coil with 25 turns of wire is wrapped on a square frame 1.80 cm on a side. Each turn has the same area, equal to that of the frame, and the total resistance of the coil is 0.35 Ω. A uniform magnetic field is applied perpendicularly to the plane of the coil. If the field changes uniformly from 0 to 0.500 T in 0.800 s, find the magnitude of the induced emf in the coil while the field is changing.

**Reasoning** The magnitude of the induced emf can be found from Faraday's law of induction, $\mathcal{E} = -N(\Delta\Phi_B/\Delta t)$. In this equation, $\Delta\Phi_B$ is the difference between the final and initial fluxes, where $\Phi_B = BA\cos\theta = BA\cos 0° = BA$.

**Solution** The area of the coil is $(0.018\,0\text{ m})^2 = 3.24 \times 10^{-4}\text{ m}^2$. The magnetic flux through the coil at $t = 0$ is zero because $B = 0$. At $t = 0.800$ s, the magnetic flux through the coil is

$$\Phi_{B,f} = BA = (0.500\text{ T})(3.24 \times 10^{-4}\text{ m}^2) = 1.62 \times 10^{-4}\text{ T}\cdot\text{m}^2$$

Therefore, the *change* in flux through the coil during the 0.800-s interval is

$$\Delta\Phi_B = \Phi_{B,f} - \Phi_{B,i} = 1.62 \times 10^{-4}\ \mathrm{T \cdot m^2}$$

Faraday's law of induction enables us to find the magnitude of the induced emf:

$$|\mathcal{E}| = N\frac{\Delta\Phi_B}{\Delta t} = (25\ \text{turns})\left(\frac{1.62 \times 10^{-4}\ \mathrm{T \cdot m^2}}{0.800\ \mathrm{s}}\right) = 5.1\ \mathrm{mV}$$

Note that $1\ \mathrm{T \cdot m^2/s} = 1(\mathrm{N \cdot m/C \cdot s})(\mathrm{m^2/s}) = 1\ \mathrm{N \cdot m/C} = 1\ \mathrm{J/C} = 1\ \mathrm{V}$.

**EXERCISE** Find the magnitude of the induced current in the coil while the field is changing.

**ANSWER** 14 mA

## 20.3 MOTIONAL emf

In Section 20.2, we considered a situation in which an emf is induced in a circuit when the magnetic field changes with time. In this section we describe a particular application of Faraday's law in which a so-called **motional emf** is produced. This is the emf induced in a conductor moving through a magnetic field.

First consider a straight conductor of length $\ell$ moving with constant velocity through a uniform magnetic field directed into the paper, as in Figure 20.10. For simplicity, we assume that the conductor is moving perpendicularly to the field. The electrons in the conductor experience a force of magnitude $F = qvB$ directed downward along the conductor. Because of this magnetic force, the free electrons move to the lower end and accumulate there, leaving a net positive charge at the upper end. As a result of this charge separation, an electric field is produced in the conductor. The charge at the ends builds up until the downward magnetic force $qvB$ is balanced by the upward electric force $qE$. At this point, charge stops flowing and the condition for equilibrium requires that

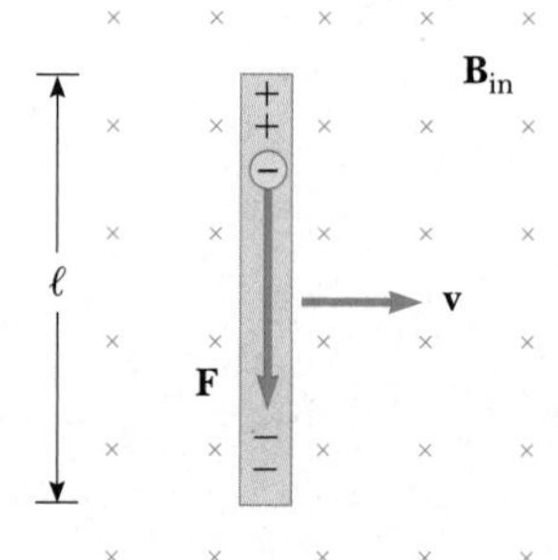

**FIGURE 20.10** A straight conductor of length $\ell$ moving with velocity **v** through a uniform magnetic field **B** directed perpendicularly to **v**. The vector **F** is the force on an electron in the conductor. An emf of $B\ell v$ is induced between the ends of the bar.

$$qE = qvB \qquad \text{or} \qquad E = vB$$

Because the electric field is uniform, the field produced in the conductor is related to the potential difference across the ends by $\Delta V = E\ell$. Thus,

$$\Delta V = E\ell = B\ell v \tag{20.3}$$

Because there is an excess of positive charge at the upper end of the conductor and an excess of negative charge at the lower end, the upper end is at a higher potential than the lower end. Thus,

**A potential difference is maintained across the conductor as long as there is motion through the field. If the motion is reversed, the polarity of the potential difference is also reversed.**

A more interesting situation occurs if the moving conductor is part of a closed conducting path. This situation is particularly useful for illustrating how a changing loop area induces a current in a closed circuit described by Faraday's law. Consider a circuit consisting of a conducting bar of length $\ell$, sliding along two fixed parallel conducting rails, as in Figure 20.11a. For simplicity, we assume that the moving bar has zero resistance and that the stationary part of the circuit has constant resistance $R$. A uniform and constant magnetic field **B** is applied perpendicularly to the plane of the circuit. As the bar is pulled to the right with velocity **v** under the influence of an applied force $\mathbf{F}_{app}$, the free charges in the bar experience a magnetic force along the length of the bar. This force in turn sets up an induced current because the charges are free to move in a closed conducting path. In this

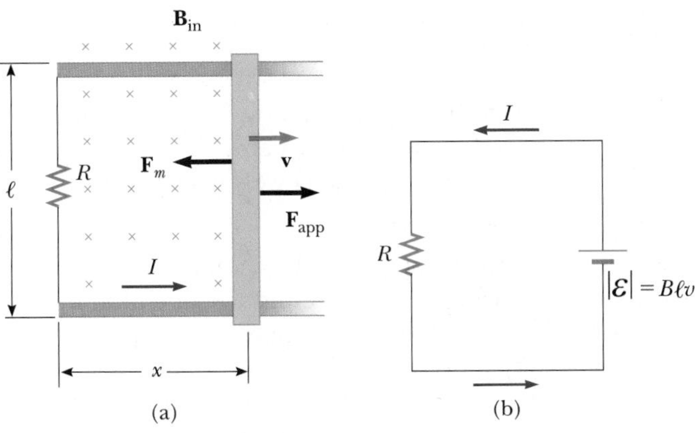


**FIGURE 20.11** (a) A conducting bar sliding with velocity **v** along two conducting rails under the action of an applied force $\mathbf{F}_{app}$. The magnetic force $\mathbf{F}_m$ opposes the motion, and a counterclockwise current is induced in the loop. (b) The equivalent circuit of (a).

case, the changing magnetic flux through the loop and the corresponding induced emf across the moving bar arise from the *change in area of the loop* as the bar moves through the magnetic field.

Let us assume that the bar moves a distance $\Delta x$ in time $\Delta t$, as shown in Figure 20.12. The increase in flux $\Delta\Phi_B$ through the loop in that time is the amount of flux that now passes through the portion of the circuit that has area $\ell\,\Delta x$:

$$\Delta\Phi_B = BA = B\ell\,\Delta x$$

Using Faraday's law and noting that there is one loop ($N = 1$), we find that the magnitude of the induced emf is

$$|\mathcal{E}| = \frac{\Delta\Phi_B}{\Delta t} = B\ell\frac{\Delta x}{\Delta t} = B\ell v \qquad [20.4]$$

This induced emf is often called a **motional emf** because it arises from the motion of a conductor through a magnetic field.

Furthermore, if the resistance of the circuit is $R$, the magnitude of the induced current in the circuit is

$$I = \frac{|\mathcal{E}|}{R} = \frac{B\ell v}{R} \qquad [20.5]$$

Figure 20.11b is the equivalent circuit diagram for this example.

## Webnote 20.1

Change the shape of a coil in a magnetic field and watch as a current is induced in the coil. Visit *http://www.lightlink.com/sergey/java/java/indcur/index.html*

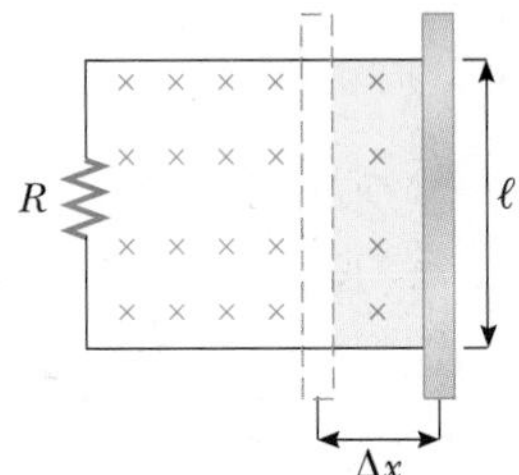


**FIGURE 20.12** As the bar moves to the right, the area of the loop increases by the amount $\ell\,\Delta x$ and the magnetic flux through the loop increases by $B\ell\,\Delta x$.

## APPLYING PHYSICS 20.2

We have discussed applying a force on the bar, which results in an induced emf in the circuit shown in Figure 20.11. Suppose we remove the external magnetic field in the diagram and replace the resistor with a high-voltage source and a switch, as in Figure 20.13. What happens when the switch is closed? Will the bar move, and does it matter which way we connect the high-voltage source?

**Explanation** Suppose the source is capable of establishing high current. The two horizontal conducting rods will create a strong magnetic field in the area between them, directed into the page. (The movable bar also creates a magnetic field, but this field cannot exert force on the bar itself.) As the current passes downward through the movable bar, it experiences a magnetic force to the right. Hence, it accelerates along the rails away from the power supply. If the polarity of the power were reversed, the magnetic field would be out of the page, the current in the bar would be upward, and the force on the bar would still be to the right. (This is the essence of a railgun.) The $BI\ell$ force exerted by a magnetic field per Equation 19.6 causes the bar to accelerate away from the voltage source. Studies have shown that it is possible to launch payloads into space with this technology. Very large accelerations can be obtained with currently available technology, with payloads being accelerated to a speed of several kilometers per second in a fraction of a second. This is a larger acceleration than humans can withstand.

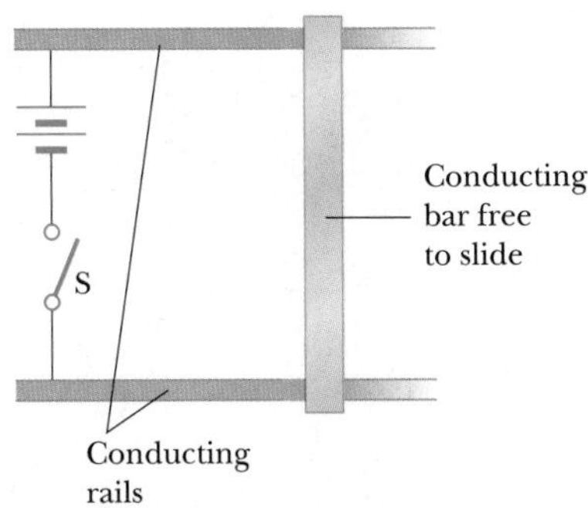


**FIGURE 20.13** (Applying Physics 20.2)

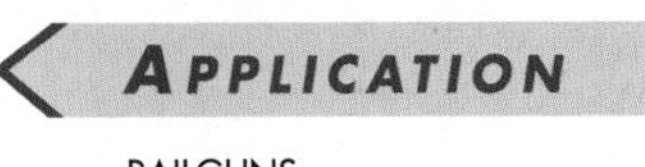


RAILGUNS

**Quick Quiz 20.2** As an airplane flies due north from Los Angeles to Seattle, it cuts through Earth's magnetic field. As a result, an emf is developed between the wing tips. Which wing tip is positively charged?

**Quick Quiz 20.3** You wish to move a rectangular loop of wire into a region of uniform magnetic field at a given speed so as to induce an emf in the loop. The plane of the loop must remain perpendicular to the magnetic field lines. In which orientation should you hold the loop while you move it into the region of magnetic field in order to generate the largest emf? (a) With the long dimension of the loop parallel to the velocity vector. (b) With the short dimension of the loop parallel to the velocity vector. (c) Either way—the emf is the same regardless of orientation.

### Example 20.2 The Electrified Airplane Wing

An airplane with a wing span of 30.0 m flies parallel to Earth's surface at a location at which the downward component of Earth's magnetic field is $0.60 \times 10^{-4}$ T. Find the difference in potential between the wing tips when the speed of the plane is 250 m/s.

**Solution** Because the plane is flying horizontally, we do not have to concern ourselves with the horizontal component of Earth's magnetic field. Thus, we find that

$$\mathcal{E} = B\ell v = (0.60 \times 10^{-4}\ \text{T})(30.0\ \text{m})(250\ \text{m/s}) = \boxed{0.45\ \text{V}}$$

### Example 20.3 Where Is the Energy Source?

**A** The sliding bar in Figure 20.11a has a length of 0.50 m and moves at 2.0 m/s in a magnetic field of magnitude 0.25 T. Find the induced voltage in the moving rod.

**Solution** We use Equation 20.3 and find that

$$\mathcal{E} = B\ell v = (0.25\ \text{T})(0.50\ \text{m})(2.0\ \text{m/s}) = \boxed{0.25\ \text{V}}$$

**B** If the resistance in the circuit is 0.50 Ω, find the current in the circuit.

**Solution** The current is found from Ohm's law to be

$$I = \frac{\mathcal{E}}{R} = \frac{0.25\ \text{V}}{0.50\ \Omega} = \boxed{0.50\ \text{A}}$$

**C** Find the amount of energy delivered to the 0.50-Ω resistor in 1 s.

**Solution** The power delivered to the resistor is

$$\mathcal{P} = I\,\Delta V = (0.50\ \text{A})(0.25\ \text{V}) = 0.13\ \text{W}$$

Because power is defined as the rate at which energy is converted in a device, the energy $W$ delivered to the resistor in 1 s is

$$W = \mathcal{P}t = (0.125\ \text{W})(1.0\ \text{s}) = \boxed{0.13\ \text{J}}$$

**D** The source of the energy calculated in part C is some external agent that keeps the bar moving at a constant speed of 2.0 m/s by exerting an applied force $F_{\text{app}}$. Find the value of $F_{\text{app}}$.

**Solution** From part C, we know that the work done by the applied force in 1 s is 0.13 J. In 1 s, the bar moves a distance of

$$d = vt = (2.0 \text{ m/s})(1.0 \text{ s}) = 2.0 \text{ m}$$

Thus, from the definition of work, we find that $W = F_{\text{app}}d$, or

$$F_{\text{app}} = \frac{W}{d} = \frac{0.13 \text{ J}}{2.0 \text{ m}} = 0.063 \text{ N}$$

**Exercise** If the rod is to move at constant speed, the applied force must be equal in magnitude to the retarding magnetic force, $BI\ell$. Show that this approach also gives $F_{\text{app}} = 0.063$ N, as found in part (d).

## 20.4 LENZ'S LAW REVISITED (THE MINUS SIGN IN FARADAY'S LAW)

To attain a better understanding of Lenz's law, let us return to the example of a bar moving to the right on two parallel rails in the presence of a uniform magnetic field directed into the paper (Fig. 20.14a). As the bar moves to the right, the magnetic flux through the circuit increases with time because the area of the loop increases. Lenz's law says that the induced current must be in a direction such that the flux *it* produces opposes the change in the external magnetic flux. Because the flux due to the external field is increasing *into* the paper, the induced current, to oppose the change, must produce a flux *out* of the paper. Hence, the induced current must be counterclockwise when the bar moves to the right. (Use right-hand rule #2 from Chapter 19 to verify this direction.) On the other hand, if the bar is moving to the left, as in Figure 20.14b, the magnetic flux through the loop decreases with time. Because the flux is into the paper, the induced current has to be clockwise to produce its own flux into the paper (which opposes the decrease in the external flux). In either case, the induced current tends to maintain the original flux through the circuit.

Now let us examine this situation from the viewpoint of energy conservation. Suppose that the bar is given a slight push to the right. In the preceding analysis, we found that this motion led to a counterclockwise current in the loop. What would happen if we assume that the current is clockwise, that is, opposite the direction required by Lenz's law? For a clockwise current $I$, the direction of the magnetic force $BI\ell$ on the sliding bar would be to the right. This force would accelerate the rod and increase its velocity. This, in turn, would cause the area of the loop to increase more rapidly, thereby increasing the induced current, which would increase the force, which would increase the current, which would. . . . In effect, the system would acquire energy with zero input energy. This is clearly inconsistent both with our experience and with the law of conservation of energy. We are forced to conclude that the current must be counterclockwise.

Consider another situation. A bar magnet is moved to the right toward a stationary loop of wire, as in Figure 20.15a. As the magnet moves, the magnetic flux through the loop increases with time. To counteract this, the induced current produces a flux to the left, as in Figure 20.15b; hence, the induced current is in the direction shown. Note that the magnetic field lines associated with the induced current oppose the motion of the magnet. Therefore, the left face of the current loop is a north pole and the right face is a south pole.

On the other hand, if the magnet moves to the left, as in Figure 20.15c, its flux through the loop, which is toward the right, decreases in time. Under these circumstances, the induced current in the loop is in a direction to set up a field directed from left to right through the loop, in an effort to maintain a constant

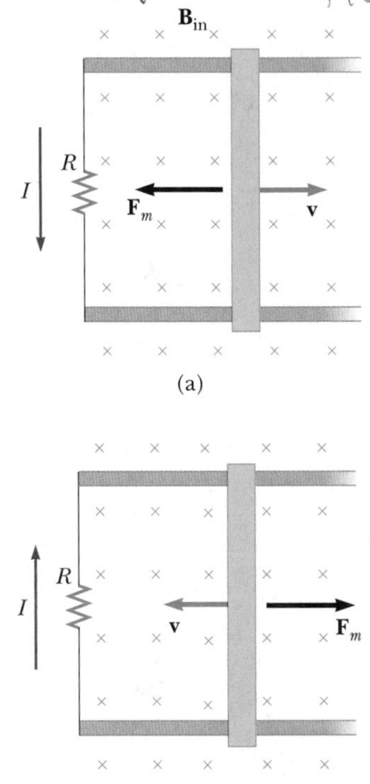

**FIGURE 20.14** (a) As the conducting bar slides on the two fixed conducting rails, the magnetic flux through the loop increases in time. By Lenz's law, the induced current must be *counterclockwise* so as to produce a counteracting flux *out of the paper.* (b) When the bar moves to the left, the induced current must be *clockwise.* Why?

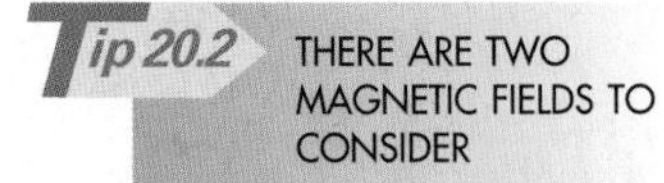
Tip 20.2
THERE ARE TWO MAGNETIC FIELDS TO CONSIDER

When applying Lenz's law, note that there are *two* magnetic fields to consider. The first is the external changing magnetic field that induces the current in a conducting loop. The second is the magnetic field produced *by* the induced current in the loop.

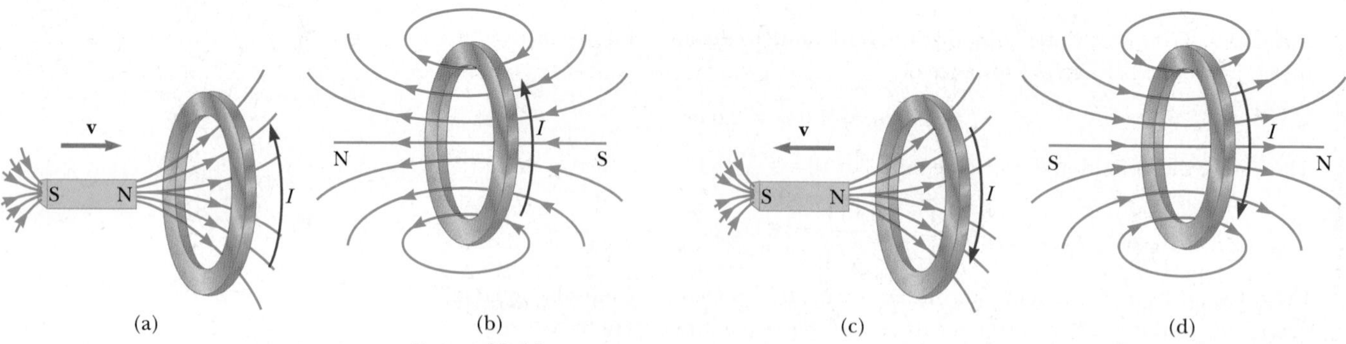


**FIGURE 20.15** (a) When the magnet is moved toward the stationary conducting loop, a current is induced in the direction shown. (b) This induced current produces its own flux to the left to counteract the increasing external flux to the right. (c) When the magnet is moved away from the stationary conducting loop, a current is induced in the direction shown. (d) This induced current produces its own flux to the right to counteract the decreasing external flux to the right.

number of flux lines. Hence, the induced current in the loop is as shown in Figure 20.15d. In this case, the left face of the loop is a south pole and the right face is a north pole.

| **Quick Quiz 20.4** | A bar magnet is falling through a loop of wire with constant velocity with the north pole entering first. Viewed from the same side of the loop as the magnet, as the north pole approaches the loop, the induced current will be in what direction? (a) clockwise (b) zero (c) counterclockwise (d) along the length of the magnet |
|---|---|

## Example 20.4 Application of Lenz's Law

A coil of wire is placed near an electromagnet, as in Figure 20.16a. Find the direction of the induced current in the coil (a) at the instant the switch is closed, (b) after the switch has been closed for several seconds, and (c) when the switch is opened.

### Reasoning and Solution

**A** When the switch is closed, the situation changes from a condition in which no lines of flux pass through the coil to one in which lines of flux pass through in the direction

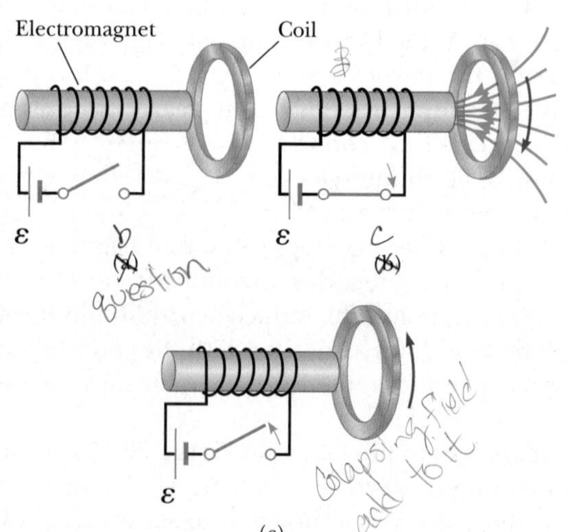


**FIGURE 20.16** (Example 20.4)

shown in Figure 20.16b. To counteract this change in the number of lines, the coil must set up a field from left to right in the figure. This requires a current directed as shown in Figure 20.16b.

**B** After the switch has been closed for several seconds, there is no change in the number of lines through the loop; hence, the induced current is zero.

**C** Opening the switch causes the magnetic field to change from a condition in which flux lines thread through the coil from right to left to a condition of zero flux. The induced current must then be as shown in Figure 20.16c, so as to set up its own field from right to left.

## Tape Recorders

One common practical use of induced currents and emfs is in the tape recorder. Many different types of tape recorders are made, but the basic principles are the same for all. A magnetic tape moves past a recording and playback head, as in Figure 20.17a. The tape is a plastic ribbon coated with iron oxide or chromium oxide.

**APPLICATION**

MAGNETIC TAPE RECORDERS

The recording process uses the fact that a current in an electromagnet produces a magnetic field. Figure 20.17b illustrates the steps in the process. A sound wave sent into a microphone is transformed into an electric current, amplified, and allowed to pass through a wire coiled around a doughnut-shaped piece of iron, which functions as the recording head. The iron ring and the wire constitute an electromagnet, in which the lines of the magnetic field are contained completely inside the iron except at the point where a slot is cut in the ring. Here the magnetic field fringes out of the iron and magnetizes the small pieces of iron oxide embedded in the tape. Thus, as the tape moves past the slot, it becomes magnetized in a pattern that reproduces both the frequency and the intensity of the sound signal entering the microphone.

To reconstruct the sound signal, the previously magnetized tape is allowed to pass through a recorder head operating in the playback mode. A second wire-wound doughnut-shaped piece of iron with a slot in it passes close to the tape, so that the varying magnetic fields on the tape produce changing field lines through the wire coil. The changing flux induces a current in the coil that corresponds to the current in the recording head that originally produced the tape. This changing electric current can be amplified and used to drive a speaker. Playback is thus an example of induction of a current by a moving magnet.

### Webnote 20.2

Take a look at audiotapes, tape decks, and tape recorders by visiting *http://www.howstuffworks.com/cassette.htm*

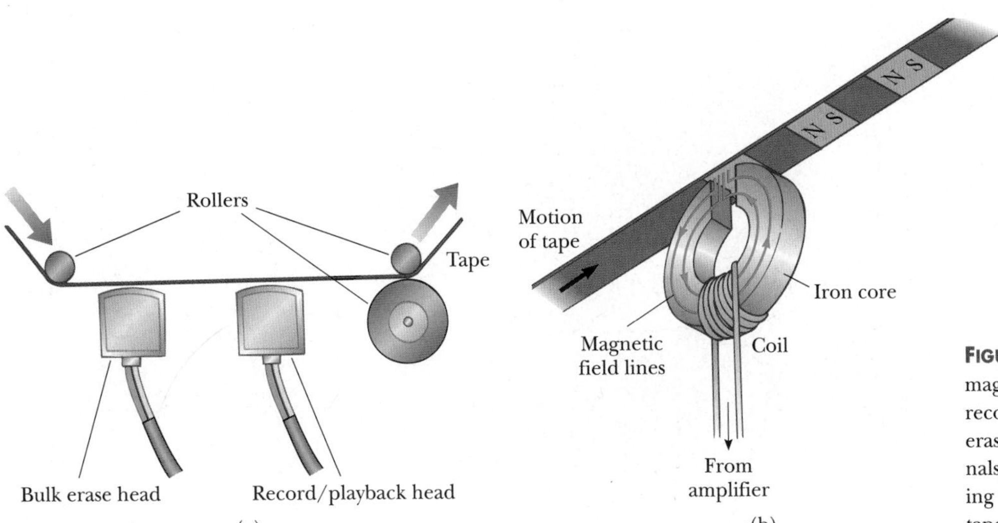


**FIGURE 20.17** (a) Major parts of a magnetic tape recorder. If a new recording is to be made, the bulk erase head wipes the tape clean of signals before recording. (b) The fringing magnetic field magnetizes the tape during recording.

## 20.5 GENERATORS

**APPLICATION**

ALTERNATING CURRENT GENERATORS

Generators and motors are important practical devices that operate on the principle of electromagnetic induction. First, let us consider the **alternating current (AC) generator,** a device that converts mechanical energy to electrical energy. In its simplest form, the AC generator consists of a wire loop rotated in a magnetic field by some external means (Fig. 20.18a). In commercial power plants, the energy required to rotate the loop can be derived from a variety of sources. For example, in a hydroelectric plant, falling water directed against the blades of a turbine produces the rotary motion; in a coal-fired plant, heat produced by burning coal is used to convert water to steam, and this steam is directed against the turbine blades. As the loop rotates, the magnetic flux through it changes with time, inducing an emf and a current in an external circuit. The ends of the loop are connected to slip rings that rotate with the loop. Connections to the external circuit are made by stationary brushes in contact with the slip rings.

We can derive an expression for the emf generated in the rotating loop by making use of the expression for motional emf, $\mathcal{E} = B\ell v$. Figure 20.19a shows a loop of wire rotating clockwise in a uniform magnetic field directed to the right. The magnetic force ($qvB$) on the charges in wires $AB$ and $CD$ is not along the lengths of the wires. (The force on the electrons in these wires is perpendicular to the wires.) Hence, an emf is generated only in wires $BC$ and $AD$. At any instant, wire $BC$ has velocity $\mathbf{v}$ at an angle $\theta$ with the magnetic field, as shown in Figure 20.19b. (Note that the component of velocity parallel to the field has no effect on the charges in the wire, whereas the component of velocity perpendicular to the field produces a magnetic force on the charges that moves electrons from $C$ to $B$.) The emf generated in wire $BC$ equals $B\ell v_\perp$, where $\ell$ is the length of the wire and $v_\perp$ is the component of velocity perpendicular to the field. An emf of $B\ell v_\perp$ is also generated in wire $DA$, and the sense of this emf is the same as in wire $BC$. Because $v_\perp = v \sin \theta$, the total induced emf is

$$\mathcal{E} = 2B\ell v_\perp = 2B\ell v \sin \theta \qquad \textbf{[20.6]}$$

If the loop rotates with a constant angular speed $\omega$, we can use the relation $\theta = \omega t$ in Equation 20.6. Furthermore, because every point on wires $BC$ and $DA$ rotates in a circle about the axis of rotation with the same angular speed $\omega$, we have $v = r\omega = (a/2)\omega$, where $a$ is the length of sides $AB$ and $CD$. Therefore, Equation 20.6 reduces to

$$\mathcal{E} = 2B\ell \left(\frac{a}{2}\right) \omega \sin \omega t = B\ell a\omega \sin \omega t$$

If a coil has $N$ turns, the emf is $N$ times as large because each loop has the same

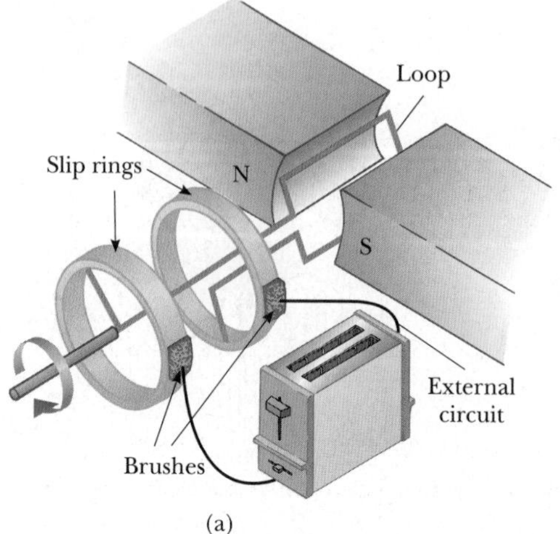


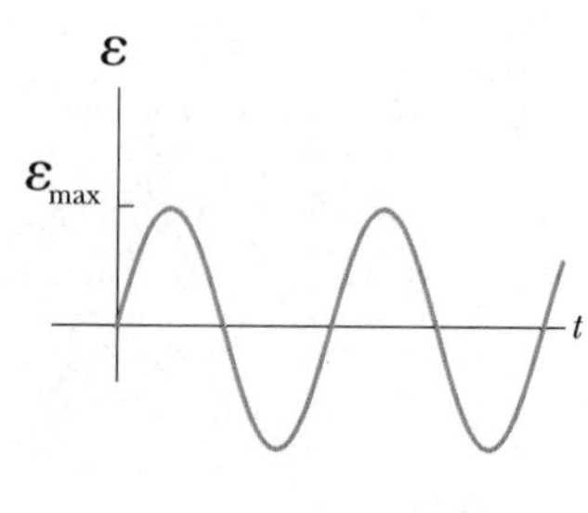


**FIGURE 20.18** (a) Diagram of an AC generator. An emf is induced in a coil, which rotates by some external means in a magnetic field. (b) Plot of the alternating emf induced in the loop versus time.

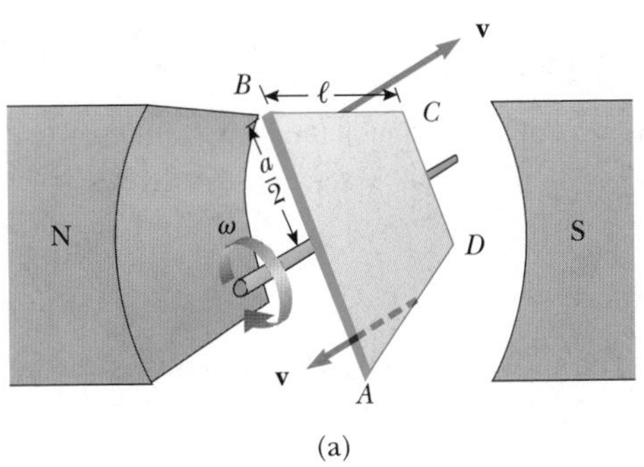


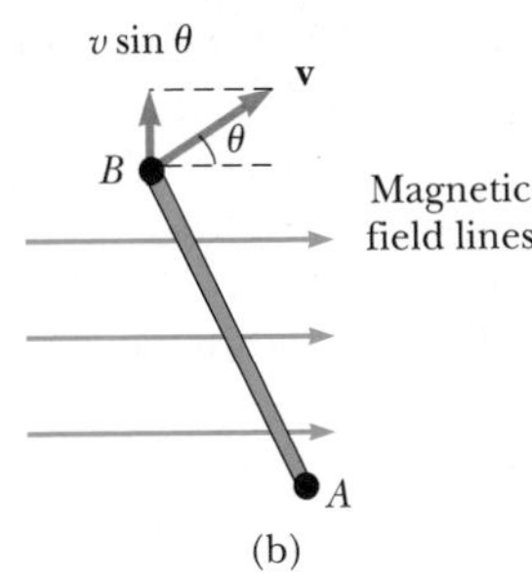


**FIGURE 20.19** (a) A loop rotating at constant angular velocity in an external magnetic field. The emf induced in the loop varies sinusoidally with time. (b) An edge view of the rotating loop.

emf induced in it. Furthermore, because the area of the loop is $A = \ell a$, the total emf is

$$\mathcal{E} = NBA\omega \sin \omega t \qquad [20.7]$$

This result shows that the emf varies sinusoidally with time, as plotted in Figure 20.18b. Note that the maximum emf has the value

$$\mathcal{E}_{\text{max}} = NBA\omega \qquad [20.8]$$

which occurs when $\omega t = 90°$ or $270°$. In other words, $\mathcal{E} = \mathcal{E}_{\text{max}}$ when the plane of the loop is parallel to the magnetic field. Furthermore, the emf is zero when $\omega t = 0$ or $180°$, that is, when the magnetic field is perpendicular to the plane of the loop. In the United States and Canada, the frequency of rotation for commercial generators is 60 Hz, whereas in some European countries 50 Hz is used. (Recall that $\omega = 2\pi f$, where $f$ is the frequency in hertz.)

Turbines turn electric generators at a hydroelectric power plant. *(Luis Castaneda/The IMAGE Bank)*

The **direct current (DC) generator** is illustrated in Figure 20.20a. The components are essentially the same as those of the AC generator, except that the contacts to the rotating loop are made by a split ring, or commutator. In this design, the output voltage always has the same polarity and the current is a pulsating direct current, as in Figure 20.20b. This can be understood by noting that the contacts to the split ring reverse their roles every half cycle. At the same time, the polarity of the induced emf reverses. Hence, the polarity of the split ring remains the same.

**APPLICATION**

DIRECT CURRENT GENERATORS

A pulsating DC current is not suitable for most applications. To produce a steady DC current, commercial DC generators use many loops and commutators distributed around the axis of rotation so that the sinusoidal pulses from the loops overlap in phase. When these pulses are superimposed, the DC output is almost free of fluctuations.

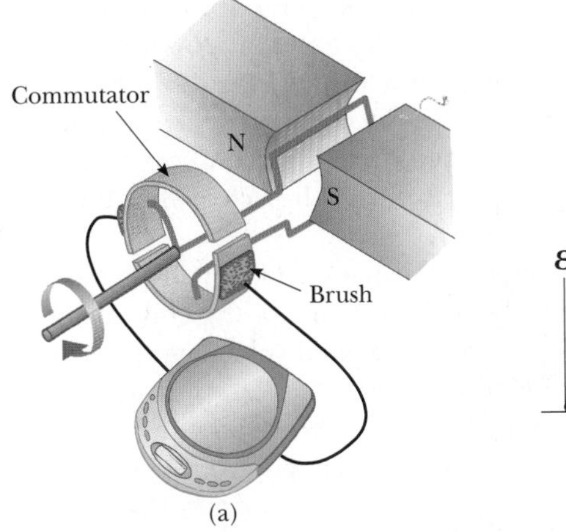


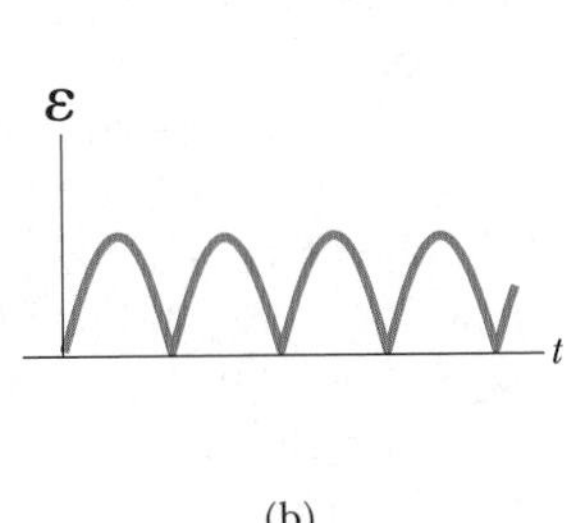


**FIGURE 20.20** (a) Diagram of a DC generator. (b) The emf fluctuates in magnitude but always has the same polarity.

## Example 20.5 Emf Induced in an AC Generator

An AC generator consists of eight turns of wire of area $A = 0.090\,0\ \text{m}^2$ with a total resistance of 12.0 Ω. The loop rotates in a magnetic field of 0.500 T at a constant frequency of 60.0 Hz.

**A** Find the maximum induced emf.

**Solution** First note that $\omega = 2\pi f = 2\pi(60.0\ \text{Hz}) = 377\ \text{rad/s}$. When we substitute the appropriate numerical values into Equation 20.8, we obtain

$$\mathcal{E}_{\text{max}} = NAB\omega = 8(0.090\,0\ \text{m}^2)(0.500\ \text{T})(377\ \text{rad/s}) = \boxed{136\ \text{V}}$$

**B** What is the maximum induced current?

**Solution** From Ohm's law and the result of part A, we find that

$$I_{\text{max}} = \frac{\mathcal{E}_{\text{max}}}{R} = \frac{136\ \text{V}}{12.0\ \Omega} = \boxed{11.3\ \text{A}}$$

**C** Determine the induced emf as a function of time.

**Solution** We can use Equation 20.7 to obtain the time variation of $\mathcal{E}$:

$$\mathcal{E} = \mathcal{E}_{\text{max}} \sin \omega t = \boxed{(136\ \text{V}) \sin 377t}$$

where $t$ is in seconds.

**EXERCISE** Determine the time variation of the induced current.

**ANSWER** $I = (11.3\ \text{A}) \sin 377t$

**APPLICATION**

MOTORS

## MOTORS AND BACK emf

Motors are devices that convert electrical energy to mechanical energy. Essentially, **a motor is a generator run in reverse.** Instead of a current being generated by a rotating loop, a current is supplied to the loop by a source of emf, and the magnetic torque on the current-carrying loop causes it to rotate.

A motor can perform useful mechanical work when a shaft connected to its rotating coil is attached to some external device. As the coil in the motor rotates, however, the changing magnetic flux through it induces an emf, which acts to reduce the current in the coil. If it increased the current, Lenz's law would be violated. The phrase **back emf** is used for an emf that tends to reduce the applied current. The back emf increases in magnitude as the rotational speed of the coil increases. We can picture this state of affairs as the equivalent circuit in Figure 20.21. For illustrative purposes, assume that the external power source attempting to supply current in the coil of the motor has a voltage of 120 V, that the coil has a resistance of 10 Ω, and that the back emf induced in the coil at this instant is 70 V. Thus, the voltage available to supply current equals the difference between the applied voltage and the back emf, 50 V in this case. It is clear that the current is reduced by the back emf.

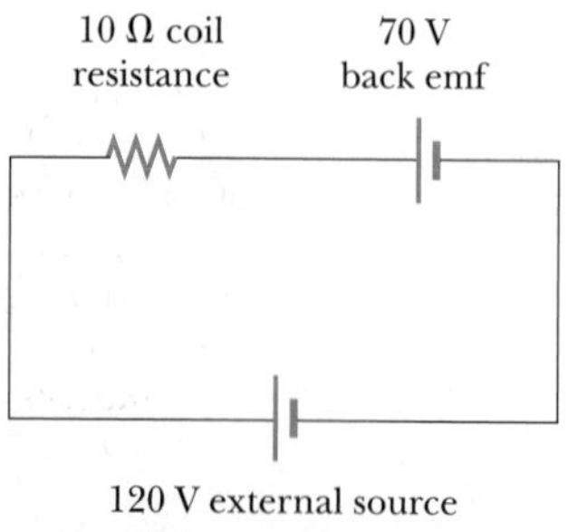

When a motor is turned on, there is no back emf initially, and the current is very large because it is limited only by the resistance of the coil. As the coil begins to rotate, the induced back emf opposes the applied voltage and the current in the coil is reduced. If the mechanical load increases, the motor slows down, which decreases the back emf. This reduction in the back emf increases the current in the coil and therefore also increases the power needed from the external voltage source. As a consequence, the power requirements for starting a motor and for running it under heavy loads are greater than those for running the motor under average loads. If the motor is allowed to run under no mechanical load, the back



**FIGURE 20.21** A motor can be represented as a resistance plus a back emf.

emf reduces the current to a value just large enough to balance energy losses by heat and friction.

## Example 20.6 The Induced Current in a Motor

A motor has coils with a resistance of 10 Ω and is supplied by a voltage of 120 V. When the motor is running at its maximum speed, the back emf is 70 V. Find the current in the coils (a) when the motor is first turned on and (b) when the motor has reached maximum speed.

**Solution**

**A** When the motor is first turned on, the back emf is zero. (The coils are motionless.) Thus, the current in the coils is a maximum and is

$$I = \frac{\mathcal{E}}{R} = \frac{120\text{ V}}{10\ \Omega} = 12\text{ A}$$

**B** At the maximum speed, the back emf has its maximum value. Thus, the effective supply voltage is now that of the external source minus the back emf, and the current is reduced to

$$I = \frac{\mathcal{E} - \mathcal{E}_{\text{back}}}{R} = \frac{120\text{ V} - 70\text{ V}}{10\ \Omega} = \frac{50\text{ V}}{10\ \Omega} = 5.0\text{ A}$$

**EXERCISE** If the current in the motor is 8.0 A at some instant, what is the back emf at this time?

**ANSWER** 40 V

## 20.6 SELF-INDUCTANCE

Consider a circuit consisting of a switch, a resistor, and a source of emf, as in Figure 20.22. When the switch is closed, the current does not immediately change from zero to its maximum value, $\mathcal{E}/R$. The law of electromagnetic induction, Faraday's law, prevents this. What happens instead is the following. As the current increases with time, the magnetic flux through the loop due to this current also increases. The increasing flux induces an emf in the circuit that opposes the change in magnetic flux. By Lenz's law, the induced emf is in the direction indicated by the dashed battery in Figure 20.22. The net potential difference across the resistor is the emf of the battery minus the opposing induced emf. As the magnitude of the current increases, the *rate* of increase lessens and hence the induced emf decreases. This opposing emf results in a gradual increase in the current. For the same reason, when the switch is opened, the current does not immediately decrease to zero. This effect is called **self-induction** because the changing flux through the circuit arises from the circuit itself. The emf that is set up in this case is called a **self-induced emf.**

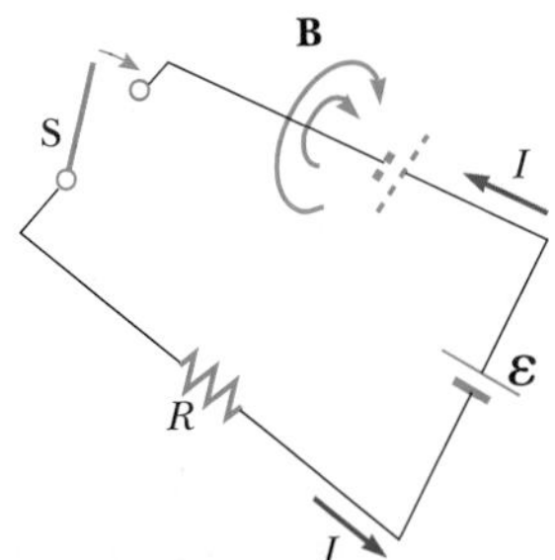

**FIGURE 20.22** After the switch in the circuit is closed, the current produces its own magnetic flux through the loop. As the current increases toward its equilibrium value, the flux changes in time and induces an emf in the loop. The battery drawn with dashed lines is a symbol for the self-induced emf.

As a second example of self-inductance, consider Figure 20.23, which shows a coil wound on a cylindrical iron core. (A practical device would have several hundred turns.) Assume that the current changes with time. When the current is in the direction shown, a magnetic field is set up inside the coil, directed from right to left. As a result, some lines of magnetic flux pass through the cross-sectional area of the coil. As the current changes with time, the flux through the coil changes and induces an emf in the coil. Application of Lenz's law shows that this induced emf has a direction so as to oppose the change in the current. That is, if the current is increasing, the induced emf is as pictured in Figure 20.23b, and if the current is decreasing, the induced emf is as shown in Figure 20.23c.

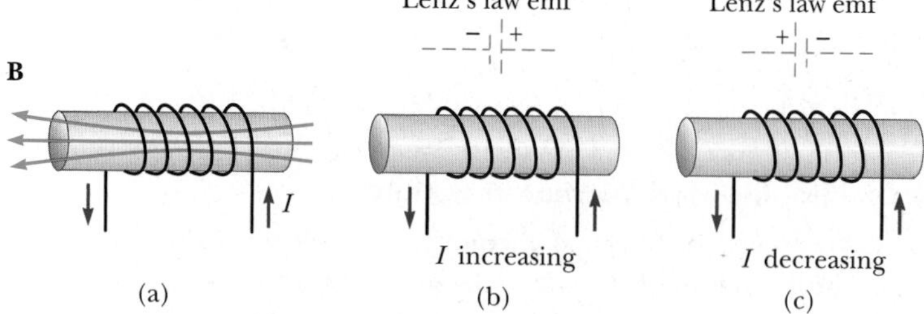


**FIGURE 20.23** (a) A current in the coil produces a magnetic field directed to the left. (b) If the current increases, the coil acts as a source of emf directed as shown by the dashed battery. (c) The induced emf in the coil changes its polarity if the current decreases.

To evaluate self-inductance quantitatively, first note that, according to Faraday's law, the induced emf is given by Equation 20.2:

$$\mathcal{E} = -N\frac{\Delta\Phi_B}{\Delta t}$$

The magnetic flux is proportional to the magnetic field, which is proportional to the current in the coil. Thus, **the self-induced emf must be proportional to the time rate of change of the current:**

Self-induced emf

$$\mathcal{E} \equiv -L\frac{\Delta I}{\Delta t} \quad [20.9]$$

where $L$ is a proportionality constant called the **inductance** of the device. The negative sign indicates that a changing current induces an emf in opposition to that change. This means that if the current is increasing ($\Delta I$ positive), the induced emf is negative, to indicate opposition to the increase in current. Likewise, if the current is decreasing ($\Delta I$ negative), the sign of the induced emf is positive to indicate that the emf is acting to oppose the decrease.

The inductance of a coil depends on the cross-sectional area of the coil and other quantities, all of which can be grouped under the general heading of geometric factors. The SI unit of inductance is the **henry** (H), which, from Equation 20.9, is equal to 1 volt-second per ampere:

$$1 \text{ H} = 1 \text{ V}\cdot\text{s/A}$$

Examples 20.7 and 20.8 discuss simple situations for which self-inductances can be calculated. In the process, it is often convenient to equate Equations 20.2 and 20.9 to find an expression for $L$:

$$N\frac{\Delta\Phi_B}{\Delta t} = L\frac{\Delta I}{\Delta t}$$

Inductance

$$L = N\frac{\Delta\Phi_B}{\Delta I} = \frac{N\Phi_B}{I} \quad [20.10]$$

## APPLYING PHYSICS 20.3

In some circuits, a spark occurs between the poles of a switch when the switch is opened. Yet, when the switch for this circuit is closed, there is no spark. Why is there this difference?

**Explanation** According to Lenz's law, induced emfs are in a direction such as to attempt to maintain the original magnetic flux when a change occurs. When the switch is opened, the sudden drop in the magnetic field in the circuit induces an emf in a direction that attempts to maintain the original current. This can cause a spark as the current bridges the air gap between the poles of the switch. The spark does not occur when the switch is closed, because the original current is zero, and the induced emf attempts to maintain it at zero.

## Example 20.7 Inductance of a Solenoid

Find the inductance of a uniformly wound solenoid with $N$ turns and length $\ell$. Assume that $\ell$ is large compared with the radius and that the core of the solenoid is air.

**Reasoning** The inductance can be found from $L = N\Phi_B/I$. The flux through each turn is $\Phi_B = BA$, and $B = \mu_0 nI$.

**Solution** We take the interior field to be uniform and given by Equation 19.15:

$$B = \mu_0 nI = \mu_0 \frac{N}{\ell} I$$

where $n = N/\ell$ is the number of turns per unit length. The flux through each turn is

$$\Phi_B = BA = \mu_0 \frac{N}{\ell} AI$$

where $A$ is the cross-sectional area of the solenoid. From this expression and Equation 20.10, we find that

$$L = \frac{N\Phi_B}{I} = \frac{\mu_0 N^2 A}{\ell} \quad (1)$$

This shows that $L$ depends on the geometric factors $\ell$ and $A$ and on $\mu_0$ and is proportional to the square of the number of turns. Because $N = n\ell$ we can also express the result in the form

$$L = \mu_0 \frac{(n\ell)^2}{\ell} A = \mu_0 n^2 A\ell = \mu_0 n^2 V \quad (2)$$

where $V = A\ell$ is the volume of the solenoid.

JOSEPH HENRY, AMERICAN PHYSICIST (1797–1878)

Henry became the first director of the Smithsonian Institution and first president of the Academy of Natural Science. He was the first to produce an electric current with a magnetic field, but he failed to publish his results as early as Faraday because of his heavy teaching duties at the Albany Academy in New York state. He improved the design of the electromagnet and constructed one of the first motors. He also discovered the phenomena of self-induction. The unit of inductance, the henry, is named in his honor. *(North Wind Picture Archives)*

## Example 20.8 Calculating Inductance and Self-Induced emf

**A** Calculate the inductance of a solenoid containing 300 turns if the length of the solenoid is 25.0 cm and its cross-sectional area is $4.00\ \text{cm}^2 = 4.00 \times 10^{-4}\ \text{m}^2$.

**Solution** Using the results to Example 20.7, we get

$$L = \frac{\mu_0 N^2 A}{\ell} = (4\pi \times 10^{-7}\ \text{T}\cdot\text{m/A}) \frac{(300)^2(4.00 \times 10^{-4}\ \text{m}^2)}{25.0 \times 10^{-2}\ \text{m}}$$

$$= 1.81 \times 10^{-4}\ \text{T}\cdot\text{m}^2/\text{A} = 0.181\ \text{mH}$$

**B** Calculate the self-induced emf in the solenoid described in part A if the current through it is decreasing at the rate of 50.0 A/s.

**Solution** Equation 20.9 can be combined with $\Delta I/\Delta t = -50.0$ A/s to give

$$\mathcal{E} = -L\frac{\Delta I}{\Delta t} = -(1.81 \times 10^{-4}\ \text{H})(-50.0\ \text{A/s}) = 9.05\ \text{mV}$$

## 20.7 *RL* CIRCUITS

A circuit element that has a large inductance, such as a closely wrapped coil of many turns, is called an **inductor.** The circuit symbol for an inductor is ——ꝏꝏ——. We shall always assume that the self-inductance of the remainder of the circuit is negligible compared with that of the inductor in the circuit.

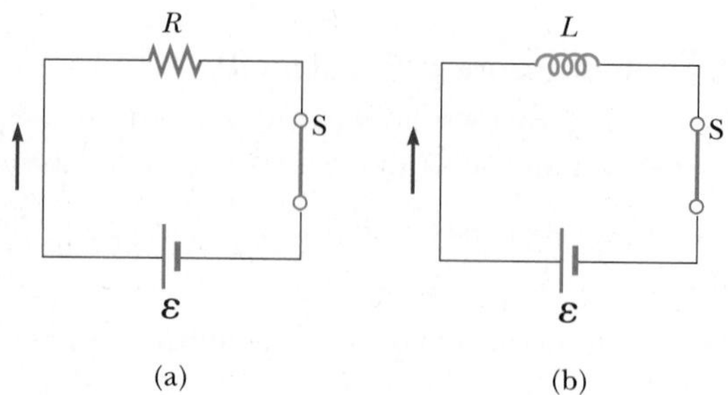


**FIGURE 20.24** A comparison of the effect of a resistor to that of an inductor in a simple circuit.

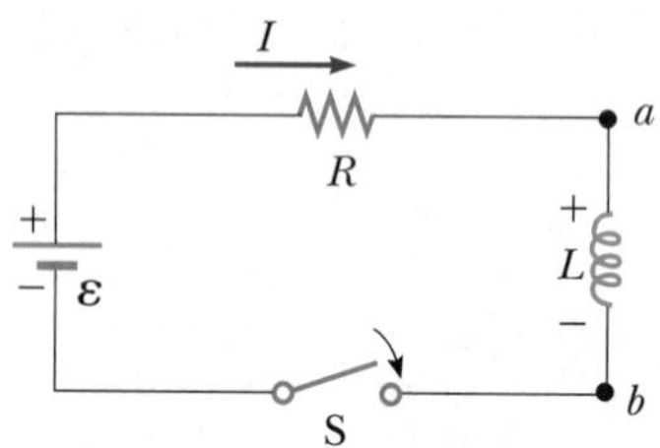


**FIGURE 20.25** A series *RL* circuit. As the current increases toward its maximum value, the inductor produces an emf that opposes the increasing current.

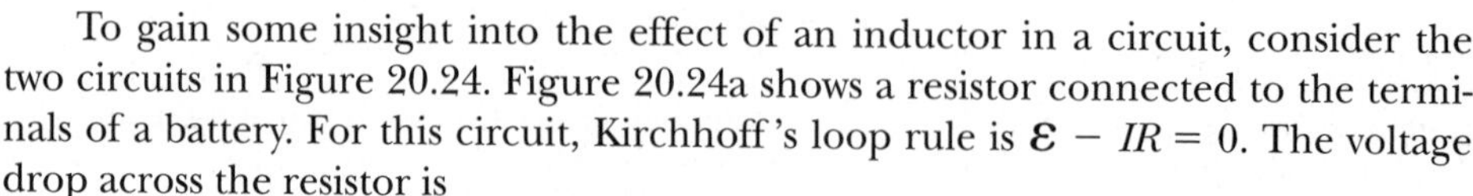

To gain some insight into the effect of an inductor in a circuit, consider the two circuits in Figure 20.24. Figure 20.24a shows a resistor connected to the terminals of a battery. For this circuit, Kirchhoff's loop rule is $\mathcal{E} - IR = 0$. The voltage drop across the resistor is

$$\Delta V_R = -IR \qquad [20.11]$$

In this case, we interpret **resistance as a measure of opposition to the current.** Now consider the circuit in Figure 20.24b, consisting of an inductor connected to the terminals of a battery. At the instant the switch in this circuit is closed, because $IR = 0$, the emf of the battery equals the back emf generated in the coil. Thus, we have

$$\mathcal{E}_L = -L\frac{\Delta I}{\Delta t} \qquad [20.12]$$

From this expression, **we can interpret *L* as a measure of opposition to the rate of change in current.**

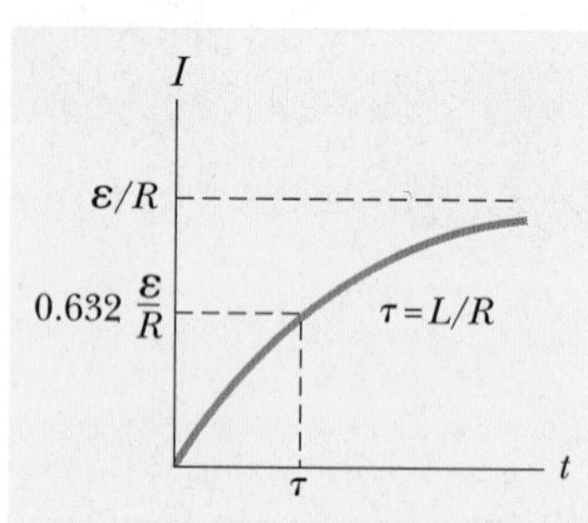


**FIGURE 20.26** Plot of current versus time for the *RL* circuit shown in Figure 20.25. The switch is closed at $t = 0$, and the current increases toward its maximum value $\mathcal{E}/R$. The time constant $\tau$ is the time it takes the current to reach 63.2% of its maximum value.

Figure 20.25 shows a circuit consisting of a resistor, inductor, and battery. Suppose the switch is closed at $t = 0$. The current begins to increase, but the inductor produces an emf that opposes the increasing current. Thus, the current is unable to change from zero to its maximum value of $\mathcal{E}/R$ instantaneously. Equation 20.12 shows that the induced emf is a maximum when the current is changing most rapidly, which occurs when the switch is first closed. As the current approaches its steady-state value, the back emf of the coil falls off because the current is changing more slowly. Finally, when the current reaches its steady-state value, the rate of change is zero and the back emf is also zero. Figure 20.26 plots current in the circuit as a function of time.[1] This plot is similar to that of the charge on a capacitor as a function of time, discussed in Chapter 18. In that case, we found it convenient to introduce a quantity called the *time constant of the circuit,* which told us something about the time required for the capacitor to approach its steady-state charge. In the same fashion, time constants are defined for circuits containing resistors and inductors. The **time constant** $\tau$ for an *RL* circuit is the time required for the current in the circuit to reach 63.2% of its final value $\mathcal{E}/R$; the time constant of an *RL* circuit is given by

Time constant for an *RL* circuit

$$\tau = \frac{L}{R} \qquad [20.13]$$

[1] The equation for the current in the circuit as a function of time may be obtained from calculus and is

$$I = \frac{\mathcal{E}}{R}(1 - e^{-Rt/L})$$

**Quick Quiz 20.5** The switch in the circuit shown in Figure 20.27 is closed and the lightbulb glows steadily. The inductor is a simple air-core solenoid. An iron rod is inserted into the interior of the solenoid, which increases the magnitude of the magnetic field in the solenoid. As the rod is inserted into the solenoid, the brightness of the lightbulb (a) increases, (b) decreases, or (c) remains the same.

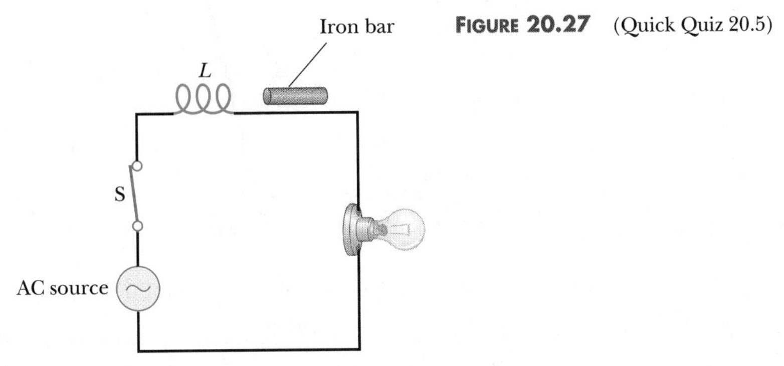


**FIGURE 20.27** (Quick Quiz 20.5)

## Example 20.9 The Time Constant for an *RL* Circuit

The circuit shown in Figure 20.25 consists of a 30-mH inductor, a 6.0-Ω resistor, and a 12-V battery. The switch is closed at $t = 0$.

**A** Find the time constant of the circuit.

**Solution** The time constant is given by Equation 20.13:

$$\tau = \frac{L}{R} = \frac{30 \times 10^{-3}\ \text{H}}{6.0\ \Omega} = \boxed{5.0\ \text{ms}}$$

**B** Find the current after one time constant has elapsed.

**Solution** After one time constant, the current in the circuit has risen to 63.2% of its final value. Thus, the current is

$$I = 0.632\frac{\mathcal{E}}{R} = (0.632)\left(\frac{12\ \text{V}}{6.0\ \Omega}\right) = \boxed{1.3\ \text{A}}$$

**EXERCISE** What is the voltage drop across the resistor (a) at $t = 0$? (b) after one time constant?

**ANSWER** (a) 0, (b) 7.6 V

## 20.8 ENERGY STORED IN A MAGNETIC FIELD

The emf induced by an inductor prevents a battery from establishing an instantaneous current in a circuit. The battery has to do work to produce a current. We can think of this needed work as energy stored by the inductor in its magnetic field. In a manner quite similar to that used in Section 16.9 to find the energy stored by a capacitor, we find that the energy stored by an inductor is

Energy stored in an inductor

$$PE_L = \tfrac{1}{2}LI^2 \qquad [20.14]$$

Note that the result is similar in form to the expression for the energy stored in a charged capacitor (Eq. 16.18):

Energy stored in a capacitor

$$PE_C = \tfrac{1}{2}C(\Delta V)^2$$

## SUMMARY

The magnetic flux $\Phi_B$ through a closed loop is defined as

$$\Phi_B \equiv BA\cos\theta \qquad [20.1]$$

where $B$ is the strength of the uniform magnetic field, $A$ is the cross-sectional area of the loop, and $\theta$ is the angle between **B** and the direction perpendicular to the plane of the loop.

**Faraday's law of induction** states that the instantaneous emf induced in a circuit equals the rate of change of magnetic flux through the circuit:

$$\mathcal{E} = -N\frac{\Delta\Phi_B}{\Delta t} \qquad [20.2]$$

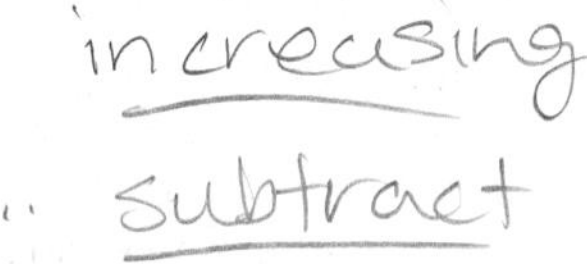

where $N$ is the number of loops in the circuit.

**Lenz's law** states that the polarity of the induced emf is such that it produces a current whose magnetic field opposes the *change* in magnetic flux through a circuit.

If a conducting bar of length $\ell$ moves through a magnetic field with a speed $v$ so that **B** is perpendicular to the bar, the emf induced in the bar, often called a **motional emf,** is

$$|\mathcal{E}| = B\ell v \qquad [20.4]$$

When a coil of wire with $N$ turns, each of area $A$, rotates with constant angular speed $\omega$ in a uniform magnetic field **B** as in Figure 20.19, the emf induced in the coil is

$$\mathcal{E} = NAB\omega\sin\omega t \qquad [20.7]$$

When the current in a coil changes with time, an emf is induced in the coil according to Faraday's law. This **self-induced emf** is defined by the expression

$$\mathcal{E} \equiv -L\frac{\Delta I}{\Delta t} \qquad [20.9]$$

where $L$ is the inductance of the coil. The SI unit for inductance is the henry (H); $1\text{ H} = 1\text{ V}\cdot\text{s/A}$.

The **inductance** of a coil can be found from the expression

$$L = \frac{N\Phi_B}{I} \qquad [20.10]$$

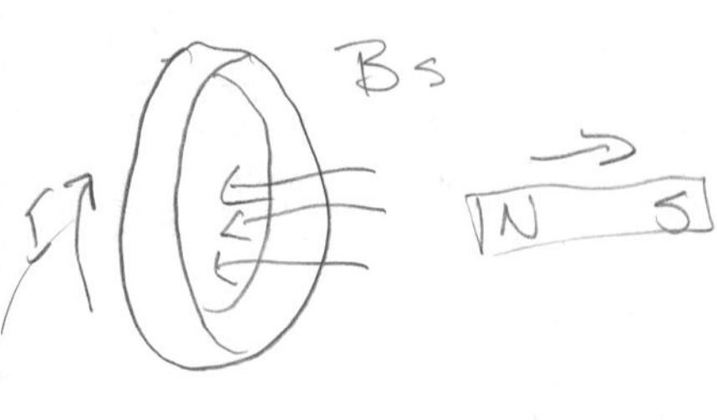

where $N$ is the number of turns on the coil, $I$ is the current in the coil, and $\Phi_B$ is the magnetic flux through the coil produced by that current.

If a resistor and inductor are connected in series to a battery and a switch is closed at $t = 0$, the current in the circuit does not rise instantly to its maximum value. After one **time constant** $\tau = L/R$ the current in the circuit is 63.2% of its final value, $\mathcal{E}/R$.

The **energy stored** in the magnetic field of an inductor carrying current $I$ is

$$PE_L = \tfrac{1}{2}LI^2 \qquad [20.14]$$

## CONCEPTUAL QUESTIONS

1. A circular loop is located in a uniform and constant magnetic field. Describe how an emf can be induced in the loop in this situation.
2. Does dropping a magnet down a copper tube produce a current in the tube? Explain.
3. A spacecraft orbiting Earth has a coil of wire in it. An astronaut measures a small current in the coil, although no battery is connected to it and there are no magnets in the spacecraft. What is causing the current?
4. A loop of wire is placed in a uniform magnetic field. For what orientation of the loop is the magnetic flux a maximum? For what orientation is the flux zero?
5. As the conducting bar in Figure Q20.5 moves to the right, an electric field is set up directed downward. If the bar were moving to the left, explain why the electric field would be upward.

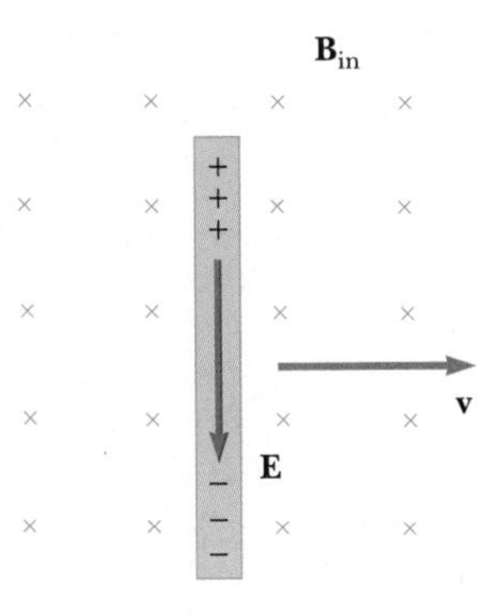


**FIGURE Q20.5** (Conceptual Questions 5 and 6)

6. As the bar in Figure Q20.5 moves perpendicular to the field, is an external force required to keep it moving with constant speed?
7. Wearing a metal bracelet in a region of strong magnetic field can be hazardous. Discuss.
8. How is electrical energy produced in dams (that is, how is the energy of motion of the water converted to ac electricity)?
9. Eddy currents are induced currents set up in a piece of metal when it moves through a nonuniform magnetic field. For example, consider the flat metal plate swinging at the end of a bar as a pendulum as shown in Figure Q20.9. At position 1, the pendulum is moving from a region where there is no magnetic field into a region where the field $\mathbf{B}_{in}$ is directed into the paper. Show that at position 1, the direction of the eddy current is counterclockwise. Also, at position 2 the pendulum is moving out of the field into a region of zero field. Show that the direction of the eddy current is clockwise in this case. Use right-hand rule #2 to show that these eddy currents lead to a magnetic force on the plate directed as shown in the figure. Because the induced eddy current always produces a retarding force when the plate enters or leaves the field, the swinging plate quickly comes to rest.

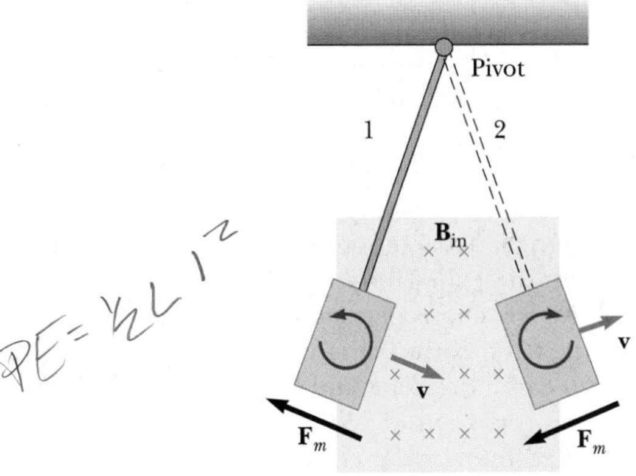


**FIGURE Q20.9**

10. Suppose you would like to steal power for your home from the electric company by placing a loop of wire near a transmission cable in order to induce an emf in the loop (an illegal procedure). Should you locate the loop so that the transmission cable passes through your loop or simply place your loop near the transmission cable? Does the orientation of the loop matter?
11. A piece of aluminum is dropped vertically downward between the poles of an electromagnet. Does the magnetic field affect the velocity of the aluminum? (*Hint:* See Conceptual Question 9.)
12. A bar magnet is dropped toward a conducting ring lying on the floor. As the magnet falls toward the ring, does it move as a freely falling object?
13. If the current in an inductor is doubled, by what factor does the stored energy change?
14. Is it possible to induce a constant emf for an infinite amount of time?
15. Why is the induced emf that appears in an inductor called a back (counter) emf?
16. A magneto is used to cause the spark in a spark plug in many lawn mowers today. A magneto consists of a permanent magnet mounted on the flywheel so that it spins past a fixed coil. Explain how this arrangement generates a large enough potential difference to cause the spark.

## PROBLEMS

1, 2, 3 = straightforward, intermediate, challenging ☐ = full solution available in Student Solutions Manual/Study Guide

**web** = solution posted at **http://info.brookscole.com/serway** = biomedical application

### Section 20.1 Induced emf and Magnetic Flux

1. A magnetic field of strength 0.30 T is directed perpendicular to a plane circular loop of wire of radius 25 cm. Find the magnetic flux through the area enclosed by this loop.
2. A circular loop with a radius of 0.200 m is placed in a uniform magnetic field of magnitude 0.850 T. The normal to

the loop makes an angle of 30.0° with respect to the direction of **B**. If the field increases to 0.950 T, what is the increase in magnetic flux through the loop?

**3.** Consider a place where Earth's magnetic field has strength of $0.520 \times 10^{-4}$ T and makes an angle of 62.0° with the horizontal. At this location a magnetic compass points toward true north. What is the magnetic flux in a rectangular loop of wire that is 15.0 by 25.0 cm and is lying on a table? What is the magnetic flux in this loop if it is mounted vertically on a north wall? On an east wall?

**4.** A long, straight wire carrying a current of 2.00 A is placed along the axis of a cylinder of radius 0.500 m and a length of 3.00 m. Determine the total magnetic flux through the cylinder.

**5.** A long, straight wire lies in the plane of a circular coil with a radius of 0.010 m. The wire carries a current of 2.0 A and is placed along a diameter of the coil. (a) What is the net flux through the coil? (b) If the wire passes through the center of the coil and is perpendicular to the plane of the coil, find the net flux through the coil.

**6.** A solenoid 4.00 cm in diameter and 20.0 cm long has 250 turns and carries a current of 15.0 A. Calculate the magnetic flux through the circular cross-sectional area of the solenoid.

**7.** A cube of edge length $\ell = 2.5$ cm is positioned as shown in Figure P20.7. There is a uniform magnetic field throughout the region with components of $B_x = +5.0$ T, $B_y = +4.0$ T, and $B_z = +3.0$ T. (a) Calculate the flux through the shaded face of the cube. (b) What is the total flux emerging from the volume enclosed by the cube (that is, total flux through all six faces)?

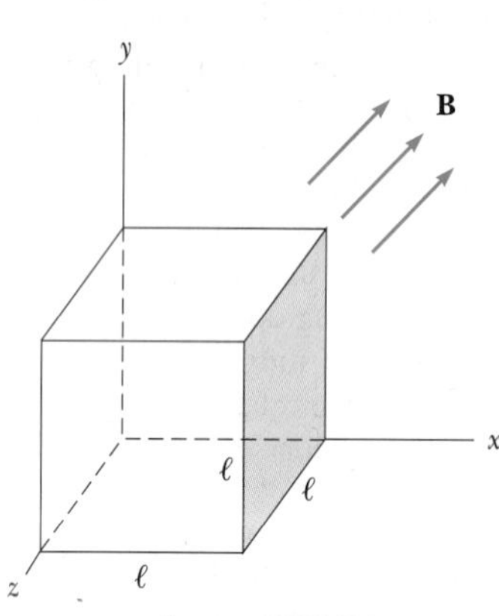


**FIGURE P20.7**

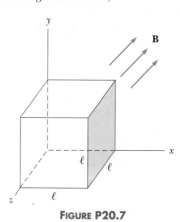

## Section 20.2 Faraday's Law of Induction

**8.** A circular coil of radius 20 cm is placed in an external magnetic field of strength 0.20 T so that the plane of the coil is perpendicular to the field. The coil is pulled out of the field in 0.30 s. Find the average induced emf during this interval.

**9.** A 25-turn circular coil of wire has a diameter of 1.00 m. It is placed with its axis along the direction of Earth's magnetic field of 50.0 μT, and then in 0.200 s it is flipped 180°. An average emf of what magnitude is generated in the coil?

**10.** The flexible loop in Figure P20.10 has a radius of 12 cm and is in a magnetic field of strength 0.15 T. The loop is grasped at points $A$ and $B$ and stretched until its area is nearly zero. If it takes 0.20 s to close the loop, find the magnitude of the average induced emf in it during this time.

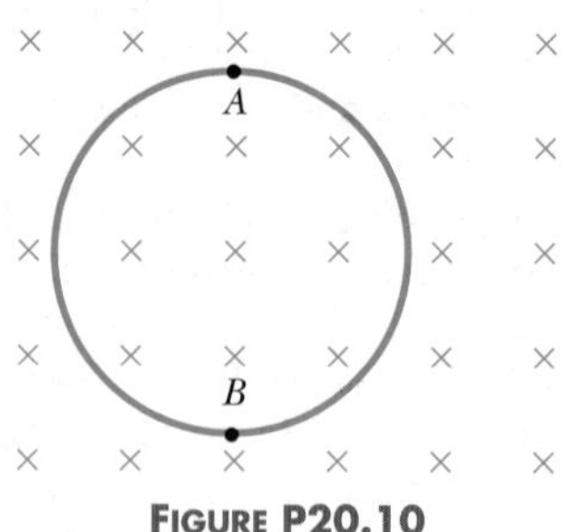


**FIGURE P20.10**

**11.** A strong electromagnet produces a uniform field of 1.60 T over a cross-sectional area of 0.200 m². We place a coil having 200 turns and a total resistance of 20.0 Ω around the electromagnet. We then smoothly decrease the current in the electromagnet until it reaches zero in 20.0 ms. What is the current induced in the coil?

**12.** A 500-turn circular-loop coil 15.0 cm in diameter is initially aligned so that its axis is parallel to Earth's magnetic field. In 2.77 ms, the coil is flipped so that its axis is perpendicular to Earth's magnetic field. If an average voltage of 0.166 V is thereby induced in the coil, what is the value of Earth's magnetic field at that location?

**13.** web The plane of a rectangular coil, 5.0 cm by 8.0 cm, is perpendicular to the direction of a magnetic field **B**. If the coil has 75 turns and a total resistance of 8.0 Ω, at what rate must the magnitude of **B** change to induce a current of 0.10 A in the windings of the coil?

**14.** A square, single-turn wire loop 1.00 cm on a side is placed inside a solenoid that has a circular cross section of radius 3.00 cm, as shown in Figure P20.14. The solenoid is 20.0 cm long and wound with 100 turns of wire. (a) If the current in the solenoid is 3.00 A, find the flux through the loop. (b) If the current in the solenoid is reduced to zero in 3.00 s, find the magnitude of the average induced emf in the loop.

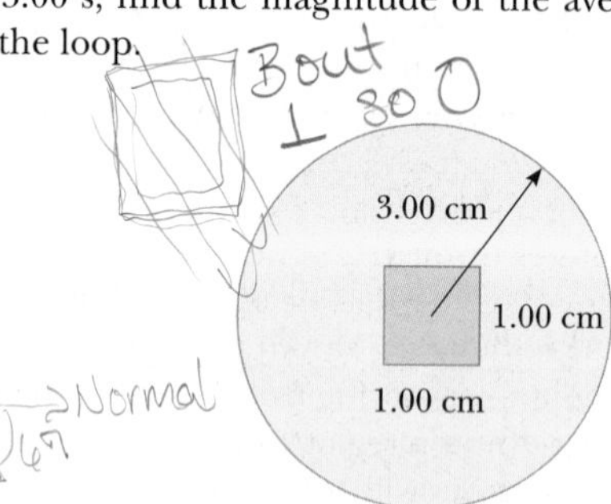


**FIGURE P20.14**

**15.** A 300-turn solenoid with a length of 20 cm and a radius of 1.5 cm carries a current of 2.0 A. A second coil of four turns is wrapped tightly about this solenoid so that it can be considered to have the same radius as the solenoid. Find (a) the change in the magnetic flux through the coil and (b) the magnitude of the average induced emf in the coil when the current in the solenoid increases to 5.0 A in a period of 0.90 s.

**16.** A circular coil, enclosing an area of 100 cm², is made of 200 turns of copper wire. The wire making up the coil has resistance of 5.0 Ω, and the ends of the wire are connected to form a closed loop. Initially, a 1.1-T uniform magnetic field points perpendicularly upward through the plane of the coil. The direction of the field then reverses so that the final magnetic field has a magnitude of 1.1 T and points downward through the coil. If the time required for the field to reverse directions is 0.10 s, what average current flows through the coil during this time?

17. The person in Figure P20.17 is girded about the chest with a breathing monitor, which is a 100-turn coil. As a breath is inhaled, the area of the coil varies from 0.120 m$^2$ to 0.124 m$^2$. Earth's magnetic field is 50.0 $\mu$T and makes an angle of 22.5° with respect to the normal to the loop. If the patient inhales a breath in 1.59 s, what is the average voltage induced in the coil during the inhalation?

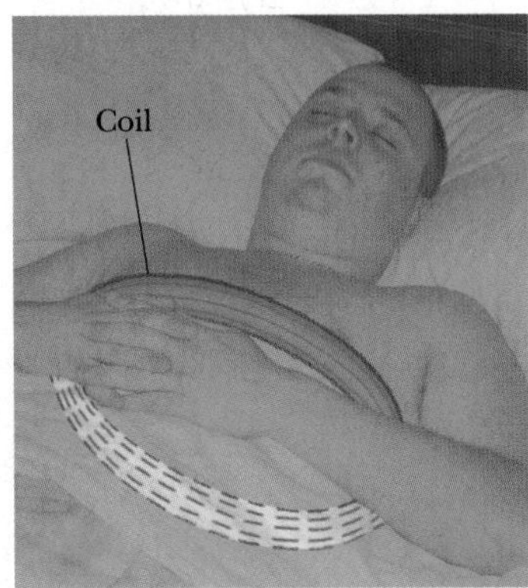


**FIGURE P20.17**

## Section 20.3 Motional emf

18. Consider the arrangement shown in Figure P20.18. Assume that $R = 6.00\ \Omega$ and $\ell = 1.20$ m, and that a uniform 2.50-T magnetic field is directed *into* the page. At what speed should the bar be moved to produce a current of 0.500 A in the resistor?

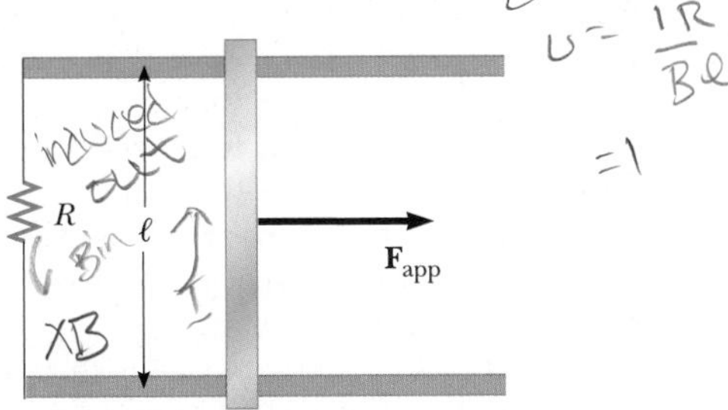


**FIGURE P20.18** (Problems 18 and 57)

19. A Boeing-747 jet with a wing span of 60.0 m is flying horizontally at a speed of 300 m/s over Phoenix, Arizona, at a location where Earth's magnetic field is 50.0 $\mu$T at 58.0° below the horizontal. What voltage is generated between the wingtips?

20. Over a region where the *vertical* component of Earth's magnetic field is 40.0 $\mu$T directed downward, a 5.00-m length of wire is held in an east-west direction and moved horizontally to the north with a speed of 10.0 m/s. Calculate the potential difference between the ends of the wire, and determine which end is positive.

21. An automobile has a vertical radio antenna 1.20 m long. The automobile travels at 65.0 km/h on a horizontal road where Earth's magnetic field is 50.0 $\mu$T directed toward the north and downward at an angle of 65.0° below the horizontal. (a) Specify the direction that the automobile should move in order to generate the maximum motional emf in the antenna, with the top of the antenna positive relative to the bottom. (b) Calculate the magnitude of this induced emf.

22. A helicopter has blades of length 3.0 m, rotating at 2.0 rev/s about a central hub. If the vertical component of Earth's magnetic field is $5.0 \times 10^{-5}$ T, what is the emf induced between the blade tip and the central hub?

## Section 20.4 Lenz's Law Revisited (The Minus Sign in Faraday's Law)

23. A bar magnet is positioned near a coil of wire as shown in Figure P20.23. What is the direction of the current through the resistor when the magnet is moved (a) to the left? (b) to the right?

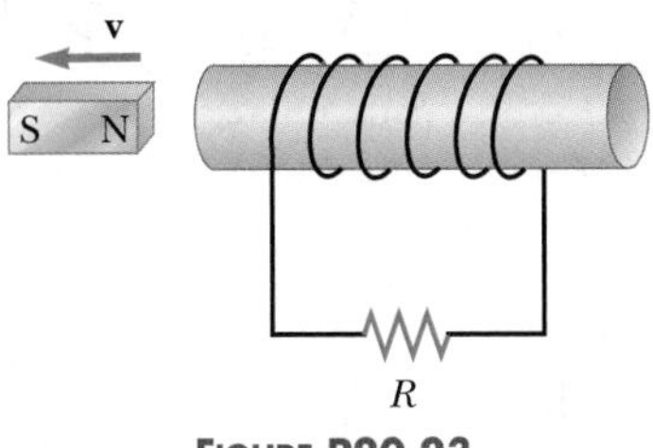


**FIGURE P20.23**

24. A bar magnet is held above the center of a wire loop in a horizontal plane, as shown in Figure P20.24. The south end of the magnet is toward the loop. The magnet is dropped. Find the direction of the current through the resistor (a) while the magnet is falling toward the loop and (b) after the magnet has passed through the loop and moves away from it.

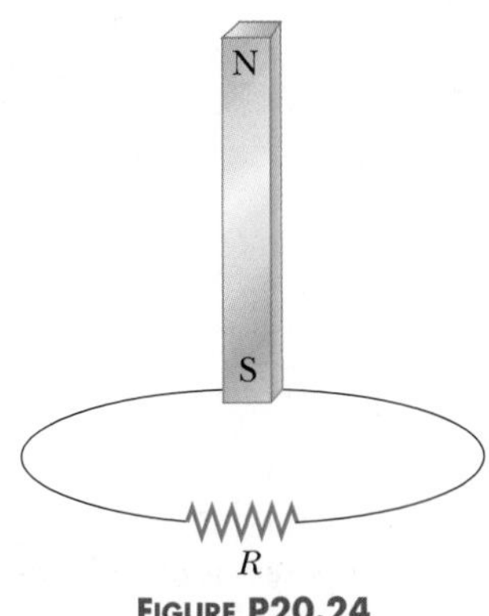


**FIGURE P20.24**

25. What is the direction of the current induced in the resistor when the current in the long, straight wire in Figure P20.25 decreases rapidly to zero?

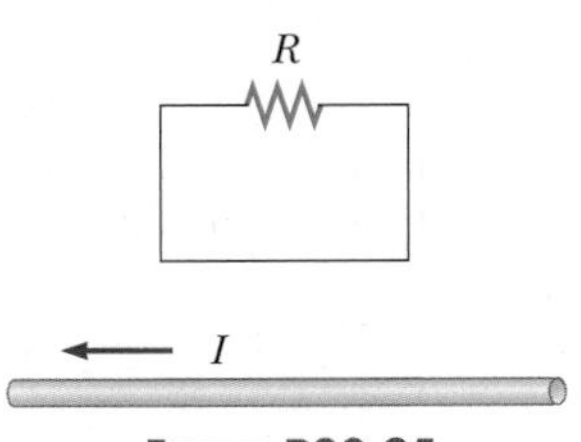


**FIGURE P20.25**

26. In Figure P20.26, what is the direction of the current induced in the resistor at the instant the switch is closed?

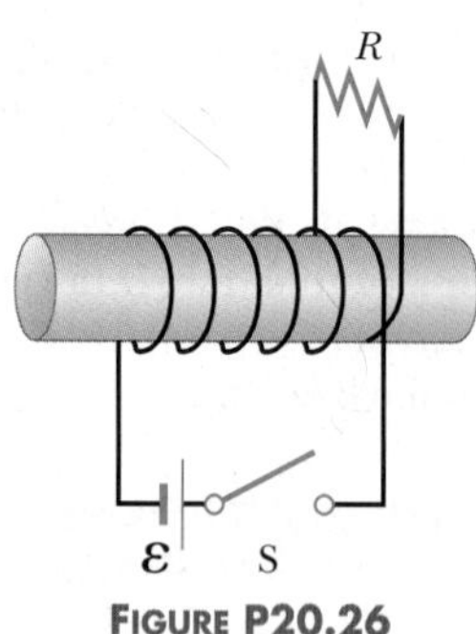

FIGURE P20.26

27. A copper bar is moved to the right while its axis is maintained in a direction perpendicular to a magnetic field, as shown in Figure P20.27. If the top of the bar becomes positive relative to the bottom, what is the direction of the magnetic field?

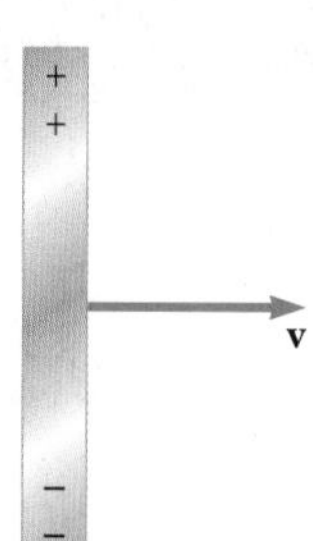

FIGURE P20.27

28. Find the direction of the current through the resistor in Figure P20.28, (a) at the instant the switch is closed, (b) after the switch has been closed for several minutes, and (c) at the instant the switch is opened.

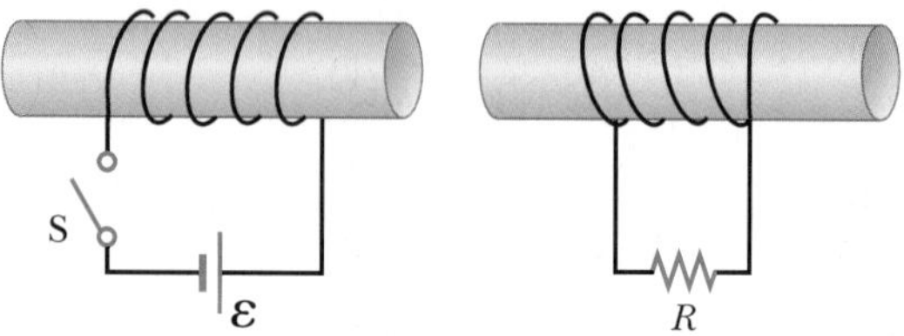

FIGURE P20.28

29. Find the direction of the current in resistor $R$ in Figure P20.29 after each of the following steps (taken in the order given). (a) The switch is closed. (b) The variable resistance in series with the battery is decreased. (c) The circuit containing resistor $R$ is moved to the left. (d) The switch is opened.

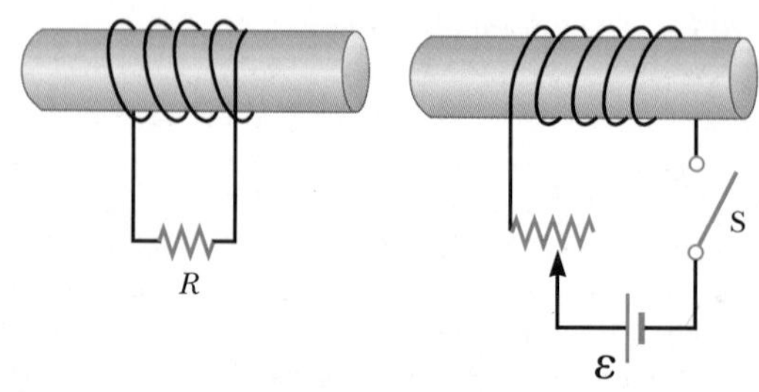

FIGURE P20.29

## Section 20.5 Generators

30. A 100-turn square wire coil of area 0.040 m$^2$ rotates about a vertical axis at 1 500 rev/min, as indicated in Figure P20.30. The horizontal component of Earth's magnetic field at the location of the loop is $2.0 \times 10^{-5}$ T. Calculate the maximum emf induced in the coil by Earth's field.

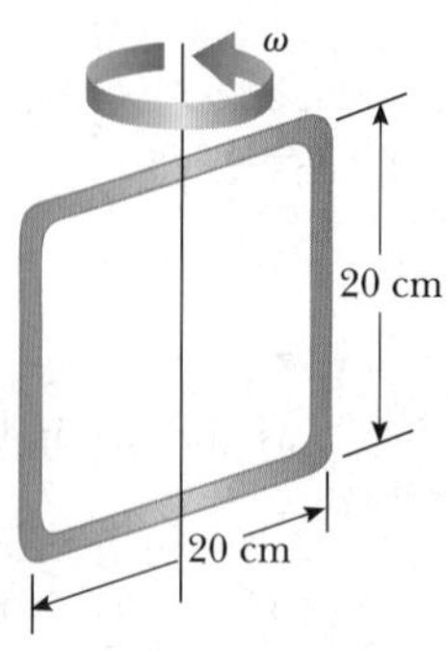

FIGURE P20.30

31. An automobile generator produces 12.0 V when turning at 500 rev/min. What potential difference will it produce at 1 200 rev/min?

32. A motor has coils with a resistance of 30 Ω and operates from a voltage of 240 V. When the motor is operating at its maximum speed, the back emf is 145 V. Find the current in the coils (a) when the motor is first turned on and (b) when the motor has reached maximum speed. (c) If the current in the motor is 6.0 A at some instant, what is the back emf at that time?

33. When the coil of a motor is rotating at maximum speed, the current in the windings is 4.0 A. When the motor is first turned on, the current in the windings is 11 A. If the motor is operated at 120 V, find (a) the resistance of the windings and (b) the back emf in the coil at maximum speed.

34. A loop of area 0.10 m$^2$ is rotating at 60 rev/s with its axis of rotation perpendicular to a 0.20-T magnetic field. (a) If there are 1 000 turns on the loop, what is the maximum voltage induced in the loop? (b) When the maximum induced voltage occurs, what is the orientation of the loop with respect to the magnetic field?

35. web In a model ac generator, a 500-turn rectangular coil, 8.0 cm by 20 cm, rotates at 120 rev/min in a uniform magnetic field of 0.60 T. (a) What is the maximum emf induced in the coil? (b) What is the instantaneous value of the emf in the coil at $t = (\pi/32)$ s? Assume that the emf is zero at $t = 0$. (c) What is the smallest value of $t$ for which the emf will have its maximum value?

## Section 20.6 Self-Inductance

36. A coiled telephone cord forms a spiral with 70.0 turns, a diameter of 1.30 cm, and an unstretched length of 60.0 cm. Determine the self-inductance of one conductor in the unstretched cord.

37. A 2.00-H inductor carries a steady current of 0.500 A. When the switch in the circuit is thrown open, the current is effectively zero after 10.0 ms. What is the average induced emf in the inductor during this time?

38. Show that the two expressions for inductance given by

$$L = \frac{N\Phi_B}{I} \quad \text{and} \quad L = \frac{-\mathcal{E}}{\Delta I/\Delta t}$$

have the same units.

39. A solenoid of radius 2.5 cm has 400 turns and a length of 20 cm. Find (a) its inductance and (b) the rate at which current must change through it to produce an emf of 75 mV.

40. An emf of 24.0 mV is induced in a 500-turn coil when the current is changing at a rate of 10.0 A/s. What is the magnetic flux through each turn of the coil at an instant when the current is 4.00 A?

### Section 20.7 *RL* Circuits

41. Show that the SI units for the inductive time constant $\tau = L/R$ are seconds.

42. An *RL* circuit with $L = 3.00$ H and an *RC* circuit with $C = 3.00\ \mu$F have the same time constant. If the two circuits have the same resistance *R*, (a) what is the value of *R*, and (b) what is this common time constant?

43. A 6.0-V battery is connected in series with a resistor and an inductor. The series circuit has a time constant of 600 $\mu$s, and the maximum current is 300 mA. What is the value of the inductance?

44. A 25-mH inductor, an 8.0-Ω resistor, and a 6.0-V battery are connected in series. The switch is closed at $t = 0$. Find the voltage drop across the resistor (a) at $t = 0$ and (b) after one time constant has passed. Also, find the voltage drop across the inductor (c) at $t = 0$ and (d) after one time constant has elapsed.

45. Calculate the resistance in an *RL* circuit in which **web** $L = 2.50$ H and the current increases to 90.0% of its final value in 3.00 s.

46. Consider the circuit in Figure P20.46, taking $\mathcal{E} = 6.00$ V, $L = 8.00$ mH, and $R = 4.00\ \Omega$. (a) What is the inductive time constant of the circuit? (b) Calculate the current in the circuit 250 $\mu$s after the switch is closed. (c) What is the value of the final steady-state current? (d) How long does it take the current to reach 80.0% of its maximum value?

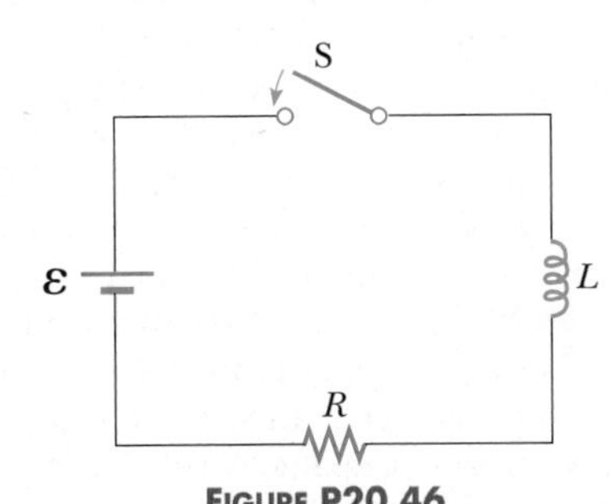


FIGURE P20.46

### 20.8 Energy Stored in a Magnetic Field

47. How much energy is stored in a 70.0-mH inductor at an instant when the current is 2.00 A?

48. An air-core solenoid with 68 turns is 8.00 cm long and has a diameter of 1.20 cm. How much energy is stored in its magnetic field when it carries a current of 0.770 A?

49. A 24-V battery is connected in series with a resistor and an inductor, where $R = 8.0\ \Omega$ and $L = 4.0$ H. Find the energy stored in the inductor (a) when the current reaches its maximum value and (b) one time constant after the switch is closed.

### ADDITIONAL PROBLEMS

50. What is the time constant for (a) the circuit shown in Figure P20.50a? (b) the circuit shown in Figure P20.50b?

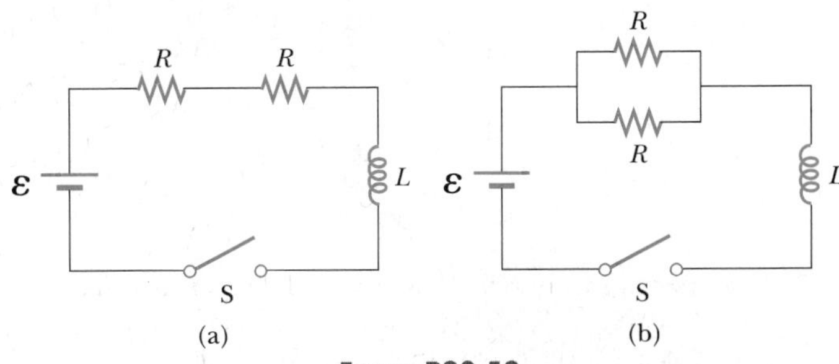


FIGURE P20.50

51. In Figure P20.51, the bar magnet is being moved toward the loop. Is $(V_a - V_b)$ positive, negative, or zero during this motion? Explain.

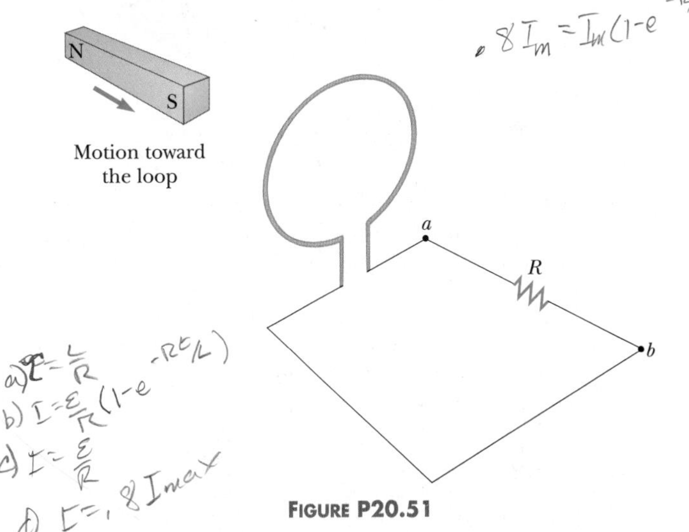


FIGURE P20.51

52. Your physics teacher asks you to help her set up a demonstration of Faraday's law for the class. The apparatus consists of a strong permanent magnet that has a field of 0.10 T, a small 10-turn coil of radius 2.0 cm cemented on a wood frame with a handle, some flexible connecting wires, and an ammeter as in Figure P20.52. The idea is to pull the coil out of the center of the magnetic field as quickly as possible and read the average current registered on the meter. The combined resistance of the coil, leads, and meter is 2.0 Ω and you must flip the coil out of the field in about 0.20 s. The ammeter you must use has a full-scale sensitivity of 1 000 $\mu$A. Will this meter be sensitive enough to show clearly the induced current?

53. An 820-turn wire coil of resistance 24.0 Ω is placed on top **web** of a 12 500-turn, 7.00-cm-long solenoid, as in Figure P20.53. Both coil and solenoid have cross-sectional areas of $1.00 \times 10^{-4}$ m$^2$. (a) How long does it take the solenoid current to reach 0.632 times its maximum value? (b) Determine the average back emf caused by the self-inductance of the solenoid during this interval. The magnetic field produced by the solenoid at the location of the coil is one-half as strong as the field at the center of the

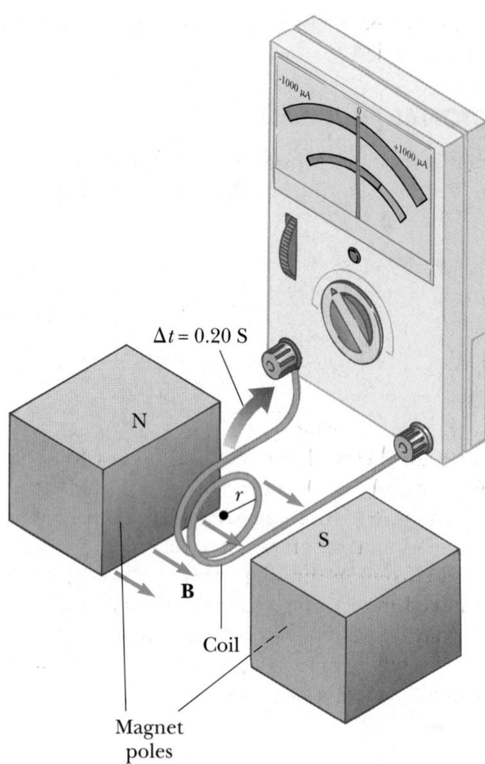


FIGURE P20.52

solenoid. (c) Determine the average rate of change in magnetic flux through each turn of the coil during this interval. (d) Find the magnitude of the average induced current in the coil.

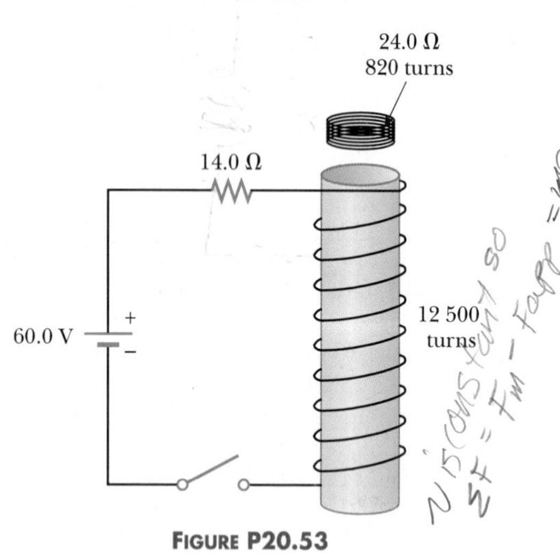


FIGURE P20.53

54. Figure P20.54 is a graph of induced emf versus time for a coil of $N$ turns rotating with angular speed $\omega$ in a uniform magnetic field directed perpendicularly to the axis of rotation of the coil. Copy this sketch (increasing the scale), and on the same set of axes show the graph of emf versus $t$ when (a) the number of turns in the coil is doubled, (b) the angular speed is doubled, and (c) the angular speed is doubled while the number of turns in the coil is halved.

55. The plane of a square loop of wire with edge length $a = 0.200$ m is perpendicular to Earth's magnetic field at a

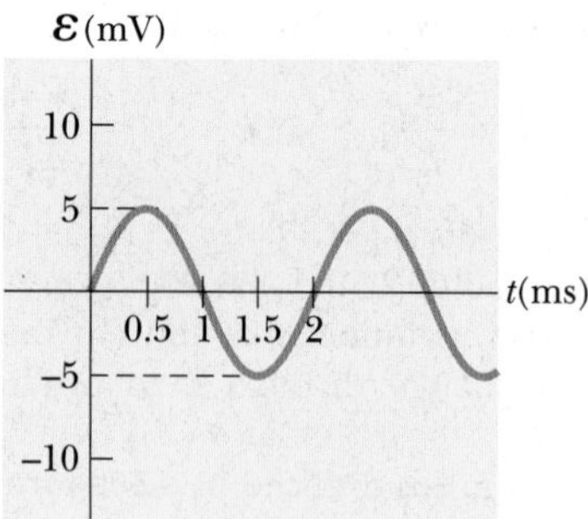


FIGURE P20.54

point where $B = 15.0\ \mu$T, as in Figure P20.55. The total resistance of the loop and the wires connecting it to the ammeter is 0.500 Ω. If the loop is suddenly collapsed by horizontal forces as shown, what total charge passes through the ammeter?

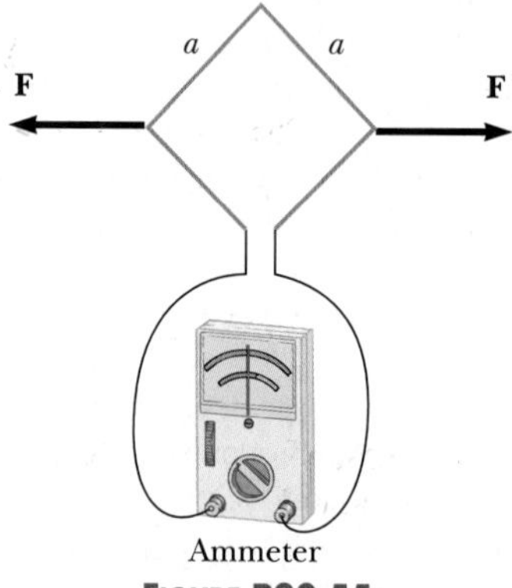


FIGURE P20.55

56. A novel method of storing electrical energy has been proposed. A huge underground superconducting coil, 1.00 km in diameter, would be fabricated. It would carry a maximum current of 50.0 kA through each winding of a 150-turn $Nb_3Sn$ solenoid. (a) If the inductance of this huge coil is 50.0 H, what is the total energy stored? (b) What is the compressive force per meter length acting between two adjacent windings 0.250 m apart? (*Hint:* Because the radius of the coil is so large, the magnetic field created by one winding and acting on an adjacent turn can be considered to be that of a long, straight wire.)

57. A conducting rod of length $\ell$ moves on two horizontal frictionless rails, as in Figure P20.18. A constant force of magnitude 1.00 N moves the bar at a uniform speed of 2.00 m/s through a magnetic field **B** that is into the page. (a) What is the current in an 8.00-Ω resistor $R$? (b) What is the rate of energy dissipation in the resistor? (c) What is the mechanical power delivered by the constant force?

58. The square loop in Figure P20.58 is made of wires with total series resistance 10.0 Ω. It is placed in a uniform 0.100-T magnetic field directed perpendicular into the plane of the paper. The loop, which is hinged at each corner, is pulled as shown until the separation between points $A$ and $B$ is 3.00 m. If this process takes 0.100 s, what is the average current generated in the loop? What is the direction of the current?

59. The bolt of lightning depicted in Figure P20.59 passes 200 m from a 100-turn coil oriented as shown. If the cur-

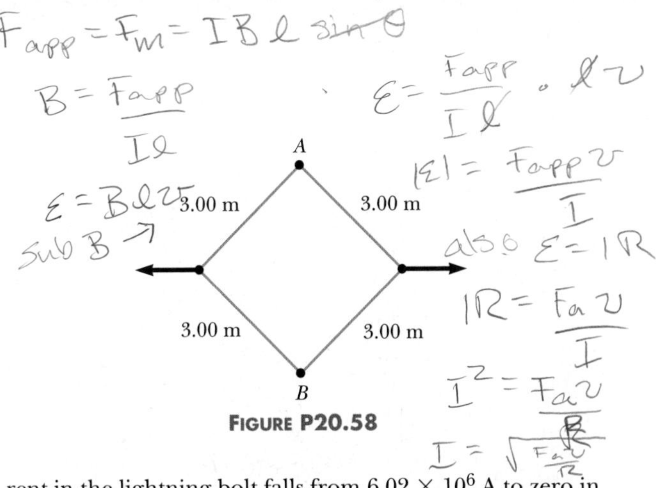


**FIGURE P20.58**

rent in the lightning bolt falls from $6.02 \times 10^6$ A to zero in 10.5 $\mu$s, what is the average voltage induced in the coil? Assume that the distance to the center of the coil determines the average magnetic field at the coil's position. Treat the lightning bolt as a long, vertical wire.

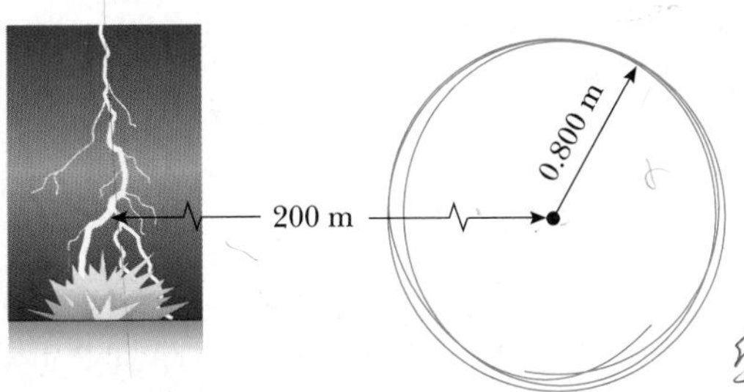


**FIGURE P20.59**

60. The wire shown in Figure P20.60 is bent in the shape of a "tent" with $\theta = 60°$ and $L = 1.5$ m, and is placed in a uniform magnetic field of 0.30 T perpendicular to the tabletop. The wire is "hinged" at points $a$ and $b$. If the tent is flattened out on the table in 0.10 s, what is the average induced emf in the wire during this time?

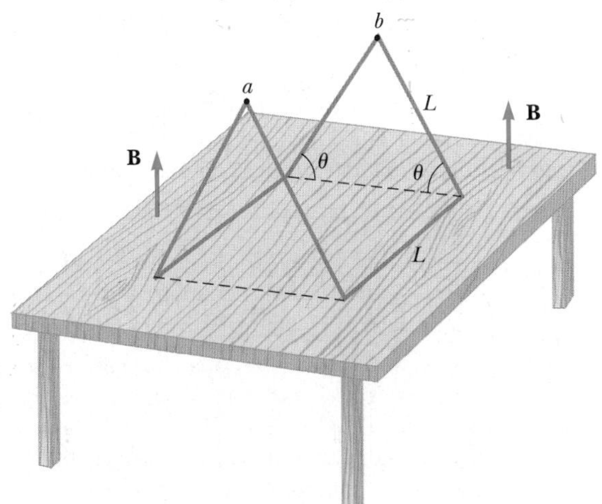


**FIGURE P20.60**

61. The magnetic field shown in Figure P20.61 has a uniform magnitude of 25.0 mT directed into the paper. The initial diameter of the kink is 2.00 cm. (a) The wire is quickly pulled taut, and the kink shrinks to a diameter of zero in 50.0 ms. Determine the average voltage induced between endpoints $A$ and $B$. Include the polarity. (b) Suppose the kink is undisturbed, but the magnetic field increases to 100 mT in $4.00 \times 10^{-3}$ s. Determine the average voltage across terminals $A$ and $B$, including polarity, during this period.

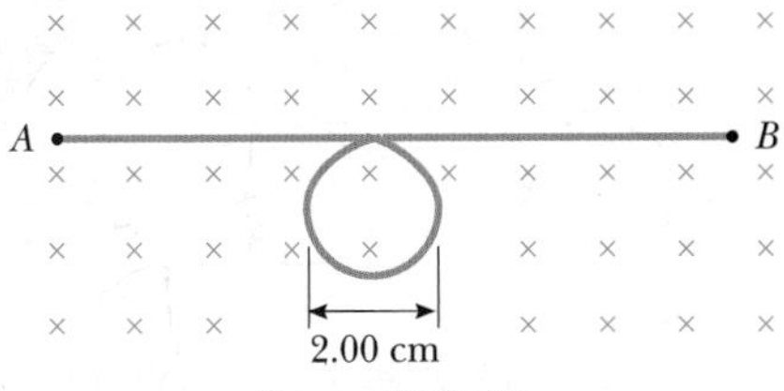


**FIGURE P20.61**

62. An aluminum ring of radius 5.00 cm and resistance $3.00 \times 10^{-4}$ Ω is placed around the top of a long air-core solenoid with 1 000 turns per meter and smaller radius 3.00 cm as in Figure P20.62. If the current in the solenoid is increasing at a constant rate of 270 A/s, what is the induced current in the ring? Assume that the magnetic field produced by the solenoid over the area at the end of the solenoid is one-half as strong as the field at the center of the solenoid. Assume that the solenoid produces a negligible field outside its cross-sectional area.

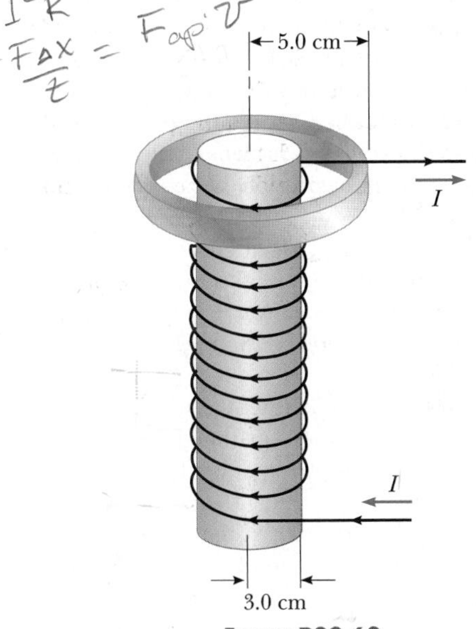


**FIGURE P20.62**

63. In Figure P20.63, the rolling axle, 1.50 m long, is pushed along horizontal rails at a constant speed $v = 3.00$ m/s. A resistor $R = 0.400$ Ω is connected to the rails at points $a$ and $b$, directly opposite each other. (The wheels make good electrical contact with the rails, and so the axle, rails, and $R$ form a closed-loop circuit. The only significant resistance in the circuit is $R$.) There is a uniform magnetic field $B = 0.800$ T vertically downward. (a) Find the induced current $I$ in the resistor. (b) What horizontal force **F** is required to keep the axle rolling at constant speed? (c) Which end of the resistor, $a$ or $b$, is at the higher electric potential? (d) After the axle rolls past the resistor, does the current in $R$ reverse direction?

64. In 1832 Faraday proposed that the apparatus shown in Figure P20.64 could be used to generate electric current

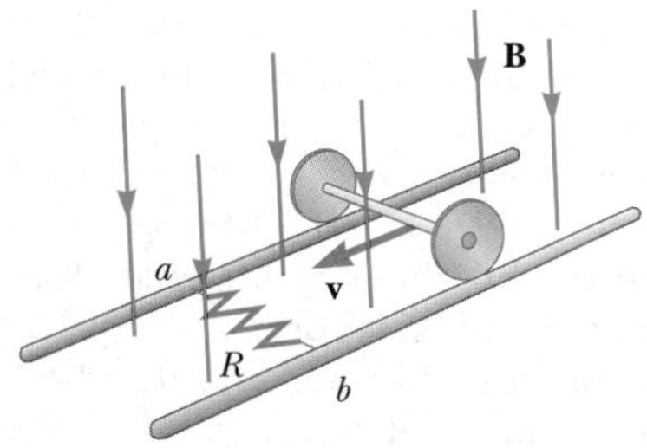

FIGURE P20.63

from the flowing water in the Thames River.[2] Two conducting plates of lengths $a$ and widths $b$ are placed facing one another on opposite sides of the river, a distance $w$ apart and immersed entirely. The flow velocity of the river is $\mathbf{v}$, and the vertical component of Earth's magnetic field is $B$. Show that the current in the load resistor $R$ is

$$I = \frac{abvB}{\rho + abR/w}$$

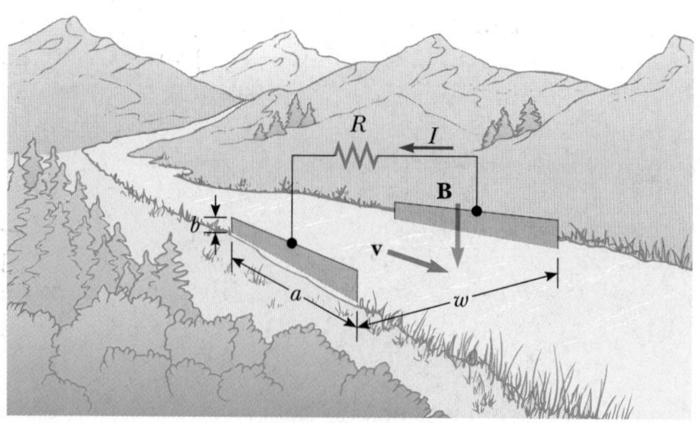

FIGURE P20.64

where $\rho$ is the resistivity of the water. (b) Calculate the short-circuit current ($R = 0$) if $a = 100$ m, $b = 5.00$ m, $v = 3.00$ m/s, $B = 50.0$ $\mu$T, and $\rho = 100$ $\Omega$ m.

## GROUP ACTIVITIES

**G.1** Experimenting with induced currents is not easy. For small magnets and small coils of wire, the resulting induced currents are so small that they are difficult to detect. Thus, you may have to try this exercise several times before you are satisfied with your results. Wind a coil of wire on a cardboard mailing tube. Use insulated wire with as small a diameter as possible because you need as many turns as possible on the coil. Connect the coil to a flashlight bulb and see if you can get it to light by moving a bar magnet into and out of the coil in rapid succession. Why does the speed of movement make a difference? If you are unsuccessful, place two magnets side by side and repeat.

After you have experimented with the bulb, ask your instructor to let you use a galvanometer as a current detector. These devices are capable of measuring very small currents and they have the added advantage of detecting the direction of the current in the circuit.

Use your equipment to observe or test the following. (a) Does the magnitude of the induced current depend on the speed of movement of the magnet? (b) Can you induce a current by holding the magnet still and moving the coil over it? (c) Does the direction of the current depend on whether the magnet is pushed in or pulled out of the coil? (d) Does the direction of the current depend on whether the inserted pole of the magnet is north or south? (e) Can you predict the direction of the current by using Lenz's law? (f) Replace your bar magnet with the electromagnet you constructed in the last chapter and repeat the observations above.

**G.2** As explained in the text, a cassette tape is made up of tiny particles of metal oxide attached to a long plastic strip. Pull a tape out of a cassette that you do not mind destroying and see if it is repelled or attracted by a refrigerator magnet. Also, try this with an expendable floppy computer disk.

**G.3** This experiment takes steady hands, a dime, and a strong magnet. After verifying that a dime is not attracted to the magnet, carefully balance the coin on its edge. (This will not work with other coins because they require too much force to topple them.) Hold one pole of the magnet within a millimeter of the face of the dime, but do not make contact with it. Now very rapidly pull the magnet straight back away from the coin. Which way does the dime tip? Does the coin fall the same way most of the time? Explain what is going on in terms of Lenz's law.

**G.4** A ramp runs from the bed of a truck down to the level ground. The ramp holds two parallel conducting rails connected at its base. A metal bar slides on the rails without friction. A magnet supplies an external magnetic field

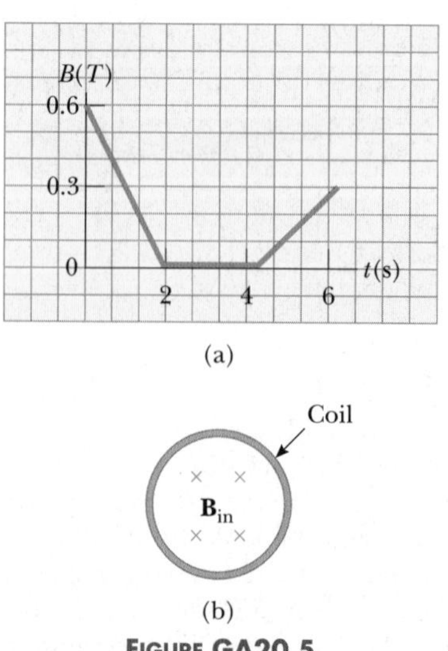

FIGURE GA20.5

[2] The idea for this problem and Figure P20.64 is from Oleg D. Jefimenko, *Electricity and Magnetism: An Introduction to the Theory of Electric and Magnetic Fields,* Star City, WV, Electret Scientific Co., 1989.

directed toward the ground. It is found that the bar slides down the ramp at a constant speed. (a) In a sketch, indicate the direction of the induced current in the bar. (b) Find the direction of the magnetic force acting on the bar. (c) What is the relationship between the forces that must be acting on the bar if it is to move down the ramp at constant speed? (d) How are the directions of the current and the force altered if the magnetic field is upward, away from the ground? Can the bar still slide at constant speed down the ramp in this case?

**G.5** A one-turn coil of wire of area 0.20 $m^2$ and resistance 0.25 Ω is in a magnetic field that varies with time as shown in Figure GA20.5a. The magnetic flux through the coil at $t = 0$ is as shown in Figure GA20.5b. (a) When is the induced current the largest? (b) When is it zero? (c) Is the induced current always in the same direction? (d) Find the direction (clockwise or counterclockwise) and magnitude of the current between $t = 0$ and $t = 2.0$ s, between $t = 2.0$ s and $t = 4.0$ s, and between $t = 4.0$ s and $t = 6.0$ s.

# 21

# Alternating Current Circuits and Electromagnetic Waves

## Chapter Outline

**Radiowaves from distant galaxies are detected by this array of radio telescopes. Everything that we know about celestial objects is determined by studying electromagnetic waves emitted by the objects. In this chapter, we will discuss the various kinds of electromagnetic waves that surround us.** *(The Image Bank/Getty Images)*

It is important to understand the basic principles of alternating-current (AC) circuits because they are so much a part of our everyday life. Every time we turn on a television set or a stereo, or any of a multitude of other electric appliances, we are calling on alternating currents to provide the power to operate them. We begin our study of AC circuits by examining the characteristics of a circuit containing a source of emf and one other circuit element: either a resistor, a capacitor, or an inductor. Then we examine what happens when these elements are connected in combination with each other. Our discussion is limited to situations in which the elements are arranged in simple series configurations.

We conclude this chapter with a discussion of **electromagnetic waves,** which are composed of fluctuating electric and magnetic fields. Electromagnetic waves in the form of visible light enable us to view the world around us; infrared waves warm our environment; radio-frequency waves carry our favorite television and radio programs; the list goes on and on.

## 21.1 RESISTORS IN AN AC CIRCUIT

An AC circuit consists of combinations of circuit elements and an AC generator or an AC source, which provides the alternating current. We have seen that the output of an AC generator is sinusoidal and varies with time according to

$$\Delta v = \Delta V_{\text{max}} \sin 2\pi f t \qquad [21.1]$$

where $\Delta v$ is the instantaneous voltage, $\Delta V_{\text{max}}$ is the maximum voltage of the AC generator, and $f$ is the frequency at which the voltage changes, measured in hertz (Hz). (Compare Eqs. 20.7 and 20.8 with Eq. 21.1.) We first consider a simple circuit consisting of a resistor and an AC source (designated by the symbol —(~)—) as in Figure 21.1. The current and the voltage across the resistor are shown in Figure 21.2.

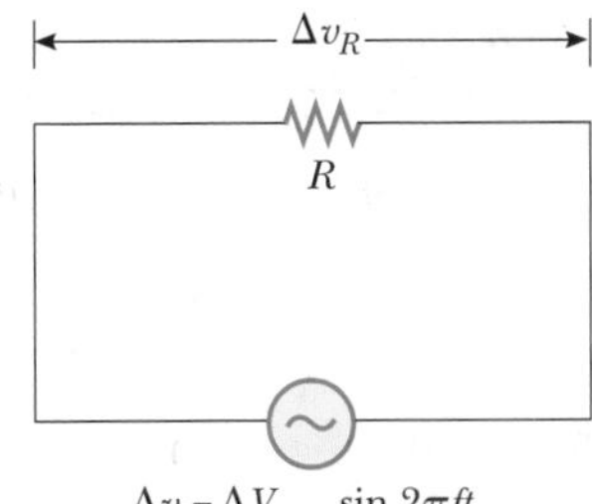


**FIGURE 21.1** A series circuit consisting of a resistor $R$ connected to an AC generator.

Let us try to explain the concept of alternating current by briefly discussing the current-versus-time curve in Figure 21.2. At point $a$ on the curve, the current has a maximum value in one direction, arbitrarily called the *positive direction.* Between points $a$ and $b$, the current is decreasing in magnitude but is still in the positive direction. At point $b$, the current is momentarily zero; it then begins to increase in the opposite (negative) direction between points $b$ and $c$. At point $c$, the current has reached its maximum value in the negative direction.

Note that the current and voltage are in step with each other because they vary identically with time. **Because the current and the voltage reach their maximum values at the same time, they are said to be in phase.** Also note that **the average value of the current over one cycle is zero.** That is, the current is maintained in one direction (the positive direction) for the same amount of time and at the same magnitude as it is in the opposite direction (the negative direction). However, the direction of the current has no effect on the behavior of the resistor in the circuit. This can be understood by realizing that collisions between electrons and the fixed atoms of the resistor result in an increase in the temperature of the resistor. Although this temperature increase depends on the magnitude of the current, it is independent of its direction.

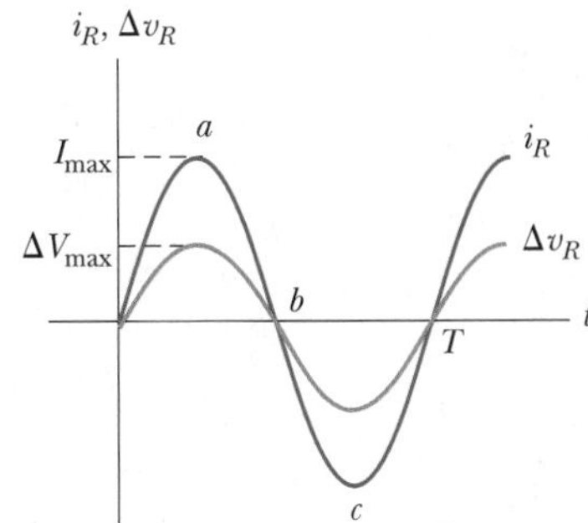


**FIGURE 21.2** A plot of current and voltage across a resistor versus time.

We can quantify this discussion by recalling that the rate at which electrical energy is dissipated in a resistor, which is the power $\mathcal{P}$, is

$$\mathcal{P} = i^2 R$$

where $i$ is the *instantaneous* current in the resistor. Because the heating effect of a current is proportional to the *square* of the current, it makes no difference whether the current is direct or alternating—that is, whether the sign associated with the current is positive or negative. However, the heating effect produced by an

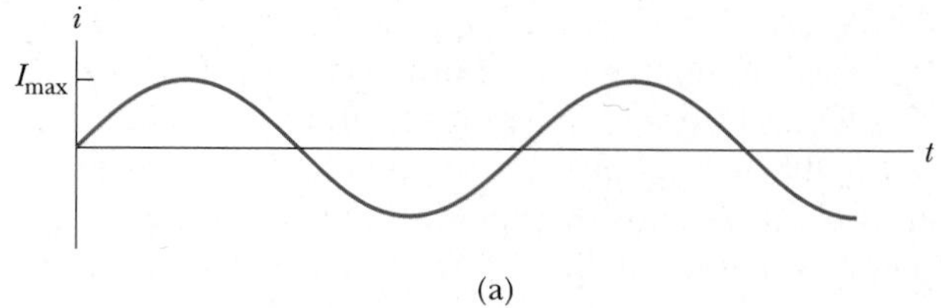


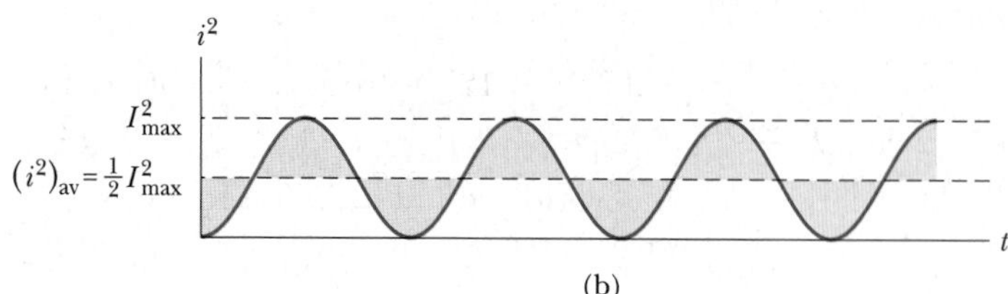


**FIGURE 21.3** (a) Plot of the current in a resistor as a function of time. (b) Plot of the current squared in a resistor as a function of time. Notice that the gray shaded regions *under* the curve and *above* the dashed lines for $I^2_{max}/2$ have the same area as the gray shaded regions *above* the curve and *below* the dashed line for $I^2_{max}/2$. Thus, the average value of $i^2$ is $I^2_{max}/2$.

alternating current with a maximum value of $I_{max}$ *is not the same* as that produced by a direct current of the same value. This is because the alternating current has this maximum value for only an instant of time during a cycle. What is important in an AC circuit is a special kind of average value of current, referred to as the rms current. The **rms current** is the direct current that would dissipate the same amount of energy in a resistor as is dissipated by the actual alternating current. To find the rms current, first we square the current, next find its average value, and then take the square root of this average value. Hence, the rms current is the square *root* of the average (*mean*) of the *square* of the current. Because $i^2$ varies as $\sin^2 2\pi ft$, one can show that the average value of $i^2$ is $\frac{1}{2}I_{max}{}^2$ (Fig. 21.3).[1] Therefore, the rms current $I_{rms}$ is related to the maximum value of the alternating current $I_{max}$ by

rms current 

$$I_{rms} = \frac{I_{max}}{\sqrt{2}} = 0.707 I_{max} \quad [21.2]$$

This equation says that an alternating current with a maximum value of 3 A produces the same heating effect in a resistor as a direct current of $(3/\sqrt{2})$ A. Thus, we can say that the average power dissipated in a resistor that carries alternating current $I$ is $\mathcal{P}_{av} = I^2_{rms}R$.

Alternating voltages are also best discussed in terms of rms voltages, with the relationship being identical to the preceding one; that is, the rms voltage $\Delta V_{rms}$ is related to the maximum value of the alternating voltage $\Delta V_{max}$ by

rms voltage 

$$\Delta V_{rms} = \frac{\Delta V_{max}}{\sqrt{2}} = 0.707 \Delta V_{max} \quad [21.3]$$

[1] The fact that $(i^2)_{av} = I^2_{max}/2$ can be shown as follows. The current in the circuit varies with time according to the expression $i = I_{max} \sin 2\pi ft$, and so $i^2 = I^2_{max} \sin^2 2\pi ft$. Therefore, we can find the average value of $i^2$ by calculating the average value of $\sin^2 2\pi ft$. Note that a graph of $\cos^2 2\pi ft$ versus time is identical to a graph of $\sin^2 2\pi ft$ versus time, except that the points are shifted on the time axis. Thus, the time average of $\sin^2 2\pi ft$ is equal to the time average of $\cos^2 2\pi ft$ when taken over one or more cycles. That is,

$$(\sin^2 2\pi ft)_{av} = (\cos^2 2\pi ft)_{av}$$

With this fact and the trigonometric identity $\sin^2 \theta + \cos^2 \theta = 1$, we get

$$(\sin^2 2\pi ft)_{av} + (\cos^2 2\pi ft)_{av} = 2(\sin^2 2\pi ft)_{av} = 1$$

$$(\sin^2 2\pi ft)_{av} = \tfrac{1}{2}$$

When this result is substituted into the expression $i^2 = I^2_{max} \sin^2 2\pi ft$, we get $(i^2)_{av} = I^2_{rms} = I^2_{max}/2$, or $I_{rms} = I_{max}/\sqrt{2}$, where $I_{rms}$ is the rms current.

**TABLE 21.1 Notation Used in This Chapter**

| | Voltage | Current |
|---|---|---|
| Instantaneous value | $\Delta v$ | $i$ |
| Maximum value | $\Delta V_{max}$ | $I_{max}$ |
| rms value | $\Delta V_{rms}$ | $I_{rms}$ |

When we speak of measuring an AC voltage of 120 V from an electric outlet, we really mean an rms voltage of 120 V. A quick calculation using Equation 21.3 shows that such an AC voltage actually has a peak value of about 170 V. In this chapter we use rms values when discussing alternating currents and voltages. One reason is that AC ammeters and voltmeters are designed to read rms values. Furthermore, if we use rms values, many of the equations for alternating current will have the same form as those used in the study of direct-current (DC) circuits. Table 21.1 summarizes the notations used in this chapter.

Consider the series circuit in Figure 21.1, consisting of a resistor connected to an AC generator. A resistor impedes the current in an AC circuit just as it does in a DC circuit. Therefore, Ohm's law is valid for an AC circuit, and we have

$$\Delta V_{R,\text{rms}} = I_{\text{rms}} R \qquad [21.4]$$

That is, **the rms voltage across a resistor is equal to the rms current in the circuit times the resistance.** This equation also applies if maximum values of current and voltage are used. That is, the maximum voltage drop across a resistor equals the maximum current in the resistor times the resistance.

**Quick Quiz 21.1** Which of the following statements might be true for a resistor connected to an AC generator? (a) $\mathcal{P}_{av} = 0$ and $i_{av} = 0$, (b) $\mathcal{P}_{av} = 0$ and $i_{av} > 0$, (c) $\mathcal{P}_{av} > 0$ and $i_{av} = 0$, (d) $\mathcal{P}_{av} > 0$ and $i_{av} > 0$.

## Example 21.1 What Is the rms Current?

An AC voltage source has an output of $\Delta v = (200\text{ V})\sin 2\pi ft$. This source is connected to a 100-Ω resistor as in Figure 21.1. Find the rms current in the resistor.

**Reasoning** Compare the expression for the voltage output just given with the general form, $\Delta v = \Delta V_{max} \sin 2\pi ft$.

**Solution** By comparison, we see that the maximum output voltage of the device is 200 V. Thus, the rms voltage output of the source is

$$\Delta V_{\text{rms}} = \frac{\Delta V_{\text{max}}}{\sqrt{2}} = \frac{200\text{ V}}{\sqrt{2}} = 141\text{ V}$$

Ohm's law can be used in purely resistive AC circuits as well as in DC circuits. The calculated rms voltage can be used with Ohm's law to find the rms current in the circuit:

$$I_{\text{rms}} = \frac{\Delta V_{\text{rms}}}{R} = \frac{141\text{ V}}{100\ \Omega} = 1.41\text{ A}$$

**EXERCISE** Find the maximum current in the circuit and the power delivered to the circuit.

**ANSWER** 2.00 A; 199 W

## 21.2 CAPACITORS IN AN AC CIRCUIT

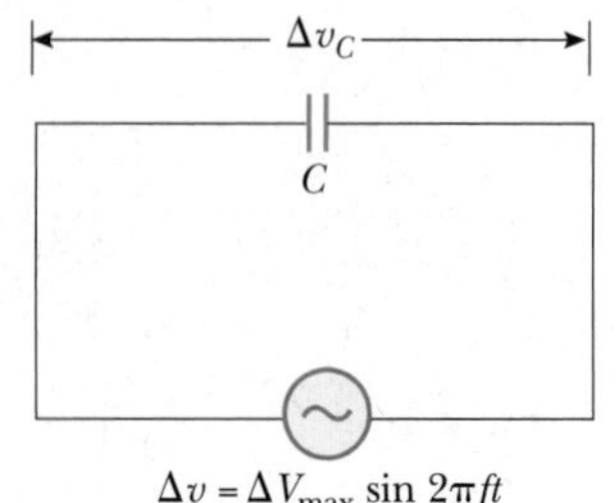


**FIGURE 21.4** A series circuit consisting of a capacitor $C$ connected to an AC generator.

To understand the effect of a capacitor on the behavior of a circuit containing an AC voltage source, let us first review what happens when a capacitor is placed in a circuit containing a DC source, such as a battery. At the instant a switch is closed in a series circuit containing a battery, a resistor, and a capacitor, there is zero charge on the plates of the capacitor. Therefore, the motion of charge through the circuit is relatively free, and initially there is a large current in the circuit. As more charge accumulates on the capacitor, the voltage across it increases, opposing the current. After some time interval, which depends on the time constant $RC$, the current approaches zero. From this, we see that a capacitor in a DC circuit limits or impedes the current so that it approaches zero after a brief time.

Now consider the simple series circuit in Figure 21.4, consisting of a capacitor connected to an AC generator. We sketch curves of current versus time and voltage versus time and then attempt to make the graphs seem reasonable. The curves are shown in Figure 21.5. First, note that the segment of the current curve from $a$ to $b$ indicates that the current starts out at a rather large value. This can be understood by recognizing that there is no charge on the capacitor at $t = 0$; as a consequence, there is nothing in the circuit except the resistance of the wires to hinder the flow of charge at this instant. However, the current decreases as the voltage across the capacitor increases from $c$ to $d$ on the voltage curve. When the voltage is at point $d$, the current reverses and begins to increase in the opposite direction (from $b$ to $e$ on the current curve). During this time, the voltage across the capacitor decreases from $d$ to $f$ because the plates are now losing the charge they accumulated earlier. The remainder of the cycle for both voltage and current is a repeat of what happened during the first half of the cycle. The current reaches a maximum value in the opposite direction at point $e$ on the current curve and then decreases as the voltage across the capacitor builds up.

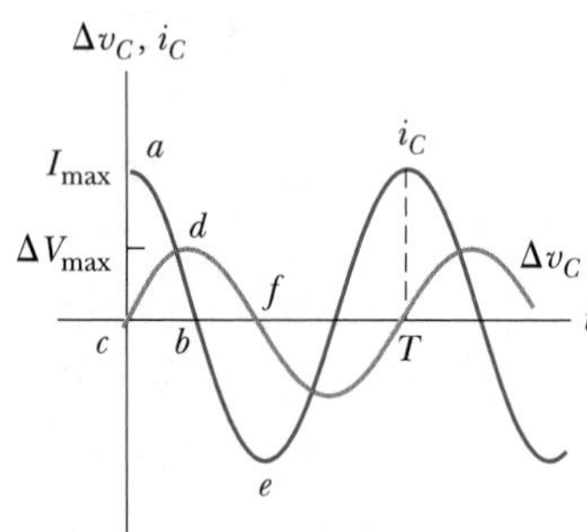


**FIGURE 21.5** Plots of current and voltage across a capacitor versus time in an AC circuit. The voltage lags the current by 90°.

Note that the current and voltage are not in step with each other, as they are in a purely resistive circuit. The curves of Figure 21.5 indicate that when an alternating voltage is applied across a capacitor, the voltage reaches its maximum value one quarter of a cycle after the current reaches its maximum value. In this situation, it is common to say that **the voltage across a capacitor always lags behind the current by 90°.**

The voltage across a capacitor lags behind the current by 90°.

The impeding effect of a capacitor on the current in an AC circuit is expressed in terms of a factor called the **capacitive reactance,** $X_C$, defined as

$$X_C \equiv \frac{1}{2\pi fC} \qquad [21.5]$$

You will be asked in Problem 7 at the end of the chapter to show that when $C$ is in farads and $f$ is in hertz, the unit of $X_C$ is the ohm.

Equation 21.5 indicates that as the frequency $f$ of the voltage source increases, the capacitive reactance $X_C$, or the impeding effect of the capacitor, decreases and therefore the current increases. This may be understood by noting that the voltage across the capacitor controls the current in the circuit. At high frequency, there is less time available to charge the capacitor, and less charge and voltage accumulate on the capacitor resulting in less opposition to charge flow and higher current. The analogy between capacitive reactance and resistance allows us to write an equation of the same form as Ohm's law to describe AC circuits containing capacitors. This equation relates the rms voltage and rms current in the circuit to the reactance as

$$\Delta V_{C,\text{rms}} = I_{\text{rms}} X_C \qquad [21.6]$$

**Example 21.2 A Purely Capacitive AC Circuit**

An 8.00-$\mu$F capacitor is connected to the terminals of an AC generator with an rms voltage of 150 V and a frequency of 60.0 Hz. Find the capacitive reactance and the rms current in the circuit.

**Solution** From Equation 21.5 and the fact that $2\pi f = 377\ \text{s}^{-1}$, we have

$$X_C = \frac{1}{2\pi fC} = \frac{1}{(377\ \text{s}^{-1})(8.00 \times 10^{-6}\ \text{F})} = 332\ \Omega$$

If we substitute this result into Equation 21.6, we find that

$$I_{\text{rms}} = \frac{\Delta V_{C,\text{rms}}}{X_C} = \frac{150\ \text{V}}{332\ \Omega} = 0.452\ \text{A}$$

**EXERCISE** If the frequency is doubled, what happens to the capacitive reactance and the current?

**ANSWER** $X_C$ is halved, and $I_{\text{rms}}$ is doubled.

## 21.3 INDUCTORS IN AN AC CIRCUIT

Now consider an AC circuit consisting only of an inductor connected to the terminals of an AC source, as in Figure 21.6. (In any real circuit, there is some resistance in the wire forming the inductive coil, but we ignore this for now.) The changing current output of the generator produces a back emf in the coil of magnitude

$$\Delta v_L = L\frac{\Delta I}{\Delta t} \qquad [21.7]$$

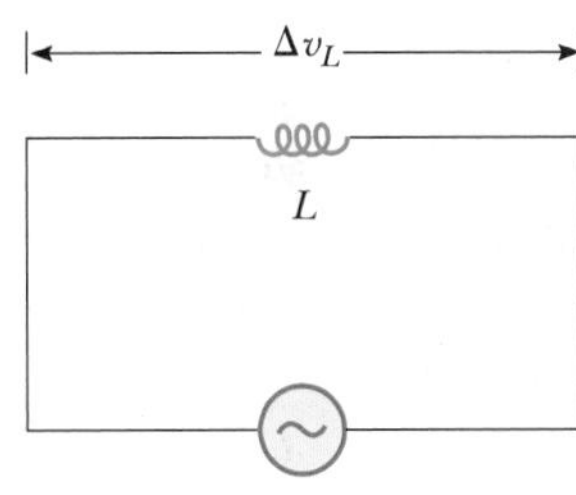


**FIGURE 21.6** A series circuit consisting of an inductor $L$ connected to an AC generator.

Thus, the current in the circuit is impeded by the back emf of the inductor. The effective resistance of the coil in an AC circuit is measured by a quantity called the **inductive reactance,** $X_L$:

$$X_L \equiv 2\pi fL \qquad [21.8]$$

Inductive reactance

You will be asked in Problem 13 at the end of the chapter to show that when $f$ is in hertz and $L$ is in henries, the unit of $X_L$ is the ohm. Note that the inductive reactance increases with increasing frequency and increasing inductance.

To understand the meaning of inductive reactance, let us compare this equation for $X_L$ with Equation 21.7. First, note from Equation 21.8 that the inductive reactance depends on the inductance $L$. This seems reasonable because the back emf (Eq. 21.7) is large for large values of $L$. Second, note that the inductive reactance depends on the frequency $f$. This, too, seems reasonable because the back emf depends on $\Delta I/\Delta t$, a quantity that is large when the current changes rapidly, as it would for high frequencies.

With inductive reactance defined in this manner, we can write an equation of the same form as Ohm's law for the voltage across the coil or inductor:

$$\Delta V_{L,\text{rms}} = I_{\text{rms}} X_L \qquad [21.9]$$

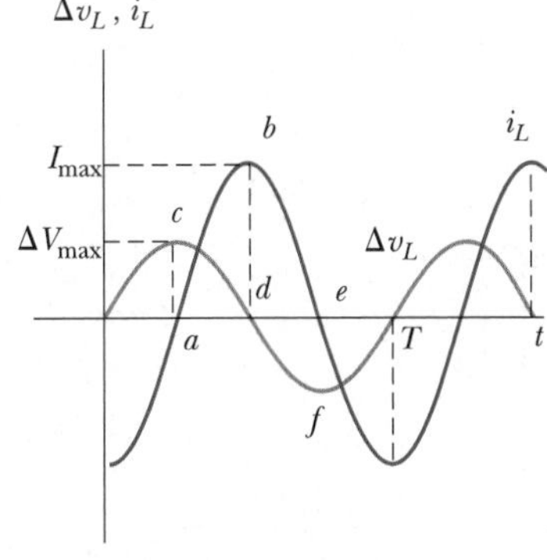


**FIGURE 21.7** Plots of current and voltage across an inductor versus time in an AC circuit. The voltage leads the current by 90°.

where $\Delta V_{L,\text{rms}}$ is the rms voltage across the coil and $I_{\text{rms}}$ is the rms current in the coil.

Figure 21.7 shows the instantaneous voltage and instantaneous current across the coil as functions of time. When a sinusoidal voltage is applied across an

The voltage across an inductor leads the current by 90°.

inductor, the voltage reaches its maximum value one quarter of an oscillation period before the current reaches its maximum value. In this situation, we say that **the voltage across an inductor always leads the current by 90°.**

To see why there is a phase relationship between voltage and current, let us examine a few points on the curves of Figure 21.7. Note that at point *a* on the current curve, the current is beginning to increase in the positive direction. At this instant, the rate of change of current, $\Delta I/\Delta t$ (the slope of the current curve), is at a maximum, and we see from Equation 21.7 that the voltage across the inductor is consequently also at a maximum at this time. As the current rises between points *a* and *b* on the curve, $\Delta I/\Delta t$ gradually decreases until it reaches zero at point *b*. As a result, the voltage across the inductor is decreasing during this same time interval, as the segment between *c* and *d* on the voltage curve indicates. Immediately after point *b*, the current begins to decrease, although it still has the same direction it had during the previous quarter cycle. As the current decreases to zero (from *b* to *e* on the curve), a voltage is again induced in the coil ($d$ to $f$), but the sense of this voltage is opposite the sense of the voltage induced between *c* and *d*. This occurs because back emfs are always directed to oppose the change in the current.

We could continue to examine other segments of the curves, but no new information would be gained because the current and voltage variations are repetitive.

**Example 21.3 A Purely Inductive AC Circuit**

In a purely inductive AC circuit (see Fig. 21.6), $L = 25.0$ mH and the rms voltage is 150 V. Find the inductive reactance and rms current in the circuit if the frequency is 60.0 Hz.

**Solution** First, note that $2\pi f = 2\pi(60.0) = 377\ \text{s}^{-1}$. Equation 21.8 then gives

$$X_L = 2\pi f L = (377\ \text{s}^{-1})(25.0 \times 10^{-3}\ \text{H}) = 9.43\ \Omega$$

Substituting this result into Equation 21.9 gives

$$I_{\text{rms}} = \frac{\Delta V_{L,\text{rms}}}{X_L} = \frac{150\ \text{V}}{9.43\ \Omega} = 15.9\ \text{A}$$

**EXERCISE** Calculate the inductive reactance and rms current in the circuit if the frequency is 6 kHz.

**ANSWER** $X_L = 943\ \Omega$; $I = 0.159$ A

## 21.4 THE *RLC* SERIES CIRCUIT

In the foregoing sections, we examined the effects of an inductor, a capacitor, and a resistor when they are connected separately across an AC voltage source. We now consider what happens when these devices are combined.

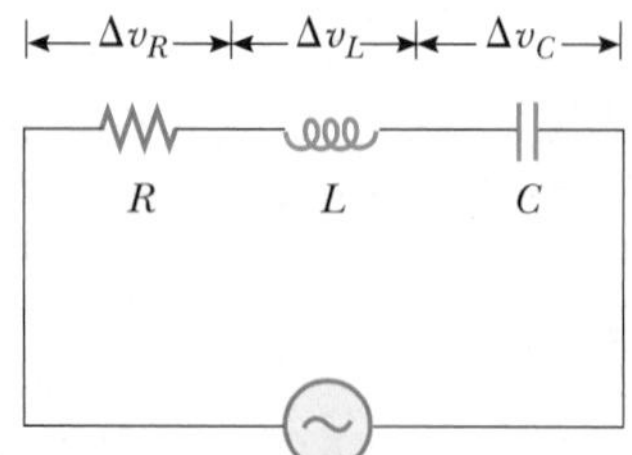


**FIGURE 21.8** A series circuit consisting of a resistor, an inductor, and a capacitor connected to an AC generator.

Figure 21.8 shows a circuit containing a resistor, an inductor, and a capacitor connected in series across an AC source that supplies a total voltage $\Delta v$ at some instant. The current in the circuit is the same at all points in the circuit at any instant and varies sinusoidally with time, as indicated in Figure 21.9a. Thus,

$$i = I_{\max} \sin 2\pi f t$$

Earlier we learned that the voltage across each element may or may not be in phase with the current. The instantaneous voltages across the three elements, shown in Figure 21.9, have the following phase relations to the instantaneous current:

1. The instantaneous voltage $\Delta v_R$ across the resistor is *in phase* with the instantaneous current. (See Fig. 21.9b.)

2. The instantaneous voltage $\Delta v_L$ across the inductor *leads* the current by 90°. (See Fig. 21.9c.)
3. The instantaneous voltage $\Delta v_C$ across the capacitor *lags behind* the current by 90°. (See Fig. 21.9d.)

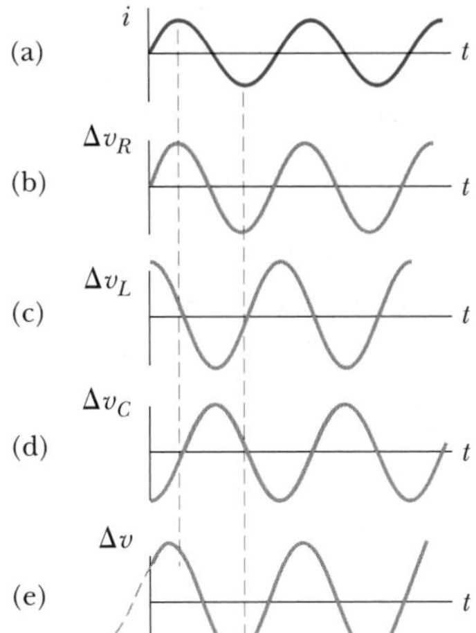


**FIGURE 21.9** Phase relations in the series *RLC* circuit shown in Figure 21.8.

The net instantaneous voltage $\Delta v$ supplied by the AC source equals the sum of the instantaneous voltages across the separate elements; that is, $\Delta v = \Delta v_R + \Delta v_C + \Delta v_L$. This does not mean, however, that the voltages measured with an AC voltmeter across *R*, *C*, and *L* sum up to the measured source voltage! The measured voltages do not sum to the measured source voltage because the voltages across *R*, *C*, and *L* all have different phases.

To account for the different phases of the voltage drops, we use a technique involving vectors. We represent the voltage across each element with a rotating vector, as in Figure 21.10. The rotating vectors are referred to as **phasors,** and the diagram is called a **phasor diagram.** This particular diagram represents the circuit voltage given by the expression $\Delta v = \Delta V_{max} \sin(2\pi ft + \phi)$, where $\Delta V_{max}$ is the maximum voltage (the magnitude or length of the rotating vector or phasor) and $\phi$ is the angle between the phasor and the $+x$ axis when $t = 0$. The phasor can be viewed as a vector of magnitude $\Delta V_{max}$ rotating at a constant frequency $f$ so that its projection along the $y$ axis is the instantaneous voltage in the circuit. Because $\phi$ is the phase angle between the voltage and current in the circuit, the phasor for the current (not shown in Fig. 21.10) lies along the $+x$ axis when $t = 0$ and is expressed by the relation $i = I_{max} \sin(2\pi ft)$.

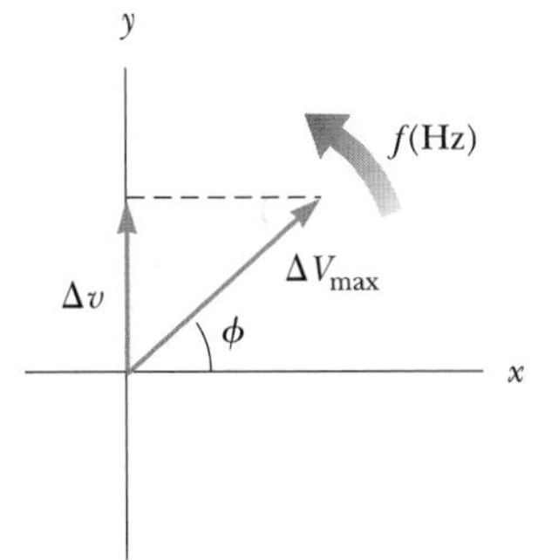


**FIGURE 21.10** A phasor diagram for the voltage in an AC circuit, where $\phi$ is the phase angle between the voltage and current and $\Delta v$ is the instantaneous voltage.

The phasor diagrams in Figure 21.11 are useful for analyzing the *series RLC* circuit. Voltages in phase with the current are represented by vectors along the $+x$ axis, and voltages out of phase with the current lie along other directions. Thus, $\Delta V_R$ is horizontal and to the right because it is in phase with the current. Likewise, $\Delta V_L$ is represented by a phasor along the $+y$ axis because it leads the current by 90°. Finally, $\Delta V_C$ is along the $-y$ axis because it lags behind the current by 90°.[2] If the phasors are added as vector quantities in order to account for the different phases of the voltages across *R*, *L*, and *C*, Figure 21.11a shows that the only $x$ component for the voltages is $\Delta V_R$, and the net $y$ component is $\Delta V_L - \Delta V_C$. We now add the phasors vectorially to find the phasor $\Delta V_{max}$ (Fig. 21.11b), where $\Delta V_{max}$ represents the maximum voltage. The right triangle in Figure 21.11b gives the

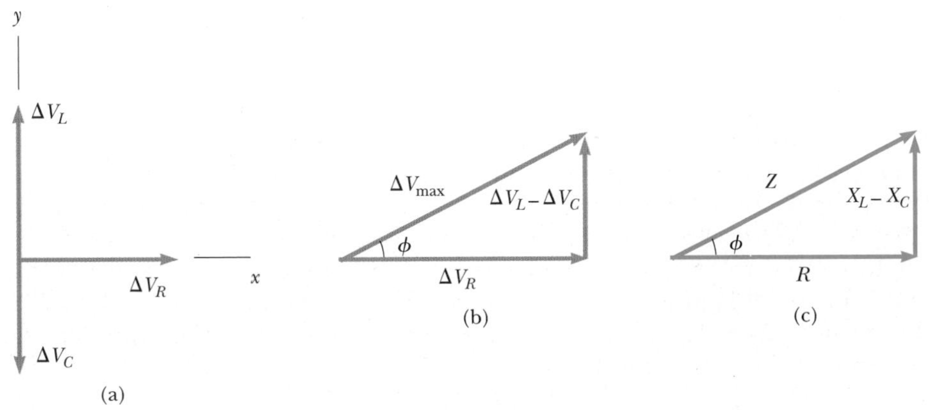


**FIGURE 21.11** (a) A phasor diagram for the *RLC* circuit. (b) Addition of the phasors as vectors gives $\Delta V_{max} = \sqrt{\Delta V_R^2 + (\Delta V_L - \Delta V_C)^2}$. (c) The reactance triangle that gives the impedance relation, $Z = \sqrt{R^2 + (X_L - X_C)^2}$.

[2] A mnemonic to help you remember the phase relationships in *RLC* circuits is "*ELI* the *ICE* man." *E* represents the voltage, *I* the current, *L* the inductance, and *C* the capacitance. Thus, the name *ELI* means that in an inductive circuit, the voltage *E* leads the current *I*. In a capacitive circuit, *ICE* means that the current leads the voltage.

following equations for the maximum voltage and the phase angle $\phi$ between the maximum voltage and the current:

$$\Delta V_{max} = \sqrt{\Delta V_R^2 + (\Delta V_L - \Delta V_C)^2} \quad [21.10]$$

$$\tan \phi = \frac{\Delta V_L - \Delta V_C}{\Delta V_R} \quad [21.11]$$

where all voltages are maximum values. Note that although we choose to use maximum voltages in our analysis, the preceding equations apply equally well to rms voltages, because the two quantities are related to each other by the same factor for all circuit elements. The result for the maximum voltage $\Delta V_{max}$ as given by Equation 21.10 reinforces the fact that **the voltages across the resistor, capacitor, and inductor are not in phase, so one cannot simply add them to get the voltage across the combination of elements or the source voltage.**

**Quick Quiz 21.2** For the circuit of Figure 21.8, is the voltage of the source equal to (a) the sum of the maximum voltages across the elements, (b) the sum of the instantaneous voltages across the elements, or (c) the sum of the rms voltages across the elements?

We can write Equation 21.10 in the form of Ohm's law, using the relations $\Delta V_R = I_{max}R$, $\Delta V_L = I_{max}X_L$, and $\Delta V_C = I_{max}X_C$, where $I_{max}$ is the maximum current in the circuit:

$$\Delta V_{max} = I_{max}\sqrt{R^2 + (X_L - X_C)^2} \quad [21.12]$$

It is convenient to define a parameter called the **impedance,** $Z$, of the circuit as

Impedance

$$Z \equiv \sqrt{R^2 + (X_L - X_C)^2} \quad [21.13]$$

so that Equation 21.12 becomes

$$\Delta V_{max} = I_{max}Z \quad [21.14]$$

Note that Equation 21.14 is in the form of Ohm's law, $\Delta V = IR$, where $R$ is replaced by the impedance in ohms. Equation 21.14 can be regarded as a generalized form of Ohm's law applied to a series AC circuit. Note that the impedance and therefore the current in an AC circuit depend on the resistance, the inductance, the capacitance, *and* the frequency (because the reactances are frequency dependent).

It is useful to represent the impedance $Z$ with a vector diagram such as the one depicted in Figure 21.11c. A right triangle is constructed; the right side is the quantity $X_L - X_C$, the base is $R$, and the hypotenuse is $Z$. Applying the Pythagorean theorem to this triangle, we see that

$$Z = \sqrt{R^2 + (X_L - X_C)^2}$$

which is consistent with Equation 21.13. Furthermore, we see from the vector diagram in Figure 21.11c that the phase angle $\phi$ between the current and the voltage obeys the relationship

Phase angle $\phi$

$$\tan \phi = \frac{X_L - X_C}{R} \quad [21.15]$$

The physical significance of the phase angle will become apparent in Section 21.5.

**TABLE 21.2 Impedance Values and Phase Angles for Various Combinations of Circuit Elements**[a]

| Circuit Elements | Impedance $Z$ | Phase Angle $\phi$ |
|---|---|---|
| $R$ | $R$ | 0° |
| $C$ | $X_C$ | −90° |
| $L$ | $X_L$ | +90° |
| $R$ $C$ | $\sqrt{R^2 + X_C^2}$ | Negative, between −90° and 0° |
| $R$ $L$ | $\sqrt{R^2 + X_L^2}$ | Positive, between 0° and 90° |
| $R$ $L$ $C$ | $\sqrt{R^2 + (X_L - X_C)^2}$ | Negative if $X_C > X_L$ Positive if $X_C < X_L$ |

[a] In each case, an AC voltage (not shown) is applied across the combination of elements (that is, across the dots).

Table 21.2 provides impedance values and phase angles for some series circuits containing different combinations of circuit elements.

Parallel alternating-current circuits are also useful in everyday applications. We shall not discuss them here, however, because their analysis is beyond the scope of this book.

### Webnote 21.1

Take a look at a real-time Java applet of voltage across each element of an *RLC* series circuit. Go to *http://www.phy.ntnu.edu.tw/java/rlc/rlc.html*

**Quick Quiz 21.3**

The switch in the circuit shown in Figure 21.12 is closed and the lightbulb glows steadily. The inductor is a simple air-core solenoid. As an iron rod is being inserted into the interior of the solenoid, the brightness of the lightbulb (a) increases, (b) decreases, or (c) remains the same.

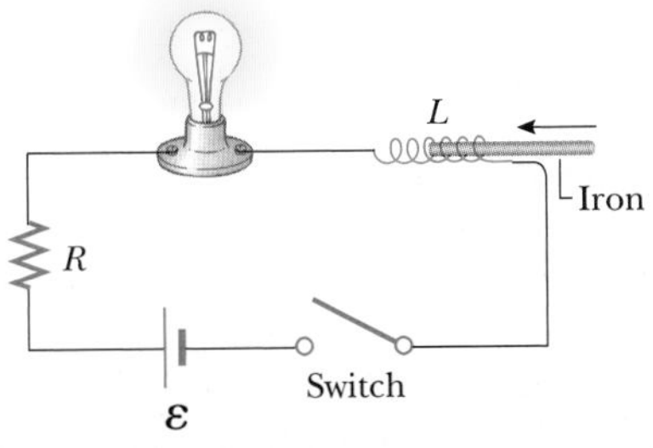


**FIGURE 21.12** (Quick Quiz 21.3)

## PROBLEM-SOLVING STRATEGY Alternating Current

The following procedures are recommended for solving alternating-current problems:

1. The first step in analyzing alternating-current circuits is to calculate as many of the unknown quantities, such as $X_L$ and $X_C$, as possible. (When you calculate $X_C$, express the capacitance in farads rather than, say, microfarads.)
2. Apply the equation $\Delta V_{\max} = I_{\max} Z$ to the portion of the circuit that is of interest. For example, if you want to know the voltage drop across the combination of an inductor and a resistor, the equation for the voltage drop reduces to $\Delta V_{\max} = I_{\max}\sqrt{R^2 + X_L{}^2}$.

**Example 21.4 Analyzing a Series *RLC* AC Circuit**

Analyze a series *RLC* AC circuit for which $R = 250\ \Omega$, $L = 0.600$ H, $C = 3.50\ \mu$F, $f = 60$ Hz, and $\Delta V_{max} = 150$ V.

**Solution** The reactances are given by $X_L = 2\pi fL = 226\ \Omega$ and $X_C = 1/2\pi fC = 758\ \Omega$. Therefore, the impedance is

$$Z = \sqrt{R^2 + (X_L - X_C)^2} = \sqrt{(250\ \Omega)^2 + (226\ \Omega - 758\ \Omega)^2} = 588\ \Omega$$

The maximum current is

$$I_{max} = \frac{\Delta V_{max}}{Z} = \frac{150\ \text{V}}{588\ \Omega} = 0.255\ \text{A}$$

The phase angle between the current and voltage is

$$\phi = \tan^{-1}\left(\frac{X_L - X_C}{R}\right) = \tan^{-1}\left(\frac{226\ \Omega - 758\ \Omega}{250\ \Omega}\right) = -64.8°$$

Because the circuit is more capacitive than inductive (that is, $X_C > X_L$), $\phi$ is negative. A negative phase angle means that the current leads the applied voltage.

The maximum voltages across the elements are

$$\Delta V_R = I_{max}R = (0.255\ \text{A})(250\ \Omega) = 63.8\ \text{V}$$

$$\Delta V_L = I_{max}X_L = (0.255\ \text{A})(226\ \Omega) = 57.6\ \text{V}$$

$$\Delta V_C = I_{max}X_C = (0.255\ \text{A})(758\ \Omega) = 193\ \text{V}$$

Note that the sum of the maximum voltages across the elements is $\Delta V_R + \Delta V_L + \Delta V_C = 314$ V, which is much greater than the maximum voltage of the generator, 150 V. As we saw in Quick Quiz 21.2, the sum of the maximum voltages is a meaningless quantity because when alternating voltages are added, *both their amplitudes and their phases* must be taken into account. We know that the maximum voltages across the various elements occur at different times. That is, the voltages must be added in a way that takes account of the different phases. When this is done, Equation 21.12 is satisfied. You should verify this result.

**NIKOLA TESLA (1856–1943)**

Tesla was born in Croatia but spent most of his professional life as an inventor in the United States. He was a key figure in the development of alternating-current electricity, high-voltage transformers, and the transport of electrical power using AC transmission lines. Tesla's viewpoint was at odds with the ideas of Edison, who committed himself to the use of direct current in power transmission. Tesla's AC approach won out. *(CORBIS/Bettmann)*

## 21.5 POWER IN AN AC CIRCUIT

No power losses are associated with capacitors and pure inductors in an AC circuit. (A pure inductor is defined as one with no resistance or capacitance.) Let us begin by analyzing the power dissipated in an AC circuit that contains only a generator and a capacitor.

When the current begins to increase in one direction in an AC circuit, charge begins to accumulate on the capacitor and a voltage drop appears across it. When this voltage reaches its maximum value, the energy stored in the capacitor is

$$PE_C = \tfrac{1}{2}C(\Delta V_{max})^2$$

However, this energy storage is only momentary. When the current reverses direction, the charge leaves the capacitor plates and returns to the voltage source. Thus, during one half of each cycle the capacitor is being charged, and during the other half the charge is being returned to the voltage source. Therefore, the average power supplied by the source is zero. In other words, **no power losses occur in a capacitor in an AC circuit.**

Similarly, the source must do work against the back emf of an inductor that is carrying a current. When the current reaches its maximum value, the energy stored in the inductor is a maximum and is given by

$$PE_L = \tfrac{1}{2}LI_{max}^2$$

When the current begins to decrease in the circuit, this stored energy is returned to the source as the inductor attempts to maintain the current in the circuit. The average power delivered to a resistor in an *RLC* circuit is

$$\mathcal{P}_{av} = I_{rms}^2 R \qquad [21.16]$$

Note **that the average power delivered by the generator is converted to internal energy in the resistor. No power loss occurs in an ideal capacitor or inductor.**

An alternative equation for the average power loss in an AC circuit can be found by substituting (from Ohm's law) $R = \Delta V_R / I_{rms}$ into Equation 21.16:

$$\mathcal{P}_{av} = I_{rms}\,\Delta V_R$$

It is convenient to refer to a voltage triangle that shows the relationship among $\Delta V_{rms}$, $\Delta V_R$, and $\Delta V_L - \Delta V_C$, such as Figure 21.11b. (Remember that Figure 21.11 was valid for *both* maximum and rms voltages.) From this figure, we see that the voltage drop across a resistor can be written in terms of the voltage of the source, $\Delta V_{rms}$:

$$\Delta V_R = \Delta V_{rms} \cos \phi$$

Hence, the average power delivered by a generator in an AC circuit is

$$\mathcal{P}_{av} = I_{rms}\,\Delta V_{rms} \cos \phi \qquad [21.17]$$

Average power

where the quantity $\cos \phi$ is called the **power factor.**

Equation 21.17 shows that the power delivered by an AC source to any circuit depends on the phase difference between the source voltage and the resulting current. This fact has many interesting applications. For example, factories often use large motors in devices such as machines, generators, and transformers that have a large inductive load due to all the windings. To deliver greater power to such devices without using excessively high voltages, factory technicians introduce capacitance in the circuits to shift the phase.

**APPLICATION**

SHIFTING PHASE TO DELIVER MORE POWER

### Example 21.5 Average Power in a *RLC* Series Circuit

Calculate the average power delivered to the series *RLC* circuit described in Example 21.4.

**Solution** First, let us calculate the rms voltage and rms current using the value of $Z$ calculated in Example 21.4:

$$\Delta V_{rms} = \frac{\Delta V_{max}}{\sqrt{2}} = \frac{150 \text{ V}}{\sqrt{2}} = 106 \text{ V}$$

$$I_{rms} = \frac{I_{max}}{\sqrt{2}} = \frac{\Delta V_{max}/Z}{\sqrt{2}} = \frac{0.255 \text{ A}}{\sqrt{2}} = 0.180 \text{ A}$$

Because $\phi = -64.8°$, the power factor, $\cos \phi$, is 0.426, and hence the average power using Equation 21.17 is

$$\mathcal{P}_{av} = I_{rms}\Delta V_{rms} \cos \phi = (0.180 \text{ A})(106 \text{ V})(0.426) = 8.12 \text{ W}$$

The same result can be obtained using Equation 21.16.

## 21.6 Resonance in a Series *RLC* Circuit

In general, the rms current in a series *RLC* circuit can be written

$$I_{rms} = \frac{\Delta V_{rms}}{Z} = \frac{\Delta V_{rms}}{\sqrt{R^2 + (X_L - X_C)^2}} \qquad [21.18]$$

Resonance frequency

From this we see that if the frequency is varied, the current has its *maximum* value when the impedance has its *minimum* value. This occurs when $X_L = X_C$. In such a circumstance, the impedance of the circuit reduces to $Z = R$. The frequency $f_0$ at which this happens is called the **resonance frequency** of the circuit. To find $f_0$, we set $X_L = X_C$, which gives, from Equations 21.5 and 21.8,

$$2\pi f_0 L = \frac{1}{2\pi f_0 C}$$

$$f_0 = \frac{1}{2\pi\sqrt{LC}} \quad [21.19]$$

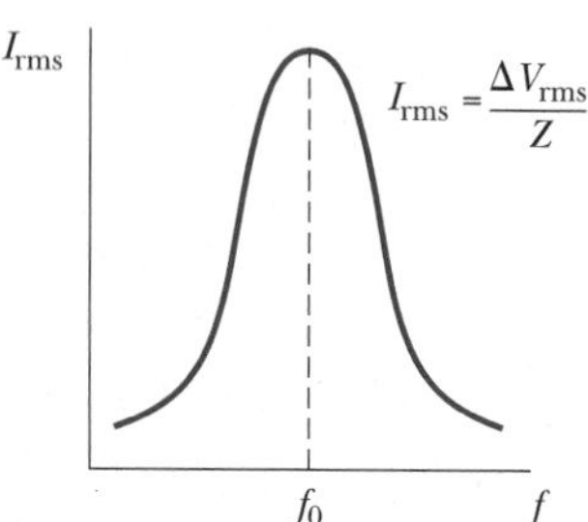


**FIGURE 21.13** A plot of current amplitude in a series *RLC* circuit versus frequency of the generator voltage. Note that the current reaches its maximum value at the resonance frequency $f_0$.

Figure 21.13 is a plot of current as a function of frequency for a circuit containing a fixed value for both the capacitance and inductance. From Equation 21.18 it must be concluded that the current would become infinite at resonance when $R = 0$. Although Equation 21.18 predicts this result, real circuits always have some resistance, which limits the value of the current.

**APPLICATION**

TUNING YOUR RADIO

The tuning circuit of a radio is an important application of a series resonance circuit. The radio is tuned to a particular station (which transmits a specific radio-frequency signal) by varying a capacitor, which changes the resonance frequency of the tuning circuit. When this resonance frequency matches that of the incoming radio wave, the current in the tuning circuit increases.

## APPLYING PHYSICS 21.1

**APPLICATION**

METAL DETECTORS IN AIRPORTS

When you walk through the doorway of an airport metal detector as in Figure 21.14, you are really walking through a coil of many turns. How might this work?

**Explanation** The metal detector is essentially a resonant circuit. The portal you step through is an inductor (a large loop of conducting wire) that is part of the circuit. The frequency of the circuit is tuned to the resonant frequency of the circuit when there is no metal in the inductor. When you walk through with metal in your pocket, you change the effective inductance of the resonance circuit, resulting in a change in the current in the circuit. This change in current is detected, and an electronic circuit causes a sound to be emitted as an alarm.

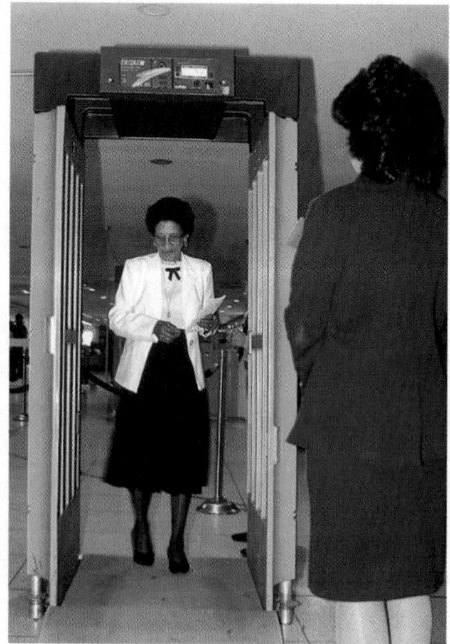

**FIGURE 21.14** (Applying Physics 21.1) An airport metal detector. *(Ryan Williams/International Stock Photography)*

## Example 21.6 The Capacitance of a Circuit in Resonance

Consider a series *RLC* circuit for which $R = 150\ \Omega$, $L = 20$ mH, $\Delta V_{rms} = 20$ V, and $2\pi f = 5.0 \times 10^3\ \text{s}^{-1}$. Determine the value of the capacitance for which the rms current is a maximum.

**Reasoning** The current is a maximum at the resonance frequency $f_0$, which should be made to match the driving frequency, $5.0 \times 10^3\ \text{s}^{-1}$.

**Solution** In this problem,

$$2\pi f_0 = 5.0 \times 10^3\ \text{s}^{-1} = \frac{1}{\sqrt{LC}}$$

$$C = \frac{1}{(25 \times 10^6\ \text{s}^{-2})L} = \frac{1}{(25 \times 10^6\ \text{s}^{-2})(20.0 \times 10^{-3}\ \text{H})} = 2.0\ \mu F$$

**EXERCISE** Calculate the maximum rms current in the circuit.

**ANSWER** 0.13 A

## 21.7 THE TRANSFORMER

In many situations it is desirable or necessary to change a small AC voltage to a larger one or vice versa. Before we examine a few such cases, let us consider the device that makes these conversions possible, the transformer.

In its simplest form, the **AC transformer** consists of two coils of wire wound around a core of soft iron, as in Figure 21.15. The coil on the left, which is connected to the input AC voltage source and has $N_1$ turns, is called the primary winding, or the *primary*. The coil on the right, which is connected to a resistor $R$ and consists of $N_2$ turns, is the *secondary*. The purpose of the common iron core is to increase the magnetic flux and to provide a medium in which nearly all the flux through one coil passes through the other.

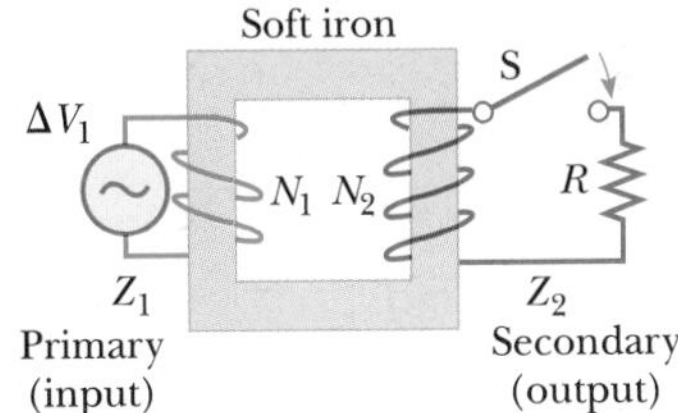


**FIGURE 21.15** An ideal transformer consists of two coils wound on the same soft iron core. An AC voltage $\Delta V_1$ is applied to the primary coil, and the output voltage $\Delta V_2$ is observed across the load resistance $R$.

When an input AC voltage $\Delta V_1$ is applied to the primary, the induced voltage across it is given by

$$\Delta V_1 = -N_1 \frac{\Delta \Phi_B}{\Delta t} \qquad [21.20]$$

where $\Phi_B$ is the magnetic flux through each turn. If we assume that no flux leaks from the iron core, then the flux through each turn of the primary equals the flux through each turn of the secondary. Hence, the voltage across the secondary coil is

$$\Delta V_2 = -N_2 \frac{\Delta \Phi_B}{\Delta t} \qquad [21.21]$$

The term $\Delta\Phi_B/\Delta t$ is common to Equations 21.20 and 21.21. Therefore, we see that

$$\Delta V_2 = \frac{N_2}{N_1} \Delta V_1 \qquad [21.22]$$

When $N_2$ is greater than $N_1$, and thus $\Delta V_2$ exceeds $\Delta V_1$, the transformer is referred to as a *step-up transformer*. When $N_2$ is less than $N_1$, making $\Delta V_2$ less than $\Delta V_1$, we speak of a *step-down transformer*.

It should be clear that a voltage is generated across the secondary only when there is a *change* in the number of flux lines passing through the secondary. Thus, the input current in the primary must change with time, which is what happens when an alternating current is used. However, when the input at the primary is a direct current, a voltage output occurs at the secondary only at the instant a switch in the primary circuit is opened or closed. Once the current in the primary reaches a steady value, the output voltage at the secondary is zero.

It may seem that a transformer is a device in which it is possible to get something for nothing. For example, a step-up transformer can change an input voltage from, say, 10 V to 100 V. This means that each 1 coulomb of charge leaving the secondary has 100 J of energy, whereas each coulomb of charge entering the primary has only 10 J of energy. However, this is not an example of a breakdown in the principle of conservation of energy, because **the power input to the primary equals the power output at the secondary;** that is,

$$I_1 \Delta V_1 = I_2 \Delta V_2 \qquad [21.23]$$

Thus, if the voltage at the secondary is ten times that at the primary, the current at the secondary is reduced by a factor of 10. Equation 21.23 assumes an **ideal transformer,** in which there are no power losses between the primary and the secondary. Real transformers typically have power efficiencies ranging from 90% to 99%. Power losses occur because of such factors as eddy currents induced in the iron core of the transformer, which dissipate energy in the form of $I^2R$ losses.

In an ideal transformer, the input power equals the output power.

When electric power is transmitted over large distances, it is economical to use a high voltage and a low current because the power lost via resistive heating in the transmission lines varies as $I^2R$. This means that if a utility company can reduce

LONG-DISTANCE ELECTRIC POWER TRANSMISSION

the current by a factor of 10, for example, the power loss is reduced by a factor of 100. In practice, the voltage is stepped up to around 230 000 V at the generating station, then stepped down to around 20 000 V at a distribution station, and finally stepped down to 120 V at the customer's utility pole.

### Example 21.7 Distributing Power to a City

A generator at a utility company produces 100 A of current at 4 000 V. The voltage is stepped up to 240 000 V by a transformer before it is sent on a high-voltage transmission line across a rural area to a city. Assume that the effective resistance of the power line is 30.0 Ω.

**A** Determine the percentage of power lost.

**Solution** From Equation 21.23, the current in the transmission line is

$$I_2 = \frac{I_1 \Delta V_1}{\Delta V_2} = \frac{(100\text{ A})(4.00 \times 10^3\text{ V})}{2.40 \times 10^5\text{ V}} = 1.67\text{ A}$$

and the power lost in the transmission line is

$$\mathcal{P}_{\text{lost}} = I_2^2 R = (1.67\text{ A})^2(30.0\ \Omega) = 83.7\text{ W}$$

The power output of the generator is

$$\mathcal{P} = I\Delta V = (100\text{ A})(4\,000\text{ V}) = 4.00 \times 10^5\text{ W}$$

From this we can find the percentage of power lost as

$$\%\text{ power lost} = \left(\frac{83.7\text{ W}}{4.00 \times 10^5\text{ W}}\right) \times 100 = 0.0209\%$$

**B** What percentage of the original power would be lost in the transmission line if the voltage were not stepped up?

**Solution** If the voltage were not stepped up, the current in the transmission line would be 100 A and the power lost in the line would be

$$\mathcal{P}_{\text{lost}} = I^2 R = (100\text{ A})^2(30.0\ \Omega) = 3.00 \times 10^5\text{ W}$$

In this case, the percentage of power lost would be

$$\%\text{ power lost} = \left(\frac{3.00 \times 10^5\text{ W}}{4.00 \times 10^5\text{ W}}\right) \times 100 = 75\%$$

This example illustrates the advantage of high-voltage transmission lines. At the city, a transformer at a substation steps the voltage back down to about 4 000 V, and this voltage is maintained across utility lines throughout the city. When the power is to be used at a home or business, a transformer on a utility pole near the establishment reduces the voltage to 240 V or 120 V.

**EXERCISE** If the transmission line is cooled so that the resistance is reduced to 5.0 Ω, how much power is lost in the line if it carries a current of 0.89 A?

**ANSWER** 4.0 W

This cylindrical step-down transformer drops the voltage from 4 000 V to 220 V for delivery to a group of residences. *(George Semple)*

## 21.8 MAXWELL'S PREDICTIONS

During the early stages of their study and development, electric and magnetic phenomena were thought to be unrelated. In 1865, however, James Clerk Maxwell (1831–1879) provided a mathematical theory that showed a close relationship between all electric and magnetic phenomena. In addition to unifying the formerly separate fields of electricity and magnetism, his brilliant theory predicted that electric and magnetic fields can move through space as waves. The theory he developed is based on the following four pieces of information:

1. Electric field lines originate on positive charges and terminate on negative charges.
2. Magnetic field lines always form closed loops—that is, they do not begin or end anywhere.
3. A varying magnetic field induces an emf and hence an electric field. This is a statement of Faraday's law (Chapter 20).
4. Magnetic fields are generated by moving charges (or currents), as summarized in Ampère's law (Chapter 19).

Let us examine these statements further in order to understand their significance and Maxwell's great contribution to the theory of electromagnetism. The first statement is a consequence of the nature of the electrostatic force between charged particles, given by Coulomb's law. It embodies the fact that **free charges (electric monopoles) exist in nature.**

The second statement—that magnetic fields form continuous loops—is exemplified by the magnetic field lines around a long, straight wire, which are closed circles, and the magnetic field lines of a bar magnet, which form closed loops.

The third statement is equivalent to Faraday's law of induction, and the fourth statement is equivalent to Ampère's law.

In one of the greatest theoretical developments of the 19th century, Maxwell used these four statements within a corresponding mathematical framework to prove that electric and magnetic fields play symmetric roles in nature. It was already known from experiments that a changing magnetic field produced an electric field according to Faraday's law. Maxwell believed that nature was symmetric, and he therefore hypothesized that a changing electric field should produce a magnetic field. This hypothesis could not be proven experimentally at the time it was developed, because the magnetic fields generated by changing electric fields are generally very weak and therefore difficult to detect.

To justify his hypothesis, Maxwell searched for other phenomena that might be explained by it. He turned his attention to the motion of rapidly oscillating (accelerating) charges, such as those in a conducting rod connected to an alternating voltage. Such charges experience accelerations and, according to Maxwell's predictions, generate changing electric and magnetic fields. The changing fields cause electromagnetic disturbances that travel through space as waves, similar to the spreading water waves created by a pebble thrown into a pool. The waves sent out by the oscillating charges are fluctuating electric and magnetic fields, and so they are called *electromagnetic waves.* From Faraday's law and from his generalization of Ampère's law, Maxwell calculated their speed to be equal to the speed of light, $c = 3 \times 10^8$ m/s. He concluded that visible light and other electromagnetic waves consist of fluctuating electric and magnetic fields traveling through empty space, with each varying field inducing the other! This was truly one of the greatest discoveries of science on a par with Newton's discovery of the laws of motion. Like Newton's laws, it had a profound influence on later scientific developments.

**JAMES CLERK MAXWELL, SCOTTISH THEORETICAL PHYSICIST (1831–1879)**

Maxwell developed the electromagnetic theory of light and the kinetic theory of gases and explained the nature of Saturn's rings and color vision. Maxwell's successful interpretation of the electromagnetic field resulted in the equations that bear his name. Formidable mathematical ability combined with great insight enabled him to lead the way in the study of electromagnetism and kinetic theory. Although it is difficult to know what influences and characteristics shape a great scientist, Maxwell already showed remarkable curiosity at the age of three. He was described in a letter by his mother as follows: "He is a very happy man, and has improved much since the weather got moderate; he has great work with doors, locks, keys, etc., and 'Show me how it doos' is never out of his mouth. He also investigates the hidden course of streams and bellwires, the way the water gets from the pond through the wall . . . and down a drain into Water Orr (River Urr) . . . As to the bells, they will not rust; he stands sentry in the kitchen . . . or he rings, and sends Bessy to see and shout to let him know, and he drags papa all over to show him the holes where the wires go through." (From *Faraday, Maxwell, and Kelvin,* D. K. C. MacDonald, Doubleday-Anchor Books, 1964.) *(North Wind Picture Archives)*

## 21.9 HERTZ'S CONFIRMATION OF MAXWELL'S PREDICTIONS

In 1887, after Maxwell's death, Heinrich Hertz (1857–1894) was the first to generate and detect electromagnetic waves in a laboratory setting. To appreciate the details of his experiment, let us re-examine the properties of an *LC* circuit. In such a circuit, a charged capacitor is connected to an inductor, as in Figure 21.16. When the switch is closed, oscillations occur in the current in the circuit and in the charge on the capacitor. If the resistance of the circuit is neglected, no energy is dissipated, and the oscillations continue.

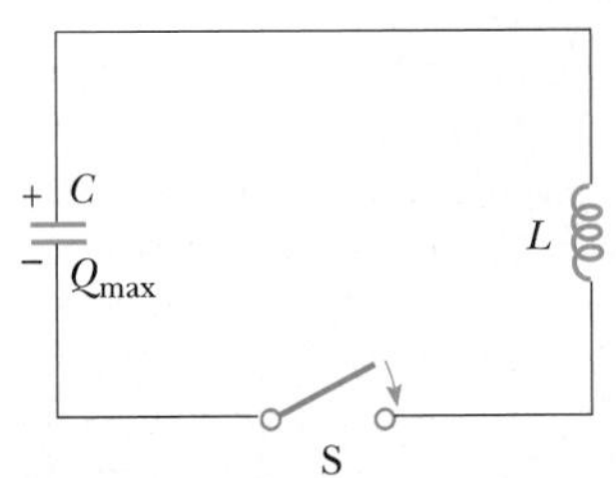


**FIGURE 21.16** A simple *LC* circuit. The capacitor has an initial charge of $Q_{max}$ and the switch is closed at $t = 0$.

**HEINRICH RUDOLF HERTZ, GERMAN PHYSICIST (1857–1894)**

Hertz made his most important discovery of radio waves in 1887. After finding that the speed of a radio wave was the same as that of light, Hertz showed that radio waves, like light waves, could be reflected, refracted, and diffracted. Hertz died of blood poisoning at the age of 36. During his short life, he made many contributions to science. The hertz, equal to one complete vibration or cycle per second, is named after him. *(Hulton-Deutsch Collection/CORBIS)*

In the following analysis, we shall neglect the resistance in the circuit. Let us assume that the capacitor has an initial charge of $Q_{\max}$ and that the switch is closed at $t = 0$. It is convenient to describe what ensues from an energy viewpoint. When the capacitor is fully charged, the total energy in the circuit is stored in the electric field of the capacitor and is equal to $Q_{\max}^2/2C$. At this time, the current is zero and so no energy is stored in the inductor. As the capacitor begins to discharge, the energy stored in its electric field decreases. At the same time, the current increases and energy equal to $LI^2/2$ is now stored in the magnetic field of the inductor. Thus, energy is transferred from the electric field of the capacitor to the magnetic field of the inductor. When the capacitor is fully discharged, it stores no energy. At this time, the current reaches its maximum value, and all of the energy is stored in the inductor. The process then repeats in the reverse direction. The energy continues to transfer between the inductor and the capacitor, corresponding to oscillations in the current and charge.

As we saw in Section 21.6, the frequency of oscillation of an $LC$ circuit is called the *resonance frequency* of the circuit and is given by

$$f_0 = \frac{1}{2\pi\sqrt{LC}}$$

The circuit Hertz used in his investigations of electromagnetic waves is similar to that just discussed and is shown schematically in Figure 21.17. An induction coil (a large coil of wire) is connected to two metal spheres with a narrow gap between them to form a capacitor. Oscillations are initiated in the circuit by short voltage pulses sent via the coil to the spheres, charging one positive, the other negative. Because $L$ and $C$ are quite small in this circuit, the frequency of oscillation is quite high, $f \approx 100$ MHz. This circuit is called a *transmitter* because it produces electromagnetic waves.

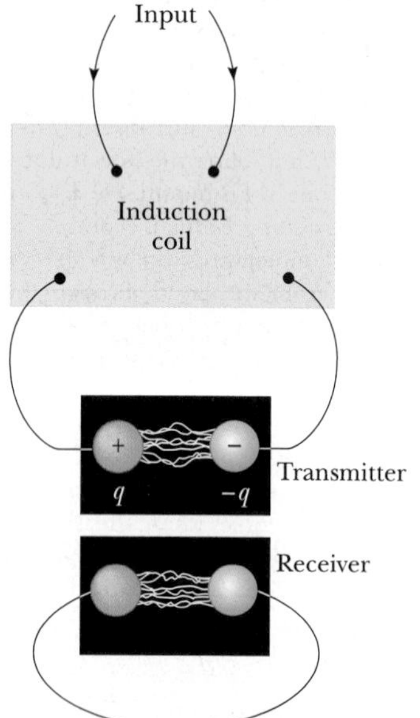


**FIGURE 21.17** A schematic diagram of Hertz's apparatus for generating and detecting electromagnetic waves. The transmitter consists of two spherical electrodes connected to an induction coil, which provides short voltage surges to the spheres, setting up oscillations in the discharge. The receiver is a nearby single loop of wire containing a second spark gap.

Several meters from the transmitter circuit, Hertz placed a second circuit, the receiver, which consisted of a single loop of wire connected to two spheres. It had its own effective inductance, capacitance, and natural frequency of oscillation. Hertz found that energy was being sent from the transmitter to the receiver when the resonance frequency of the receiver was adjusted to match that of the transmitter. The energy transfer was detected when the voltage across the spheres in the receiver circuit became high enough to produce ionization in the air, which caused sparks to appear in the air gap separating the spheres. Hertz's experiment is analogous to the mechanical phenomenon in which a tuning fork picks up the vibrations from another, identical tuning fork.

Hertz hypothesized that the energy transferred from the transmitter to the receiver is carried in the form of waves, which are now known to be electromagnetic waves. In a series of experiments, he also showed that the radiation generated by the transmitter exhibits wave properties: interference, diffraction, reflection, refraction, and polarization. As you will see shortly, all of these properties are exhibited by light. Thus, it became evident that these electromagnetic waves had the same known properties of light waves (already known at the time) and differed only in frequency and wavelength. Hertz effectively confirmed Maxwell's theory by showing that Maxwell's mysterious electromagnetic waves existed and had all the properties of light waves.

Perhaps the most convincing experiment Hertz performed was the measurement of the speed of waves from the transmitter, accomplished as follows. Waves of known frequency from the transmitter were reflected from a metal sheet so that an interference pattern was set up, much like the standing wave pattern on a stretched string. As we saw in our discussion of standing waves, the distance between nodes is $\lambda/2$, so Hertz was able to determine the wavelength $\lambda$. Using the relationship $v = \lambda f$, he found that $v$ was close to $3 \times 10^8$ m/s, the known speed of visible light. Hertz's experiments thus provided the first evidence in support of Maxwell's theory.

## 21.10 PRODUCTION OF ELECTROMAGNETIC WAVES BY AN ANTENNA

In the previous section, we found that the energy stored in an *LC* circuit is continually transferred between the electric field of the capacitor and the magnetic field of the inductor. However, this energy transfer continues for prolonged periods of time only when the changes occur slowly. If the current alternates rapidly, the circuit loses some of its energy in the form of electromagnetic waves. In fact, electromagnetic waves are radiated by *any* circuit carrying an alternating current. The fundamental mechanism responsible for this radiation is the acceleration of a charged particle. Whenever a charged particle undergoes an acceleration, it must radiate energy.

An accelerating charge radiates energy.

An alternating voltage applied to the wires of an antenna forces an electric charge in the antenna to oscillate. This is a common technique for accelerating charged particles and is the source of the radio waves emitted by the broadcast antenna of a radio station.

RADIO-WAVE TRANSMISSION

Figure 21.18 illustrates the production of an electromagnetic wave by oscillating electric charges in an antenna. Two metal rods are connected to an AC source, which causes charges to oscillate between the two rods. The output voltage of the generator is sinusoidal. At $t = 0$, the upper rod is given a maximum positive charge and the bottom rod an equal negative charge, as in Figure 21.18a. The electric field near the antenna at this instant is also shown in Figure 21.18a. As the charges oscillate, the rods become less charged, the field near the rods decreases in strength, and the downward-directed maximum electric field produced at $t = 0$ moves away from the rod. When the charges are neutralized, as in Figure 21.18b, the electric field has dropped to zero. This occurs after an interval equal to one quarter of the period of oscillation. Continuing in this fashion, the upper rod soon obtains a maximum negative charge and the lower rod becomes positive, as in Figure 21.19c, resulting in an electric field directed upward. This occurs after an interval equal to one-half the period of oscillation. The oscillations continue as indicated in Figure 21.18d. Note that the electric field near the antenna oscillates in phase with the charge distribution. That is, the field points down when the upper rod is positive and up when the upper rod is negative. Furthermore, the magnitude of the field at any instant depends on the amount of charge on the rods at that instant.

ACCELERATED CHARGES PRODUCE EM WAVES

Stationary charges produce only electric fields, while charges in uniform motion (i.e., constant velocity) produce electric and magnetic fields, but no electromagnetic waves. In contrast, accelerated charges produce electromagnetic waves as well as electric and magnetic fields.

As the charges continue to oscillate (and accelerate) between the rods, the electric field set up by the charges moves away from the antenna in all directions at the speed of light. Figure 21.18 shows the electric field pattern on one side of the antenna at certain times during the oscillation cycle. As you can see, one cycle of charge oscillation produces one full wavelength in the electric field pattern.

Because the oscillating charges create a current in the rods, a magnetic field is also generated when the current in the rods is upward, as shown in Figure 21.19. The magnetic field lines circle the antenna (right-hand rule #2) and are perpendicular to the electric field at all points. As the current changes with time, the

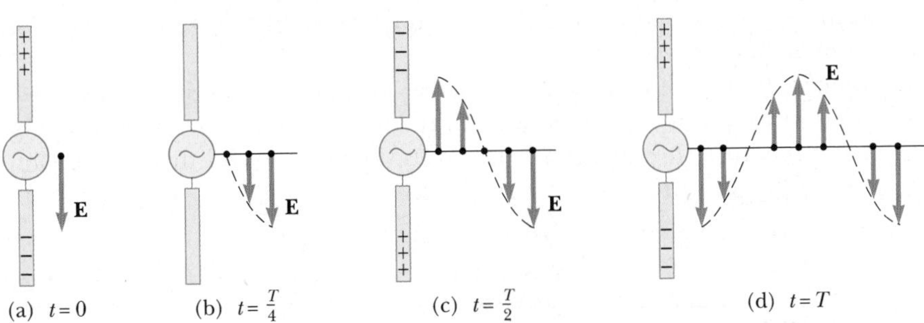


**FIGURE 21.18** An electric field set up by oscillating charges in an antenna. The field moves away from the antenna at the speed of light.

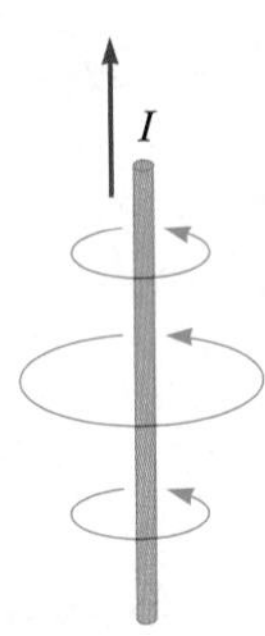


**FIGURE 21.19** Magnetic field lines around an antenna carrying a changing current.

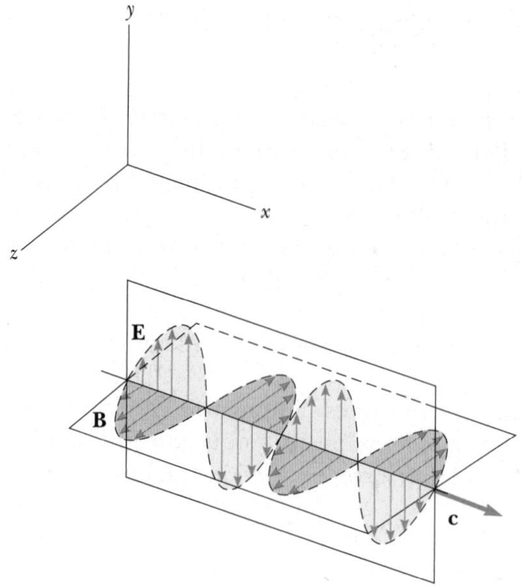

**FIGURE 21.20** An electromagnetic wave sent out by oscillating charges in an antenna, represented at one instant of time and far from the antenna. Note that the electric field is perpendicular to the magnetic field, and both are perpendicular to the direction of wave propagation.

magnetic field lines spread out from the antenna. At great distances from the antenna, the strengths of the electric and magnetic fields become very weak. However, at these distances it is necessary to take into account the facts that (1) a changing magnetic field produces an electric field and (2) a changing electric field produces a magnetic field, as predicted by Maxwell. These induced electric and magnetic fields are in phase: At any point, the two fields reach their maximum values at the same instant. This is illustrated at one instant of time in Figure 21.20. Note that (1) **E** and **B** fields are perpendicular to each other, and (2) both fields are perpendicular to the direction of motion of the wave. This second property is characteristic of transverse waves. Hence, we see that **an electromagnetic wave is a transverse wave.**

**Webnote 21.2**
Watch the propagation of an electromagnetic wave through the real-time Java applet at *http://www.phy.ntnu.edu.tw/~hwang/emWave/emWave.html*

## 21.11 PROPERTIES OF ELECTROMAGNETIC WAVES

We have seen that Maxwell's detailed analysis predicted the existence and properties of electromagnetic waves. We have already examined some of those properties. In this section we summarize what we know about electromagnetic waves thus far and consider some additional properties. In our discussion here and in future sections, we shall often make reference to a type of wave called a **plane wave.** A plane electromagnetic wave is a wave traveling from a very distant source. Figure 21.20 pictures such a wave at a given instant of time. In this case, the oscillations of the electric and magnetic fields take place in planes perpendicular to the $x$ axis and thus to the direction of travel for the wave. Because the electric and magnetic fields are perpendicular to the direction of travel of the wave, electromagnetic waves are transverse waves. In Figure 21.20, the electric field **E** is in the $y$ direction and the magnetic field **B** is in the $z$ direction.

Electromagnetic waves travel with the speed of light. In fact, it can be shown that the speed of an electromagnetic wave is related to the permeability and permittivity of the medium through which it travels. Maxwell found this relationship for free space to be

Speed of light

$$c = \frac{1}{\sqrt{\mu_0 \epsilon_0}} \quad [21.24]$$

where $c$ is the speed of light, $\mu_0 = 4\pi \times 10^{-7}\ \text{N} \cdot \text{s}^2/\text{C}^2$ is the permeability constant of vacuum, and $\epsilon_0 = 8.854\ 19 \times 10^{-12}\ \text{C}^2/\text{N}\cdot\text{m}^2$ is the permittivity of free space. Substituting these values into Equation 21.24, we find that

$$c = 2.997\ 92 \times 10^8\ \text{m/s} \qquad [21.25]$$

Because electromagnetic waves travel at a speed that is precisely the same as the speed of light in vacuum, one is led to believe (correctly) that **light is an electromagnetic wave.**

Maxwell also proved that the ratio of the electric to the magnetic field in an electromagnetic wave equals the speed of light. That is,

$$\frac{E}{B} = c \qquad [21.26]$$

Electromagnetic waves carry energy as they travel through space, and this energy can be transferred to objects placed in their paths. The average rate at which energy passes through an area perpendicular to the direction of travel of a wave, or the average power per unit area, is given by

$$\text{Average power per unit area} = \frac{E_{\text{max}}B_{\text{max}}}{2\mu_0} \qquad [21.27]$$

where $E_{\text{max}}$ and $B_{\text{max}}$ are the *maximum* values of $E$ and $B$. As in Chapter 14, we call this quantity the *intensity* of the wave. From Equation 21.26, we see that $E_{\text{max}} = cB_{\text{max}} = B_{\text{max}}/\sqrt{\mu_0\epsilon_0}$. Therefore, Equation 21.27 can also be expressed as

$$\text{Average power per unit area} = \frac{E_{\text{max}}^2}{2\mu_0 c} = \frac{c}{2\mu_0}B_{\text{max}}^2 \qquad [21.28]$$

Note that in these expressions we use the *average* power per unit area. A detailed analysis would show that the energy carried by an electromagnetic wave is shared equally by the electric and magnetic fields.

Light is an electromagnetic wave and transports energy and momentum.

Electromagnetic waves transport linear momentum as well as energy. Hence it follows that pressure is exerted on a surface when an electromagnetic wave impinges on it. In what follows, we assume that the electromagnetic wave transports a total energy $U$ to a surface in a time $t$. If the surface absorbs all the incident energy $U$ in this time, Maxwell showed that the total momentum $\mathbf{p}$ delivered to this surface has a magnitude

$$p = \frac{U}{c} \quad \text{(complete absorption)} \qquad [21.29]$$

If the surface is a perfect reflector, then the momentum delivered in a time $t$ for normal incidence is twice that given by Equation 21.29. That is, a momentum $U/c$ is delivered first by the incident wave and then again by the reflected wave, in analogy with a ball colliding elastically with a wall. Therefore,

$$p = \frac{2U}{c} \quad \text{(complete reflection)} \qquad [21.30]$$

Although radiation pressures are very small (about $5 \times 10^{-6}\ \text{N/m}^2$ for direct sunlight), they have been measured with a device such as the one shown in Figure 21.21. Light is allowed to strike a mirror and a black disk that are connected to each other by a horizontal bar suspended from a fine fiber. Light striking the black disk is completely absorbed, so *all* of the momentum of the light is transferred to the disk. Light striking the mirror head on is totally reflected; hence, the momentum transfer to the mirror is twice that transmitted to the disk. As a result, the horizontal bar supporting the disks twists counterclockwise as seen from above. The bar comes to equilibrium at some angle under the action of the torques caused by radiation pressure and the twisting of the fiber. The radiation pressure can be determined by measuring the angle at which equilibrium occurs. The apparatus

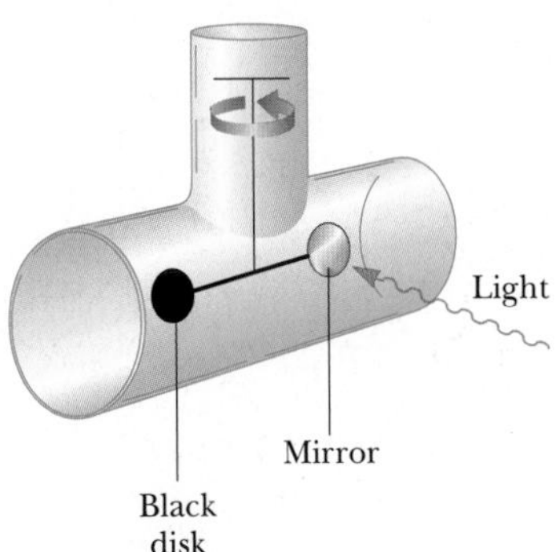

**FIGURE 21.21** An apparatus for measuring the radiation pressure of light. In practice, the system is contained in a high vacuum.

must be placed in a high vacuum to eliminate the effects of air currents. It is interesting to note that similar experiments demonstrate that electromagnetic waves carry angular momentum as well.

In summary, electromagnetic waves traveling through free space have the following properties:

1. Electromagnetic waves travel at the speed of light.
2. Electromagnetic waves are transverse waves, because the electric and magnetic fields are perpendicular to the direction of propagation of the wave and to each other.
3. The ratio of the electric field to the magnetic field in an electromagnetic wave equals the speed of light.
4. Electromagnetic waves carry both energy and momentum, which can be delivered to a surface.

## APPLYING PHYSICS 21.2

In the interplanetary space in the Solar System, there is a large amount of dust. Although interplanetary dust can in theory have a variety of sizes—from molecular size upward—there are very few dust particles smaller than about 0.2 $\mu$m in our Solar System. Why? (*Hint:* The Solar System originally contained dust particles of all sizes.)

**Explanation** Dust particles in the solar system are subject to two forces—the gravitational force toward the Sun and the force from radiation pressure, which is away from the Sun. The gravitational force is proportional to the cube of the radius of a spherical dust particle, because it is proportional to the mass ($\rho V$) of the particle. The radiation pressure is proportional to the square of the radius, because it depends on the cross-sectional area of the particle. For large particles, the gravitational force is larger than the force of radiation pressure and the weak attraction to the Sun causes such particles to move slowly toward the Sun. For small particles, less than about 0.2 $\mu$m, the larger force from radiation pressure sweeps these particles out of the Solar System.

**Quick Quiz 21.4** In an apparatus such as that in Figure 21.21, suppose the black disk is replaced by one with half the radius. Which of the following are different after the disk is replaced: (a) radiation pressure on the disk, (b) radiation force on the disk, (c) radiation momentum delivered to the disk in a given time interval?

### Example 21.8 Solar Energy

Assume that the Sun delivers an average power per unit area of about 1 000 W/m$^2$ to Earth's surface. Calculate the total power incident on a roof 8.00 m by 20.0 m. Assume that the radiation is incident *normal* to the roof (the Sun is directly overhead).

**Solution** The power per unit area, or light intensity, is 1 000 W/m$^2$. For normal incidence, the power supplied to the roof is

$$\mathcal{P} = (1\,000 \text{ W/m}^2)(8.00 \times 20.0 \text{ m}^2) = \boxed{1.60 \times 10^5 \text{ W}}$$

Note that if this power could *all* be converted to electric power, it would be more than enough for the average home. Unfortunately, solar energy is not easily harnessed, and the prospects for large-scale conversion are not as bright as they may appear from this

(Example 21.8) A solar home in Oregon. *(John Neal/Photo Researchers, Inc.)*

**APPLICATION**

SOLAR-POWERED HOMES

simple calculation. For example, the conversion efficiency from solar to electrical energy is far less than 100%; 10% is typical for photovoltaic cells. Roof systems for using solar energy to raise the temperature of water with efficiencies of around 50% have been built. However, other practical problems must be considered, such as overcast days, geographic location, and energy storage.

**EXERCISE** How much solar energy (in joules) is incident on the roof in 1.00 h?

**ANSWER** $5.76 \times 10^8$ J

## Example 21.9 A High-Intensity Laser Beam

The intensity of light ranges from about 1 W/m$^2$ for a candle to about 30 MW/m$^2$ for a modest-size laser. (A laser is a high-intensity light source that concentrates several watts of power into a very narrow beam producing MW/m$^2$ intensity. See Chapter 28.) A particular laser produces a power of 4.0 W in a beam 0.40 mm in diameter. (a) What is its average intensity? (b) Find the peak electric field of the laser light. (c) Find the peak magnetic field of the laser light. (d) If the laser beam is aimed upward to levitate a 20-$\mu$m-diameter sphere (this has actually been done), what is the maximum mass and density of this sphere? Assume the sphere is perfectly reflecting.

### Solution

**A** Intensity is power per unit area carried by the beam:

$$\text{Power per unit area} = \frac{\mathcal{P}}{\pi r^2} = \frac{4.0\text{ W}}{(3.14)(0.20 \times 10^{-3}\text{ m})^2} = 32\text{ MW/m}^2$$

**B** Since the average power per unit area $= E_{\text{max}}^2/2\mu_0 c$,

$$\begin{aligned} E_{\text{max}} &= \sqrt{(2\mu_0 c)(\text{average power per unit area})} \\ &= \sqrt{(8\pi \times 10^{-7}\text{ Ns}^2/C^2)(3.0 \times 10^8\text{ m/s})(32 \times 10^6\text{ W/m}^2)} \\ &= 1.6 \times 10^5\text{ V/m} \end{aligned}$$

**C** $E_{\text{max}}$ and $B_{\text{max}}$ are not independent in an electromagnetic wave, but are related by $B_{\text{max}} = E_{\text{max}}/c$:

$$B_{\text{max}} = \frac{E_{\text{max}}}{c} = \frac{1.6 \times 10^5\text{ V/m}}{3.0 \times 10^8\text{ m/s}} = 5.3 \times 10^{-4}\text{ T}$$

**D** To suspend a 20-$\mu$m-diameter sphere, the upward electromagnetic force $F_{EM}$ must balance the sphere's weight.

$$F_{EM} = mg \qquad \text{or} \qquad m = \frac{F_{EM}}{g}$$

Because $F_{EM} = \Delta p/\Delta t$, and Equation 21.30 states that the momentum $p$ delivered to a perfectly reflecting object is given by $p = 2U/c$ where $U$ is the electromagnetic energy incident on the object, $\Delta p/\Delta t = 2U/c\Delta t$. Finally, we find for the mass

$$m = \frac{2U/\Delta t}{gc}$$

$U/\Delta t$ is the energy per unit time incident on the sphere and is given by the product of the laser intensity and the cross-sectional area of the sphere ($A = \pi r^2$).

$$\frac{U}{\Delta t} = \text{intensity} \cdot \pi r^2 = (3.2 \times 10^7\ \text{W/m}^2)(3.14)(1.0 \times 10^{-10}\ \text{m}^2) = 0.010\ \text{W}$$

$$= 0.010\ \text{J/s}$$

The maximum mass that can be supported by this laser beam is

$$m = \frac{2U/\Delta t}{gc} = \frac{0.020\ \text{J/s}}{(9.8\ \text{m/s}^2)(3.0 \times 10^8\ \text{m/s})} = 6.8 \times 10^{-12}\ \text{kg}$$

The maximum density is given by

$$\rho_{\text{max}} = \frac{\text{mass}}{\text{volume}} = \frac{m}{(4/3)\pi r^3} = \frac{6.8 \times 10^{-12}\ \text{kg}}{(4/3)(3.14)(1.0 \times 10^{-5}\ \text{m})^3} = 1.6 \times 10^3\ \text{kg/m}^3$$

which is about the density of bone.

## 21.12 THE SPECTRUM OF ELECTROMAGNETIC WAVES

We have seen that all electromagnetic waves travel in vacuum with the speed of light, $c$. These waves transport energy and momentum from some source to a receiver. In 1887, Hertz successfully generated and detected the radio-frequency electromagnetic waves predicted by Maxwell. Maxwell himself had recognized as electromagnetic waves both visible light and the infrared radiation discovered in 1800 by William Herschel. It is now known that other forms of electromagnetic waves exist that are distinguished by their frequencies and wavelengths.

Because all electromagnetic waves travel through vacuum with a speed $c$, their frequency $f$ and wavelength $\lambda$ are related by the important expression

$$c = f\lambda \qquad \textbf{[21.31]}$$

The types of electromagnetic waves are presented in Figure 21.22. Note the wide and overlapping range of frequencies and wavelengths. For instance, an AM radio wave with a frequency of 5.00 MHz (a typical value) has a wavelength of

$$\lambda = \frac{c}{f} = \frac{3.00 \times 10^8\ \text{m/s}}{5.00 \times 10^6\ \text{s}^{-1}} = 60.0\ \text{m}$$

The following abbreviations are often used to designate short wavelengths and distances:

$$1\ \text{micrometer}\ (\mu\text{m}) = 10^{-6}\ \text{m}$$

$$1\ \text{nanometer}\ (\text{nm}) = 10^{-9}\ \text{m}$$

$$1\ \text{angstrom}\ (\text{Å}) = 10^{-10}\ \text{m}$$

The wavelengths of visible light, for example, range from 0.4 $\mu$m to 0.7 $\mu$m, or 400 nm to 700 nm, or 4 000 Å to 7 000 Å.

Brief descriptions of these wave types follow, in order of decreasing wavelength. There is no sharp division between one kind of electromagnetic wave and the next. Note that all forms of radiation are produced by accelerating charges.

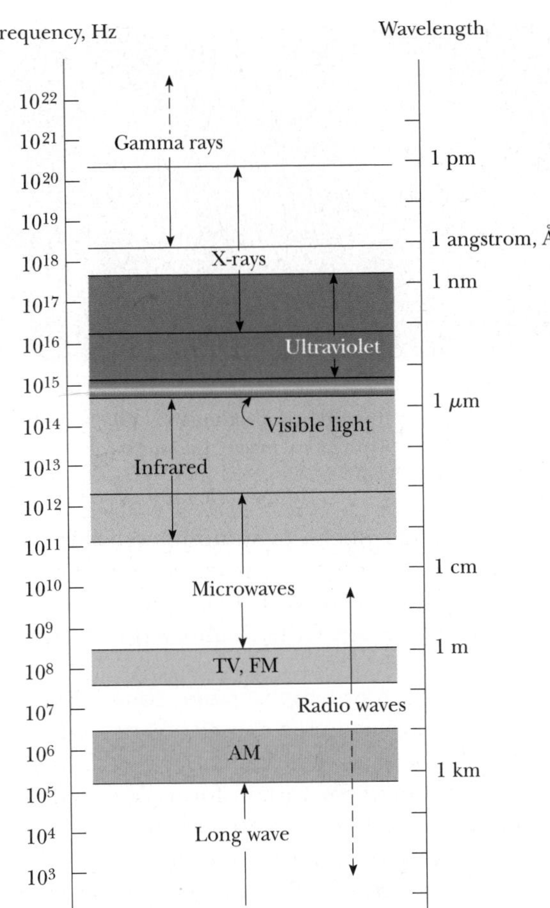


**FIGURE 21.22** The electromagnetic spectrum. Note the overlap between one type of wave and the next. There is no sharp division between the types of waves.

**Radio waves,** which were discussed in Section 21.10, are the result of charges accelerating through conducting wires. They are, of course, used in radio and television communication systems.

**Microwaves** (short-wavelength radio waves) have wavelengths ranging between about 1 mm and 30 cm and are generated by electronic devices. Their short wavelengths make them well suited for the radar systems used in aircraft navigation and for the study of atomic and molecular properties of matter. Microwave ovens are an interesting domestic application of these waves. It has been suggested that solar energy might be harnessed by beaming microwaves to Earth from a solar collector in space.

**Infrared waves** (sometimes and incorrectly called "heat waves"), produced by hot objects and molecules, have wavelengths ranging from about 1 mm to the longest wavelength of visible light, $7 \times 10^{-7}$ m. They are readily absorbed by most materials. The infrared energy absorbed by a substance causes it to get warmer. This is because the energy agitates the atoms of the object, increasing their vibrational or translational motion, and the result is a temperature rise. Infrared radiation has many practical and scientific applications, including physical therapy, infrared photography, and the study of the vibrations of atoms.

**Visible light,** the most familiar form of electromagnetic waves, may be defined as the part of the spectrum that is detected by the human eye. Light is produced by the rearrangement of electrons in atoms and molecules. The wavelengths of visible light are classified as colors ranging from violet ($\lambda \approx 4 \times 10^{-7}$ m) to red ($\lambda \approx 7 \times 10^{-7}$ m). The eye's sensitivity is a function of wavelength and is greatest at a wavelength of about $5.6 \times 10^{-7}$ m (yellow-green).

**Ultraviolet (UV) light** covers wavelengths ranging from about $4 \times 10^{-7}$ m (400 nm) down to $6 \times 10^{-10}$ m (0.6 nm). The Sun is an important source of ultraviolet light (which is the main cause of suntans). Most of the ultraviolet light from

### Webnote 21.3

For much more detail on the radio spectrum and the electromagnetic spectrum in general, go to *http://www.howstuffworks.com/radio-spectrum.htm*

### Webnote 21.4

Microwave ovens produce microwaves at a frequency of 2.45 GHz. How can they heat food from the inside out? Find out at *http://www.howstuffworks.com/microwave.htm*

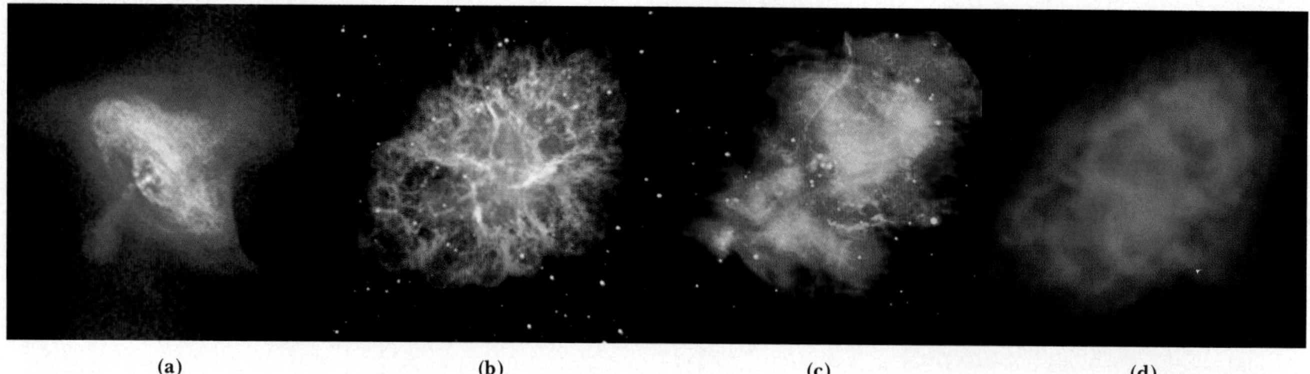

(a) (b) (c) (d)

**FIGURE 21.23** Observations in different parts of the electromagnetic spectrum show different features of the Crab Nebula. (a) X-ray image *(NASA/CXC/SAO)* (b) Optical image *(Palomar Observatory)* (c) Infrared image *(WM Keck Observatory)* (d) Radio image *(VLA/NRAO)*.

the Sun is absorbed by atoms in the upper atmosphere, or stratosphere. This is fortunate, because UV light in large quantities has harmful effects on humans. One important constituent of the stratosphere is ozone ($O_3$) from reactions of oxygen with ultraviolet radiation. This ozone shield causes lethal high-energy ultraviolet radiation to warm the stratosphere.

**X-rays** are electromagnetic waves with wavelengths from about $10^{-8}$ m (10 nm) down to $10^{-13}$ m ($10^{-4}$ nm). The most common source of x-rays is the acceleration of high-energy electrons bombarding a metal target. X-rays are used as a diagnostic tool in medicine and as a treatment for certain forms of cancer. Because x-rays easily penetrate and damage or destroy living tissues and organisms, care must be taken to avoid unnecessary exposure and overexposure.

**Gamma rays,** electromagnetic waves emitted by radioactive nuclei, have wavelengths ranging from about $10^{-10}$ m to less than $10^{-14}$ m. They are highly penetrating and cause serious damage when absorbed by living tissues. As a consequence, those working near such radiation must be protected by garments containing heavily absorbing materials, such as layers of lead.

When astronomers observe the same celestial object using detectors sensitive to different regions of the electromagnetic spectrum, striking variations in the object's features can be seen. Figure 21.23 shows images of the Crab Nebula made in four different wavelength ranges. The Crab Nebula is the remnant of a supernova explosion that was seen from Earth in 1054 A.D. (Compare Fig. 8.28).

## APPLYING PHYSICS 21.3

The center of sensitivity of our eyes coincides with the center of the wavelength distribution of the Sun. Is this an amazing coincidence?

**Explanation** This is not a coincidence—it is the result of biological evolution. Humans have evolved with vision most sensitive to wavelengths that are strongest from the Sun. It is an interesting conjecture to imagine aliens from another planet, with a Sun of a different temperature, arriving at Earth. Their eyes would have a center of sensitivity at different wavelengths than ours. How would their vision of the Earth compare to ours?

## 21.13 THE DOPPLER EFFECT FOR ELECTROMAGNETIC WAVES

As we saw in Section 14.6, sound waves exhibit the Doppler effect when the observer, the source, or both are moving relative to the medium of propagation. Recall that in the Doppler effect, the observed frequency of the wave is larger or smaller than the frequency emitted by the source of the wave.

A Doppler effect also occurs for electromagnetic waves, but it differs from the Doppler effect for sound waves in two ways. First, in the Doppler effect for sound waves, motion relative to the medium is most important because sound waves require a medium in which to propagate. In contrast, the medium of propagation plays no role in the Doppler effect for electromagnetic waves because the waves require no medium in which to propagate. Second, the speed of sound that appears in the equation for the Doppler effect for sound depends on the reference frame in which it is measured. In contrast, as we shall see in Chapter 26, the speed of electromagnetic waves has the same value in all coordinate systems that are either at rest or moving at constant velocity with respect to one another.

The single equation that describes the Doppler effect for electromagnetic waves is given by the approximate expression

$$f' = f\left(1 \pm \frac{u}{c}\right) \quad \text{if } u \ll c \qquad \textbf{[21.32]}$$

where $f'$ is the observed frequency, $f$ is the frequency emitted by the source, $c$ is the speed of light in vacuum, and $u$ is the *relative* speed of the observer and source. Note that Equation 21.32 is valid only if $u$ is much smaller than $c$. Furthermore, this result can also be used for sound as long as the relative velocity of the source and observer is much less than the velocity of sound. The positive sign in Equation 21.32 must be used when the source and observer are moving toward one another, and the negative sign must be used when they are moving away from each other. Thus, we anticipate an increase in the observed frequency if the source and observer are approaching each other, and a decrease in observed frequency if the source and observer recede from each other.

Astronomers have made important discoveries using Doppler observations on light reaching Earth from distant stars and galaxies. Such measurements have shown that most distant galaxies are moving away from the Earth. Thus, the Universe is expanding. This is called a *red shift* because the observed wavelengths are shifted toward the red portion (longest wavelength) of the visible spectrum. Furthermore, measurements show that the speed of a galaxy increases with increasing distance from the Earth. More recent Doppler effect measurements made with the Hubble Space Telescope have shown that a galaxy labeled M87 is rotating—one edge moving toward us and the other moving away. Its measured speed of rotation was used to identify a supermassive black hole located at the center of this galaxy.

## SUMMARY

If an AC circuit consists of a generator and a resistor, the current in the circuit is in phase with the voltage. That is, the current and voltage reach their maximum values at the same time.

In discussions of voltages and currents in AC circuits, **rms values** of voltages are usually used. One reason is that AC ammeters and voltmeters are designed to read rms values. The rms values of currents and voltage ($I_{\text{rms}}$ and $\Delta V_{\text{rms}}$) are related to the maximum values of these quantities ($I_{\text{max}}$ and $\Delta V_{\text{max}}$) as follows:

$$I_{\text{rms}} = \frac{I_{\text{max}}}{\sqrt{2}} \qquad \Delta V_{\text{rms}} = \frac{\Delta V_{\text{max}}}{\sqrt{2}} \qquad \textbf{[21.2, 21.3]}$$

The rms voltage across a resistor is related to the rms current in the resistor by **Ohm's law:**

$$\Delta V_{R,\text{rms}} = I_{\text{rms}} R \qquad \textbf{[21.4]}$$

If an AC circuit consists of a generator and a capacitor, the voltage lags behind the current by 90°. That is, the voltage reaches its maximum value one quarter of a period after the current reaches its maximum value.

The impeding effect of a capacitor on current in an AC circuit is given by the **capacitive reactance** $X_C$, defined as

$$X_C \equiv \frac{1}{2\pi f C} \quad [21.5]$$

where $f$ is the frequency of the AC voltage source.

The rms voltage across and the rms current in a capacitor are related by

$$\Delta V_{C,\text{rms}} = I_{\text{rms}} X_C \quad [21.6]$$

If an AC circuit consists of a generator and an inductor, the voltage leads the current by 90°. That is, the voltage reaches its maximum value one quarter of a period before the current reaches its maximum value.

The effective impedance of a coil in an AC circuit is measured by a quantity called the **inductive reactance** $X_L$, defined as

$$X_L \equiv 2\pi f L \quad [21.8]$$

The rms voltage across a coil is related to the rms current in the coil by

$$\Delta V_{L,\text{rms}} = I_{\text{rms}} X_L \quad [21.9]$$

In an *RLC* series AC circuit, the maximum applied voltage $\Delta V$ is related to the maximum voltages across the resistor ($\Delta V_R$), capacitor ($\Delta V_C$), and inductor ($\Delta V_L$) by

$$\Delta V_{\text{max}} = \sqrt{\Delta V_R^2 + (\Delta V_L - \Delta V_C)^2} \quad [21.10]$$

If an AC circuit contains a resistor, an inductor, and a capacitor connected in series, the limit they place on the current is given by the **impedance** $\mathbf{Z}$ of the circuit, defined as

$$Z \equiv \sqrt{R^2 + (X_L - X_C)^2} \quad [21.13]$$

The relationship between the maximum voltage supplied to an *RLC* circuit and the maximum current in the circuit that is the same in every element is

$$\Delta V_{\text{max}} = I_{\text{max}} Z \quad [21.14]$$

In an *RLC* series AC circuit, the applied rms voltage and current are out of phase. The **phase angle** $\phi$ between the current and voltage is given by

$$\tan \phi = \frac{X_L - X_C}{R} \quad [21.15]$$

The **average power** delivered by the voltage source in an *RLC* series AC circuit is

$$\mathcal{P}_{\text{av}} = I_{\text{rms}} \Delta V_{\text{rms}} \cos \phi \quad [21.17]$$

where the constant $\cos \phi$ is called the **power factor.**

**Electromagnetic waves** were predicted by James Clerk Maxwell and later experimentally confirmed by Heinrich Hertz. These waves have the following properties:

1. Electromagnetic waves are transverse waves, because the electric and magnetic fields are perpendicular to the direction of travel.
2. Electromagnetic waves travel with the speed of light.
3. The ratio of the electric field to the magnetic field at a given point in an electromagnetic wave equals the speed of light. That is,

$$\frac{E}{B} = c \quad [21.26]$$

4. Electromagnetic waves carry energy as they travel through space. The average power per unit area is

$$\frac{E_{\text{max}} B_{\text{max}}}{2\mu_0} = \frac{E_{\text{max}}^2}{2\mu_0 c} = \frac{c}{2\mu_0} B_{\text{max}}^2 \quad [21.27, 21.28]$$

where $E_{\text{max}}$ and $B_{\text{max}}$ are the maximum values of the electric and magnetic fields.

5. Electromagnetic waves transport linear and angular momentum as well as energy. The speed $c$, frequency $f$, and wavelength $\lambda$ of an electromagnetic wave are related by

$$c = f\lambda \qquad [21.31]$$

The **electromagnetic spectrum** includes waves covering a broad range of frequencies and wavelengths. These waves have a variety of applications and characteristics, depending on their frequencies or wavelengths.

## CONCEPTUAL QUESTIONS

1. Before the advent of cable television and satellite dishes, homeowners either mounted a television antenna on the roof or used "rabbit ears" atop their sets (see Fig. Q21.1). Certain orientations of the receiving antenna on a television set gave better reception than others. Furthermore, the best orientation varied from station to station. Explain.

**FIGURE Q21.1** *(George Semple)*

2. What is the impedance of an *RLC* circuit at the resonance frequency?
3. When a DC voltage is applied to a transformer, the primary coil sometimes will overheat and burn. Why?
4. Why are the primary and secondary coils of a transformer wrapped on an iron core that passes through both coils?
5. Receiving radio antennas can be in the form of conducting lines or loops. What should the orientation of each of these antennas be relative to a broadcasting antenna that is vertical?
6. If the fundamental source of a sound wave is a vibrating object, what is the fundamental source of an electromagnetic wave?
7. In radio transmission, a radio wave serves as a carrier wave and the sound signal is superimposed on the carrier wave. In amplitude modulation (AM) radio, the amplitude of the carrier wave varies according to the sound wave. The Navy sometimes uses flashing lights to send Morse code between neighboring ships, a process that has similarities to radio broadcasting. Is this AM or FM? What is the carrier frequency? What is the signal frequency? What is the broadcasting antenna? What is the receiving antenna?
8. When light (or other electromagnetic radiation) travels across a given region, what is it that moves?
9. In space sailing, which is a proposed alternative method to reach the planets, a spacecraft carries a very large sail that experiences a force due to radiation pressure from the Sun. Should the sail be absorptive or reflective to be most effective?
10. How can the average value of an alternating current be zero yet the square root of the average squared value not be zero?
11. Suppose a creature from another planet had eyes that were sensitive to infrared radiation. Describe what it would see if it looked around the room that you are now in. That is, what would be bright and what would be dim?
12. Why should an infrared photograph of a person look different from a photograph taken with visible light?
13. Radio stations often advertise "instant news." If what they mean is that you hear the news at the instant they speak it, is their claim true? About how long would it take for a message to travel across this country by radio waves, assuming that these waves could travel this great distance and still be detected?
14. Would an inductor and a capacitor used together in an AC circuit dissipate any energy?
15. Does a wire connected to a battery emit an electromagnetic wave?
16. Do all current-carrying conductors emit electromagnetic waves? Explain.
17. If the resistance in an *RLC* circuit is doubled but the capacitance and inductance are unchanged, how will the resonance frequency change?
18. If the resistance in an *RLC* circuit remains the same but the capacitance and inductance are each doubled, how will the resonance frequency change?

## PROBLEMS

1, 2, 3 = straightforward, intermediate, challenging ☐ = full solution available in Student Solutions Manual/Study Guide

**web** = solution posted at **http://info.brookscole.com/serway** = biomedical application

### Section 21.1 Resistors in an AC Circuit

1. An rms voltage of 100 V is applied to a purely resistive load of 5.00 Ω. Find (a) the maximum voltage applied, (b) the rms current supplied, (c) the maximum current supplied, and (d) the power dissipated.
2. (a) What is the resistance of a lightbulb that uses an average power of 75.0 W when connected to a 60-Hz power source with a peak voltage of 170 V? (b) What is the resistance of a 100-W bulb?
3. An AC power supply produces a maximum voltage of $\Delta V_{max}$ = 100 V. This power supply is connected to a 24.0-Ω resistor, and the current and resistor voltage are measured with an ideal AC ammeter and an ideal AC voltmeter, as shown in Figure P21.3. What does each meter read? Recall that an ideal ammeter has zero resistance and an ideal voltmeter has infinite resistance.

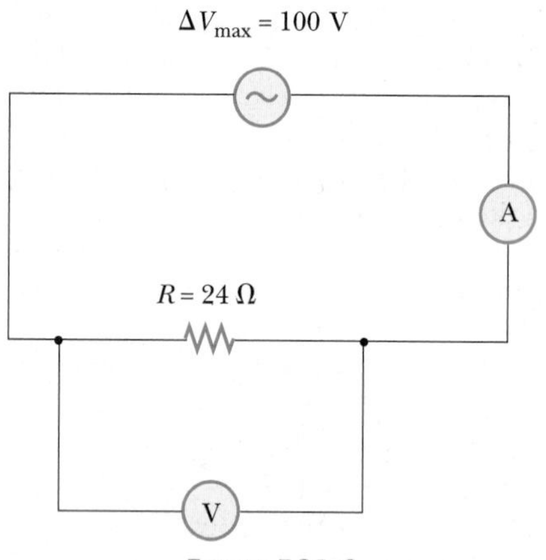


**FIGURE P21.3**

4. Figure P21.4 shows three lamps connected to a 120-V AC (rms) household supply voltage. Lamps 1 and 2 have 150-W bulbs; lamp 3 has a 100-W bulb. Find the rms current and resistance of each bulb.

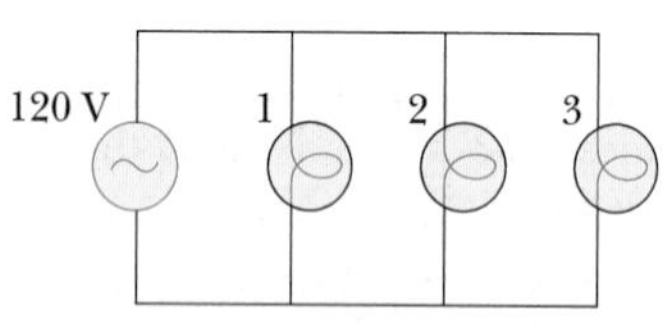


**FIGURE P21.4**

5. An audio amplifier, represented by the AC source and the resistor $R$ in Figure P21.5, delivers alternating voltages at audio frequencies to the speaker. If the source puts out an alternating voltage of 15.0 V (rms), resistance $R$ is 8.20 Ω, and the speaker is equivalent to a resistance of 10.4 Ω, what is the time-averaged power delivered to the speaker?

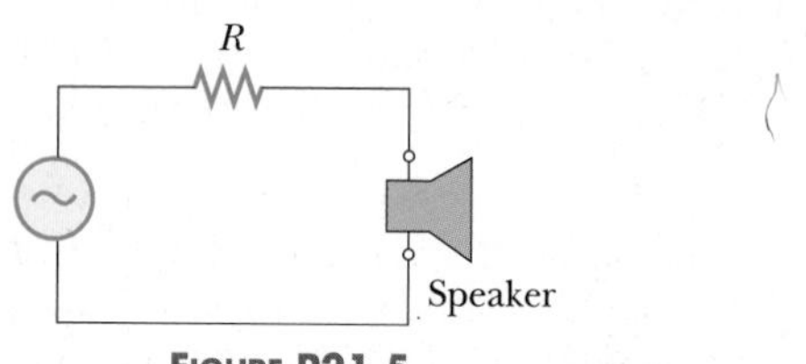


**FIGURE P21.5**

6. An AC voltage source has an output of $\Delta v = 150 \sin 377t$. Find (a) the rms voltage output, (b) the frequency of the source, and (c) the voltage at $t = 1/120$ s. (d) Find the maximum current in the circuit when the generator is connected to a 50.0-Ω resistor.

### Section 21.2 Capacitors in an AC Circuit

7. Show that the SI unit of capacitive reactance, $X_C$, is the ohm.
8. What maximum current is delivered by a 2.20-μF capacitor when connected across (a) a North American outlet having $\Delta V_{rms}$ = 120 V, $f$ = 60.0 Hz; and (b) a European outlet having $\Delta V_{rms}$ = 240 V, $f$ = 50.0 Hz?
9. When a 4.0-μF capacitor is connected to a generator whose rms output is 30 V, the current in the circuit is observed to be 0.30 A. What is the frequency of the source?
10. What maximum current is delivered by an AC generator with $\Delta V_{max}$ = 48.0 V and $f$ = 90.0 Hz when connected across a 3.70-μF capacitor?
11. **web** What value of capacitor must be inserted in a 60-Hz circuit in series with a generator of 170 V maximum output voltage to produce an rms current output of 0.75 A?
12. The generator in a purely capacitive AC circuit has an angular frequency of 120π rad/s. If $\Delta V_{max}$ = 140 V and $C$ = 6.00 μF, what is the rms current in the circuit?

### Section 21.3 Inductors in an AC Circuit

13. Show that the inductive reactance $X_L$ has SI units of ohms.
14. In a purely inductive AC circuit, as in Figure 21.6, $\Delta V_{max}$ = 100 V. (a) If the maximum current is 7.50 A at 50.0 Hz, calculate the inductance $L$. (b) At what angular frequency ω is the maximum current 2.50 A?
15. An inductor has a 54.0-Ω reactance at 60.0 Hz. What will be the *maximum* current if this inductor is connected to a 50.0-Hz source that produces a 100-V rms voltage?
16. A 2.40-μF capacitor is connected across an alternating voltage with an rms value of 9.00 V. The rms current in the circuit is 25.0 mA. (a) What is the source frequency? (b) If the capacitor is replaced by an ideal coil with an inductance of 0.160 H, what is the rms current in the coil?
17. Determine the maximum magnetic flux through an inductor connected to a standard outlet ($\Delta V_{rms}$ = 120 V, $f$ = 60.0 Hz).

### Section 21.4 The *RLC* Series Circuit

18. An inductor ($L$ = 400 mH), a capacitor ($C$ = 4.43 μF), and a resistor ($R$ = 500 Ω) are connected in series. A 50.0-Hz AC generator connected in series to these elements produces a maximum current of 250 mA in the circuit. (a) Calculate the required maximum voltage $\Delta V_{max}$. (b) Determine the phase angle by which the current leads or lags the applied voltage.
19. A 10.0-μF capacitor and a 2.00-H inductor are connected in series with a 60.0-Hz source whose rms output is 100 V. Find (a) the rms current in the circuit, (b) the voltage drop across the inductor, (c) the voltage drop across the capacitor, and (d) the phase angle for the circuit. (e) Sketch the phasor diagram for this circuit.

**20.** A series AC circuit contains the following components: $R = 150\ \Omega$, $L = 250$ mH, $C = 2.00\ \mu$F, and a generator with $\Delta V_{max} = 210$ V operating at 50.0 Hz. Calculate the (a) inductive reactance, (b) capacitive reactance, (c) impedance, (d) maximum current, and (e) phase angle between the current and the generator voltage.

**21.** A resistor ($R = 900\ \Omega$), a capacitor ($C = 0.25\ \mu$F), and an inductor ($L = 2.5$ H) are connected in series across a 240-Hz AC source for which $\Delta V_{max} = 140$ V. Calculate the (a) impedance of the circuit, (b) maximum current delivered by the source, and (c) phase angle between the current and voltage. (d) Is the current leading or lagging behind the voltage?

**22.** A sinusoidal voltage $\Delta v(t) = (40.0\text{ V})\sin(100t)$ is applied to a series *RLC* circuit with $L = 160$ mH, $C = 99.0\ \mu$F, and $R = 68.0\ \Omega$. (a) What is the impedance of the circuit? (b) What is the maximum current?

23. A 60.0-$\Omega$ resistor, a 3.00-$\mu$F capacitor, and a 0.400-H inductor are connected in series to a 90.0-V, 60.0-Hz source. Find (a) the voltage drop across the *LC* combination and (b) the voltage drop across the *RC* combination.

24. A 50.0-$\Omega$ resistor is connected in series with a 15.0-$\mu$F capacitor and a 60.0-Hz, 120-V source. (a) Find the current in the circuit. (b) What is the value of the inductor that must be inserted in the circuit to reduce the current to one-half that found in (a)?

25. A person is working near the secondary of a transformer, as shown in Figure P21.25. The primary voltage is 120 V (rms) at 60.0 Hz. The capacitance $C_s$, which is the stray capacitance between the hand and the secondary winding, is 20.0 pF. Assuming the person has a body resistance to ground of $R_b = 50.0$ k$\Omega$, determine the rms voltage across the body. (*Hint:* Redraw the circuit with the secondary of the transformer as a simple AC source.)

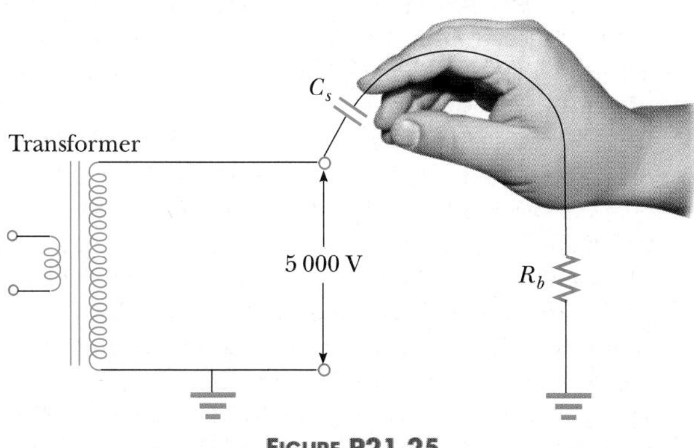


**FIGURE P21.25**

**26.** A coil of resistance 35.0 $\Omega$ and inductance 20.5 H is in series with a capacitor and a 200-V (rms), 100-Hz source. The rms current in the circuit is 4.00 A. (a) Calculate the capacitance in the circuit. (b) What is $\Delta V_{rms}$ across the coil?

**27.** web An AC source with a maximum voltage of 150 V and $f = 50.0$ Hz is connected between points *a* and *d* in Figure P21.27. Calculate the rms voltages between points (a) *a* and *b*, (b) *b* and *c*, (c) *c* and *d*, (d) *b* and *d*.

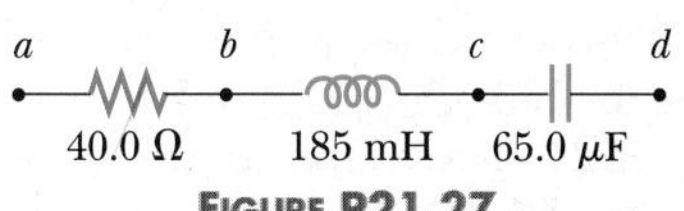


**FIGURE P21.27**

## Section 21.5 Power in an AC Circuit

**28.** A 50.0-$\Omega$ resistor is connected to a 30.0-$\mu$F capacitor and to a 60.0-Hz, 100-V (rms) source. (a) Find the power factor and the average power delivered to the circuit. (b) Repeat part (a) when the capacitor is replaced with a 0.300-H inductor.

**29.** A multimeter in an *RL* circuit records an rms current of 0.500 A and a 60.0-Hz rms generator voltage of 104 V. A wattmeter shows that the average power delivered to the resistor is 10.0 W. Determine (a) the impedance in the circuit, (b) the resistance *R*, and (c) the inductance *L*.

30. In a certain *RLC* circuit, the rms current is 6.0 A, the rms voltage is 240 V, and the current leads the voltage by 53°. (a) What is the total resistance of the circuit? (b) Calculate the total reactance $X_L - X_C$. (c) Find the average power delivered to the circuit.

**31.** An inductor and a resistor are connected in series. When connected to a 60-Hz, 90-V source, the voltage drop across the resistor is found to be 50 V and the power delivered to the circuit is 14 W. Find (a) the value of the resistance and (b) the value of the inductance.

32. An AC voltage of the form $\Delta v = (100\text{ V})\sin(1\,000t)$ is applied to a series *RLC* circuit. If $R = 400\ \Omega$, $C = 5.00\ \mu$F, and $L = 0.500$ H, find the average power delivered to the circuit.

## Section 21.6 Resonance in a Series *RLC* Circuit

**33.** An *RLC* circuit is used to tune a radio to an FM station broadcasting at 88.9 MHz. The resistance in the circuit is 12.0 $\Omega$ and the capacitance is 1.40 pF. What inductance should be present in the circuit?

**34.** A resonant circuit in a radio receiver is tuned to a certain station when the inductor has a value of 0.200 mH and the capacitor has a value of 30.0 pF. Find the frequency of the radio station and the wavelength sent out by the station.

**35.** The AM band extends from approximately 500 kHz to 1 600 kHz. If a 2.0-$\mu$H inductor is used in a tuning circuit for a radio, what are the extremes that a capacitor must reach in order to cover the complete band of frequencies?

36. A series circuit contains a 3.00-H inductor, a 3.00-$\mu$F capacitor, and a 30.0-$\Omega$ resistor connected to a 120-V (rms) source of variable frequency. Find the power delivered to the circuit when the frequency of the source is (a) the resonance frequency, (b) one-half the resonance frequency, (c) one-fourth the resonance frequency, (d) two times the resonance frequency, (e) four times the resonance frequency. From your calculations, can you draw a conclusion about the frequency at which the maximum power is delivered to the circuit?

**37.** A 10.0-$\Omega$ resistor, 10.0-mH inductor, and a 100-$\mu$F capacitor are connected in series to a 50.0-V (rms) source having variable frequency. Find the energy delivered to the circuit during one period if the operating frequency is twice the resonance frequency.

## Section 21.7 The Transformer

**38.** A step-down transformer is used for recharging the batteries of portable devices such as tape players. The turns ratio inside the transformer is 13:1 and it is used with 120 V (rms) household service. If a particular ideal transformer draws 0.350 A from the house outlet, what are (a) the voltage and (b) the current supplied to a tape player from the transformer? (c) How much power is delivered?

**39.** A step-up transformer is designed to have an output voltage of 2 200 V (rms) when the primary is connected across a 110-V (rms) source. (a) If there are 80 turns on the primary winding, how many turns are required on the secondary? (b) If a load resistor across the secondary draws a current of 1.5 A, what is the current in the primary, assuming ideal conditions?

**40.** A transformer has $N_1 = 350$ turns and $N_2 = 2\,000$ turns. If the input voltage is $\Delta v(t) = (170\text{ V})\cos \omega t$, what rms voltage is developed across the secondary coil?

**41.** A transformer on a pole near a factory steps the voltage down from 3 600 V to 120 V. The transformer is to deliver 1 000 kW to the factory at 90% efficiency. Find (a) the power delivered to the primary, (b) the current in the primary, and (c) the current in the secondary.

**42.** A transmission line that has a resistance per unit length of $4.50 \times 10^{-4}\ \Omega/\text{m}$ is to be used to transmit 5.00 MW over 400 miles ($6.44 \times 10^5$ m). The output voltage of the generator is 4.50 kV. (a) What is the line loss if a transformer is used to step up the voltage to 500 kV? (b) What fraction of the input power is lost to the line under these circumstances? (c) What difficulties would be encountered on attempting to transmit the 5.00 MW at the generator voltage of 4.50 kV?

## Section 21.10 Production of Electromagnetic Waves by an Antenna

## Section 21.11 Properties of Electromagnetic Waves

**43.** The U.S. Navy has long proposed the construction of extremely low-frequency (ELF) communications systems; such waves could penetrate the oceans to reach distant submarines. Calculate the length of a quarter-wavelength antenna for a transmitter generating ELF waves of frequency 75 Hz. How practical is this?

**44.** Experimenters at the National Institute of Standards and Technology have made precise measurements of the speed of light using the fact that, in vacuum, the speed of electromagnetic waves is $c = 1/\sqrt{\mu_0 \epsilon_0}$, where the constants $\mu_0 = 4\pi \times 10^{-7}\ \text{N}\cdot\text{s}^2/\text{C}^2$ and $\epsilon_0 = 8.854 \times 10^{-12}\ \text{C}^2/\text{N}\cdot\text{m}^2$. What value (to four significant figures) does this give for the speed of light in vacuum?

**45.** An electromagnetic wave in vacuum has an electric field amplitude of 220 V/m. Calculate the amplitude of the corresponding magnetic field.

**46.** A particular electromagnetic wave traveling in vacuum has a magnetic field amplitude of $1.5 \times 10^{-7}$ T. Find (a) the electric field amplitude and (b) the average power per unit area associated with the wave.

**47.** A microwave oven is powered by an electron tube called a magnetron that generates electromagnetic waves of frequency 2.45 GHz. The microwaves enter the oven and are reflected by the walls. The standing wave pattern produced in the oven can cook food unevenly, with hot spots in the food at antinodes and cool spots at nodes, so a turntable is often used to rotate the food and distribute the energy. If a microwave oven is used with a cooking dish in a fixed position, the antinodes can appear as burn marks on foods such as carrot strips or cheese. The separation distance between the burns is measured to be 6.00 cm. Calculate the speed of the microwaves from these data.

**48.** Assume that the solar radiation incident on Earth is 1 340 W/m$^2$ (at the top of Earth's atmosphere). Calculate the total power radiated by the Sun, taking the average separation between Earth and the Sun to be $1.49 \times 10^{11}$ m.

**49.** web The Sun delivers an average power of 1 340 W/m$^2$ to the top of Earth's atmosphere. Find the magnitudes of $\mathbf{E}_{max}$ and $\mathbf{B}_{max}$ for the electromagnetic waves at the top of the atmosphere.

## Section 21.12 The Spectrum of Electromagnetic Waves

**50.** The eye is most sensitive to light of wavelength $5.50 \times 10^{-7}$ m, which is in the green-yellow region of the visible electromagnetic spectrum. What is the frequency of this light?

**51.** What are the wavelength ranges in the (a) AM radio band (540–1 600 kHz) and (b) the FM radio band (88–108 MHz)?

**52.** A diathermy machine, used in physiotherapy, generates electromagnetic radiation that gives the effect of "deep heat" when absorbed in tissue. One assigned frequency for diathermy is 27.33 MHz. What is the wavelength of this radiation?

**53.** A radar pulse returns to the receiver after a total travel time of $4.00 \times 10^{-4}$ s. How far away is the object that reflected the wave?

**54.** An important news announcement is transmitted by radio waves to people who are 100 km away, sitting next to their radios, and by sound waves to people sitting across the newsroom, 3.0 m from the newscaster. Who receives the news first? Explain. Take the speed of sound in air to be 343 m/s.

## ADDITIONAL PROBLEMS

**55.** An AC adapter for a telephone answering unit uses a transformer to reduce the line voltage of 120 V (rms) to a voltage of 9.0 V. The rms current delivered to the answering system is 400 mA. (a) If the primary (input) coil in the transformer in the adapter has 240 turns, how many turns are there on the secondary (output) coil? (b) What is the rms power delivered to the transformer? Assume an ideal transformer.

**56.** The primary coil of a certain transformer has an inductance of 2.50 H and a resistance of 80.0 Ω. (a) If the primary coil is connected to an AC source with a frequency of 60.0 Hz and a voltage of 110 V (rms), what is the rms current in the primary? (b) If the primary is connected to 110 V DC, what is the current in the primary? (c) In each case, compare the power delivered to the coil.

**57.** As a way of determining the inductance of a coil used in a research project, a student first connects the coil to a 12.0-V battery and measures a current of 0.630 A. The student then connects the coil to a 24.0-V (rms), 60.0-Hz generator and measures an rms current of 0.570 A. What is the inductance?

**58.** The intensity of solar radiation at the top of Earth's atmosphere is 1 340 W/m$^3$. Assuming that 60% of the incoming solar energy reaches Earth's surface and assuming that you absorb 50% of the incident energy, make an order-of-magnitude estimate of the amount of solar energy you absorb in a 60-minute sunbath.

**59.** A 200-Ω resistor is connected in series with a 5.0-$\mu$F capacitor and a 60-Hz, 120-V rms line. If electrical energy costs \$0.080/kWh, how much does it cost to leave this circuit connected for 24 h?

**60.** A series *RLC* circuit has a resonance frequency of $2\,000/\pi$ Hz. When it is operating at a frequency of

$\omega > \omega_0$, $X_L = 12\ \Omega$ and $X_C = 8.0\ \Omega$. Calculate the values of $L$ and $C$ for the circuit.

61. Two connections allow contact with two circuit elements in series inside a box, but it is not known whether the circuit elements are *R*, *L*, or *C*. In an attempt to find what is inside the box, you make some measurements, with the following results. When a 3.0-V DC power supply is connected across the terminals, there is a direct current of 300 mA in the circuit. When a 3.0-V, 60-Hz source is connected, the current becomes 200 mA. (a) What are the two elements in the box? (b) What are their values of *R*, *L*, or *C*?

web

62. (a) What capacitance will resonate with a one-turn loop of inductance 400 pH to give a radar wave of wavelength 3.0 cm? (b) If the capacitor has square parallel plates separated by 1.0 mm of air, what should the edge length of the plates be? (c) What is the common reactance of the loop and capacitor at resonance?

63. A dish antenna with a diameter of 20.0 m receives (at normal incidence) a radio signal from a distant source, as shown in Figure P21.63. The radio signal is a continuous sinusoidal wave with amplitude $E_{\text{max}} = 0.20\ \mu\text{V/m}$. Assume the antenna absorbs all the radiation that falls on the dish. (a) What is the amplitude of the magnetic field in this wave? (b) What is the intensity of the radiation received by this antenna? (c) What is the power received by the antenna?

64. A particular inductor has appreciable resistance. When the inductor is connected to a 12-V battery, the current in the inductor is 3.0 A. When it is connected to an AC source with an rms output of 12 V and a frequency of 60 Hz, the current drops to 2.0 A. What are (a) the impedance at 60 Hz and (b) the inductance of the inductor?

**FIGURE P21.63**

65. A possible means of space flight is to place a perfectly reflecting aluminized sheet into Earth's orbit and to use the light from the Sun to push this solar sail. Suppose such a sail, of area $6.00 \times 10^4\ \text{m}^2$ and mass 6 000 kg, is placed in orbit facing the Sun. (a) What force is exerted on the sail? (b) What is the sail's acceleration? (c) How long does it take for this sail to reach the Moon, $3.84 \times 10^8$ m away? Ignore all gravitational effects, and assume a solar intensity of 1 340 W/m$^2$. [*Hint:* The radiation pressure by a reflected wave is given by 2(average power per unit area)/*c*.]

66. Suppose you wish to use a transformer as an impedance-matching device between an audio amplifier that has an output impedance of 8.0 kΩ and a speaker that has an input impedance of 8.0 Ω. What should be the ratio of primary to secondary turns on the transformer?

67. Compute the average energy content of a liter of sunlight as it reaches the top of Earth's atmosphere, where its intensity is 1 340 W/m$^2$.

## GROUP ACTIVITIES

**G.1** For this observation, you will need some items that can be found at many electronics stores. You will need a bicolored LED (light-emitting diode), a resistor of about 100 Ω, 2 m of flexible wire, and a step-down transformer with an output of 3 to 6 V. Use the wire to connect the LED and the resistor in series with the transformer. A bicolored LED is designed such that it emits a red color when the current in the LED is in one direction and green when the current reverses. When connected to an AC source, the LED is yellow. Why?

Hold the wires and whirl the LED in a circular path. In a darkened room, you will see red and green bars at equally spaced intervals along the path of the LED. Why?

As you continue to whirl the LED in a circular path, have your partner count the number of green bars in the circle then measure the time it takes for the LED to travel ten times around the circular path. Based on this information, determine the time it takes for the color of the LED to change from green to red to green. You should obtain an answer of (1/60) s. Why?

**G.2** Rotate a portable radio (with a telescoping antenna) about a horizontal axis while it is tuned to a weak station. Such an antenna detects the varying electric field produced by the station. What can you determine about the direction of the electric field produced by the transmitter?

Now turn on your radio to a nearby station and experiment with shielding the radio from incoming waves. Is the reception affected by surrounding the radio by aluminum foil? By plastic wrap? Use any other material you have available. What kinds of material block the signal? Why?

**G.3** An *RLC* circuit is brought to resonance. (a) What is the impedance of the circuit at resonance? (b) When will the current be greatest—at resonance, at 10% below the resonant frequency, or at 10% above the resonant frequency? (c) Check your guesses by examining a circuit with $\Delta V = 12$ V, $R = 15\ \Omega$, $C = 75\ \mu$F, and $L = 200$ mH.

**G.4** A transformer is to be used to provide power for a computer disk drive that needs 6.0 V instead of the 120 V from the wall outlet. (a) Should the power supply have more turns in the primary as compared to the secondary, or the reverse? (b) Will the larger current be in the primary or the secondary? (c) Assume the number of turns in the primary is 400 and that it delivers 4.0 A at the 6.0-V output. Do the calculations to confirm your answers.

**G.5** Consider an *RLC* circuit with $R = 25\ \Omega$, $L = 6.0$ mH, and $C = 25\ \mu$F. This circuit is connected to a 10-V, 600-Hz AC source. (a) Is the sum of the voltage drops across *R*, *L*, and *C* equal to 10 V? (b) Which is greatest, the power delivered to the resistor, the capacitor, or the inductor? (c) Do the calculations to verify your predictions.

# Light and Optics

Scientists have long been intrigued by the nature of light, and philosophers have argued endlessly concerning the proper definition and perception of light. It is important to understand the nature of this basic ingredient of life on Earth. Plants convert light energy from the Sun to chemical energy through photosynthesis. Light is the means by which we transmit and receive information from objects around us and throughout the Universe.

The ancient Greeks believed that vision worked by means of "fire" emanating from the eye, which then interacted with objects and returned to the eye carrying object properties with it. Around 1 000 A.D., it was proposed that light consisted of tiny particles that were emitted by a light source. These particles stimulated the perception of vision on striking the observer's eye. Newton used this particle model to explain the reflection and refraction of light. In 1670 one of Newton's contemporaries, the Dutch scientist Christian Huygens, succeeded in explaining many properties of light by proposing that light was wave-like. In 1801 Thomas Young gave strong support to the wave theory by showing that light beams can interfere with one another. In 1865 Maxwell developed a brilliant theory that electromagnetic waves travel with the speed of light (Chapter 21). By that time, the wave theory of light seemed to be on firm ground.

However, at the beginning of the 20th century, Max Planck introduced the notion that vibrating electrons in the wall of a glowing solid had quantized (discrete) energies in order to explain the light spectrum emitted by hot objects. Albert Einstein extended this idea to a full-fledged particle theory of light to explain the photoelectric effect (electrons emitted by a metal exposed to light). We shall discuss those and other modern topics in the last part of this book.

Today, scientists view light as having a dual nature. Experiments can be devised that will display either its particle-like or its wave-like nature. All we can say about light is that whenever we design an experiment to determine its wave-like nature, we will find it, and whenever we look for its particle-like nature, we will find that. Nature prevents us from testing both qualities at the same time.

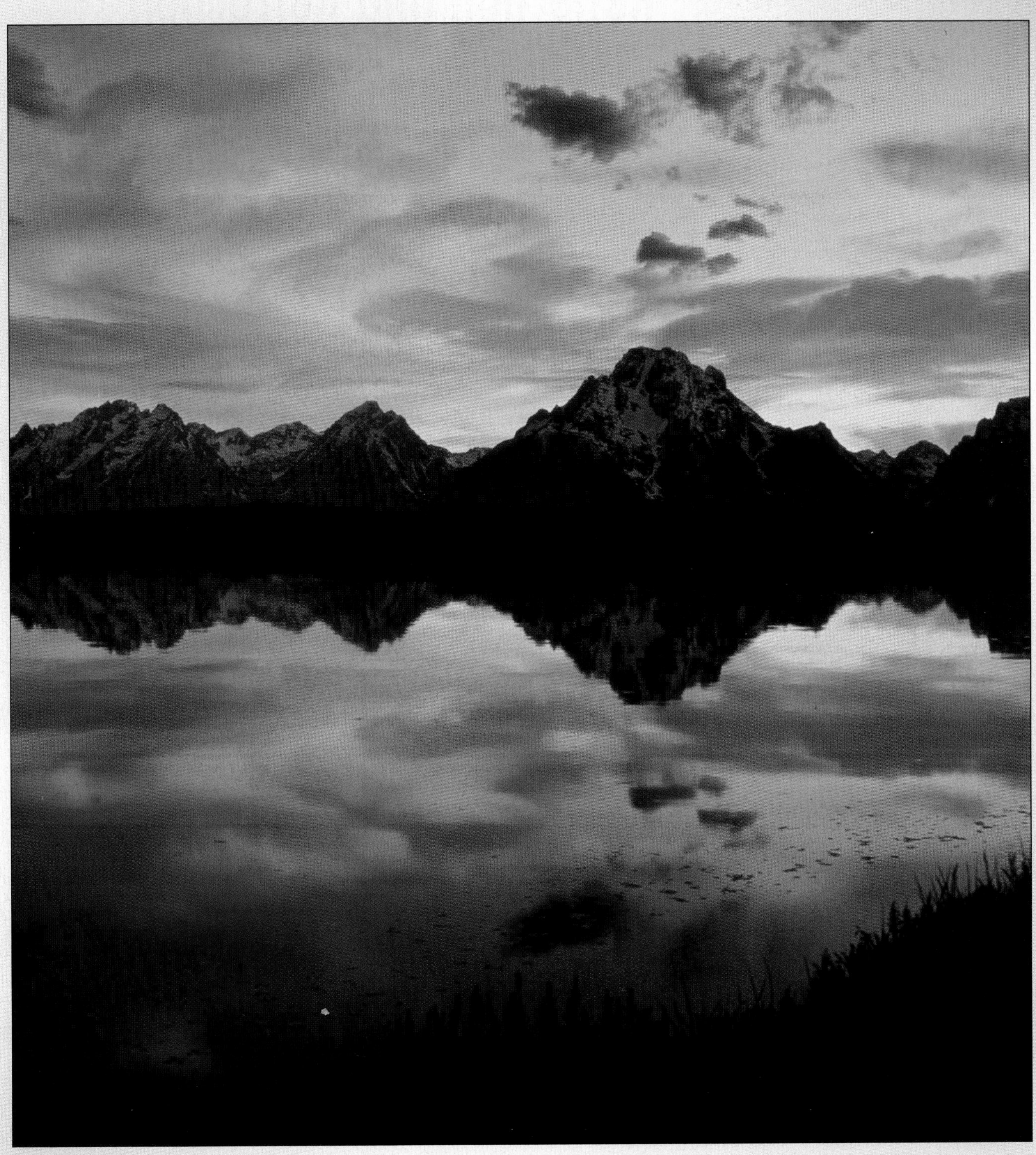

This striking photograph shows the Grand Tetons in Wyoming reflected in a lake which acts like a mirror. Note that the image of the mountains has the same size as the mountains.
*(David Muench/CORBIS)*

# 22

# Reflection and Refraction of Light

## Chapter Outline

**The prism in this photograph demonstrates both refraction and reflection. A beam of white light at the top left is refracted and strikes the right face of the prism. At this face, some of the light is refracted once more, exiting the prism as a spectrum of light with different colors. The spectrum arises from the fact that the incident white light contains different wavelengths that refract at different angles. Part of the light is reflected at the right face and exits the prism at the bottom left as white light.** *(David Parker/Science Photo Library/Photo Researchers, Inc.)*

In this part of the book, we concentrate on the aspects of light that are best understood through the wave model. First, we discuss the reflection of light at the boundary between two media and the refraction (bending) of light as it travels from one medium into another. We use these ideas to study the refraction of light as it passes through lenses and the reflection of light from mirrored surfaces. Finally, we describe how lenses and mirrors can be used to view objects with telescopes and microscopes and how lenses are used in photography.

## 22.1 THE NATURE OF LIGHT

Until the beginning of the 19th century, light was considered to be a stream of particles, emitted by a light source, that stimulated the sense of sight on entering the eye. The chief architect of the particle theory of light was Newton. With this theory he provided simple explanations of some known experimental facts concerning the nature of light—namely, the laws of reflection and refraction.

Most scientists accepted Newton's particle theory of light. However, during Newton's lifetime another theory was proposed. In 1678 a Dutch physicist and astronomer, Christian Huygens (1629–1695), showed that a wave theory of light could also explain the laws of reflection and refraction. The wave theory did not receive immediate acceptance for several reasons. All the waves known at the time (sound, water, and so on) traveled through some sort of medium, but light from the Sun could travel to the Earth through empty space. Furthermore, it was argued that if light were some form of wave, it would bend around obstacles; hence, we should be able to see around corners. It is now known that light does indeed bend around the edges of objects. This phenomenon, known as *diffraction,* is not easy to observe because light waves have such short wavelengths. Even though experimental evidence for the diffraction of light was discovered by Francesco Grimaldi (1618–1663) around 1660, for more than a century most scientists rejected the wave theory and adhered to Newton's particle theory. This was also due to Newton's great reputation as a scientist.

**CHRISTIAN HUYGENS, DUTCH PHYSICIST AND ASTRONOMER (1629–1695)**

Huygens is best known for his contributions to the fields of optics and dynamics. To Huygens, light was a vibratory motion in the ether, spreading out and producing the sensation of light when impinging on the eye. On the basis of this theory, he deduced the laws of reflection and refraction and explained the phenomenon of double refraction. *(Courtesy of Rijksmuseum voor de Geschiedenis der Natuurwetenschappen. Courtesy AIP Niels Bohr Library)*

The first clear demonstration of the wave nature of light was provided in 1801 by Thomas Young (1773–1829), who showed that under appropriate conditions, light exhibits interference behavior. That is, light waves emitted by a single source and traveling along two different paths can arrive at some point, combine, and cancel each other by destructive interference. Such behavior could not be explained at that time by a particle model, because scientists could not imagine how two or more particles could come together and cancel one another.

The most important development concerning the theory of light was the work of Maxwell, who in 1865 predicted that light was a form of high-frequency electromagnetic wave (Chapter 21). His theory predicted that these waves should have a speed of $3 \times 10^8$ m/s, in agreement with the measured value.

Although the classical theory of electricity and magnetism explained most known properties of light, some subsequent experiments could not be explained by the assumption that light was a wave. The most striking of these was the *photoelectric effect* (which we shall examine more closely in Chapter 27), discovered by Hertz. Hertz found that clean metal surfaces emit charges when exposed to ultraviolet light.

In 1905 Einstein published a paper that formulated the theory of light quanta ("particles") and explained the photoelectric effect. He reached the conclusion that light is composed of corpuscles, or discontinuous quanta of energy. These corpuscles or quanta are now called *photons* to emphasize their particle-like nature. According to Einstein's theory, the energy of a photon is proportional to the frequency of the electromagnetic wave:

$$E = hf \qquad [22.1]$$

Energy of a photon

where $h = 6.63 \times 10^{-34}$ J·s is *Planck's constant.* This theory retains some features of both the wave and particle theories of light. As we shall discuss later, the photoelectric effect is the result of energy transfer from a single photon to an electron in the metal. That is, the electron interacts with one photon of light as if the electron had been struck by a particle. Yet the photon has wave-like characteristics, as implied by the fact that a frequency is used in its definition.

In view of these developments, light must be regarded as having a *dual nature.* That is, **in some cases light acts as a wave and in others it acts as a particle.** Classical electromagnetic wave theory provides adequate explanations of light propagation and of the effects of interference, whereas the photoelectric effect and other experiments involving the interaction of light with matter are best explained by assuming that light is a particle. However, the question "Is light a wave or a particle?" is inappropriate; sometimes it acts as one, sometimes as the other. Fortunately, it never acts as both in the same experiment.

## 22.2 THE RAY APPROXIMATION IN GEOMETRIC OPTICS

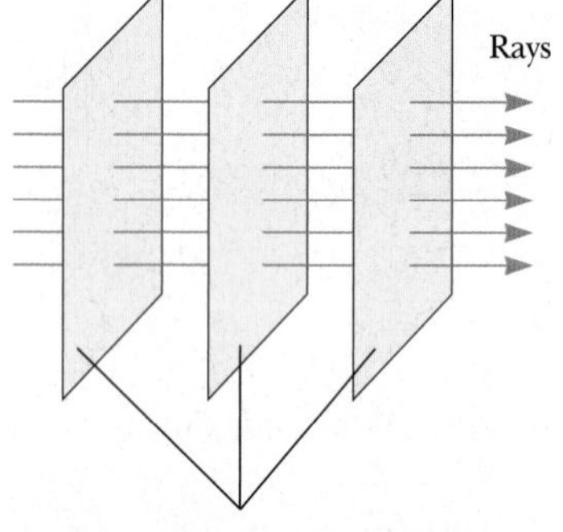


**FIGURE 22.1** A plane wave traveling to the right. Note that the rays, corresponding to the direction of wave motion, are straight lines perpendicular to the wave fronts.

In studying geometric optics here and in future chapters, we shall make use of an important property of light that can be understood based on common experience. **Light travels in a straight-line path in a homogeneous medium until it encounters a boundary between two different materials.** As we shall see, when light strikes a boundary it either is reflected from that boundary or passes into the material on the other side of the boundary or partially does both.

Based on this observation, we will use what is called the **ray approximation** to represent beams of light. As shown in Figure 22.1, a ray of light is an imaginary line drawn along the direction of travel of the light beam. For example, a beam of sunlight passing through a darkened room traces out the path of a light ray. We will also have occasion to refer to wave fronts of light. A **wave front** is a surface passing through the points of a wave that have the same phase and amplitude. For instance, the wave fronts in Figure 22.1 could be surfaces passing through the crests of waves. You should note that the rays, corresponding to the direction of wave motion, are straight lines perpendicular to the wave fronts. Also, note that when light rays travel in parallel paths, the wave fronts are planes perpendicular to the rays.

## 22.3 REFLECTION AND REFRACTION

### REFLECTION OF LIGHT

When a light ray traveling in a transparent medium encounters a boundary leading into a second medium, part of the incident ray is reflected back into the first medium. Figure 22.2a shows several rays of a beam of light incident on a smooth, mirror-like, reflecting surface. The reflected rays are parallel to each other, as indicated in the figure. Reflection of light from such a smooth surface is called **specular reflection.** On the other hand, if the reflecting surface is rough, as in Figure 22.2b, the surface reflects the rays in a variety of directions. Reflection from any rough surface is known as **diffuse reflection.** A surface behaves as a smooth surface as long as the surface variations are small compared with the wavelength of the incident light. Figures 22.2c and 22.2d are photographs of specular and diffuse reflection of laser light.

**APPLICATION**

SEEING THE ROAD ON A RAINY NIGHT

For instance, consider the two types of reflection from a road surface that one sees while driving at night. When the road is dry, light from oncoming vehicles is scattered off the road in different directions (diffuse reflection) and the road is quite visible. On a rainy night, when the road is wet, the road irregularities are filled with water. Because the wet surface is quite smooth, the light undergoes

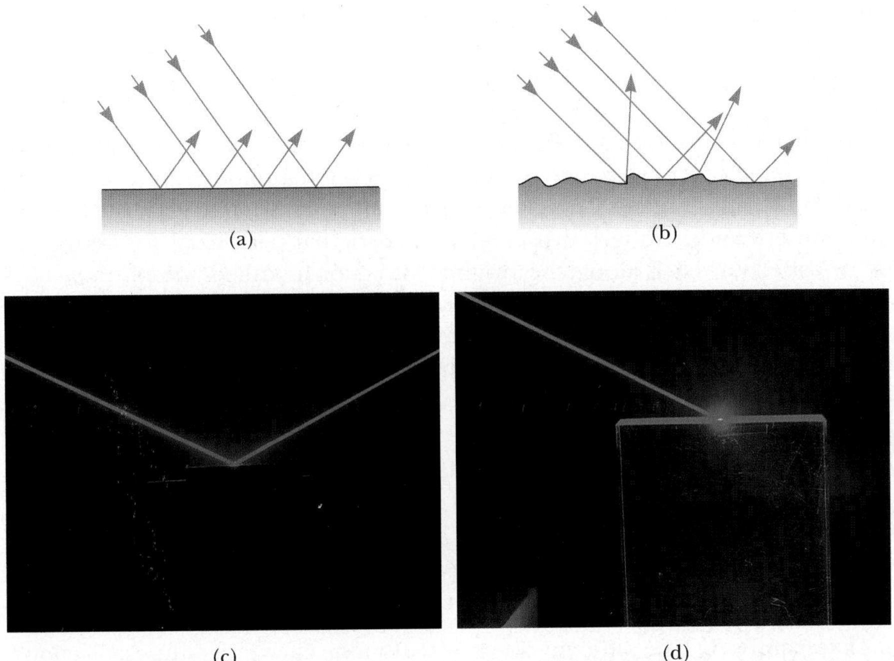


**FIGURE 22.2** A schematic representation of (a) specular reflection, where the reflected rays are all parallel to each other, and (b) diffuse reflection, where the reflected rays travel in random directions. (c, d) Photographs of specular and diffuse reflection, using laser light. *(Photographs courtesy of Henry Leap and Jim Lehman)*

**FIGURE 22.3** (Quick Quiz 22.1) *(Photos by Charles D. Winters)*

specular reflection. This means that the light is reflected straight ahead, and the driver of a car sees only what is directly in front. Light from the side never reaches the driver's eye. In this book we concern ourselves only with specular reflection, and we use the term *reflection* to mean specular reflection.

**Quick Quiz 22.1** Which part of Figure 22.3, (a) or (b), shows specular reflection of light from the roadway?

Consider a light ray traveling in air and incident at some angle on a flat, smooth surface, as in Figure 22.4. The incident and reflected rays make angles $\theta_1$ and $\theta_1'$, respectively, **with a line perpendicular to the surface** at the point where the incident ray strikes the surface. We call this line the *normal* to the surface. Experiments

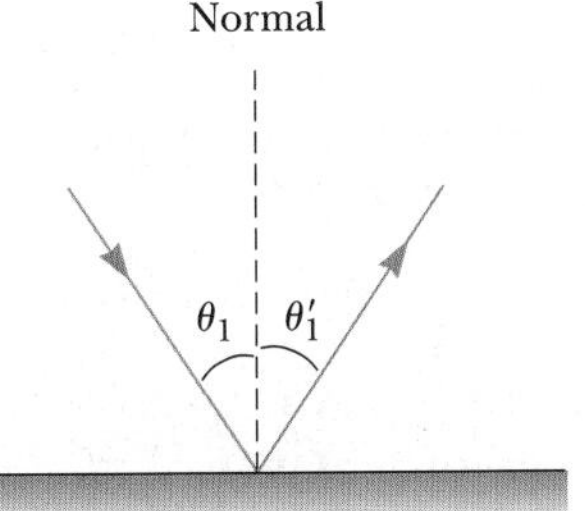


**FIGURE 22.4** According to the law of reflection, $\theta_1 = \theta_1'$.

show that the angle of reflection equals *the angle of incidence;* that is,

$$\theta_1' = \theta_1 \tag{22.2}$$

**APPLICATION**

RED EYES IN FLASH PHOTOGRAPHS

You may have noticed a common occurrence in photographs of individuals—their eyes appear to be glowing red. This occurs when a photographic flash device is used and the flash unit is very close to the camera lens. Light from the flash unit enters the eye and is reflected back along its original path from the retina. This type of reflection back along the original direction is called *retroreflection.* If the flash unit and lens are close together, this retroreflected light can enter the lens. Most of the light reflected from the retina is red, due to the blood vessels at the back of the eye, giving the red-eye effect in the photograph.

## APPLYING PHYSICS 22.1

**APPLICATION**

THE COLORS OF WATER RIPPLES AT SUNSET

An observer on the west-facing beach of a large lake is watching the beginning of a sunset. The water is very smooth except for some areas with small ripples. The observer notices that some areas of the water are blue and some are pink. Why does the water appear to be different colors in different areas?

**Explanation** The different colors arise from specular and diffuse reflection. The smooth areas of the water will specularly reflect the light from the west, which is the pink light from the sunset. The areas with small ripples will reflect the light diffusely. Thus, light from all parts of the sky will be reflected into the observer's eyes. Because most of the sky is still blue at the beginning of the sunset, these areas will appear to be blue.

## APPLYING PHYSICS 22.2

When looking outdoors through a glass window at night, you sometimes see a double image of yourself. Why?

**Explanation** Reflection occurs whenever there is an interface between two different media. For the glass in the window, there are two such surfaces. The first is the inner surface of the glass, and the second is the outer surface. Each of these interfaces results in an image.

## Example 22.1 The Double-Reflecting Light Ray

Two mirrors make an angle of 120° with each other, as in Figure 22.5. A ray is incident on mirror $M_1$ at an angle of 65° to the normal. Find the angle the ray makes with the normal to $M_2$ after it is reflected from both mirrors.

**Reasoning and Solution** From the law of reflection, we see that the first reflected ray also makes an angle of 65° with the normal to $M_1$. It follows that this same ray makes an angle of 90° − 65°, or 25°, with the horizontal. From the triangle made by the first reflected ray and the two mirrors, we see that the first reflected ray makes an angle of 35° with $M_2$ (because the sum of the interior angles of any triangle is 180°). This means that this ray makes an angle of 55° with the normal to $M_2$. From the law of reflection, it follows that the second reflected ray makes an angle of 55° with the normal to $M_2$.

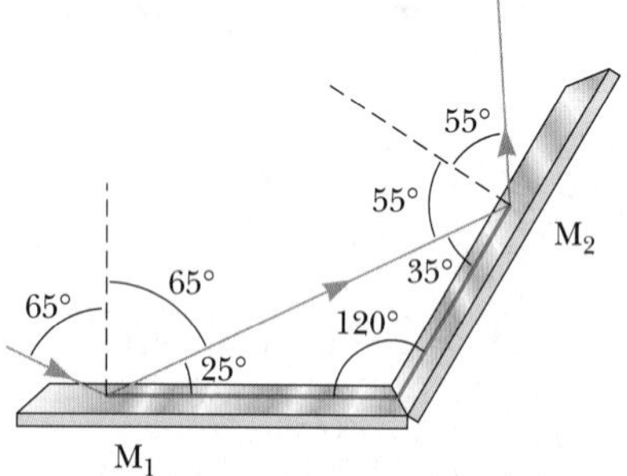


**FIGURE 22.5** (Example 22.1) Mirrors $M_1$ and $M_2$ make an angle of 120° with each other.

### REFRACTION OF LIGHT

When a ray of light traveling through a transparent medium encounters a boundary leading into another transparent medium, as in Figure 22.6a, part of the ray is reflected and part enters the second medium. The ray that enters the second

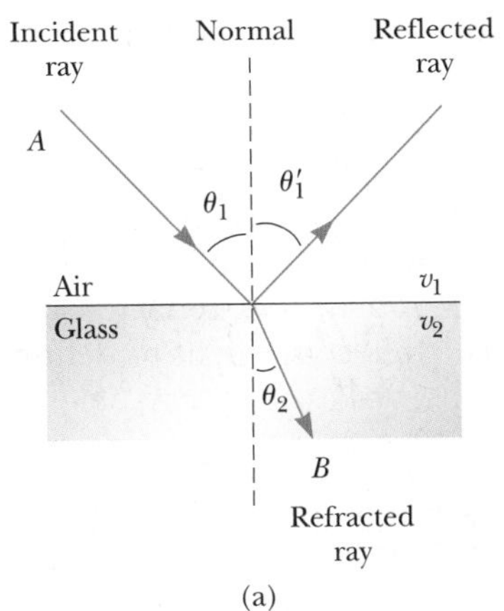


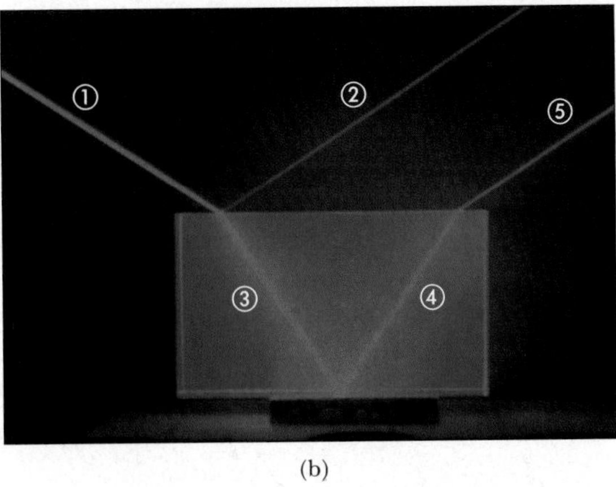


**FIGURE 22.6** (a) A ray obliquely incident on an air-glass interface. The refracted ray is bent toward the normal because $v_2 < v_1$. (b) Light incident on the Lucite block bends both when it enters the block and when it leaves the block. *(Courtesy of Henry Leap and Jim Lehman)*

medium is bent at the boundary and is said to be *refracted.* The incident ray, the reflected ray, the refracted ray, and the normal at the point of incidence all lie in the same plane. The **angle of refraction,** $\theta_2$, in Figure 22.6a depends on the properties of the two media and on the angle of incidence, through the relationship

$$\frac{\sin \theta_2}{\sin \theta_1} = \frac{v_2}{v_1} = \text{constant} \qquad [22.3]$$

where $v_1$ is the speed of light in medium 1 and $v_2$ is the speed of light in medium 2. Note that the angle of refraction is also measured with respect to the normal. In Section 22.7 we shall derive the laws of reflection and refraction using Huygens's principle.

Experiment shows that **the path of a light ray through a refracting surface is reversible.** For example, the ray in Figure 22.6a travels from point *A* to point *B*. If the ray originated at *B*, it would follow the same path to reach point *A*, but the reflected ray would be in the glass.

**Quick Quiz 22.2** If beam 1 is the incoming beam in Figure 22.6b, which of the other four beams are reflected and which are refracted?

When light moves from a material in which its speed is high to a material in which its speed is lower, the angle of refraction $\theta_2$ is less than the angle of incidence. The refracted ray therefore bends toward the normal, as shown in Figure 22.7a. If the ray moves from a material in which it travels slowly to a material in

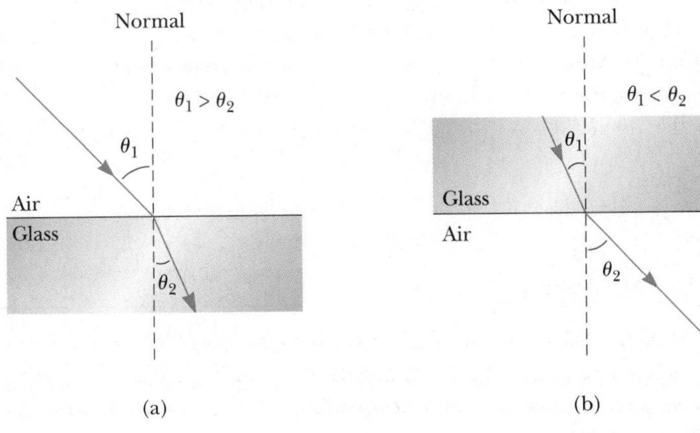


**FIGURE 22.7** (a) When the light beam moves from air into glass, its path is bent toward the normal. (b) When the beam moves from glass into air, its path is bent away from the normal.

which it travels more rapidly, $\theta_2$ is greater than $\theta_1$, and so the ray bends away from the normal, as shown in Figure 22.7b.

## 22.4 THE LAW OF REFRACTION

When light passes from one transparent medium to another, it is refracted because the speed of light is different in the two media.[1] It is convenient to define the **index of refraction,** $n$, of a medium as the ratio

Index of refraction 

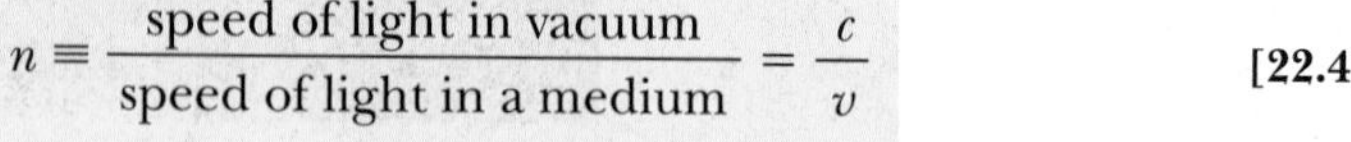

$$n \equiv \frac{\text{speed of light in vacuum}}{\text{speed of light in a medium}} = \frac{c}{v} \qquad [22.4]$$

**Tip 22.1 AN INVERSE RELATIONSHIP**

The index of refraction is *inversely* proportional to the wave speed. Therefore, as the wave speed $v$ decreases, the index of refraction $n$ *increases.*

From this definition, we see that the index of refraction is a dimensionless number that is greater than or equal to unity because $v$ is always less than $c$. Furthermore, $n$ is equal to unity for vacuum. Table 22.1 lists the indices of refraction for various substances.

**As light travels from one medium to another, its frequency does not change.** To see why, consider Figure 22.8. Wave fronts pass an observer at point $A$ in medium 1 with a certain frequency and are incident on the boundary between medium 1 and medium 2. The frequency at which the wave fronts pass an observer at point $B$ in medium 2 must equal the frequency at which they arrive at point $A$. If this were not the case, the wave fronts would either pile up at the boundary or be destroyed or created at the boundary. Because this does not occur, the frequency must remain the same as a light ray passes from one medium into another.

**Tip 22.2 THE FREQUENCY REMAINS THE SAME**

Note that the *frequency* of a wave does *not* change as the wave passes from one medium to another. Both the wave speed and wavelength *do* change, but the frequency remains constant.

Therefore, because the relation $v = f\lambda$ must be valid in both media and because $f_1 = f_2 = f$, we see that

$$v_1 = f\lambda_1 \qquad \text{and} \qquad v_2 = f\lambda_2$$

Because $v_1 \neq v_2$, it follows that $\lambda_1 \neq \lambda_2$. A relationship between index of refraction and wavelength can be obtained by dividing these two equations and making use of the definition of index of refraction given by Equation 22.4:

$$\frac{\lambda_1}{\lambda_2} = \frac{v_1}{v_2} = \frac{c/n_1}{c/n_2} = \frac{n_2}{n_1} \qquad [22.5]$$

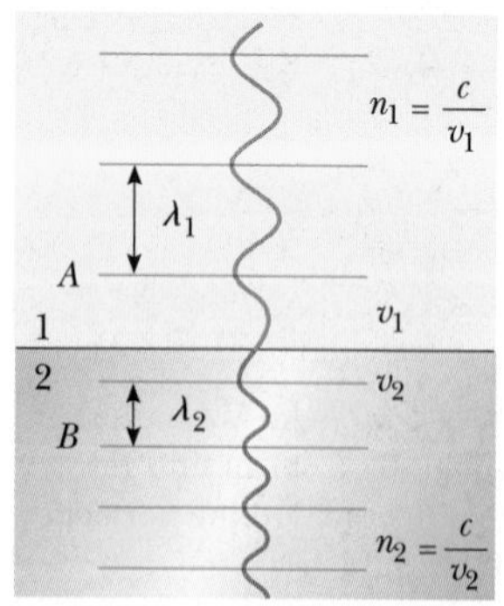


**FIGURE 22.8** As the wave moves from medium 1 to medium 2, its wavelength changes but its frequency remains constant.

**TABLE 22.1** Indices of Refraction for Various Substances, Measured with Light of Vacuum Wavelength $\lambda_0 = 589$ nm

| Substance | Index of Refraction | Substance | Index of Refraction |
|---|---|---|---|
| **Solids at 20°C** | | **Liquids at 20°C** | |
| Diamond (C) | 2.419 | Benzene | 1.501 |
| Fluorite ($CaF_2$) | 1.434 | Carbon disulfide | 1.628 |
| Fused quartz ($SiO_2$) | 1.458 | Carbon tetrachloride | 1.461 |
| Glass, crown | 1.52 | Ethyl alcohol | 1.361 |
| Glass, flint | 1.66 | Glycerine | 1.473 |
| Ice ($H_2O$) (at 0°C) | 1.309 | Water | 1.333 |
| Polystyrene | 1.49 | | |
| Sodium chloride (NaCl) | 1.544 | **Gases at 0°C, 1 atm** | |
| Zircon | 1.923 | Air | 1.000 293 |
| | | Carbon dioxide | 1.000 45 |

[1] The speed of light varies between media because the time lags caused by absorption and reemission of light as it travels from atom to atom depends on the particular electronic structure of the atoms constituting each material.

which gives

$$\lambda_1 n_1 = \lambda_2 n_2 \qquad [22.6]$$

Let medium 1 be the vacuum so that $n_1 = 1$. It follows from Equation 22.6 that the index of refraction of any medium can be expressed as the ratio

$$n = \frac{\lambda_0}{\lambda_n} \qquad [22.7]$$

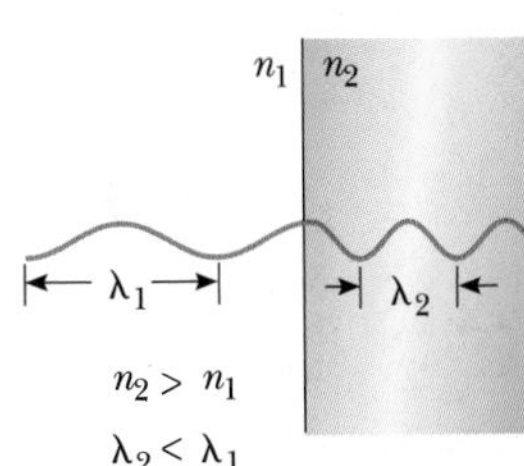


**FIGURE 22.9** A schematic diagram of the *reduction* in wavelength when light travels from a medium with a low index of refraction to one with a higher index of refraction.

where $\lambda_0$ is the wavelength of light in vacuum and $\lambda_n$ is the wavelength in the medium whose index of refraction is $n$. Figure 22.9 is a schematic representation of this reduction in wavelength when light passes from vacuum into a transparent medium.

We are now in a position to express Equation 22.3 in an alternative form. If we substitute Equation 22.5 into Equation 22.3, we get

$$n_1 \sin \theta_1 = n_2 \sin \theta_2 \qquad [22.8]$$

Snell's law of refraction

The experimental discovery of this relationship is usually credited to Willebord Snell (1591–1627) and is therefore known as **Snell's law of refraction.**

**Quick Quiz 22.3** A material has an index of refraction that increases continuously from top to bottom. Of the three paths shown in Figure 22.10, which path will a light ray follow as it passes through the material.

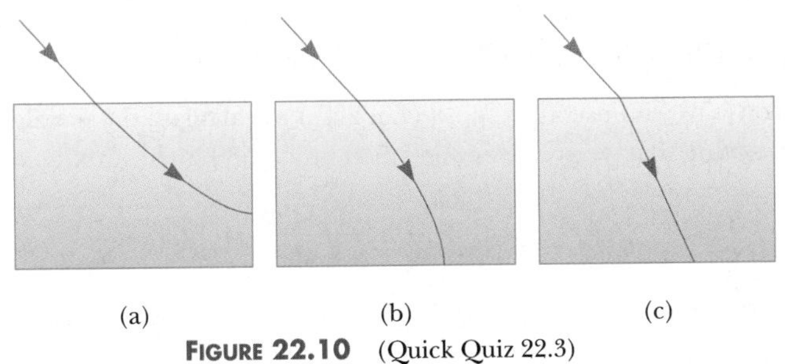


**FIGURE 22.10** (Quick Quiz 22.3)

**Quick Quiz 22.4** As light travels from vacuum ($n = 1$) to a medium such as glass ($n > 1$), which of the following properties remains the same: (a) wavelength, (b) wave speed, or (c) frequency?

## Example 22.2 Angle of Refraction for Glass

A light ray of wavelength 589 nm (produced by a sodium lamp) traveling through air is incident on a smooth, flat slab of crown glass at an angle of 30.0° to the normal, as sketched in Figure 22.11. Find the angle of refraction, $\theta_2$.

**Solution** Snell's law (Eq. 22.8) can be rearranged as

$$\sin \theta_2 = \frac{n_1}{n_2} \sin \theta_1$$

From Table 22.1, we find that $n_1 = 1.00$ for air and $n_2 = 1.52$ for crown glass. Therefore, the unknown refraction angle is determined by

$$\sin \theta_2 = \left(\frac{1.00}{1.52}\right)(\sin 30.0°) = 0.329$$

$$\theta_2 = \sin^{-1}(0.329) = 19.2°$$

We see that the ray is bent *toward* the normal, as expected.

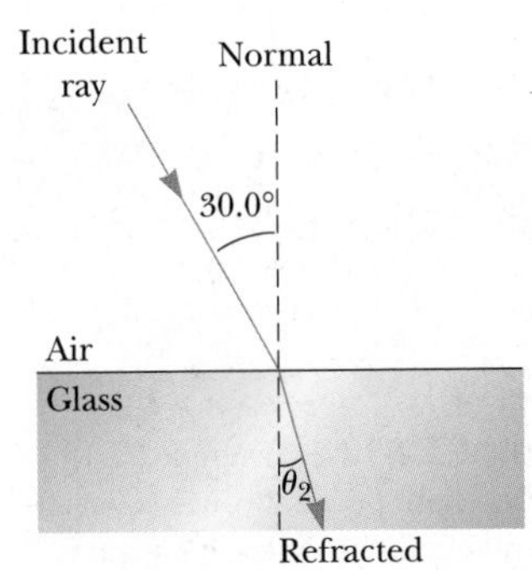


**FIGURE 22.11** (Example 22.2) Refraction of light by glass.

**EXERCISE** If the light ray moves from inside the glass toward the glass-air interface at an angle of 30.0° to the normal, determine the angle of refraction.

**ANSWER** 49.5° *away* from the normal

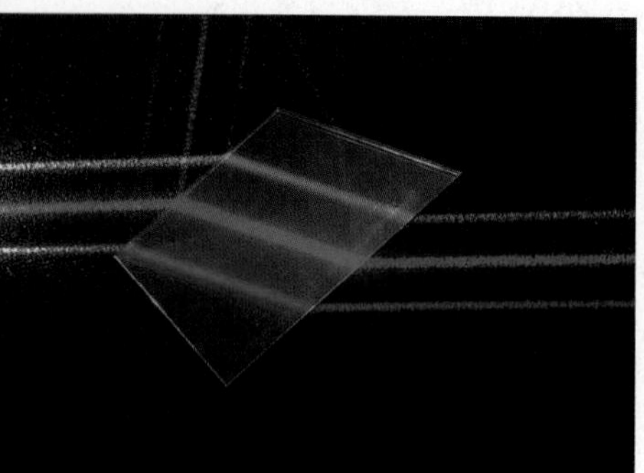

A transparent block has refracted three incident light beams, which exit parallel to each other but are offset from their original direction of travel. Note the partial reflection at both interfaces. *(© Leonard Lessin/Peter Arnold, Inc.)*

## Example 22.3 The Speed of Light in Fused Quartz

Light of wavelength 589 nm in vacuum passes through a piece of fused quartz of index of refraction $n = 1.458$.

**A** Find the speed of light in fused quartz.

**Solution** The speed of light in fused quartz can be obtained from Equation 22.4:

$$v = \frac{c}{n} = \frac{3.00 \times 10^8 \text{ m/s}}{1.458} = 2.06 \times 10^8 \text{ m/s}$$

It is interesting to note that the speed of light in vacuum, $3.00 \times 10^8$ m/s, is an upper limit for the speed of material objects. In our treatment of relativity in Chapter 26, we shall find that this upper limit is consistent with experimental observations. However, it is possible for a particle moving in a medium to have a speed that exceeds the speed of light in that medium. For example, it is theoretically possible for a particle to travel through fused quartz at a speed greater than $2.06 \times 10^8$ m/s, but it must have a speed less than $3.00 \times 10^8$ m/s.

**B** What is the wavelength of this light in fused quartz?

**Solution** We can use $\lambda_n = \lambda_0/n$ (Eq. 22.7) to calculate the wavelength in fused quartz, noting that we are given $\lambda_0 = 589 \text{ nm} = 589 \times 10^{-9}$ m:

$$\lambda_n = \frac{\lambda_0}{n} = \frac{589 \text{ nm}}{1.458} = 404 \text{ nm}$$

**EXERCISE** Find the frequency of the light passing through the fused quartz.

**ANSWER** $5.09 \times 10^{14}$ Hz

### Webnote 22.1

Scientists have been able to slow down the speed of light to as low as 38 mph! See how this was achieved at *http://ajanta.sci.ccny.cuny.edu/~jupiter/pub/sciinfo/slowlight.html*

## Example 22.4 Light Passing Through a Slab

A light beam traveling through a transparent medium of index of refraction $n_1$ passes through a thick transparent slab with parallel faces and index of refraction $n_2$ (Fig. 22.12). Show that the emerging beam is parallel to the incident beam.

**Reasoning** To solve this problem, it is necessary to apply Snell's law twice, once at the upper surface and once at the lower surface. The two equations will be related because the angle of refraction at the upper surface equals the angle of incidence at the lower surface. The ray passing through the slab makes equal angles with the normal at the entry and exit points. This procedure will enable us to compare angles $\theta_1$ and $\theta_3$.

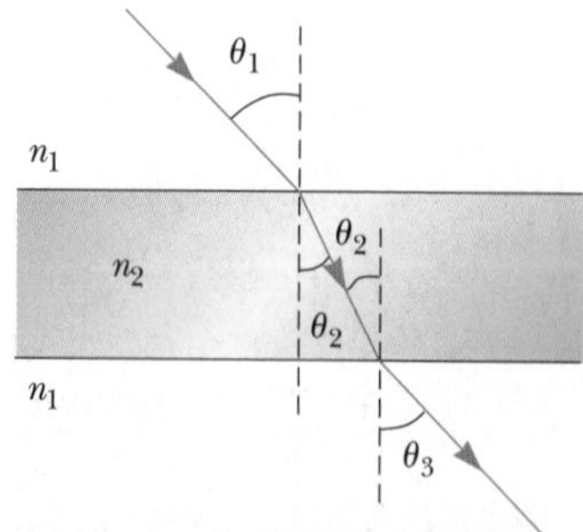


**FIGURE 22.12** (Example 22.4) When light passes through a flat slab of material, the emerging beam is parallel to the incident beam, and therefore $\theta_1 = \theta_3$.

**Solution** First, let us apply Snell's law to the upper surface:

$$\sin \theta_2 = \frac{n_1}{n_2} \sin \theta_1 \quad (1)$$

Applying Snell's law to the lower surface gives

$$\sin \theta_3 = \frac{n_2}{n_1} \sin \theta_2 \quad (2)$$

Substituting (1) into (2) gives

$$\sin \theta_3 = \frac{n_2}{n_1}\left(\frac{n_1}{n_2}\sin \theta_1\right) = \sin \theta_1$$

Thus, $\theta_3 = \theta_1$, and so the slab does not alter the direction of the beam as shown by the dashed line in Figure 22.12. It does, however, produce a lateral displacement of the beam.

## Example 22.5 Refraction of Laser Light in a DVD (Digital Video Disk)

A DVD is a video recording consisting of a spiral track of digital information about 1.0 $\mu$m wide (see Fig. 22.13a). The digital information consists of a series of pits that are "read" by a laser beam sharply focused on a track in the information layer. If the width of the laser beam $a$ at the information layer must equal 1.0 $\mu$m to distinguish individual tracks, and the width of the beam $w$ as it enters the plastic is 0.70 mm, find the angle $\theta_1$ at which the conical beam should enter the plastic (see Fig. 22.13b). Assume the plastic has a thickness $t = 1.2$ mm and an index of refraction $n = 1.55$. Note that this system is relatively immune to small dust particles degrading the video quality, because particles would have to be as large as 0.70 mm to obscure the beam at the point where it enters the plastic.

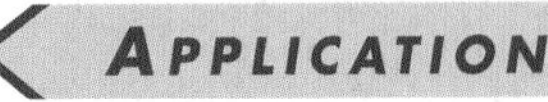


READING DATA ON A DVD

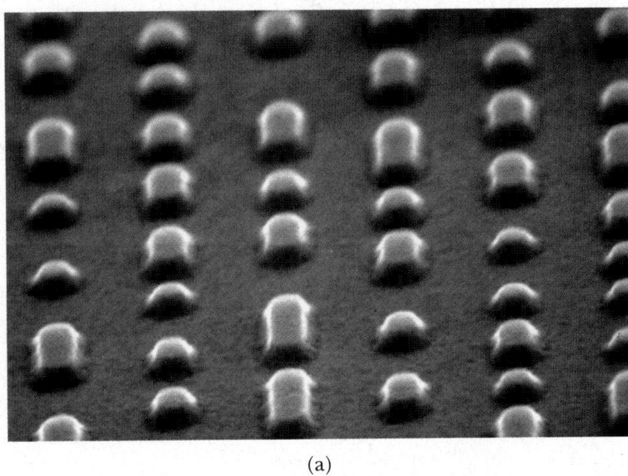

(a)

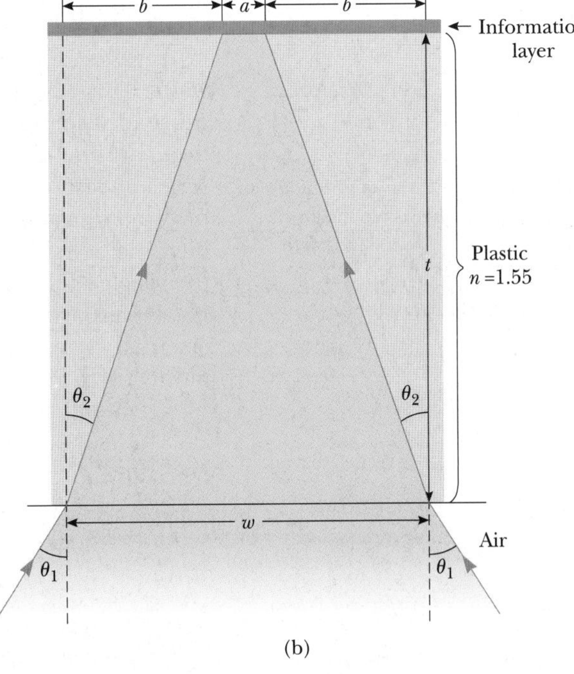


(b)

**FIGURE 22.13** (a) A micrograph of a DVD surface showing tracks and pits along each track. *(Courtesy of Sony Disc Manufacturing)* (b) Cross section of a cone-shaped laser beam used to read a DVD.

**Solution** From Figure 22.13b we see that

$$w = 2b + a$$

$$b = \frac{w - a}{2} = \frac{699\ \mu\text{m}}{2} = 349.5\ \mu\text{m}$$

Also,

$$\tan \theta_2 = \frac{b}{t} = \frac{349.5\ \mu\text{m}}{1200\ \mu\text{m}}; \qquad \theta_2 = 16.2°$$

Finally we can use Snell's law to find $\theta_1$.

$$n_1 \sin \theta_1 = n_2 \sin \theta_2$$

$$\sin \theta_1 = \frac{n_2 \sin \theta_2}{n_1} = \frac{1.55 \sin 16.2°}{1.00} = 0.433$$

$$\theta_1 = 26°$$

## 22.5 DISPERSION AND PRISMS

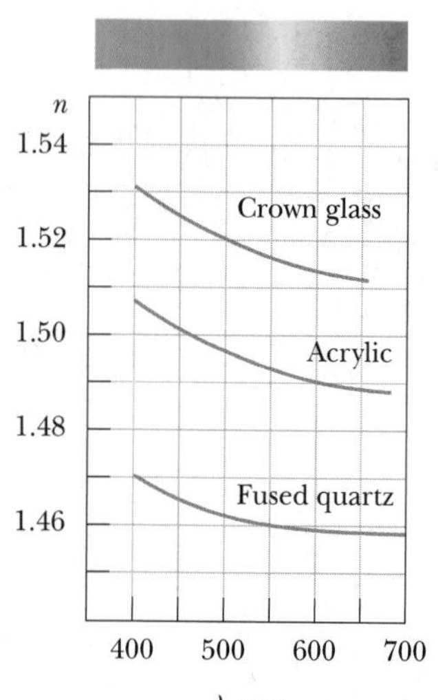


**FIGURE 22.14** Variations of index of refraction in the visible spectrum with respect to vacuum wavelength for three materials.

In Table 22.1, we presented the index of refraction values for various materials. If we make careful measurements, however, we find that the index of refraction in anything but vacuum depends on the wavelength of light. The dependence of the index of refraction on wavelength is called **dispersion.** Figure 22.14 is a graphical representation of this variation in index of refraction with wavelength. Because $n$ is a function of wavelength, Snell's law indicates that **the angle of refraction when light enters a material depends on the wavelength of the light.** As seen in Figure 22.14, the index of refraction for a material usually decreases with increasing wavelength. This means that violet light ($\lambda \cong 400$ nm) refracts more than red light ($\lambda \cong 650$ nm) when passing from air into a material.

To understand the effects of dispersion on light, consider what happens when light strikes a prism, as in Figure 22.15a. A ray of light of a single wavelength that is incident on the prism from the left emerges bent away from its original direction of travel by an angle $\delta$, called the **angle of deviation.** Now suppose a beam of white light (a combination of all visible wavelengths) is incident on a prism. Because of dispersion, the different colors refract through different angles of deviation, and the rays that emerge from the second face of the prism spread out in a series of colors known as a visible **spectrum,** as shown in Figure 22.16. These colors, in order of decreasing wavelength, are red, orange, yellow, green, blue, and violet. Violet light deviates the most, red light the least, and the remaining colors in the visible spectrum fall between these extremes.

Prisms are often used in an instrument known as a **prism spectrometer,** the essential elements of which are shown in Figure 22.17a. This instrument is commonly used to study the wavelengths emitted by a light source, such as a sodium

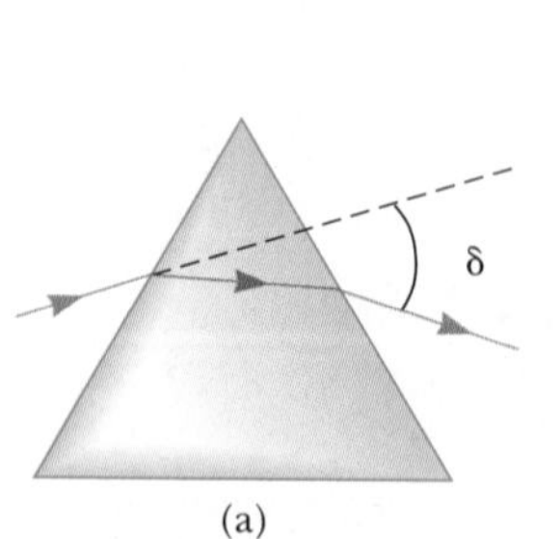


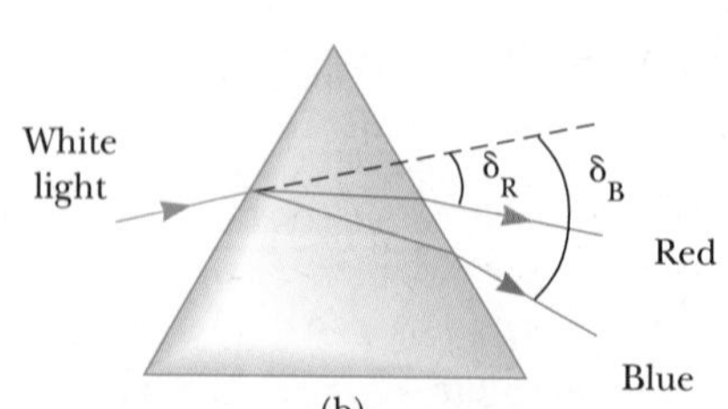


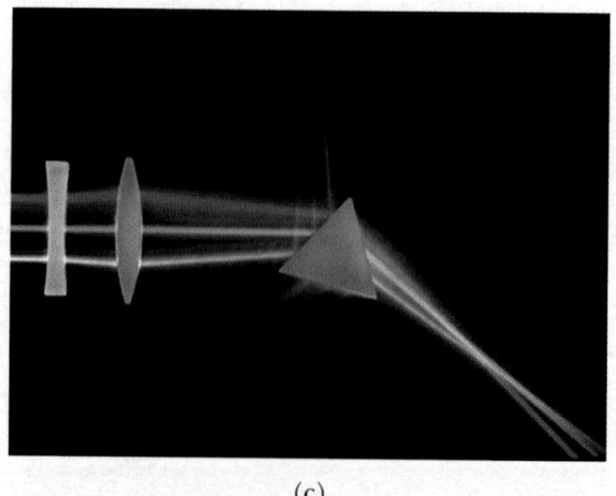


**FIGURE 22.15** (a) A prism refracts a light ray and deviates the light through the angle $\delta$. (b) When light is incident on a prism, the blue light is bent more than the red. (c) Light of different colors passes through a prism and two lenses. Note that as the light passes through the prism, different wavelengths are refracted at different angles. *(David Parker/SPL/Photo Researchers)*

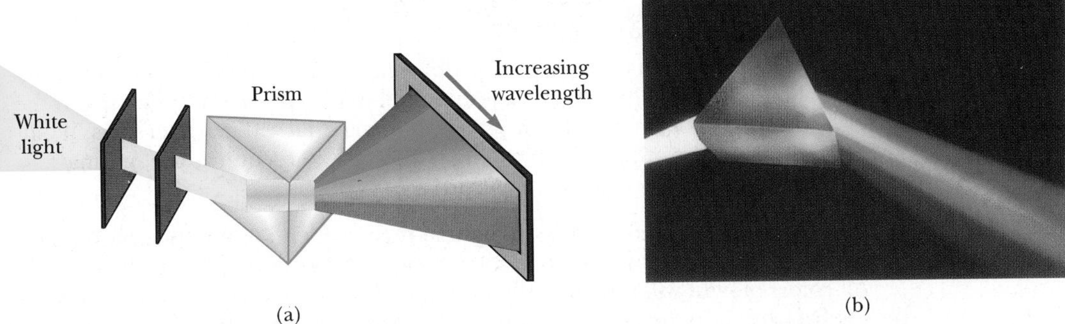


**FIGURE 22.16** (a) Dispersion of white light by a prism. Since $n$ varies with wavelength, the prism disperses the white light into its various spectral components. (b) Different colors of light that pass through a prism are refracted at different angles because the index of refraction of the glass depends on wavelength. The violet light bends the most; red light bends the least. *(b, courtesy of Bausch & Lomb)*

vapor lamp. Light from the source is sent through a narrow, adjustable slit and lens to produce a parallel, or collimated, beam. The light then passes through the prism and is dispersed into a spectrum. The refracted light is observed through a telescope. The experimenter sees different colored images of the slit through the eyepiece of the telescope. The telescope can be moved or the prism can be rotated in order to view the various wavelengths, which have different angles of deviation. Figure 22.17b shows one type of prism spectrometer used in undergraduate laboratories.

**APPLICATION**

IDENTIFYING GASES USING A SPECTROMETER

All hot, low-pressure gases emit their own characteristic spectra. Thus, one use of a prism spectrometer is to identify gases. For example, sodium emits only two wavelengths in the visible spectrum: two closely spaced yellow lines. (The bright line-like images of the slit seen in a spectroscope are called *spectral lines.*) A gas emitting these and only these colors can thus be identified as sodium. Likewise, mercury vapor has its own characteristic spectrum, consisting of four prominent wavelengths—orange, green, blue, and violet lines—along with some wavelengths of lower intensity. The particular wavelengths emitted by a gas serve as "fingerprints" of that gas. Spectral analysis, which is the measurement of the wavelengths emitted or absorbed by a substance, has proven to be a powerful general tool in many areas: Chemists and biologists use infrared spectroscopy to identify molecules, astronomers use visible light spectroscopy to identify elements on distant stars, and geologists use spectral analysis to identify minerals.

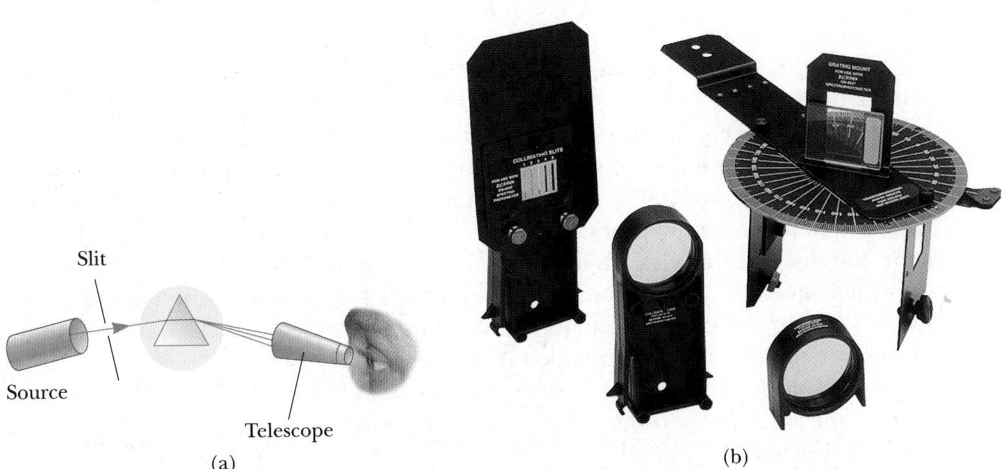


**FIGURE 22.17** (a) A diagram of a prism spectrometer. The colors in the spectrum are viewed through a telescope. (b) A prism spectrometer with interchangeable components. *(Courtesy of PASCO Scientific)*

## APPLYING PHYSICS 22.3

When a beam of light enters a glass prism, which has nonparallel sides, the rainbow of color exiting the prism is a testimonial to the dispersion occurring in the glass. Suppose a beam of light enters a slab of material with parallel sides. When the beam exits the other side, traveling in the same direction as the original beam, is there any evidence of dispersion?

**Explanation** Due to dispersion, light at the violet end of the spectrum will exhibit a larger angle of refraction on entering the glass than light at the red end. All colors of light will return to the original direction of propagation on refracting back out into the air. Thus, the outgoing beam will be white. But the net shift in the position of the violet light along the edge of the slab will be larger than that of the red light. Thus, one edge of the outgoing beam will have a bluish tinge (it will appear blue rather than violet, because the eye is not very sensitive to violet light), whereas the other edge will have a reddish tinge. This effect is indicated in Figure 22.18. The colored edges of the outgoing beam of white light are evidence of dispersion.

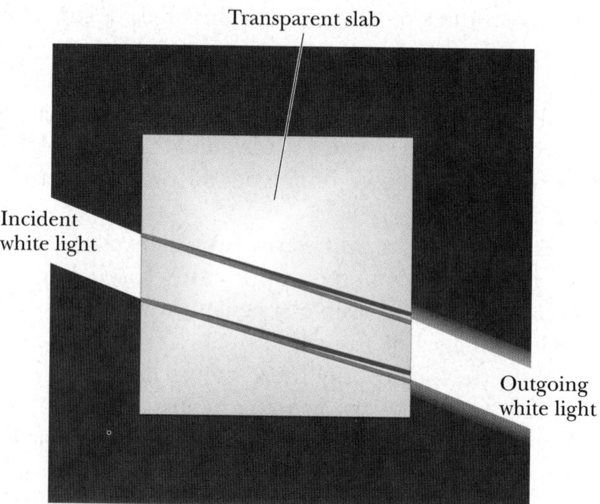


**FIGURE 22.18** (Applying Physics 22.3)

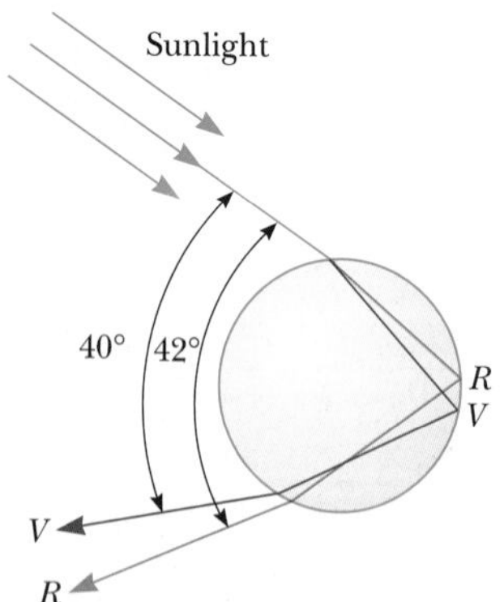


**FIGURE 22.19** Refraction of sunlight by a spherical raindrop.

**Webnote 22.2**
See what makes the bow in a rainbow. Go to
*http://www.unidata.ucar.edu/staff/blynds/rnbw.html*

## 22.6 THE RAINBOW

The dispersion of light into a spectrum is demonstrated most vividly in nature through the formation of a rainbow, often seen by an observer positioned between the Sun and a rain shower. To understand how a rainbow is formed, consider Figure 22.19. A ray of light passing overhead strikes a drop of water in the atmosphere and is refracted and reflected as follows. It is first refracted at the front surface of the drop, with the violet light deviating the most and the red light the least. At the back surface of the drop, the light is reflected and returns to the front surface, where it again undergoes refraction as it moves from water into air. The rays leave the drop so that the angle between the incident white light and the returning violet ray is 40°, and the angle between the white light and the returning red ray is 42° (Fig. 22.19). This small angular difference between the returning rays causes us to see the bow as explained in the next paragraph.

Now consider an observer viewing a rainbow, as in Figure 22.20a. If a raindrop high in the sky is being observed, the red light returning from the drop can reach the observer because it is deviated the most, but the violet light passes over the observer because it is deviated the least. Hence, the observer sees this drop as being red. Similarly, a drop lower in the sky would direct violet light toward the observer and appear to be violet. (The red light from this drop would strike the ground

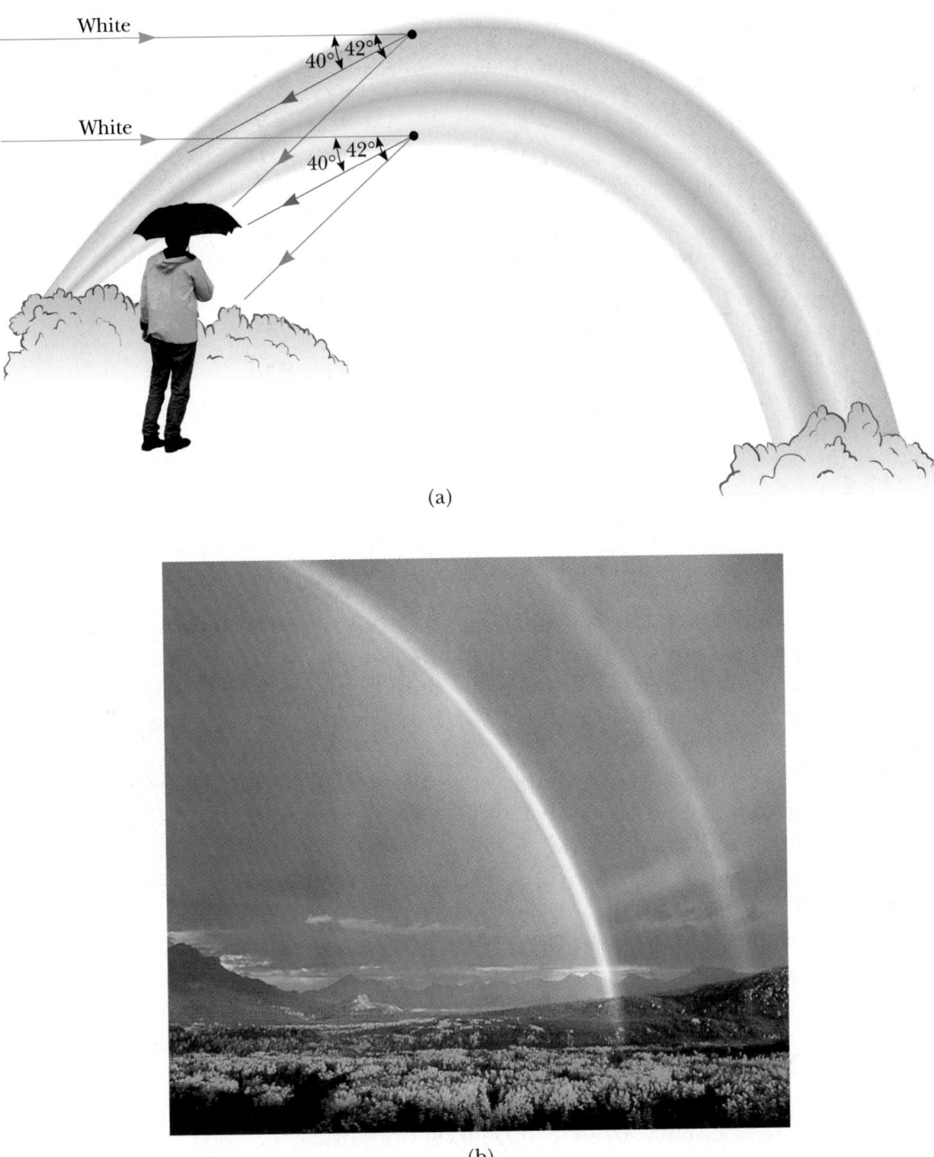


**FIGURE 22.20** (a) The formation of a rainbow. (b) This photograph of a rainbow shows a distinct secondary rainbow with the colors reversed. *(Mark D. Phillips/Photo Researchers, Inc.)*

and not be seen.) The remaining colors of the spectrum would reach the observer from raindrops lying between these two extreme positions. Figure 22.20b shows a beautiful rainbow.

## 22.7 HUYGENS'S PRINCIPLE

The laws of reflection and refraction can be deduced using a geometric method proposed by Huygens in 1678. Huygens assumed that light is a form of wave motion rather than a stream of particles. He had no knowledge of the nature of light or of its electromagnetic character. Nevertheless, his simplified wave model is adequate for understanding many practical aspects of the propagation of light.

Huygens's principle is a geometric construction for determining at some instant the position of a new wave front from the knowledge of the wave front that preceded it. (A wave front is a surface passing through those points of a wave that have the same phase and amplitude. For instance, a wave front could be a surface passing through the crests of waves.) In Huygens's construction, **all points on a**

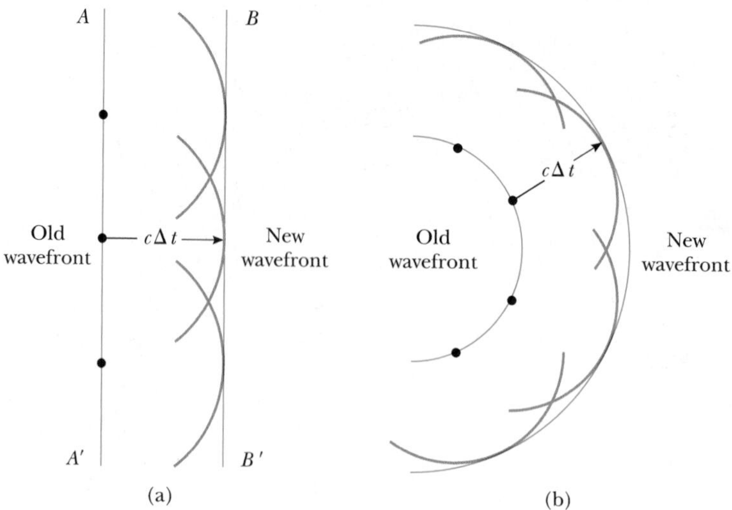


**FIGURE 22.21** Huygens's constructions for (a) a plane wave propagating to the right and (b) a spherical wave.

Huygens's principle

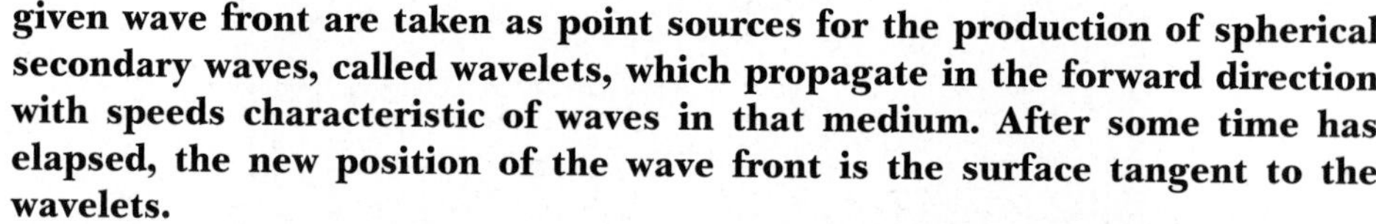
**given wave front are taken as point sources for the production of spherical secondary waves, called wavelets, which propagate in the forward direction with speeds characteristic of waves in that medium. After some time has elapsed, the new position of the wave front is the surface tangent to the wavelets.**

Figure 22.21 illustrates two simple examples of Huygens's construction. First, consider a plane wave moving through free space, as in Figure 22.21a. At $t = 0$, the wave front is indicated by the plane labeled $AA'$. In Huygens's construction, each point on this wave front is considered a point source. For clarity, only a few points on $AA'$ are shown. With these points as sources for the wavelets, we draw circles each of radius $c\,\Delta t$, where $c$ is the speed of light in vacuum and $\Delta t$ is the period of propagation from one wave front to the next. The surface drawn tangent to these wavelets is the plane $BB'$, which is parallel to $AA'$. In a similar manner, Figure 22.21b shows Huygens's construction for an outgoing spherical wave.

Figure 22.22 shows a convincing demonstration of Huygens's principle. Plane waves coming from far off shore emerge from the openings between the barriers as two-dimensional circular waves propagating outward.

**FIGURE 22.22** This photograph of the beach at Tel Aviv, Israel, shows Huygens wavelets radiating from each opening between breakwalls. Note how the beach has been shaped by the wave action. *(Courtesy of Sabina Zigman/Benjamin Cardozo High School, and by permission of PHYSICS TEACHER, vol. 37, January 1999, p. 55)*

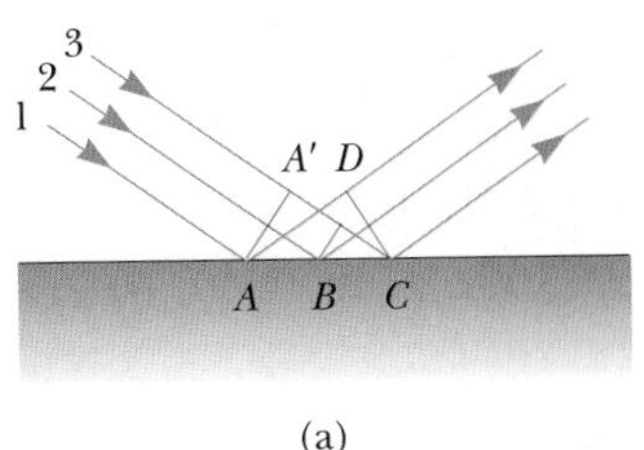


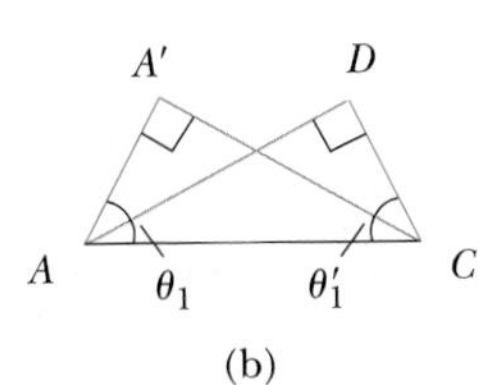


**FIGURE 22.23** (a) Huygens's construction for proving the law of reflection. (b) Triangle *ADC* is identical to triangle *AA′C*.

## HUYGENS'S PRINCIPLE APPLIED TO REFLECTION AND REFRACTION

The laws of reflection and refraction were stated earlier in this chapter without proof. We shall now derive these laws using Huygens's principle. For the law of reflection, refer to Figure 22.23a. The line $AA'$ represents a wave front of the incident light. As ray 3 travels from $A'$ to $C$, ray 1 reflects from $A$ and produces a spherical wavelet of radius $AD$. (Recall that the radius of a Huygens wavelet is $v\Delta t$.) Because the two wavelets having radii $A'C$ and $AD$ are in the same medium, they have the same speed, $v$, and thus $AD = A'C$. Meanwhile, the spherical wavelet centered at $B$ has spread only half as far as the one centered at $A$, because ray 2 strikes the surface later than ray 1.

From Huygens's principle, we find that the reflected wave front is $CD$, a line tangent to all the outgoing spherical wavelets. The remainder of our analysis depends on geometry, as summarized in Figure 22.23b. Note that the right triangles $ADC$ and $AA'C$ are congruent because they have the same hypotenuse, $AC$, and because $AD = A'C$. From Figure 22.23b we have

$$\sin\theta_1 = \frac{A'C}{AC} \quad \text{and} \quad \sin\theta_1' = \frac{AD}{AC}$$

Thus,

$$\sin\theta = \sin\theta_1'$$

$$\theta_1 = \theta_1'$$

which is the law of reflection.

Now let us use Huygens's principle and Figure 22.24a to derive Snell's law of refraction. Note that in the time interval $\Delta t$, ray 1 moves from $A$ to $B$ and ray 2 moves from $A'$ to $C$. The radius of the outgoing spherical wavelet centered at $A$ is equal to $v_2\,\Delta t$. The distance $A'C$ is equal to $v_1\,\Delta t$. Geometric considerations show

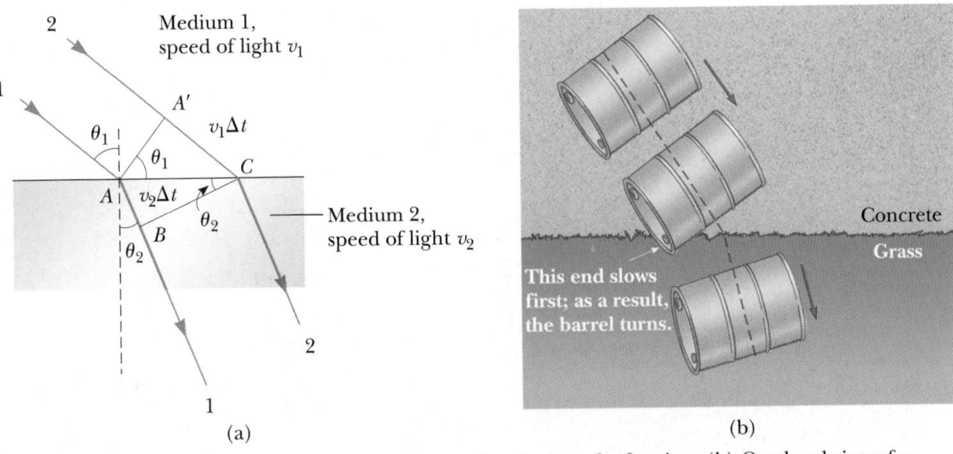


**FIGURE 22.24** (a) Huygens's construction for proving the law of refraction. (b) Overhead view of a barrel rolling from concrete onto grass.

that angle $A'AC$ equals $\theta_1$ and angle $ACB$ equals $\theta_2$. From triangles $AA'C$ and $ACB$ we find that

$$\sin\theta_1 = \frac{v_1\,\Delta t}{AC} \quad \text{and} \quad \sin\theta_2 = \frac{v_2\,\Delta t}{AC}$$

If we divide these two equations, we get

$$\frac{\sin\theta_1}{\sin\theta_2} = \frac{v_1}{v_2}$$

But from Equation 22.4 we know that $v_1 = c/n_1$ and $v_2 = c/n_2$. Therefore,

$$\frac{\sin\theta_1}{\sin\theta_2} = \frac{c/n_1}{c/n_2} = \frac{n_2}{n_1}$$

$$n_1 \sin\theta_1 = n_2 \sin\theta_2$$

which is the law of refraction.

A mechanical analog of refraction is shown in Figure 22.24b. When the left end of the rolling barrel reaches the grass, it slows down, while the right end remains on the concrete and moves at its original speed. This difference in speeds causes the barrel to pivot, and this changes the direction of travel.

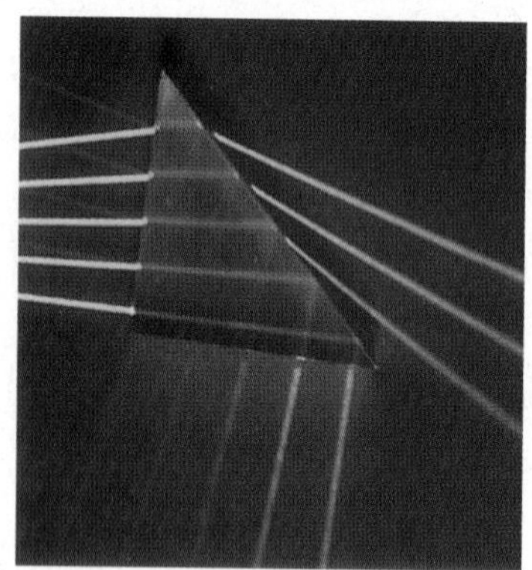

This photograph shows nonparallel light rays entering a glass prism. The bottom two rays undergo total internal reflection at the longest side of the prism. The top three rays are refracted at the longest side as they leave the prism. *(Courtesy of Henry Leap and Jim Lehman)*

## 22.8 TOTAL INTERNAL REFLECTION

An interesting effect called *total internal reflection* can occur when light attempts to move from a medium with a *high* index of refraction to one with a *lower* index of refraction. Consider a light beam traveling in medium 1 and meeting the boundary between medium 1 and medium 2, where $n_1$ is greater than $n_2$ (Fig. 22.25). Possible directions of the beam are indicated by rays 1 through 5. Note that the refracted rays are bent away from the normal because $n_1$ is greater than $n_2$. At some particular angle of incidence $\theta_c$, called the **critical angle,** the refracted light ray moves parallel to the boundary so that $\theta_2 = 90°$ (Fig. 22.25b). *For angles of incidence greater than* $\theta_c$, the beam is entirely reflected at the boundary, as is ray 5 in Figure 22.25a. This ray is reflected at the boundary as though it had struck a perfectly reflecting surface. It and all rays like it obey the law of reflection; that is, the angle of incidence equals the angle of reflection.

We can use Snell's law to find the critical angle. When $\theta_1 = \theta_c$, $\theta_2 = 90°$ and Snell's law (Eq. 22.8) gives

$$n_1 \sin\theta_c = n_2 \sin 90° = n_2$$

$$\sin\theta_c = \frac{n_2}{n_1} \quad \text{for } n_1 > n_2 \qquad [22.9]$$

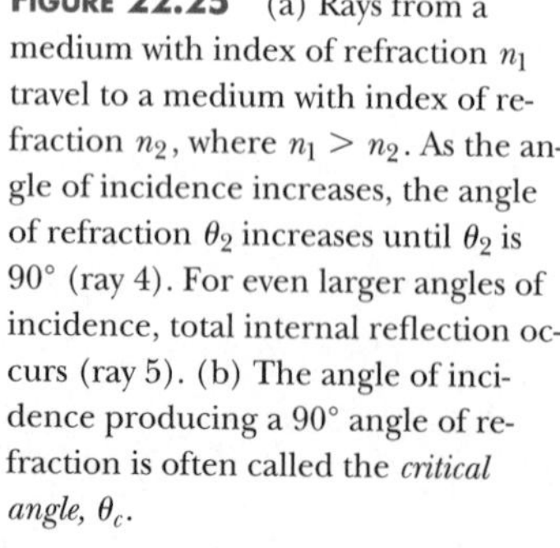

**FIGURE 22.25** (a) Rays from a medium with index of refraction $n_1$ travel to a medium with index of refraction $n_2$, where $n_1 > n_2$. As the angle of incidence increases, the angle of refraction $\theta_2$ increases until $\theta_2$ is 90° (ray 4). For even larger angles of incidence, total internal reflection occurs (ray 5). (b) The angle of incidence producing a 90° angle of refraction is often called the *critical angle*, $\theta_c$.

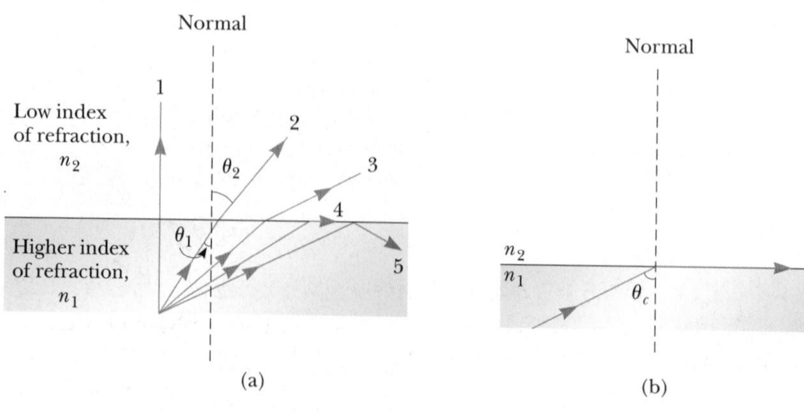

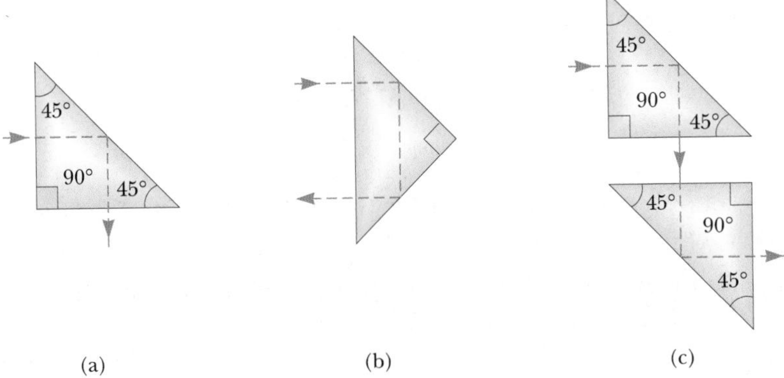


**FIGURE 22.26** Internal reflection in a prism. (a) The ray is deviated by 90°. (b) The direction of the ray is reversed. (c) Two prisms used as a periscope.

Note that this equation can be used only when $n_1$ is greater than $n_2$. That is, **total internal reflection occurs only when light attempts to move from a medium of high index of refraction to a medium of lower index of refraction.** If $n_1$ were less than $n_2$, Equation 22.9 would give $\sin\theta_c > 1$, which is an absurd result because the sine of an angle can never be greater than unity.

When medium 2 is air, the critical angle is small for substances with large indices of refraction, such as diamond, where $n = 2.42$ and $\theta_c = 24.0°$. By comparison, for crown glass, $n = 1.52$ and $\theta_c = 41.0°$. This property, combined with proper faceting, causes a diamond to sparkle brilliantly.

One can use a prism and the phenomenon of total internal reflection to alter the direction of travel of a light beam. Figure 22.26 illustrates two such possibilities. In one case the light beam is deflected by 90.0° (Fig. 22.26a), and in the second case the path of the beam is reversed (Fig. 22.26b). A common application of total internal reflection is a submarine periscope. In this device, two prisms are arranged, as in Figure 22.26c so that an incident beam of light follows the path shown and the user can "see around corners."

**APPLICATION**

SUBMARINE PERISCOPES

## APPLYING PHYSICS 22.4

A beam of white light is incident on the curved edge of a semicircular piece of glass, as shown in Figure 22.27. The light enters the curved surface along the normal, so it shows no refraction. It encounters the straight side at the center of curvature of the curved side and refracts into the air. The incoming beam is moved clockwise (so that the angle $\theta$ increases) such that the beam always enters along the normal to the curved side and encounters the straight side at the center of curvature of the curved side. As the refracted beam approaches a direction parallel to the straight side, it becomes redder. Why?

**Explanation** When the outgoing beam approaches the direction parallel to the straight side, the incident angle is approaching the critical angle for total internal reflection. Dispersion occurs as the light passes out of the glass. The index of refraction for light at the violet end of the visible spectrum is larger than at the red end. Thus, as the outgoing beam approaches the straight side, the violet light experiences total internal reflection first, followed by the other colors. The red light is the last to experience total internal reflection, so just before the outgoing light disappears, it is composed of light from the red end of the visible spectrum.

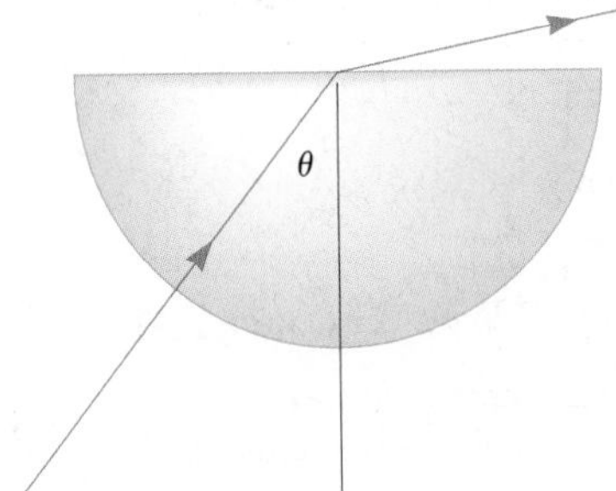


**FIGURE 22.27** (Applying Physics 22.4)

### Example 22.6 A View from the Fish's Eye

**A** Find the critical angle for a water-air boundary if the index of refraction of water is 1.33.

### Webnote 22.3

See what total internal reflection would look like beneath the surface of a pond. Go to *http://www.lightlink.com/sergey/java/java/totintrefl/index.html*

**Solution** Applying Equation 22.9, we find the critical angle to be

$$\sin \theta_c = \frac{n_2}{n_1} = \frac{1.00}{1.33} = 0.752$$

$$\theta_c = 48.8°$$

**B** Use the results of (part A) to predict what a fish will see if it looks up toward the water surface at angles of 40.0°, 48.8°, and 60.0°.

**Reasoning and Solution** Because the path of a light ray is reversible, the fish can see out of the water if it looks toward the surface at an angle less than the critical angle. Thus, at 40.0°, the fish can see into the air above the water. At an angle of 48.8°, the critical angle for water, the light that reaches the fish has to skim along the water surface before being refracted to the fish's eye. At angles greater than the critical angle, the light reaching the fish comes via internal reflection at the surface. Thus, beyond 60.0°, the fish sees a reflection of some object on the bottom of the pool.

This finger tip is being illuminated by a bundle of optical fibers. The light source is not shown in the photograph. (*© Novastock/Photo Researchers, Inc.*)

## FIBER OPTICS

Another interesting application of total internal reflection is the use of solid glass or transparent plastic rods to "pipe" light from one place to another. As indicated in Figure 22.28, light is confined to traveling within the rods, even around gentle curves, as a result of successive internal reflections. Such a light pipe can be quite flexible if thin fibers are used rather than thick rods. If a bundle of parallel fibers is used to construct an optical transmission line, images can be transferred from one point to another.

**APPLICATION**

FIBER OPTICS IN MEDICAL DIAGNOSIS AND SURGERY

Very little light intensity is lost in these fibers as a result of reflections on the sides. Any loss of intensity is due essentially to reflections from the two ends and absorption by the fiber material. Fiber-optic devices are particularly useful for viewing images produced at inaccessible locations. Physicians often use fiber-optic cables to aid in the diagnosis and correction of certain medical problems without the intrusion of major surgery. For example, a fiber-optic cable can be threaded through the esophagus and into the stomach to look for ulcers. In this application, the cable actually consists of two fiber-optic lines, one to transmit a beam of light into the stomach for illumination and the other to allow this light to be transmitted out of the stomach. The resulting image can, in some cases, be viewed directly by the physician but most often is displayed on a television monitor or captured on film. In a similar fashion, the cables can be used to examine the colon or to do repair work without the need for large incisions.

**APPLICATION**

FIBER OPTICS IN TELECOMMUNICATIONS

Most important, the field of fiber optics has revolutionized the entire communications industry. Billions of kilometers of optical fiber have been installed in the United States to carry high-speed Internet traffic, radio and television signals, and telephone calls. The fibers can carry much higher volumes of telephone calls and other forms of communication than electrical wires because of the higher frequency of the infrared light used to carry the information on optical fibers. Optical fibers are also preferable to copper wires because they are insulators and do not pick up stray electric and magnetic fields or electronic "noise."

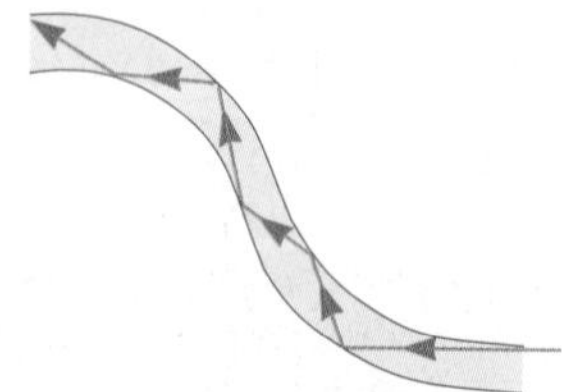

**FIGURE 22.28** Light travels in a curved transparent rod by multiple internal reflections.

### APPLYING PHYSICS 22.5

An optical fiber consists of a transparent core, surrounded by cladding, which is a material with a lower index of refraction than the core. There is a cone of angles, called the *acceptance cone,* at the entrance to the fiber. Incoming light at angles within this cone will be transmitted through the fiber, whereas light entering the core from angles outside the cone will not be transmitted. Figure 22.29 shows a light ray entering the fiber just within the acceptance cone and experiencing total internal reflection at the interface between the core and the cladding. If it is technologically difficult to produce light entering the fiber from a small range of angles, how could you adjust the indices of refraction of the core and cladding to increase the size of the acceptance cone—would you design them to be farther apart or closer together?

**APPLICATION**

DESIGN OF AN OPTICAL FIBER

**Explanation** The acceptance cone would become larger if the critical angle ($\theta_c$ in the diagram) could be made smaller. This can be done by making the index of refraction of the cladding material smaller, so that the indices of refraction of the core and cladding material would be farther apart.

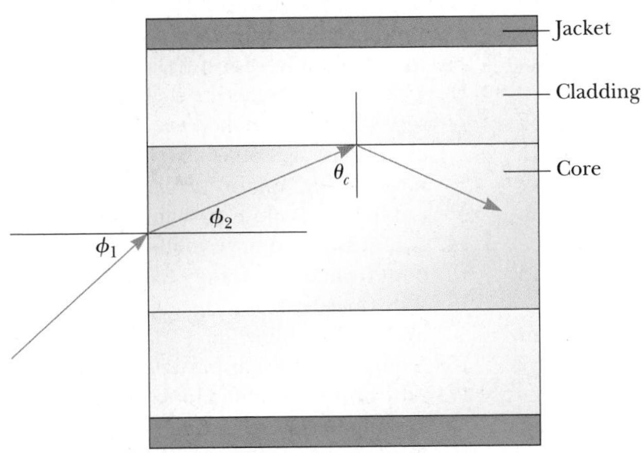


**FIGURE 22.29** (Applying Physics 22.5)

## SUMMARY

The **index of refraction** of a material, $n$, is defined as

$$n \equiv \frac{c}{v} \qquad [22.4]$$

where $c$ is the speed of light in a vacuum and $v$ is the speed of light in the material. The index of refraction of a material is also

$$n = \frac{\lambda_0}{\lambda_n} \qquad [22.7]$$

where $\lambda_0$ is the wavelength of the light in vacuum and $\lambda_n$ is its wavelength in the material.

The **law of reflection** states that a wave reflects from a surface so that the *angle of reflection* $\theta_1'$ equals the *angle of incidence* $\theta_1$.

The **law of refraction,** or **Snell's law,** states that

$$n_1 \sin \theta_1 = n_2 \sin \theta_2 \qquad [22.8]$$

where $n_1$ and $n_2$ are the indices of refraction in the two media. The incident ray, the reflected ray, the refracted ray, and the normal to the surface all lie in the same plane.

**Huygens's principle** states that all points on a wave front are point sources for the production of spherical secondary waves called wavelets. These wavelets propagate forward at a speed characteristic of waves in a particular medium. After some time has elapsed, the new position of the wave front is the surface tangent to the wavelets.

Total internal reflection can occur when light attempts to move from a material with a high index of refraction to one with a lower index of refraction. The *maximum angle of incidence* $\theta_c$ for which light can move from a medium with index $n_1$ into a medium with index $n_2$, where $n_1$ is greater than $n_2$, is called the **critical angle** and is given by

$$\sin \theta_c = \frac{n_2}{n_1} \quad \text{for } n_1 > n_2 \qquad [22.9]$$

## CONCEPTUAL QUESTIONS

1. Under certain circumstances, sound can be heard from extremely far away. This frequently happens over a body of water, where the air near the water surface is cooler than the air at higher altitudes. Explain how the refraction of sound waves could increase the distance over which sound can be heard.
2. What are some reasons that most ceilings are made of white textured material?
3. The color of an object is said to depend on wavelength. So, if you view colored objects under water, in which the wavelength of the light will be different, does the color change?
4. How is it possible that a complete circle of a rainbow can sometimes be seen from an airplane?
5. Pass white light through a prism to produce a multicolored spectrum. Let that spectrum fall on a piece of cardboard that has a slit in it such that only green light passes through. Now let this green light fall on a second prism. Describe the light that comes out of the second prism.
6. Why does the arc of a rainbow appear with red on top and violet on the bottom?
7. A scientific supply catalog advertises a material having an index of refraction of 0.85. Is this a good product to buy? Why or why not?
8. Under what conditions is a mirage formed? On a hot day, what are we seeing when we observe a mirage water puddle on the road?
9. In dispersive materials, the angle of refraction for a light ray depends on the wavelength of the light. Does the angle of reflection from the surface of the material depend on the wavelength? Why or why not?
10. A type of mirage called a *pingo* is observed often in Alaska. These occur when the light from a small hill passes to an observer by a path that takes the light over a body of water warmer than the air. What is seen is the hill and an inverted image directly below it. Explain how these mirages are formed.
11. Explain why a diamond loses most of its sparkle when submerged in carbon disulfide.
12. Suppose you are told that only two colors of light (X and Y) are sent through a glass prism and that X is bent more than Y. Which color travels more slowly in the prism?
13. The level of water in a clear, colorless glass is easily observed with the naked eye. The level of liquid helium in a clear glass vessel is extremely difficult to see with the naked eye. Explain. (*Hint:* The index of refraction of liquid helium is close to that of air.)
14. Is it possible to have total internal reflection for light incident from air on water? Explain.
15. Why does a diamond show flashes of color when observed under white light?
16. Explain why an oar partially in water appears to be bent.
17. Why do astronomers looking at distant galaxies talk about looking backward in time?
18. If a beam of light with a given cross section enters a new medium, the cross section of the refracted beam is different from the incident beam. Is it larger or smaller, or is there no definite direction to the change?

## PROBLEMS

1, 2, 3 = straightforward, intermediate, challenging □ = full solution available in Student Solutions Manual/Study Guide

**web** = solution posted at **http://info.brookscole.com/serway** = biomedical application

### Section 22.1 The Nature of Light

1. During the Apollo XI Moon landing, a retroreflecting panel was erected on the Moon's surface. The speed of light may be found by measuring the time it takes a laser beam to travel from Earth, reflect from the panel, and return to Earth. If this interval is measured to be 2.51 s, what is the measured speed of light? Take the center-to-center distance from Earth to the Moon to be $3.84 \times 10^8$ m. Assume the Moon is directly overhead and do not neglect the sizes of Earth and the Moon.
2. Figure P22.2 shows the apparatus used by Armand H. L. Fizeau (1819–1896) to measure the speed of light. The basic idea is to measure the total time it takes light to travel from some point to a distant mirror and back. If $d$ is the distance between the light source and the mirror and if the transit time for one round trip is $t$, then the speed of light is $c = 2d/t$. To measure the transit time, Fizeau used a rotating toothed wheel, which converts an otherwise continuous beam of light to a series of light pulses. The rotation of the wheel controls what an observer at the light source sees. For example, assume that the toothed wheel of the Fizeau experiment has 360 teeth and is rotating at a speed of 27.5 rev/s when the light from the source is extinguished—that is, when a burst of light passing through opening $A$ in Figure P22.2 is blocked by tooth $B$ on return. If the distance to the mirror is 7 500 m, find the speed of light.

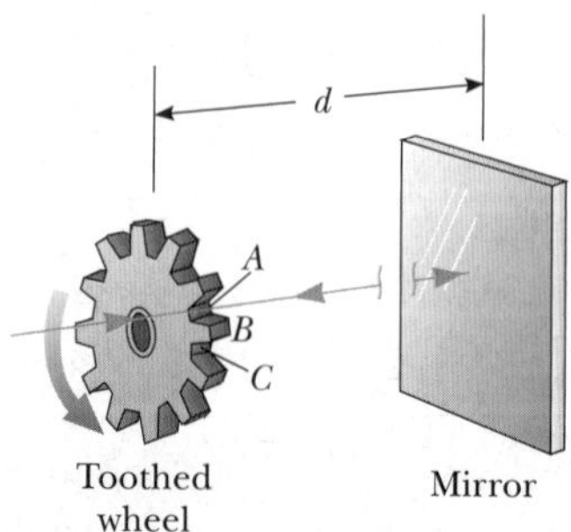


**FIGURE P22.2** (Problems 2 and 3)

3. In an experiment to measure the speed of light using the apparatus of Fizeau described in the preceding problem, the distance between light source and mirror was 11.45 km and the wheel had 720 notches. The experimen-

tally determined value of $c$ was $2.998 \times 10^8$ m/s. Calculate the minimum angular speed of the wheel for this experiment.

4. Albert A. Michelson very carefully measured the speed of light using an alternative version of the technique developed by Fizeau. (See Problem 2.) Figure P22.4 shows the approach he used. Light was reflected from one face of a rotating eight-sided mirror toward a stationary mirror 35.0 km away. At certain rates of rotation, the returning beam of light was directed toward the eye of an observer as shown. (a) What minimum angular speed must the rotating mirror have in order that side A will have rotated to position B, causing the light to be reflected to the eye? (b) What is the next-higher angular velocity that will enable the source of light to be seen?

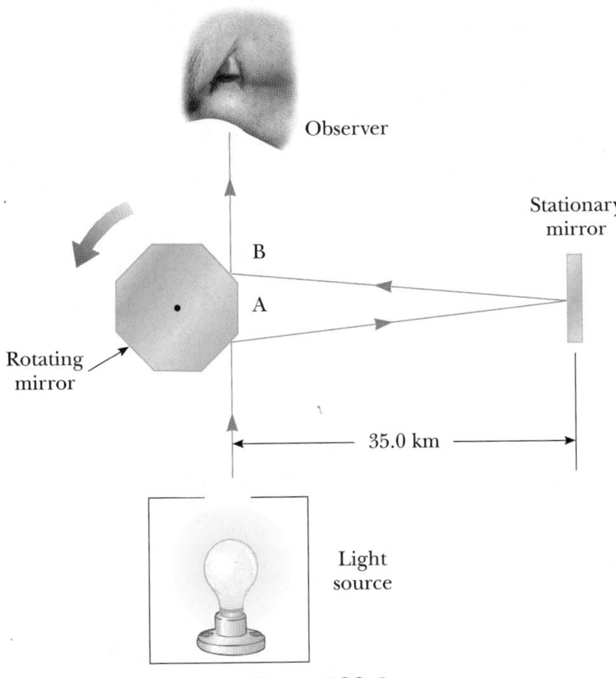


**FIGURE P22.4**

5. Figure P22.5 shows an apparatus used to measure the speed distribution of gas molecules. It consists of two slotted rotating disks separated by a distance $d$, with the slots displaced by the angle $\theta$. Suppose the speed of light is measured by sending a light beam toward the left disk of this apparatus. (a) Show that a light beam will be seen in the detector (that is, will make it through both slots) only if its speed is given by $c = \omega d/\theta$, where $\omega$ is the angular speed of the disks and $\theta$ is measured in radians. (b) What is the measured speed of light if the distance between the two slotted rotating disks is 2.500 m, the slot in the second disk is displaced 1/60 of one degree from the slot in the first disk, and the disks are rotating at 5 555 rev/s?

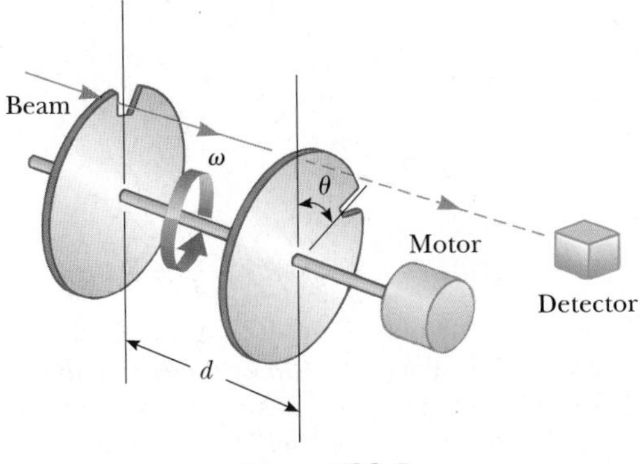


**FIGURE P22.5**

## Section 22.3 Reflection and Refraction

## Section 22.4 The Law of Refraction

6. The two mirrors in Figure P22.6 meet at a right angle. The beam of light in the vertical plane $P$ strikes mirror 1 as shown. (a) Determine the distance the reflected light beam travels before striking mirror 2. (b) In what direction does the light beam travel after being reflected from mirror 2?

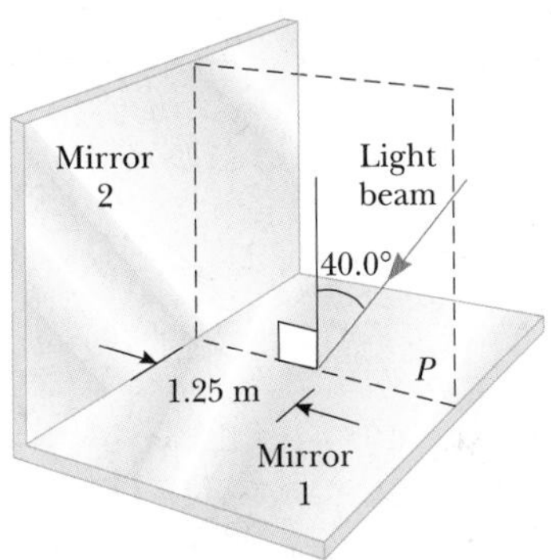


**FIGURE P22.6**

7. An underwater scuba diver sees the Sun at an apparent angle of 45.0° from the vertical. What is the actual direction of the Sun?
8. Light is incident normally on a 1.00-cm layer of water that lies on top of a flat Lucite plate with a thickness of 0.500 cm. How much more time is required for light to pass through this double layer than is required to traverse the same distance in air ($n_{\text{Lucite}} = 1.59$)?
9. A laser beam is incident at an angle of 30.0° to the vertical onto a solution of corn syrup in water. If the beam is refracted to 19.24° to the vertical, (a) what is the index of refraction of the syrup solution? Suppose the light is red, with vacuum wavelength 632.8 nm. Find its (b) wavelength, (c) frequency, and (d) speed in the solution.
10. Find the speeds of light in (a) flint glass, (b) water, and (c) zircon.
11. Light of wavelength $\lambda_0$ in vacuum has a wavelength of 438 nm in water and a wavelength of 390 nm in benzene. (a) What is the wavelength $\lambda_0$ of this light in vacuum? (b) Using only the given wavelengths, determine the ratio of the index of refraction of benzene to that of water.
12. Light of wavelength 436 nm in air enters a fishbowl filled with water, then exits through the crown-glass wall of the container. Find the wavelengths of the light (a) in the water and (b) in the glass.
13. A ray of light is incident on the surface of a block of clear ice at an angle of 40.0° with the normal. Part of the light is reflected and part is refracted. Find the angle between the reflected and refracted light.
14. A narrow beam of sodium yellow light ($\lambda_0 = 589$ nm) is incident from air on a smooth surface of water at an angle

of $\theta_1 = 35.0°$. Determine the angle of refraction $\theta_2$ and the wavelength of the light in water.

**15.** A beam of light, traveling in air, strikes the surface of mineral oil at an angle of 23.1° with the normal to the surface. If the light travels at $2.17 \times 10^8$ m/s through the oil, what is the angle of refraction?

**16.** A flashlight on the bottom of a 4.00-m-deep swimming pool sends a ray upward and at an angle so that the ray strikes the surface of the water 2.00 m from the point directly above the flashlight. What angle (in air) does the emerging ray make with the water's surface?

**17.** How many times will the incident beam shown in Figure P22.17 be reflected by each of the parallel mirrors?

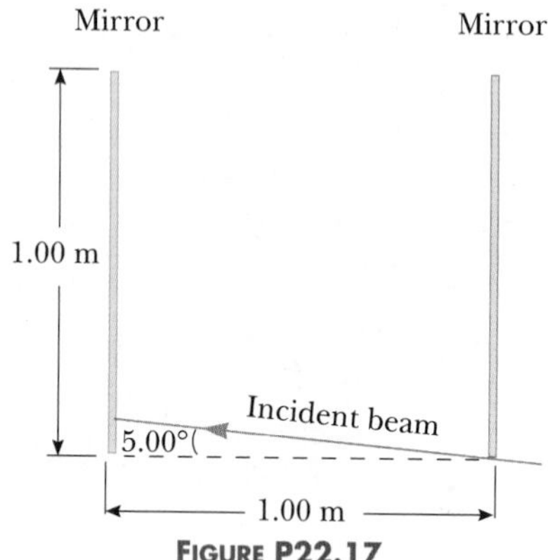


**FIGURE P22.17**

**18.** A ray of light strikes a flat, 2.00-cm-thick block of glass ($n = 1.50$) at an angle of 30.0° with the normal (Fig. P22.18). Trace the light beam through the glass and find the angles of incidence and refraction at each surface.

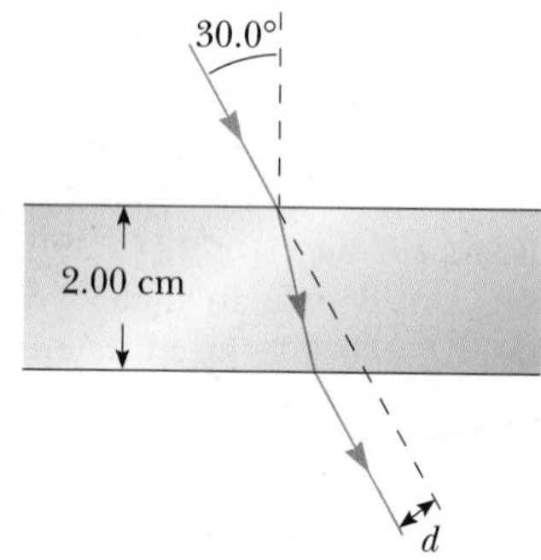


**FIGURE P22.18** (Problems 18, 19, and 20)

**19.** **web** When the light ray in Problem 18 passes through the glass block, it is shifted laterally by a distance $d$ (Fig. P22.18). Find the value of $d$.

**20.** Find the time required for the light to pass through the glass block described in Problem 19.

**21.** The light beam shown in Figure P22.21 makes an angle of 20.0° with the normal line $NN'$ in the linseed oil. Determine the angles $\theta$ and $\theta'$. (The refractive index for linseed oil is 1.48.)

**22.** A submarine is 300 m horizontally out from the shore and 100 m beneath the surface of the water. A laser beam is sent from the sub so that it strikes the surface of the water at a point 210 m from the shore. If the beam just strikes the top of a building standing directly at the water's edge, find the height of the building.

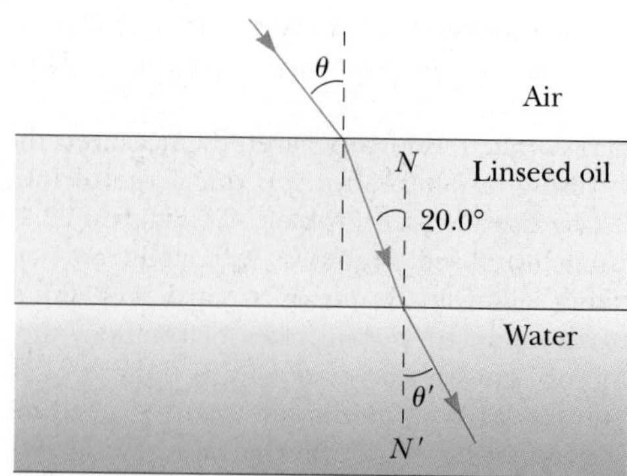


**FIGURE P22.21**

**23.** Two light pulses are emitted simultaneously from a source. The pulses take parallel paths to a detector 6.20 m away, but one moves through air and the other through a block of ice. Determine the difference in the pulses' times of arrival at the detector.

**24.** A narrow beam of ultrasonic waves reflects off the liver tumor in Figure P22.24. If the speed of the wave is 10.0% less in the liver than in the surrounding medium, determine the depth of the tumor.

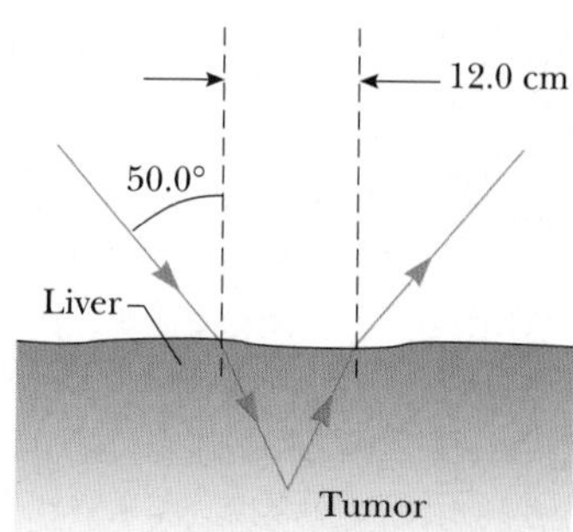


**FIGURE P22.24**

**25.** A beam of light both reflects and refracts at the surface between air and glass as shown in Figure P22.25. If the index of refraction of the glass is $n_g$, find the angle of incidence $\theta_1$ in the air that would result in the reflected ray and the refracted ray being perpendicular to each other. [*Hint:* Remember the identity $\sin(90° - \theta) = \cos\theta$.]

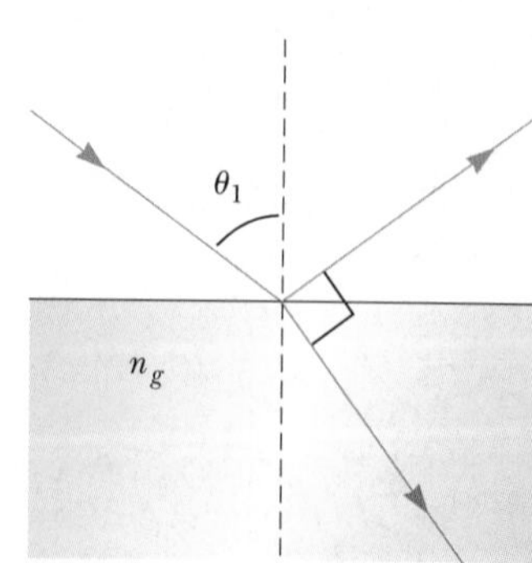


**FIGURE P22.25**

**26.** Three sheets of plastic have unknown indices of refraction. Sheet 1 is placed on top of sheet 2, and a laser beam is directed onto the sheets from above so that it strikes the interface at an angle of 26.5° with the normal. The refracted beam in sheet 2 makes an angle of 31.7° with the normal. The experiment is repeated with sheet 3 on top of sheet 2 and, with the same angle of incidence, the refracted beam makes an angle of 36.7° with the normal. If the experiment is repeated again with sheet 1 on top of sheet 3, what is the expected angle of refraction in sheet 3? Assume the same angle of incidence.

27. An opaque cylindrical tank with an open top has a diameter of 3.00 m and is completely filled with water. When the afternoon Sun reaches an angle of 28.0° above the horizon, sunlight ceases to illuminate the bottom of the tank. How deep is the tank?

28. A cylindrical cistern, constructed below ground level, is 3.0 m in diameter and 2.0 m deep and is filled to the brim with a liquid whose index of refraction is 1.5. A small object rests on the bottom of the cistern at its center. How far from the edge of the cistern can a girl whose eyes are 1.2 m from the ground stand and still see the object?

## Section 22.5 Dispersion and Prisms

29. The index of refraction for red light in water is 1.331, and that for blue light is 1.340. If a ray of white light enters the water at an angle of incidence of 83.00°, what are the underwater angles of refraction for the blue and red components of the light?

30. A certain kind of glass has an index of refraction of 1.650 for blue light of wavelength 430 nm and an index of 1.615 for red light of wavelength 680 nm. If a beam containing these two colors is incident at an angle of 30.00° on a piece of this glass, what is the angle between the two beams inside the glass?

31. A ray of light strikes the midpoint of one face of an equiangular (60°-60°-60°) glass prism ($n = 1.5$) at an angle of incidence of 30°. (a) Trace the path of the light ray through the glass, and find the angles of incidence and refraction at each surface. (b) If a small fraction of light is also reflected at each surface, find the angles of reflection at these surfaces.

32. The index of refraction for violet light in silica flint glass is 1.66 and that for red light is 1.62. What is the angular dispersion of visible light passing through an equilateral prism of apex angle 60.0° if the angle of incidence is 50.0°? (See Fig. P22.32.)

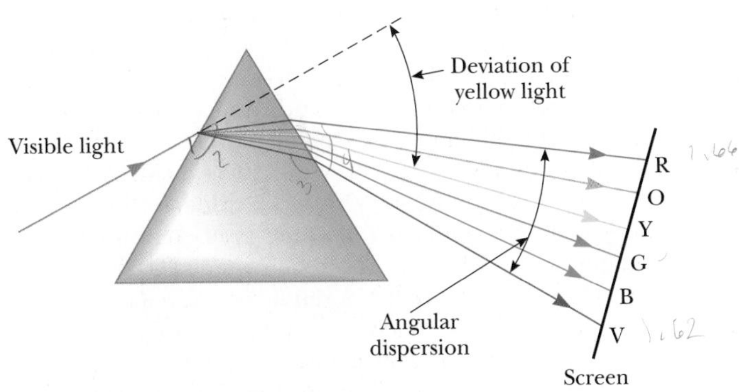


FIGURE P22.32

## Section 22.8 Total Internal Reflection

33. Calculate the critical angles for the following materials when surrounded by air: (a) zircon, (b) fluorite, (c) ice. Assume that $\lambda = 589$ nm.

34. For 589-nm light, calculate the critical angle for the following materials surrounded by air: (a) diamond and (b) flint glass.

35. Repeat Problem 34 when the materials are surrounded by water.

36. web A beam of light is incident from air on the surface of a liquid. If the angle of incidence is 30.0° and the angle of refraction is 22.0°, find the critical angle for the liquid when surrounded by air.

37. A light pipe consists of a central strand of material surrounded by an outer coating. The interior portion of the pipe has an index of refraction of 1.60. If all rays striking the interior walls of the pipe with incident angles greater than 59.5° are subject to total internal reflection, what is the index of refraction of the coating?

38. Determine the maximum angle $\theta$ for which the light rays incident on the end of the light pipe in Figure P22.38 are subject to total internal reflection along the walls of the pipe. Assume that the light pipe has an index of refraction of 1.36 and that the outside medium is air.

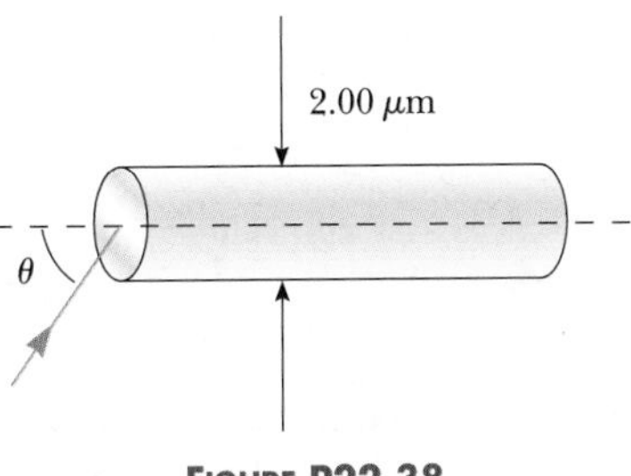


FIGURE P22.38

39. Consider a common mirage formed by super-heated air just above a roadway. A truck driver whose eyes are 2.00 m above the road, where $n = 1.000\ 3$, looks forward. She has the illusion of seeing a patch of water ahead on the road, where her line of sight makes an angle of 1.20° below the horizontal. Find the index of refraction of the air just above the road surface. (*Hint:* Treat this as a problem in total internal reflection.)

40. A jewel thief hides a diamond by placing it on the bottom of a public swimming pool. He places a circular raft on the surface of the water directly above and centered on the diamond, as shown in Figure P22.40. If the surface of the water is calm and the pool is 2.00 m deep, find the minimum diameter of the raft that would prevent the diamond from being seen.

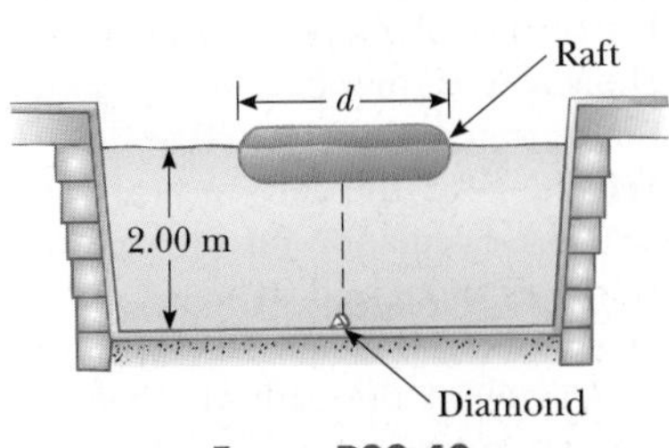


FIGURE P22.40

41. A room contains air in which the speed of sound is 343 m/s. The walls of the room are made of concrete, in which the speed of sound is 1 850 m/s. (a) Find the critical angle for total internal reflection of sound at the

concrete-air boundary. (b) In which medium must the sound be traveling in order to undergo total internal reflection? (c) "A bare concrete wall is a highly efficient mirror for sound." Give evidence for or against this statement.

42. A light ray is incident normally to the long face (the hypotenuse) of a 45°-45°-90° prism surrounded by air, as shown in Figure 22.26b. Calculate the minimum index of refraction of the prism for which the ray will follow the path shown.

43. The light beam in Figure P22.43 strikes surface 2 at the critical angle. Determine the angle of incidence $\theta_i$.

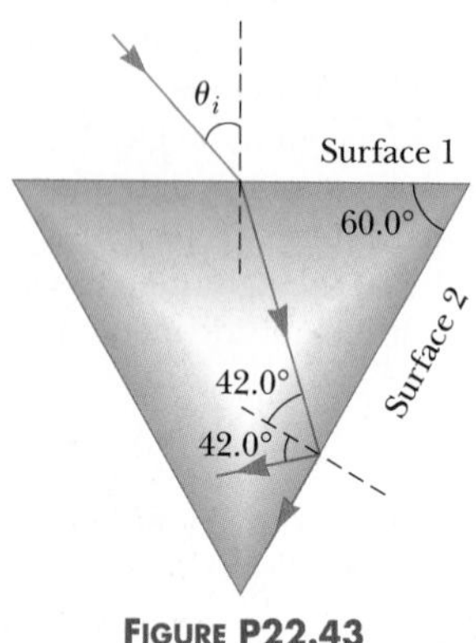


FIGURE P22.43

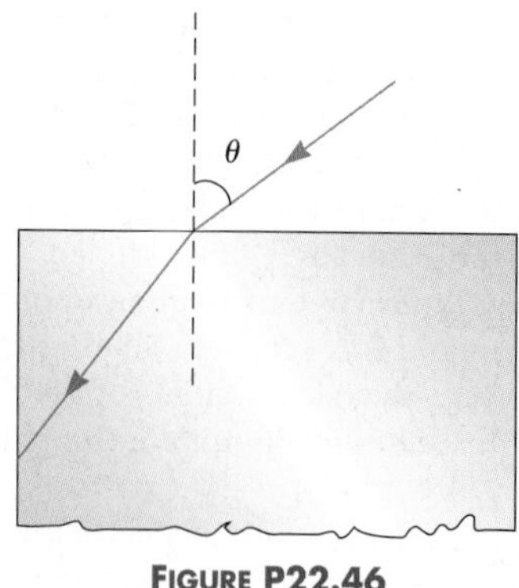


FIGURE P22.46

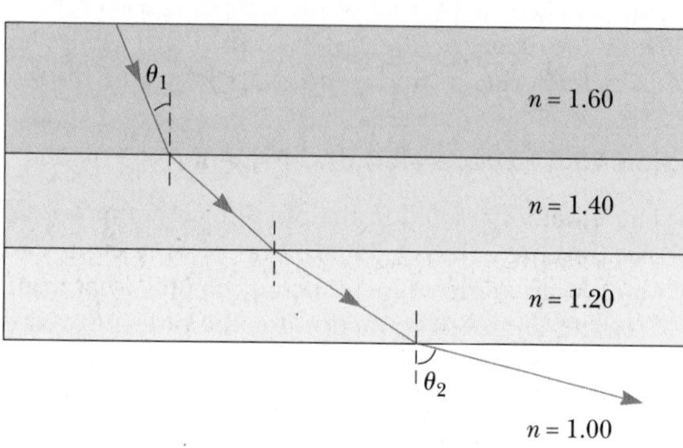


FIGURE P22.47

## ADDITIONAL PROBLEMS

44. (a) Consider a horizontal interface between air above and glass with an index of 1.55 below. Draw a light ray incident from the air at an angle of incidence of 30.0°. Determine the angles of the reflected and refracted rays and show them on the diagram. (b) Suppose instead that the light ray is incident from the glass at an angle of incidence of 30.0°. Determine the angles of the reflected and refracted rays and show all three rays on a new diagram. (c) For rays incident from the air onto the air-glass surface, determine and tabulate the angles of reflection and refraction for all the angles of incidence at 10.0° intervals from 0° to 90.0°. (d) Do the same for light rays traveling up to the interface through the glass.

45. web A layer of ice, having parallel sides, floats on water. If light is incident on the upper surface of the ice at an angle of incidence of 30.0°, what is the angle of refraction in the water?

46. A light ray of wavelength 589 nm is incident at an angle $\theta$ on the top surface of a block of polystyrene surrounded by air, as shown in Figure P22.46. (a) Find the maximum value of $\theta$ for which the refracted ray will undergo total internal reflection at the left vertical face of the block. (b) Repeat the calculation for the case in which the polystyrene block is immersed in water. (c) What happens if the block is immersed in carbon disulfide?

47. Figure P22.47 shows the path of a beam of light through several layers of different indices of refraction. (a) If $\theta_1 = 30.0°$, what is the angle $\theta_2$ of the emerging beam? (b) What must the incident angle $\theta_1$ be in order to have total internal reflection at the surface between the $n = 1.20$ medium and the $n = 1.00$ medium?

48. The walls of a prison cell are perpendicular to the four cardinal compass directions. On the first day of spring, light from the rising Sun enters a rectangular window in the eastern wall. The light traverses 2.37 m horizontally to shine perpendicularly on the wall opposite the window. A prisoner observes the patch of light moving across this western wall and for the first time forms his own understanding of the rotation of the Earth. (a) With what speed does the illuminated rectangle move? (b) The prisoner holds a small square mirror flat against the wall at one corner of the rectangle of light. The mirror reflects light back to a spot on the eastern wall close beside the window. How fast does the smaller square of light move across that wall? (c) Seen from a latitude of 40.0° north, the rising Sun moves through the sky along a line making a 50.0° angle with the southeastern horizon. In what direction does the rectangular patch of light on the western wall of the prisoner's cell move? (d) In what direction does the smaller square of light on the eastern wall move?

49. As shown in Figure P22.49, a light ray is incident normally on one face of a 30°-60°-90° block of dense flint glass (a prism) that is immersed in water. (a) Determine the exit angle $\theta_4$ of the ray. (b) A substance is dissolved in the water to increase the index of refraction. At what value of $n_2$ does total internal reflection cease at point $P$?

50. A narrow beam of light is incident from air onto a glass surface with index of refraction 1.56. Find the angle of incidence for which the corresponding angle of refraction is one half the angle of incidence. (*Hint:* You might want to use the trigonometric identity $\sin 2\theta = 2 \sin \theta \cos \theta$.)

51. One technique to measure the angle of a prism is shown in Figure P22.51. A parallel beam of light is directed on

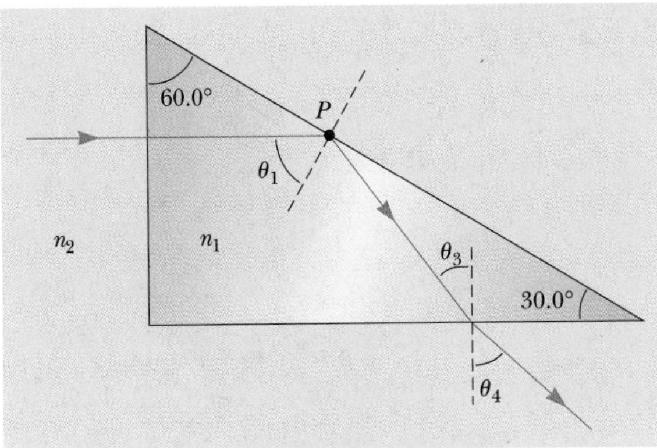


FIGURE P22.49

the apex of the prism so that the beam reflects from opposite faces of the prism. Show that the angular separation of the two reflected beams is given by $B = 2A$.

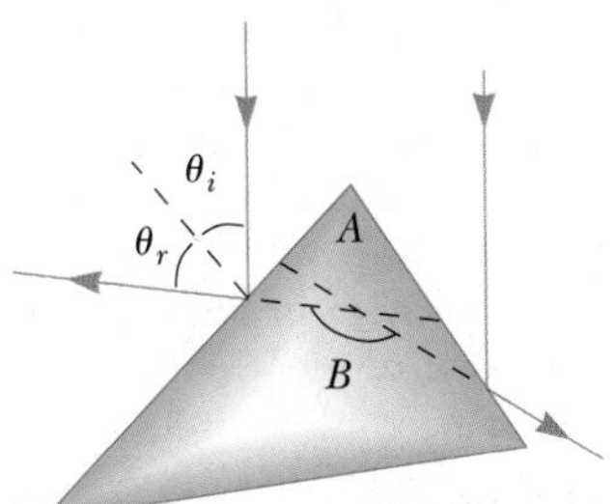


FIGURE P22.51

52. A 4.00-m-long pole stands vertically in a lake having a depth of 2.00 m. When the Sun is 40.0° above the horizontal, determine the length of the pole's shadow on the bottom of the lake. Take the index of refraction for water to be 1.33.

53. A piece of wire is bent through an angle $\theta$. The bent wire is partially submerged in benzene (index of refraction = 1.50) so that to a person looking along the dry part, the wire appears to be straight and makes an angle of 30.0° with the horizontal. Determine the value of $\theta$.

54. When you look through a window, by how much time is the light you see delayed by having to go through glass instead of air? Make an order-of-magnitude estimate on the basis of data you specify. By how many wavelengths is it delayed?

55. **web** A transparent cylinder of radius $R = 2.00$ m has a mirrored surface on its right half, as shown in Figure P22.55. A light ray traveling in air is incident on the left side of the cylinder. The incident light ray and the exiting light ray are parallel and $d = 2.00$ m. Determine the index of refraction of the material.

56. A laser beam strikes one end of a slab of material, as in Figure P22.56. The index of refraction of the slab is 1.48. Determine the number of internal reflections of the beam before it emerges from the opposite end of the slab.

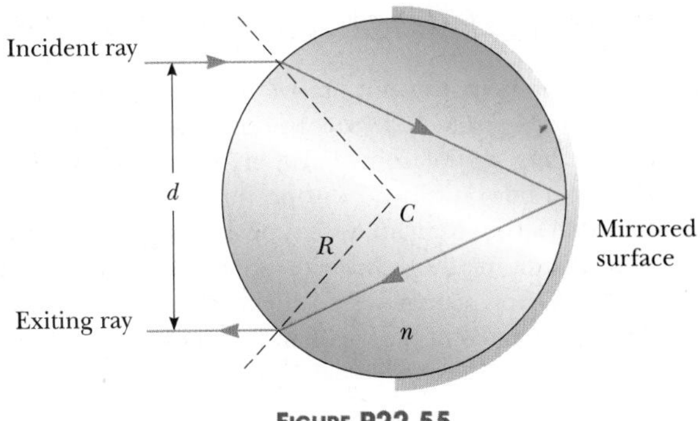


FIGURE P22.55

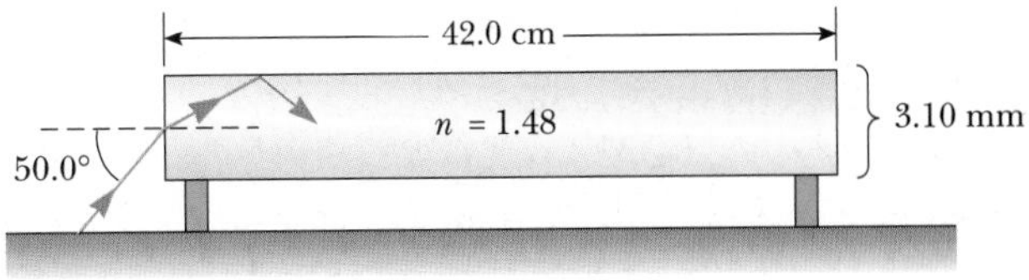

42.0 cm

$n = 1.48$

3.10 mm

50.0°

FIGURE P22.56

57. For this problem, refer to Figure 22.15. For various angles of incidence, it can be shown that the deviation angle $\delta$ is a minimum when the ray passes through the glass so that the interior ray is parallel to the base of the prism. A measurement of this minimum angle of deviation enables one to find the index of refraction of the prism material. Show that $n$ is given by the expression

$$n = \frac{\sin\left[\frac{1}{2}(A + \delta_{\min})\right]}{\sin\left(\dfrac{A}{2}\right)}$$

where $A$ is the apex angle of the prism.

58. A hiker stands on a mountain peak near sunset and observes a rainbow caused by water droplets in the air about 8.00 km away. The valley is 2.00 km below the mountain peak and entirely flat. What fraction of the complete circular arc of the rainbow is visible to the hiker?

59. A light ray is incident on a prism and refracted at the first surface, as shown in Figure P22.59. Let $\phi$ represent the apex angle of the prism and $n$ its index of refraction. Find, in terms of $n$ and $\phi$, the smallest allowed value of the angle of incidence at the first surface for which the refracted ray will not undergo internal reflection at the second surface.

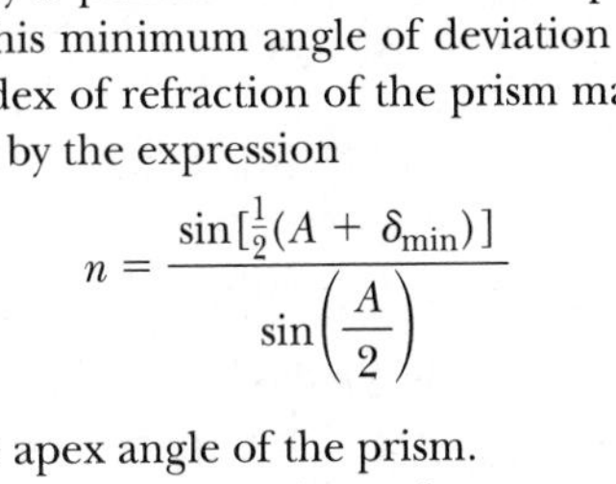

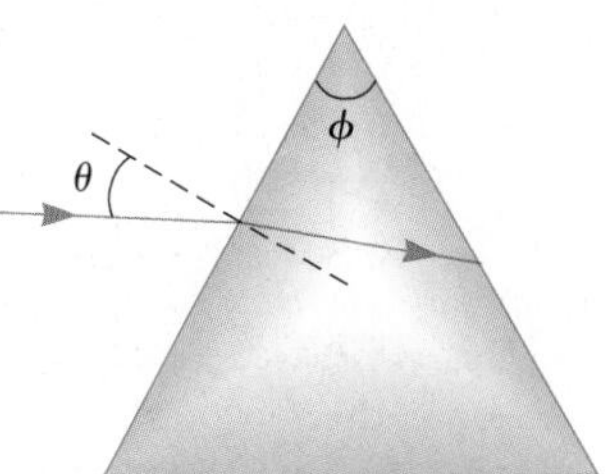


FIGURE P22.59

## GROUP ACTIVITIES

**G.1** Tape a coin to the bottom of a large opaque bowl, as shown in Figure GA22.1a. Stand over the bowl so that you are looking at the coin and then move backwards away from the bowl until you can no longer see the coin over the bowl's rim. Remain at that position and have your co-investigator fill the bowl with water, as shown in Figure GA22.1b. You can now see the coin again because the light is refracted at the water-air interface.

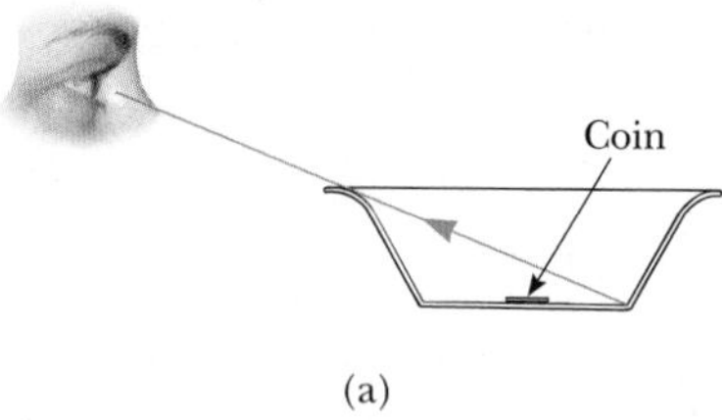


(a)

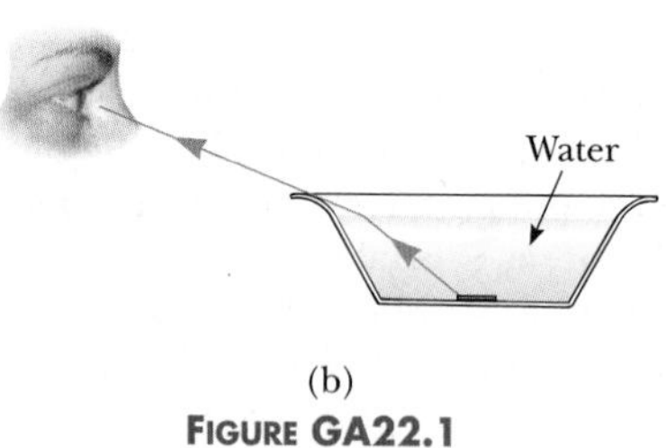


(b)

**FIGURE GA22.1**

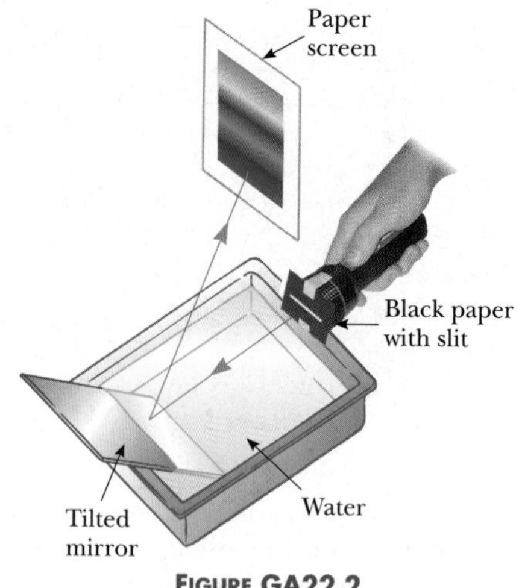


**FIGURE GA22.2**

**G.2** Tape a piece of black paper to the end of a flashlight and cut a narrow slit in the middle of the paper, as shown in Figure GA22.2. Lean a flat mirror against one end of a tray partially filled with water. Shine your flashlight on that part of the mirror that is under water, and hold a sheet of white paper such that the reflected light shines on the paper. You should observe a spectrum of colors on the paper as the light is dispersed when it travels from air into water, and then from water into air. According to your observations, which color is bent the most? Which is bent the least?

**G.3** Create an artificial rainbow by standing with your back to the Sun and spraying water into the air with a hose. You should cover the end of the hose slightly with a finger or use a nozzle so that the water is broken up into tiny droplets. You should be able to form a rainbow in this way. Which color is on the outside of the arc? The inside?

If the droplets are close to you in the experiment above, you may be able to see two rainbows, one for each eye. Close one eye and only one bow is seen.

With the Sun high in the sky, stand on a ladder and spray the hose toward the ground. In this case, you should be able to form a complete circle rainbow.

**G.4** You can demonstrate that light rays travel in straight-line paths by observing the shadows formed by large and small light sources. Figure GA22.4 shows the shadows formed on a screen by a baseball when the light from a lightbulb (a large source) falls on it. Use the figure to explain why the shadow is dark at location A and less dark at B. Replace the light source with a small source such as a high-intensity lamp with a small clear bulb. (A substitute for a high-intensity lamp is an ordinary lightbulb with a piece of cardboard close in front of it with a hole about the size of a penny punched in the cardboard. The hole serves as the small source of light.) What kind of shadow is formed in this case? Why?

**G.5** A ray of light is moving from a material having a high index of refraction into a material with a lower index of refraction. (a) Is the ray bent toward the normal or away from it? (b) If the wavelength is 600 nm in the high-index-of-refraction material, is it greater, smaller, or the

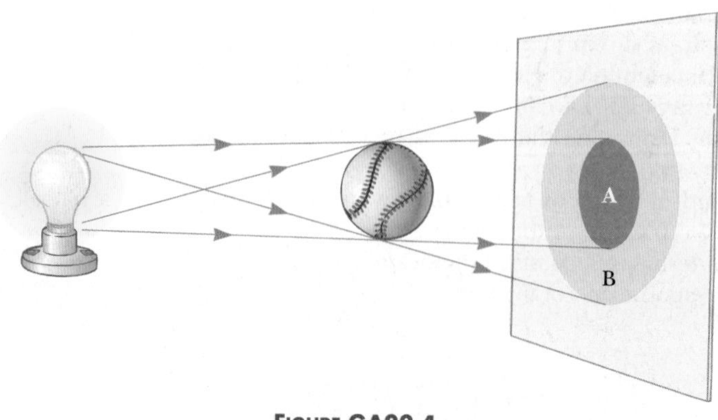


**FIGURE GA22.4**

same in the lower-index material? (c) How does the frequency change as the light moves between the two materials? Does it increase, decrease, or remain the same? (d) Assume the high-index material has $n = 1.55$, the lower index is 1.33, and the angle of incidence in the higher material is, say, 30°. Do the calculations to see if they agree with your predictions.

**G.6** Light attempts to move from cubic zirconia ($n = 1.61$) to air. In a nearby experiment, light is attempting to move from cubic zirconia to water. Finally, in a third trial, light is attempting to move from cubic zirconia to a material with an index of refraction of 1.62. (a) Without doing any calculations, in which case is the angle of refraction the largest? (b) Without calculation, can there be situations in the preceding cases in which the light will not emerge from the cubic zirconia? Explain. (c) Verify your predictions by doing the calculations with an angle of incidence of 40°.

**G.7** Light containing three wavelengths—400 nm, 500 nm, and 600 nm—is incident from air on a block of crown glass at an angle of 30°. Are all colors refracted alike or is one color bent more than the others? If you think one is bent more than the others, which one is it? Explain your answer. Calculate the angle of refraction in each case to verify your answer.

# 23 Mirrors and Lenses

## Chapter Outline



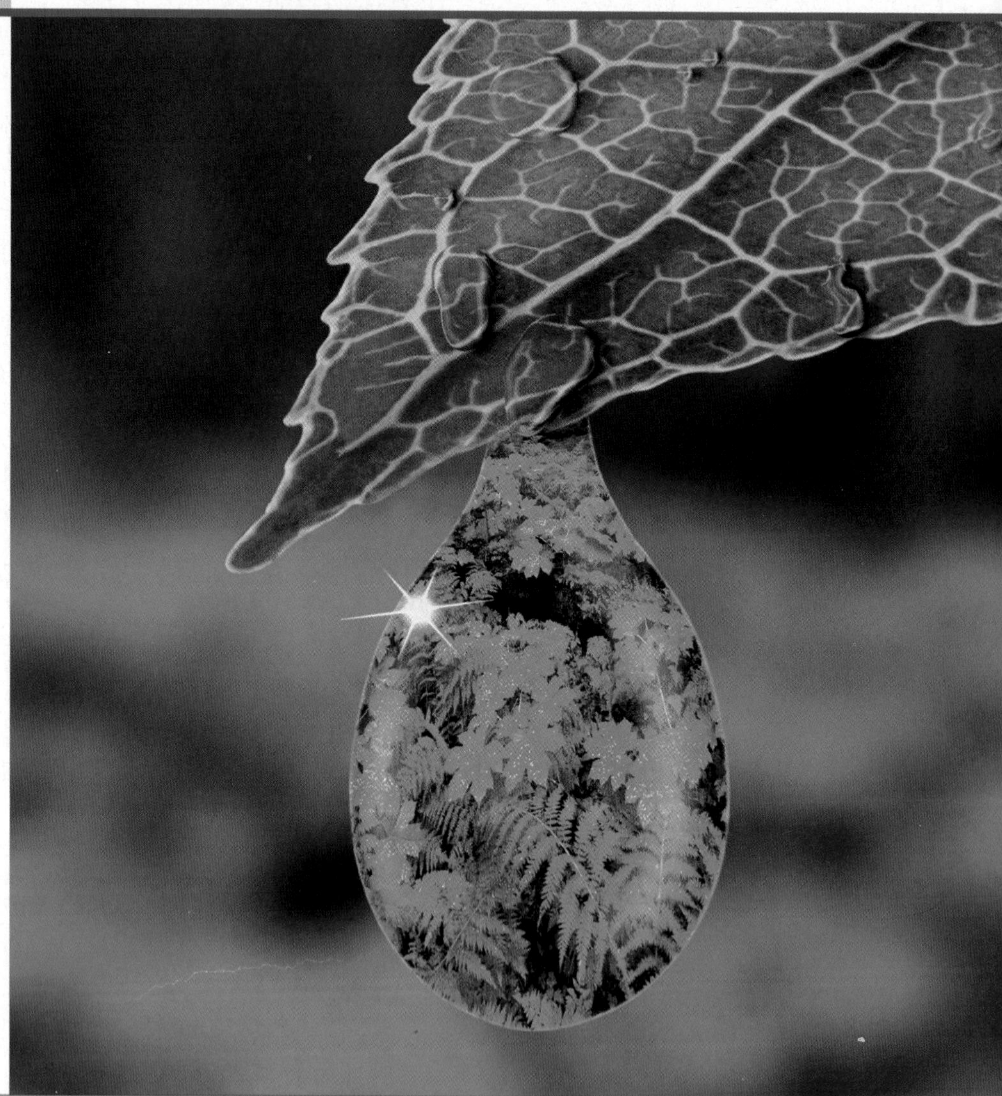

**This beautiful photograph shows a raindrop suspended from a leaf. The raindrop acts as a lens. It refracts light twice to produce a real, inverted image of the foliage beyond.** *(Don Hammond/CORBIS)*

This chapter is concerned with the formation of images when plane and spherical light waves fall on plane and spherical surfaces. Images can be formed by either reflection or refraction. Mirrors and lenses form images in both ways. In our study of mirrors and lenses, we continue to use the ray approximation and to assume that light travels in straight lines (in other words, we ignore diffraction).

## 23.1 FLAT MIRRORS

We begin our investigation by examining the simplest possible mirror, the flat mirror. Consider a point source of light placed at $O$ in Figure 23.1, a distance $p$ in front of a flat mirror. The distance $p$ is called the **object distance.** Light rays leave the source and are reflected from the mirror. After reflection, the rays diverge (spread apart), but they appear to the viewer to come from a point $I$ behind the mirror. Point $I$ is called the **image** of the object at $O$. Regardless of the system under study, **images are formed at the point at which rays of light actually intersect or at which they appear to originate.** Because the rays in Figure 23.1 appear to originate at $I$, which is a distance $q$ behind the mirror, this is the location of the image. The distance $q$ is called the **image distance.**

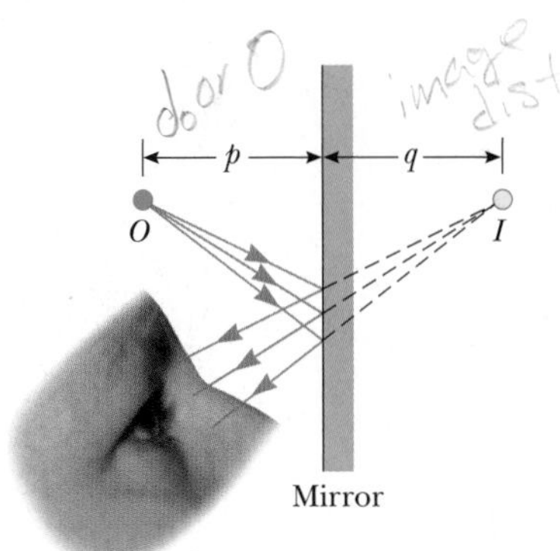


**FIGURE 23.1** An image formed by reflection from a flat mirror. The image point $I$ is behind the mirror at distance $q$, which is equal in magnitude to the object distance $p$.

Images are classified as real or virtual. A *real image* is one in which light actually passes through the image point; a *virtual image* is one in which the light does not pass through the image point but appears to come (diverge) from that point. The image formed by the flat mirror in Figure 23.1 is a virtual image. In fact, the images seen in flat mirrors are always virtual (for real objects). Real images can be displayed on a screen (as at a movie), but virtual images cannot.

We shall examine some of the properties of the images formed by flat mirrors by using the simple geometric techniques shown in Figure 23.2. To find out where an image is formed, it is always necessary to follow at least two rays of light as they reflect from the mirror. One of those rays starts at $P$, follows a horizontal path $PQ$ to the mirror, and reflects back on itself. The second ray follows the oblique path $PR$ and reflects as shown. An observer to the left of the mirror would trace the two reflected rays back to the point from which they appear to have originated—that is, point $P'$. A continuation of this process for points other than $P$ on the object would result in a virtual image (drawn as a yellow arrow) to the right of the mirror. Because triangles $PQR$ and $P'QR$ are identical, $PQ = P'Q$. Hence, we conclude that **the image formed by an object placed in front of a flat mirror is as far behind the mirror as the object is in front of the mirror.** Geometry also shows that the object height $h$ equals the image height $h'$. **Lateral magnification** $M$ is defined as follows:

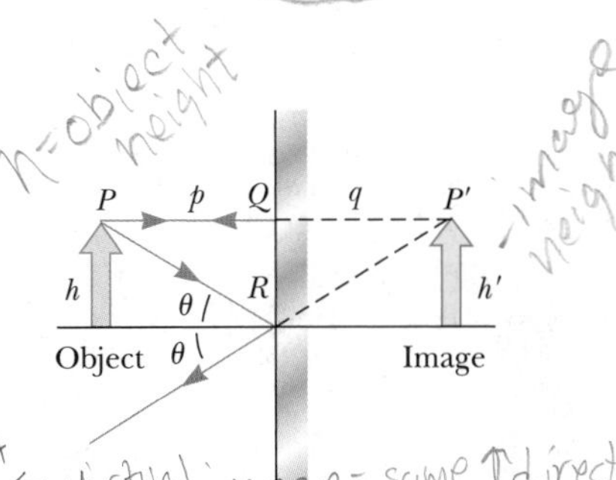


**FIGURE 23.2** A geometric construction to locate the image of an object placed in front of a flat mirror. Because the triangles $PQR$ and $P'QR$ are identical, $p = |q|$ and $h = h'$.

$$M \equiv \frac{\text{image height}}{\text{object height}} = \frac{h'}{h} \qquad [23.1]$$

This is a general definition of the lateral magnification of any type of mirror. For a flat mirror, $M = 1$ because $h' = h$.

In summary, the image formed by a flat mirror has the following properties:

1. **The image is as far behind the mirror as the object is in front.**
2. **The image is unmagnified, virtual, and upright. (By *upright* we mean that if the object arrow points upward, as in Figure 23.2, so does the image arrow. The opposite of an upright image is an inverted image.)**

**Tip 23.1 MAGNIFICATION ≠ ENLARGEMENT**

Note that the word *magnification* used in optics does not always mean *enlargement* because the image could be smaller than the object.

Finally, note that a flat mirror produces an image having an *apparent* left-right reversal. This reversal can be seen by standing in front of a mirror and raising your right hand. The image you see raises its left hand. Likewise, your hair appears to be parted on the opposite side and a mole on your right cheek appears to be on your left cheek.

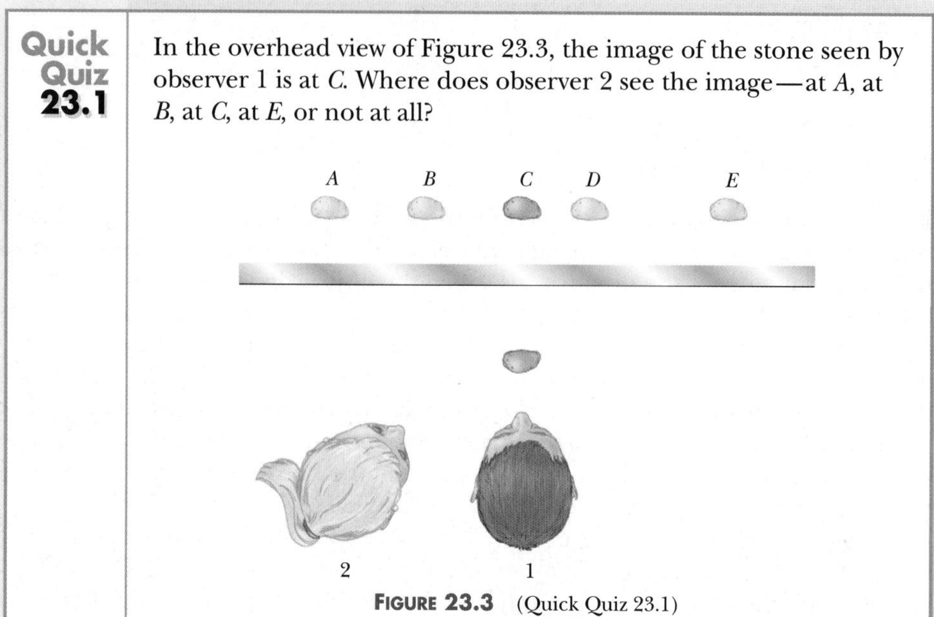

Quick Quiz 23.1
In the overhead view of Figure 23.3, the image of the stone seen by observer 1 is at *C*. Where does observer 2 see the image—at *A*, at *B*, at *C*, at *E*, or not at all?
A
B
C
D
E
2
1


**FIGURE 23.3** (Quick Quiz 23.1)

## Example 23.1 "I Can See Myself!"

A 1.80-m tall man stands in front of a mirror in hopes of seeing his full height, no more and no less. If his eyes are 0.10 m from the top of his head, what is the minimum height of the mirror?

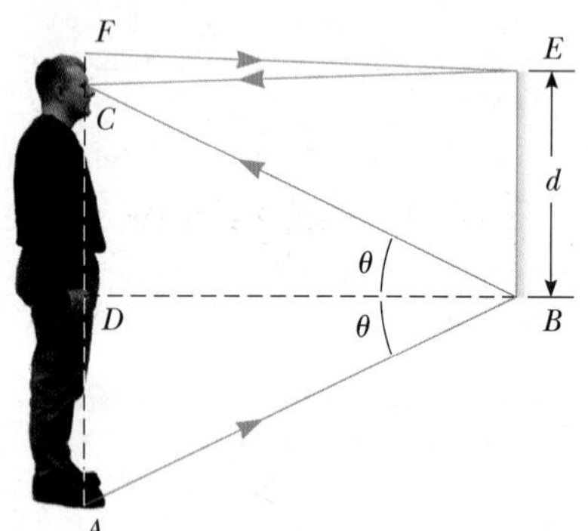


**FIGURE 23.4** (Example 23.1)

**Reasoning and Solution** Figure 23.4 shows two rays of light originating at the extremes of the body, reflecting from the mirror, and entering the eye of the viewer. The ray from the feet just strikes the bottom of the mirror, so if the mirror were longer, it would be too long; if shorter, the ray would not be reflected. The angle of incidence and the angle of reflection are equal, labeled $\theta$. The two triangles, *ABD* and *DBC* are identical because they are right triangles with a common side (*DB*) and two identical angles ($\theta$). Thus, we have

$$AD = DC = \tfrac{1}{2}AC = \tfrac{1}{2}(1.80\text{ m} - 0.10\text{ m}) = 0.85\text{ m}$$

Furthermore, because $\tfrac{1}{2}CF = \tfrac{1}{2}(0.10\text{ m}) = 0.05\text{ m}$, we find that

$$d = FA - AD - \tfrac{1}{2}CF = 1.80\text{ m} - 0.85\text{ m} - 0.05\text{ m} = 0.90\text{ m}$$

Thus, the mirror must be exactly equal to half the height of the man in order for him to see only his full height and nothing more or less.

**EXERCISE** Is the result of this problem still valid if the man moves farther away from the mirror?

**APPLICATION**

DAY AND NIGHT SETTINGS FOR REARVIEW MIRRORS

Most rearview mirrors in cars have a day setting and a night setting. The night setting greatly diminishes the intensity of the image so that lights from trailing cars will not blind the driver. To understand how such a mirror works, consider Figure 23.5. The mirror is a wedge of glass with a reflecting metallic coating on the back side. When the mirror is in the day setting, as in Figure 23.5a, light from an object

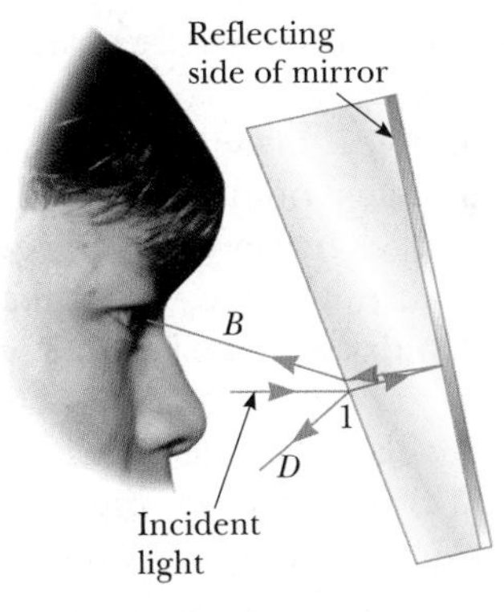


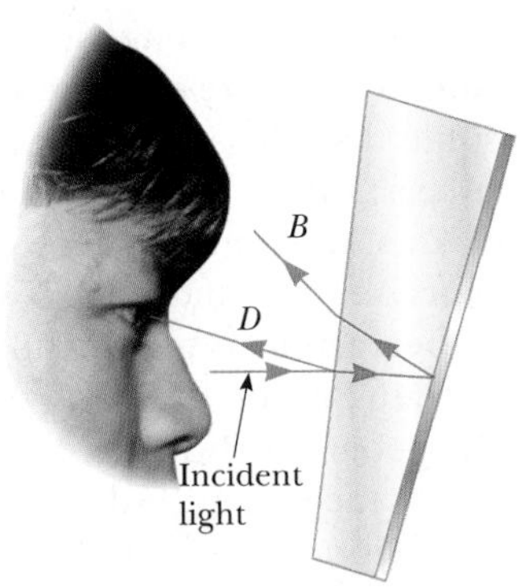


**FIGURE 23.5** Cross-sectional views of a rearview mirror. (a) With the day setting, the silvered back surface of the mirror reflects a bright ray $B$ into the driver's eyes. (b) With the night setting, the glass of the unsilvered front surface of the mirror reflects a dim ray $D$ into the driver's eyes.

behind the car strikes the mirror at point 1. Most of the light enters the wedge, is refracted, and reflects from the back of the mirror to return to the front surface, where it is refracted again as it re-enters the air as ray $B$ (for *bright*). In addition, a small portion of the light is reflected at the front surface, as indicated by ray $D$ (for *dim*). This dim reflected light is responsible for the image observed when the mirror is in the night setting, as in Figure 23.5b. In this case, the wedge is rotated so that the path followed by the bright light (ray $B$) does not lead to the eye. Instead, the dim light reflected from the front surface travels to the eye, and the brightness of trailing headlights does not become a hazard.

## APPLYING PHYSICS 23.1

The professor in the box shown in Figure 23.6 appears to be balancing himself on a few fingers with both of his feet elevated from the floor. The professor can maintain this position for a long time, and he appears to defy gravity. How do you suppose this illusion was created?

**Explanation** This is one example of an optical illusion, used by magicians, that makes use of a mirror. The box that the professor is standing in is a cubical frame that contains a flat vertical mirror through a diagonal plane. The professor straddles the mirror so that the foot you see is in front of the mirror and the other foot is behind the mirror where you cannot see it. When he raises the foot that you see in front of the mirror, the reflection of this foot also rises, so he appears to float in air.

**FIGURE 23.6** (Applying Physics 23.1) *(Courtesy of Henry Leap and Jim Lehman)*

## 23.2 IMAGES FORMED BY SPHERICAL MIRRORS

### CONCAVE MIRRORS

A **spherical mirror,** as its name implies, has the shape of a segment of a sphere. Figure 23.7 shows a spherical mirror with light reflecting from its silvered inner, concave surface; this is called a **concave mirror.** The mirror has radius of curvature $R$, and its center of curvature is at point $C$. Point $V$ is the center of the spherical segment, and a line drawn from $C$ to $V$ is called the **principal axis** of the mirror.

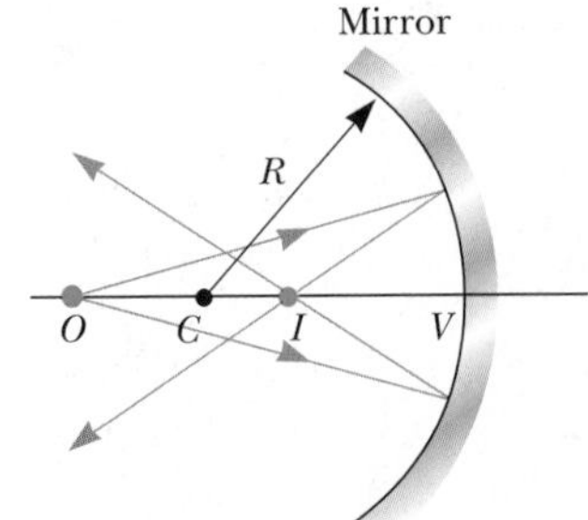


**FIGURE 23.7** A point object placed at $O$, outside the center of curvature of a concave spherical mirror, forms a real image at $I$ as shown. If the rays diverge from $O$ at small angles, they all reflect through the same image point.

Now consider a point source of light placed at point $O$ in Figure 23.7, on the principal axis and outside point $C$. Several diverging rays originating at $O$ are shown. After reflecting from the mirror, these rays converge to meet at $I$, called the **image point.** The rays then continue and diverge from $I$ as if there were an

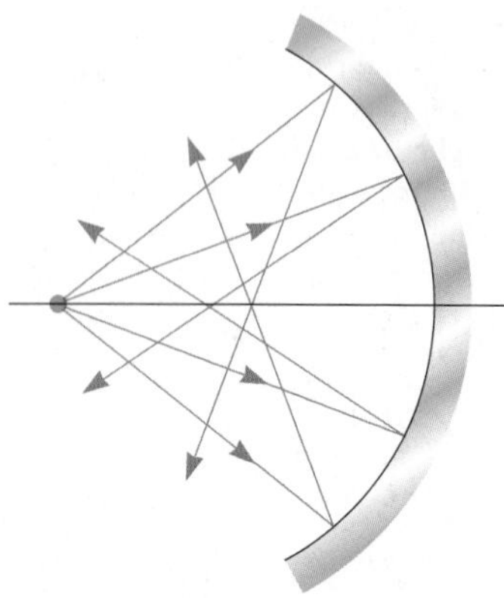

**FIGURE 23.8** Rays at large angles from the horizontal axis reflect from a spherical concave mirror to intersect the principal axis at different points, resulting in a blurred image. This is called *spherical aberration.*

object there. As a result, a real image is formed. **Whenever reflected light actually passes through a point, the image formed there is real.**

We often assume that all rays that diverge from the object make small angles with the principal axis. All such rays reflect through the image point, as in Figure 23.7. Rays that make a large angle with the principal axis, as in Figure 23.8, converge to other points on the principal axis, producing a blurred image. This effect, called **spherical aberration,** is present to some extent with any spherical mirror and will be discussed in Section 23.7.

We can use the geometry shown in Figure 23.9 to calculate the image distance $q$ from the object distance $p$ and radius of curvature $R$. By convention, these distances are measured from point $V$. Figure 23.9 shows two rays of light leaving the tip of the object. One ray passes through the center of curvature $C$ of the mirror, hitting the mirror head on (perpendicularly to the mirror surface) and reflecting back on itself. The second ray strikes the mirror at point $V$ and reflects as shown, obeying the law of reflection. The image of the tip of the arrow is at the point at which the two rays intersect. From the largest triangle in Figure 23.9, we see that $\tan\theta = h/p$; the light blue triangle gives $\tan\theta = -h'/q$. The negative sign has been introduced to satisfy our convention that $h'$ is negative when the image is inverted with respect to the object, as it is here. Thus, from Equation 23.1 and these results, we find that the magnification of the mirror is

$$M = \frac{h'}{h} = -\frac{q}{p} \tag{23.2}$$

We also note, from two other triangles in the figure, that

$$\tan\alpha = \frac{h}{p-R} \quad \text{and} \quad \tan\alpha = -\frac{h'}{R-q}$$

from which we find that

$$\frac{h'}{h} = -\frac{R-q}{p-R} \tag{23.3}$$

If we compare Equation 23.2 to Equation 23.3, we see that

$$\frac{R-q}{p-R} = \frac{q}{p}$$

Simple algebra reduces this to

Mirror equation

$$\frac{1}{p} + \frac{1}{q} = \frac{2}{R} \tag{23.4}$$

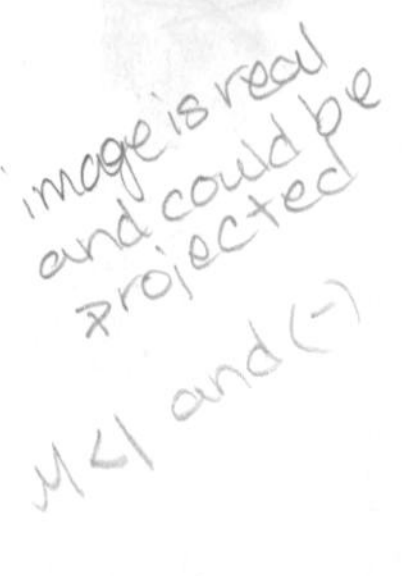


This expression is called the **mirror equation.**

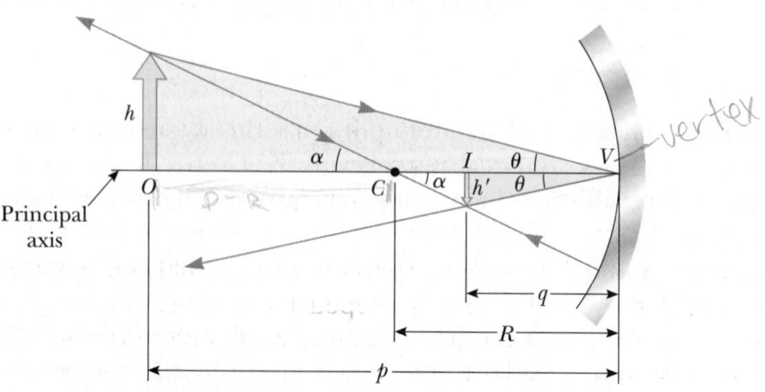


**FIGURE 23.9** The image formed by a spherical concave mirror, where the object at $O$ lies outside the center of curvature, $C$.

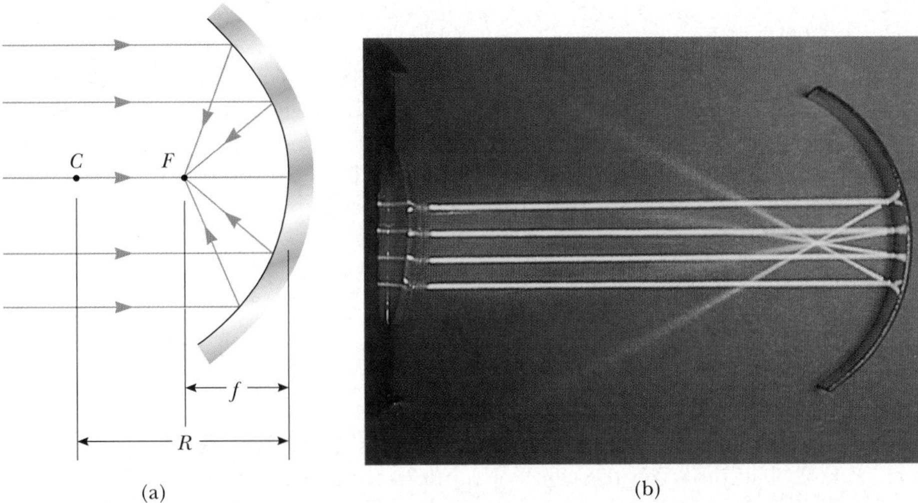


**FIGURE 23.10** (a) Light rays from a distant object ($p = \infty$) reflect from a concave mirror through the focal point $F$. In this case, the image distance $q = R/2 = f$, where $f$ is the focal length of the mirror. (b) A photograph of the reflection of parallel rays from a concave mirror. *(Courtesy of Jim Lehman, James Madison University)*

If the object is very far from the mirror—that is, if the object distance $p$ is great enough compared with $R$ that $p$ can be said to approach infinity—then $1/p \approx 0$, and we see from Equation 23.4 that $q \approx R/2$. In other words, when the object is very far from the mirror, **the image point is halfway between the center of curvature and the center of the mirror,** as in Figure 23.10a. The incoming rays are essentially parallel in this figure because the source is assumed to be very far from the mirror. In this special case we call the image point the **focal point,** $F$, and the image distance the **focal length,** $f$, where

$$f = \frac{R}{2} \qquad [23.5]$$

Focal length

The mirror equation can therefore be expressed in terms of the focal length:

$$\frac{1}{p} + \frac{1}{q} = \frac{1}{f} \qquad [23.6]$$

Note that rays from objects at infinity are always focused at the focal point.

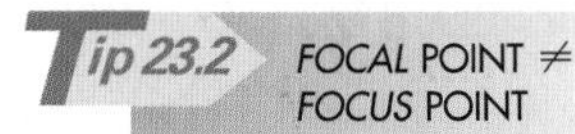


The focal point *is not* the point at which light rays focus to form an image. The focal point of a mirror is determined *solely* by its curvature—it does not depend on the location of any object.

## 23.3 CONVEX MIRRORS AND SIGN CONVENTIONS

Figure 23.11 shows the formation of an image by a **convex mirror,** which is silvered so that light is reflected from the outer, convex surface. This is sometimes called a **diverging mirror** because the rays from any point on the object diverge after reflection as though they were coming from some point behind the mirror. The image in Figure 23.11 is virtual rather than real because it lies behind the mirror at the point at which the reflected rays appear to originate. In general, as shown in the figure, the image formed by a convex mirror is upright, virtual, and smaller than the object.

We shall not derive any equations for convex spherical mirrors. If we did, we would find that the equations developed for concave mirrors can be used with convex mirrors if particular sign conventions are used. Let us call the region in which light rays move the *front side* of the mirror, and the other side, where virtual images are formed, the *back side.* For example, in Figures 23.9 and 23.11, the side to the left of the mirror is the front side, and the side to the right is the back side.

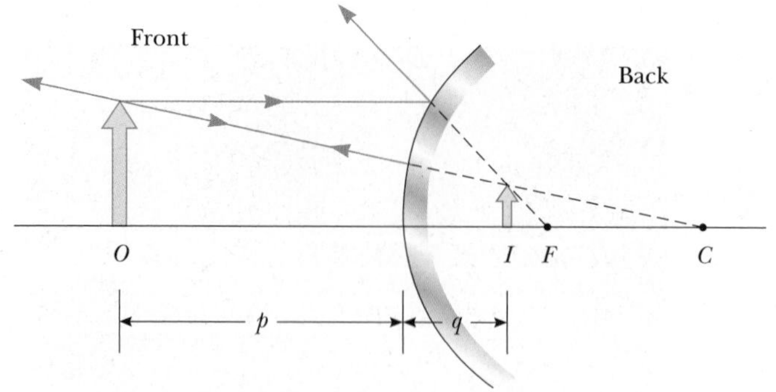


**FIGURE 23.11** Formation of an image by a spherical convex mirror. Note that the image is virtual and upright.

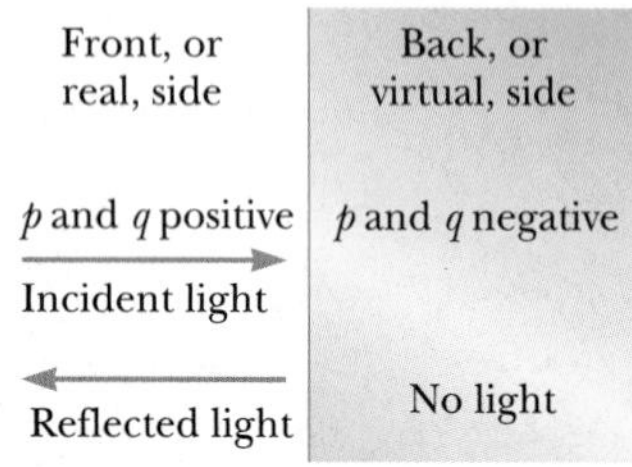


Convex or concave mirror

**FIGURE 23.12** A diagram describing the signs of $p$ and $q$ for convex and concave mirrors.

Figure 23.12 is helpful for understanding the rules for object and image distances, and Table 23.1 summarizes the sign conventions for all the necessary quantities.

## RAY DIAGRAMS FOR MIRRORS

We can determine conveniently the positions and sizes of images formed by mirrors by constructing *ray diagrams* similar to the ones we have been using. This kind of graphical construction tells us the overall nature of the image and can be used to check parameters calculated from the mirror and magnification equations. To make a ray diagram, one needs to know the position of the object and the location of the center of curvature. To locate the image, three rays are constructed (rather than just the two we have been constructing so far), as shown by the examples in Figure 23.13. All three rays start from the same object point; for these examples the tip of the arrow was chosen. For the concave mirrors in Figure 23.13a and b, the rays are drawn as follows:

1. **Ray 1 is drawn parallel to the principal axis and is reflected back through the focal point *F*.**
2. **Ray 2 is drawn through the focal point. Thus, it is reflected parallel to the principal axis.**
3. **Ray 3 is drawn through the center of curvature *C* and is reflected back on itself.**

Note that rays actually go in all directions from the object; we choose to follow those moving in a direction that simplifies our drawing.

The intersection of any *two* of these rays at a point locates the image. The third ray serves as a check of construction. The image point obtained in this fashion must always agree with the value of $q$ calculated from the mirror formula.

In the case of a concave mirror, note what happens as the object is moved closer to the mirror. The real, inverted image in Figure 23.13a moves to the left as the object approaches the focal point. When the object is at the focal point, the

**TABLE 23.1** Sign Conventions for Mirrors

| Quantity | Positive When | Negative When |
|---|---|---|
| Object location ($p$) | Object is in front of mirror | Object is behind mirror |
| Image location ($q$) | Image is behind mirror | Image is in front of mirror |
| Image height ($h'$) | Image is upright | Image is inverted |
| Focal length ($f$) and radius ($R$) | Mirror is concave | Mirror is convex |
| Magnification ($M$) | Image is upright | Image is inverted |

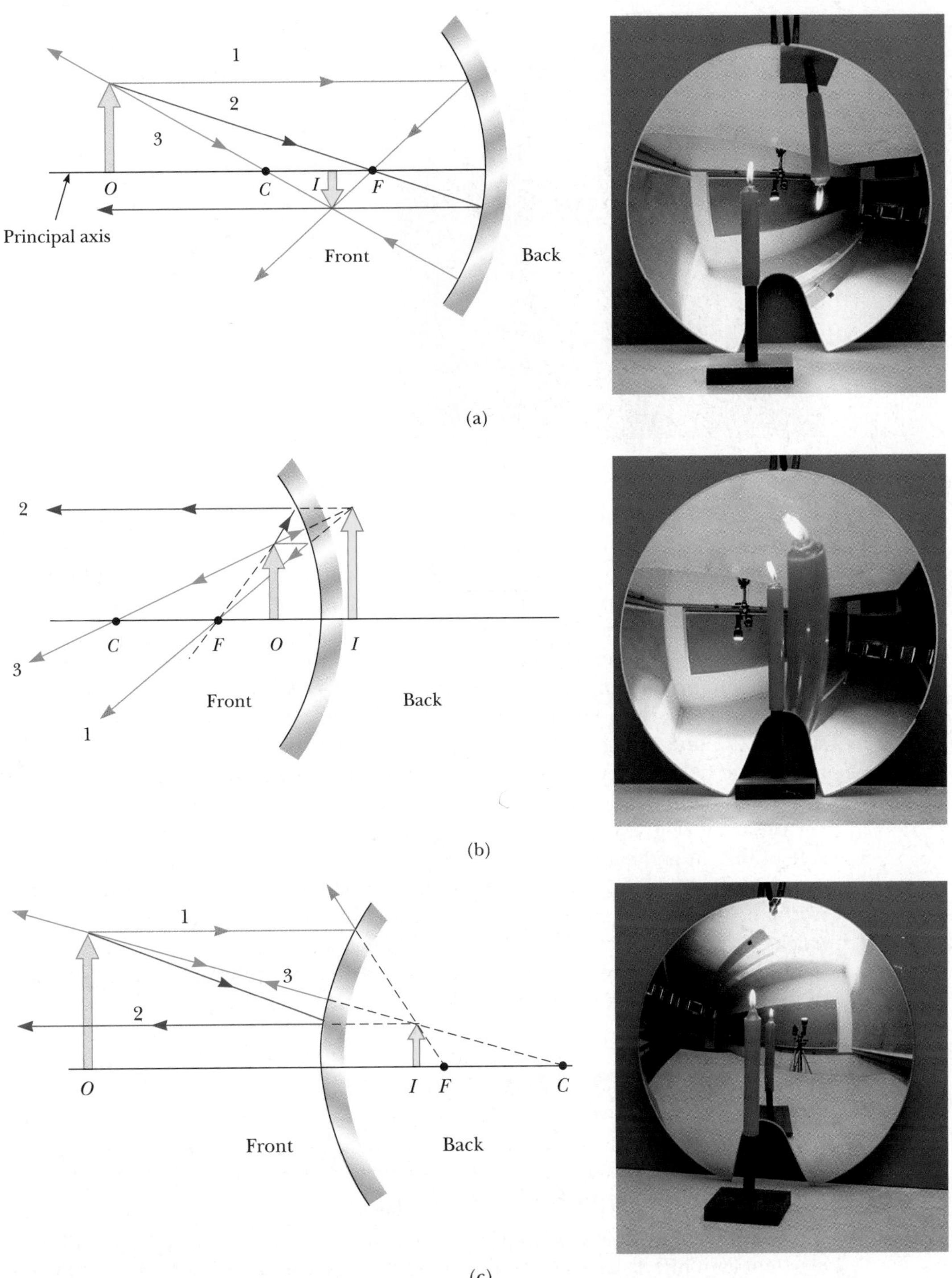


**FIGURE 23.13** Ray diagrams for spherical mirrors, and corresponding photographs of the images of candles. (a) When an object is outside the center of curvature of a concave mirror, the image is real, inverted, and reduced in size. (b) When an object is between a concave mirror and the focal point, the image is virtual, upright, and magnified. (c) When an object is in front of a convex mirror, the image is virtual, upright, and reduced in size. *(Photos courtesy of David Rogers)*

image is infinitely far to the left. However, when the object lies between the focal point and the mirror surface, as in Figure 23.13b, the image is virtual and upright.

With the convex mirror shown in Figure 23.13c, the image of a real object is always virtual and upright. As the object distance increases, the virtual image

**FIGURE 23.14** A convex sideview mirror on a vehicle produces an upright image that is smaller than the object. *(© Junebug Clark 1988/Photo Researchers, Inc.)*

shrinks and approaches the focal point as $p$ approaches infinity. You should construct a ray diagram to verify this.

The image-forming characteristics of curved mirrors obviously determine their uses. For example, suppose you want to design a mirror that will help people shave or apply cosmetics, such as the one in Figure 23.13b. That is, you need a concave mirror that puts the user inside the focal point. In such a situation, the image is upright and greatly enlarged. In contrast, suppose that the primary purpose of a mirror is to observe a large field of view, in which case you need a convex mirror such as the one in Figure 23.13c. The diminished size of the image means that a fairly large field of view is seen in the mirror. Mirrors such as this are often placed in stores to help employees watch for shoplifting. A second use is as a sideview mirror on a car (Fig. 23.14). This kind of mirror is usually placed on the passenger side of the car and carries the warning "Objects are closer than they appear." Without this warning, a driver might think she is looking into a flat mirror, which does not alter the size of the image. Thus, she could be fooled into believing that a truck is far away because it looks small, when it is actually a large semi very close behind her but diminished in size because of the image formation characteristics of the convex mirror.

## APPLYING PHYSICS 23.2

For a concave mirror, a virtual image can be anywhere behind the mirror. For a convex mirror, however, there is a maximum distance at which the image can exist behind the mirror. Why?

**Explanation** Let us consider the concave mirror first and imagine two different light rays leaving a tiny object and striking the mirror. If the object is at the focal point, the light rays reflecting from the mirror will be parallel to the mirror axis. They can be interpreted as forming a virtual image infinitely far away behind the mirror. As the object is brought closer to the mirror, the reflected rays will diverge through larger and larger angles, resulting in their extensions converging closer and closer to the back of the mirror. When the object is brought right up to the mirror, the image is right behind the mirror. When the object is much closer to the mirror than the focal length, the mirror acts like a flat mirror, and the image is just as far behind the mirror as the object is in front of it. Thus, the image can be anywhere from infinitely far away to right at the surface of the mirror. For the convex mirror, an object at infinity produces a virtual image at the focal point. As the object is brought closer, the reflected rays diverge more sharply and the image moves closer to the mirror. Thus, the virtual image is restricted to the region between the mirror and the focal point.

## APPLYING PHYSICS 23.3

Large trucks often have a sign on the back saying, "If you can't see my mirror, I can't see you." Explain this sign.

**Explanation** The trucking companies are making use of the principle of reversibility of light rays. In order for an image of you to be formed in the driver's mirror, there must be a pathway for rays of light to reach the mirror, allowing the driver to see your image. If you can't see the mirror, obviously there is no such pathway.

## Example 23.2 Images Formed by a Concave Mirror

Assume that a certain concave spherical mirror has a focal length of 10.0 cm. Locate the images for object distances of (a) 25.0 cm, (b) 10.0 cm, and (c) 5.00 cm. Describe the image in each case.

### Solution

**A** For an object distance of 25.0 cm, we find the image distance using the mirror equation:

$$\frac{1}{p} + \frac{1}{q} = \frac{1}{f}$$

$$\frac{1}{25.0 \text{ cm}} + \frac{1}{q} = \frac{1}{10.0 \text{ cm}}$$

$$q = 16.7 \text{ cm}$$

The magnification is given by Equation 23.2:

$$M = -\frac{q}{p} = -\frac{16.7 \text{ cm}}{25.0 \text{ cm}} = -0.667$$

Thus, the image is smaller than the object. Furthermore, the image is inverted because $M$ is negative. Finally, because $q$ is positive, the image is on the front side of the mirror and is real. This situation is pictured in Figure 23.13a.

**B** When the object distance is 10.0 cm, the object is at the focal point. Substituting the values $p = 10.0$ cm and $f = 10.0$ cm into the mirror equation, we find that

$$\frac{1}{10.0 \text{ cm}} + \frac{1}{q} = \frac{1}{10.0 \text{ cm}}$$

$$q = \infty$$

Thus, we see that rays of light originating from an object at the focal point of a concave mirror are reflected so that the image is formed an infinite distance from the mirror—that is, the rays travel parallel to one another after reflection. Furthermore, because the image distance is infinite, the magnification $M$ is infinite. In other words, light diverging from a point of zero height forms an image of height $2R$ and so $M$ is infinite.

**C** When the object is at 5.00 cm, inside the focal point of the mirror, the mirror equation gives

$$\frac{1}{5.00 \text{ cm}} + \frac{1}{q} = \frac{1}{10.0 \text{ cm}}$$

$$q = -10.0 \text{ cm}$$

That is, the image is virtual because it is behind the mirror. The magnification is

$$M = -\frac{q}{p} = -\left(\frac{-10.0 \text{ cm}}{5.00 \text{ cm}}\right) = 2.00$$

We see that the image height is magnified by a factor of 2, and the positive sign indicates that the image is upright (Fig. 23.13b).

Note the characteristics of an image formed by a concave spherical mirror. When the object is outside the focal point, the image is inverted and real; at the focal point, the image is formed at infinity; inside the focal point, the image is upright and virtual.

**EXERCISE** If the object distance is 20.0 cm, find the image distance and the magnification of the mirror.

**ANSWER** $q = 20.0$ cm, $M = -1.00$

## Example 23.3 Images Formed by a Convex Mirror

An object 3.00 cm high is placed 20.0 cm from a convex mirror with a focal length of 8.00 cm. Find (a) the position of the final image and (b) the magnification of the mirror.

**Solution**

**A** Because the mirror is convex, its focal length is negative. To find the image position, we use the mirror equation:

$$\frac{1}{p} + \frac{1}{q} = \frac{1}{f}$$

$$\frac{1}{20.0 \text{ cm}} + \frac{1}{q} = \frac{1}{-8.00 \text{ cm}}$$

$$q = -5.71 \text{ cm}$$

The negative value of $q$ indicates that the image is virtual, or behind the mirror, as in Figure 23.13c.

**B** The magnification of the mirror is

$$M = -\frac{q}{p} = -\left(\frac{-5.71 \text{ cm}}{20.0 \text{ cm}}\right) = 0.286$$

The image is upright because $M$ is positive.

**EXERCISE** Find the height of the image.

**ANSWER** 0.857 cm

### Webnote 23.1

To see how objects are formed into virtual images for a convex (diverging) mirror, go to *http://www.lightlink.com/sergey/java/java/dmirr/index.html*

## Example 23.4 An Enlarged Image

When a woman stands with her face 40.0 cm from a cosmetic mirror, the upright image is twice as tall as her face. What is the focal length of the mirror?

**Reasoning** Most of the problems we have encountered so far have been simple applications of the mirror equation. However, to find $f$ in this example, we must first find $q$, the image distance. Because the problem states that the image is upright, the magnification must be positive (in this case, $M = +2$), and because $M = -q/p$, we can determine $q$.

**Solution** The magnification equation gives us a relationship between the object and image distances:

$$M = -\frac{q}{p} = 2$$

$$q = -2p = -2(40.0 \text{ cm}) = -80.0 \text{ cm}$$

First, note that a virtual image is formed because the woman is able to see her upright image in the mirror. This explains why the image distance is negative. Substitute $q = -80.0$ cm into the mirror equation to obtain

$$\frac{1}{40.0\text{ cm}} - \frac{1}{80.0\text{ cm}} = \frac{1}{f}$$

$$f = 80.0\text{ cm}$$

The positive sign for the focal length indicates that the mirror is concave, a fact that we already knew because the mirror magnified the object. (A convex mirror would have produced a diminished image.)

## 23.4 IMAGES FORMED BY REFRACTION

In this section we describe how images are formed by refraction at a spherical surface. Consider two transparent media with indices of refraction $n_1$ and $n_2$, where the boundary between the two media is a spherical surface of radius $R$ (Fig. 23.15). Let us assume that the medium to the right has a higher index of refraction than the one to the left; that is, $n_2 > n_1$. This would be the case for light entering a curved piece of glass from air or for light entering the water in a fishbowl from air. The rays originating at the object location $O$ are refracted at the spherical surface and then converge to the image point $I$. We can begin with Snell's law of refraction and use simple geometric techniques to show that the object distance, image distance, and radius of curvature are related by the equation

$$\frac{n_1}{p} + \frac{n_2}{q} = \frac{n_2 - n_1}{R} \qquad [23.7]$$

Furthermore, the magnification of a refracting surface is

$$M = \frac{h'}{h} = -\frac{n_1 q}{n_2 p} \qquad [23.8]$$

As with mirrors, we must use a sign convention if we are to apply these equations to a variety of circumstances. First note that real images are formed on the side of the surface *opposite* the side from which the light comes. This is in contrast with mirrors, where real images are formed on the *same* side of the reflecting surface. We define the side of the surface in which light rays originate as the front side. The other side is called the back side. Because of the difference in location of real images, the refraction sign conventions for $q$ and $R$ are the opposite of those for reflection. For example, $p$, $q$, and $R$ are all positive in Figure 23.15. The sign convention for spherical refracting surfaces is summarized in Table 23.2.

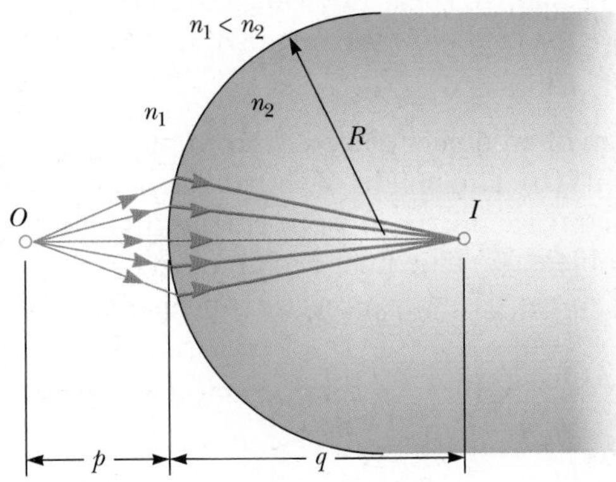


**FIGURE 23.15** An image formed by refraction at a spherical surface. Rays making small angles with the principal axis diverge from a point object at $O$ and pass through the image point $I$.

**TABLE 23.2 Sign Conventions for Refracting Surfaces**

| Quantity | Positive When | Negative When |
|---|---|---|
| Object location ($p$) | Object is in front of surface | Object is in back of surface |
| Image location ($q$) | Image is in back of surface | Image is in front of surface |
| Image height ($h'$) | Image is upright | Image is inverted |
| Radius ($R$) | Center of curvature is in back of surface | Center of curvature is in front of surface |

## APPLYING PHYSICS 23.4

**APPLICATION**

OPENING YOUR EYES UNDERWATER

Why does a person with normal vision see a blurry image if the eyes are opened underwater without goggles or a diving mask?

**Explanation** The eye presents a spherical refraction surface. The eye normally functions so that light entering from the air is refracted to form an image in the retina located at the back of the eyeball. The difference in index of refraction between water and the eye is smaller than the difference in index of refraction between air and the eye. Thus, light entering the eye from the water does not experience as much refraction as does light entering from the air, and the image is formed behind the retina. A diving mask or swimming goggles have no optical action of their own; they are simply flat pieces of glass or plastic in a rubber mount. However, they provide a region of air adjacent to the eyes, so that the correct refraction relationship is established, and images will be in focus.

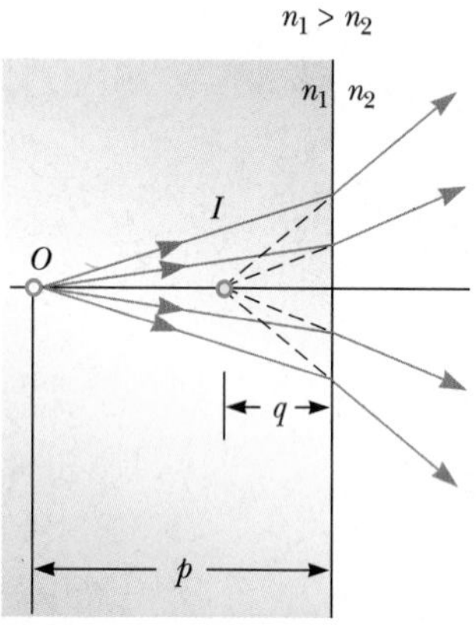

$n_1 > n_2$

**FIGURE 23.16** The image formed by a flat refracting surface is virtual; that is, it forms to the left of the refracting surface. Note that if the light rays are reversed in direction, we have the situation described in Example 22.6.

### FLAT REFRACTING SURFACES

If the refracting surface is flat, then $R$ approaches infinity and Equation 23.7 reduces to

$$\frac{n_1}{p} = -\frac{n_2}{q}$$

$$q = -\frac{n_2}{n_1}p \qquad [23.9]$$

From Equation 23.9 we see that the sign of $q$ is opposite that of $p$. Thus, **the image formed by a flat refracting surface is on the same side of the surface as the object.** This is illustrated in Figure 23.16 for the situation in which $n_1$ is greater than $n_2$, where a virtual image is formed between the object and the surface. Note that the refracted ray bends *away* from the normal in this case, because $n_1 > n_2$.

**Quick Quiz 23.2** A person spear fishing from a boat sees a fish located 3 m from the boat at an apparent depth of 1 m. To spear the fish, should the person aim (a) at, (b) above, or (c) below the image of the fish?

**Quick Quiz 23.3** True or false? (a) The image of an object placed in front of a concave mirror is always upright. (b) The height of the image of an object placed in front of a concave mirror must be smaller than or equal to the height of the object. (c) The image of an object placed in front of a convex mirror is always upright and smaller than the object.

## Example 23.5 Gaze into the Crystal Ball

A coin 2.00 cm in diameter is embedded in a solid glass ball of radius 30.0 cm (Fig. 23.17). The index of refraction of the ball is 1.50, and the coin is 20.0 cm from the surface. Find the position and height of the image of the coin.

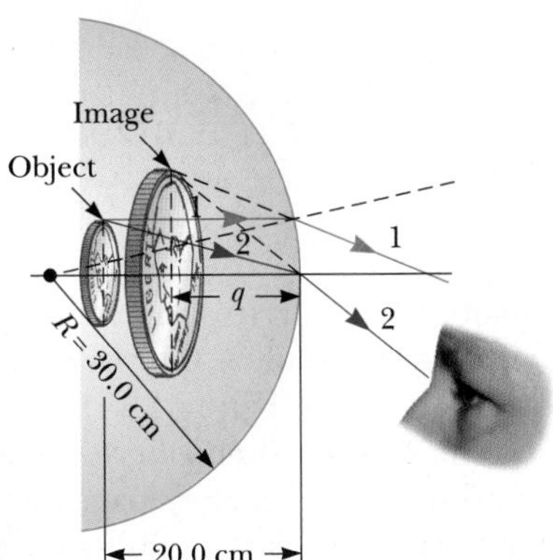


**FIGURE 23.17** (Example 23.5) A coin embedded in a glass ball forms a virtual image between the coin and the glass surface.

**Solution** Because they are moving from a medium of high index of refraction to a medium of lower index of refraction, the rays originating at the coin are refracted away from the normal at the surface and diverge outward. The image is formed in the glass and is virtual. Applying Equation 23.7 and taking $n_1 = 1.50$, $n_2 = 1.00$, $p = 20.0$ cm, and $R = -30.0$ cm, we get

$$\frac{n_1}{p} + \frac{n_2}{q} = \frac{n_2 - n_1}{R}$$

$$\frac{1.50}{20.0 \text{ cm}} + \frac{1.00}{q} = \frac{1.00 - 1.50}{-30.0 \text{ cm}}$$

$$q = -17.1 \text{ cm}$$

The negative sign indicates that the image is in the same medium as the object (the side of incident light), in agreement with our ray diagram, and therefore must be virtual.

To find the image height, we first use Equation 23.8 for the magnification:

$$M = -\frac{n_1 q}{n_2 p} = -\frac{1.50(-17.1 \text{ cm})}{1.00(20.0 \text{ cm})} = \frac{h'}{h}$$

Therefore,

$$h' = 1.28h = (1.28)(2.00 \text{ cm}) = 2.56 \text{ cm}$$

The positive value for $M$ indicates an upright image.

## Example 23.6 The One That Got Away

A small fish is swimming at a depth of $d$ below the surface of a pond (Fig. 23.18). What is the *apparent depth* of the fish as viewed from directly overhead?

**Reasoning** In this example, the refracting surface is flat, and so $R$ is infinite. Hence, we can use Equation 23.9 to determine the location of the image, which is the apparent location of the fish.

**Solution** The facts that $n_1 = 1.33$ for water and $p = d$ give us

$$q = -\frac{n_2}{n_1}p = -\frac{1}{1.33}d = -0.752d$$

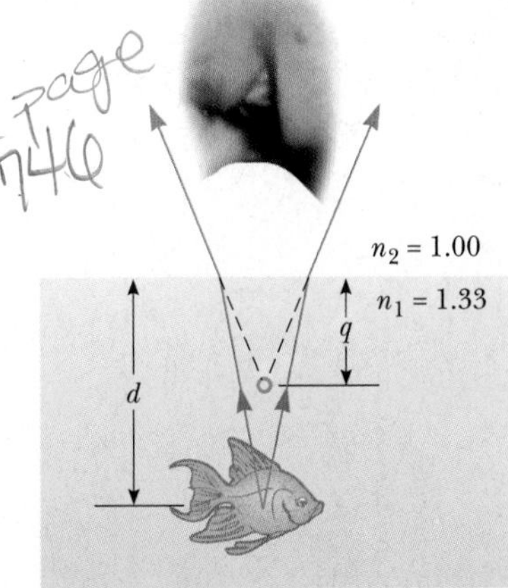


**FIGURE 23.18** (Example 23.6) The apparent depth $q$ of the fish is less than the true depth $d$.

Again, because $q$ is negative, the image is virtual, as indicated in Figure 23.18. The apparent depth is three-fourths the actual depth. For instance, if $d = 4.0$ m, $q = -3.0$ m.

**EXERCISE** If the fish is 12 cm long, how long is its image?

**ANSWER** 12 cm

## 23.5 ATMOSPHERIC REFRACTION

Images formed by refraction in our atmosphere lead to some interesting results. In this section we look at two examples. A situation that occurs daily is the visibility of the Sun at dusk even though it has passed below the horizon. Figure 23.19 shows why this occurs. Rays of light from the Sun strike the Earth's atmosphere (represented by the shaded area around the Earth) and are bent as they pass into a medium that has an index of refraction different from that of the almost empty space in which they have been traveling. The bending in this situation differs somewhat from the bending we have considered previously in that it is gradual and continuous as the light moves through the atmosphere toward an observer at point $O$. This is because the light moves through layers of air that have a continuously changing index of refraction. When the rays reach the observer, the eye follows them back along the direction from which they appear to have come (indicated by the dashed path in the figure). The end result is that the Sun is seen to be above the horizon even after it has fallen below it.

The **mirage** is another phenomenon of nature produced by refraction in the atmosphere. A mirage can be observed when the ground is so hot that the air directly above it is warmer than the air at higher elevations. The desert is a region in which such circumstances prevail, but mirages are also seen on heated roadways during the summer. The layers of air at different heights above the Earth have different densities and different refractive indices. The effect this can have is pictured in Figure 23.20a. In this situation the observer sees the sky and a tree in two different ways. One group of light rays reaches the observer by the straight-line path $A$, and the eye traces these rays back to see the tree in the normal fashion. In addition, a second group of rays travels along the curved path $B$. These rays are directed toward the ground and are then bent as a result of refraction. As a consequence, the observer also sees an inverted image of the tree and sky background as he traces these rays back to the point at which they appear to have originated. Because an upright image and an inverted image are seen when the image of a tree is observed in a reflecting pool of water, the observer unconsciously calls on this past experience and concludes that the sky background is reflected by a pool of water in front of the tree.

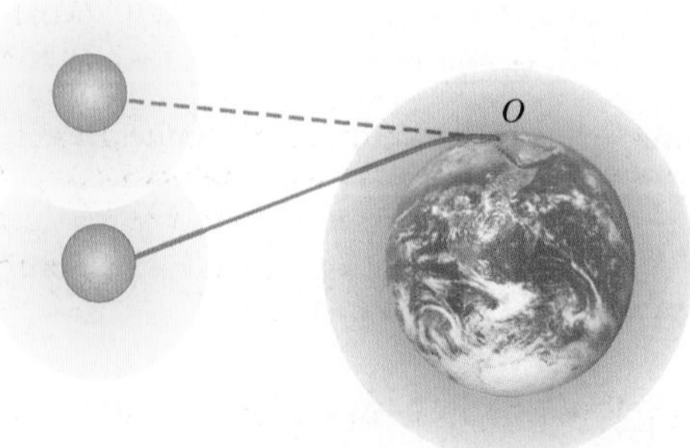


**FIGURE 23.19** Because light is refracted by the Earth's atmosphere, an observer at $O$ sees the Sun even though it has fallen below the horizon.

(a)

(b)

**FIGURE 23.20** (a) A mirage is produced by the bending of light rays in the atmosphere when there are large temperature differences between the ground and the air. (b) *(John M. Dunay IV, Fundamental Photographs, NYC)*

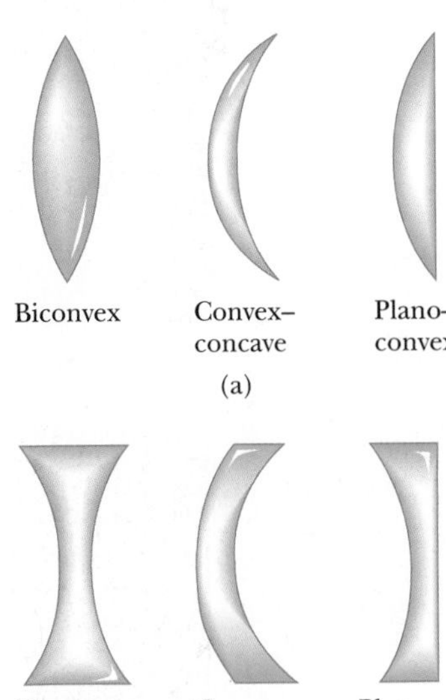


**FIGURE 23.21** Lens shapes. (a) Converging lenses have positive focal lengths and are thickest at the middle. (b) Diverging lenses have negative focal lengths and are thickest at the edges.

## 23.6 THIN LENSES

A typical **thin lens** consists of a piece of glass or plastic, ground so that each of its two refracting surfaces is a segment of either a sphere or a plane. Lenses are commonly used to form images by refraction in optical instruments, such as cameras, telescopes, and microscopes. The equation that relates object and image distances for a lens is virtually identical to the mirror equation derived earlier, and the method used to derive it is also similar.

Figure 23.21 shows some representative shapes of lenses. Notice that we have placed these lenses in two groups. Those in Figure 23.21a are thicker at the center than at the rim, and those in Figure 23.21b are thinner at the center than at the rim. The lenses in the first group are examples of **converging lenses,** and those in the second group are **diverging lenses.** The reason for these names will become apparent shortly.

As we did for mirrors, it is convenient to define a point called the **focal point** for a lens. For example, in Figure 23.22a, a group of rays parallel to the axis passes through the focal point $F$ after being converged by the lens. The distance from the focal point to the lens is called the **focal length,** $f$. **The focal length is the image distance that corresponds to an infinite object distance.** Recall that we are considering the lens to be very thin. As a result, it makes no difference whether we take the focal length to be the distance from the focal point to the surface of the lens or the distance from the focal point to the center of the lens, because the difference between these two lengths is negligible. A thin lens has *two* focal points, as illustrated in Figure 23.22, corresponding to parallel rays traveling from the left and from the right.

Rays parallel to the axis diverge after passing through a lens of biconcave shape in Figure 23.22b. In this case, the focal point is defined to be the point at which the diverged rays appear to originate, labeled $F$ in the figure. Figures 23.22a and 23.22b indicate why the names *converging* and *diverging* are applied to these lenses.

Consider a ray of light passing through the center of a lens, labeled ray 1 in Figure 23.23. For a thin lens, a ray passing through the center is undeviated. Ray 2 in Figure 23.23 is parallel to the principal axis of the lens (the horizontal axis passing through $O$), and as a result it passes through the focal point $F$ after refraction. Rays 1 and 2 intersect at the point which is the tip of the image arrow.

We first note that the tangent of the angle $\alpha$ can be found by using the blue and gold shaded triangles in Figure 23.23:

$$\tan \alpha = \frac{h}{p} \qquad \text{or} \qquad \tan \alpha = -\frac{h'}{q}$$

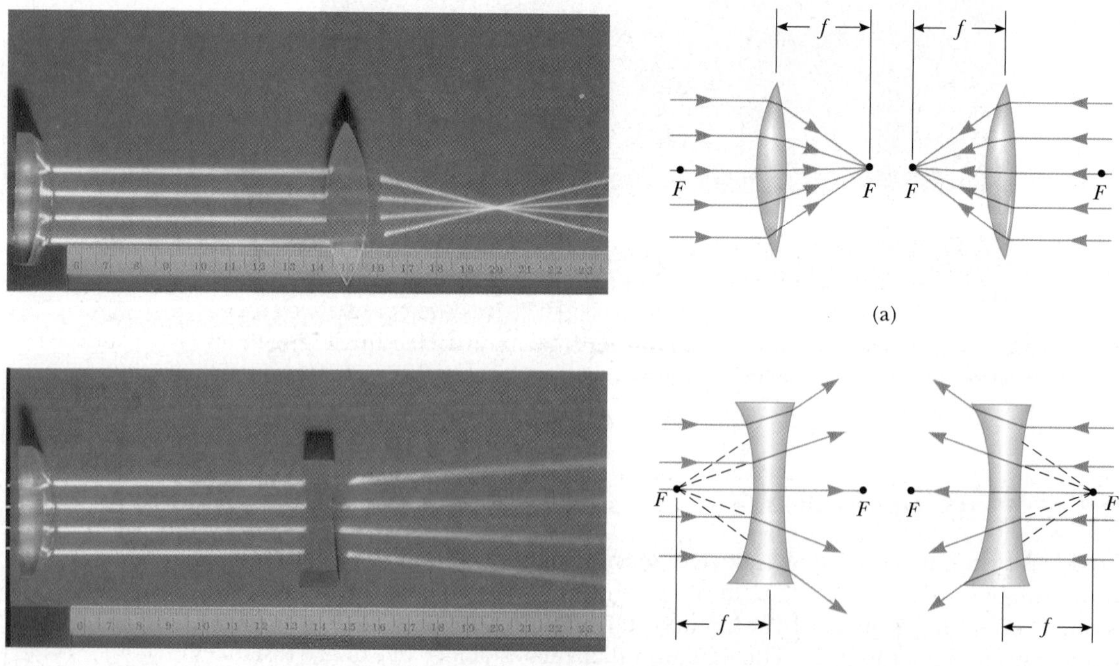


**FIGURE 23.22** (*Left*) Photographs of the effects of converging and diverging lenses on parallel rays. (*Courtesy of Jim Lehman, James Madison University*) (*Right*) The focal points of (a) the biconvex lens and (b) the biconcave lens.

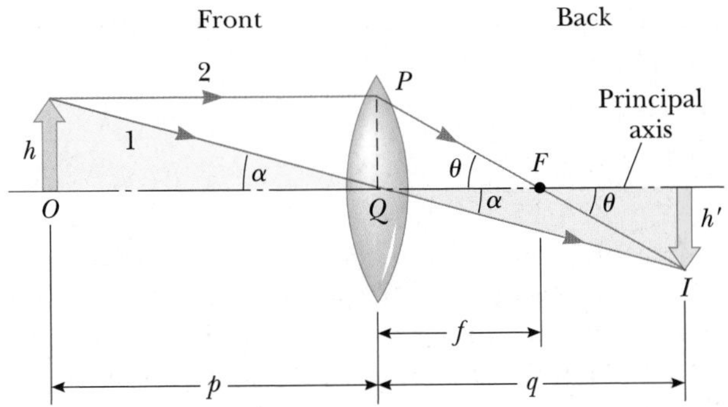


**FIGURE 23.23** A geometric construction for developing the thin-lens equation.

From this we find that

$$M = \frac{h'}{h} = -\frac{q}{p} \qquad [23.10]$$

Thus, the equation for magnification by a lens is the same as the equation for magnification by a mirror. We also note from Figure 23.23 that the tangent of $\theta$ is

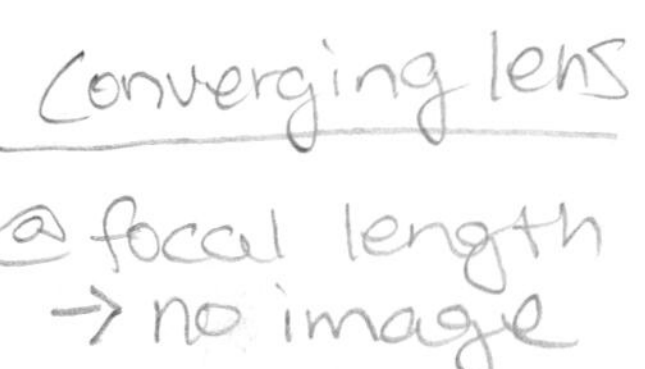


$$\tan\theta = \frac{PQ}{f} \qquad \text{or} \qquad \tan\theta = -\frac{h'}{q-f}$$

However, the height $PQ$ used in the first of these equations is the same as $h$, the height of the object. Therefore,

$$\frac{h}{f} = -\frac{h'}{q-f}$$

$$\frac{h'}{h} = -\frac{q-f}{f}$$

Using this in combination with Equation 23.10 gives

$$\frac{q}{p} = \frac{q-f}{f}$$

which reduces to

$$\frac{1}{p} + \frac{1}{q} = \frac{1}{f} \qquad [23.11]$$

Thin-lens equation

This equation, called the **thin-lens equation,** can be used with both converging and diverging lenses if we adhere to a set of sign conventions. Figure 23.24 is useful for obtaining the signs of $p$ and $q$, and Table 23.3 gives the complete sign conventions for lenses. Note that **a converging lens has a positive focal length** under this convention, and **a diverging lens has a negative focal length.** Hence the names *positive* and *negative* are often given to these lenses.

| Front | Back |
|---|---|
| $p$ positive<br>$q$ negative | $p$ negative<br>$q$ positive |
| Incident light | Refracted light |

**FIGURE 23.24** A diagram for obtaining the signs of $p$ and $q$ for a thin lens or a refracting surface.

**TABLE 23.3** Sign Conventions for Thin Lenses

| Quantity | Positive When | Negative When |
|---|---|---|
| Object location ($p$) | Object is in front of lens | Object is in back of lens |
| Image location ($q$) | Image is in back of lens | Image is in front of lens |
| Image height ($h'$) | Image is upright | Image is inverted |
| $R_1$ and $R_2$ | Center of curvature is in back of lens | Center of curvature is in front of lens |
| Focal length ($f$) | Converging lens | Diverging lens |

The focal length for a lens in air is related to the curvatures of its front and back surfaces and to the index of refraction $n$ of the lens material by

Lens maker's equation

$$\frac{1}{f} = (n - 1)\left(\frac{1}{R_1} - \frac{1}{R_2}\right) \qquad [23.12]$$

where $R_1$ is the radius of curvature of the front surface of the lens and $R_2$ is the radius of curvature of the back surface. (As with mirrors, we arbitrarily call the side from which the light approaches the *front* of the lens.) Table 23.3 gives the sign conventions for $R_1$ and $R_2$. Equation 23.12 enables us to calculate the focal length from the known properties of the lens. It is called the **lens maker's equation.**

## Ray Diagrams for Thin Lenses

Ray diagrams are essential for understanding the overall image formation by a thin lens or a system of lenses. They should also help clarify the sign conventions we have already discussed. Figure 23.25 illustrates this method for three single-lens situations. To locate the image formed by a converging lens (Fig. 23.25a and b), the following three rays are drawn from the top of the object:

1. The first ray is drawn parallel to the principal axis. After being refracted by the lens, this ray passes through (or appears to come from) one of the focal points.
2. The second ray is drawn through the center of the lens. This ray continues in a straight line.
3. The third ray is drawn through the other focal point and emerges from the lens parallel to the principal axis.

A similar construction is used to locate the image formed by a diverging lens, as shown in Figure 23.25c. The point of intersection of *any two* of the rays in these diagrams can be used to locate the image. The third ray serves as a check on construction.

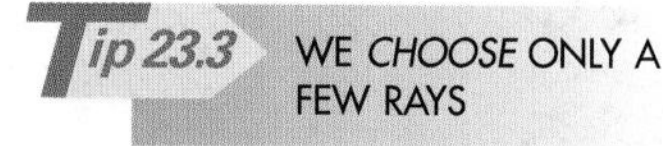

Tip 23.3
WE CHOOSE ONLY A FEW RAYS


Note that although our ray diagrams in Figure 23.25 only show three rays leaving an object, there are an infinite number of rays that can be drawn between the object and its image.

For the converging lens in Figure 23.25a, where the object is *outside* the front focal point ($p > f$), the ray diagram shows the image is real and inverted. When the real object is *inside* the front focal point ($p < f$), as in Figure 23.25b, the image is virtual and upright. For the diverging lens of Figure 23.25c, the image is virtual and upright.

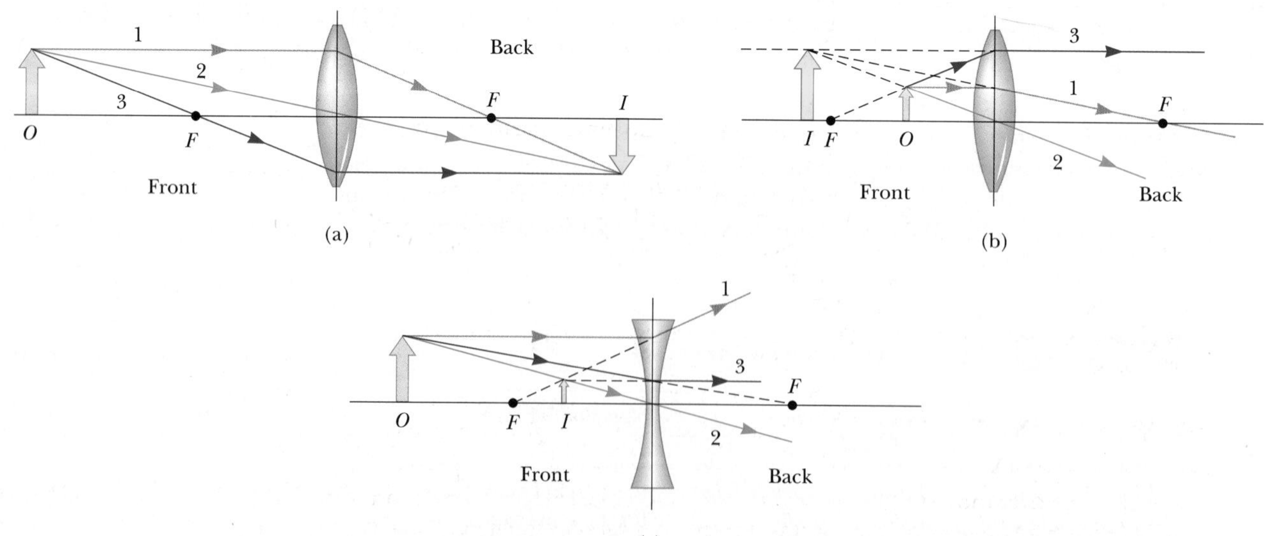


**FIGURE 23.25** Ray diagrams for locating the image of an object. (a) The object is outside the focal point of a converging lens. (b) The object is inside the focal point of a converging lens. (c) The object is outside the focal point of a diverging lens.

**Quick Quiz 23.4** A plastic sandwich bag filled with water can act as a crude converging lens in air. If the bag is filled with air and placed under water, is the effective lens (a) converging or (b) diverging?

**Quick Quiz 23.5** In Figure 23.25a, the blue object arrow is replaced by one that is much taller than the lens. How many rays from the object will strike the lens?

**Quick Quiz 23.6** An object is placed to the left of a converging lens. Which of the following statements are true and which are false? (a) The image is always to the right of the lens. (b) The image can be upright or inverted. (c) The image is always smaller or the same size as the object. Justify your answers with ray diagrams.

## PROBLEM-SOLVING STRATEGY Lenses and Mirrors

Your success or failure in working lens and mirror problems will be determined largely by whether or not you make sign errors when substituting into the lens and mirror equations. The only way to ensure that you don't make sign errors is to become adept at using the sign conventions. The best way to do this is to work a multitude of problems on your own and to make confirming ray diagrams. Watching an instructor or reading the example problems is no substitute for practice.

## APPLYING PHYSICS 23.5

**APPLICATION**

VISION AND DIVING MASKS

Diving masks often have a lens built into the glass for divers who do not have perfect vision. This allows the individual to dive without the necessity of glasses, because the lenses in the faceplate perform the necessary refraction to produce clear vision. Normal glasses have lenses that are curved on both the front and rear surfaces. The lenses in a diving mask faceplate often only have curved surfaces on the inside of the glass. Why is this design desirable?

**Explanation** The main reason for curving only the inner surface of the lenses in the diving mask faceplate is so that the diver can see clearly while underwater and in the air. If there were curved surfaces on both the front and the back of the diving lens, there would be two refractions. The lens could be designed so that these two refractions would give clear vision while the diver is in air. When the diver goes underwater, however, the refraction between the water and the glass at the first interface is now different, because the index of refraction of water is different from that of air. Thus, the vision will not be clear underwater.

## APPLYING PHYSICS 23.6

Consider a glass plano-convex lens, one that is flat on one side and convex on the other. You project three laser beams through it, as shown in Figure 23.26a, and measure the focal length $f$—that is, the distance from the lens to the point at which the three beams cross. Now you hold the flat side of the lens against the glass of an aquarium filled with water, as in Figure 23.26b. When you shine the

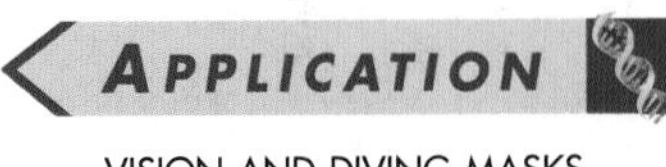


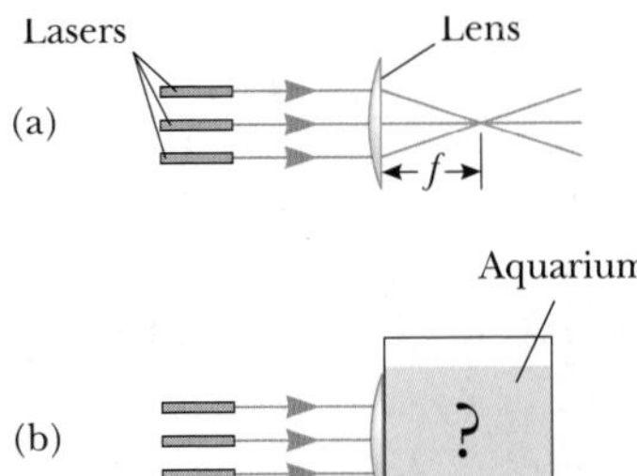


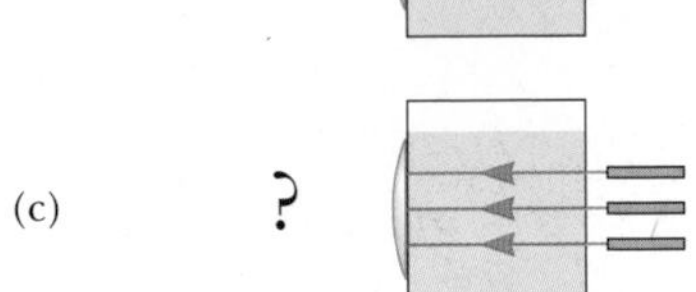


**FIGURE 23.26** (Applying Physics 23.6)

laser beams through the lens from the outside of the aquarium, will the beams cross at a point closer to the lens than before, farther away, or at the same distance? What if you direct the laser beams through the lens from the other side of the aquarium, as in Figure 23.26c? What happens in this case?

**Explanation** The laser beam going through the center of the lens is undeviated by the lens in all three cases. We need to look at what happens to the outer beams. In Figure 23.26a, these two laser beams will refract toward the normal as they pass from the air into the glass of the lens. Thus, they will be deviated *toward* the central beam. As they reach the other side, and pass from the glass back into the air, they will deviate away from the normal, which causes additional deviation toward the central beam. As a result, all three beams cross at the focal point. When the flat side of the lens is held against the glass side of the aquarium as in Figure 23.26b, the outgoing laser beams pass through the glass of the aquarium without additional refraction, but then enter *water.* The change in index of refraction in going from glass to water is much smaller than in the previous case, and the deviation toward the central beam is less. As a result, the crossing point for the three beams is farther from the lens. When the laser beams are directed through the water and then through the lens, as in Figure 23.26c, we must consider two factors. First, the focal length of a lens is independent of which way the light passes through it. Second, in the situation shown, the laser beams experience no refraction on entering the flat side of the lens, because they strike the glass at normal incidence. Thus, all of the refraction occurs at the interface of the air with the curved side of the lens. This is exactly the same situation as if the laser beams had struck the flat side of the lens while it was in air. Thus, there is no effect of the water in the aquarium. The laser beams cross at the same distance from the lens as they did in the first diagram. This third situation is very similar to the curvature of a lens on the inside of a diving mask, discussed previously.

## Example 23.7 Images Formed by a Converging Lens

A converging lens of focal length 10.0 cm forms images of objects placed (a) 30.0 cm, (b) 10.0 cm, and (c) 5.00 cm from the lens. In each case, construct a ray diagram, find the image distance, and describe the image.

**Solution**

**A** First we construct a ray diagram as shown in Figure 23.27a. The diagram shows that a real, inverted, smaller image is formed on the far side of the lens. The thin-lens equation, Equation 23.11, can be used to find the image distance:

$$\frac{1}{p} + \frac{1}{q} = \frac{1}{f}$$

$$\frac{1}{30.0 \text{ cm}} + \frac{1}{q} = \frac{1}{10.0 \text{ cm}}$$

$$q = +15.0 \text{ cm}$$

The positive sign for the image distance tells us that the image is real and on the back side of the lens. The magnification of the lens is

$$M = -\frac{q}{p} = -\frac{15.0 \text{ cm}}{30.0 \text{ cm}} = -0.500$$

Thus, the image is reduced in height by one half, and the negative sign for $M$ tells us that the image is inverted.

**B** No calculation is necessary for this case because we know that when the object is placed at the focal point, the image is formed at infinity. This is readily verified by substituting $p = 10.0$ cm into the lens equation.

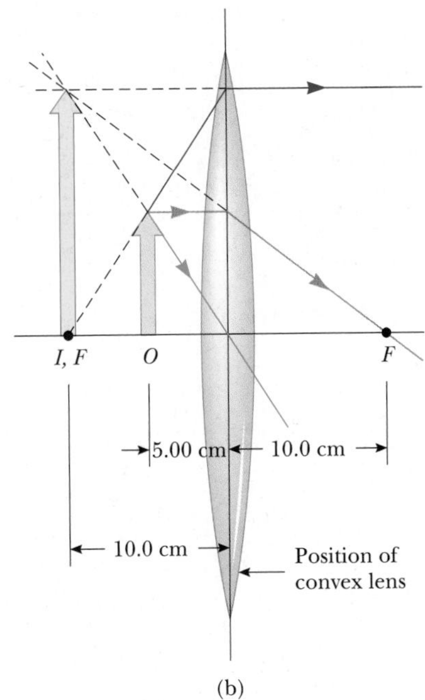

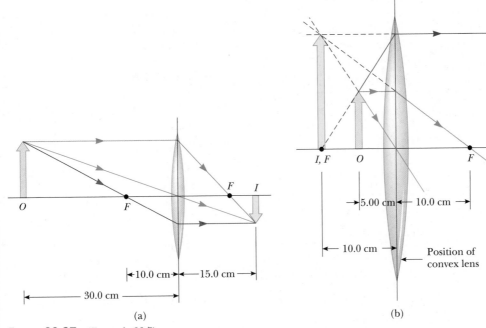

(a) (b)

**FIGURE 23.27** (Example 23.7)

**C** We now move inside the focal point. The ray diagram in Figure 23.27b immediately shows that in this case the lens is being used as a magnifying glass; that is, the image is magnified, erect, on the same side as the object, and virtual. Because the object distance is 5.00 cm, the lens equation gives

$$\frac{1}{5.00 \text{ cm}} + \frac{1}{q} = \frac{1}{10.0 \text{ cm}}$$

$$q = \boxed{-10.0 \text{ cm}}$$

$$M = -\frac{q}{p} = -\left(\frac{-10.0 \text{ cm}}{5.00 \text{ cm}}\right) = \boxed{+2.00}$$

The negative image distance tells us that the image is virtual and formed on the side of the lens from which the light is incident, the front side. The image is enlarged, and the positive sign for $M$ tells us that the image is upright.

## Example 23.8 The Case of a Diverging Lens

Repeat the problem of Example 23.7 for a *diverging* lens of focal length 10.0 cm.

### Solution

**A** We begin by constructing a ray diagram as in Figure 23.28a taking the object distance to be 30.0 cm. The diagram shows that the image is virtual, smaller than the object, and upright. Let us now apply the lens equation with $p = 30.0$ cm:

$$\frac{1}{p} + \frac{1}{q} = \frac{1}{f}$$

$$\frac{1}{30.0 \text{ cm}} + \frac{1}{q} = -\frac{1}{10.0 \text{ cm}}$$

$$q = \boxed{-7.50 \text{ cm}}$$

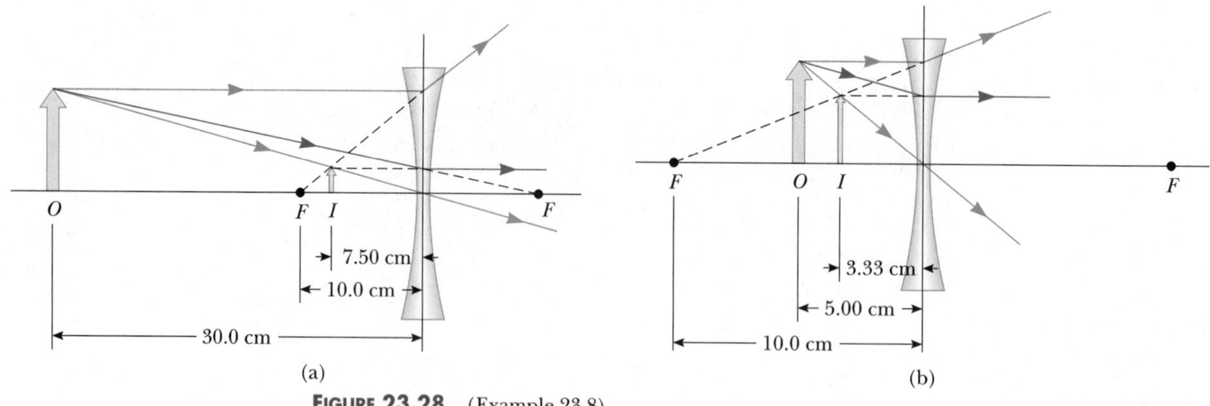


**FIGURE 23.28** (Example 23.8)

The magnification is

$$M = -\frac{q}{p} = -\left(\frac{-7.50 \text{ cm}}{30.0 \text{ cm}}\right) = +0.250$$

This result confirms that the image is virtual, smaller than the object, and upright.

**B** When the object is at the focal point, $p = 10.0$ cm, we have

$$\frac{1}{10.0 \text{ cm}} + \frac{1}{q} = -\frac{1}{10.0 \text{ cm}}$$

$$q = -5.00 \text{ cm}$$

$$M = -\frac{q}{p} = -\left(\frac{-5.00 \text{ cm}}{10.0 \text{ cm}}\right) = +0.500$$

**C** When the object is inside the focal point, at $p = 5.00$ cm, the ray diagram in Figure 23.28b shows that we have a virtual image that is smaller than the object and upright. In this case, the lens equation gives

$$\frac{1}{5.00 \text{ cm}} + \frac{1}{q} = -\frac{1}{10.0 \text{ cm}}$$

$$q = -3.33 \text{ cm}$$

$$M = -\left(\frac{-3.33 \text{ cm}}{5.00 \text{ cm}}\right) = +0.667$$

This confirms that the image is virtual, smaller than the object, and upright.

## COMBINATIONS OF THIN LENSES

If two thin lenses are used to form an image, the system can be treated in the following manner. First, the image produced by the first lens is calculated as though the second lens were not present. The light then approaches the second lens *as if* it had come from the image formed by the first lens. Hence, **the image formed by the first lens is treated as the object for the second lens.** The image formed by the second lens is the final image of the system. If the image formed by the first lens lies on the back side of the second lens, then the image is treated as a virtual object for the second lens (that is, $p$ is negative). The same procedure can be extended to a system of three or more lenses. The overall magnification of a system of thin lenses is the *product* of the magnifications of the separate lenses.

### Webnote 23.2

For a more advanced version of lens and mirror combinations, go to *http://webphysics.ph.msstate.edu/javamirror/ntnujava/Lens/lens_e.html*

## Example 23.9 Two Lenses in a Row

Two converging lenses are placed 20.0 cm apart, as shown in Figure 23.29a. If the first lens has a focal length of 10.0 cm and the second has a focal length of 20.0 cm, locate the final image formed of an object 30.0 cm in front of the first lens. Find the magnification of the system.

**Reasoning** We apply the thin-lens equation to both lenses. The image formed by the first lens is treated as the object for the second lens. Also, we use the fact that the total magnification of the system is the product of the magnifications produced by the separate lenses.

**Solution** First we make ray diagrams roughly to scale to see where the image from the first lens falls and how it acts as the object for the second lens. (See Fig. 23.29b.) The location of the image formed by the first lens is found via the thin-lens equation:

$$\frac{1}{30.0 \text{ cm}} + \frac{1}{q} = \frac{1}{10.0 \text{ cm}}$$

$$q = \boxed{+15.0 \text{ cm}}$$

The magnification of this lens is

$$M_1 = -\frac{q}{p} = -\frac{15.0 \text{ cm}}{30.0 \text{ cm}} = \boxed{-0.500}$$

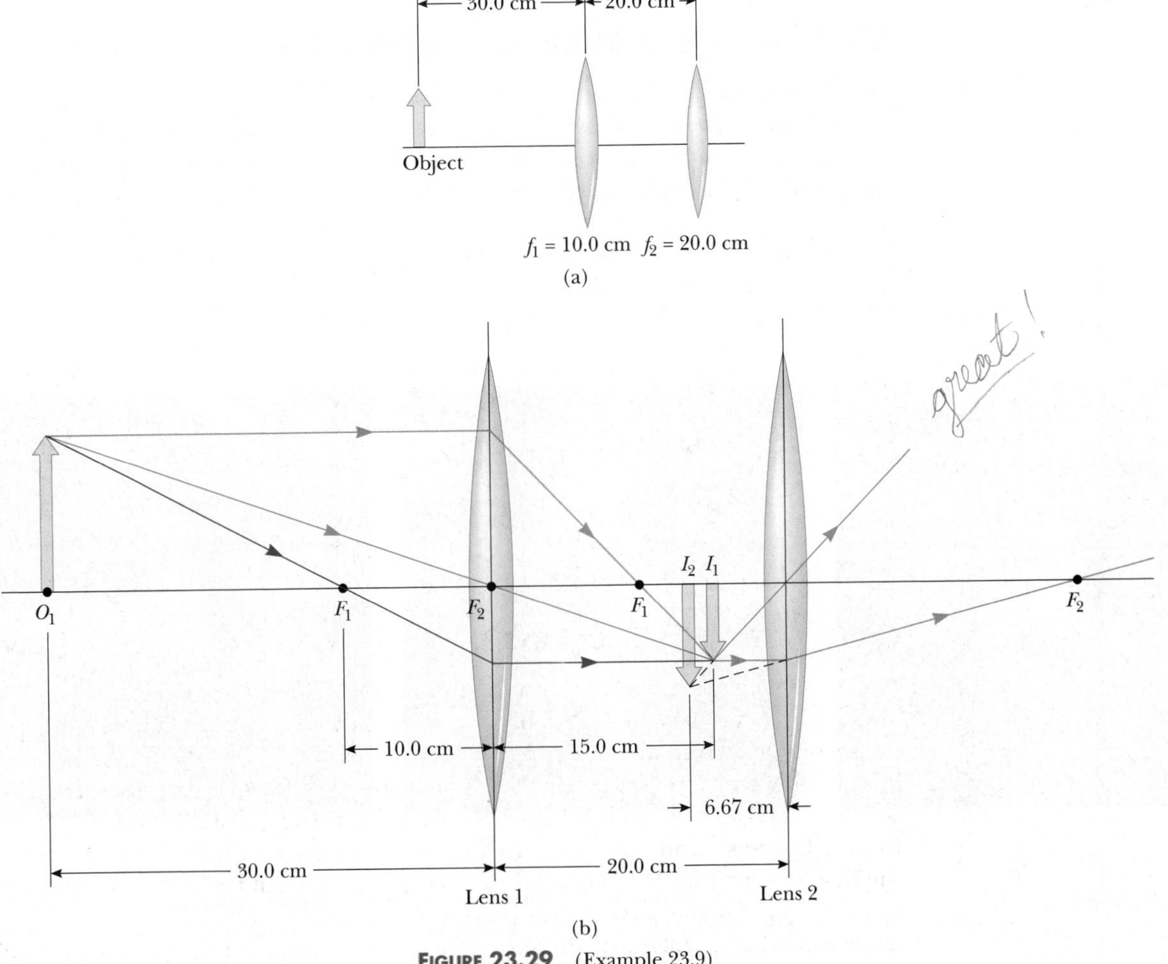


**FIGURE 23.29** (Example 23.9)

The image formed by this lens becomes the object for the second lens. Thus, the object distance for the second lens is 20.0 cm − 15.0 cm = 5.00 cm. We again apply the thin-lens equation to find the location of the final image.

$$\frac{1}{5.00 \text{ cm}} + \frac{1}{q} = \frac{1}{20.0 \text{ cm}}$$

$$q = -6.67 \text{ cm}$$

The magnification of the second lens is

$$M_2 = -\frac{q}{p} = -\frac{(-6.67 \text{ cm})}{5.00 \text{ cm}} = +1.33$$

Thus, the final image is 6.67 cm to the left of the second lens, and the overall magnification of the system is

$$M = M_1 M_2 = (-0.500)(1.33) = -0.667$$

The negative sign indicates that the final image is inverted with respect to the initial object.

**EXERCISE** If the two lenses in Figure 23.29 are separated by 10.0 cm, locate the final image and find the magnification of the system.

**ANSWER** 4.00 cm behind the second lens; $M = -0.400$

## 23.7 LENS AND MIRROR ABERRATIONS

One of the basic problems of systems containing mirrors and lenses is the imperfect quality of the images, which is largely the result of defects in shape and form. The simple theory of mirrors and lenses assumes that rays make small angles with the principal axis and that all rays reaching the lens or mirror from a point source are focused at a single point, producing a sharp image. Clearly, this is not always true in the real world. Where the approximations used in this theory do not hold, imperfect images are formed.

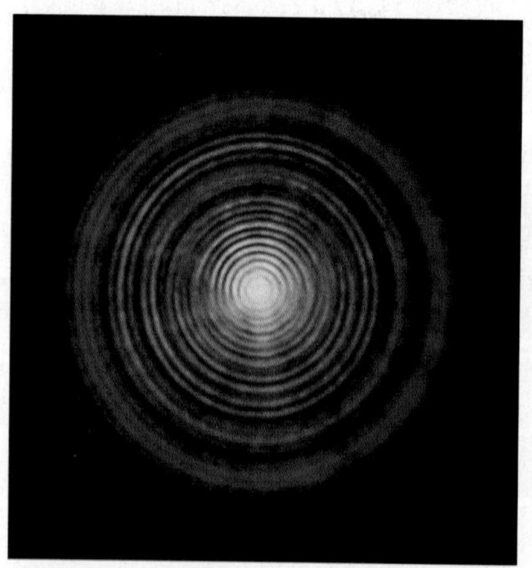

(a)

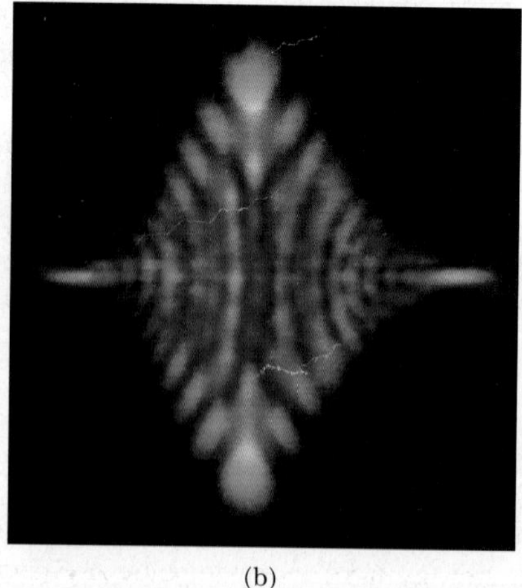

(b)

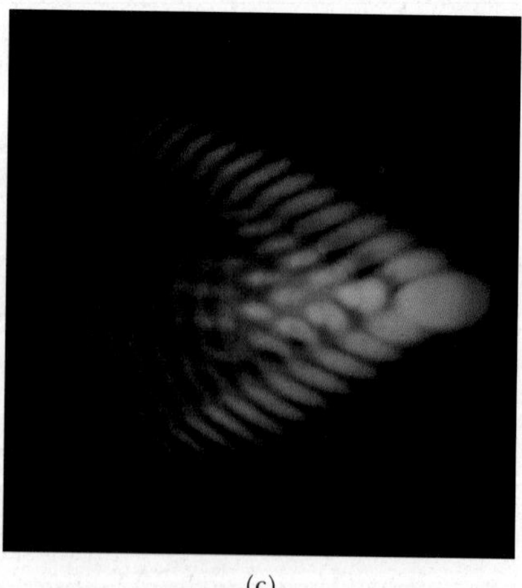

(c)

**FIGURE 23.30** Lenses can produce varied forms of aberrations, as shown by these blurred photographic images of a point source. (a) Spherical aberration occurs when light passing through the lens at different distances from the principal axis is focused at different points. (b) Astigmatism is an aberration that occurs when the object is not on the principal axis of the lens. (c) Coma. This aberration occurs when light passing through the lens far from the principal axis focuses at a different part of the focal plane from light passing near the center of the lens. *(Photos by Norman Goldberg)*

If one wishes to analyze image formation precisely, it is necessary to trace each ray, using Snell's law, at each refracting surface. This procedure shows that there is no single-point image; instead, the image is blurred. The departures of real (imperfect) images from the ideal predicted by the simple theory are called **aberrations.** Two common types of aberrations are spherical aberration and chromatic aberration. Photographs of three forms of lens aberrations are shown in Figure 23.30.

## SPHERICAL ABERRATION

Spherical aberration results from the fact that the focal points of light rays far from the principal axis of a spherical lens (or mirror) are different from the focal points of rays with the same wavelength passing near the axis. Figure 23.31 illustrates spherical aberration for parallel rays passing through a converging lens. Rays near the middle of the lens are imaged farther from the lens than rays at the edges. Hence, there is no single focal length for a spherical lens.

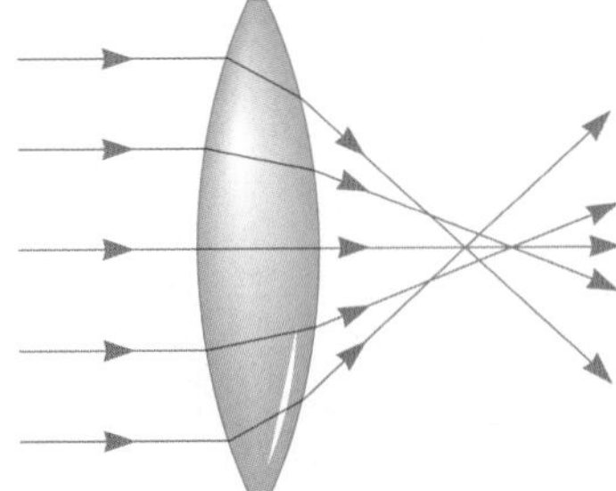

**FIGURE 23.31** Spherical aberration produced by a converging lens. Does a diverging lens produce spherical aberration? (Angles are greatly exaggerated for clarity.)

Most cameras are equipped with an adjustable aperture to control the light intensity and, when possible, reduce spherical aberration. (An aperture is an opening that controls the amount of light transmitted through the lens.) As the aperture size is reduced, sharper images are produced, because only the central portion of the lens is exposed to the incident light when the aperture is very small. At the same time, however, progressively less light is imaged. To compensate for this loss, a longer exposure time is used. An example of the results obtained with small apertures is the sharp image produced by a "pinhole" camera, with an aperture size of approximately 0.1 mm.

In the case of mirrors used for very distant objects, one can eliminate, or at least minimize, spherical aberration by employing a parabolic rather than spherical surface. Parabolic surfaces are not used in many applications, however, because they are very expensive to make with high-quality optics. Parallel light rays incident on such a surface focus at a common point. Parabolic reflecting surfaces are used in many astronomical telescopes to enhance the image quality. They are also used in flashlights, in which a nearly parallel light beam is produced from a small lamp placed at the focus of the reflecting surface.

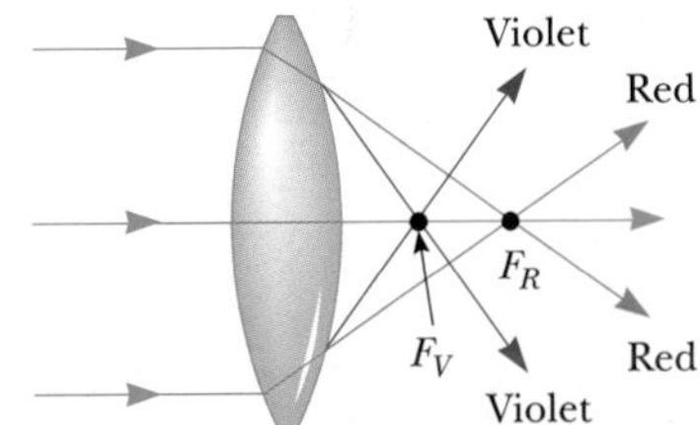


**FIGURE 23.32** Chromatic aberration produced by a converging lens. Rays of different wavelengths focus at different points. (Angles are greatly exaggerated for clarity.)

## CHROMATIC ABERRATION

The fact that different wavelengths of light refracted by a lens focus at different points gives rise to chromatic aberration. In Chapter 22 we described how the index of refraction of a material varies with wavelength. When white light passes through a lens, one finds, for example, that violet light rays are refracted more than red light rays (Fig. 23.32); thus, the focal length for red light is greater than that for violet light. Other wavelengths (not shown in Fig. 23.32) would have intermediate focal points. The chromatic aberration for a diverging lens is opposite that for a converging lens. Chromatic aberration can be greatly reduced by the use of a combination of converging and diverging lenses.

### Webnote 23.3

To learn more about optical aberrations, visit *http://nikon.topica.ne.jp/bi_e/encyclo/ad.htm*

## SUMMARY

Images are formed where rays of light intersect or where they appear to originate. A **real image** is one in which light intersects, or passes through, an image point. A **virtual image** is one in which the light does not pass through the image point but appears to diverge from that point.

The image formed by a flat mirror has the following properties:

1. The image is as far behind the mirror as the object is in front.
2. The image is unmagnified, virtual, and upright.

The **magnification** $M$ of a mirror is defined as the ratio of **image height** $h'$ to **object height** $h$, which is the negative of the ratio of image distance $q$ to object distance $p$:

$$M = \frac{h'}{h} = -\frac{q}{p} \quad [23.2]$$

The **object distance** and **image distance** for a spherical mirror of radius $R$ are related by the **mirror equation:**

$$\frac{1}{p} + \frac{1}{q} = \frac{1}{f} \quad [23.6]$$

where $f = R/2$ is the **focal length** of the mirror.

An image can be formed by refraction at a spherical surface of radius $R$. The object and image distances for refraction from such a surface are related by

$$\frac{n_1}{p} + \frac{n_2}{q} = \frac{n_2 - n_1}{R} \quad [23.7]$$

The **magnification of a refracting surface** is

$$M = \frac{h'}{h} = -\frac{n_1 q}{n_2 p} \quad [23.8]$$

where the object is located in the medium with index of refraction $n_1$ and the image is formed in the medium with index of refraction $n_2$.

The **magnification for a thin lens** is

$$M = \frac{h'}{h} = -\frac{q}{p} \quad [23.10]$$

and the object and image distances are related by the **thin-lens equation:**

$$\frac{1}{p} + \frac{1}{q} = \frac{1}{f} \quad [23.11]$$

**Aberrations** are responsible for the formation of imperfect images by lenses and mirrors. **Spherical aberration** results from the fact that the focal points of light rays far from the principal axis of a spherical lens or mirror are different from those of rays passing through the center. **Chromatic aberration** arises from the fact that light rays of different wavelengths focus at different points when refracted by a lens.

## CONCEPTUAL QUESTIONS

1. Tape a picture of yourself on a bathroom mirror. Stand several centimeters away from the mirror. Can you focus your eyes on *both* the picture taped to the mirror *and* your image in the mirror *at the same time?* So, where is the image of yourself?
2. One method for determining the position of an image, either real or virtual, is by means of *parallax.* If a finger or another object is placed at the position of the image, as shown in Figure Q23.2, and the finger and the image are viewed simultaneously (the image is viewed through the lens if it is virtual), the finger and image have the same parallax; that is, if the image is viewed from different positions, it will appear to move along with the finger. Use this method to locate the image formed by a lens. Explain why this method works.

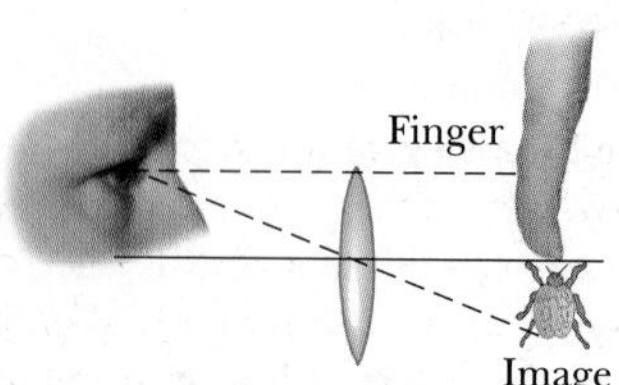


**FIGURE Q23.2**

3. A flat mirror creates a virtual image of your face. Suppose the flat mirror is combined with another optical element. Can a flat mirror form a real image in such a combination?
4. Explain why a mirror cannot give rise to chromatic aberration.
5. You are taking a picture of yourself with a camera that uses an ultrasonic range finder to measure the distance to the object. When you take a picture of yourself in a mirror with this camera, your image is out of focus. Why?
6. A solar furnace can be constructed by using a concave mirror to reflect and focus sunlight into a furnace enclosure. What factors in the design of the reflecting mirror will guarantee that very high temperatures can be achieved?

7. A virtual image is often described as one through which the light rays do not actually travel, as they do for a real image. Can a virtual image be photographed?

8. What is wrong with the caption of the cartoon shown in Figure Q23.8?

**FIGURE Q23.8** "Most mirrors reverse left and right. This one reverses top and bottom."

9. Suppose you want to use a converging lens to project the image of two trees onto a screen. One tree is distance $x$ from the lens; the other is at $2x$, as in Figure Q23.9. You adjust the screen so that the near tree is in focus. If you now want the far tree to be in focus, do you move the screen toward or away from the lens?

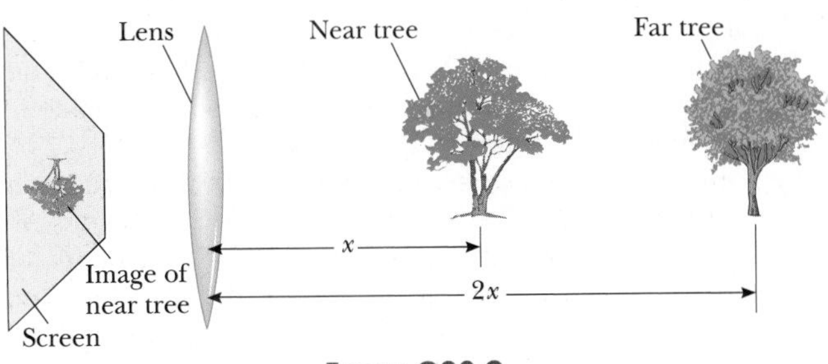


**FIGURE Q23.9**

10. Why does a clear stream always appear to be shallower than it actually is?

11. Can a converging lens be made to diverge light if placed in a liquid? How about a converging mirror?

12. A common mirage is formed when the air gets gradually cooler as the height above the ground increases. What might happen if the air grows gradually warmer as the height is increased? This often happens over bodies of water or snow-covered ground; the effect is called *looming.*

13. In a Jules Verne novel, a piece of ice is shaped into a magnifying lens to focus sunlight to start a fire. Is this possible?

14. Lenses used in eyeglasses, whether converging or diverging, are always designed such that the middle of the lens curves away from the eye. Why?

15. Why does the focal length of a mirror not depend on the mirror material whereas the focal length of a lens does depend on the lens material?

16. If a cylinder of solid glass or clear plastic is placed above the words LEAD OXIDE and viewed from the side, as shown in Figure Q23.16, the word LEAD appears inverted but the word OXIDE does not. Explain.

**FIGURE Q23.16** *(Richard Megna/Fundamental Photographs, NYC)*

17. What is the focal length of a flat mirror? Is that value consistent with the mirror equation?

## PROBLEMS

1, 2, 3 = straightforward, intermediate, challenging ☐ = full solution available in Student Solutions Manual/Study Guide

**web** = solution posted at **http://info.brookscole.com/serway** = biomedical application

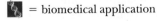

### Section 23.1 Flat Mirrors

1. Does your bathroom mirror show you older or younger than your actual age? Compute an order-of-magnitude estimate for the age difference, based on data that you specify.

2. Use Figure 23.2 to give a geometric proof that the virtual image formed by a plane mirror is the same distance behind the mirror as the object is in front of it.

3. A person walks into a room that has, on opposite walls, two plane mirrors producing multiple images. When the person is 5.00 ft from the mirror on the left wall and 10.0 ft

from the mirror on the right wall, find the distances from the person to the first three images seen in the left-hand mirror.

4. In a church choir loft, two parallel walls are 5.30 m apart. The singers stand against the north wall. The organist faces the south wall, sitting 0.800 m away from it. So that she can see the choir, a flat mirror 0.600 m wide is mounted on the south wall, straight in front of the organist. What width of the north wall can she see? (*Hint:* Draw a top-view diagram to justify your answer.)

## Section 23.2 Images Formed by Spherical Mirrors

## Section 23.3 Convex Mirrors and Sign Conventions

In the following problems, algebraic signs are not given. We leave it to you to determine the correct sign to use with each quantity, based on an analysis of the problem and the sign conventions in Table 23.1.

5. At an intersection of hospital hallways, a convex mirror is mounted high on a wall to help people avoid collisions. The mirror has a radius of curvature of 0.550 m. Locate and describe the image of a patient 10.0 m from the mirror. Determine the magnification.

6. A spherical Christmas tree ornament is 6.00 cm in diameter. What is the magnification of an object placed 10.0 cm away from the ornament?

7. A concave spherical mirror has a radius of curvature of 20.0 cm. Locate the images for object distances of (a) 40.0 cm, (b) 20.0 cm, and (c) 10.0 cm. In each case, state whether the image is real or virtual and upright or inverted, and find the magnification.

8. A dentist uses a mirror to examine a tooth. The tooth is 1.00 cm in front of the mirror, and the image is formed 10.0 cm behind the mirror. Determine (a) the mirror's radius of curvature and (b) the magnification of the image.

9. A large church has a niche in one wall. On the floor plan it appears as a semicircular indentation of radius 2.50 m. A worshiper stands on the center line of the niche, 2.00 m out from its deepest point, and whispers a prayer. Where is the sound concentrated after reflection from the back wall of the niche?

10. A convex mirror has a focal length of 20.0 cm. Determine the object location for which the image will be one-half as tall as the object.

11. A 2.00-cm-high object is placed 3.00 cm in front of a concave mirror. If the image is 5.00 cm high and virtual, what is the focal length of the mirror?

12. A 2.00-cm-high object is placed 10.0 cm in front of a mirror. What type of mirror and what radius of curvature are needed to create an upright image that is 4.00 cm high?

13. web A concave makeup mirror is designed so that a person 25 cm in front of it sees an upright image magnified by a factor of 2. What is the radius of curvature of the mirror?

14. A concave mirror has a focal length of 40.0 cm. Determine the object position for which the resulting image is upright and four times the size of the object.

15. A man standing 1.52 m in front of a shaving mirror produces an inverted image 18.0 cm in front of it. How close to the mirror should he stand if he wants to form an upright image of his chin that is twice the chin's actual size?

16. A convex spherical mirror with a radius of curvature of 10.0 cm produces a virtual image one-third the size of the real object. Where is the object?

17. A child holds a candy bar 10.0 cm in front of a convex mirror and notices that the image is only one-half the size of the candy bar. What is the radius of curvature of the mirror?

18. It is observed that the size of a *real* image formed by a concave mirror is four times the size of the object when the object is 30.0 cm in front of the mirror. What is the radius of curvature of this mirror?

19. A spherical mirror is to be used to form an image, five times as tall as an object, on a screen positioned 5.0 m from the mirror. (a) Describe the type of mirror required. (b) Where should the mirror be positioned relative to the object?

20. A ball is dropped from rest 3.00 m directly above the vertex of a concave mirror having a radius of 1.00 m and lying in a horizontal plane. (a) Describe the motion of the ball's image in the mirror. (b) At what time do the ball and its image coincide?

## Section 23.4 Images Formed by Refraction

21. A cubical block of ice 50.0 cm on an edge is placed on a level floor over a speck of dust. Locate the image of the speck, when viewed from directly above, if the index of refraction of ice is 1.309.

22. The top of a swimming pool is at ground level. If the pool is 2 m deep, how far below ground level does the bottom of the pool appear to be located when (a) the pool is completely filled with water? (b) the pool is filled halfway with water?

23. web A paperweight is made of a solid glass hemisphere of index of refraction 1.50. The radius of the circular cross section is 4.0 cm. The hemisphere is placed on its flat surface with the center directly over a 2.5-mm-long line drawn on a sheet of paper. What length of line is seen by someone looking vertically down on the hemisphere?

24. A flint glass plate ($n = 1.66$) rests on the bottom of an aquarium tank. The plate is 8.00 cm thick (vertical dimension) and covered with water ($n = 1.33$) to a depth of 12.0 cm. Calculate the apparent thickness of the plate as viewed from above the water. (Assume nearly normal incidence.)

25. One end of a long glass rod ($n = 1.50$) is formed into the shape of a convex surface of radius 8.00 cm. An object is positioned in air along the axis of the rod. Find the image position that corresponds to each of the following object positions: (a) 20.0 cm, (b) 8.00 cm, (c) 4.00 cm, (d) 2.00 cm.

26. A goldfish is swimming at 2.00 cm/s toward the front wall of a rectangular aquarium. What is the apparent speed of the fish measured by an observer looking in from outside the front wall of the tank? The index of refraction of water is 1.333.

## Section 23.6 Thin Lenses

27. A contact lens is made of plastic with an index of refraction of 1.50. The lens has an outer radius of curvature of +2.00 cm and an inner radius of curvature of +2.50 cm. What is the focal length of the lens?

28. The left face of a biconvex lens has a radius of curvature of 12.0 cm, and the right face has a radius of curvature of 18.0 cm. The index of refraction of the glass is 1.44. (a) Calculate the focal length of the lens. (b) Calculate the focal length if the radii of curvature of the two faces are interchanged.

29. A converging lens has a focal length of 20.0 cm. Locate the images for object distances of (a) 40.0 cm, (b) 20.0 cm, and (c) 10.0 cm. For each case, state whether the image is real or virtual and upright or inverted, and find the magnification.

30. Where must an object be placed to have unit magnification ($|M| = 1.00$) (a) for a converging lens of focal length 12.0 cm? (b) for a diverging lens of focal length 12.0 cm?

31. A diverging lens has a focal length of 20.0 cm. Locate the images for object distances of (a) 40.0 cm, (b) 20.0 cm, and (c) 10.0 cm. For each case, state whether the image is real or virtual and upright or inverted, and find the magnification.

32. A convex lens of focal length 15.0 cm is used as a magnifying glass. At what distance from a postage stamp should you hold this lens to get a magnification of +2.00?

33. A transparent photographic slide is placed in front of a converging lens with a focal length of 2.44 cm. The lens forms an image of the slide 12.9 cm from the slide. How far is the lens from the slide if the image is (a) real? (b) virtual?

34. The nickel's image in Figure P23.34 has twice the diameter of the nickel when the lens is 2.84 cm from the nickel. Determine the focal length of the lens.

**FIGURE P23.34**

35. A certain LCD projector contains a single thin lens. An object 24.0 mm high is to be projected so that its image fills a screen 1.80 m high. The object-to-screen distance is 3.00 m. (a) Determine the focal length of the projection lens. (b) How far from the object should the lens of the projector be placed in order to form the image on the screen?

36. An object's distance from a converging lens is ten times the focal length. How far is the image from the focal point? Express the answer as a fraction of the focal length.

37. web A diverging lens is to be used to produce a virtual image one-third as tall as the object. Where should the object be placed?

38. An object is 5.00 m to the left of a flat screen. A converging lens for which the focal length is $f = 0.800$ m is placed between object and screen. (a) Show that there are two lens positions that form an image on the screen, and determine how far these positions are from the object. (b) How do the two images differ from each other?

39. A converging lens is placed 30.0 cm to the right of a diverging lens of focal length 10.0 cm. A beam of parallel light enters the diverging lens from the left, and the beam is again parallel when it emerges from the converging lens. Calculate the focal length of the converging lens.

40. An object is placed 20.0 cm to the left of a converging lens of focal length 25.0 cm. A diverging lens of focal length 10.0 cm is 25.0 cm to the right of the converging lens. Find the position and magnification of the final image.

41. Two converging lenses, each of focal length 15.0 cm, are placed 40.0 cm apart, and an object is placed 30.0 cm in front of the first. Where is the final image formed, and what is the magnification of the system?

42. Object $O_1$ is 15.0 cm to the left of a converging lens of 10.0-cm focal length. A second lens is positioned 10.0 cm to the right of the first lens and is observed to form a final image at the position of the original object $O_1$. (a) What is the focal length of the second lens? (b) What is the overall magnification of this system? (c) What is the nature (i.e., real or virtual, upright or inverted) of the final image?

43. A 1.00-cm-high object is placed 4.00 cm to the left of a converging lens of focal length 8.00 cm. A diverging lens of focal length −16.00 cm is 6.00 cm to the right of the converging lens. Find the position and height of the final image. Is the image inverted or upright? Real or virtual?

44. Two converging lenses having focal lengths of 10.0 cm and 20.0 cm are placed 50.0 cm apart, as shown in Figure P23.44. The final image is to be located between the lenses, at the position indicated. (a) How far to the left of the first lens should the object be positioned? (b) What is the overall magnification? (c) Is the final image upright or inverted?

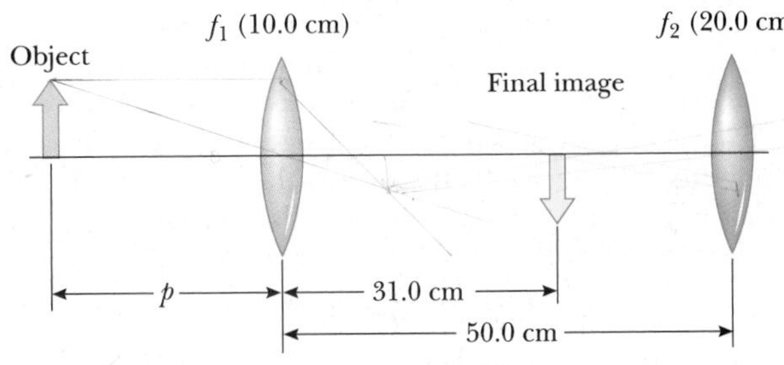


**FIGURE P23.44**

45. Lens $L_1$ in Figure P23.45 has a focal length of 15.0 cm and is located a fixed distance in front of the film plane of a camera. Lens $L_2$ has a focal length of 13.0 cm, and its distance $d$ from the film plane can be varied from 5.00 cm to 10.0 cm. Determine the range of distances for which objects can be focused on the film.

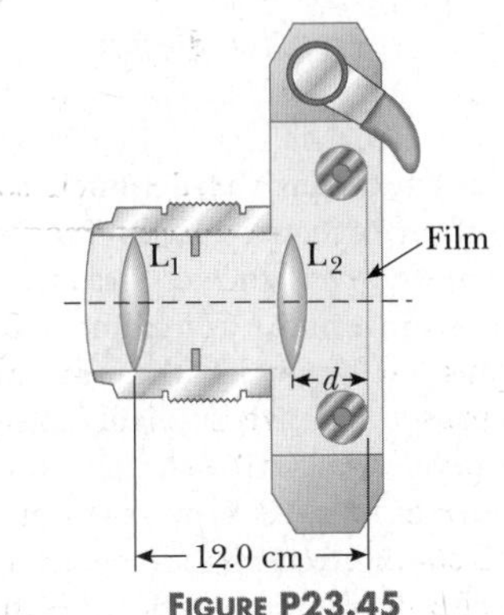


**FIGURE P23.45**

46. Consider two thin lenses, one of focal length $f_1$ and the other of focal length $f_2$, placed in contact with each other as shown in Figure P23.46. Apply the thin-lens equation to each of these lenses and combine these results to show that this combination of lenses behaves like a thin lens having a focal length $f$ given by $1/f = 1/f_1 + 1/f_2$. Assume that the thicknesses of the lenses can be ignored in comparison to the other distances involved.

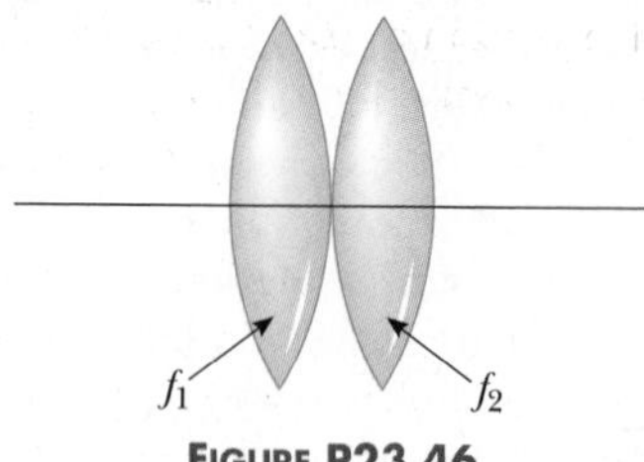


**FIGURE P23.46**

## ADDITIONAL PROBLEMS

47. An object placed 10.0 cm from a concave spherical mirror produces a real image 8.00 cm from the mirror. If the object is moved to a new position 20.0 cm from the mirror, what is the position of the image? Is the final image real or virtual?

48. An object is placed 12 cm to the left of a diverging lens of focal length −6.0 cm. A converging lens of focal length 12 cm is placed a distance of $d$ to the right of the diverging lens. Find the distance $d$ that places the final image at infinity.

49. The distance between an object and its upright image is 20.0 cm. If the magnification is 0.500, what is the focal length of the lens being used to form the image?

50. The object in Figure P23.50 is midway between the lens and the mirror. The mirror's radius of curvature is 20.0 cm, and the lens has a focal length of −16.7 cm. Considering only the light that leaves the object and travels first toward the mirror, locate the final image formed by this system. Is this image real or virtual? Is it upright or inverted? What is the overall magnification?

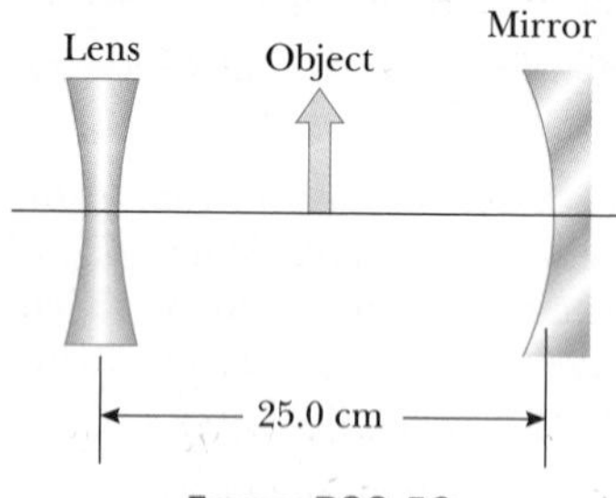


**FIGURE P23.50**

51. web The lens and mirror in Figure P23.51 are separated by 1.00 m and have focal lengths of +80.0 cm and −50.0 cm, respectively. If an object is placed 1.00 m to the left of the lens, locate the final image. State whether the image is upright or inverted, and determine the overall magnification.

52. A diverging lens ($n = 1.50$) is shaped like that in Figure 23.25c. The radius of the first surface is 15.0 cm and that of the second surface is 10.0 cm. (a) Find the focal length of the lens. Determine the positions of the images for object distances of (b) infinity, (c) $3|f|$, (d) $|f|$, and (e) $|f|/2$.

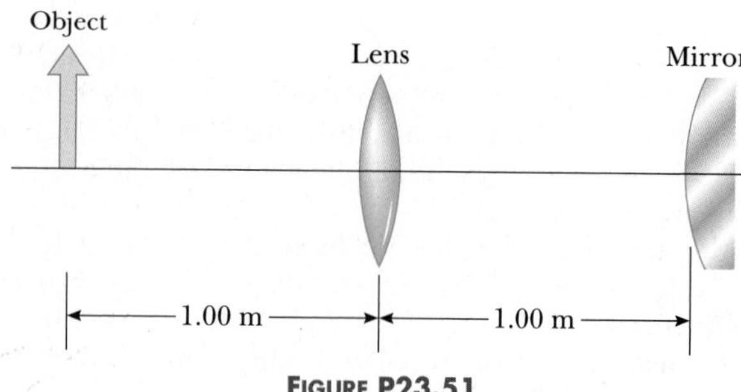


**FIGURE P23.51**

53. A parallel beam of light enters a glass hemisphere perpendicular to the flat face, as shown in Figure P23.53. The radius is $R = 6.00$ cm, and the index of refraction is $n = 1.560$. Determine the point at which the beam is focused. (Assume paraxial rays; that is, all rays are located close to the principal axis.)

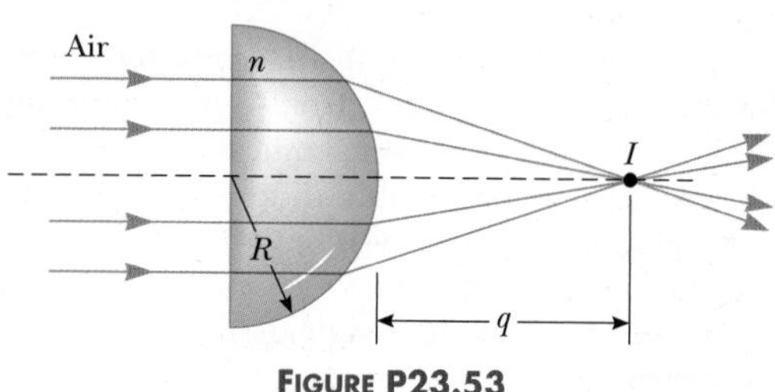


**FIGURE P23.53**

54. A converging lens of focal length 20.0 cm is separated by 50.0 cm from a converging lens of focal length 5.00 cm. (a) Find the position of the final image of an object placed 40.0 cm in front of the first lens. (b) If the height of the object is 2.00 cm, what is the height of the final image? Is the image real or virtual? (c) If the two lenses are now placed in contact with each other and the object is 5.00 cm in front of this combination, where will the image be located? (See Problem 46.)

55. To work this problem, use the fact that the image formed by the first surface becomes the object for the second surface. Figure P23.55 shows a piece of glass with index of refraction of 1.50. The ends are hemispheres with radii 2.00 cm and 4.00 cm, and the centers of the hemispherical ends are separated by a distance of 8.00 cm. A point object is in air, 1.00 cm from the left end of the glass. Locate the image of the object due to refraction at the two spherical surfaces.

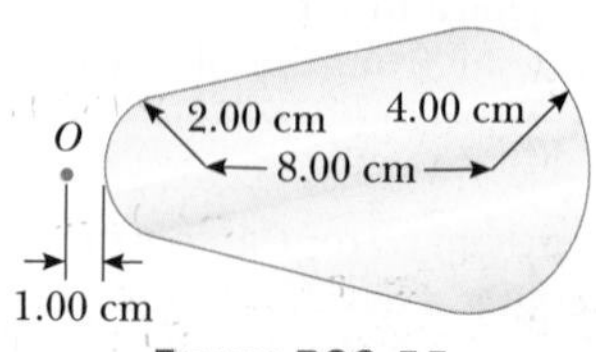


**FIGURE P23.55**

56. In a darkened room, a burning candle is placed 1.50 m from a white wall. A lens is placed between candle and wall at a location that causes a larger, inverted image to form on the wall. When the lens is moved 90.0 cm toward the wall, another image of the candle is formed. Find (a) the two object distances that produce the images and (b) the focal length of the lens. (c) Characterize the second image.

57. An object 2.00 cm high is placed 40.0 cm to the left of a converging lens having a focal length of 30.0 cm. A diverging lens having a focal length of − 20.0 cm is placed 110 cm to the right of the converging lens. (a) Determine the final position and magnification of the final image. (b) Is the image upright or inverted? (c) Repeat parts (a) and (b) for the case where the second lens is a converging lens having a focal length of + 20.0 cm.

58. A "floating strawberry" illusion consists of two parabolic mirrors, each with a focal length of 7.5 cm, facing each other so that their centers are 7.5 cm apart (Fig. P23.58). If a strawberry is placed on the lower mirror, an image of the strawberry forms at the small opening at the center of the top mirror. Show that the final image forms at that location, and describe its characteristics. (*Note:* A flashlight beam shone on these *images* has a very startling effect. Even at a glancing angle, the incoming light beam is seemingly reflected off the *images* of the strawberry! Do you understand why?)

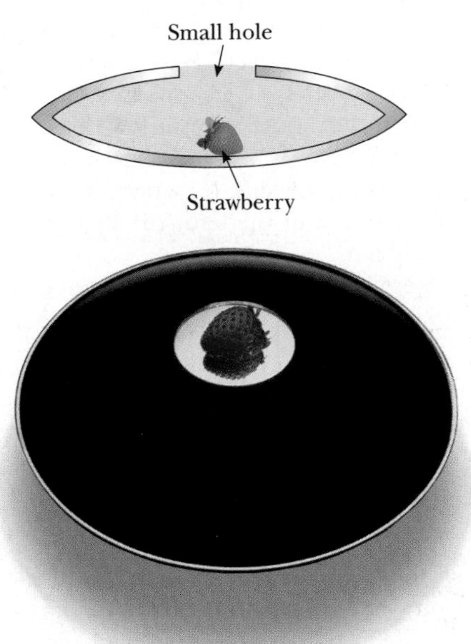


**FIGURE P23.58** *(Photo © Michael Levin/Opti-Gone Associates)*

59. Figure P23.59 shows a converging lens with radii $R_1 = 9.00$ cm and $R_2 = -11.00$ cm in front of a concave spherical mirror of radius $R = 8.00$ cm. The focal points ($F_1$ and $F_2$) for the thin lens and the center of curvature ($C$) of the mirror are also shown. (a) If the focal points $F_1$ and $F_2$ are 5.00 cm from the vertex of the thin lens, determine the index of refraction for the lens. (b) If the lens and mirror are 20.0 cm apart, and an object is placed 8.00 cm to the left of the lens, determine the position of the final image and its magnification as seen by the eye in the figure. (c) Is the final image inverted or upright? Explain.

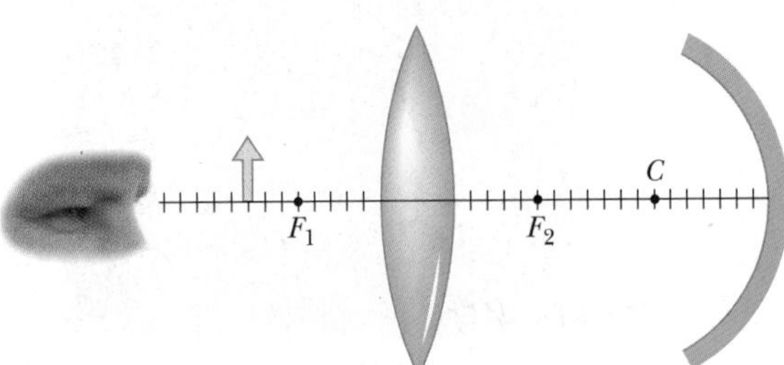


**FIGURE P23.59**

60. Find the object distances (in terms of $f$) for a thin converging lens of focal length $f$ if (a) the image is real and the image distance is four times the focal length; (b) the image is virtual and the image distance is three times the focal length. (c) Calculate the magnification of the lens for case (a) and (b).

61. The lens maker's equation for a lens with index $n_1$ immersed in a medium with index $n_2$ takes the form

$$\frac{1}{f} = \left(\frac{n_1}{n_2} - 1\right)\left(\frac{1}{R_1} - \frac{1}{R_2}\right)$$

A thin diverging glass lens (index = 1.50) with $R_1 = -3.00$ m and $R_2 = -6.00$ m is surrounded by air. An arrow is placed 10.0 m to the left of the lens. (a) Determine the position of the image. Repeat part (a) with the arrow and lens immersed in (b) water (index = 1.33); (c) a medium with an index of refraction of 2.00. (d) How can a lens that is diverging in air be changed into a converging lens?

## GROUP ACTIVITIES

**G.1** This experiment will enable you to examine some of the properties of images formed by flat mirrors. As shown in Figure GA23.1, place a clear plastic sheet between two small candles of the same height as in the figure. Light one candle placed about 6 inches from the sheet and observe the reflected flame from the front side. Move the unlit candle until it also appears to be lit. The candles should be approximately equidistant from the sheet at this point. Why? When you place your finger on the wick of the unlit candle, you will get the illusion that your finger is burning. Explain your observation.

A similar experiment can be performed with a flat mirror and two pencils. Hold one of the pencils in front of the mirror at a distance of about 10 inches from the mirror and at a position such that about half of the image appears in the mirror. Move the second pencil back and forth behind the mirror until it appears to align perfectly with the image in the mirror. When alignment is achieved, measure the distance from the mirror to the location of the second pencil. This pencil is located at the apparent position of the virtual image. Why? The two pencils should be at equal distances from the mirror. Why?

**G.2** Fill a clear glass tumbler with water and place a pencil or straw into the tumbler. Now observe the pencil from the side at an angle of about 45° to the surface, and note that the line of the portion of the pencil under water is not parallel with the line of the portion in air. That is, the

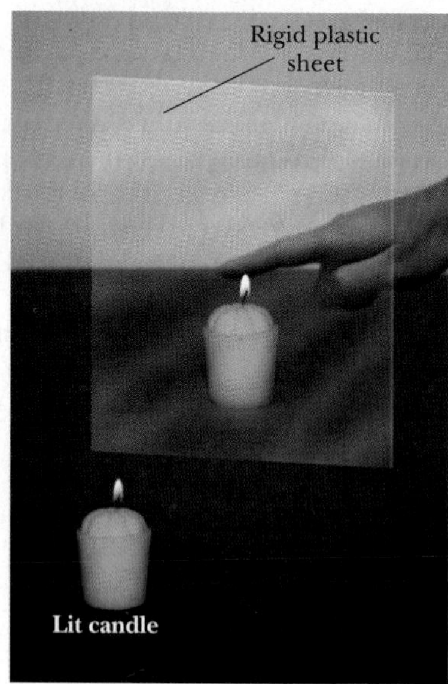


**FIGURE GA23.1**

pencil appears to be bent at the point where it enters the water. Use the techniques of Section 23.4 to explain this observation.

**FIGURE GA23.2**

**G.3** View yourself in a full-length mirror. Standing close to the mirror, place one piece of tape at the top of the image of your head and another piece of tape at the very bottom of the image of your feet. Now step back a few meters and observe your image. How big is it relative to the original size? How does the distance between the pieces of tape compare with your actual height?

Move to a position in front of the mirror such that you can see a full image of yourself with the top of your head just level with the top of the mirror. Have your co-worker gradually block off the lower portion of the mirror with a sheet or newspaper page until you can see your complete image but no more. Measure the length of the mirror and compare this measurement to your height. How do the two compare?

**G.4** Compare the images formed of your face when you look first at the front side and then at the back side of a shiny soupspoon. For each side observe the change in the image of your face as you move closer to and farther away from the spoon.

**G.5** An object is placed 25 cm in front of a concave makeup mirror that has a focal length of 40 cm. (a) The sign of the focal length is ________ (plus or minus). (b) The sign of the object distance is ________. (c) The sign of the image distance will be ________. (d) Use the mirror equation to find the image distance. (e) Repeat for the case in which the object is 5.0 m away.

**G.6** Describe in words what happens to the following rays of light when they strike a converging lens: (a) a ray parallel to the optic axis, (b) a ray passing through the focal point before striking the lens, (c) a ray passing through the center of curvature of the lens. (d) Repeat for the case of a diverging lens.

**G.7** An object is placed 25 cm in front of a converging lens that has a focal length of 10 cm. (a) The sign of the focal length to be used in the lens equation is ________. (b) The sign of the object distance is ________. (c) The sign of the image distance will be ________. (d) Verify your answer to (c) by a calculation. (d) Repeat for the case when the lens is diverging with a focal length of 15 cm.

**G.8** Draw a sketch of you and your image in a plane mirror when you are walking away from it with a velocity **v**. Indicate on your diagram the velocity vectors for you and your image. Keep the relative lengths of the two velocities as you believe they will be. Explain your answer. (b) Repeat for when you are walking toward the mirror. Explain why you have the vectors drawn as you do.

# Wave Optics

# 24

**Peacock feathers appear to be brilliant in color, especially in the blues and greens. However, these colors are not caused by pigments in the feathers. This chapter describes diffraction, the optical process responsible for this bird's magnificent display.** *(Terry Qing/FPG International, Inc.)*

## Chapter Outline

Thus far our discussion of light has been concerned with what happens when light passes through a lens or reflects from a mirror. Because explanations of such phenomena rely on a geometric analysis of light rays, that part of optics is often called geometric optics. We now expand our study of light into an area called *wave optics*. The three primary topics we examine in this chapter are interference, diffraction, and polarization. These phenomena cannot be adequately explained with ray optics, but the wave theory leads us to satisfying descriptions.

## 24.1 CONDITIONS FOR INTERFERENCE

When you blew bubbles as a child, did you notice the brilliant colors playing on the surface of the bubbles? Have you ever wondered what causes the vivid rainbow of colors reflected from the oil films floating in parking lot puddles on rainy days? Or, have you noticed that good-quality lenses used in cameras have a violet hue? All of these effects are caused by interference in thin films, a subject we shall discuss in this chapter. However, before we can examine interference, we must pause to consider the conditions necessary for it to occur.

In our discussion of interference of mechanical waves in Chapter 13, we found that two waves can add together either constructively or destructively. In constructive interference, the amplitude of the resultant wave is greater than that of either of the individual waves, whereas in destructive interference, the resultant amplitude is less than that of either individual wave. Light waves also interfere with each other. Fundamentally, all interference associated with light waves arises when the electromagnetic fields that constitute the individual waves combine.

Interference effects in light waves are not easy to observe because of the short wavelengths involved (about $4 \times 10^{-7}$ m to about $7 \times 10^{-7}$ m). For sustained interference between two sources of light to be observed, the following conditions must be met:

Conditions for interference

1. The sources must be **coherent**—that is, they must maintain a constant phase with respect to each other.
2. The waves must have identical wavelengths.

Let us examine the characteristics of coherent sources. Two sources (producing two traveling waves) are needed to create interference. To produce a stable interference pattern, the individual waves must maintain a constant phase with one another. When this situation prevails, the sources are said to be coherent. The sound waves emitted by two side-by-side loudspeakers driven by a single amplifier can produce interference because the two speakers respond to the amplifier in the same way at the same time—that is, they are in phase.

If two light sources are placed side by side, however, no interference effects are observed, because the light waves from one source are emitted independently of the waves from the other source. Hence, the emissions from the two sources do not maintain a constant phase relationship with each other during the time of observation. An ordinary light source undergoes random changes about once every $10^{-8}$ s. Therefore, the conditions for constructive interference, destructive interference, and intermediate states have durations on the order of $10^{-8}$ s. The result is that no interference effects are observed, because the eye cannot follow such short-term changes. Such light sources are said to be **incoherent.**

An older method for producing two coherent light sources is to pass light from a single wavelength (monochromatic) source through a narrow slit and then allow this light to fall on a screen containing two narrow slits. The first slit is needed to insure that light comes from a tiny region of the light source in which excited atoms emit coherently. The light emerging from the two slits is coherent because a single source produces the original light beam and the slits serve only to separate the original beam into two parts (which is exactly what was done to the sound signal just mentioned). Any random change in the light emitted by the source will occur in the two separate beams at the same time, and interference effects can be observed.

Currently it is much more common to use a laser as a coherent source to demonstrate interference. A laser produces an intense, coherent, monochromatic beam over a width of several millimeters. This means that the laser can be used to illuminate multiple slits directly, and that interference effects can be observed easily in a fully lighted room. The principles of operation of lasers are explained in Chapter 28.

## 24.2 YOUNG'S DOUBLE-SLIT INTERFERENCE

Thomas Young first demonstrated interference in light waves from two sources in 1801. Figure 24.1a is a diagram of the apparatus used in this experiment. (Young used pinholes rather than slits in his original experiments.) Light is incident on a screen in which there is a narrow slit $S_0$. The light waves emerging from this slit arrive at a second screen that contains two narrow, parallel slits $S_1$ and $S_2$. These slits serve as a pair of coherent light sources because waves emerging from them originate from the same wave front and therefore are always in phase. The light from the two slits produces a visible pattern on screen C consisting of a series of bright and dark parallel bands called **fringes** (Fig. 24.1b). When the light from slits $S_1$ and $S_2$ arrives at a point on the screen so that constructive interference occurs at that location, a bright fringe appears. When the light from the two slits combines destructively at any location on the screen, a dark fringe results.

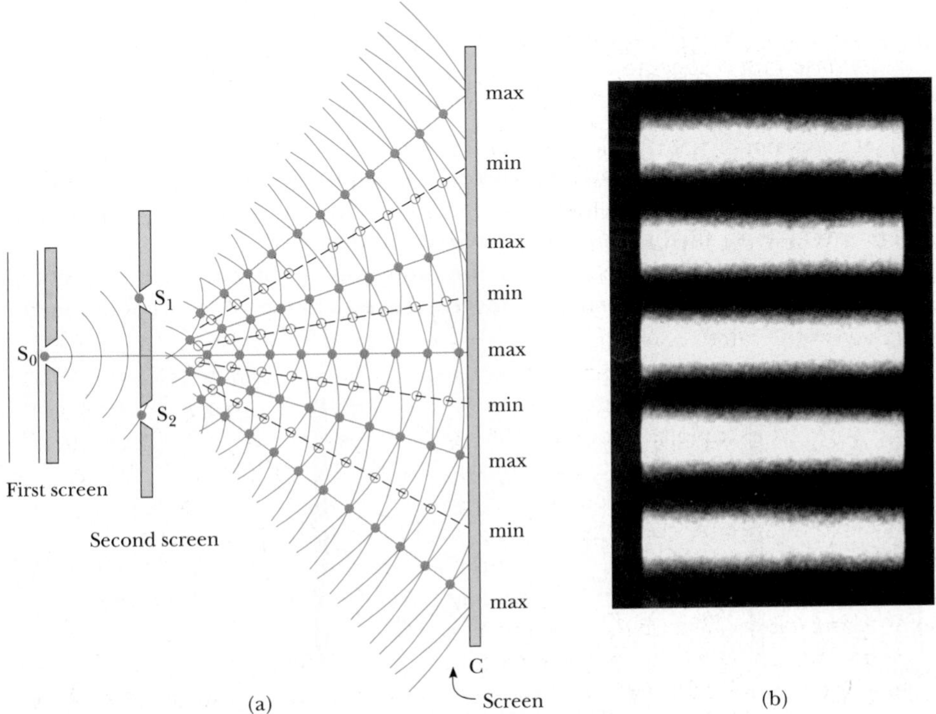


**FIGURE 24.1** (a) A diagram of Young's double-slit experiment. The narrow slits act as sources of waves. Slits $S_1$ and $S_2$ behave as coherent sources that produce an interference pattern on screen C. (This drawing is not to scale.) (b) The fringe pattern formed on screen C could look like this.

**FIGURE 24.2** An interference pattern involving water waves is produced by two vibrating sources at the water's surface. The pattern is analogous to that observed in Young's double-slit experiment. Note the regions of constructive and destructive interference. *(Richard Megna, Fundamental Photographs)*

Figure 24.2 is a photograph of an interference pattern produced by two coherent vibrating sources in a water tank.

Figure 24.3 is a diagram of some of the ways in which the two waves can combine at screen C. In Figure 24.3a, the two waves, which leave the two slits in phase, strike the screen at the central point *P*. Because these waves travel equal distances, they arrive in phase at *P*, and as a result constructive interference occurs there and a bright fringe is observed. In Figure 24.3b, the two light waves again start in phase, but the upper wave has to travel one wavelength farther to reach point *Q* on the screen. Because the upper wave falls behind the lower one by exactly one wavelength, the two waves still arrive in phase at *Q*, and so a second bright fringe appears at that location. Now consider point *R*, midway between *P* and *Q* in Figure 24.3c. At that point, the upper wave has fallen half a wavelength behind the lower wave. This means that the trough of the bottom wave overlaps the crest of the upper wave, giving rise to destructive interference at *R*. As a consequence, a dark fringe can be observed at *R*.

We can describe Young's experiment quantitatively with the help of Figure 24.4. Consider point *P* on the viewing screen; the screen is positioned a perpendicular distance $L$ from the screen containing slits $S_1$ and $S_2$, which are separated by distance $d$, and $r_1$ and $r_2$ are the distances the secondary waves travel from slit to screen. Let us assume the waves emerging from $S_1$ and $S_2$ have the same constant frequency, the same amplitude, and start out in phase. The light intensity on the screen at *P* is the resultant of the light from both slits. Note that a wave from the lower slit travels farther than a wave from the upper slit by the amount $d \sin \theta$. This distance is called the **path difference** $\delta$ (lowercase Greek delta), where

Path difference 

$$\delta = r_2 - r_1 = d \sin \theta \qquad [24.1]$$

This equation assumes that the two waves travel in parallel lines, which is approximately true, because $L$ is much greater than $d$. As noted earlier, the value of this path difference determines whether the two waves are in phase when they arrive at *P*. If the path difference is either zero or some integral multiple of the wavelength, the two waves are in phase at *P* and constructive interference results. Therefore, the condition for bright fringes, or **constructive interference,** at *P* is

Condition for constructive interference (two slits)

$$\delta = d \sin \theta_{\text{bright}} = m\lambda \qquad m = 0, \pm 1, \pm 2, \ldots \qquad [24.2]$$

The number $m$ is called the **order number.** The central bright fringe at $\theta_{\text{bright}} = 0$ ($m = 0$) is called the *zeroth-order maximum.* The first maximum on either side, where $m = \pm 1$, is called the *first-order maximum,* and so forth.

When $\delta$ is an odd multiple of $\lambda/2$, the two waves arriving at *P* are 180° out of phase and give rise to **destructive interference.** Therefore, the condition for

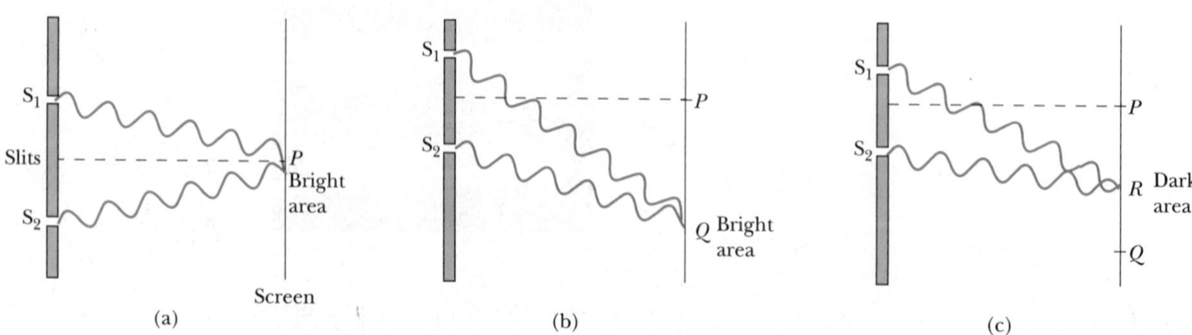


**FIGURE 24.3** (a) Constructive interference occurs at *P* when the waves combine. (b) Constructive interference also occurs at *Q*. (c) Destructive interference occurs at *R* when the wave from the upper slit falls half a wavelength behind the wave from the lower slit. (These figures are not drawn to scale.)

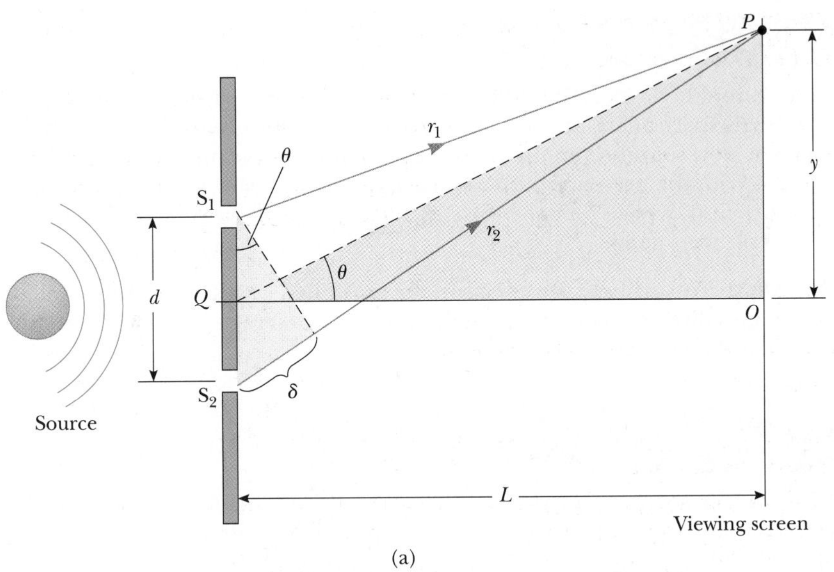


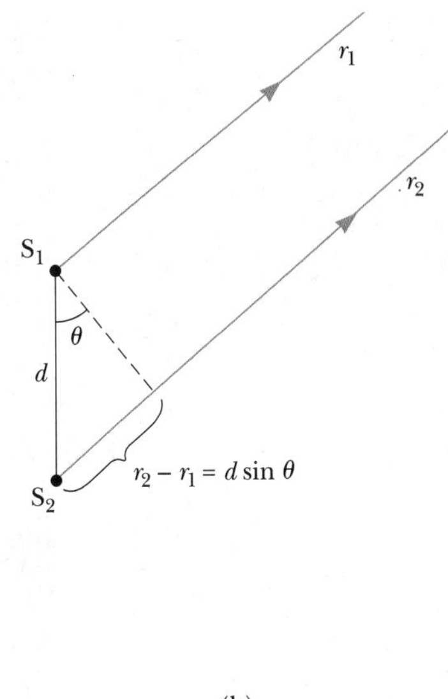


**FIGURE 24.4** A geometric construction that describes Young's double-slit experiment. The path difference between the two rays is $r_2 - r_1 = d \sin \theta$. (This figure is not drawn to scale.)

dark fringes at *P* is

$$\delta = d \sin \theta_{\text{dark}} = (m + \tfrac{1}{2})\lambda \qquad m = 0, \pm 1, \pm 2, \ldots \qquad [24.3]$$

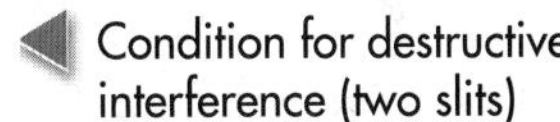
Condition for destructive interference (two slits)

If $m = 0$ in this equation, the path difference is $\delta = \lambda/2$, which is the condition for the location of the first dark fringe on either side of the central (bright) maximum. Likewise, if $m = 1$, $\delta = 3\lambda/2$, which is the condition for the second dark fringe on each side, and so forth.

It is useful to obtain expressions for the positions of the bright and dark fringes measured vertically from *O* to *P*. In addition to our assumption that $L >> d$, we assume that $d >> \lambda$. These can be valid assumptions because in practice, *L* is often of the order of 1 m, *d* a fraction of a millimeter, and $\lambda$ a fraction of a micrometer for visible light. Under these conditions, $\theta$ is small; we can thus use the approximation $\sin \theta \cong \tan \theta$. Then, from triangle *OPQ* in Figure 24.4, we see that

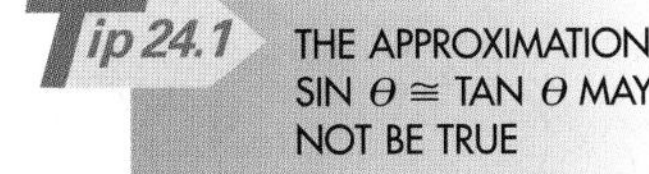
**Tip 24.1 THE APPROXIMATION SIN θ ≅ TAN θ MAY NOT BE TRUE**

The small-angle approximation $\sin \theta \cong \tan \theta$ is true to three-digit precision only for angles less than about 4°.

$$y = L \tan \theta \approx L \sin \theta \qquad [24.4]$$

Solving Equation 24.2 for $\sin \theta$ and substituting the result into Equation 24.4, we see that the positions of the *bright fringes,* measured from *O* are

$$y_{\text{bright}} = \frac{\lambda L}{d} m \qquad m = 0, \pm 1, \pm 2, \ldots \qquad [24.5]$$

Using Equations 24.3 and 24.4, we find that the *dark fringes* are located at

$$y_{\text{dark}} = \frac{\lambda L}{d}(m + \tfrac{1}{2}) \qquad m = 0, \pm 1, \pm 2, \ldots \qquad [24.6]$$

As we shall demonstrate in Example 24.1, Young's double-slit experiment provides a method for measuring the wavelength of light. In fact, Young used this technique to do just that. In addition, his experiment gave the wave model of light a great deal of credibility. It was inconceivable that particles of light coming through the slits could cancel each other in a way that would explain the dark fringes.

As the waves pass through a narrow gap, they spread out (diffract), and the interference of two waveforms is manifested in cross-patterned areas. *(John S. Shelton)*

## APPLYING PHYSICS 24.1

Consider a double-slit experiment in which a laser beam is passed through a pair of very closely spaced slits, and a clear interference pattern is displayed on a distant screen. Now suppose you place smoke particles between the double slit and the screen. With the presence of the smoke particles, will you see the effects of interference in the space between the slits and the screen, or will you only see the effects on the screen?

**Explanation** You will see the effects in the area filled with smoke. There will be bright lines directed toward the bright areas on the screen and dark lines directed toward the dark areas on the screen.

**Webnote 24.1**

See the Young's two-slit diffraction pattern by visiting *http://vsg.tripod.com/interfer.htm*

## APPLYING PHYSICS 24.2

Suppose you are watching television by means of an antenna rather than a cable system. If an airplane flies near your location, you will sometimes notice wavering ghost images in the television picture. What might cause this?

**APPLICATION**

TELEVISION SIGNAL INTERFERENCE

**Explanation** Your television antenna receives two signals—the direct signal from the transmitting antenna and a signal reflected from the surface of the airplane. As the airplane changes position, there are times when these two signals are in phase and other times when they are out of phase. As a result, the intensity of the combined signal received at your antenna varies. The wavering of the ghost images of the picture is evidence for this variation.

### Example 24.1 Measuring the Wavelength of a Light Source

A screen is separated from a double-slit source by 1.2 m. The distance between the two slits is 0.030 mm. The second-order bright fringe ($m = 2$) is measured to be 4.5 cm from the centerline. Determine (a) the wavelength of the light and (b) the distance between adjacent bright fringes.

**Reasoning and Solution**

**A** We can use Equation 24.5 with $m = 2$, $y_2 = 4.5 \times 10^{-2}$ m, $L = 1.2$ m, and $d = 3.0 \times 10^{-5}$ m:

$$\lambda = \frac{y_2 d}{mL} = \frac{(4.5 \times 10^{-2}\text{ m})(3.0 \times 10^{-5}\text{ m})}{2(1.2\text{ m})} = 5.6 \times 10^{-7}\text{ m} = \boxed{560\text{ nm}}$$

**B** Because the positions of the bright fringes are given by Equation 24.5, we see that the distance between *any* adjacent bright fringes (say, those characterized by $m$ and $m + 1$) is

$$\Delta y = y_{m+1} - y_m = \frac{\lambda L}{d}(m+1) - \frac{\lambda L}{d}m = \frac{\lambda L}{d}$$

$$= \frac{(5.6 \times 10^{-7}\text{ m})(1.2\text{ m})}{3.0 \times 10^{-5}\text{ m}} = \boxed{2.2\text{ cm}}$$

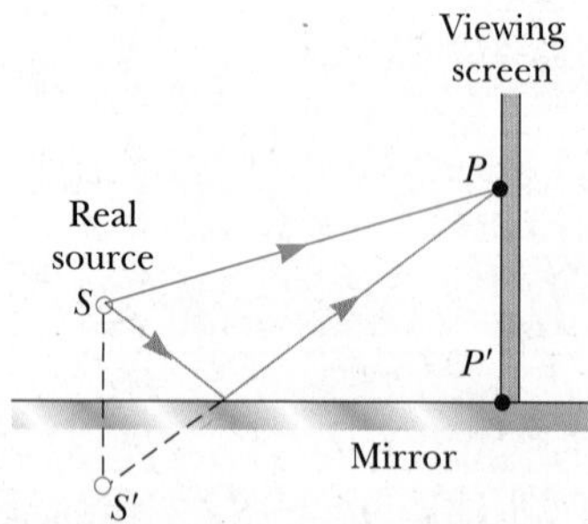


**FIGURE 24.5** Lloyd's mirror. An interference pattern is produced on a screen at $P$ as a result of the combination of the direct ray (blue) and the reflected ray (brown). The reflected ray undergoes a phase change of 180°.

## 24.3 CHANGE OF PHASE DUE TO REFLECTION

Young's method of producing two coherent light sources involves illuminating a pair of slits with a single source. Another simple, although ingenious, arrangement for producing an interference pattern with a single light source is known as *Lloyd's mirror.* A light source is placed at point $S$, close to a mirror, as illustrated in Figure 24.5. Waves can reach the viewing point $P$ either by the direct path $SP$ or by the

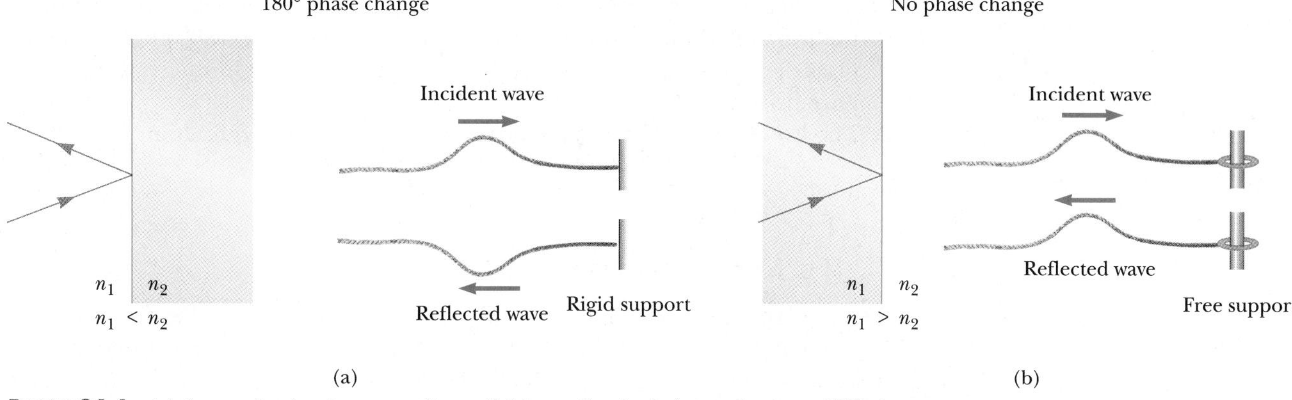


**FIGURE 24.6** (a) A ray reflecting from a medium of higher refractive index undergoes a 180° phase change. The right side shows the analogy with a reflected pulse on a string. (b) A ray reflecting from a medium of lower refractive index undergoes no phase change.

path involving reflection from the mirror. The reflected ray can be treated as a ray originating at the source $S'$ behind the mirror. Source $S'$, which is the image of $S$, can be considered a virtual source.

At points far from the source, we would expect an interference pattern due to waves from $S$ and $S'$, just as is observed for two real coherent sources. An interference pattern is indeed observed. However, the positions of the dark and bright fringes are *reversed* relative to the pattern of two real coherent sources (Young's experiment). This is because the coherent sources $S$ and $S'$ differ in phase by 180°. This 180° phase change is produced by reflection.

To illustrate this further, consider point $P'$, at which the mirror meets the screen. This point is equidistant from $S$ and $S'$. If path difference alone were responsible for the phase difference, we would expect to see a bright fringe at $P'$ (because the path difference is zero for this point), corresponding to the central fringe of the two-slit interference pattern. Instead, we observe a *dark* fringe at $P'$ because of the 180° phase change produced by reflection. In general, an electromagnetic wave undergoes a phase change of 180° upon reflection from a medium of higher index of refraction than the one in which it was traveling.

It is useful to draw an analogy between reflected light waves and the reflections of a transverse wave on a stretched string when the wave meets a boundary, as in Figure 24.6. The reflected pulse on a string undergoes a phase change of 180° when it is reflected from the boundary of a denser string, or from a rigid barrier, and no phase change when it is reflected from the boundary of a less dense string. Similarly, an electromagnetic wave undergoes a 180° phase change when reflected from the boundary of a medium of higher index of refraction than the one in which it has been traveling. There is no phase change when the wave is reflected from a boundary leading to a medium of lower index of refraction. The transmitted wave that crosses the boundary also undergoes no phase change.

## 24.4 INTERFERENCE IN THIN FILMS

Interference effects are commonly observed in thin films, such as soap bubbles and thin layers of oil on water. The varied colors observed with incoherent white light result from the interference of waves reflected from the opposite surfaces of the film.

Consider a film of uniform thickness $t$ and index of refraction $n$, as in Figure 24.7. Let us assume that the light rays traveling in air are nearly normal to the two surfaces of the film. To determine whether the reflected rays interfere constructively or destructively, we must first note the following facts:

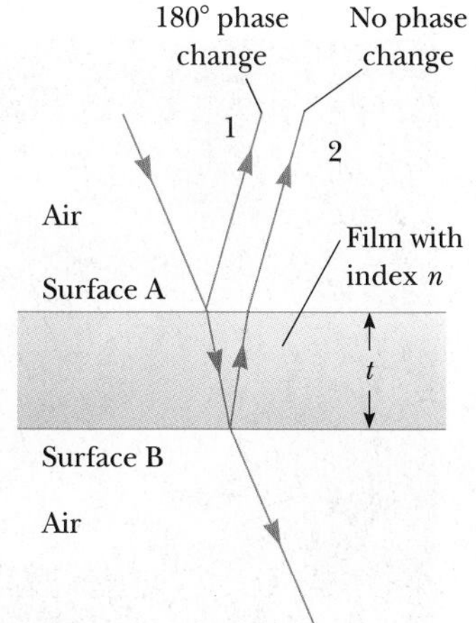


**FIGURE 24.7** Interference observed in light reflected from a thin film is due to a combination of rays reflected from the upper and lower surfaces.

1. An electromagnetic wave traveling from a medium of index of refraction $n_1$ toward a medium of index of refraction $n_2$ undergoes a 180° phase change on reflection when $n_2 > n_1$. There is no phase change in the reflected wave if $n_2 < n_1$.
2. The wavelength of light $\lambda_n$ in a medium with index of refraction $n$ is

$$\lambda_n = \frac{\lambda}{n} \qquad [24.7]$$

where $\lambda$ is the wavelength of light in vacuum.

We apply these rules to the film of Figure 24.7. According to the first rule, ray 1, which is reflected from the upper surface *A*, undergoes a phase change of 180° with respect to the incident wave. Ray 2, which is reflected from the lower surface *B*, undergoes no phase change with respect to the incident wave. Therefore, ray 1 is 180° out of phase with respect to ray 2, a situation that is equivalent to a path difference of $\lambda_n/2$. However, we must also consider the fact that ray 2 travels an extra distance of $2t$ before the waves recombine. For example, if $2t = \lambda_n/2$, rays 1 and 2 recombine in phase and constructive interference results. In general, the condition for *constructive interference* is

$$2t = (m + \tfrac{1}{2})\lambda_n \qquad m = 0, 1, 2, \ldots \qquad [24.8]$$

This condition takes into account two factors: (a) the difference in optical path length for the two rays (the term $m\lambda_n$) and (b) the 180° phase change upon reflec-

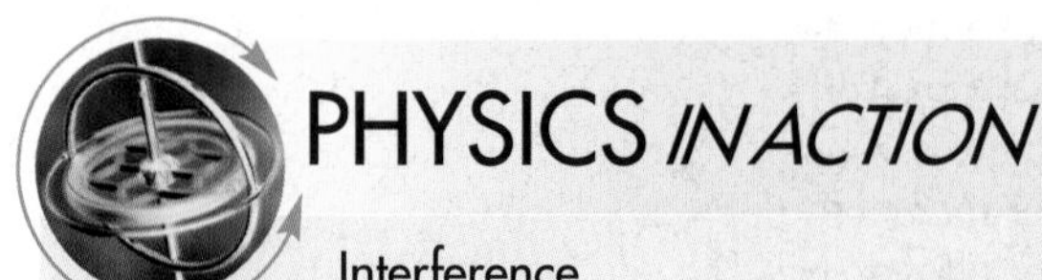

## PHYSICS *IN ACTION*

### Interference

The image on the left shows a layer of soap-film bubbles on water. The colors, produced just before the bubbles burst, are due to interference between light rays reflected from the front and back of the thin film of soap making up the bubble. The color depends on the thickness of the film, ranging from black where the film is at its thinnest to magenta where it is thickest.

On the right, a thin film of oil on water displays interference, evidenced by the pattern of colors when white light is incident on the film. Variations in the film's thickness in the vicinity of the blade produces the intersecting color pattern.

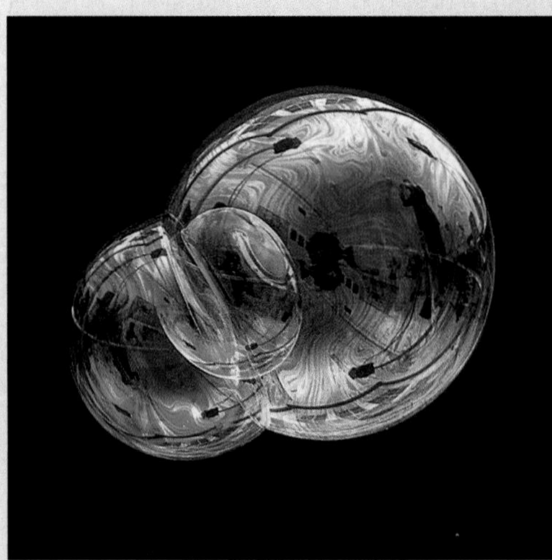

*(Dr. Jeremy Burgess/Science Photo Library)*

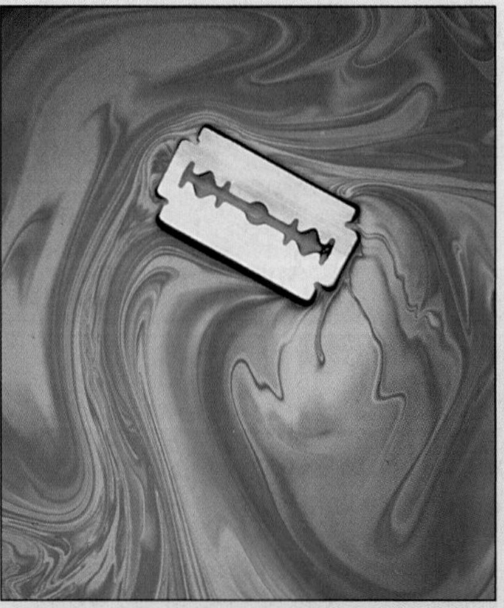

*(Peter Aprahamian/Science Photo Library/Photo Researchers, Inc.)*

tion (the term $\lambda_n/2$). Because $\lambda_n = \lambda/n$, we can write Equation 24.8 in the form

$$2nt = (m + \tfrac{1}{2})\lambda \qquad m = 0, 1, 2, \ldots \qquad [24.9]$$

Condition for constructive interference (thin film)

If the extra distance $2t$ traveled by ray 2 is a multiple of $\lambda_n$, the two waves combine out of phase and destructive interference results. The general equation for *destructive interference* is

$$2nt = m\lambda \qquad m = 0, 1, 2, \ldots \qquad [24.10]$$

Condition for destructive interference (thin film)

It is important to realize that two factors influence interference: (1) possible phase reversals on reflection and (2) differences in travel distance. The foregoing conditions for constructive and destructive interference are valid when the medium above the top surface of the film is the same as the medium below the bottom surface. The surrounding medium may have a refractive index less than or greater than that of the film. In either case, the rays reflected from the two surfaces will be out of phase by 180°. If the film is placed between two *different* media, one of lower refractive index than the film and one of higher refractive index, the conditions for constructive and destructive interference are reversed. In this case, either there is a phase change of 180° for both ray 1 reflecting from surface *A* and ray 2 reflecting from surface *B*, or there is no phase change for either ray; hence, the net change in relative phase due to the reflections is *zero*.

**Tip 24.2 BE CAREFUL WITH THIN FILMS**

Be sure to include *both* effects—path length and phase change—when analyzing an interference pattern from a thin film.

## NEWTON'S RINGS

Another method for observing interference of light waves is to place a planoconvex lens on top of a flat glass surface, as in Figure 24.8a. With this arrangement, the air film between the glass surfaces varies in thickness from zero at the point of contact to some value *t* at *P*. If the radius of curvature *R* of the lens is very large compared with the distance *r*, and if the system is viewed from above using light of wavelength $\lambda$, a pattern of light and dark rings is observed (Fig. 24.8b). These circular fringes are called **Newton's rings** after their discoverer. Newton's particle model of light could not explain the origin of the rings.

The interference is due to the combination of ray 1, reflected from the plate, with ray 2, reflected from the lower surface of the lens. Ray 1 undergoes a phase change of 180° on reflection, because it is reflected from a boundary leading into a medium of higher refractive index, whereas ray 2 undergoes no phase change. Hence, the conditions for constructive and destructive interference are given by

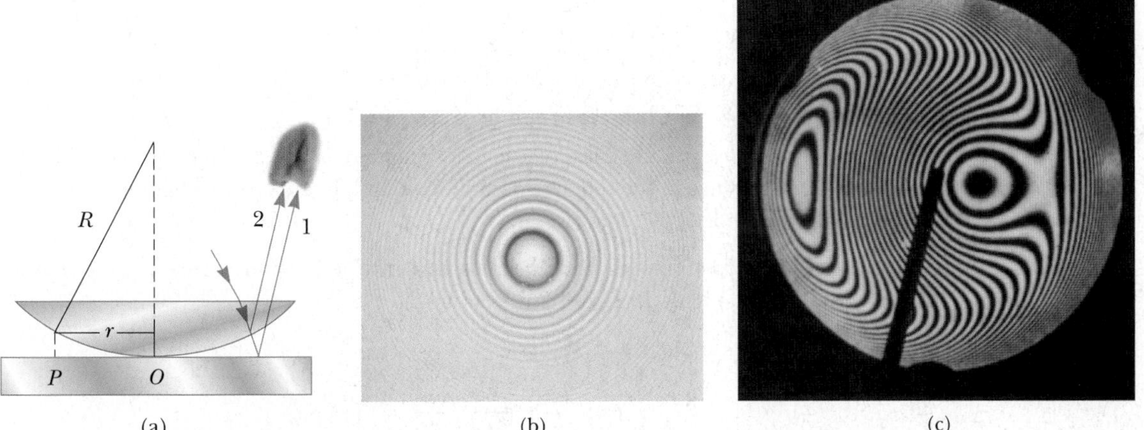


**FIGURE 24.8** (a) The combination of rays reflected from the glass plate and the curved surface of the lens gives rise to an interference pattern known as Newton's rings. (b) A photograph of Newton's rings. *(Richard Megna, Fundamental Photographs, NYC)* (c) This asymmetric interference pattern indicates imperfections in the lens. *(From Physical Science Study Committee,* College Physics, *Lexington, Mass., D. C. Heath and Co., 1968)*

Equations 24.9 and 24.10, respectively, with $n = 1$, because the "film" is air. Here again, we might guess that the contact point $O$ would be bright, corresponding to constructive interference. Instead, it is dark, as seen in Figure 24.8b, because ray 1, reflected from the plate, undergoes a 180° phase change with respect to ray 2. Using the geometry shown in Figure 24.8a, we can obtain expressions for the radii of the bright and dark bands in terms of the radius of curvature $R$, and vacuum wavelength $\lambda$. For example, the dark rings have radii of $r \approx \sqrt{m\lambda R/n}$. In Problem 64 at the end of the chapter, you will be asked to supply the details.

**APPLICATION**

CHECKING FOR IMPERFECTIONS IN OPTICAL LENSES

One of the important uses of Newton's rings is in the testing of optical lenses. A circular pattern like that in Figure 24.8b is achieved only when the lens is ground to a perfectly spherical curvature. Variations from such symmetry might produce a pattern like that in Figure 24.8c. These variations give an indication of how the lens must be ground and polished to remove the imperfections.

### PROBLEM-SOLVING STRATEGY — Thin-Film Interference

The following features should be kept in mind when you work thin-film interference problems:

1. Identify the thin film causing the interference.
2. The type of interference—constructive or destructive—that occurs is determined by the phase relationship between the portion of the wave reflected at the upper surface of the film and the portion reflected at the lower surface.
3. Phase differences between the two portions of the wave have two causes: (a) differences in the distances traveled by the two portions and (b) phase changes occurring on reflection. *Both* causes must be considered when you are determining whether constructive or destructive interference occurs.
4. The interference is constructive if the path difference between the two waves is an integral multiple of $\lambda$, and destructive if the equivalent path difference is $\lambda/2$, $3\lambda/2$, $5\lambda/2$, and so forth. However, the conditions for constructive and destructive interference are reversed if one of the waves undergoes a phase change on reflection.

### Example 24.2 Interference in a Soap Film

Calculate the minimum thickness of a soap-bubble film ($n = 1.33$) that will result in constructive interference in the reflected light if the film is illuminated by light with a wavelength in free space of 602 nm.

**Reasoning** The minimum film thickness for constructive interference corresponds to $m = 0$ in Equation 24.9. This gives $2nt = \lambda/2$.

**Solution** Because $2nt = \lambda/2$, we have

$$t = \frac{\lambda}{4n} = \frac{602 \text{ nm}}{(4)(1.33)} = 113 \text{ nm}$$

**EXERCISE** What other film thicknesses will produce constructive interference?

**ANSWER** 338 nm, 564 nm, 789 nm, and so on

## Example 24.3 Nonreflective Coatings for Solar Cells

Semiconductors such as silicon are used to fabricate solar cells—devices that generate electric energy when exposed to sunlight. Solar cells are often coated with a transparent thin film, such as silicon monoxide (SiO; $n = 1.45$) to minimize reflective losses (Fig. 24.9). A silicon solar cell ($n = 3.50$) is coated with a thin film of silicon monoxide for this purpose. Assuming normal incidence, determine the minimum thickness of the film that will produce the least reflection at a wavelength of 552 nm.

**APPLICATION**

NONREFLECTIVE COATINGS FOR SOLAR CELLS AND OPTICAL LENSES

**Reasoning** Reflection is least when rays 1 and 2 in Figure 24.9 meet the condition of destructive interference. Note that both rays undergo 180° phase changes on reflection. Hence, the net change in phase due to reflection is zero, and the condition for a reflection *minimum* is a path difference of $\lambda_n/2$; therefore, $2t = \lambda_n/2 = \lambda/2n$.

**Solution** Because $2t = \lambda/2n$, the required thickness is

$$t = \frac{\lambda}{4n} = \frac{552 \text{ nm}}{(4)(1.45)} = 95.2 \text{ nm}$$

Typically, such coatings reduce the reflective loss from 30% (with no coating) to 10% (with coating), thereby increasing the cell's efficiency, because more light is available to create charge carriers in the cell. In reality, the coating is never perfectly nonreflecting, because the required thickness is wavelength-dependent and the incident light covers a wide range of wavelengths.

Glass lenses used in cameras and other optical instruments are usually coated with a transparent thin film, such as magnesium fluoride ($MgF_2$), to reduce or eliminate unwanted reflection. More important, such coatings enhance the transmission of light through the lenses.

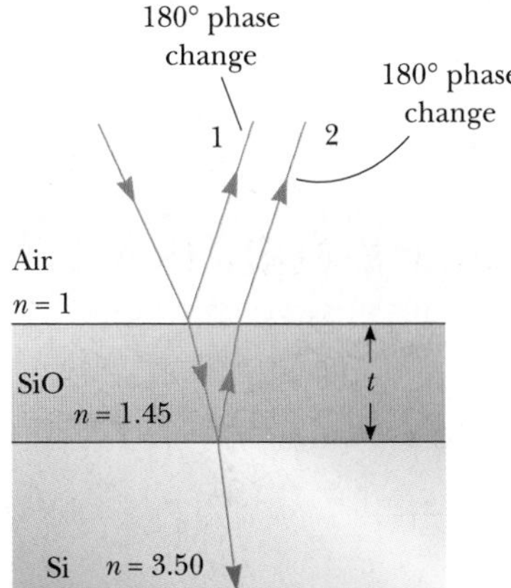


**FIGURE 24.9** (Example 24.3) Reflective losses from a silicon solar cell are minimized by coating it with a thin film of silicon monoxide, SiO.

## Example 24.4 Interference in a Wedge-Shaped Film

A thin, wedge-shaped film of refractive index $n$ is illuminated with monochromatic light of wavelength $\lambda$, as illustrated in Figure 24.10a. Describe the interference pattern observed in this case.

**Reasoning and Solution** The interference pattern is that of a thin film of variable thickness consisting of air. Hence, the pattern is a series of alternating bright and dark

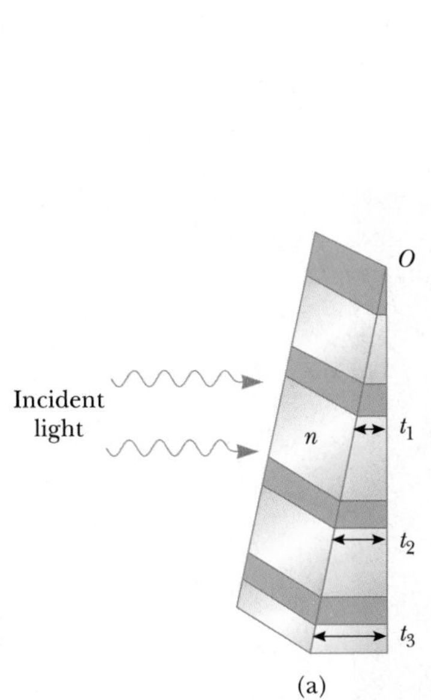

**FIGURE 24.10** (Example 24.4) (a) Interference bands in reflected light can be observed by illuminating a wedge-shaped film with monochromatic light. The dark areas correspond to positions of destructive interference. (b) Interference in a vertical film of variable thickness. The top of the film appears darkest where the film is thinnest. *(Richard Megna, Fundamental Photographs)*

parallel bands. A dark band corresponding to destructive interference appears at point $O$, the apex, because the upper reflected ray undergoes a 180° phase change and the lower one does not. According to Equation 24.10, other dark bands appear when $2nt = m\lambda$, so that $t_1 = \lambda/2n$, $t_2 = \lambda/n$, $t_3 = 3\lambda/2n$, and so on. Similarly, bright bands are observed when the thickness satisfies the condition $2nt = (m + \frac{1}{2})\lambda$, corresponding to thicknesses of $\lambda/4n$, $3\lambda/4n$, $5\lambda/4n$, and so on. If white light is used, bands of different colors are observed at different points, corresponding to the different wavelengths of light present. This is shown in the soap film in Figure 24.10b.

## 24.5 Using Interference to Read CDs and DVDs

**APPLICATION**

THE PHYSICS OF CDS AND DVDS

Compact disks (CDs) and digital video disks (DVDs) have revolutionized the computer and entertainment industries by providing fast access; high-density storage of text, graphics, and movies; and high-quality sound recordings. The data on these disks are stored digitally—that is, as a series of zeros and ones—and these zeros and ones are read by laser light reflected from the disk. Strong reflections (constructive interference) from the disk are chosen to represent zeros and weak reflections (destructive interference) represent ones.

To see in more detail how thin-film interference plays a crucial role in reading CDs, consider Figure 24.11. This shows a photomicrograph of several CD tracks, which consist of a sequence of pits (when viewed from the top or label side of the disk) of varying length formed in a reflecting-metal information layer. A cross-sectional view of a CD as shown in Figure 24.12 reveals that the pits appear as bumps to the laser beam, which shines on the metallic layer through a clear plastic coating from below. As the disk rotates, the laser reflects off the sequence of bumps and lower areas into a photodetector, which converts the fluctuating reflected light intensity into an electrical string of zeros and ones. To make the light fluctuations more pronounced and easier to detect, the pit depth $t$ is made equal to one quarter of a wavelength of the laser light in the plastic. When the beam hits a rising or falling bump edge, part of the beam reflects from the top of the bump and part from the lower adjacent area, ensuring destructive interference and very low intensity when the reflected beams combine at the detector. Thus bump edges are read as ones, and flat bump tops and intervening flat plains are read as zeros.

Example 24.5 determines the pit depth for a standard CD using an infrared laser of wavelength 780 nm. DVDs use shorter-wavelength lasers of 635 nm, and the track separation, pit depth, and minimum pit length are all smaller. This allows a DVD to store about 30 times more information than a CD.

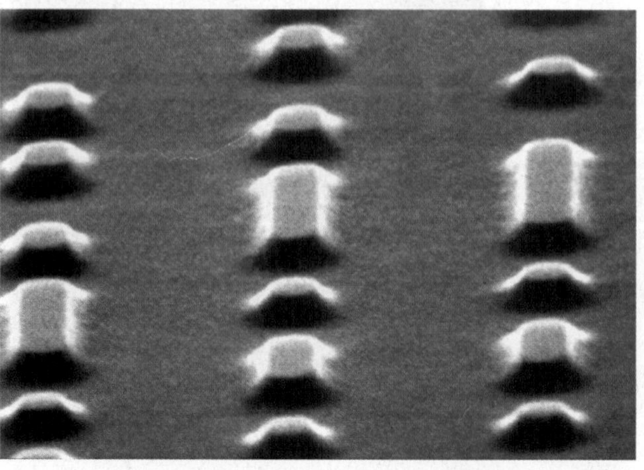

**FIGURE 24.11** A photomicrograph of adjacent tracks on a compact disc (CD). The information encoded in these pits and smooth areas is read by a laser beam. *(Courtesy of Sony Disc Manufacturing)*

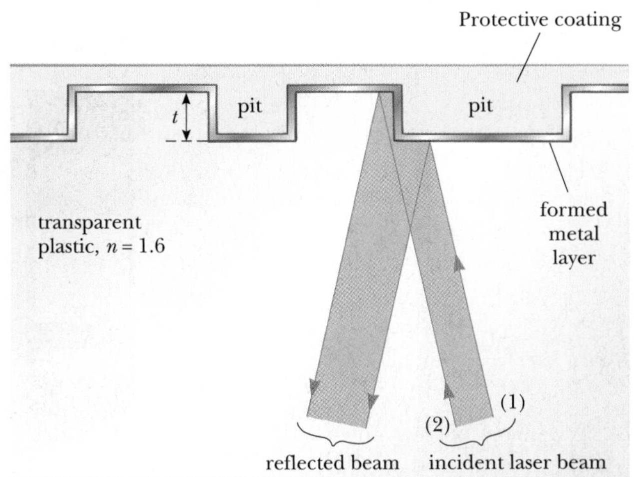


**FIGURE 24.12** Cross section of a CD showing metallic pits of depth $t$ and a laser beam detecting a pit edge.

### Example 24.5 Pit Depth in a CD

Find the pit depth in a CD, which has a plastic transparent layer with index of refraction of 1.6 and is designed for use in a CD player, using a laser that has a wavelength of 780 nm in air.

**Reasoning and Solution** Rays 1 and 2 undergo a 180° phase change on reflection from the metal layer, which acts like a mirror. (See Fig. 24.12.) Thus, there is no phase difference between rays 1 and 2 because of reflection, but there is a phase difference caused by the extra distance $2t$ traveled by ray 2. The wavelength of light in a substance with index of refraction $n$ is $\lambda/n$, where $\lambda$ is the wavelength in air. Hence, for destructive interference of rays 1 and 2, we know that $2t = \lambda/2n$, or

$$t = \frac{\lambda}{4n} = \frac{780 \text{ nm}}{(4)(1.60)} = 1.2 \times 10^2 \text{ nm}$$

## 24.6 DIFFRACTION

Suppose a light beam is incident on two slits, as in Young's double-slit experiment. If the light truly traveled in straight-line paths after passing through the slits, as in Figure 24.13a, the waves would not overlap and no interference pattern would be seen. Instead, Huygens's principle requires that the waves spread out from the slits, as shown in Figure 24.13b. In other words, the light deviates from a straight-line path and enters the region that would otherwise be shadowed. This spreading out of light from its initial line of travel is called **diffraction.**

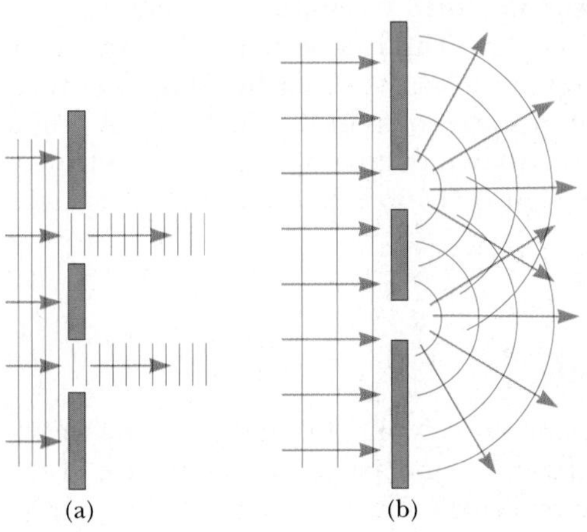


**FIGURE 24.13** (a) If light did not spread out after passing through the slits, no interference would occur. (b) The light from the two slits overlaps as it spreads out, filling the expected shadowed regions with light and producing interference fringes.

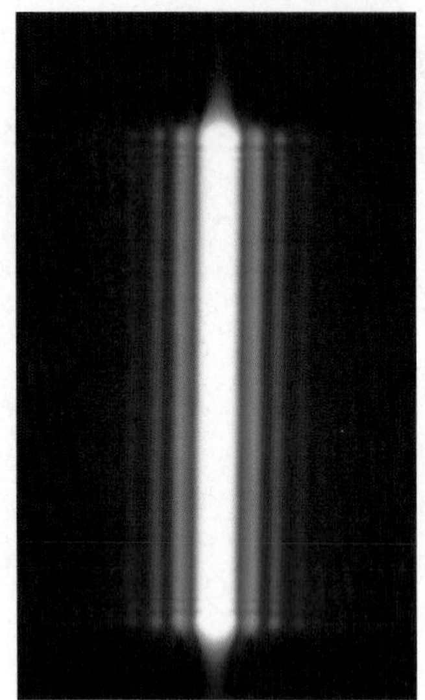

**FIGURE 24.14** The diffraction pattern that appears on a screen when light passes through a narrow, vertical slit. The pattern consists of a broad central band and a series of less intense and narrower side bands.

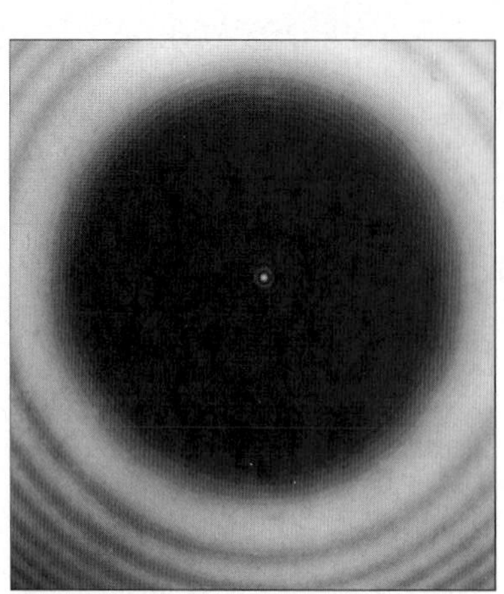

**FIGURE 24.15** The diffraction pattern of a penny placed midway between the screen and the source. *(Courtesy of P. M. Rinard, from* Am. J. Phys., *44:70, 1976)*

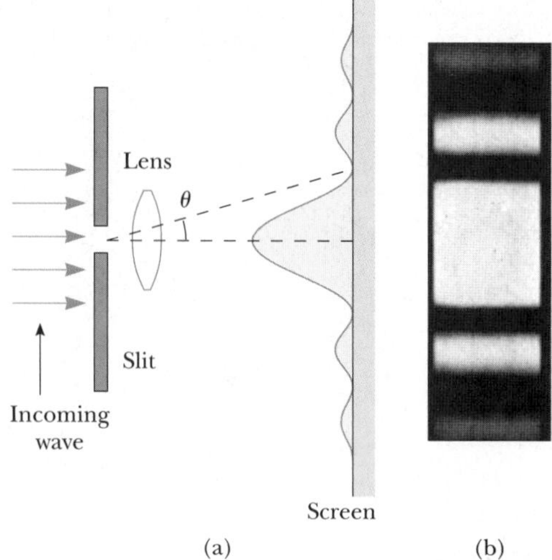


**FIGURE 24.16** (a) The Fraunhofer diffraction pattern of a single slit. The parallel rays are brought into focus on the screen with a converging lens. The pattern consists of a central bright region flanked by much weaker maxima. (This drawing is not to scale.) (b) A photograph of a single-slit Fraunhofer diffraction pattern. *(From M. Cagnet, M. Francon, and J. C. Thierr,* Atlas of Optical Phenomena, *Berlin, Springer-Verlag, 1962, plate 18)*

In general, diffraction occurs when waves pass through small openings, around obstacles, or by sharp edges. For example, when a single narrow slit is placed between a distant light source (or a laser beam) and a screen, the light produces a diffraction pattern like that in Figure 24.14. The pattern consists of a broad, intense central band flanked by a series of narrower, less intense secondary bands (called **secondary maxima**) and a series of dark bands, or **minima.** This cannot be explained within the framework of geometric optics, which says that light rays traveling in straight lines should cast a sharp image of the slit on the screen.

Figure 24.15 shows the diffraction pattern and shadow of a penny. The pattern consists of the shadow, a bright spot at its center, and a series of bright and dark circular bands of light near the edge of the shadow. The bright spot at the center (called the *Fresnel bright spot*) is explained by Augustin Fresnel's wave theory of light, which predicts constructive interference at this point for certain locations of the penny. In contrast, from the viewpoint of geometric optics, the center of the pattern would be completely screened by the penny, and so we would never observe a central bright spot.

One type of diffraction, called **Fraunhofer diffraction,** occurs when the rays leave the diffracting object in parallel directions. This can be achieved experimentally either by placing the observing screen far from the slit or by using a converging lens to focus the parallel rays on a nearby screen, as in Figure 24.16a. A bright fringe is observed along the axis at $\theta = 0$, with alternating dark and bright fringes on each side of the central bright fringe. Figure 24.16b is a photograph of a single-slit Fraunhofer diffraction pattern.

## 24.7 SINGLE-SLIT DIFFRACTION

Until now we have assumed that slits are line sources of light. In this section we determine how their finite widths are the basis for understanding the nature of the Fraunhofer diffraction pattern produced by a single slit.

We can deduce some important features of this problem by examining waves coming from various portions of the slit, as shown in Figure 24.17. According to Huygens's principle, **each portion of the slit acts as a source of waves. Hence, light from one portion of the slit can interfere with light from another portion,** and the resultant intensity on the screen depends on the direction $\theta$.

To analyze the diffraction pattern, it is convenient to divide the slit into halves, as in Figure 24.17. All the waves that originate at the slit are in phase. Consider waves 1 and 3, which originate at the bottom and center of the slit, respectively. Wave 1 travels farther than wave 3 by an amount equal to the path difference $(a/2)\sin\theta$, where $a$ is the width of the slit. Similarly, the path difference between waves 3 and 5 is also $(a/2)\sin\theta$. If this path difference is exactly half of a wavelength (corresponding to a phase difference of 180°), the two waves cancel each other and destructive interference results. This is true, in fact, for any two waves that originate at points separated by half the slit width, because the phase difference between two such points is 180°. Therefore, waves from the upper half of the slit interfere *destructively* with waves from the lower half of the slit when

$$\frac{a}{2}\sin\theta = \frac{\lambda}{2} \quad \text{or when} \quad \sin\theta = \frac{\lambda}{a}$$

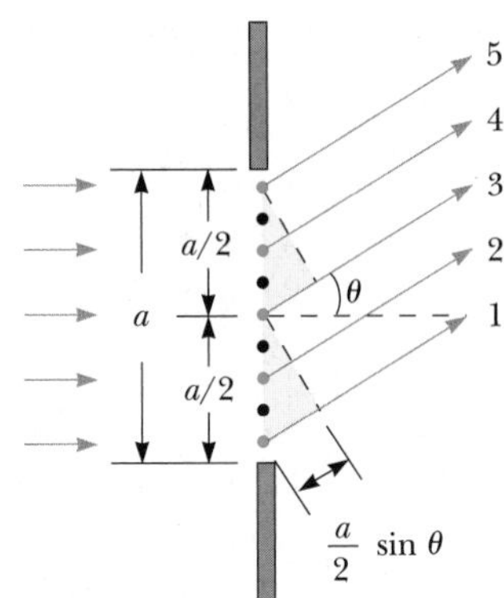


**FIGURE 24.17** Diffraction of light by a narrow slit of width $a$. Each portion of the slit acts as a point source of waves. The path difference between rays 1 and 3 or between rays 2 and 4 is equal to $(a/2)\sin\theta$. (This drawing is not to scale, and the rays are assumed to converge at a distant point.)

If we divide the slit into four parts rather than two, and use similar reasoning, we find that the screen is also dark when

$$\sin\theta = \frac{2\lambda}{a}$$

Likewise, we can divide the slit into six parts and show that darkness occurs on the screen when

$$\sin\theta = \frac{3\lambda}{a}$$

### Webnote 24.2

To see the diffraction pattern of a single slit as a function of the incident wavelength, explore *http://www.lightlink.com/sergey/java/java/slitdiffr/index.html*

Therefore, the general condition for **destructive interference** for a single slit of width $a$ is

$$\sin\theta_{\text{dark}} = m\frac{\lambda}{a} \qquad m = \pm 1, \pm 2, \pm 3, \ldots \qquad [24.11]$$

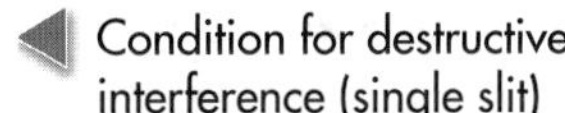


Equation 24.11 gives the values of $\theta$ for which the diffraction pattern has zero intensity—that is, a dark fringe is formed. However, this equation tells us nothing about the variation in intensity along the screen. The general features of the intensity distribution along the screen are shown in Figure 24.18. A broad central bright fringe is observed, flanked by much weaker bright fringes alternating with dark fringes. The various dark fringes (points of zero intensity) occur at the values of $\theta$ that satisfy Equation 24.11. The points of constructive interference lie approximately halfway between the dark fringes. Note that the central bright fringe is twice as wide as the weaker maxima having $m > 1$.

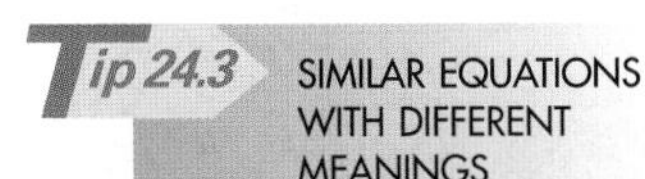


Although Equations 24.2 and 24.11 have the same form, they have different meanings. Equation 24.2 describes the *bright* regions in a two-slit interference pattern, whereas Equation 24.11 describes the *dark* regions in a single-slit interference pattern.

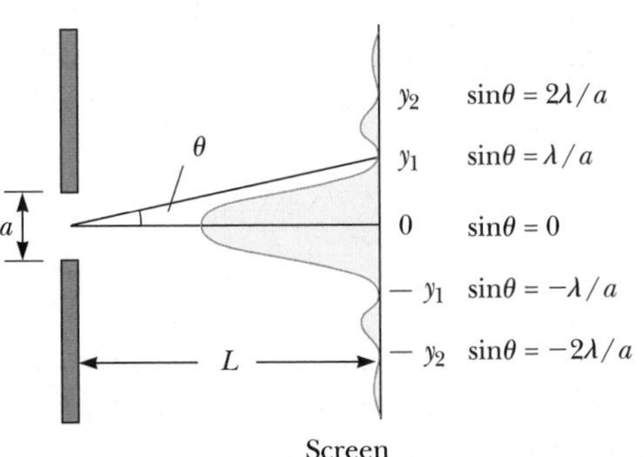


**FIGURE 24.18** Positions of the minima for the Fraunhofer diffraction pattern of a single slit of width $a$. (This is not to scale.)

**Quick Quiz 24.1** In a single-slit diffraction experiment, as the width of the slit is made smaller, the width of the central maximum of the diffraction pattern becomes (a) smaller, (b) larger, or (c) remains the same.

## APPLYING PHYSICS 24.3

If a classroom door is open even just a small amount, you can hear sounds coming from the hallway. Yet you cannot see what is going on in the hallway. Why is there this difference?

**Explanation** The space between the slightly open door and the wall is acting as a single slit for waves. Sound waves have wavelengths larger than the slit width, so sound is effectively diffracted by the opening and the central maximum spreads throughout the room. Light wavelengths are much smaller than the slit width, so there is virtually no diffraction for the light. You must have a direct line of sight to detect the light waves.

### Example 24.6 Where Are the Dark Fringes?

Light of wavelength 580 nm is incident on a slit of width 0.30 mm. The observing screen is placed 2.0 m from the slit. Find the positions of the first dark fringes and the width of the central bright fringe.

**Solution** The first dark fringes that flank the central bright fringe correspond to $m = \pm 1$ in Equation 24.11:

$$\sin \theta = \pm \frac{\lambda}{a} = \pm \frac{5.8 \times 10^{-7}\text{ m}}{0.30 \times 10^{-3}\text{ m}} = \pm 1.9 \times 10^{-3}$$

From the triangle in Figure 24.16, we see that $\tan \theta = y_1/L$. Because $\theta$ is very small, we can use the approximation $\sin \theta \approx \tan \theta$, so that $\sin \theta \approx y_1/L$. Therefore, the positions of the first minima, measured from central axis, are

$$y_1 \approx L \sin \theta = \pm L \frac{\lambda}{a} = \pm 3.9 \times 10^{-3}\text{ m}$$

The positive and negative signs correspond to the first dark fringes on either side of the central bright fringe. Hence, the width of the central bright fringe is given by $2|y_1| = 7.8 \times 10^{-3}\text{ m} = 7.8\text{ mm}$. Note that this value is much greater than the width of the slit. However, as the width of the slit is *increased,* the diffraction pattern *narrows,* corresponding to smaller values of $\theta$. In fact, for large values of $a$, the maxima and minima are so closely spaced that the only observable pattern is a large central bright area resembling the geometric image of the slit. Because the width of the geometric image increases as the slit width increases, the narrowest image occurs when the geometric and diffraction widths are equal.

**EXERCISE** Determine the width of the first-order bright fringe.

**ANSWER** 3.9 mm

## 24.8 THE DIFFRACTION GRATING

The diffraction grating, a very useful device for analyzing light sources, consists of many equally spaced parallel slits. A grating can be made by scratching parallel lines on a glass plate with a precision machining technique. The scratches act to scatter the light and so behave as slits. A typical grating contains several thousand lines per centimeter. For example, a grating ruled with 5 000 lines/cm has a slit spacing $d$ equal to the inverse of this number; hence, $d = (1/5\,000)$ cm $= 2 \times 10^{-4}$ cm.

Figure 24.19 is a diagram of a section of a plane diffraction grating. A plane wave is incident from the left, normal to the plane of the grating. The intensity of the pattern on the screen is the result of the combined effects of interference and diffraction. Each slit produces diffraction, and the diffracted beams in turn interfere with one another to produce the pattern. Moreover, each slit acts as a source of waves, and all waves start at the slits in phase. However, for some arbitrary direction $\theta$ measured from the horizontal, the waves must travel *different* path lengths before reaching a particular point $P$ on the screen. From Figure 24.19, note that the path difference between waves from any two adjacent slits is $d \sin \theta$. If this path difference equals one wavelength or some integral multiple of a wavelength, waves from all slits will be in phase at $P$ and a bright line will be observed at this point. Therefore, the condition for **maxima** in the interference pattern at the angle $\theta$ is

$$d \sin \theta_{\text{bright}} = m\lambda \qquad m = 0, 1, 2, \ldots \qquad \textbf{[24.12]}$$

Condition for maxima in the interference pattern of a diffraction grating

Light emerging from a slit at an angle other than that for a maximum interferes with nearly complete destruction with light from some other slit on the grating. All such pairs will result in little or no transmission in that direction, as illustrated in Figure 24.20.

Equation 24.12 can be used to calculate the wavelength from the grating spacing and the angle of deviation, $\theta$. The integer $m$ is the **order number** of the diffraction pattern. If the incident radiation contains several wavelengths, each wavelength deviates through a specific angle, which can be found from Equation 24.12. All wavelengths are focused at $\theta = 0$, corresponding to $m = 0$. This is called the *zeroth-order maximum.* The *first-order maximum,* corresponding to $m = 1$, is observed at an angle that satisfies the relationship $\sin \theta = \lambda/d$; the *second-order maximum,* corresponding to $m = 2$, is observed at a larger angle $\theta$, and so on. Figure 24.20 is a sketch of the intensity distribution for some of the orders produced by a diffrac-

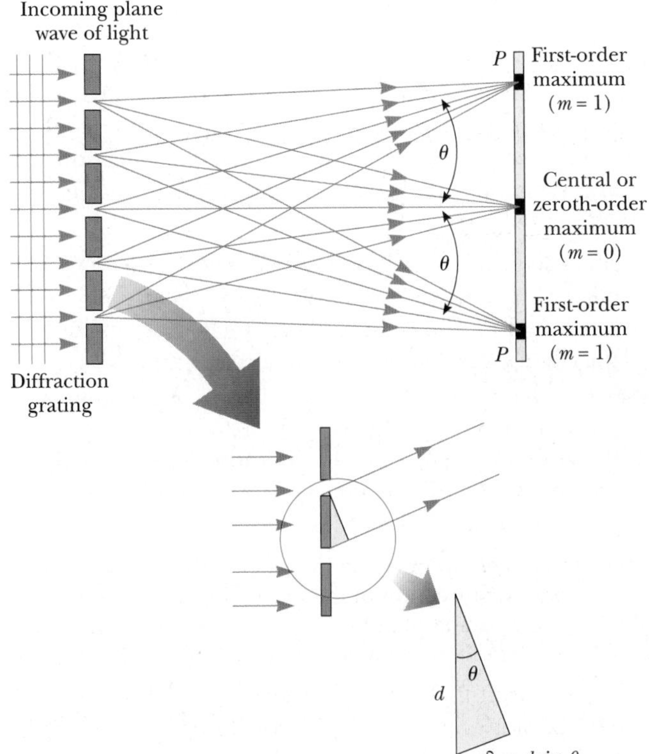


**FIGURE 24.19** A side view of a diffraction grating. The slit separation is $d$, and the path difference between adjacent slits is $d \sin \theta$.

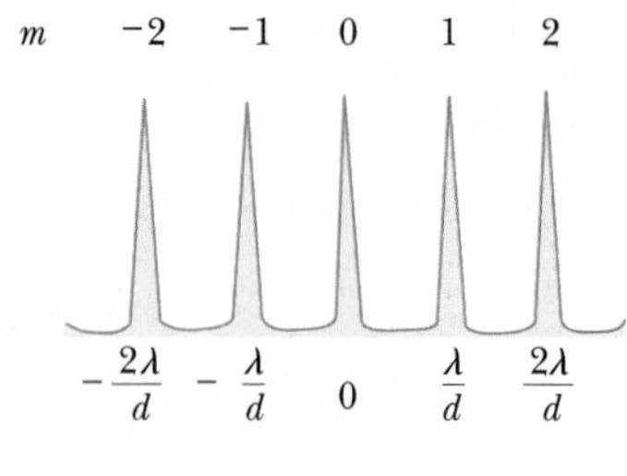


**FIGURE 24.20** Intensity versus sin $\theta$ for the diffraction grating. The zeroth-, first-, and second-order principal maxima are shown.

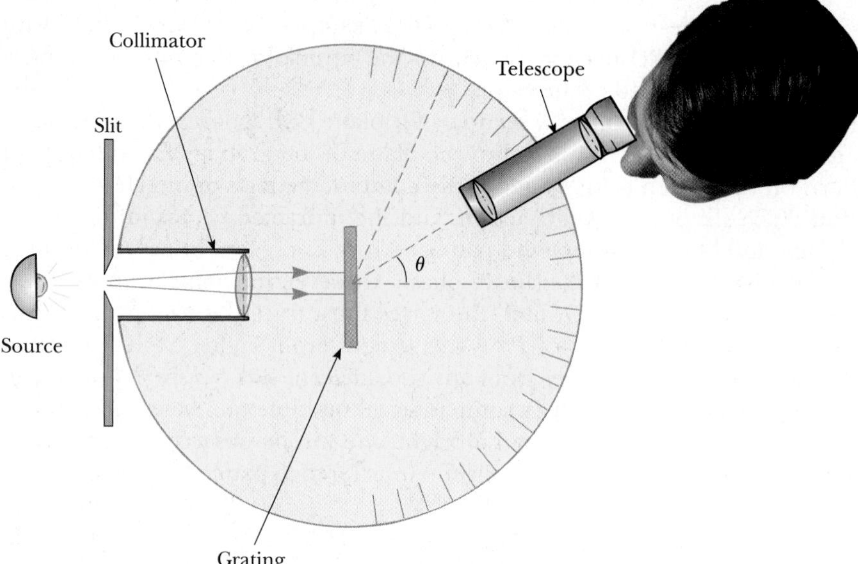


**FIGURE 24.21** A diagram of a diffraction grating spectrometer. The collimated beam incident on the grating is diffracted into the various orders at the angles $\theta$ that satisfy the equation $d \sin \theta = m\lambda$, where $m = 0, 1, 2, \ldots$ .

tion grating. Note the sharpness of the principal maxima and the broad range of the dark areas. This is in contrast to the broad, bright fringes characteristic of the two-slit interference pattern.

A simple arrangement that can be used to measure the angles in a diffraction pattern is shown in Figure 24.21. This is a form of diffraction-grating spectrometer. The light to be analyzed passes through a slit and is formed into a parallel beam by a lens. The light then strikes the grating at a 90° angle. The diffracted light leaves the grating at angles that satisfy Equation 24.12. A telescope is used to view the image of the slit. The wavelength can be determined by measuring the angles at which the images of the slit appear for the various orders.

**Quick Quiz 24.2**

If laser light is reflected from a phonograph record or a compact disc, a diffraction pattern appears. This occurs because both devices contain parallel tracks of information that act as a reflection diffraction grating. Which device, record or compact disc, results in diffraction maxima that are farther apart?

## APPLYING PHYSICS 24.4

White light enters through an opening in an opaque box, exits through an opening on the other side of the box, and a spectrum of colors appears on the wall. How would you determine whether the box *contains* a prism or a diffraction grating?

**Explanation** The determination could be made by noticing the order of the colors in the spectrum relative to the direction of the original beam of white light. For a prism, in which the separation of light is a result of dispersion, the violet light will be refracted more than the red light. Hence, the order of the spectrum from a prism will be from red, closest to the original direction, to violet. For a diffraction grating, the angle of diffraction increases with wavelength. Thus, the spectrum from the diffraction grating will have colors in the order of violet, closest to the original direction, to red. Furthermore, the diffraction grating will produce *two* first-order spectra on either side of the grating, whereas the prism will only produce a single spectrum.

## APPLYING PHYSICS 24.5

White light reflected from the surface of a compact disc has a multicolored appearance, as shown in Figure 24.22. Furthermore, the observation depends on the orientation of the disc relative to the eye and the position of the light source. Explain how this works.

**Explanation** The surface of a compact disc has a spiral-shaped track (with a spacing of approximately 1 $\mu$m) that acts as a reflection grating. The light scattered by these closely spaced parallel tracks interferes constructively in certain directions that depend on the wavelength and on the direction of the incident light. Any one section of the disc serves as a diffraction grating for white light, sending beams of constructive interference for different colors in different directions. The different colors you see when viewing one section of the disc change as the light source, the disc, or you move to change the angles of incidence or diffraction.

**FIGURE 24.22** (Applying Physics 24.5) Compact discs act as diffraction gratings when observed under white light. *(© Kristen Brochmann/Fundamental Photographs)*

## USE OF DIFFRACTION GRATING IN CD TRACKING

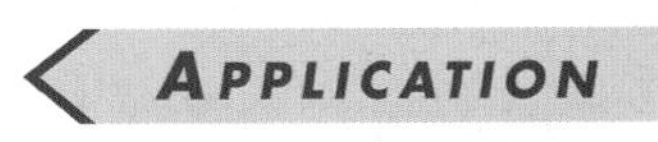


TRACKING INFORMATION ON A CD

If a CD player is to reproduce sound faithfully, the laser beam must follow the spiral track of information perfectly. However, sometimes the laser beam can drift off track, and without a feedback procedure to let the player know this is happening, the fidelity of the music would be greatly reduced.

Figure 24.23 shows how a diffraction grating is used in a three-beam method to keep the beam on track. The central maximum of the diffraction pattern is

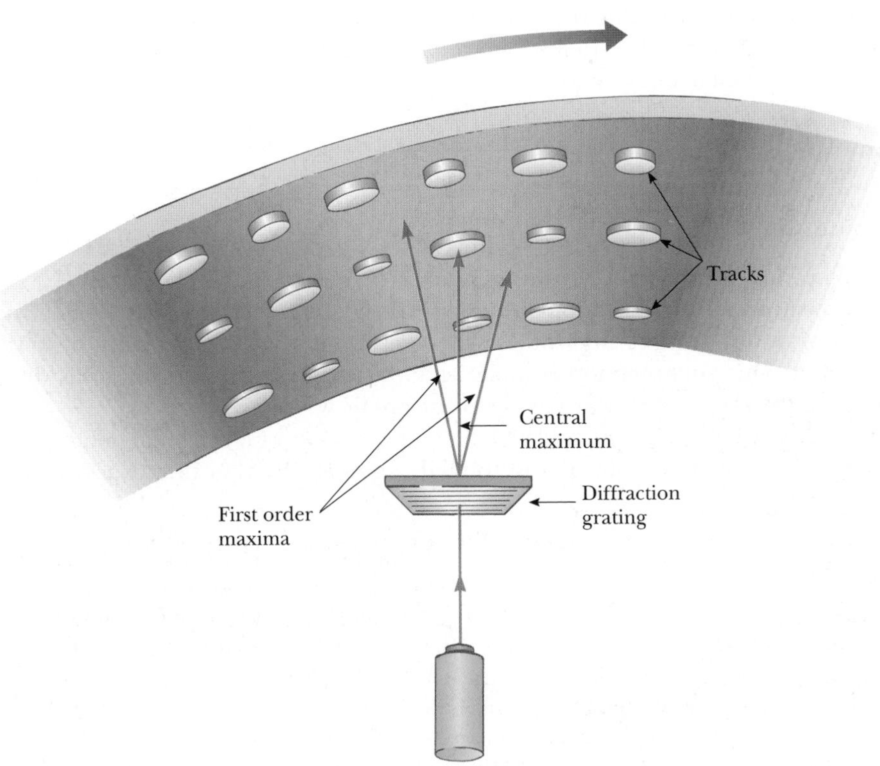


**FIGURE 24.23** The laser beam in a CD player is able to follow the spiral track by using three beams produced with a diffraction grating.

used to read the information on the CD track, and the two first-order maxima are used for steering. The grating is designed so that the first-order maxima fall on the smooth surfaces on either side of the information track. Both of these reflected beams have their own detectors, and because both beams are reflected from smooth surfaces, they should both have the same strong intensity when they are detected. If the central beam wanders off the track however, one of the steering beams will begin to strike bumps on the information track and the amount of light reflected will decrease. This information is then used by electronic circuits to drive the main beam back to its desired location.

### Example 24.7 The Orders of a Diffraction Grating

Monochromatic light from a helium-neon laser ($\lambda = 632.8$ nm) is incident normally on a diffraction grating containing 6 000 lines/cm. Find the angles at which we would observe the first-order maximum, the second-order maximum, and so forth.

**Solution** First we must calculate the slit separation, which is the inverse of the number of lines per centimeter:

$$d = \frac{1}{6\,000}\text{ cm} = 1.667 \times 10^{-4}\text{ cm} = 1\,667\text{ nm}$$

For the first-order maximum ($m = 1$), we get

$$\sin\theta_1 = \frac{\lambda}{d} = \frac{632.8\text{ nm}}{1\,667\text{ nm}} = 0.379\,6$$

$$\theta_1 = 22.31°$$

For $m = 2$, we find that

$$\sin\theta_2 = \frac{2\lambda}{d} = \frac{2(632.8\text{ nm})}{1\,667\text{ nm}} = 0.759\,2$$

$$\theta_2 = 49.39°$$

However, for $m = 3$, we find that $\sin\theta_3 = 1.139$. Because $\sin\theta$ cannot exceed unity, this is not a realistic solution. Hence, only zeroth-, first-, and second-order maxima would be observed in this situation.

## 24.9 POLARIZATION OF LIGHT WAVES

In Chapter 21 we described the transverse nature of electromagnetic waves. Figure 24.24 shows that the electric and magnetic field vectors associated with an electromagnetic wave are at right angles to each other and also to the direction of wave propagation. The phenomenon of polarization, described in this section, is firm evidence of the transverse nature of electromagnetic waves.

An ordinary beam of light consists of a large number of electromagnetic waves emitted by the atoms or molecules of the light source. The vibrating charges asso-

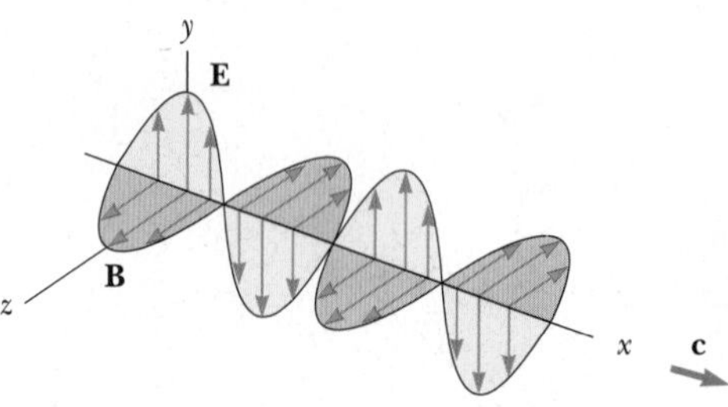


**FIGURE 24.24** A diagram of a polarized electromagnetic wave propagating in the $x$ direction. The electric field vector **E** vibrates in the $xy$ plane, and the magnetic field vector **B** vibrates in the $xz$ plane.

ciated with the atoms act as tiny antenna. Each atom produces a wave with its own orientation of **E**, as in Figure 24.24, corresponding to the direction of atomic vibration. However, because all directions of vibration are possible, the resultant electromagnetic wave is a superposition of waves produced by the individual atomic sources. The result is an **unpolarized** light wave, represented schematically in Figure 24.25a. The direction of wave propagation in this figure is perpendicular to the page. Note that *all* directions of the electric field vector are equally probable and lie in a plane (such as the plane of this page) perpendicular to the direction of propagation.

A wave is said to be **linearly polarized** if the resultant electric field **E** vibrates in the same direction *at all times* at a particular point, as in Figure 24.25b. (Sometimes such a wave is described as *plane-polarized* or simply *polarized.*) The wave in Figure 24.24 is an example of a wave linearly polarized in the $y$ direction. As the wave propagates in the $x$ direction, **E** is always in the $y$ direction. The plane formed by **E** and the direction of propagation is called the *plane of polarization* of the wave. In Figure 24.24, the plane of polarization is the $xy$ plane.

It is possible to obtain a linearly polarized beam from an unpolarized beam by removing all waves from the beam except those whose electric field vectors oscillate in a single plane. We now discuss three processes for doing this: (1) selective absorption, (2) reflection, and (3) scattering.

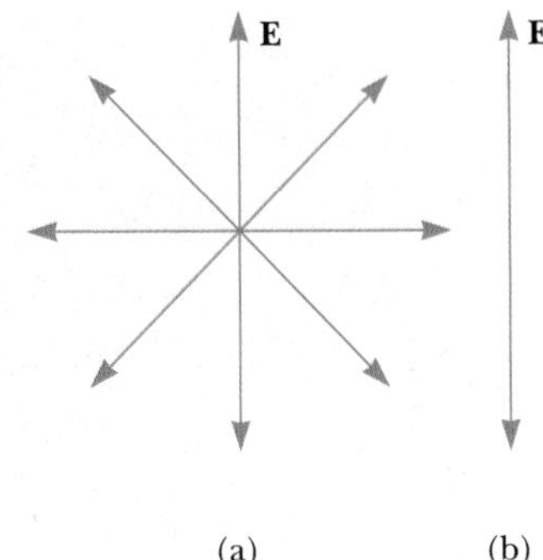


**FIGURE 24.25** (a) An unpolarized light beam viewed along the direction of propagation (perpendicular to the page). The transverse electric field vector can vibrate in any direction with equal probability. (b) A linearly polarized light beam with the electric field vector vibrating in the vertical direction.

## POLARIZATION BY SELECTIVE ABSORPTION

The most common technique for polarizing light is to use a material that transmits waves whose electric field vectors vibrate in a plane parallel to a certain direction and absorbs those waves whose electric field vectors vibrate in directions perpendicular to that direction.

In 1932, E. H. Land discovered a material, which he called **polaroid,** that polarizes light through selective absorption by oriented molecules. This material is fabricated in thin sheets of long-chain hydrocarbons, which are stretched during manufacture so that the molecules align. After a sheet is dipped into a solution containing iodine, the molecules become good electrical conductors. However, the conduction takes place primarily along the hydrocarbon chains, because the valence electrons of the molecules can move easily only along the chains. (Recall that valence electrons are "free" electrons that can readily move through the conductor.) As a result, the molecules readily *absorb* light whose electric field vector is parallel to their lengths, and *transmit* light whose electric field vector is perpendicular to their lengths. It is common to refer to the direction perpendicular to the molecular chains as the **transmission axis.** In an ideal polarizer, all light with **E** parallel to the transmission axis is transmitted, and all light with **E** perpendicular to the transmission axis is absorbed.

### Webnote 24.3

Polarized light has helped dermatologic surgeons estimate the size (area) of skin cancers. Explore *http://omlc.ogi.edu/news/feb98/polarization/index.html*

Let us now describe the intensity of light that passes through a polarizing material. In Figure 24.26, an unpolarized light beam is incident on the first polarizing sheet, called the **polarizer,** where the transmission axis is as indicated. The light that is passing through this sheet is polarized vertically, and the transmitted elec-

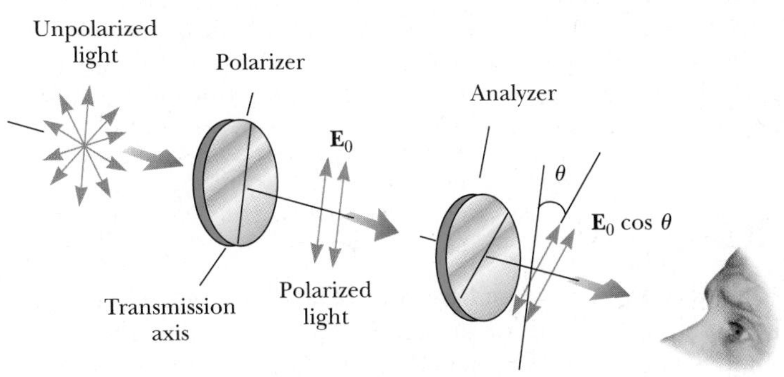


**FIGURE 24.26** Two polarizing sheets whose transmission axes make an angle $\theta$ with each other. Only a fraction of the polarized light incident on the analyzer is transmitted.

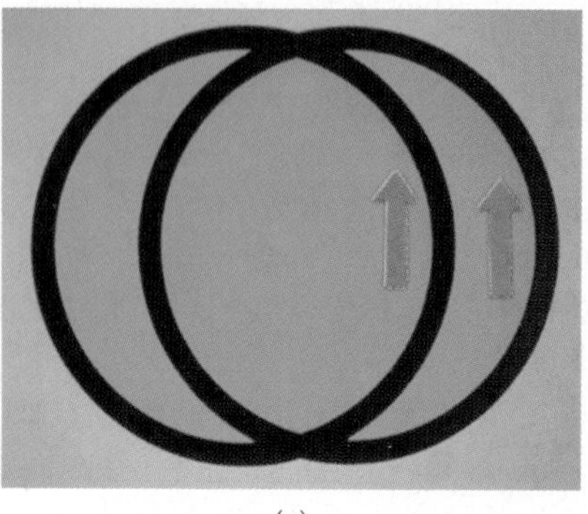
(a)

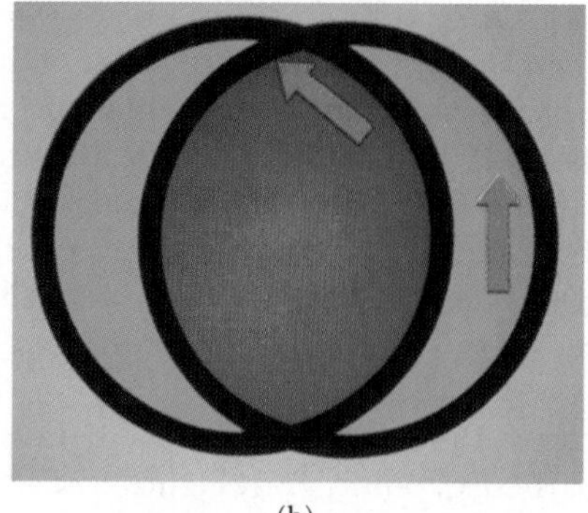
(b)

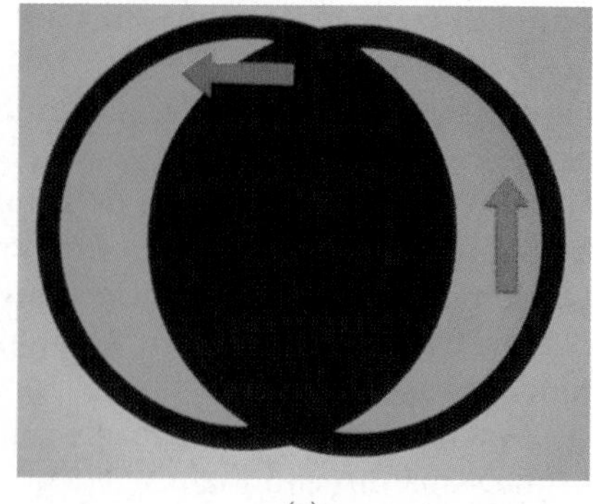
(c)

**FIGURE 24.27** The intensity of light transmitted through two polarizers depends on the relative orientations of their transmission axes. (a) The transmitted light has *maximum* intensity when the transmission axes are *aligned* with each other. (b) The transmitted light intensity diminishes when the transmission axes are at an angle of 45° with each other. (c) The transmitted light intensity is a *minimum* when the transmission axes are at *right angles* to each other. *(Photos courtesy of Henry Leap)*

tric field vector is $\mathbf{E}_0$. A second polarizing sheet, called the **analyzer,** intercepts this beam with its transmission axis at an angle of $\theta$ to the axis of the polarizer. The component of $\mathbf{E}_0$ that is perpendicular to the axis of the analyzer is completely absorbed. The component of $\mathbf{E}_0$ parallel to the analyzer axis, $E_0 \cos \theta$, is allowed to pass through the analyzer. Because the intensity of the transmitted beam varies as the *square* of its amplitude $E$, we conclude that the intensity of the (polarized) beam transmitted through the analyzer varies as

$$I = I_0 \cos^2 \theta \qquad [24.13]$$

where $I_0$ is the intensity of the polarized wave incident on the analyzer. This expression, known as **Malus's law,** applies to any two polarizing materials whose transmission axes are at an angle of $\theta$ to each other. From this expression, note that the transmitted intensity is a maximum when the transmission axes are parallel ($\theta = 0$ or 180°) and zero (complete absorption by the analyzer) when the transmission axes are perpendicular to each other. This variation in transmitted intensity through a pair of polarizing sheets is illustrated in Figure 24.27.

When unpolarized light of intensity $I_0$ is sent through a single ideal polarizer, the transmitted linearly polarized light has intensity $I_0/2$. This fact follows from Malus's law because the average value of $\cos^2 \theta$ is one half.

## APPLYING PHYSICS 24.6

**APPLICATION**

POLARIZING MICROWAVES

A polarizer for microwaves can be made as a grid of parallel metal wires about a centimeter apart. Is the electric field vector for microwaves transmitted through this polarizer parallel to, or perpendicular to, the metal wires?

**Explanation** Electric field vectors parallel to the metal wires cause electrons in the metal to oscillate parallel to the wires. Thus, the energy from the waves with these electric field vectors is transferred to the metal by accelerating these electrons and is eventually transformed to internal energy through the metal's resistance. Waves with electric field vectors perpendicular to the metal wires are not able to accelerate electrons and therefore pass through. Thus, the electric field polarization is perpendicular to the metal wires.

### POLARIZATION BY REFLECTION

When an unpolarized light beam is reflected from a surface, the reflected light is completely polarized, partially polarized, or unpolarized, depending on the angle of incidence. If the angle of incidence is either 0° or 90° (a normal or grazing an-

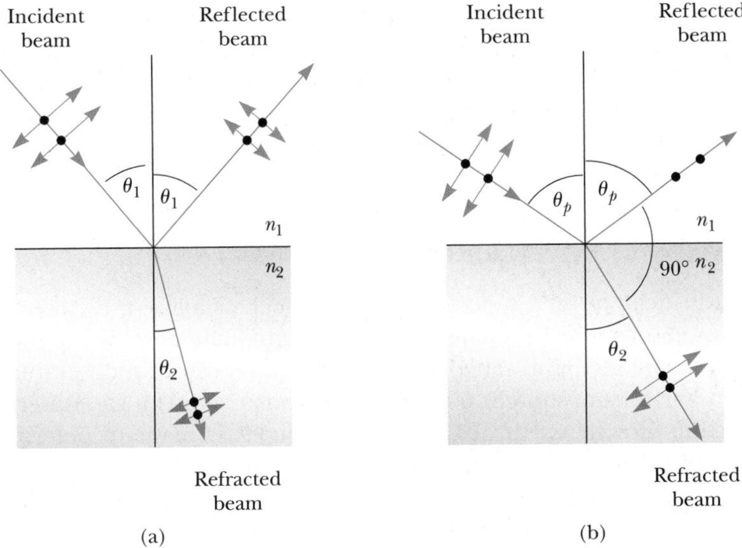


**FIGURE 24.28** (a) When unpolarized light is incident on a reflecting surface, the reflected and refracted beams are partially polarized. (b) The reflected beam is completely polarized when the angle of incidence equals the polarizing angle $\theta_p$, satisfying the equation $n = \tan\theta_p$.

gle), the reflected beam is unpolarized. However, for angles of incidence between 0° and 90°, the reflected light is polarized to some extent. For one particular angle of incidence, the reflected beam is completely polarized. Let us now investigate that special angle.

Suppose an unpolarized light beam is incident on a surface, as in Figure 24.28a. The beam can be described by two electric field components, one parallel to the surface (represented by dots) and the other perpendicular to the first component and to the direction of propagation (represented by brown arrows). It is found that the parallel component reflects more strongly than the other components, and this results in a partially polarized beam. Furthermore, the refracted beam is also partially polarized.

Now suppose that the angle of incidence $\theta_1$ is varied until the angle between the reflected and refracted beams is 90° (Fig. 24.28b). At this particular angle of incidence, the reflected beam is completely polarized, with its electric field vector parallel to the surface, whereas the refracted beam is partially polarized. The angle of incidence at which this occurs is called the **polarizing angle** $\theta_p$.

An expression relating the polarizing angle to the index of refraction of the reflecting surface can be obtained by use of Figure 24.28b. From this figure we see that, at the polarizing angle, $\theta_p + 90° + \theta_2 = 180°$, so that $\theta_2 = 90° - \theta_p$. Using Snell's law and taking $n_1 = n_{air} = 1.00$ and $n_2 = n$,

$$n = \frac{\sin\theta_1}{\sin\theta_2} = \frac{\sin\theta_p}{\sin\theta_2}$$

Because $\sin\theta_2 = \sin(90° - \theta_p) = \cos\theta_p$, the expression for $n$ can be written

$$n = \frac{\sin\theta_p}{\cos\theta_p} = \tan\theta_p \qquad [24.14]$$

Brewster's law

This expression is called **Brewster's law,** and the polarizing angle $\theta_p$ is sometimes called **Brewster's angle** after its discoverer, Sir David Brewster (1781–1868). For example, Brewster's angle for crown glass ($n = 1.52$) has the value $\theta_p = \tan^{-1}(1.52) = 56.7°$. Because $n$ varies with wavelength for a given substance, Brewster's angle is also a function of wavelength.

POLAROID SUNGLASSES

Polarization by reflection is a common phenomenon. Sunlight reflected from water, glass, or snow is partially polarized. If the surface is horizontal, the electric

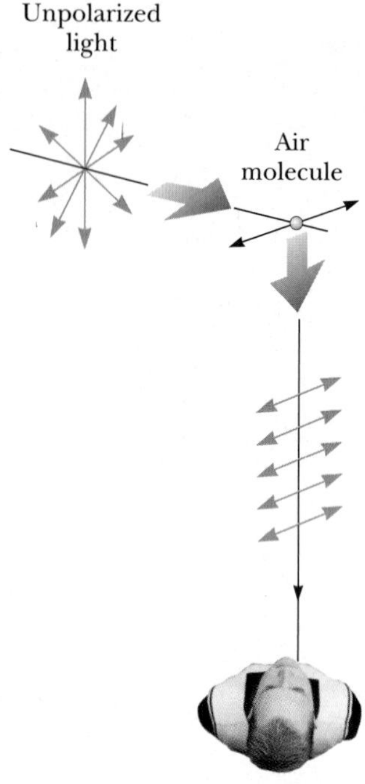


**FIGURE 24.29** The scattering of unpolarized sunlight by air molecules. The light observed at right angles is linearly polarized because the vibrating molecule has a horizontal component of vibration.

field vector of the reflected light has a strong horizontal component. Sunglasses made of polarizing material reduce the glare, which *is* the reflected light. The transmission axes of the lenses are oriented vertically to absorb the strong horizontal component of the reflected light. Because the reflected light is mostly polarized, most of the glare can be eliminated without removing most of the normal light.

## POLARIZATION BY SCATTERING

When light is incident on a system of particles, such as a gas, the electrons in the medium can absorb and reradiate part of the light. The absorption and reradiation of light by the medium, called **scattering,** is what causes sunlight reaching an observer on Earth from straight overhead to be polarized. You can observe this effect by looking directly up through a pair of sunglasses made of polarizing glass. Less light passes through at certain orientations of the lenses than at others.

Figure 24.29 illustrates how the sunlight becomes polarized. The left side of the figure shows an incident unpolarized beam of sunlight on the verge of striking an air molecule. When the beam strikes the air molecule, it sets the electrons of the molecule into vibration. These vibrating charges act like those in an antenna except that they vibrate in a complicated pattern. The horizontal part of the electric field vector in the incident wave causes the charges to vibrate horizontally, and the vertical part of the vector simultaneously causes them to vibrate vertically. A horizontally polarized wave is emitted by the electrons as a result of their horizontal motion, and a vertically polarized wave is emitted parallel to Earth as a result of their vertical motion.

Scientists have found that bees and homing pigeons use the polarization of sunlight as a navigational aid.

## OPTICAL ACTIVITY

Many important practical applications of polarized light involve the use of certain materials that display the property of **optical activity.** A substance is said to be optically active if it rotates the plane of polarization of transmitted light. Suppose unpolarized light is incident on a polarizer from the left, as in Figure 24.30a. The transmitted light is polarized vertically, as shown. If this light is then incident on an analyzer with its axis perpendicular to that of the polarizer, no light emerges from it. If an optically active material is placed between the polarizer and analyzer, as in Figure 24.30b, the material causes the direction of the polarized beam to ro-

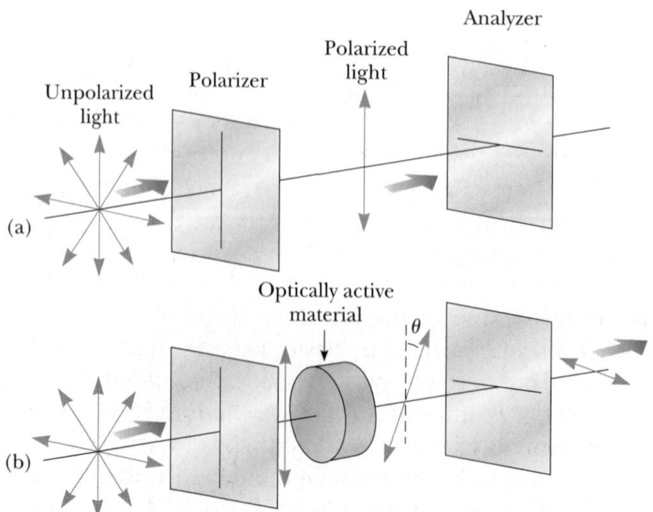


**FIGURE 24.30** (a) When crossed polarizers are used, none of the polarized light can pass through the analyzer. (b) An optically active material rotates the direction of polarization through the angle $\theta$, enabling some of this light to pass through the analyzer.

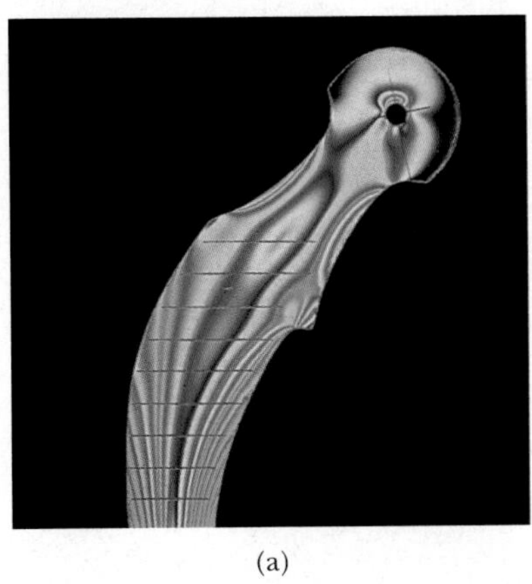

(a)

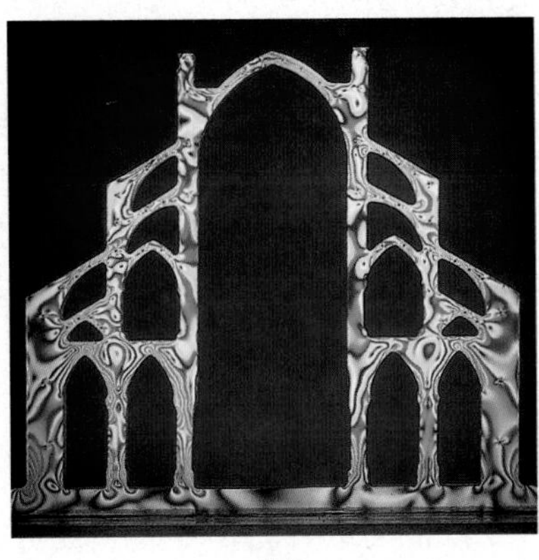

(b)

**FIGURE 24.31** (a) Strain distribution in a plastic model of a hip replacement used in a medical research laboratory. The pattern is produced when the plastic model is placed between a polarizer and analyzer oriented perpendicular to each other. *(Sepp Seitz 1981)* (b) A plastic model of an arch structure under load conditions observed between perpendicular polarizers. Such patterns are useful in the optimum design of architectural components. *(Peter Aprahamian/Science Photo Library)*

tate through the angle $\theta$. As a result, some light is able to pass through the analyzer. The angle through which the light is rotated by the material can be found by rotating the polarizer until the light is again extinguished. It is found that the angle of rotation depends on the length of the sample and, if the substance is in solution, on the concentration. One optically active material is a solution of common sugar, dextrose. A standard method for determining the concentration of sugar solutions is to measure the rotation produced by a fixed length of the solution.

**APPLICATION**

FINDING THE CONCENTRATIONS OF SOLUTIONS USING THEIR OPTICAL ACTIVITY

Optical activity occurs in a material because of an asymmetry in the shape of its constituent molecules. For example, some proteins are optically active because of their spiral shapes. Other materials, such as glass and plastic, become optically active when placed under stress. If polarized light is passed through an unstressed piece of plastic and then through an analyzer with an axis perpendicular to that of the polarizer, none of the polarized light is transmitted. However, if the plastic is placed under stress, the regions of greatest stress produce the largest angles of rotation of polarized light. Hence, we observe a series of light and dark bands in the transmitted light. Engineers often use this procedure in the design of structures ranging from bridges to small tools. A plastic model is built and analyzed under different load conditions to determine positions of potential weakness and failure under stress. If the design is poor, patterns of light and dark bands will indicate the points of greatest weakness, and the design can be corrected at an early stage. Figure 24.31 shows examples of stress patterns in plastic.

## LIQUID CRYSTALS

**APPLICATION**

LIQUID CRYSTAL DISPLAYS (LCDS)

An effect similar to rotation of the plane of polarization is used to create the familiar displays on pocket calculators, wristwatches, laptop computers, and so forth. The properties of a unique substance called a liquid crystal make these displays (called LCDs for *liquid crystal displays*) possible. As its name implies, a **liquid crystal** is a substance with properties intermediate between those of a crystalline solid and those of a liquid—that is, the molecules of the substance are more orderly than those in a liquid but less than those in a pure crystalline solid. The forces that hold the molecules together in such a state are just barely strong enough to enable the substance to maintain a definite shape, so it is reasonable to call it a solid. However, small inputs of mechanical or electrical energy can disrupt these weak bonds and make the substance flow, rotate, or twist.

To see how liquid crystals are used to create a display, consider Figure 24.32a. The liquid crystal is placed between two glass plates in the pattern shown, and electrical contacts, indicated by the thin lines, are made to the liquid crystal. When a voltage is applied across any segment in the display, that segment turns dark. In this fashion, any number between 0 and 9 can be formed by the pattern, depending on the voltages applied to the seven segments.

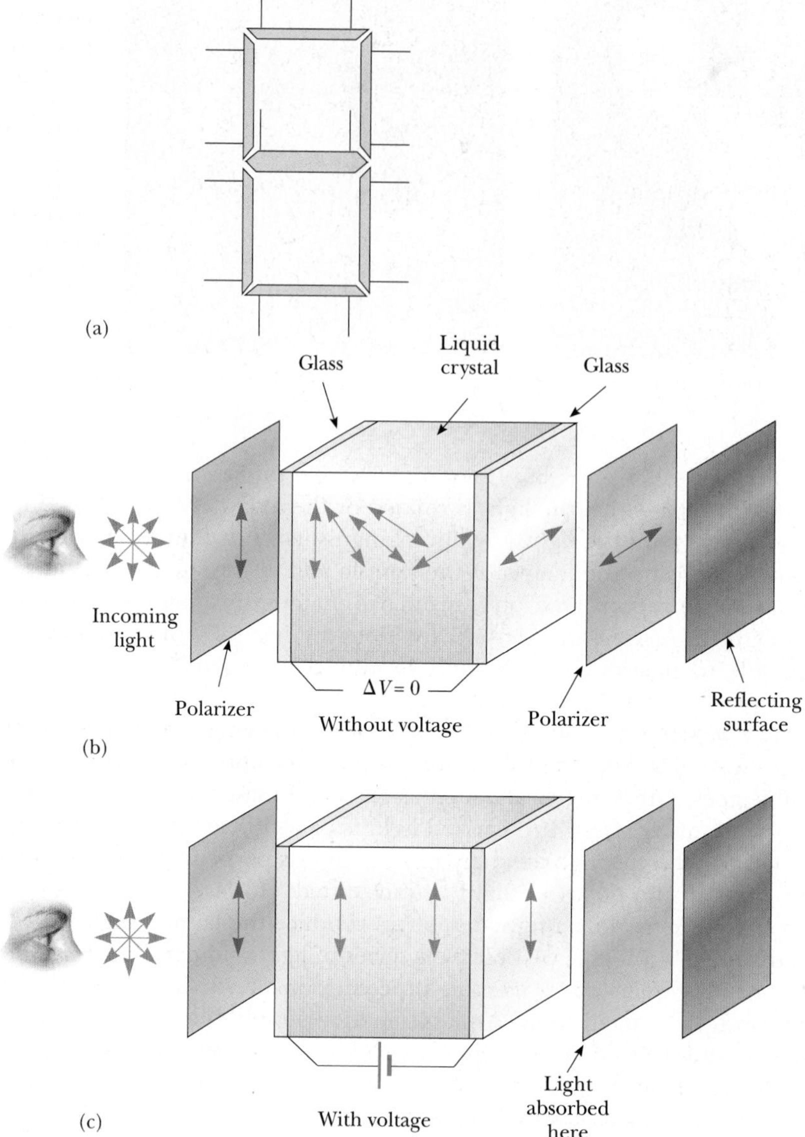


**FIGURE 24.32** (a) The light-segment pattern of a liquid crystal display. (b) Rotation of a polarized light beam by a liquid crystal when the applied voltage is zero. (c) Molecules of the liquid crystal align with the electric field when a voltage is applied.

To see why a segment can be changed from dark to light by the application of a voltage, consider Figure 24.32b, which shows the basic construction of a portion of the display. The liquid crystal is placed between two glass substrates that are packaged between two pieces of Polaroid material with their transmission axes perpendicular. A reflecting surface is placed behind one of the pieces of Polaroid. First consider what happens when light falls on this package and no voltages are applied to the liquid crystal, as shown in Figure 24.32b. Incoming light is polarized by the polarizer on the left and then falls on the liquid crystal. As the light passes through the crystal, its plane of polarization is rotated by 90°, allowing it to pass through the polarizer on the right. It reflects from the reflecting surface and retraces its path through the crystal. Thus, an observer to the left of the crystal sees the segment as being bright. When a voltage is applied as in Figure 24.32c, the molecules of the liquid crystal do not rotate the plane of polarization of the light. In this case, the light is absorbed by the polarizer on the right and none is reflected back to the observer to the left of the crystal. Thus, the observer sees this segment as black. Changing the applied voltage to the crystal in a precise pattern and at precise times can make the pattern tick off the seconds on a watch, display a letter on a computer display, and so forth.

## SUMMARY

**Interference** occurs when two or more light waves overlap at a given point. A sustained interference pattern is observed if (1) the sources are coherent (that is, they maintain a constant phase relationship with one another); (2) the sources have identical wavelengths; and (3) the superposition principle is applicable.

In **Young's double-slit experiment,** two slits separated by distance $d$ are illuminated by a single-wavelength light source. An interference pattern consisting of bright and dark fringes is observed on a screen a distance $L$ from the slits. The condition for **bright fringes** (constructive interference) is

$$d \sin \theta_{\text{bright}} = m\lambda \qquad m = 0, \pm 1, \pm 2, \ldots \qquad [24.2]$$

The number $m$ is called the **order number** of the fringe. The condition for **dark fringes** (destructive interference) is

$$d \sin \theta_{\text{dark}} = (m + \tfrac{1}{2})\lambda \qquad m = 0, \pm 1, \pm 2, \ldots \qquad [24.3]$$

An electromagnetic wave undergoes a phase change of 180° on reflection from a medium with an index of refraction higher than that of the medium in which the wave is traveling.

The wavelength $\lambda_n$ of light in a medium with index of refraction $n$ is

$$\lambda_n = \frac{\lambda}{n} \qquad [24.7]$$

where $\lambda$ is the wavelength of the light in free space.

The **diffraction pattern** produced by a single slit on a distant screen consists of a central bright maximum flanked by less-bright fringes alternating with dark regions. The angles $\theta$ at which the diffraction pattern has zero intensity (regions of destructive interference) are described by

$$\sin \theta_{\text{dark}} = m\frac{\lambda}{a} \qquad m = \pm 1, \pm 2, \pm 3, \ldots \qquad [24.11]$$

where $a$ is the width of the slit and $\lambda$ is the wavelength of the light incident on the slit.

A **diffraction grating** consists of many equally spaced, identical slits. The condition for **maximum intensity** in the interference pattern of a diffraction grating is

$$d \sin \theta_{\text{bright}} = m\lambda \qquad m = 0, 1, 2, \ldots \qquad [24.12]$$

where $d$ is the spacing between adjacent slits and $m$ is the order number of the diffraction pattern.

Unpolarized light can be polarized by selective absorption, reflection, and scattering.

In general, light reflected from an amorphous material, such as glass, is partially polarized. Reflected light is completely polarized with its electric field parallel to the surface when the angle of incidence produces a 90° angle between the reflected and refracted beams. This angle of incidence, called the **polarizing angle** $\theta_p$, satisfies **Brewster's law,** given by

$$n = \tan \theta_p \qquad [24.14]$$

where $n$ is the index of refraction of the reflecting medium.

## CONCEPTUAL QUESTIONS

1. Your automobile has two headlights. What sort of interference pattern do you expect to see from them? Why?
2. Holding your hand at arm's length, you can readily block sunlight from your eyes. Why can you not block sound from your ears this way?
3. Consider a dark fringe in an interference pattern, at which almost no light energy is arriving. Light from both slits is arriving at this point, but the waves are canceling. Where does the energy go?
4. If Young's double-slit experiment were performed under water, how would the observed interference pattern be affected?
5. In a laboratory accident, you spill two liquids onto water, neither of which mixes with the water. They both form

thin films on the water surface. You notice, as the films become very thin as they spread, that one film becomes bright and the other black in reflected light. Why might this be?

6. If white light is used in Young's double-slit experiment rather than monochromatic light, how does the interference pattern change?
7. In our discussion of thin-film interference, we looked at light *reflecting* from a thin film. Consider one light ray, the direct ray, that transmits through the film without reflecting. Consider a second ray, the reflected ray, that transmits through the first surface, reflects back to the second, reflects again from the first, and then transmits out into the air, parallel to the direct ray. For normal incidence, how thick must the film be, in terms of the wavelength of the light, for the outgoing rays to interfere destructively? Is it the same thickness as for reflected destructive interference?
8. What is the necessary condition on path-length difference between two waves that interfere (a) constructively and (b) destructively? Assume that the wave sources are coherent.
9. A lens with outer radius of curvature $R$ and index of refraction $n$ rests on a flat glass plate, and the combination is illuminated from white light from above. Is there a dark spot or a light spot at the center of the lens? What does it mean if the observed rings are noncircular?
10. Often fingerprints left on a piece of glass such as a window pane show colored spectra like that from a diffraction grating. Why?
11. In our everyday experience, radio waves are polarized, but light is not. Why?
12. Suppose reflected white light is used to observe a thin, transparent coating on glass as the coating material is gradually deposited by evaporation in a vacuum. Describe possible color changes that might occur during the process of building up the thickness of the coating.
13. Would it be possible to place a nonreflective coating on an airplane to cancel radar waves of wavelength 3 cm?
14. Certain sunglasses use a polarizing material to reduce the intensity of light reflected from shiny surfaces, such as water or the hood of a car. What orientation of the transmission axis should the material have to be most effective?
15. Why is it so much easier to perform interference experiments with a laser than with an ordinary light source?
16. A simple way to observe an interference pattern is to look at a distant light source through a stretched handkerchief or an open umbrella. Explain how this works.
17. Although we can hear around corners, we cannot see around corners. How can you explain this in view of the fact that sound and light are both waves?
18. Can a sound wave be polarized? Explain.
19. When you receive a chest x-ray at a hospital, the x-rays pass through a series of parallel ribs in your chest. Do the ribs act as a diffraction grating for x-rays?

## PROBLEMS

1, 2, 3 = straightforward, intermediate, challenging ☐ = full solution available in Student Solutions Manual/Study Guide

web = solution posted at **http://info.brookscole.com/serway** = biomedical application

### Section 24.2 Young's Double-Slit Interference

1. A laser beam ($\lambda$ = 632.8 nm) is incident on two slits 0.200 mm apart. How far apart are the bright interference fringes on a screen 5.00 m away from the double slits?
2. In a Young's double-slit experiment, a set of parallel slits with a separation of 0.100 mm is illuminated by light having a wavelength of 589 nm and the interference pattern observed on a screen 4.00 m from the slits. (a) What is the difference in path lengths from each of the slits to the screen location of a third-order bright fringe? (b) What is the difference in path lengths from the two slits to the screen location of the third dark fringe away from the center of the pattern?
3. A pair of narrow, parallel slits separated by 0.250 mm are illuminated by the green component from a mercury vapor lamp ($\lambda$ = 546.1 nm). The interference pattern is observed on a screen 1.20 m from the plane of the parallel slits. Calculate the distance (a) from the central maximum to the first bright region on either side of the central maximum and (b) between the first and second dark bands in the interference pattern.
4. Light of wavelength 460 nm falls on two slits spaced 0.300 mm apart. What is the required distance from the slit to a screen if the spacing between the first and second dark fringes is to be 4.00 mm?
5. In a location where the speed of sound is 354 m/s, a 2 000-Hz sound wave impinges on two slits 30.0 cm apart. (a) At what angle is the first maximum located? (b) If the sound wave is replaced by 3.00-cm microwaves, what slit separation gives the same angle for the first maximum? (c) If the slit separation is 1.00 $\mu$m, what frequency light gives the same first maximum angle?
6. White light spans the wavelength range between about 400 nm and 700 nm. If white light passes through two slits 0.30 mm apart and falls on a screen 1.5 m from the slits, find the distance between the first-order violet and the first-order red fringes.
7. Two radio antennas separated by 300 m, as shown in Figure P24.7, simultaneously transmit identical signals of the same wavelength. A radio in a car traveling due north receives the signals. (a) If the car is at the position of the second maximum, what is the wavelength of the signals? (b) How much farther must the car travel to encounter the next minimum in reception? (*Hint:* Determine the path difference between the two signals at the two locations of the car.)
8. Light of wavelength 575 nm falls on a double slit, and the first bright fringe is seen at an angle of 16.5°. Find the distance between the two slits.
9. Waves from a radio station have a wavelength of 300 m.

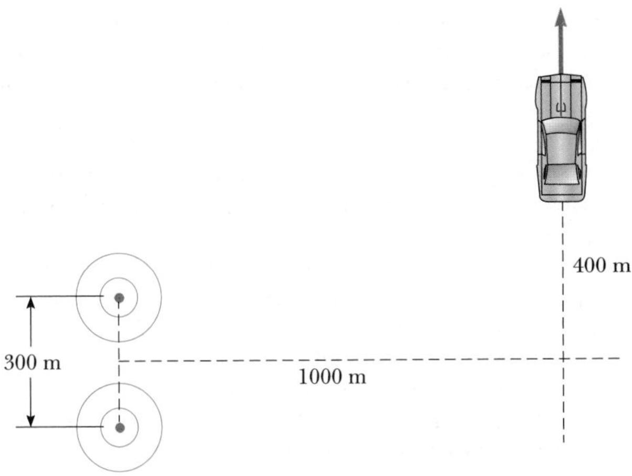


FIGURE P24.7

They travel by two paths to a home receiver 20.0 km from the transmitter. One path is a direct path, and the second is by reflection from a mountain directly behind the home receiver. What is the minimum distance from the mountain to the receiver that produces destructive interference at the receiver? (Assume that no phase change occurs on reflection from the mountain.)

10. A Young's interference experiment is performed with blue-green argon laser light. The separation between the slits is 0.500 mm, and the interference pattern on a screen 3.30 m away shows the first maximum 3.40 mm from the center of the pattern. What is the wavelength of argon laser light?

11. A riverside warehouse has two open doors, as in Figure P24.11. Its interior is lined with a sound-absorbing material. A boat on the river sounds its horn. To person A the sound is loud and clear. To person B the sound is barely audible. The principal wavelength of the sound waves is 3.00 m. Assuming person B is at the position of the first minimum, determine the distance between the doors, center to center.

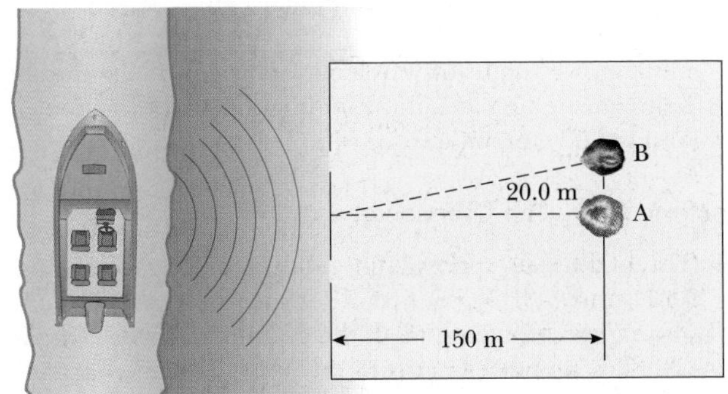


FIGURE P24.11

12. web The waves from a radio station can reach a home receiver by two different paths. One is a straight-line path from the transmitter to the home, a distance of 30.0 km. The second path is by reflection from a storm cloud. Assume that this reflection takes place at a point midway between receiver and transmitter. If the wavelength broadcast by the radio station is 400 m, find the minimum height of the storm cloud that will produce destructive interference between the direct and reflected beams. (Assume no phase changes on reflection.)

13. Radio waves from a star, of wavelength 250 m, reach a radio telescope by two separate paths as shown in Figure P24.13. One is a direct path to the receiver, which is situated on the edge of a cliff by the ocean. The second is by reflection off the water. The first minimum of destructive interference occurs when the star is 25.0° above the horizon. Find the height of the cliff. (Assume no phase change on reflection.)

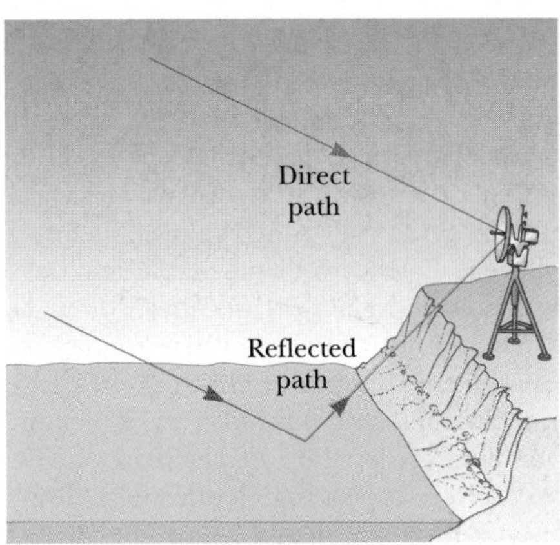


FIGURE P24.13

## Section 24.3 Change of Phase Due to Reflection

## Section 24.4 Interference in Thin Films

14. A soap bubble ($n = 1.33$) is floating in air. If the thickness of the bubble wall is 115 nm, what is the wavelength of the visible light that is most strongly reflected?

15. Suppose the film shown in Figure 24.7 has an index of refraction of 1.36 and is surrounded by air on both sides. Find the minimum thickness that will produce constructive interference in the reflected light when the film is illuminated by light of wavelength 500 nm.

16. A thin film of oil ($n = 1.25$) is located on a smooth, wet pavement. When viewed perpendicular to the pavement, the film appears to be predominantly red (640 nm) and has no blue color (512 nm). How thick is the oil film?

17. A thin layer of liquid methylene iodide ($n = 1.756$) is sandwiched between two flat parallel plates of glass ($n = 1.50$). What must be the thickness of the liquid layer if normally incident light with $\lambda = 600$ nm in air is to be strongly reflected?

18. A transparent oil of index of refraction 1.29 spills on the surface of water (index of refraction 1.33), producing a maximum of reflection with normally incident orange light (wavelength 600 nm in air). Assuming the maximum occurs in the first order, determine the thickness of the oil slick.

19. A possible means for making an airplane invisible to radar is to coat the plane with an antireflective polymer. If radar waves have a wavelength of 3.00 cm and the index of refraction of the polymer is $n = 1.50$, how thick would you make the coating?

20. A beam of light of wavelength 580 nm passes through two closely spaced glass plates, as shown in Figure P24.20. For what minimum nonzero value of the plate separation $d$

will the transmitted light be bright? This arrangement is often used to measure the wavelength of light and is called a Fabry-Perot interferometer.

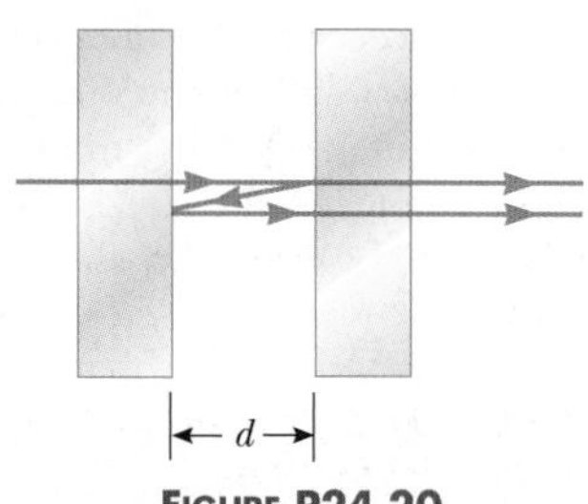


**FIGURE P24.20**

21. Astronomers observe the chromosphere of the Sun with a filter that passes the red hydrogen spectral line of wavelength 656.3 nm, called the $H_\alpha$ line. The filter consists of a transparent dielectric of thickness $d$ held between two partially aluminized glass plates. The filter is held at a constant temperature. (a) Find the minimum value of $d$ that will produce maximum transmission of perpendicular $H_\alpha$ light, if the dielectric has index of refraction 1.378. (b) If the temperature of the filter increases above the normal value, thereby increasing its thickness, what happens to the transmitted wavelength? (c) The dielectric will also pass what near-visible wavelength? One of the glass plates is colored red to absorb this light.

22. Two rectangular optically flat plates ($n = 1.52$) are in contact along one end and are separated along the other end by a 2.00-$\mu$m-thick spacer (Fig. P24.22). The top plate is illuminated by monochromatic light of wavelength 546.1 nm. Calculate the number of dark parallel bands crossing the top plate (including the dark band at zero thickness along the edge of contact between the two plates).

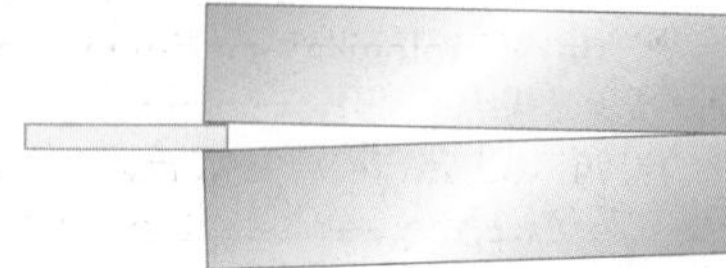

**FIGURE P24.22** (Problems 22 and 23)

23. An air wedge is formed between two glass plates separated at one edge by a very fine wire as in Figure P24.22. When the wedge is illuminated from above by 600-nm light, 30 dark fringes are observed. Calculate the radius of the wire.

24. A planoconvex lens with radius of curvature $R = 3.0$ m is in contact with a flat plate of glass. A light source and the observer's eye are both close to the normal, as shown in Figure 24.8. The radius of the 50th bright Newton's ring is found to be 9.8 mm. What is the wavelength of the light produced by the source?

25. A planoconvex lens rests with its curved side on a flat glass surface and is illuminated from above by light of wavelength 500 nm (see Fig. 24.8). A dark spot is observed at the center, surrounded by 19 concentric dark rings (with bright rings in between). How much thicker is the air wedge at the position of the 19th dark ring than at the center?

26. Nonreflective coatings on camera lenses reduce the loss of light at the surfaces of multi-lens systems and prevent internal reflections that might mar the image. Find the minimum thickness of a layer of magnesium fluoride ($n = 1.38$) on flint glass ($n = 1.66$) that will cause destructive interference of reflected light of wavelength 550 nm near the middle of the visible spectrum.

27. **web** A thin film of $MgF_2$ ($n = 1.38$) with thickness $1.00 \times 10^{-5}$ cm is used to coat a camera lens. Are any wavelengths in the visible spectrum intensified in the reflected light?

28. A flat piece of glass is supported horizontally above the flat end of a 10.0-cm-long metal rod that has its lower end rigidly fixed. The thin film of air between the rod and glass is observed to be bright when illuminated by light of wavelength 500 nm. As the temperature is slowly increased by 25.0°C, the film changes from bright to dark and back to bright 200 times. What is the coefficient of linear expansion of the metal?

## Section 24.7 Single-Slit Diffraction

29. Helium-neon laser light ($\lambda = 632.8$ nm) is sent through a 0.300-mm-wide single slit. What is the width of the central maximum on a screen 1.00 m from the slit?

30. Light of wavelength 600 nm falls on a 0.40-mm-wide slit and forms a diffraction pattern on a screen 1.5 m away. (a) Find the position of the first dark band on each side of the central maximum. (b) Find the width of the central maximum.

31. Light of wavelength 587.5 nm illuminates a single 0.75-mm-wide slit. (a) At what distance from the slit should a screen be placed if the first minimum in the diffraction pattern is to be 0.85 mm from the central maximum? (b) Calculate the width of the central maximum.

32. Microwaves of wavelength 5.00 cm enter a long, narrow window in a building that is otherwise essentially opaque to the microwaves. If the window is 36.0 cm wide, what is the distance from the central maximum to the first-order minimum along a wall 6.50 m from the window?

33. A slit of width 0.50 mm is illuminated with light of wavelength 500 nm, and a screen is placed 120 cm in front of the slit. Find the widths of the first and second maxima on each side of the central maximum.

34. A screen is placed 50.0 cm from a single slit, which is illuminated with light of wavelength 680 nm. If the distance between the first and third minima in the diffraction pattern is 3.00 mm, what is the width of the slit?

## Section 24.8 The Diffraction Grating

35. Three discrete spectral lines occur at angles of 10.1°, 13.7°, and 14.8° in the first-order spectrum of a diffraction grating spectrometer. (a) If the grating has 3 660 slits/cm, what are the wavelengths of the light? (b) At what angles are these lines found in the second-order spectra?

36. Intense white light is incident on a diffraction grating that has 600 lines/mm. (a) What is the highest order in which the complete visible spectrum can be seen using this grating? (b) What is the angular separation between the violet edge (400 nm) and the red edge (700 nm) of the first-order spectrum produced by this grating?

37. The hydrogen spectrum has a red line at 656 nm and a violet line at 434 nm. What is the angular separation between these two spectral lines obtained with a diffraction grating that has 4 500 lines/cm?

38. A grating with 1 500 slits per centimeter is illuminated with light of wavelength 500 nm. (a) What is the highest-order number that can be observed with this grating? (b) Repeat for a grating of 15 000 slits per centimeter.

39. A light source emits two major spectral lines, an orange line of wavelength 610 nm and a blue-green line of wavelength 480 nm. If the spectrum is resolved by a diffraction grating having 5 000 lines/cm and viewed on a screen 2.00 m from the grating, what is the distance (in centimeters) between the two spectral lines in the second-order spectrum?

40. White light is spread out into its spectral components by a diffraction grating. If the grating has 2 000 lines per centimeter, at what angle does red light of wavelength 640 nm appear in the first-order spectrum?

41. Sunlight is incident on a diffraction grating that has 2 750 lines/cm. The second-order spectrum over the visible range (400–700 nm) is to be limited to 1.75 cm along a screen that is distance $L$ from the grating. What is the required value of $L$?

42. Light containing two different wavelengths passes through a diffraction grating with 1 200 slits/cm. On a screen 15.0 cm from the grating, the third-order maximum of the shorter wavelength falls midway between the central maximum and the first side maximum for the longer wavelength. If the neighboring maxima of the longer wavelength are 8.44 mm apart on the screen, what are the wavelengths in the light? (*Hint:* Use the small-angle approximation.)

43. web A beam of 541-nm light is incident on a diffraction grating that has 400 lines/mm. (a) Determine the angle of the second-order ray. (b) If the entire apparatus is immersed in water, determine the new second-order angle of diffraction. (c) Show that the two diffracted rays of parts (a) and (b) are related through the law of refraction.

44. Light from an helium-neon laser ($\lambda = 632.8$ nm) is incident on a single slit. What is the maximum width for which no diffraction minima are observed? (*Hint:* Values of $\sin\theta > 1$ are not possible.)

## Section 24.9 Polarization of Light Waves

45. The angle of incidence of a light beam in air onto a reflecting surface is continuously variable. The reflected ray is found to be completely polarized when the angle of incidence is 48.0°. (a) What is the index of refraction of the reflecting material? (b) If some of the incident light (at an angle of 48°) passes into the material below the surface, what is the angle of refraction?

46. Unpolarized light passes through two polaroid sheets. The axis of the first is vertical, and that of the second is at 30.0° to the vertical. What fraction of the initial light is transmitted?

47. The index of refraction of a glass plate is 1.52. What is the Brewster's angle when the plate is (a) in air? (b) in water? (See Problem 51.)

48. At what angle above the horizon is the Sun if light from it is completely polarized upon reflection from water?

49. A light beam is incident on heavy flint glass ($n = 1.65$) at the polarizing angle. Calculate the angle of refraction for the transmitted ray.

50. The critical angle for total internal reflection for sapphire surrounded by air is 34.4°. Calculate the Brewster's angle for sapphire if the light is incident from the air.

51. Equation 24.14 assumes that the incident light is in air. If the light is incident from a medium of index $n_1$ on a medium of index $n_2$, follow the procedure used to derive Equation 24.14 to show that $\tan\theta_p = n_2/n_1$.

52. Plane-polarized light is incident on a single polarizing disk with the direction of $E_0$ parallel to the direction of the transmission axis. Through what angle should the disk be rotated so that the intensity in the transmitted beam is reduced by a factor of (a) 3.00, (b) 5.00, (c) 10.0?

53. Three polarizing plates whose planes are parallel are centered on a common axis. The directions of the transmission axes relative to the common vertical direction are shown in Figure P24.53. A linearly polarized beam of light with the plane of polarization parallel to the vertical reference direction is incident from the left on the first disk with intensity $I_i = 10.0$ units (arbitrary). Calculate the transmitted intensity $I_f$ when $\theta_1 = 20.0°$, $\theta_2 = 40.0°$, and $\theta_3 = 60.0°$. (*Hint:* Make repeated use of Malus's law.)

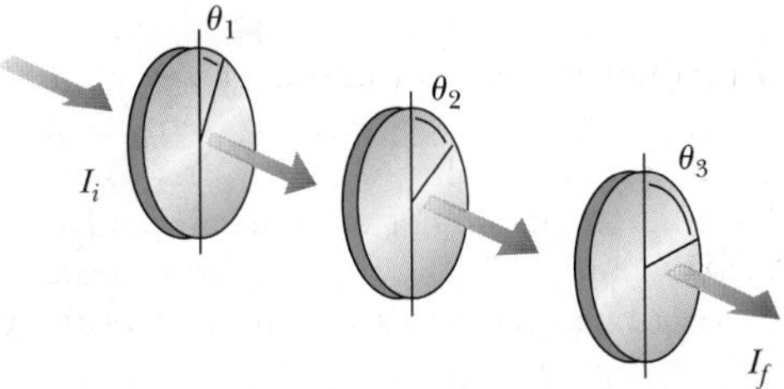


**FIGURE P24.53** (Problems 53 and 62)

54. web Light of intensity $I_0$ and polarized parallel to the transmission axis of a polarizer, is incident on an analyzer. (a) If the transmission axis of the analyzer makes an angle of 45° with the axis of the polarizer, what is the intensity of the transmitted light? (b) What should the angle between the transmission axes be to make $I/I_0 = 1/3$?

55. Light with a wavelength in vacuum of 546.1 nm falls perpendicularly on a biological specimen that is 1.000 $\mu$m thick. The light splits into two beams polarized at right angles, for which the indices of refraction are 1.320 and 1.333. (a) Calculate the wavelength of each component of the light while it is traversing the specimen. (b) Calculate the phase difference between the two beams when they emerge from the specimen.

## ADDITIONAL PROBLEMS

56. A beam containing light of wavelengths $\lambda_1$ and $\lambda_2$ is incident on a set of parallel slits. In the interference pattern, the fourth bright line of the $\lambda_1$ light occurs at the same position as the fifth bright line of the $\lambda_2$ light. If $\lambda_1$ is known to be 540 nm, what is the value of $\lambda_2$?

57. Light of wavelength 546 nm (the intense green line from a mercury source) produces a Young's interference pattern in which the second minimum from the central maximum is along a direction that makes an angle of 18.0 min of arc with the axis through the central maximum. What is the distance between the parallel slits?

58. The two speakers of a boom box are 35.0 cm apart. A single oscillator makes the speakers vibrate in phase at a frequency of 2.00 kHz. At what angles, measured from the perpendicular bisector of the line joining the speakers, would a distant observer hear maximum sound intensity?

minimum sound intensity? (Take the speed of sound as 340 m/s.)

59. Interference effects are produced at point $P$ on a screen as a result of direct rays from a 500-nm source and reflected rays off the mirror, as in Figure P24.59. If the source is 100 m to the left of the screen, and 1.00 cm above the mirror, find the distance $y$ (in millimeters) to the first dark band above the mirror.

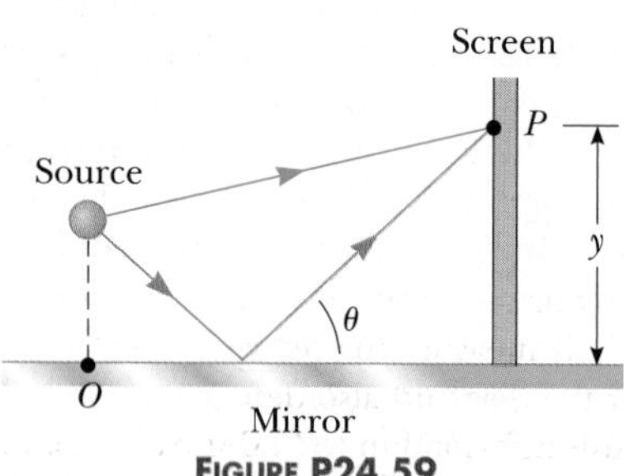


**FIGURE P24.59**

60. Many cells are transparent and colorless. Structures of great interest in biology and medicine can be practically invisible to ordinary microscopy. An *interference microscope* reveals a difference in refractive index as a shift in interference fringes, to indicate the size and shape of cell structures. The idea is exemplified in the following problem: An air wedge is formed between two glass plates in contact along one edge and slightly separated at the opposite edge. When the plates are illuminated with monochromatic light from above, the reflected light has 85 dark fringes. Calculate the number of dark fringes that appear if water ($n = 1.33$) replaces the air between the plates.

**61.** When a monochromatic beam of light is incident from air at an angle of 37.0° with the normal on the surface of a glass block, it is observed that the refracted ray is directed at 22.0° with the normal. What angle of incidence from air would result in total polarization of the reflected beam?

62. Three polarizers, centered on a common axis and with their planes parallel to each other, have transmission axes oriented at angles of $\theta_1$, $\theta_2$, and $\theta_3$ from the vertical as shown in Figure P24.53. Light of intensity $I_i$, polarized with its plane of polarization oriented vertically, is incident from the left on the first polarizer. What is the ratio $I_f/I_i$ of the final transmitted intensity to the incident intensity if (a) $\theta_1 = 45°$, $\theta_2 = 90°$, and $\theta_3 = 0°$? (b) $\theta_1 = 0°$, $\theta_2 = 45°$, and $\theta_3 = 90°$?

63. Figure P24.63 shows a radio wave transmitter and a receiver, both $h = 50.0$ m above the ground and $d = 600$ m apart. The receiver can receive signals both directly, from the transmitter, and indirectly, from signals that bounce off the ground. If the ground is level between the transmitter and receiver and a $\lambda/2$ phase shift occurs upon reflection, determine the longest wavelengths that interfere (a) constructively and (b) destructively.

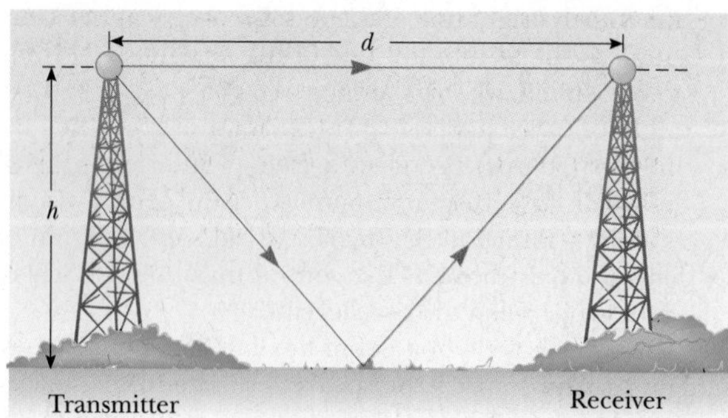


**FIGURE P24.63**

64. A planoconvex lens (flat on one side, convex on the other) with index of refraction $n$ rests with its curved side (radius of curvature $R$) on a flat glass surface of the same index of refraction with a film of index $n_{film}$ between them. The lens is illuminated from above by light of wavelength $\lambda$. Show that the dark Newton rings have radii of

$$r \approx \sqrt{m\lambda R/n_{film}}$$

where $m$ is an integer.

65. When a liquid is introduced into the air space between the lens and the plate in a Newton's rings apparatus, the diameter of the tenth ring changes from 1.50 to 1.31 cm. Find the index of refraction of the liquid. (See Problem 64.)

66. (a) If light is incident at an angle $\theta$ from a medium of index $n_1$ on a medium of index $n_2$ so that the angle between the reflected ray and refracted ray is $\beta$, show that

$$\tan\theta = \frac{n_2 \sin\beta}{n_1 - n_2\cos\beta}$$

*Hint:* Use the following identity.

$$\sin(A + B) = \sin A\cos B + \cos A\sin B$$

(b) Show that the foregoing equation for $\tan\theta$ reduces to Brewster's law when $\beta = 90°$, $n_1 = 1$, and $n_2 = n$.

67. A diffraction pattern is produced on a screen 140 cm from a single slit, using monochromatic light of wavelength 500 nm. The distance from the center of the central maximum to the first-order maximum is 3.00 mm. Calculate the slit width. (*Hint:* Assume that the first-order maximum is halfway between the first- and second-order minima.)

68. A glass plate ($n = 1.61$) is covered with a thin, uniform layer of oil ($n = 1.20$). A light beam of variable wavelength is normally incident from air onto the oil surface. Observation of the reflected beam shows destructive interference at 500 nm and constructive interference at 750 nm. From this information, calculate the thickness of the oil film.

69. The condition for constructive interference by reflection from a thin film in air, as developed in Section 24.4, assumes nearly normal incidence. (a) Show that for large angles of incidence, the condition for constructive interference of light reflecting from a thin film of thickness $t$, index of refraction $n$, and surrounded by air may be written as

$$2nt\cos\theta_2 = (m + \tfrac{1}{2})\lambda$$

where $\theta_2$ is the angle of refraction. (b) Calculate the minimum thickness for constructive interference if sodium light ($\lambda = 590$ nm) is incident at an angle of 30.0° on a film with index of refraction 1.38.

70. Figure P24.70 illustrates the formation of an interference pattern by the Lloyd's mirror method. Light from source $S$ reaches the screen via two different pathways. One is a direct path, and the second is by reflection from a horizontal mirror. The interference is as though light from two

different sources $S$ and $S'$ had interfered as in the Young's double-slit arrangement. Assume that the actual source $S$ and the virtual source $S'$ are in a plane 25 cm to the left of the mirror, and the screen is a distance of $L = 120$ cm to the right of this plane. Source $S$ is a distance of $h = 2.5$ mm above the top surface of the mirror, and the light is monochromatic with $\lambda = 620$ nm. Determine the distance of the first bright fringe above the surface of the mirror.

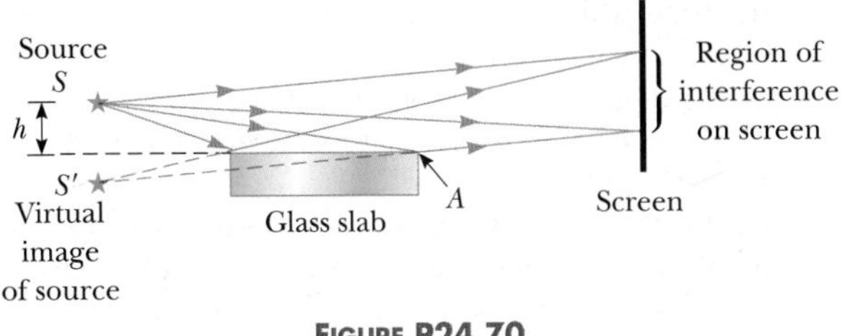


**FIGURE P24.70**

## GROUP ACTIVITIES

**G.1** A simple way to observe interference effects is to look at a distant light source through a stretched handkerchief or an opened umbrella. You will observe the light passing through the threads of the cloth spreads out to form streaks. Explain why the streaks look as they do.

**G.2** Place a clear dish or plate on a black surface such as a sheet of black construction paper. Now add a thin layer of water to the glass and place a few drops of kerosene or light machine oil on the water. Darken the room and shine a flashlight from an angle, as in Figure GA24.2. Note the interference pattern of various colors you observe under the white light. How does the pattern change if you cover the flashlight with a sheet of red, blue, or green cellophane, which acts as a filter?

As an extension of the above, observe the colors that appear to swirl on the surface of a soap bubble. What color do you see just before a bubble bursts?

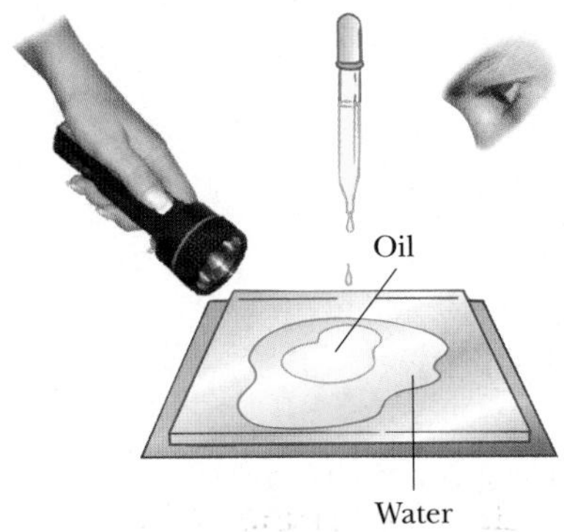


**FIGURE GA24.2**

**G.3** Make a V with your index and middle fingers. Hold your hand up very close to your eye so that you are looking between your two fingers toward a distant light source. Now bring the fingers together until there is only a very tiny slit between them. You should be able to see a series of parallel lines. Although the lines appear to be located in the narrow space between your fingers, what you are actually seeing is a diffraction pattern cast upon your retina.

**G.4** Stand a couple of meters from a lightbulb. Facing away from the light, hold a compact disc about 10 cm from your eye and tilt it until the reflection of the bulb is located in the hole at the disc's center. You should see spectra radiating out from the center, with violet on the inside and red on the outside. Now move the disc away from your eye until the violet band is at the outer edge. Carefully measure the distance from your eye to the center of the disc and also determine the radius of the disc. Use this information to find the angle $\theta$ to the first-order maximum for violet light. Now use $d \sin \theta = m\lambda$ to determine the spacing between the grooves of the disc. The industry standard is 1.6 $\mu$m. How close did you come? While you are observing the spectrum from a CD, note that the color of the light from a given point changes with the viewing angle. Explain this in terms of changes in the path length. It is of interest that the blues and blue-greens in peacock feathers and butterflies are caused by diffraction off finely aligned structures in feathers and wings (see chapter opener photo).

**G.5** (a) Devise a way to use a protractor, desklamp, and polarizing sunglasses to measure Brewster's angle for the glass in a window. From this, determine the index of refraction of the glass. (b) Put on a pair of polarizing sunglasses and close one eye. Hold up a lens of a second pair of polarizing glasses in front of your open eye so that light must pass through a lens of each pair before entering your eye. Now rotate the second pair of glasses around. You will note that the light reaching your eye is considerably reduced at some orientations and will pass freely at others. (c) On a sunny day, rotate your polarizing sunglasses in front of your eye and observe light reflection from a window or the surface of water. Note the change in the amount of light entering your eye for various orientations of the glasses. (d) For a final observation concerning polarized light, rotate a pair of polarizing sunglasses as you observe various areas of the sky. From what direction do you find the light to be most highly polarized?

**G.6** (a) In case 1, light from a laser passes through a double slit that has a separation $d$. The light then forms an interference pattern on a screen a distance $L$ away. In case 2, the slit is replaced by a slit with a separation $2d$. In which case is the distance from the central maximum to the next maximum the greatest? (b) Check your logic by considering light of wavelength $\lambda = 694.3$ nm, the distance $d = 0.100$ mm, and $L = 2.50$ m.

**G.7** Light of a certain wavelength passes through a double-slit arrangement that has a separation $d$ and forms a pattern on a screen a distance $L$ away. (a) If light having a wavelength 10% smaller now shines on the slits, is the distance from the central maximum to the next maximum larger or smaller? (b) Based on your logic for case (a), what will happen when visible light containing all wavelengths falls on the slits? (c) Let the first light have a wavelength of $\lambda = 694.3$ nm, the second wavelength to be 10% smaller, $d = 0.100$ mm and $L = 2.50$ m to check your reasoning.

**G.8** Light of wavelength $\lambda$ in air shines on an oil film of index $n_o$ floating on water where $n_w > n_o$. (a) What is the wavelength of the light in the oil? (b) Is there a phase change upon reflection at the oil-water boundary? (c) Use $\lambda = 450$ nm and $n_o = 1.25$ to find the minimum thickness for constructive interference to occur at the top of the oil film.

**G.9** Light in air that is reflected from a water surface is found to be completely polarized at an angle $\theta$. (a) If the light is instead reflected from a glass coffee table, will the new angle for complete polarization be larger or smaller? (b) Use $n_g = 1.54$ for glass and $n_w = 1.33$ for water to check your prediction.

**G.10** Light of a certain wavelength passes through a diffraction grating that has a slit separation $d$ and forms a pattern on a screen. (a) If light having a wavelength 10% smaller now shines on the grating, are the bright lines separated by larger, smaller, or the same angles? (b) Based on your logic for case (a), what will happen when visible light containing all wavelengths falls on the slits? (c) Let the first light have a wavelength $\lambda = 694.3$ nm, take the second wavelength to be 10% smaller, assume $d = 1\ 500$ nm, and find the diffraction angles.

Chapter 23

A cylindrical glass rod in air has an index refraction n = 1.52. One end is ground to a hemispherical surface w/ radius of R = 2 cm. A small object is located 8 cm to the left of the vertex

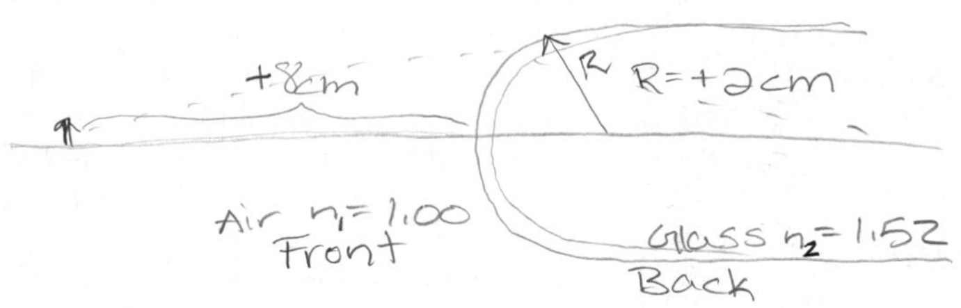


a) Find the image distance.

$$\frac{n_1}{p} + \frac{n_2}{q} = \frac{n_2 - n_1}{R}$$

$$\frac{n_2}{q} = \frac{n_2 - n_1}{R} - \frac{n_1}{p} \qquad \frac{1.52}{q} = \frac{1.52 - 1}{2} - \frac{1}{8} = 1.35$$

$$q = \frac{1.52}{1.35} = +11.259\ \text{cm}$$

Image real

$$M = \frac{-n_1 q}{n_2 p} = -\frac{1 \cdot 11.259}{1.52 \cdot 8} = -.926$$ inverted/real

In #20 $\frac{n_2}{q} = \frac{n_2 - n_1}{R} - \frac{n_1}{p}$ $\qquad \frac{1.52}{q} = \frac{1.52 - 1.33}{2} - \frac{1.33}{8}$

$q = -21.33 \qquad M = +2.33$

# Optical Instruments

# 25

## Chapter Outline

**A normal lens for a camera might have a focal length of 50 mm. A telephoto lens might have a focal length of 200 mm, resulting in a larger image on the film.** *(Courtesy of Canon, USA)*

We use devices made from lenses, mirrors, or other optical components every time we put on a pair of eyeglasses or contact lenses, take a photograph, look at the sky through a telescope, and so on. In this chapter we examine how these and other optical instruments work. For the most part, our analyses will involve the laws of reflection and refraction and the procedures of geometric optics. However, to explain certain phenomena we must use the wave nature of light.

## 25.1 THE CAMERA

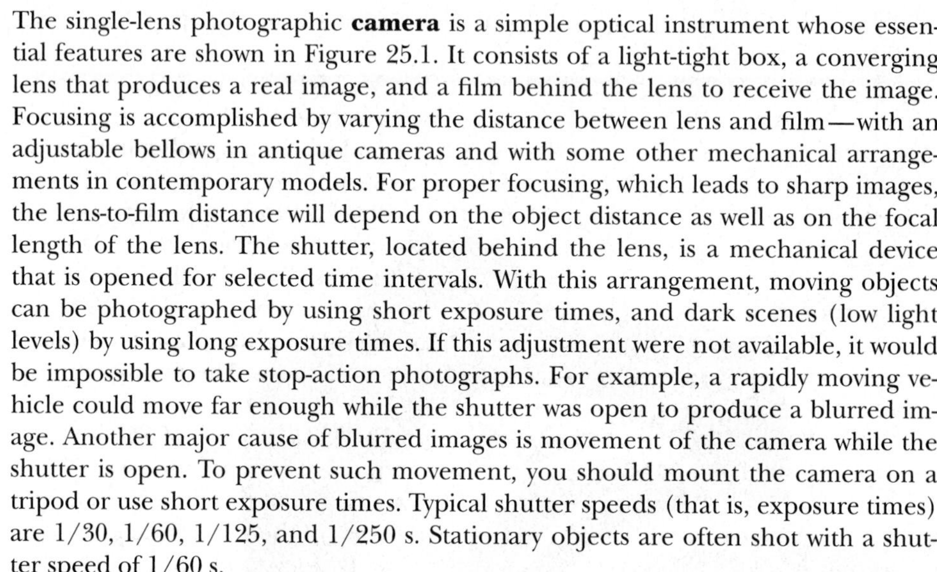

The single-lens photographic **camera** is a simple optical instrument whose essential features are shown in Figure 25.1. It consists of a light-tight box, a converging lens that produces a real image, and a film behind the lens to receive the image. Focusing is accomplished by varying the distance between lens and film—with an adjustable bellows in antique cameras and with some other mechanical arrangements in contemporary models. For proper focusing, which leads to sharp images, the lens-to-film distance will depend on the object distance as well as on the focal length of the lens. The shutter, located behind the lens, is a mechanical device that is opened for selected time intervals. With this arrangement, moving objects can be photographed by using short exposure times, and dark scenes (low light levels) by using long exposure times. If this adjustment were not available, it would be impossible to take stop-action photographs. For example, a rapidly moving vehicle could move far enough while the shutter was open to produce a blurred image. Another major cause of blurred images is movement of the camera while the shutter is open. To prevent such movement, you should mount the camera on a tripod or use short exposure times. Typical shutter speeds (that is, exposure times) are 1/30, 1/60, 1/125, and 1/250 s. Stationary objects are often shot with a shutter speed of 1/60 s.

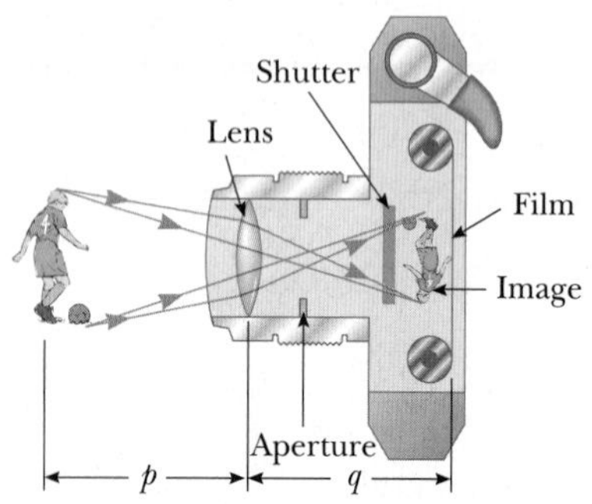


**FIGURE 25.1** A cross-sectional view of a simple camera.

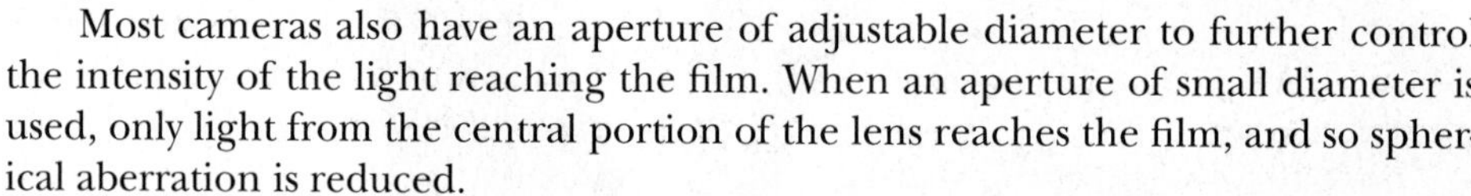

Most cameras also have an aperture of adjustable diameter to further control the intensity of the light reaching the film. When an aperture of small diameter is used, only light from the central portion of the lens reaches the film, and so spherical aberration is reduced.

The intensity $I$ of the light reaching the film is proportional to the area of the lens. Because this area is proportional to the square of the lens diameter $D$, the intensity is also proportional to $D^2$. Light intensity is a measure of the rate at which energy is received by the film per unit area of the image. Because the area of the image is proportional to $q^2$, and $q \approx f$ (when $p >> f$, so that $p$ can be approximated as infinite), we conclude that the intensity is also proportional to $1/f^2$, so that $I \propto D^2/f^2$. The brightness of the image formed on the film depends on the light intensity, and so we see that the image brightness depends on both the focal length and diameter $D$ of the lens.

The ratio $f/D$ is called the ***f*-number** of a lens:

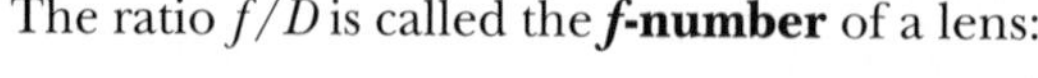

$$f\text{-number} \equiv \frac{f}{D} \qquad [25.1]$$

The *f*-number is often given as a description of the lens "speed." The lens with a low *f*-number is a "fast" lens. Extremely fast lenses, which have an *f*-number as low as approximately 1.2, are expensive because it is very difficult to keep aberrations acceptably small with light rays passing through a large area of the lens. Camera lenses are often marked with a range of *f*-numbers such as 1.4, 2, 2.8, 4, 5.6, 8, 11, . . . . Any one of these settings can be selected by adjusting the aperture, which changes the value of $D$. Increasing the setting from one *f*-number to the next higher value (for example, from 2.8 to 4) decreases the area of the aperture by a factor of 2. The lowest *f*-number setting on a camera corresponds to the aperture wide open and the maximum possible lens area in use.

Simple cameras usually have a fixed focal length and fixed aperture size, with an *f*-number of about 11. This high value for the *f*-number allows for a large **depth of field.** This means that objects at a wide range of distances from the lens form reasonably sharp images on the film. In other words, the camera does not have to be focused. Most cameras with variable *f*-numbers adjust them automatically.

## 25.2 THE EYE

Like a camera, a normal eye focuses light and produces a sharp image. However, the mechanisms by which the eye controls the amount of light admitted and adjusts to produce correctly focused images are far more complex, intricate, and effective than those in even the most sophisticated camera. In all respects, the eye is a physiological wonder.

Figure 25.2 shows the essential parts of the eye. Light entering the eye passes through a transparent structure called the *cornea,* behind which are a clear liquid (the *aqueous humor*), a variable aperture (the *pupil,* which is an opening in the *iris*), and the *crystalline lens.* Most of the refraction occurs at the outer surface of the eye, at which the cornea is covered with a film of tears. Relatively little refraction occurs in the crystalline lens because the aqueous humor in contact with the lens has an average index of refraction close to that of the lens. The iris, which is the colored portion of the eye, is a muscular diaphragm that controls pupil size. The iris regulates the amount of light entering the eye by dilating the pupil in low-light conditions and contracting the pupil in high-light conditions. The *f*-number range of the eye is from about 2.8 to 16.

The cornea-lens system focuses light onto the back surface of the eye, the *retina,* which consists of millions of sensitive receptors called *rods* and *cones.* When stimulated by light, these structures send impulses via the optic nerve to the brain, which converts these impulses into our conscious view of the world. The process by which the brain performs this conversion is not well understood and is the subject of much speculation and research. Unlike film in a camera, the rods and cones chemically adjust their sensitivity according to the prevailing light conditions. This adjustment, which takes about 15 minutes, is responsible for the experience of

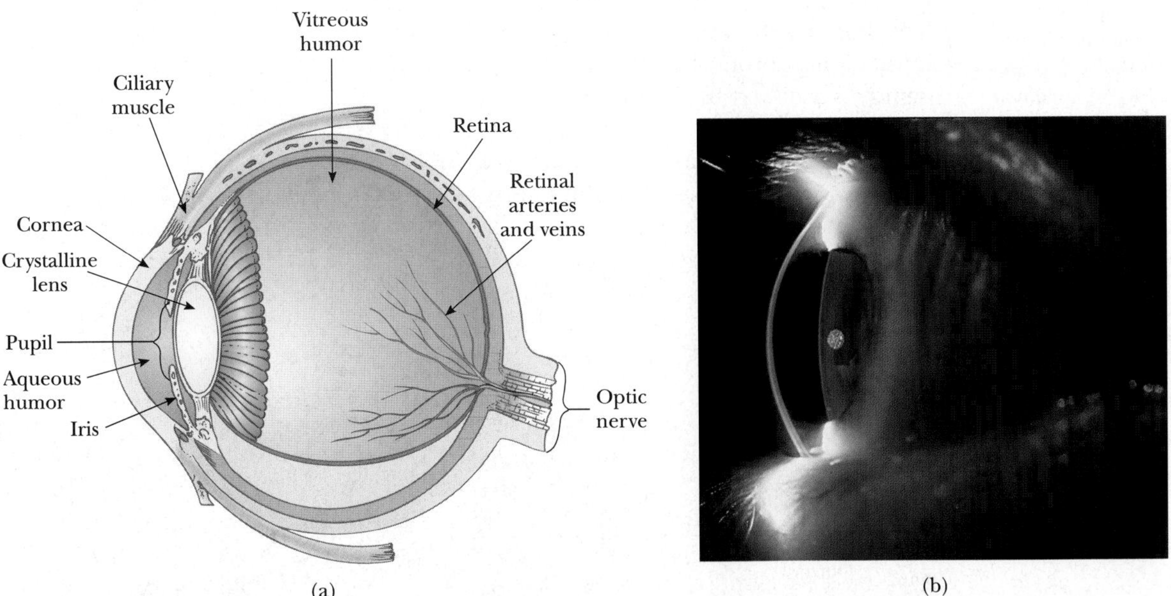


**FIGURE 25.2** (a) Essential parts of the eye. Can you correlate the essential parts of the eye with those of the simple camera in Figure 25.1? (b) Close-up photograph of the human cornea. *(From Lennart Nilson, in collaboration with Jan Lindberg,* Behold Man: A Photographic Journey of Discovery Inside the Body, *Boston, Little, Brown & Co., 1974)*

"getting used to the dark" in such places as movie theaters. Iris aperture control, which takes less than 1 second, helps protect the retina from overload in the adjustment process.

The eye focuses on an object by varying the shape of the pliable crystalline lens through an amazing process called **accommodation.** An important component in accommodation is the *ciliary muscle,* which is situated in a circle around the rim of the lens. Thin filaments, called *zonules,* run from this muscle to the edge of the lens. When the eye is focused on a distant object, the ciliary muscle is relaxed, tightening the zonules that attach the ciliary muscle to the edge of the lens. The force of the zonules causes the lens to flatten, increasing its focal length. For an object distance of infinity, the focal length of the eye is equal to the fixed distance between lens and retina, about 1.7 cm. The eye focuses on nearby objects by tensing the ciliary muscle, which relaxes the zonules. This action allows the lens to bulge a bit and its focal length decreases, resulting in the image being focused on the retina. All these lens adjustments take place so swiftly that we are not even aware of the change. In this respect, even the finest electronic camera is a toy compared with the eye.

There is a limit to accommodation because objects that are very close to the eye produce blurred images. The **near point** is the closest distance for which the lens can accommodate to focus light on the retina. This distance usually increases with age and has an average value of 25 cm. Typically, at age 10 the near point of the eye is about 18 cm. This increases to about 25 cm at age 20, to 50 cm at age 40, and to 500 cm or greater at age 60. The **far point** of the eye represents the largest distance for which the lens of the relaxed eye can focus light on the retina. A person with normal vision is able to see very distant objects, such as the Moon, and thus has a far point at infinity.

## CONDITIONS OF THE EYE

When the eye suffers a mismatch between the focusing power of the lens-cornea system and the length of the eye such that light rays reach the retina before they converge to form an image, as in Figure 25.3a, the condition is known as **farsight-**

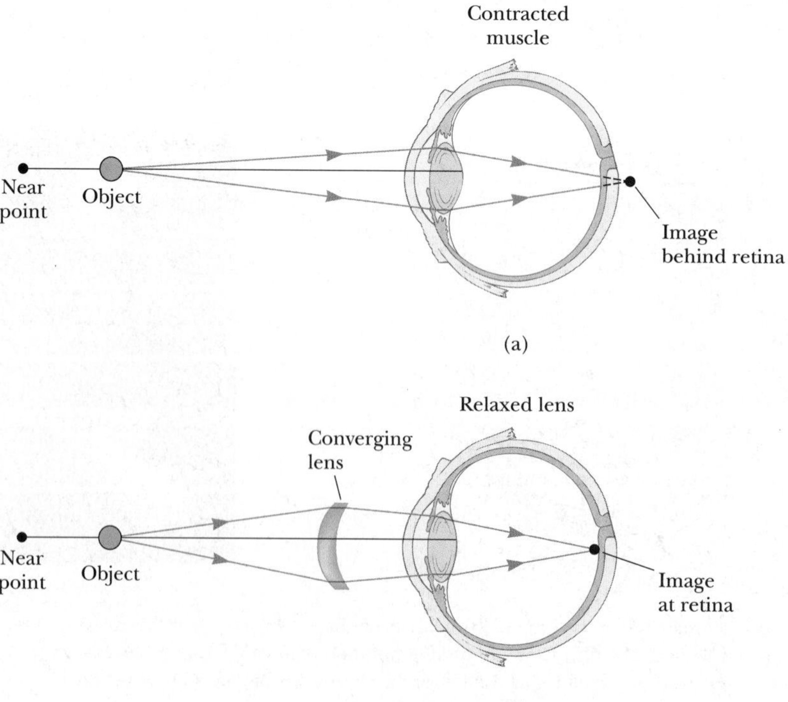


**FIGURE 25.3** (a) A farsighted eye is slightly shorter than normal; hence, the image of a nearby object focuses *behind* the retina. (b) The condition can be corrected with a converging lens. (The object is assumed to be very small in these figures.)

**edness** (or *hyperopia*). A farsighted person can usually see faraway objects clearly but not nearby objects. Although the near point of a normal eye is approximately 25 cm, the near point of a farsighted person is much farther than this. The eye of a farsighted person tries to focus by accommodation, that is, by shortening its focal length. This works for distant objects, but because the focal length of the farsighted eye is longer than normal, the light from nearby objects cannot be brought to a sharp focus before it reaches the retina, causing a blurred image. There is insufficient refracting power in the cornea and lens to satisfactorily focus the light from all but distant objects. The condition can be corrected by placing a converging lens in front of the eye, as in Figure 25.3b. The lens refracts the incoming rays more toward the principal axis before entering the eye, allowing them to converge and focus on the retina.

**Nearsightedness** (or *myopia*) is another mismatch condition in which a person is able to focus on nearby objects but not faraway objects. In the case of *axial myopia,* nearsightedness is caused by the lens being too far from the retina. It is also possible to have *refractive myopia,* in which case the lens-cornea system is too powerful for the normal length of the eye. The far point of the nearsighted eye is not at infinity and may be less than a meter. The maximum focal length of the nearsighted eye is insufficient to produce a sharp image on the retina, and rays from a distant object converge to a focus in front of the retina. They then continue past that point, diverging before they finally reach the retina and causing blurred vision (Fig. 25.4a). Nearsightedness can be corrected with a diverging lens, as shown in Figure 25.4b. The lens refracts the rays away from the principal axis before they enter the eye, a modification that allows them to focus on the retina.

CORRECTING FOR DEFECTS WITH OPTICAL LENSES

Beginning with middle age, most people lose some of their accommodation ability as the ciliary muscle weakens and the lens hardens. Unlike farsightedness, which is a mismatch of focusing power and eye length, **presbyopia** (literally, "old-age vision") is due to a reduction in accommodation ability. The cornea and lens do not have sufficient focusing power to bring nearby objects into focus on the retina. The symptoms are the same as with farsightedness, and the condition can be corrected with converging lenses.

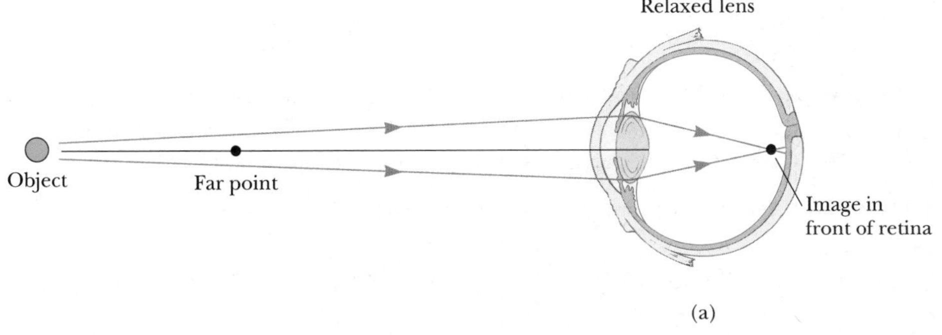


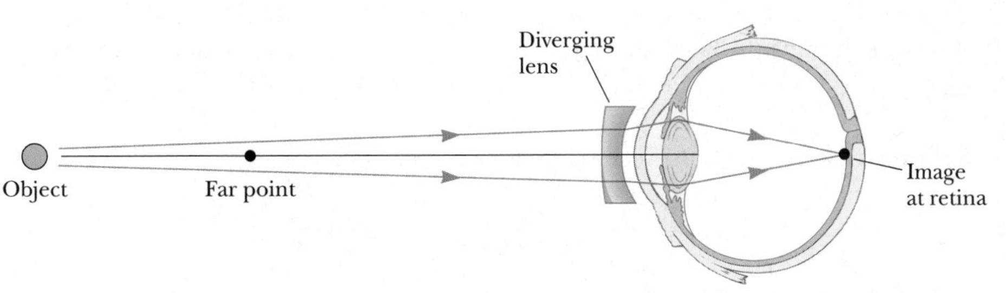


**FIGURE 25.4** (a) A nearsighted eye is slightly longer than normal; hence, the image of a distant object focuses *in front* of the retina. (b) The condition can be corrected with a diverging lens. (The object is assumed to be very small in these figures.)

In the eye defect known as **astigmatism,** light from a point source produces a line image on the retina. This condition arises when either the cornea or the lens or both are not perfectly symmetric. Astigmatism can be corrected with lenses having different curvatures in two mutually perpendicular directions.

Optometrists and ophthalmologists usually prescribe lenses measured in **diopters:**

**The power $\mathcal{P}$ of a lens in diopters equals the inverse of the focal length in meters: $\mathcal{P} = 1/f$.**

For example, a converging lens of focal length +20 cm has a power of +5.0 diopters, and a diverging lens of focal length −40 cm has a power of −2.5 diopters.

## APPLYING PHYSICS 25.1

A classic science fiction story, *The Invisible Man,* tells of a person who becomes invisible by changing the index of refraction of his body to that of air. This story has been criticized by students who know how the eye works; they claim the invisible man would be unable to see. On the basis of your knowledge of the eye, could he see or not?

**Explanation** He would not be able to see. In order for the eye to "see" an object, incoming light must be refracted at the cornea and lens to form an image on the retina. If the cornea and lens have the same index of refraction as air, refraction cannot occur, and an image would not be formed.

## Example 25.1 Prescribing a Lens

The near point of an eye is 50.0 cm.

**A** What focal length must a corrective lens have to enable the eye to see clearly an object 25.0 cm away?

**Reasoning** The thin-lens equation (Eq. 23.11) enables us to solve this problem. We have placed an object at 25.0 cm, and we want the lens to form an image at the closest point that the eye can see clearly. This corresponds to the near point, 50.0 cm.

**Solution** Applying the thin-lens equation, we have

$$\frac{1}{25.0 \text{ cm}} + \frac{1}{(-50.0 \text{ cm})} = \frac{1}{f}$$

$$f = 50.0 \text{ cm}$$

Why did we use a negative sign for the image distance? Notice that the focal length is positive, indicating the need for a converging lens to correct farsightedness such as this.

**B** What is the power of this lens?

**Solution** The power is the reciprocal of the focal length in meters:

$$\mathcal{P} = \frac{1}{f} = \frac{1}{0.500 \text{ m}} = 2.00 \text{ diopters}$$

### Example 25.2 A Case of Nearsightedness

A particular nearsighted person cannot see objects clearly when they are beyond 25 cm (the far point of the eye). What focal length should the prescribed lens have to correct this problem?

**Solution** The purpose of the lens in this instance is to "move" an object from infinity to a distance where it can be seen clearly. This is accomplished by having the lens produce an image at the far point. From the thin-lens equation, we have

$$\frac{1}{p}+\frac{1}{q}=\frac{1}{\infty}+\frac{1}{(-25\text{ cm})}=\frac{1}{f}$$

$$f= -25\text{ cm}$$

Why did we use a negative sign for the image distance? As you should have suspected, the lens must be diverging (one with a negative focal length) to correct nearsightedness.

**EXERCISE** What is the power of this lens?

**ANSWER** $-0.40$ diopters

**Webnote 25.1**

To study how corrective laser eye surgery is related to the structures of the eye, visit *http://www.medscape.com/viewarticle/200310_4*

## 25.3 THE SIMPLE MAGNIFIER

The **simple magnifier** is one of the most basic of all optical instruments because it consists only of a single converging lens. As the name implies, this device is used to increase the apparent size of an object. Suppose an object is viewed at some distance $p$ from the eye, as in Figure 25.5. Clearly, the size of the image formed at the retina depends on the angle $\theta$ subtended by the object at the eye. As the object moves closer to the eye, $\theta$ increases and a larger image is observed. However, a normal eye cannot focus on an object closer than about 25 cm, the near point (Fig. 25.6a). (Try it!) Therefore, $\theta$ is maximum at the near point.

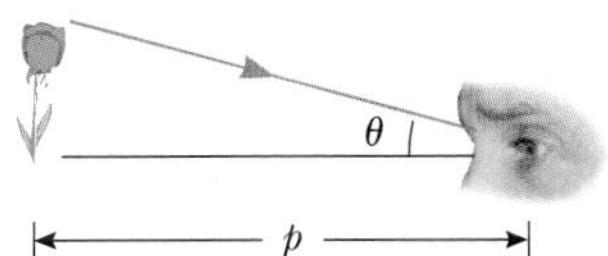


**FIGURE 25.5** The size of the image formed on the retina depends on the angle $\theta$ subtended at the eye.

To further increase the apparent angular size of an object, a converging lens can be placed in front of the eye with the object positioned at point $O$, just inside the focal point of the lens, as in Figure 25.6b. At this location, the lens forms a virtual, upright, and enlarged image, as shown. The lens allows the object to be

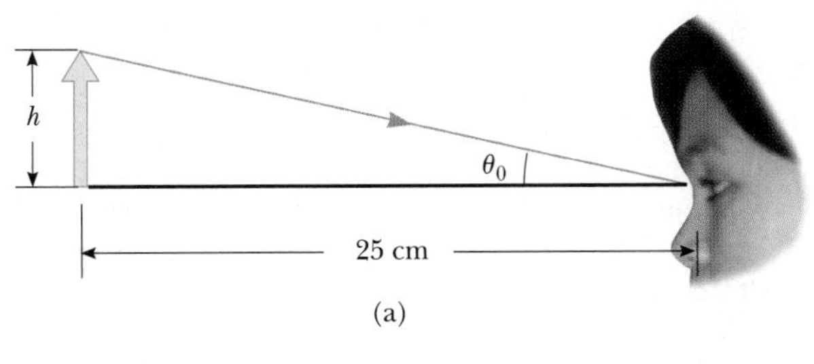


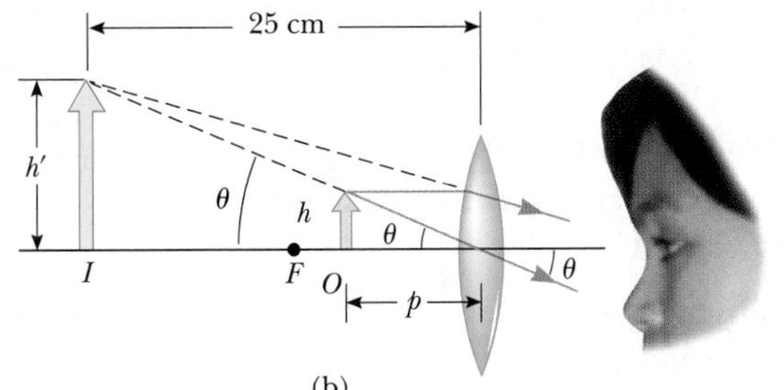


**FIGURE 25.6** (a) An object placed at the near point ($p$ = 25 cm) subtends an angle of $\theta_0 \approx h/25$ at the eye. (b) An object placed near the focal point of a converging lens produces a magnified image, which subtends an angle of $\theta \approx h'/25$ at the eye. Note that in this situation, $q = -25$ cm.

viewed closer to the eye than is possible without the lens. We define the **angular magnification** $m$ as the ratio of the angle subtended by the object when the lens is in use (angle $\theta$ in Fig. 25.6b) to the angle subtended by the object placed at the near point with no lens in use (angle $\theta_0$ in Fig. 25.6a):

Angular magnification with the object at the near point

$$m \equiv \frac{\theta}{\theta_0} \qquad [25.2]$$

The angular magnification is a maximum when the image formed by the lens is at the near point of the eye—that is, when $q = -25$ cm (see Fig. 25.6b). The object distance corresponding to this image distance can be calculated from the thin-lens equation:

$$\frac{1}{p} + \frac{1}{(-25\text{ cm})} = \frac{1}{f}$$
$$p = \frac{25f}{25 + f} \qquad [25.3]$$

where $f$ is the focal length of the magnifier in centimeters. From Figures 25.6a and 25.6b, the small-angle approximation gives

$$\tan\theta_0 \approx \theta_0 \approx \frac{h}{25} \quad \text{and} \quad \tan\theta \approx \theta \approx \frac{h}{p} \qquad [25.4]$$

Thus, Equation 25.2 becomes

$$m_{\text{max}} = \frac{\theta}{\theta_0} = \frac{h/p}{h/25} = \frac{25}{p} = \frac{25}{25f/(25+f)}$$
$$= 1 + \frac{25\text{ cm}}{f} \qquad [25.5]$$

The maximum angular magnification given by Equation 25.5 is the ratio of the angular size seen with the lens to the angular size seen without the lens, with the object at the near point of the eye. Although the normal eye can focus on an image formed anywhere between the near point and infinity, it is most relaxed when the image is at infinity (Sec. 25.2). For the image formed by the magnifying lens to appear at infinity, the object must be placed at the focal point of the lens—that is, $p = f$. In this case, Equation 25.4 becomes

$$\theta_0 \approx \frac{h}{25} \quad \text{and} \quad \theta \approx \frac{h}{f}$$

and the angular magnification is

$$m = \frac{\theta}{\theta_0} = \frac{25\text{ cm}}{f} \qquad [25.6]$$

With a single lens, it is possible to achieve angular magnifications up to about 4 without serious aberrations. Magnifications up to about 20 can be achieved by using one or two additional lenses to correct for aberrations.

## Example 25.3 Maximum Angular Magnification of a Lens

What is the maximum angular magnification of a lens with a focal length of 10.0 cm, and what is the angular magnification of this lens when the eye is relaxed?

**Reasoning** The maximum angular magnification occurs when the image formed by the lens is at the near point of the eye. Under these circumstances, Equation 25.5 gives us the maximum angular magnification.

**Solution** Using Equation 25.5, we have

$$m_{\text{max}} = 1 + \frac{25 \text{ cm}}{f} = 1 + \frac{25 \text{ cm}}{10.0 \text{ cm}} = 3.5$$

When the eye is relaxed, the image is at infinity. In this case, we use Equation 25.6:

$$m = \frac{25 \text{ cm}}{f} = \frac{25 \text{ cm}}{10.0 \text{ cm}} = 2.5$$

## 25.4 THE COMPOUND MICROSCOPE

A simple magnifier provides only limited assistance with inspection of the minute details of an object. Greater magnification can be achieved by combining two lenses in a device called a compound microscope, a diagram of which is shown in Figure 25.7a. It consists of two lenses: an objective with a very short focal length $f_o$ (where $f_o < 1$ cm), and an ocular lens, or eyepiece, with a focal length $f_e$ of a few centimeters. The two lenses are separated by distance $L$ which is much greater than either $f_o$ or $f_e$.

As you read the discussion that follows, note that the basic approach used to analyze the image formation properties of a microscope is that of two lenses in a row—that is, the image formed by the first becomes the object for the second. The object $O$ placed just outside the focal length of the objective, forms a real, inverted image at $I_1$, which is at or just inside the focal point of the eyepiece. This image is real and much enlarged. (For clarity, the enlargement of $I_1$ is not shown in Fig. 25.7a.) The eyepiece, which serves as a simple magnifier, uses the image at $I_1$ as its object and produces an image at $I_2$. The image seen by the eye at $I_2$ is virtual, inverted, and very much enlarged.

The lateral magnification $M_1$ of the first image is $-q_1/p_1$. Note in Figure 25.7a that $q_1$ is approximately equal to $L$. This occurs because the object is placed close to the focal point of the objective lens, which ensures that the image formed will be far from the objective lens. Furthermore, because the object is very close to the focal point of the objective lens, $p_1 \approx f_o$. This gives a lateral magnification of

$$M_1 = -\frac{q_1}{p_1} \approx -\frac{L}{f_o}$$

for the objective. The angular magnification of the eyepiece for an object (corresponding to the image at $I_1$) placed at the focal point is found from Equation 25.6 to be

$$m_e = \frac{25 \text{ cm}}{f_e}$$

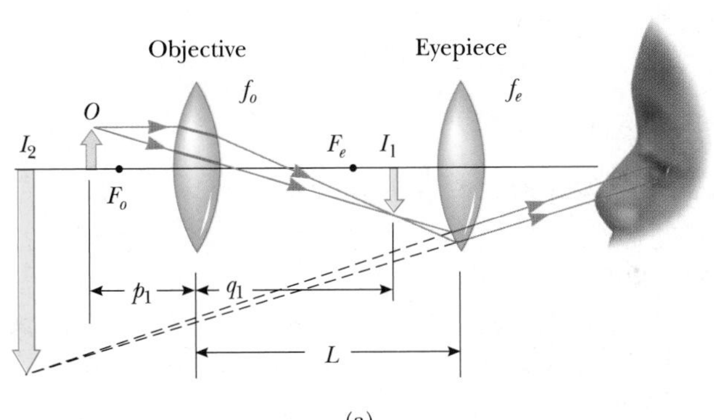

(a)

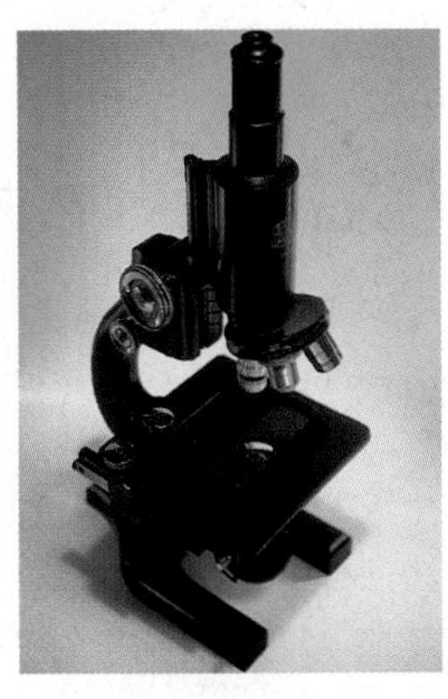
(b)

**FIGURE 25.7** (a) A diagram of a compound microscope, which consists of an objective and an eyepiece, or ocular lens. (b) A compound microscope. The three-objective turret allows the user to switch to several different powers of magnification. Combinations of eyepieces with different focal lengths and different objectives can produce a wide range of magnifications. *(Henry Leap and Jim Lehman)*

The overall magnification of the compound microscope is defined as the product of the lateral and angular magnifications:

Magnification of a microscope

$$m = M_1 m_e = -\frac{L}{f_o}\left(\frac{25\text{ cm}}{f_e}\right) \quad [25.7]$$

The negative sign indicates that the image is inverted with respect to the object.

The microscope has extended our vision into the previously unknown realm of incredibly small objects, and the capabilities of this instrument have increased steadily with improved techniques in precision grinding of lenses. A question that is often asked about microscopes is, "With extreme patience and care, would it be possible to construct a microscope that would enable us to see an atom?" The answer to this question is no, as long as visible light is used to illuminate the object. The reason is that, in order to be seen, the object under a microscope must be at least as large as a wavelength of light. An atom is many times smaller than the wavelength of visible light, and so its mysteries must be probed via other techniques.

The wavelength dependence of the "seeing" ability of a wave can be illustrated by water waves set up in a bathtub in the following manner. Imagine that you vibrate your hand in the water until waves with a wavelength of about 6 in. are moving along the surface. If you fix a small object, such as a toothpick, in the path of the waves, you will find that the waves are not appreciably disturbed by the toothpick but continue along their path. Now suppose you fix a larger object, such as a toy sailboat, in the path of the waves. In this case, the waves are considerably "disturbed" by the object. The toothpick was much smaller than the wavelength of the waves, and as a result the waves did not "see" it. The toy sailboat, in contrast, is about the same size as the wavelength of the waves and hence creates a disturbance. Light waves behave in this same general way. The ability of an optical microscope to view an object depends on the size of the object relative to the wavelength of the light used to observe it. Hence, it will never be possible to observe atoms or molecules with such a microscope, because their dimensions are so small ($\approx 0.1$ nm) relative to the wavelength of the light ($\approx 500$ nm).

**Webnote 25.2**

How do you effectively light and use a compound microscope? This web site will help you learn these skills. Visit *http://www.microscopy-uk.org.uk/mag/art98/incid1.html*

## Example 25.4 Magnifications of a Microscope

A certain microscope has two interchangeable objectives. One has a focal length of 20.0 mm, and the other has a focal length of 2.0 mm. Also available are two eyepieces of focal lengths 2.5 cm and 5.0 cm. If the length of the microscope is 18 cm, what magnifications are possible?

**Reasoning and Solution** The solution consists of applying Equation 25.7 to four different combinations of lenses. For the combination of the two long focal lengths, we have

$$m = -\frac{L}{f_o}\left(\frac{25\text{ cm}}{f_e}\right) = -\frac{18}{2.0}\left(\frac{25}{5.0}\right) = -45$$

The combination of the 20.0-mm objective and the 2.5-cm eyepiece gives

$$m = -\frac{18}{2.0}\left(\frac{25}{2.5}\right) = -90$$

The 2.0-mm and 5.0-cm combination produces

$$m = -\frac{18}{0.20}\left(\frac{25}{5.0}\right) = -450$$

Finally, the two short focal lengths give

$$m = -\frac{18}{0.20}\left(\frac{25}{2.5}\right) = -900$$

## 25.5 THE TELESCOPE

There are two fundamentally different types of telescopes, both designed to help us view distant objects such as the planets in our Solar System. These two types are (1) the **refracting telescope,** which uses a combination of lenses to form an image, and (2) the **reflecting telescope,** which uses a curved mirror and a lens to form an image. Once again, we will be able to analyze the telescope by considering it to be a system of two optical elements in a row. The basic technique followed is that the image formed by the first element becomes the object for the second.

Let us first consider the refracting telescope. In this device, two lenses are arranged so that the objective forms a real, inverted image of the distant object very near the focal point of the eyepiece (Fig. 25.8a). Furthermore, the image at $I_1$ is formed at the focal point of the objective because the object is essentially at infinity. Hence, the two lenses are separated by the distance $f_o + f_e$, which corresponds to the length of the telescope's tube. The eyepiece finally forms, at $I_2$, an enlarged, inverted image of the image at $I_1$.

The angular magnification of the telescope is given by $\theta/\theta_o$, where $\theta_o$ is the angle subtended by the object at the objective and $\theta$ is the angle subtended by the final image. From the triangles in Figure 25.8a, and for small angles, we have

$$\theta \approx \frac{h'}{f_e} \quad \text{and} \quad \theta_o \approx \frac{h'}{f_o}$$

Therefore, the angular magnification of the telescope can be expressed as

$$m = \frac{\theta}{\theta_o} = \frac{h'/f_e}{h'/f_o} = \frac{f_o}{f_e} \qquad [25.8]$$

Angular magnification of a telescope

This says that the angular magnification of a telescope equals the ratio of the objective focal length to the eyepiece focal length. Here again, the angular magnification is the ratio of the angular size seen with the telescope to the angular size seen with the unaided eye.

In some applications—for instance, the observation of relatively nearby objects such as the Sun, Moon, or planets—angular magnification is important.

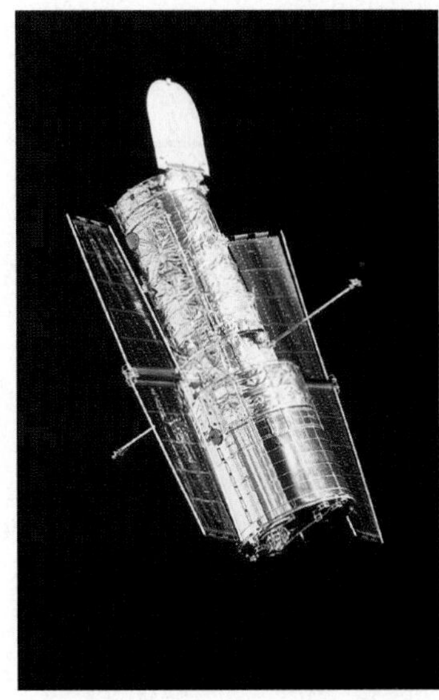

The Hubble Space telescope now enables us to see both further into space and back in time than possible before its launch. *(NASA)*

### Webnote 25.3

How does the Hubble Space Telescope operate? Take a fascinating look to see how it is built. Visit *http://www.howstuffworks.com/hubble.htm*
Another good Hubble site is *http://hubble.stsci.edu/gallery*

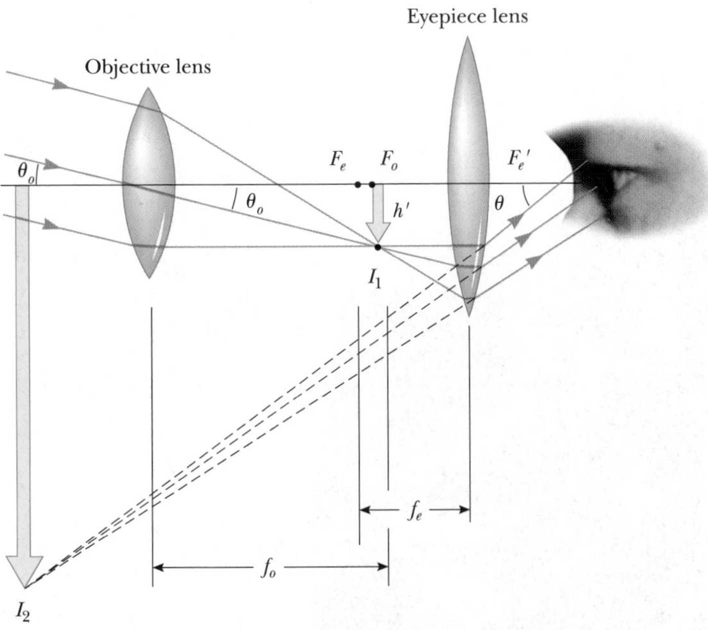


(a)

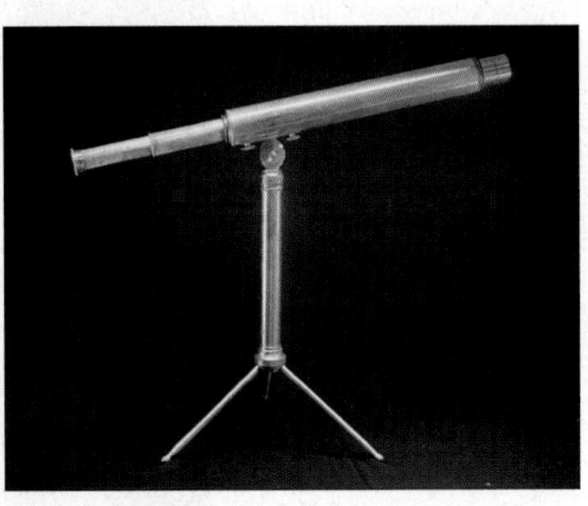

(b)

**FIGURE 25.8** (a) A diagram of a refracting telescope, with the object at infinity. (b) A photograph of a refracting telescope. *(Photo courtesy of Henry Leap and Jim Lehman)*

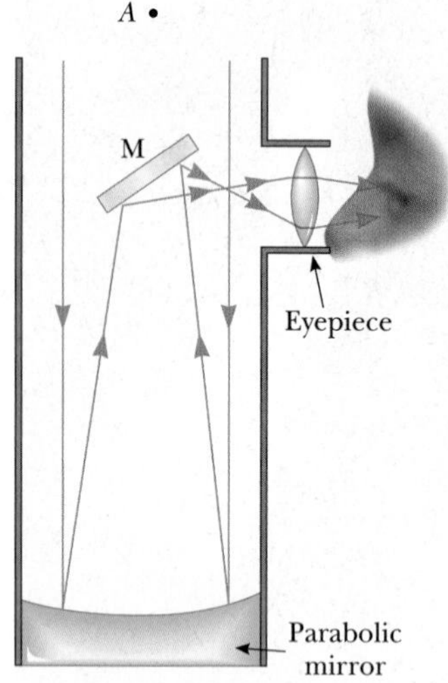


**FIGURE 25.9** A reflecting telescope with a Newtonian focus.

Stars, in contrast, are so far away that they always appear as small points of light, regardless of how much angular magnification is used. The large research telescopes used to study very distant objects must have great diameters to gather as much light as possible. It is difficult and expensive to manufacture such large lenses for refracting telescopes. In addition, the heaviness of large lenses leads to sagging, which is another source of aberration.

These problems can be partially overcome by replacing the objective lens with a reflecting, concave mirror, which usually has a parabolic shape to avoid spherical aberration. Figure 25.9 shows the design for a typical reflecting telescope. Incoming light rays pass down the barrel of the telescope and are reflected by a parabolic mirror at the base. These rays converge toward point $A$ in the figure, where an image would be formed on a photographic plate or another detector. However, before this image is formed, a small flat mirror at point $M$ reflects the light toward an opening in the side of the tube that passes into an eyepiece. This design is said to have a *Newtonian focus,* after its developer. Note that in the reflecting telescope the light never passes through glass (except the small eyepiece). As a result, problems associated with chromatic aberration are virtually eliminated.

The largest optical telescopes in the world are the two 10-m-diameter Keck reflectors on Mauna Kea in Hawaii. The largest single-mirrored reflecting telescope in the United States is the 5-m-diameter instrument on Mount Palomar in California (see Figure 25.10). In contrast, the largest refracting telescope in the world, at the Yerkes Observatory in Williams Bay, Wisconsin, has a diameter of only 1 m.

### Example 25.5 Angular Magnification of a Reflecting Telescope

A reflecting telescope has an 8-in.-diameter objective mirror with a focal length of 1 500 mm. What is the angular magnification of this telescope when an eyepiece with an 18-mm focal length is used?

**Solution** The equation for finding the angular magnification of a reflector is the same as that for a refractor. Thus, Equation 25.8 gives

$$m = \frac{f_o}{f_e} = \frac{1\,500 \text{ mm}}{18 \text{ mm}} = 83$$

**FIGURE 25.10** The Hale telescope at Mount Palomar Observatory. Just before taking the elevator up to the prime-focus cage, a first-time observer is always told, "Good viewing! And, if you should fall, try to miss the mirror." *(Courtesy of Palomar Observatory/California Institute of Technology)*

## 25.6 RESOLUTION OF SINGLE-SLIT AND CIRCULAR APERTURES

The ability of an optical system such as the eye, a microscope, or telescope to distinguish between closely spaced objects is limited because of the wave nature of light. To understand this difficulty, consider Figure 25.11, which shows two light sources far from a narrow slit of width $a$. The sources can be taken as two point sources $S_1$ and $S_2$ that are *not* coherent. For example, they could be two distant stars. If no diffraction occurred, we would observe two distinct bright spots (or images) on the screen at the right in the figure. However, because of diffraction, each source is imaged as a bright central region flanked by weaker bright and dark rings. What is observed on the screen is the sum of two diffraction patterns, one from $S_1$ and the other from $S_2$.

If the two sources are separated so that their central maxima do not overlap, as in Figure 25.11a, their images can be distinguished and are said to be *resolved*. If the sources are close together, however, as in Figure 25.11b, the two central maxima may overlap and the images are *not resolved*. To decide whether two images are resolved, the following condition is often applied to their diffraction patterns:

**When the central maximum of one image falls on the first minimum of another image, the images are said to be just resolved. This limiting condition of resolution is known as Rayleigh's criterion.**

Rayleigh's criterion

Figure 25.12 shows diffraction patterns in three situations. The images are just resolved when their angular separation satisfies Rayleigh's criterion (Fig. 25.12a). As the objects are brought closer together, their images are barely resolved (Fig. 25.12b). Finally, when the sources are very close to each other, their images are not resolved (Fig. 25.12c).

From Rayleigh's criterion, we can determine the minimum angular separation $\theta_{min}$ subtended by the source at the slit so that the images will be just resolved. In Chapter 24 we found that the first minimum in a single-slit diffraction pattern occurs at the angle that satisfies the relationship

$$\sin \theta = \frac{\lambda}{a}$$

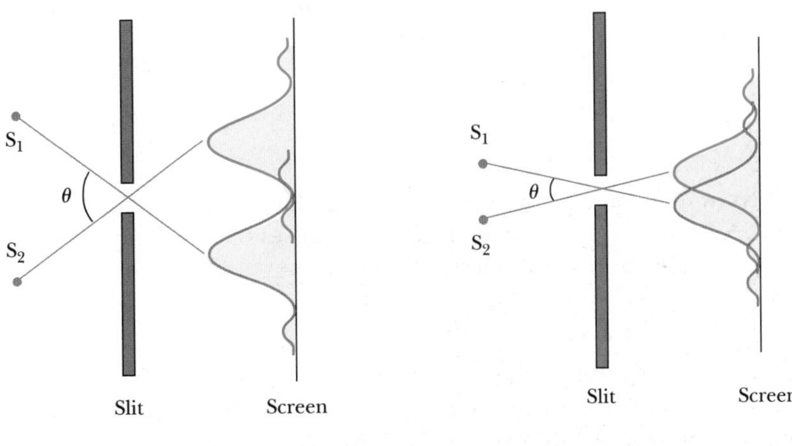


**FIGURE 25.11** Each of two point sources at some distance from a small aperture produces a diffraction pattern. (a) The angle subtended by the sources at the aperture is large enough so that the diffraction patterns are distinguishable. (b) The angle subtended by the sources is so small that the diffraction patterns are not distinguishable. (Note that the angles are greatly exaggerated.)

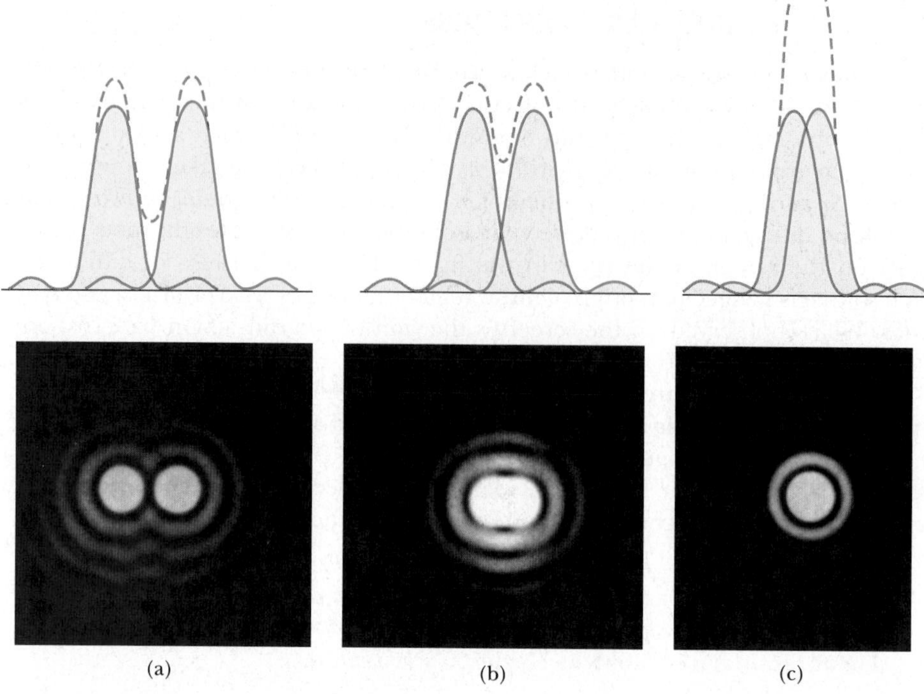


**FIGURE 25.12** The diffraction patterns of two point sources (solid curves) and the resultant pattern (dashed curve) for three angular separations of the sources. (a) The sources are separated such that their patterns are just resolved. (b) The sources are closer together, and their patterns are barely resolved. (c) The sources are so close together that their patterns are not resolved. *(From M. Cagnet, M. Francon, and J. C. Thierr,* Atlas of Optical Phenomena, *Berlin, Springer-Verlag, 1962, plate 16)*

where $a$ is the width of the slit. According to Rayleigh's criterion, this expression gives the smallest angular separation for which the two images can be resolved. Because $\lambda \ll a$ in most situations, $\sin\theta$ is small and we can use the approximation $\sin\theta \approx \theta$. Therefore, the limiting angle of resolution for a slit of width $a$ is

Limiting angle for a slit 

$$\theta_{\min} \approx \frac{\lambda}{a} \tag{25.9}$$

where $\theta_{\min}$ is in radians. Hence, the angle subtended by the two sources at the slit must be *greater* than $\lambda/a$ if the images are to be resolved.

Limiting angle for a circular aperture

Many optical systems use circular apertures rather than slits. The diffraction pattern of a circular aperture (Fig. 25.13) consists of a central, circular bright region surrounded by progressively fainter rings. Analysis shows that the limiting angle of resolution of the circular aperture is

$$\theta_{\min} = 1.22\frac{\lambda}{D} \tag{25.10}$$

where $D$ is the diameter of the aperture. Note that Equation 25.10 is similar to Equation 25.9 except for the factor 1.22, which arises from a complex mathematical analysis of diffraction from a circular aperture.

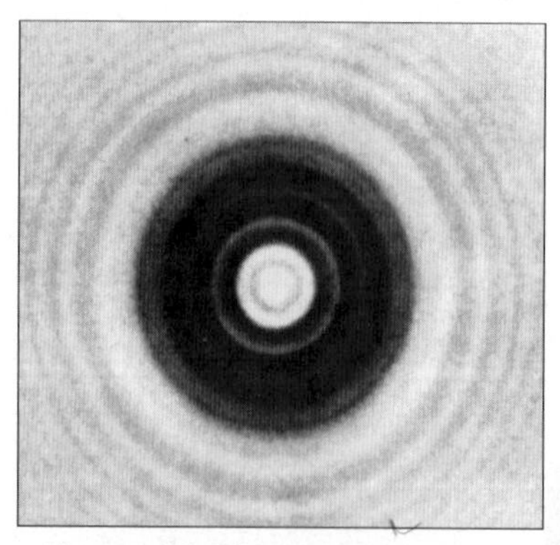

**FIGURE 25.13** The diffraction pattern of a circular aperture consists of a central bright disk surrounded by concentric bright and dark rings. *(From M. Cagnet, M. Francon, and J. C. Thierr,* Atlas of Optical Phenomena, *Berlin, Springer-Verlag, 1962, plate 34)*

**Quick Quiz 25.1** Suppose you are observing a binary star with a telescope and are having difficulty resolving the two stars. You decide to use a colored filter to help you. Should you choose a blue filter or a red filter?

## APPLYING PHYSICS 25.2

Cats' eyes have vertical pupils when in dim light. Which would cats be most successful at resolving at night—headlights on a distant car or vertically separated running lights on a distant boat's mast having the same separation?

**Explanation** The effective slit width in the vertical direction of the cat's eye is larger than that in the horizontal direction. Thus, it has more resolving power for lights separated in the vertical direction and would be more effective at resolving the mast lights on the boat.

## Example 25.6 Limiting Resolution of a Microscope

Sodium light of wavelength 589 nm is used to view an object under a microscope. The aperture of the objective has a diameter of 0.90 cm.

**A** Find the limiting angle of resolution.

**Solution** From Equation 25.10, we find the limiting angle of resolution to be

$$\theta_{\text{min}} = 1.22\left(\frac{589 \times 10^{-9}\text{ m}}{0.90 \times 10^{-2}\text{ m}}\right) = 8.0 \times 10^{-5}\text{ rad}$$

This means that any two points on the object subtending an angle of less than $8.0 \times 10^{-5}$ rad at the objective cannot be distinguished in the image.

**B** Using visible light of any wavelength you desire, what is the maximum limit of resolution for this microscope?

**Solution** To obtain the maximum resolution, we have to use the shortest wavelength available in the visible spectrum. Violet light of wavelength 400 nm gives us a limiting angle of resolution of

$$\theta_{\text{min}} = 1.22\left(\frac{400 \times 10^{-9}\text{ m}}{0.90 \times 10^{-2}\text{ m}}\right) = 5.4 \times 10^{-5}\text{ rad}$$

**C** Suppose water of index of refraction 1.33 filled the space between the object and the objective. What effect would this have on the resolving power of the microscope?

**Solution** In this case, the wavelength of the sodium light in the water is found by $\lambda_w = \lambda_a/n$ (Chapter 22). Thus, we have

$$\lambda_w = \frac{\lambda_a}{n} = \frac{589\text{ nm}}{1.33} = 443\text{ nm}$$

The limiting angle of resolution at this wavelength is

$$\theta_{\text{min}} = 1.22\left(\frac{443 \times 10^{-9}\text{ m}}{0.90 \times 10^{-2}\text{ m}}\right) = 6.0 \times 10^{-5}\text{ rad}$$

## Example 25.7 Resolution of a Telescope

The Hale telescope at Mount Palomar has a diameter of 200 in. What is its limiting angle of resolution at a wavelength of 600 nm?

**Solution** Because $D = 200$ in. $= 5.08$ m and $\lambda = 6.00 \times 10^{-7}$ m, Equation 25.10 gives

$$\theta_{\text{min}} = 1.22\frac{\lambda}{D} = 1.22\left(\frac{6.00 \times 10^{-7}\text{ m}}{5.08\text{ m}}\right) = 1.44 \times 10^{-7}\text{ rad}$$

Therefore, any two stars that subtend an angle greater than or equal to this value could be resolved if the air above the telescope were perfectly steady.

It is interesting to compare this value with the resolution of a large radio telescope, such as the system at Arecibo, Puerto Rico, which has a diameter of 1 000 ft (305 m). This telescope detects radio waves at a wavelength of 0.75 m. The corresponding minimum angle of resolution is calculated to be $3.0 \times 10^{-3}$ rad (10 min 19 s of arc), which is more than 10 000 times larger than the calculated minimum angle for the Hale telescope.

### Example 25.8 Comparing Two Telescopes

Two telescopes have the following properties:

| Telescope | Diameter of Objective (in.) | Focal Length of Objective (mm) | Focal Length of Eyepiece (mm) |
|---|---|---|---|
| A | 6.00 | 1 000 | 6.00 |
| B | 8.00 | 1 250 | 25.00 |

**A** Which has the better resolving power?

**Solution** The telescope with the larger objective has the greater ability to discriminate between objects that are close together, and this is telescope B.

**B** Which has the greater light-gathering ability?

**Solution** The telescope with the larger objective can collect more light. Hence, B is again the choice.

**C** Which produces a greater magnification?

**Solution** The magnification of telescope A is

$$m_A = \frac{f_o}{f_e} = \frac{1\,000 \text{ mm}}{6.00 \text{ mm}} = 167$$

That of telescope B is

$$m_B = \frac{1\,250 \text{ mm}}{25.0 \text{ mm}} = 50.0$$

Thus, telescope A has the greater magnification.

## RESOLVING POWER OF THE DIFFRACTION GRATING

The diffraction grating, which we studied in Chapter 24, is most useful for making accurate wavelength measurements. Like the prism, it can be used to disperse a spectrum into its components. Of the two devices, the grating is more precise if we want to distinguish between two closely spaced wavelengths. We say that the grating spectrometer has a higher *resolution* than the prism spectrometer. If $\lambda_1$ and $\lambda_2$ are two nearly equal wavelengths between which the spectrometer can just barely distinguish, the **resolving power** $R$ of the grating is defined as

$$R \equiv \frac{\lambda}{\lambda_2 - \lambda_1} = \frac{\lambda}{\Delta\lambda} \qquad [25.11]$$

where $\lambda \approx \lambda_1 \approx \lambda_2$ and $\Delta\lambda = \lambda_2 - \lambda_1$. Thus, we see that a grating with a high resolving power can distinguish small differences in wavelength. Furthermore, if $N$ lines of the grating are illuminated, it can be shown that the resolving power in the $m$th-order diffraction equals the product $Nm$:

$$R = Nm \qquad [25.12]$$

Resolving power of a grating

Thus, the resolving power increases with order number. Furthermore, $R$ is large for a grating with a great number of illuminated slits. Note that for $m = 0$, $R = 0$, which signifies that *all wavelengths are indistinguishable* for the zeroth-order maximum (all wavelengths fall at the same point on the screen). However, consider the second-order diffraction pattern of a grating that has 5 000 rulings illuminated by the light source. The resolving power of such a grating in second order is $R = 5\ 000 \times 2 = 10\ 000$. Therefore, the *minimum* wavelength separation between two spectral lines that can be just resolved, assuming a mean wavelength of 600 nm, is calculated from Equation 25.12 to be $\Delta\lambda = \lambda/R = 6 \times 10^{-2}$ nm. For the third-order principal maximum, $R = 15\ 000$ and $\Delta\lambda = 4 \times 10^{-2}$ nm, and so on.

### Example 25.9 Resolving the Sodium Spectral Lines

Two bright lines in the spectrum of sodium have wavelengths of 589.00 nm and 589.59 nm.

**A** What must the resolving power of a grating be in order to distinguish these wavelengths?

**Solution** From Equation 25.11, we find that

$$R = \frac{\lambda}{\Delta\lambda} = \frac{589\text{ nm}}{589.59\text{ nm} - 589.00\text{ nm}} = \frac{589}{0.59} = 998$$

**B** To resolve these lines in the second-order spectrum, how many lines of the grating must be illuminated?

**Solution** From Equation 25.12 and the result of Part A, we find that

$$N = \frac{R}{m} = \frac{998}{2} = 499\text{ lines}$$

## 25.7 THE MICHELSON INTERFEROMETER

The camera and the telescope are examples of commonly used optical instruments. In contrast, the Michelson interferometer is an optical instrument that is unfamiliar to most people. It has great scientific importance, however. Invented by the American physicist A. A. Michelson (1852–1931), it is an ingenious device that splits a light beam into two parts and then recombines them to form an interference pattern. The interferometer is used to make accurate length measurements.

Figure 25.14 is a diagram of an interferometer. A beam of light provided by a monochromatic source is split into two rays by a partially silvered mirror, M, inclined at an angle of 45° relative to the incident light beam. One ray is reflected vertically upward to mirror $M_1$, and the other ray is transmitted horizontally through mirror M to mirror $M_2$. Hence, the two rays travel separate paths, $L_1$ and $L_2$. After reflecting from mirrors $M_1$ and $M_2$, the two rays eventually recombine to produce an interference pattern, which can be viewed through a telescope. The glass plate P equal in thickness to mirror M is placed in the path of the horizontal ray to ensure that the two rays travel the same distance through glass.

The interference pattern for the two rays is determined by the difference in their path lengths. When the two rays are viewed as shown, the image of $M_2$ is at

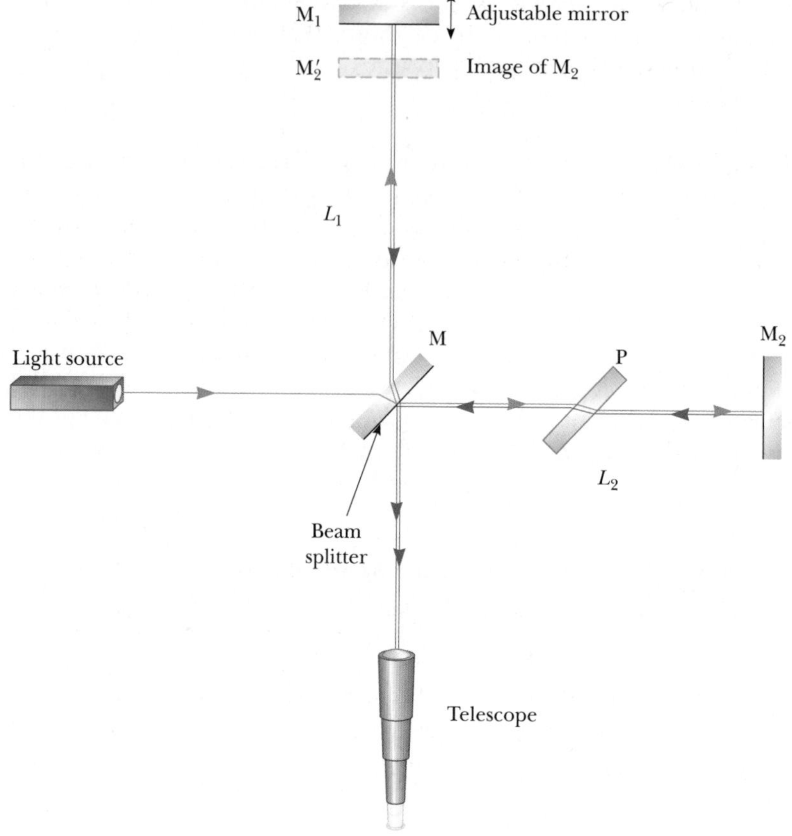


**FIGURE 25.14** A diagram of the Michelson interferometer. A single beam is split into two rays by the half-silvered mirror, M. The path difference between the two rays is varied with the adjustable mirror, $M_1$.

$M_2'$ parallel to $M_1$. Hence, the space between $M_2'$ and $M_1$ forms the equivalent of a parallel air film. The effective thickness of the air film is varied by using a finely threaded screw to move mirror $M_1$ in the direction indicated by the arrows in Figure 25.14. If one of the mirrors is tipped slightly with respect to the other, the thin film between the two is wedge-shaped, and an interference pattern consisting of parallel fringes is set up, as described in Example 24.4.

Now suppose we focus on one of the dark lines with the crosshairs of a telescope. As the mirror $M_1$ is moved to lengthen the path $L_1$, the thickness of the wedge increases. When the thickness increases by $\lambda/4$, the destructive interference that initially produced the dark fringe has changed to constructive interference, and we observe a bright fringe at the location of the crosshairs. The term *fringe shift* is used to describe the change in a fringe from dark to light or light to dark. Thus, successive light and dark fringes are formed each time $M_1$ is moved a distance of $\lambda/4$. The wavelength of light can be measured by counting the number of fringe shifts for a measured displacement of $M_1$. Conversely, if the wavelength is accurately known (as with a laser beam), the mirror displacement can be determined to within a fraction of the wavelength. Because the interferometer can measure displacements precisely, it is often used to make highly accurate measurements of the dimensions of mechanical components.

If the mirrors are perfectly aligned, rather than tipped with respect to one another, the path difference differs slightly for different angles of view. This results in an interference pattern that resembles Newton's rings. The pattern can be used in a fashion similar to that for tipped mirrors. One mirror concentrates on the center spot in the interference pattern. For example, suppose the spot is initially dark, indicating that destructive interference is occurring. If $M_1$ is now moved by a distance of $\lambda/4$, this central spot changes to a light region, corresponding to a fringe shift.

## SUMMARY

The light-concentrating power of a lens of focal length $f$ and diameter $D$ is determined by the ***f*-number,** defined as

$$f\text{-number} \equiv \frac{f}{D} \qquad [25.1]$$

The smaller the $f$-number of a lens, the brighter the image formed.

**Hyperopia** (farsightedness) is a defect of the eye that occurs either when the eyeball is too short or when the ciliary muscle cannot change the shape of the lens enough to form a properly focused image. **Myopia** (nearsightedness) occurs either when the eye is longer than normal or when the maximum focal length of the lens is insufficient to produce a clearly focused image on the retina.

The **power** of a lens in **diopters** is the inverse of the focal length in meters.

The **angular magnification of a lens** is defined as

$$m \equiv \frac{\theta}{\theta_0} \qquad [25.2]$$

where $\theta$ is the angle subtended by an object at the eye with a lens in use and $\theta_0$ is the angle subtended by the object when it is placed at the near point of the eye and no lens is used. The **maximum angular magnification of a lens** is

$$m_{\text{max}} = 1 + \frac{25 \text{ cm}}{f} \qquad [25.5]$$

When the eye is relaxed, the angular magnification is

$$m = \frac{25 \text{ cm}}{f} \qquad [25.6]$$

The overall **magnification of a compound microscope** of length $L$ is the product of the magnification produced by the objective, of focal length $f_o$, and the magnification produced by the eyepiece, of focal length $f_e$:

$$m = -\frac{L}{f_o}\left(\frac{25 \text{ cm}}{f_e}\right) \qquad [25.7]$$

The **angular magnification of a telescope** is

$$m = \frac{f_o}{f_e} \qquad [25.8]$$

where $f_o$ is the focal length of the objective and $f_e$ is the focal length of the eyepiece.

Two images are said to be **just resolved** when the central maximum of the diffraction pattern for one image falls on the first minimum of the other image. This limiting condition of resolution is known as **Rayleigh's criterion.** The limiting angle of resolution for a **slit** of width $a$ is

$$\theta_{\text{min}} \approx \frac{\lambda}{a} \qquad [25.9]$$

The limiting angle of resolution of a **circular aperture** is

$$\theta_{\text{min}} = 1.22\frac{\lambda}{D} \qquad [25.10]$$

where $D$ is the diameter of the aperture.

The **resolving power** of a diffraction grating in the $m$th order is

$$R = Nm \qquad [25.12]$$

where $N$ is the number of illuminated rulings on the grating.

## CONCEPTUAL QUESTIONS

1. A lens is used to examine an object across the room. Is the lens probably being used as a simple magnifier?
2. Why is it difficult or impossible to focus a microscope on an object across the room?
3. The optic nerve and the brain invert the image formed on the retina. Why do we not see everything upsidedown?
4. Why is it difficult or impossible to focus a telescope on an object very close to you?
5. Suppose you are observing the interference pattern formed by a Michelson interferometer in a laboratory, and a joking colleague holds a lit match in the light path of one arm of the interferometer. Will this have an effect on the interference pattern?
6. Compare and contrast the eye and a camera. What parts of the camera correspond to the iris, the retina, and the cornea of the eye?
7. Large telescopes are usually reflecting rather than refracting. List some reasons for this choice.
8. If you want to use a converging lens to set fire to a piece of paper, why should the light source be farther from the lens than its focal point?
9. Explain why it is theoretically impossible to see an object as small as an atom regardless of the quality of the light microscope being used.
10. Which is most important in the use of a camera photoflash unit, the intensity of the light (the energy per unit area per unit time), or the product of the intensity and the time of the flash, assuming the time is less than the shutter speed?

## PROBLEMS

1, 2, 3 = straightforward, intermediate, challenging ☐ = full solution available in Student Solutions Manual/Study Guide

**web** = solution posted at **http://info.brookscole.com/serway** = biomedical application

### Section 25.1 The Camera

1. A camera used by a professional photographer to shoot portraits has a focal length of 25.0 cm. The photographer takes a portrait of a person 1.50 m in front of the camera. Where is the image formed, and what is the lateral magnification?
2. The lens of a certain 35-mm camera (where 35-mm is the width of the film strip) has a focal length of 55 mm and a speed (an f-number) of f/1.8. Determine the diameter of the lens.
3. The image area of a typical 35-mm slide is 23.5 mm by 35.0 mm. If a camera has a 55.0-mm focal length lens, will the full image this camera forms of the constellation Orion, which is 20° across, fit on a 35-mm slide?
4. The full Moon is photographed using a camera with a 120-mm focal length lens. Determine the diameter of the Moon's image on the film. (*Note:* The radius of the Moon is $1.74 \times 10^6$ m, and the Earth-Moon distance is $3.84 \times 10^8$ m.)
5. **web** A camera is being used with the correct exposure at $f/4$ and a shutter speed of 1/32 s. In order to "stop" a fast-moving subject, the shutter speed is changed to 1/256 s. Find the new *f*-stop that should be used to maintain satisfactory exposure, assuming no change in lighting conditions.
6. (a) Use conceptual arguments to show that the intensity of light (energy per unit area per unit time) reaching the film in a camera is proportional to the square of the reciprocal of the *f*-number, as

$$I \propto \frac{1}{(f/D)^2}$$

(b) The correct exposure time for a camera set to $f/1.8$ is (1/500) s. Calculate the correct exposure time if the *f*-number is changed to $f/4$ under the same lighting conditions.

7. A certain type of film requires an exposure time of 0.010 s with an $f/11$ lens setting. Another type of film requires twice the light energy to produce the same level of exposure. What *f*-stop does the second type of film need with the 0.010-s exposure time?
8. Assume that the camera in Figure 25.1 has a fixed focal length of 65.0 mm and is adjusted to properly focus the image of a distant object. How far and in what direction must the lens be moved to focus the image of an object that is 2.00 m away?

### Section 25.2 The Eye

9. A retired bank president can easily read the fine print of the financial page when the newspaper is held no closer than arm's length, 60.0 cm from the eye. What should be the focal length of an eyeglass lens that allows her to read at the more comfortable distance of 24.0 cm?
10. A person has far points 84.4 cm from the right eye and 122 cm from the left eye. Write a prescription for the powers of the corrective lenses.
11. The accommodation limits for Nearsighted Nick's eyes are 18.0 cm and 80.0 cm. When he wears his glasses, he is able to see faraway objects clearly. At what minimum distance is he able to see objects clearly?
12. The near point of an eye is 100 cm. A corrective lens is to be used to allow this eye to clearly focus on objects 25.0 cm in front of it. (a) What should be the focal length of this lens? (b) What is the power of the needed corrective lens?
13. An individual is nearsighted; his near point is 13.0 cm and his far point is 50.0 cm. (a) What lens power is needed to correct his nearsightedness? (b) When the lenses are in use, what is this person's near point?
14. A person sees clearly wearing eyeglasses that have a power of −4.00 diopters and sit 2.00 cm in front of the eyes. If the person wants to switch to contact lenses, which are placed directly on the eyes, what lens power should be prescribed?
15. An artificial lens is implanted in a person's eye to replace a diseased lens. The distance between the artificial lens and the retina is 2.80 cm. In the absence of the lens, an image

of a distant object (Formed by refraction at the cornea) falls 2.53 cm behind the retina. The lens is designed to put the image of the distant object on the retina. What is the power of the implanted lens? (*Hint:* Consider the image formed by the cornea as a virtual object.)

16. A person is to be fitted with bifocals. She can see clearly when the object is between 30 cm and 1.5 m from the eye. (a) The upper portions of the bifocals (Fig. P25.16) should be designed to enable her to see distant objects clearly. What power should they have? (b) The lower portions of the bifocals should enable her to see objects comfortably at 25 cm. What power should they have?

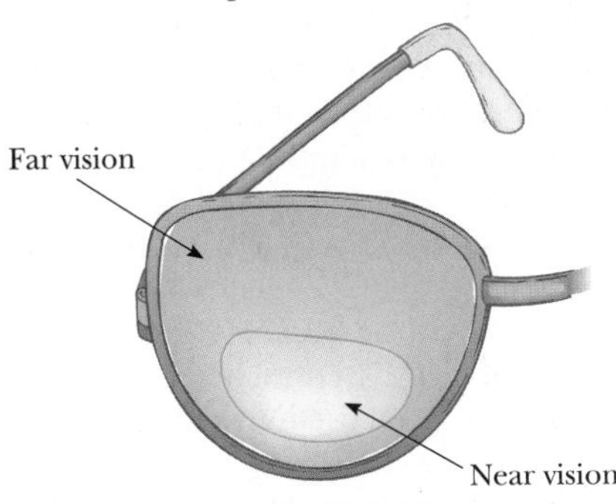


**FIGURE P25.16**

## Section 25.3 The Simple Magnifier

17. A philatelist examines the printing detail on a stamp using a biconvex lens of focal length 10.0 cm as a simple magnifier. The lens is held close to the eye, and the lens-to-object distance is adjusted so that the virtual image is formed at the normal near point (25.0 cm). Calculate the magnification.

18. A lens having a focal length of 25 cm is used as a simple magnifier. (a) What is the angular magnification obtained when the image is formed at the normal near point ($q = -25$ cm)? (b) What is the angular magnification produced by this lens when the eye is relaxed?

19. A biology student uses a simple magnifier to examine the structural features of the wing of an insect. The wing is held 3.50 cm in front of the lens, and the image is formed 25.0 cm from the eye. (a) What is the focal length of the lens? (b) What angular magnification is achieved?

20. A lens that has a focal length of 5.00 cm is used as a magnifying glass. (a) To obtain maximum magnification, where should the object be placed? (b) What is the magnification?

21. A leaf of length $h$ is positioned 71.0 cm in front of a converging lens with a focal length of 39.0 cm. An observer views the image of the leaf from a position 1.26 m behind the lens, as shown in Figure P25.21. (a) What is the magnitude of the lateral magnification (ratio of image size to object size) produced by the lens? (b) What angular magnification is achieved by viewing the image of the leaf rather than viewing the leaf directly?

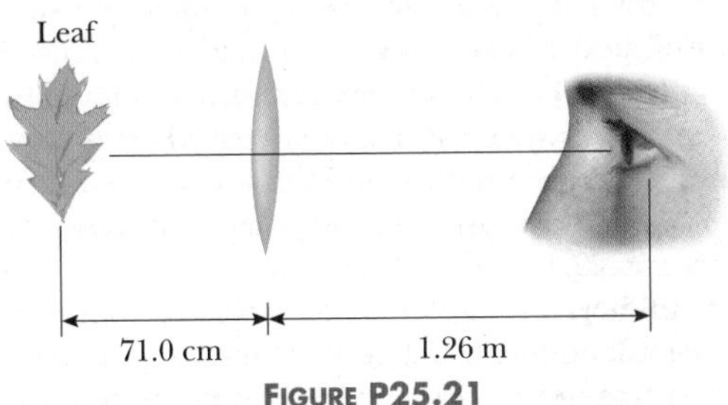


**FIGURE P25.21**

## Section 25.4 The Compound Microscope

## Section 25.5 The Telescope

22. The distance between eyepiece and objective lens in a certain compound microscope is 23.0 cm. The focal length of the eyepiece is 2.50 cm, and that of the objective is 0.400 cm. What is the overall magnification of the microscope?

23. The desired overall magnification of a compound microscope is 140×. The objective alone produces a lateral magnification of 12×. Determine the required focal length of the eyepiece.

24. A microscope has an objective lens with a focal length of 16.22 mm and an eyepiece with a focal length of 9.50 mm. With the length of the barrel set at 29.0 cm, the diameter of a red blood cell's image subtends an angle of 1.43 mrad with the eye. If the final image distance is 29.0 cm from the eyepiece, what is the actual diameter of the red blood cell?

25. web The length of a microscope tube is 15.0 cm. The focal length of the objective is 1.00 cm, and the focal length of the eyepiece is 2.50 cm. What is the magnification of the microscope, assuming it is adjusted so that the eye is relaxed?

26. The Yerkes refracting telescope has a 1.00-m-diameter objective lens of focal length 20.0 m. Assume it is used with an eyepiece of focal length 2.50 cm. (a) Determine the angular magnification of the planet Mars as seen through this telescope. (b) Are the Martian polar caps right side up or upside down?

27. The lenses of an astronomical telescope are 92 cm apart when adjusted for viewing a distant object with minimum eyestrain. The angular magnification produced by the telescope is 45. Compute the focal length of each lens.

28. An elderly sailor is shipwrecked on a desert island but manages to save his eyeglasses. The lens for one eye has a power of +1.20 diopters, and the other lens has a power of +9.00 diopters. (a) What is the magnifying power of the telescope he can construct with these lenses? (b) How far apart are the lenses when the telescope is adjusted for minimum eyestrain?

29. Astronomers often take photographs with the objective lens or mirror of a telescope alone, without an eyepiece. (a) Show that the image size $h'$ for this telescope is given by $h' = fh/(f - p)$ where $h$ is the object size, $f$ the objective focal length, and $p$ the object distance. (b) Simplify the expression in part (a) if the object distance is much greater than objective focal length. (c) The "wingspan" of the International Space Station is 108.6 m, the overall width of its solar panel configuration. When it is orbiting at an altitude of 407 km, find the width of the image formed by a telescope objective of focal length 4.00 m.

30. Galileo devised a simple terrestrial telescope that produces an upright image. It consists of a converging objective lens and a diverging eyepiece at opposite ends of the telescope tube. For distant objects, the tube length is the objective focal length less the absolute value of the eyepiece focal length. (a) Does the user of the telescope see a real or virtual image? (b) Where is the final image? (c) If a telescope is to be constructed with a tube of length 10.0 cm and a magnification of 3.00, what are the focal lengths of the objective and eyepiece?

31. A person decides to use an old pair of eyeglasses to make some optical instruments. He knows that the near point in his left eye is 50.0 cm and the near point in his right eye is 100 cm. (a) What is the maximum angular magnification he can produce in a telescope? (b) If he places the lenses 10.0 cm apart, what is the maximum overall magnification he can produce in a microscope? (Go back to basics and use the thin-lens equation to solve part (b).)

### Section 25.6 Resolution of Single-Slit and Circular Apertures

32. If the distance from Earth to the Moon is $3.8 \times 10^8$ m, what diameter would be required for a telescope objective to resolve a Moon crater 300 m in diameter? Assume a wavelength of 500 nm.

33. A converging lens with a diameter of 30.0 cm forms an image of a satellite passing overhead. The satellite has two green lights (wavelength 500 nm) spaced 1.00 m apart. If the lights can just be resolved according to the Rayleigh criterion, what is the altitude of the satellite?

34. The pupil of a cat's eye narrows to a vertical slit of width 0.500 mm in daylight. What is the angular resolution for a pair of horizontally separated mice? (Use 500-nm light in your calculation.)

35. To increase the resolving power of a microscope, the object and the objective are immersed in oil ($n = 1.5$). If the limiting angle of resolution without the oil is 0.60 $\mu$rad, what is the limiting angle of resolution with the oil? (*Hint:* The oil changes the wavelength of the light.)

36. Two motorcycles, separated laterally by 2.00 m, are approaching an observer holding an infrared detector that is sensitive to radiation of wavelength 885 nm. What aperture diameter is required in the detector if the two headlights are to be resolved at a distance of 10.0 km?

37. A helium-neon laser emits light that has a wavelength of 632.8 nm. The circular aperture through which the beam emerges has a diameter of 0.500 cm. Estimate the diameter of the beam 10.0 km from the laser.

38. A spy satellite circles Earth at an altitude of 200 km and carries out surveillance with a special high-resolution telescopic camera having a lens diameter of 35 cm. If the angular resolution of this camera is limited by diffraction, estimate the separation of two small objects on Earth's surface that are just resolved in yellow-green light ($\lambda = 550$ nm).

39. Suppose a 5.00-m-diameter telescope is constructed on the Moon, where the absence of atmospheric distortion permits excellent viewing. If observations are made using 500-nm light, what minimum separation between two objects could just be resolved on Mars at closest approach (when Mars is $8.0 \times 10^7$ km from the Moon)?

### Section 25.7 The Michelson Interferometer

40. Light of wavelength 550 nm is used to calibrate a Michelson interferometer. By use of a micrometer screw, the platform on which one mirror is mounted is moved 0.180 mm. How many fringe shifts are counted?

41. An interferometer is used to measure the length of a bacterium. The wavelength of the light used is 650 nm. As one arm of the interferometer is moved from one end of the cell to the other, 310 fringe shifts are counted. How long is the bacterium?

42. Mirror $M_1$ in Figure 25.14 is displaced a distance $\Delta L$. During this displacement, 250 fringe shifts are counted. The light being used has a wavelength of 632.8 nm. Calculate the displacement $\Delta L$.

43. Monochromatic light is beamed into a Michelson interferometer. The movable mirror is displaced 0.382 mm, causing the central spot in the interferometer pattern to change from bright to dark and back to bright 1 700 times. Determine the wavelength of the light. What color is it?

44. The Michelson interferometer can be used to measure the index of refraction of a gas by placing an evacuated transparent tube in the light path along one arm of the device. Fringe shifts occur as the gas is slowly added to the tube. Assume that 600-nm light is used, the tube is 5.00 cm long, and that 160 fringe shifts occur as the pressure of the gas in the tube increases to atmospheric pressure. What is the index of refraction of the gas? (*Hint:* The fringe shifts occur because the wavelength of the light changes inside the gas-filled tube.)

45. web The light path in one arm of a Michelson interferometer includes a transparent cell that is 5.00 cm long. How many fringe shifts would be observed if all the air were evacuated from the cell? The wavelength of the light source is 590 nm and the refractive index of air is 1.000 29. (See the Hint in Problem 44.)

46. The $H_\alpha$ line in hydrogen has a wavelength of 656.20 nm. This line differs in wavelength from the corresponding spectral line in deuterium (the heavy stable isotope of hydrogen) by 0.18 nm. (a) Determine the minimum number of lines a grating must have to resolve these two wavelengths in the first order. (b) Repeat part (a) for the second order.

47. A 15.0-cm-long grating has 6 000 slits per centimeter. Can two lines of wavelengths 600.000 nm and 600.003 nm be separated using this grating? Explain.

### ADDITIONAL PROBLEMS

48. A person with a nearsighted eye has near and far points of 16 cm and 25 cm, respectively. (a) Assuming a lens is placed 2.0 cm from the eye, what power must the lens have to correct this condition? (b) Contact lenses placed directly on the cornea are used to correct the eye in this example. What is the power of the lens required in this case, and what is the new near point? (*Hint:* The contact lens and the eyeglass lens require slightly different powers because they are at different distances from the eye.)

49. The near point of an eye is 75.0 cm. (a) What should be the power of a corrective lens prescribed to enable the eye to see an object clearly at 25.0 cm? (b) If, using the corrective lens, the user can see an object clearly at 26.0 cm but not 25.0 cm, by how many diopters did the lens grinder miss the prescription?

50. A compound microscope has an objective of focal length 0.300 cm and an eyepiece of focal length 2.50 cm. If an object is 3.40 mm from the objective, what is the magnification? (*Hint:* Use the lens equation for the objective.)

51. web A cataract-impaired lens in an eye can be surgically removed and replaced by a manufactured lens. The focal length required for the new lens is determined by the lens-to-retina distance, which is measured by a sonar-like device, and by the requirement that the implant provide for correct distant vision. (a) If the distance from lens to retina is 22.4 mm, calculate the power of the implanted lens in diopters. (b) Since there is no accommodation and the implant allows for correct distant vision, a corrective lens for close work or reading must be used. Assume a

reading distance of 33.0 cm and calculate the power of the lens in the reading glasses.

52. Estimate the minimum angle subtended at the eye of a hawk flying at an altitude of 50 m to recognize a mouse on the ground.

53. The wavelengths of the sodium spectrum are $\lambda_1 =$ 589.00 nm and $\lambda_2 =$ 589.59 nm. Determine the minimum number of lines in a grating that will allow resolution of the sodium spectrum in (a) the first order and (b) the third order.

54. The text discusses the astronomical telescope. Another type is the Galilean telescope, in which an objective lens gathers light (Fig. P25.54), and tends to form an image at point *A*. An eyepiece, consisting of a diverging lens, intercepts the light before it comes to a focus and forms a virtual image at point *B*. When adjusted for minimum eyestrain, point *B* is an infinite distance in front of the lens and parallel rays emerge from the lens, as in Figure P25.54b. An opera glass, which is a Galilean telescope, is used to view a 30.0-cm-tall singer's head that is 40.0 m from the objective lens. The focal length of the objective is + 8.00 cm, and that of the eyepiece is − 2.00 cm. The telescope is adjusted so parallel rays enter the eye. Compute (a) the size of the real image that would have been formed by the objective, (b) the virtual object distance for the diverging lens, (c) the distance between the lenses, and (d) the overall angular magnification.

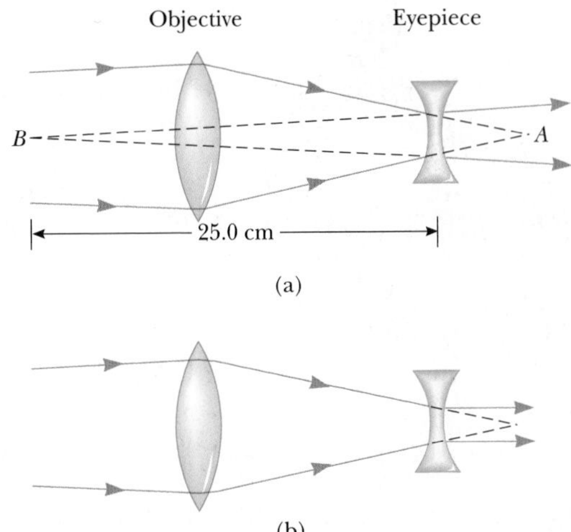


**FIGURE P25.54**

55. A laboratory (astronomical) telescope is used to view a scale that is 300 cm from the objective, which has a focal length of 20.0 cm; the eyepiece has a focal length of 2.00 cm. Calculate the angular magnification when the telescope is adjusted for minimum eyestrain. (*Note:* The object is not at infinity, and so the simple expression $m = f_o/f_e$ is not sufficiently accurate for this problem. Also, assume small angles so that $\tan\theta \approx \theta$.)

56. If the aqueous humor of the eye has an index of refraction of 1.34 and the distance from the vertex of the cornea to the retina is 2.00 cm, what is the radius of curvature of the cornea for which distant objects will be focused on the retina? (For simplicity, assume that all refraction occurs in the aqueous humor.)

## GROUP ACTIVITIES

**G.1** (a) Move this book toward your face until the letters just begin to blur. The distance from the book to your eye is your near point. (b) On a sheet of paper make a dot near the center and about 3 inches to the left and right place an x. With one eye shut and while looking at the dot, slowly move the paper toward your eye. You will notice at some distance from your eye, one of the x's will disappear. This is the location of the blind spot of your eye, the point where the optic nerve enters the eye. (c) Stand before a mirror in a darkened room for a few minutes. Then, turn on a light in the room and observe your pupils in the mirror as they change size. Such dark adaptation also takes place at the rods and cones as they chemically adjust their sensitivity. This adjustment takes 15–30 minutes as you have noted when you enter a darkened movie theater. Iris aperture control takes less than a second and helps protect the retina from overload.

**G.2** On a sunny day, hold a magnifying glass above a nonflammable surface, such as a sidewalk, so that the image of the Sun forms a round spot of light on the surface. Note where the spot formed by the lens is most distinct, or smallest. Use a ruler to measure the distance between the glass and the image. The distance is equal to the focal length of the lens.

**G.3** Hold a pair of prescription glasses about 12 cm from your eye, and look at different objects through the lenses. Try this with different types of glasses, such as those for farsightedness and nearsightedness, and describe what effects the differences have on the image you see. If you have bifocals, how do the images produced by the top and bottom portions of the bifocal lens compare?

**G.4** If you have never experimented with a 35-mm camera with adjustable *f*-numbers and shutter speeds, use up a couple of rolls of film to see what happens. Take several shots of the same object with different settings for these two variables. (You should record your *f*-numbers and shutter speeds for each photograph.) Explain any differences you see in the final images in terms of the settings used.

**G.5** (a) A patient has a near point of 1.25 m. Is she nearsighted or farsighted? (b) Should the corrective lens be converging or diverging? (c) For a comfortable reading distance of 25.0 cm from her eye, what diopter lens should be prescribed? (d) Does the power of the lens change if you assume that the lens is always placed 2.00 cm from the eye?

**G.6** A lens with a certain power is used as a simple magnifier. (a) If the power of the lens is doubled, does the angular magnification increase or decrease? (b) What assumptions are made when you use the equation $m = 25\text{ cm}/f$? (c) What is the angular magnification of a lens having a focal length of 5.0 cm?

**G.7** A telescope uses an objective lens of focal length 40 cm and an eyepiece of 2.0 cm. (a) Without doing a calculation, how will the view through the telescope change if the eyepiece is changed to one having a focal length of 4.0 cm? (b) Do the calculations to verify your prediction. (c) Does your answer depend on whether you are using a refracting or a reflecting telescope?

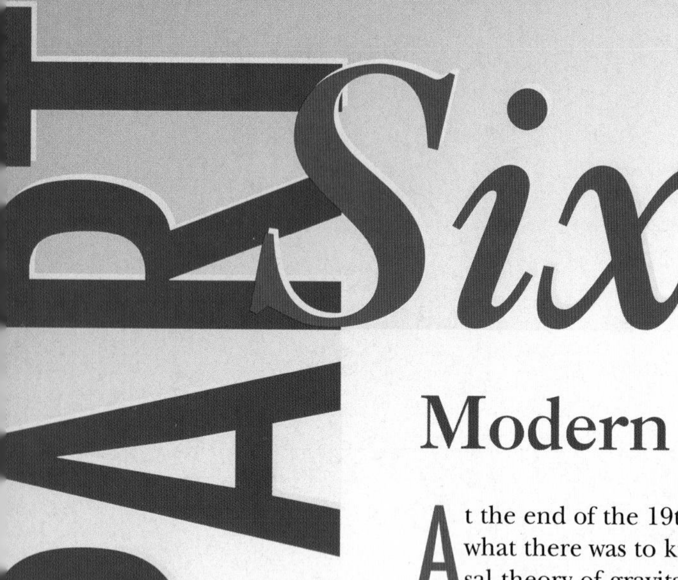


# Modern Physics

At the end of the 19th century, scientists believed that they had learned most of what there was to know about physics. Newton's laws of motion and his universal theory of gravitation, Maxwell's theoretical work in unifying electricity and magnetism, and the laws of thermodynamics and kinetic theory were highly successful in explaining a wide variety of phenomena.

However, at the turn of the 20th century, a major revolution shook the world of physics. In 1900 Planck provided the basic ideas that led to the formulation of the quantum theory, and in 1905 Einstein formulated his brilliant special theory of relativity. The excitement of the times is captured in Einstein's own words: "It was a marvelous time to be alive." Both ideas were to have a profound effect on our understanding of nature. Within a few decades, these two theories inspired new developments in the fields of atomic physics, nuclear physics, and condensed-matter physics.

Our discussion of modern physics will begin with a treatment of the special theory of relativity in Chapter 26. The theory provides us with a new and deeper view of physical laws. In Chapter 27 we shall discuss various developments in quantum theory, which provides us with a successful model for understanding electrons, atoms, and molecules. The last three chapters of the text are concerned with applications of quantum theory. Chapter 28 discusses the structure and properties of atoms using concepts from quantum mechanics. Chapter 29 is concerned with the structure and properties of the atomic nucleus. Chapter 30 discusses practical applications of nuclear physics and concludes with a discussion of elementary particles.

Keep in mind that although modern physics was developed during the 20th century and has led to a multitude of important technological achievements, the story is still incomplete as we begin this 21st century. Surprising discoveries will continue to be made during our lifetimes, and many of these discoveries will deepen or refine our understanding of nature and the world around us. It is still "a marvelous time to be alive."

The 20th century saw the development of large accelerators capable of producing high-energy collisions between atoms and the subatomic particles within them. Charged particles resulting from these collisions follow curved tracks similar to the ones shown above when the detector is placed in a strong magnetic field. *(Science VU/FNL/DOE)*

# 26 Relativity

## Chapter Outline



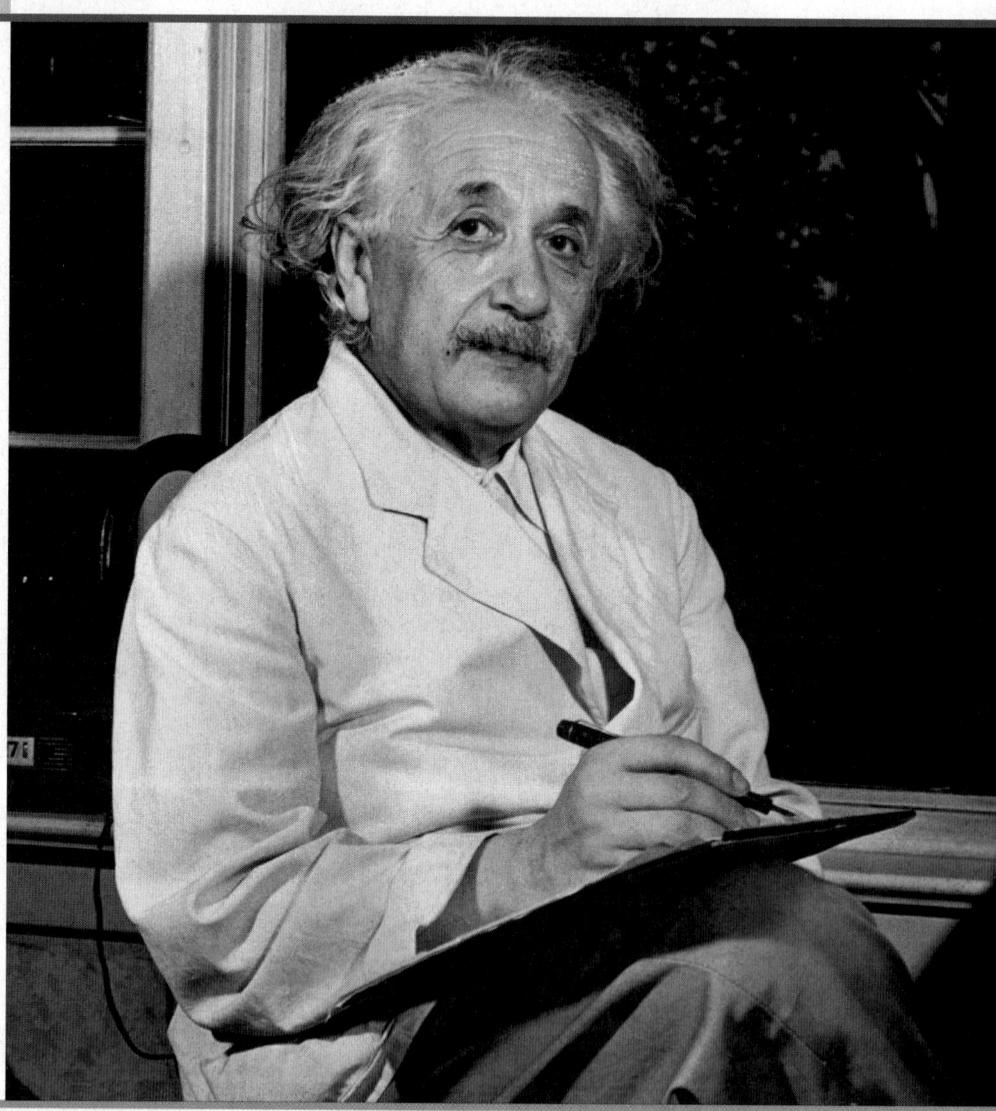

**Albert Einstein, one of the greatest theoretical physicists of all times, revolutionized the world of modern physics. He is quoted as saying, "I have no particular talent. I am merely inquisitive." Although he is recognized primarily for his monumental intellectual achievements, he was also deeply concerned with the social impact of scientific discovery.** *(© Bettmann/CORBIS)*

Most of our everyday experiences and observations have to do with objects that move at speeds much less than the speed of light. Newtonian mechanics was formulated to describe the motion of such objects, and this formalism is quite successful in describing a wide range of phenomena that occur at low speeds. It fails, however, when applied to particles whose speeds approach that of light.

This chapter is concerned with the theory of special relativity. Although the concepts of special relativity often violate our common sense, the theory correctly predicts the results of experiments involving speeds near the speed of light.

## 26.1 INTRODUCTION

Experimentally, the predictions of Newtonian theory can be tested at high speeds by accelerating electrons or other charged particles through a large electric potential difference. For example, it is possible to accelerate an electron to a speed of $0.99c$ (where $c$ is the speed of light) by using a potential difference of several million volts. According to Newtonian mechanics, if the potential difference is increased by a factor of 4, the electron's kinetic energy is four times greater and its speed should double to $1.98c$. However, experiments show that the speed of the electron—as well as the speed of any other particle in the Universe—always remains *less* than the speed of light, regardless of the size of the accelerating voltage. Because it places no upper limit on the speed, Newtonian mechanics is contrary to modern experimental results and is clearly a limited theory.

In 1905, at the age of 26, Einstein published his special theory of relativity. Regarding the theory, Einstein wrote,

> The relativity theory arose from necessity, from serious and deep contradictions in the old theory from which there seemed no escape. The strength of the new theory lies in the consistency and simplicity with which it solves all these difficulties, using only a few very convincing assumptions.[1]

Although Einstein made many other important contributions to science, his theory of relativity alone represents one of the greatest intellectual achievements of all time. With this theory, experimental observations can be correctly predicted over the range of speeds from $v = 0$ to speeds approaching the speed of light. Newtonian mechanics, which was accepted for more than 200 years, is in fact a specialized case of Einstein's theory. This chapter introduces the special theory of relativity, with emphasis on some of the consequences of the theory. A brief discussion of general relativity and some of its consequences is presented in Section 26.11.

As we shall see, the special theory of relativity is based on two postulates:

1. **The laws of physics are the same in all coordinate systems either at rest or moving at constant velocity with respect to one another.**
2. **The speed of light in a vacuum has the same value, $3 \times 10^8$ m/s, regardless of the velocity of the observer or the velocity of the source emitting the light.**

Special relativity covers such phenomena as the slowing down of moving clocks and the contraction of lengths in reference frames moving with constant velocity relative to an observer. We also discuss the relativistic forms of momentum and energy, as well as some consequences of the famous mass-energy formula, $E = mc^2$.

[1] A. Einstein and L. Infeld, *The Evolution of Physics,* New York, Simon and Schuster, 1961.

**APPLYING PHYSICS 26.1**

Imagine a very powerful lighthouse with a rotating beacon. Imagine also drawing a horizontal circle around the lighthouse, with the lighthouse at the center. Along the circumference of the circle, the light beam lights up a portion of the circle and the lit portion of the circle moves around the circle at a certain tangential speed. If we now imagine a circle twice as big in radius, the tangential speed of the lit portion is faster, because it must travel a larger circumference in the time of one rotation of the light source. Imagine that we continue to make the circle larger and larger, eventually moving it out into space. The tangential speed of the lit portion will keep increasing. Is it possible that the tangential speed could become larger than the speed of light? Would this violate a principle of special relativity?

**Explanation** For a large-enough circle, it is possible that the tangential speed of the lit portion of the circle could be larger than the speed of light. This does not violate a principle of special relativity, however, because no matter or information is traveling faster than the speed of light.

## 26.2 THE PRINCIPLE OF GALILEAN RELATIVITY

In order to describe a physical event, it is necessary to choose a *frame of reference.* For example, when you perform an experiment in a laboratory, you select a coordinate system, or frame of reference, that is at rest with respect to the laboratory. However, suppose an observer in a passing car moving at a constant velocity with respect to the lab were to observe your experiment. Would the observations made by the moving observer differ dramatically from yours? That is, if you found Newton's first law to be valid in your frame of reference, would the moving observer agree with you? According to the principle of Galilean relativity, **the laws of mechanics must be the same in all inertial frames of reference.** Inertial frames of reference are those reference frames in which Newton's laws are valid. Practically, such frames are those in which objects subjected to no forces move in straight lines at constant speed—thus the name *inertial frame* because objects observed from these frames obey Newton's first law, the law of inertia. For the situation just described, the laboratory coordinate system and the coordinate system of the moving car are both inertial frames of reference. As a consequence, if the laws of mechanics are found to be true in the lab, the person in the car must also observe the same laws.[2]

Let us describe a hypothetical observation to illustrate the equivalence of the laws of mechanics in different inertial frames. Consider an airplane in flight, moving with a constant velocity, as in Figure 26.1a. If a passenger in the airplane throws a ball straight up in the air, the passenger observes that the ball moves in a vertical path. The motion of the ball is precisely the same as it would be if the ball were thrown while at rest on Earth. The law of gravity and the equations of motion under constant acceleration are obeyed whether the airplane is at rest or in uniform motion. Now consider the same experiment when viewed by another observer at rest on Earth. This stationary observer views the path of the ball in the plane to be a parabola, as in Figure 26.1b. Furthermore, according to this observer, the ball has a velocity to the right equal to the velocity of the plane. Although the two observers disagree on the shape of the ball's path, both agree that the motion of the ball obeys the law of gravity and Newton's laws of motion and even agree on how long the ball is in the air. Thus, we draw the following important conclusion: **There is no preferred frame of reference for describing the laws of mechanics.**

[2] What is an example of a noninertial frame? A frame undergoing translational acceleration or a frame rapidly rotating with respect to the two inertial frames mentioned above.

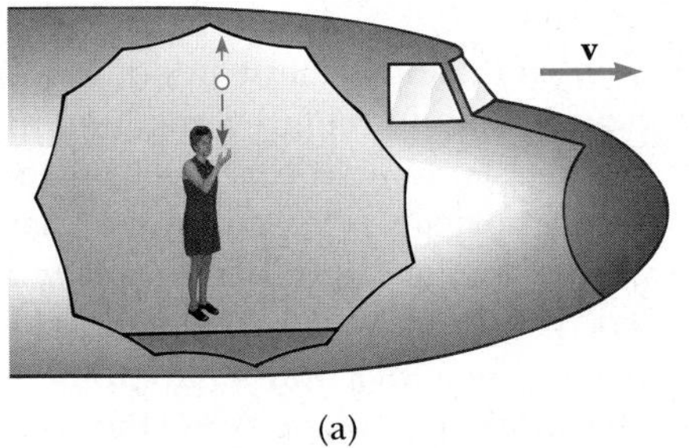


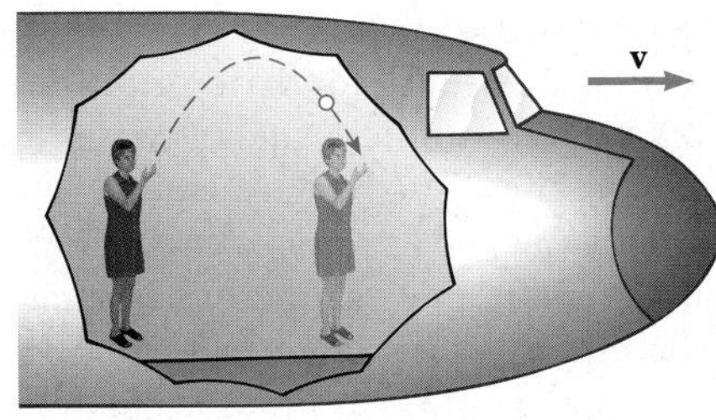

**FIGURE 26.1** (a) The observer on the airplane sees the ball move in a vertical path when thrown upward. (b) The Earth observer views the path of the ball to be a parabola.

## 26.3 THE SPEED OF LIGHT

It is quite natural to ask whether the concept of Galilean relativity in mechanics also applies to experiments in electricity, magnetism, optics, and other areas. Experiments indicate that the answer is no. For example, if we assume that the laws of electricity and magnetism are the same in all inertial frames, a paradox concerning the speed of light immediately arises. This can be understood by recalling that according to electromagnetic theory, the speed of light always has the fixed value of $2.997\,924\,58 \times 10^8$ m/s in free space. But this is in direct contradiction to common sense. For example, suppose a light pulse is sent out by an observer in a boxcar moving with a velocity **v** (Fig. 26.2). The light pulse has a velocity **c** relative to observer S′ in the boxcar. According to Galilean relativity, the velocity of the pulse relative to the stationary observer S outside the boxcar should be **c** + **v**. This obviously contradicts Einstein's theory, which postulates that the velocity of the light pulse is the same for all observers.

In order to resolve this paradox, we must conclude either that (1) the addition law for velocities is incorrect or that (2) the laws of electricity and magnetism are not the same in all inertial frames. Assume that the second conclusion is true; then a preferred reference frame must exist in which the speed of light has the value $c$, but in any other reference frame the speed of light must have a value that is greater or less than $c$. It is useful to draw an analogy with sound waves, which propagate through a medium such as air. The speed of sound in air is about 330 m/s when measured in a reference frame in which the air is stationary. However, the speed of sound is greater or less than this value when measured from a reference frame that is moving with respect to the air.

In the case of light signals (electromagnetic waves), recall that electromagnetic theory predicted that such waves must propagate through free space with a

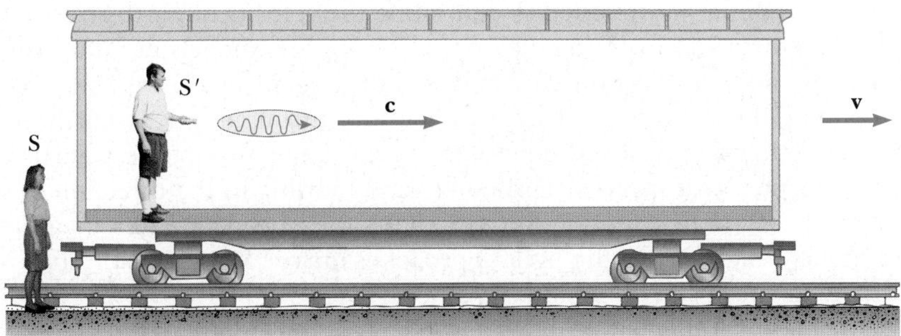


**FIGURE 26.2** A pulse of light is sent out by a person in a moving boxcar. According to Newtonian relativity, the speed of the pulse should be **c** + **v** relative to a stationary observer.

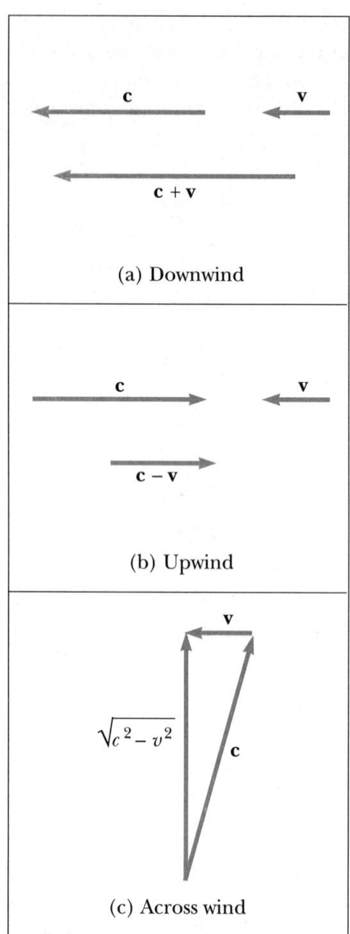


**FIGURE 26.3** If the speed of the ether wind relative to Earth is $v$, and $c$ is the speed of light relative to the ether, the speed of light relative to Earth is (a) $c + v$ in the downwind direction, (b) $c - v$ in the upwind direction, and (c) $(c^2 - v^2)^{1/2}$ in the direction perpendicular to the wind.

speed equal to the speed of light. However, the theory does not require the presence of a medium for wave propagation. This is in contrast to other types of waves that we have studied, such as water and sound waves, that do require a medium to support the disturbances. In the 19th century, physicists thought that electromagnetic waves also required a medium in order to propagate. They proposed that such a medium existed, and they gave it the wonderful name **luminiferous ether.** The ether was assumed to be present everywhere, even in empty space, and light waves were viewed as ether oscillations. Furthermore, the ether would have to be a massless but rigid medium with no effect on the motion of planets or other objects. These are strange concepts indeed. In addition, it was found that the troublesome laws of electricity and magnetism would take on their simplest forms in a special frame of reference at *rest* with respect to the ether. This frame was called the *absolute frame.* The laws of electricity and magnetism would be valid in this absolute frame, but they would have to be modified in any reference frame moving with respect to the absolute frame.

As a result of the importance attached to the ether and this absolute frame, it became of considerable interest in physics to prove by experiment that they existed. Since it was considered likely that Earth was in motion through the ether, from the view of an experimenter on Earth, there was an "ether wind" blowing through his lab. A direct method for detecting the ether wind would use an apparatus fixed to Earth to measure the wind's influence on the speed of light. If $v$ is the speed of the ether relative to Earth, then the speed of light should have its maximum value $c + v$ when propagating downwind, as shown in Figure 26.3a. Likewise, the speed of light should have its minimum value $c - v$ when propagating upwind, as in Figure 26.3b, and an intermediate value $(c^2 - v^2)^{1/2}$ in the direction perpendicular to the ether wind, as in Figure 26.3c. If the Sun is assumed to be at rest in the ether, then the velocity of the ether wind would be equal to the orbital velocity of Earth around the Sun, which has a magnitude of approximately $3 \times 10^4$ m/s. Because $c = 3 \times 10^8$ m/s, it should be possible to detect a change in speed of about 1 part in $10^4$ for measurements in the upwind or downwind directions. However, as we shall see in the next section, all attempts to detect such changes and establish the existence of the ether (and hence the absolute frame) proved futile!

In conclusion, we see that the second hypothesis in our introduction to this section is false—and we now believe that **the laws of electricity and magnetism are the same in all inertial frames.** It is the simple classical addition laws for velocities that are incorrect and must be modified, as shown in Section 26.8.

## 26.4 THE MICHELSON–MORLEY EXPERIMENT

The most famous experiment designed to detect small changes in the speed of light was first performed in 1881 by Michelson and later repeated under various conditions by Michelson and Edward W. Morley (1838–1923). We state at the outset that the outcome of the experiment contradicted the ether hypothesis.

The experiment was designed to determine the velocity of Earth relative to the hypothetical ether. The experimental tool used was the Michelson interferometer, which was discussed in Section 25.7 and is shown again in Figure 26.4. Arm 2 is aligned along the direction of Earth's motion through space. Earth moving through the ether at speed $v$ is equivalent to the ether flowing past Earth in the opposite direction with speed $v$. This ether wind blowing in the direction opposite the direction of Earth's motion should cause the speed of light measured in Earth's frame to be $c - v$ as the light approaches mirror $M_2$ and $c + v$ after reflection, where $c$ is the speed of light in the ether frame.

The two beams reflected from $M_1$ and $M_2$ recombine, and an interference pattern consisting of alternating dark and bright fringes is formed. The interference pattern was observed while the interferometer was rotated through an angle of 90°. This rotation supposedly would change the speed of the ether wind along the direction of arm 1. The effect of this rotation should have been to cause the fringe pattern to shift slightly but measurably, but measurements failed to show any change in the interference pattern! The Michelson–Morley experiment was repeated at different times of the year when the ether wind was expected to change direction, but the results were always the same: **No fringe shift of the magnitude required was ever observed.**

The negative results of the Michelson–Morley experiment not only contradicted the ether hypothesis but also showed that it was impossible to measure the absolute velocity of Earth with respect to the ether frame. However, as we shall see in the next section, Einstein suggested a postulate in the special theory of relativity that places quite a different interpretation on these negative results. In later years, when more was known about the nature of light, the idea of an ether that permeates all of space was relegated to the ash heap of worn-out concepts. **Light is now understood to be an electromagnetic wave, which requires no medium for its propagation.** As a result, the idea of an ether in which these waves could travel became unnecessary.

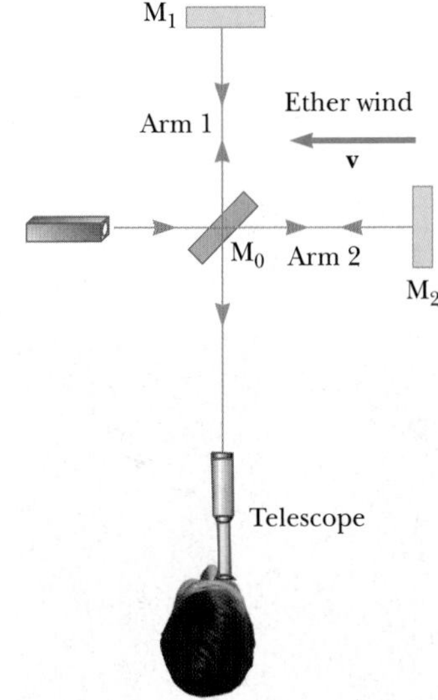


**FIGURE 26.4** According to the ether wind theory, the speed of light should be $c - v$ as the beam approaches mirror $M_2$ and $c + v$ after reflection.

## DETAILS OF THE MICHELSON–MORLEY EXPERIMENT

As we mentioned earlier, the Michelson–Morley experiment was designed to detect the motion of Earth with respect to the ether. Before we examine the details of this historical experiment, it is instructive to first consider a race between two airplanes, as shown in Figure 26.5a. One airplane flies from point $O$ to point $A$ perpendicular to the direction of the wind, and the second airplane flies from point $O$ to point $B$ parallel to the wind. We shall assume that they start at $O$ at the same time, travel the same distance $L$ with the same cruising speed $c$ with respect to the wind, and return to $O$. Which airplane will win the race? In order to answer this question, we calculate the time of flight for both airplanes.

First, consider the airplane that moves along path I parallel to the wind. As it moves to the right, its speed is enhanced by the wind, and its speed with respect to Earth is $c + v$. As it moves to the left on its return journey, it must fly opposite the wind; hence its speed with respect to Earth is $c - v$. Thus, the times of flight to the right and to the left are, respectively,

$$t_R = \frac{L}{c + v} \quad \text{and} \quad t_L = \frac{L}{c - v}$$

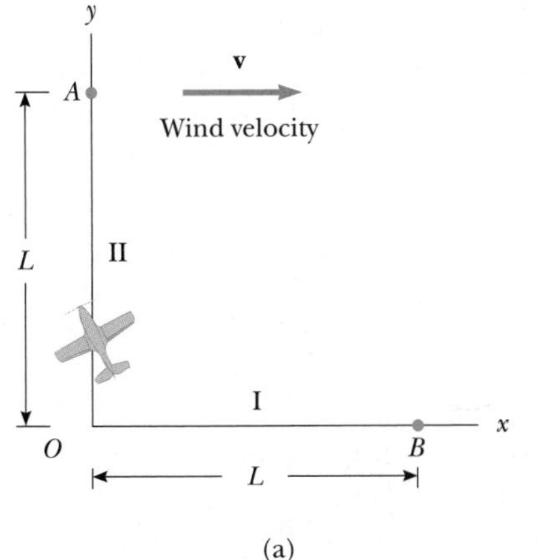


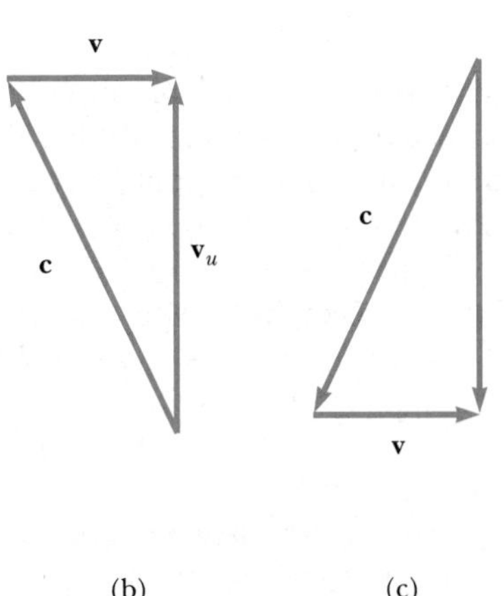


(a) (b) (c)

**FIGURE 26.5** (a) If an airplane wishes to travel from $O$ to $A$ with a wind blowing to the right, it must head into the wind at some angle. (b) Vector diagram for determining the airplane's direction for the trip from $O$ to $A$. (c) Vector diagram for determining its direction for the trip from $A$ to $O$.

and the total time of flight for the airplane moving along path I is

$$t_1 = t_R + t_L = \frac{L}{c+v} + \frac{L}{c-v} = \frac{2Lc}{c^2 - v^2} \quad \textbf{[26.1]}$$

$$= \frac{2L}{c\left(1 - \frac{v^2}{c^2}\right)}$$

**ALBERT A. MICHELSON, GERMAN AMERICAN PHYSICIST (1852–1931)**

Michelson spent much of his life making accurate measurements of the speed of light. In 1907 he was the first American to be awarded the Nobel prize, which he received for his work in optics. His most famous experiment, conducted with Edward Morley in 1887, implied that it was impossible to measure the absolute velocity of Earth with respect to the ether. *(AIP Emilio Segre Visual Archives, Michelson Collection)*

Now consider the airplane flying along path II. If the pilot aims the airplane directly toward point *A*, it will be blown off course by the wind and will not reach its destination. To compensate for the wind, the pilot must point the airplane into the wind at some angle, as shown in Figure 26.5a. This angle must be selected so that the vector sum of **c** and **v** leads to a velocity vector pointed directly toward *A*. The resultant vector diagram is shown in Figure 26.5b, where $\mathbf{v}_u$ is the velocity of the airplane with respect to the ground as it moves from *O* to *A*. From the Pythagorean theorem, the magnitude of the vector $\mathbf{v}_u$ is

$$v_u = \sqrt{c^2 - v^2} = c\sqrt{1 - \frac{v^2}{c^2}}$$

Likewise, on the return trip from *A* to *O*, the pilot must again head into the wind so that the airplane's velocity $\mathbf{v}_d$ with respect to Earth will be directed toward *O*, as shown in Figure 26.5c. From this figure, we see that

$$v_d = \sqrt{c^2 - v^2} = c\sqrt{1 - \frac{v^2}{c^2}}$$

Thus, the total time of flight for the trip along path II is

$$t_2 = \frac{L}{v_u} + \frac{L}{v_d} = \frac{L}{c\sqrt{1 - \frac{v^2}{c^2}}} + \frac{L}{c\sqrt{1 - \frac{v^2}{c^2}}} \quad \textbf{[26.2]}$$

$$= \frac{2L}{c\sqrt{1 - \frac{v^2}{c^2}}}$$

Comparing Equations 26.1 and 26.2, we see that the airplane flying along path II wins the race. The difference in flight times is given by

$$\Delta t = t_1 - t_2 = \frac{2L}{c}\left[\frac{1}{\left(1 - \frac{v^2}{c^2}\right)} - \frac{1}{\sqrt{1 - \frac{v^2}{c^2}}}\right]$$

This expression can be simplified by noting that the ratio $v/c$ of wind speed to plane speed is usually much smaller than 1 and by using the following binomial expansions in $v/c$ after dropping all terms higher than second order:

$$\left(1 - \frac{v^2}{c^2}\right)^{-1} \approx 1 + \frac{v^2}{c^2}$$

and

$$\left(1 - \frac{v^2}{c^2}\right)^{-1/2} \approx 1 + \frac{1}{2}\frac{v^2}{c^2}$$

Therefore, the difference in flight times is

$$\Delta t \approx \frac{Lv^2}{c^3} \quad \text{for} \quad v/c \ll 1 \quad \textbf{[26.3]}$$

The analogy between this airplane race and the Michelson–Morley experiment is shown in Figure 26.6a. Two beams of light travel along two arms of an

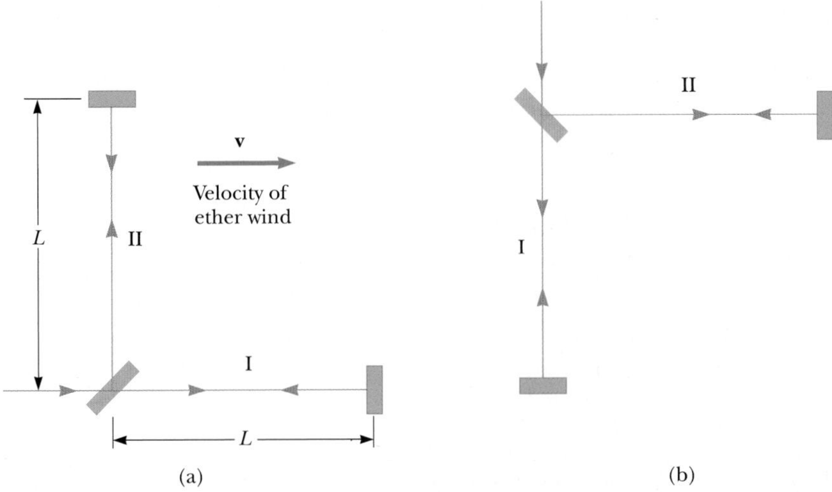


**FIGURE 26.6** (a) Top view of the Michelson–Morley interferometer, where **v** is the velocity of the ether and $L$ is the length of each arm. (b) When the interferometer is rotated by 90°, the role of each arm is reversed.

interferometer. In this case, the "wind" is the ether blowing across Earth from left to right as Earth moves through the ether from right to left. Because the speed of Earth in its orbital path is approximately $3 \times 10^4$ m/s, it is reasonable to use this value for the speed of the ether wind. Also notice that in this case $v/c \approx 1 \times 10^{-4} \ll 1$. The two light beams start out in phase and return to form an interference pattern. Let us assume that the interferometer is adjusted for parallel fringes and that a telescope is focused on one of these fringes. The time difference between the two light beams gives rise to a phase difference between the beams, producing an interference pattern when they combine at the position of the telescope. The difference in the pattern is detected by rotating the interferometer through 90° in a horizontal plane so that the two beams exchange roles (Fig. 26.6b). This results in a net time shift of twice the time difference given by Equation 26.3. Thus, the net time difference is

$$\Delta t_{\text{net}} = 2\,\Delta t = \frac{2Lv^2}{c^3} \qquad [26.4]$$

The corresponding path difference is

$$\Delta d = c\,\Delta t_{\text{net}} = \frac{2Lv^2}{c^2} \qquad [26.5]$$

In the first experiments by Michelson and Morley, each light beam was reflected by the mirrors many times to give an increased effective path length $L$ of about 11 meters. Using this value and taking $v$ to be equal to $3 \times 10^4$ m/s gives a path difference of

$$\Delta d = \frac{2(11\text{ m})(3.0 \times 10^4\text{ m/s})^2}{(3.0 \times 10^8\text{ m/s})^2} = 2.2 \times 10^{-7}\text{ m}$$

This extra travel distance should produce a noticeable shift in the fringe pattern. Specifically, calculations show that if the pattern is viewed while the interferometer is rotated through 90°, a shift of about 0.4 fringes should be observed. The instrument used by Michelson and Morley was capable of detecting a shift in the fringe pattern as small as 0.01 fringes. However, *it detected no shift whatsoever in the fringe pattern.* Since then, the experiment has been repeated many times by different scientists under a wide variety of conditions and no fringe shift has ever been detected. Thus, it was concluded that the motion of Earth with respect to the ether cannot be detected.

Many efforts were made to explain the null results of the Michelson–Morley experiment and to save the ether frame concept and the Galilean addition law for the velocity of light. All proposals resulting from these efforts have been shown to be wrong. No experiment in the history of physics has been the subject of such valiant efforts to explain the absence of an expected result as was the Michelson–Morley experiment. The stage was set for Einstein, who solved the problem in 1905 with his special theory of relativity.

## 26.5 EINSTEIN'S PRINCIPLE OF RELATIVITY

In the previous section we noted the serious contradiction between the Galilean addition law for velocities and the fact that the speed of light is the same for all observers. In 1905 Albert Einstein proposed a theory that resolved this contradiction but at the same time completely altered our notions of space and time. He based his special theory of relativity on two postulates:

Postulates of relativity

1. **The principle of relativity: All the laws of physics are the same in all inertial frames.**
2. **The constancy of the speed of light: The speed of light in a vacuum has the same value, $c = 2.997\ 924\ 58 \times 10^8$ m/s, in all inertial reference frames, regardless of the velocity of the observer or the velocity of the source emitting the light.**

**ALBERT EINSTEIN, GERMAN AMERICAN PHYSICIST (1879–1955)**

Einstein, one of the greatest physicists of all times, was born in Ulm, Germany. In 1905, at the age of 26, he published four scientific papers that revolutionized physics. Two of these papers were concerned with what is now considered his most important contribution, the special theory of relativity. In 1916 Einstein published his work on the general theory of relativity. The most dramatic prediction of this theory is the degree to which light is deflected by a gravitational field. Measurements made by astronomers on bright stars in the vicinity of the eclipsed Sun in 1919 confirmed Einstein's prediction, and as a result Einstein became a world celebrity. Einstein was deeply disturbed by the development of quantum mechanics in the 1920s despite his own role as a scientific revolutionary. In particular, he could never accept the probabilistic view of events in nature that is a central feature of quantum theory. The last few decades of his life were devoted to an unsuccessful search for a unified theory that would combine gravitation and electromagnetism. *(AIP Niels Bohr Library)*

The first postulate asserts that *all* the laws of physics are the same in all reference frames moving with constant velocity relative to each other. This postulate is a sweeping generalization of the principle of Galilean relativity, which refers only to the laws of mechanics. From an experimental point of view, Einstein's principle of relativity means that any kind of experiment—mechanical, thermal, optical, or electrical—performed in a laboratory at rest must give the same result when performed in a laboratory moving at a constant speed past the first one. Hence no preferred inertial reference frame exists, and it is impossible to detect absolute motion.

Postulate 2 was a brilliant theoretical insight on Einstein's part in 1905, and it has since been confirmed experimentally in many ways. Perhaps the most direct demonstration involves measuring the speed of photons emitted by particles traveling at 99.99% of the speed of light. The measured photon speed in this case agrees to five significant figures with the speed of light in empty space.

The null result of the Michelson–Morley experiment can be readily understood within the framework of Einstein's theory. According to his principle of relativity, the premises of the Michelson–Morley experiment were incorrect. In the process of trying to explain the expected results, we stated that when light traveled against the ether wind its speed was $c - v$. However, if the state of motion of the observer or of the source has no influence on the value found for the speed of light, one always measures the value to be $c$. Likewise, the light makes the return trip after reflection from the mirror at a speed of $c$, not at a speed of $c + v$. Thus, the motion of Earth does not influence the fringe pattern observed in the Michelson–Morley experiment and a null result should be expected.

If we accept Einstein's theory of relativity, we must conclude that uniform relative motion is unimportant when measuring the speed of light. At the same time, we must alter our common-sense notions of space and time and be prepared for some rather bizarre consequences.

## 26.6 CONSEQUENCES OF SPECIAL RELATIVITY

Almost everyone who has dabbled even superficially in science is aware of some of the startling predictions that arise because of Einstein's approach to relative motion. As we examine some of the consequences of relativity in this section, we shall find that they conflict with some of our basic notions of space and time. We shall restrict our discussion to the concepts of length, time, and simultaneity, which are quite different in relativistic mechanics from what they are in Newtonian mechanics. For example, in relativistic mechanics, the distance between two points and the time interval between two events depend on the frame of reference in which they are measured. That is, **in relativistic mechanics, there is no such thing as absolute length or absolute time.** Furthermore, **events at different locations that are observed to occur simultaneously in one frame are not observed to be simultaneous in another frame moving uniformly past the first.**

Absolute length and absolute time intervals are meaningless in relativity.

## SIMULTANEITY AND THE RELATIVITY OF TIME

A basic premise of Newtonian mechanics is that a universal time scale exists that is the same for all observers. In fact, Newton wrote, "Absolute, true, and mathematical time, of itself, and from its own nature, flows equably without relation to anything external." Newton and his followers simply took simultaneity for granted. In his special theory of relativity, Einstein abandoned this assumption.

Einstein devised the following thought experiment to illustrate this point. A boxcar moves with uniform velocity, and two lightning bolts strike its ends, as in Figure 26.7a, leaving marks on the boxcar and the ground. The marks on the boxcar are labeled $A'$ and $B'$, and those on the ground are labeled $A$ and $B$. An observer at $O'$ moving with the boxcar is midway between $A'$ and $B'$, and an observer on the ground at $O$ is midway between $A$ and $B$. The events recorded by the observers are the striking of the boxcar by the two lightning bolts.

The light signals recording the instant at which the two bolts struck reach observer $O$ at the same time, as indicated in Figure 26.7b. This observer realizes that the signals have traveled at the same speed over equal distances, and so rightly concludes that the events at $A$ and $B$ occurred simultaneously. Now consider the same events as viewed by observer $O'$. By the time the signals have reached observer $O$, observer $O'$ has moved, as indicated in Figure 26.7b. Thus, the signal from $B'$ has already swept past $O'$, but the signal from $A'$ has not yet reached $O'$. In other words, $O'$ sees the signal from $B'$ before seeing the signal from $A'$. According to Einstein, *the two observers must find that light travels at the same speed.* Therefore, observer $O'$ concludes that the lightning struck the front of the boxcar before it struck the back.

### Tip 26.1 WHO'S RIGHT?

Which person is correct concerning the simultaneity of the two events? Both are correct, because the principle of relativity states that no inertial frame of reference is preferred. Although the two observers may reach different conclusions, both are correct in their own reference frame. Any uniformly moving frame of reference can be used to describe events and do physics.

This thought experiment clearly demonstrates that the two events that appear to be simultaneous to observer $O$ do not appear to be simultaneous to observer $O'$. In other words,

**two events that are simultaneous in one reference frame are in general not simultaneous in a second frame moving relative to the first. That is, simultaneity is not an absolute concept but rather one that depends on the state of motion of the observer.**

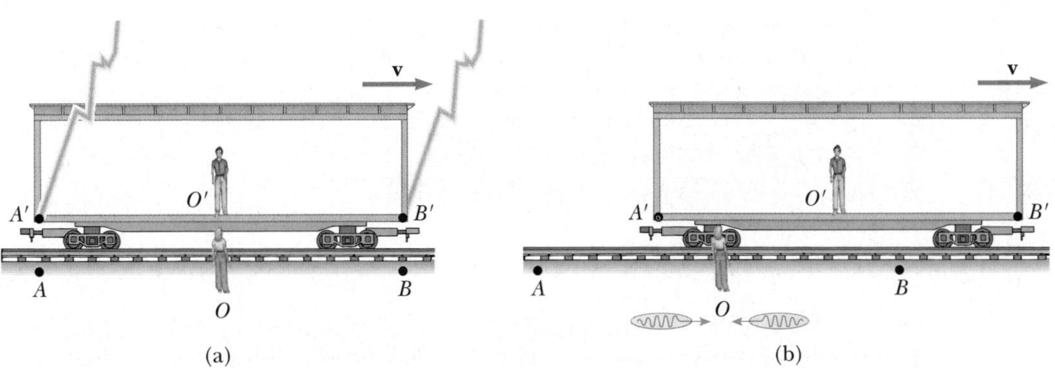


**FIGURE 26.7** Two lightning bolts strike the ends of a moving boxcar. (a) The events appear to be simultaneous to the stationary observer at $O$, who is midway between $A$ and $B$. (b) The events do not appear to be simultaneous to the observer at $O'$, who claims that the front of the train is struck *before* the rear.

At this point, you might wonder which observer is right concerning the two events. The answer is that *both* are correct because the principle of relativity states that **there is no preferred inertial frame of reference.** Although the two observers reach different conclusions, both are correct in their own reference frames because the concept of simultaneity is not absolute. In fact, this is the central point of relativity. Any inertial frame of reference can be used to describe events and do physics.

## TIME DILATION

We can illustrate the fact that observers in different inertial frames may measure different time intervals between a pair of events by considering a vehicle moving to the right with a speed $v$ as in Figure 26.8a. A mirror is fixed to the ceiling of the vehicle, and an observer $O'$ at rest in this system holds a laser a distance $d$ below the mirror. At some instant, the laser emits a pulse of light directed toward the mirror (event 1), and at some later time after reflecting from the mirror, the pulse arrives back at the laser (event 2). Observer $O'$ carries a clock and uses it to measure the time interval, $\Delta t_p$, between these two events which she views to occur at the same place. (The subscript $p$ stands for *proper,* as we shall see in a moment.) Because the light pulse has a speed $c$, the time it takes it to travel from point $A$ to the mirror and back to point $A$ is

$$\Delta t_p = \frac{\text{distance traveled}}{\text{speed}} = \frac{2d}{c} \qquad [26.6]$$

The time interval $\Delta t_p$ measured by $O'$ requires only a single clock located at the same place as the laser in this frame.

Now consider the same set of events as viewed by $O$ in a second frame, as shown in Fig. 26.8b. According to this observer, the mirror and laser are moving to the right with a speed $v$, and as a result the sequence of events appears different. By the time the light from the laser reaches the mirror, the mirror has moved to the right a distance $v\,\Delta t/2$, where $\Delta t$ is the time it takes the light pulse to travel from point $A$ to the mirror and back to point $A$ as measured by $O$. In other words, $O$ concludes that, because of the motion of the vehicle, if the light is to hit the mirror, it must leave the laser at an angle with respect to the vertical direction. Comparing Figures 26.8a and 26.8b, we see that the light must travel farther in (b) than in (a). (Note that neither observer "knows" that he or she is moving; each is at rest in his or her own inertial frame.)

According to the second postulate of the special theory of relativity, both observers must measure $c$ for the speed of light. Because the light travels farther in

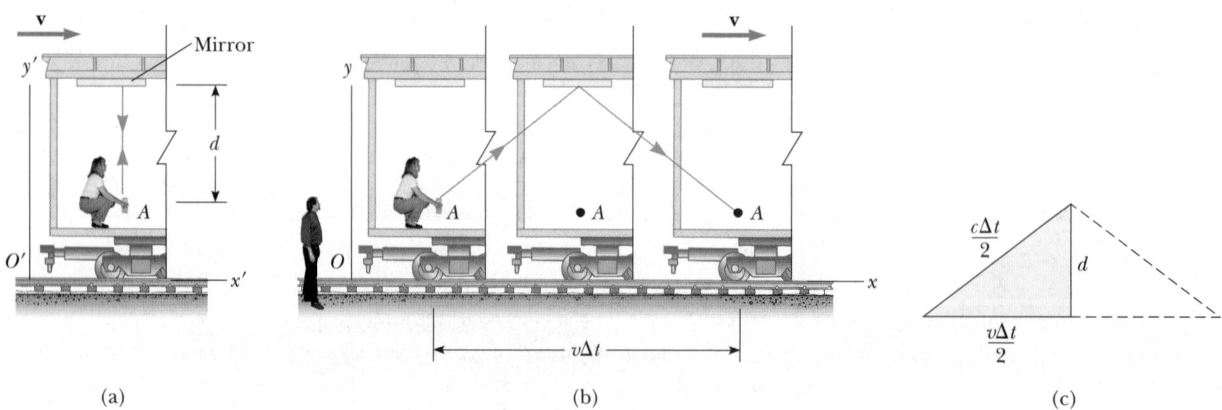


**FIGURE 26.8** (a) A mirror is fixed to a moving vehicle, and a light pulse leaves $O'$ at rest in the vehicle. (b) Relative to a stationary observer on Earth, the mirror and $O'$ move with a speed $v$. Note that the distance the pulse travels is greater than $2d$ as measured by the stationary observer. (c) The right triangle for calculating the relationship between $\Delta t$ and $\Delta t_p$.

the frame of $O$, it follows that the time interval $\Delta t$ measured by $O$ is longer than the time interval $\Delta t_p$ measured by $O'$. To obtain a relationship between these two time intervals, it is convenient to use the right triangle shown in Figure 26.8c. The Pythagorean theorem gives

$$\left(\frac{c\,\Delta t}{2}\right)^2 = \left(\frac{v\,\Delta t}{2}\right)^2 + d^2$$

Solving for $\Delta t$ gives

$$\Delta t = \frac{2d}{\sqrt{c^2 - v^2}} = \frac{2d}{c\sqrt{1 - v^2/c^2}}$$

Because $\Delta t_p = 2d/c$, we can express this result as

$$\Delta t = \frac{\Delta t_p}{\sqrt{1 - v^2/c^2}} = \gamma\,\Delta t_p \qquad [26.7]$$

Time dilation

where

$$\gamma = \frac{1}{\sqrt{1 - v^2/c^2}} \qquad [26.8]$$

Because $\gamma$ is always greater than unity, Equation 26.7 says that **the time interval $\Delta t$ between two events measured by an observer moving with respect to a clock[3] is longer than the time interval $\Delta t_p$ between the same two events measured by an observer at rest with respect to the clock.** Thus, $\Delta t > \Delta t_p$ and the proper time interval is expanded or dilated by the factor $\gamma$. Hence this effort is known as **time dilation.**

For example, suppose the observer who is at rest with respect to the clock measures the time required for the light flash to leave the laser and return. Let us assume that the measured time interval $\Delta t_p$ in this frame of reference is 1 second. (This would require a very tall vehicle.) Now let us find the time interval as measured by observer $O$ moving with respect to the same clock. If observer $O$ is traveling at half the speed of light ($v = 0.500c$), then $\gamma = 1.15$, and according to Equation 26.7 $\Delta t = \gamma\,\Delta t_p = 1.15(1.00\text{ s}) = 1.15\text{ s}$. Thus, when observer $O'$ claims that 1.00 s has passed, observer $O$ claims that 1.15 s has passed. Observer $O$ considers the clock of $O'$ to be reading too low a value for the elapsed time between events and says that the clock of $O'$ is "runn ing slow." From this we may conclude that

**A clock moving past an observer at speed $v$ runs more slowly than an identical clock at rest with respect to the observer by a factor of $\gamma^{-1}$.**

A clock in motion runs more slowly than an identical stationary clock.

The time interval $\Delta t_p$ in Equations 26.6 and 26.7 is called the **proper time.** In general, **proper time is the time interval between two events as measured by an observer who sees two events occur at the same position.**

Although you may have realized it by now, it is important to spell out that relativity is a scientific democracy—the view of $O'$ that $O$ is really the one moving with speed $v$ to the left and that $O$'s clock is running more slowly is just as valid. The principle of relativity requires that the views of two observers in uniform relative motion must be equally valid and capable of being checked experimentally.

We have seen that moving clocks run slow by a factor of $\gamma^{-1}$. This is true for ordinary mechanical clocks as well as for the light clock just described. In fact, we can generalize these results by stating that all physical processes, including chemical and biological ones, slow down relative to a clock when those processes occur

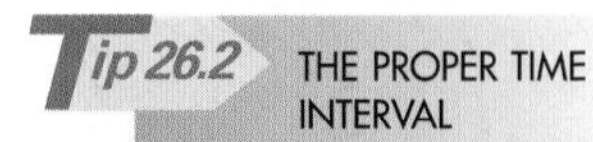

THE PROPER TIME INTERVAL

You must be able to correctly identify the observer who measures the proper time interval. The proper time interval between two events is the time interval measured by an observer for whom the two events take place at the same position.

[3] Actually Figure 26.8 shows the clock moving and not the observer, but this is equivalent to the observer $O$ moving to the left with velocity v with respect to the clock.

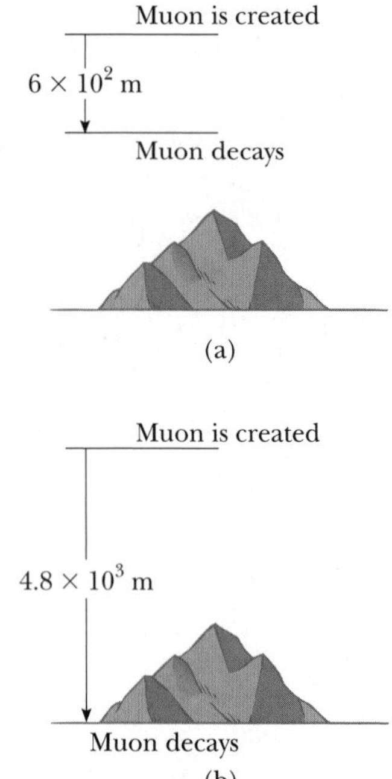


**FIGURE 26.9** (a) The muons travel only about 600 m as measured in their reference frame, in which their lifetime is about 2.2 $\mu$s. (b) Because of time dilation, the muons' lifetime is longer as measured by the observer on Earth. Muons traveling with a speed of $0.99c$ travel a distance of about 4 800 m as measured by an observer on Earth.

in a frame moving with respect to the clock. For example, the heartbeat of an astronaut moving through space would keep time with a clock inside the spaceship. Both the astronaut's clock and heartbeat would be slowed down relative to a clock back on Earth (although the astronaut would have no sensation of life slowing down in the spaceship).

Time dilation is a very real phenomenon that has been verified by various experiments involving the ticking of natural clocks. An interesting example of time dilation involves the observation of *muons,* unstable elementary particles that have a charge equal to that of the electron and a mass 207 times that of the electron. Muons can be produced by the collision of cosmic radiation with atoms high in the atmosphere. These particles have a lifetime of $\Delta t_p = 2.2\ \mu$s when measured in a reference frame at rest with respect to them. If we take 2.2 $\mu$s as the average lifetime of a muon and assume that their speed is close to the speed of light, we find that these particles can travel only about 600 m before they decay (Fig. 26.9a). Hence, they could never reach Earth from the upper atmosphere where they are produced. However, experiments show that a large number of muons *do* reach Earth, and the phenomenon of time dilation explains how. Relative to an observer on Earth, the muons have a lifetime equal to $\gamma\, \Delta t_p$, where $\Delta t_p = 2.2\ \mu$s is the lifetime in a frame of reference traveling with the muons (the "proper" lifetime). For example, for $v = 0.99c$, $\gamma \approx 7.1$ and $\gamma\, \Delta t_p \approx 16\ \mu$s. Hence, the average distance traveled as measured by an observer on Earth is $\gamma\, v\, \Delta t_p \approx 4\,800$ m, as indicated in Figure 26.9b. Hence, muons can reach Earth's surface.

In 1976 experiments with muons were conducted at the laboratory of the European Council for Nuclear Research (CERN) in Geneva. Muons were injected into a large storage ring, reaching speeds of about $0.9994c$. Electrons produced by the decaying muons were detected by counters around the ring, enabling scientists to measure the decay rate and hence the lifetime of the muons. The lifetime of the moving muons was measured to be about 30 times as long as that of stationary muons to within two parts in a thousand, in agreement with the prediction of relativity.

| **Quick Quiz 26.1** | Imagine that you are an astronaut who is being paid according to the time spent traveling in space as measured by a clock on Earth. You take a long voyage traveling at a speed near that of light. Upon your return to Earth, your paycheck will be (a) smaller than if you had remained on Earth, (b) larger than if you had remained on Earth, or (c) the same as if you had remained on Earth. |
|---|---|

## APPLYING PHYSICS 26.2

Suppose a student explains time dilation with the following argument: "If you start running at $0.99c$ away from a clock at 12:00, you would not see the time change, because the light from the clock representing 12:01 would never reach you." What is the flaw in this argument?

**Explanation** This argument infers that the velocity of light relative to the runner is approximately zero—"the light . . . would never reach you." This is a Galilean relativity point of view, in which the relative velocity is a simple subtraction of running velocity from the light velocity. From the point of view of special relativity, one of the fundamental postulates is that the speed of light is the same for all observers, including one running away from the light source at the speed of light. Thus, the light from 12:01 will move toward the runner at the speed of light. Section 26.8 gives the remarkable relativistic formulas for velocity addition that always give $c$ for the speed of light.

## Example 26.1 What Is the Period of the Pendulum?

The period of a pendulum is measured to be 3.0 s in the inertial frame of the pendulum. What is the period when measured by an observer moving at a speed of $0.95c$ with respect to the pendulum?

**Reasoning and Solution** In this case, the proper time is 3.0 s. We can use Equation 26.7 to calculate the period measured by the moving observer:

$$T = \gamma T_p = \frac{1}{\sqrt{1 - \frac{(0.95c)^2}{c^2}}} T_p = (3.2)(3.0 \text{ s}) = 9.6 \text{ s}$$

That is, the observer moving at a speed of $0.95c$ observes that the pendulum slows down.

### THE TWIN PARADOX

An intriguing consequence of time dilation is the so-called twin paradox (Fig. 26.10). Consider an experiment involving a set of twins named Speedo and Goslo. When they are 20 years old, Speedo, the more adventuresome of the two, sets out on an epic journey to Planet X, located 20 lightyears from Earth. Furthermore, his spaceship is capable of reaching a speed of $0.95c$ relative to the inertial frame of his twin brother back home. After reaching Planet X, Speedo becomes homesick and immediately returns to Earth at the same speed, $0.95c$. Upon his return, Speedo is shocked to discover that Goslo has aged $2D/v = 2(20y)/(0.95y/y)$ 42 years and is now 62 years old. Speedo, on the other hand, has aged only 13 years.

The space traveler ages more slowly than his twin who remains on Earth.

At this point, it is fair to raise the following question—which twin is the traveler and which is really younger as a result of this experiment? From Goslo's frame of reference, he was at rest while his brother traveled at a high speed. From Speedo's perspective, it is he who was at rest while Goslo was on the high-speed space journey. According to Speedo, he himself remained stationary while Goslo and Earth raced away from him on a 6.5-year journey and then headed back for another 6.5 years. This leads to an apparent contradiction. Which twin has aged the most?

To resolve this apparent paradox, recall that the special theory of relativity deals with inertial frames of reference moving relative to each other at uniform speed. However, the trip in our current problem is not symmetrical. Speedo, the space traveler, must experience a series of accelerations during his journey. As a re-

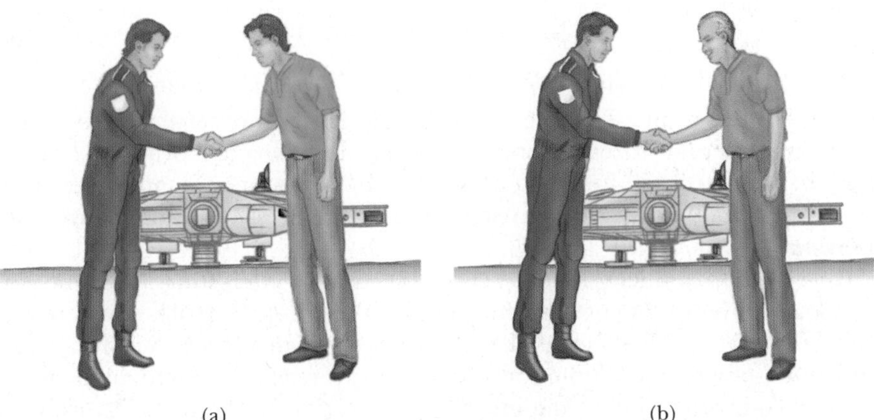

(a) (b)

**FIGURE 26.10** (a) As the twins depart, they are the same age. (b) When Speedo returns from his journey to Planet X, he is younger than his twin Goslo who remained on Earth.

**Tip 26.3 THE PROPER LENGTH**

You must be able to correctly identify the observer who measures the proper length. The proper length between two points in space is the length measured by an observer at rest with respect to the length. Very often, the proper time interval and the proper length are not measured by the same observer.

sult, his speed is not always uniform and consequently he is not in an inertial frame at all times. He cannot be regarded as always being at rest while Goslo is in uniform motion because to do so would be an incorrect application of the special theory of relativity. Therefore there is no paradox. During each passing year noted by Goslo, slightly less than four months elapsed for Speedo.

The conclusion that Speedo is in a noninertial frame is inescapable. Each twin observes the other as accelerating, but it is Speedo who actually undergoes dynamical acceleration due to the real forces acting on him. The time required to accelerate and decelerate Speedo's spaceship may be made very small by using large rockets, so Speedo can claim that he spends most of his time traveling to Planet X at $0.95c$ in an inertial frame. However, Speedo must slow down, reverse his motion, and return to Earth in an altogether different inertial frame. At the very best, Speedo is in two different inertial frames during his journey. Only Goslo, who is in a single inertial frame, can apply the simple time-dilation formula to Speedo's trip. Thus, Goslo finds that instead of aging 42 years, Speedo ages only $(1 - v^2/c^2)^{1/2}(42 \text{ years}) = 13$ years. Of this 13 years, Speedo spends 6.5 years traveling to Planet X and 6.5 years returning, for a total travel time of 13 years, in agreement with our earlier statement.

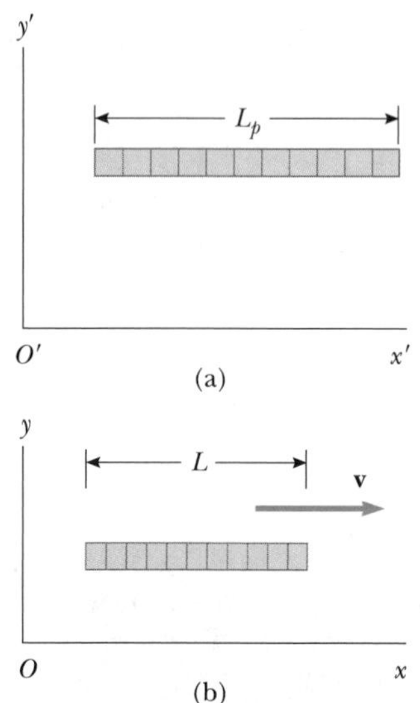


**FIGURE 26.11** A meter stick moves to the right with a speed $v$. (a) The meter stick as viewed by an observer at rest with respect to the meter stick. (b) The meter stick as seen by an observer moving with a speed $v$ with respect to the meter stick. The moving meter stick is always measured to be *shorter* than in its own rest frame by a factor of $\sqrt{1 - v^2/c^2}$.

## LENGTH CONTRACTION

The measured distance between two points depends on the frame of reference of the observer. The **proper length,** $L_p$, of an object is **the length of the object measured by someone at rest relative to the object.** The length of an object measured in a reference frame that is moving with respect to the object is always less than the proper length. This effect is known as **length contraction.**

To understand length contraction quantitatively, consider a spaceship traveling with a speed $v$ from one star to another. There are two observers: one on Earth and the other in the spaceship. The observer at rest on Earth (and also assumed to be at rest with respect to the two stars) measures the distance between the stars to be $L_p$. According to this observer, the time it takes the spaceship to complete the voyage is $\Delta t = L_p/v$. Because of time dilation, the space traveler using the spaceship clock measures a smaller time of travel: $\Delta t_p = \Delta t/\gamma$. The space traveler claims to be at rest and sees the destination star moving toward the spaceship with speed $v$. Because the spaceship reaches the star in the time $\Delta t_p$, the traveler concludes that the distance $L$ between the stars is shorter than $L_p$. The distance measured by the space traveler is

$$L = v\,\Delta t_p = v\frac{\Delta t}{\gamma}$$

Because $L_p = v\,\Delta t$, we see that

Length contraction

$$L = \frac{L_p}{\gamma} = L_p\sqrt{1 - v^2/c^2} \qquad [26.9]$$

According to this result, illustrated in Figure 26.11, if an observer at rest with respect to an object measures its length to be $L_p$, an observer moving at a speed $v$ relative to the object will find it to be shorter than its proper length by the factor $\sqrt{1 - v^2/c^2}$. Note that **length contraction takes place only along the direction of motion.**

Time dilation and length contraction effects have interesting applications for future space travel to distant stars. In order for the star to be reached in a fraction of human lifetime, the trip must be taken at very high speeds. According to an Earth-bound observer, the time for a spacecraft to reach the destination star will be dilated compared to the time interval measured by the travelers. Thus, as was discussed in the treatment of the twin paradox, the travelers will be younger than their twins when they return to Earth. Thus, by the time the travelers reach the star, they have aged by some number of years, while their partners back on Earth will have aged a larger number of years, the exact ratio depending on the speed of the spacecraft. At a spacecraft speed of $0.94c$, this ratio is about 3:1.

**Quick Quiz 26.2** You are packing for a trip to another star, to which you will be traveling at $0.99c$. Should you buy smaller sizes of your clothing, because you will be skinnier on the trip? Can you sleep in a smaller cabin than usual, because you will be shorter when you lie down? Explain your answers.

**Quick Quiz 26.3** You are observing a rocket moving away from you. Compared to its length when it was at rest on the ground, you will measure its length to be (a) shorter, (b) longer, or (c) the same. Now you see a clock through a window on the rocket. Compared to the passage of time measured by the watch on your wrist, you observe that the passage of time on the rocket's clock is (d) faster, (e) slower, or (f) the same. Answer the same questions if the rocket turns around and comes toward you.

### Example 26.2 The Contraction of a Spaceship

A spaceship is measured to be 120 m long while it is at rest with respect to an observer. If this spaceship now flies past the observer with a speed of $0.99c$, what length will the observer measure for the spaceship?

**Solution** From Equation 26.9, the length measured by the observer is

$$L = L_p\sqrt{1 - v^2/c^2} = (120\text{ m})\sqrt{1 - \frac{(0.99c)^2}{c^2}} = 17\text{ m}$$

**EXERCISE** If the ship moves past the observer with a speed of $0.01000c$, what length will the observer measure?

**ANSWER** 119.994 m

### Example 26.3 How High Is the Spaceship?

An observer on Earth sees a spaceship at an altitude of 435 m moving downward toward Earth with a speed of $0.970c$. What is the distance from the spaceship to the ground as measured by an observer in the spaceship?

**Solution** The moving observer in the spaceship finds the altitude to be

$$L = L_p\sqrt{1 - v^2/c^2} = (435\text{ m})\sqrt{1 - \frac{(0.970c)^2}{c^2}} = 106\text{ m}$$

### Example 26.4 The Triangular Spaceship

A spaceship in the form of a triangle flies by an observer with a speed of $0.95c$. When the spaceship is at rest (Fig. 26.12a), the distances $x$ and $y$ are found to be 52 m and 25 m, respectively. What is the shape of the spaceship as seen by an observer at rest when the spaceship is in motion along the direction shown in Figure 26.12b?

y

x

(a)

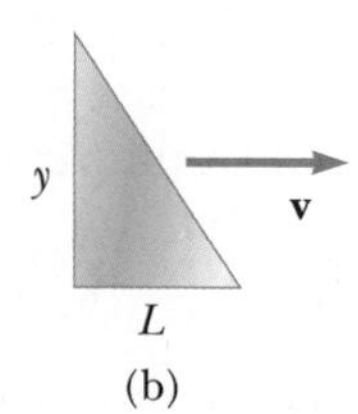


(b)

**FIGURE 26.12** (Example 26.4) (a) When the spaceship is at rest, its shape is as shown. (b) The spaceship appears to look like this when it moves to the right with a speed $v$. Note that only its $x$ dimension is contracted in this case.

**Solution** The observer sees the horizontal length of the spaceship to be contracted to a length of

$$L = L_p\sqrt{1 - v^2/c^2} = (52 \text{ m})\sqrt{1 - \frac{(0.95c)^2}{c^2}} = 16 \text{ m}$$

The 25-m vertical height is unchanged because it is perpendicular to the direction of relative motion between the observer and the spaceship. Figure 26.12b represents the shape of the spaceship as seen by the observer at rest.

## 26.7 RELATIVISTIC MOMENTUM

In order to describe properly the motion of particles within the framework of special relativity, we must generalize Newton's laws of motion and the definitions of momentum and energy. As we shall see, these generalized definitions reduce to the classical (nonrelativistic) definitions when $v$ is much less than $c$.

First, recall that conservation of momentum states that when two objects collide, the total momentum of the system remains constant, assuming that the objects are isolated (that is, they interact only with each other). However, analyzing such collisions from rapidly moving inertial frames, it is found that momentum is not conserved if the classical definition of momentum, $p = mv$, is used. In order to have momentum conservation in all inertial frames, even those moving at an appreciable fraction of $c$, the definition of momentum must be modified as follows:

Momentum

$$p \equiv \frac{mv}{\sqrt{1 - v^2/c^2}} = \gamma mv \qquad [26.10]$$

where $v$ is the speed of the particle and $m$ is its mass as measured by an observer at rest with respect to the mass. Note that when $v$ is much less than $c$, the denominator of Equation 26.10 approaches unity, so that $p$ approaches $mv$. Therefore, the relativistic equation for momentum reduces to the classical expression when $v$ is small compared with $c$.

**Webnote 26.1**

The Earth's upper atmosphere is a very violent setting of high-energy cosmic rays colliding with atoms. See what takes place at *http://www.sciam.com/specialissues/0398cosmos/0398croninbox3.html*

**Webnote 26.2**

AMANDA is a detector being constructed at the South Pole whose purpose is to observe high-energy neutrinos from astrophysical point sources. Visit the project home page and take a look at photos of the science measurement site at *http://amanda.berkeley.edu*

### Example 26.5 The Relativistic Momentum of an Electron

An electron, which has a mass of $9.11 \times 10^{-31}$ kg, moves with a speed of $0.75c$. Find its relativistic momentum and compare this value to the momentum calculated from the classical expression.

**Solution** From Equation 26.10, with $v = 0.75c$, we have

$$p = \frac{mv}{\sqrt{1 - v^2/c^2}}$$

$$= \frac{(9.11 \times 10^{-31} \text{ kg})(0.75 \times 3 \times 10^8 \text{ m/s})}{\sqrt{1 - (0.75c)^2/c^2}} = 3.1 \times 10^{-22} \text{ kg} \cdot \text{m/s}$$

The classical expression gives

$$\text{Momentum} = mv = 2.1 \times 10^{-22}\ \text{kg} \cdot \text{m/s}$$

The (correct) relativistic result is 50% greater than the classical result!

## 26.8 RELATIVISTIC ADDITION OF VELOCITIES

Imagine a motorcycle rider moving with a speed of $0.80c$ past a stationary observer, as shown in Figure 26.13. If the rider tosses a ball in the forward direction with a speed of $0.70c$ relative to himself, what is the speed of the ball as seen by the stationary observer at the side of the road? Common sense and the ideas of Newtonian relativity say that the speed should be the sum of the two speeds, or $1.50c$. This answer must be incorrect because it contradicts the assertion that no material object can travel faster than the speed of light.

Einstein resolved this dilemma by deriving an equation for the relativistic addition of velocities. For one-dimensional motion, this equation is

$$v_{ab} = \frac{v_{ad} + v_{db}}{1 + \dfrac{v_{ad}v_{db}}{c^2}} \qquad [26.11]$$

Velocity addition

where $v_{ad}$ is the speed of some object a with respect to a moving frame d, and $v_{db}$ is the speed of the moving frame d with respect to frame b. The left side of this equation and the numerator on the right are like the equations of Galilean relativity discussed in Chapter 3, and the evaluation of subscripts is applied in the same fashion as discussed in Section 3.6. The denominator of Equation 26.11 is a correction to Galilean relativity based on length contraction and time dilation. Let us apply this equation to the case of the speedy motorcycle rider and the stationary observer.

We are given

$v_{bm}$ = the velocity of the ball with respect to the motorcycle = $0.70c$

$v_{mo}$ = the velocity of the motorcycle with respect to the stationary observer = $0.80c$,

and we want to find

$v_{bo}$ = the velocity of the ball with respect to the stationary observer.

**FIGURE 26.13** A motorcycle moves past a stationary observer with a speed of $0.80c$; the motorcyclist throws a ball in the direction of motion with a speed of $0.70c$ relative to himself.

The speed of light is the speed limit of the Universe.

Thus,

$$v_{bo} = \frac{v_{bm} + v_{mo}}{1 + \frac{v_{bm}v_{mo}}{c^2}} = \frac{0.70c + 0.80c}{1 + \frac{(0.70c)(0.80c)}{c^2}} = 0.96c$$

### Example 26.6 Measuring the Speed of a Light Beam

Suppose that the motorcyclist moving with a speed of $0.80c$ turns on a beam of light that moves away from the motorcycle with a speed of $c$ in the same direction as the moving motorcycle. What speed would the stationary observer measure for the beam of light?

**Solution** In this case, we have

$v_{lm}$ = the velocity of the light with respect to the motorcycle = $c$

$v_{mo}$ = the velocity of the motorcycle with respect to the stationary observer = $0.80c$

and we want

$v_{lo}$ = the velocity of the light with respect to the stationary observer

Thus,

$$v_{lo} = \frac{v_{lm} + v_{mo}}{1 + \frac{v_{lm}v_{mo}}{c^2}} = \frac{c + 0.80c}{1 + \frac{(c)(0.80c)}{c^2}} = c$$

This is consistent with the statement made earlier that **all observers measure the speed of light to be *c* regardless of the motion of the source of light.**

## 26.9 Relativistic Energy and the Equivalence of Mass and Energy

We have seen that the definition of momentum required generalization to make it compatible with the principle of relativity. Likewise, the definition of kinetic energy requires modification in relativistic mechanics. Einstein found that the correct expression for the **kinetic energy** of an object is

Kinetic energy ▶

$$KE = \gamma mc^2 - mc^2 \qquad [26.12]$$

The constant term $mc^2$ in Equation 26.12, which is independent of the speed of the object, is called the **rest energy,** $E_R$, of the object.

Rest energy ▶

$$E_R = mc^2 \qquad [26.13]$$

The term $\gamma mc^2$ in Equation 26.12 depends on the object's speed and is the sum of the kinetic and rest energies. We define $\gamma mc^2$ to be the **total energy,** $E$; that is,

$$\text{Total energy} = \text{kinetic energy} + \text{rest energy}$$

or using Equation 26.12, gives

$$E = KE + mc^2 = \gamma mc^2 \qquad [26.14]$$

Because $\gamma = (1 - v^2/c^2)^{-1/2}$, we can also express $E$ as

Total energy ▶

$$E = \frac{mc^2}{\sqrt{1 - v^2/c^2}} \qquad [26.15]$$

This is Einstein's famous mass-energy equivalence equation.[4] The relation $E = \gamma mc^2 = KE + mc^2$ shows the amazing result that **a particle has energy by virtue of its mass alone; that is, a stationary particle with zero kinetic energy has an energy proportional to its mass.** Furthermore, this result shows that a small mass corresponds to an enormous amount of energy because the proportionality constant between mass and energy is a large quantity—$c^2 = 9 \times 10^{16}\ m^2/s^2$. The equation $E_R = mc^2$, as Einstein first suggested, indicates that the mass of a particle may be completely convertible to energy, and the reverse—pure energy, say electromagnetic energy, may be converted to particles—is also true. This is indeed the case as has been shown in the laboratory many times. For example, the coming together of a slow-moving electron and its antiparticle, the positron, a particle with the same mass $m_e$ as the electron but opposite charge, results in the disappearance of both particles and the appearance of a burst of electromagnetic energy in the amount $2m_ec^2$. The reverse process is also fairly easily observed in the lab: A high-energy pulse of electromagnetic energy, a gamma ray, disappears near an atom and an electron-positron pair is created with nearly 100% conversion of the gamma ray's energy into mass. Such a pair-production process is shown in the bubble-chamber photo of Figure 26.14. We shall discuss pair production and annihilation in more detail in Section 26.10.

**FIGURE 26.14** Bubble-chamber photograph of electron (green) and positron (red) tracks produced by energetic gamma rays. The highly curved tracks at the top are due to the electron and positron in an electron-positron pair bending in opposite directions in the magnetic field. *(Lawrence Berkeley Laboratory/Science Photo Library/Photo Researchers, Inc.)*

On a larger scale, nuclear power plants produce energy by the fission of uranium, which involves the conversion of a small amount of the mass of the uranium into energy. The Sun, too, converts mass into energy and continually loses mass in pouring out a tremendous amount of electromagnetic energy in all directions.

It is extremely interesting that while we have been talking about the interconversion of mass and energy for particles, the expression $E = mc^2$ is universal and applies to all objects, processes, and systems: A hot object has slightly more mass and is slightly more difficult to accelerate than an identical cold object because it has more thermal energy; a stretched spring has more elastic potential energy and more mass than an identical unstretched spring. A key point, however, is that these mass changes are often far too small to measure. Our best bet for measuring mass changes is in nuclear transformations, where a measurable fraction of the mass is converted to energy.

## ENERGY AND RELATIVISTIC MOMENTUM

In many situations, the momentum or energy of a particle is measured rather than its speed. It is therefore useful to have an expression relating the total energy $E$ to the relativistic momentum $p$. This is accomplished by using the expressions $E = \gamma mc^2$ and $p = \gamma mv$. By squaring these equations and subtracting, we can eliminate $v$. The result, after some algebra, is

$$E^2 = p^2c^2 + (mc^2)^2 \qquad \textbf{[26.16]}$$

When the particle is at rest, $p = 0$ and so $E = E_R = mc^2$. That is, the total energy equals the rest energy. For the case of particles that have zero mass, such as photons (massless, chargeless particles of light), we set $m = 0$ in Equation 26.16, and we see that

$$E = pc \qquad \textbf{[26.17]}$$

This equation is an exact expression relating energy and momentum for photons, which always travel at the speed of light.

When dealing with subatomic particles, it is convenient to express their energy in electron volts (eV), because the particles are given energy by acceleration

[4] Although this doesn't look exactly like the famous equation $E = mc^2$, it used to be common to write $m = \gamma m_0$ (Einstein himself wrote it that way), where $m$ is the effective mass of an object moving at speed $v$ and $m_0$ is the mass of that object measured by an observer at rest with respect to the object. Then our $E = \gamma mc^2$ becomes the familiar $E = mc^2$. It is currently unfashionable to use $m = \gamma m_0$.

through an electrostatic potential difference. The conversion factor is

$$1 \text{ eV} = 1.60 \times 10^{-19} \text{ J}$$

For example, the mass of an electron is $9.11 \times 10^{-31}$ kg. Hence, the rest energy of the electron is

$$m_e c^2 = (9.11 \times 10^{-31} \text{ kg})(3.00 \times 10^8 \text{ m/s})^2 = 8.20 \times 10^{-14} \text{ J}$$

Converting this to eV, we have

$$m_e c^2 = (8.20 \times 10^{-14} \text{ J})(1 \text{ eV}/1.60 \times 10^{-19} \text{ J}) = 0.511 \text{ MeV}$$

Because we frequently use the expression $E = \gamma mc^2$ in nuclear physics and $m$ is usually in atomic mass units, u, it is useful to have the conversion factor $1 \text{ u} = 939.494 \text{ MeV}/c^2$. For example, this makes it easy to find the rest energy in MeV of a nucleus with mass of 235.043 924 u:

$$E_R = mc^2 = (235.043\,924 \text{ u})(931.494 \text{ MeV/u}c^2)(c^2) = 2.189\,42 \times 10^5 \text{ MeV}$$

**Quick Quiz 26.4**

A photon is reflected from a mirror. **True or false:** (a) Because a photon has a zero mass, it does not exert a force on the mirror. (b) Although the photon has energy, it cannot transfer any energy to the surface because it has zero mass. (c) The photon carries momentum, and when it reflects off the mirror, it undergoes a change in momentum and exerts a force on the mirror. (d) Although the photon carries momentum, its change in momentum is zero when it reflects from the mirror, so it cannot exert a force on the mirror.

### Example 26.7 The Energy Contained in a Baseball

If a 0.50-kg baseball could be converted completely to energy of forms other than mass, how much energy of other forms would be released?

**Solution** The energy equivalent of the baseball is found from Equation 26.12 (with $KE = 0$):

$$E_R = mc^2 = (0.50 \text{ kg})(3.0 \times 10^8 \text{ m/s})^2 = 4.5 \times 10^{16} \text{ J}$$

This is enough energy to keep a 100-W lightbulb burning for approximately 10 million years. However, it is generally impossible to achieve complete conversion from mass to energy of other forms. For example, mass is converted to energy in nuclear power plants, but only a small fraction of the mass actually undergoes conversion.

### Example 26.8 The Energy of a Speedy Electron

An electron moves with a speed of $v = 0.850c$. Find its total energy and kinetic energy in electron volts.

**Solution** The fact that the rest energy of an electron is 0.511 MeV, along with Equation 26.15, gives

$$E = \frac{m_e c^2}{\sqrt{1 - v^2/c^2}} = \frac{0.511 \text{ MeV}}{\sqrt{1 - \frac{(0.850c)^2}{c^2}}}$$

$$= 1.90(0.511 \text{ MeV}) = 0.970 \text{ MeV}$$

The kinetic energy is obtained by subtracting the rest energy from the total energy:

$$KE = E - m_e c^2 = 0.970 \text{ MeV} - 0.511 \text{ MeV} = 0.459 \text{ MeV}$$

## Example 26.9 The Conversion of Mass to Kinetic Energy in Uranium Fission

The fission or splitting of uranium was discovered in 1938 by Otto Hahn and is a process in which about 0.1% of the mass is converted to kinetic energy of the fission fragments. The fission of $^{235}_{92}\text{U}$ begins with the absorption of a slow-moving neutron that produces an unstable nucleus of $^{236}\text{U}$. The $^{236}\text{U}$ nucleus then quickly decays into two heavy fragments, moving at high speed, as well as several neutrons. Most of the kinetic energy released in such a fission is carried off by the two large fragments. (a) For the typical fission process

$$^1_0\text{n} + ^{235}_{92}\text{U} \rightarrow ^{141}_{56}\text{Ba} + ^{92}_{36}\text{Kr} + 3^1_0\text{n}$$

calculate the kinetic energy in MeV carried off by the fission fragments. (b) Assuming the barium and krypton fragments split the kinetic energy equally between themselves leaving nothing for the three neutrons, find the speed of each fragment. The atomic masses involved are given next in atomic mass units.

$^1_0\text{n} = 1.008\ 665$ $^{235}_{92}\text{U} = 235.043\ 924$ $^{141}_{56}\text{Ba} = 140.903\ 496$ $^{92}_{36}\text{Kr} = 91.907\ 720$

### Solution

**A** We use the conservation of relativistic energy to find the kinetic energy released:

$$(KE + mc^2)_{\text{initial}} = (KE + mc^2)_{\text{final}}$$

$$0 + m_n c^2 + m_U c^2 = m_{Ba} c^2 + m_{Kr} c^2 + 3m_n c^2 + KE_{\text{final}}$$

where we assumed that the initial kinetic energy was zero. Solving for $KE_{\text{final}}$,

$$KE_{\text{final}} = [(m_n + m_U) - (m_{Ba} + m_{Kr} + 3m_n)]c^2$$

$$KE_{\text{final}} = [(1.008\ 665 \text{ u} + 235.043\ 924 \text{ u}) - (140.903\ 496 \text{ u} + 91.907\ 936 \text{ u} + 3(1.008\ 665 \text{ u}))]c^2$$

$$= (0.215\ 162 \text{ u})c^2 = (0.215\ 162 \text{ u})(931.494 \text{ MeV/u} \cdot c^2)(c^2) = \boxed{200.422 \text{ MeV}}$$

This means that a little over 0.2 u of the original mass is converted to kinetic energy in this fission.

**B** According to relativity, the total energy $E$ of a particle of mass $m$ moving at velocity $v$ is given by

$$E = \frac{mc^2}{\sqrt{1 - v^2/c^2}}$$

Solving for $v/c$, we find

$$\frac{v}{c} = \sqrt{1 - (mc^2)^2/E^2}$$

Applying this expression for $v/c$ to the barium fragment,

$$m_{Ba}c^2 = (140.903\ 496 \text{ u})(931.49 \text{ MeV/u} \cdot c^2)(c^2) = 1.312\ 5 \times 10^5 \text{ MeV}$$

$$E_{Ba} = KE_{Ba} + m_{Ba}c^2 = 200.42 \text{ MeV}/2 + 1.312\ 5 \times 10^5 \text{ MeV} = 1.313\ 5 \times 10^5 \text{ MeV}$$

$$\frac{v_{Ba}}{c} = \sqrt{1 - (1.312\ 5 \times 10^5 \text{ MeV})^2/(1.313\ 5 \times 10^5 \text{ MeV})^2} = 0.039\ 013$$

Thus $v_{Ba} = 0.039\ 013c \approx \boxed{1.2 \times 10^7 \text{ m/s.}}$ In a similar fashion we find $v_{Kr} = 0.0483\ 39c \approx \boxed{1.5 \times 10^7 \text{ m/s.}}$

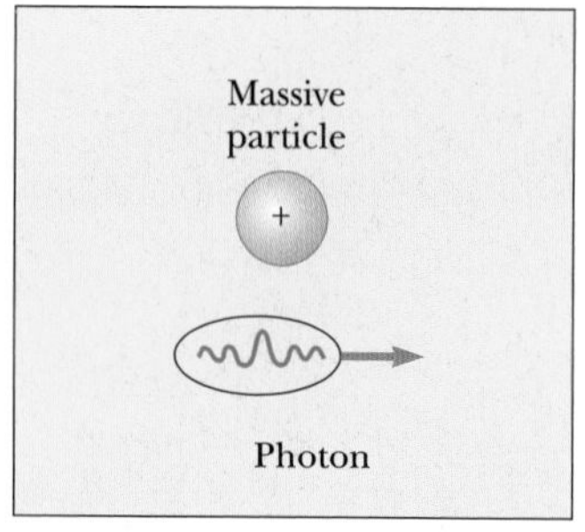


Before

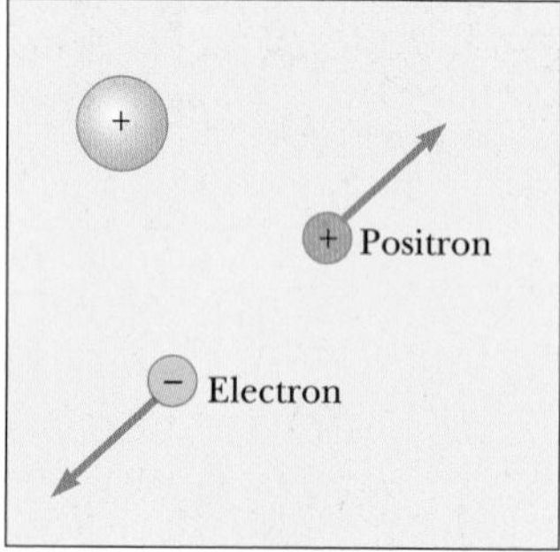


After

**FIGURE 26.15** Representation of the process of pair production.

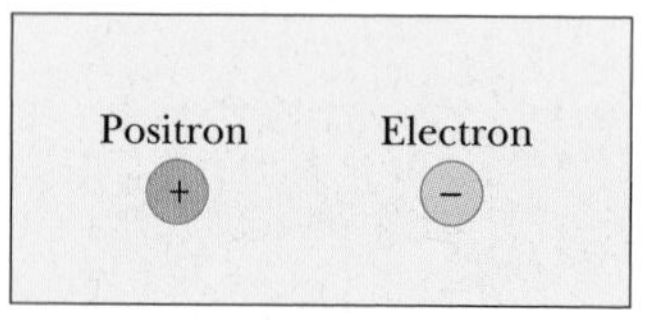


Before

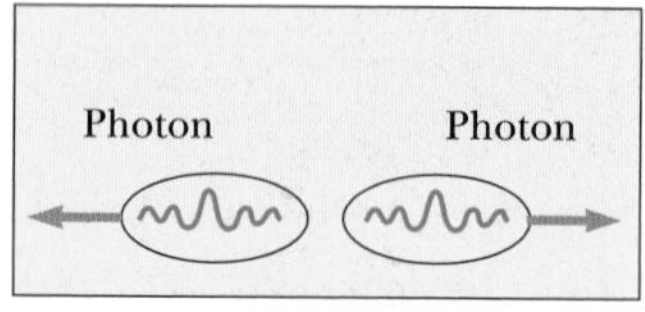


After

**FIGURE 26.16** Representation of the process of pair annihilation.

## 26.10 PAIR PRODUCTION AND ANNIHILATION

We shall now describe a process in which the energy of a photon is converted completely into mass. This is a striking verification of the equivalence of mass and other forms of energy as predicted by Einstein's theory of relativity.

A common process in which a photon creates matter is called **pair production,** illustrated in Figure 26.15. In this process, an electron and a positron are simultaneously produced, while the photon disappears. (Note that the positron is a positively charged particle having the same mass as an electron. The positron is often called the *antiparticle* of the electron.) In order for pair production to occur, energy, momentum, and charge must all be conserved during the process. It is impossible for a photon to produce a single electron because the photon has zero charge, and charge would not be conserved in the process.

As we explain in more detail in Chapter 27, the energy of a photon having a frequency $f$ is given by $E = hf$, where $h$ is Planck's constant. The *minimum* energy that a photon must have to produce an electron-positron pair can be found using conservation of energy by equating the photon energy, $hf_{\text{min}}$, to the total rest energy of the pair. That is,

$$hf_{\text{min}} = 2m_ec^2 \qquad [26.18]$$

Because the energy of an electron is $m_ec^2 = 0.51$ MeV, the minimum energy required for pair production is 1.02 MeV.

Pair production cannot occur in a vacuum, but can only take place in the presence of a massive particle such as an atomic nucleus. The massive particle must participate in the interaction in order that energy and momentum be conserved simultaneously.

**Pair annihilation** is a process in which an electron-positron pair produces two photons, the inverse of pair production. Figure 26.16 is one example of pair annihilation in which an electron and positron initially at rest combine with each other, disappear, and create two photons. Because the initial momentum of the pair is zero, it is impossible to produce a single photon. Momentum can be conserved only if two photons moving in opposite directions, both with the same energy and magnitude of momentum, are produced. We shall discuss particles and their antiparticles further in Chapter 30.

## 26.11 GENERAL RELATIVITY

Up to this point, we have sidestepped a curious puzzle. Mass has two seemingly different properties: a *gravitational attraction* for other masses and an *inertial* property that resists acceleration. To designate these two attributes, we use the subscripts $g$ and $i$ and write

$$\text{Gravitational property} \qquad F_g = G\frac{m_g m'_g}{r^2}$$

$$\text{Inertial property} \qquad F_i = m_i a$$

The value for the gravitational constant $G$ was chosen to make the magnitudes of $m_g$ and $m_i$ numerically equal. Regardless of how $G$ is chosen, however, the strict proportionality of $m_g$ and $m_i$ has been established experimentally to an extremely high degree: a few parts in $10^{12}$. Thus, it appears that gravitational mass and inertial mass may indeed be exactly proportional.

But why? They seem to involve two entirely different concepts: a force of mutual gravitational attraction between two masses and the resistance of a single mass to being accelerated. This question, which puzzled Newton and many other physicists over the years, was answered when Einstein published his theory of gravita-

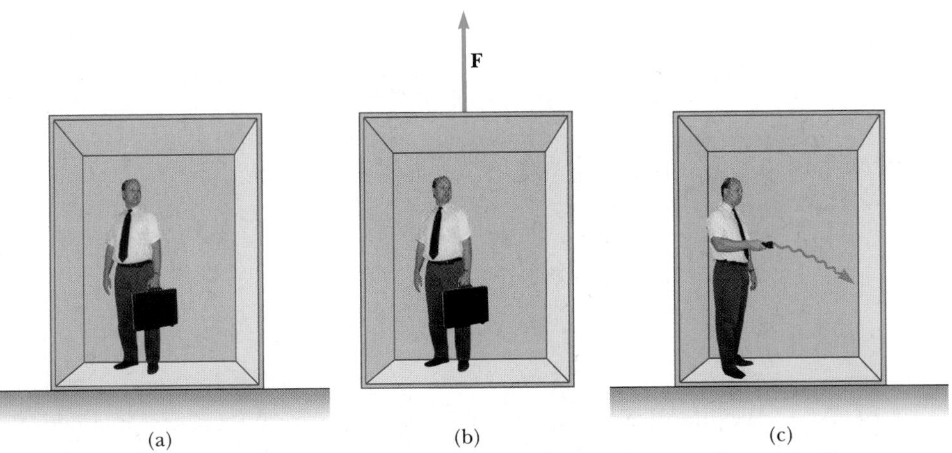


**FIGURE 26.17** (a) The observer is at rest in a uniform gravitational field $g$. (b) The observer is in a region in which gravity is negligible, but the frame of reference is accelerated by an external force **F** that produces an acceleration $g$. According to Einstein, the frames of reference in parts (a) and (b) are equivalent in every way. No local experiment could distinguish any difference between the two frames. (c) If parts (a) and (b) are truly equivalent, as Einstein proposed, then a ray of light would bend in a gravitational field.

tion, known as *general relativity,* in 1916. Because it is a mathematically complex theory, we merely offer a hint of its elegance and insight.

In Einstein's view, the remarkable coincidence that $m_g$ and $m_i$ were exactly proportional was evidence for a very intimate and basic connection between the two concepts. He pointed out that no mechanical experiment (such as dropping a mass) could distinguish between the two situations illustrated in Figures 26.17a and 26.17b. In each case, a mass released by the observer undergoes a downward acceleration of $g$ relative to the floor.

Einstein carried this idea further and proposed that *no* experiment, mechanical or otherwise, could distinguish between the two cases. This extension to include all phenomena (not just mechanical ones) has interesting consequences. For example, suppose that a light pulse is sent horizontally across the box, as in Figure 26.17c. The trajectory of the light pulse bends downward as the box accelerates upward to meet it. Einstein proposed that a beam of light should also be bent downward by a gravitational field. (No such bending is predicted in Newton's theory of gravitation.)

The two postulates of Einstein's **general relativity** are as follows:

1. All the laws of nature have the same form for observers in any frame of reference, whether accelerated or not.
2. In the vicinity of any given point, a gravitational field is equivalent to an accelerated frame of reference without a gravitational field. (This is the *principle of equivalence.*)

The second postulate implies that gravitational mass and inertial mass are completely equivalent, not just proportional. What were thought to be two different types of mass are actually identical.

One interesting effect predicted by general relativity is that time scales are altered by gravity. A clock in the presence of gravity runs more slowly than one where gravity is negligible. As a consequence, the frequencies of radiation emitted by atoms in the presence of a strong gravitational field are shifted to lower frequencies when compared with the same emissions in a weak field. This gravitational shift has been detected in spectral lines emitted by atoms in massive stars. It has also been verified on Earth by comparing the frequencies of gamma rays emitted from nuclei separated vertically by about 20 m.

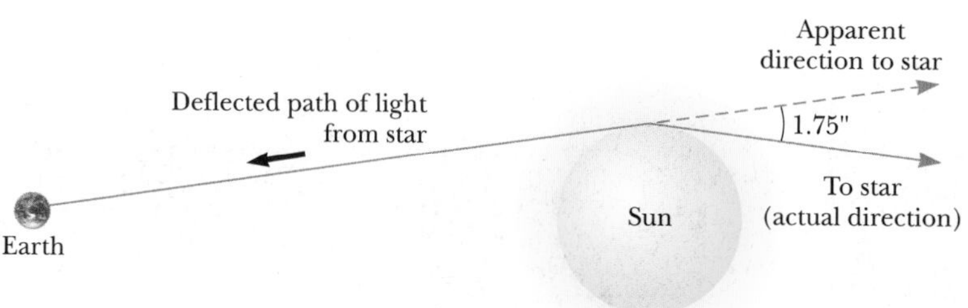


**FIGURE 26.18** Deflection of starlight passing near the Sun. Because of this effect, the Sun and other remote objects can act as a *gravitational lens.* In his general theory of relativity, Einstein calculated that starlight just grazing the Sun's surface should be deflected by an angle of 1.75″.

**Quick Quiz 26.5** Two identical clocks are in the same house, one upstairs in a bedroom and the other downstairs in the kitchen. Which statement is correct? (a) The clock in the kitchen runs more slowly than the clock in the bedroom. (b) The clock in the bedroom runs more slowly than the clock in the kitchen. (c) Both clocks keep the same time.

The second postulate suggests that a gravitational field may be "transformed away" at any point if we choose an appropriate accelerated frame of reference—a freely falling one. Einstein developed an ingenious method of describing the acceleration necessary to make the gravitational field "disappear." He specified a certain quantity, the *curvature of space-time,* that describes the gravitational effect at every point. In fact, the curvature of space-time completely replaces Newton's gravitational theory. According to Einstein, there is no such thing as a gravitational force. Rather, the presence of a mass causes a curvature of space-time in the vicinity of the mass, and this curvature dictates the space-time path that all freely moving objects must follow. In 1979, John Wheeler summarized Einstein's general theory of relativity in a single sentence: "Mass one tells space-time how to curve; curved space-time tells mass two how to move."

One important test of the general theory of relativity is the prediction that a light ray passing near the Sun should be deflected by the curved space-time created by the Sun's mass. This prediction was confirmed by astronomers during a total solar eclipse shortly following World War I when they detected starlight being bent near the Sun (Fig. 26.18). When this discovery was announced, Einstein became an international celebrity.

If the concentration of mass becomes very great, as is believed to occur when a large star exhausts its nuclear fuel and collapses to a very small volume, a **black hole** may form. Here the curvature of space-time is so extreme that, within a certain distance from the center of the black hole, all matter and light become trapped.

### Webnote 26.3

Have you ever wondered what it would look like to travel to a black hole? Learn more about the fascinating topic of black holes by visiting *http://antwrp.gsfc.nasa.gov/htmltest/rjn_bht.html*

### APPLYING PHYSICS 26.3

Atomic clocks are extremely accurate; in fact, an error of 1 second in 3 million years is typical. This error can be described as about one part in $10^{14}$. On the other hand, the atomic clock in Boulder, Colorado, is often 15 ns faster than the one in Washington after only one day. This is an error of about one part in $6 \times 10^{12}$, which is about 17 times larger than the previously expressed error. If atomic clocks are so accurate, why does a clock in Boulder not remain in synchronism with one in Washington? (*Hint:* Denver, near Boulder, is known as the Mile High City.)

**Explanation** According to the general theory of relativity, a clock's rate depends on gravity—clocks run more slowly in strong gravitational fields. Washington is at an elevation very close to sea level, whereas Boulder is about a mile higher in altitude. This will result in a weaker gravitational field at Boulder than at Washington. As a result, an atomic clock runs more rapidly in Boulder than in Washington.

## SUMMARY

The two basic postulates of the **special theory of relativity** are as follows:

1. The laws of physics are the same in all inertial frames of reference.
2. The speed of light is the same for all inertial observers, independent of their motion or of the motion of the source of light.

Some of the consequences of the special theory of relativity are as follows:

1. Clocks in motion relative to an observer slow down. This is known as **time dilation.** The relationship between time intervals in the moving and at-rest systems is

$$\Delta t = \gamma \, \Delta t_p \qquad [26.7]$$

where $\Delta t$ is the time interval measured in the system in relative motion with respect to the clock,

$$\gamma = 1/\sqrt{1 - v^2/c^2} \qquad [26.8]$$

and $\Delta t_p$ is the proper time interval measured in the system moving with the clock.

2. The length of an object in motion is *contracted* in the direction of motion. The equation for **length contraction** is

$$L = L_p\sqrt{1 - v^2/c^2} \qquad [26.9]$$

where $L$ is the length measured by an observer in motion relative to the object and $L_p$ is the proper length measured by an observer for whom the object is at rest.

3. Events that are simultaneous for one observer are not simultaneous for another observer in motion relative to the first.

The relativistic expression for the **momentum** of a particle moving with a velocity $v$ is

$$p \equiv \frac{mv}{\sqrt{1 - v^2/c^2}} = \gamma mv \qquad [26.10]$$

The relativistic expression for the addition of velocities is

$$v_{\text{ab}} = \frac{v_{\text{ad}} + v_{\text{db}}}{1 + \dfrac{v_{\text{ad}}v_{\text{db}}}{c^2}} \qquad [26.11]$$

where $v_{\text{ab}}$ is the velocity of object a with respect to object b, $v_{\text{ad}}$ is the velocity of object a with respect to object d, and so forth.

The relativistic expression for the **kinetic energy** of an object is

$$KE = \gamma mc^2 - mc^2 \qquad [26.12]$$

where $mc^2$ is the **rest energy,** $E_R$, of the object.

The **total energy** of a particle is

$$E = \frac{mc^2}{\sqrt{1 - v^2/c^2}} \qquad [26.15]$$

This is Einstein's famous mass-energy equivalence equation.

The relativistic momentum is related to the total energy through the equation

$$E^2 = p^2c^2 + (mc^2)^2 \qquad [26.16]$$

**Pair production** is a process in which the energy of a photon is converted into mass. In this process, the photon disappears as an electron-positron pair is created. Likewise, the energy of an electron-positron pair can be converted into electromagnetic radiation by the process of **pair annihilation.**

## CONCEPTUAL QUESTIONS

1. Why must you be careful with such phrases as *at the same time* and *right now* when dealing with very high speeds?
2. Without looking outside the laboratory, is it possible to conduct a laboratory experiment to tell if, relative to the stars, you are at rest? Moving at constant velocity? Moving at constant acceleration?
3. You are in a speedboat on a lake. You see ahead of you a wave front, caused by the previous passage of another boat, moving away from you. You accelerate, catch up with, and pass the wave front. Is this scenario possible if you are in a rocket and you detect a wave front of light ahead of you?
4. What two speed measurements will two observers in relative motion always agree on?
5. In the *Principia,* Newton suggested that the "fixed" stars constitute an absolute frame of reference. Criticize this suggestion.
6. A rocket is moving with half the speed of light. Discuss the value of the speed of a light pulse emitted from the nose of the rocket as measured by (a) the rocket captain, (b) an observer who sees the rocket approaching, (c) an observer who sees the rocket receding.
7. Some distant star-like objects, called quasars, are receding from us at half the speed of light (or greater). What is the speed of the light we receive from these quasars?
8. It is said that Einstein, in his teenage years, asked the question, "What would I see in a mirror if I carried it in my hands and ran at a speed near that of light?" How would you answer this question?
9. List some ways our day-to-day lives would change if the speed of light were only 50 m/s.
10. Two identically constructed clocks are synchronized. One is put in orbit around Earth while the other remains on Earth. Which clock runs more slowly? When the moving clock returns to Earth, will the two clocks still be synchronized?
11. As discussed in Chapter 22, the speed of light in a transparent material medium is less than the speed of light in a vacuum. Is this a violation of relativity? Discuss.
12. Imagine an astronaut on a trip to Sirius, which lies 8 lightyears from Earth. Upon arrival at Sirius, the astronaut finds that the trip lasted 6 years. If the trip was made at a constant speed of $0.8c$, how can the 8-lightyear distance be reconciled with the 6-year duration?
13. Explain why it is necessary, when defining length, to specify that the positions of the ends of a rod are to be measured simultaneously.
14. The equation $E = mc^2$ is often given in popular descriptions of Einstein's theory of relativity. Is this expression strictly correct? For example, does it accurately account for the kinetic energy of a moving mass?

## PROBLEMS

1, 2, 3 = straightforward, intermediate, challenging ☐ = full solution available in Student Solutions Manual/Study Guide

**web** = solution posted at **http://info.brookscole.com/serway** [biomedical icon] = biomedical application

### Section 26.4 The Michelson–Morley Experiment

1. Two airplanes fly paths I and II as specified in Figure 26.5a. Both planes have airspeeds of 100 m/s and fly a distance $L = 200$ km. The wind blows at 20.0 m/s in the direction shown in the figure. Find (a) the time of flight to each city, (b) the time to return, and (c) the difference in total flight times.
2. In one version of the Michelson–Morley experiment, the lengths $L$ in Figure 26.6 were 28 m. Take $v$ to be $3.0 \times 10^4$ m/s and find the time difference caused by rotation of the interferometer and (b) the expected fringe shift, assuming that the light used has a wavelength of 550 nm.

### Section 26.6 Consequences of Special Relativity

3. A deep-space probe moves away from Earth with a speed of $0.80c$. An antenna on the probe requires 3.0 s, probe time, to rotate through 1.0 rev. How much time is required for 1.0 rev according to an observer on Earth?
4. How fast must a meter stick be moving if its length is observed to shrink to 0.500 m?
5. At what speed does a clock have to move in order to be seen to run at a rate that is one-half the rate of a clock at rest?
6. An astronaut at rest on Earth has a heartbeat rate of 70 beats/min. When the astronaut is traveling in a spaceship at $0.90c$, what will this rate be as measured by (a) an observer also in the ship and (b) an observer at rest on Earth?
7. The average lifetime of a pi meson in its own frame of reference (i.e., the proper lifetime) is $2.6 \times 10^{-8}$ s. If the meson moves with a speed of $0.98c$, what is (a) its mean lifetime as measured by an observer on Earth and (b) the average distance it travels before decaying as measured by an observer on Earth? (c) What distance would it travel if time dilation did not occur?
8. An astronaut is traveling in a space vehicle that has a speed of $0.500c$ relative to Earth. The astronaut measures his pulse rate at 75.0 per minute. Signals generated by the

astronaut's pulse are radioed to Earth when the vehicle is moving perpendicularly to a line that connects the vehicle with an Earth observer. What pulse rate does Earth observer measure? What would be the pulse rate if the speed of the space vehicle were increased to $0.990c$?

9. web A muon formed high in Earth's atmosphere travels at a speed $v = 0.99c$ for a distance of 4.6 km before it decays into an electron, a neutrino, and an antineutrino $(\mu^- \rightarrow e^- + \nu + \bar{\nu})$. (a) How long does the muon live, as measured in its reference frame? (b) How far does the muon travel, as measured in its frame?

10. A box is cubical with sides of proper lengths $L_1 = L_2 = L_3$, as shown in Figure P26.10, when viewed in its own rest frame. If this block moves parallel to one of its edges with a speed of $0.80c$ past an observer, (a) what shape does it appear to have to this observer, and (b) what is the length of each side as measured by this observer?

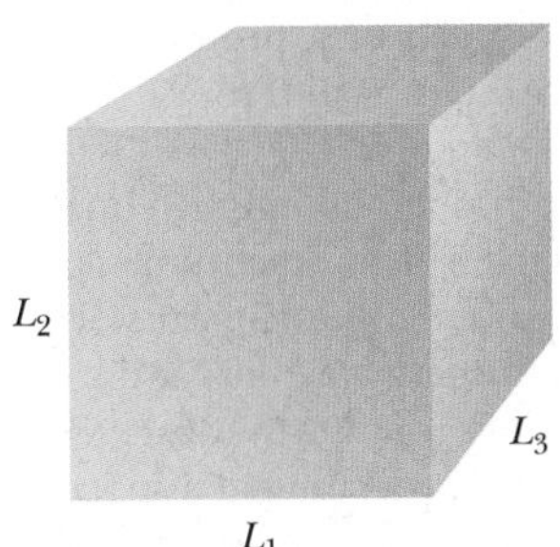

**FIGURE P26.10**

11. The proper length of one spaceship is three times that of another. The two spaceships are traveling in the same direction and, while both are passing overhead, an Earth observer measures the two spaceships to be the same length. If the slower spaceship is moving with a speed of $0.350c$, determine the speed of the faster spaceship.

12. An atomic clock moves at 1 000 km/h for one hour as measured by an identical clock on Earth. How many nanoseconds slow will the moving clock be at the end of the one-hour interval? (*Hint:* The following approximation will prove helpful: $\sqrt{1-x} \approx 1 - (x/2)$ for $x \ll 1$.)

13. A supertrain of proper length 100 m travels at a speed of $0.95c$ as it passes through a tunnel having proper length 50 m. As seen by a trackside observer, is the train ever completely within the tunnel? If so, by how much?

14. An observer moving at a speed of $0.995c$ relative to a rod (Fig. P26.14) measures its length to be 2.00 m and sees its length to be oriented at 30.0° with respect to the direction of motion. What is the proper length of the rod? (b) What is the orientation angle in a reference frame moving with the rod?

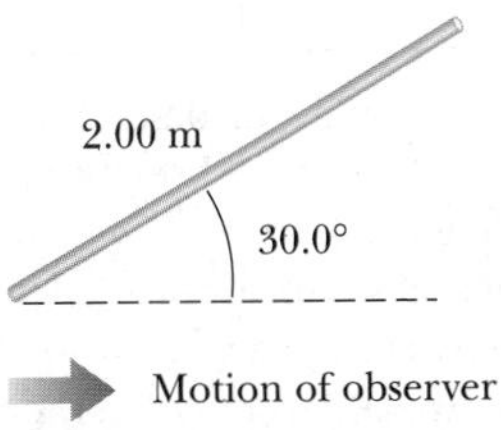

**FIGURE P26.14** View of rod as seen by an observer moving to the right.

15. In 1963 when Mercury astronaut Gordon Cooper orbited Earth 22 times, the press stated that for each orbit he aged 2 millionths of a second less than if he had remained on Earth. (a) Assuming that he was 160 km above Earth in a circular orbit, determine the time difference between someone on Earth and the orbiting astronaut for the 22 orbits. You will need to use the approximation

$$\frac{1}{\sqrt{1-x}} \approx 1 + \frac{x}{2} \qquad \text{for } x \ll 1$$

(b) Did the press report accurate information? Explain.

16. An interstellar space probe is launched from Earth. After a brief period of acceleration, it moves with a constant velocity, 70.0% of the speed of light. Its nuclear-powered batteries supply the energy to keep its data transmitter active continuously. The batteries have a lifetime of 15.0 years as measured in a rest frame. (a) How long do the batteries on the space probe last as measured by mission control on Earth? (b) How far is the probe from Earth when its batteries fail, as measured by mission control? (c) How far is the probe from Earth, as measured by its built-in trip odometer, when its batteries fail? (d) For what total time after launch is data received from the probe by mission control? Note that radio waves travel at the speed of light and fill the space between the probe and Earth at the time of battery failure.

## Section 26.7 Relativistic Momentum

17. An electron has a speed of $v = 0.90c$. At what speed will a proton have a momentum equal to that of the electron?

18. Calculate the momentum of an electron moving with a speed of (a) $0.010c$, (b) $0.50c$, (c) $0.90c$.

19. An unstable particle at rest breaks up into two fragments of *unequal mass*. The mass of the lighter fragment is $2.50 \times 10^{-28}$ kg, and that of the heavier fragment is $1.67 \times 10^{-27}$ kg. If the lighter fragment has a speed of $0.893c$ after the breakup, what is the speed of the heavier fragment?

20. The nonrelativistic expression for the momentum of a particle, $p = mv$, can be used if $v \ll c$. For what speed does the use of this formula give an error in the momentum of (a) 1.00% and (b) 10.0%?

## Section 26.8 Relativistic Addition of Velocities

21. An electron moves to the right with a speed of $0.90c$ relative to the laboratory frame. A proton moves to the left with a speed of $0.70c$ relative to the electron. Find the speed of the proton relative to the laboratory frame.

22. Spaceship R is moving to the right at a speed of $0.70c$ with respect to Earth. A second spaceship, L, moves to the left at the same speed with respect to Earth. What is the speed of L with respect to R?

23. A Klingon space ship moves away from Earth at a speed of $0.800c$ (Fig. P26.23). The Starship *Enterprise* pursues at a speed of $0.900c$ relative to Earth. Observers on Earth see the *Enterprise* overtaking the Klingon ship at a relative speed of $0.100c$. With what speed is the *Enterprise* overtaking the Klingon ship as seen by the crew of the *Enterprise*?

24. A spaceship travels at $0.750c$ relative to Earth. If the spaceship fires a small rocket in the forward direction, how fast

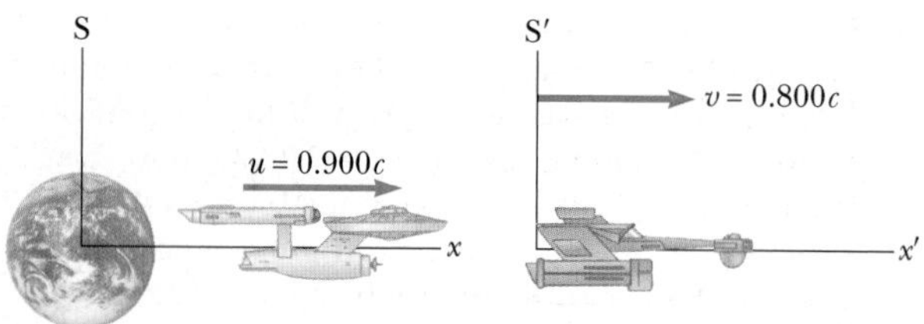


FIGURE P26.23

(relative to the ship) must it be fired for it to travel at $0.950c$ relative to Earth?

25. web A rocket moves with a velocity of $0.92c$ to the right with respect to a stationary observer A. An observer B moving relative to observer A finds that the rocket is moving with a velocity of $0.95c$ to the left. What is the velocity of observer B relative to observer A? (*Hint:* Consider observer B's velocity in the frame of reference of the rocket.)

26. A pulsar is a stellar object that emits light in short bursts. Suppose a pulsar with a speed of $0.950c$ approaches Earth, and a rocket with a speed of $0.995c$ heads toward the pulsar (both speeds measured in Earth's frame of reference). If the pulsar emits 10.0 pulses per second in its own frame of reference, at what rate are the pulses emitted in the rocket's frame of reference?

27. Spaceship I, which contains students taking a physics exam, approaches Earth with a speed of $0.60c$, while spaceship II, which contains instructors proctoring the exam, moves away from Earth at $0.28c$, as in Figure P26.27. If the instructors in spaceship II stop the exam after 50 min have passed on *their clock,* how long does the exam last as measured by (a) the students? (b) an observer on Earth?

FIGURE P26.27

## Section 26.9 Relativistic Energy and the Equivalence of Mass and Energy

28. A proton moves with a speed of $0.950c$. Calculate its (a) rest energy, (b) total energy, and (c) kinetic energy.

29. What is the speed of a particle whose kinetic energy is equal to its own rest energy?

30. A proton in a high-energy accelerator is given a kinetic energy of 50.0 GeV. Determine (a) the momentum and (b) the speed of the proton.

31. In a color television tube, electrons are accelerated through a potential difference of 20 000 V. With what speed do the electrons strike the screen?

32. Determine the energy required to accelerate an electron from (a) $0.500c$ to $0.900c$ and (b) $0.900c$ to $0.990c$.

33. A cube of steel has a volume of $1.00\ \text{cm}^3$ and a mass of 8.00 g when at rest on Earth. If this cube is now given a speed of $v = 0.900c$, what is its density as measured by a stationary observer? Note that relativistic density is $E_R/c^2V$.

34. An unstable particle with a mass equal to $3.34 \times 10^{-27}$ kg is initially at rest. The particle decays into two fragments that fly off with velocities of $0.987c$ and $-0.868c$. Find the masses of the fragments. (*Hint:* Conserve both mass-energy and momentum.)

## Section 26.10 Pair Production and Annihilation

35. How much total kinetic energy will an electron-positron pair have if produced by a photon of energy 3.00 MeV?

36. If an electron-positron pair with a total kinetic energy of 2.50 MeV is produced, find (a) the energy of the photon that produced the pair and (b) its frequency.

37. web Two photons are produced when a proton and an antiproton annihilate each other. What is the minimum frequency and corresponding wavelength of each photon?

38. An electron moving at a speed of $0.60c$ collides head on with a positron also moving at $0.60c$. Determine the energy and momentum of each photon produced in the process.

## ADDITIONAL PROBLEMS

39. What is the speed of a proton that has been accelerated from rest through a difference of potential of (a) 500 V and (b) $5.00 \times 10^8$ V?

40. An electron has a total energy equal to 5 times its rest energy. (a) What is its momentum? (b) Repeat for a proton.

41. What is the momentum (in units of MeV/$c$) of an electron with a kinetic energy of 1.00 MeV?

42. An astronomer on Earth observes a meteoroid in the southern sky approaching at a speed of $0.800c$. At the time of its discovery, the meteoroid is 20.0 lightyears from Earth. Calculate (a) the time required for the meteoroid to reach Earth as measured by the Earthbound astronomer, (b) this time as measured by a tourist on the meteoroid, and (c) the distance to Earth as measured by the tourist.

43. Ted and Mary are playing a game of catch in frame S′, which is moving with a speed of $0.60c$; Jim in frame S is watching (Fig. P26.43). Ted throws the ball to Mary with a speed of $0.80c$ (according to Ted) and their separation (measured in S′) is $1.8 \times 10^{12}$ m. (a) According to Mary, how fast is the ball moving? (b) According to Mary, how long will it take the ball to reach her? (c) According to Jim, how far apart are Ted and Mary, and how fast is the ball moving?

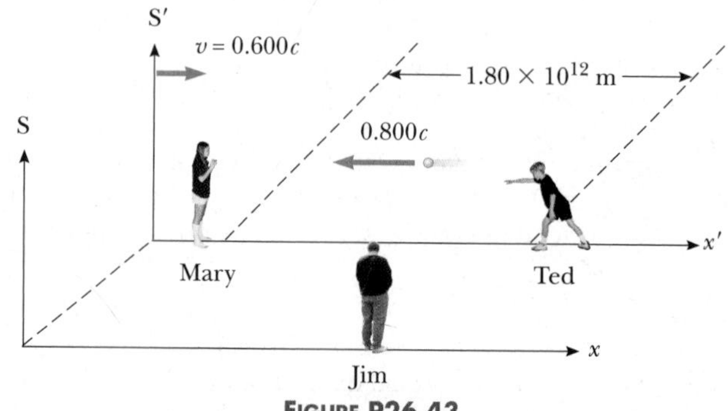


FIGURE P26.43

44. An alarm clock is set to sound in 10 h. At $t = 0$, the clock is placed in a spaceship moving with a speed of $0.75c$ (relative to Earth). What distance, as determined by an Earth observer, does the spaceship travel before the alarm clock sounds?

45. A radioactive nucleus moves with a speed of $v$ relative to a laboratory observer. The nucleus emits an electron in the positive $x$ direction with a speed of $0.70c$ relative to the decaying nucleus, and a speed of $0.85c$ in the $+x$ direction relative to the laboratory observer. What is the value of $v$?

46. A certain quasar recedes from Earth at $v = 0.870c$. A jet of material ejected from the quasar back toward Earth moves at $0.550c$ relative to the quasar. Find the speed of the ejected material relative to Earth.

47. An astronaut wishes to visit the Andromeda galaxy, making a one-way trip that will take 30.0 y in the spaceship's frame of reference. Assume that the galaxy is 2.00 million lightyears away and that the speed is constant. (a) How fast must the astronaut travel relative to Earth? (b) What will be the kinetic energy of the spacecraft, whose mass is $1.00 \times 10^6$ kg? (c) What is the cost of this energy if it is purchased at a typical consumer price for electric energy, 13.0 cents per kWh? The following approximation will prove useful:

$$\frac{1}{\sqrt{1+x}} \approx 1 - \frac{x}{2} \qquad \text{for } x \ll 1$$

48. The cosmic rays of highest energy are protons that have kinetic energy on the order of $10^{13}$ MeV. (a) How long would it take a proton of this energy to travel across the Milky Way galaxy, having a diameter ~$10^5$ lightyears, as measured in the proton's frame? (b) From the point of view of the proton, how many kilometers across is the galaxy?

49. A spaceship of proper length 300 m takes 0.75 $\mu$s to pass an Earth observer. Determine the speed of this spaceship as measured by Earth observer.

50. Find the kinetic energy of a 78.0-kg spacecraft launched out of the solar system with a speed of 106 km/s by using (a) the classical equation $KE = \frac{1}{2}mv^2$ and (b) the relativistic equation. You will need to use the approximation

$$\frac{1}{\sqrt{1-x}} \approx 1 + \frac{x}{2} \qquad \text{for } x \ll 1$$

51. web An alien spaceship traveling at $0.600c$ toward Earth launches a landing craft with an advance guard of purchasing agents. The lander travels in the same direction with a velocity $0.800c$ relative to the spaceship. As observed on Earth, the spaceship is 0.200 lightyears from Earth when the lander is launched. (a) With what velocity is the lander observed to be approaching by observers on Earth? (b) What is the distance to Earth at the time of lander launch, as observed by the aliens on the mother ship? (c) How long does it take the lander to reach Earth as observed by the aliens on the mother ship? (d) If the lander has a mass of $4.00 \times 10^5$ kg, what is its kinetic energy as observed in Earth's reference frame?

52. (a) Show that a potential difference of $1.02 \times 10^6$ V would be sufficient to give an electron a speed equal to twice the speed of light if Newtonian mechanics remained valid at high speeds. (b) What speed would an electron actually acquire in falling through a potential difference of $1.02 \times 10^6$ V?

53. The muon is an unstable particle that spontaneously decays into an electron and two neutrinos. In a reference frame in which the muons are stationary, if the number of muons at $t = 0$ is $N_0$, the number at time $t$ is given by $N = N_0 e^{-t/\tau}$, where $\tau$ is the mean lifetime, equal to 2.2 $\mu$s. Suppose that the muons move at a speed of $0.95c$ and that there are $5.0 \times 10^4$ muons at $t = 0$. (a) What is the observed lifetime of the muons? (b) How many muons remain after traveling a distance of 3.0 km?

54. An observer in a rocket moves toward a mirror at speed $v$ relative to the reference frame labeled by S in Figure P26.54. The mirror is stationary with respect to S. A light pulse emitted by the rocket travels toward the mirror and is reflected back to the rocket. The front of the rocket is a distance $d$ from the mirror (as measured by observers in S) at the moment the light pulse leaves the rocket. What is the total travel time of the pulse as measured by observers in (a) the S frame and (b) the front of the rocket?

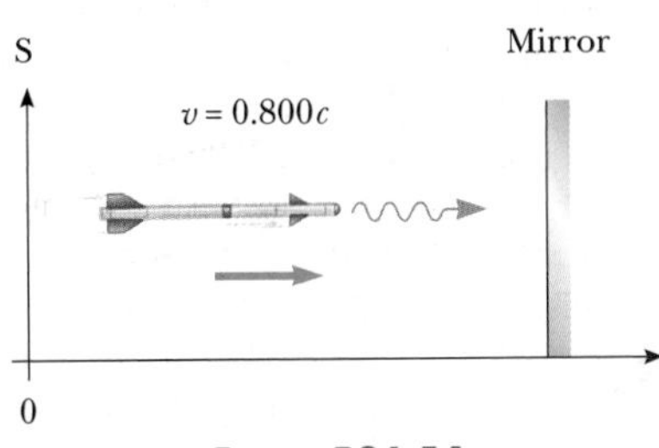


**FIGURE P26.54**

55. A physics professor on Earth gives an exam to her students who are on a rocket ship traveling at a speed of $v$ with respect to Earth. The moment the ship passes the professor, she signals the start of the exam. If she wishes her students to have $T_0$ (rocket time) to complete the exam, show that she should wait a time of

$$T = T_0\sqrt{\frac{1 - v/c}{1 + v/c}}$$

(Earth time) before sending a light signal telling them to stop. (*Hint:* Remember that it takes some time for the second light signal to travel from the professor to the students.)

56. Imagine that the entire Sun collapses to a sphere of radius $R_g$ such that the work required to remove a small mass $m$ from the surface would be equal to its rest energy, $mc^2$. This radius is called the *gravitational radius* for the Sun. Find $R_g$. (It is believed that the ultimate fate of very massive stars is to collapse beyond their gravitational radii into black holes.)

57. A rod of length $L_0$ moves with a speed of $v$ along the horizontal direction. The rod makes an angle of $\theta_0$ with respect to the axis of a coordinate system moving with the rod. (a) Show that the length of the rod as measured by a stationary observer is given by

$$L = L_0\left[1 - \left(\frac{v^2}{c^2}\right)\cos^2\theta_0\right]^{1/2}$$

(b) Show that the angle the rod makes with the axis as seen by the stationary observer is given by the expression $\tan\theta = \gamma\tan\theta_0$. These results show that the rod is both contracted and rotated. (Take the lower end of the rod to be at the origin of the moving coordinate system.)

## GROUP ACTIVITIES

**G.1** How would your life change if the speed of light were 55 mi/h? As you and your co-worker go about your daily activities, make notes about how events or occurrences would be changed. See which of the two of you can find the largest number of ways in which your day-to-day life would be affected.

**G.2** Consider two travelers on the surface of Earth walking directly toward the North Pole but from different starting locations. How is the fact that they are getting closer together as they go north similar to the curvature of space discussed in the general theory of relativity?

**G.3** Find a comfortable location on campus to observe life going on about you. Take note of all the motion going on about you and describe how this motion would look in a system moving at a constant velocity with respect to you.

**G.4** Two rockets are of length $L$ and each carries an antenna also of length $L$ that is supported in a direction perpendicular to the length of the rocket. Two pilots, A and B, are in the rockets, which are initially at rest with respect to one another. Rocket A takes off and travels at a speed of $0.400c$ away from B. Answer the following questions for observers A and B either with greater than $L$, less than $L$, or equal to $L$.

(a) The length of A as seen by pilot A is ________.
(b) The length of A as seen by pilot B is ________.
(c) The length of B as seen by pilot B is ________.
(d) The length of B as seen by pilot A is ________.
(e) The length of the antenna of A as seen by pilot B is ________.
(f) The length of the antenna of B as seen by pilot A is ________.
(g) Take $L$ to be 100 m to find precise values for all of the above answers.

**G.5** Two rockets carry identical clocks. When they are at rest with respect to each other, pilot A on ship A and pilot B on ship B agree that the second hand on each other's watch takes precisely time $\Delta T$ to go around the dial of the watch once. Rocket A takes off and travels at a speed of $0.500c$ away from B. Answer the following questions for observers A and B either with greater than $\Delta T$, less than $\Delta T$, or equal to $\Delta T$.

(a) The interval on A as seen by pilot A is ________.
(b) The interval on A as seen by pilot B is ________.
(c) The interval on B as seen by pilot B is ________.
(d) The interval on B as seen by pilot A is ________.
(e) Find the time intervals when $\Delta T$ is equal to 1 min.

**G.6** An alien spacecraft fires torpedoes forward with a speed of $0.75c$ relative to the spacecraft. The spacecraft has a speed of $0.30c$ relative to Earth. In such a case, will the torpedoes travel at $1.05c$ with respect to Earth? If you do not agree with the speed of $1.05c$, find the speed at which the torpedoes do travel with respect to Earth.

# Quantum Physics

# 27

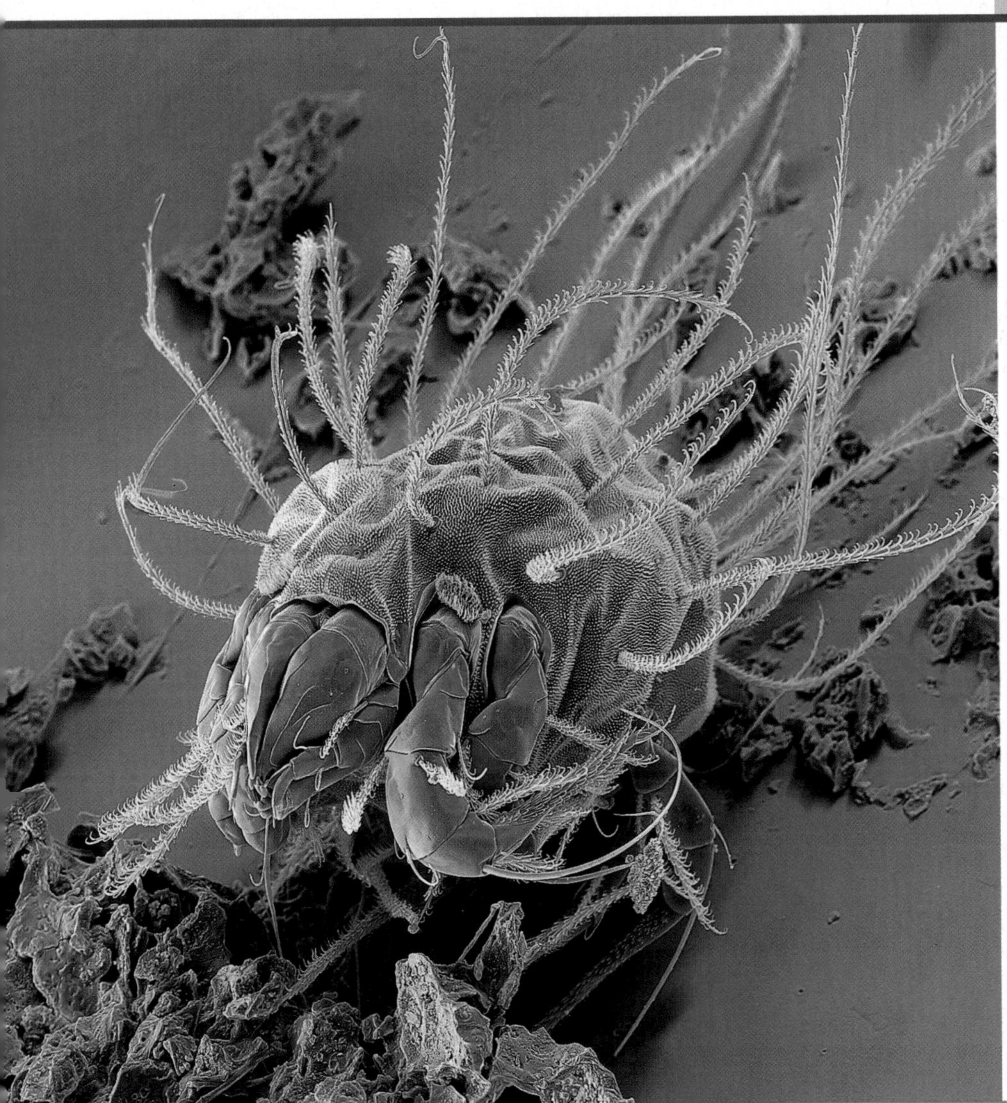

## Chapter Outline



**Color-enhanced scanning electron micrograph of the storage mite *Lepidoglyphus destructor*. These common mites grow up to 0.75 mm and feed on molds, flour and rice. They thrive at 25°C and high humidity, and can trigger allergies.** *(© Eye of Science/Science Source/Photo Researchers, Inc.)*

Although many problems were resolved by the theory of relativity in the early part of the 20th century, many other problems remained unsolved. Attempts to explain the behavior of matter on the atomic level with the laws of classical physics were consistently unsuccessful. Various phenomena, such as the electromagnetic radiation emitted by a heated object (blackbody radiation), the emission of electrons by illuminated metals (photoelectric effect), and the emission of sharp spectral lines by gas atoms in an electric discharge tube, could not be understood within the framework of classical physics. Between 1900 and 1930, however, a modern version of mechanics called *quantum mechanics* or *wave mechanics* was highly successful in explaining the behavior of atoms, molecules, and nuclei. Moreover, quantum mechanics reduces to classical physics when applied to macroscopic systems. As with relativity, the quantum theory requires a modification of our everyday ideas concerning the physical world.

The earliest and most basic ideas of quantum theory were introduced by Planck, and most of the subsequent mathematical developments, interpretations, and improvements were made by a number of distinguished physicists, including Einstein, Bohr, Schrödinger, de Broglie, Heisenberg, Born, and Dirac. An extensive study of quantum theory is certainly beyond the scope of this book. This chapter is simply an introduction to the underlying ideas of quantum theory and the wave-particle nature of matter. We also discuss some simple applications of quantum theory, including the photoelectric effect, the Compton effect, and x-rays.

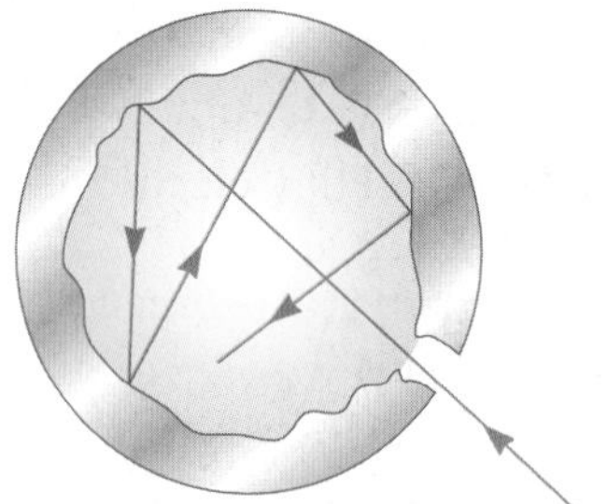

**FIGURE 27.1** The opening in the cavity of a body is a good approximation of a blackbody. As light enters the cavity through the small opening, part is reflected and part is absorbed on each reflection from the interior walls. After many reflections, essentially all of the incident energy is absorbed.

## 27.1 BLACKBODY RADIATION AND PLANCK'S HYPOTHESIS

An object at any temperature is known to emit electromagnetic radiation that is sometimes referred to as **thermal radiation.** Stefan's law, which we discussed in Section 11.7, describes the total power radiated. The spectrum of the radiation depends on the temperature and properties of the object. At low temperatures, the wavelengths of the thermal radiation are mainly in the infrared region and hence not observable by the eye. As the temperature of an object increases, the object eventually begins to glow red. At sufficiently high temperatures, it appears to be white, as in the glow of the hot tungsten filament of a lightbulb. A careful study of thermal radiation shows that it consists of a continuous distribution of wavelengths from the infrared, visible, and ultraviolet portions of the spectrum.

From a classical viewpoint, thermal radiation originates from accelerated charged particles near the surface of an object; such charges emit radiation much as small antennas do. The thermally agitated charges can have a distribution of frequencies, which accounts for the continuous spectrum of radiation emitted by the object. By the end of the 19th century, it had become apparent that the classical theory of thermal radiation was inadequate. The basic problem was in understanding the observed distribution of energy as a function of wavelength in the radiation emitted by a blackbody. By definition, a blackbody is an ideal system that absorbs *all* radiation incident on it. A good approximation of a blackbody is a small hole leading to the inside of a hollow object, as shown in Figure 27.1. The nature of the radiation emitted through the small hole leading to the cavity depends *only on the temperature* of the cavity walls, and not at all on the material composition of the object, its shape, or other factors.

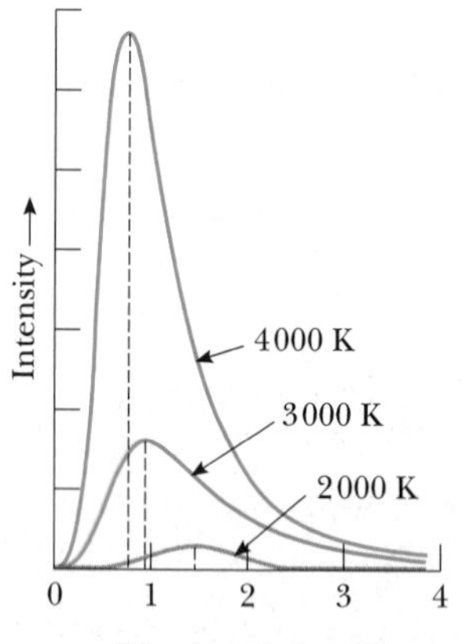


**FIGURE 27.2** Intensity of blackbody radiation versus wavelength at three different temperatures. Note that the total radiation emitted (the area under the curve) increases with increasing temperature.

Experimental data for the distribution of energy in blackbody radiation at three temperatures are shown in Figure 27.2. The radiated energy varies with wavelength and temperature. As the temperature of the blackbody increases, the total amount of energy (area under the curve) it emits increases. Also, with increasing temperature, the peak of the distribution shifts to shorter wavelengths. This shift was found to obey the following relationship, called **Wien's displacement law:**

$$\lambda_{max} T = 0.2898 \times 10^{-2}\ \mathrm{m \cdot K} \qquad [27.1]$$

where $\lambda_{max}$ is the wavelength at which the curve peaks and $T$ is the absolute temperature of the object emitting the radiation.

## APPLYING PHYSICS 27.1

If you look carefully at stars in the night sky, you can distinguish three main colors—red, white, and blue. What is the reason for these colors?

**Explanation** These colors are a result of the thermal radiation from the surfaces of the stars. A relatively cool star, with a surface temperature of 3 000 K, has a radiation curve like the middle curve in Figure 27.2. Notice that the peak in this curve is above the visible wavelengths, 0.4 $\mu$m–0.7 $\mu$m. Thus, significantly more radiation is emitted within the visible range at the red end than the blue end. As a result, the star appears to be reddish in color, similar to the red glow from the burner of an electric stove.

A hotter star has a radiation curve more like the upper curve in Figure 27.2. In this case, the star emits significant radiation throughout the visible range and the combination of all colors causes the star to look white. This is the case with our own Sun, with a surface temperature of 5 800 K. For very hot stars, the peak can be shifted so far below the visible range that significantly more blue radiation is emitted than red, so that the star appears bluish in color.

**APPLICATION**

COLOR OF THE STARS

Attempts to use classical ideas to explain the shape of the curves shown in Figure 27.2 failed. Figure 27.3 shows an experimental plot of the blackbody radiation spectrum (red curve) together with the theoretical picture of what this curve should look like based on classical theories (blue curve). At long wavelengths, classical theory is in good agreement with the experimental data. At short wavelengths, however, major disagreement exists between classical theory and experiment. As $\lambda$ approaches zero, classical theory predicts that the amount of energy being radiated should increase. In fact, theory predicts that the intensity should be infinite. This is contrary to the experimental data, which show that as $\lambda$ approaches zero, the amount of energy carried by short-wavelength radiation also approaches zero. This contradiction is called the **ultraviolet catastrophe** because theory and experiment disagree strongly in the short-wavelength ultraviolet region of the spectrum.

In 1900 Planck developed a formula for blackbody radiation that was in complete agreement with experiments at all wavelengths, yielding an intensity versus wavelength distribution that matches the red line in Figure 27.3. Planck's original theoretical approach is rather abstract in that it involves arguments based on entropy and thermodynamics. We shall present arguments that are easier to visualize physically, while attempting to convey the spirit and revolutionary impact of Planck's original work.

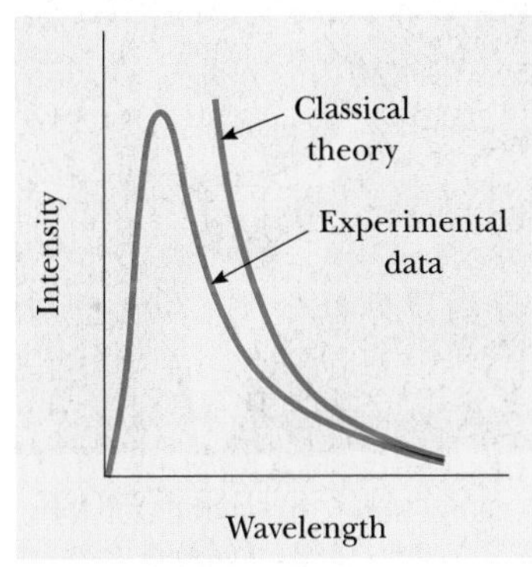


**FIGURE 27.3** Comparison of experimental data with the classical theory of blackbody radiation. Planck's theory perfectly matches the experimental data.

Planck hypothesized that blackbody radiation was produced by submicroscopic charged oscillators, which he called *resonators.* He assumed that the walls of a glowing cavity were composed of literally billions of these resonators whose exact nature was unknown. In his theory, Planck made a bold and controversial assumption concerning the nature of the oscillating molecular charges in the blackbody:

### Webnote 27.1

Physics 2000 from the University of Colorado at Boulder is an interactive journey through modern physics using applets, animations, and cartoons. Go to *http://www.colorado.edu/physics/2000/index.pl*

The resonators were allowed to have only certain discrete energies, $E_n$, given by

$$E_n = nhf \qquad [27.2]$$

where $n$ is a positive integer called a **quantum number,** $f$ is the frequency of vibra-

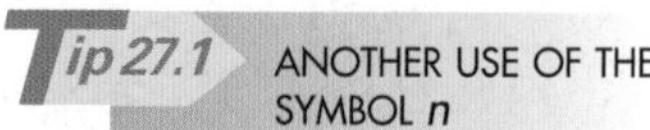


The symbol $n$ in this chapter represents an integer quantum number that is used to identify a particular state of a system. Do not confuse this with the same symbol $n$ used for the index of refraction (which was not an integer) in our study of optics.

tion of the resonator, and $h$ is a constant, known as **Planck's constant,** which has the value

$$h = 6.626 \times 10^{-34}\ \text{J} \cdot \text{s} \qquad [27.3]$$

Because the energy of each resonator can have only discrete values given by Equation 27.2, we say the energy is *quantized.* Each discrete energy value represents a different *quantum state,* with each value of $n$ representing a specific quantum state. (When the resonator is in the $n = 1$ quantum state, its energy is $hf$; when it is in the $n = 2$ quantum state, its energy is $2hf$; and so on.)

The key point in Planck's theory is the radical assumption of quantized energy states. This development marked the birth of the quantum theory. When Planck presented his theory, most scientists (including Planck!) did not consider the quantum concept to be realistic since it violated several centuries of belief that energy was a continuous quantity. Hence, Planck and others continued to search for a classical explanation of blackbody radiation. However, subsequent developments showed that a theory based on the quantum concept (rather than on classical concepts) had to be used to explain a number of other phenomena at the atomic level.

MAX PLANCK, GERMAN PHYSICIST (1858–1947)

Planck was born in Germany. His first son was killed in World War I, and his second son was executed by the Nazis for conspiring to assassinate Hitler. During an air raid on Berlin, he lost his home and valuable library. Planck introduced the concept of a "quantum of action" (Planck's constant, $h$) in an attempt to explain the spectral distribution of blackbody radiation, which laid the foundations for quantum theory. In 1918 he was awarded the Nobel prize for this discovery of the quantized nature of energy. The work leading to the "lucky" blackbody radiation formula was described by Planck in his Nobel prize acceptance speech (1920): "But even if the radiation formula proved to be perfectly correct, it would after all have been only an interpolation formula found by lucky guesswork and, thus, would have left us rather unsatisfied." *(© Bettmann/CORBIS)*

### Example 27.1 Thermal Radiation from the Human Body

The temperature of the skin is approximately 35°C. At what wavelength does the radiation emitted from the skin reach its peak?

**Solution** From Wien's displacement law, Equation 27.1, we have

$$\lambda_{\text{max}} T = 0.2898 \times 10^{-2}\ \text{m} \cdot \text{K}$$

Solving for $\lambda_{\text{max}}$, noting that 35°C corresponds to an absolute temperature of 308 K, we find

$$\lambda_{\text{max}} = \frac{0.2898 \times 10^{-2}\ \text{m} \cdot \text{K}}{308\ \text{K}} = 940\ \mu\text{m}$$

This radiation is in the infrared region of the spectrum.

**Exercise** (a) Find the wavelength corresponding to the peak of the radiation curve for the heating element of an electric oven at a temperature of $1.20 \times 10^3$ K. (Note that although this radiation peak lies in the infrared, there is enough visible radiation at this temperature to give the element a red glow.) (b) Calculate the wavelength corresponding to the peak of the radiation curve for an object whose temperature is $5.00 \times 10^3$ K, an approximate temperature for the surface of the Sun.

**Answer** (a) 2.42 $\mu$m; (b) 580 nm in the visible region

### Example 27.2 The Quantized Macroscopic Oscillator

A 2.0-kg object is attached to a massless spring of force constant $k = 25$ N/m. The spring is stretched 0.40 m from its equilibrium position and released.

**A** Find the total energy and frequency of oscillation according to classical calculations.

**Solution** The total energy of a simple harmonic oscillator having an amplitude $A$ is $kA^2/2$. Therefore,

$$E = \tfrac{1}{2}kA^2 = \tfrac{1}{2}(25\ \text{N/m})(0.40\ \text{m})^2 = 2.0\ \text{J}$$

The frequency of oscillation is

$$f = \frac{1}{2\pi}\sqrt{\frac{k}{m}} = \frac{1}{2\pi}\sqrt{\frac{25\text{ N/m}}{2.0\text{ kg}}} = 0.56\text{ Hz}$$

**B** Assume that Planck's law of energy quantization applies to any oscillator (atomic or large-scale) and find the quantum number $n$ for this system.

**Solution** If the energy is quantized, we have $E_n = nhf$, and from the result of Part A, we have

$$E_n = nhf = n(6.63 \times 10^{-34}\text{ J}\cdot\text{s})(0.56\text{ Hz}) = 2.0\text{ J}$$

Therefore,

$$n = 5.4 \times 10^{33}$$

and we see that macroscopic systems have huge quantum numbers.

**C** How much energy would be carried away in a one-quantum change?

**Solution** The energy carried away in a one-quantum change of energy is $\Delta E = E_{n+1} - E_n = hf$, so

$$\Delta E = hf = (6.63 \times 10^{-34}\text{ J}\cdot\text{s})(0.56\text{ Hz}) = 3.7 \times 10^{-34}\text{ J}$$

The energy carried away by a one-quantum change is such a small fraction of the total energy of the oscillator that we could not expect to measure it. Thus, the energy of an object-spring system decreases by such small quantum transitions that the decrease in energy of this macroscopic oscillator appears to be continuous.

## 27.2 THE PHOTOELECTRIC EFFECT AND THE PARTICLE THEORY OF LIGHT

In the latter part of the 19th century, experiments showed that when light is incident on certain metallic surfaces, electrons are emitted from the surfaces. This phenomenon is known as the **photoelectric effect,** and the emitted electrons are called **photoelectrons.** The first discovery of this phenomenon was made by Hertz, who was also the first to produce the electromagnetic waves predicted by Maxwell.

Figure 27.4 is a schematic diagram of a photoelectric effect apparatus. An evacuated glass tube known as a photocell contains a metal plate E connected to the negative terminal of a variable power supply. Another metal plate C is maintained at a positive potential by the power supply. When the tube is kept in the dark, the ammeter reads zero, indicating that there is no current in the circuit. However, when plate E is illuminated by light having a wavelength shorter than some particular wavelength that depends on the material used to make plate E, a current is detected by the ammeter, indicating a flow of charges across the gap between E and C. This current arises from photoelectrons emitted from the negative plate (the emitter) and collected at the positive plate (the collector).

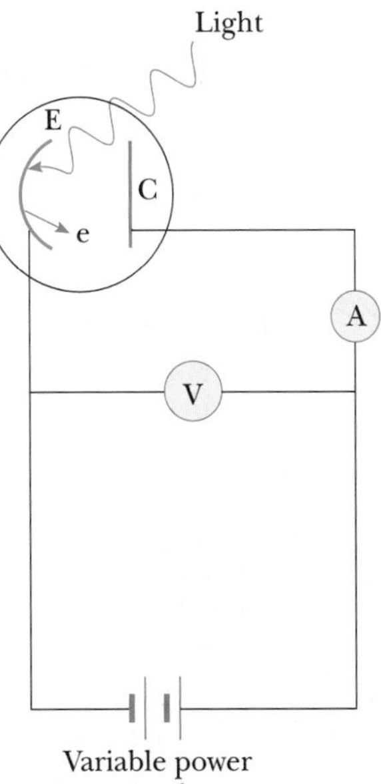


**FIGURE 27.4** Circuit diagram for observing the photoelectric effect. When light strikes plate E, photoelectrons are ejected from the plate. Electrons collected at C and passing through the ammeter constitute a current in the circuit.

Figure 27.5 is a plot of the photoelectric current versus the potential difference (voltage) $\Delta V$ between E and C for two light intensities. At large values of $\Delta V$, the current reaches a maximum value. In addition, the current increases as the incident light intensity increases, as you might expect. Finally, when $\Delta V$ is negative—that is, when the power supply in the circuit is reversed to make E positive and C negative—the current drops to a low value because most of the emitted photoelectrons are repelled by the now negative plate C. In this situation, only those electrons having a kinetic energy greater than the magnitude of $e\,\Delta V$ reach C, where $e$ is the charge on the electron.

When $\Delta V$ is equal to or more negative than $-\Delta V_s$, the **stopping potential,** no electrons reach C and the current is zero. The stopping potential is *independent* of the radiation intensity. The maximum kinetic energy of the photoelectrons is

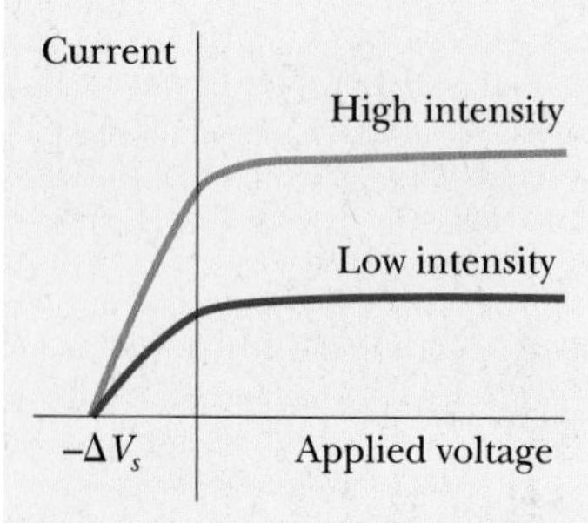


**FIGURE 27.5** Photoelectric current versus applied voltage for two light intensities. The current increases with intensity but reaches a saturation level for large values of $\Delta V$. At voltages equal to or less than $-\Delta V_s$, the current is zero.

related to the stopping potential through the relationship

$$KE_{\text{max}} = e\,\Delta V_s \qquad [27.4]$$

Several features of the photoelectric effect cannot be explained with classical physics or with the wave theory of light:

- No electrons are emitted if the incident light frequency falls below some **cutoff frequency,** $f_c$, that is characteristic of the material being illuminated. This is inconsistent with the wave theory, which predicts that the photoelectric effect should occur at any frequency, provided the light intensity is sufficiently high.
- The maximum kinetic energy of the photoelectrons is independent of light intensity. According to wave theory, light of higher intensity should carry more energy into the metal per unit time and therefore eject photoelectrons having higher kinetic energies.
- The maximum kinetic energy of the photoelectrons increases with increasing light frequency. The wave theory predicts no relationship between photoelectron energy and incident light frequency.
- Electrons are emitted from the surface almost instantaneously (less than $10^{-9}$ s after the surface is illuminated), even at low light intensities. Classically, we expect the photoelectrons to require some time to absorb the incident radiation before they acquire enough kinetic energy to escape from the metal.

A successful explanation of the photoelectric effect was given by Einstein in 1905, the same year he published his special theory of relativity. As part of a general paper on electromagnetic radiation, for which he received the Nobel prize in 1921, Einstein extended Planck's concept of quantization to electromagnetic waves. He suggested that a tiny packet of light energy or **photon** would be emitted when a quantized oscillator made a jump from an energy state $E_n = nhf$ to the next lower state $E_{n-1} = (n-1)hf$. Conservation of energy would require the decrease in oscillator energy, $hf$, to be equal to the photon's energy, $E$, so that

$$E = hf \qquad [27.5]$$

$h$ is Planck's constant and $f$ is the frequency of the light which is equal to the frequency of the Planckian oscillator.

The key point here is that the light energy lost by the emitter, $hf$, stays sharply localized in a tiny packet or particle called a photon. In Einstein's model, a photon is so localized that it can give *all* its energy $hf$ to a single electron in the metal. According to Einstein, the maximum kinetic energy for these liberated photoelectrons is

$$KE_{\text{max}} = hf - \phi \qquad [27.6]$$

where $\phi$ is called the **work function** of the metal. The work function represents the minimum energy with which an electron is bound in the metal and is on the order of a few electron volts. Table 27.1 lists work functions for various metals.

With the photon theory of light, we can explain the previously mentioned features of the photoelectric effect that cannot be understood using concepts of classical physics:

- That the effect is not observed below a certain cutoff frequency follows from the fact that the photon energy must be greater than or equal to $\phi$. If the energy of the incoming photon does not satisfy this condition, the electrons are never ejected from the surface, regardless of light intensity.
- That $KE_{\text{max}}$ is independent of the light intensity can be understood with the following argument. If the light intensity is doubled, the number of photons is doubled, which doubles the number of photoelectrons emitted. However, their maximum kinetic energy, which equals $hf - \phi$, depends only on the light frequency and the work function, not on the light intensity.
- That $KE_{\text{max}}$ increases with increasing frequency is easily understood with Equation 27.6.

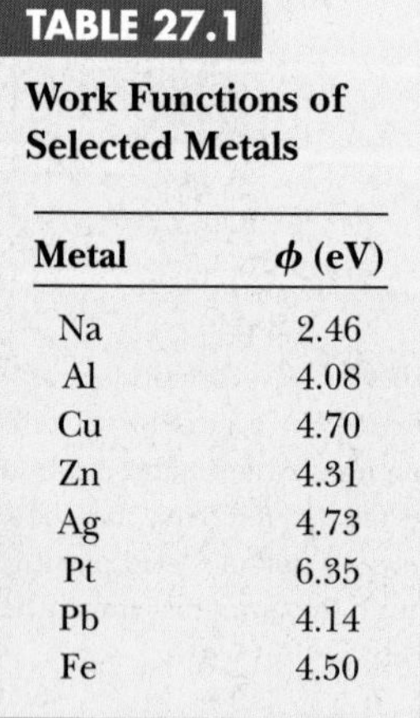
**TABLE 27.1**

**Work Functions of Selected Metals**

| Metal | $\phi$ (eV) |
|---|---|
| Na | 2.46 |
| Al | 4.08 |
| Cu | 4.70 |
| Zn | 4.31 |
| Ag | 4.73 |
| Pt | 6.35 |
| Pb | 4.14 |
| Fe | 4.50 |

- That the electrons are emitted almost instantaneously is consistent with the particle theory of light, in which the incident energy arrives at the surface in small spatial packets and there is a one-to-one interaction between photons and photoelectrons. In this interaction, the photon's energy is imparted to an electron that then has enough energy to leave the metal. This is in contrast to the wave theory, in which the incident energy is distributed uniformly over a large area of the metal surface.

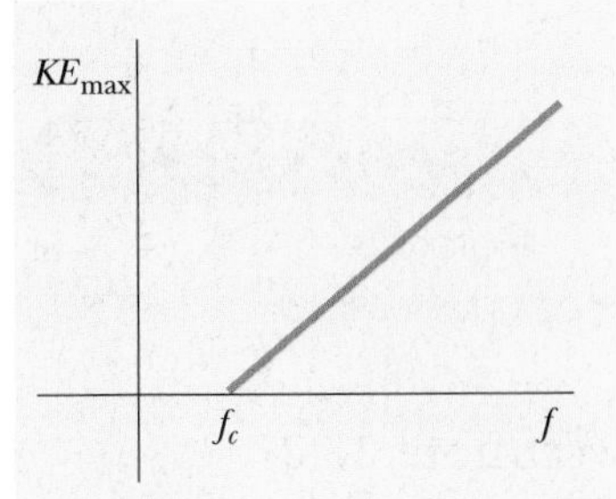


**FIGURE 27.6** A sketch of $KE_{max}$ versus frequency of incident light for photoelectrons in a typical photoelectric effect experiment. Photons with frequency less than $f_c$ do not have sufficient energy to eject an electron from the metal.

Experimental observation of a linear relationship between $f$ and $KE_{max}$ would be a final confirmation of Einstein's theory. Indeed, such a linear relationship is observed, as sketched in Figure 27.6. The intercept on the horizontal axis gives the cutoff frequency below which no photoelectrons are emitted, regardless of light intensity. The cutoff frequency is related to the work function through the relationship $f_c = \phi/h$. This cutoff frequency corresponds to a **cutoff wavelength** of

$$\lambda_c = \frac{c}{f_c} = \frac{c}{\phi/h} = \frac{hc}{\phi} \qquad [27.7]$$

where $c$ is the speed of light. Wavelengths *greater* than $\lambda_c$ incident on a material with a work function of $\phi$ do not result in the emission of photoelectrons.

## Example 27.3 The Energy of a "Yellow" Photon

Yellow light with a frequency of approximately $6.0 \times 10^{14}$ Hz is the predominant frequency in sunlight. What is the energy carried by a photon (a quantum of light) having this frequency?

**Solution** The energy carried by a photon having this frequency is given by Equation 27.5:

$$E = hf = (6.63 \times 10^{-34}\,\text{J}\cdot\text{s})(6.0 \times 10^{14}\,\text{Hz}) = 4.0 \times 10^{-19}\,\text{J} = 2.5\,\text{eV}$$

## Example 27.4 The Photoelectric Effect for Sodium

A sodium surface is illuminated with light of wavelength 0.300 $\mu$m. The work function for sodium is 2.46 eV. Find (a) the maximum kinetic energy of the ejected photoelectrons and (b) the cutoff wavelength for sodium.

**Solution**

**A** The energy of each photon of the illuminating light beam is

$$E = hf = \frac{hc}{\lambda} = \frac{(6.63 \times 10^{-34}\,\text{J}\cdot\text{s})(3.00 \times 10^{8}\,\text{m/s})}{3.00 \times 10^{-7}\,\text{m}}$$

$$= 6.63 \times 10^{-19}\,\text{J} = \frac{6.63 \times 10^{-19}\,\text{J}}{1.60 \times 10^{-19}\,\text{J/eV}} = 4.14\,\text{eV}$$

where we have used the conversion 1 eV = $1.6 \times 10^{-19}$ J. Using Equation 27.6 gives

$$KE_{max} = hf - \phi = 4.14\,\text{eV} - 2.46\,\text{eV} = 1.68\,\text{eV}$$

**B** The cutoff wavelength can be calculated from Equation 27.7 after we convert $\phi$ from electron volts to joules:

$$\phi = 2.46\,\text{eV} = (2.46\,\text{eV})(1.6 \times 10^{-19}\,\text{J/eV}) = 3.94 \times 10^{-19}\,\text{J}$$

$$\lambda_c = \frac{hc}{\phi} = \frac{(6.63 \times 10^{-34}\,\text{J}\cdot\text{s})(3.00 \times 10^{8}\,\text{m/s})}{3.94 \times 10^{-19}\,\text{J}}$$

$$= 5.05 \times 10^{-7}\,\text{m} = 505\,\text{nm}$$

This wavelength is in the yellow-green region of the visible spectrum.

**APPLICATION**

PHOTOCELLS

### Webnote 27.2

To see an applet of the photoelectric effect, go to *http://home.a-city.de/walter.fendt/physengl/photoeffect.htm*

### PHOTOCELLS

The photoelectric effect has many interesting applications using a device called the *photocell.* The photocell shown in Figure 27.4 acts much like a switch in an electric circuit with associated active elements in that it produces a current in the circuit when light of sufficiently high frequency falls on the cell, but it does not allow a current in the dark. Many practical devices in our everyday lives make use of photocells. For example, a use familiar to everyone is that of turning streetlights on at night and off in the morning. A photoelectric control unit in the base of the light activates a switch to turn off the streetlight when ambient light strikes it. Many garage-door systems and elevators use a light beam and a photocell as a safety feature in their design. When the light beam strikes the photocell, the electric current generated is sufficiently large to maintain a closed circuit. When an object or person blocks the light beam, the current is interrupted, which signals the door to open.

## 27.3 X-RAYS

In 1895 at the University of Wurzburg, Wilhelm Roentgen (1845–1923) was studying electrical discharges in low-pressure gases when he noted that a fluorescent screen glowed even when placed several meters from the gas discharge tube and even when black cardboard was placed between the tube and the screen. He concluded that the effect was caused by a mysterious type of radiation, which he called **x-rays** because of their unknown nature. Subsequent study showed that these rays traveled at or near the speed of light and that they could not be deflected by either electric or magnetic fields. This last fact indicated that x-rays did not consist of beams of charged particles, although the possibility that they were beams of uncharged particles remained.

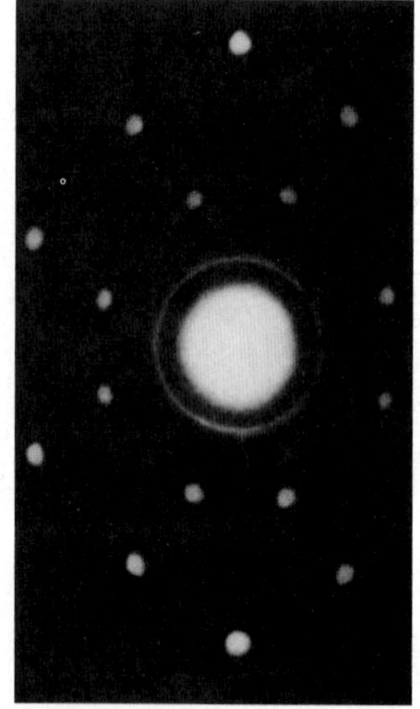

**FIGURE 27.7** X-ray diffraction pattern of NaCl.

In 1912 Max von Laue (1879–1960) suggested that if x-rays were electromagnetic waves with very short wavelengths, one should be able to diffract x-rays by using the regular atomic spacings of a crystal lattice as a diffraction grating, just as visible light is diffracted by a ruled grating. Shortly thereafter, researchers demonstrated that such a diffraction pattern could be observed, similar to that shown in Figure 27.7 for NaCl. The wavelengths of the x-rays were then determined from the diffraction data and the known values of the spacing between atoms in the crystal. X-ray diffraction has proved to be an invaluable technique for understanding the structure of matter. We shall discuss this subject in more detail in the next section.

Typical x-ray wavelengths are about 0.1 nm, which is on the order of the atomic spacing in a solid. We now know that x-rays are a part of the electromagnetic spectrum, characterized by frequencies higher than those of ultraviolet radiation and having the ability to penetrate most materials with relative ease.

X-rays are produced when high-speed electrons are suddenly slowed down, for example, when a metal target is struck by electrons that have been accelerated through a potential difference of several thousand volts. Figure 27.8a shows a schematic diagram of an x-ray tube. A current in the filament causes electrons to be emitted, and these freed electrons are accelerated toward a dense metal target, such as tungsten, which is held at a higher potential than the filament.

Figure 27.9 represents a plot of x-ray intensity versus wavelength for the spectrum of radiation emitted by an x-ray tube. Note that the spectrum has two distinct components. One component is a continuous broad spectrum that depends on the voltage applied to the tube. Superimposed on this component is a series of sharp, intense lines that depend on the nature of the target material. The accelerating voltage must exceed a certain value, called the **threshold voltage,** in order to observe these sharp lines, which represent radiation emitted by the target atoms as their electrons undergo rearrangements. We shall discuss this further in Chap-

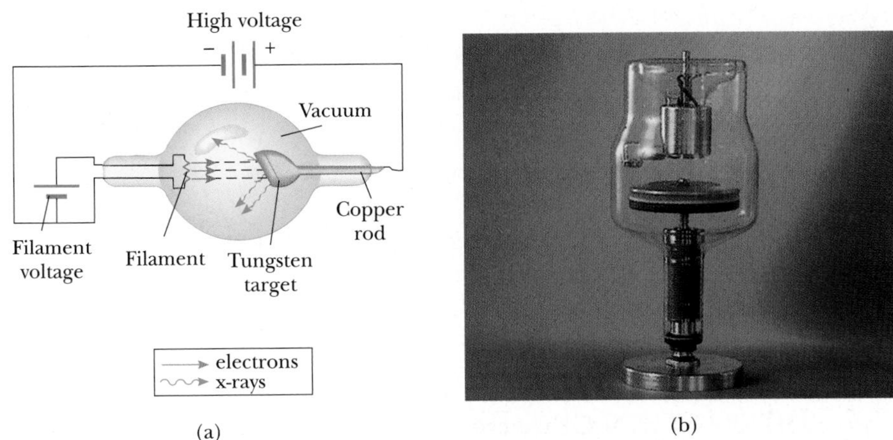


**FIGURE 27.8** (a) Diagram of an x-ray tube. (b) Photograph of an x-ray tube. *(Courtesy of GE Medical Systems)*

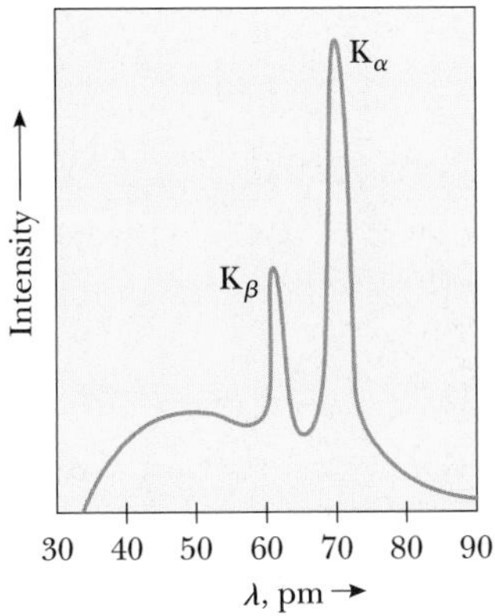


**FIGURE 27.9** The x-ray spectrum of a metal target consists of a broad continuous spectrum plus a number of sharp lines, which are due to *characteristic x-rays.* The data shown were obtained when 35-keV electrons bombarded a molybdenum target. Note that 1 pm = $10^{-12}$ m = $10^{-3}$ nm.

ter 28. The continuous radiation is sometimes called **bremsstrahlung,** a German word meaning "braking radiation." The term arises from the nature of the mechanism responsible for the radiation. That is, electrons emit radiation when they undergo an acceleration inside the target.

Figure 27.10 illustrates how x-rays are produced when an electron passes near a charged target nucleus. As an electron passes close to a positively charged nucleus contained in the target material, it is deflected from its path because of its electrical attraction to the nucleus, and hence experiences an acceleration. An analysis from classical physics shows that any charged particle will emit electromagnetic radiation when it is accelerated. (An example of this is the production of electromagnetic waves by accelerated charges in a radio antenna, as described in Chapter 21.) According to quantum theory, this radiation must appear in the form of photons. Because the radiated photon shown in Figure 27.10 carries energy, the electron must lose kinetic energy because of its encounter with the target nucleus. Let us consider an extreme example in which the electron loses all of its energy in a single collision. In this case, the initial energy of the electron ($e\,\Delta V$) is transformed completely into the energy of the photon ($hf_{\text{max}}$). In equation form, we have

$$e\,\Delta V = hf_{\text{max}} = \frac{hc}{\lambda_{\text{min}}} \qquad [27.8]$$

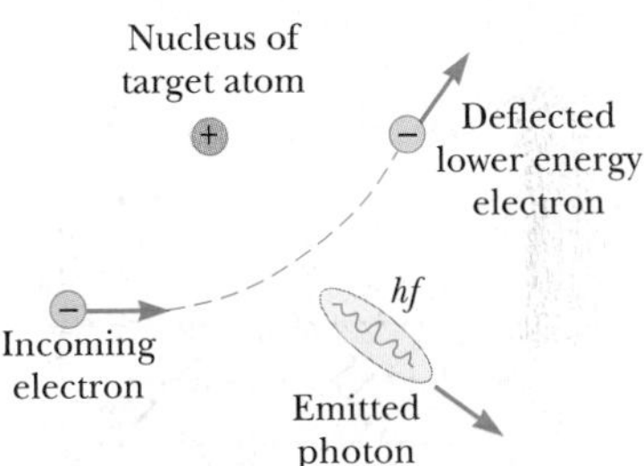


**FIGURE 27.10** An electron passing near a charged target atom experiences an acceleration, and a photon is emitted in the process.

where $e\,\Delta V$ is the energy of the electron after it has been accelerated through a potential difference of $\Delta V$ volts, and $e$ is the charge on the electron. This says that the shortest-wavelength radiation that can be produced is

$$\lambda_{\text{min}} = \frac{hc}{e\,\Delta V} \qquad [27.9]$$

The reason that all the radiation produced does not have this particular wavelength is because many of the electrons are not stopped in a single collision. This results in the production of the continuous spectrum of wavelengths.

**APPLICATION**

USING X-RAYS TO STUDY THE WORK OF MASTER PAINTERS

Interesting insights into the process of painting and revising a masterpiece are being revealed by x-rays. Long-wavelength x-rays are absorbed in varying degrees by some paints, such as those having lead, cadmium, chromium, or cobalt as a base. The x-ray interactions with the paints give contrast because the different elements in the paints have different electron densities. Also, thicker layers will absorb more than thin layers. To examine a painting by an old master, a film is placed behind it while it is x-rayed from the front. Ghost outlines of earlier paintings and earlier forms of the final masterpiece are sometimes revealed when the film is developed.

## Example 27.5 The Minimum X-Ray Wavelength

Calculate the minimum wavelength produced when electrons are accelerated through a potential difference of 100 000 V, a not-uncommon voltage for an x-ray tube.

**Solution** From Equation 27.9, we have

$$\lambda_{\min} = \frac{(6.63 \times 10^{-34}\ \text{J}\cdot\text{s})(3.00 \times 10^{8}\ \text{m/s})}{(1.60 \times 10^{-19}\ \text{C})(10^{5}\ \text{V})} = 1.24 \times 10^{-11}\ \text{m}$$

## 27.4 DIFFRACTION OF X-RAYS BY CRYSTALS

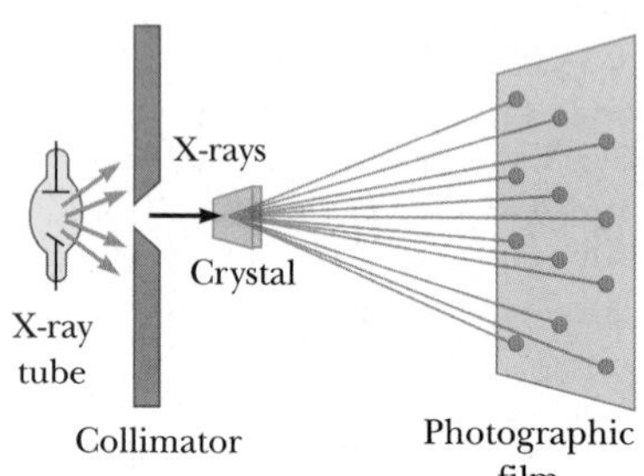


**FIGURE 27.11** Schematic diagram of the technique used to observe the diffraction of x-rays by a single crystal. The array of spots formed on the film by the diffracted beams is called a Laue pattern. (See Figure 27.7.)

In Chapter 24 we described how a diffraction grating can be used to measure the wavelength of light. In principle, the wavelength of *any* electromagnetic wave can be measured if a grating having a suitable line spacing can be found. The spacing between lines must be approximately equal to the wavelength of the radiation to be measured. X-rays are electromagnetic waves with wavelengths on the order of 0.1 nm. It would be impossible to construct a grating with such a small spacing. However, as noted in the previous section, Max von Laue suggested that the regular array of atoms in a crystal could act as a three-dimensional grating for observing the diffraction of x-rays.

One experimental arrangement for observing x-ray diffraction is shown in Figure 27.11. A narrow beam of x-rays with a continuous wavelength range is incident on a crystal such as sodium chloride. The diffracted radiation is very intense in certain directions, corresponding to constructive interference from waves reflected from layers of atoms in the crystal. The diffracted radiation is detected by a photographic film and forms an array of spots known as a Laue pattern. The crystal structure is determined by analyzing the positions and intensities of the various spots in the pattern.

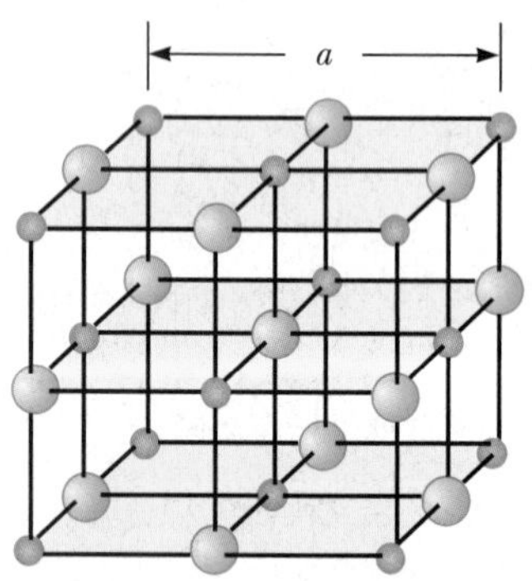


**FIGURE 27.12** A model of the cubic crystalline structure of sodium chloride. The blue spheres represent the $Cl^-$ ions, and the red spheres represent the $Na^+$ ions. The length of the cube edge is $a = 0.563$ nm.

The arrangement of atoms in a crystal of NaCl is shown in Figure 27.12. The smaller red spheres represent $Na^+$ ions and the larger blue spheres represent $Cl^-$ ions. The spacing between successive $Na^+$ (or $Cl^-$) ions in this cubic structure, denoted by the symbol $a$ in Figure 27.12, is approximately 0.563 nm.

A careful examination of the NaCl structure shows that the ions lie in various planes. The shaded areas in Figure 27.12 represent one example in which the atoms lie in equally spaced planes. Now suppose an x-ray beam is incident at grazing angle $\theta$ on one of the planes, as in Figure 27.13. The beam can be reflected from both the upper and lower plane of atoms. However, the geometric construction in Figure 27.13 shows that the beam reflected from the lower surface travels farther than the beam reflected from the upper surface by a distance of $2d \sin \theta$. The two portions of the reflected beam will combine to produce constructive in-

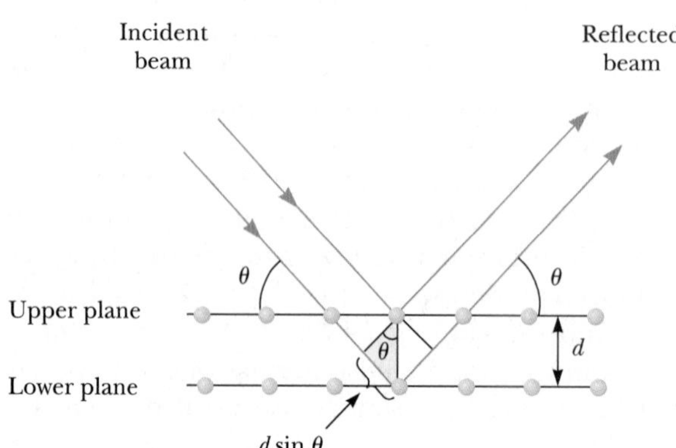


**FIGURE 27.13** A two-dimensional description of the reflection of an x-ray beam from two parallel crystalline planes separated by a distance $d$. The beam reflected from the lower plane travels farther than the one reflected from the upper plane by an amount equal to $2d \sin \theta$.

terference when this path difference equals some integral multiple of the wavelength λ. The condition for constructive interference is given by

$$2d \sin \theta = m\lambda \qquad (m = 1, 2, 3, \ldots) \qquad [27.10]$$

Bragg's law

This condition is known as **Bragg's law** after W. L. Bragg (1890–1971), who first derived the relationship. If the wavelength and diffraction angle are measured, Equation 27.10 can be used to calculate the spacing between atomic planes.

The method of x-ray diffraction to determine crystalline structures was thoroughly developed in England by W. H. Bragg and his son W. L. Bragg, who shared a Nobel prize in 1915 for their work. Since then, thousands of crystalline structures have been investigated. Most important, the technique of x-ray diffraction has been and is being used to determine the atomic arrangement of complex organic molecules such as proteins. Proteins are large molecules containing thousands of atoms that help to regulate and speed up chemical life processes in cells. Some proteins are amazing catalysts, speeding up the slow room-temperature reactions in cells by 17 orders of magnitude. In order to understand this incredible biochemical reactivity, it is important to determine the structure of these intricate molecules.

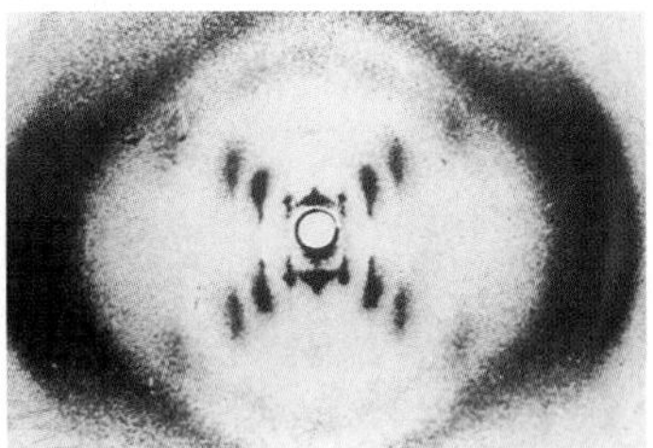

**FIGURE 27.14** An x-ray diffraction photograph of DNA taken by Rosalind Franklin. The cross pattern of spots was a clue that DNA has a helical structure. *(Science Source/Photo Researchers, Inc.)*

The main technique used to determine the molecular structure of proteins, DNA, and RNA is x-ray diffraction using x-rays of wavelength of about 1.0 A. This allows the experimenter to "see" individual atoms that are separated by about this distance in molecules. Because the biochemical x-ray diffraction sample is prepared in crystal form, the *geometry* (position of the bright spots in space) of the diffraction pattern is determined by the regular three-dimensional crystal lattice arrangement of molecules in the sample. The *intensities* of the bright diffraction spots are determined by the atoms and their electronic distributions in the fundamental building block of the crystal, the unit cell. Using complicated computational techniques, investigators can essentially deduce the molecular structure by matching the observed intensities of diffracted beams with a series of assumed atomic positions, resulting in the atomic structure and electron density of the molecule. Figure 27.14 shows a classic x-ray diffraction image of DNA made by Rosalind Franklin in 1952. This and similar x-ray diffraction photos played an important role in the determination of the double-helix structure of DNA by F. H. C. Crick and J. D. Watson in 1953. A model of the famous DNA double helix is shown in Figure 27.15.

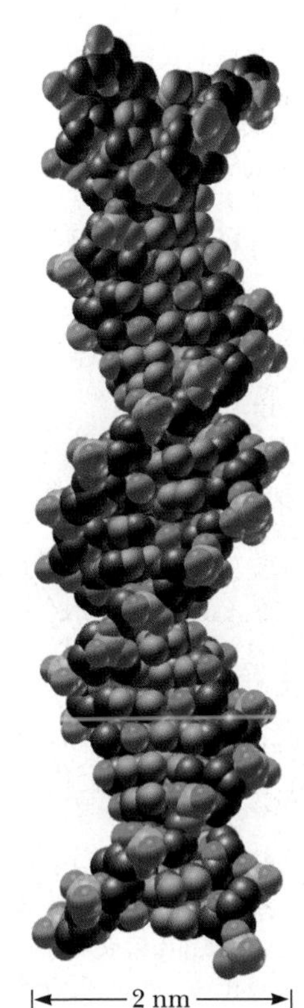


**FIGURE 27.15** The double-helix structure of DNA.

## Example 27.6 X-Ray Diffraction from Calcite

If the spacing between certain planes in a crystal of calcite ($CaCO_3$) is 0.314 nm, find the grazing angles at which first- and third-order interference will occur for x-rays of wavelength 0.070 nm.

**Solution** For first-order interference, the value of $m$ in Equation 27.10 is 1. Thus, the grazing angle corresponding to this order of interference is found as follows:

$$\sin \theta = \frac{m\lambda}{2d} = \frac{(0.0700 \text{ nm})}{2(0.314 \text{ nm})} = 0.111$$

$$\theta = 6.37°$$

In third-order interference, $m = 3$, and we find

$$\sin \theta = \frac{m\lambda}{2d} = \frac{3(0.0700 \text{ nm})}{2(0.314 \text{ nm})} = 0.334$$

$$\theta = 19.5°$$

### Webnote 27.3

To learn much more about x-ray crystallography, its techniques, and the equipment used for measurement, explore *http://www-structure.llnl.gov/xray/101index.html*

## 27.5 THE COMPTON EFFECT

Further justification for the photon nature of light came from an experiment conducted by Arthur H. Compton in 1923. In his experiment, Compton directed an x-ray beam of wavelength $\lambda_0$ toward a block of graphite. He found that the scattered x-rays had a slightly longer wavelength $\lambda$ than the incident x-rays, and hence the energies of the scattered rays were lower. The amount of energy reduction depended on the angle at which the x-rays were scattered. The change in wavelength, $\Delta\lambda$, between a scattered x-ray and an incident x-ray is called the **Compton shift.**

In order to explain this effect, Compton assumed that if a photon behaves like a particle, its collision with other particles is similar to that between two billiard balls. Hence, the x-ray photon carries both measurable *energy and momentum* and these two quantities must be conserved in a collision. If the incident photon collides with an electron initially at rest, as in Figure 27.16, the photon transfers some of its energy and momentum to the electron. As a consequence, the energy and frequency of the scattered photon are lowered and its wavelength increases. Applying relativistic energy and momentum conservation to the collision described in Figure 27.16, the shift in wavelength of the scattered photon is given by

The Compton shift formula

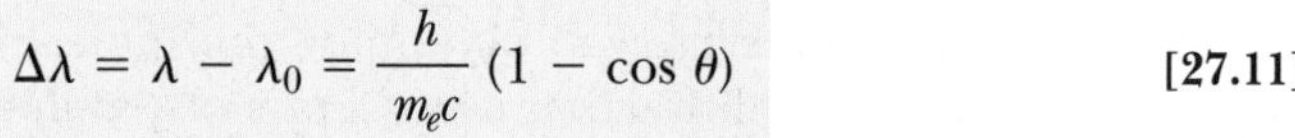

$$\Delta\lambda = \lambda - \lambda_0 = \frac{h}{m_e c}(1 - \cos\theta) \qquad [27.11]$$

where $m_e$ is the mass of the electron and $\theta$ is the angle between the directions of the scattered and incident photons. The quantity $h/m_e c$ is called the **Compton wavelength** and has a value of $h/m_e c = 0.002\,43$ nm. Note that the Compton wavelength is very small relative to the wavelengths of visible light, and hence the shift in wavelength would be difficult to detect if visible light were used. Furthermore, note that the Compton shift depends on the scattering angle $\theta$ and not on the wavelength. Experimental results for x-rays scattered from various targets obey Equation 27.11 and strongly support the photon concept.

**ARTHUR HOLLY COMPTON, AMERICAN PHYSICIST (1892–1962)**

Compton was born in Wooster, Ohio, and attended Wooster College and Princeton University. He became director of the laboratory at the University of Chicago where experimental work concerned with sustained chain reactions was conducted. This work was of central importance to the construction of the first atomic bomb. His discovery of the Compton effect and his work with cosmic rays led to his sharing the 1927 Nobel prize in physics with Charles Wilson. *(Courtesy of AIP Niels Bohr Library)*

An x-ray photon is scattered by an electron. The frequency of the scattered photon relative to that of the incident photon (a) increases, (b) decreases, or (c) remains the same.

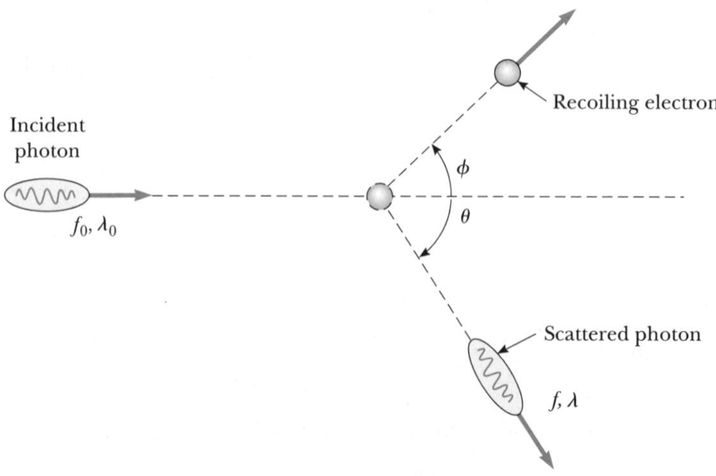


**FIGURE 27.16** Diagram representing Compton scattering of a photon by an electron. The scattered photon has less energy (or longer wavelength) than the incident photon.

**Quick Quiz 27.2** A photon of energy $E_0$ strikes a free electron, with the scattered photon of energy $E$ moving in the direction opposite that of the incident photon. In this Compton effect interaction, the resulting kinetic energy of the electron is (a) $E_0$, (b) $E$, (c) $E_0 - E$, (d) $E_0 + E$, (e) none of the above.

**Quick Quiz 27.3** A photon of energy $E_0$ strikes a free electron with the scattered photon of energy $E$ moving in the direction opposite that of the incident photon. In this Compton effect interaction, the resulting momentum of the electron is (a) $E_0/c$, (b) $< E_0/c$, (c) $> E_0/c$, (d) $(E_0 - E)/c$, (e) $(E - E_0)/c$.

## APPLYING PHYSICS 27.2

The Compton effect involves a change in wavelength as photons are scattered through different angles. Suppose we illuminate a piece of material with a beam of light and then view the material from different angles relative to the beam of light. Will we see a color change corresponding to the change in wavelength of the scattered light?

**Explanation** There will be a wavelength change for visible light scattered by the material, but the change will be far too small to detect as a color change. The largest possible wavelength change, at 180° scattering, will be twice the Compton wavelength, about 0.005 nm. This represents a change of less than 0.001% of the wavelength of red light. The Compton effect is only detectable for wavelengths that are very short to begin with, so that the Compton wavelength is an appreciable fraction of the incident wavelength. As a result, the usual radiation for observing the Compton effect is in the x-ray range of the electromagnetic spectrum.

## Example 27.7 Compton Scattering at 45°

X-rays of wavelength $\lambda_0 = 0.200\ 000$ nm are scattered from a block of material. The scattered x-rays are observed at an angle of 45.0° to the incident beam. Calculate the wavelength of the x-rays scattered at this angle.

**Solution** The shift in wavelength of the scattered x-rays is given by Equation 27.11. Taking $\theta = 45.0°$, we find that

$$\begin{aligned}\Delta\lambda &= \frac{h}{m_e c}(1 - \cos\theta)\\ &= \frac{6.626 \times 10^{-34}\ \text{J}\cdot\text{s}}{(9.11 \times 10^{-31}\ \text{kg})(3.00 \times 10^{8}\ \text{m/s})}(1 - \cos 45.0°)\\ &= 7.10 \times 10^{-13}\ \text{m} = 0.000\ 710\ \text{nm}\end{aligned}$$

Hence, the wavelength of the scattered x-ray at this angle is

$$\lambda = \Delta\lambda + \lambda_0 = 0.200\ 710\ \text{nm}$$

**EXERCISE** Find the fraction of energy lost by the photon in this collision.

**ANSWER** Fraction $= \Delta E/E = 0.003\ 54$

## 27.6 PHOTONS AND ELECTROMAGNETIC WAVES

Phenomena such as the photoelectric effect and the Compton effect offer iron-clad evidence that when light (or other forms of electromagnetic radiation) and matter interact, the light behaves as if it were composed of particles having energy $hf$ and momentum $h/\lambda$. An obvious question that arises at this point is, "How can light be considered a photon (in other words, a particle) when we know it is a wave?" On the one hand, we describe light in terms of photons having energy and momentum. On the other hand, we must also recognize that light and other electromagnetic waves exhibit interference and diffraction effects that are consistent only with a wave interpretation.

Which model is correct? Is light a wave or a particle? The answer depends on the phenomenon being observed. Some experiments can be better explained with the photon concept, whereas others are best described with a wave model. The end result is that we must use both models and admit that the true nature of light is not describable in terms of a single classical picture. We can say that

Light has a dual nature

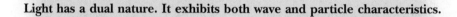

**Light has a dual nature. It exhibits both wave and particle characteristics.**

To understand why photons are compatible with electromagnetic waves, consider 2.5-MHz radio waves as an example. The energy of a photon having this frequency is only about $10^{-8}$ eV, too small to allow the photon to be detected. A sensitive radio receiver might require as many as $10^{10}$ of these photons to produce a detectable signal. Such a large number of photons would appear, on the average, as a continuous wave. With so many photons reaching the detector every second, it is unlikely that any graininess would appear in the detected signal. That is, with 2.5-MHz waves, we would not be able to detect the individual photons striking the antenna.

Now consider what happens as we go to higher frequencies. In the visible region, it is possible to observe both the particle characteristics and the wave characteristics of light. As we mentioned earlier, a light beam shows interference phenomena (thus, it is a wave) and at the same time can produce photoelectrons (thus, it is a particle). At even higher frequencies, the momentum and energy of the photons increase. Consequently, the particle nature of light becomes more evident than its wave nature. For example, absorption of an x-ray photon is easily detected as a single event, but wave effects are difficult to observe.

LOUIS DE BROGLIE, FRENCH PHYSICIST (1892–1987)

De Broglie was born in Dieppe, France. At the Sorbonne in Paris, he studied history in preparation for what he hoped to be a career in the diplomatic service. The world of science is lucky that he changed his career path to become a theoretical physicist. De Broglie was awarded the Nobel prize in 1929 for his discovery of the wave nature of electrons. "It would seem that the basic idea of quantum theory is the impossibility of imaging an isolated quantity of energy without associating with it a certain frequency." *(AIP Niels Bohr Library)*

## 27.7 THE WAVE PROPERTIES OF PARTICLES

The "magic" duality of light mentioned in the last section is not the only duality we have encountered in this course. The mass-energy duality mentioned in Chapter 26 is probably about as easy or difficult to accept as the duality of light. All these dualities should make it easy to accept that *matter* has a dual nature as well!

In 1924, in his doctoral dissertation, Louis de Broglie postulated that **because photons have wave and particle characteristics, perhaps all forms of matter have both properties.** This was a highly revolutionary idea with no experimental confirmation at that time. According to de Broglie, electrons, just like light, have a dual particle-wave nature. Let us follow de Broglie's argument for the wavelength of a material particle.

In Chapter 26 we found that the relationship between energy and momentum for a photon, which has a rest energy of zero, is $p = E/c$. We also know from Equation 27.5 that the energy of a photon is

Energy of a photon

$$E = hf = \frac{hc}{\lambda} \qquad [27.12]$$

Thus, the momentum of a photon can be expressed as

$$p = \frac{E}{c} = \frac{hc}{c\lambda} = \frac{h}{\lambda} \quad [27.13]$$

Momentum of a photon

From this equation we see that the photon wavelength can be specified by its momentum, or $\lambda = h/p$. De Broglie suggested that all material particles of momentum $p$ should have a characteristic wavelength $\lambda = h/p$. Because the momentum of a particle of mass $m$ and speed $v$ is $mv = p$, the **de Broglie wavelength** of a particle is

$$\lambda = \frac{h}{p} = \frac{h}{mv} \quad [27.14]$$

De Broglie's hypothesis

Furthermore, in analogy with photons, de Broglie postulated that the frequencies of matter waves (that is, waves associated with particles having nonzero rest energy) obey the Einstein relationship $E = hf$, so that

$$f = \frac{E}{h} \quad [27.15]$$

Frequency of matter waves

The dual nature of matter is quite apparent in Equations 27.14 and 27.15 because each contains both particle concepts ($mv$ and $E$) and wave concepts ($\lambda$ and $f$). The fact that these relationships had been established experimentally for photons made the de Broglie hypothesis that much easier to accept.

## THE DAVISSON-GERMER EXPERIMENT

De Broglie's proposal in 1923 that matter exhibits both wave and particle properties was first regarded as pure speculation. If particles such as electrons had wave-like properties, then under the correct conditions they should exhibit diffraction effects. In 1927, three years after de Broglie published his work, C. J. Davisson (1881–1958) and L. H. Germer (1896–1971) of the United States succeeded in measuring the wavelength of electrons. Their important discovery provided the first experimental confirmation of the matter waves proposed by de Broglie.

Interestingly, the intent of the initial Davisson–Germer experiment was not to confirm the de Broglie hypothesis. In fact, their discovery was made by accident (as is often the case). The experiment involved the scattering of low-energy electrons (about 54 eV) from a nickel target in a vacuum. During one experiment, the nickel surface was badly oxidized because of an accidental break in the vacuum system. After the nickel target was heated in a flowing stream of hydrogen to remove the oxide coating, electrons scattered by it exhibited intensity maxima and minima at specific angles. The experimenters finally realized that the nickel had formed large crystalline regions upon heating and that the regularly spaced planes of atoms in the crystalline regions served as a diffraction grating for electron matter waves (see Section 27.4).

Shortly thereafter, Davisson and Germer performed more extensive diffraction measurements on electrons scattered from single-crystal targets. Their results showed conclusively the wave nature of electrons and confirmed the de Broglie relation $\lambda = h/p$. In the same year, G. P. Thomson (1892–1975) of Scotland also observed electron diffraction patterns by passing electrons through very thin gold foils. Diffraction patterns have since been observed for helium atoms, hydrogen atoms, and neutrons. Hence, the universal nature of matter waves has been established in various ways.

**Quick Quiz 27.4** A nonrelativistic electron and a nonrelativistic proton are moving and have the same de Broglie wavelength. Which of the following are also the same for the two particles: (a) speed, (b) kinetic energy, (c) momentum, (d) frequency?

**Quick Quiz 27.5** We have seen two wavelengths assigned to the electron, the Compton wavelength and the de Broglie wavelength. Which is an actual *physical* wavelength associated with the electron: (a) the Compton wavelength, (b) the de Broglie wavelength, (c) both wavelengths, (d) neither wavelength?

### Example 27.8 The Wavelength of an Electron

Calculate the de Broglie wavelength for an electron ($m_e = 9.11 \times 10^{-31}$ kg) moving with a speed of $1.00 \times 10^7$ m/s.

**Solution** Equation 27.14 gives

$$\lambda = \frac{h}{m_e v} = \frac{6.63 \times 10^{-34}\ \text{J}\cdot\text{s}}{(9.11 \times 10^{-31}\ \text{kg})(1.00 \times 10^7\ \text{m/s})} = 7.28 \times 10^{-11}\ \text{m}$$

This wavelength corresponds to that of x-rays in the electromagnetic spectrum.

**EXERCISE** Find the de Broglie wavelength of a proton ($m_p = 1.67 \times 10^{-27}$ kg) moving with a speed of $1.00 \times 10^7$ m/s.

**ANSWER** $3.97 \times 10^{-14}$ m

### Example 27.9 The Wavelength of a Baseball

A baseball of mass 0.145 kg is thrown with a speed of 40.0 m/s. What is the de Broglie wavelength of the ball?

**Solution** From Equation 27.14, we have

$$\lambda = \frac{h}{mv} = \frac{6.63 \times 10^{-34}\ \text{J}\cdot\text{s}}{(0.145\ \text{kg})(40.0\ \text{m/s})} = 1.14 \times 10^{-34}\ \text{m}$$

This wavelength is much smaller than any aperture through which the baseball could possibly pass. This means that we could not observe diffraction effects, and as a result the wave properties of large-scale objects cannot be observed.

## APPLICATION: THE ELECTRON MICROSCOPE

**APPLICATION**
ELECTRON MICROSCOPES

A practical device that relies on the wave characteristics of electrons is the **electron microscope.** A *transmission* electron microscope, used for viewing flat, thin samples, is shown in Figure 27.17. In many respects it is similar to an optical microscope, but the electron microscope has a much greater resolving power because it can accelerate electrons to very high kinetic energies, giving them very short wavelengths. No microscope can resolve details that are significantly smaller than the wavelength of the radiation used to illuminate the object. Typically, the wavelengths of electrons are about 100 times smaller than those of the visible light used

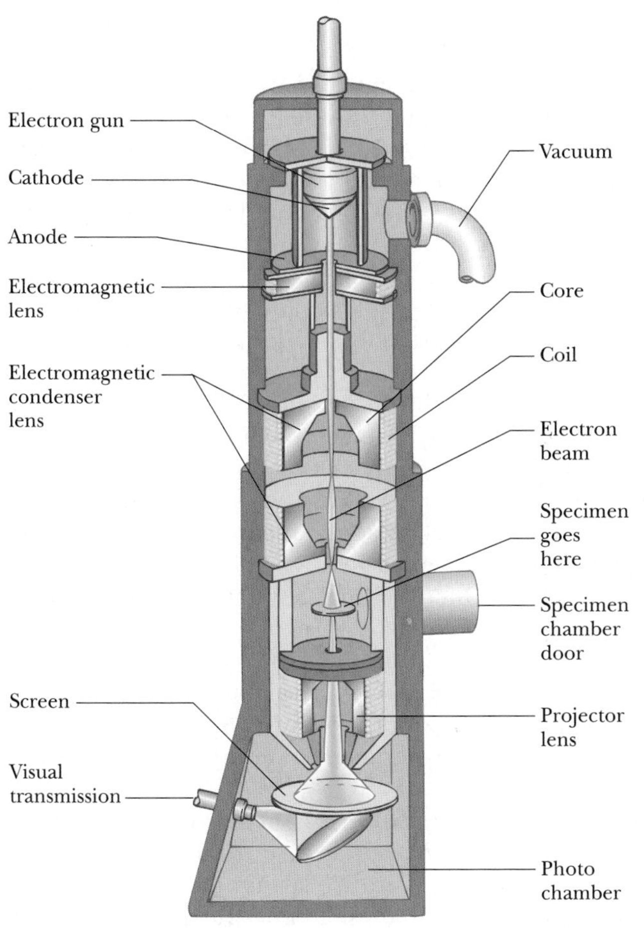


(a)

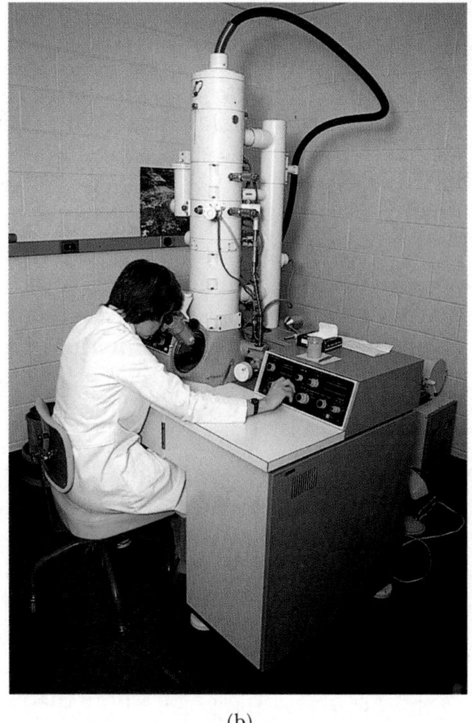

(b)

**FIGURE 27.17** (a) Diagram of a transmission electron microscope for viewing a thin, sectioned sample. The "lenses" that control the electron beam are magnetic deflection coils. (b) An electron microscope. *(W. Ormerod/Visuals Unlimited)*

in optical microscopes. (Radiation of the same wavelength as the electrons in an electron microscope is in the x-ray region of the spectrum.)

The electron beam in an electron microscope is controlled by electrostatic or magnetic deflection, which acts on the electrons to focus the beam to an image. Rather than examining the image through an eyepiece as in an optical microscope, the viewer looks at an image formed on a fluorescent screen. (The viewing screen must be fluorescent because otherwise the image produced would not be visible.)

## APPLYING PHYSICS 27.3

Electron microscopes (Fig. 27.17) take advantage of the wave nature of particles. Electrons are accelerated to high speeds, giving them a short de Broglie wavelength. Imagine an electron microscope using electrons with a de Broglie wavelength of 0.2 nm. Why don't we design a microscope using 0.2-nm *photons* to do the same thing?

**Explanation** Because electrons are charged particles, they interact electrically with the sample in the microscope and scatter according to the shape and density of various portions of the sample, providing a means of viewing the sample. Photons of wavelength 0.2 nm are uncharged and in the x-ray region of the spectrum. They tend to simply pass through the thin sample without interacting.

### Webnote 27.4

To see how an electron microscope works, visit *http://www.mos.org/sln/sem*

**ERWIN SCHRÖDINGER, AUSTRIAN THEORETICAL PHYSICIST (1887–1961)**

Schrödinger is best known as the creator of wave mechanics. He demonstrated the mathematical equivalence between wave mechanics and the more abstract matrix mechanics developed by Heisenberg. In 1933 Schrödinger left Germany and eventually settled at the Dublin Institute of Advanced Study, where he spent 17 happy, creative years working on problems in general relativity, cosmology, and the application of quantum physics to biology. In 1956 he returned home to Austria and to his beloved Tirolean mountains, where he died in 1961. *(AIP Emilio Segré Visual Archives)*

## 27.8 THE WAVE FUNCTION

De Broglie's revolutionary idea that particles should have a wave nature soon moved out of the realm of skepticism to the point where it was viewed as a necessary concept in understanding the subatomic world. In 1926 the Austrian-German physicist Erwin Schrödinger proposed a wave equation that described the manner in which matter waves change in space and time. The Schrödinger wave equation represents a key element in the theory of quantum mechanics. It is as important in quantum mechanics as Newton's laws in classical mechanics. Schrödinger's equation has been successfully applied to the hydrogen atom and to many other microscopic systems. Its importance in most aspects of modern physics cannot be overemphasized.

We shall not go through a mathematical derivation of Schrödinger's wave equation, nor shall we even state the equation here because it involves mathematical operations beyond the scope of this textbook, but we do wish to discuss the general features of the solution to this equation. When we attempt to solve the Schrödinger equation, the basic entity we seek to determine is a quantity $\Psi$, called the **wave function.** Each particle is represented by a wave function $\Psi$ that depends both on its position and on time. Once $\Psi$ is found, what information about the particle does it give? To answer this question, let us consider an analogy with light.

In Chapter 24 we discussed Young's double-slit experiment and explained experimental observations of the interference pattern solely in terms of the wave nature of light. Let us now discuss this same experiment in terms of both the wave and particle nature of light.

First, recall from Chapter 21 that the intensity of a light beam is proportional to the square of the electric field strength $E$ associated with the beam. That is, $I \propto E^2$. According to the wave model of light, there are certain points on the viewing screen where the net electric field is zero as a result of destructive interference of waves from the two slits. Because $E$ is zero at these points, the intensity is also zero, and the screen is dark at these locations. Likewise, at points on the screen at which constructive interference occurs, $E$ is large, as is the intensity; hence these locations are bright.

Now consider the same experiment when light is viewed as having a particle nature. The number of photons reaching a point on the screen per second increases as the intensity (brightness) increases. Thus, the number of photons that strikes a unit area on the screen each second is proportional to the square of the electric field, or $N \propto E^2$. Now let us consider the behavior of a single photon. What will be the fate of the photon as it moves through the slits in Young's experiment? From a probabilistic point of view, a photon has a high probability of striking the screen at a point at which the intensity (and $E^2$) is high and a low probability of striking the screen where the intensity is low.

When describing particles rather than photons, $\Psi$ rather than $E$ plays the role of the amplitude. Using an analogy with the description of light, we make the following interpretation of $\Psi$ for particles: If $\Psi$ is a wave function used to describe a single particle, the value of $\Psi^2$ at some location at a given time is proportional to the probability of finding the particle at that location at that time.

## 27.9 THE UNCERTAINTY PRINCIPLE

If you were to measure the position and speed of a particle at any instant, you would always be faced with experimental uncertainties in your measurements. According to classical mechanics, no fundamental barrier to an ultimate refinement of the apparatus or experimental procedures exists. In other words, it is possible, in principle, to make such measurements with arbitrarily small uncertainty. Quantum theory predicts, however, that such a barrier does exist. In 1927 Werner

Heisenberg (1901–1976) introduced this notion, which is now known as the **uncertainty principle:**

**If a measurement of position of a particle is made with precision $\Delta x$ and a simultaneous measurement of linear momentum is made with precision $\Delta p_x$, then the product of the two uncertainties can never be smaller than $h/4\pi$:**

$$\Delta x \, \Delta p_x \geq \frac{h}{4\pi} \qquad [27.16]$$

Uncertainty principle

In other words, **it is physically impossible to measure simultaneously the exact position and exact linear momentum of a particle.** If $\Delta x$ is very small, then $\Delta p_x$ is large, and vice versa.

To understand the physical origin of the uncertainty principle, consider the following thought experiment introduced by Heisenberg. Suppose you wish to measure the position and linear momentum of an electron as accurately as possible. You might be able to do this by viewing the electron with a powerful light microscope. For you to see the electron and thus determine its location, at least one photon of light must bounce off the electron, as shown in Figure 27.18a, and pass through the microscope into your eye, as shown in Figure 27.18b. When it strikes the electron, however, the photon transfers some unknown amount of its momentum to the electron. Thus, in the process of locating the electron very accurately (that is, by making $\Delta x$ very small), the light that enables you to succeed in your measurement changes the electron's momentum to some undeterminable extent (making $\Delta p_x$ very large.)

Let us analyze the collision by first noting that the incoming photon has a momentum of $h/\lambda$. As a result of the collision, the photon transfers part or all of its momentum along the $x$ axis to the electron. Thus, the *uncertainty* in the electron's momentum after the collision is as great as the momentum of the incoming photon: $\Delta p_x = h/\lambda$. Furthermore, because the photon also has wave properties, we expect to be able to determine the electron's position to within one wavelength of the light being used to view it, so $\Delta x = \lambda$. Multiplying these two uncertainties gives

$$\Delta x \, \Delta p_x = \lambda \left(\frac{h}{\lambda}\right) = h$$

**WERNER HEISENBERG, GERMAN THEORETICAL PHYSICIST (1901–1976)**

Heisenberg obtained his Ph.D. in 1923 at the University of Munich, where he studied under Arnold Sommerfeld. While physicists such as de Broglie and Schrödinger tried to develop physical models of the atom, Heisenberg developed an abstract mathematical model called *matrix mechanics* to explain the wavelengths of spectral lines. The more successful wave mechanics of Schrödinger, announced a few months after Heisenberg's model, was shown to be equivalent to Heisenberg's approach. Heisenberg made many other significant contributions to physics, including his famous uncertainty principle, for which he received the Nobel prize in 1932; the prediction of two forms of molecular hydrogen; and theoretical models of the nucleus. *(Courtesy of the University of Hamburg)*

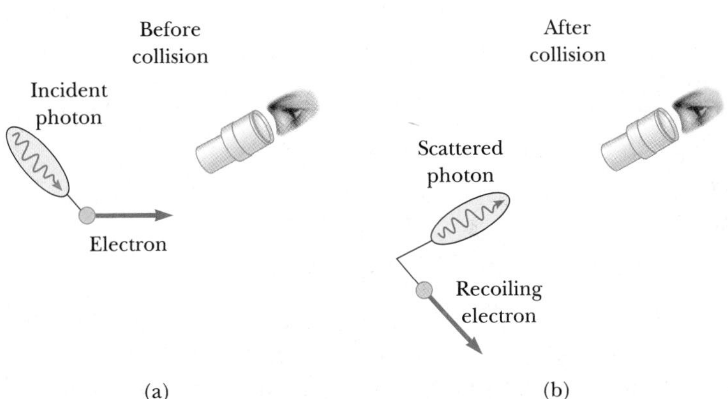


**FIGURE 27.18** A thought experiment for viewing an electron with a powerful microscope. (a) The electron is viewed before colliding with the photon. (b) The electron recoils (is disturbed) as the result of the collision with the photon.

The value $h$ represents the minimum in the products of the uncertainties. Because the uncertainty can always be greater than this minimum, we have

$$\Delta x \, \Delta p_x \geq h$$

Apart from the numerical factor $1/4\pi$ introduced by Heisenberg's more precise analysis, this agrees with Equation 27.16.

Another form of the uncertainty relationship sets a limit on the accuracy with which the energy of a system, $\Delta E$, can be measured in a finite time interval $\Delta t$:

$$\Delta E \, \Delta t \geq \frac{h}{4\pi} \qquad [27.17]$$

It can be inferred from this relationship that the energy of a particle cannot be measured with complete precision in a very short interval of time. Thus, when an electron is viewed as a particle, the uncertainty principle tells us that (a) its position and velocity cannot both be known precisely at the same time and (b) its energy can be uncertain for a period of time given by $\Delta t = h/(4\pi \, \Delta E)$.

## APPLYING PHYSICS 27.4

A common, but erroneous, description of the absolute zero of temperature is "that temperature at which all molecular motion ceases." How can the uncertainty principle be used to argue against this description?

**Explanation** Let us imagine molecules in a piece of material. The molecules are confined within the material, so there is a fixed uncertainty in their position along one axis, $\Delta x$, which is the size of the piece of material along this axis. If the molecular motion were to cease at absolute zero, we would be claiming that the molecule velocity is zero, with no uncertainty, $\Delta v = 0$. The product of zero uncertainty in velocity and a nonzero uncertainty in position will be zero, violating the uncertainty principle. Thus, even at absolute zero, there must be some molecular motion.

## Example 27.10 Locating an Electron

The speed of an electron is measured to be $5.00 \times 10^3$ m/s to an accuracy of 0.003 00%. Find the uncertainty in determining the position of this electron.

**Solution** The momentum of the electron is

$$p = m_e v = (9.11 \times 10^{-31}\ \text{kg})(5.00 \times 10^3\ \text{m/s}) = 4.56 \times 10^{-27}\ \text{kg} \cdot \text{m/s}$$

Because the uncertainty in $p$ is 0.00 300% of this value, we get

$$\Delta p = 0.000\,0300p = (0.000\,0300)(4.56 \times 10^{-27}\ \text{kg} \cdot \text{m/s})$$
$$= 1.37 \times 10^{-31}\ \text{kg} \cdot \text{m/s}$$

The uncertainty in position can now be calculated by using this value of $\Delta p$ and Equation 27.16:

$$\Delta x \, \Delta p_x \geq \frac{h}{4\pi}$$

$$\Delta x \geq \frac{h}{4\pi \, \Delta p_x} = \frac{6.626 \times 10^{-34}\ \text{J} \cdot \text{s}}{4\pi(1.37 \times 10^{-31}\ \text{kg} \cdot \text{m/s})}$$
$$= 0.384 \times 10^{-3}\ \text{m} = 0.384\ \text{mm}$$

**Webnote 27.5**

See an applet that shows you the conjugate effect of defining position and momentum. Visit *http://www2.adnc.com/~topquark/quantum/heisenbergmain.html*

**Example 27.11 Excited States of Atoms**

As we shall see in the next chapter, electrons in atoms can be found in higher states of energy called **excited states** for short periods of time. If the average time that an electron exists in one of these states is $1.00 \times 10^{-8}$ s, what is the minimum uncertainty in energy of the excited state?

**Solution** From the uncertainty principle in the form of Equation 27.17, we find that the minimum uncertainty in energy is

$$\Delta E = \frac{h}{4\pi\,\Delta t} = \frac{(6.63 \times 10^{-34}\,\text{J}\cdot\text{s})}{4\pi(1.00 \times 10^{-8}\,\text{s})} = 5.28 \times 10^{-27}\,\text{J} = 3.30 \times 10^{-8}\,\text{eV}$$

## 27.10 THE SCANNING TUNNELING MICROSCOPE[1]

One of the basic phenomena of quantum mechanics—tunneling—is at the heart of a very practical device, the scanning tunneling microscope, or STM, which enables us to get highly detailed images of surfaces with resolution comparable to the size of a single atom.

**APPLICATION**

SCANNING TUNNELING MICROSCOPES

Figure 27.19 is an image of a ring of 48 iron atoms located on a copper surface. Note the high quality of the STM image and the recognizable ring of iron atoms. What makes this image so remarkable is that at its resolution the size of the smallest detail that can be discerned is about 0.2 nm. For an ordinary microscope, the resolution is limited by the wavelength of the waves used to make the image. Thus, an optical microscope has a resolution no better than 200 nm, about half the wavelength of visible light, and so could never show the detail displayed in Figure 27.19. Electron microscopes can have a resolution of 0.2 nm by using electron waves of this wavelength, given by the de Broglie formula $\lambda = h/p$. The electron momentum $p$ required to give this wavelength is 10 000 eV/$c$, corresponding to an electron speed of 2% of the speed of light. Electrons traveling at this speed would penetrate into the interior of the sample in Figure 27.19 and so could not give us information about individual surface atoms.

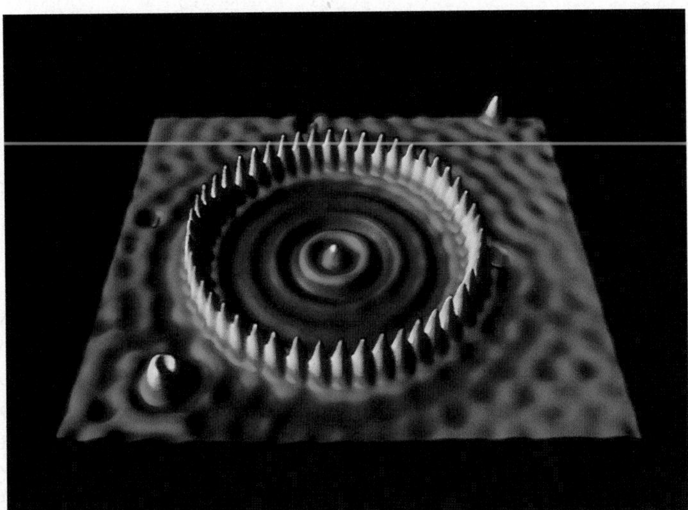

**FIGURE 27.19** This is a photograph of a "quantum-corral" consisting of a ring of 48 iron atoms located on a copper surface. The diameter of the ring is 143 nm, and the photograph was obtained using a low-temperature scanning tunneling microscope (STM). *(IBM Corporation Research Division)*

[1] This section was written by Roger A. Freedman and Paul K. Hansma, University of California, Santa Barbara.

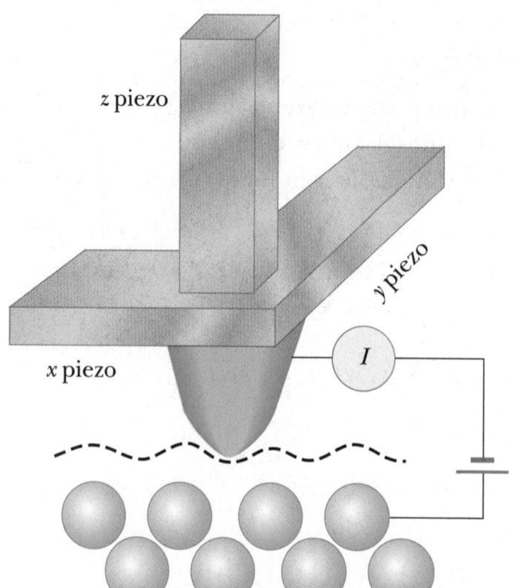


**FIGURE 27.20** A schematic view of an STM. The tip, shown as a rounded cone, is mounted on a piezoelectric *x, y, z* scanner. A scan of the tip over the sample can reveal contours of the surface down to the atomic level. An STM image is composed of a series of scans displaced laterally from each other. *(Based on a drawing from P. K. Hansma, V. B. Elings, O. Marti, and C. Bracker,* Science 242:209, 1988, *Copyright 1988 by the AAAS.)*

The STM achieves its very fine resolution by using the basic idea shown in Figure 27.20. A conducting probe with a sharp tip is brought near the surface to be studied. Because it is attracted to the positive ions in the surface, an electron in the surface has a lower total energy than an electron in the empty space between surface and tip. The same thing is true for an electron in the probe tip, which is attracted to the positive ions in the tip. In Newtonian mechanics, this means that electrons cannot move between surface and tip because they lack the energy to escape either material. Because the electrons obey quantum mechanics, however, they can "tunnel" across the barrier of empty space. By applying a voltage between surface and tip, the electrons can be made to tunnel preferentially from surface to tip. In this way the tip samples the distribution of electrons just above the surface.

Because of the nature of tunneling, the STM is very sensitive to the distance $z$ from tip to surface. The reason is that in the empty space between tip and surface, the electron wave function falls off exponentially with a decay length of order 0.1 nm—that is, the wave function decreases by $1/e$ over that distance. For distances $z$ greater than 1 nm (that is, beyond a few atomic diameters), essentially no tunneling takes place. This exponential behavior causes the current of electrons tunneling from surface to tip to depend very strongly on $z$. This sensitivity is the basis of the operation of the STM: By monitoring the tunneling current as the tip is scanned over the surface, scientists obtain a sensitive measure of the topography of the electron distribution on the surface. The result of this scan is used to make images like that in Figure 27.19. In this way, the STM can measure the height of surface features to within 0.001 nm, approximately 1/100 of an atomic diameter!

### Webnote 27.6

Take a fascinating look at an exquisite gallery of STM photographs. Jump to *http://www.almaden.ibm.com/vis/stm/gallery.html*

The STM has, however, one serious limitation: It depends on electrical conductivity of the sample and the tip. Unfortunately, most materials are not electrically conductive at their surface. Even metals such as aluminum are covered with nonconductive oxides. A newer microscope, the atomic force microscope, or AFM, overcomes this limitation. It measures the force between a tip and the sample rather than an electrical current. This force depends very strongly on the tip-sample separation, just as the electron tunneling current does for the STM. Thus the AFM has comparable sensitivity for measuring topography and has become widely used for technological applications.

### Webnote 27.7

Learn more about an atomic force microscope. Explore *http://stm2.nrl.navy.mil/how-afm/how-afm.html*

### Webnote 27.8

See some exciting visuals from an atomic force microscope. Go to *http://www.physics.purdue.edu/nanophys/afm.html*

Perhaps the most remarkable thing about the STM is that its operation is based on a quantum-mechanical phenomenon—tunneling—that was well understood in the 1920s, even though the first STM was not built until the 1980s. What other applications of quantum mechanics may yet be waiting to be discovered?

## SUMMARY

The characteristics of **blackbody radiation** cannot be explained using classical concepts. The peak of a blackbody radiation curve is given by **Wien's displacement law:**

$$\lambda_{\max} T = 0.2898 \times 10^{-2}\ \mathrm{m \cdot K} \qquad [27.1]$$

where $\lambda_{\max}$ is the wavelength at which the curve peaks and $T$ is the absolute temperature of the object emitting the radiation.

Planck first introduced the quantum concept when he assumed that the subatomic oscillators responsible for blackbody radiation could have only discrete amounts of energy given by

$$E_n = nhf \qquad [27.2]$$

where $n$ is a positive integer called a **quantum number** and $f$ is the frequency of vibration of the resonator.

The **photoelectric effect** is a process whereby electrons are ejected from a metal surface when light is incident on that surface. Einstein provided a successful explanation of this effect by extending Planck's quantum hypothesis to electromagnetic waves. In this model, light is viewed as a stream of particles called photons, each with energy $E = hf$, where $f$ is the light frequency and $h$ is **Planck's constant.** The maximum kinetic energy of the ejected photoelectrons is

$$KE_{\max} = hf - \phi \qquad [27.6]$$

where $\phi$ is the **work function** of the metal.

**X-rays** are produced when high-speed electrons are suddenly decelerated. When electrons have been accelerated through a voltage $V$, the shortest-wavelength radiation that can be produced is

$$\lambda_{\min} = \frac{hc}{e\Delta V} \qquad [27.9]$$

The regular array of atoms in a crystal can act as a diffraction grating for x-rays and for electrons. The condition for constructive interference of the diffracted rays is given by **Bragg's law:**

$$2d \sin\theta = m\lambda \qquad (m = 1, 2, 3, \ldots) \qquad [27.10]$$

X-rays from an incident beam are scattered at various angles by electrons in a target such as carbon. In such a scattering event, a shift in wavelength is observed for the scattered x-rays. This phenomenon is known as the **Compton shift.** Conservation of momentum and energy applied to a photon-electron collision yields the following expression for the shift in wavelength of the scattered x-rays:

$$\Delta\lambda = \lambda - \lambda_0 = \frac{h}{m_e c}(1 - \cos\theta) \qquad [27.11]$$

where $m_e$ is the mass of the electron, $c$ is the speed of light, and $\theta$ is the scattering angle.

De Broglie proposed that all matter has both a particle and a wave nature. The **de Broglie wavelength** of any particle of mass $m$ and speed $v$ is

$$\lambda = \frac{h}{p} = \frac{h}{mv} \qquad [27.14]$$

De Broglie also proposed that the frequencies of the waves associated with particles obey the Einstein relationship, $E = hf$.

In the theory of **quantum mechanics,** each particle is described by a quantity $\Psi$ called the **wave function.** The probability of finding the particle at a particular point at some instant is proportional to $\Psi^2$. Quantum mechanics is very successful in describing the behavior of atomic and molecular systems.

According to Heisenberg's **uncertainty principle,** it is impossible to measure simultaneously the exact position and exact momentum of a particle. If $\Delta x$ is the uncertainty in the measured position and $\Delta p_x$ the uncertainty in the momentum, the product $\Delta x \, \Delta p_x$ is given by

$$\Delta x \, \Delta p_x \geq \frac{h}{4\pi} \qquad [27.16]$$

Also,

$$\Delta E \, \Delta t \geq \frac{h}{4\pi} \qquad [27.17]$$

where $\Delta E$ is the uncertainty in the energy of the particle and $\Delta t$ is the uncertainty in the time it takes to measure the energy.

## CONCEPTUAL QUESTIONS

1. If you observe objects inside a very hot kiln, it is difficult to discern the shapes of the objects. Why?
2. Why is an electron microscope more suitable than an optical microscope for "seeing" objects of an atomic size?
3. Are blackbodies black?
4. Why is it impossible to simultaneously measure the position and velocity of a particle with infinite accuracy?
5. All objects radiate energy. Why, then, are we not able to see all objects in a dark room?
6. Is light a wave or a particle? Support your answer by citing specific experimental evidence.
7. A student claims that he is going to eject electrons from a piece of metal by placing a radio transmitter antenna adjacent to the metal and sending a strong AM radio signal into the antenna. The work function of a metal is typically a few electron volts. Will this work?
8. In view of Heisenberg's uncertainty principle, how can we predict such things as maximum height and range when describing something like a baseball?
9. In the photoelectric effect, explain why the stopping potential depends on the frequency of the light but not on the intensity.
10. Which has more energy, a photon of ultraviolet radiation or a photon of yellow light?
11. Why does the existence of a cutoff frequency in the photoelectric effect favor a particle theory of light rather than a wave theory?
12. What effect, if any, would you expect the temperature of a material to have on the ease with which electrons can be ejected from it in the photoelectric effect?
13. If photons that cause photoelectrons to be released all have the same energy, how can the photoelectrons have different energies?
14. The brightest star in the constellation Lyra is the bluish star Vega, whereas the brightest star in Bootes is the reddish star Arcturus. How do you account for the difference in color of the two stars?
15. If the photoelectric effect is observed for one metal, can you conclude that the effect will also be observed for another metal under the same conditions? Explain.
16. Discuss whether the behavior of an electron is mainly wave-like or particle-like in each of the following situations: (a) traversing a circular orbit in a magnetic field; (b) absorbing a photon and being photoelectrically ejected from the surface of a metal; (c) forming an interference pattern.
17. If a photon is deflected via the Compton effect, can its wavelength ever become shorter?
18. In an adult, 13% of the bone marrow is in the head. Discuss this in relation to dental x-rays.

## PROBLEMS

1, 2, 3 = straightforward, intermediate, challenging ☐ = full solution available in Student Solutions Manual/Study Guide

**web** = solution posted at **http://info.brookscole.com/serway** [icon] = biomedical application

### Section 27.1 Blackbody Radiation and Planck's Hypothesis

### Section 27.2 The Photoelectric Effect and the Particle Theory of Light

1. (a) What is the surface temperature of Betelgeuse, a red giant star in the constellation of Orion, which radiates with a peak wavelength of about 970 nm? (b) Rigel, a bluish-white star in Orion, radiates with a peak wavelength of 145 nm. Find the temperature of Rigel's surface.
2. (a) Lightning produces a maximum air temperature on the order of $10^4$ K, while (b) a nuclear explosion produces a temperature on the order of $10^7$ K. Use Wien's displacement law to find the order of magnitude of the wavelength of the thermally produced photons radiated with greatest intensity by each of these sources. Name the part of the electromagnetic spectrum where you would expect each to radiate most strongly.
3. (a) Assuming that the tungsten filament of a lightbulb is a blackbody, determine its peak wavelength if its temperature is 2 900 K. (b) Why does your answer to part (a) suggest that more energy from a lightbulb goes into heat than into light?
4. Calculate the energy, in electron volts, of a photon whose frequency is (a) 620 THz, (b) 3.10 GHz, (c) 46.0 MHz. (d) Determine the corresponding wavelengths for these

photons and state the classification of each on the electromagnetic spectrum.

**5.** Calculate the energy in electron volts of a photon having a wavelength in (a) the microwave range, 5.00 cm; (b) the visible light range, 500 nm; and (c) the x-ray range, 5.00 nm.

**6.** A sodium-vapor lamp has a power output of 1 000 W. Using 589.3 nm as the average wavelength of this source, calculate the number of photons emitted per second.

**7.** An FM radio transmitter has a power output of 150 kW and operates at a frequency of 99.7 MHz. How many photons per second does the transmitter emit?

**8.** The threshold of dark-adapted (scotopic) vision is $4.0 \times 10^{-11}$ W/m$^2$ at a central wavelength of 500 nm. If light with this intensity and wavelength enters the eye when the pupil is open to its maximum diameter of 8.5 mm, how many photons per second enter the eye?

**9.** A 1.5-kg mass vibrates at an amplitude of 3.0 cm on the end of a spring of spring constant 20 N/m. (a) If the energy of the spring is quantized, find its quantum number. (b) If $n$ changes by 1, find the fractional change in energy of the spring.

**10.** A 0.50-kg mass falls from a height of 3.0 m. If all of the energy of this mass could be converted to visible light of wavelength $5.0 \times 10^{-7}$ m, how many photons would be produced?

**11.** web When light of wavelength 350 nm falls on a potassium surface, electrons are emitted that have a maximum kinetic energy of 1.31 eV. Find (a) the work function of potassium, (b) the cutoff wavelength, and (c) the frequency corresponding to the cutoff wavelength.

**12.** Electrons are ejected from a metallic surface with speeds ranging up to $4.6 \times 10^5$ m/s when light with a wavelength of $\lambda = 625$ nm is used. (a) What is the work function of the surface? (b) What is the cutoff frequency for this surface?

**13.** Molybdenum has a work function of 4.20 eV. (a) Find the cutoff wavelength and threshold frequency for the photoelectric effect. (b) Calculate the stopping potential if the incident light has a wavelength of 180 nm.

**14.** Lithium, beryllium, and mercury have work functions of 2.30 eV, 3.90 eV, and 4.50 eV, respectively. If 400-nm light is incident on each of these metals, determine (a) which metals exhibit the photoelectric effect and (b) the maximum kinetic energy for the photoelectrons in each case.

**15.** From the scattering of sunlight, Thomson calculated that the classical radius of the electron has a value of $2.82 \times 10^{-15}$ m. If sunlight having an intensity of 500 W/m$^2$ falls on a disk with this radius, find the time required to accumulate 1.00 eV of energy. Assume that light is a classical wave and that the light striking the disk is completely absorbed. How does your value compare with the observation that photoelectrons are promptly (within $10^{-9}$ s) emitted?

16. An isolated copper sphere of radius 5.00 cm, initially uncharged, is illuminated by ultraviolet light of wavelength 200 nm. What charge will the photoelectric effect induce on the sphere? The work function for copper is 4.70 eV.

17. When light of wavelength 254 nm falls on cesium, the required stopping potential is 3.00 V. If light of wavelength 436 nm is used, the stopping potential is 0.900 V. Use this information to plot a graph like that shown in Figure 27.6, and from the graph determine the cutoff frequency for cesium and its work function.

18. Ultraviolet light is incident normally on the surface of a certain substance. The binding energy of the electrons in this substance is 3.44 eV. The incident light has an intensity of 0.055 W/m$^2$. The electrons are photoelectrically emitted with a maximum speed of $4.2 \times 10^5$ m/s. How many electrons are emitted from a square centimeter of the surface each second? Assume that the absorption of every photon ejects an electron.

## Section 27.3 X-Rays

**19.** The extremes of the x-ray portion of the electromagnetic spectrum range from approximately $1.0 \times 10^{-8}$ m to $1.0 \times 10^{-13}$ m. Find the minimum accelerating voltages required to produce wavelengths at these two extremes.

**20.** Calculate the minimum wavelength x-ray that can be produced when a target is struck by an electron that has been accelerated through a potential difference of (a) 15.0 kV, (b) 100 kV.

**21.** What minimum accelerating voltage would be required to produce an x-ray with a wavelength of 0.0300 nm?

## Section 27.4 Diffraction of X-Rays by Crystals

**22.** A monochromatic x-ray beam is incident on a NaCl crystal surface where $d = 0.353$ nm. The second-order maximum in the reflected beam is found when the angle between the incident beam and the surface is 20.5°. Determine the wavelength of the x-rays.

**23.** Potassium iodide has an interplanar spacing of $d = 0.296$ nm. A monochromatic x-ray beam shows a first-order diffraction maximum when the grazing angle is 7.6°. Calculate the x-ray wavelength.

**24.** The spacing between certain planes in a crystal is known to be 0.30 nm. Find the smallest grazing angle at which constructive interference will occur for wavelength 0.070 nm.

**25.** web X-rays of wavelength 0.140 nm are reflected from a certain crystal, and the first-order maximum occurs at an angle of 14.4°. What value does this give for the interplanar spacing of this crystal?

## Section 27.5 The Compton Effect

**26.** X-rays are scattered from electrons in a carbon target. The measured wavelength shift is $1.50 \times 10^{-3}$ nm. Calculate the scattering angle.

**27.** Calculate the energy and momentum of a photon of wavelength 700 nm.

**28.** A beam of 0.68-nm photons undergoes Compton scattering from free electrons. What are the energy and momentum of the photons that emerge at a 45° angle with respect to the incident beam?

**29.** A 0.001 6-nm photon scatters from a free electron. For what (photon) scattering angle will the recoiling electron and scattered photon have the same kinetic energy?

30. X-rays with an energy of 300 keV undergo Compton scattering from a target. If the scattered rays are deflected at 37.0° relative to the direction of the incident rays, find (a) the Compton shift at this angle, (b) the energy of the scattered x-ray, and (c) the kinetic energy of the recoiling electron.

31. A 0.110-nm photon collides with a stationary electron. After the collision, the electron moves forward and the photon recoils backward. Find the momentum and kinetic energy of the electron.

32. After a 0.800 nm x-ray photon scatters from a free electron, the electron recoils with a speed equal to $1.40 \times 10^6$ m/s. (a) What was the Compton shift in the photon's wavelength? (b) Through what angle was the photon scattered?

33. A 0.45-nm x-ray photon is deflected through a 23° angle after scattering from a free electron. (a) What is the kinetic energy of the recoiling electron? (b) What is its speed?

## Section 27.7 The Wave Properties of Particles

34. Calculate the de Broglie wavelength for an electron that has kinetic energy (a) 50.0 eV and (b) 50.0 keV (ignore relativistic effects).

35. (a) If the wavelength of an electron is equal to $5.00 \times 10^{-7}$ m, how fast is it moving? (b) If the electron has a speed of $1.00 \times 10^7$ m/s, what is its wavelength?

36. Through what potential difference would an electron have to be accelerated from rest to give it a de Broglie wavelength of $1.0 \times 10^{-10}$ m?

37. The nucleus of an atom is on the order of $10^{-14}$ m in diameter. For an electron to be confined to a nucleus, its de Broglie wavelength would have to be of this order of magnitude or smaller. (a) What would be the kinetic energy of an electron confined to this region? (b) On the basis of this result, would you expect to find an electron in a nucleus? Explain.

38. After learning about de Broglie's hypothesis that particles of momentum $p$ have wave characteristics with wavelength $\lambda = h/p$, an 80-kg student has grown concerned about being diffracted when passing through a 75-cm-wide doorway. Assume that significant diffraction occurs when the width of the diffraction aperture is less than 10 times the wavelength of the wave being diffracted. (a) Determine the order of magnitude of the maximum speed at which the student can pass through the doorway in order to be significantly diffracted. (b) With that speed, how long will it take the student to pass through a doorway in a wall 15 cm thick? Compare your order-of-magnitude result to the currently accepted age of the Universe, which is $4 \times 10^{17}$ s. (c) Should this student worry about being diffracted?

39. De Broglie postulated that the relationship $\lambda = h/p$ is
web valid for relativistic particles. What is the de Broglie wavelength for a (relativistic) electron whose kinetic energy is 3.00 MeV?

40. At what speed must an electron move so that its de Broglie wavelength equals its Compton wavelength? (*Hint:* This electron is relativistic.)

41. The resolving power of a microscope is proportional to the wavelength used. A resolution of approximately $1.0 \times 10^{-11}$ m (0.010 nm) would be required in order to "see" an atom. (a) If electrons were used (electron microscope), what minimum kinetic energy would be required for the electrons? (b) If photons were used, what minimum photon energy would be needed to obtain $1.0 \times 10^{-11}$ m resolution?

## Section 27.9 The Uncertainty Principle

42. A 50.0-g ball moves at 30.0 m/s. If its speed is measured to an accuracy of 0.10%, what is the minimum uncertainty in its position?

43. A 0.50-kg block rests on the icy surface of a frozen pond, which we can assume to be frictionless. If the location of the block is measured to a precision of 0.50 cm, what speed must the block acquire because of the measurement process?

44. Suppose Fuzzy, a quantum-mechanical duck, lives in a world in which $h = 2\pi$ J·s. Fuzzy has a mass of 2.00 kg and is initially known to be within a pond 1.00 m wide. (a) What is the minimum uncertainty in his speed? (b) Assuming this uncertainty in speed to prevail for 5.00 s, determine the uncertainty in position after this time.

45. Suppose optical radiation ($\lambda = 5.00 \times 10^{-7}$ m) is used to determine the position of an electron to within the wavelength of the light. What will be the resulting uncertainty in the electron's velocity?

46. (a) Show that the kinetic energy of a nonrelativistic particle can be written in terms of its momentum as $KE = p^2/2m$. (b) Use the results of (a) to find the minimum kinetic energy of a proton confined within a nucleus having a diameter of $1.0 \times 10^{-15}$ m.

## ADDITIONAL PROBLEMS

47. Figure P27.47 shows the spectrum of light emitted by a firefly. Determine the temperature of a blackbody that would emit radiation peaked at the same frequency. Based on your result, would you say firefly radiation is blackbody radiation?

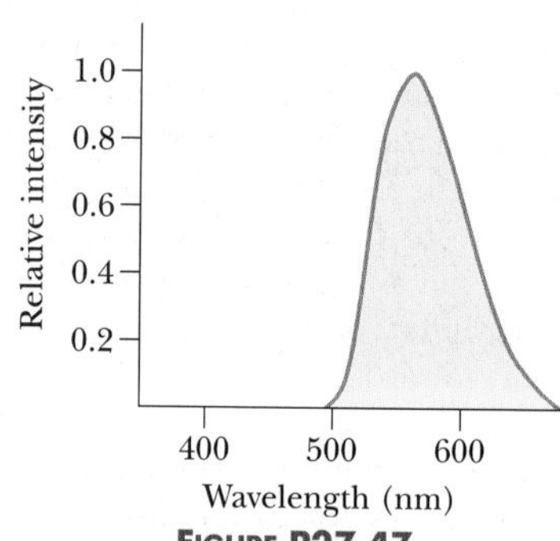


**FIGURE P27.47**

48. An x-ray tube is operated at 50 000 V. (a) Find the minimum wavelength of the radiation emitted by this tube. (b) If this radiation is directed at a crystal, the first-order maximum in the reflected radiation occurs when the grazing angle is 2.50°. What is the spacing between reflecting planes in the crystal?

49. The spacing between planes of nickel atoms in a nickel crystal is 0.352 nm. At what angle does a second-order Bragg reflection occur in nickel for 11.3-keV x-rays?

50. Johnny Jumper's favorite trick is to step out of his 16th story window and fall 50.0 m into a pool. A news reporter takes a picture of 75.0-kg Johnny just before he makes a splash, using an exposure time of 5.00 ms. Find (a) Johnny's de Broglie wavelength at this moment, (b) the uncertainty of his kinetic energy measurement during such a period of time, and (c) the percent error caused by such an uncertainty.

51. Photons of wavelength 450 nm are incident on a metal. The most energetic electrons ejected from the metal are bent into a circular arc of radius 20.0 cm by a magnetic field with a magnitude of $2.00 \times 10^{-5}$ T. What is the work function of the metal?

52. A 200-MeV photon is scattered at 40.0° by a free proton initially at rest. Find the energy (in MeV) of the scattered photon.

53. web A light source of wavelength $\lambda$ illuminates a metal and ejects photoelectrons with a maximum kinetic energy of 1.00 eV. A second light source of wavelength $\lambda/2$ ejects photoelectrons with a maximum kinetic energy of 4.00 eV. What is the work function of the metal?

54. Red light of wavelength 670 nm produces photoelectrons from a certain photoemissive material. Green light of wavelength 520 nm produces photoelectrons from the same material with 1.50 times the maximum kinetic energy. What is the material's work function?

55. How fast must an electron be moving if all its kinetic energy is lost to a single x-ray photon (a) at the long-wavelength end of the x-ray electromagnetic spectrum with a wavelength of $1.00 \times 10^{-8}$ m; (b) at the short-wavelength end of the x-ray electromagnetic spectrum with a wavelength of $1.00 \times 10^{-13}$ m?

56. Show that if an electron were confined inside an atomic nucleus of diameter $2.0 \times 10^{-15}$ m, it would have to be moving relativistically, while a proton confined to the same nucleus can be moving at less than one-tenth the speed of light.

57. A photon strikes a metal with a work function of $\phi$ and produces a photoelectron with a de Broglie wavelength equal to the wavelength of the original photon. (a) Show that the energy of this photon must have been given by

$$E = \frac{\phi(m_ec^2 - \phi/2)}{m_ec^2 - \phi}$$

where $m_e$ is the mass of the electron. (*Hint:* Begin with conservation of energy, $E + m_ec^2 = \phi + \sqrt{(pc)^2 + (m_ec^2)^2}$.) (b) If one of these photons strikes platinum ($\phi = 6.35$ eV), determine the resulting maximum speed of the photoelectron.

58. In a Compton scattering event, the scattered photon has an energy of 120.0 keV and the recoiling electron has a kinetic energy of 40.0 keV. Find (a) the wavelength of the incident photon, (b) the angle $\theta$ at which the photon is scattered, and (c) the recoil angle of the electron. (*Hint:* Conserve both mass-energy and relativistic momentum.)

59. A woman on a ladder drops small pellets toward a spot on the floor. (a) Show that according to the uncertainty principle, the average miss distance must be at least

$$\Delta x = \left(\frac{h}{2\pi m}\right)^{1/2}\left(\frac{H}{2g}\right)^{1/4}$$

where $H$ is the initial height of each pellet above the floor and $m$ is the mass of each pellet. (b) If $H = 2.00$ m and $m = 0.500$ g, what is $\Delta x$?

60. Show that the speed of a particle having de Broglie wavelength $\lambda$ and Compton wavelength $\lambda_C = h/(mc)$ is

$$v = \frac{c}{\sqrt{1 + (\lambda/\lambda_C)^2}}$$

## GROUP ACTIVITIES

**G.1** Use a black marker or pieces of dark electrical tape to make a very dark area on the outside of a shoebox. Poke a hole in the center of the dark area with a pencil. Now put a lid on the box and compare the blackness of the hole with the blackness of the surrounding dark area. Based on your observation explain why the radiation emitted from the hole is like that emitted from a blackbody.

**G.2** On a clear night, go outdoors far from city lights and find the constellation Orion. Your instructor should be able to furnish you with a star chart to assist you in locating this grouping of stars. Look very carefully at the color of the two stars Betelgeuse and Rigel. Can you tell which star is hotter? Orion is visible from November through April in the evening sky. Thus, if Orion is not visible, compare two of the brightest stars you can see, such as Vega in the constellation Lyra and Arcturus in Bootes.

**G.3** (a) If a pulse of blue light and a pulse of red light both carry energy $E$, which pulse contains more photons? (b) If the wavelength of the red light is 690 nm, the wavelength of the blue light is 420 nm, and the energy $E$ is 1 000 eV, find the number of photons in each beam to verify your prediction of part (a).

**G.4** (a) The threshold frequency for the photoelectric effect for a material is $f_0$. Are electrons emitted from the material when light incident on the material is of frequency (i) greater than $f_0$? (ii) less than $f_0$? (iii) equal to $f_0$? (b) The work function for zinc is 4.31 eV. What is the lowest frequency of light that releases photoelectrons when incident on the material?

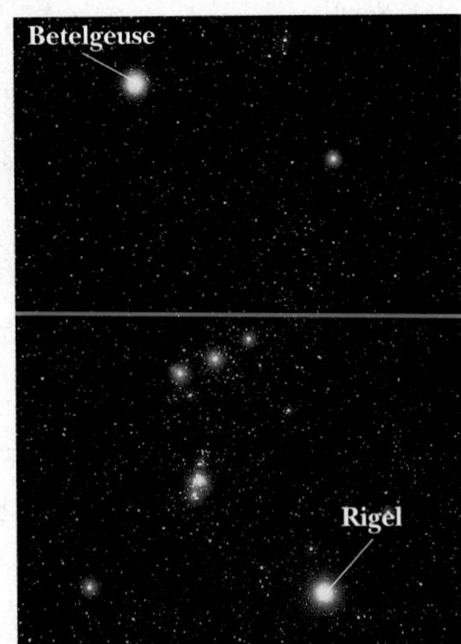


**FIGURE GA27.2** (*John Chumack/Photo Researchers, Inc.*)

**G.5** Light sometimes acts like a wave and sometimes like a particle. For the following situations, which best describes the behavior of light? Defend your answers. (a) The photoelectric effect. (b) The Compton effect. (c) Young's double-slit experiment.

# 28 Atomic Physics

## Chapter Outline



**"Neon lights", commonly used in advertising signs, consist of thin glass tubes filled with various gases such as neon and helium. The gas atoms are excited to higher energy levels by electric discharge through the tube. When the electrons in these excited levels return to lower energy levels, the atoms emit light having a wavelength (color) that depends on the type of gas in the tube. For example, a tube filled with neon produces a red-orange color, while helium produces pink.** *(Terry Gleason, Visuals Unlimited)*

A large portion of this chapter is concerned with the study of the hydrogen atom. Although the hydrogen atom is the simplest atomic system, it is especially important for several reasons:

- The quantum numbers used to characterize the allowed states of hydrogen can also be used to describe (approximately) the allowed states of more complex atoms. This enables us to understand the periodic table of the elements, one of the great triumphs of quantum mechanics.
- The hydrogen atom is an ideal system for performing precise comparisons of theory and experiment and for improving our overall understanding of atomic structure.
- Much of what we know about the hydrogen atom with its single electron can be extended to such single-electron ions as $He^{+}$ and $Li^{2+}$.

In this chapter we first discuss the Bohr model of hydrogen, which helps us understand many features of hydrogen but fails to explain the finer details of atomic structure. Next we examine the hydrogen atom from the viewpoint of quantum mechanics and the quantum numbers used to characterize various atomic states. In addition, we examine the physical significance of the quantum numbers and the effect of a magnetic field on certain quantum states. The Pauli exclusion principle is also presented. This physical principle is extremely important in understanding the properties of complex atoms and the arrangement of elements in the periodic table. Finally, we apply our knowledge of atomic structure to describe the mechanisms involved in the production of x-rays, the operation of lasers, and the behavior of solid-state devices such as diodes and transistors.

## 28.1 EARLY MODELS OF THE ATOM

The model of the atom in the days of Newton was a tiny, hard, indestructible sphere. Although this model was a good basis for the kinetic theory of gases, new models had to be devised when later experiments revealed the electronic nature of atoms. J. J. Thomson (1856–1940) suggested a model of the atom as a volume of positive charge with electrons embedded throughout the volume, much like the seeds in a watermelon (Fig. 28.1).

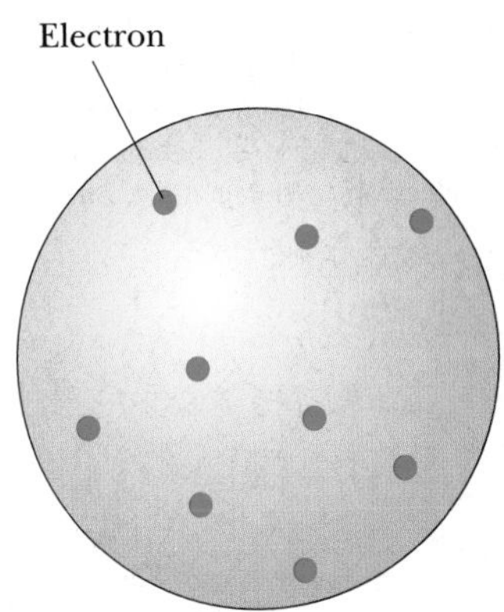


**FIGURE 28.1** Thomson's model of the atom, with the electrons embedded inside the positive charge like seeds in a watermelon.

In 1911 Ernest Rutherford (1871–1937) and his students Hans Geiger and Ernest Marsden performed a critical experiment showing that Thomson's model could not be correct. In this experiment, a beam of positively charged **alpha particles** was projected against a thin metal foil, as in Figure 28.2a. The results of the experiment astounded scientists. Most of the alpha particles passed through the foil as if it were empty space, but a few particles deflected from their original direc-

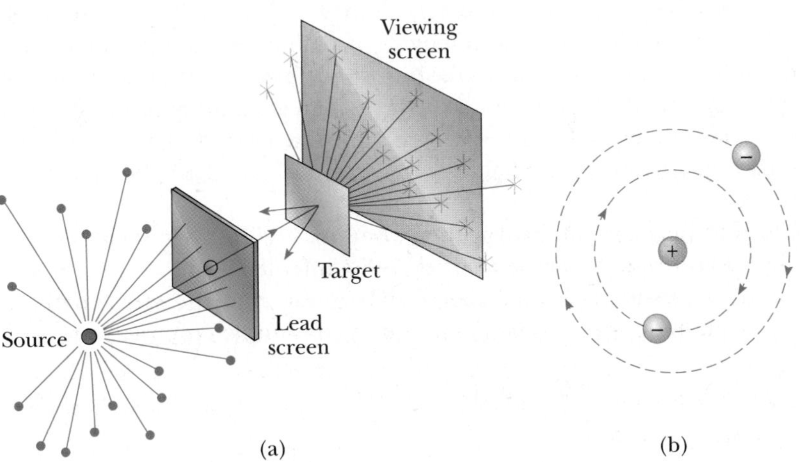


**FIGURE 28.2** (a) Geiger and Marsden's technique for observing the scattering of alpha particles from a thin foil target. The source is a naturally occurring radioactive substance, such as radium. (b) Rutherford's planetary model of the atom.

SIR JOSEPH JOHN THOMSON, ENGLISH PHYSICIST (1856–1940)

Thomson, usually considered the discoverer of the electron, opened up the field of subatomic particle physics with his extensive work on the deflection of cathode rays (electrons) in an electric field. Thomson received the 1906 Nobel prize for his discovery of the electron. *(Stock Montage, Inc.)*

tion of travel were scattered through large angles. Some particles even deflected backward, reversing their direction of travel. When Geiger informed Rutherford of these results, Rutherford wrote, "It was quite the most incredible event that has ever happened to me in my life. It was almost as incredible as if you fired a 15-inch [artillery] shell at a piece of tissue paper and it came back and hit you."

Such large deflections were not expected on the basis of Thomson's model. According to this model, a positively charged alpha particle would never come close enough to a large positive charge to cause large-angle deflections. Rutherford explained these astounding results by assuming that the positive charge in an atom was concentrated in a region that was small relative to the size of the atom. He called this concentration of positive charge the **nucleus** of the atom. Any electrons belonging to the atom were assumed to be in the relatively large volume outside the nucleus. In order to explain why electrons in this outer region of the atom were not pulled into the nucleus, Rutherford viewed them as moving in orbits about the positively charged nucleus in the same manner as the planets orbit the Sun, as shown in Figure 28.2b. Alpha particles themselves were later identified as the nuclei of helium atoms.

There are two basic difficulties with Rutherford's planetary model. As we shall see in the next section, an atom emits certain discrete characteristic frequencies of electromagnetic radiation and no others; the Rutherford model is unable to explain this phenomenon. A second difficulty is that Rutherford's electrons are undergoing a centripetal acceleration. According to Maxwell's theory of electromagnetism, centripetally accelerated charges revolving with frequency $f$ should radiate electromagnetic waves of the same frequency. Unfortunately, this classical model falls apart when applied to the atom. As the electron radiates energy, the radius of its orbit steadily decreases and its frequency of revolution increases. This leads to an ever-increasing frequency of emitted radiation and a rapid collapse of the atom as the electron spirals into the nucleus.

## 28.2 ATOMIC SPECTRA

As you may have already learned in chemistry, the hydrogen atom is the simplest atomic system, and an especially important one to understand. Much of what we know about the hydrogen atom (which consists of one proton and one electron) can be extended directly to other single-electron ions such as $He^+$ and $Li^{2+}$. Furthermore, a thorough understanding of the physics underlying the hydrogen atom can then be used to describe more complex atoms and the periodic table of the elements.

Suppose an evacuated glass tube is filled with hydrogen (or some other gas) at low pressure. If a voltage applied between metal electrodes in the tube is great enough to produce an electric current in the gas, the tube emits light whose color is characteristic of the gas in the tube. (This is how a neon sign works.) When the emitted light is analyzed with a spectrometer, a series of discrete bright lines is observed, each line having a different wavelength, or color. Such a series of spectral lines is commonly referred to as an **emission spectrum.** The wavelengths contained in a given line spectrum are characteristic of the element emitting the light (Fig. 28.3a). Because no two elements emit the same line spectrum, this phenomenon represents a marvelous and reliable technique for identifying elements in a gaseous substance.

The emission spectrum of hydrogen shown in Figure 28.4 includes four prominent lines that occur at wavelengths of 656.3 nm, 486.1 nm, 434.1 nm, and 410.2 nm. In 1885 Johann Balmer (1825–1898) found that the wavelengths of these and less prominent lines can be described by the simple empirical equation:

Balmer series ▶

$$\frac{1}{\lambda} = R_H\left(\frac{1}{2^2} - \frac{1}{n^2}\right) \qquad [28.1]$$

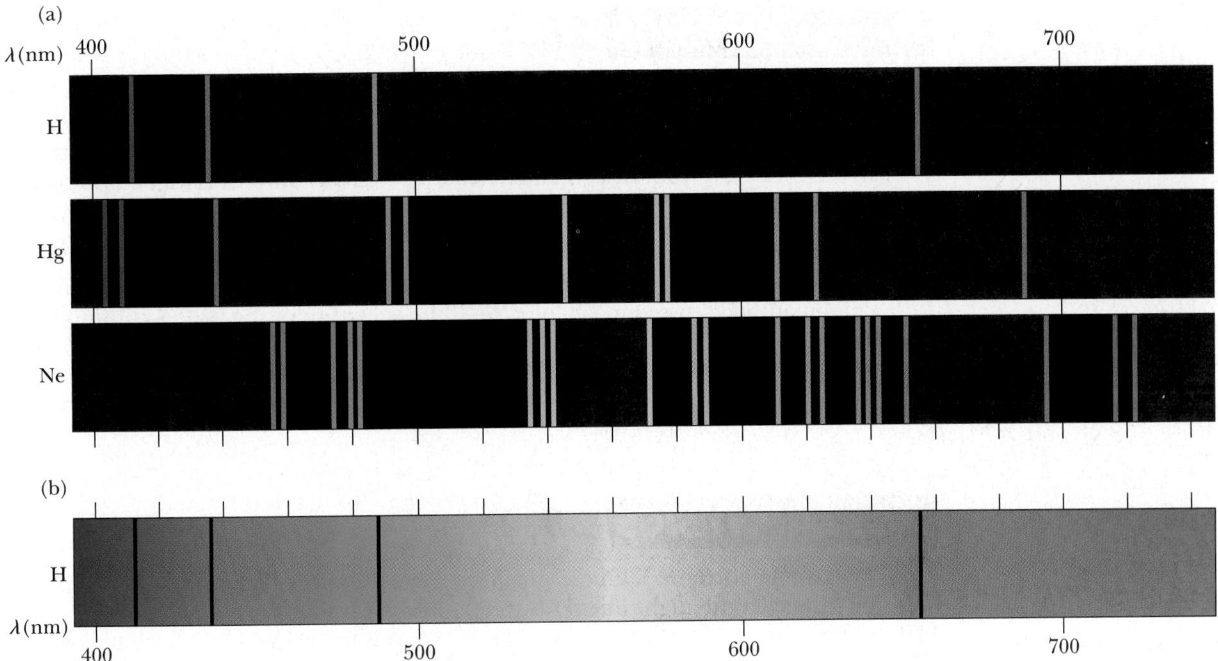


**FIGURE 28.3** Visible spectra. (a) Line spectra produced by emission in the visible range for the elements hydrogen, mercury, and neon. (b) The absorption spectrum for hydrogen. The dark absorption lines occur at the same wavelengths as the emission lines for hydrogen shown in (a).

where $n$ has integral values of 3, 4, 5, . . . , and $R_H$ is a constant, called the **Rydberg constant.** If the wavelength is in meters, $R_H$ has the value

$$R_H = 1.097\,373\,2 \times 10^7 \text{ m}^{-1} \qquad [28.2]$$

The first line in the Balmer series, at 656.3 nm, corresponds to $n = 3$ in Equation 28.1; the line at 486.1 nm corresponds to $n = 4$; and so on. In addition to this Balmer series of spectral lines, a Lyman series was subsequently discovered in the far ultraviolet, with the radiated wavelengths described by a similar equation.

In addition to emitting light at specific wavelengths, an element can also absorb light at specific wavelengths. The spectral lines corresponding to this process form what is known as an **absorption spectrum.** An absorption spectrum can be obtained by passing a continuous radiation spectrum (one containing all wavelengths) through a vapor of the element being analyzed. The absorption spectrum consists of a series of dark lines superimposed on the otherwise bright continuous spectrum. Each line in the absorption spectrum of a given element coincides with a line in the emission spectrum of the element. That is, if hydrogen is the absorbing vapor, dark lines will appear at the visible wavelengths 656.3 nm, 486.1 nm, 434.1 nm, and 410.2 nm, as shown in Figures 28.3b and 28.4.

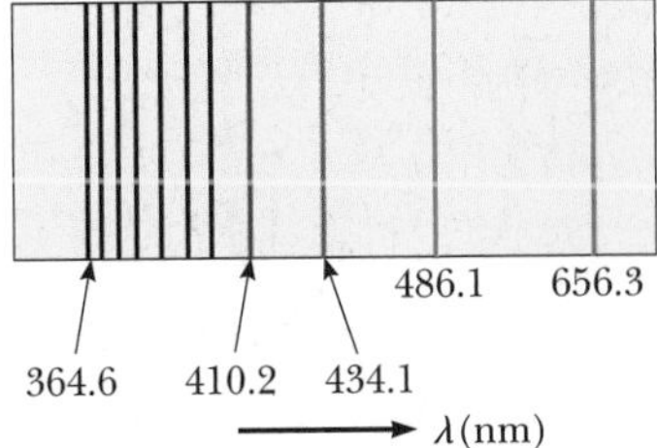


**FIGURE 28.4** A series of spectral lines for atomic hydrogen. The prominent labeled lines are part of the Balmer series.

The absorption spectrum of an element has many practical applications. For example, the continuous spectrum of radiation emitted by the Sun must pass through the cooler gases of the solar atmosphere before reaching Earth. The various absorption lines observed in the solar spectrum have been used to identify elements in the solar atmosphere. It is interesting to note that when the solar spectrum was first studied, some lines were found that did not correspond to any known element. A new element had been discovered! Because the Greek word for Sun is *helios,* the new element was named *helium.* It was later identified in underground gases on Earth. Scientists are able to examine the light from stars other than our Sun in this fashion, but elements other than those present on Earth have never been detected.

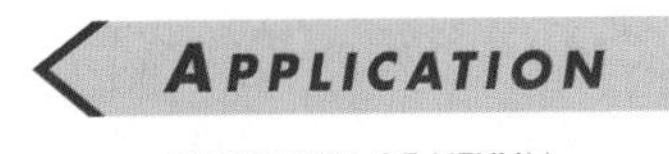


DISCOVERY OF HELIUM

**APPLYING PHYSICS 28.1**

You are observing a yellow candle flame, and your laboratory partner claims that the light from the flame originates from excited sodium atoms in the flame. You disagree, claiming that the candle flame is hot, so the radiation must be thermal in origin. Before your disagreement leads to fisticuffs, how can you determine who is correct?

**Explanation** A simple determination can be made by observing the light from the candle flame through a spectrometer, which is a slit and diffraction grating combination discussed in Chapter 25. If the spectrum of the light is continuous, then it is most likely thermal in origin. If the spectrum shows discrete lines, it is atomic in origin. The results of the experiment show that the light is indeed thermal in origin and originates from random molecular motion in the candle flame.

**APPLYING PHYSICS 28.2**

At extreme northern latitudes, the aurora borealis provides a beautiful and colorful display in the nighttime sky. A similar display occurs near the southern polar region and is called the aurora australis. What is the origin of the various colors of radiation seen in the auroras?

**APPLICATION**

THE NORTHERN AND SOUTHERN LIGHTS

**Explanation** The auroras are due to high-speed particles interacting with Earth's magnetic field and entering the atmosphere, as described on the back cover of this book. When these particles collide with molecules in the atmosphere, they excite the molecules in a way similar to the voltage in the spectrum tubes discussed earlier in this section. In response, the molecules emit colors of light according to the characteristic spectrum of their atomic constituents. For our atmosphere, the primary constituents are nitrogen and oxygen, which provide the red, blue, and green colors of the auroras.

## 28.3 THE BOHR THEORY OF HYDROGEN

At the beginning of the 20th century, scientists were perplexed by the failure of classical physics to explain the characteristics of spectra. Why did atoms of a given element emit only certain lines? Furthermore, why did the atoms absorb only those wavelengths that they emitted? In 1913 Bohr provided an explanation of atomic spectra that includes some features of the currently accepted theory. Using the simplest atom, hydrogen, Bohr developed a model of what he thought must be the atom's structure in an attempt to explain why the atom was stable. His model of the hydrogen atom contains some classical features as well as some revolutionary postulates that could not be justified within the framework of classical physics. The basic assumptions of the Bohr theory as it applies to the hydrogen atom are as follows:

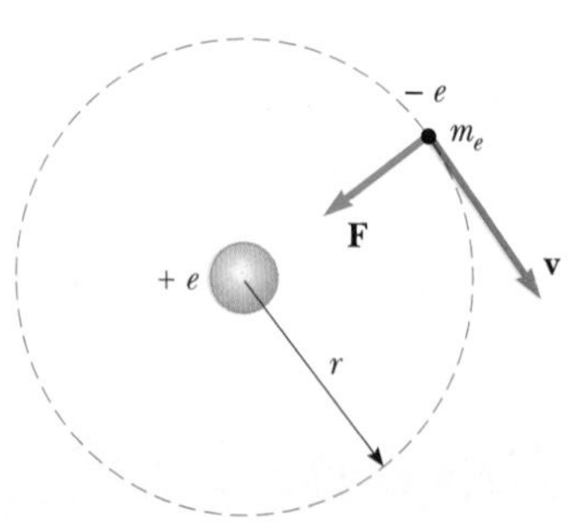


**FIGURE 28.5** Bohr's model of the hydrogen atom. In this model, the orbiting electron is allowed only in specific orbits of discrete radius.

1. The electron moves in circular orbits about the proton under the influence of the Coulomb force of attraction, as in Figure 28.5. In this case, the Coulomb force produces the electron's centripetal acceleration.
2. Only certain electron orbits are stable. These are orbits in which the hydrogen atom does not emit energy in the form of electromagnetic radiation. Hence, the total energy of the atom remains constant, and classical mechanics can be used to describe the electron's motion.
3. Radiation is emitted by the hydrogen atom when the electron "jumps" from a more energetic initial state to a lower state. The "jump" cannot be visualized or treated classically. In particular, the frequency, $f$, of the radiation emitted in the jump is related to the change in the atom's energy and is

*generally not the same as the frequency of the electron's orbital motion.* The frequency of the emitted radiation is given by

$$E_i - E_f = hf \quad [28.3]$$

where $E_i$ is the energy of the initial state, $E_f$ is the energy of the final state, $h$ is Planck's constant, and $E_i > E_f$.

4. The size of the allowed electron orbits is determined by a condition imposed on the electron's orbital angular momentum: The allowed orbits are those for which the electron's orbital angular momentum about the nucleus is an integral multiple of $\hbar$ (pronounced "$h$ bar"), where $\hbar = h/2\pi$.

$$m_e vr = n\hbar \qquad n = 1, 2, 3, \ldots \quad [28.4]$$

With these assumptions, we can calculate the allowed energies and emission wavelengths of the hydrogen atom. We shall use the model pictured in Figure 28.5, in which the electron travels in a circular orbit of radius $r$ with an orbit speed $v$.

The electrical potential energy of the atom is

$$PE = k_e \frac{q_1 q_2}{r} = k_e \frac{(-e)(e)}{r} = -k_e \frac{e^2}{r}$$

where $k_e$ is the Coulomb constant. Assuming the nucleus is at rest the total energy $E$ of the atom is the sum of the kinetic and potential energy:

$$E = KE + PE = \tfrac{1}{2} m_e v^2 - k_e \frac{e^2}{r} \quad [28.5]$$

Let us apply Newton's second law to the electron. We know that the electric force of attraction on the electron, $k_e e^2/r^2$, must equal $m_e a_r$, where $a_r = v^2/r$ is the centripetal acceleration of the electron. Thus,

$$k_e \frac{e^2}{r^2} = m_e \frac{v^2}{r} \quad [28.6]$$

From this equation, we see that the kinetic energy of the electron is

$$\tfrac{1}{2} mv^2 = \frac{k_e e^2}{2r} \quad [28.7]$$

We can combine this result with Equation 28.5 and express the energy of the atom as

$$E = -\frac{k_e e^2}{2r} \quad [28.8]$$

Energy of the hydrogen atom

where the negative value of the energy indicates that the electron is bound to the proton.

An expression for $r$ is obtained by solving Equations 28.4 and 28.6 for $v$ and equating the results:

$$v^2 = \frac{n^2 \hbar^2}{m_e^2 r^2} = \frac{k_e e^2}{m_e r}$$

$$r_n = \frac{n^2 \hbar^2}{m_e k_e e^2} \qquad n = 1, 2, 3, \ldots \quad [28.9]$$

The radii of the Bohr orbits are quantized

This equation is based on the assumption that the **electron can exist only in certain allowed orbits determined by the integer $n$.**

The orbit with the smallest radius, called the **Bohr radius,** $a_0$, corresponds to $n = 1$ and has the value

$$a_0 = \frac{\hbar^2}{m_e k_e e^2} = 0.052\,9 \text{ nm} \quad [28.10]$$

**NIELS BOHR, DANISH PHYSICIST (1885–1962)**

Bohr was an active participant in the early development of quantum mechanics and provided much of its philosophical framework. During the 1920s and 1930s, Bohr headed the Institute for Advanced Studies in Copenhagen. The institute was a magnet for many of the world's best physicists and provided a forum for the exchange of ideas. When Bohr visited the United States in 1939 to attend a scientific conference, he brought news that the fission of uranium had been observed by Hahn and Strassman in Berlin. The results were the foundations of the atomic bomb developed in the United States during World War II. Bohr was awarded the 1922 Nobel prize for his investigation of the structure of atoms and of the radiation emanating from them. *(Princeton University/Courtesy of AIP Emilio Segre Visual Archives)*

A general expression for the radius of any orbit in the hydrogen atom is obtained by substituting Equation 28.10 into Equation 28.9:

$$r_n = n^2 a_0 = n^2(0.052\,9 \text{ nm}) \qquad [28.11]$$

The first three Bohr orbits for hydrogen are shown in Figure 28.6.

Equation 28.9 may be substituted into Equation 28.8 to give the following expression for the energies of the quantum states:

$$E_n = -\frac{m_e k_e^2 e^4}{2\hbar^2}\left(\frac{1}{n^2}\right) \qquad n = 1, 2, 3, \ldots \qquad [28.12]$$

If we insert numerical values into Equation 28.12, we find

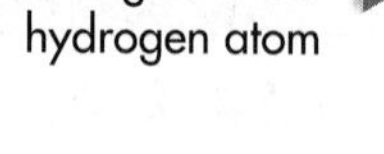

Allowed energies of the hydrogen atom

$$E_n = -\frac{13.6}{n^2} \text{ eV} \qquad [28.13]$$

The lowest energy state, or **ground state,** corresponds to $n = 1$ and has an energy $E_1 = -m_e k_e^2 e^4/2\hbar^2 = -13.6$ eV. The next state, corresponding to $n = 2$, has an energy $E_2 = E_1/4 = -3.40$ eV, and so on. An energy level diagram showing the energies (horizontal lines) of these stationary states and the corresponding quantum numbers is shown in Figure 28.7. The uppermost level shown, corresponding to $E = 0$ and $n \rightarrow \infty$, represents the state for which the electron is completely removed from the atom. In this case, both the electron's *KE* and *PE* are zero, which means that the electron is at rest infinitely far away from the proton. The minimum energy required to ionize the atom, that is, to completely remove the electron, is called the **ionization energy.** The ionization energy for hydrogen is 13.6 eV.

**Tip 28.1 ENERGY DEPENDS ON N ONLY FOR HYDROGEN**

According to Equation 28.13, the energy depends only on the quantum number $n$. Note that this is only true for the hydrogen atom. For more complicated atoms, the energy levels depend primarily on $n$, but also on other quantum numbers.

Equations 28.3 and 28.12 and the third Bohr postulate show that if the electron jumps from one orbit, whose quantum number is $n_i$, to a second orbit, whose quantum number is $n_f$, it emits a photon of frequency $f$, given by

$$f = \frac{E_i - E_f}{h} = \frac{m_e k_e^2 e^4}{4\pi\hbar^3}\left(\frac{1}{n_f^2} - \frac{1}{n_i^2}\right) \qquad [28.14]$$

where $n_f < n_i$.

Finally, to compare this result with the empirical formulas for the various spectral series, we use the fact that for light, $\lambda f = c$ and Equation 28.14 to get

$$\frac{1}{\lambda} = \frac{f}{c} = \frac{m_e k_e^2 e^4}{4\pi c\hbar^3}\left(\frac{1}{n_f^2} - \frac{1}{n_i^2}\right) \qquad [28.15]$$

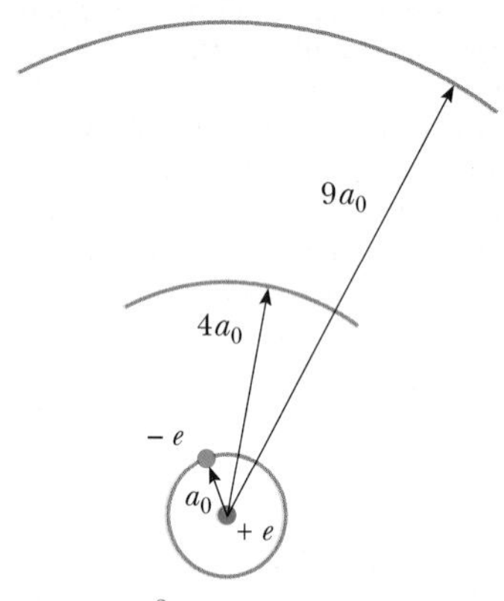


**FIGURE 28.6** The first three circular orbits predicted by the Bohr model of the hydrogen atom.

A comparison of this result with Equation 28.1 gives the following expression for the Rydberg constant:

$$R_H = \frac{m_e k_e^2 e^4}{4\pi c\hbar^3} \qquad [28.16]$$

If we insert the known values of $m_e$, $k_e$, $e$, $c$, and $\hbar$ into this expression, the resulting theoretical value for $R_H$ is found to be in excellent agreement with the value determined experimentally for the Rydberg constant. When Bohr demonstrated this agreement, it was recognized as a major accomplishment of his theory.

In order to compare Equation 28.15 with spectroscopic data, it is convenient to express it in the form

$$\frac{1}{\lambda} = R_H\left(\frac{1}{n_f^2} - \frac{1}{n_i^2}\right) \qquad [28.17]$$

We can use this expression to evaluate the wavelengths for the various series in the hydrogen spectrum. For example, in the Balmer series, $n_f = 2$ and $n_i = 3, 4, 5, \ldots$

(Eq. 28.1). For the Lyman series, we take $n_f = 1$ and $n_i = 2, 3, 4, \ldots$ The energy level diagram for hydrogen, shown in Figure 28.7, indicates the origin of the spectral lines described previously. The transitions between levels are represented by vertical arrows. Note that whenever a transition occurs between a state designated by $n_i$ to one designated by $n_f$ (where $n_i > n_f$), a photon with a frequency of $(E_i - E_f)/h$ is emitted. This can be interpreted as follows. The lines in the visible part of the hydrogen spectrum arise when the electron jumps from the third, fourth, or even higher orbit to the second orbit. Likewise, the lines of the Lyman series (in the ultraviolet) arise when the electron jumps from the second, third, or even higher orbit to the innermost ($n_f = 1$) orbit. Hence, the Bohr theory successfully predicts the wavelengths of all observed spectral lines of hydrogen.

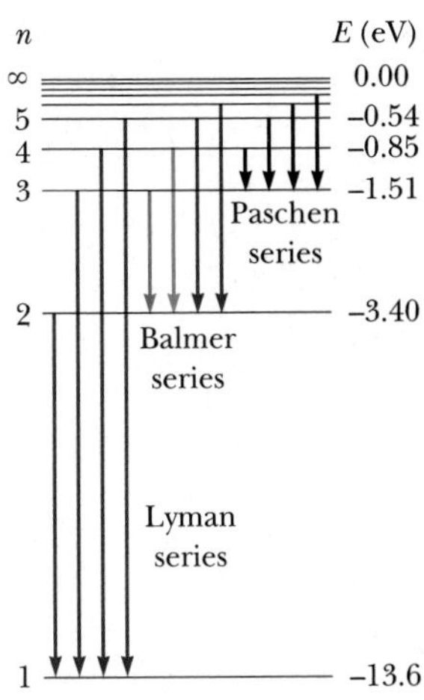


**FIGURE 28.7** An energy level diagram for hydrogen. In such diagrams the discrete allowed energies are plotted on the vertical axis. Nothing is plotted on the horizontal axis, but the horizontal extent of the diagram is made large enough to show allowed transitions. Note that the quantum numbers are given on the left and the energies (in eV) are on the right.

## APPLYING PHYSICS 28.3

According to the Bohr model of the hydrogen atom, the electron in the ground state moves in a circular orbit of radius $0.529 \times 10^{-10}$ m, and the speed of the electron in this state is $2.2 \times 10^6$ m/s. How could Bohr's model be so successful initially when its specific orbital radii seem to contradict the "fuzziness" demanded by the Heisenberg uncertainty principle?

**Explanation** We have no answer. The Bohr theory works much better than it should. The theory pictures an electron as having definite position and momentum, ignoring quantum uncertainty. The theory proceeds from Newton's second law, but Schrödinger's equation describes the wave motion of the electron. Scattering experiments show that the electron does not lie on a flat circle, but fills a sphere around the nucleus.

## Example 28.1 An Electronic Transition in Hydrogen

The electron in the hydrogen atom makes a transition from the $n = 2$ energy state to the ground state (corresponding to $n = 1$). Find the wavelength and frequency of the emitted photon.

**Solution** We can use Equation 28.17 directly to obtain $\lambda$, with $n_i = 2$ and $n_f = 1$:

$$\frac{1}{\lambda} = R_H\left(\frac{1}{n_f^2} - \frac{1}{n_i^2}\right)$$

$$\frac{1}{\lambda} = R_H\left(\frac{1}{1^2} - \frac{1}{2^2}\right) = \frac{3R_H}{4}$$

$$\lambda = \frac{4}{3R_H} = \frac{4}{3(1.097 \times 10^7\ \text{m}^{-1})} = 1.215 \times 10^{-7}\ \text{m} = \boxed{121.5\ \text{nm}}$$

This wavelength lies in the ultraviolet region.

Because $c = f\lambda$, the frequency of the photon is

$$f = \frac{c}{\lambda} = \frac{3.00 \times 10^8\ \text{m/s}}{1.215 \times 10^{-7}\ \text{m}} = \boxed{2.47 \times 10^{15}\ \text{Hz}}$$

**EXERCISE** What is the wavelength of the photon emitted by hydrogen when the electron makes a transition from the $n = 3$ state to the $n = 1$ state?

**ANSWER** $\frac{9}{8R_H} = 102.6$ nm

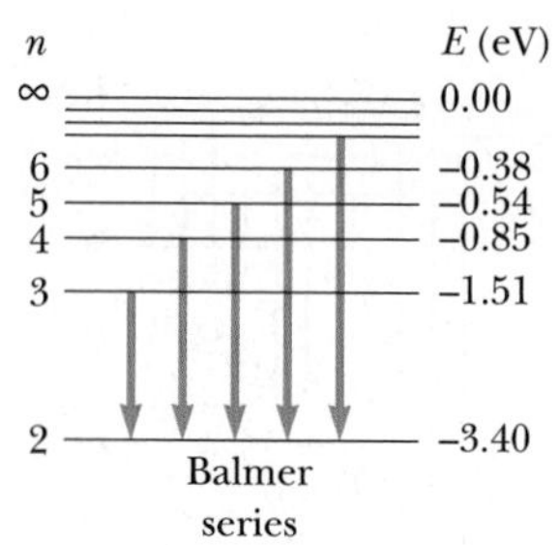


**FIGURE 28.8** (Example 28.2) Transitions responsible for the Balmer series for the hydrogen atom. All transitions terminate at the $n = 2$ level.

## Example 28.2 The Balmer Series for Hydrogen

The Balmer series for the hydrogen atom corresponds to electronic transitions that terminate in the state of quantum number $n = 2$, as shown in Figure 28.8.

**A** Find the longest-wavelength photon emitted and determine its energy.

**Solution** The longest-wavelength photon in the Balmer series results from the transition from $n = 3$ to $n = 2$. Using Equation 28.17 gives

$$\frac{1}{\lambda} = R_H\left(\frac{1}{n_f^2} - \frac{1}{n_i^2}\right)$$

$$\frac{1}{\lambda_{max}} = R_H\left(\frac{1}{2^2} - \frac{1}{3^2}\right) = \frac{5}{36}R_H$$

$$\lambda_{max} = \frac{36}{5R_H} = \frac{36}{5(1.097 \times 10^7 \text{ m}^{-1})} = \boxed{656.3 \text{ nm}}$$

This wavelength is in the red region of the visible spectrum.

The energy of this photon is

$$E_{photon} = hf = \frac{hc}{\lambda_{max}}$$

$$= \frac{(6.626 \times 10^{-34} \text{ J}\cdot\text{s})(3.00 \times 10^8 \text{ m/s})}{656.3 \times 10^{-9} \text{ m}}$$

$$= 3.03 \times 10^{-19} \text{ J} = \boxed{1.89 \text{ eV}}$$

We could also obtain the energy of the photon by using Equation 28.3 in the form $hf = E_3 - E_2$, where $E_2$ and $E_3$ are the energy levels of the hydrogen atom, calculated from Equation 28.13. Note that this is the lowest-energy photon in this series, because it involves the smallest energy change.

**B** Find the shortest-wavelength photon emitted in the Balmer series.

**Solution** The shortest-wavelength photon in the Balmer series is emitted when the electron makes a transition from $n = \infty$ to $n = 2$. Therefore,

$$\frac{1}{\lambda_{min}} = R_H\left(\frac{1}{2^2} - \frac{1}{\infty}\right) = \frac{R_H}{4}$$

$$\lambda_{min} = \frac{4}{R_H} = \frac{4}{1.097 \times 10^7 \text{ m}^{-1}} = \boxed{364.6 \text{ nm}}$$

This wavelength is in the ultraviolet region and corresponds to the series limit.

## BOHR'S CORRESPONDENCE PRINCIPLE

In our study of relativity in Chapter 26, we found that Newtonian mechanics cannot be used to describe phenomena that occur at speeds approaching the speed of light. Newtonian mechanics is a special case of relativistic mechanics and is usable only when $v$ is much smaller than $c$. Similarly, **quantum mechanics is in agreement with classical physics when the energy differences between quantized levels are very small.** This principle, first set forth by Bohr, is called the **correspondence principle.**

For example, consider the hydrogen atom with $n > 10\,000$. For such large values of $n$, the energy differences between adjacent levels approach zero and the levels are nearly continuous as Equation 28.13 shows. As a consequence, the classical model is reasonably accurate in describing the system for large values of $n$. According to the classical model, the frequency of the light emitted by the atom is equal to the frequency of revolution of the electron in its orbit about the nucleus. Calculations show that for $n > 10\,000$, the frequency of the emitted light is the same as the electron's frequency of revolution to within 0.015%.

## 28.4 MODIFICATION OF THE BOHR THEORY

The Bohr theory of the hydrogen atom was a tremendous success in certain areas because it explains several features of the hydrogen spectrum that previously defied explanation. It accounts for the Balmer series and other series; it predicts a value for the Rydberg constant that is in excellent agreement with the experimental value; it gives an expression for the radius of the atom; and it predicts the energy levels of hydrogen. Although these successes are important to scientists, it is perhaps even more significant that the Bohr theory gives us a model of what the atom looks like and how it behaves. Once a basic model is constructed, refinements and modifications can be made to enlarge on the concept and to explain finer details.

The analysis used in the Bohr theory is also successful when applied to *hydrogen-like* atoms. An atom is said to be hydrogen-like when it contains only one electron. Examples are singly ionized helium, doubly ionized lithium, triply ionized beryllium, and so forth. The results of the Bohr theory for hydrogen can be extended to hydrogen-like atoms by substituting $Ze^2$ for $e^2$ in the hydrogen equations, where $Z$ is the atomic number of the element. For example, Equations 28.12 and 28.15 become

$$E_n = -\frac{m_e k_e^2 Z^2 e^4}{2\hbar^2}\left(\frac{1}{n^2}\right) \qquad n = 1, 2, 3, \ldots \tag{28.18}$$

and

$$\frac{1}{\lambda} = \frac{m_e k_e^2 Z^2 e^4}{4\pi c\hbar^3}\left(\frac{1}{n_f^2} - \frac{1}{n_i^2}\right) \tag{28.19}$$

Although many attempts have been made to extend the Bohr theory to more complex multi-electron atoms, the results have thus far been unsuccessful. Even today, only approximate methods are available for treating multi-electron atoms.

Within a few months following the publication of Bohr's theory, Arnold Sommerfeld (1868–1951) extended the results to include elliptical orbits. We shall examine his model briefly because much of the nomenclature used in this treatment is still in use today. Bohr's concept of quantization of angular momentum led to the **principal quantum number** $n$, which determines the energy of the allowed states of hydrogen. Sommerfeld's theory retained $n$, but also introduced a new quantum number $\ell$ called the **orbital quantum number,** where the value of $\ell$ ranges from 0 to $n - 1$ in integer steps. According to this model, an electron in any one of the allowed energy states of a hydrogen atom may move in any one of a number of orbits corresponding to different $\ell$ values. For each value of $n$ there are $n$ possible orbits corresponding to different $\ell$ values. Because $n = 1$ and $\ell = 0$ for the first energy level (ground state), there is only one possible orbit for this state. The second energy level, with $n = 2$, has two possible orbits corresponding to $\ell = 0$ and $\ell = 1$. The third energy level, with $n = 3$, has three possible orbits corresponding to $\ell = 0$, $\ell = 1$, and $\ell = 2$.

For historical reasons, **all states with the same principal quantum number *n* are said to form a shell.** Shells are identified by the letters K, L, M, . . . , which designate the states for which $n = 1, 2, 3, \ldots$ . Likewise, **the states with given values of *n* and $\ell$ are said to form a subshell.** The letters *s, p, d, f, g,* . . . are used to designate the states for which $\ell = 0, 1, 2, 3, 4, \ldots$ . These notations are summarized in Table 28.1.

States that violate the rules given in Table 28.1 cannot exist. For instance, the $2d$ state, which would have $n = 2$ and $\ell = 2$, cannot exist because the highest al-

**TABLE 28.1** Shell and Subshell Notations

| $n$ | Shell Symbol | $\ell$ | Subshell Symbol |
|---|---|---|---|
| 1 | K | 0 | *s* |
| 2 | L | 1 | *p* |
| 3 | M | 2 | *d* |
| 4 | N | 3 | *f* |
| 5 | O | 4 | *g* |
| 6 | P | 5 | *h* |
| ⋮ | | ⋮ | |

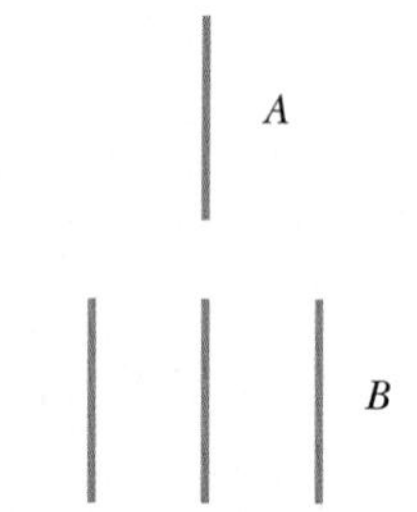


**FIGURE 28.9** A single line ($A$) can split into three separate lines ($B$) in a magnetic field.

lowed value of $\ell$ is $n - 1$, or 1 in this case. Thus, for $n = 2$, $2s$ and $2p$ are allowed subshells but $2d$, $2f$, . . . are not. For $n = 3$, the allowed states are $3s$, $3p$, and $3d$. The maximum number of electrons allowed in any given subshell is $2(2\ell + 1)$. For example, the $p$ subshell ($\ell = 1$) is filled when it contains six electrons. This fact will be important to us later when we discuss the *Pauli exclusion principle.*

Another modification of the Bohr theory arose when it was discovered that the spectral lines of a gas are split into several closely spaced lines when the gas is placed in a strong magnetic field. (This is called the *Zeeman effect,* after its discoverer.) Figure 28.9 shows a single spectral line being split into three closely spaced lines. This observation indicates that the energy of an electron is slightly modified when the atom is immersed in a magnetic field. In order to explain this observation, a new quantum number $m_\ell$ called the **orbital magnetic quantum number,** was introduced. The theory is in accord with experimental results when $m_\ell$ is restricted to values ranging from $-\ell$ to $+\ell$, in integer steps.

Finally, very high resolution spectrometers revealed that spectral lines of gases are in fact two very closely spaced lines even in the absence of an external magnetic field. This splitting was referred to as **fine structure.** In 1925 Samuel Goudsmit and George Uhlenbeck introduced the idea of an electron spinning about its own axis to explain the origin of fine structure. The results of their work introduced yet another quantum number $m_s$ called the **spin magnetic quantum number.** We shall save further discussion of this quantum number for a later section. It is interesting to note that each of the new concepts introduced into the original Bohr theory added new quantum numbers and improved the original model. However, the most profound step forward in our understanding of atomic structure came with the development of quantum mechanics.

## Example 28.3 Singly Ionized Helium

Singly ionized helium, $He^+$, a hydrogen-like system, has one electron in the $1s$ orbit when the atom is in its ground state.

**A** Find the energy of the system in the ground state.

**Solution** From Equation 28.18, the energy of a hydrogen-like system whose principal quantum number is $n$ is given by

$$E_n = -\frac{m_e k_e^2 Z^2 e^4}{2\hbar^2}\left(\frac{1}{n^2}\right)$$

This can be expressed in eV units as

$$E_n = -\frac{Z^2(13.6)}{n^2}\ \text{eV}$$

Because $Z = 2$ for helium, and $n = 1$ in the ground state, we have

$$E_1 = -4(13.6)\ \text{eV} = \boxed{-54.4\ \text{eV}}$$

**B** Find the radius of the ground-state orbit.

**Solution** The radius of the ground-state orbit can be found with the help of Equation 28.9. This equation must be modified in the case of a hydrogen-like atom by substituting $Ze^2$ for $e^2$ to obtain

$$r_n = \frac{n^2\hbar^2}{m_e k_e Z e^2} = \frac{n^2}{Z}(0.052\,9\ \text{nm})$$

For our case, $n = 1$ and $Z = 2$, and the result is

$$r_1 = \boxed{0.026\,5\ \text{nm}}$$

## 28.5 DE BROGLIE WAVES AND THE HYDROGEN ATOM

One of the postulates made by Bohr in his theory of the hydrogen atom was that angular momentum of the electron is quantized in units of $\hbar$ or

$$m_e vr = n\hbar$$

For more than a decade following Bohr's publication, no one was able to explain why the angular momentum of the electron was restricted to these discrete values. Finally, de Broglie gave a direct physical way of interpreting this condition. He assumed that an electron orbit would be stable (allowed) only if it contained an integral number of electron wavelengths. Figure 28.10a demonstrates this point when three complete wavelengths are contained in one circumference of the orbit. Similar patterns can be drawn for orbits containing one wavelength, two wavelengths, four wavelengths, five wavelengths, and so forth. This situation is analogous to that of standing waves on a string, discussed in Chapter 14. There we found that strings have preferred (resonant) frequencies of vibration. Figure 28.10b shows a standing-wave pattern containing three wavelengths for a string fixed at each end. Now imagine that the vibrating string is removed from its supports at $A$ and $B$ and bent into a circular shape that brings points $A$ and $B$ together. The end result is a pattern such as the one shown in Figure 28.10a.

In general, the condition for a de Broglie standing wave in an electron orbit is that the circumference must contain an integral number of electron wavelengths. We can express this condition as

$$2\pi r = n\lambda \qquad n = 1, 2, 3, \ldots$$

Because the de Broglie wavelength of an electron is $\lambda = h/m_e v$, we can write the preceding equation as $2\pi r = nh/m_e v$, or

$$m_e vr = n\hbar$$

This is precisely the quantization of angular momentum condition imposed by Bohr in his original theory of hydrogen.

The electron orbit shown in Figure 28.10a contains three complete wavelengths and corresponds to the case in which the principal quantum number $n = 3$. The orbit with one complete wavelength in its circumference corresponds to the first Bohr orbit, $n = 1$; the orbit with two complete wavelengths corresponds to the second Bohr orbit, $n = 2$, and so forth.

By applying the wave theory of matter to electrons in atoms, de Broglie was able to explain the appearance of integers in the Bohr theory as a natural consequence of standing-wave patterns. This was the first convincing argument that the wave nature of matter was at the heart of the behavior of atomic systems. Although the analysis provided by de Broglie was a promising first step, gigantic strides were made subsequently with the development of Schrödinger's wave equation and its application to atomic systems.

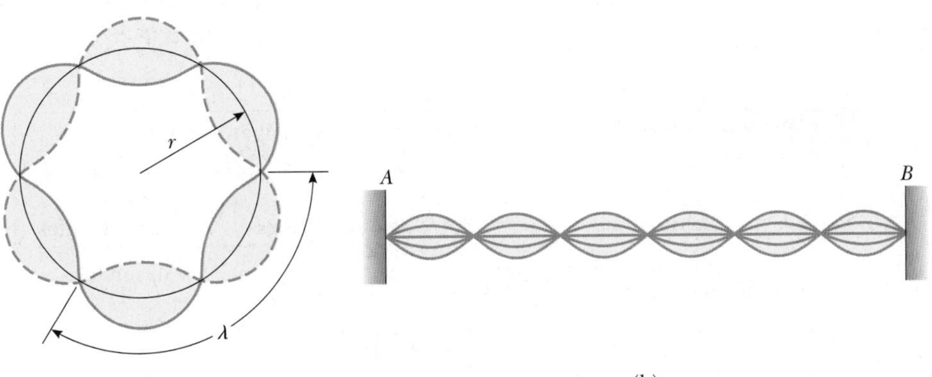


**FIGURE 28.10** (a) Standing-wave pattern for an electron wave in a stable orbit of hydrogen. There are three full wavelengths in this orbit. (b) Standing-wave pattern for a vibrating stretched string fixed at its ends. This pattern has three full wavelengths.

| Quick Quiz 28.1 | In an analysis relating Bohr's theory to the de Broglie wavelength of electrons, when an electron moves from the $n = 1$ level to the $n = 3$ level, the circumference of its orbit becomes 9 times greater. This occurs because (a) there are 3 times as many wavelengths in the new orbit, (b) there are 3 times as many wavelengths and each wavelength is 3 times as long, (c) the wavelength of the electron becomes 9 times as long, or (d) the electron is moving 9 times as fast. |
|---|---|

## 28.6 QUANTUM MECHANICS AND THE HYDROGEN ATOM

One of the first great achievements of quantum mechanics was the solution of the wave equation for the hydrogen atom. We shall not attempt to carry out this solution. Rather, we will simply describe its properties and some of its implications with regard to atomic structure.

According to quantum mechanics, the energies of the allowed states are in exact agreement with the values obtained by the Bohr theory (Eq. 28.12), when the allowed energies depend only on the principal quantum number $n$.

In addition to the principal quantum number, two other quantum numbers emerge from the solution of the wave equation, $\ell$ and $m_\ell$. The quantum number $\ell$ is called the **orbital quantum number,** and $m_\ell$ is called the **orbital magnetic quantum number.** As pointed out in Section 28.4, these quantum numbers had already appeared in modifications made to the Bohr theory. The significance of quantum mechanics is that these quantum numbers and the restrictions placed on their values arise directly from mathematics and not from any ad hoc assumptions to make the theory consistent with experimental observation. Because we shall need to make use of the various quantum numbers in the next sections, we repeat the allowed ranges of their values here.

**The values of $n$ can range from 1 to $\infty$ in integer steps. The values of $\ell$ can range from 0 to $n - 1$ in integer steps. The values of $m_\ell$ can range from $-\ell$ to $\ell$ in integer steps.**

For example, if $n = 1$, only $\ell = 0$ and $m_\ell = 0$ are permitted. If $n = 2$, the value of $\ell$ may be 0 or 1; if $\ell = 0$, then $m_\ell = 0$, but if $\ell = 1$, then $m_\ell$ may be 1, 0, or $-1$. Table 28.2 summarizes the rules for determining the allowed values of $\ell$ and $m_\ell$ for a given value of $n$.

States that violate the rules given in Table 28.2 cannot exist. For instance, one state that cannot exist is the $2d$ state, which would have $n = 2$ and $\ell = 2$. This state

**TABLE 28.2 Three Quantum Numbers for the Hydrogen Atom**

| Quantum Number | Name | Allowed Values | Number of Allowed States |
|---|---|---|---|
| $n$ | Principal quantum number | $1, 2, 3, \ldots$ | Any number |
| $\ell$ | Orbital quantum number | $0, 1, 2, \ldots, n-1$ | $n$ |
| $m_\ell$ | Orbital magnetic quantum number | $-\ell, -\ell+1, \ldots, 0, \ldots, \ell-1, \ell$ | $2\ell + 1$ |

is not allowed because the highest allowed value of $\ell$ is $n - 1$, or 1 in this case. Thus, for $n = 2$, $2s$ and $2p$ are allowed states but $2d$, $2f$, . . . are not. For $n = 3$, the allowed states are $3s$, $3p$, and $3d$.

### Example 28.4 The $n = 2$ Level of Hydrogen

Determine the number of states in the hydrogen atom for $n = 2$ and calculate the energies of these states.

**Solution** For $n = 2$, $\ell$ can have the values 0 and 1. For $\ell = 0$, $m_\ell$ can only be 0; for $\ell = 1$, $m_\ell$ can be $-1$, 0, or 1. Hence we have one state designated as the $2s$ state associated with the quantum numbers $n = 2$, $\ell = 0$, and $m_\ell = 0$, and three states designated as $2p$ states, for which the quantum numbers are $n = 2$, $\ell = 1$, $m_\ell = -1$; $n = 2$, $\ell = 1$, $m_\ell = 0$; and $n = 2$, $\ell = 1$, $m_\ell = 1$.

Because all these states have the same principal quantum number, $n = 2$, they also have the same energy, which can be calculated using Equation 28.13, $E_n = -(13.6/n^2)\text{eV}$. For $n = 2$, this gives

$$E_2 = -\frac{13.6}{2^2}\text{ eV} = -3.40\text{ eV}$$

**EXERCISE** How many possible states are there for the $n = 3$ level of hydrogen? For the $n = 4$ level?

**ANSWER** 9 states with different values of $\ell$ or $m_\ell$ for $n = 3$, and 16 states for $n = 4$

**Quick Quiz 28.2** How many possible orbital states are there for (a) the $n = 3$ level of hydrogen? (b) the $n = 4$ level?

**Quick Quiz 28.3** When the principal quantum number is $n = 5$, how many different values of (a) $\ell$ and (b) $m_\ell$ are possible?

## 28.7 THE SPIN MAGNETIC QUANTUM NUMBER

Example 28.4 was presented to give you some practice in manipulating quantum numbers, but as we shall see in this section, there actually are *eight* states corresponding to $n = 2$ for hydrogen, not four. This happens because another quantum number, $m_s$, the **spin magnetic quantum number,** has to be introduced to explain the splitting of each level into two.

As pointed out in Section 28.4, the need for this new quantum number first came about because of an unusual feature in the spectra of certain gases, such as sodium vapor. Close examination of one of the prominent lines of sodium shows that it is, in fact, two very closely spaced lines. The wavelengths of these lines occur in the yellow region at 589.0 nm and 589.6 nm. In 1925, when this doublet was first noticed, atomic theory could not explain it. To resolve the dilemma, Samuel Goudsmit and George Uhlenbeck, following a suggestion by the Austrian physicist Wolfgang Pauli, proposed that a fourth quantum number, called the *spin quantum number,* be introduced to describe any atomic level.

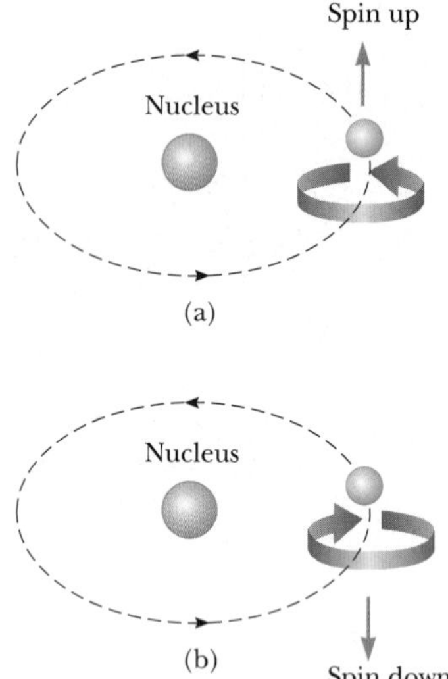


**FIGURE 28.11** As an electron moves in its orbit about the nucleus, its spin can be either (a) up or (b) down.

In order to describe the spin quantum number, it is convenient (but technically incorrect) to think of the electron as spinning on its axis as it orbits the nucleus, just as Earth spins on its axis as it orbits the Sun. Strangely, there are only two ways in which the electron can spin as it orbits the nucleus, as shown in Figure 28.11. If the direction of spin is as shown in Figure 28.11a, the electron is said to have "spin up." If the direction of spin is reversed, as in Figure 28.11b, the electron is said to have "spin down." The energy of the electron is slightly different for the two spin directions, and this energy difference accounts for the sodium doublet. The quantum numbers associated with electron spin are $m_s = \frac{1}{2}$ for the spin-up state and $m_s = -\frac{1}{2}$ for the spin-down state. As we shall see in Example 28.5, this new quantum number doubles the number of allowed states specified by the quantum numbers $n$, $\ell$, and $m_\ell$.

Any classical description of electron spin is technically incorrect because quantum mechanics tells us that, because the electron cannot be precisely located in space, it cannot be considered to be a spinning solid object, as pictured in Figure 28.11. In spite of this conceptual difficulty, all experimental evidence supports the fact that an electron does have some intrinsic property that can be described by the spin magnetic quantum number.

**Tip 28.2 THE ELECTRON IS NOT SPINNING**

The electron is *not* physically spinning. Electron spin is a purely quantum effect that gives the electron an angular momentum as if it were physically spinning.

### Example 28.5 The Quantum Numbers for the 2*p* Subshell

List the quantum numbers for electrons in the 2*p* subshell.

**Solution** For this subshell, $n = 2$ and $\ell = 1$. The magnetic quantum number can have the values $-1$, 0, 1, and the spin quantum number is always $+\frac{1}{2}$ or $-\frac{1}{2}$. Thus, the six possibilities are as follows:

| $n$ | $\ell$ | $m_\ell$ | $m_s$ |
|---|---|---|---|
| 2 | 1 | $-1$ | $-\frac{1}{2}$ |
| 2 | 1 | $-1$ | $\frac{1}{2}$ |
| 2 | 1 | 0 | $-\frac{1}{2}$ |
| 2 | 1 | 0 | $\frac{1}{2}$ |
| 2 | 1 | 1 | $-\frac{1}{2}$ |
| 2 | 1 | 1 | $\frac{1}{2}$ |

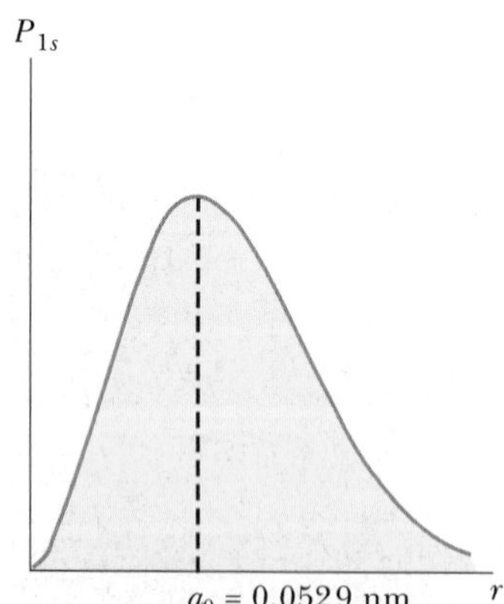


**FIGURE 28.12** The probability of finding the electron versus distance from the nucleus for the hydrogen atom in the 1*s* (ground) state. Note that the probability has its maximum value when $r$ equals the first Bohr radius, $a_0$.

## 28.8 ELECTRON CLOUDS

The solution of the wave equation, discussed in Section 27.10, yields a wave function $\Psi$ that depends on the quantum numbers $n$, $\ell$, and $m_\ell$. Let us assume that we have found a wave function $\Psi$ and see what it may tell us about the hydrogen atom. We choose a value of $n = 1$ for the principal quantum number, which corresponds to the lowest energy state for hydrogen. For $n = 1$, the restrictions placed on the remaining quantum numbers are that $\ell = 0$ and $m_\ell = 0$.

The quantity $\Psi^2$ has great physical significance because it is proportional to the probability of finding the electron at a given position. Figure 28.12 gives the probability of finding the electron at various radial distances from the nucleus in the 1*s* state of hydrogen. Some useful and surprising information can be extracted from this curve. First, the curve peaks at a value of $r = 0.052\,9$ nm, the Bohr value of the radius of the first electron orbit in hydrogen. This means that there is a maximum probability of finding the electron in a small interval centered at this distance from the nucleus. However, as the curve indicates, there is also a probability of finding the electron in a small interval centered at any other distance from the nucleus. In other words, the electron is not confined to a particular orbital distance from the nucleus, as assumed in the Bohr model. The electron may be found at various distances from the nucleus, but **the probability of finding it at a distance corresponding to the first Bohr orbit is a maximum.** Quantum mechanics also predicts that the wave function for the hydrogen atom in the ground state is spherically symmetric; hence, the electron can be found in a spherical region surrounding the nucleus. This is in contrast to the Bohr theory, which confines the position of the electron to points in a plane. The quantum mechanical result is often interpreted by viewing the electron as a cloud surrounding the nucleus. An attempt at picturing this cloud-like behavior is shown in Figure 28.13. The densest regions of the cloud represent those locations where the electron is most likely to be found.

If a similar analysis is carried out for the $n = 2$, $\ell = 0$ state of hydrogen, a peak of the probability curve is found at $4a_0$. Likewise, for the $n = 3$, $\ell = 0$ state, the curve peaks at $9a_0$. Thus, quantum mechanics predicts a most-probable electron distance to the nucleus that is in agreement with the location predicted by the Bohr theory.

**FIGURE 28.13** The spherical electron cloud for the hydrogen atom in its 1*s* state.

## 28.9 THE EXCLUSION PRINCIPLE AND THE PERIODIC TABLE

Earlier we found that the state of an electron in an atom is specified by four quantum numbers: $n$, $\ell$, $m_\ell$, and $m_s$. For example, an electron in the ground state of hydrogen could have quantum numbers of $n = 1$, $\ell = 0$, $m_\ell = 0$, $m_s = \frac{1}{2}$. As it turns out, the state of an electron in any other atom may also be specified by this same set of quantum numbers. In fact, these four quantum numbers can be used to describe all the electronic states of an atom regardless of the number of electrons in its structure.

An obvious question that arises here is, How many electrons in an atom can have a particular set of quantum numbers? This important question was answered by Pauli in 1925 in a powerful statement known as the **Pauli exclusion principle:**

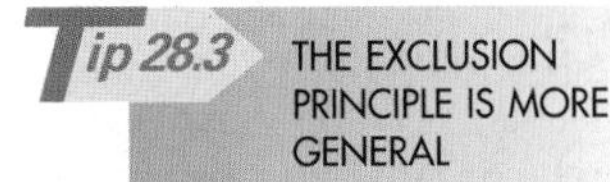

Tip 28.3
THE EXCLUSION PRINCIPLE IS MORE GENERAL


The exclusion principle stated here is a limited form of the more general exclusion principle, which states that no two *fermions* (particles with spin $\frac{1}{2}, \frac{3}{2}, \ldots$) can be in the same quantum state.

**No two electrons in an atom can ever be in the same quantum state; that is, no two electrons in the same atom can have exactly the same values for the set of quantum numbers $n$, $\ell$, $m_\ell$, and $m_s$.**

The Pauli exclusion principle

The Pauli exclusion principle explains the electronic structure of complex atoms as a succession of filled levels with different quantum numbers increasing in energy, where the outermost electrons are primarily responsible for the chemical properties of the element. It is interesting to note that if this principle were not valid, every electron would end up in the lowest energy state of the atom and the chemical behavior of the elements would be grossly different. Nature as we know it would not exist—and *we* would not exist to wonder about it!

As a general rule, the order that electrons fill an atom's subshell is as follows. Once one subshell is filled, the next electron goes into the vacant subshell that is lowest in energy. We can understand this principle by recognizing that if the atom were not in the lowest energy state available to it, it would radiate energy until it reached this state. A subshell is filled when it contains $2(2\ell + 1)$ electrons. This rule is based on the analysis of quantum numbers to be described later. Following this rule, shells and subshells can contain a number of electrons according to the pattern given in Table 28.3.

The exclusion principle can be illustrated by examining the electronic arrangement in a few of the lighter atoms.

*Hydrogen* has only one electron, which, in its ground state, can be described by either of two sets of quantum numbers: 1, 0, 0, $\frac{1}{2}$ or 1, 0, 0, $-\frac{1}{2}$. The electronic

Wolfgang Pauli and Niels Bohr watch a spinning top. *(Courtesy of AIP Niels Bohr Library, Margarethe Bohr Collection)*

**WOLFGANG PAULI (1900–1958)**

An extremely talented Austrian theoretical physicist who made important contributions in many areas of modern physics. Pauli gained public recognition at the age of 21 with a masterful review article on relativity, which is still considered one of the finest and most comprehensive introductions to the subject. Other major contributions were the discovery of the exclusion principle, the explanation of the connection between particle spin and statistics, and theories of relativistic quantum electrodynamics, the neutrino hypothesis, and the hypothesis of nuclear spin. *(CERN/Courtesy of AIP Emilio Segre Visual Archives)*

**TABLE 28.3 Number of Electrons in Filled Subshells and Shells**

| Shell | Subshell | Number of Electrons in Filled Subshell | Number of Electrons in Filled Shell |
|---|---|---|---|
| K ($n = 1$) | $s(\ell = 0)$ | 2 | 2 |
| L ($n = 2$) | $s(\ell = 0)$ | 2 | 8 |
| | $p(\ell = 1)$ | 6 | |
| M($n = 3$) | $s(\ell = 0)$ | 2 | 18 |
| | $p(\ell = 1)$ | 6 | |
| | $d(\ell = 2)$ | 10 | |
| N ($n = 4$) | $s(\ell = 0)$ | 2 | 32 |
| | $p(\ell = 1)$ | 6 | |
| | $d(\ell = 2)$ | 10 | |
| | $f(\ell = 3)$ | 14 | |

configuration of this atom is often designated as $1s^1$. The notation $1s$ refers to a state for which $n = 1$ and $\ell = 0$, and the superscript indicates that one electron is present in this level.

Neutral *helium* has two electrons. In the ground state, the quantum numbers for these two electrons are 1, 0, 0, $\frac{1}{2}$ and 1, 0, 0, $-\frac{1}{2}$. No other possible combinations of quantum numbers exist for this level, and we say that the K shell is filled. The helium electronic configuration is designated as $1s^2$.

Neutral *lithium* has three electrons. In the ground state, two of these are in the $1s$ subshell and the third is in the $2s$ subshell because it is lower in energy than the $2p$ subshell. Hence, the electronic configuration for lithium is $1s^22s^1$.

A list of electronic ground-state configurations for a number of atoms is provided in Table 28.4. In 1871 Dmitri Mendeleev (1834–1907), a Russian chemist, arranged the elements known at that time in a table according to their atomic masses and chemical similarities. The first table Mendeleev proposed contained many blank spaces, and he boldly stated that the gaps were there only because those elements had not yet been discovered. By noting the column in which these missing elements should be located, he was able to make rough predictions about their chemical properties. Within 20 years of this announcement, these elements were indeed discovered.

The elements in our current version of the periodic table are still arranged so that all those in a vertical column have similar chemical properties. For example, consider the elements in the last column: He (helium), Ne (neon), Ar (argon), Kr (krypton), Xe (xenon), and Rn (radon). The outstanding characteristic of these elements is that they do not normally take part in chemical reactions; that is, they do not join with other atoms to form molecules, and are therefore classified as inert. Because of this aloofness, they are referred to as the *noble gases.* We can partially understand their behavior by looking at the electronic configurations shown in Table 28.4. The element helium has the electronic configuration $1s^2$. In other words, one shell is filled. The electrons in this filled shell are considerably separated in energy from the next available level, the $2s$ level.

The electronic configuration for neon is $1s^22s^22p^6$. Again, the outer shell is filled and the difference in energy between the $2p$ level and the $3s$ level is large. Argon has the configuration $1s^22s^22p^63s^23p^6$. Here, the $3p$ subshell is filled and a wide gap in energy exists between the $3p$ subshell and the $3d$ subshell. Through all the noble gases, the pattern remains the same. A noble gas is formed when either a shell or a subshell is filled and there is a large gap in energy before the next possible level is encountered.

**TABLE 28.4 Electronic Configuration of Some Elements**

| Z | Symbol | Ground-State Configuration | Ionization Energy (eV) | Z | Symbol | Ground-State Configuration | Ionization Energy (eV) |
|---|---|---|---|---|---|---|---|
| 1 | H | $1s^1$ | 13.595 | 19 | K | [Ar] $4s^1$ | 4.339 |
| 2 | He | $1s^2$ | 24.581 | 20 | Ca | $4s^2$ | 6.111 |
| | | | | 21 | Sc | $3d4s^2$ | 6.54 |
| 3 | Li | [He] $2s^1$ | 5.390 | 22 | Ti | $3d^24s^2$ | 6.83 |
| 4 | Be | $2s^2$ | 9.320 | 23 | V | $3d^34s^2$ | 6.74 |
| 5 | B | $2s^22p^1$ | 8.296 | 24 | Cr | $3d^54s^1$ | 6.76 |
| 6 | C | $2s^22p^2$ | 11.256 | 25 | Mn | $3d^54s^2$ | 7.432 |
| 7 | N | $2s^22p^3$ | 14.545 | 26 | Fe | $3d^64s^2$ | 7.87 |
| 8 | O | $2s^22p^4$ | 13.614 | 27 | Co | $3d^74s^2$ | 7.86 |
| 9 | F | $2s^22p^5$ | 17.418 | 28 | Ni | $3d^84s^2$ | 7.633 |
| 10 | Ne | $2s^22p^6$ | 21.559 | 29 | Cu | $3d^{10}4s^1$ | 7.724 |
| | | | | 30 | Zn | $3d^{10}4s^2$ | 9.391 |
| 11 | Na | [Ne] $3s^1$ | 5.138 | 31 | Ga | $3d^{10}4s^24p^1$ | 6.00 |
| 12 | Mg | $3s^2$ | 7.644 | 32 | Ge | $3d^{10}4s^24p^2$ | 7.88 |
| 13 | Al | $3s^23p^1$ | 5.984 | 33 | As | $3d^{10}4s^24p^3$ | 9.81 |
| 14 | Si | $3s^23p^2$ | 8.149 | 34 | Se | $3d^{10}4s^24p^4$ | 9.75 |
| 15 | P | $3s^23p^3$ | 10.484 | 35 | Br | $3d^{10}4s^24p^5$ | 11.84 |
| 16 | S | $3s^23p^4$ | 10.357 | 36 | Kr | $3d^{10}4s^24p^6$ | 13.996 |
| 17 | Cl | $3s^23p^5$ | 13.01 | | | | |
| 18 | Ar | $3s^23p^6$ | 15.755 | | | | |

*Note:* The bracket notation is used as a shorthand method to avoid repetition in indicating inner-shell electrons. Thus, [He] represents $1s^2$, [Ne] represents $1s^22s^22p^6$, [Ar] represents $1s^22s^22p^63s^23p^6$, and so on.

The elements in the first column of the periodic table are called the *alkali metals* and are very active chemically. Referring to Table 28.4, we can understand why these elements interact so strongly with other elements. All of these alkali metals have a single outer electron in an *s* subshell. This electron is shielded from the nucleus by all the electrons in the inner shells. Thus, it is only loosely bound to the atom and can readily be accepted by other atoms that bind it more tightly to form molecules.

All the elements in the seventh column of the periodic table (called the *halogens*) are also very active chemically. Note that all these elements lack one electron in a subshell. As a consequence, they readily accept electrons from other atoms to form molecules.

**Quick Quiz 28.4**

Krypton (atomic number 36) has how many electrons in its next to outer shell ($n = 3$)?
(a) 2 (b) 4 (c) 8 (d) 18

## APPLYING PHYSICS 28.4

As we move from left to right across one row of the periodic table, the effective size of the atoms first decreases and then increases. What would cause this behavior?

**Explanation** If we begin at the left side of the periodic table and move toward the middle, the nuclear charge increases. As a result, the Coulomb attraction between the nucleus and the electrons increases, and the electrons are pulled into an average position that is closer to the nucleus. From the middle of the row to the right side, the increasing number of electrons being placed in proximity to each other results in a mutual repulsion that increases the average distance from the nucleus and causes the atomic size to grow.

## 28.10 CHARACTERISTIC X-RAYS

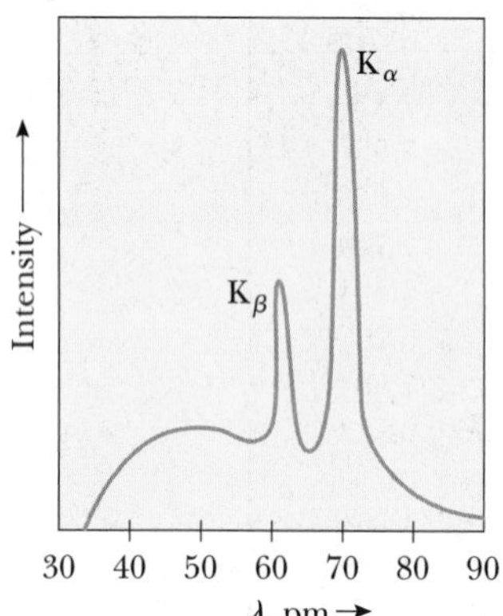


**FIGURE 28.14** The x-ray spectrum of a metal target consists of broad continuous spectrum plus a number of sharp lines that are due to *characteristic x-rays.* The data shown were obtained when 35-keV electrons bombarded a molybdenum target. (*Note:* 1 pm = $10^{-12}$ m = 0.001 nm.)

When a metal target is bombarded with high-energy electrons, x-rays are emitted. The x-ray spectrum typically consists of a broad continuous band and a series of intense sharp lines that are dependent on the type of metal used for the target, as shown in Figure 28.14. These discrete lines, called **characteristic x-rays,** were discovered in 1908, but their origin remained unexplained until the details of atomic structure were developed.

The first step in the production of characteristic x-rays occurs when a bombarding electron collides with an electron in an inner shell of a target atom with sufficient energy to remove the electron from the atom. The vacancy created in the shell is filled when an electron in a higher level drops down into the lower energy level containing the vacancy. The time it takes for this to happen is very short, less than $10^{-9}$ s. This transition is accompanied by the emission of a photon whose energy equals the difference in energy between the two levels. Typically, the energy of such transitions is greater than 1 000 eV, and the emitted x-ray photons have wavelengths in the range of 0.01–1 nm.

Let us assume that the incoming electron has dislodged an atomic electron from the innermost shell, the K shell. If the vacancy is filled by an electron dropping from the next higher shell, the L shell, the photon emitted in the process is referred to as the $K_\alpha$ line on the curve of Figure 28.14. If the vacancy is filled by an electron dropping from the M shell, the line produced is called the $K_\beta$ line.

Other characteristic x-ray lines are formed when electrons drop from upper levels to vacancies other than those in the K shell. For example, L lines are produced when vacancies in the L shell are filled by electrons dropping from higher shells. An $L_\alpha$ line is produced as an electron drops from the M shell to the L shell, and an $L_\beta$ line is produced by a transition from the N shell to the L shell.

We can estimate the energy of the emitted x-rays as follows. Consider two electrons in the K shell of an atom whose atomic number is $Z$. Each electron partially shields the other from the charge of the nucleus, $Ze$, and so each is subject to an effective nuclear charge of $Z_{\text{eff}} = (Z - 1)e$. We can now use a modified form of Equation 28.18 to estimate the energy of either electron in the K shell (with $n = 1$):

$$E_{\text{K}} = -m_e Z_{\text{eff}}^2 \frac{k_e^2 e^4}{2\hbar^2} = -Z_{\text{eff}}^2 E_0$$

where $E_0$ is the ground-state energy. Substituting $Z_{\text{eff}} = Z - 1$ gives

$$E_{\text{K}} = -(Z - 1)^2(13.6 \text{ eV}) \qquad [28.20]$$

As Example 28.6 shows, we can estimate the energy of an electron in an L or M shell in a similar fashion. Taking the energy difference between these two levels, we can then calculate the energy and wavelength of the emitted photon.

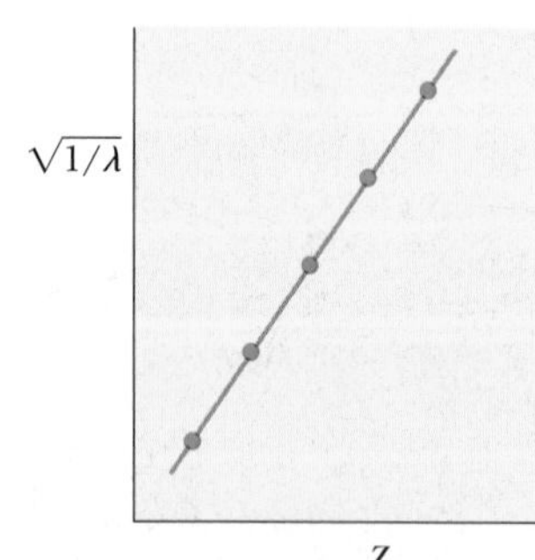


**FIGURE 28.15** A Moseley plot. A straight line is obtained when $\sqrt{1/\lambda}$ is plotted versus $Z$ for the $K_\alpha$ x-ray lines of a number of elements.

In 1914 Henry G. J. Moseley plotted the $Z$ values for a number of elements versus $\sqrt{1/\lambda}$, where $\lambda$ is the wavelength of the $K_\alpha$ line for each element. He found that such a plot produced a straight line, as in Figure 28.15. This is consistent with our rough calculations of the energy levels based on Equation 28.20. From this plot, Moseley was able to determine the $Z$ values of other elements, providing a periodic chart in excellent agreement with the known chemical properties of the elements.

## Example 28.6 Estimating the Energy of an x-Ray

Estimate the energy of the characteristic x-ray emitted from a tungsten target when an electron drops from an M shell ($n = 3$ state) to a vacancy in the K shell ($n = 1$ state).

**Solution** The atomic number for tungsten is $Z = 74$. Using Equation 28.20, we see that the energy of the electron in the K shell state is approximately

$$E_K = -(74 - 1)^2(13.6 \text{ eV}) = -72\,500 \text{ eV}$$

The electron in the M shell ($n = 3$) is subject to an effective nuclear charge that depends on the number of electrons in the $n = 1$ and $n = 2$ states, which shield the nucleus. Because there are eight electrons in the $n = 2$ state and one electron in the $n = 1$ state, roughly nine electrons shield the nucleus, and so $Z_{\text{eff}} = Z - 9$. Hence, the energy of an electron in the M shell ($n = 3$), following Equation 28.20, is equal to

$$E_M = -Z_{\text{eff}}^2 E_3 = -(Z-9)^2 \frac{E_0}{3^2} = -(74-9)^2 \frac{(13.6 \text{ eV})}{9} = -6\,380 \text{ eV}$$

where $E_3$ is the energy of an electron in the $n = 3$ level of the hydrogen atom. Therefore, the emitted x-ray has an energy equal to $E_M - E_K = -6\,380 \text{ eV} - (-72\,500 \text{ eV}) =$ $66\,100 \text{ eV}$. Note that this energy difference is also equal to $hf$, where $hf = hc/\lambda$, and where $\lambda$ is the wavelength of the emitted x-ray.

**EXERCISE** Calculate the wavelength of the emitted x-ray for this transition.

**ANSWER** 0.018 8 nm

## 28.11 ATOMIC TRANSITIONS

We have seen that an atom emits radiation only at certain frequencies that correspond to the energy separation between the various allowed states. Consider an atom with many allowed energy states, labeled $E_1$, $E_2$, $E_3$, . . . , as in Figure 28.16. When light is incident on the atom, only those photons whose energy, $hf$, matches the energy separation, $\Delta E$, between two levels can be absorbed by the atom. A diagram representing this **stimulated absorption process** is shown in Figure 28.17. At ordinary temperatures, most of the atoms in a sample are in the ground state. If a vessel containing many atoms of a gas is illuminated with a light beam containing all possible photon frequencies (that is, a continuous spectrum), only those photons of energies $E_2 - E_1$, $E_3 - E_1$, $E_4 - E_1$, and so on, can be ab-

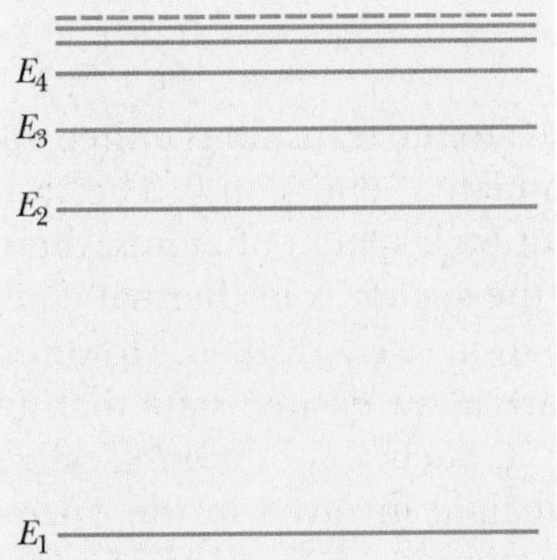


**FIGURE 28.16** Energy level diagram of an atom with various allowed states. The lowest energy state, $E_1$, is the ground state. All others are excited states.

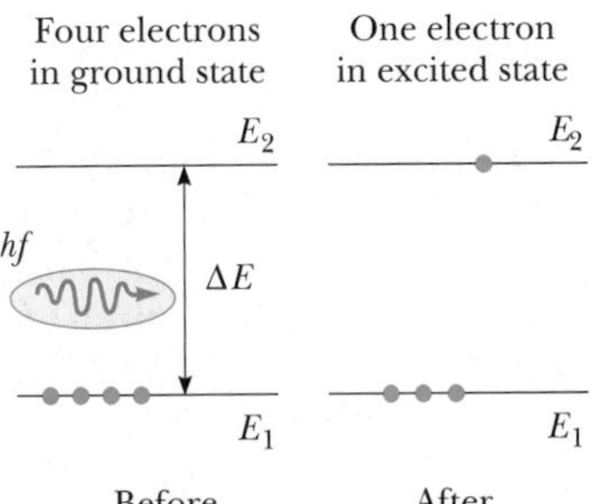


**FIGURE 28.17** *Stimulated absorption of a photon by an atom.* The blue dots represent electrons in the various states. One electron is transferred from the ground state to the excited state when the atom absorbs one photon whose energy is $hf = E_2 - E_1$.

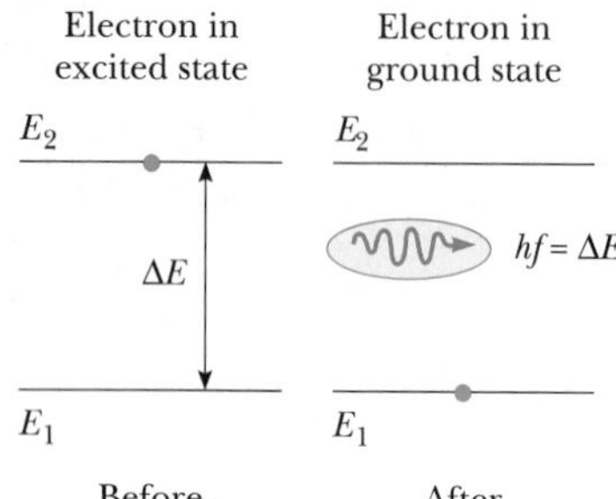


**FIGURE 28.18** *Spontaneous emission* of a photon by an atom that is initially in the excited state $E_2$. When the electron falls to the ground state, the atom emits a photon whose energy is $hf = E_2 - E_1$.

sorbed. As a result of this absorption, some atoms are raised to various allowed higher energy levels, called **excited states.**

Once an atom is in an excited state, there is a constant probability that it will jump back to a lower level by emitting a photon, as shown in Figure 28.18. This process is known as **spontaneous emission.** Typically, an atom remains in an excited state for only about $10^{-8}$ s.

A third process that is important in lasers, **stimulated emission,** was predicted by Einstein in 1917. Suppose an atom is in the excited state $E_2$, as in Figure 28.19, and a photon with energy $hf = E_2 - E_1$ is incident on it. The incoming photon increases the probability that the excited atom will return to the ground state and thereby emit a second photon having the same energy $hf$. Note that two identical photons result from stimulated emission—the incident photon and the emitted photon. *The emitted photon is exactly in phase with the incident photon.* These photons can stimulate other atoms to emit photons in a chain of similar processes. The many photons produced in this fashion are the source of the intense, coherent (in phase) light in a laser.

## APPLYING PHYSICS 28.5

A physics student is watching a meteor shower in the early morning hours. She notices that the streaks of light from the meteoroids entering the very high regions of the atmosphere last for as long as 2 or 3 seconds before fading. She also notices a lightning storm off in the distance. The streaks of light from the lightning fade away almost immediately after the flash, certainly in much less than 1 second. Both lightning and meteors cause the air to turn into a plasma because of the very high temperatures generated. The light is given off when the stripped electrons in the plasma recombine with the ionized atoms. Why would this light last longer for meteors than for lightning?

**Explanation** To answer this question, let us examine the phrase—". . . the streaks of light from the meteoroids entering the very high regions of the atmosphere." In the very high regions of the atmosphere, the pressure is very low. Thus, the density is very low, and the atoms of the gas are relatively far apart. Thus, after the air is ionized by the passing meteoroid, the probability of freed electrons finding an ionized atom with which to recombine is relatively low. As a result, the recombination process occurs over a relatively long time, measured in seconds. However, lightning occurs in the lower regions of the atmosphere (the troposphere) in which the pressure and density are relatively high. After the ionization by the lightning flash, the electrons and ionized atoms are much closer together than in the upper atmosphere. The probability of a recombination is much higher, and the time for the recombination to occur is much shorter.

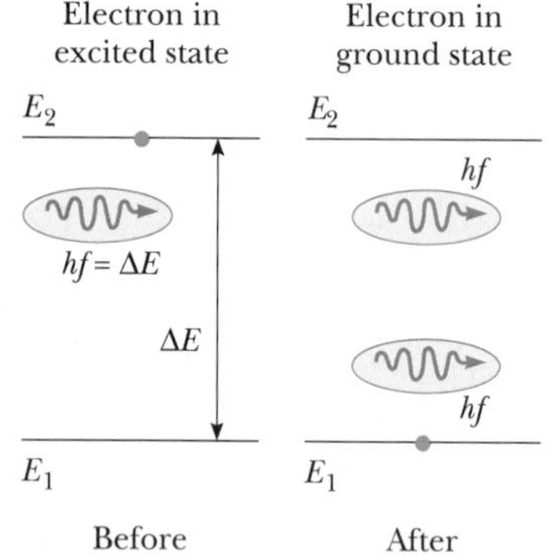


**FIGURE 28.19** *Stimulated emission* of a photon by an incoming photon of energy $hf$. Initially, the atom is in the excited state. The incoming photon stimulates the atom to emit a second photon of energy $hf = E_2 - E_1$, which is in phase with the first.

## 28.12 LASERS AND HOLOGRAPHY

We have described how an incident photon can cause atomic transitions either upward (stimulated absorption) or downward (stimulated emission). The two processes are equally probable. When light is incident on a system of atoms, there is usually a net absorption of energy because, when the system is in thermal equilibrium, many more atoms are in the ground state than in excited states. However, if the situation can be inverted so that more atoms are in an excited state than in the ground state, a net emission of photons can result. Such a condition is called **population inversion.** This is the fundamental principle involved in the operation of a laser, an acronym for *l*ight *a*mplification by *s*timulated *e*mission of *r*adiation. The amplification corresponds to a buildup of photons in the system as the result of a chain reaction of events. To achieve laser action, the following three conditions must be satisfied:

**Webnote 28.1**

Visualize how a photon interacts with an atom. Go to *http://www.lightlink.com/sergey/java/java/atomphoton/index.html*

1. The system must be in a state of population inversion (that is, more atoms in an excited state than in the ground state).

2. The excited state of the system must be a *metastable state,* which means its lifetime must be long compared with the usually short lifetimes of excited states. When that is the case, stimulated emission will occur before spontaneous emission.
3. The emitted photons must be confined in the system long enough to allow them to stimulate further emission from other excited atoms. This is achieved by using reflecting mirrors at the ends of the system. One end is totally reflecting, and the other is slightly transparent to allow the laser beam to escape.

One device that exhibits stimulated emission of radiation is the helium-neon gas laser. Figure 28.20 is an energy level diagram for the neon atom in this system. The mixture of helium and neon is confined to a glass tube sealed at the ends by mirrors. A high voltage applied to the tube causes electrons to sweep through the tube, colliding with the atoms of the gas and raising them into excited states. Neon atoms are excited to state $E_3^*$ through this process and also as a result of collisions with excited helium atoms. When a neon atom makes a transition to state $E_2$, it stimulates emission by neighboring excited atoms. This results in the production of coherent light at a wavelength of 632.8 nm. Figure 28.21a summarizes the steps in the production of a laser beam.

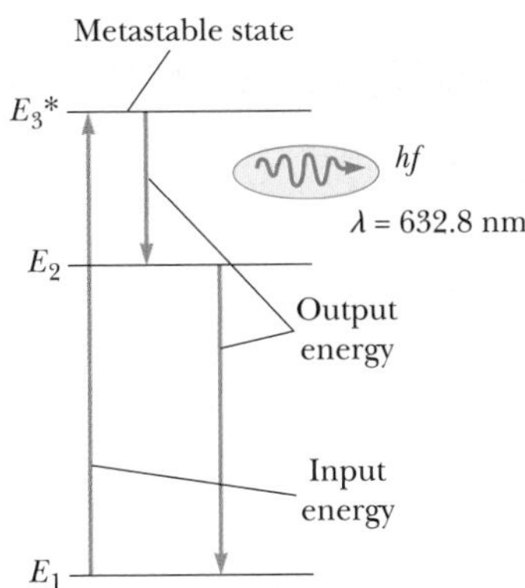


**FIGURE 28.20** Energy level diagram for the neon atom, which emits photons at a wavelength of 632.8 nm through stimulated emission. The photon at this wavelength arises from the transition $E_3^* \rightarrow E_2$. This is the source of coherent light in the helium-neon gas laser.

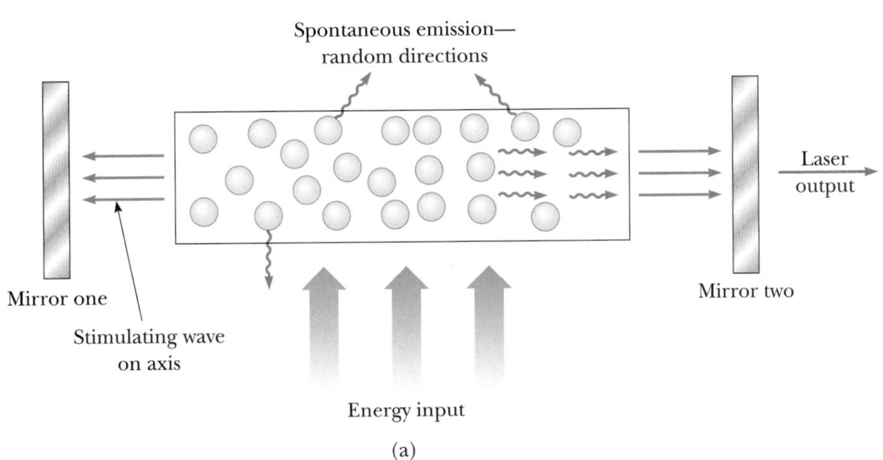


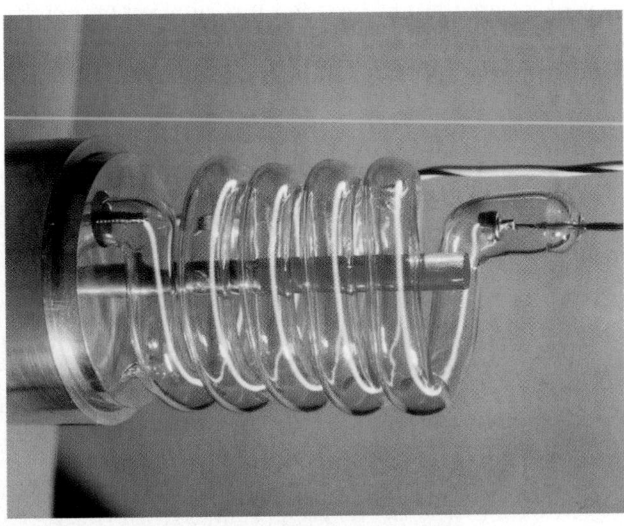

(b)

**FIGURE 28.21** (a) Steps in the production of a laser beam. The tube contains atoms, which represent the active medium. An external source of energy (optical, electrical, etc.) is needed to "pump" the atoms to excited energy states. The parallel end mirrors provide the feedback of the stimulating wave. (b) Photograph of the first ruby laser showing the flash lamp surrounding the ruby rod. *(Courtesy of HRL Laboratories LLC, Malibu, CA)*

Scientist checking the performance of an experimental laser-cutting device mounted on a robot arm. The laser is being used to cut through a metal plate. *(Philippe Plailly/Photo Researchers)*

**APPLICATION**

LASER TECHNOLOGY

Since the development of the first laser in 1960, laser technology has experienced tremendous growth. Lasers that cover wavelengths in the infrared, visible, and ultraviolet regions are now available. Applications include surgical "welding" of detached retinas, "lasik" surgery, precision surveying and length measurement, a potential source for inducing nuclear fusion reactions, precision cutting of metals and other materials, and telephone communication along optical fibers. These and other applications are possible because of the unique characteristics of laser light. In addition to being highly monochromatic and coherent, laser light is also highly directional and can be sharply focused to produce regions of extremely intense light energy.

### HOLOGRAPHY

**APPLICATION**

HOLOGRAPHY

**Webnote 28.2**

You can learn more about holograms, and your instructor can produce holograms on a realistic budget by viewing *http://members.aol.com/gakall/holopg.html*

**Webnote 28.3**

Take a look at the popular *Lasik* procedure. Go to *http://www.lasik.md/lasiksection/normaleye.htm*

One interesting application of the laser is holography, the production of three-dimensional images of objects. Figure 28.22a shows how a hologram is made. Light from the laser is split into two parts by a half-silvered mirror at *B*. One part of the beam reflects off the object to be photographed and strikes an ordinary photographic film. The other half of the beam is diverged by lens $L_2$, reflects from mirrors $M_1$ and $M_2$, and finally strikes the film. The two beams overlap to form an extremely complicated interference pattern on the film, one that can be produced only if the phase relationship of the two waves is constant throughout the exposure of the film. This condition is met through the use of light from a laser because such light is coherent. The hologram records not only the intensity of the light scattered from the object (as in a conventional photograph) but also the phase difference between the reference beam and the beam scattered from the object. Because of this phase difference, an interference pattern is formed that produces an image with full three-dimensional perspective.

A hologram is best viewed by allowing coherent light to pass through the developed film while looking back along the direction from which the beam comes. Figure 28.22b is a photograph of a hologram made using a cylindrical film.

## 28.13 ENERGY BANDS IN SOLIDS

In this section we trace the changes that occur in the discrete energy levels of isolated atoms when atoms group together and form a solid. We find that in solids, the discrete levels of isolated atoms broaden into allowed energy bands separated by forbidden gaps. The separation and electron population of the highest bands determines whether a given solid is a conductor, insulator, or semiconductor.

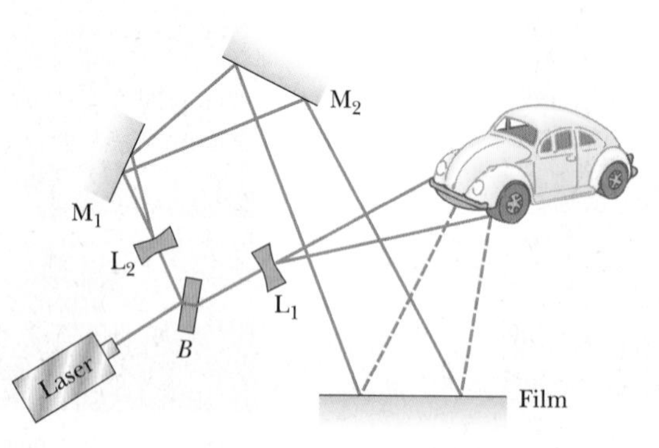


(a)

(b)

**FIGURE 28.22** (a) Experimental arrangement for producing a hologram. (b) Photograph of a hologram made using a cylindrical film. Note the detail of the Volkswagen image. *(Courtesy of Central Scientific Company)*

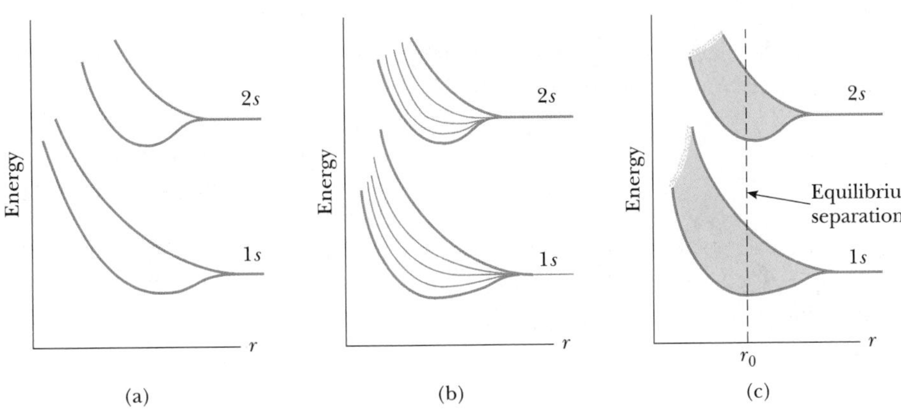


**FIGURE 28.23** (a) Splitting of the 1*s* and 2*s* states when two atoms are brought together. (b) Splitting of the 1*s* and 2*s* states when five atoms are brought close together. (c) Formation of energy bands when a large number of atoms are assembled to form a solid.

Consider two identical atoms, initially widely separated, that are brought closer and closer together. If two identical atoms are very far apart, they do not interact, and their electronic energy levels can be considered to be those of isolated atoms. Hence, the energy levels are exactly the same. As the atoms come close together, they essentially become one quantum system, and the Pauli exclusion principle demands that the two electrons be in different quantum states for this single system. The exclusion principle manifests itself as a changing or splitting of electron energy levels that were identical in the widely separated atoms as shown in Figure 28.23a. Figure 28.23b shows that with five atoms, each energy level in the isolated atom splits into five different, more closely spaced levels.

If we extend this argument to the large number of atoms found in solids (of the order of $10^{23}$ atoms/cm$^3$), we obtain a large number of levels so closely spaced that they may be regarded as a continuous **band** of energy levels, as in Figure 28.23c. An electron can have any energy within an allowed energy band, but cannot have an energy in the **band gap** or the region between allowed bands. In practice we are only interested in the band structure of a solid at some equilibrium separation of its atoms $r_0$, and so we remove the distance scale on the $r$ axis and simply plot the allowed energy bands of a solid as a series of horizontal bands as shown in Figure 28.24 for sodium.

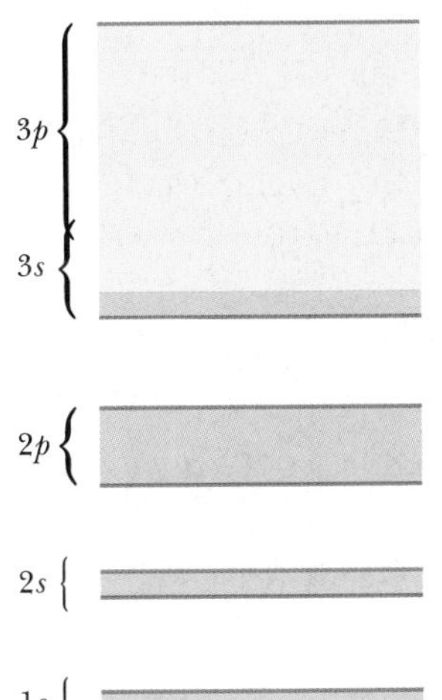


**FIGURE 28.24** Energy bands of sodium. Note the energy gaps (white regions) between the allowed bands; electrons cannot occupy states that lie in these forbidden gaps. Blue represents energy bands occupied by the sodium electrons when the atom is in its ground state. Gold represents energy bands that are empty. Note that the 3*s* and 3*p* levels broaden so much that they overlap.

## CONDUCTORS AND INSULATORS

Figure 28.24 shows that the band structure of a particular solid is quite complicated with individual atomic levels broadening by varying amounts and some levels (3*s* and 3*p*) broadening so much that they overlap. Nevertheless, it is possible to gain a qualitative understanding of whether a solid is a conductor, insulator, or semiconductor by considering only the structure of the upper one or two energy bands, and whether they are occupied by electrons.

Deciding whether an energy band is empty (unoccupied by electrons), partially filled, or full is carried out in basically the same way as we decided on the energy level population of atoms: We distribute the total number of electrons from the lowest energy levels up in a way consistent with the exclusion principle. Although we omit the details of this process here, one important case is shown in Figure 28.25a, where the highest-energy occupied band is only partially full. The other important case, where the highest occupied band is completely full, is shown in Figure 28.25b. Notice that this figure also shows that the highest filled band is the **valence band** and the next higher empty band is the **conduction band.** The energy band gap, which varies with the type of material, is also indicated as the energy difference $E_g$ between the top of the valence band and the bottom of the conduction band.

With these ideas and definitions we are now in a position to understand what determines, quantum mechanically, whether a solid will be a conductor or an insu-

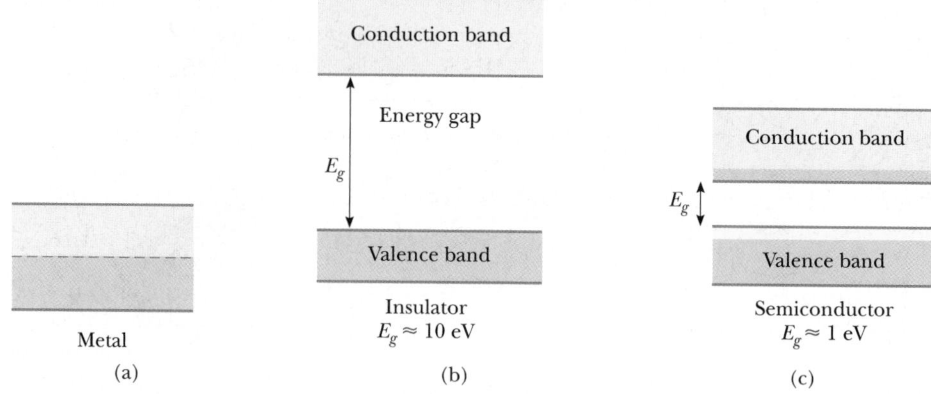


**FIGURE 28.25** (a) Half-filled band of a metal, an electrical conductor. (b) An electrical insulator at $T = 0$ K has a filled valence band and an empty conduction band. (c) Band structure of a semiconductor at ordinary temperatures ($T \approx 300$ K). The energy gap is much smaller than in an insulator, and many electrons occupy states in the conduction band.

lator. When a modest voltage is applied to a good conductor, the electrons accelerate and gain energy. In quantum terms, electron energies increase *if there are higher unoccupied energy levels for electrons to jump to.* For example, electrons near the top of the partially filled band in sodium need to gain very little energy from the applied voltage to reach one of the nearby, closely spaced, empty states. Thus, it is easy for a small voltage to kick electrons into higher energy states, and charge flows easily in sodium, an excellent conductor.

Now consider the case of a material where the highest occupied band is completely full of electrons and a band gap separates this filled valence band from the vacant conduction band, as in Figure 28.25b. A typical case might be that of diamond (carbon) where the band gap is about 10 eV. When a voltage is applied, electrons cannot easily gain energy because there are no vacant states nearby in energy to which electrons can make transitions. Because the only empty band is the conduction band, an electron must gain an amount of energy at least equal to the band gap for it to move through the solid. This large amount of energy can't be supplied by a modest applied voltage, and so no charge flows and diamond is a good insulator. In summary then, a conductor has a highest-energy occupied band that is *partially filled,* and an insulator has a highest-energy occupied band that is *completely filled* with a large energy gap between its valence and conduction bands.

## SEMICONDUCTORS

Thus far, we have completely ignored the influence of temperature on the electronic populations of energy bands. Recalling that the average thermal energy of a particle at temperature $T$ is $3k_B T/2$, we find that an electron at room temperature has an average energy of about 0.04 eV. Because this energy is about 100 times smaller than the band gap in a typical insulator, very few electrons have enough random thermal energy to jump the energy gap in an insulator and contribute to conduction. However, things are different for a semiconductor. As we see in Figure 28.25c, a **semiconductor** is a material with a small band gap of about 1 eV whose conductivity results from appreciable thermal excitation of electrons across the band gap into the conduction band at room temperature. The most commonly used semiconductors are silicon and gallium arsenide with band gaps of 1.14 eV and 1.43 eV, respectively, at 300 K. As you might expect, the resistivity of semiconductors usually decreases with increasing temperature, because $k_B T$ becomes a larger fraction of the band gap energy.

It is interesting that the electrons in the conduction band of a semiconductor do not carry the entire current when a voltage is applied, as Figure 28.25c shows. (We are tempted to say that conduction electrons do not constitute the "whole" story.) The missing electrons in the valence band, shown as a narrow white band in the figure, provide a few empty states called **holes** for valence band electrons to fill; so some electrons in the valence band gain energy and move toward a positive electrode and thus also carry the current. Since the valence band electrons that fill holes leave behind other holes, it is equally valid and more common to view the conduction process in the valence band as a flow of positive holes toward the negative electrode applied to a semiconductor. Thus, a pure semiconductor, such as silicon, can be viewed in a symmetric way: Silicon has equal numbers of mobile electrons in the conduction band and holes in the valence band. Furthermore, when an external voltage is applied to the semiconductor, electrons move toward the positive electrode, and holes move toward the negative electrode. This symmetric current process in a semiconductor is summarized in Figure 28.26. In the next section we shall look at the concepts of electron and hole in a simpler, more graphic way as the presence or absence of an outer-shell electron at a particular location in a crystal lattice.

When small amounts of impurities are added to a semiconductor (about one impurity atom per $10^7$ silicon atoms), both the band structure of the semiconductor and its resistivity are modified. The process of adding impurities, called **doping,** is important in making devices that have well-defined regions of different resistivity. For example, when an atom containing five outer-shell electrons, such as arsenic, is added to a semiconductor such as silicon, four of the arsenic electrons form shared bonds with atoms of the semiconductor and one is left over. This extra electron is nearly free of its parent atom and has an energy level that lies in the energy gap, just below the conduction band. Such a pentavalent atom in effect donates an electron to the structure and hence is referred to as a **donor atom.** Because the spacing between the energy level of the electron of the donor atom and the bottom of the conduction band is very small (typically, about 0.05 eV), only a small amount of thermal energy is needed to cause this electron to move into the conduction band. (Recall that the average thermal energy of an electron at room temperature is $3k_BT/2 \approx 0.04$ eV). Semiconductors doped with donor atoms are called ***n*-type semiconductors** because the charge carriers are electrons, the charge of which is **n**egative.

If a semiconductor is doped with atoms containing three outer-shell electrons, such as aluminum, the three electrons form shared bonds with neighboring semiconductor atoms, leaving an electron deficiency—a hole—where the fourth bond would be if an impurity-atom electron was available to form it. The energy level of this hole lies in the energy gap, just above the valence band. An electron from the valence band has enough energy at room temperature to fill this impurity level, leaving behind a hole in the valence band. Because a trivalent atom in effect accepts an electron from the valence band, such impurities are referred to as **acceptor atoms.** A semiconductor doped with acceptor impurities is known as a ***p*-type semiconductor** because the majority of charge carriers are **p**ositively charged holes.

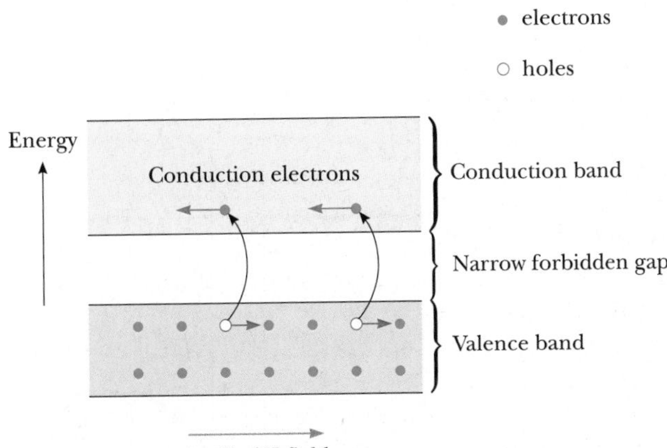


**FIGURE 28.26** Movement of charges (holes and electrons) in a semiconductor. The electrons move in the direction opposite the direction of the external electric field, and the holes move in the direction of the field.

## 28.14 SEMICONDUCTOR DEVICES

### THE *P-N* JUNCTION

Now let us consider what happens when a *p*-semiconductor is joined to an *n*-semiconductor to form a *p-n* junction. The junction consists of the three distinct regions shown in Figure 28.27a: a *p* region, a depletion region, and an *n* region.

The depletion region, which extends several micrometers to either side of the center of the junction, may be visualized as arising when the two halves of the junction are brought together. Mobile donor electrons from the *n* side nearest the junction (blue area in Figure 28.27a) diffuse to the *p* side, leaving behind immobile positive ions. At the same time, holes from the *p* side nearest the junction diffuse to the *n* side and leave behind a region (red area in Figure 28.27a) of fixed negative ions. Thus, the depletion region is so named because it is depleted of mobile charge carriers.

The depletion region contains an internal electric field (arising from the charges of the fixed ions) on the order of $10^4$–$10^6$ V/cm. This field sweeps mobile charge out of the depletion region and keeps it truly depleted. This internal electric field creates an internal potential difference $\Delta V_0$ that prevents further diffusion of holes and electrons across the junction and thereby ensures zero current through the junction when no external potential difference is applied.

Perhaps the most notable feature of the *p-n* junction is its ability to pass current in only one direction. Such *diode* action is easiest to understand in terms of the potential difference graph shown in Figure 28.27c. If an external voltage $\Delta V$ is applied to the junction such that the *p* side is connected to the positive terminal of a voltage source as in Figure 28.27a, the internal potential difference $\Delta V_0$ across the junction is decreased, resulting in a current that increases exponentially with increasing forward voltage, or *forward bias*. For *reverse bias* (where the *n* side of the junction is connected to the positive terminal of a voltage source), the internal potential difference $\Delta V_0$ increases with increasing reverse bias. This results in a very small reverse current that quickly reaches a saturation value $I_0$. The current-

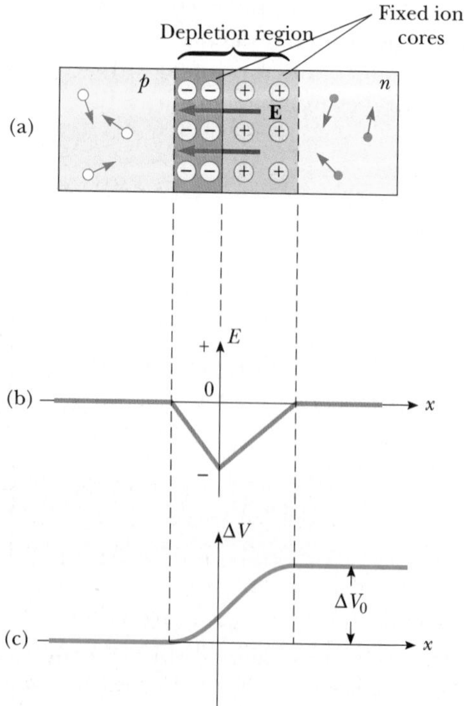


**FIGURE 28.27** (a) Physical arrangement of a *p-n* junction. (b) Internal electric field versus *x* for the *p-n* junction. (c) Internal electric potential $\Delta V$ versus *x* for the *p-n* junction. $\Delta V_0$ represents the potential difference across the junction in the absence of an applied electric field.

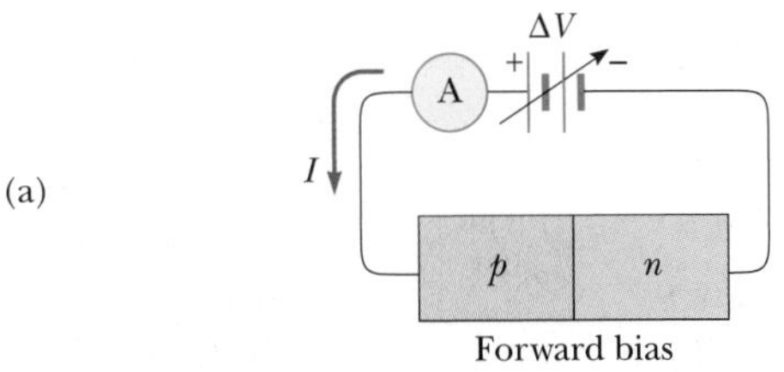


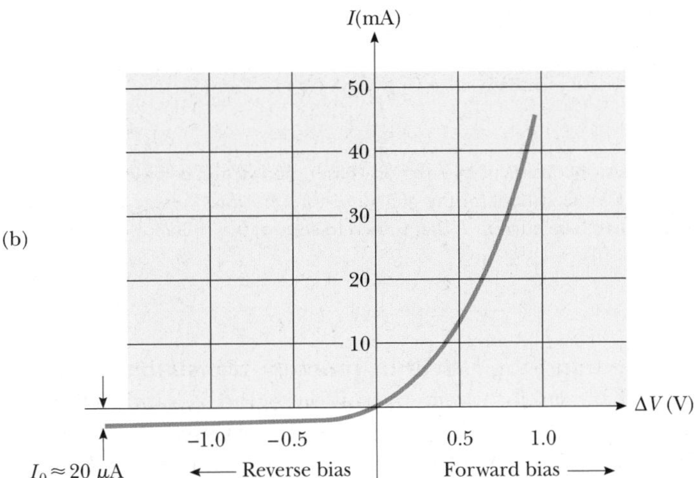


**FIGURE 28.28** (a) Schematic of a *p-n* junction under forward bias. (b) The characteristic curve for a real *p-n* junction.

voltage relationship for an ideal diode is

$$I = I_0(e^{q\Delta V/k_B T} - 1) \quad [28.21]$$

where $e$ is Euler's number, $q$ is the electron charge, $k_B$ is Boltzmann's constant, and $T$ is the temperature in kelvins. Figure 28.28 shows an $I$-$\Delta V$ plot characteristic of a real *p-n* junction along with a schematic of such a device under forward bias.

The most common use of the semiconductor diode is as a rectifier, a device that changes 120-V AC voltage supplied by the power company to, say the 12-V DC voltage needed by your music keyboard. We can understand how a diode rectifies a current by considering Figure 28.29a, which shows a diode connected in series with a resistor and an AC source. Because appreciable current can pass through the diode only in one direction, the alternating current in the resistor is reduced to the form shown in Figure 28.29b. The diode is said to act as a **half-wave rectifier** because there is current in the circuit during only half of each cycle.

Figure 28.30a shows a circuit that lowers the AC voltage to 12 V with a step-down transformer and then rectifies both halves of the 12 V AC. Such a rectifier is called a **full-wave rectifier** and when combined with a step-down transformer is the most common DC power supply around the home today. A capacitor added in parallel with the load will yield an even steadier DC voltage.

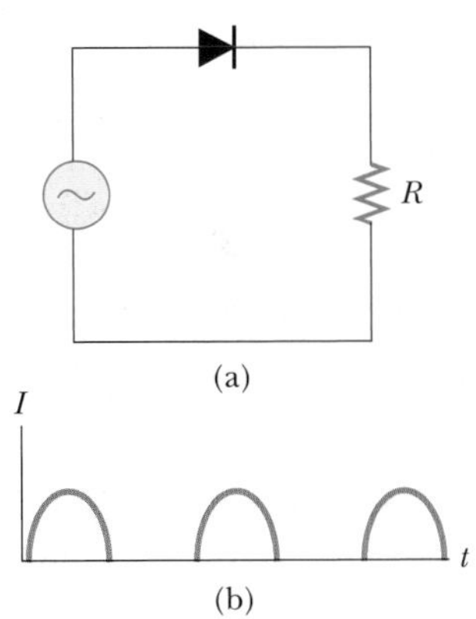


**FIGURE 28.29** (a) A diode in series with a resistor allows current to pass in only one direction. (b) The current versus time for the circuit in (a).

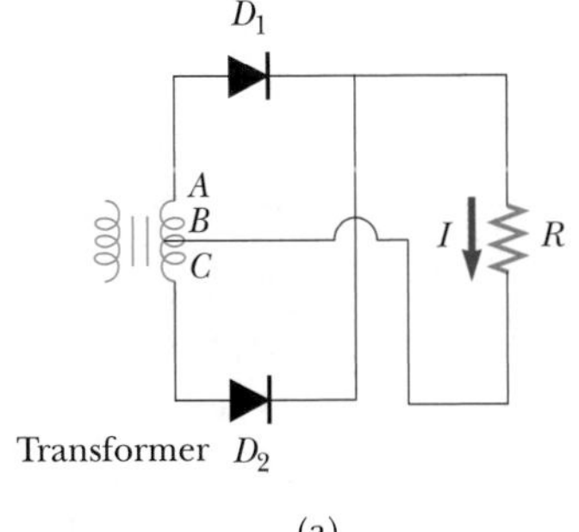


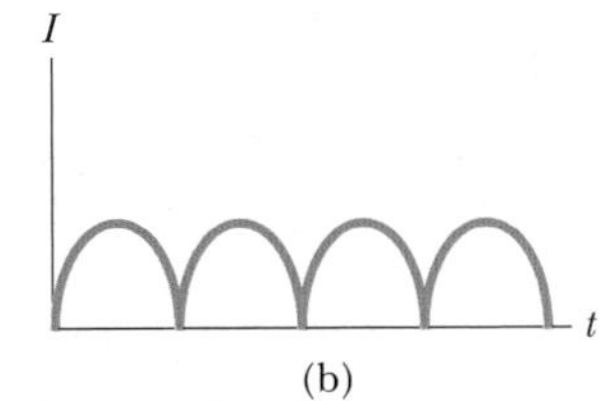


**FIGURE 28.30** (a) A full-wave rectifier circuit. (b) The current versus time in the resistor *R*.

## THE JUNCTION TRANSISTOR

The invention of the transistor by John Bardeen (1908–1991), Walter Brattain (1902–1987), and William Shockley (1910–1989) in 1948 totally revolutionized the world of electronics. For this work, these three men shared a Nobel prize in 1956. By 1960, the transistor had replaced the vacuum tube in many electronic applications. The advent of the transistor created a multitrillion dollar industry that produced such popular devices as pocket radios, handheld calculators, computers, television receivers, and electronic games. In this section we explain how a transistor acts as an amplifier to boost the tiny voltages and currents generated in a microphone to the ear-splitting levels required to drive a speaker.

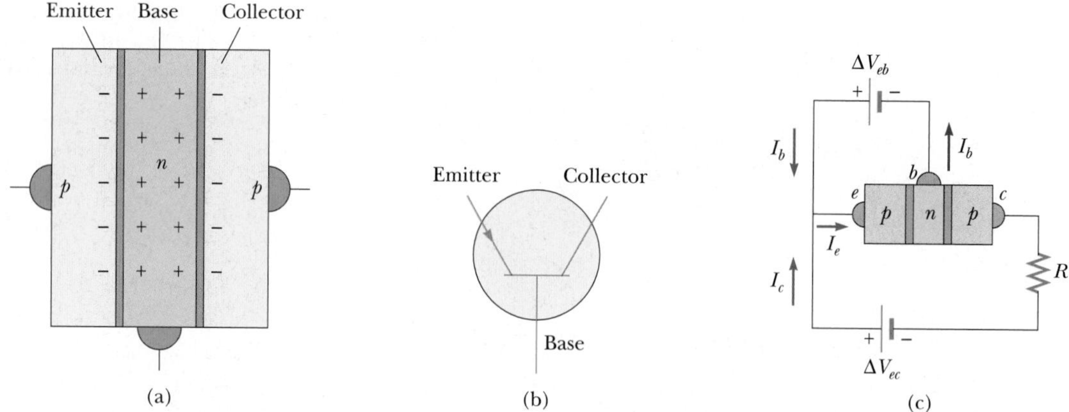


**FIGURE 28.31** (a) The *pnp* transistor consists of an *n* region (base) sandwiched between two *p* regions (emitter and collector). (b) Circuit symbol for the *pnp* transistor. (c) A bias voltage $\Delta V_{eb}$ applied to the base as shown produces a small base current $I_b$ that is used to control the collector current $I_c$ in a *pnp* transistor.

One simple form of the transistor, called the **junction transistor,** consists of a semiconducting material in which a very narrow *n* region is sandwiched between two *p* regions. This configuration is called a ***pnp* transistor.** Another configuration is the ***npn* transistor,** which consists of a *p* region sandwiched between two *n* regions. Because the operation of the two transistors is essentially the same, we describe only the *pnp* transistor. The structure of the *pnp* transistor, together with its circuit symbol, is shown in Figure 28.31. The outer regions are called the **emitter** and **collector,** and the narrow central region is called the **base.** The configuration contains two junctions: the emitter-base interface and the collector-base interface.

Suppose a voltage is applied to the transistor so that the emitter is at a higher electric potential than the collector. (This is accomplished with the battery labeled $\Delta V_{ec}$ in Figure 28.31c.) If we think of the transistor as two diodes back to back, we see that the emitter-base junction is forward-biased and the base-collector junction is reverse-biased. The emitter is heavily doped relative to the base, and as a result nearly all the current consists of holes moving across the emitter-base junction. Most of these holes do not recombine with electrons in the base because the base is very narrow. Instead, they are accelerated across the reverse-biased base-collector junction, producing the collector current $I_c$ in Figure 28.31c.

Although only a small percentage of holes recombine in the base, those that do limit the emitter current to a small value because positive charge carriers accumulating in the base prevent holes from flowing in. In order to prevent this current limitation, some of the positive charge on the base must be drawn off; this is accomplished by connecting the base to the battery labeled $\Delta V_{eb}$ in Figure 28.31c. Those positive charges that are not swept across the base-collector junction leave the base through this added pathway. **This base current $I_b$ is very small, but a small change in it can significantly change the collector current $I_c$.** If the transistor is properly biased, the collector (output) current is directly proportional to the base (input) current, and the transistor acts as a current amplifier. This condition may be written

$$I_c = \beta I_b$$

where $\beta$, the *current gain* factor, is typically in the range of 10–100. Thus, the transistor may be used to amplify a small signal. The small voltage to be amplified is placed in series with the battery $\Delta V_{eb}$. The input signal produces a small variation in the base current, resulting in a large change in the collector current and hence a large change in the voltage across the output resistor.

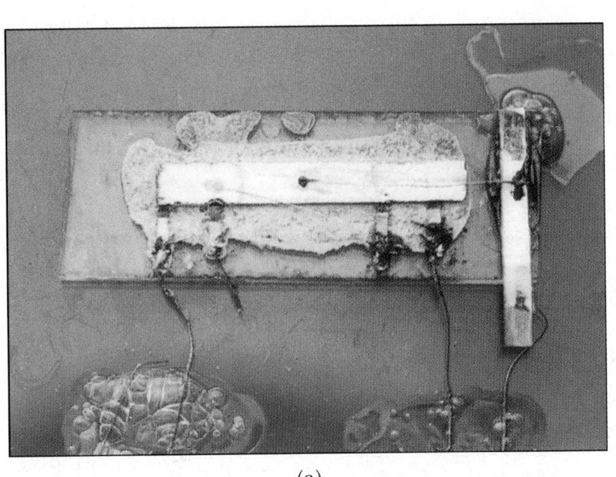

(a)

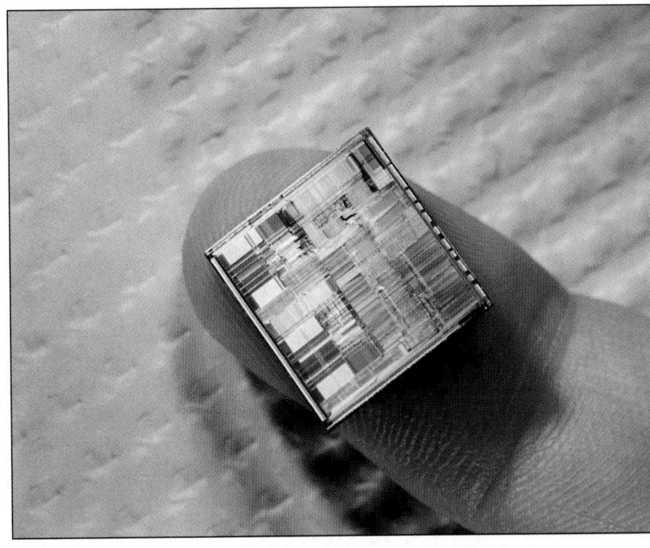

(b)

**FIGURE 28.32** (a) Jack Kilby's first integrated circuit tested on September 12, 1958. *(Courtesy of Texas Instruments, Inc.)* (b) Integrated circuits continue to shrink in size and price while simultaneously growing in capability. *(Courtesy of Intel Corporation)*

## THE INTEGRATED CIRCUIT

Invented independently by Jack Kilby at Texas Instruments in late 1958 and by Robert Noyce at Fairchild Camera and Instrument in early 1959, the integrated circuit has been justly called "the most remarkable technology ever to hit mankind." Kilby's first device is shown in Figure 28.32a. Integrated circuits have indeed started a "second industrial revolution" and are at the heart of computers, watches, cameras, automobiles, aircraft, robots, space vehicles, and all sorts of communication and switching networks.

In simplest terms, an **integrated circuit** is a collection of interconnected transistors, diodes, resistors, and capacitors fabricated on a single piece of silicon, known as a chip. State-of-the-art chips easily contain several million components in a 1-$cm^2$ area, with the number of components per square inch having doubled every year since the integrated circuit was invented.

Integrated circuits were invented partly to solve the interconnection problem spawned by the transistor. In the era of vacuum tubes, power and size considerations of individual components set significant limits on the number of components that could be interconnected in a given circuit. With the advent of the tiny, low-power, highly reliable transistor, design limits on the number of components disappeared and were replaced by the problem of wiring together hundreds of thousands of components. The magnitude of this problem can be appreciated when we consider that second-generation computers (consisting of discrete transistors rather than integrated circuits) contained several hundred thousand components requiring more than a million hand-soldered joints to be made and tested.

In addition to solving the interconnection problem, integrated circuits possess the advantages of miniaturization and fast response, two attributes critical for high-speed computers. The fast response results from the miniaturization and close packing of components, because the response time of a circuit depends on the time it takes for electrical signals traveling at about the speed of light to pass from one component to another. This time is clearly reduced by packing components closely.

## SUMMARY


The **Bohr model** of the atom is successful in describing the spectra of atomic hydrogen and hydrogen-like ions. One of the basic assumptions of the model is that the electron can exist only in certain orbits such that its angular momentum, $mvr$, is an integral multiple of $\hbar$, where $\hbar$ is Planck's constant divided by $2\pi$. Assuming circular orbits and a Coulomb force of attraction between electron and proton, the energies of the quantum states for hydrogen are

$$E_n = -\frac{m_e k_e^2 e^4}{2\hbar^2}\left(\frac{1}{n^2}\right) \qquad n = 1, 2, 3, \ldots \qquad [28.12]$$

where $k_e$ is the Coulomb constant, $e$ is the charge on the electron, and $n$ is an integer called a **quantum number.**

If the electron in the hydrogen atom jumps from an orbit whose quantum number is $n_i$ to an orbit whose quantum number is $n_f$, it emits a photon of frequency $f$, given by

$$f = \frac{m_e k_e^2 e^4}{4\pi\hbar^3}\left(\frac{1}{n_f^2} - \frac{1}{n_i^2}\right) \qquad [28.14]$$

Bohr's **correspondence principle** states that quantum mechanics is in agreement with classical physics when the quantum numbers for a system are very large.

One of the many successes of quantum mechanics is that the quantum numbers $n$, $\ell$, and $m_\ell$ associated with atomic structure arise directly from the mathematics of the theory. The quantum number $n$ is called the **principal quantum number,** $\ell$ is the **orbital quantum number,** and $m_\ell$ is the **orbital magnetic quantum number.** In addition, a fourth quantum number, called the **spin magnetic quantum number,** $m_s$, is needed to explain certain features of atomic structure.

An understanding of the periodic table of the elements became possible when Pauli formulated the **exclusion principle,** which states that no two electrons in an atom can ever be in the same quantum state; that is, no two electrons in the same atom can have the same values for the set of quantum numbers, $n$, $\ell$, $m_\ell$, and $m_s$.

**Characteristic x-rays** are produced when a bombarding electron collides with an electron in an inner shell of an atom with sufficient energy to remove the electron from the atom. The vacancy thus created is filled when an electron from a higher level drops down into the level containing the vacancy.

**Lasers** are monochromatic, coherent light sources that work on the principle of **stimulated emission** of radiation from a system of atoms.

In solids, the discrete energy levels of isolated atoms broaden into **bands,** and the structure and electron population of the bands determine whether a substance is a conductor, insulator, or semiconductor. By **doping** semiconductors, specific regions with known resistivity and positive or negative majority charge carriers are created, and devices such as diodes, transistors, and integrated circuits may be fabricated. These solid state devices have revolutionized life in the 20th century by their reliability, low power consumption, speed of response, and their virtually limitless interconnection capability in chips.


## CONCEPTUAL QUESTIONS

1. In the hydrogen atom, the quantum number $n$ can increase without limit. Because of this, does the frequency of possible spectral lines from hydrogen also increase without limit?
2. Does the light emitted by a neon sign constitute a continuous spectrum or only a few colors? Defend your answer.
3. In an x-ray tube, if the energy with which the electrons strike the metal target is increased, the wavelengths of the characteristic x-rays do not change. Why not?
4. Must an atom first be ionized before it can emit light? Discuss.
5. Is it possible for a spectrum from an x-ray tube to show the continuous spectrum of x-rays without the presence of the characteristic x-rays?
6. Suppose that the electron in the hydrogen atom obeyed classical mechanics rather than quantum mechanics. Why should such a hypothetical atom emit a continuous spectrum rather than the observed line spectrum?

7. When a hologram is produced, the system (including light source, object, beam splitter, and so on) must be held motionless within a quarter of the light's wavelength. Why?
8. If matter has a wave nature, why is this not observable in our daily experiences?
9. Discuss some consequences of the exclusion principle.
10. Can the electron in the ground state of hydrogen absorb a photon of energy less than 13.6 eV? Can it absorb a photon of energy greater than 13.6 eV? Explain.
11. Why do lithium, potassium, and sodium exhibit similar chemical properties?
12. List some ways in which quantum mechanics altered our view of the atom pictured by the Bohr theory.
13. It is easy to understand how two electrons (one spin-up, one spin-down) can fill the $1s$ shell for a helium atom. How is it possible that eight more electrons can fit into the $2s$, $2p$ level to complete the $1s^22s^22p^6$ shell for a neon atom?
14. The ionization energies for Li, Na, K, Rb, and Cs are 5.390, 5.138, 4.339, 4.176, and 3.893 eV, respectively. Explain why this pattern of decrease should be expected in terms of the atomic structures.
15. How can the total energy of an atom be negative? For that matter, how can any energy be negative?
16. A 0.1-mW laser might damage your eye if you look directly at it, yet there is no harm in looking at a bare 100-W lightbulb. Why?
17. Why is stimulated emission so important in the operation of a laser?

## PROBLEMS

1, 2, 3 = straightforward, intermediate, challenging □ = full solution available in Student Solutions Manual/Study Guide

**web** = solution posted at **http://info.brookscole.com/serway** = biomedical application

### Section 28.1 Early Models of the Atom

### Section 28.2 Atomic Spectra

1. Use Equation 28.1 to calculate the wavelength of the first three lines in the Balmer series for hydrogen.
2. Show that the wavelengths for the Balmer series satisfy the equation

$$\lambda = \frac{364.5n^2}{n^2 - 4}\text{ nm} \qquad \text{where } n = 3, 4, 5, \ldots$$

3. The "size" of the *atom* in Rutherford's model is about $1.0 \times 10^{-10}$ m. (a) Determine the attractive electrical force between an electron and a proton separated by this distance. (b) Determine (in eV) the electrical potential energy of the atom.
4. The "size" of the *nucleus* in Rutherford's model of the atom is about 1.0 fm $= 1.0 \times 10^{-15}$ m. (a) Determine the repulsive electrical force between two protons separated by this distance. (b) Determine (in MeV) the electrical potential energy of the pair of protons.
5. **web** The "size" of the atom in Rutherford's model is about $1.0 \times 10^{-10}$ m. (a) Determine the speed of an electron moving in a circle around the proton using the attractive electrical force between an electron and a proton separated by this distance. (b) Does this speed suggest that Einsteinian relativity must be considered when studying the atom? (c) Compute the de Broglie wavelength of the electron as it moves about the proton. (d) Does this wavelength suggest that wave effects, such as diffraction and interference, must be considered when studying the atom?
6. In a Rutherford scattering experiment, an $\alpha$-particle (charge $= +2e$) heads directly toward a gold nucleus (charge $= +79e$). The $\alpha$-particle has a kinetic energy of 5.0 MeV when very far ($r \to \infty$) from the nucleus. Assuming the gold nucleus to be fixed in space, determine the distance of closest approach. (*Hint:* Use conservation of energy with $PE = k_eq_1q_2/r$.)

### Section 28.3 The Bohr Theory of Hydrogen

7. A hydrogen atom is in its first excited state ($n = 2$). Using the Bohr theory of the atom, calculate (a) the radius of the orbit, (b) the linear momentum of the electron, (c) the angular momentum of the electron, (d) the kinetic energy, (e) the potential energy, and (f) the total energy.
8. For a hydrogen atom in its ground state, use the Bohr model to compute (a) the orbital speed of the electron, (b) the kinetic energy of the electron, and (c) the electrical potential energy of the atom.
9. Show that the speed of the electron in the $n$th Bohr orbit in hydrogen is given by

$$v_n = \frac{k_e e^2}{n\hbar}$$

10. A photon is emitted as a hydrogen atom undergoes a transition from the $n = 6$ state to the $n = 2$ state. Calculate (a) the energy, (b) the wavelength, and (c) the frequency of the emitted photon.
11. Calculate the Coulomb force of attraction on the electron when it is in the ground state of the hydrogen atom.
12. Four possible transitions for a hydrogen atom are listed below:

I. $n_i = 2$; $n_f = 5$ II. $n_i = 5$; $n_f = 3$
III. $n_i = 7$; $n_f = 4$ IV. $n_i = 4$; $n_f = 7$

(a) Which transition will emit the shortest-wavelength photon? (b) For which transition will the atom gain the most energy? (c) For which transition(s) does the atom lose energy?
13. What is the energy of the photon that, when absorbed by a hydrogen atom, could cause (a) an electronic transition from the $n = 3$ state to the $n = 5$ state and (b) an electronic transition from the $n = 5$ state to the $n = 7$ state?
14. A hydrogen atom initially in its ground state ($n = 1$) absorbs a photon and ends up in the state for which $n = 3$. (a) What is the energy of the absorbed photon? (b) If the

atom eventually returns to the ground state, what photon energies could the atom emit?

15. Determine both the longest and the shortest wavelengths in (a) the Lyman series ($n_f = 1$) and (b) the Paschen series ($n_f = 3$) of hydrogen.

16. How much energy is required to ionize hydrogen when it is in (a) the ground state and (b) the state for which $n = 3$?

17. A monochromatic beam of light is absorbed by a collection of ground-state hydrogen atoms in such a way that six different wavelengths are observed when the hydrogen relaxes back to the ground state. What is the wavelength of the incident beam?

18. A particle of charge $q$ and mass $m$, moving with a constant speed $v$, perpendicular to a constant magnetic field, $B$, follows a circular path. If the angular momentum about the center of this circle is quantized so that $mvr = n\hbar$, show that the allowed radii for the particle are

$$r_n = \sqrt{\frac{n\hbar}{qB}} \qquad \text{where } n = 1, 2, 3, \ldots$$

19. (a) If an electron makes a transition from the $n = 4$ Bohr orbit to the $n = 2$ orbit, determine the wavelength of the photon created in the process. (b) Assuming that the atom was initially at rest, determine the recoil speed of the hydrogen atom when this photon is emitted.

20. An electron is in the first Bohr orbit of hydrogen. Find (a) the speed of the electron, (b) the time required for the electron to circle the nucleus, and (c) the current in amperes corresponding to the motion of the electron.

21. Analyze the Earth-Sun system following the Bohr model, where the gravitational force between Earth (mass $m$) and the Sun (mass $M$) replaces the Coulomb force between the electron and proton (so that $F = GMm/r^2$ and $PE = -GMm/r$). Show that (a) Earth's total energy in an orbit of radius $r$ is given by $E = -GMm/2r$, (b) the radius of the $n$th orbit is given by $r_n = r_0 n^2$, where $r_0 = \hbar^2/GMm^2 = 2.32 \times 10^{-138}$ m, and (c) the energy of the $n$th orbit is given by $E_n = -E_0/n^2$, where $E_0 = G^2M^2m^3/2\hbar^2 = 1.71 \times 10^{182}$ J. (d) Using the Earth-Sun orbit radius of $r = 1.49 \times 10^{11}$ m, determine the value of the quantum number $n$. (e) Should you expect to observe quantum effects in the Earth-Sun system?

22. An electron is in the $n$th Bohr orbit of the hydrogen atom. (a) Show that the period of the electron is $T = t_0 n^3$, and determine the numerical value of $t_0$. (b) On the average, an electron remains in the $n = 2$ orbit for about 10 $\mu$s before it jumps down to the $n = 1$ (ground-state) orbit. How many revolutions does the electron make before it jumps to the ground state? (c) If one revolution of the electron is defined as an "electron-year" (analogous to an Earth-year being one revolution of Earth around the Sun), does the electron in the $n = 2$ orbit "live" very long? Explain. (d) How does the above calculation support the "electron cloud" concept?

23. Consider a hydrogen atom. (a) Calculate the frequency **web** $f$ of the $n = 2 \rightarrow n = 1$ transition and compare with the frequency $f_{orb}$ of the electron orbital motion in the $n = 2$ state. (b) Make the same calculation for the $n = 10\ 000 \rightarrow n = 9\ 999$ transition. Comment on the results.

24. Two hydrogen atoms collide head-on and end up with zero kinetic energy. Each then emits a 121.6-nm photon ($n = 2$ to $n = 1$ transition). At what speed were the atoms moving before the collision?

25. Two hydrogen atoms, both initially in the ground state, undergo a head-on collision. If both atoms are to be excited to the $n = 2$ level in this collision, what is the minimum speed each atom can have before the collision?

26. (a) Calculate the angular momentum of the Moon due to its orbital motion about Earth. In your calculation, use $3.84 \times 10^8$ m as the average Earth-Moon distance and $2.36 \times 10^6$ s as the period of the Moon in its orbit. (b) If the angular momentum of the moon obeys Bohr's quantization rule ($L = n\hbar$), determine the value of the quantum number, $n$. (c) By what fraction would the Earth-Moon radius have to be increased to increase the quantum number by 1?

### Section 28.4 Modification of the Bohr Theory

### Section 28.5 de Broglie Waves and the Hydrogen Atom

27. (a) Find the energy of the electron in the ground state of doubly ionized lithium, which has an atomic number $Z = 3$. (b) Find the radius of its ground-state orbit.

28. (a) Construct an energy level diagram for the $He^+$ ion, for which $Z = 2$. (b) What is the ionization energy for $He^+$?

29. The orbital radii of a hydrogen-like atom is given by the equation $r = n^2\hbar^2/Zm_ek_ee^2$. What is the radius of the first Bohr orbit in (a) $He^+$, (b) $Li^{2+}$, and (c) $Be^{3+}$?

30. Construct an energy level diagram like that in Figure 28.7 for doubly ionized lithium ($Li^+$), for which $Z = 3$.

31. Determine the wavelength of an electron in the third excited orbit of the hydrogen atom, with $n = 4$.

32. Using the concept of standing waves, de Broglie was able to derive Bohr's stationary orbit postulate. He assumed that a confined electron could exist only in states where its de Broglie waves form standing-wave patterns, as in Figure 28.10a. Consider a particle confined in a box of length $L$ to be equivalent to a string of length $L$ and fixed at both ends. Apply de Broglie's concept to show that (a) the linear momentum of this particle is quantized with $p = mv = nh/2L$ *and* (b) the allowed states correspond to particle energies of $E_n = n^2E_0$ where $E_0 = h^2/(8mL^2)$.

### Section 28.6 Quantum Mechanics and the Hydrogen Atom

### Section 28.7 The Spin Magnetic Quantum Number

33. List the possible sets of quantum numbers for electrons in the $3p$ subshell.

34. List the possible sets of quantum numbers for electrons in the $3d$ subshell.

35. The $\rho$-meson has a charge of $-e$, a spin quantum number of 1, and a mass 1 507 times that of the electron. If the electrons in atoms were replaced by $\rho$-mesons, list the possible sets of quantum numbers for $\rho$-mesons in the $3d$ subshell.

### Section 28.9 The Exclusion Principle and the Periodic Table

36. (a) Write out the electronic configuration of the ground state for oxygen ($Z = 8$). (b) Write out values for the set of

quantum numbers $n$, $\ell$, $m_\ell$, and $m_s$ for each of the electrons in oxygen.

**37.** Two electrons in the same atom both have $n = 3$ and $\ell = 1$. List the quantum numbers for the possible states of the atom. (a) How many states are possible? (b) How many states would be possible if the exclusion principle did not apply to the atom?

**38.** How many different sets of quantum numbers are possible for an electron for which (a) $n = 1$, (b) $n = 2$, (c) $n = 3$, (d) $n = 4$, and (e) $n = 5$? Check your results to show that they agree with the general rule that the number of different sets of quantum numbers is equal to $2n^2$.

**39.** **web** Zirconium ($Z = 40$) has two electrons in an incomplete $d$ subshell. (a) What are the values of $n$ and $\ell$ for each electron? (b) What are all possible values of $m_\ell$ and $m_s$? (c) What is the electron configuration in the ground state of zirconium?

## Section 28.10 Characteristic X-Rays

**40.** The K shell ionization energy of copper is 8 979 eV. The L shell ionization energy is 951 eV. Determine the wavelength of the $K_\alpha$ emission line of copper. What must the minimum voltage be on an x-ray tube with a copper target in order to see the $K_\alpha$ line?

**41.** The $K_\alpha$ x-ray is emitted when an electron undergoes a transition from the L shell ($n = 2$) to the K shell ($n = 1$). Use the method illustrated in Example 28.7 to calculate the wavelength of the $K_\alpha$ x-ray from a nickel target ($Z = 28$).

**42.** When an electron drops from the M shell ($n = 3$) to a vacancy in the K shell ($n = 1$), the measured wavelength of the emitted x-ray is found to be 0.101 nm. Identify the element.

**43.** The K series of the discrete spectrum of tungsten contains wavelengths of 0.018 5 nm, 0.020 9 nm, and 0.021 5 nm. The K shell ionization energy is 69.5 keV. Determine the ionization energies of the L, M, and N shells. Sketch the transitions that produce the above wavelengths.

## ADDITIONAL PROBLEMS

**44.** In a hydrogen atom, what is the principal quantum number of the electron orbit with a radius closest to 1.0 $\mu$m?

**45.** (a) How much energy is required to cause an electron in hydrogen to move from the $n = 1$ state to the $n = 2$ state? (b) If the electrons gain this energy by collision between hydrogen atoms in a high-temperature gas, find the minimum temperature of the heated hydrogen gas. The thermal energy of the heated atoms is given by $3k_BT/2$, where $k_B$ is the Boltzmann constant.

**46.** A pulsed ruby laser emits light at 694.3 nm. For a 14.0-ps pulse containing 3.00 J of energy, find (a) the physical length of the pulse as it travels through space and (b) the number of photons in it. (c) If the beam has a circular cross section of 0.600-cm diameter, find the number of photons per cubic millimeter.

**47.** The Lyman series for a (new!?) one-electron atom is observed in a distant galaxy. The wavelengths of the first four lines and the short-wavelength limit of this Lyman series are given by the energy level diagram in Figure P28.47. Based on this information, calculate (a) the energies of the ground state and first four excited states for this one-electron atom, and (b) the longest-wavelength (alpha) lines and the short-wavelength series limit in the Balmer series for this atom.

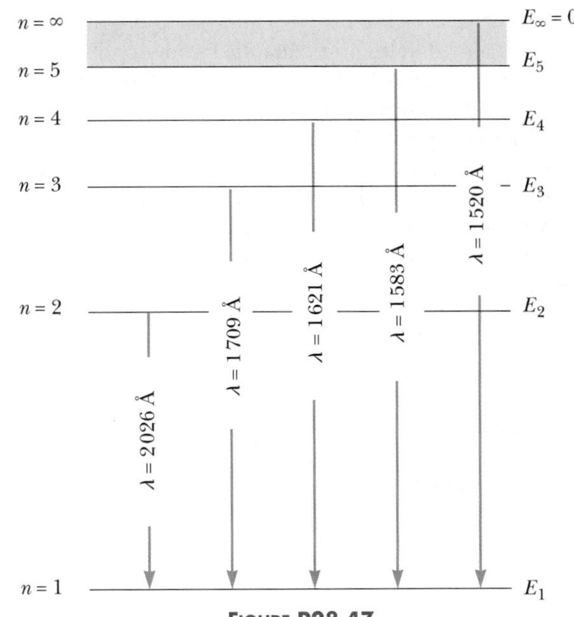

**FIGURE P28.47**

**48.** A dimensionless number that often appears in atomic physics is the fine-structure constant $\alpha = k_e e^2/\hbar c$, where $k_e$ is the Coulomb constant. (a) Obtain a numerical value for $1/\alpha$. (b) In terms of $\alpha$, what is the ratio of the Bohr radius $a_0$ to the Compton wavelength $\lambda_C = h/m_e c$? (c) In terms of $\alpha$, what is the ratio of the reciprocal of the Rydberg constant $1/R_H$ to the Bohr radius?

**49.** Mercury's ionization energy is 10.39 eV. The three longest wavelengths of the absorption spectrum of mercury are 253.7 nm, 185.0 nm, and 158.5 nm. (a) Construct an energy level diagram for mercury. (b) Indicate all emission lines that can occur when an electron is raised to the third level above the ground state. (c) Disregarding recoil of the mercury atom, determine the minimum speed an electron must have in order to make an inelastic collision with a mercury atom in its ground state.

**50.** Suppose the ionization energy of an atom is 4.100 eV. In this same atom, we observe emission lines with wavelengths 310.0 nm, 400.0 nm, and 1 378 nm. Use this information to construct the energy level diagram with the least number of levels. Assume the higher energy levels are closer together.

**51.** A laser used in eye surgery emits a 3.00-mJ pulse in 1.00 ns, focused to a spot 30.0 $\mu$m in diameter on the retina. (a) Find (in SI units) the power per unit area at the retina. (This quantity is called the *irradiance*.) (b) What energy is delivered per pulse to an area of molecular size—say, a circular area 0.600 nm in diameter?

**52.** An electron has a de Broglie wavelength equal to the diameter of a hydrogen atom in its ground state. (a) What is the kinetic energy of the electron? (b) How does this energy compare with the ground-state energy of the hydrogen atom?

**53.** Use Bohr's model of the hydrogen atom to show that when the atom makes a transition from the state $n$ to the state $n - 1$, the frequency of the emitted light is given by

$$f = \frac{2\pi^2 m k_e^2 e^4}{h^3}\left[\frac{2n - 1}{(n - 1)^2 n^2}\right]$$

54. Calculate the classical frequency for the light emitted by an atom. To do so, note that the frequency of revolution is $v/2\pi r$, where $r$ is the Bohr radius. Show that as $n$ approaches infinity in the equation of the preceding problem, the expression given there varies as $1/n^3$ and reduces to the classical frequency. (This is an example of the correspondence principle, which requires that the classical and quantum models agree for large values of $n$.)

55. web A $\pi$-meson ($\pi^-$) of charge $-e$ and mass 273 times greater than that of the electron is captured by a helium nucleus ($Z = +2$) as shown in Figure P28.55. (a) Draw an energy level diagram (in units of eV) for this "Bohr-type" atom up to the first six energy levels. (b) When the $\pi$-meson makes a transition between two orbits, a photon is emitted that Compton scatters off a free electron initially at rest, producing a scattered photon of wavelength $\lambda' = 0.089\ 929\ 3$ nm at an angle of $\theta = 42.68°$, as shown on the right-hand side of Figure P28.55. Between which two orbits did the $\pi$-meson make a transition?

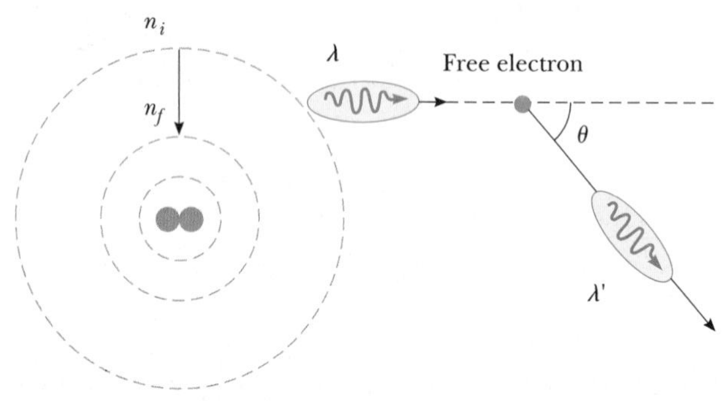


**FIGURE P28.55**

56. When a muon with charge $-e$ is captured by a proton, the resulting bound system forms a "muonic atom," which is the same as hydrogen except with a muon (of mass 207 times the mass of an electron) replacing the electron. For this "muonic atom" determine: (a) the Bohr radius and (b) the three lowest energy levels.

57. In this problem you will estimate the classical lifetime of the hydrogen atom. An accelerating charge loses electromagnetic energy at a rate given by $\mathcal{P} = -2k_e q^2 a^2/(3c^3)$, where $k_e$ is the Coulomb constant, $q$ is the charge of the particle, $a$ is its acceleration, and $c$ is the speed of light in a vacuum. Assuming that the electron is one Bohr radius (0.052 9 nm) from the center of the hydrogen atom, (a) determine its acceleration. (b) Show that $\mathcal{P}$ has units of energy per unit time and determine the rate of energy loss. (c) Calculate the kinetic energy of the electron and determine how long it will take for all of this energy to be converted into electromagnetic waves, assuming that the rate calculated in part (b) remains constant throughout the electron's motion.

## GROUP ACTIVITIES

**G.1** With your partner not looking, use modeling clay to build one or more mounds on top of a table. Place a piece of cardboard over your mound(s), and assign your partner the task of determining the size, shape, and number of mounds without looking. He is to do this by rolling marbles at the unseen mounds and observing how they emerge. This experiment models the Rutherford scattering experiment.

**G.2** Your instructor can probably lend you a small plastic diffraction grating that will enable you to examine the spectra of different light sources. You can use this grating to examine a source by holding the grating very close to your eye with the grooves vertical and noting the spectrum appearing off to the side when you look at a small or distant light source. You should look at light sources such as sodium vapor lights and mercury vapor lights used in many parking lots, neon lights used in many signs, black lights, ordinary incandescent light bulbs, and so forth.

**G.3** Assume that a quantum mechanical oscillator behaves in a fashion similar to a simple harmonic motion oscillator in our large-scale world. That is, its acceleration and velocity vary in the same way. Assume the oscillation takes place along a segment of the $x$ axis and sketch a graph of the probability of finding the oscillator versus distance from the origin. Where does the probability distribution peak? Where would it have the smallest amplitude?

**G.4** Photons are emitted when an electron drops from the $n = 3$ and $n = 4$ levels to the $n = 2$ level in the hydrogen atom. (a) In which case is the wavelength of the emitted photon greater? (b) In which case is the energy of the emitted photon greater? (c) Use the Bohr theory to justify your answers.

**G.5** What are the quantum numbers that lead to the minimum and maximum wavelengths in the Balmer series for hydrogen? (b) Use the Bohr theory to find these wavelengths. Do your results agree with your predictions?

**G.6** Consider the following quantum numbers, $n$, $\ell$, $m_\ell$, and $m_s$. (a) Which of these are integers and which are fractional? (b) Which of the above are always positive and which can be negative? (c) If $n = 2$, the largest value of $\ell$ is ________. (d) If $\ell = 1$, the values of $m_\ell$ range from ________ to ________.

# Nuclear Physics

# 29

## Chapter Outline



**Aerial view of a nuclear power plant that generates electrical power. Energy is generated in such plants from the process of nuclear fission, in which a heavy nucleus such as $^{235}U$ splits into two smaller nuclei. The fission process and nuclear reactors are discussed in more detail in Chapter 30.** *(Courtesy of Public Service Electric and Gas Company)*

**ERNEST RUTHERFORD, NEW ZEALAND PHYSICIST (1871 – 1937)**

Rutherford was awarded the Nobel Prize in 1908 for discovering that atoms can be broken apart by alpha rays, and for studying radioactivity. "On consideration, I realized that this scattering backward must be the result of a single collision, and when I made calculations I saw that it was impossible to get anything of that order of magnitude unless you took a system in which the greater part of the mass of the atom was concentrated in a minute nucleus. It was then that I had the idea of an atom with a minute massive center carrying a charge." *(North Wind Picture Archives)*

In 1896, the year that marks the birth of nuclear physics, Henri Becquerel (1852–1908) discovered radioactivity in uranium compounds. A great deal of activity followed this discovery as researchers attempted to understand and characterize the radiation that we now know to be emitted by radioactive nuclei. Pioneering work by Rutherford showed that the radiation was of three types, which he called *alpha, beta,* and *gamma rays.* These types are classified according to the nature of their electric charge and according to their ability to penetrate matter. Later experiments showed that alpha rays are helium nuclei, beta rays are electrons, and gamma rays are high-energy photons.

In 1911 Rutherford and his students Geiger and Marsden performed a number of important scattering experiments involving alpha particles. These experiments established that the nucleus of an atom can be regarded as essentially a point mass and point charge and that most of the atomic mass is contained in the nucleus. Furthermore, such studies demonstrated a wholly new type of force, the *nuclear force,* which is predominant at distances of less than about $10^{-14}$ m and zero at greater distances.

Other milestones in the development of nuclear physics include

- the first observations of nuclear reactions by Rutherford and coworkers in 1919 in which naturally occurring alpha particles bombarded nitrogen nuclei to produce oxygen,
- the first use of artificially accelerated protons to produce nuclear reactions by Cockcroft and Walton in 1932,
- the discovery of the neutron by Chadwick in 1932,
- the discovery of artificial radioactivity by Joliot and Irene Curie in 1933,
- the discovery of nuclear fission by Hahn and Strassman in 1938,
- the development of the first controlled fission reactor by Fermi and his collaborators in 1942.

In this chapter we discuss the properties and structure of the atomic nucleus. We start by describing the basic properties of nuclei and follow with a discussion of the phenomenon of radioactivity. Finally, we explore nuclear reactions and the various processes by which nuclei decay.

**Webnote 29.1**

See a timeline of the atomic and nuclear age from 1942 to 1998. Go to *http://www.pbs.org/wgbh/amex/three/timeline/index.html*

## 29.1 SOME PROPERTIES OF NUCLEI

All nuclei are composed of two types of particles: protons and neutrons. The only exception is the ordinary hydrogen nucleus, which is a single proton. In describing some of the properties of nuclei, such as their charge, mass, and radius, we make use of the following quantities:

- the **atomic number** $Z$, which equals the number of protons in the nucleus,
- the **neutron number** $N$, which equals the number of neutrons in the nucleus,
- the **mass number** $A$, which equals the number of nucleons in the nucleus. (*Nucleon* is a generic term used to refer to either a proton or a neutron.)

We use the symbol ${}^{A}_{Z}\mathrm{X}$ to represent nuclei, where X represents the chemical symbol for the element. For example, ${}^{27}_{13}\mathrm{Al}$ has the mass number 27 and the atomic number 13; therefore, it contains 13 protons and 14 neutrons. When no confusion is likely to arise, we often omit the subscript $Z$ because the chemical symbol can always be used to determine $Z$.

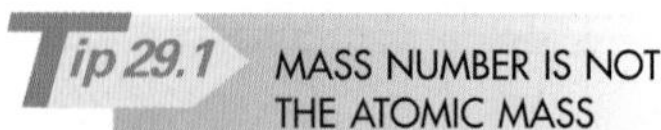

**Tip 29.1 MASS NUMBER IS NOT THE ATOMIC MASS**

Do not confuse the mass number $A$ with the atomic mass. Mass number is an integer that specifies an isotope and has no units—it is simply equal to the number of nucleons. Atomic mass is generally not an integer. It is an average of the masses of the isotopes of a given element and has units of u.

The nuclei of all atoms of a particular element must contain the same number of protons, but they may contain different numbers of neutrons. Nuclei that are related in this way are called **isotopes. The isotopes of an element have the same Z value but different *N* and *A* values.** The natural abundances of isotopes can differ substantially. For example, ${}^{11}_{6}\mathrm{C}$, ${}^{12}_{6}\mathrm{C}$, ${}^{13}_{6}\mathrm{C}$, and ${}^{14}_{6}\mathrm{C}$ are four isotopes of carbon. The natural abundance of the ${}^{12}_{6}\mathrm{C}$ isotope is about 98.9%, whereas that of

**TABLE 29.1 Mass of the Proton, Neutron, and Electron in Various Units**

| | Mass | | |
|---|---|---|---|
| Particle | kg | u | MeV/$c^2$ |
| Proton | $1.672\,6 \times 10^{-27}$ | 1.007 276 | 938.28 |
| Neutron | $1.675\,0 \times 10^{-27}$ | 1.008 665 | 939.57 |
| Electron | $9.109 \times 10^{-31}$ | $5.486 \times 10^{-4}$ | 0.511 |

the ${}^{13}_{6}C$ isotope is only about 1.1%. Some isotopes do not occur naturally but can be produced in the laboratory through nuclear reactions. Even the simplest element, hydrogen, has isotopes: ${}^{1}_{1}H$, hydrogen; ${}^{2}_{1}H$, deuterium; and ${}^{3}_{1}H$, tritium.

## CHARGE AND MASS

The proton carries a single positive charge, $+e$, the electron carries a single negative charge, $-e$, where $e = 1.602\,177\,33 \times 10^{-19}$ C, and the neutron is electrically neutral. Because the neutron has no charge, it is difficult to detect. The proton is about 1 836 times as massive as the electron, and the masses of the proton and the neutron are almost equal (Table 29.1).

It is convenient to define, for atomic masses, the **unified mass unit** u in such a way that the mass of one atom of the isotope ${}^{12}C$ is exactly 12 u, where $1\text{ u} = 1.660\,559 \times 10^{-27}$ kg. The proton and neutron each have a mass of about 1 u, and the electron has a mass that is only a small fraction of an atomic mass unit.

Definition of the unified mass unit, u

Because the rest energy of a particle is given by $E_R = mc^2$, it is often convenient to express the particle's mass in terms of its energy equivalent. For one atomic mass unit, we have an energy equivalent of

$$\begin{aligned} E_R = mc^2 &= (1.660\,559 \times 10^{-27}\text{ kg})(2.997\,92 \times 10^8\text{ m/s})^2 \\ &= 1.492\,431 \times 10^{-10}\text{ J} = 931.494\text{ MeV} \end{aligned}$$

Nuclear physicists often express *mass* in terms of the unit MeV/$c^2$ in calculations, where

$$1\text{ u} = 931.494\text{ MeV}/c^2$$

## THE SIZE OF NUCLEI

The size and structure of nuclei were first investigated in the scattering experiments of Rutherford, discussed in Section 28.1. Using the principle of conservation of energy, Rutherford found an expression for how close an alpha particle moving directly toward the nucleus can come to the nucleus before being turned around by Coulomb repulsion.

In such a head-on collision, the kinetic energy of the incoming alpha particle must be converted completely to electrical potential energy when the particle stops at the point of closest approach and turns around (Fig. 29.1). If we equate the initial kinetic energy of the alpha particle to the maximum electrical potential energy of the system (alpha particle + target nucleus), we have

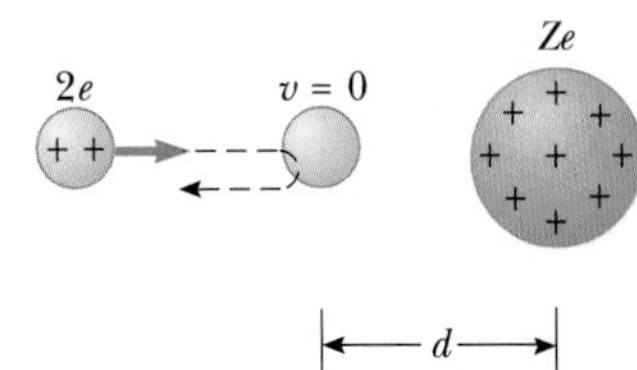


**FIGURE 29.1** An alpha particle on a head-on collision course with a nucleus of charge *Ze*. Because of the Coulomb repulsion between the like charges, the alpha particle will stop instantaneously at a distance *d* from the nucleus.

$$\tfrac{1}{2}mv^2 = k_e\frac{q_1 q_2}{r} = k_e\frac{(2e)(Ze)}{d}$$

where $d$ is the distance of closest approach. Solving for $d$, we get

$$d = \frac{4k_e Ze^2}{mv^2}$$

From this expression, Rutherford found that alpha particles approached to within $3.2 \times 10^{-14}$ m of a nucleus when the foil was made of gold. Thus, the radius of the gold nucleus must be less than this value. For silver atoms, the distance of closest approach was $2 \times 10^{-14}$ m. From these results, Rutherford concluded that the positive charge in an atom is concentrated in a small sphere, which he called the nucleus, whose radius is no greater than about $10^{-14}$ m. Because such small lengths are common in nuclear physics, a convenient unit of length is the *femtometer* (fm), sometimes called the **fermi,** defined as

$$1 \text{ fm} \equiv 10^{-15} \text{ m}$$

Since the time of Rutherford's scattering experiments, a multitude of other experiments have shown that most nuclei are approximately spherical and have an average radius given by

$$r = r_0 A^{1/3} \qquad [29.1]$$

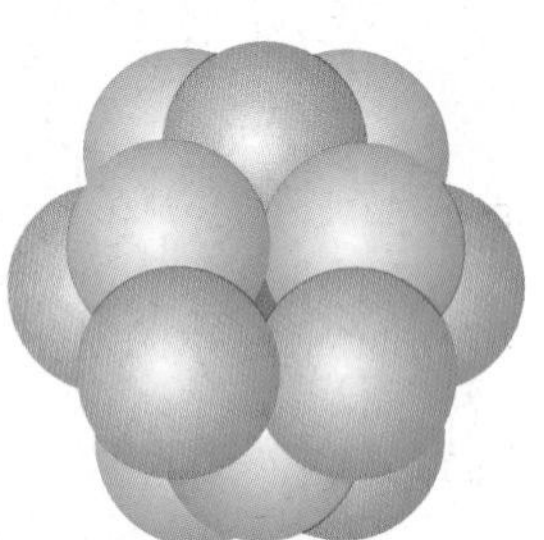

**FIGURE 29.2** A nucleus can be visualized as a cluster of tightly packed spheres in which each sphere is a nucleon.

where $A$ is the total number of nucleons and $r_0$ is a constant equal to $1.2 \times 10^{-15}$ m. Because the volume of a sphere is proportional to the cube of its radius, it follows from Equation 29.1 that the volume of a nucleus (assumed to be spherical) is directly proportional to $A$, the total number of nucleons. This suggests that **all nuclei have nearly the same density.** Nucleons combine to form a nucleus *as though* they were tightly packed spheres (Fig. 29.2).

## Example 29.1 Nuclear Volume and Density

**A** Find an approximate expression for the mass of a nucleus of mass number $A$.

**Solution** The mass of the proton is approximately equal to that of the neutron. Thus, if the mass of one of these particles is $m$, the mass of the nucleus is approximately $Am$.

**B** Find an expression for the volume of this nucleus in terms of the mass number.

**Solution** Assuming the nucleus is spherical and using Equation 29.1, we find that the volume is

$$V = \tfrac{4}{3}\pi r^3 = \tfrac{4}{3}\pi r_0{}^3 A$$

**C** Find a numerical value for its density.

**Solution** The nuclear density can be found as follows:

$$\rho_n = \frac{\text{mass}}{\text{volume}} = \frac{Am}{\frac{4}{3}\pi r_0{}^3 A} = \frac{3m}{4\pi r_0{}^3}$$

The fact that $A$ cancels out of our expression for nuclear density proves our statement above that all nuclei have roughly the same density.

Taking $r_0 = 1.2 \times 10^{-15}$ m and $m = 1.67 \times 10^{-27}$ kg, we find that

$$\rho_n = \frac{3(1.67 \times 10^{-27} \text{ kg})}{4\pi(1.2 \times 10^{-15} \text{ m})^3} = 2.3 \times 10^{17} \text{ kg/m}^3$$

Note that the nuclear density is about $2.3 \times 10^{14}$ times greater than the density of water ($1 \times 10^3$ kg/m$^3$)!

## NUCLEAR STABILITY

Given that the nucleus consists of a closely packed collection of protons and neutrons, you might be surprised that it can exist. The very large repulsive electrostatic forces between protons should cause the nucleus to fly apart. However, nuclei are stable because of the presence of another, short-range (about 2 fm) force, the

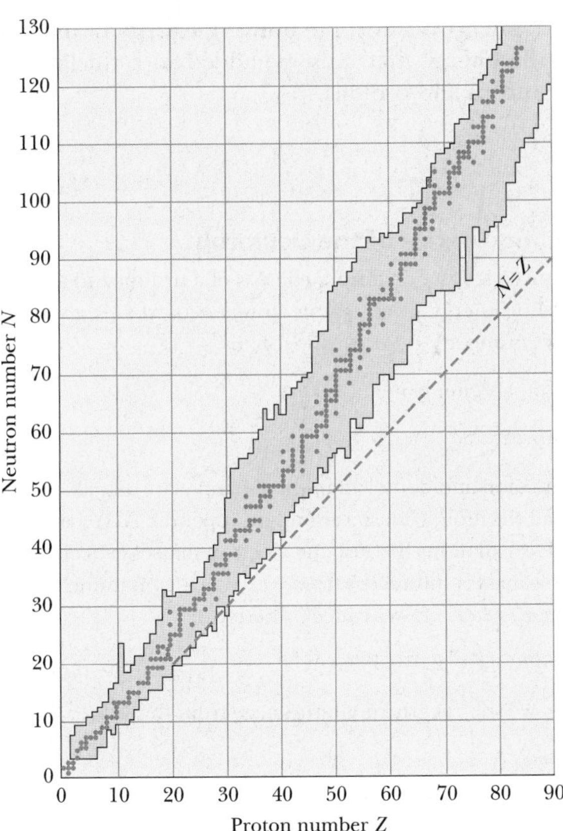


**FIGURE 29.3** A plot of the neutron number $N$ versus the proton number $Z$ for the stable nuclei (solid points). They are centered on the so-called line of stability. The dashed straight line corresponds to the condition $N = Z$. The shaded area shows radioactive (unstable) nuclei.

MARIA GOEPPERT-MAYER, GERMAN PHYSICIST (1906–1972)

Goeppert-Mayer was born and educated in Germany. She is best known for her develpment of the shell model of the nucleus, published in 1950. A similar model was simultaneously developed by Hans Jensen, a German scientist. Maria Goeppert-Mayer and Hans Jensen were awarded the Nobel prize in physics in 1963 for their extraordinary work in understanding the structure of the nucleus. *(Courtesy of Louise Barker/AIP Niels Bohr Library)*

**nuclear force.** This is an attractive force that acts between all nuclear particles. The protons attract each other via the nuclear force, and at the same time they repel each other through the Coulomb force. The attractive nuclear force also acts between pairs of neutrons and between neutrons and protons.

The nuclear attractive force is stronger than the Coulomb repulsive force within the nucleus (at short ranges). If this were not the case, stable nuclei would not exist. Moreover, the strong nuclear force is nearly independent of charge. In other words, the nuclear forces associated with the proton-proton, proton-neutron, and neutron-neutron interactions are approximately the same.

There are about 260 stable nuclei; hundreds of others have been observed but are unstable. A plot of $N$ versus $Z$ for a number of stable nuclei is given in Figure 29.3. Note that light nuclei are most stable if they contain equal numbers of protons and neutrons—that is, if $N = Z$—but heavy nuclei are more stable if $N > Z$. This can be partially understood by recognizing that, as the number of protons increases, the strength of the Coulomb force increases, which tends to break the nucleus apart. As a result, more neutrons are needed to keep the nucleus stable, because neutrons experience only the attractive nuclear forces. In effect, the additional neutrons "dilute" the nuclear charge density. Eventually, when $Z = 83$, the repulsive forces between protons cannot be compensated by the addition of more neutrons. Elements that contain more than 83 protons do not have stable nuclei, but decay or disintegrate into other particles in various amounts of time. The masses of several stable particles are given in Table 29.1. The masses and some other properties of selected isotopes are provided in Appendix B.

### Webnote 29.2

A detailed, "clickable" chart of the nuclide's extending information in Figure 29.3 can be found at *http://www2.bnl.gov/ton*

## 29.2 BINDING ENERGY

The total mass of a nucleus is always less than the sum of the masses of its nucleons. Because mass is another manifestation of energy, **the total energy of the bound system (the nucleus) is less than the combined energy of the sepa-**

**rated nucleons.** This difference in energy is called the **binding energy** of the nucleus and can be thought of as the energy that must be added to a nucleus to break it apart into its separated neutrons and protons.

### Example 29.2 The Binding Energy of the Deuteron

The nucleus of the deuterium atom, called the deuteron, consists of a proton and a neutron. Calculate the deuteron's binding energy, given that its atomic mass, that is, *the mass of a deuterium nucleus plus an electron,* is measured to be 2.014 102 u.

**Solution** We know that the proton and neutron masses are

$$m_p = 1.007\ 825\ \text{u} \qquad m_n = 1.008\ 665\ \text{u}$$

Note that the masses used for the proton and deuteron in this example are actually those of the neutral atoms, as are all the atomic masses found in Appendix B. We are able to use atomic masses for these calculations because the electron masses cancel. That is, when $m_d$ containing one electron mass is subtracted from $(m_p + m_n)$ containing one electron mass in the line below, the electron masses cancel. Therefore,

$$m_p + m_n = 2.016\ 490\ \text{u}$$

To calculate the mass difference, we subtract the deuteron mass from this value:

$$\begin{aligned}\Delta m &= (m_p + m_n) - m_d \\ &= 2.016\ 490\ \text{u} - 2.014\ 102\ \text{u} = 0.002\ 388\ \text{u}\end{aligned}$$

Because 1 u corresponds to an equivalent energy of 931.494 MeV (that is, $1\ \text{u} \cdot c^2 = 931.494$ MeV), the mass difference corresponds to the binding energy

$$E_b = (0.002\ 388\ \text{u})(931.494\ \text{MeV/u}) = \boxed{2.224\ \text{MeV}}$$

This result tells us that, to separate a deuteron into a proton and a neutron, it is necessary to add 2.224 MeV of energy to the deuteron to overcome the attractive nuclear force between proton and neutron. One way of supplying the deuteron with this energy is by bombarding it with energetic particles.

If the binding energy of a nucleus were zero, the nucleus would separate into its constituent protons and neutrons without the addition of any energy—that is, it would spontaneously break apart.

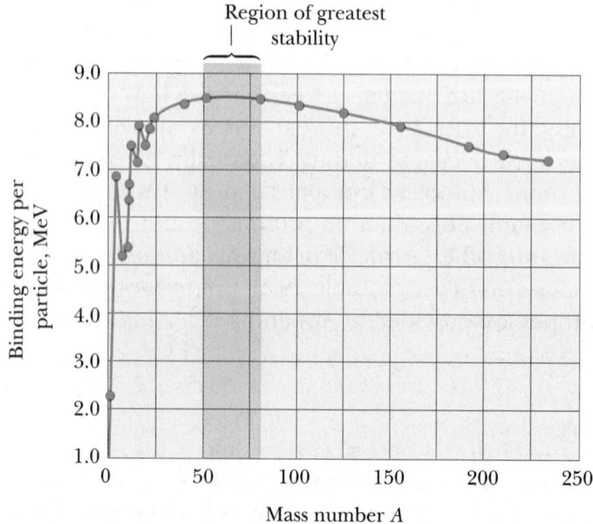


**FIGURE 29.4** A plot of the binding energy per nucleon versus the mass number $A$ for nuclei that are along the line of stability shown in Figure 29.3.

It is interesting to examine a plot of binding energy per nucleon, $E_b/A$, as a function of mass number for various stable nuclei (Fig. 29.4). Except for the lighter nuclei, the average binding energy per nucleon is about 8 MeV. Note that the curve peaks in the vicinity of $A = 60$. That is, nuclei with mass numbers greater or less than 60 are not as strongly bound as those near the middle of the periodic table. As we shall see later, this fact allows energy to be released in fission and fusion reactions. The curve is slowly varying for $A > 40$, which suggests that the nuclear force saturates. In other words, a particular nucleon can interact with only a limited number of other nucleons, which can be viewed as the "nearest neighbors" in the close-packed structure illustrated in Figure 29.2.

## APPLYING PHYSICS 29.1

Figure 29.4 shows a graph of the amount of energy required to remove a nucleon from the nucleus. The figure indicates that an approximately constant amount of energy is necessary to remove a nucleon (above $A = 40$), whereas we saw in Chapter 28 that widely varying amounts of energy are required to remove an electron from the atom. Why does this difference occur?

**Explanation** In the case of Figure 29.4, the approximately constant value of the nuclear binding energy is a result of the short-range nature of the nuclear force. A given nucleon interacts only with its few nearest neighbors, rather than with all of the nucleons in the nucleus. Thus, no matter how many nucleons are present in the nucleus, pulling any one nucleon out involves separating it only from its nearest neighbors. The energy to do this, therefore, is approximately independent of how many nucleons are present. For clearest comparison with the electron, think of averaging the energies required to strip all of the electrons out of a particular atom, from the outermost valence electron to the innermost K shell electron. This average increases steeply with increasing atomic number. The electrical force binding the electrons to the nucleus in an atom is a long-range force. An electron in the atom interacts with all the protons in the nucleus. As the nuclear charge increases, the attraction between the nucleus and the electrons becomes stronger. Therefore, as the nuclear charge increases, more energy is necessary to remove an average electron.

## 29.3 RADIOACTIVITY

In 1896 Becquerel accidentally discovered that uranium salt crystals emit an invisible radiation that can darken a photographic plate even if the plate is covered to exclude light. After several such observations under controlled conditions, he concluded that the radiation emitted by the crystals was of a new type, one that required no external stimulation. This spontaneous emission of radiation was soon called **radioactivity.** Subsequent experiments by other scientists showed that other substances were also radioactive.

**MARIE CURIE, POLISH SCIENTIST (1867–1934)**

In 1903 Marie Curie shared the Nobel prize in physics with her husband, Pierre, and with Becquerel for their studies of radioactive substances. In 1911 she was awarded a second Nobel prize in chemistry for the discovery of radium and polonium. Marie Curie died of leukemia caused by years of exposure to radioactive substances. "I persist in believing that the ideas that then guided us are the only ones which can lead to the true social progress. We cannot hope to build a better world without improving the individual. Toward this end, each of us must work toward his own highest development, accepting at the same time his share of responsibility in the general life of humanity." *(FPG/Getty)*

The most significant investigations of this type were conducted by Marie and Pierre Curie. After several years of careful and laborious chemical separation processes on tons of pitchblende, a radioactive ore, the Curies reported the discovery of two previously unknown elements, both of which were radioactive. These were named polonium and radium. Subsequent experiments, including Rutherford's famous work on alpha-particle scattering, suggested that radioactivity was the result of the decay, or disintegration, of unstable nuclei.

Three types of radiation can be emitted by a radioactive substance: alpha ($\alpha$) particles, in which the emitted particles are $^4_2$He nuclei; beta ($\beta$) particles, in which the emitted particles are either electrons or positrons; and gamma ($\gamma$) rays, in which the emitted "rays" are high-energy photons. A **positron** is a particle similar to the electron in all respects except that it has a charge of $+e$. (The positron is

The hands and numbers of this luminous watch contain minute amounts of radium salt. The radioactive decay of radium causes the phosphors to glow in the dark. *(© Richard Megna 1990, Fundamental Photographs)*

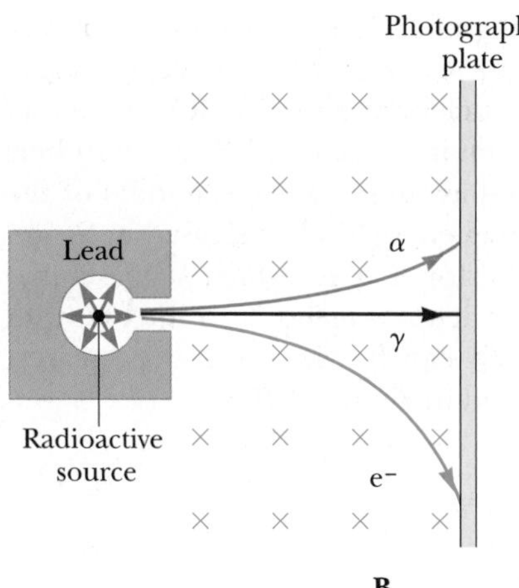


**FIGURE 29.5** The radiation from a radioactive source, such as radium, can be separated into three components using a magnetic field to deflect the charged particles. The photographic plate at the right records the events.

said to be the **antiparticle** of the electron.) The symbol $e^-$ is used to designate an electron and $e^+$ designates a positron.

It is possible to distinguish these three forms of radiation using the scheme shown in Figure 29.5. The radiation from a radioactive sample is directed into a region with a magnetic field, and the beam splits into three components, two bending in opposite directions and the third not changing direction. From this simple observation, it can be concluded that the radiation of the undeflected beam carries no charge (the gamma ray), the component deflected upward contains positively charged particles (alpha particles), and the component deflected downward contains negatively charged particles ($e^-$). If the beam includes a positron ($e^+$), it is deflected upward.

The three types of radiation have quite different penetrating powers. Alpha particles barely penetrate a sheet of paper, beta particles can penetrate a few millimeters of aluminum, and gamma rays can penetrate several centimeters of lead.

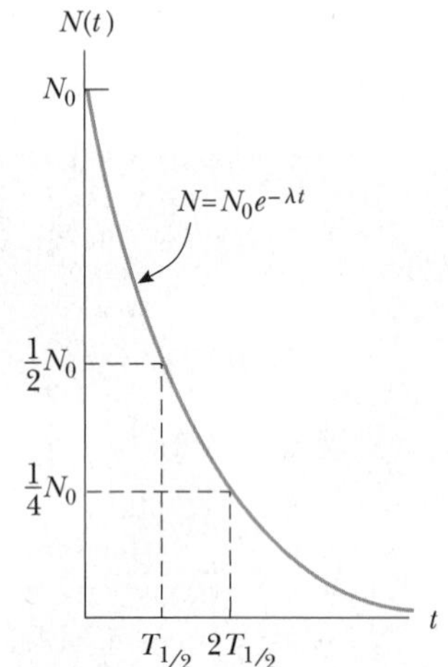


**FIGURE 29.6** Plot of the exponential decay law for radioactive nuclei. The vertical axis represents the number of radioactive nuclei present at any time $t$ and the horizontal axis is time. The parameter $T_{1/2}$ is the half-life of the sample.

## THE DECAY CONSTANT AND HALF-LIFE

If a radioactive sample contains $N$ radioactive nuclei at some instant, it is found that the number of nuclei, $\Delta N$, that decay in a small time interval $\Delta t$ is proportional to $N$:

$$\frac{\Delta N}{\Delta t} \propto N$$

or

$$\Delta N = -\lambda N \Delta t \qquad [29.2]$$

where $\lambda$ is a constant called the **decay constant.** The negative sign signifies that $N$ decreases with time; that is, $\Delta N$ is negative. The value of $\lambda$ for any isotope determines the rate at which that isotope will decay. **The decay rate, or activity, *R*, of a sample is defined as the number of decays** per second. From Equation 29.2, we see that the decay rate is

Decay rate

$$R = \left|\frac{\Delta N}{\Delta t}\right| = \lambda N \qquad [29.3]$$

Thus, we see that isotopes with a large $\lambda$ value decay at a rapid rate and those with a small $\lambda$ value decay slowly.

A general decay curve for a radioactive sample is shown in Figure 29.6. It can be shown from Equation 29.2 (using calculus) that the number of nuclei present varies with time according to the expression

Exponential decay

$$N = N_0 e^{-\lambda t} \qquad [29.4]$$

where $N$ is the number of radioactive nuclei present at time $t$, $N_0$ is the number present at time $t = 0$, and $e = 2.718\ldots$ is Euler's constant. Processes that obey Equation 29.4 are sometimes said to undergo exponential decay.[1]

Another parameter that is useful for characterizing radioactive decay is the **half-life, $T_{1/2}$. The half-life of a radioactive substance is the time it takes for half of a given number of radioactive nuclei to decay.** Setting $N = N_0/2$ and $t = T_{1/2}$ in Equation 29.4 gives

$$\frac{N_0}{2} = N_0 e^{-\lambda T_{1/2}}$$

Writing this in the form $e^{\lambda T_{1/2}} = 2$ and taking the natural logarithm of both sides, we get

$$T_{1/2} = \frac{\ln 2}{\lambda} = \frac{0.693}{\lambda} \qquad [29.5]$$

Half-life

This is a convenient expression relating the half-life to the decay constant. Note that after an elapsed time of one half-life, $N_0/2$ radioactive nuclei remain (by definition); after two half-lives, half of these will have decayed and $N_0/4$ radioactive nuclei will be left; after three half-lives, $N_0/8$ will be left; and so on.

The unit of activity $R$ is the **curie** (Ci), defined as

$$1 \text{ Ci} \equiv 3.7 \times 10^{10} \text{ decays/s} \qquad [29.6]$$

This unit was selected as the original activity unit because it is the approximate activity of 1 g of radium. The SI unit of activity is the **becquerel** (Bq):

$$1 \text{ Bq} = 1 \text{ decay/s} \qquad [29.7]$$

Therefore, 1 Ci = $3.7 \times 10^{10}$ Bq. The most commonly used units of activity are the millicurie ($10^{-3}$ Ci) and the microcurie ($10^{-6}$ Ci).

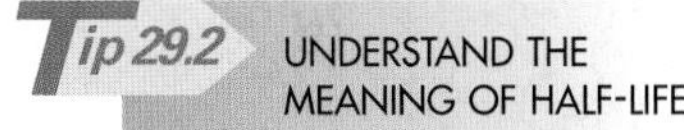

Tip 29.2
UNDERSTAND THE MEANING OF HALF-LIFE


Be sure that you understand the concept of the half-life of a substance. It is simply the time it takes half of a given number of nuclei to decay. Note that not all of the nuclei have decayed after two half-lives.

What fraction of a radioactive sample has decayed after two half-lives have elapsed? (a) $\frac{1}{4}$ (b) $\frac{1}{2}$ (c) $\frac{3}{4}$ (d) not enough information to say

## APPLYING PHYSICS 29.2

The isotope $^{14}_{6}$C is radioactive and has a half-life of 5 730 years. If you start with a sample of 1 000 carbon-14 nuclei, how many will still be around in 17 190 years?

**Explanation** In 5 730 years, half the sample will have decayed, leaving 500 radioactive $^{14}_{6}$C nuclei. In another 5 730 years (for a total elapsed time of 11 460 years), the number will be reduced to 250 nuclei. After another 5 730 years (total time 17 190 years), 125 nuclei remain.

These numbers represent ideal circumstances. Radioactive decay is an averaging process over a very large number of atoms, and the actual outcome depends on statistics. Our original sample in this example contained only 1 000 nuclei, certainly not a very large number. Thus, if we were actually to count the number remaining after one half-life for this small sample, it probably would not be exactly 500.

[1] Other examples of exponential decays were discussed in Chapter 18 in connection with *RC* circuits, and in Chapter 20 in connection with *RL* circuits.

## Example 29.3 The Activity of Radium

The half-life of the radioactive nucleus $^{226}_{86}\text{Ra}$ is $1.6 \times 10^3$ yr. If a sample contains $3.0 \times 10^{16}$ such nuclei, determine the activity at this time.

**Solution** First, let us convert the half-life to seconds:

$$T_{1/2} = (1.6 \times 10^3 \text{ yr})(3.16 \times 10^7 \text{ s/yr}) = 5.0 \times 10^{10} \text{ s}$$

Now we can use this value in Equation 29.5 to get the decay constant:

$$\lambda = \frac{0.693}{T_{1/2}} = \frac{0.693}{5.0 \times 10^{10} \text{ s}} = 1.4 \times 10^{-11} \text{ s}^{-1}$$

We can calculate the activity of the sample at $t = 0$ using $R_0 = \lambda N_0$, where $R_0$ is the decay rate at $t = 0$ and $N_0$ is the number of radioactive nuclei present at $t = 0$:

$$R_0 = \lambda N_0 = (1.4 \times 10^{-11} \text{ s}^{-1})(3.0 \times 10^{16}) = 4.1 \times 10^5 \text{ decays/s}$$

Because $1 \text{ Ci} = 3.7 \times 10^{10}$ decays/s, the activity, or decay rate, at $t = 0$ is

$$R_0 = 11.1\ \mu\text{Ci}$$

## Example 29.4 The Activity of Radon Gas

Radon, $^{222}_{86}\text{Rn}$, is a radioactive gas that can be trapped in the basement of homes, and its presence in high concentrations is a known health hazard. Radon has a half-life of 3.83 days. A gas sample contains $4.0 \times 10^8$ radon atoms initially.

**A** How many atoms will remain after 12 days have passed if no more radon leaks in?

**Reasoning** First note that 12 days corresponds to about 3.1 half-lives. In three half-lives, the number of radon atoms is reduced by a factor $2^3 = 8$. Therefore, the number of remaining radon atoms is approximately $4.0 \times 10^8/8 = 5.0 \times 10^7$.

**Solution** A more precise answer is obtained by first finding the decay constant from Equation 29.5:

$$\lambda = \frac{0.693}{T_{1/2}} = \frac{0.693}{3.83 \text{ days}} = 0.181 \text{ days}^{-1}$$

We now use Equation 29.4, taking $N_0 = 4.0 \times 10^8$ and the value of $\lambda$ just found to obtain the number $N$ remaining after 12 days:

$$N = N_0 e^{-\lambda t} = (4.0 \times 10^8 \text{ atoms}) e^{-(0.181 \text{ days}^{-1})(12 \text{ days})} = 4.6 \times 10^7 \text{ atoms}$$

This is very close to our original estimate of $5.0 \times 10^7$ atoms.

**B** What is the initial activity of the radon sample?

**Solution** First, we must express the decay constant in units of $\text{s}^{-1}$.

$$\lambda = \frac{0.693}{(3.83 \text{ days})(8.64 \times 10^4 \text{ s/day})} = 2.09 \times 10^{-6} \text{ s}^{-1}$$

From Equation 29.3 and the above value of $\lambda$, we find that the initial activity is

$$R = \lambda N_0 = (2.09 \times 10^{-6} \text{ s}^{-1})(4.0 \times 10^8) = 840 \text{ decays/s} = 840 \text{ Bq}$$

**EXERCISE** Find the activity of the radon sample after 12 days have elapsed.

**ANSWER** 95 Bq

The activity of a newly discovered radioactive isotope reduces to 96% of its original value in an interval of 2 hours. What is its half-life? (a) 10.2 h (b) 34.0 h (c) 44.0 h (d) 68.6 h

## 29.4 THE DECAY PROCESSES

As stated in the previous section, radioactive nuclei spontaneously decay via alpha, beta, and gamma decay. Let us discuss these processes in more detail.

### ALPHA DECAY

If a nucleus emits an alpha particle ($^4_2$He), it loses two protons and two neutrons. Therefore, the neutron number $N$ of a single nucleus decreases by 2, $Z$ decreases by 2, and $A$ decreases by 4. The decay can be written symbolically as

$$^A_Z\text{X} \longrightarrow {}^{A-4}_{Z-2}\text{Y} + {}^4_2\text{He} \qquad [29.8]$$

where X is called the **parent nucleus** and Y the **daughter nucleus.** As examples, $^{238}$U and $^{226}$Ra are both alpha emitters and decay according to the schemes

$$^{238}_{92}\text{U} \longrightarrow {}^{234}_{90}\text{Th} + {}^4_2\text{He} \qquad [29.9]$$

$$^{226}_{88}\text{Ra} \longrightarrow {}^{222}_{86}\text{Rn} + {}^4_2\text{He} \qquad [29.10]$$

The half-life for $^{238}$U decay is $4.47 \times 10^9$ yr, and the half-life for $^{226}$Ra decay is $1.60 \times 10^3$ yr. In both cases, note that the $A$ of the daughter nucleus is 4 less than that of the parent nucleus. Likewise, $Z$ is reduced by 2.

The decay of $^{226}$Ra is shown in Figure 29.7. When one element changes into another, as happens in alpha decay, the process is called **spontaneous decay** or transmutation. As a general rule, (1) the sum of the mass numbers $A$ must be the same on both sides of the equation, and (2) the sum of the atomic numbers $Z$ must be the same on both sides of the equation.

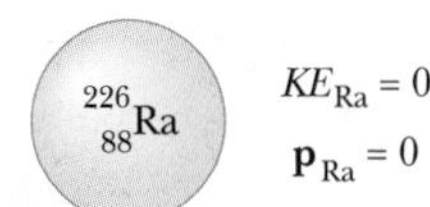


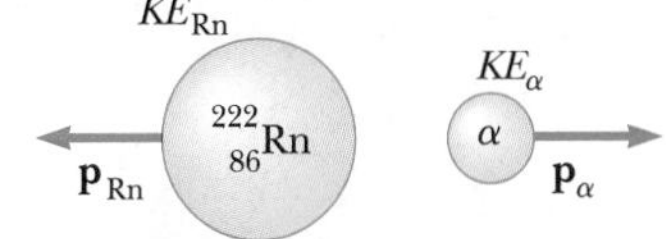


**FIGURE 29.7** Alpha decay of radium. The radium nucleus is initially at rest. After the decay, the radon nucleus has kinetic energy $KE_{\text{Rn}}$, and momentum $\mathbf{p}_{\text{Rn}}$, and the alpha particle has kinetic energy $KE_\alpha$ and momentum $\mathbf{p}_\alpha$.

In order for alpha emission to occur, the mass of the parent must be greater than the combined mass of the daughter and the alpha particle. In the decay process, this excess mass is converted into energy of other forms and appears in the form of kinetic energy in the daughter nucleus and the alpha particle. Most of the kinetic energy is carried away by the alpha particle because it is much less massive than the daughter nucleus. This can be understood by first noting that a particle's kinetic energy and momentum, $p$, are related as follows:

$$KE = \frac{p^2}{2m}$$

Because momentum is conserved, the two particles emitted in the decay of a nucleus at rest must have equal, but oppositely directed, momenta. Thus, the lighter particle has more kinetic energy than the more massive particle.

If a nucleus such as $^{226}$Ra that is initially at rest undergoes alpha decay, which of the following statements is true? (a) The alpha particle has more kinetic energy than the daughter nucleus. (b) The daughter nucleus has more kinetic energy than the alpha particle. (c) The daughter nucleus and the alpha particle have the same kinetic energy.

## APPLYING PHYSICS 29.3

In comparing alpha decay energies from a number of radioactive nuclides, it is found that the half-life of the decay goes down as the energy of the decay goes up. Why is this?

**Explanation** It should seem reasonable that the higher the energy of the alpha particle, the more likely it is to escape the confines of the nucleus. The higher probability of escape translates to a faster rate of decay, which appears as a shorter half-life.

### Example 29.5 The Energy Liberated When Radium Decays

We showed that the $^{226}_{88}\text{Ra}$ nucleus undergoes alpha decay to $^{222}_{86}\text{Rn}$ (Eq. 29.10). Calculate the amount of energy liberated in this decay. Take the mass of $^{226}_{88}\text{Ra}$ to be 226.025 402 u, that of $^{222}_{86}\text{Rn}$ to be 222.017 571 u and that of $^{4}_{2}\text{He}$ to be 4.002 602 u as found in Appendix B.

**Solution** After decay, the mass of the daughter $m_d$ plus the mass of the alpha particle $m_\alpha$ is

$$m_d + m_\alpha = 222.017\,571 \text{ u} + 4.002\,602 \text{ u} = 226.020\,173 \text{ u}$$

Thus, calling the mass of the parent nucleus $M_p$, we find that the mass lost during decay is

$$\Delta m = M_p - (m_d + m_\alpha) = 226.025\,402 \text{ u} - 226.020\,173 \text{ u} = 0.005\,229 \text{ u}$$

Using the relationship that 1 u corresponds to 931.494 MeV, we find that the energy liberated is

$$E = (0.005\,229 \text{ u})(931.494 \text{ MeV/u}) = 4.87 \text{ MeV}$$

ENRICO FERMI, ITALIAN PHYSICIST (1901–1954)

Fermi was awarded the Nobel prize in 1938 for producing the transuranic elements by neutron irradiation and for his discovery of nuclear reactions brought about by slow neutrons. He made many other outstanding contributions to physics, including his theory of beta decay, the free electron theory of metals, and the development of the world's first fission reactor in 1942. Fermi was truly a gifted theoretical and experimental physicist. He was also well known for his ability to present physics in a clear and exciting manner. "Whatever Nature has in store for mankind, unpleasant as it may be, men must accept, for ignorance is never better than knowledge." *(National Accelerator Laboratory)*

## BETA DECAY

When a radioactive nucleus undergoes beta decay, the daughter nucleus has the same number of nucleons as the parent nucleus, but the atomic number is changed by 1:

$$^{A}_{Z}\text{X} \longrightarrow {}^{A}_{Z+1}\text{Y} + \text{e}^- \qquad [29.11]$$

$$^{A}_{Z}\text{X} \longrightarrow {}^{A}_{Z-1}\text{Y} + \text{e}^+ \qquad [29.12]$$

Again, note that the nucleon number and total charge are both conserved in these decays. However, as we shall see shortly, these processes are not described completely by these expressions. A typical beta decay event is

$$^{14}_{6}\text{C} \longrightarrow {}^{14}_{7}\text{N} + \text{e}^- \qquad [29.13]$$

The emission of electrons from a *nucleus* is surprising because, in all our previous discussions, we stated that the nucleus is composed of protons and neutrons only. This apparent discrepancy can be explained by noting that the emitted electron is created in the nucleus by a process in which a neutron is transformed into a proton. This can be represented by the equation

$$^{1}_{0}\text{n} \longrightarrow {}^{1}_{1}\text{p} + \text{e}^- \qquad [29.14]$$

Let us consider the energy of the system of Equation 29.13 before and after decay. As with alpha decay, energy must be conserved in beta decay. The following example illustrates how to calculate the amount of energy released in the beta decay of $^{14}_{6}\text{C}$.

## Example 29.6 The Beta Decay of Carbon-14

Find the energy liberated in the beta decay of $^{14}_{6}C$ to $^{14}_{7}N$ as represented by Equation 29.13. Equation 29.13 refers to nuclei, whereas Appendix B shows masses of neutral atoms. Adding six electrons to both sides of Equation 29.13 gives

$$^{14}_{6}\text{C atom} \longrightarrow {}^{14}_{7}\text{N atom}$$

**Solution** We find from Appendix B that $^{14}_{6}C$ has a mass of 14.003 242 u and $^{14}_{7}N$ has a mass of 14.003 074 u. Here, the mass difference between the initial and final states is

$$\Delta m = 14.003\ 242\ \text{u} - 14.003\ 074\ \text{u} = 0.000\ 168\ \text{u}$$

This corresponds to an energy release of

$$E = (0.000\ 168\ \text{u})(931.494\ \text{MeV/u}) = 0.156\ \text{MeV}$$

From Example 29.6, we see that the energy released in the beta decay of $^{14}C$ is approximately 0.16 MeV. As with alpha decay, we expect the electron to carry away virtually all of this as kinetic energy because apparently it is the lightest particle produced in the decay. However, as Figure 29.8 shows, only a small number of electrons have this maximum kinetic energy, represented as $KE_{\text{max}}$ on the graph; most of the electrons emitted have kinetic energies lower than this predicted value. If the daughter nucleus and the electron are not carrying away this liberated energy, then the energy conservation requirement leads to the question, What accounts for the missing energy? As an additional complication, further analysis of beta decay shows that the principles of conservation of both angular momentum and linear momentum appear to be violated!

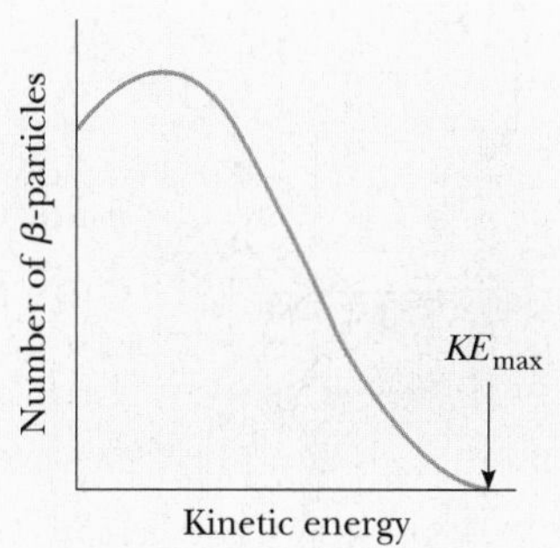


**FIGURE 29.8** A typical beta-decay spectrum.

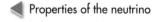

In 1930 Pauli proposed that a third particle must be present to carry away the "missing" energy and to conserve momentum. Enrico Fermi later developed a complete theory of beta decay and named this particle the **neutrino** ("little neutral one") because it had to be electrically neutral and have little or no mass. Although it eluded detection for many years, the neutrino ($\nu$) was finally detected experimentally in 1956. The neutrino has the following properties:

Properties of the neutrino

- Zero electric charge
- A mass much smaller than that of the electron, but probably not zero (Recent experiments suggest that the neutrino definitely has mass, but the value is uncertain—perhaps less than 1 eV/$c^2$.)
- A spin of $\frac{1}{2}$
- Very weak interaction with matter, making it quite difficult to detect

### Webnote 29.3

Visit a neutrino measurement facility in Japan. Check out *http://neutrino.kek.jp*

Thus, with the introduction of the neutrino, we are now able to represent the beta decay process of Equation 29.13 in its correct form:

$$^{14}_{6}C \longrightarrow {}^{14}_{7}N + e^- + \bar{\nu} \qquad [29.15]$$

where the bar in the symbol $\bar{\nu}$ indicates an **antineutrino.** To explain what an antineutrino is, let us first consider the following decay:

$$^{12}_{7}N \longrightarrow {}^{12}_{6}C + e^+ + \nu \qquad [29.16]$$

Here we see that when $^{12}N$ decays into $^{12}C$, a particle is produced that is identical to the electron except that it has a positive charge of $+e$. This particle is called a **positron.** Because it is like the electron in all respects except charge, the positron is said to be the **antiparticle** to the electron. We shall discuss antiparticles further in Chapter 30. For now, suffice it to say that **in beta decay, an electron and an antineutrino are emitted or a positron and a neutrino are emitted.**

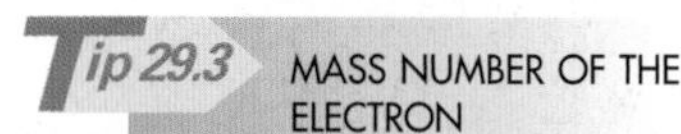


Another notation that is sometimes used for an electron is $^{\ 0}_{-1}e$. This notation does not imply that the electron has zero rest energy. The mass of the electron is much smaller than that of the lightest nucleon, so we can approximate it as zero when we study nuclear decays and reactions.

## GAMMA DECAY

Very often a nucleus that undergoes radioactive decay is left in an excited energy state. The nucleus can then undergo a second decay to a lower energy state, perhaps to the ground state, by emitting one or more high-energy photons. The process is very similar to the emission of light by an atom. An atom emits radiation to release some extra energy when an electron "jumps" from a state of high energy to a state of lower energy. Likewise, the nucleus uses essentially the same method to release any extra energy it may have following a decay or some other nuclear event. In nuclear de-excitation, the "jumps" that release energy are made by protons or neutrons in the nucleus as they move from a higher energy level to a lower level. The photons emitted in such a de-excitation process are called **gamma rays,** which have very high energy relative to the energy of visible light.

A nucleus may reach an excited state as the result of a violent collision with another particle. However, it is more common for a nucleus to be in an excited state as a result of alpha or beta decay. The following sequence of events represents a typical situation in which gamma decay occurs:

$$^{12}_{5}\text{B} \longrightarrow {}^{12}_{6}\text{C}^* + e^- + \bar{\nu} \quad [29.17]$$

$$^{12}_{6}\text{C}^* \longrightarrow {}^{12}_{6}\text{C} + \gamma \quad [29.18]$$

Equation 29.17 represents a beta decay in which $^{12}$B decays to $^{12}$C* where the asterisk indicates that the carbon nucleus is left in an excited state following the decay. The excited carbon nucleus then decays to the ground state by emitting a gamma ray, as indicated by Equation 29.18. Note that gamma emission does not result in any change in either $Z$ or $A$.

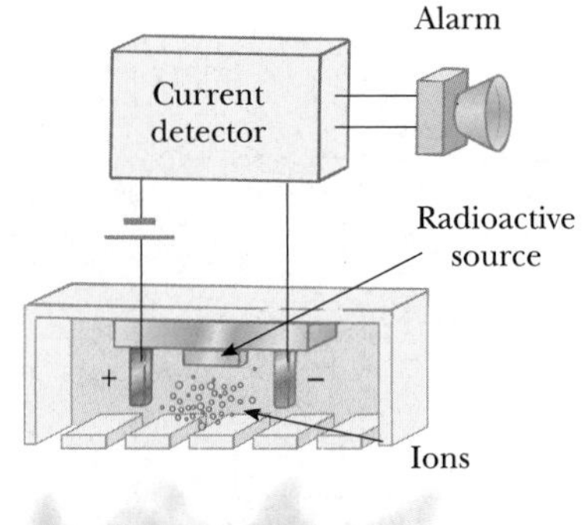


**FIGURE 29.9** An ionization-type smoke detector. Smoke entering the chamber reduces the detected current, causing the alarm to sound.

## PRACTICAL USES OF RADIOACTIVITY

### *Carbon Dating*

The beta decay of $^{14}$C given by Equation 29.15 is commonly used to date organic samples. Cosmic rays (high-energy particles from outer space) in the upper atmosphere cause nuclear reactions that create $^{14}$C from $^{14}$N. In fact, the ratio of $^{14}$C to $^{12}$C (by numbers of nuclei) in the carbon dioxide molecules of our atmosphere has a constant value of about $1.3 \times 10^{-12}$ as determined by measuring carbon ratios in tree rings. All living organisms have the same ratio of $^{14}$C to $^{12}$C because they continuously exchange carbon dioxide with their surroundings. When an organism dies, however, it no longer absorbs $^{14}$C from the atmosphere, and so the ratio of $^{14}$C to $^{12}$C decreases as the result of the beta decay of $^{14}$C. It is therefore possible to determine the age of a material by measuring its activity per unit mass as a result of the decay of $^{14}$C. Using carbon dating, samples of wood, charcoal, bone, and shell have been identified as having lived from 1 000 to 25 000 years ago. This knowledge has helped researchers to reconstruct the history of living organisms—including humans—during this time span.

**Webnote 29.4**
Learn a lot more about $^{14}$C dating by exploring
*http://abcnews.go.com/sections/science/DailyNews/carbon0220.html*

**APPLICATION**
CARBON DATING OF THE DEAD SEA SCROLLS

A particularly interesting example is the dating of the Dead Sea Scrolls. This group of manuscripts was first discovered by a young Bedouin boy in a cave at Qumran near the Dead Sea in 1947. Translation showed them to be religious documents, including most of the books of the Old Testament. Because of their historical and religious significance, scholars wanted to know their age. Carbon dating applied to fragments of the scrolls and to the material in which they were wrapped established that they were about 1 950 years old. The scrolls are now stored at the Israel Museum in Jerusalem.

**APPLICATION**
SMOKE DETECTORS

**Smoke Detectors** Smoke detectors are frequently used in homes and industry for fire protection. Most of the common ones are the ionization type that use radioactive materials (see Fig. 29.9). A smoke detector consists of an ionization chamber, a sensitive current detector, and an alarm. A weak radioactive source ionizes the air in the chamber of the detector, which creates charged particles. A voltage is maintained between the plates inside the chamber, setting up a small but detectable current in the external circuit. As long as the current is maintained, the alarm is deactivated. However, if smoke drifts into the chamber, the ions become attached to the smoke particles. These heavier particles do not drift as readily as do the lighter ions, which causes a decrease in the detector current. The external circuit senses this decrease in current and sets off the alarm.

**Radon Detecting** Radioactivity can also affect our daily lives in harmful ways. Soon after the discovery of radium by the Curies, it was found that the air in contact with radium compounds becomes radioactive. It was shown that this radioactivity came from the radium itself, and the product was therefore called "radium emanation." Rutherford and Soddy succeeded in condensing this "emanation," confirming that it is a real substance—the inert, gaseous element now called **radon,** Rn. We now know that the air in uranium mines is radioactive because of the presence of radon gas. The mines must therefore be well ventilated to help protect miners. The fear of radon pollution has now moved from uranium mines into our own homes (see Example 29.4). Because certain types of rock, soil, brick, and concrete contain small quantities of radium, some of the resulting radon gas finds its way into our homes and other buildings. The most serious problems arise from leakage of radon from the ground into the structure. One practical remedy is to exhaust the air through a pipe just above the underlying soil or gravel directly to the outdoors by means of a small fan or blower.

**APPLICATION**

RADON POLLUTION

**Webnote 29.5**

Read more on the risks associated with radon gas. Go to *http://www.howstuffworks.com/radon.htm*

## APPLYING PHYSICS 29.4

In 1991 a German tourist discovered the well-preserved remains of the Iceman trapped in a glacier in the Italian Alps (see Fig. 29.10). Radioactive dating of a sample of Iceman revealed an age of 5 300 years. Why did scientists date the sample using the isotope $^{14}$C, rather than $^{11}$C, a beta emitter with a half-life of 20.4 min?

**Explanation** Carbon-14 has a long half-life of 5 730 years, so the fraction of $^{14}$C nuclei remaining after one half-life is high enough to measure accurate changes in the sample's activity. The $^{11}$C isotope, which has a very short half-life, is not useful because its activity decreases to a vanishingly small value over the age of the sample, making it impossible to detect.

**APPLICATION**

RADIOACTIVE DATING OF ICEMAN

**FIGURE 29.10** (Applying Physics 29.4) The body of an ancient man (dubbed the Iceman) was exposed by a melting glacier in the Alps. *(Hanny Paul/Gamma Liaison)*

If a sample to be dated is not very old, say about 50 years, then you should select the isotope of some other element whose half-life is comparable with the age of the sample. For example, if the sample contained hydrogen, you could measure the activity of $^3H$ (tritium), a beta emitter of half-life 12.3 years. As a general rule, the expected age of the sample should be long enough to measure a change in activity, but not so long that its activity cannot be detected.

## APPLYING PHYSICS 29.5

A wooden coffin is found that contains a skeleton holding a gold statue. Which of the three objects (coffin, skeleton, and gold statue) can be carbon dated to find out how old it is?

**Explanation** Only the coffin and the skeleton can be carbon dated. These two items exchanged air with the environment during their lifetimes, resulting in a fixed ratio of $^{14}C$ to $^{12}C$. Once the human and the tree died, the amount of $^{14}C$ began to decrease due to radioactive decay. The gold in the statue was never alive and did not have an uptake of carbon; therefore, there is no $^{14}C$ to detect.

## Example 29.7 Should We Report This to Homicide?

A 50.0-g sample of carbon is taken from the pelvis bone of a skeleton and is found to have a $^{14}C$ decay rate of 200.0 decays/min. It is known that carbon from a living organism has a decay rate of 15.0 decays/min · g and that $^{14}C$ has a half-life of 5 730 yr = $3.01 \times 10^9$ min. Find the age of the skeleton.

**Solution** Let us start with Equation 29.4

$$N = N_0 e^{-\lambda t}$$

and multiply both sides by $\lambda$ to get

$$\lambda N = \lambda N_0 e^{-\lambda t}$$

But from Equation 29.3 we see that this is equivalent to

$$R = R_0 e^{-\lambda t}$$

or

$$\frac{R}{R_0} = e^{-\lambda t}$$

where $R$ is the present activity and $R_0$ was the activity when the skeleton was a part of a living organism. We can solve for the time by taking the natural log of both sides of this equation.

$$\ln\left(\frac{R}{R_0}\right) = \ln(e^{-\lambda t}) = -\lambda t$$

$$t = -\frac{\ln\left(\frac{R}{R_0}\right)}{\lambda}$$

Because we are given the decay rate and mass of the sample, we can find $R_0$ as

$$R_0 = \left(15.0\,\frac{\text{decays}}{\text{min}\cdot\text{g}}\right)(50.0\text{ g}) = 750\,\frac{\text{decays}}{\text{min}}$$

The decay constant is found from Equation 29.5 as

$$\lambda = \frac{0.693}{T_{1/2}} = \frac{0.693}{3.01 \times 10^9\text{ min}} = 2.30 \times 10^{-10}\text{ min}^{-1}$$

Thus, we make the following substitutions:

$$t = -\frac{\ln\left(\frac{R}{R_0}\right)}{\lambda} = -\frac{\ln\left(\frac{200.0 \text{ decays/min}}{750 \text{ decays/min}}\right)}{2.30 \times 10^{-10} \text{ min}^{-1}} = \frac{1.32}{2.30 \times 10^{-1}\text{min}^{-1}}$$

$$= 5.74 \times 10^9 \text{ min} = \boxed{10\,900 \text{ yr}}$$

## CARBON-14 AND THE SHROUD OF TURIN

**APPLICATION**

$^{14}C$ DATING OF THE SHROUD OF TURIN

Since the Middle Ages, many people have marveled at a 14-foot-long, yellowing piece of linen found in Turin, Italy, purported to be the burial shroud of Jesus Christ (Fig. 29.11). The cloth bears a remarkable, full-size likeness of a crucified body, with wounds on the head that could have been caused by a crown of thorns, and another in the side that could have been the cause of death. Skepticism over the authenticity of the shroud has existed since its first public showing in 1354; in fact, a French bishop declared it to be a fraud at the time. Because of its controversial nature, religious bodies have taken a neutral stance on its authenticity.

In 1978 the bishop of Turin allowed the cloth to be subjected to scientific analysis, but notably missing from these tests was a $^{14}C$ dating. The reason for this omission was that, at the time, carbon-dating techniques required a piece of cloth about the size of a handkerchief. By 1988 the process had been refined to the point that pieces as small as 1 in$^2$ were sufficient, and at that time permission was granted to allow the dating to proceed. Three labs were selected for the testing, and each was given four pieces of material. One of these was a piece of the shroud, and the other three pieces were control pieces similar in appearance to the shroud.

The testing procedure consisted of burning the cloth to produce carbon dioxide, which was then converted chemically to graphite. The graphite sample was subjected to $^{14}C$ analysis, and in the end all three labs agreed amazingly well on the age of the shroud. The average of their results gave a date for the cloth of A.D. 1320 ± 60 years, with an assurance that the cloth could not be older than A.D. 1200. Carbon-14 dating has thus unraveled the most important mystery concerning the shroud, but others remain. For example, investigators have not yet been able to explain how the image was imprinted.

**FIGURE 29.11** The Shroud of Turin as it appears in a photographic negative image. *(Santi Visali/The IMAGE Bank)*

## 29.5 NATURAL RADIOACTIVITY

Radioactive nuclei are generally classified into two groups: (1) unstable nuclei found in nature, which give rise to what is called **natural radioactivity,** and (2) nuclei produced in the laboratory through nuclear reactions, which exhibit **artificial radioactivity.**

Three series of naturally occurring radioactive nuclei exist (Table 29.2). Each series starts with a specific long-lived radioactive isotope whose half-life exceeds

**TABLE 29.2** The Four Radioactive Series

| Series | Starting Isotope | Half-life (yr) | Stable End Product |
|---|---|---|---|
| Uranium | $^{238}_{92}U$ | $4.47 \times 10^9$ | $^{206}_{82}Pb$ |
| Actinium | $^{235}_{92}U$ | $7.04 \times 10^8$ | $^{207}_{82}Pb$ |
| Thorium | $^{232}_{90}Th$ | $1.41 \times 10^{10}$ | $^{208}_{82}Pb$ |
| Neptunium | $^{237}_{93}Np$ | $2.14 \times 10^6$ | $^{209}_{83}Bi$ |

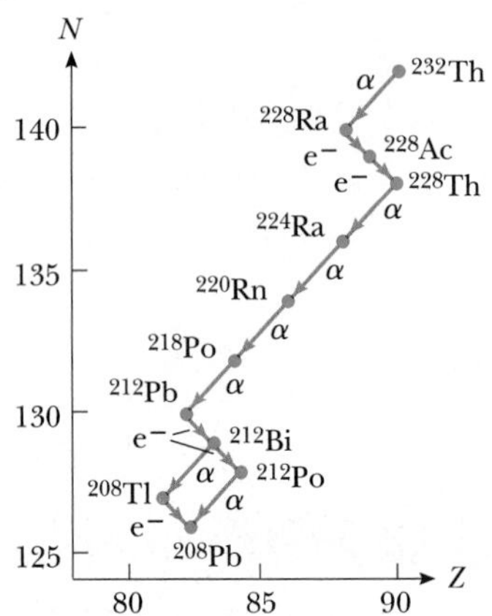


**FIGURE 29.12** Decay series beginning with $^{232}$Th.

that of any of its descendants. The fourth series in Table 29.2 begins with $^{237}$Np, a transuranic element (one having an atomic number greater than that of uranium) not found in nature. This element has a half-life of "only" $2.14 \times 10^6$ yr.

The two uranium series are somewhat more complex than the $^{232}$Th series (Fig. 29.12). Also, several other naturally occurring radioactive isotopes, such as $^{14}$C and $^{40}$K, are not part of either decay series.

Natural radioactivity constantly supplies our environment with radioactive elements that would otherwise have disappeared long ago. For example, because the Solar System is about $5 \times 10^9$ yr old, the supply of $^{226}$Ra (whose half-life is only 1 600 yr) would have been depleted by radioactive decay long ago if it were not for the decay series that starts with $^{238}$U, with a half-life of $4.47 \times 10^9$ yr.

## 29.6 NUCLEAR REACTIONS

It is possible to change the structure of nuclei by bombarding them with energetic particles. Such changes are called **nuclear reactions.** Rutherford was the first to observe nuclear reactions, using naturally occurring radioactive sources for the bombarding particles. He found that protons were released when alpha particles were allowed to collide with nitrogen atoms. The process can be represented symbolically as

$$^{4}_{2}\text{He} + ^{14}_{7}\text{N} \longrightarrow \text{X} + ^{1}_{1}\text{H} \qquad [29.19]$$

This equation says that an alpha particle ($^{4}_{2}$He) strikes a nitrogen nucleus and produces an unknown product nucleus (X) and a proton ($^{1}_{1}$H). Balancing atomic numbers and mass numbers, as we did for radioactive decay, enables us to conclude that the unknown is characterized as $^{17}_{8}$X. Because the element with atomic number 8 is oxygen, we see that the reaction is

$$^{4}_{2}\text{He} + ^{14}_{7}\text{N} \longrightarrow ^{17}_{8}\text{O} + ^{1}_{1}\text{H} \qquad [29.20]$$

This nuclear reaction starts with two stable isotopes, helium and nitrogen, and produces two different stable isotopes, hydrogen and oxygen.

Since the time of Rutherford, thousands of nuclear reactions have been observed, particularly following the development of charged-particle accelerators in the 1930s. With today's advanced technology in particle accelerators and particle accelerators, it is possible to achieve particle energies of at least 1 000 GeV = 1 TeV. These high-energy particles are used to create new particles whose properties are helping us solve the mysteries of the nucleus.

Which of the following are possible reactions?

(a) $^{1}_{0}\text{n} + ^{235}_{92}\text{U} \rightarrow ^{140}_{54}\text{Xe} + ^{94}_{38}\text{Sr} + 2\,(^{1}_{0}\text{n})$

(b) $^{1}_{0}\text{n} + ^{235}_{92}\text{U} \rightarrow ^{132}_{50}\text{Sn} + ^{101}_{42}\text{Mo} + 3\,(^{1}_{0}\text{n})$

(c) $^{1}_{0}\text{n} + ^{239}_{94}\text{Pu} \rightarrow ^{127}_{53}\text{I} + ^{93}_{41}\text{Nb} + 3\,(^{1}_{0}\text{n})$

### Example 29.8 The Discovery of the Neutron

A nuclear reaction of significant note occurred in 1932 when Chadwick, in England, bombarded a beryllium target with alpha particles. Analysis of the experiment indicated that the following reaction occurred:

$$^{4}_{2}\text{He} + ^{9}_{4}\text{Be} \longrightarrow ^{12}_{6}\text{C} + \text{X}$$

What is X in this reaction?

**Solution** Balancing mass numbers and atomic numbers, we see that the unknown particle must be represented as ${}^{1}_{0}X$, that is, a particle having a mass of 1 and zero charge.

Hence, the particle X is the neutron, ${}^{1}_{0}n$. This experiment was the first to provide positive proof of the existence of neutrons.

## Example 29.9 Synthetic Elements

**A** A beam of neutrons is directed at a target of ${}^{238}_{92}U$. The reaction products are a gamma ray and another isotope. What is the isotope?

**Solution**

Balancing input with output gives

$$ {}^{1}_{0}n + {}^{238}_{92}U \longrightarrow {}^{239}_{92}U + \gamma $$

**B** Isotope ${}^{239}U$ is radioactive and undergoes beta decay. Write the equation symbolizing this decay and identify the resulting isotope.

**Solution** The decay of ${}^{239}U$ by beta emission is

$$ {}^{239}_{92}U \longrightarrow {}^{239}_{93}Np + e^- + \bar{\nu} $$

**C** The isotope ${}^{239}Np$ is also radioactive and decays by beta emission. What is the end product?

**Solution** The decay of ${}^{239}Np$ by beta emission gives

$$ {}^{239}_{93}Np \longrightarrow {}^{239}_{94}Pu + e^- + \bar{\nu} $$

**D** What is the significance of these reactions?

**Solution** The interesting feature of these reactions is the fact that uranium is the element with the greatest number of protons, 92, that exists in nature in any appreciable amount. The reactions in parts A, B, and C do occur occasionally in nature; hence, minute traces of neptunium and plutonium are present. In 1940, however, researchers bombarded uranium with neutrons to produce plutonium and neptunium by the steps given previously. These two elements were thus the first elements made in the laboratory, and by bombarding them with neutrons and other particles, the list of synthetic elements has been extended to include those up to atomic number 112.

## *Q* VALUES

We have just examined some nuclear reactions for which mass numbers and atomic numbers must be balanced in the equations. We shall now consider the energy involved in these reactions, because energy is another important quantity that must be conserved.

Let us illustrate this procedure by analyzing the following nuclear reaction:

$$ {}^{2}_{1}H + {}^{14}_{7}N \longrightarrow {}^{12}_{6}C + {}^{4}_{2}He \qquad [29.21] $$

The total mass on the left side of the equation is the sum of the mass of ${}^{2}_{1}H$ (2.014 102 u) and the mass of ${}^{14}_{7}N$ (14.003 074 u), which equals 16.017 176 u. Similarly, the mass on the right side of the equation is the sum of the mass of ${}^{12}_{6}C$ (12.000 000 u) plus the mass of ${}^{4}_{2}He$ (4.002 602 u), for a total of 16.002 602 u. Thus, the total mass before the reaction is greater than the total mass after the reaction. The mass difference in this reaction is equal to 16.017 176 u − 16.002 602 u = 0.014 574 u. This "lost" mass is converted to the kinetic energy of the nuclei present after the reaction. In energy units, 0.014 574 u is equivalent to 13.576 MeV of kinetic energy carried away by the carbon and helium nuclei.

The energy required to balance the equation is called the $Q$ value of the reaction. In Equation 29.21 the $Q$ value is 13.576 MeV. Nuclear reactions in which there is a release of energy—that is, positive $Q$ values—are said to be **exothermic reactions.**

The energy balance sheet is not complete, however. We must also consider the kinetic energy of the incident particle before the collision. As an example, let us assume that the deuteron in Equation 29.21 has a kinetic energy of 5 MeV. Adding this to our $Q$ value, we find that the carbon and helium nuclei have a total kinetic energy of 18.576 MeV following the reaction.

Now consider the reaction

$$^{4}_{2}\text{He} + ^{14}_{7}\text{N} \longrightarrow ^{17}_{8}\text{O} + ^{1}_{1}\text{H} \qquad [29.22]$$

Before the reaction, the total mass is the sum of the masses of the alpha particle and the nitrogen nucleus: 4.002 602 u + 14.003 074 u = 18.005 676 u. After the reaction, the total mass is the sum of the masses of the oxygen nucleus and the proton: 16.999 133 u + 1.007 825 u = 18.006 958 u. In this case, the total mass after the reaction is *greater* than the total mass before the reaction. The mass deficit is 0.001 282 u, equivalent to an energy deficit of 1.194 MeV. This deficit is expressed by the negative $Q$ value of the reaction, − 1.194 MeV. Reactions with negative $Q$ values are called **endothermic reactions.** Such reactions will not take place unless the incoming particle has at least enough kinetic energy to overcome the energy deficit.

At first it might appear that the reaction in Equation 29.22 could take place if the incoming alpha particle had a kinetic energy of 1.194 MeV. In practice, however, the alpha particle must have more energy than this. If it had an energy of only 1.194 MeV, energy would be conserved but careful analysis would show that momentum was not. This can easily be understood by recognizing that the incoming alpha particle has some momentum before the reaction. However, if its kinetic energy were only 1.194 MeV, the products (oxygen and a proton) would be created with zero kinetic energy and, thus, zero momentum. It can be shown that, in order to conserve both energy and momentum, the incoming particle must have a minimum kinetic energy given by

$$KE_{\min} = \left(1 + \frac{m}{M}\right)|Q| \qquad [29.23]$$

where $m$ is the mass of the incident particle, $M$ is the mass of the target, and the absolute value of the $Q$ value is used. For the reaction given by Equation 29.22, we find

$$KE_{\min} = \left(1 + \frac{4.002\ 602}{14.003\ 074}\right)|-1.194\ \text{MeV}| = 1.535\ \text{MeV}$$

This minimum value of the kinetic energy of the incoming particle is called the **threshold energy.** The nuclear reaction shown in Equation 29.22 will not occur if the incoming alpha particle has a kinetic energy of less than 1.535 MeV, but can occur if its kinetic energy is equal to or greater than 1.535 MeV.

**Webnote 29.6**

Take a step-by-step tour of an atom smasher. Start by going to *http://www.howstuffworks.com/atom-smasher.htm*

If the $Q$ value of an endothermic reaction is − 2.17 MeV, the minimum kinetic energy needed in the reactant nuclei if the reaction is to occur must be (a) equal to 2.17 MeV, (b) greater than 2.17 MeV, (c) less than 2.17 MeV, or (d) precisely half of 2.17 MeV.

## 29.7 MEDICAL APPLICATIONS OF RADIATION

### RADIATION DAMAGE IN MATTER

Radiation absorbed by matter can cause severe damage. The degree and type of damage depend on several factors, including the type and energy of the radiation and the properties of the absorbing material. Radiation damage in biological organisms is primarily due to ionization effects in cells. The normal function of a cell may be disrupted when highly reactive ions or radicals are formed as the result of ionizing radiation. For example, hydrogen and hydroxyl radicals produced from water molecules can induce chemical reactions that may break bonds in proteins and other vital molecules. Large acute doses of radiation are especially dangerous because damage to a great number of molecules in a cell can cause death of the cell. Also, cells that do survive the radiation may become defective, which can lead to cancer.

In biological systems, it is common to separate radiation damage into two categories: somatic damage and genetic damage. **Somatic damage** is radiation damage to any cells except the reproductive cells. Such damage can lead to cancer at high radiation levels or seriously alter the characteristics of specific organisms. **Genetic damage** affects only reproductive cells. Damage to the genes in reproductive cells can lead to defective offspring. Clearly, we must be concerned about the effect of diagnostic treatments, such as x-rays and other forms of radiation exposure.

Several units are used to quantify radiation exposure and dose. The **roentgen** (R) is defined as **that amount of ionizing radiation that will produce $2.08 \times 10^9$ ion pairs in 1 $cm^3$ of air under standard conditions.** Equivalently, the roentgen is **that amount of radiation that deposits $8.76 \times 10^{-3}$ J of energy into 1 kg of air.**

For most applications, the roentgen has been replaced by the **rad** (which is an acronym for *r*adiation *a*bsorbed *d*ose), defined as follows: **1 rad is that amount of radiation that deposits $10^{-2}$ J of energy into 1 kg of absorbing material.**

Although the rad is a perfectly good physical unit, it is not the best unit for measuring the degree of biological damage produced by radiation. This is because the degree of biological damage depends not only on the dose but also on the type of radiation. For example, a given dose of alpha particles causes about 10 times more biological damage than an equal dose of x-rays. The **RBE** (*r*elative *b*iological *e*ffectiveness) factor is defined as **the number of rad of x-radiation or gamma radiation that produces the same biological damage as 1 rad of the radiation being used.** The RBE factors for different types of radiation are given in Table 29.3. Note that the values are only approximate because they vary with particle energy and form of damage.

Finally, the **rem** (*r*oentgen *e*quivalent in *m*an) is defined as the product of the dose in rad and the RBE factor:

$$\text{Dose in rem} = \text{dose in rad} \times \text{RBE}$$

**TABLE 29.3** RBE Factors for Several Types of Radiation

| Radiation | RBE Factor |
|---|---|
| x-Rays and gamma rays | 1.0 |
| Beta particles | 1.0–1.7 |
| Alpha particles | 10–20 |
| Slow neutrons | 4–5 |
| Fast neutrons and protons | 10 |
| Heavy ions | 20 |

According to this definition, 1 rem of any two radiations produces the same amount of biological damage. From Table 29.3, we see that a dose of 1 rad of fast neutrons represents an effective dose of 10 rem and that 1 rad of x-radiation is equivalent to a dose of 1 rem.

OCCUPATIONAL RADIATION EXPOSURE LIMITS

Low-level radiation from natural sources, such as cosmic rays and radioactive rocks and soil, delivers to each of us a dose of about 0.13 rem/yr. The upper limit of radiation dose recommended by the U.S. government (apart from background radiation and exposure related to medical procedures) is 0.5 rem/yr. Many occupations involve higher levels of radiation exposure, and for individuals in these occupations an upper limit of 5 rem/yr has been set for whole-body exposure. Higher upper limits are permissible for certain parts of the body, such as the hands and forearms. An acute whole-body dose of 400–500 rem results in a mortality rate of about 50%. The most dangerous form of exposure is ingestion or inhalation of radioactive isotopes, especially those elements the body retains and concentrates, such as $^{90}Sr$. In some cases, a dose of 1 000 rem can result from ingesting 1 mCi of radioactive material.

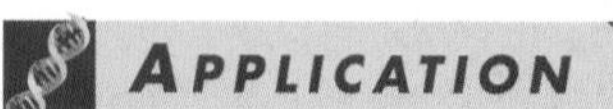


IRRADIATION OF FOOD AND MEDICAL EQUIPMENT

Sterilizing objects by exposing them to radiation has been going on for at least 25 years, but in recent years, the methods used have become safer to use and more economical. Most bacteria, worms, and insects are easily destroyed by exposure to gamma radiation from radioactive cobalt. There is no intake of radioactive nuclei by an organism in such sterilizing processes, as there is in the use of radioactive tracers. The process is very effective in destroying Trichinella worms in pork, salmonella bacteria in chickens, insect eggs in wheat, and surface bacteria on fruit and vegetables that can lead to rapid spoilage. Recently, this procedure has been expanded to include the sterilization of medical equipment while in its protective covering. Surgical gloves, sponges, sutures, and so forth are irradiated while packaged. Also, bone, cartilage, and skin used for grafting is often irradiated to reduce the chance for infection.

## TRACING

APPLICATION

RADIOACTIVE TRACERS IN MEDICINE

Radioactive particles can be used to trace chemicals participating in various reactions. One of the most valuable uses of radioactive tracers is in medicine. For example, $^{131}I$ is an artificially produced isotope of iodine (the natural, nonradioactive isotope is $^{127}I$). Iodine, a necessary nutrient for our bodies, is obtained largely through the intake of seafood and iodized salt. The thyroid gland plays a major role in the distribution of iodine throughout the body. In order to evaluate the performance of the thyroid, the patient drinks a small amount of radioactive sodium iodide. Two hours later, the amount of iodine in the thyroid gland is determined by measuring the radiation intensity at the neck area.

A medical application of the use of radioactive tracers occurring in emergency situations is that of locating a hemorrhage inside the body. Often the location of the site cannot easily be determined, but radioactive chromium can identify the location with a high degree of precision. Chromium is taken up by red blood cells and carried uniformly throughout the body. However, the blood will be dumped at a hemorrhage site, and the radioactivity of that region will increase markedly.

RADIOACTIVE TRACERS IN AGRICULTURAL RESEARCH

The tracer technique is also useful in agricultural research. Suppose the best method of fertilizing a plant is to be determined. A certain material in the fertilizer, such as nitrogen, can be tagged with one of its radioactive isotopes. The fertilizer is then sprayed on one group of plants, sprinkled on the ground for a second group, and raked into the soil for a third. A Geiger counter is then used to track the nitrogen through the three types of plants.

Tracing techniques are as wide-ranging as human ingenuity can devise. Present applications range from checking the absorption of fluorine by teeth to checking contamination of food-processing equipment by cleansers to monitoring deterioration inside an automobile engine. In the latter case, a radioactive material is used in the manufacture of the pistons, and the oil is checked for radioactivity to determine the amount of wear on the pistons.

## COMPUTED AXIAL TOMOGRAPHY (CAT SCANS)

The normal x-ray of a human body has two primary disadvantages when used as a source of clinical diagnosis. First, it is difficult to distinguish between various types of tissue in the body because they all have similar x-ray absorption properties. Second, a conventional x-ray absorption picture is indicative of the average amount of absorption along a particular direction in the body, leading to somewhat obscured pictures. To overcome these problems, a device called a CAT scanner was developed in England in 1973; it is capable of producing pictures of much greater clarity and detail than were previously obtainable.

The operation of a CAT scanner can be understood by considering the following hypothetical experiment. Suppose a box consists of four compartments, labeled A, B, C, and D, as in Figure 29.13a. Each compartment has a different amount of absorbing material from any other compartment. What set of experimental procedures will enable us to determine the relative amounts of material in each compartment? The following steps outline one method that will provide this information. First, a beam of x-rays is passed through compartments A and C, as Figure 29.13b. The intensity of the exiting radiation is reduced by absorption by some number that we assign as 8. (The number 8 could mean, for example, that the intensity of the exiting beam is reduced by eight tenths of 1% from its initial value.) Because we do not know which of the compartments, A or C, was responsible for this reduction in intensity, half the loss is assigned to each compartment, as in Figure 29.13c. Next, a beam of x-rays is passed through compartments B and D, as in Figure 29.13b. The reduction in intensity for this beam is 10, and again we assign half the loss to each compartment. We now redirect the x-ray source so that it sends one beam through compartments A and B and another through compartments C and D, as in Figure 29.13d, and again measure the absorption. Suppose the absorption through compartments A and B in this experiment is measured to be 7 units. On the basis of our first experiment, we would have guessed it would be 9 units, 4 by compartment A, and 5 by compartment B. Thus, we have reduced the guessed absorption for each compartment by 1 unit so that the sum is 7 rather than 9, to give the numbers shown in Figure 29.13e. Likewise, when the beam is passed through compartments C and D, as in Figure 29.13d, we may find the total absorption to be 11 as compared to our first experiment of 9. In this case, we add

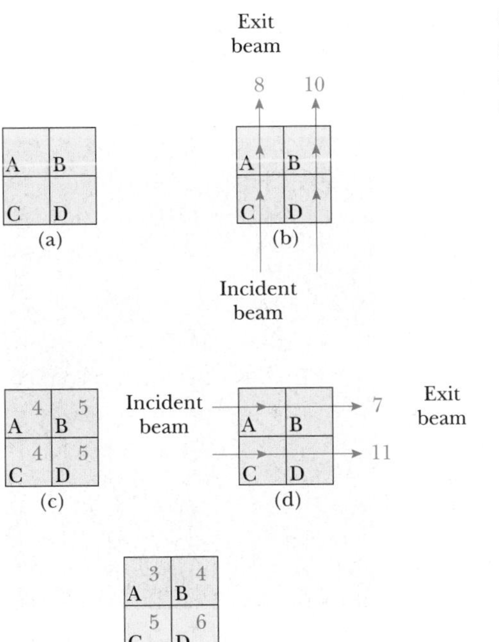


**FIGURE 29.13** An experimental procedure for determining the relative amounts of x-ray absorption by four different compartments in a box.

1 unit of absorption to each compartment to give a sum of 11 as in Figure 29.13e. This somewhat crude procedure could be improved by measuring the absorption along other paths. However, these simple measurements are sufficient to enable us to conclude that compartment D contains the most absorbing material and A the least. A visual representation of these results can be obtained by assigning to each compartment a shade of gray corresponding to the particular number associated with the absorption. In our example, compartment D would be very dark and compartment A would be very light.

CAT SCANS

The steps outlined above are representative of how a CAT scanner produces images of the human body. A thin slice of the body is subdivided into perhaps 10 000 compartments, rather than 4 compartments as in our simple example. The function of the CAT scanner is to determine the relative absorption in each of these 10 000 compartments and to display a picture of its calculations in various shades of gray. Note that CAT stands for **computed axial tomography.** The term *axial* is used because the slice of the body to be analyzed corresponds to a plane perpendicular to the head-to-toe axis. *Tomos* is the Greek word for slice and *graph* is the Greek word for picture. In a typical diagnosis, the patient is placed in the position shown in Figure 29.14 and a narrow beam of x-rays is sent through the plane of interest. The emerging x-rays are detected and measured by photomultiplier tubes behind the patient. The x-ray tube is then rotated a few degrees, and the intensity is recorded again. An extensive amount of information is obtained by rotating the beam through 180° at intervals of about 1° per measurement, resulting in a set of numbers assigned to each of the 10 000 "compartments" in the slice. These numbers are then converted by the computer to a photograph in various shades of gray for this segment of the body.

A brain scan of a patient can now be made in about 2 s, and a full-body scan requires about 6 s. The final result is a picture containing much greater quantitative information and clarity than a conventional x-ray photograph. Because CAT scanners use x-rays, which are an ionizing form of radiation, the technique presents a modest health risk to the patient being diagnosed.

## MAGNETIC RESONANCE IMAGING (MRI)

MAGNETIC RESONANCE IMAGING (MRI)

At the heart of magnetic resonance imaging (MRI) is the fact that when a nucleus having a magnetic moment is placed in an external magnetic field, its moment precesses about the magnetic field with a frequency that is proportional to the field. For example, a proton, whose spin is $\frac{1}{2}$, can occupy one of two energy states when placed in an external magnetic field. The lower-energy state corresponds to

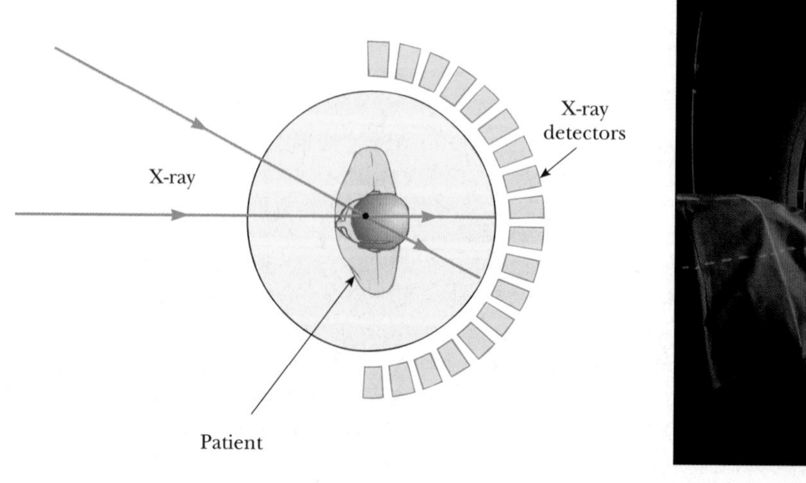

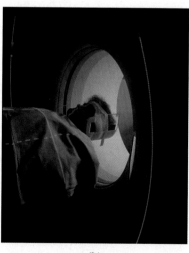

**FIGURE 29.14** (a) CAT scanner detector assembly. (b) Photograph of a patient undergoing a CAT scan in a hospital. *(Jay Freis/The Image Bank)*

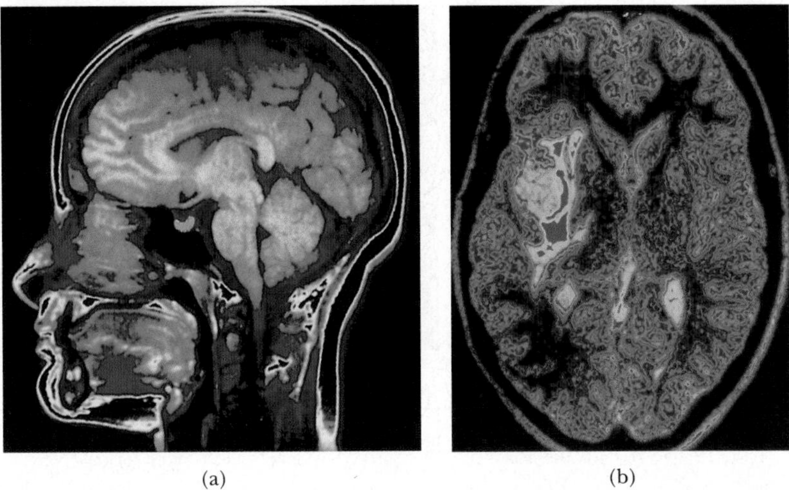

(a) (b)

**FIGURE 29.15** Computer-enhanced MRI images of (a) a normal human brain with the pituitary gland highlighted and (b) a human brain with a glioma tumor. *(Scott Camazine/Science Source/Photo Researchers, Inc.)*

the case in which the spin is aligned with the field, where as the higher-energy state corresponds to the case in which the spin is opposite the field. Transitions between these two states can be observed using a technique known as **nuclear magnetic resonance.** A DC magnetic field is applied to align the magnetic moments, and a second, weak oscillating magnetic field is applied perpendicular to the DC field. When the frequency of the oscillating field is adjusted to match the precessional frequency of the magnetic moments, the nuclei "flip" between the two spin states. These transitions result in a net absorption of energy by the spin system, which can be detected electronically.

In MRI, image reconstruction is obtained using spatially varying magnetic fields and a procedure for encoding each point in the sample being imaged. Some MRI images taken on a human head are shown in Figure 29.15. In practice, a computer-controlled pulse sequencing technique is used to produce signals that are captured by a suitable processing device. This signal is then subjected to appropriate mathematical manipulations to provide data for the final image. The main advantage of MRI over other imaging techniques in medical diagnostics is that it causes minimal damage to cellular structures. Photons associated with the radio frequency signals used in MRI have energies of only about $10^{-7}$ eV. Because molecular bond strengths are much larger (of the order of 1 eV), the rf photons cause little cellular damage. In comparison, x-rays or gamma-rays have energies ranging from $10^4$ to $10^6$ eV and can cause considerable cellular damage.

**Webnote 29.7**

Learn about other medical applications of nuclear physics by going to *http://www.howstuffworks.com/nuclear-medicine.htm*

## 29.8 RADIATION DETECTORS

Although most medical applications of radiation require instruments to make quantitative measurements of radioactive intensity, we have not yet explained how such instruments operate. Various devices have been developed to detect the energetic particles emitted when a radioactive nucleus decays. The **Geiger counter** (Fig. 29.16) is perhaps the most common form of device used to detect radioactivity. It can be considered the prototype of all counters that use the ionization of a medium as the basic detection process. It consists of a thin wire electrode aligned along the central axis of a cylindrical metallic tube filled with a gas at low pressure. The wire is maintained at a high positive voltage of about 1 000 V relative to the tube. When an energetic charged particle or gamma-ray photon enters the tube through a thin window at one end, some of the gas atoms are ionized. The electrons removed from these atoms are attracted toward the wire electrode, and in the process they ionize other atoms in their path. This sequential ionization results in an *avalanche* of electrons that produces a current pulse. After the pulse has been amplified, it can either be used to trigger an electronic counter or delivered

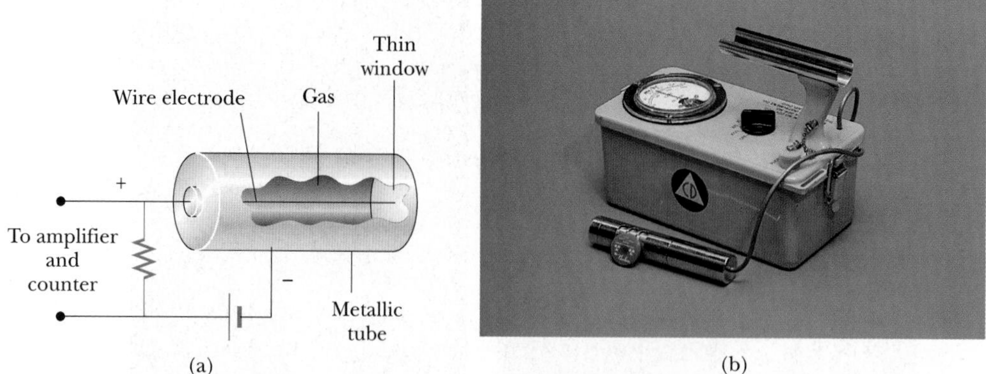


**FIGURE 29.16** (a) Diagram of a Geiger counter. The voltage between the wire electrode and the metallic tube is usually about 1 000 V. (b) A Geiger counter. *(David Rogers)*

to a loudspeaker that clicks each time a particle is detected. Although a Geiger counter reliably detects the presence and quantity of radiation, it cannot be used to measure the energy of the detected radiation.

A **semiconductor diode detector** is essentially a reverse-biased *p-n* junction. As an energetic particle passes through the junction, it produces electron-hole pairs that are separated by the internal electric field. This movement of electrons and holes creates a brief pulse of current that is measured with an electronic counter. In a typical device, the duration of the pulse is $10^{-8}$ s.

A **scintillation counter** typically uses a solid or liquid material whose atoms are easily excited by radiation. The excited atoms then emit visible-light photons when they return to their ground state. Common materials used as scintillators are transparent crystals of sodium iodide and certain plastics. If the scintillator material is attached to one end of a device called a **photomultiplier** (PM) tube as shown in Figure 29.17, the photons emitted by the scintillator can be converted to an electrical signal. The PM tube consists of numerous electrodes, called *dynodes,* whose electric potentials increase in succession along the length of the tube. Between the top of the tube and the scintillator material is a plate called a photocathode. When photons leaving the scintillator hit this plate, electrons are emitted because of the photoelectric effect. As one of these emitted electrons strikes the first

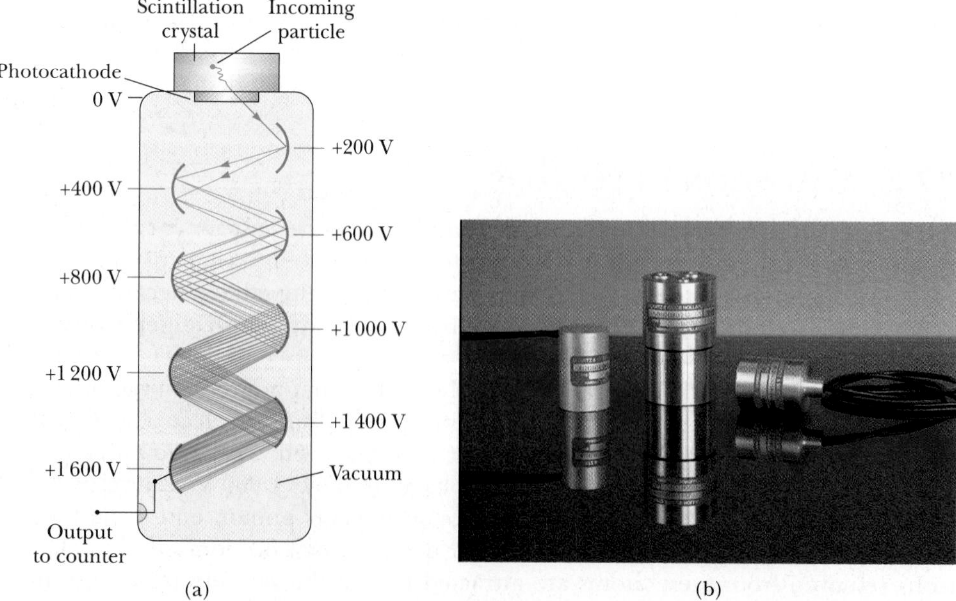


**FIGURE 29.17** (a) Diagram of a scintillation counter connected to a photomultiplier tube. (b) The sodium iodide in these scintillation crystals flashes when an energetic particle passes through it, something like the way the atmosphere flashes when a meteor passes through it.

dynode, the electron has sufficient kinetic energy to eject several other electrons from the dynode surface. When these electrons are accelerated to the second dynode, many more electrons are ejected, and thus a multiplication process occurs. The end result is 1 million or more electrons striking the last dynode. Hence, one particle striking the scintillator produces a sizable electrical pulse at the PM output, and this pulse is sent to an electronic counter.

Both the scintillator and the semiconductor diode detector are much more sensitive than a Geiger counter, mainly because of the higher mass density of the detecting medium. Both can also be used to measure particle energy from the height of the pulses produced.

**Track detectors** are various devices used to view the tracks or paths of charged particles directly. High-energy particles produced in particle accelerators may have energies ranging from $10^9$ to $10^{12}$ eV. Thus, they cannot be stopped and cannot have their energy measured with the small detectors already mentioned. Instead, the energy and momentum of these energetic particles are found from the curvature of their path in a magnetic field of known magnitude and direction.

A **photographic emulsion** is the simplest example of a track detector. A charged particle ionizes the atoms in an emulsion layer. The path of the particle corresponds to a family of points at which chemical changes have occurred in the emulsion. When the emulsion is developed, the particle's track becomes visible.

A **cloud chamber** contains a gas that has been supercooled to just below its usual condensation point. An energetic charged particle passing through ionizes the gas along its path. The ions serve as centers for condensation of the supercooled gas. The track can be seen with the naked eye and can be photographed. A magnetic field can be applied to determine the charges of the radioactive particles, as well as their momentum and energy.

A device called a **bubble chamber,** invented in 1952 by D. Glaser, uses a liquid (usually liquid hydrogen) maintained near its boiling point. Ions produced by incoming charged particles leave bubble tracks, which can be photographed (Fig. 29.18). Because the density of the liquid in a bubble chamber is much higher than the density of the gas in a cloud chamber, the bubble chamber has a much higher sensitivity.

A **wire chamber** consists of thousands of closely spaced parallel wires that collect the electrons created by a passing ionizing particle. A second grid, with wires

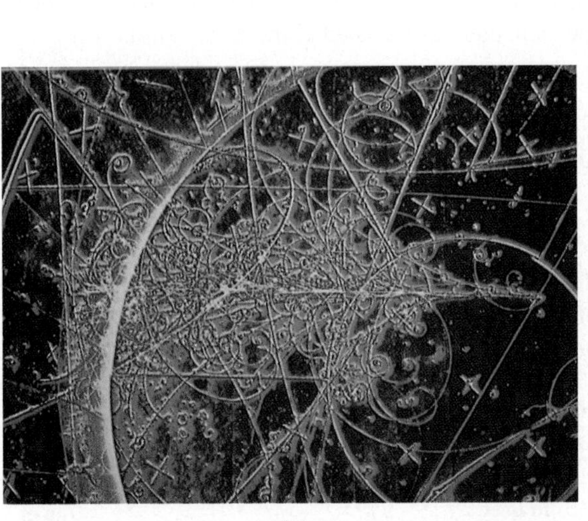

(a)

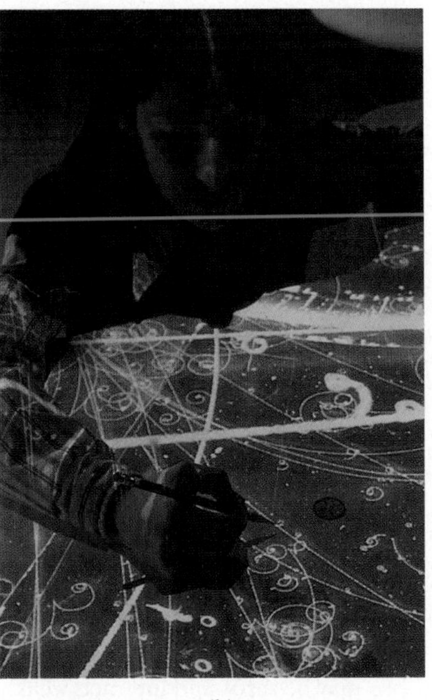

(b)

**FIGURE 29.18** (a) Artificially colored bubble-chamber photograph showing tracks of particles that have passed through the chamber. *(Photo Researchers, Inc./Science Photo Library)* (b) This research scientist is studying a photograph of particle tracks made in a bubble chamber at Fermilab.

### Webnote 29.8

Nuclear aircraft? Nuclear automobiles? Nuclear excavation? These were some of the ideas that were proposed following World War II. To learn why one of them failed, go to *http://www.megazone.org/ANP*

perpendicular to the first, allows the *x,y* position of the particle in the plane of the two sets of wires to be determined. Finally several such *x,y* grids arranged parallel to each other along the *z* axis can be used to determine the particle's track in three dimensions. Wire chambers form a part of most detectors used at high-energy accelerator labs, and provide electronic readout to a computer for rapid track reconstruction and display.

## SUMMARY

Nuclei are represented symbolically as $^A_Z\text{X}$, where X represents the chemical symbol for the element. The quantity $A$ is the **mass number,** which equals the total number of nucleons (neutrons plus protons) in the nucleus. The quantity $Z$ is the **atomic number,** which equals the number of protons in the nucleus. Nuclei that contain the same number of protons but different numbers of neutrons are called **isotopes.** In other words, isotopes have the same $Z$ value but different $A$ values.

Most nuclei are approximately spherical, with an average radius given by

$$r = r_0 A^{1/3} \qquad [29.1]$$

where $A$ is the mass number and $r_0$ is a constant equal to $1.2 \times 10^{-15}$ m.

The total mass of a nucleus is always less than the sum of the masses of its individual nucleons. This mass difference, $\Delta m$, multiplied by $c^2$ gives the **binding energy** of the nucleus.

The spontaneous emission of radiation by certain nuclei is called **radioactivity.** There are three processes by which a radioactive substance can decay: alpha ($\alpha$) decay, in which the emitted particles are $^4_2\text{He}$ nuclei; beta ($\beta$) decay, in which the emitted particles are electrons or positrons; and gamma ($\gamma$) decay, in which the emitted particles are high-energy photons.

The **decay rate,** or **activity,** $R$, of a sample is given by

$$R = \left|\frac{\Delta N}{\Delta t}\right| = \lambda N \qquad [29.3]$$

where $N$ is the number of radioactive nuclei at some instant and $\lambda$ is a constant for a given substance called the **decay constant.**

Nuclei in a radioactive substance decay in such a way that the number of nuclei present varies with time according to the expression

$$N = N_0 e^{-\lambda t} \qquad [29.4]$$

where $N$ is the number of radioactive nuclei present at time $t$, $N_0$ is the number at time $t = 0$, and $e = 2.718. \ . \ . \ .$

The **half-life,** $T_{1/2}$, of a radioactive substance is the time required for half of a given number of radioactive nuclei to decay. The half-life is related to the decay constant as

$$T_{1/2} = \frac{0.693}{\lambda} \qquad [29.5]$$

If a nucleus decays by alpha emission, it loses two protons and two neutrons. A typical alpha decay is

$$^{238}_{92}\text{U} \longrightarrow {}^{234}_{90}\text{Th} + {}^4_2\text{He} \qquad [29.9]$$

Note that in this decay, as in all radioactive decay processes, the sum of the $Z$ values on the left equals the sum of the $Z$ values on the right; the same is true for the $A$ values.

A typical beta decay is

$$^{14}_6\text{C} \longrightarrow {}^{14}_7\text{N} + \text{e}^- + \bar{\nu} \qquad [29.15]$$

When a nucleus beta decays, an **antineutrino** is emitted along with an electron, or a **neutrino** along with a position. A neutrino has zero electric charge and a small mass (which may be zero) and interacts weakly with matter.

Nuclei are often in an excited state following radioactive decay, and release their extra energy by emitting a high-energy photon called a **gamma ray** ($\gamma$). A typical gamma-ray emission is

$$^{12}_6\text{C}^* \longrightarrow {}^{12}_6\text{C} + \gamma \qquad [29.18]$$

where the asterisk indicates that the carbon nucleus was in an excited state before gamma emission.

**Nuclear reactions** can occur when a bombarding particle strikes another nucleus. A typical nuclear reaction is

$$^{4}_{2}\text{He} + ^{14}_{7}\text{N} \longrightarrow ^{17}_{8}\text{O} + ^{1}_{1}\text{H} \qquad [29.20]$$

In this reaction an alpha particle strikes a nitrogen nucleus, producing an oxygen nucleus and a proton. As in radioactive decay, atomic numbers and mass numbers balance on the two sides of the arrow.

Nuclear reactions in which energy is released are said to be **exothermic reactions** and are characterized by positive $Q$ values. Reactions with negative $Q$ values, called **endothermic reactions,** cannot occur unless the incoming particle has at least enough kinetic energy to overcome the energy deficit. In order to conserve both energy and momentum, the incoming particle must have a minimum kinetic energy, called the **threshold energy,** given by

$$KE_{\text{min}} = \left(1 + \frac{m}{M}\right)|Q| \qquad [29.23]$$

where $m$ is the mass of the incident particle and $M$ is the mass of the target atom.

## CONCEPTUAL QUESTIONS

1. Isotopes of a given element have many different properties, such as mass, but the same chemical properties. Why is this?
2. Physicists have attempted to measure the solar neutrinos associated with nuclear processes in the Sun's interior. These studies involve placing a large volume of detecting material in a deep mine shaft to reduce the background from cosmic rays. Explain why neutrinos are not similarly reduced.
3. A student claims that a heavy form of hydrogen decays by alpha emission. How do you respond?
4. Explain the main differences between alpha, beta, and gamma rays.
5. In beta decay, the energy of the electron or positron emitted from the nucleus lies somewhere in a relatively large range of possibilities. In alpha decay, however, the alpha-particle energy can only have discrete values. Why is there this difference?
6. If film is kept in a box, alpha particles from a radioactive source outside the box cannot expose the film but beta particles can. Explain.
7. In positron decay, a proton in the nucleus becomes a neutron, and the positive charge is carried away by the positron. But a neutron has a larger rest energy than a proton. How is this possible?
8. An alpha particle has twice the charge of a beta particle. Why does the former deflect less than the latter when passing between electrically charged plates, assuming they both have the same speed?
9. Can carbon-14 dating be used to measure the age of a stone?
10. Pick any beta-decay process and show that the neutrino must have zero charge.
11. Why do heavier elements require more neutrons in order to maintain stability?
12. Suppose it could be shown that cosmic-ray intensity was much greater 10 000 years ago. How would this affect the ages we assign to ancient samples of once-living matter?
13. If no more people were to be born, the law of population growth would strongly resemble the radioactive decay law. Discuss this statement.
14. Why is carbon dating unable to provide accurate estimates of very old materials?
15. Two samples of the same radioactive nuclide are prepared. Sample A has twice the intial activity of sample B. How does the half-life of A compare with the half-life of B? After each has passed through five half-lives, what is the ratio of their activities?
16. Which number determines the chemical characteristics of an atom: $A$, $N$, or $Z$?
17. Compare and contrast a photon and a neutrino.
18. Why do you think neutrons are more effective at starting a nuclear reaction than alpha particles or protons?

## PROBLEMS

1, 2, 3 = straightforward, intermediate, challenging ☐ = full solution available in Student Solutions Manual/Study Guide

**web** = solution posted at **http://info.brookscole.com/serway** = biomedical application

*Note:* Table 29.4 will be useful for many of these problems. A more complete list of atomic masses is given in Appendix B.

### Section 29.1 Some Properties of Nuclei

1. Compare the nuclear radii of the following nuclides: $^{2}_{1}\text{H}$, $^{60}_{27}\text{Co}$ $^{197}_{79}\text{Au}$ $^{239}_{94}\text{Pu}$
2. What is the order of magnitude of the number of protons in your body? Of the number of neutrons? Of the number of electrons?
3. Using the result of Example 29.1, find the radius of a sphere of nuclear matter that would have a mass equal to that of Earth. (Earth has a mass of $5.98 \times 10^{24}$ kg and equatorial radius of $6.38 \times 10^{6}$ m.)

**TABLE 29.4 Some Atomic Masses**

| Element | Atomic Mass (u) |
|---|---|
| $(_{-1}^{0}e)$ | 0.000 549 |
| $(_{1}^{0}n)$ | 1.008 665 |
| $_{1}^{1}H$ | 1.007 825 |
| $_{1}^{2}H$ | 2.014 102 |
| $_{2}^{4}He$ | 4.002 602 |
| $_{3}^{7}Li$ | 7.016 003 |
| $_{4}^{9}Be$ | 9.012 174 |
| $_{5}^{10}B$ | 10.012 936 |
| $_{6}^{12}C$ | 12.000 000 |
| $_{6}^{13}C$ | 13.003 355 |
| $_{7}^{14}N$ | 14.003 074 |
| $_{7}^{15}N$ | 15.000 108 |
| $_{8}^{15}O$ | 15.003 065 |
| $_{8}^{17}O$ | 16.999 131 |
| $_{8}^{18}O$ | 17.999 160 |
| $_{9}^{18}F$ | 18.000 937 |
| $_{10}^{20}Ne$ | 19.992 435 |
| $_{11}^{23}Na$ | 22.989 770 |
| $_{12}^{23}Mg$ | 22.994 127 |
| $_{13}^{27}Al$ | 26.981 538 |
| $_{15}^{30}P$ | 29.978 310 |
| $_{20}^{40}Ca$ | 39.962 591 |
| $_{20}^{42}Ca$ | 41.958 63 |
| $_{20}^{43}Ca$ | 42.958 770 |
| $_{26}^{56}Fe$ | 55.934 940 |
| $_{30}^{64}Zn$ | 63.929 144 |
| $_{29}^{64}Cu$ | 63.929 599 |
| $_{41}^{93}Nb$ | 92.906 376 8 |
| $_{79}^{197}Au$ | 196.966 543 |
| $_{80}^{202}Hg$ | 201.970 617 |
| $_{84}^{216}Po$ | 216.001 790 |
| $_{86}^{220}Rn$ | 220.011 401 |
| $_{90}^{234}Th$ | 234.043 583 |
| $_{92}^{238}U$ | 238.050 784 |

4. Consider the hydrogen atom to be a sphere of radius equal to the Bohr radius, $0.53 \times 10^{-10}$ m, and calculate the approximate value of the ratio of the nuclear density to the atomic density.
5. An alpha particle ($Z = 2$, mass $6.64 \times 10^{-27}$ kg) approaches to within $1.00 \times 10^{-14}$ m of a carbon nucleus ($Z = 6$). What are (a) the maximum Coulomb force on the alpha particle, (b) the acceleration of the alpha particle at this point, and (c) the potential energy of the alpha particle at this point?
6. Singly ionized carbon is accelerated through 1 000 V and passed into a mass spectrometer to determine the isotopes present (see Chapter 19). The magnetic field strength in the spectrometer is 0.200 T. (a) Determine the orbit radii for the $^{12}C$ and $^{13}C$ isotopes as they pass through the field. (b) Show that the ratio of radii may be written in the form

$$\frac{r_1}{r_2} = \sqrt{\frac{m_1}{m_2}}$$

and verify that your radii in part (a) agree with this.
7. web (a) Find the speed an alpha particle requires to come within $3.2 \times 10^{-14}$ m of a gold nucleus. (b) Find the energy of the alpha particle in MeV.
8. Find the nucleus that has a radius approximately equal to one half of the radius of uranium $_{92}^{238}U$.

### Section 29.2 Binding Energy

9. Calculate the average binding energy per nucleon of $_{41}^{93}Nb$ and $_{79}^{197}Au$.
10. Calculate the binding energy per nucleon for (a) $^2H$, (b) $^4He$, (c) $^{56}Fe$, and (d) $^{238}U$.
11. A pair of nuclei for which $Z_1 = N_2$ and $Z_2 = N_1$ are called *mirror isobars* (the atomic and neutron numbers are interchangeable). Binding energy measurements on such pairs can be used to obtain evidence of the charge independence of nuclear forces. Charge independence means that the proton-proton, proton-neutron, and neutron-neutron nuclear forces are approximately equal. Calculate the difference in binding energy for the two mirror nuclei, $_{8}^{15}O$ and $_{7}^{15}N$.
12. The peak of the stability curve occurs at $^{56}Fe$. Elements up to iron are produced in the cores of massive stars by exothermic fusion reactions. This is the fundamental reason that iron and lighter elements are much more common in the Universe than elements with higher mass numbers. Show that $^{56}Fe$ has a higher binding energy per nucleon than its neighbors $^{55}Mn$ and $^{59}Co$. Compare your results with Figure 29.4.
13. Two nuclei having the same mass number are known as *isobars*. Calculate the difference in binding energy per nucleon for the isobars $_{11}^{23}Na$ and $_{12}^{23}Mg$. How do you account for the difference?
14. Compare the average binding energy per nucleon of $_{12}^{24}Mg$ and $_{37}^{85}Rb$.

### Section 29.3 Radioactivity

15. The half-life of an isotope of phosphorus is 14 days. If a sample contains $3.0 \times 10^{16}$ such nuclei, determine its activity. Express your answer in curies.
16. A drug tagged with $_{43}^{99}Tc$ (half-life = 6.05 h) is prepared for a patient. If the original activity of the sample was $1.1 \times 10^4$ Bq, what is its activity after it has sat on the shelf for 2.0 h?
17. The half-life of $^{131}I$ is 8.04 days. (a) Calculate the decay constant for this isotope. (b) Find the number of $^{131}I$ nuclei necessary to produce a sample with an activity of 0.50 $\mu$Ci.
18. The half-life of $^{131}I$ is 8.04 days. On a certain day, the activity of an $^{131}I$ sample is 6.40 mCi. What is its activity 40.2 days later?
19. A radioactive sample contains 3.50 $\mu$g of pure $^{11}C$, which has a half-life of 20.4 min. (a) How many moles of $^{11}C$ is present initially? (b) Determine the number of nuclei present initially. What is the activity of the sample (c) initially and (d) after 8.00 h?

20. How much time elapses before 90.0% of the radioactivity of a sample of $^{72}_{33}As$ disappears, as measured by its activity? The half-life of $^{72}_{33}As$ is 26 h.

21. Many smoke detectors use small quantities of the isotope $^{241}Am$ in their operation. The half-life of $^{241}Am$ is 432 yr. How long will it take for the activity of this material to decrease to $1.00 \times 10^{-3}$ of the original activity?

22. After a plant or animal dies, its $^{14}C$ content decreases with a half-life of 5 730 yr. If an archaeologist finds an ancient firepit containing partially consumed firewood, and the $^{14}C$ content of the wood is only 12.5% that of an equal carbon sample from a present-day tree, what is the age of the ancient site?

23. web A freshly prepared sample of a certain radioactive isotope has an activity of 10.0 mCi. After 4.00 h, the activity is 8.00 mCi. (a) Find the decay constant and half-life of the isotope. (b) How many atoms of the isotope were contained in the freshly prepared sample? (c) What is the sample's activity 30 h after it is prepared?

24. A building has become accidentally contaminated with radioactivity. The longest-lived material in the building is strontium-90 (the atomic mass of $^{90}_{38}Sr$ is 89.907 7). If the building initially contained 5.0 kg of this substance and the safe level is less than 10.0 counts/min, how long will the building be unsafe?

### Section 29.4 The Decay Processes

25. Complete the following radioactive decay formulas:

$$^{212}_{83}Bi \longrightarrow ? + ^{4}_{2}He$$
$$^{95}_{36}Kr \longrightarrow ? + e^-$$
$$? \longrightarrow ^{4}_{2}He + ^{140}_{58}Ce$$

26. Complete the following radioactive decay formulas:
(a) $^{12}_{5}B \rightarrow ? + e^-$
(b) $^{234}_{90}Th \rightarrow ^{230}_{88}Ra + ?$
(c) $? \rightarrow ^{14}_{7}N + e^-$

27. Complete the following radioactive decay formulas:
(a) $? \rightarrow e^+ + ^{40}_{19}K$
(b) $^{94}_{44}Ru \rightarrow ^{4}_{2}He + ?$
(c) $^{144}_{60}Nd \rightarrow ? + ^{140}_{58}Ce$

28. Figure P29.28 shows the steps by which $^{235}_{92}U$ decays to $^{207}_{82}Pb$. Enter the correct isotope symbol in each square.

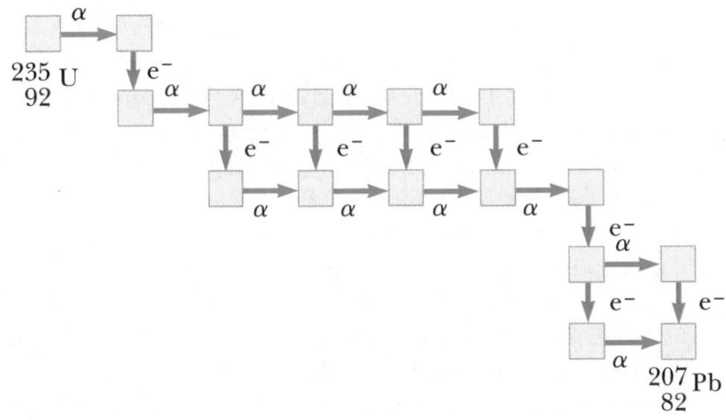


**FIGURE P29.28**

29. The mass of $^{56}Fe$ is 55.934 9 u and the mass of $^{56}Co$ is 55.939 9 u. Which isotope decays into the other, and by what process?

30. Find the energy released in the alpha decay of $^{238}_{92}U$. The following mass value will be useful: $^{234}_{90}Th$ has a mass of 234.043 583 u.

31. A nucleus of mass 228 u, initially at rest, undergoes alpha decay. If the alpha particle emitted has a kinetic energy of 4.00 MeV, what is the kinetic energy of the recoiling daughter nucleus?

32. $^{66}_{28}Ni$ (mass = 65.929 1 u) undergoes beta decay to $^{66}_{29}Cu$ (mass = 65.928 9 u). (a) Write the complete decay formula for this process. (b) Find the maximum kinetic energy of the emerging electrons.

33. An $^{3}H$ nucleus beta-decays into $^{3}He$ by creating an electron and an antineutrino according to the reaction

$$^{3}_{1}H \longrightarrow ^{3}_{2}He + e^- + \bar{\nu}$$

Use Appendix B to determine the total energy released in this reaction.

34. A piece of charcoal used for cooking is found at the remains of an ancient campsite. A 1.00-kg sample of carbon from the wood has an activity of $2.00 \times 10^3$ decays per minute. Find the age of the charcoal. (*Hint:* Living material has an activity of 15.0 decays/minute per gram of carbon present.)

35. A wooden artifact is found in an ancient tomb. Its $^{14}C$ activity is measured to be 60.0% of that in a fresh sample of wood from the same region. Assuming the same amount of $^{14}C$ was initially present in the wood from which the artifact was made, determine the age of the artifact.

36. A living specimen in equilibrium with the atmosphere contains one atom of $^{14}C$ (half-life = 5 730 yr) for every $7.70 \times 10^{11}$ stable carbon atoms. An archeological sample of wood (cellulose, $C_{12}H_{22}O_{11}$) contains 21.0 mg of carbon. When the sample is placed inside a shielded beta counter with 88.0% counting efficiency, 837 counts are accumulated in one week. Assuming that the cosmic-ray flux and Earth's atmosphere have not changed appreciably since the sample was formed, find the age of the sample.

### Section 29.6 Nuclear Reactions

37. web The first known reaction in which the product nucleus was radioactive (achieved in 1934) was one in which $^{27}_{13}Al$ was bombarded with alpha particles. Produced in the reaction were a neutron and a product nucleus. (a) What was the product nucleus? (b) Find the *Q* value of the reaction.

38. Complete the following nuclear reactions:
(a) $? + ^{14}_{7}N \rightarrow ^{1}_{1}H + ^{17}_{8}O$
(b) $^{7}_{3}Li + ^{1}_{1}H \rightarrow ^{4}_{2}He + ?$

39. Identify the unknown particles X and X′ in the following nuclear reactions:
(a) $X + ^{4}_{2}He \rightarrow ^{24}_{12}Mg + ^{1}_{0}n$
(b) $^{235}_{92}U + ^{1}_{0}n \rightarrow ^{90}_{38}Sr + X + 2\,^{1}_{0}n$
(c) $2\,^{1}_{1}H \rightarrow ^{2}_{1}H + X + X'$

40. The first nuclear reaction utilizing particle accelerators was performed by Cockcroft and Walton. Accelerated protons were used to bombard lithium nuclei, producing the following reaction:

$$^{1}_{1}H + ^{7}_{3}Li \longrightarrow ^{4}_{2}He + ^{4}_{2}He$$

Since the masses of the particles involved in the reaction were well known, these results were used to obtain an early proof of the Einstein mass-energy relation. Calculate the *Q* value of the reaction.

41. (a) Suppose $^{10}_{5}B$ is struck by an alpha particle, releasing a proton and a product nucleus in the reaction. What is the product nucleus? (b) An alpha particle and a product nucleus are produced when $^{13}_{6}C$ is struck by a proton. What is the product nucleus?

42. (a) Determine the product of the reaction $^{7}_{3}Li + ^{4}_{2}He \rightarrow ? + n$. (b) What is the $Q$ value of the reaction?

43. A beam of 6.61-MeV protons is incident on a target of $^{27}_{13}Al$. Those that collide produce the reaction

$$p + ^{27}_{13}Al \longrightarrow ^{27}_{14}Si + n$$

($^{27}_{14}Si$ has a mass of 26.986 721 u.) Neglect any recoil of the product nucleus and determine the kinetic energy of the emerging neutrons.

44. Find the threshold energy that the incident neutron must have to produce the reaction:

$$^{1}_{0}n + ^{4}_{2}He \longrightarrow ^{2}_{1}H + ^{3}_{1}H$$

45. When $^{18}O$ is struck by a proton, $^{18}F$ and another particle are produced. (a) What is the other particle? (b) This reaction has a $Q$ value of $-2.453$ MeV, and the atomic mass of $^{18}O$ is 17.999 160 u. What is the atomic mass of $^{18}F$?

## Section 29.7 Medical Applications of Radiation

46. In terms of biological damage, how many rad of heavy ions are equivalent to 100 rad of x-rays?

47. A person whose mass is 75.0 kg is exposed to a whole-body dose of 25.0 rad. How many joules of energy are deposited in the person's body?

48. A 200-rad dose of radiation is administered to a patient in an effort to combat a cancerous growth. Assuming all of the energy deposited is absorbed by the growth, (a) calculate the amount of energy delivered per unit mass. (b) Assuming the growth has a mass of 0.25 kg and a specific heat equal to that of water, calculate its temperature rise.

49. A "clever" technician decides to heat some water for his coffee with an x-ray machine. If the machine produces 10 rad/s, how long will it take to raise the temperature of a cup of water by 50°C? Ignore heat losses during this time.

50. An x-ray technician works 5 days per week, 50 weeks per year. Assume that the technician takes an average of eight x-rays per day and receives a dose of 5.0 rem/yr as a result. (a) Estimate the dose in rem per x-ray taken. (b) How does this result compare with the amount of low-level background radiation the technician is exposed to?

51. web A patient swallows a radiopharmaceutical tagged with phosphorus-32 ($^{32}_{15}P$), a $\beta^-$ emitter with a half-life of 14.3 days. The average kinetic energy of the emitted electrons is 700 keV. If the initial activity of the sample is 1.31 MBq, determine (a) the number of electrons emitted in a 10-day period, (b) the total energy deposited in the body during the 10 days, and (c) the absorbed dose if the electrons are completely absorbed in 100 g of tissue.

52. A particular radioactive source produces 100 mrad of 2-MeV gamma rays per hour at a distance of 1.0 m. (a) How long could a person stand at this distance before accumulating an intolerable dose of 1 rem? (b) Assuming the gamma radiation is emitted uniformly in all directions, at what distance would a person receive a dose of 10 mrad/h from this source?

## ADDITIONAL PROBLEMS

53. A 200.0-mCi sample of a radioactive isotope is purchased by a medical supply house. If the sample has a half-life of 14.0 days, how long will it keep before its activity is reduced to 20.0 mCi?

54. One method for producing neutrons for experimental use is to bombard $^{7}_{3}Li$ with protons. The neutrons are emitted according to the following reaction:

$$^{1}_{1}H + ^{7}_{3}Li \longrightarrow ^{7}_{4}Be + ^{1}_{0}n$$

What is the minimum kinetic energy the incident proton must have if this reaction is to occur?

55. Deuterons that have been accelerated are used to bombard other deuterium nuclei, resulting in the reaction

$$^{2}_{1}H + ^{2}_{1}H \longrightarrow ^{3}_{2}He + ^{1}_{0}n$$

Does this reaction require a threshold energy? If so, what is its value?

56. Consider a radioactive sample. Determine the ratio of the number of atoms decaying during the first half of its half-life to the number of atoms decaying during the second half of its half-life.

57. A by-product of some fission reactors is the isotope $^{239}_{94}Pu$, an alpha emitter having a half-life of 24 120 yr:

$$^{239}_{94}Pu \longrightarrow ^{235}_{92}U + \alpha$$

Consider a sample of 1.00 kg of pure $^{239}_{94}Pu$ at $t = 0$. Calculate (a) the number of $^{239}_{94}Pu$ nuclei present at $t = 0$ and (b) the initial activity in the sample. (c) How long does the sample have to be stored if a "safe" activity level is 0.100 Bq?

58. (a) Find the radius of the $^{12}_{6}C$ nucleus. (b) Find the force of repulsion between a proton at the surface of a $^{12}_{6}C$ nucleus and the remaining five protons. (c) How much work (in MeV) has to be done to overcome this electrostatic repulsion in order to put the last proton into the nucleus? (d) Repeat (a), (b), and (c) for $^{238}_{92}U$.

59. In a piece of rock from the Moon, the $^{87}Rb$ content is assayed to be $1.82 \times 10^{10}$ atoms per gram of material, and the $^{87}Sr$ content is found to be $1.07 \times 10^{9}$ atoms per gram. (The relevant decay is $^{87}Rb \rightarrow ^{87}Sr + e^-$. The half-life of the decay is $4.8 \times 10^{10}$ yr.) (a) Determine the age of the rock. (b) Could the material in the rock actually be much older? What assumption is implicit in using the radioactive dating method?

60. Many radioisotopes have important industrial, medical, and research applications. One of these is $^{60}Co$, which has a half-life of 5.2 yr and decays by the emission of a beta particle (energy 0.31 MeV) and two gamma photons (energies 1.17 MeV and 1.33 MeV). A scientist wishes to prepare a $^{60}Co$ sealed source that will have an activity of at least 10 Ci after 30 months of use. What is the minimum initial mass of $^{60}Co$ required?

61. A medical laboratory stock solution is prepared with an initial activity due to $^{24}Na$ of 2.5 mCi/mL, and 10.0 mL of the stock solution is diluted at $t_0 = 0$ to a working solution whose total volume is 250 mL. After 48 h, a 5.0-mL sample of the working solution is monitored with a counter. What is the measured activity? (Note that 1 mL = 1 milliliter.)

62. A theory of nuclear astrophysics is that all the heavy elements like uranium are formed in supernova explosions of massive stars, which immediately release the elements into space. If we assume that at the time of explosion there

were equal amounts of $^{235}U$ and $^{238}U$, how long ago were the elements that formed our Earth released, given that the present $^{235}U/^{238}U$ ratio is 0.007? (The half-lives of $^{235}U$ and $^{238}U$ are $0.70 \times 10^9$ yr and $4.47 \times 10^9$ yr, respectively.)

63. A fission reactor is hit by a nuclear weapon, causing $5.0 \times 10^6$ Ci of $^{90}Sr$ ($T_{1/2} = 28.7$ yr) to evaporate into the air. The $^{90}Sr$ falls out over an area of $10^4$ km$^2$. How long will it take the activity of the $^{90}Sr$ to reach the agriculturally "safe" level of 2.0 $\mu$Ci/m$^2$?

64. After the sudden release of radioactivity from the Chernobyl nuclear reactor accident in 1986, the radioactivity of milk in Poland rose to 2 000 Bq/L due to iodine-131, with a half-life of 8.04 days. Radioactive iodine is particularly hazardous, because the thyroid gland concentrates iodine. The Chernobyl accident caused a measurable increase in thyroid cancers among children in Belarus. (a) For comparison, find the activity of milk due to potassium. Assume that 1 L of milk contains 2.00 g of potassium, of which 0.011 7% is the isotope $^{40}K$ which has a half-life of $1.28 \times 10^9$ yr. (b) After what time would the activity due to iodine fall below that due to potassium?

65. During the manufacture of a steel engine component, radioactive iron ($^{59}Fe$) is included in the total mass of 0.20 kg. The component is placed in a test engine when the activity due to the isotope is 20.0 $\mu$Ci. After a 1 000-h test period, oil is removed from the engine and found to contain enough $^{59}Fe$ to produce 800 disintegrations/min per liter of oil. The total volume of oil in the engine is 6.5 L. Calculate the total mass worn from the engine component per hour of operation. (The half-life for $^{59}Fe$ is 45.1 days.)

66. After determining that the Sun has existed for hundreds of millions of years, but before the discovery of nuclear physics, scientists could not explain why the Sun has continued to burn for such a long time. For example, if it were a coal fire, it would have burned up in about 3 000 yr. Assume that the Sun, whose mass is 1.99 $10^{30}$ kg, originally consisted entirely of hydrogen and that its total power output is $3.76 \times 10^{26}$ W. (a) If the energy-generating mechanism of the Sun is the transforming of hydrogen into helium via the net reaction

$$4\,{}^1_1H + 2e^- \longrightarrow {}^4_2He + 2\nu + \gamma$$

calculate the energy (in joules) given off by this reaction. (b) Determine how many hydrogen atoms constitute the Sun. Take the mass of one hydrogen atom to be $1.67 \times 10^{-27}$ kg. (c) Assuming that the total power output remains constant, after what time will all the hydrogen be converted into helium, making the Sun die? The actual projected lifetime of the Sun is about 10 billion years, because only the hydrogen in a relatively small core is available as a fuel. Only in the core are temperatures and densities high enough for the fusion reaction to be self-sustaining.

## GROUP ACTIVITIES

**G.1** This experiment will take a little longer to do than most that we have suggested, but the time spent is worthwhile to help you understand the concept of half-life. Obtain a box of sugar cubes and with a pencil make a mark on one side of each of about 200 cubes. Each of these cubes will represent the nucleus of a radioactive substance. Thus, at $t = 0$, you have 200 undecayed nuclei. Now, put the 200 marked cubes in a box and roll them out on a table, just as you would roll dice. Next, count and remove any cubes that have landed marked-side up. These cubes represent nuclei that emitted radiation during the roll. They are no longer radioactive and thus do not participate in the rest of the action. Record the number of undecayed cubes remaining as the number of undecayed nuclei at $t = 1$ roll.

Continue rolling, counting, and removing until you have completed 12 to 15 rolls. By then, you should have only a few cubes remaining. Plot a graph of undecayed cubes versus the roll number and from this determine the "half-roll" of the cubes.

**G.2** Use a nail to punch a hole in the bottom of a large tin can. Hold the can beneath a faucet and adjust the water flow from the faucet to a fine constant stream. Although water flows from the hole at the bottom, you will note that the level of the water in the can rises. As it does so, however, the flow of water leaving the can increases due to increased water pressure caused by the greater depth of water. Unless the flow of water is too great, an equilibrium point will be reached at which the amount of water flowing out of the can each second exactly equals the amount flowing in each second. When this happens, the level of water in the can is constant. As noted in the text, carbon-14 is continually being produced in the atmosphere and is also continually disappearing as it decays into nitrogen. What is the analogy between water entering the can, remaining in the can, and flowing out of the can and the behavior of carbon-14 in the atmosphere?

**G.3** (a) Describe what happens to the number of protons and neutrons in a nucleus when the nucleus undergoes alpha decay. (b) Repeat for a beta decay. (c) In alpha decay how does the mass of the parent nuclei compare to the sum of the masses of the daughter plus alpha? Is the parent mass greater, less, equal to, or no relation to the sum of the daughter plus alpha? (d) Use the decay relationship ${}^{230}_{90}Th \rightarrow {}^{226}_{98}Ra + {}^4_2He$ to check your answer to part (c). The mass of $^{230}Th$ is 230.033 131 u and the mass of $^{226}Ra$ is 226.025 406 u.

**G.4** (a) Assume the energy released in a beta decay is 0.150 MeV. If the daughter nucleus is found to have an energy of 0.100 MeV, then for energy conservation, is it true that the beta must have an energy of 0.050 MeV? (b) If you find that the beta actually has only an energy of 0.025 MeV, what happens to the remaining energy? (c) Correct the following beta decay reaction to agree with your answers to the above: ${}^{14}_6C \rightarrow {}^{14}_7N + e^-$.

**G.5** (a) In a radioactive decay process, which of the following are reduced by one half after one half-life, (i) the number of nuclei, (ii) the activity, or (iii) the decay constant? (b) If the half-life of a given material is 14 days, what is the ratio of the number of particles remaining after 10 days to the original number? What is the ratio of the activity after 10 days to the initial activity?

# 30

# Nuclear Energy and Elementary Particles

## Chapter Outline



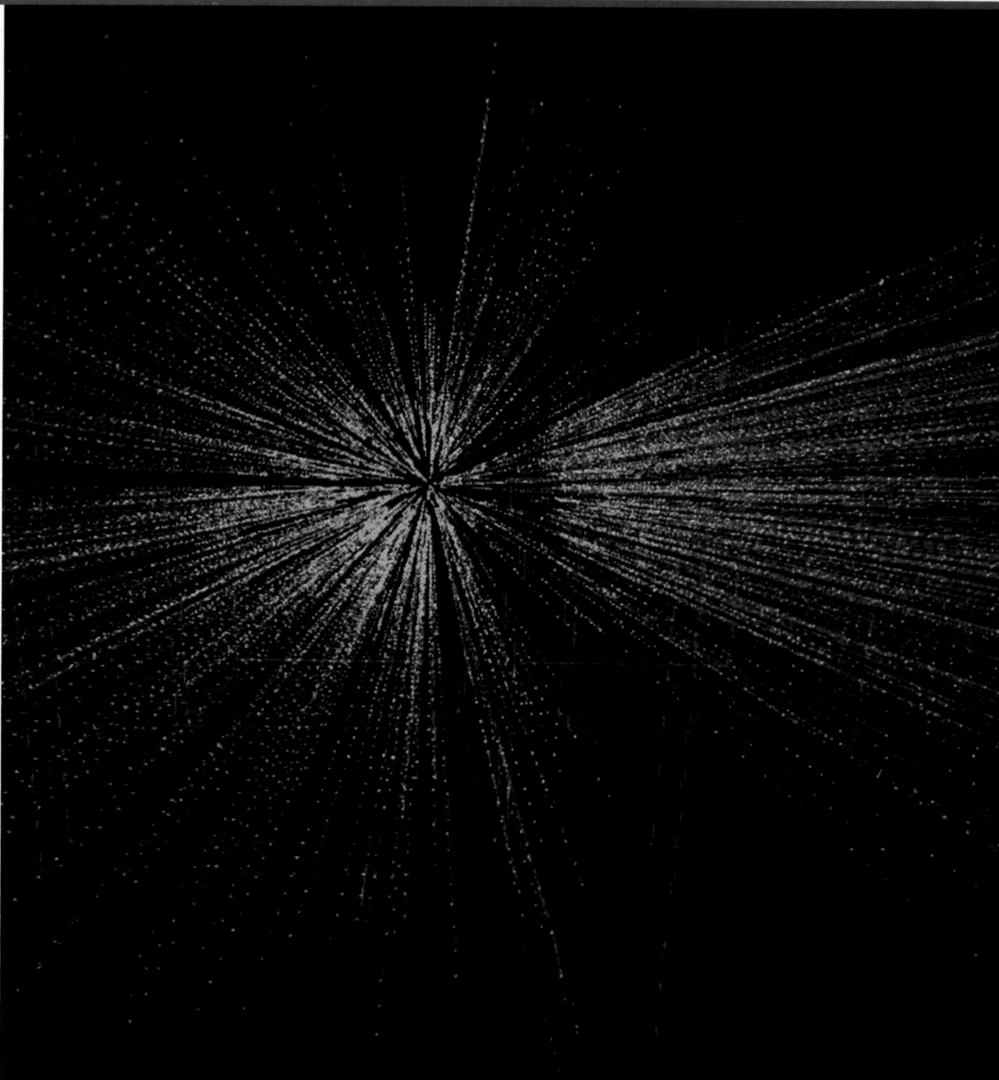

**In this image from the NA49 experiment at CERN, hundreds of subatomic particles are created in the collision of high-energy lead nuclei with a lead target. The aim of the experiment was to create extreme densities in matter to break down the strong force that normally locks quarks within protons and neutrons.** *(Courtesy of CERN)*

In this concluding chapter we discuss the two means by which energy can be derived from nuclear reactions. These two techniques are fission, in which a nucleus of large mass number splits, or fissions, into two smaller nuclei, and fusion, in which two light nuclei fuse to form a heavier nucleus. In either case, large amounts of energy are released that can be used destructively, through bombs, or constructively, through the production of electric power.

We end our study of physics by examining the known subatomic particles and the fundamental interactions that govern their behavior. We also discuss the current theory of elementary particles, which states that all matter in nature is constructed from only two families of particles, quarks and leptons. Finally, we describe how clarifications of such models might help us understand the evolution of the Universe.

## 30.1 NUCLEAR FISSION

**Nuclear fission** occurs when a heavy nucleus, such as $^{235}U$, splits, or fissions, into two smaller nuclei. In such a reaction, **the total mass of the products is less than the original mass of the heavy nucleus.**

**Nuclear fission** was first observed in 1939 by Otto Hahn and Fritz Strassman, following some basic studies by Fermi. After bombarding uranium ($Z = 92$) with neutrons, Hahn and Strassman discovered among the reaction products two medium-mass elements, barium and lanthanum. Shortly thereafter, Lisa Meitner and Otto Frisch explained what had happened. The uranium nucleus had split into two nearly equal fragments after absorbing a neutron. Such an occurrence was of considerable interest to physicists attempting to understand the nucleus, but it was to have even more far-reaching consequences. Measurements showed that about 200 MeV of energy is released in each fission event, and this fact was to affect the course of human history.

The fission of $^{235}U$ by slow (low energy) neutrons can be represented by the equation

$$^{1}_{0}n + ^{235}_{92}U \longrightarrow ^{236}_{92}U^* \longrightarrow X + Y + \text{neutrons} \qquad [30.1]$$

where $^{236}U^*$ is an intermediate state that lasts only for about $10^{-12}$ s before splitting into X and Y. The resulting nuclei, X and Y, are called **fission fragments.** Many combinations of X and Y satisfy the requirements of conservation of energy and charge. In the fission of uranium, there are about 90 different daughter nuclei that can be formed. The process also results in the production of several (typically two or three) neutrons per fission event. On the average, 2.47 neutrons are released per event.

A typical reaction of this type is

$$^{1}_{0}n + ^{235}_{92}U \longrightarrow ^{141}_{56}Ba + ^{92}_{36}Kr + 3\,^{1}_{0}n \qquad [30.2]$$

The fission fragments, barium and krypton, and the released neutrons have a great deal of kinetic energy following the fission event.

The breakup of the uranium nucleus can be likened to what happens to a drop of water when excess energy is added to it. All of the atoms in the drop have energy, but not enough to break up the drop. However, if enough energy is added to set the drop vibrating, it will undergo elongation and compression until the amplitude of vibration becomes large enough to cause the drop to break apart. In the uranium nucleus, a similar process occurs (Fig. 30.1). The sequence of events is as follows:

Sequence of events in a nuclear fission process

1. The $^{235}U$ nucleus captures a thermal (slow-moving) neutron.
2. This capture results in the formation of $^{236}U^*$, and the excess energy of this nucleus causes it to undergo violent oscillations.
3. The $^{236}U^*$ nucleus becomes highly elongated, and the force of repulsion

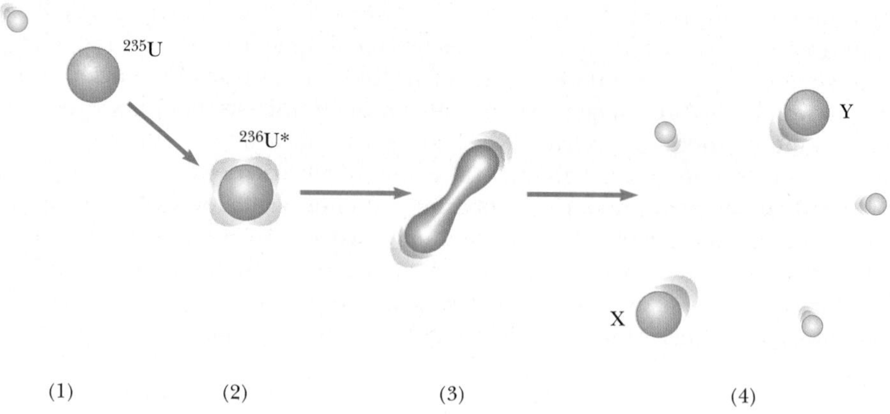


**FIGURE 30.1** The stages involved in a nuclear fission event as described by the liquid-drop model of the nucleus.

between protons in the two halves of the dumbbell shape tends to increase the distortion.

4. The nucleus splits into two fragments, emitting several neutrons in the process.

Let us estimate the disintegration energy, $Q$, released in a typical fission process. From Figure 29.4 we see that the binding energy per nucleon is about 7.2 MeV for heavy nuclei (those having a mass number of approximately 240) and about 8.2 MeV for nuclei of intermediate mass. This means that the nucleons in the fission fragments are more tightly bound and therefore have less mass than the nucleons in the original heavy nucleus. This decrease in mass per nucleon appears as released energy when fission occurs. The amount of energy released is (8.2–7.2) MeV per nucleon. Assuming a total of 240 nucleons, we find that the energy released per fission event is

$$Q = (240 \text{ nucleons})(8.2 \text{ MeV/nucleon} - 7.2 \text{ MeV/nucleon}) = 240 \text{ MeV}$$

This is indeed a very large amount of energy relative to the amount released in chemical processes. For example, the energy released in the combustion of one molecule of the octane used in gasoline engines is about one hundred-millionth the energy released in a single fission event!

## APPLYING PHYSICS 30.1

If a heavy nucleus were to fission into just two product nuclei, they would be very unstable. Why is this?

**Explanation** According to Figure 29.3, the ratio of the number of neutrons to the number of protons increases with $Z$. As a result, when a heavy nucleus splits in a fission reaction to two lighter nuclei, the lighter nuclei tend to have too many neutrons. This leads to instability, as the nucleus returns to the curve in Figure 29.3 by decay processes that reduce the number of neutrons.

## Example 30.1 The Fission of Uranium

Two other possible ways by which $^{235}U$ can undergo fission when bombarded with a neutron are (1) by the release of $^{140}Xe$ and $^{94}Sr$ as fission fragments and (2) by the release of $^{132}Sn$ and $^{101}Mo$ as fission fragments. In each case, neutrons are also released. Find the number of neutrons released in each of these events.

**Solution** By balancing mass numbers and atomic numbers, we find that these reactions can be written

$$^{1}_{0}n + ^{235}_{92}U \longrightarrow ^{140}_{54}Xe + ^{94}_{38}Sr + 2\,^{1}_{0}n$$

$$^{1}_{0}n + ^{235}_{92}U \longrightarrow ^{132}_{50}Sn + ^{101}_{42}Mo + 3\,^{1}_{0}n$$

Thus, two neutrons are released in the first event and three in the second.

**Quick Quiz 30.1** In the first atomic bomb, the energy released was equivalent to about 30 kilotons of TNT, where a ton of TNT releases an energy of $4.0 \times 10^9$ J. The amount of mass converted into energy in this event is nearest to (a) 1 $\mu$g, (b) 1 mg, (c) 1 g, (d) 1 kg, (e) 20 kilotons

## Example 30.2 The Energy Released in the Fission of $^{235}U$

Calculate the total energy released if 1.00 kg of $^{235}U$ undergoes fission, taking the disintegration energy per event to be $Q$ = 208 MeV (a more accurate value than the estimate given previously).

**Solution** We need to know the number of nuclei in 1.00 kg of uranium. Because $A$ = 235, the number of nuclei is

$$N = \left(\frac{6.02 \times 10^{23} \text{ nuclei/mol}}{235 \text{ g/mol}}\right)(1.00 \times 10^3 \text{ g}) = 2.56 \times 10^{24} \text{ nuclei}$$

Hence the disintegration energy is

$$E = NQ = (2.56 \times 10^{24} \text{ nuclei})\left(208 \frac{\text{MeV}}{\text{nucleus}}\right) = 5.32 \times 10^{26} \text{ MeV}$$

Because 1 MeV is equivalent to $4.45 \times 10^{-20}$ kWh, $E = 2.37 \times 10^7$ kWh. This is enough energy to keep a 100-W lightbulb burning for about 30 000 years.

## 30.2 NUCLEAR REACTORS

We have seen that neutrons are emitted when $^{235}U$ undergoes fission. These neutrons can in turn trigger other nuclei to undergo fission, with the possibility of a chain reaction (Fig. 30.2). Calculations show that if the chain reaction is not controlled (that is, if it does not proceed slowly), it can result in a violent explosion, with the release of an enormous amount of energy, even from only 1 g of $^{235}U$. If the energy in 1 kg of $^{235}U$ were released, it would equal that released by the detonation of about 20 000 tons of TNT! This, of course, is the principle behind the first nuclear bomb, an uncontrolled fission reaction.

A nuclear reactor is a system designed to maintain what is called a **self-sustained chain reaction.** This important process was first achieved in 1942 by a group led by Fermi at the University of Chicago, with natural uranium as the fuel. Most reactors in operation today also use uranium as fuel. Natural uranium contains only about 0.7% of the $^{235}U$ isotope, with the remaining 99.3% being the

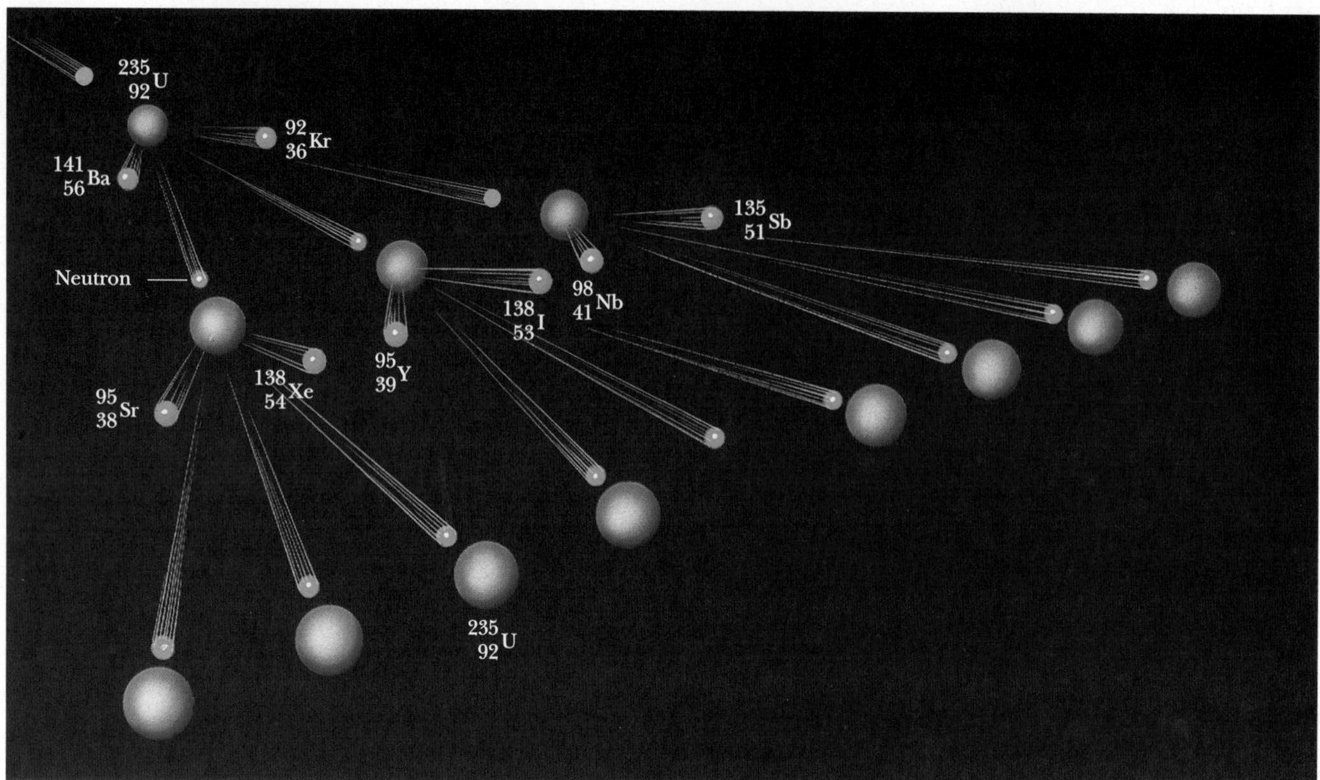


**FIGURE 30.2** A nuclear chain reaction initiated by capture of a neutron.

$^{238}$U isotope. This is important to the operation of a reactor because $^{238}$U almost never undergoes fission. Instead, it tends to absorb neutrons, producing neptunium and plutonium. For this reason, reactor fuels must be artificially enriched so that they contain several percent of the $^{235}$U isotope.

Earlier, we mentioned that an average of about 2.5 neutrons are emitted in each fission event of $^{235}$U. In order to achieve a self-sustained chain reaction, one of these neutrons must be captured by another $^{235}$U nucleus and cause it to un-

Painting of the world's first reactor. Because of wartime secrecy, there are no photographs of the completed reactor. The reactor was composed of layers of graphite interspersed with uranium. A self-sustained chain reaction was first achieved on December 2, 1942. Word of the success was telephoned immediately to Washington with this message: "The Italian navigator has landed in the New World and found the natives very friendly." The historic event took place in an improvised laboratory in the racquet court under the west stands of the University of Chicago's Stagg Field; the Italian navigator was Fermi. *(Courtesy of Chicago Historical Society)*

dergo fission. A useful parameter for describing the level of reactor operation is the **reproduction constant *K*, defined as the average number of neutrons from each fission event that will cause another event.** As we have seen, $K$ can have a maximum value of 2.5 in the fission of uranium. However, in practice $K$ is less than this because of several factors, which we shall discuss in a moment.

A self-sustained chain reaction is achieved when $K = 1$. Under this condition, the reactor is said to be **critical.** When $K$ is less than unity, the reactor is subcritical and the reaction dies out. When $K$ is greater than unity, the reactor is said to be supercritical, and a runaway reaction occurs. In a nuclear reactor used to furnish power to a utility company, it is necessary to maintain a $K$ value close to unity.

The basic design of a nuclear reactor is shown in Figure 30.3. The fuel elements consist of enriched uranium. Now let us look at the function of the remaining parts of the reactor and some aspects of its design.

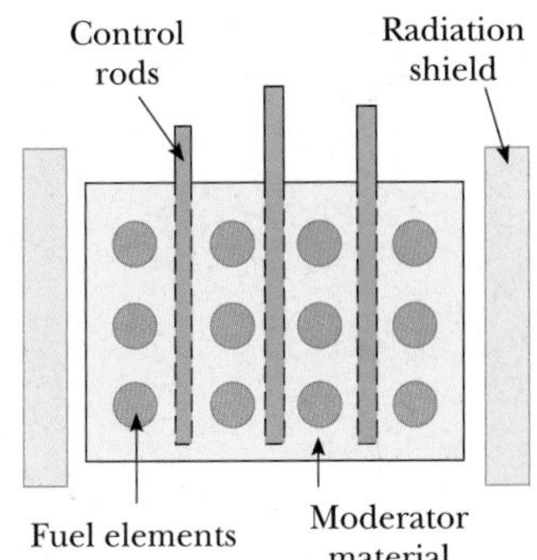


**FIGURE 30.3** Cross section of a reactor core surrounded by a radiation shield.

## NEUTRON LEAKAGE

In any reactor, a fraction of the neutrons produced in fission will leak out of the core before inducing other fission events. If the fraction leaking out is too large, the reactor will not operate. The percentage lost is large if the reactor is very small because leakage is a function of the ratio of surface area to volume. Therefore, a critical requirement of reactor design is choosing the correct surface-area-to-volume ratio so that a sustained reaction can be achieved.

## REGULATING NEUTRON ENERGIES

The neutrons released in fission events are very energetic, with kinetic energies of about 2 MeV. It is found that slow neutrons are far more likely than fast neutrons to produce fission events in $^{235}$U. Furthermore, $^{238}$U does not absorb slow neutrons. Therefore, in order for the chain reaction to continue, the neutrons must be slowed down. This is accomplished by surrounding the fuel with a **moderator** substance.

To understand how neutrons are slowed down, consider a collision between a light object and a very massive one. In such an event, the light object rebounds from the collision with most of its original kinetic energy. However, if the collision is between objects with nearly the same masses, the incoming projectile transfers a large percentage of its kinetic energy to the target. In the first nuclear reactor ever constructed, Fermi placed bricks of graphite (carbon) between the fuel elements. Carbon nuclei are about 12 times more massive than neutrons, but after about 100 collisions with carbon nuclei, a neutron is slowed sufficiently to increase its likelihood of fission with $^{235}$U. In this design the carbon is the moderator; most modern reactors use heavy water ($D_2O$) as the moderator.

## NEUTRON CAPTURE

In the process of being slowed down, the neutrons may be captured by nuclei that do not undergo fission. The most common event of this type is neutron capture by $^{238}$U. The probability of neutron capture by $^{238}$U is very high when the neutrons have high kinetic energies and very low when they have low kinetic energies. Thus, the slowing down of the neutrons by the moderator serves the dual purpose of making them available for reaction with $^{235}$U and decreasing their chances of being captured by $^{238}$U.

## CONTROL OF POWER LEVEL

**APPLICATION**

NUCLEAR REACTOR DESIGN

It is possible for a reactor to reach the critical stage ($K = 1$) after all neutron losses described previously are minimized. However, a method of control is needed to adjust $K$ to a value near unity. If $K$ were to rise above this value, the heat produced

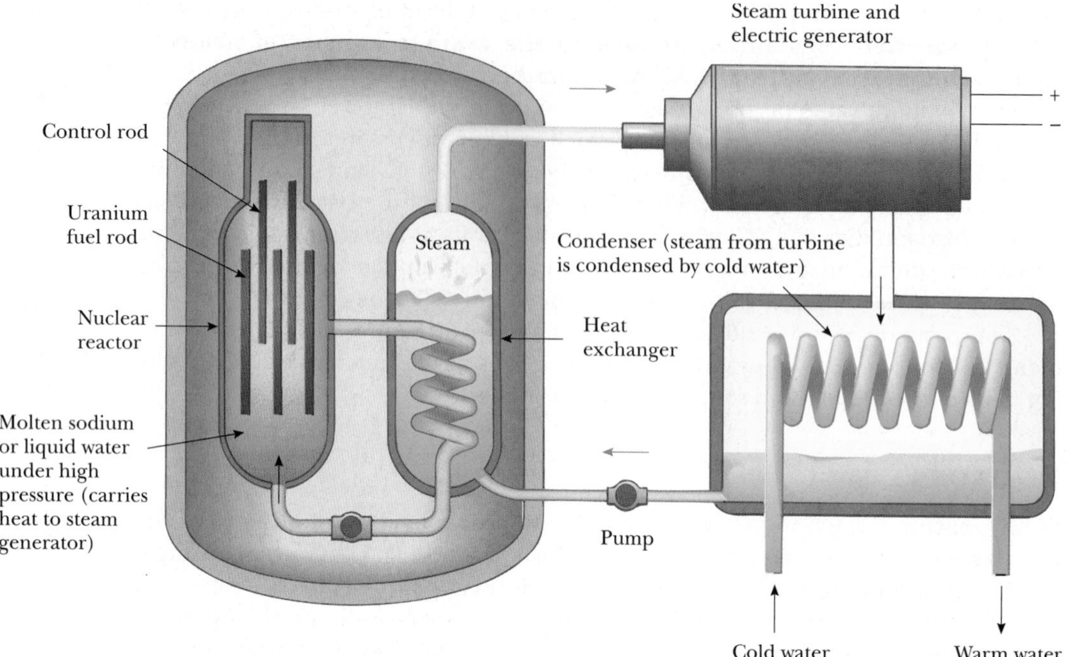


**FIGURE 30.4** Main components of a pressurized-water reactor.

in the runaway reaction would melt the reactor. To control the power level, control rods are inserted into the reactor core (see Fig. 30.3). These rods are made of materials such as cadmium that are very efficient in absorbing neutrons. By adjusting the number and position of these control rods in the reactor core, the $K$ value can be varied and any power level within the design range of the reactor can be achieved.

A diagram of a pressurized-water reactor is shown in Figure 30.4. This type of reactor is commonly used in electric power plants in the United States. Fission events in the reactor core supply heat to the water contained in the primary (closed) system, which is maintained at high pressure to keep it from boiling. This water also serves as the moderator. The hot water is pumped through a heat exchanger, and the heat is transferred to the water contained in the secondary system. There the hot water is converted to steam, which drives a turbine-generator to create electric power. Note that the water in the secondary system is isolated from the water in the primary system in order to prevent contamination of the secondary water and steam by radioactive nuclei from the reactor core.

## REACTOR SAFETY[1]

The safety aspects of nuclear power reactors are often sensationalized by the media and misunderstood by the public. The 1979 near-disaster of Three Mile Island in Pennsylvania and the accident at the Chernobyl reactor in the Ukraine rightfully focused attention on reactor safety. Yet the safety record in the United States is enviable. The records show no fatalities attributed to commercial nuclear power generation in the history of the United States nuclear industry.

Commercial reactors achieve safety through careful design and rigid operating procedures. Radiation exposure and the potential health risks associated with such exposure are controlled by three layers of containment. The fuel and radioactive fission products are contained inside the reactor vessel. Should this vessel rupture, the reactor building acts as a second containment structure to prevent

[1] The authors are grateful to Professor Gene Skluzacek of the University of Nebraska at Omaha for rewriting this section on reactor safety.

radioactive material from contaminating the environment. Finally, the reactor facilities must be in a remote location to protect the general public from exposure should radiation escape the reactor building.

According to the Oak Ridge National Laboratory Review, "the health risk of living within 8 km (5 miles) of a nuclear reactor for 50 years is no greater than the risk of smoking 1.4 cigarettes, drinking 0.5 liters of wine, traveling 240 km by car, flying 9 600 km by jet, or having one chest x-ray in a hospital. Each of these activities is estimated to increase a person's chances of dying in any given year by one in a million."

Another potential danger in nuclear reactor operations is the possibility that the water flow could be interrupted. Even if the nuclear fission chain reaction were stopped immediately, residual heat could build up in the reactor to the point of melting the fuel elements. The molten reactor core would melt to the bottom of the reactor vessel and conceivably melt its way into the ground below—the so-called "China syndrome." Although it might appear that this deep underground burial site would be an ideal safe haven for a radioactive blob, there would be danger of a steam explosion should the molten mass encounter water. This nonnuclear explosion could spread radioactive material to the areas surrounding the power plant. To prevent such an unlikely chain of events, nuclear reactors are designed with emergency core-cooling systems, requiring no power, that automatically flood the reactor with water in the event of coolant loss. The emergency cooling water moderates heat buildup in the core, which in turn prevents the occurrence of melting.

A continuing concern in nuclear fission reactors is the safe disposal of radioactive material when the reactor core is replaced. This waste material contains long-lived, highly radioactive isotopes and must be stored over long periods of time in such a way that there is no chance of environmental contamination. At present, sealing radioactive wastes in waterproof containers and burying them in deep salt mines seems to be the most promising solution.

Transportation of reactor fuel and reactor wastes poses additional safety risks. However, neither the waste nor the fuel of nuclear power reactors can be used to construct a nuclear bomb.

Accidents during transportation of nuclear fuel could expose the public to harmful levels of radiation. The Department of Energy requires stringent crash tests on all containers used to transport nuclear materials. Container manufacturers must demonstrate that their containers will not rupture even in high-speed collisions.

The safety issues associated with nuclear power reactors are complex and often emotional. All sources of energy have associated risks. In each case, we must weigh the risks against the benefits and the availability of the energy source.

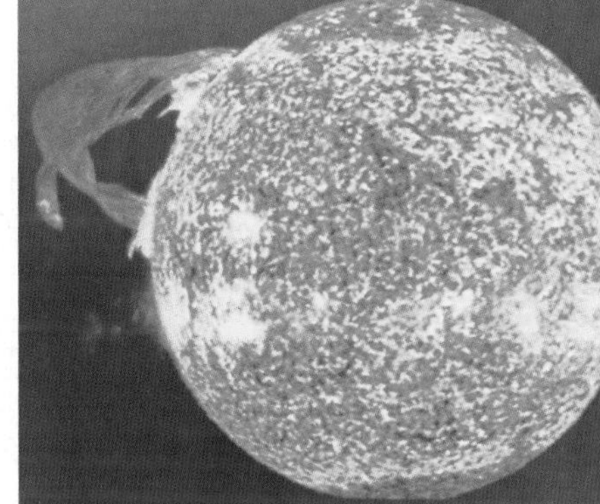

This photograph of the Sun, taken on December 19, 1973, during the third and final manned Skylab mission, shows one of the most spectacular solar flares ever recorded, spanning more than 588 000 km (365 000 mi) across the solar surface. The last picture, taken some 17 hours earlier, showed this feature as a large quiescent prominence on the eastern side of the Sun. The flare gives the distinct impression of a twisted sheet in the process of unwinding itself. In this photograph the solar poles are distinguished by a relative absence of granulation and a much darker tone than the central portions of the disk. Several active regions are seen on the eastern side of the disk. The photograph was taken in the light of ionized helium by the extreme ultraviolet spectroheliograph instrument of the U.S. Naval Research Laboratory. *(NASA)*

## 30.3 NUCLEAR FUSION

Figure 29.4 shows that the binding energy for light nuclei (those having a mass number lower than 20) is much smaller than the binding energy for heavier nuclei. This suggests a possible process that is the reverse of fission. **When two light nuclei combine to form a heavier nucleus, the process is called nuclear fusion.** Because the mass of the final nucleus is less than the masses of the original nuclei, there is a loss of mass accompanied by a release of energy. Although fusion power plants have not yet been developed, a worldwide effort is under way to harness the energy from fusion reactions in the laboratory. Later we shall discuss the possibilities and advantages of this process for generating electric power.

### FUSION IN THE SUN

All stars generate their energy through fusion processes. About 90% of the stars, including the Sun, fuse hydrogen, whereas some older stars fuse helium or other heavier elements. Stars are born in regions of space containing vast clouds of dust and gas. Recent mathematical models of these clouds indicate that star formation is triggered by shock waves passing through a cloud. These shock waves are similar to sonic booms and are produced by events such as the explosion of a nearby star, called a *supernova explosion*. The shock wave compresses certain regions of the cloud, causing these regions to collapse under their own gravity. As the gas falls inward toward the center, the atoms gain speed, which causes the temperature of the gas to rise. Two conditions must be met before fusion reactions in the star can sustain its energy needs: (1) The temperature must be high enough (about $10^7$ K for hydrogen) to allow the kinetic energy of the positively charged hydrogen nuclei to overcome their mutual Coulomb repulsion as they collide, and (2) the density of nuclei must be high enough to ensure a high rate of collision.

When fusion reactions occur at the core of a star, the energy liberated eventually becomes sufficient to prevent further collapse of the star under its own gravity. The star then continues to live out the remainder of its life under a balance between the inward force of gravity pulling it toward collapse and the outward force due to thermal effects and radiation pressure.

The **proton-proton cycle** is a series of three nuclear reactions that are believed to be the stages in the liberation of energy in the Sun and other stars rich in hydrogen. An overall view of the proton-proton cycle is that four protons combine to form an alpha particle and two positrons, with the release of 25 MeV of energy in the process.

The specific steps in the proton-proton cycle are

$$\begin{aligned} {}^1_1\text{H} + {}^1_1\text{H} &\longrightarrow {}^2_1\text{H} + e^+ + \nu \\ {}^1_1\text{H} + {}^2_1\text{H} &\longrightarrow {}^3_2\text{He} + \gamma \end{aligned} \qquad [30.3]$$

This second reaction is followed by either hydrogen-helium fusion or helium-helium fusion:

$${}^1_1\text{H} + {}^3_2\text{He} \longrightarrow {}^4_2\text{He} + e^+ + \nu$$

or

$${}^3_2\text{He} + {}^3_2\text{He} \longrightarrow {}^4_2\text{He} + {}^1_1\text{H} + {}^1_1\text{H}$$

The energy liberated is carried primarily by gamma rays, positrons, and neutrinos, as can be seen from the reactions. The gamma rays are soon absorbed by the dense gas, thus raising its temperature. The positrons combine with electrons to produce gamma rays, which in turn are also absorbed by the gas within a few centimeters. The neutrinos, however, almost never interact with matter; hence they escape from the star, carrying about 2% of the generated energy with them. These energy-liberating fusion reactions are called **thermonuclear fusion reactions.** The hydrogen (fusion) bomb, first exploded in 1952, is an example of an uncontrolled thermonuclear fusion reaction.

## FUSION REACTORS

**APPLICATION**

FUSION REACTORS

The enormous amount of energy released in fusion reactions suggests the possibility of harnessing this energy for useful purposes on Earth. A great deal of effort is focused on developing a sustained and controllable thermonuclear reactor—a fusion power reactor. Controlled fusion is often called the ultimate energy source because of the availability of its fuel source: water. For example, if deuterium were used as the fuel, 0.06 g of it could be extracted from 1 gal of water at a cost of about four cents. Such rates would make the fuel costs of even an inefficient reactor almost insignificant. An additional advantage of fusion reactors is that comparatively few radioactive by-products are formed. As noted in Equation 30.3, the end product of the fusion of hydrogen nuclei is safe, nonradioactive helium. Unfortunately, a thermonuclear reactor that can deliver a net power output over a reasonable time interval is not yet a reality, and many difficulties must be solved before a successful device is constructed.

We have seen that the Sun's energy is based, in part, on a set of reactions in which ordinary hydrogen is converted to helium. Unfortunately, the proton-proton interaction is not suitable for use in a fusion reactor because the event requires very high pressures and densities. The process works in the Sun only because of the extremely high density of protons in the Sun's interior. In fact, even at the densities and temperatures that exist at the center of the Sun, the average proton takes 14 billion years to react!

The fusion reactions that appear most promising in the construction of a fusion power reactor involve deuterium and tritium, which are isotopes of hydrogen. These reactions are

$$\begin{aligned} {}^2_1\text{H} + {}^2_1\text{H} &\longrightarrow {}^3_2\text{He} + {}^1_0\text{n} \qquad Q = 3.27 \text{ MeV} \\ {}^2_1\text{H} + {}^2_1\text{H} &\longrightarrow {}^3_1\text{H} + {}^1_1\text{H} \qquad Q = 4.03 \text{ MeV} \\ {}^2_1\text{H} + {}^3_1\text{H} &\longrightarrow {}^4_2\text{He} + {}^1_0\text{n} \qquad Q = 17.59 \text{ MeV} \end{aligned} \tag{30.4}$$

where the $Q$ values refer to the amount of energy released per reaction. As noted earlier, deuterium is available in almost unlimited quantities from our lakes and oceans and is very inexpensive to extract. Tritium, however, is radioactive ($T_{1/2} = 12.3$ yr) and undergoes beta decay to $^3$He. For this reason, tritium does not occur naturally to any great extent and must be artificially produced.

One of the major problems in obtaining energy from nuclear fusion is the fact that the Coulomb repulsion force between two charged nuclei must be overcome before they can fuse. The fundamental challenge is to give the two nuclei enough kinetic energy to overcome this repulsive force. This can be accomplished by heating the fuel to extremely high temperatures (about $10^8$ K, far greater than the interior temperature of the Sun). As you might expect, such high temperatures are not easy to obtain in a laboratory or a power plant. At these high temperatures the atoms are ionized, and the system consists of a collection of electrons and nuclei, commonly referred to as a *plasma*.

In addition to the high temperature requirements, two other critical factors determine whether or not a thermonuclear reactor will be successful: **plasma ion density,** $n$, and **plasma confinement time,** $\tau$, the time the interacting ions are maintained at a temperature equal to or greater than that required for the reaction to proceed successfully. The density and confinement time must both be large enough to ensure that more fusion energy will be released than is required to heat the plasma.

**Lawson's criterion** states that a net power output in a fusion reactor is possible under the following conditions:

$$\begin{aligned} n\tau &\geq 10^{14} \text{ s/cm}^3 \qquad \text{Deuterium-tritium interaction} \\ n\tau &\geq 10^{16} \text{ s/cm}^3 \qquad \text{Deuterium-deuterium interaction} \end{aligned} \tag{30.5}$$

Lawson's criterion

The problem of plasma confinement time has yet to be solved. How can a plasma be confined at a temperature of $10^8$ K for times on the order of 1 s? The basic plasma-confinement technique under investigation is discussed following Example 30.3.

## Example 30.3 The Deuterium-Deuterium Reaction

Find the energy released in the deuterium-deuterium reaction

$${}^2_1\text{H} + {}^2_1\text{H} \longrightarrow {}^3_1\text{H} + {}^1_1\text{H}$$

**Solution** The mass of the ${}^2_1$H atom is 2.014 102 u. Thus, the total mass before the reaction is 4.028 204 u. After the reaction, the sum of the masses is equal to 3.016 049 u + 1.007 825 u = 4.023 874 u. Thus, the excess mass is 0.004 33 u. In energy units, this is equivalent to 4.03 MeV.

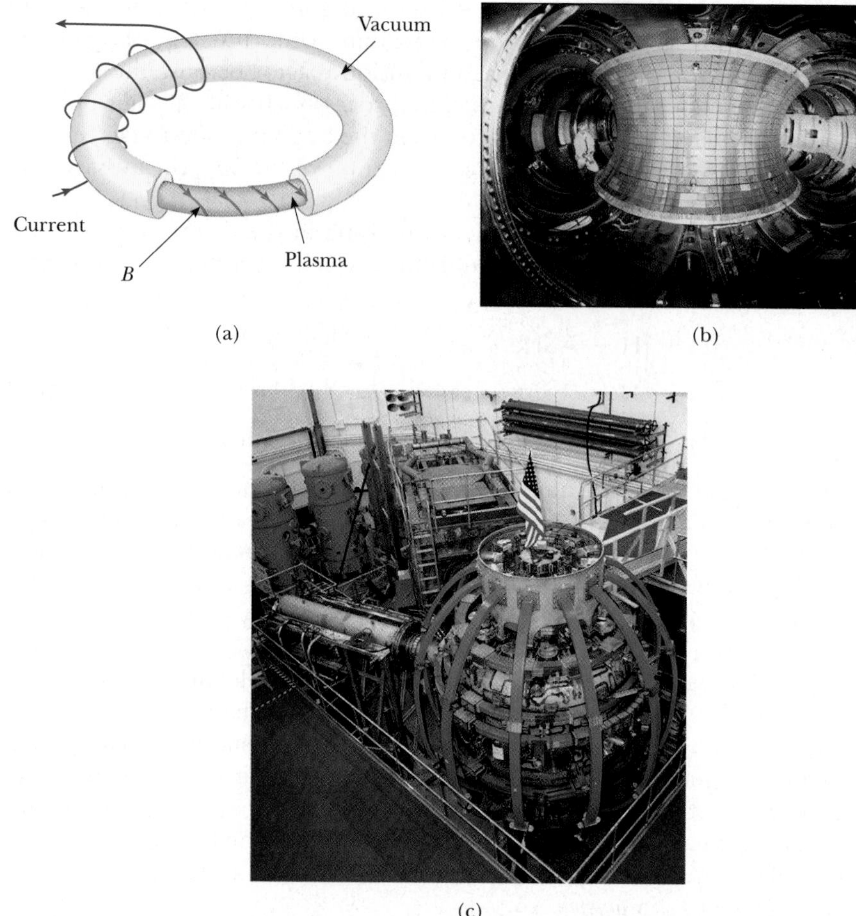


**FIGURE 30.5** (a) Diagram of a tokamak used in the magnetic confinement scheme. The plasma is trapped within the spiraling magnetic field lines as shown. (b) Interior view of the Tokamak Fusion Test Reactor (TFTR) vacuum vessel located at the Princeton Plasma Physics Laboratory, Princeton University, New Jersey. *(Courtesy of Princeton Plasma Physics Laboratory)* (c) The National Spherical Torus Experiment (NSTX) that began operation in March 1999. *(Courtesy of Princeton University)*

## MAGNETIC FIELD CONFINEMENT

Most fusion experiments use magnetic field confinement to contain a plasma. One device, called a **tokamak,** has a doughnut-shaped geometry (a toroid), as shown in Figure 30.5a. This device, first developed in the former Soviet Union, uses a combination of two magnetic fields to confine the plasma inside the doughnut. A strong magnetic field is produced by the current in the windings, and a weaker magnetic field is produced by the current in the toroid. The resulting magnetic field lines are helical, as in Figure 30.5a. In this configuration, the field lines spiral around the plasma and prevent it from touching the walls of the vacuum chamber.

In order for the plasma to reach ignition temperature, some form of auxiliary heating is necessary. A successful and efficient auxiliary heating technique that has been used recently is the injection of a beam of energetic neutral particles into the plasma.

**Webnote 30.1**
Want to operate a tokamak on your own? Do it here with a Java applet.
*http://ippex.pppl.gov/tokamak/default.htm*

When it was in operation, the Tokamak Fusion Test Reactor (TFTR) at Princeton reported central ion temperatures of 510 million degrees Celsius, more than 30 times hotter than the center of the Sun. TFTR $n\tau$ values for the D-T reaction were well above $10^{13}$ s/cm$^3$ and close to the value required by Lawson's criterion. In 1991, reaction rates of $6 \times 10^{17}$ D-T fusions per second were reached in the JET tokamak at Abington, England.

One of the new generations of fusion experiments is the National Spherical Torus Experiment (NSTX) shown in Figure 30.5c. Rather than the donut-shaped plasma of a tokamak, the NSTX produces a spherical plasma that has a hole through its center. The major advantage of the spherical configuration is its ability to confine the plasma at a higher pressure in a given magnetic field. This approach could lead to the development of smaller and more economical fusion reactors.

An international collaboration involving four major fusion programs is currently working on building a fusion reactor called ITER (International Thermonuclear Experimental Reactor). This facility will address the remaining technological and scientific issues concerning the feasibility of fusion power. The design is completed, and site and construction negotiations are under way. If the planned device works as expected, the Lawson number for ITER will be about six times greater than the current record holder, the JT-60U tokamak in Japan.

**Webnote 30.2**

One means of creating high plasma-ion density is through magnetic field confinement. Another approach is to use laser fusion. Find out how laser fusion can produce electricity at *http://www.llnl.gov/nif/library/ife.pdf* (*Note:* You will need Acrobat Reader to view this file.)

## 30.4 Elementary Particles

The word "atom" comes from the Greek word *atomos,* meaning "indivisible." At one time atoms were thought to be the indivisible constituents of matter; that is, they were regarded as elementary particles. Discoveries in the early part of the 20th century revealed that the atom is not elementary, but has as its constituents protons, neutrons, and electrons. Until 1932 physicists viewed these three constituent particles as elementary because, with the exception of the free neutron, they are very stable. The theory soon fell apart, however, and beginning in 1937, many new particles were discovered in experiments involving high-energy collisions between known particles. These new particles are characteristically unstable and have very short half-lives, ranging between $10^{-6}$ s and $10^{-23}$ s. So far more than 300 of them have been cataloged.

Until the 1960s, physicists were bewildered by the large number and variety of subatomic particles being discovered. They wondered if the particles were like animals in a zoo or if a pattern was emerging that would provide a better understanding of the elaborate structure in the subnuclear world. In the last 30 years, physicists have made tremendous advances in our knowledge of the structure of matter by recognizing that all particles (with the exception of electrons, photons, and a few others) are made of smaller particles called *quarks.* Thus, protons and neutrons, for example, are not truly elementary but are systems of tightly bound quarks. The quark model has reduced the bewildering array of particles to a manageable number and has successfully predicted new quark combinations that were subsequently found in many experiments.

## 30.5 The Fundamental Forces in Nature

The key to understanding the properties of elementary particles is to be able to describe the forces between them. All particles in nature are subject to four fundamental forces: strong, electromagnetic, weak, and gravitational.

The **strong force** is responsible for the tight binding of quarks to form neutrons and protons and for the nuclear force, a sort of residual strong force, binding neutrons and protons into nuclei. This force represents the "glue" that holds the nucleons together and is the strongest of all the fundamental forces. It is very short-ranged and is negligible for separations greater than about $10^{-15}$ m (the approximate size of the nucleus). The **electromagnetic force,** which is about $10^{-2}$ times the strength of the strong force, is responsible for the binding of atoms and molecules. It is a long-range force that decreases in strength as the inverse square of the separation between interacting particles. The **weak force** is a short-range nuclear force that tends to produce instability in certain nuclei. It is responsible for beta decay, and its strength is only about $10^{-6}$ times that of the strong force. (As we shall discuss later, scientists now believe that the weak and electromagnetic

**TABLE 30.1 Particle Interactions**

| Interaction (Force) | Relative Strength[a] | Range of Force | Mediating Field Particle |
|---|---|---|---|
| Strong | 1 | Short ($\sim$1 fm) | Gluon |
| Electromagnetic | $10^{-2}$ | Long ($\propto 1/r^2$) | Photon |
| Weak | $10^{-6}$ | Short ($\sim 10^{-3}$ fm) | $W^{\pm}$ and $Z^0$ bosons |
| Gravitational | $10^{-43}$ | Long ($\propto 1/r^2$) | Graviton |

[a] For two quarks separated by $3 \times 10^{-17}$ m

forces are two manifestations of a single force called the *electroweak* force). Finally, the **gravitational force** is a long-range force with a strength of only about $10^{-43}$ times that of the strong force. Although this familiar interaction is the force that holds the planets, stars, and galaxies together, its effect on elementary particles is negligible. Thus, the gravitational force is the weakest of all the fundamental forces.

Modern physics often describes the forces between particles in terms of the actions of field particles or quanta. In the case of the familiar electromagnetic interaction, the field particles are photons. In the language of modern physics, it can be said that the electromagnetic force is *mediated* (carried) by photons, which are the quanta of the electromagnetic field. Likewise, the strong force is mediated by field particles called *gluons,* the weak force is mediated by particles called the W and Z *bosons,* and the gravitational force is mediated by quanta of the gravitational field called *gravitons.* All of these field quanta have been detected except for the graviton, which may never be found directly because of the weakness of the gravitational field. These interactions, their ranges, and their relative strengths are summarized in Table 30.1.

PAUL ADRIEN MAURICE DIRAC (1902–1984)

Winner of the Nobel prize for physics in 1933. *(Courtesy AIP Emilio Segre Visual Archives)*

## 30.6 POSITRONS AND OTHER ANTIPARTICLES

In the 1920s, the theoretical physicist Paul Adrien Maurice Dirac (1902–1984) developed a version of quantum mechanics that incorporated special relativity. Dirac's theory successfully explained the origin of the electron's spin and its magnetic moment. But it had one major problem: Its relativistic wave equation required solutions corresponding to negative energy states even for free electrons. If negative energy states existed, we would expect a normal free electron in a state of positive energy to make a rapid transition to one of these lower states, emitting a photon in the process. Thus, normal electrons would not exist, and we would be left with a universe of photons and electrons locked up in negative energy states.

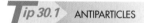

**Tip 30.1 ANTIPARTICLES**

An antiparticle is not identified solely on the basis of opposite charge. For example, even neutral particles have antiparticles.

Dirac circumvented this difficulty by postulating that all negative energy states are normally filled. The electrons that occupy the negative energy states are said to be in the "Dirac sea" and are not directly observable when all negative energy states are filled. However, if one of these negative energy states is vacant, leaving a hole in the sea of filled states, the hole can react to external forces and therefore can be observed as the electron's positive antiparticle. The general and profound implication of Dirac's theory is that **for every particle, there is an antiparticle.** The antiparticle has the same mass as the particle, but the opposite charge. For example, the electron's antiparticle, the *positron,* has a mass of 0.511 MeV/$c^2$ and a positive charge of $1.6 \times 10^{-19}$ C. As noted in Chapter 29, we usually designate an antiparticle with a bar over the symbol for the particle. Thus, $\overline{\text{p}}$ denotes the antiproton and $\overline{\nu}$ the antineutrino. In this book, the notation $e^+$ is preferred for the positron.

### Webnote 30.3

In 1933, Dirac shared the Nobel Prize in Physics with Erwin Schroedinger. Read about it at *http://www.nobel.se/physics/laureates/1933*

The positron was discovered by Carl Anderson in 1932, and in 1936 he was awarded the Nobel prize for his achievement. Anderson discovered the positron while examining tracks created by electron-like particles of positive charge in a cloud chamber. (These early experiments used cosmic rays—mostly energetic protons passing through interstellar space—to initiate high-energy reactions on the order of several GeV.) In order to discriminate between positive and negative charges, the cloud chamber was placed in a magnetic field, causing moving charges to follow curved paths. Anderson noted that some of the electron-like tracks deflected in a direction corresponding to a positively charged particle.

Since Anderson's initial discovery, the positron has been observed in a number of experiments. Perhaps the most common process for producing positrons is pair production, which was introduced in Chapter 26. In this process, a gamma ray with sufficiently high energy collides with a nucleus, creating an electron-positron pair. Because the total rest energy of the electron-positron pair is $2m_ec^2 = 1.02$ MeV, the gamma ray must have at least this much energy to create an electron-positron pair.

**HIDEKI YUKAWA, JAPANESE PHYSICIST (1907–1981)**

Yukawa was awarded the Nobel Prize in 1949 for predicting the existence of mesons. This photograph of Yukawa at work was taken in 1950 in his office at Columbia University. *(UPI/Corbis-Bettman)*

Practically every known elementary particle has a distinct antiparticle. Among the exceptions are the photon and the neutral pion ($\pi^0$), which are their own antiparticles. Following the construction of high-energy accelerators in the 1950s, many of these antiparticles were discovered. They included the antiproton $\overline{\text{p}}$ discovered by Emilio Segré and Owen Chamberlain in 1955 and the antineutron $\overline{\text{n}}$ discovered shortly thereafter.

The process of electron-positron annihilation is used in the medical diagnostic technique of positron emission tomography (PET). The patient is injected with glucose solution containing a radioactive substance that decays by positron emission. Examples of such substances are oxygen-15, nitrogen-13, carbon-11, and fluorine-18. The radioactive material is carried to the brain. When a decay occurs, the emitted positron annihilates with an electron in the brain tissue, resulting in two gamma-ray photons. With the assistance of a computer, we can display an image of the sites in the brain at which the glucose accumulates.

POSITRON EMISSION TOMOGRAPHY (PET SCANNING)

The images from a PET scan can indicate a wide variety of disorders in the brain, including Alzheimer's disease. In addition, because glucose metabolizes more rapidly in active areas of the brain, the PET scan can indicate which areas of the brain are involved in various processes such as language, music, and vision.

## 30.7 MESONS AND THE BEGINNING OF PARTICLE PHYSICS

Physicists in the mid-1930s had a fairly simple view of the structure of matter. The building blocks were the proton, the electron, and the neutron. Three other particles were known or postulated at the time: the photon, the neutrino, and the positron. These six particles were considered the fundamental constituents of matter. Although the accepted picture of the world was marvelously simple, no one was able to provide an answer to the following important question: Because the many protons in proximity in any nucleus should strongly repel each other due to their like charges, what is the nature of the force that holds the nucleus together? Scientists recognized that this mysterious nuclear force must be much stronger than anything encountered up to that time.

The first theory to explain the nature of the nuclear force was proposed in 1935 by the Japanese physicist Hideki Yukawa (1907–1981), an effort that later earned him the Nobel prize. In order to understand Yukawa's theory, it is useful to first note that **two atoms can form a covalent chemical bond by the exchange of electrons.** Similarly, in the modern view of electromagnetic interactions, **charged particles interact by exchanging a photon.** Yukawa used this same idea to explain the nuclear force by proposing a new particle that is exchanged by nucleons in the nucleus to produce the strong force. Furthermore, he established that the range of the force is inversely proportional to the mass of this particle, and predicted that the mass would be about 200 times the mass of the electron. Because the new particle would have a mass between that of the electron and the proton, it was called a **meson** (from the Greek *meso,* meaning "middle").

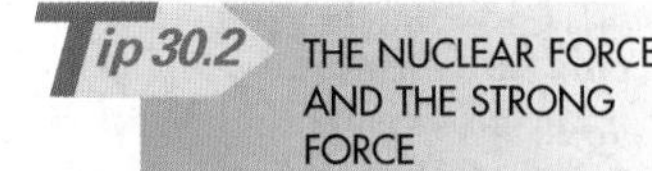

THE NUCLEAR FORCE AND THE STRONG FORCE

The nuclear force discussed in Chapter 29 was originally called the *strong force.* Once the quark theory was established, however, the phrase *strong force* was reserved for the force between quarks. We shall follow this convention—the strong force is between quarks and the nuclear force is between nucleons.

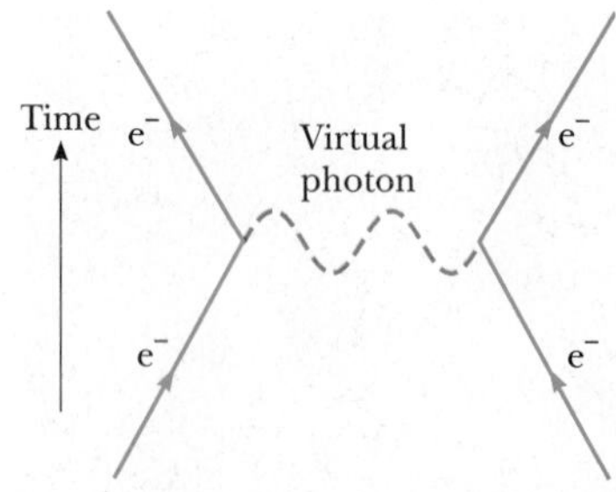


**FIGURE 30.6** Feynman diagram representing a photon mediating the electromagnetic force between two electrons.

In an effort to substantiate Yukawa's predictions, physicists began looking for the meson by studying cosmic rays that enter Earth's atmosphere. In 1937 Carl Anderson and his collaborators discovered a particle whose mass was 106 MeV/$c^2$, about 207 times the mass of the electron. However, subsequent experiments showed that the particle interacted very weakly with matter, and hence could not be the carrier of the nuclear force. This puzzling situation inspired several theoreticians to propose that there are two mesons with slightly different masses, an idea that was confirmed in 1947 with the discovery of the pi meson ($\pi$), or simply *pion*, by Cecil Frank Powell (1903–1969) and Guiseppe P. S. Occhialini (1907–1993). The particle discovered earlier by Anderson in 1937, the one thought to be a meson, is not really a meson. Instead, it takes part in weak and electromagnetic interactions only, and it is now called the *muon* ($\mu$).

The pion comes in three varieties, corresponding to three charge states: $\pi^+$, $\pi^-$, and $\pi^0$. The $\pi^+$ and $\pi^-$ particles have masses of 139.6 MeV/$c^2$, and the $\pi^0$ has a mass of 135.0 MeV/$c^2$. Pions and muons are very unstable particles. For example, the $\pi^-$, which has a lifetime of about $2.6 \times 10^{-8}$ s, decays into a muon and an antineutrino. The muon, with a lifetime of 2.2 $\mu$s, then decays into an electron, a neutrino, and an antineutrino. The sequence of decays is

$$\begin{aligned} \pi^- &\longrightarrow \mu^- + \overline{\nu} \\ \mu^- &\longrightarrow e^- + \nu + \overline{\nu} \end{aligned} \qquad [30.6]$$

## Webnote 30.4

For a good summary on Feynman diagrams, take a look at *http://www2.slac.stanford.edu/vvc/theory/feynman.html*

The interaction between two particles can be understood in general with a simple diagram called a *Feynman diagram,* developed by Richard P. Feynman (1918–1988). Figure 30.6 is a Feynman diagram for the electromagnetic interaction between two electrons. In this simple case, a photon is the field particle that mediates the electromagnetic force between the electrons. The photon transfers energy and momentum from one electron to the other in this interaction. Such a photon, called a *virtual photon,* can never be detected directly because it is absorbed by the second electron very shortly after being emitted by the first electron. The existence of a virtual photon would violate the law of conservation of energy, but because of the uncertainty principle and its very short lifetime $\Delta t$, the photon's excess energy is less than the uncertainty in its energy, given by $\Delta E \approx \hbar/\Delta t$.

**RICHARD FEYNMAN, AMERICAN PHYSICIST (1918–1988)**

With his son, Carl, in 1965. Feynman, together with Julian S. Schwinger and Shinichiro Tomonaga, won the 1965 Nobel prize for physics for fundamental work in the principles of quantum electrodynamics. His many important contributions to physics include the invention of simple diagrams to represent particle interactions graphically, the theory of the weak interaction of subatomic particles, a reformulation of quantum mechanics, the theory of superfluid helium, and his contribution to physics education through the magnificent three-volume text *The Feynman Lectures on Physics. (UPI Telephotos)*

Now consider the pion exchange between a proton and a neutron via the nuclear force (Fig. 30.7). We can reason that the energy $\Delta E$ needed to create a pion of mass $m_\pi$ is given by $\Delta E = m_\pi c^2$. Again, the existence of the pion is allowed in spite of conservation of energy if this energy is surrendered in a short enough time $\Delta t$, the time it takes the pion to transfer from one nucleon to the other. From the uncertainty principle, $\Delta E\,\Delta t \approx \hbar$, we get

$$\Delta t \approx \frac{\hbar}{\Delta E} = \frac{\hbar}{m_\pi c^2} \qquad [30.7]$$

Because the pion cannot travel faster than the speed of light, the maximum distance $d$ it can travel in a time $\Delta t$ is $c\,\Delta t$. Using Equation 30.7 and $d = c\,\Delta t$, we find this maximum distance to be

$$d \approx \frac{\hbar}{m_\pi c} \qquad [30.8]$$

The measured range of the nuclear force is about $1.5 \times 10^{-15}$ m. Using this value for $d$ in Equation 30.8, the rest energy of the pion is calculated to be

$$\begin{aligned} m_\pi c^2 &\approx \frac{\hbar c}{d} = \frac{(1.05 \times 10^{-34}\ \text{J}\cdot\text{s})(3.00 \times 10^8\ \text{m/s})}{1.5 \times 10^{-15}\ \text{m}} \\ &= 2.1 \times 10^{-11}\ \text{J} \cong 130\ \text{MeV} \end{aligned}$$

This corresponds to a mass of 130 MeV/$c^2$ (about 250 times the mass of the electron), which is in good agreement with the observed mass of the pion.

The concept we have just described is quite revolutionary. In effect, it says that a proton can change into a proton plus a pion, as long as it returns to its original state in a very short time. High-energy physicists often say that a nucleon undergoes "fluctuations" as it emits and absorbs pions. As we have seen, these fluctuations are a consequence of a combination of quantum mechanics (through the uncertainty principle) and special relativity (through Einstein's energy-mass relation $E = mc^2$).

This section has dealt with Yukawa's early theory of particles that mediate the nuclear force, namely the pions, and the mediators of the electromagnetic force, the photons. Although Yukawa's model led to the modern view, it has been superseded by the more basic quark-gluon theory as explained in Sections 30.12 and 30.13.

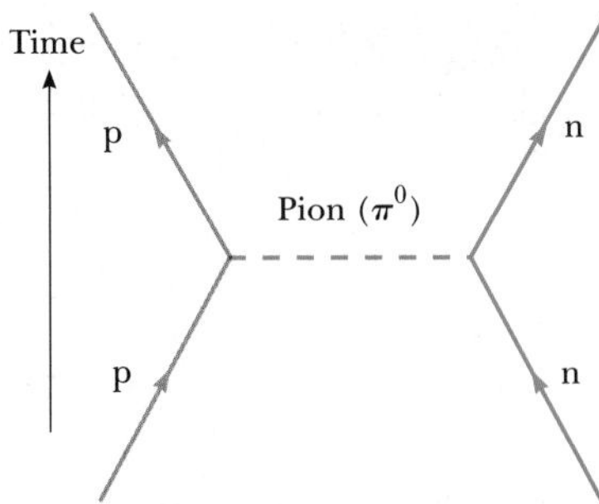


**FIGURE 30.7** Feynman diagram representing a proton interacting with a neutron via the strong force. In this case, the pion mediates the nuclear force.

## 30.8 CLASSIFICATION OF PARTICLES

### HADRONS

All particles other than photons can be classified into two broad categories, hadrons and leptons, according to their interactions. Particles that interact through the strong force are called *hadrons.* There are two classes of hadrons, known as *mesons* and *baryons,* distinguished by their masses and spins. All mesons are known to decay finally into electrons, positrons, neutrinos, and photons. The pion is the lightest of known mesons, with a mass of about 140 MeV/$c^2$ and a spin of 0. Another is the K meson, with a mass of about 500 MeV/$c^2$ and spin 0.

**Webnote 30.5**

Take a look at particle physics in terms of the Big Bang Theory by visiting *http://hepwww.rl.ac.uk/Pub/Phil/ppintro/ppintro.html*

Baryons have masses equal to or greater than the proton mass (the name *baryon* means "heavy" in Greek), and their spin is always a noninteger value ($\frac{1}{2}$ or $\frac{3}{2}$). Protons and neutrons are baryons, as are many other particles. With the exception of the proton, all baryons decay in such a way that the end products include a proton. For example, the baryon called the $\Xi$ hyperon first decays to a $\Lambda^0$ in about $10^{-10}$ s. The $\Lambda^0$ then decays to a proton and a $\pi^-$ in about $3 \times 10^{-10}$ s.

Today it is believed that hadrons are composed of quarks. (We shall have more to say about the quark model later.) Some of the important properties of hadrons are listed in Table 30.2.

### LEPTONS

Leptons (from the Greek *leptos* meaning "small" or "light") are a group of particles that participate in the weak interaction. All leptons have a spin of $\frac{1}{2}$. Included in this group are electrons, muons, and neutrinos, which are less massive than the lightest hadron. Although hadrons have size and structure, leptons appear to be truly elementary, with no structure (that is, point-like).

Quite unlike hadrons, the number of known leptons is small. Currently, scientists believe there only are six leptons (each having an antiparticle)—the electron, the muon, the tau, and a neutrino associated with each:

$$\begin{pmatrix} e^- \\ \nu_e \end{pmatrix} \qquad \begin{pmatrix} \mu^- \\ \nu_\mu \end{pmatrix} \qquad \begin{pmatrix} \tau^- \\ \nu_\tau \end{pmatrix}$$

The tau lepton, discovered in 1975, has a mass about twice that of the proton.

Although neutrinos have masses of about zero, there is strong indirect evidence that the electron neutrino has a finite mass of about 3 eV/$c^2$ or a mass 1/180 000 of the electron mass. As we shall see, a firm knowledge of the neutrino's mass could have great significance in cosmological models and the future of the Universe.

**Webnote 30.6**

Protons showering from outer space can create muons in the atmosphere. Understand more about this effect by visiting *http://www2.slac.stanford.edu/vvc/cosmic_rays.html*

**TABLE 30.2** **Some Particles and Their Properties**

| Category | Particle Name | Symbol | Anti-particle | Mass ($MeV/c^2$) | $B$ | $L_e$ | $L_\mu$ | $L_\tau$ | $S$ | Lifetime(s) | Principal Decay Modes[a] |
|---|---|---|---|---|---|---|---|---|---|---|---|
| **Leptons** | Electron | $e^-$ | $e^+$ | 0.511 | 0 | +1 | 0 | 0 | 0 | Stable | |
| | Electron–neutrino | $\nu_e$ | $\overline{\nu}_e$ | $<7\ eV/c^2$ | 0 | +1 | 0 | 0 | 0 | Stable | |
| | Muon | $\mu^-$ | $\mu^+$ | 105.7 | 0 | 0 | +1 | 0 | 0 | $2.20 \times 10^{-6}$ | $e^-\overline{\nu}_e\nu_\mu$ |
| | Muon–neutrino | $\nu_\mu$ | $\overline{\nu}_\mu$ | $<0.3$ | 0 | 0 | +1 | 0 | 0 | Stable | |
| | Tau | $\tau^-$ | $\tau^+$ | 1 784 | 0 | 0 | 0 | +1 | 0 | $<4 \times 10^{-13}$ | $\mu^-\overline{\nu}_\mu\nu_\tau$, $e^-\overline{\nu}_e\nu_\tau$ |
| | Tau–neutrino | $\nu_\tau$ | $\overline{\nu}_\tau$ | $<30$ | 0 | 0 | 0 | +1 | 0 | Stable | |
| **Hadrons** | | | | | | | | | | | |
| **Mesons** | Pion | $\pi^+$ | $\pi^-$ | 139.6 | 0 | 0 | 0 | 0 | 0 | $2.60 \times 10^{-8}$ | $\mu^+\nu_\mu$ |
| | | $\pi^0$ | Self | 135.0 | 0 | 0 | 0 | 0 | 0 | $0.83 \times 10^{-16}$ | $2\gamma$ |
| | Kaon | $K^+$ | $K^-$ | 493.7 | 0 | 0 | 0 | 0 | +1 | $1.24 \times 10^{-8}$ | $\mu^+\nu_\mu$, $\pi^+\pi^0$ |
| | | $K_S^0$ | $\overline{K}_S^0$ | 497.7 | 0 | 0 | 0 | 0 | +1 | $0.89 \times 10^{-10}$ | $\pi^+\pi^-$, $2\pi^0$ |
| | | $K_L^0$ | $\overline{K}_L^0$ | 497.7 | 0 | 0 | 0 | 0 | +1 | $5.2 \times 10^{-8}$ | $\pi^{\pm}e^{\mp}\overline{\nu}_e$, $3\pi^0$ $\pi^{\pm}\mu^{\mp}\overline{\nu}_\mu$ |
| | Eta | $\eta$ | Self | 548.8 | 0 | 0 | 0 | 0 | 0 | $<10^{-18}$ | $2\gamma$, $3\pi$ |
| | | $\eta'$ | Self | 958 | 0 | 0 | 0 | 0 | 0 | $2.2 \times 10^{-21}$ | $\eta\pi^+\pi^-$ |
| **Baryons** | Proton | p | $\overline{p}$ | 938.3 | +1 | 0 | 0 | 0 | 0 | Stable | |
| | Neutron | n | $\overline{n}$ | 939.6 | +1 | 0 | 0 | 0 | 0 | 920 | $pe^-\overline{\nu}_e$ |
| | Lambda | $\Lambda^0$ | $\overline{\Lambda}^0$ | 1 115.6 | +1 | 0 | 0 | 0 | −1 | $2.6 \times 10^{-10}$ | $p\pi^-$, $n\pi^0$ |
| | Sigma | $\Sigma^+$ | $\overline{\Sigma}^-$ | 1 189.4 | +1 | 0 | 0 | 0 | −1 | $0.80 \times 10^{-10}$ | $p\pi^0$, $n\pi^+$ |
| | | $\Sigma^0$ | $\overline{\Sigma}^0$ | 1 192.5 | +1 | 0 | 0 | 0 | −1 | $6 \times 10^{-20}$ | $\Lambda^0\gamma$ |
| | | $\Sigma^-$ | $\overline{\Sigma}^+$ | 1 197.3 | +1 | 0 | 0 | 0 | −1 | $1.5 \times 10^{-10}$ | $n\pi^-$ |
| | Xi | $\Xi^0$ | $\overline{\Xi}^0$ | 1 315 | +1 | 0 | 0 | 0 | −2 | $2.9 \times 10^{-10}$ | $\Lambda^0\pi^0$ |
| | | $\Xi^-$ | $\Xi^+$ | 1 321 | +1 | 0 | 0 | 0 | −2 | $1.64 \times 10^{-10}$ | $\Lambda^0\pi^-$ |
| | Omega | $\Omega^-$ | $\Omega^+$ | 1 672 | +1 | 0 | 0 | 0 | −3 | $0.82 \times 10^{-10}$ | $\Xi^0\pi^0$, $\Lambda^0K^-$ |

[a] Notations in this column such as $p\pi^-$, $n\pi^0$ mean two possible decay modes. In this case, the two possible decays are $\Lambda^0 \rightarrow p + \pi^-$ and $\Lambda^0 \rightarrow n + \pi^0$.

## 30.9 CONSERVATION LAWS

A number of conservation laws are important in the study of elementary particles. Although the two described here have no theoretical foundation, they are supported by abundant empirical evidence.

### BARYON NUMBER

Conservation of baryon number

The law of conservation of baryon number tells us that whenever a baryon is created in a reaction or decay, an antibaryon is also created. This can be quantified by assigning a baryon number: $B = +1$ for all baryons, $B = -1$ for all antibaryons, and $B = 0$ for all other particles. Thus, the **law of conservation of baryon number** states that whenever a nuclear reaction or decay occurs, the sum of the baryon numbers before the process must equal the sum of the baryon numbers after the process.

Note that if baryon number is absolutely conserved, the proton must be absolutely stable. If it were not for the law of conservation of baryon number, the proton could decay into a positron and a neutral pion. However, such a decay has never been observed. At present, we can only say that the proton has a half-life of at least $10^{31}$ years (the estimated age of the Universe is about $10^{10}$ years). In one recent version of a so-called grand unified theory or (GUT), physicists have predicted that the proton is actually unstable. According to this theory, the baryon number (sometimes called the *baryonic charge*) is not absolutely conserved, whereas electric charge is always conserved.

## Example 30.4 Checking Baryon Numbers

Determine whether or not each of the following reactions can occur based on the law of conservation of baryon number.

$$\mathrm{p} + \mathrm{n} \longrightarrow \mathrm{p} + \mathrm{p} + \mathrm{n} + \overline{\mathrm{p}} \qquad (1)$$

$$\mathrm{p} + \mathrm{n} \longrightarrow \mathrm{p} + \mathrm{p} + \overline{\mathrm{p}} \qquad (2)$$

**Solution** Recall that $B = +1$ for baryons and $B = -1$ for antibaryons. Hence, the left side of (1) gives a total baryon number of $1 + 1 = 2$. The right side gives a total baryon number of $1 + 1 + 1 + (-1) = 2$. Thus, reaction (1) can occur provided the incoming proton has sufficient energy.

The left side of (2) gives a total baryon number of $1 + 1 = 2$. However, the right side gives $1 + 1 + (-1) = 1$. Because baryon number is not conserved, reaction (2) cannot occur.

## LEPTON NUMBER

There are three conservation laws involving lepton numbers, one for each variety of lepton. The **law of conservation of electron-lepton number** states that the sum of the electron-lepton numbers before a reaction or decay must equal the sum of the electron-lepton numbers after the reaction or decay. The electron and the electron neutrino are assigned a positive electron-lepton number, $L_e = +1$; the antileptons $e^+$ and $\overline{\nu}_e$ are assigned the electron-lepton number $L_e = -1$; and all other particles have $L_e = 0$. For example, consider the decay of the neutron

Conservation of lepton number

$$\mathrm{n} \longrightarrow \mathrm{p}^+ + \mathrm{e}^- + \overline{\nu}_e$$

Neutron decay

Before the decay, the electron-lepton number is $L_e = 0$; after the decay it is $0 + 1 + (-1) = 0$. Thus, the electron-lepton number is conserved. It is important to recognize that the baryon number must also be conserved. This can easily be seen by noting that before the decay $B = +1$, whereas after the decay $B = +1 + 0 + 0 = +1$.

Similarly, when a decay involves muons, the muon-lepton number $L_\mu$ is conserved. The $\mu^-$ and the $\nu_\mu$ are assigned $L_\mu = +1$, the antimuons $\mu^+$ and $\overline{\nu}_\mu$ are assigned $L_\mu = -1$, and all other particles have $L_\mu = 0$. Finally, the $\tau$-lepton number $L_\tau$ is conserved, and similar assignments can be made for the $\tau$ lepton and its neutrino.

## Example 30.5 Checking Lepton Numbers

Determine which of the following decay schemes can occur on the basis of conservation of lepton number.

$$\mu^- \longrightarrow \mathrm{e}^- + \overline{\nu}_e + \nu_\mu \qquad (1)$$

$$\pi^+ \longrightarrow \mu^+ + \nu_\mu + \nu_e \qquad (2)$$

**Solution** Because decay 1 involves both a muon and an electron, $L_\mu$ and $L_e$ must both be conserved. Before the decay, $L_\mu = +1$ and $L_e = 0$. After the decay, $L_\mu = 0 + 0 + 1 = +1$, and $L_e = +1 - 1 + 0 = 0$. Thus, both numbers are conserved, and on this basis the decay mode is possible.

Before decay 2 occurs, $L_\mu = 0$ and $L_e = 0$. After the decay, $L_\mu = -1 + 1 + 0 = 0$, but $L_e = +1$. Thus, the decay is not possible because the electron-lepton number is not conserved.

**EXERCISE** Determine whether the decay $\mu^- \rightarrow \mathrm{e}^- + \overline{\nu}_e$ can occur.

**ANSWER** No. The muon-lepton number is +1 before the decay and 0 after.

**Quick Quiz 30.2** Which of the following reactions cannot occur?
(a) $p + p \rightarrow p + p + \bar{p}$
(b) $n \rightarrow p + e^- + \bar{\nu}_e$
(c) $\mu^- \rightarrow e^- + \bar{\nu}_e + \nu_\mu$
(d) $\pi^- \rightarrow \mu^- + \bar{\nu}_\mu$

**Quick Quiz 30.3** Which of the following reactions cannot occur?
(a) $p + \bar{p} \rightarrow 2\gamma$
(b) $\gamma + p \rightarrow n + \pi^0$
(c) $\pi^0 + n \rightarrow K^+ + \Sigma^-$
(d) $\pi^+ + p \rightarrow K^+ + \Sigma^+$

**Quick Quiz 30.4** Suppose a claim is made that the decay of a neutron is given by $n \rightarrow p^+ + e^-$. Which of the following conservation laws are violated by this proposed decay scheme? (a) energy, (b) linear momentum, (c) spin angular momentum, (d) electric charge, (e) lepton number, (f) baryon number.

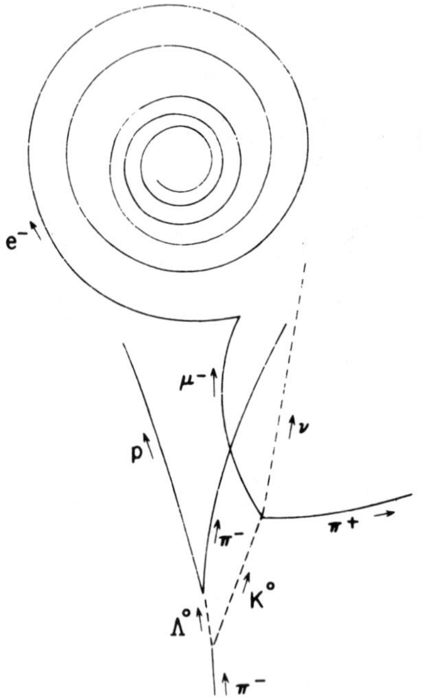

**FIGURE 30.8** This drawing represents tracks of many events obtained by analyzing a bubble-chamber photograph. The strange particles $\Lambda^0$ and $K^0$ are formed (at the bottom) as the $\pi^-$ interacts with a proton according to $\pi^- + p \rightarrow \Lambda^0 + K^0$. (Note that the neutral particles leave no tracks, as indicated by the dashed lines.) The $\Lambda^0$ and $K^0$ then decay according to $\Lambda^0 \rightarrow \pi^- + p$ and $K^0 \rightarrow \pi + \mu^- + \nu_\mu$. *(Courtesy Lawrence Berkeley Laboratory, University of California)*

## 30.10 STRANGE PARTICLES AND STRANGENESS

Many particles discovered in the 1950s were produced by the nuclear interaction of pions with protons and neutrons in the atmosphere. A group of these particles, namely the K, Λ, and Σ, were found to exhibit unusual properties in their production and decay, and hence were called *strange particles.*

One unusual property of strange particles is that they are always produced in pairs. For example, when a pion collides with a proton, two neutral strange particles are produced with high probability (Fig. 30.8) following the reaction

$$\pi^- + p^+ \longrightarrow K^0 + \Lambda^0$$

On the other hand, the reaction $\pi^- + p^+ \rightarrow K^0 + n$ never occurs, even though no known conservation laws are violated and the energy of the pion is sufficient to initiate the reaction.

The second peculiar feature of strange particles is that although they are produced by the strong interaction at a high rate, they do not decay into particles that interact via the strong force at a very high rate. Instead, they decay very slowly, which is characteristic of the weak interaction. Their half-lives are in the range $10^{-10}$ s to $10^{-8}$ s; most other particles that interact via the strong force have lifetimes on the order of $10^{-23}$ s.

To explain these unusual properties of strange particles, physicists introduced a law called *conservation of strangeness,* together with a new quantum number $S$ called **strangeness.** The strangeness numbers for some particles are given in Table 30.2. The production of strange particles in pairs is explained by assigning $S = +1$ to one of the particles and $S = -1$ to the other. All nonstrange particles are assigned strangeness $S = 0$. The **law of conservation of strangeness** states that whenever a nuclear reaction or decay occurs, the sum of the strangeness numbers before the process must equal the sum of the strangeness numbers after the process.

Conservation of strangeness number

We can explain the slow decay of strange particles by assuming that the strong and electromagnetic interactions obey the law of conservation of strangeness, whereas the weak interaction does not. Because the decay reaction involves the loss of one strange particle, it violates strangeness conservation and hence proceeds slowly via the weak interaction.

**Webnote 30.7**

Does a 600-ton bubble chamber sound big? Find out more by exploring the Imaging Cosmic and Rare Underground Signal (ICARUS) in Italy. Visit *http://www.aquila.infn.it/icarus*

## APPLYING PHYSICS 30.2

A student claims to have observed a decay of an electron into two neutrinos, traveling in opposite directions. What conservation laws would be violated by this decay?

**Explanation** Several conservation laws are violated. Conservation of electric charge is violated because the negative charge of the electron has disappeared. Conservation of electron-lepton number is also violated, because there is one lepton before the decay and two afterward. If both neutrinos were electron-neutrinos, electron-lepton number conservation would be violated in the final state. However, if one of the product neutrinos were other than an electron-neutrino, then another lepton conservation law would be violated, because no other leptons were in the initial state. Other conservation laws are obeyed by this decay. Energy can be conserved—the rest energy of the electron appears as the kinetic energy (and possibly some small rest energy) of the neutrinos. The opposite directions of the velocities of the two neutrinos allows for conservation of momentum. Conservation of baryon number and conservation of other lepton numbers are also upheld in this decay.

### Example 30.6 Is Strangeness Conserved?

**A** Determine whether the following reaction occurs on the basis of conservation of strangeness.

$$\pi^0 + \mathrm{n} \longrightarrow \mathrm{K}^+ + \Sigma^-$$

**Solution** The initial state has a total strangeness of $S = 0 + 0 = 0$. Because the strangeness of the $\mathrm{K}^+$ is $S = +1$, and the strangeness of the $\Sigma^-$ is $S = -1$, the total strangeness of the final state is $+1 - 1 = 0$. Thus, strangeness is conserved and the reaction is allowed.

**B** Show that the following reaction does not conserve strangeness.

$$\pi^- + \mathrm{p} \longrightarrow \pi^- + \Sigma^+$$

**Solution** The initial state has strangeness $S = 0 + 0 = 0$, and the final state has strangeness $S = 0 + (-1) = -1$. Thus, strangeness is not conserved.

**EXERCISE** Show that the observed reaction $\mathrm{p}^+ + \pi^- \rightarrow \mathrm{K}^0 + \Lambda^0$ obeys the law of conservation of strangeness.

## 30.11 THE EIGHTFOLD WAY

As we have seen, quantities such as spin, baryon number, lepton number, and strangeness are labels we associate with particles. Many classification schemes that group particles into families based on such labels have been proposed. First, consider the first eight baryons listed in Table 30.2, all having a spin of $\frac{1}{2}$. The family consists of the proton, the neutron, and six other particles. If we plot their strangeness versus their charge using a sloping coordinate system, as in Figure 30.9a, a fascinating pattern emerges. Six of the baryons form a hexagon, and the remaining two are at the hexagon's center. (Particles with spin quantum number $\frac{1}{2}$ or $\frac{3}{2}$ are called fermions.)

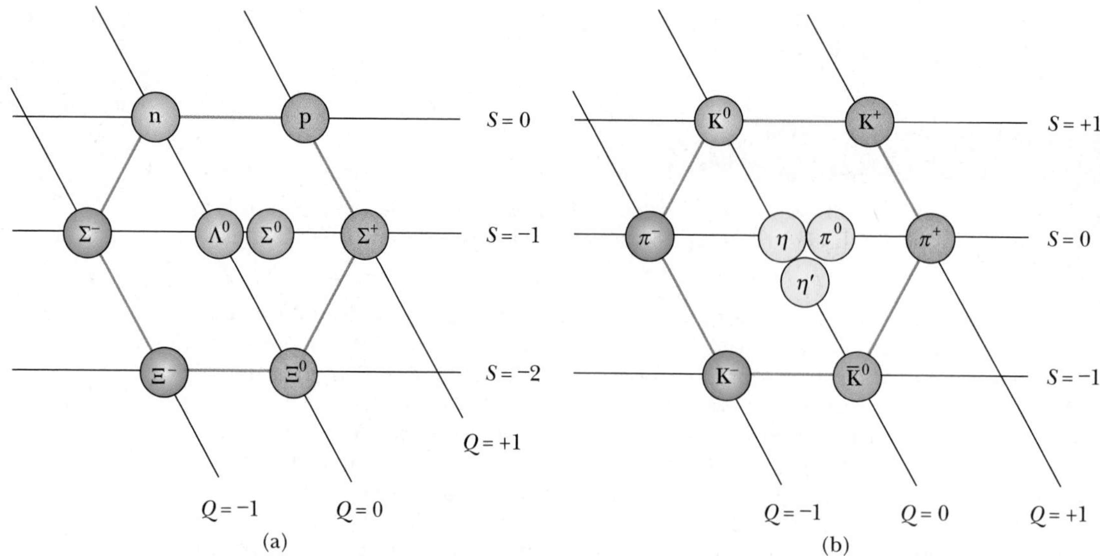


**FIGURE 30.9** (a) The hexagonal eightfold-way pattern for the eight spin-$\frac{1}{2}$ baryons. This strangeness versus charge plot uses a horizontal axis for the strangeness values $S$, but a sloping axis for the charge number $Q$. (b) The eightfold-way pattern for the nine spin-0 mesons.

Now consider the family of mesons listed in Table 30.2 with spins of 0 (Particles with spin quantum number 0 or 1 are called bosons.) If we count both particles and antiparticles, there are nine such mesons. Figure 30.9b is a plot of strangeness versus charge for this family. Again, a fascinating hexagonal pattern emerges. In this case, the particles on the perimeter of the hexagon lie opposite their antiparticles, and the remaining three (which form their own antiparticles) are at its center. These and related symmetric patterns, called the **eightfold way,** were proposed independently in 1961 by Murray Gell-Mann and Yuval Ne'eman.

The groups of baryons and mesons can be displayed in many other symmetric patterns within the framework of the eightfold way. For example, the family of spin-$\frac{3}{2}$ baryons contains ten particles arranged in a pattern such as the tenpins in a bowling alley. After the pattern was proposed, one of the particles was missing—it had yet to be discovered. Gell-Mann predicted that the missing particle, which he called the *omega minus* ($\Omega^-$), should have a spin of $\frac{3}{2}$, a charge of $-1$, a strangeness of $-3$, and a mass of about 1 680 MeV/$c^2$. Shortly thereafter, in 1964, scientists at the Brookhaven National Laboratory found the missing particle through careful analyses of bubble chamber photographs, and confirmed all its predicted properties.

The patterns of the eightfold way in the field of particle physics have much in common with the periodic table. Whenever a vacancy (a missing particle or element) occurs in the organized patterns, experimentalists have a guide for their investigations.

**MURRAY GELL-MANN, AMERICAN PHYSICIST (B. 1929)**

Gell-Mann was awarded the Nobel prize in 1969 for his theoretical studies dealing with subatomic particles. *(Photo courtesy of Michael R. Dressler)*

## 30.12 QUARKS

As we have noted, leptons appear to be truly elementary particles because they have no measurable size or internal structure, are limited in number, and do not seem to break down into smaller units. Hadrons, on the other hand, are complex particles with size and structure. Furthermore, we know that hadrons decay into other hadrons and are many in number. Table 30.2 lists only those hadrons that are stable against hadronic decay; hundreds of others have been discovered. These facts strongly suggest that hadrons cannot be truly elementary but have some substructure.

**TABLE 30.3** Properties of Quarks and Antiquarks

| Quarks | | | | | | | | |
|---|---|---|---|---|---|---|---|---|
| **Name** | **Symbol** | **Spin** | **Charge** | **Baryon Number** | **Strangeness** | **Charm** | **Bottomness** | **Topness** |
| Up | u | $\frac{1}{2}$ | $+\frac{2}{3}e$ | $\frac{1}{3}$ | 0 | 0 | 0 | 0 |
| Down | d | $\frac{1}{2}$ | $-\frac{1}{3}e$ | $\frac{1}{3}$ | 0 | 0 | 0 | 0 |
| Strange | s | $\frac{1}{2}$ | $-\frac{1}{3}e$ | $\frac{1}{3}$ | −1 | 0 | 0 | 0 |
| Charmed | c | $\frac{1}{2}$ | $+\frac{2}{3}e$ | $\frac{1}{3}$ | 0 | +1 | 0 | 0 |
| Bottom | b | $\frac{1}{2}$ | $-\frac{1}{3}e$ | $\frac{1}{3}$ | 0 | 0 | +1 | 0 |
| Top | t | $\frac{1}{2}$ | $+\frac{2}{3}e$ | $\frac{1}{3}$ | 0 | 0 | 0 | +1 |
| **Antiquarks** | | | | | | | | |
| **Name** | **Symbol** | **Spin** | **Charge** | **Baryon Number** | **Strangeness** | **Charm** | **Bottomness** | **Topness** |
| Anti-up | $\overline{u}$ | $\frac{1}{2}$ | $-\frac{2}{3}e$ | $-\frac{1}{3}$ | 0 | 0 | 0 | 0 |
| Anti-down | $\overline{d}$ | $\frac{1}{2}$ | $+\frac{1}{3}e$ | $-\frac{1}{3}$ | 0 | 0 | 0 | 0 |
| Anti-strange | $\overline{s}$ | $\frac{1}{2}$ | $+\frac{1}{3}e$ | $-\frac{1}{3}$ | +1 | 0 | 0 | 0 |
| Anti-charmed | $\overline{c}$ | $\frac{1}{2}$ | $-\frac{2}{3}e$ | $-\frac{1}{3}$ | 0 | −1 | 0 | 0 |
| Anti-bottom | $\overline{b}$ | $\frac{1}{2}$ | $+\frac{1}{3}e$ | $-\frac{1}{3}$ | 0 | 0 | −1 | 0 |
| Anti-top | $\overline{t}$ | $\frac{1}{2}$ | $-\frac{2}{3}e$ | $-\frac{1}{3}$ | 0 | 0 | 0 | −1 |

## THE ORIGINAL QUARK MODEL

In 1963 Gell-Mann and George Zweig independently proposed that hadrons have a more elementary substructure. According to their model, all hadrons are composite systems of two or three fundamental constitutents called **quarks,** which rhymes with "forks." Gell-Mann borrowed the word *quark* from the passage "Three quarks for Muster Mark" in James Joyce's book *Finnegan's Wake.* In the original model, there were three types of quarks designated by the symbols *u, d,* and *s.* These were given the arbitrary names *up, down,* and *sideways* (or, now more commonly, *strange*).

An unusual property of quarks is that they have fractional electronic charges, as shown—along with other properties—in Table 30.3. Associated with each quark is an antiquark of opposite charge, baryon number, and strangeness. The compositions of all hadrons known when Gell-Mann and Zweig presented their models could be completely specified by three simple rules:

1. Mesons consist of one quark and one antiquark, giving them a baryon number of 0, as required.
2. Baryons consist of three quarks.
3. Antibaryons consist of three antiquarks.

Table 30.4 lists the quark compositions of several mesons and baryons. Note that just two of the quarks, u and d, are contained in all hadrons encountered in ordinary matter (protons and neutrons). The third quark, s, is needed only to construct strange particles with a strangeness of either +1 or −1. Figure 30.10 shows the quark compositions of several particles.

**TABLE 30.4** Quark Composition of Several Hadrons

| Particle | Quark Composition |
|---|---|
| **Mesons** | |
| $\pi^+$ | $u\overline{d}$ |
| $\pi^-$ | $\overline{u}d$ |
| $K^+$ | $u\overline{s}$ |
| $K^-$ | $\overline{u}s$ |
| $K^0$ | $d\overline{s}$ |
| **Baryons** | |
| p | uud |
| n | udd |
| $\Lambda^0$ | uds |
| $\Sigma^+$ | uus |
| $\Sigma^0$ | uds |
| $\Sigma^-$ | dds |
| $\Xi^0$ | uss |
| $\Xi^-$ | dss |
| $\Omega^-$ | sss |

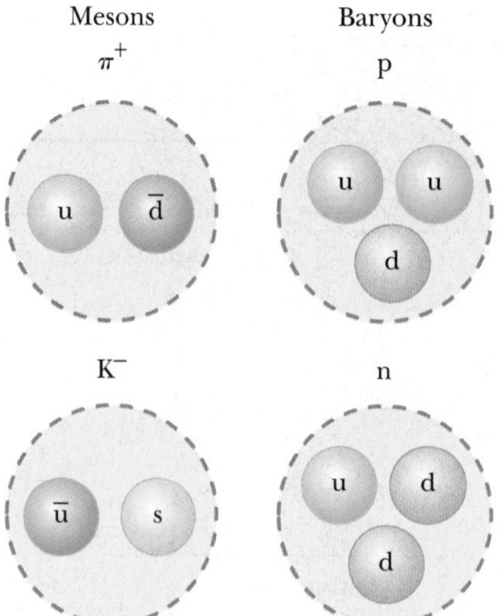


**FIGURE 30.10** Quark compositions of two mesons and two baryons. Note that the mesons on the left contain two quarks, and the baryons on the right contain three quarks.

## APPLYING PHYSICS 30.3

We have seen a law of conservation of lepton number and a law of conservation of baryon number. Why isn't there a law of conservation of meson number?

**Explanation** We can argue this from the point of view of creating particle-antiparticle pairs from available energy. If energy is converted to rest energy of a lepton-antilepton pair, then there is no net change in lepton number, because the lepton has a number of + 1 and the antilepton has a number of − 1. Energy could also be transformed into rest energy of a baryon-antibaryon pair. The baryon has baryon number + 1, the antibaryon − 1, and there is no net change in baryon number.

But now suppose energy is transformed into rest energy of a quark-antiquark pair. By definition in quark theory, a quark antiquark pair is a meson. Thus, we have created a meson from energy—there was no meson before; now there is. Thus, there is no conservation of meson number. With more energy, we can create more mesons, with no restriction from a conservation law other than that of energy.

## CHARM AND OTHER RECENT DEVELOPMENTS

Although the original quark model was highly successful in classifying particles into families, there were some discrepancies between predictions of the model and certain experimental decay rates. As a consequence, a fourth quark was proposed by several physicists in 1967. The fourth quark, designated by c, was given a property called **charm.** A charmed quark would have the charge $+2e/3$, but its charm would distinguish it from the other three quarks. The new quark would have a charm of $C = +1$, its antiquark would have a charm of $C = -1$, and all other quarks would have $C = 0$, as indicated in Table 30.3. Charm, like strangeness, would be conserved in strong and electromagnetic interactions but not in weak interactions.

In 1974 a new heavy meson called the $J/\psi$ particle (or simply $\psi$) was discovered independently by a group led by Burton Richter at the Stanford Linear Accelerator (SLAC) and another group led by Samuel Ting at the Brookhaven National Laboratory. Richter and Ting were awarded the Nobel prize in 1976 for this work. The $J/\psi$ particle does not fit into the three-quark model, but has the properties of a combination of a charmed quark and its antiquark ($c\bar{c}$). It is much heavier than the other known mesons ($\sim 3\ 100\ \text{MeV}/c^2$) and its lifetime is much longer than those of other particles that decay via the strong force. In 1975 researchers at Stan-

**TABLE 30.5 The Fundamental Particles and Some of Their Properties**

| Quarks | | |
|---|---|---|
| **Particle** | **Rest Energy** | **Charge** |
| u | 360 MeV | $+\frac{2}{3}e$ |
| d | 360 MeV | $-\frac{1}{3}e$ |
| c | 1 500 MeV | $+\frac{2}{3}e$ |
| s | 540 MeV | $-\frac{1}{3}e$ |
| t | 173 GeV | $+\frac{2}{3}e$ |
| b | 5 GeV | $-\frac{1}{3}e$ |

| Leptons | | |
|---|---|---|
| **Particle** | **Rest Energy** | **Charge** |
| $e^-$ | 511 keV | $-e$ |
| $\mu^-$ | 107 MeV | $-e$ |
| $\tau^-$ | 1 784 MeV | $-e$ |
| $\nu_e$ | $< 30$ eV | 0 |
| $\nu_\mu$ | $< 0.5$ MeV | 0 |
| $\nu_\tau$ | $< 250$ MeV | 0 |

ford University reported strong evidence for the tau ($\tau$) lepton, with a mass of 1 784 MeV/$c^2$. Such discoveries led to more elaborate quark models and the proposal of two new quarks, named *top* (t) and *bottom* (b). (Some physicists prefer the whimsical names *truth* and *beauty*.) To distinguish these quarks from the old ones, quantum numbers called *topness* and *bottomness* were assigned to these new particles and are included in Table 30.3. In 1977 researchers at the Fermi National Laboratory, under the direction of Leon Lederman, reported the discovery of a very massive new meson $\Upsilon$ whose composition is considered to be $b\bar{b}$. In March of 1995, researchers at Fermilab announced the discovery of the top quark (supposedly the last of the quarks to be found) having mass 173 GeV/$c^2$.

You are probably wondering whether or not such discoveries will ever end. How many "building blocks" of matter really exist? At the present, physicists believe that the fundamental particles in nature include six quarks and six leptons (together with their antiparticles). Some of the properties of these particles are given in Table 30.5.

Despite extensive experimental efforts, no isolated quark has ever been observed. Physicists now believe that quarks are permanently confined inside ordinary particles because of an exceptionally strong force that prevents them from escaping. This force, called the color force (which we discuss in Section 30.13), increases with separation distance (similar to the force of a spring). The great strength of the force between quarks has been described by one author as follows:[2]

**Webnote 30.8**

Further investigate the importance of the search for the top quark by exploring *http://www.duke.edu/vertices/update/fall94/quark.html*

> **Quarks are slaves of their own color charge, . . . bound like prisoners of a chain gang. . . . Any locksmith can break the chain between two prisoners, but no locksmith is expert enough to break the gluon chains between quarks. Quarks remain slaves forever.**

[2] Harald Fritzsch, *Quarks, The Stuff of Matter,* London: Allen Lane, 1983.

## 30.13 COLORED QUARKS

Shortly after the theory of quarks was proposed, scientists recognized that certain particles had quark compositions that were in violation of the Pauli exclusion principle. Because all quarks have spins of $\frac{1}{2}$, they are expected to follow the exclusion principle. One example of a particle that violates the exclusion principle is the $\Omega^-$ (sss) baryon that contains three s quarks having parallel spins, giving it a total spin of $\frac{3}{2}$. Other examples of baryons that have identical quarks with parallel spins are the $\Delta^{++}$ (uuu) and the $\Delta^-$ (ddd). To resolve this problem, Moo-Young Han and Yoichiro Nambu suggested in 1965 that quarks possess a new property called **color or color charge.** This "charge" property is similar in many respects to electric charge except that it occurs in three varieties labeled *red, green,* and *blue!* (The antiquarks are labled *antired, antigreen,* and *antiblue.*) To satisfy the exclusion principle, all three quarks in a baryon must have different colors. Just as a combination of actual colors of light can produce the neutral color white, a combination of three quarks with different colors is also white, or colorless. A meson consists of a quark of one color and an antiquark of the corresponding anticolor. The result is that baryons and mesons are always colorless (or white).

When we use the word color to describe a quark, it has nothing to do with visual sensation from light. It is simply a convenient name for a property analgous to electric charge.

Although the concept of color in the quark model was originally conceived to satisfy the exclusion principle, it also provided a better theory for explaining certain experimental results. For example, the modified theory correctly predicts the lifetime of the $\pi^0$ meson. The theory of how quarks interact with each other by means of color charge is called **quantum chromodynamics,** or QCD, to parallel quantum electrodynamics (the theory of interaction between electric charges). In QCD, the quark is said to carry a **color charge,** in analogy to electric charge. The strong force between quarks is often called the **color force.**

### Webnote 30.9

For more introductory remarks on quantum chromodynamics, take a look at
*http://www.infoplease.com/ce6/sci/A0840717.html*

The strong force between quarks is carried by massless particles called **gluons** (analogous to photons for the electromagnetic force). According to QCD, there are eight gluons, all with color charge. When a quark emits or absorbs a gluon, its color changes. For example, a blue quark that emits a gluon may become a red quark, and the red quark that absorbs this gluon becomes a blue quark. The color force between quarks is analogous to the electric force between charges; like colors repel and opposite colors attract. Therefore, two red quarks repel each other, but a red quark will be attracted to an antired quark. The attraction between quarks of opposite color to form a meson ($q\bar{q}$) is indicated in Figure 30.11a. Differently colored quarks also attract each other, but with less intensity than opposite colors of quark and antiquark. For example, a cluster of red, blue, and green quarks all attract each other to form baryons as indicated in Figure 30.11b. Thus, every baryon contains three quarks of three different colors.

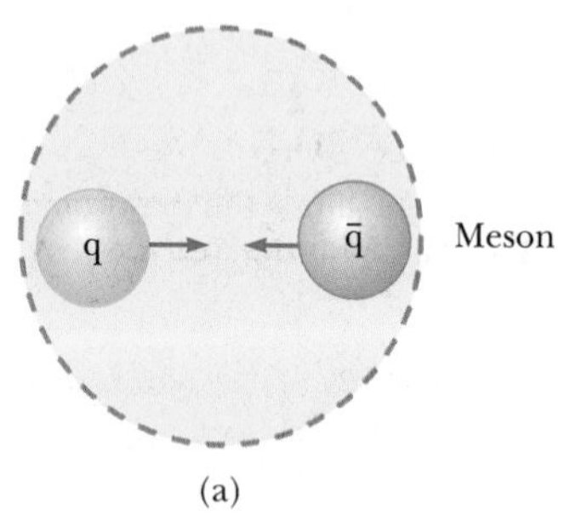


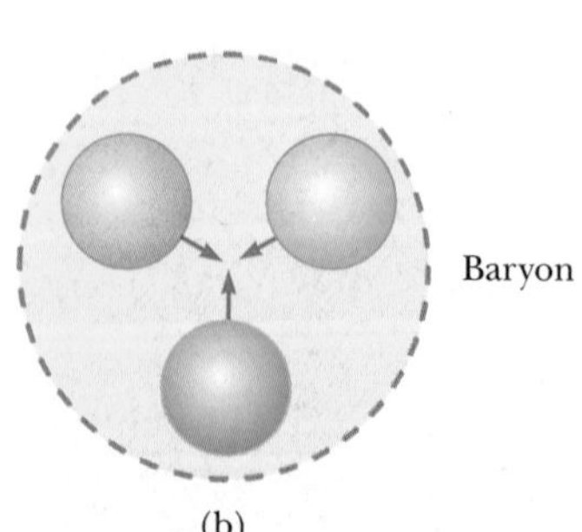


**FIGURE 30.11** (a) A red quark is attracted to an antired quark. This forms a meson whose quark structure is ($q\bar{q}$). (b) Three different colored quarks attract each other to form a baryon.

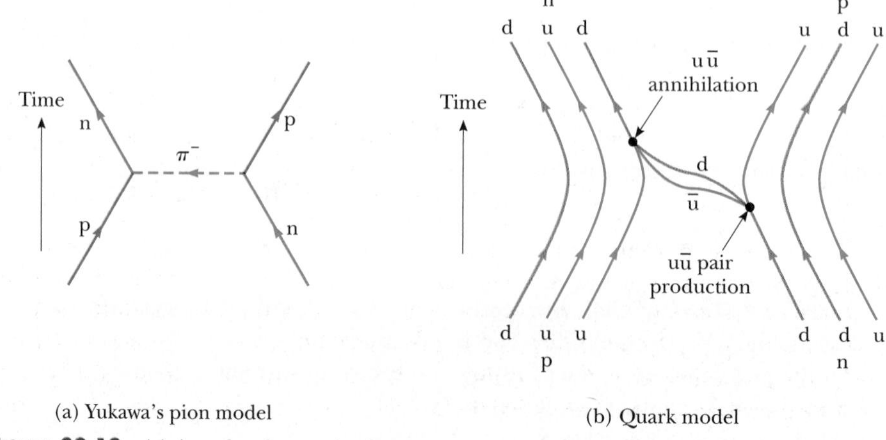


**FIGURE 30.12** (a) A nuclear interaction between a proton and a neutron explained in terms of Yukawa's pion exchange model. (b) The same interaction explained in terms of quarks and gluons. Note that the exchanged $\bar{u}d$ quark pair makes up a $\pi^-$ meson.

Although the color force between two color-neutral hadrons (like a proton and a neutron) is negligible at large separations, the strong color force between their constituent quarks does not exactly cancel at small separations of about 1 fm. **This residual strong force is in fact the nuclear force that binds protons and neutrons to form nuclei.** It is similar to the residual electromagnetic force that binds neutral atoms into molecules. According to QCD, a more basic explanation of nuclear force can be given in terms of quarks and gluons, as shown in Figure 30.12, which shows contrasting Feynman diagrams of the same process. Each quark within the neutron and proton is continually emitting and absorbing virtual gluons and creating and annihilating virtual ($q\overline{q}$) pairs. When the neutron and proton approach within 1 fm of each other, these virtual gluons and quarks can be exchanged between the two nucleons, and such exchanges produce the nuclear force. Figure 30.12b depicts one likely possibility or contribution to the process shown in Figure 30.12a. A down quark emits a virtual gluon (represented by a wavy line), which creates a $u\overline{u}$ pair. Both the recoiling d quark and the $\overline{u}$ are transmitted to the proton where the $\overline{u}$ annihilates a proton u quark (with the creation of a gluon) and the d is captured.

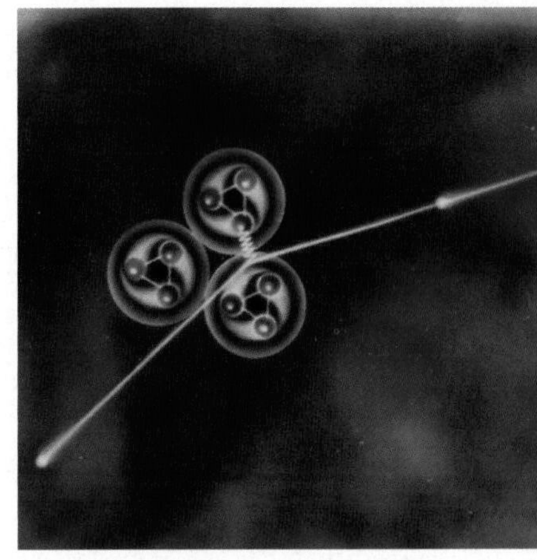

An artist's version of a high-energy particle colliding with a nucleus. The quark structure of the nucleus is indicated by the small colored spheres inside the nucleus. *(Courtesy of Janie Martz/CEBAF)*

## 30.14 Electroweak Theory and the Standard Model

Recall that the weak interaction is an extremely short-range force having an interaction distance of approximately $10^{-18}$ m (Table 30.1). Such a short-range interaction implies that the quantized particles that carry the weak field (the spin-1 $W^+$, $W^-$, and $Z^0$ bosons) are quite massive as is indeed the case. These amazing bosons can be thought of as structureless, point-like particles as massive as krypton atoms! The weak interaction is responsible for the decay of the c, s, b, and t quarks into lighter, more stable u and d quarks as well as the decay of the massive $\mu$ and $\tau$ leptons into (lighter) electrons. Thus, **the weak interaction is very important because it governs the stability of the basic matter particles.**

A mysterious feature of the weak interaction is its lack of symmetry, especially compared to the high degree of symmetry shown by the strong, electromagnetic, and gravitational interactions. For example, the weak interaction, unlike the strong interaction, is not symmetric under mirror reflection or charge exchange. (*Mirror reflection* means that all the quantities in a given particle reaction are exchanged as in a mirror reflection—left for right, an inward motion toward the mirror for an outward motion. *Charge exchange* means that all the electric charges in a particle reaction are converted to their opposites—all positives to negatives, and vice versa.) When we say that the weak interaction is not symmetric, we mean that the reaction with all quantities changed occurs less frequently than the direct reaction. For example, the decay of the $K^0$, which is governed by the weak interaction, is not symmetric under charge exchange because $K^0 \rightarrow \pi^- + e^+ + \nu_e$ occurs much more frequently than $K^0 \rightarrow \pi^+ + e^- + \overline{\nu}_e$.

In 1979, Sheldon Glashow, Abdus Salam, and Steven Weinberg won a Nobel prize for developing a theory called the **electroweak theory** that unified the electromagnetic and weak interactions. This theory postulates that the weak and electromagnetic interactions have the same strength at very high particle energies. Thus, the two interactions are viewed as two different manifestations of a single unifying electroweak interaction. The photon and the three massive bosons ($W^\pm$ and $Z^0$) play a key role in the electroweak theory. The theory makes many concrete predictions, but perhaps the most spectacular is the prediction of the masses of the W and Z particles at about 82 GeV/$c^2$ and 93 GeV/$c^2$, respectively. A 1984 Nobel prize was awarded to Carlo Rubbia and Simon van der Meer for their work leading to the discovery of these particles at just these energies at the CERN Laboratory in Geneva, Switzerland.

The combination of the electroweak theory and QCD for the strong interaction form what is referred to in high-energy physics as the **Standard Model.** Al-

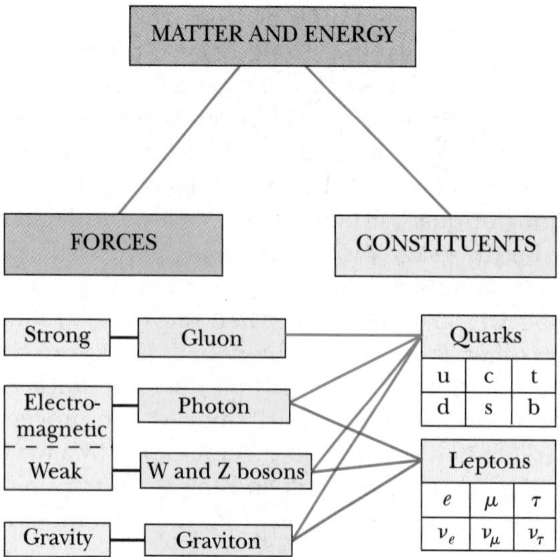


**FIGURE 30.13** The Standard Model of particle physics.

### Webnote 30.10

Visit the European Laboratory for Particle Physics in Geneva, Switzerland for more on accelerators and the invention of the World Wide Web. Go to the "General Public" link at *http://www.CERN.ch*

### Webnote 30.11

Take a look at a full-view chart of the particles in the Standard Model of particle physics. Visit *http://www.particleadventure.org/particleadventure*

A view from inside the LEP (Large Electron-Positron Collider) tunnel, which is 27 km in circumference. *(Courtesy of CERN)*

though the details of the Standard Model are complex, its essential ingredients can be summarized with the help of Figure 30.13. The strong force, mediated by gluons, holds quarks together to form composite particles such as protons, neutrons, and mesons. Leptons participate only in the electromagnetic and weak interactions. The electromagnetic force is mediated by photons, and the weak force is mediated by W and Z bosons. Note that all fundamental forces are mediated by bosons (particles with spin 1) whose properties are given, to a large extent, by symmetries involved in the theories.

However, the Standard Model does not answer all questions. A major question is why the photon has no mass, whereas the W and Z bosons do. Because of this mass difference, the electromagnetic and weak forces are quite distinct at low energies, but become similar in nature at very high energies, where the rest energies of the W and Z bosons are insignificant fractions of their total energies. This behavior as we go from high to low energies, called **symmetry breaking,** does not answer the question of the origin of particle masses. To resolve this problem, a hypothetical particle called the **Higgs boson** has been proposed that provides a mechanism for breaking the electroweak symmetry and bestowing different particle masses on different particles. The Standard Model, including the Higgs mechanism, provides a logically consistent explanation of the massive nature of the W and Z bosons. Unfortunately, the Higgs boson has not yet been found, but physicists know that its mass should be less than 1 TeV/$c^2$ ($10^{12}$ eV).

In order to determine whether the Higgs boson exists, two quarks of at least 1 TeV of energy must collide, but calculations show that this requires injecting 40 TeV of energy within the volume of a proton. Scientists are convinced that because of the limited energy available in conventional accelerators using fixed targets, it is necessary to build colliding-beam accelerators called **colliders.** The concept of colliders is straightforward. Particles with equal masses and kinetic energies, traveling in opposite directions in an accelerator ring, collide head-on to produce the required reaction and the formation of new particles. Because the total momentum of the interacting particles is zero, all of their kinetic energy is available for the reaction. The Large Electron-Positron collider (LEP) at CERN near Geneva, Switzerland, and the Stanford Linear Collider in California collide both electrons and positrons. The Super Proton Synchrotron at CERN accelerates protons and antiprotons to energies of 270 GeV, and the world's highest-energy proton acclerator, the Tevatron, at the Fermi National Laboratory in Illinois produces protons at almost 1 000 GeV (or 1 TeV). CERN has started construction of the Large Hadron Collider (LHC), a proton-proton collider that will provide a center of mass energy of 14 TeV and allow an exploration of Higgs-boson physics.

The accelerator is being constructed in the same 27-km circumference tunnel as CERN's Large Electron-Positron collider, and construction is expected to be completed in 2005.

Following the success of the electroweak theory, scientists attempted to combine it with QCD in a **grand unification theory** known as GUT. In this model, the electroweak force was merged with the strong color force to form a grand unified force. One version of the theory considers leptons and quarks as members of the same family that are able to change into each other by exchanging an appropriate particle. Many GUT theories predict that protons are unstable, and will decay with a lifetime of about $10^{31}$ years. This is far greater than the age of the Universe, and as yet proton decays have not been observed.

## APPLYING PHYSICS 30.4

Consider a car making a head-on collision with an identical car moving in the opposite direction at the same speed. Compare that collision to one in which one of the cars collides with the second car at rest. In which collision is there the larger transformation of kinetic energy to other forms during the collision? How does this relate to producing exotic particles in collisions?

**Explanation** In the head-on collision with both cars moving, conservation of momentum causes most, if not all, of the kinetic energy to be transformed to other forms. In the collision between a moving car and a stationary car, the cars are still moving after the collision, in the direction of the moving car, but with reduced speed. Thus, only part of the kinetic energy is transformed to other forms. This suggests the advantage of using colliding beams to produce exotic particles, as opposed to firing a beam into a stationary target. When particles moving in opposite directions collide, all of the kinetic energy is available for transformation into other forms—in this case, the creation of new particles. When a beam is fired into a stationary target, only part of the energy is available for transformation, so higher-mass particles cannot be created.

## 30.15 THE COSMIC CONNECTION

In this section we describe one of the most fascinating theories in all of science—the Big Bang theory of the creation of the Universe—and the experimental evidence that supports it. This theory of cosmology states that the Universe had a beginning, and that this beginning was so cataclysmic that it is impossible to look back beyond it. According to the theory, the Universe erupted from an infinitely dense singularity about 15–20 billion years ago. The first few minutes after the Big Bang saw such extremes of energy that it is believed that all four interactions of physics were unified and all matter was contained in an undifferentiated "quark soup."

The evolution of the four fundamental forces from the Big Bang to the present is shown in Figure 30.14. During the first $10^{-43}$ s (the ultra-hot epoch, $T \sim 10^{32}$ K), it is presumed that the strong, electroweak, and gravitational forces were joined to form a completely unified force. In the first $10^{-35}$ s following the Big Bang (the hot epoch, $T \sim 10^{29}$ K), gravity broke free of this unification and the strong and electroweak forces remained as one, described by a grand unification theory. This was a period when particle energies were so great ($> 10^{16}$ GeV) that very massive particles as well as quarks, leptons, and their antiparticles existed. Then, after $10^{-35}$ s, the Universe rapidly expanded and cooled (the warm epoch, $T \sim 10^{29}$ to $10^{15}$ K), the strong and electroweak forces parted company, and the grand unification scheme was broken. As the Universe continued to cool, the electroweak force split into the weak force and the electromagnetic force about $10^{-10}$ s after the Big Bang.

**GEORGE GAMOW (1904–1968)**

Gamow and two of his students, Ralph Alpher and Robert Herman, were the first to take the first half-hour of the Universe seriously. In a mostly overlooked paper published in 1948, they made truly remarkable cosmological predictions. They correctly calculated the abundances of hydrogen and helium after the first half-hour (75% H and 25% He) and predicted that radiation from the Big Bang should still be present and have an apparent temperature of about 5 K. *(Courtesy of AIP Emilio Segre Visual Archives)*

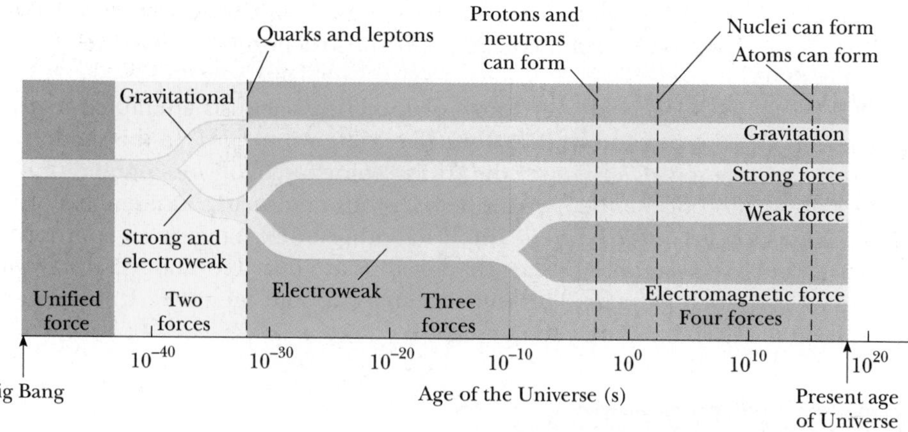


**FIGURE 30.14** A brief history of the Universe from the Big Bang to the present. The four forces became distinguishable during the first microsecond. Following this, all the quarks combined to form particles that interact via the strong force. The leptons remained separate, however, and exist as individually observable particles to this day.

After a few minutes, protons condensed out of the hot soup. For half an hour the Universe underwent thermonuclear detonation, exploding as a hydrogen bomb and producing most of the helium nuclei now present. The Universe continued to expand, and its temperature dropped. Until about 700 000 years after the Big Bang, the Universe was dominated by radiation. Energetic radiation prevented matter from forming single hydrogen atoms because collisions would instantly ionize any atoms that might form. Photons experienced continuous Compton scattering from the vast number of free electrons, resulting in a Universe that was opaque to radiation. By the time the Universe was about 700 000 years old, it had expanded and cooled to about 3 000 K, and protons could bind to electrons to form neutral hydrogen atoms. Because the energies of the atoms were quantized, far more wavelengths of radiation were not absorbed by atoms than were, and the Universe suddenly became transparent to photons. Radiation no longer dominated the Universe, and clumps of neutral matter steadily grew—first atoms, followed by molecules, gas clouds, stars, and finally galaxies.

## OBSERVATION OF RADIATION FROM THE PRIMORDIAL FIREBALL

In 1965 Arno A. Penzias (b. 1933) and Robert W. Wilson (b. 1936) of Bell Laboratories made an amazing discovery while testing a sensitive microwave receiver. A pesky signal producing a faint background hiss was interfering with their satellite communications experiments. In spite of their valiant efforts, the signal remained. Ultimately it became clear that they were observing microwave background radiation (at a wavelength of 7.35 cm) representing the leftover "glow" from the Big Bang.

The microwave horn that served as their receiving antenna is shown in Figure 30.15. The intensity of the detected signal remained unchanged as the antenna was pointed in different directions. The fact that the radiation had equal strengths in all directions suggested that the entire Universe was the source of this radiation. Evicting a flock of pigeons from the 20-foot horn and cooling the microwave detector both failed to remove the signal. Through a casual conversation, Penzias and Wilson discovered that a group at Princeton had predicted the residual radiation from the Big Bang and were planning an experiment to confirm the theory. The excitement in the scientific community was high when Penzias and Wilson announced that they had already observed an excess microwave background compatible with a 3-K blackbody source.

**FIGURE 30.15** Robert W. Wilson *(left)* and Arno A. Penzias *(right)* with Bell Telephone Laboratories' horn-reflector antenna. *(AT&T Bell Laboratories)*

Because Penzias and Wilson made their measurements at a single wavelength, they did not completely confirm the radiation as 3-K blackbody radiation. Subsequent experiments by other groups added intensity data at different wavelengths as shown in Figure 30.16. The results confirm that the radiation is that of a black body at 2.9 K. This figure is perhaps the most clear-cut evidence for the Big Bang

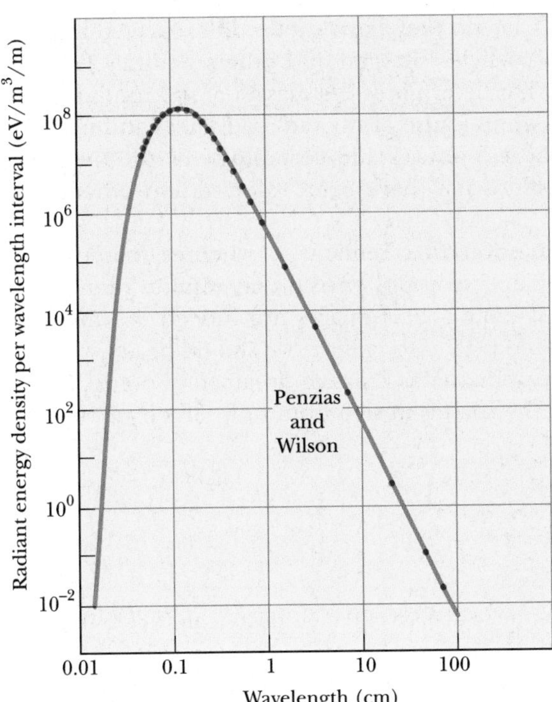


**FIGURE 30.16** Theoretical blackbody (red curve) and measured radiation spectra (black points) of the Big Bang. Most of the data were collected from the Cosmic Background Explorer (COBE) satellite. The data of Wilson and Penzias are indicated.

theory. The 1978 Nobel prize in physics was awarded to Penzias and Wilson for this most important discovery.

The discovery of the cosmic background radiation produced a problem, however—the radiation was too uniform. Scientists believed there had to be slight fluctuations in this background in order for such objects as galaxies to form. In 1989 NASA launched a satellite called COBE (KOH-bee), for Cosmic Background Explorer, to study this radiation in greater detail. In 1992 George Smoot (b. 1945) at the Lawrence Berkeley Laboratory, based on the data collected, found that the background was not perfectly uniform but instead contained irregularities corresponding to temperature variations of 0.000 3 K. It is these small variations that provided nucleation sites for the formation of the galaxies and other objects we now see in the sky.

## 30.16 PROBLEMS AND PERSPECTIVES

While particle physicists have been exploring the realm of the very small, cosmologists have been exploring cosmic history back to the first microsecond of the Big Bang. Observation of the events that occur when two particles collide in an accelerator is essential in reconstructing the early moments in cosmic history. Perhaps the key to understanding the early Universe is to first understand the world of elementary particles. Cosmologists and particle physicists find that they have many common goals and are now joining efforts to attempt to study the physical world at its most fundamental level.

Our understanding of physics at short and long distances is far from complete. Particle physics is faced with many questions. Why is there so little antimatter in the Universe? Do neutrinos have a small mass, and if so, how much do they contribute to the "dark matter" holding the Universe together gravitationally? How can we understand the latest astronomical measurements, which show that the expansion of the Universe is accelerating and that there may be a kind of "antigravity force" acting between widely separated galaxies? Is it possible to unify the strong and electroweak theories in a logical and consistent manner? Why do quarks and leptons form three similar but distinct families? Are muons the same as electrons (apart from their different masses), or do they have subtle differences that have not been detected? Why are some particles charged and others neutral? Why do quarks carry a fractional charge? What determines the masses of the fundamental particles? The questions go on and on. Because of the rapid advances and new discoveries in the related fields of particle physics and cosmology, by the time you read this book some of these questions may have been resolved and others may have emerged.

An important and obvious question that remains is whether leptons and quarks have a substructure. If they do, we could envision an infinite number of deeper structure levels. However, if leptons and quarks are indeed the ultimate constituents of matter, as physicists today tend to believe, we should be able to construct a final theory of the structure of matter as Einstein dreamed of doing. In the view of many physicists, the end of the road is in sight, but how long it will take to reach that goal is anyone's guess.

## SUMMARY

In **nuclear fission** and **nuclear fusion,** the total mass of the products is always less than the original mass of the reactants. Nuclear fission occurs when a heavy nucleus splits, or fissions, into two smaller nuclei. In nuclear fusion, two light nuclei combine to form a heavier nucleus.

A **nuclear reactor** is a system designed to maintain a self-sustaining chain reaction. Nuclear reactors using controlled fission events are currently being used to generate electric power.

Controlled fusion events offer the hope of plentiful supplies of energy in the future. The nuclear fusion reactor is considered by many scientists to be the ultimate energy source because its fuel is water. **Lawson's criterion** states that a fusion reactor will provide a net output power if the product of the plasma ion density, $n$, and the plasma confinement time, $\tau$, satisfies the following relationships:

$$n\tau \geq 10^{14}\ \mathrm{s/cm^3} \quad \text{Deuterium-tritium interaction}$$

$$n\tau \geq 10^{16}\ \mathrm{s/cm^3} \quad \text{Deuterium-deuterium interaction} \qquad [30.5]$$

Four fundamental forces in nature: **strong** (hadronic), **electromagnetic, weak,** and **gravitational.** The strong force is the force between nucleons that keeps the nucleus together. The weak force is responsible for beta decay. The electromagnetic and weak forces are now considered to be manifestations of a single force called the **electroweak** force.

Every fundamental interaction is said to be mediated by the exchange of field particles. The electromagnetic interaction is mediated by the photon; the weak interaction is mediated by the $W^{\pm}$ and $Z^0$ bosons; the gravitational interaction is mediated by gravitons; and the strong interaction is mediated by gluons.

An antiparticle and a particle have the same mass but opposite charge, and other properties may also have opposite values, such as lepton number and baryon number. It is possible to produce particle-antiparticle pairs in nuclear reactions if the available energy is greater than $2mc^2$, where $m$ is the mass of the particle (or antiparticle).

Particles other than photons are classified as hadrons or leptons. **Hadrons** interact primarily through the strong force. They have size and structure and hence are not elementary particles. There are two types of hadrons, *baryons* and *mesons*. Mesons have a baryon number of 0 and have either zero or integer spin. Baryons, which generally are the most massive particles, have nonzero baryon numbers and spins of $\frac{1}{2}$ or $\frac{3}{2}$. The neutron and proton are examples of baryons.

**Leptons** have no structure or size and are considered truly elementary particles. Leptons interact only through the weak and electromagnetic forces. There are six leptons: the electron, $e^-$; the muon, $\mu^-$; the tau, $\tau^-$; and their associated neutrinos, $\nu_e$, $\nu_\mu$, and $\nu_\tau$.

In all reactions and decays, quantities such as energy, linear momentum, angular momentum, electric charge, baryon number, and lepton number are strictly conserved. Certain particles have properties called **strangeness** and **charm.** These unusual properties are conserved only in those reactions and decays that occur via the strong force.

Recent theories postulate that all hadrons are composed of smaller units known as **quarks,** which have fractional electric charges and baryon numbers of $\frac{1}{3}$ and come in six "flavors": up, down, strange, charmed, top, and bottom. Each baryon contains three quarks, and each meson contains one quark and one antiquark.

According to the theory of **quantum chromodynamics,** quarks have a property called **color,** and the strong force between quarks is referred to as the **color force.**

Observation of background microwave radiation by Penzias and Wilson strongly confirmed that the Universe started with a Big Bang about 15 billion years ago. The background radiation is equivalent to that of a blackbody at a temperature of about 3 K.

## CONCEPTUAL QUESTIONS

1. If high-energy electrons with the de Broglie wavelengths smaller than the size of the nucleus are scattered from nuclei, the behavior of the electrons is consistent with scattering from very dense structures much smaller in size than the nucleus—quarks. How is this similar to another classic experiment that detected small structures in atoms?
2. What factors make a fusion reaction difficult to achieve?
3. Doubly charged baryons are known to exist. Why are there no doubly charged mesons?
4. Why would a fusion reactor produce less radioactive waste than a fission reactor?
5. Atoms did not exist until hundreds of thousands of years after the Big Bang. Why?
6. Particles known as resonances have very short half-lives, of the order of $10^{-23}$ s. Would you guess they are hadrons or leptons?
7. Describe the quark model of hadrons, including the properties of quarks.
8. In the theory of quantum chromodynamics, quarks come in three colors. How would you justify the statement that "all baryons and mesons are colorless"?
9. Describe the properties of baryons and mesons and the important differences between them.
10. Identify the particle decays in Table 30.2 that occur by the electromagnetic interaction. Justify your answer.
11. Kaons all decay into final states that contain no protons or neutrons. What is the baryon number of kaons?
12. When an electron and a positron meet at low speeds in free space, why are *two* 0.511-MeV gamma rays produced, rather than *one* gamma ray with an energy of 1.02 MeV?
13. Two protons in a nucleus interact via the strong interaction. Are they also subject to the weak interaction?
14. Why is a neutron stable inside the nucleus? (In free space, the neutron decays in 900 s.)
15. An antibaryon interacts with a meson. Can a baryon be produced in such an interaction? Explain.
16. Why is water a better shield against neutrons than lead or steel?
17. How many quarks are there in (a) a baryon, (b) an antibaryon, (c) a meson, (d) an antimeson? How do you account for the fact that baryons have half-integral spins and mesons have spins of 0 or 1? (*Hint:* Quarks have spin $\frac{1}{2}$.)
18. Which of the four fundamental forces have effects on an electron? On a proton? On a neutron?
19. List similarities and differences between nuclear-powered electric generating stations and coal-powered ones.

## PROBLEMS

1, 2, 3 = straightforward, intermediate, challenging ☐ = full solution available in Student Solutions Manual/Study Guide

web = solution posted at **http://info.brookscole.com/serway** [icon] = biomedical application

### Section 30.1 Nuclear Fission

### Section 30.2 Nuclear Reactors

1. When $^{235}$U absorbs a neutron, one possible fission reaction produces $^{141}$Ba and $^{92}$Kr as products. Write down this reaction. How many neutrons are released?
2. Find the energy released in the fission reaction

$$\text{n} + {}^{235}_{92}\text{U} \longrightarrow {}^{98}_{40}\text{Zr} + {}^{135}_{52}\text{Te} + 3\,\text{n}$$

The atomic masses of the fission products are: ${}^{98}_{40}$Zr, 97.912 0 u; ${}^{135}_{52}$Te, 134.908 7 u.

3. Find the energy released in the following fission reaction:

$${}^{1}_{0}\text{n} + {}^{235}_{92}\text{U} \longrightarrow {}^{88}_{38}\text{Sr} + {}^{136}_{54}\text{Xe} + 12\,{}^{1}_{0}\text{n}$$

4. Strontium-90 is a particularly dangerous fission product of $^{235}$U because it is radioactive and it substitutes for calcium in bones. What other direct fission products would accompany it in the neutron-induced fission of $^{235}$U? (*Note:* This reaction may release two, three, or four free neutrons.)
5. Assume that ordinary soil contains natural uranium in amounts of 1 part per million by mass. (a) How much uranium is in the top 1.00 meter of soil on a one acre (43 560 ft$^2$) plot of ground, assuming the specific gravity of soil is 4.00? (b) How much of the isotope $^{235}$U, appropriate for nuclear reactor fuel, is in this soil? (*Hint:* See Appendix B for the percent abundance of ${}^{235}_{92}$U.)
6. A typical nuclear fission power plant produces about 1.00 GW of electrical power. Assume that the plant has an

overall efficiency of 40.0% and that each fission produces 200 MeV of thermal energy. Calculate the mass of $^{235}U$ consumed each day.

7. Suppose that the water exerts an average frictional drag of 1.0 × $10^5$ N on a nuclear-powered ship. How far can the ship travel per kilogram of fuel if the fuel consists of enriched uranium containing 1.7% of the fissionable isotope $^{235}U$, and the ship's engine has a efficiency of 20%? (Assume 208 MeV released per fission event.)

8. The first atomic bomb released an energy equivalent to 20 kilotons of TNT. If 1 ton of TNT releases about 4.0 × $10^9$ J, how much uranium was fissioned in this bomb? (Assume 208 MeV released per fission.)

9. An all-electric home uses approximately 2 000 kWh of electric energy per month. How much uranium-235 would be required to provide this house with its energy needs for 1 year? (Assume 100% conversion efficiency and 208 MeV released per fission.)

## Section 30.3 Nuclear Fusion

10. Find the energy released in the fusion reaction

$$^1_1H + ^2_1H \longrightarrow ^3_2He + \gamma$$

11. When a star has exhausted its hydrogen fuel, it may fuse other nuclear fuels. At temperatures above 1.0 × $10^8$ K, helium fusion can occur. Write the equation for the processes described below. (a) Two alpha particles fuse to produce a nucleus A and a gamma ray. What is nucleus A? (b) Nucleus A absorbs an alpha particle to produce a nucleus B and a gamma ray. What is nucleus B? (c) Find the total energy released in the sequence of reactions given in (a) and (b).

12. Another series of nuclear reactions that can produce energy in the interior of stars is the cycle described below. This process is most efficient when the central temperature in a star is above 1.6 × $10^7$ K. Because the temperature at the center of the Sun is only 1.5 × $10^7$ K, the cycle below produces less than 10% of the Sun's energy. (a) A high-energy proton is absorbed by $^{12}C$. Another nucleus, A, is produced in the reaction, along with a gamma ray. Identify nucleus A. (b) Nucleus A decays through positron emission to form nucleus B. Identify nucleus B. (c) Nucleus B absorbs a proton to produce nucleus C and a gamma ray. Identify nucleus C. (d) Nucleus C absorbs a proton to produce nucleus D and a gamma ray. Identify nucleus D. (e) Nucleus D decays through positron emission to produce nucleus E. Identify nucleus E. (f) Nucleus E absorbs a proton to produce nucleus F plus an alpha particle. What is nucleus F? (*Note:* If nucleus F is not $^{12}C$, the nucleus you started with, you have made an error and should review the sequence of events.)

13. If an all-electric home uses approximately 2 000 kWh of electric energy per month, how many fusion events described by the reaction $^2_1H + ^3_1H \rightarrow ^4_2He + ^1_0n$ would be required to keep this house running for one year?

14. To understand why plasma containment is necessary, consider the rate at which an unconfined plasma would be lost. (a) Estimate the rms speed of deuterons in a plasma at 4.00 × $10^8$ K. (b) Estimate the order of magnitude of the time such a plasma would remain in a 10-cm cube if no steps were taken to contain it.

15. The oceans have a volume of 317 million cubic miles and contain 1.32 × $10^{21}$ kg of water. Of all the hydrogen nuclei in this water, 0.0300% of the mass is deuterium. (a) If all of these deuterium nuclei were fused to helium via the first reaction in Equation 30.4, determine the total amount of energy that could be released. (b) Present world electric power consumption is about 7.00 × $10^{12}$ W. If consumption were 100 times greater, how many years would the energy supply calculated in (a) last?

## Section 30.6 Positrons and Other Antiparticles

16. Two photons are produced when a proton and an antiproton annihilate each other. What is the minimum frequency and corresponding wavelength of each photon?

17. A photon produces a proton-antiproton pair according to the reaction $\gamma \rightarrow p + \bar{p}$. What is the minimum possible frequency of the photon? What is its wavelength?

18. A photon with an energy of 2.09 GeV creates a proton-antiproton pair in which the proton has a kinetic energy of 95.0 MeV. What is the kinetic energy of the antiproton?

## Section 30.7 Mesons and the Beginning of Particle Physics

19. When a high-energy proton or pion traveling near the speed of light collides with a nucleus, it travels an average distance of 3.0 × $10^{-15}$ m before interacting. From this information, estimate the order of magnitude of the time for the strong interaction to occur.

20. Calculate the order of magnitude of the range of the force that might be produced by the virtual exchange of a proton.

21. One of the mediators of the weak interaction is the $Z^0$ boson, whose mass is 93 GeV/$c^2$. Use this information to find the order of magnitude of the range of the weak interaction.

22. If a $\pi^0$ at rest decays into two $\gamma$'s, what is the energy of each of the $\gamma$'s?

## Section 30.9 Conservation Laws

## Section 30.10 Strange Particles and Strangeness

23. Each of the following reactions is forbidden. Determine a conservation law that is violated for each reaction.
(a) $p + \bar{p} \rightarrow \mu^+ + e^-$
(b) $\pi^- + p \rightarrow p + \pi^+$
(c) $p + p \rightarrow p + \pi^+$
(d) $p + p \rightarrow p + p + n$
(e) $\gamma + p \rightarrow n + \pi^0$

24. (a) Show that baryon number and charge are conserved in the following reactions of a pion with a proton.

$$\pi^+ + p \longrightarrow K^+ + \Sigma^+ \quad (1)$$

$$\pi^+ + p \longrightarrow \pi^+ + \Sigma^+ \quad (2)$$

(b) The first reaction is observed, but the second never occurs. Explain these observations.

25. Identify the unknown particle on the left side of the reaction

$$? + p \longrightarrow n + \mu^+$$

26. Determine the type of neutrino or antineutrino involved in each of the following processes:
(a) $\pi^+ \rightarrow \pi^0 + e^+ + ?$
(b) $? + p \rightarrow \mu^- + p + \pi^+$
(c) $\Lambda^0 \rightarrow p + \mu^- + ?$
(d) $\tau^+ \rightarrow \mu^+ + ? + ?$

**27.** The following reactions or decays involve one or more neutrinos. Supply the missing neutrinos.
(a) $\pi^- \to \mu^- + ?$ (d) $? + n \to p + e^-$
(b) $K^+ \to \mu^+ + ?$ (e) $? + n \to p + \mu^-$
(c) $? + p \to n + e^+$ (f) $\mu^- \to e^- + ? + ?$

**28.** Determine which of the reactions below can occur. For those that cannot occur, determine the conservation law (or laws) that each violates.
(a) $p \to \pi^+ + \pi^0$ (d) $\pi^+ \to \mu^+ + \nu_\mu$
(b) $p + p \to p + p + \pi^0$ (e) $n \to p + e^- + \bar{\nu}_e$
(c) $p + p \to p + \pi^+$ (f) $\pi^+ \to \pi^+ + n$

**29.** Which of the following processes are allowed by the strong interaction, the electromagnetic interaction, the weak interaction, or no interaction at all?
(a) $\pi^- + p \to 2\eta^0$ (d) $\Omega^- \to \Xi^- + \pi^0$
(b) $K^- + n \to \Lambda^0 + \pi^-$ (e) $\eta^0 \to 2\gamma$
(c) $K^- \to \pi^- + \pi^0$

**30.** A $K^0$ particle at rest decays into a $\pi^+$ and a $\pi^-$. What will be the speed of each of the pions? The mass of the $K^0$ is 497.7 MeV/$c^2$ and the mass of each pion is 139.6 MeV/$c^2$.

**31.** Determine whether or not strangeness is conserved in the following decays and reactions.
(a) $\Lambda^0 \to p + \pi^-$ (d) $\pi^- + p \to \pi^- + \Sigma^+$
(b) $\pi^- + p \to \Lambda^0 + K^0$ (e) $\Xi^- \to \Lambda^0 + \pi^-$
(c) $\bar{p} + p \to \bar{\Lambda}^0 + \Lambda^0$ (f) $\Xi^0 \to p + \pi^-$

**32.** Fill in the missing particle. Assume that (a) occurs via the strong interaction and that (b) and (c) involve the weak interaction.
(a) $K^+ + p \to$ ___ $+ p$
(b) $\Omega^- \to$ ___ $+ \pi^-$
(c) $K^+ \to$ ___ $+ \mu^+ + \nu_\mu$

**33.** Identify the conserved quantities in the following processes:
(a) $\Xi^- \to \Lambda^0 + \mu^- + \nu_\mu$ (d) $\Sigma^0 \to \Lambda^0 + \gamma$
(b) $K^0 \to 2\pi^0$ (e) $e^+ + e^- \to \mu^+ + \mu^-$
(c) $K^- + p \to \Sigma^0 + n$ (f) $\bar{p} + n \to \bar{\Lambda}^0 + \Sigma^-$

## Section 30.12 Quarks

## Section 30.13 Colored Quarks

**34.** The quark composition of the proton is uud, whereas that of the neutron in udd. Show that the charge, baryon number, and strangeness of these particles equal the sums of these numbers for their quark constituents.

**35.** Find the number of electrons, and of each species of quarks, in 1 liter of water.

**36.** The quark compositions of the $K^0$ and $\Lambda^0$ particles are $d\bar{s}$ and uds, respectively. Show that the charge, baryon number, and strangeness of these particles equal the sums of these numbers for the quark constituents.

**37.** Identify the particles corresponding to the following quark states: (a) suu; (b) $\bar{u}d$; (c) $\bar{s}d$; (d) ssd.

**38.** What is the electrical charge of the baryons with the quark compositions (a) $\bar{u}\bar{u}\bar{d}$ and (b) $\bar{u}\bar{d}\bar{d}$? What are these baryons called?

**39.** Analyze the first three of the following reactions at the quark level and show that each conserves the net number of each type quark. In the last reaction, identify the mystery particle.
(a) $\pi^- + p \to K^0 + \Lambda^0$
(b) $\pi^+ + p \to K^+ + \Sigma^+$
(c) $K^- + p \to K^+ + K^0 + \Omega^-$
(d) $p + p \to K^0 + p + \pi^+ + ?$

**40.** Imagine binding energies could be neglected. Find the masses of the u and d quarks from the masses of the proton and neutron.

## ADDITIONAL PROBLEMS

**41.** A $\Sigma^0$ particle traveling through matter strikes a proton. A $\Sigma^+$ and a gamma ray emerge, as well as a third particle. Use the quark model of each to determine the identity of the third particle.

**42.** It was stated in the text that the reaction $\pi^- + p^+ \to K^0 + \Lambda^0$ occurs with high probability, whereas the reaction $\pi^- + p^+ \to K^0 + n$ never occurs. Analyze these reactions at the quark level and show that the first conserves the net number of each type of quark while the second reaction does not.

**43.** Two protons approach each other with equal and opposite velocities. What is the minimum kinetic energy of each of the protons if they are to produce a $\pi^+$ meson at rest in the reaction,

$$p + p \longrightarrow p + n + \pi^+$$

**44.** Name at least one conservation law that prevents each of the following reactions:
(a) $\pi^- + p \to \Sigma^+ + \pi^0$
(b) $\mu^- \to \pi^- + \nu_e$
(c) $p \to \pi^+ + \pi^+ + \pi^-$

**45.** Find the energy released in the fusion reaction

$${}^1_1H + {}^3_2He \longrightarrow {}^4_2He + e^+ + \nu$$

**46.** Occasionally, high-energy muons collide with electrons and produce two neutrinos according to the reaction $\mu^+ + e^- \to 2\nu$. What kind of neutrinos are these?

**47.** Each of the following decays is forbidden. For each process, determine a conservation law that is violated.
(a) $\mu^- \to e^- + \gamma$ (d) $p \to e^+ + \pi^0$
(b) $n \to p + e^- + \nu_e$ (e) $\Xi^0 \to n + \pi^0$
(c) $\Lambda^0 \to p + \pi^0$

**48.** Two protons approach each other head on, each with 70.4 MeV of kinetic energy, and engage in a reaction in which a proton and positive pion emerge at rest. What third particle, obviously uncharged and therefore difficult to detect, must have been created?

**49.** The atomic bomb dropped on Hiroshima on August 6, 1945, released $5 \times 10^{13}$ J of energy (equivalent to that from 12 000 tons of TNT). Estimate (a) the number of ${}^{235}_{92}U$ nuclei fissioned, and (b) the mass of this ${}^{235}_{92}U$.

**50.** A 2.0-MeV neutron is emitted in a fission reactor. If it loses one half of its kinetic energy in each collision with a moderator atom, how many collisions must it undergo in order to achieve thermal energy (0.039 eV)?

**51.** If baryon number were not conserved, then one possible mechanism by which a proton could decay would be

$$p \longrightarrow e^+ + \gamma$$

(a) Show that this reaction violates conservation of baryon number. (b) Assuming that this reaction occurs, and that the proton is initially at rest, determine the energy and momentum of the photon after the reaction. (*Hint:* Recall that energy and momentum must be conserved in the reaction.) (c) Determine the speed of the positron after the reaction.

52. Classical general relativity views the space-time manifold as a deterministic structure completely well defined down to arbitrarily small distances. On the other hand, quantum general relativity forbids distances smaller than the Planck length given by $L = (\hbar G/c^3)^{1/2}$. (a) Calculate the value of $L$. The answer suggests that after the Big Bang (when all the known Universe was reduced to a singularity), nothing could be observed until that singularity grew larger than the Planck length, $L$. Because the size of the singularity grew at the speed of light, we can infer that during the time it took for light to travel the Planck length, no observations were possible. (b) Determine this time (known as the Planck time, $T$) and compare it to the ultra-hot epoch discussed in the text. (c) Does this suggest that we may never know what happened between the time $t = 0$ and the time $t = T$?

## GROUP ACTIVITIES

**G.1** (a) A typical chemical reaction is one in which a water molecule is formed by combining hydrogen and oxygen. In such a reaction about 2.5 eV of energy is released. Compare this reaction to ${}^1_0\text{n} + {}^{235}_{92}\text{U} \rightarrow {}^{136}_{53}\text{I} + {}^{98}_{39}\text{Y} + 2\,{}^1_0\text{n}$ (Take the mass of ${}^{136}\text{I}$ to be 135.914 650 u and the mass of ${}^{98}\text{Y}$ to be 97.922 300 u.) Would you expect the energy released in this reaction to be much greater, much less, or about the same as the chemical reaction? (b) Find the actual energy released in this reaction.

**G.2** (a) Explain in a couple of sentences what is meant by the conservation of baryons. (b) Use conservation of baryons to show that a proton cannot decay into a positron and a neutral pion.

**G.3** (a) Explain in a couple of sentences what is meant by conservation of strangeness. (b) Use conservation of strangeness to show that a reaction in which a negative pion strikes a proton producing a neutral kaon and a neutron cannot occur.

"QUARKS. NEUTRINOS. MESONS. ALL THOSE DAMN PARTICLES YOU CAN'T SEE. THAT'S WHAT DROVE ME TO DRINK. BUT NOW I CAN SEE THEM."

# Appendix A

# Mathematical Review

## A.1 Mathematical Notation

Many mathematical symbols are used throughout this book. You are no doubt familiar with some, such as the symbol $=$ to denote the equality of two quantities.

The symbol $\propto$ denotes a proportionality. For example, $y \propto x^2$ means that $y$ is proportional to the square of $x$.

The symbol $<$ means *is less than,* and $>$ means *is greater than.* For example, $x > y$ means $x$ is greater than $y$.

The symbol $\ll$ means *is much less than,* and $\gg$ means *is much greater than.*

The symbol $\approx$ indicates that two quantities are *approximately equal* to each other.

The symbol $\equiv$ means *is defined as.* This is a stronger statement than a simple $=$.

It is convenient to use the notation $\Delta x$ (read as "delta $x$") to indicate the *change in the quantity* $x$. (Note that $\Delta x$ does not mean "the product of $\Delta$ and $x$.") For example, suppose that a person out for a morning stroll starts measuring her distance away from home when she is 10 m from her doorway. She then moves along a straight-line path and stops strolling 50 m from the door. Her change in position during the walk is $\Delta x = 50\text{ m} - 10\text{ m} = 40\text{ m}$ or, in symbolic form,

$$\Delta x = x_f - x_i$$

In this equation $x_f$ is the *final position* and $x_i$ is the *initial position.*

We often have occasion to add several quantities. A useful abbreviation for representing such a sum is the Greek letter $\Sigma$ (capital sigma). Suppose we wish to add a set of five numbers represented by $x_1$, $x_2$, $x_3$, $x_4$, and $x_5$. In the abbreviated notation, we would write the sum as

$$x_1 + x_2 + x_3 + x_4 + x_5 = \sum_{i=1}^{5} x_i$$

where the subscript $i$ on $x$ represents any one of the numbers in the set. For example, if there are five masses in a system, $m_1$, $m_2$, $m_3$, $m_4$, and $m_5$, the total mass of the system $M = m_1 + m_2 + m_3 + m_4 + m_5$ could be expressed as

$$M = \sum_{i=1}^{5} m_i$$

Finally, the magnitude of a quantity $x$, written $|x|$, is simply the absolute value of that quantity. The sign of $|x|$ is always positive, regardless of the sign of $x$. For example, if $x = -5$, $|x| = 5$; if $x = 8$, $|x| = 8$.

## A.2 Scientific Notation

Many quantities that scientists deal with often have very large or very small values. For example, the speed of light is about 300 000 000 m/s and the ink required to make the dot over an *i* in this textbook has a mass of about 0.000 000 001 kg. Obviously, it is cumbersome to read, write, and keep track of numbers such as these. We avoid this problem by using a method dealing with powers of the number 10:

$$10^0 = 1$$
$$10^1 = 10$$
$$10^2 = 10 \times 10 = 100$$
$$10^3 = 10 \times 10 \times 10 = 1\,000$$
$$10^4 = 10 \times 10 \times 10 \times 10 = 10\,000$$
$$10^5 = 10 \times 10 \times 10 \times 10 \times 10 = 100\,000$$

and so on. The number of zeros corresponds to the power to which 10 is raised, called the **exponent** of 10. For example, the speed of light, 300 000 000 m/s, can be expressed as $3 \times 10^8$ m/s.

For numbers less than one, we note the following:

$$10^{-1} = \frac{1}{10} = 0.1$$
$$10^{-2} = \frac{1}{10 \times 10} = 0.01$$
$$10^{-3} = \frac{1}{10 \times 10 \times 10} = 0.001$$
$$10^{-4} = \frac{1}{10 \times 10 \times 10 \times 10} = 0.000\,1$$
$$10^{-5} = \frac{1}{10 \times 10 \times 10 \times 10 \times 10} = 0.000\,01$$

In these cases, the number of places the decimal point is to the left of the digit 1 equals the value of the (negative) exponent. Numbers that are expressed as some power of 10 multiplied by another number between 1 and 10 are said to be in **scientific notation.** For example, the scientific notation for 5 943 000 000 is $5.943 \times 10^9$ and that for 0.000 083 2 is $8.32 \times 10^{-5}$.

When numbers expressed in scientific notation are being multiplied, the following general rule is very useful:

$$10^n \times 10^m = 10^{n+m} \qquad \textbf{[A.1]}$$

where $n$ and $m$ can be *any* numbers (not necessarily integers). For example, $10^2 \times 10^5 = 10^7$. The rule also applies if one of the exponents is negative. For example, $10^3 \times 10^{-8} = 10^{-5}$.

When dividing numbers expressed in scientific notation, note that

$$\frac{10^n}{10^m} = 10^n \times 10^{-m} = 10^{n-m} \qquad \textbf{[A.2]}$$

## Exercises

With help from the above rules, verify the answers to the following:

1. $86\,400 = 8.64 \times 10^4$
2. $9\,816\,762.5 = 9.816\,762\,5 \times 10^6$
3. $0.000\,000\,039\,8 = 3.98 \times 10^{-8}$
4. $(4.0 \times 10^8)(9.0 \times 10^9) = 3.6 \times 10^{18}$
5. $(3.0 \times 10^7)(6.0 \times 10^{-12}) = 1.8 \times 10^{-4}$
6. $\dfrac{75 \times 10^{-11}}{5.0 \times 10^{-3}} = 1.5 \times 10^{-7}$
7. $\dfrac{(3 \times 10^6)(8 \times 10^{-2})}{(2 \times 10^{17})(6 \times 10^5)} = 2 \times 10^{-18}$

# A.3 ALGEBRA

## A. SOME BASIC RULES

When algebraic operations are performed, the laws of arithmetic apply. Symbols such as $x$, $y$, and $z$ are frequently used to represent quantities that are not specified, what are called the **unknowns.**

First, consider the equation

$$8x = 32$$

If we wish to solve for $x$, we can divide (or multiply) each side of the equation by the same factor without destroying the equality. In this case, if we divide both sides by 8, we have

$$\frac{8x}{8} = \frac{32}{8}$$

$$x = 4$$

Next consider the equation

$$x + 2 = 8$$

In this type of expression, we can add or subtract the same quantity from each side. If we subtract 2 from each side, we get

$$x + 2 - 2 = 8 - 2$$

$$x = 6$$

In general, if $x + a = b$, then $x = b - a$.

Now consider the equation

$$\frac{x}{5} = 9$$

If we multiply each side by 5, we are left with $x$ on the left by itself and 45 on the right:

$$\left(\frac{x}{5}\right)(5) = 9 \times 5$$

$$x = 45$$

In all cases, **whatever operation is performed on the left side of the equality must also be performed on the right side.**

The following rules for multiplying, dividing, adding, and subtracting fractions should be recalled, where $a$, $b$, and $c$ are three numbers:

| | Rule | Example |
|---|---|---|
| **Multiplying** | $\left(\frac{a}{b}\right)\left(\frac{c}{d}\right) = \frac{ac}{bd}$ | $\left(\frac{2}{3}\right)\left(\frac{4}{5}\right) = \frac{8}{15}$ |
| **Dividing** | $\frac{(a/b)}{(c/d)} = \frac{ad}{bc}$ | $\frac{2/3}{4/5} = \frac{(2)(5)}{(4)(3)} = \frac{10}{12}$ |
| **Adding** | $\frac{a}{b} \pm \frac{c}{d} = \frac{ad \pm bc}{bd}$ | $\frac{2}{3} - \frac{4}{5} = \frac{(2)(5) - (4)(3)}{(3)(5)} = -\frac{2}{15}$ |

## Exercises

In the following exercises, solve for $x$:

### Answers

1. $a = \dfrac{1}{1 + x}$ $\quad x = \dfrac{1 - a}{a}$
2. $3x - 5 = 13$ $\quad x = 6$
3. $ax - 5 = bx + 2$ $\quad x = \dfrac{7}{a - b}$
4. $\dfrac{5}{2x + 6} = \dfrac{3}{4x + 8}$ $\quad x = -\dfrac{11}{7}$

## B. POWERS

When powers of a given quantity $x$ are multiplied, the following rule applies:

$$x^n x^m = x^{n+m} \qquad \text{[A.3]}$$

For example, $x^2x^4 = x^{2+4} = x^6$.

When dividing the powers of a given quantity, note that

$$\frac{x^n}{x^m} = x^{n-m} \qquad \text{[A.4]}$$

For example, $x^8/x^2 = x^{8-2} = x^6$.

A power that is a fraction, such as $\frac{1}{3}$, corresponds to a root as follows:

$$x^{1/n} = \sqrt[n]{x} \qquad \text{[A.5]}$$

For example, $4^{1/3} = \sqrt[3]{4} = 1.587\ 4$. (A scientific calculator is useful for such calculations.)

Finally, any quantity $x^n$ that is raised to the $m$th power is

$$(x^n)^m = x^{nm} \qquad \text{[A.6]}$$

Table A.1 summarizes the rules of exponents.

**TABLE A.1**

**Rules of Exponents**

| |
|---|
| $x^0 = 1$ |
| $x^1 = x$ |
| $x^n x^m = x^{n+m}$ |
| $x^n/x^m = x^{n-m}$ |
| $x^{1/n} = \sqrt[n]{x}$ |
| $(x^n)^m = x^{nm}$ |

### Exercises

Verify the following:

1. $3^2 \times 3^3 = 243$
2. $x^5x^{-8} = x^{-3}$
3. $x^{10}/x^{-5} = x^{15}$
4. $5^{1/3} = 1.709\ 975$ (Use your calculator.)
5. $60^{1/4} = 2.783\ 158$ (Use your calculator.)
6. $(x^4)^3 = x^{12}$

## C. FACTORING

Some useful formulas for factoring an equation are

$$ax + ay + az = a(x + y + z) \qquad \text{common factor}$$

$$a^2 + 2ab + b^2 = (a + b)^2 \qquad \text{perfect square}$$

$$a^2 - b^2 = (a + b)(a - b) \qquad \text{differences of squares}$$

## D. QUADRATIC EQUATIONS

The general form of a quadratic equation is

$$ax^2 + bx + c = 0 \qquad \text{[A.7]}$$

where $x$ is the unknown quantity and $a$, $b$, and $c$ are numerical factors referred to as **coefficients** of the equation. This equation has two roots, given by

$$x = \frac{-b \pm \sqrt{b^2 - 4ac}}{2a} \qquad \text{[A.8]}$$

If $b^2 \geq 4ac$, the roots will be real.

## Example

The equation $x^2 + 5x + 4 = 0$ has the following roots corresponding to the two signs of the square root term:

$$x = \frac{-5 \pm \sqrt{5^2 - (4)(1)(4)}}{2(1)} = \frac{-5 \pm \sqrt{9}}{2} = \frac{-5 \pm 3}{2}$$

that is,

$$x_+ = \frac{-5 + 3}{2} = -1 \qquad x_- = \frac{-5 - 3}{2} = -4$$

where $x_+$ refers to the root corresponding to the positive sign and $x_-$ refers to the root corresponding to the negative sign.

### Exercises

Solve the following quadratic equations:

| | Answers | |
|---|---|---|
| **1.** $x^2 + 2x - 3 = 0$ | $x_+ = 1$ | $x_- = -3$ |
| **2.** $2x^2 - 5x + 2 = 0$ | $x_+ = 2$ | $x_- = 1/2$ |
| **3.** $2x^2 - 4x - 9 = 0$ | $x_+ = 1 + \sqrt{22}/2$ | $x_- = 1 - \sqrt{22}/2$ |

## E. Linear Equations

A linear equation has the general form

$$y = ax + b \qquad \text{[A.9]}$$

where $a$ and $b$ are constants. This equation is referred to as being linear because the graph of $y$ versus $x$ is a straight line, as shown in Figure A.1. The constant $b$, called the **intercept,** represents the value of $y$ at which the straight line intersects the $y$ axis. The constant $a$ is equal to the **slope** of the straight line. If any two points on the straight line are specified by the coordinates $(x_1, y_1)$ and $(x_2, y_2)$, as in Figure A.1, then the slope of the straight line can be expressed

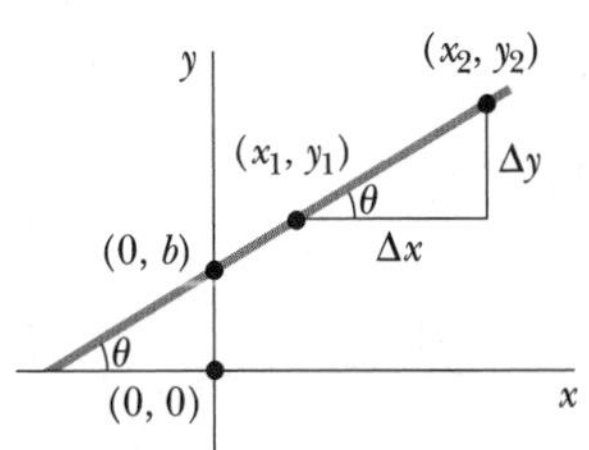

**Figure A.1**

$$\text{Slope} = \frac{y_2 - y_1}{x_2 - x_1} = \frac{\Delta y}{\Delta x} \qquad \text{[A.10]}$$

Note that $a$ and $b$ can have either positive or negative values. If $a > 0$, the straight line has a positive slope, as in Figure A.1. If $a < 0$, the straight line has a *negative* slope. In Figure A.1, both $a$ and $b$ are positive. Three other possible situations are shown in Figure A.2: $a > 0$, $b < 0$; $a < 0$, $b > 0$; and $a < 0$, $b < 0$.

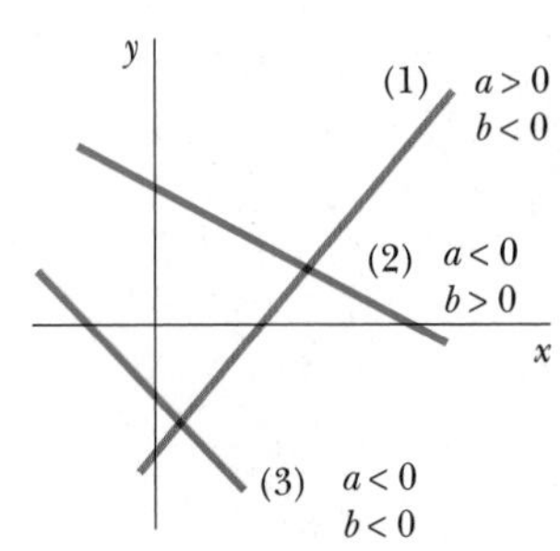

**Figure A.2**

### Exercises

1. Draw graphs of the following straight lines:
   **(a)** $y = 5x + 3$ **(b)** $y = -2x + 4$ **(c)** $y = -3x - 6$
2. Find the slopes of the straight lines described in Exercise 1.
   Answers: **(a)** 5 **(b)** $-2$ **(c)** $-3$

3. Find the slopes of the straight lines that pass through the following sets of points: **(a)** $(0, -4)$ and $(4, 2)$, **(b)** $(0, 0)$ and $(2, -5)$, and **(c)** $(-5, 2)$ and $(4, -2)$
Answers: **(a)** $\frac{3}{2}$ **(b)** $-\frac{5}{2}$ **(c)** $-\frac{4}{9}$

## F. Solving Simultaneous Linear Equations

Consider an equation such as $3x + 5y = 15$, which has two unknowns, $x$ and $y$. Such an equation does not have a unique solution. That is, $(x = 0,\ y = 3)$, $(x = 5, y = 0)$ and $(x = 2, y = \frac{9}{5})$ are all solutions to this equation.

If a problem has two unknowns, a unique solution is possible only if we have *two* independent equations. In general, if a problem has $n$ unknowns, its solution requires $n$ independent equations. In order to solve two simultaneous equations involving two unknowns, $x$ and $y$, we solve one of the equations for $x$ in terms of $y$ and substitute this expression into the other equation.

### Example

Solve the following two simultaneous equations:

$$\textbf{(1)}\quad 5x + y = -8 \qquad \textbf{(2)}\quad 2x - 2y = 4$$

**Solution** From (2), we find that $x = y + 2$. Substitution of this into (1) gives

$$5(y + 2) + y = -8$$

$$6y = -18$$

$$y = -3$$

$$x = y + 2 = -1$$

**Alternate Solution** Multiply each term in (1) by the factor 2 and add the result to (2):

$$10x + 2y = -16$$

$$\underline{2x - 2y = 4}$$

$$12x = -12$$

$$x = -1$$

$$y = x - 2 = -3$$

Two linear equations with two unknowns can also be solved by a graphical method. If the straight lines corresponding to the two equations are plotted in a conventional coordinate system, the intersection of the two lines represents the solution. For example, consider the two equations

$$x - y = 2$$

$$x - 2y = -1$$

These are plotted in Figure A.3. The intersection of the two lines has the coordinates $x = 5$, $y = 3$. This represents the solution to the equations. You should check this solution by the analytical technique discussed above.

### Exercises

Solve the following pairs of simultaneous equations involving two unknowns:

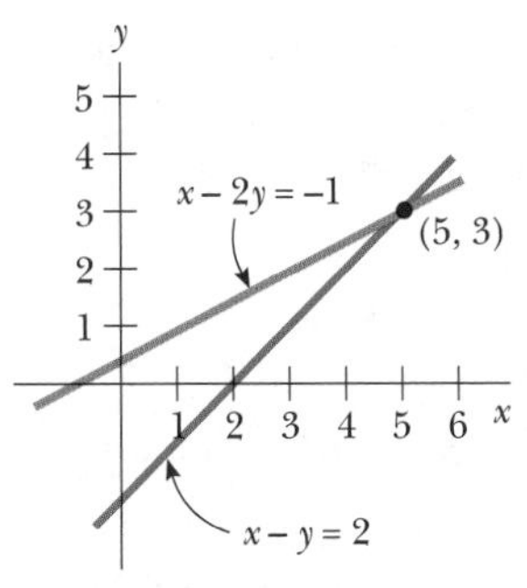


**FIGURE A.3**

### Answers

| | | |
|---|---|---|
| **1.** | $x + y = 8$<br>$x - y = 2$ | $x = 5, y = 3$ |
| **2.** | $98 - T = 10a$<br>$T - 49 = 5a$ | $T = 65, a = 3.27$ |
| **3.** | $6x + 2y = 6$<br>$8x - 4y = 28$ | $x = 2, y = -3$ |

## G. Logarithms

Suppose that a quantity $x$ is expressed as a power of some quantity $a$:

$$x = a^y \qquad \text{[A.11]}$$

The number $a$ is called the **base** number. The **logarithm** of $x$ with respect to the base $a$ is equal to the exponent to which the base must be raised in order to satisfy the expression $x = a^y$:

$$y = \log_a x \qquad \text{[A.12]}$$

Conversely, the **antilogarithm** of $y$ is the number $x$:

$$x = \text{antilog}_a\, y \qquad \text{[A.13]}$$

In practice, the two bases most often used are base 10, called the *common* logarithm base, and base $e = 2.718 \ldots$, called the *natural* logarithm base. When common logarithms are used,

$$y = \log_{10} x \qquad (\text{or } x = 10^y) \qquad \text{[A.14]}$$

When natural logarithms are used,

$$y = \ln_e x \qquad (\text{or } x = e^y) \qquad \text{[A.15]}$$

For example, $\log_{10} 52 = 1.716$, so that $\text{antilog}_{10}\, 1.716 = 10^{1.716} = 52$. Likewise, $\ln_e 52 = 3.951$, so $\text{antiln}_e\, 3.951 = e^{3.951} = 52$.

In general, note that you can convert between base 10 and base $e$ with the equality

$$\ln_e x = (2.302\,585)\log_{10} x \qquad \text{[A.16]}$$

Finally, some useful properties of logarithms are

$$\log(ab) = \log a + \log b \qquad \ln e = 1$$

$$\log(a/b) = \log a - \log b \qquad \ln e^a = a$$

$$\log(a^n) = n \log a \qquad \ln\left(\frac{1}{a}\right) = -\ln a$$

## A.4 GEOMETRY

Table A.2 gives the areas and volumes for several geometric shapes used throughout this text:

**TABLE A.2 Useful Information for Geometry**

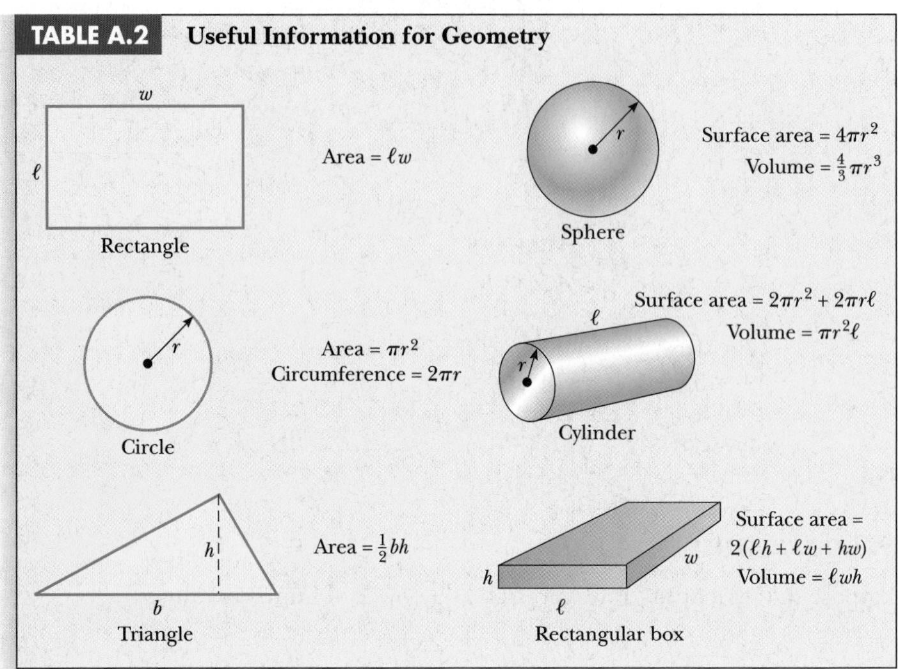


## A.5 TRIGONOMETRY

Some of the most basic facts concerning trigonometry are presented in Chapter 1, and we encourage you to study the material presented there if you are having trouble with this branch of mathematics. In addition to the discussion of Chapter 1, certain useful trig identities that can be of value to you follow.

$$\sin^2\theta + \cos^2\theta = 1$$

$$\sin\theta = \cos(90° - \theta)$$

$$\cos\theta = \sin(90° - \theta)$$

$$\sin 2\theta = 2\sin\theta\cos\theta$$

$$\cos 2\theta = \cos^2\theta - \sin^2\theta$$

$$\sin(\theta \pm \phi) = \sin\theta\cos\phi \pm \cos\theta\sin\phi$$

$$\cos(\theta \pm \phi) = \cos\theta\cos\phi \mp \sin\theta\sin\phi$$

The following relationships apply to *any* triangle, as shown in Figure A.4:

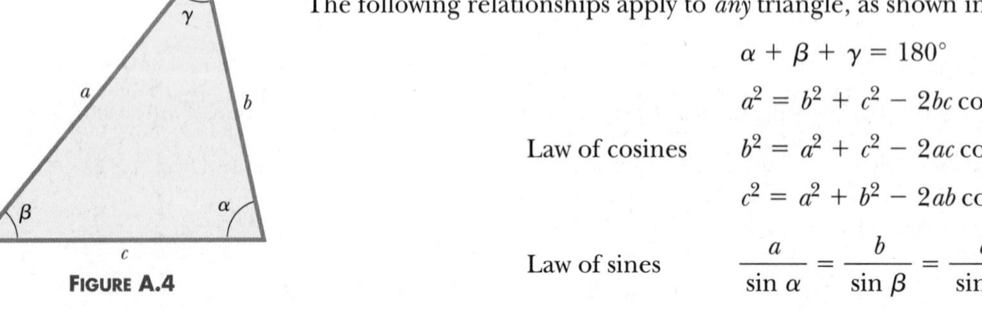

$$\alpha + \beta + \gamma = 180°$$

Law of cosines

$$a^2 = b^2 + c^2 - 2bc\cos\alpha$$

$$b^2 = a^2 + c^2 - 2ac\cos\beta$$

$$c^2 = a^2 + b^2 - 2ab\cos\gamma$$

Law of sines

$$\frac{a}{\sin\alpha} = \frac{b}{\sin\beta} = \frac{c}{\sin\gamma}$$

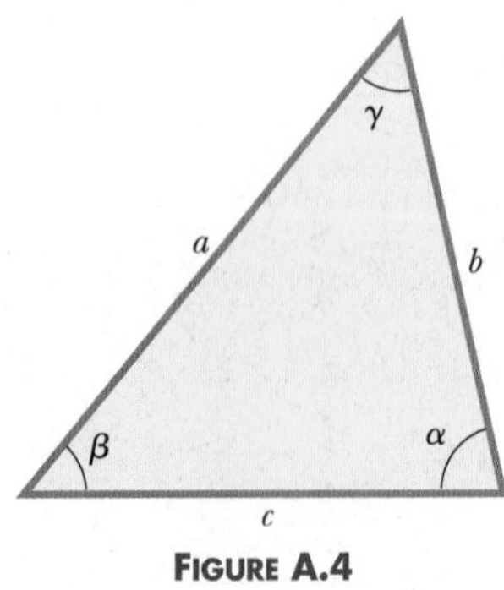


**FIGURE A.4**

# Appendix B

# An Abbreviated Table of Isotopes

| Atomic Number, $Z$ | Element | Symbol | Chemical Atomic Mass (u) | Mass Number (* Indicates Radioactive) $A$ | Atomic Mass (u)[a] | Percentage Abundance | Half-Life (if Radioactive) $T_{1/2}$ |
|---|---|---|---|---|---|---|---|
| 0 | (Neutron) | n | | 1* | 1.008 665 | | 10.4 min |
| 1 | Hydrogen | H | 1.007 9 | 1 | 1.007 825 | 99.985 | |
| | Deuterium | D | | 2 | 2.014 102 | 0.015 | |
| | Tritium | T | | 3* | 3.016 049 | | 12.33 yr |
| 2 | Helium | He | 4.002 60 | 3 | 3.016 029 | 0.000 14 | |
| | | | | 4 | 4.002 602 | 99.999 86 | |
| 3 | Lithium | Li | 6.941 | 6 | 6.015 121 | 7.5 | |
| | | | | 7 | 7.016 003 | 92.5 | |
| 4 | Beryllium | Be | 9.012 2 | 7* | 7.016 928 | | 53.3 days |
| | | | | 9 | 9.012 174 | 100 | |
| 5 | Boron | B | 10.81 | 10 | 10.012 936 | 19.9 | |
| | | | | 11 | 11.009 305 | 80.1 | |
| 6 | Carbon | C | 12.011 | 11* | 11.011 433 | | 20.4 min |
| | | | | 12 | 12.000 000 | 98.90 | |
| | | | | 13 | 13.003 355 | 1.10 | |
| | | | | 14* | 14.003 242 | | 5730 yr |
| 7 | Nitrogen | N | 14.006 7 | 13* | 13.005 738 | | 9.96 min |
| | | | | 14 | 14.003 074 | 99.63 | |
| | | | | 15 | 15.000 108 | 0.37 | |
| 8 | Oxygen | O | 15.999 4 | 15* | 15.003 065 | | 122 s |
| | | | | 16 | 15.994 915 | 99.761 | |
| | | | | 18 | 17.999 160 | 0.20 | |
| 9 | Fluorine | F | 18.998 40 | 19 | 18.998 404 | 100 | |
| 10 | Neon | Ne | 20.180 | 20 | 19.992 435 | 90.48 | |
| | | | | 22 | 21.991 383 | 9.25 | |
| 11 | Sodium | Na | 22.989 87 | 22* | 21.994 434 | | 2.61 yr |
| | | | | 23 | 22.989 770 | 100 | |
| | | | | 24* | 23.990 961 | | 14.96 h |
| 12 | Magnesium | Mg | 24.305 | 24 | 23.985 042 | 78.99 | |
| | | | | 25 | 24.985 838 | 10.00 | |
| | | | | 26 | 25.982 594 | 11.01 | |
| 13 | Aluminum | Al | 26.981 54 | 27 | 26.981 538 | 100 | |
| 14 | Silicon | Si | 28.086 | 28 | 27.976 927 | 92.23 | |
| 15 | Phosphorus | P | 30.973 76 | 31 | 30.973 762 | 100 | |
| | | | | 32* | 31.973 908 | | 14.26 days |

*(Table continues)*

[a] The masses in the sixth column are atomic masses, which include the mass of $Z$ electrons. Data are from the National Nuclear Data Center, Brookhaven National Laboratory, prepared by Jagdish K. Tuli, July 1990. The data are based on experimental results reported in *Nuclear Data Sheets* and *Nuclear Physics* and also from *Chart of the Nuclides,* 14th ed. Atomic masses are based on those by A. H. Wapstra, G. Audi, and R. Hoekstra. Isotopic abundances are based on those by N. E. Holden.

| Atomic Number, *Z* | Element | Symbol | Chemical Atomic Mass (u) | Mass Number (* Indicates Radioactive) *A* | Atomic Mass (u)[a] | Percentage Abundance | Half-Life (if Radioactive) $T_{1/2}$ |
|---|---|---|---|---|---|---|---|
| 16 | Sulfur | S | 32.066 | 32 | 31.972 071 | 95.02 | |
| | | | | 35* | 34.969 033 | | 87.5 days |
| 17 | Chlorine | Cl | 35.453 | 35 | 34.968 853 | 75.77 | |
| | | | | 37 | 36.965 903 | 24.23 | |
| 18 | Argon | Ar | 39.948 | 40 | 39.962 384 | 99.600 | |
| 19 | Potassium | K | 39.098 3 | 39 | 38.963 708 | 93.258 1 | |
| | | | | 40* | 39.964 000 | 0.0117 | $1.28 \times 10^9$ yr |
| 20 | Calcium | Ca | 40.08 | 40 | 39.962 591 | 96.941 | |
| 21 | Scandium | Sc | 44.955 9 | 45 | 44.955 911 | 100 | |
| 22 | Titanium | Ti | 47.88 | 48 | 47.947 947 | 73.8 | |
| 23 | Vanadium | V | 50.941 5 | 51 | 50.943 962 | 99.75 | |
| 24 | Chromium | Cr | 51.996 | 52 | 51.940 511 | 83.79 | |
| 25 | Manganese | Mn | 54.938 05 | 55 | 54.938 048 | 100 | |
| 26 | Iron | Fe | 55.847 | 56 | 55.934 940 | 91.72 | |
| 27 | Cobalt | Co | 58.933 20 | 59 | 58.933 198 | 100 | |
| | | | | 60* | 59.933 820 | | 5.27 yr |
| 28 | Nickel | Ni | 58.693 | 58 | 57.935 346 | 68.077 | |
| | | | | 60 | 59.930 789 | 26.223 | |
| 29 | Copper | Cu | 63.54 | 63 | 62.929 599 | 69.17 | |
| | | | | 65 | 64.927 791 | 30.83 | |
| 30 | Zinc | Zn | 65.39 | 64 | 63.929 144 | 48.6 | |
| | | | | 66 | 65.926 035 | 27.9 | |
| | | | | 68 | 67.924 845 | 18.8 | |
| 31 | Gallium | Ga | 69.723 | 69 | 68.925 580 | 60.108 | |
| | | | | 71 | 70.924 703 | 39.892 | |
| 32 | Germanium | Ge | 72.61 | 70 | 69.924 250 | 21.23 | |
| | | | | 72 | 71.922 079 | 27.66 | |
| | | | | 74 | 73.921 177 | 35.94 | |
| 33 | Arsenic | As | 74.921 6 | 75 | 74.921 594 | 100 | |
| 34 | Selenium | Se | 78.96 | 78 | 77.917 307 | 23.78 | |
| | | | | 80 | 79.916 519 | 49.61 | |
| 35 | Bromine | Br | 79.904 | 79 | 78.918 336 | 50.69 | |
| | | | | 81 | 80.916 287 | 49.31 | |
| 36 | Krypton | Kr | 83.80 | 82 | 81.913 481 | 11.6 | |
| | | | | 83 | 82.914 136 | 11.5 | |
| | | | | 84 | 83.911 508 | 57.0 | |
| | | | | 86 | 85.910 615 | 17.3 | |
| 37 | Rubidium | Rb | 85.468 | 85 | 84.911 793 | 72.17 | |
| | | | | 87* | 86.909 186 | 27.83 | $4.75 \times 10^{10}$ yr |
| 38 | Strontium | Sr | 87.62 | 86 | 85.909 266 | 9.86 | |
| | | | | 88 | 87.905 618 | 82.58 | |
| | | | | 90* | 89.907 737 | | 29.1 yr |
| 39 | Yttrium | Y | 88.905 8 | 89 | 88.905 847 | 100 | |
| 40 | Zirconium | Zr | 91.224 | 90 | 89.904 702 | 51.45 | |
| | | | | 91 | 90.905 643 | 11.22 | |
| | | | | 92 | 91.905 038 | 17.15 | |
| | | | | 94 | 93.906 314 | 17.38 | |
| 41 | Niobium | Nb | 92.906 4 | 93 | 92.906 376 | 100 | |
| 42 | Molybdenum | Mo | 95.94 | 92 | 91.906 807 | 14.84 | |
| | | | | 95 | 94.905 841 | 15.92 | |
| | | | | 96 | 95.904 678 | 16.68 | |
| | | | | 98 | 97.905 407 | 24.13 | |
| 43 | Technetium | Tc | | 98* | 97.907 215 | | $4.2 \times 10^6$ yr |
| | | | | 99* | 98.906 254 | | $2.1 \times 10^6$ yr |

*(Table continues)*

| Atomic Number, $Z$ | Element | Symbol | Chemical Atomic Mass (u) | Mass Number (* Indicates Radioactive) $A$ | Atomic Mass (u)[a] | Percentage Abundance | Half-Life (if Radioactive) $T_{1/2}$ |
|---|---|---|---|---|---|---|---|
| 44 | Ruthenium | Ru | 101.07 | 99 | 98.905 939 | 12.7 | |
| | | | | 100 | 99.904 219 | 12.6 | |
| | | | | 101 | 100.905 558 | 17.1 | |
| | | | | 102 | 101.904 348 | 31.6 | |
| | | | | 104 | 103.905 428 | 18.6 | |
| 45 | Rhodium | Rh | 102.905 5 | 103 | 102.905 502 | 100 | |
| 46 | Palladium | Pd | 106.42 | 104 | 103.904 033 | 11.14 | |
| | | | | 105 | 104.905 082 | 22.33 | |
| | | | | 106 | 105.903 481 | 27.33 | |
| | | | | 108 | 107.903 893 | 26.46 | |
| | | | | 110 | 109.905 158 | 11.72 | |
| 47 | Silver | Ag | 107.868 | 107 | 106.905 091 | 51.84 | |
| | | | | 109 | 108.904 754 | 48.16 | |
| 48 | Cadmium | Cd | 112.41 | 110 | 109.903 004 | 12.49 | |
| | | | | 111 | 110.904 182 | 12.80 | |
| | | | | 112 | 111.902 760 | 24.13 | |
| | | | | 113* | 112.904 401 | 12.22 | $9.3 \times 10^{15}$ yr |
| | | | | 114 | 113.903 359 | 28.73 | |
| 49 | Indium | In | 114.82 | 115* | 114.903 876 | 95.7 | $4.4 \times 10^{14}$ yr |
| 50 | Tin | Sn | 118.71 | 116 | 115.901 743 | 14.53 | |
| | | | | 118 | 117.901 605 | 24.22 | |
| | | | | 120 | 119.902 197 | 32.59 | |
| 51 | Antimony | Sb | 121.76 | 121 | 120.903 820 | 57.36 | |
| | | | | 123 | 122.904 215 | 42.64 | |
| 52 | Tellurium | Te | 127.60 | 126 | 125.903 309 | 18.93 | |
| | | | | 128* | 127.904 463 | 31.70 | $>8 \times 10^{24}$ yr |
| | | | | 130* | 129.906 228 | 33.87 | $\leq 1.25 \times 10^{21}$ yr |
| 53 | Iodine | I | 126.904 5 | 127 | 126.904 474 | 100 | |
| | | | | 129* | 128.904 984 | | $1.6 \times 10^{7}$ yr |
| | | | | 131* | 130.906 118 | | 8.04 days |
| 54 | Xenon | Xe | 131.29 | 129 | 128.904 779 | 26.4 | |
| | | | | 131 | 130.905 069 | 21.2 | |
| | | | | 132 | 131.904 141 | 26.9 | |
| | | | | 134 | 133.905 394 | 10.4 | |
| | | | | 136* | 135.907 215 | 8.9 | $\geq 2.36 \times 10^{21}$ yr |
| 55 | Cesium | Cs | 132.905 4 | 133 | 132.905 436 | 100 | |
| 56 | Barium | Ba | 137.33 | 137 | 136.905 816 | 11.23 | |
| | | | | 138 | 137.905 236 | 71.70 | |
| | | | | 144* | 143.922 673 | | 11.9 s |
| 57 | Lanthanum | La | 138.905 | 139 | 138.906 346 | 99.909 8 | |
| 58 | Cerium | Ce | 140.12 | 140 | 139.905 434 | 88.43 | |
| | | | | 142* | 141.909 241 | 11.13 | $>5 \times 10^{16}$ yr |
| 59 | Praseodymium | Pr | 140.907 6 | 141 | 140.907 647 | 100 | |
| 60 | Neodymium | Nd | 144.24 | 142 | 141.907 718 | 27.13 | |
| | | | | 144* | 143.910 082 | 23.80 | $2.3 \times 10^{15}$ yr |
| | | | | 146 | 145.913 113 | 17.19 | |
| 61 | Promethium | Pm | | 145* | 144.912 745 | | 17.7 yr |
| 62 | Samarium | Sm | 150.36 | 147* | 146.914 894 | 15.0 | $1.06 \times 10^{11}$ yr |
| | | | | 149* | 148.917 180 | 13.8 | $>2 \times 10^{15}$ yr |
| | | | | 152 | 151.919 728 | 26.7 | |
| | | | | 154 | 153.922 206 | 22.7 | |
| 63 | Europium | Eu | 151.96 | 151 | 150.919 846 | 47.8 | |
| | | | | 153 | 152.921 226 | 52.2 | |

*(Table continues)*

| Atomic Number, Z | Element | Symbol | Chemical Atomic Mass (u) | Mass Number (* Indicates Radioactive) A | Atomic Mass (u)[a] | Percentage Abundance | Half-Life (if Radioactive) $T_{1/2}$ |
|---|---|---|---|---|---|---|---|
| 64 | Gadolinium | Gd | 157.25 | 156 | 155.922 119 | 20.47 | |
| | | | | 158 | 157.924 099 | 24.84 | |
| | | | | 160 | 159.927 050 | 21.86 | |
| 65 | Terbium | Tb | 158.925 3 | 159 | 158.925 345 | 100 | |
| 66 | Dysprosium | Dy | 162.50 | 162 | 161.926 796 | 25.5 | |
| | | | | 163 | 162.928 729 | 24.9 | |
| | | | | 164 | 163.929 172 | 28.2 | |
| 67 | Holmium | Ho | 164.930 3 | 165 | 164.930 316 | 100 | |
| 68 | Erbium | Er | 167.26 | 166 | 165.930 292 | 33.6 | |
| | | | | 167 | 166.932 047 | 22.95 | |
| | | | | 168 | 167.932 369 | 27.8 | |
| 69 | Thulium | Tm | 168.934 2 | 169 | 168.934 213 | 100 | |
| 70 | Ytterbium | Yb | 173.04 | 172 | 171.936 380 | 21.9 | |
| | | | | 173 | 172.938 209 | 16.12 | |
| | | | | 174 | 173.938 861 | 31.8 | |
| 71 | Lutecium | Lu | 174.967 | 175 | 174.940 772 | 97.41 | |
| 72 | Hafnium | Hf | 178.49 | 177 | 176.943 218 | 18.606 | |
| | | | | 178 | 177.943 697 | 27.297 | |
| | | | | 179 | 178.945 813 | 13.629 | |
| | | | | 180 | 179.946 547 | 35.100 | |
| 73 | Tantalum | Ta | 180.947 9 | 181 | 180.947 993 | 99.988 | |
| 74 | Tungsten | W | 183.85 | 182 | 181.948 202 | 26.3 | |
| | | | | 183 | 182.950 221 | 14.28 | |
| | | | | 184 | 183.950 929 | 30.7 | |
| | | | | 186 | 185.954 358 | 28.6 | |
| 75 | Rhenium | Re | 186.207 | 185 | 184.952 951 | 37.40 | |
| | | | | 187* | 186.955 746 | 62.60 | $4.4 \times 10^{10}$ yr |
| 76 | Osmium | Os | 190.2 | 188 | 187.955 832 | 13.3 | |
| | | | | 189 | 188.958 139 | 16.1 | |
| | | | | 190 | 189.958 439 | 26.4 | |
| | | | | 191* | 190.960 94 | | 15.4 days |
| | | | | 192 | 191.961 468 | 41.0 | |
| 77 | Iridium | Ir | 192.2 | 191 | 190.960 585 | 37.3 | |
| | | | | 193 | 192.962 916 | 62.7 | |
| 78 | Platinum | Pt | 195.08 | 194 | 193.962 655 | 32.9 | |
| | | | | 195 | 194.964 765 | 33.8 | |
| | | | | 196 | 195.964 926 | 25.3 | |
| 79 | Gold | Au | 196.966 5 | 197 | 196.966 543 | 100 | |
| 80 | Mercury | Hg | 200.59 | 199 | 198.968 253 | 16.87 | |
| | | | | 200 | 199.968 299 | 23.10 | |
| | | | | 201 | 200.970 276 | 13.10 | |
| | | | | 202 | 201.970 617 | 29.86 | |
| 81 | Thallium | Tl | 204.383 | 203 | 292.972 320 | 29.524 | |
| | | | | 205 | 204.974 400 | 70.476 | |
| | | (Th C″) | | 208* | 207.981 992 | | 3.053 min |
| | | | | 210* | 209.990 069 | | 1.3 min |
| 82 | Lead | Pb | 207.2 | 204* | 203.973 020 | 1.4 | $\geq 1.4 \times 10^{17}$ yr |
| | | | | 206 | 205.974 440 | 24.1 | |
| | | | | 207 | 206.975 871 | 22.1 | |
| | | | | 208 | 207.976 627 | 52.4 | |
| | | (Ra D) | | 210* | 209.984 163 | | 22.3 yr |
| | | (Ac B) | | 211* | 210.988 734 | | 36.1 min |
| | | (Th B) | | 212* | 211.991 872 | | 10.64 h |
| | | (Ra B) | | 214* | 213.999 798 | | 26.8 min |

*(Table continues)*

| Atomic Number, $Z$ | Element | Symbol | Chemical Atomic Mass (u) | Mass Number (* Indicates Radioactive) $A$ | Atomic Mass (u)[a] | Percentage Abundance | Half-Life (if Radioactive) $T_{1/2}$ |
|---|---|---|---|---|---|---|---|
| 83 | Bismuth | Bi | 208.980 3 | 209 | 208.980 374 | 100 | |
| | | (Th C) | | 211* | 210.987 254 | | 2.14 min |
| 84 | Polonium | Po | | | | | |
| | | (Ra F) | | 210* | 209.982 848 | | 138.38 days |
| | | (Ra C′) | | 214* | 213.995 177 | | 164 μs |
| 85 | Astatine | At | | 218* | 218.008 685 | | 1.6 s |
| 86 | Radon | Rn | | 222* | 222.017 571 | | 3.823 days |
| 87 | Francium | Fr | | | | | |
| | | (Ac K) | | 223* | 223.019 733 | | 22 min |
| 88 | Radium | Ra | | 226* | 226.025 402 | | 1600 yr |
| | | (Ms $Th_1$) | | 228* | 228.031 064 | | 5.75 yr |
| 89 | Actinium | Ac | | 227* | 227.027 749 | | 21.77 yr |
| 90 | Thorium | Th | 232.038 1 | | | | |
| | | (Rd Th) | | 228* | 228.028 716 | | 1.913 yr |
| | | | | 232* | 232.038 051 | 100 | $1.40 \times 10^{10}$ yr |
| 91 | Protactinium | Pa | | 231* | 231.035 880 | | 32.760 yr |
| 92 | Uranium | U | 238.028 9 | 232* | 232.037 131 | | 69 yr |
| | | | | 233* | 233.039 630 | | $1.59 \times 10^{5}$ yr |
| | | (Ac U) | | 235* | 235.043 924 | 0.720 | $7.04 \times 10^{8}$ yr |
| | | | | 236* | 236.045 562 | | $2.34 \times 10^{7}$ yr |
| | | (UI) | | 238* | 238.050 784 | 99.274 5 | $4.47 \times 10^{9}$ yr |
| 93 | Neptunium | Np | | 237* | 237.048 168 | | $2.14 \times 10^{6}$ yr |
| 94 | Plutonium | Pu | | 239* | 239.052 157 | | $2.412 \times 10^{3}$ yr |
| | | | | 242* | 242.058 737 | | $3.73 \times 10^{5}$ yr |
| | | | | 244* | 244.064 200 | | $8.1 \times 10^{7}$ yr |

Calculator

$a_1$ $a_2$ $a_3$ $b$

$1I_1 - 1I_2 - 1I_3 = 0$

$24I_1 - 20I_2 + 0I_3 = 24$

⇒ Simult

⇒ # of equations (3)

# Appendix C

## Some Useful Tables

**TABLE C.1 Mathematical Symbols Used in the Text and Their Meaning**

| Symbol | Meaning |
|---|---|
| $=$ | is equal to |
| $\neq$ | is not equal to |
| $\equiv$ | is defined as |
| $\propto$ | is proportional to |
| $>$ | is greater than |
| $<$ | is less than |
| $\gg$ | is much greater than |
| $\ll$ | is much less than |
| $\approx$ | is approximately equal to |
| $\sim$ | is on the order of magnitude of |
| $\Delta x$ | change in $x$ or uncertainty in $x$ |
| $\Sigma x_i$ | sum of all quantities $x_i$ |
| $\lvert x \rvert$ | absolute value of $x$ (always a positive quantity) |

**TABLE C.2 Standard Symbols for Units**

| Symbol | Unit | Symbol | Unit |
|---|---|---|---|
| A | ampere | kcal | kilocalorie |
| Å | angstrom | kg | kilogram |
| atm | atmosphere | km | kilometer |
| Bq | bequerel | kmol | kilomole |
| Btu | British thermal unit | L | liter |
| C | coulomb | lb | pound |
| °C | degree Celsius | ly | lightyear |
| cal | calorie | m | meter |
| cm | centimeter | min | minute |
| Ci | curie | mol | mole |
| d | day | N | newton |
| deg | degree (angle) | nm | nanometer |
| eV | electronvolt | Pa | pascal |
| °F | degree Fahrenheit | rad | radian |
| F | farad | rev | revolution |
| ft | foot | s | second |
| G | Gauss | T | tesla |
| g | gram | u | atomic mass unit |
| H | henry | V | volt |
| h | hour | W | watt |
| hp | horsepower | Wb | weber |
| Hz | hertz | yr | year |
| in. | inch | $\mu$m | micrometer |
| J | joule | $\Omega$ | ohm |
| K | kelvin | | |

**TABLE C.3 The Greek Alphabet**

| | | | | | |
|---|---|---|---|---|---|
| Alpha | A | $\alpha$ | Nu | N | $\nu$ |
| Beta | B | $\beta$ | Xi | $\Xi$ | $\xi$ |
| Gamma | $\Gamma$ | $\gamma$ | Omicron | O | $o$ |
| Delta | $\Delta$ | $\delta$ | Pi | $\Pi$ | $\pi$ |
| Epsilon | E | $\epsilon$ | Rho | P | $\rho$ |
| Zeta | Z | $\zeta$ | Sigma | $\Sigma$ | $\sigma$ |
| Eta | H | $\eta$ | Tau | T | $\tau$ |
| Theta | $\Theta$ | $\theta$ | Upsilon | $\Upsilon$ | $\upsilon$ |
| Iota | I | $\iota$ | Phi | $\Phi$ | $\phi$ |
| Kappa | K | $\kappa$ | Chi | X | $\chi$ |
| Lambda | $\Lambda$ | $\lambda$ | Psi | $\Psi$ | $\psi$ |
| Mu | M | $\mu$ | Omega | $\Omega$ | $\omega$ |

**TABLE C.4 Physical Data Often Used[a]**

| | |
|---|---|
| Average Earth-Moon distance | $3.84 \times 10^8$ m |
| Average Earth-Sun distance | $1.496 \times 10^{11}$ m |
| Equatorial radius of Earth | $6.38 \times 10^6$ m |
| Density of air (20°C and 1 atm) | 1.20 kg/m$^3$ |
| Density of water (20°C and 1 atm) | $1.00 \times 10^3$ kg/m$^3$ |
| Free-fall acceleration | 9.80 m/s$^2$ |
| Mass of Earth | $5.98 \times 10^{24}$ kg |
| Mass of Moon | $7.36 \times 10^{22}$ kg |
| Mass of Sun | $1.99 \times 10^{30}$ kg |
| Standard atmospheric pressure | $1.013 \times 10^5$ Pa |

[a] These are the values of the constants as used in the text.

**TABLE C.5 Some Fundamental Constants[a]**

| Quantity | Symbol | Value[b] |
|---|---|---|
| Atomic mass unit | u | $1.660\,540\,2(10) \times 10^{-27}$ kg<br>931.494 32(28) MeV/$c^2$ |
| Avogadro's number | $N_A$ | $6.022\,136\,7(36) \times 10^{23}$ (mol)$^{-1}$ |
| Bohr radius | $a_0 = \dfrac{\hbar^2}{m_e e^2 k_e}$ | $0.529\,177\,249(24) \times 10^{-10}$ m |
| Boltzmann's constant | $k_B = R/N_A$ | $1.380\,658(12) \times 10^{-23}$ J/K |
| Compton wavelength | $\lambda_C = \dfrac{h}{m_e c}$ | $2.426\,310\,58(22) \times 10^{-12}$ m |
| Coulomb's law force constant | $k_e = \dfrac{1}{4\pi\epsilon_0}$ | $8.987\,551\,787 \times 10^9$ N·m$^2$/C$^2$ (exact) |
| Electron mass | $m_e$ | $9.109\,389\,7(54) \times 10^{-31}$ kg<br>$5.485\,799\,03(13) \times 10^{-4}$ u<br>0.510 999 06(15) MeV/$c^2$ |
| Electron volt | eV | $1.602\,177\,33(49) \times 10^{-19}$ J |
| Elementary charge | $e$ | $1.602\,177\,33(49) \times 10^{-19}$ C |
| Gas constant | $R$ | 8.314 510(70) J/K·mol |
| Gravitational constant | $G$ | $6.672\,59(85) \times 10^{-11}$ N·m$^2$/kg$^2$ |
| Hydrogen ionization energy | $-E_1 = \dfrac{m_e e^4 k_e^2}{2\hbar^2} = \dfrac{e^2 k_e}{2a_0}$ | 13.605 698(40) eV |
| Neutron mass | $m_n$ | $1.674\,928\,6(10) \times 10^{-27}$ kg<br>1.008 664 904(14) u<br>939.565 63(28) MeV/$c^2$ |
| Permeability of free space | $\mu_0$ | $4\pi \times 10^{-7}$ T·m/A (exact) |
| Permittivity of free space | $\epsilon_0 = 1/\mu_0 c^2$ | $8.854\,187\,817 \times 10^{-12}$ C$^2$/N·m$^2$ (exact) |
| Planck's constant | $h$<br>$\hbar = h/2\pi$ | $6.626\,075(40) \times 10^{-34}$ J·s<br>$1.054\,572\,66(63) \times 10^{-34}$ J·s |
| Proton mass | $m_p$ | $1.672\,623(10) \times 10^{-27}$ kg<br>1.007 276 470(12) u<br>938.272 3(28) MeV/$c^2$ |
| Rydberg constant | $R_H$ | $1.097\,373\,153\,4(13) \times 10^7$ m$^{-1}$ |
| Speed of light in vacuum | $c$ | $2.997\,924\,58 \times 10^8$ m/s (exact) |

[a] These constants are the values recommended in 1986 by CODATA, based on a least-squares adjustment of data from different measurements. For a more complete list, see Cohen, E. Richard, and Barry N. Taylor, *Rev. Mod. Phys.* **59:**1121, 1987.

[b] The numbers in parentheses for the values below represent the uncertainties in the last two digits.

# Appendix D

## SI Units

**TABLE D.1 SI Base Units**

| Base Quantity | SI Base Unit Name | Symbol |
|---|---|---|
| Length | meter | m |
| Mass | kilogram | kg |
| Time | second | s |
| Electric current | ampere | A |
| Temperature | kelvin | K |
| Amount of substance | mole | mol |
| Luminous intensity | candela | cd |

**TABLE D.2 Derived SI Units**

| Quantity | Name | Symbol | Expression in Terms of Base Units | Expression in Terms of Other SI Units |
|---|---|---|---|---|
| Plane angle | radian | rad | m/m | |
| Frequency | hertz | Hz | $s^{-1}$ | |
| Force | newton | N | $kg \cdot m/s^2$ | J/m |
| Pressure | pascal | Pa | $kg/m \cdot s^2$ | $N/m^2$ |
| Energy: work | joule | J | $kg \cdot m^2/s^2$ | $N \cdot m$ |
| Power | watt | W | $kg \cdot m^2/s^3$ | J/s |
| Electric charge | coulomb | C | $A \cdot s$ | |
| Electric potential (emf) | volt | V | $kg \cdot m^2/A \cdot s^3$ | W/A, J/C |
| Capacitance | farad | F | $A^2 \cdot s^4/kg \cdot m^2$ | C/V |
| Electric resistance | ohm | Ω | $kg \cdot m^2/A^2 \cdot s^3$ | V/A |
| Magnetic flux | weber | Wb | $kg \cdot m^2/A \cdot s^2$ | $V \cdot s$, $T \cdot m^2$ |
| Magnetic field intensity | tesla | T | $kg/A \cdot s^2$ | $Wb/m^2$ |
| Inductance | henry | H | $kg \cdot m^2/A^2 \cdot s^2$ | Wb/A |

# Answers to Quick Quizzes, Selected Questions and Problems

## CHAPTER 15

### Quick Quizzes

1. d
2. b
3. c
4. b
5. a
6. c and d
7. *A*, *B*, and *C*
8. **(b)** and **(d)**

### Conceptual Questions

1. Because of lower moisture content during winter, the air acts as a better insulator and allows larger static charges to build up on objects during winter than during summer. As a result, shocks from static electricity are more severe in winter months than in summer months.
3. The configuration shown is inherently unstable. The negative charges repel each other. If there is any slight rotation of one of the rods, the repulsion can result in further rotation away from this configuration. There are three conceivable final configurations shown below. Configuration (a) is stable—if the positive upper ends are pushed toward each other, they will repel and move the system back to the original configuration. Configuration (b) is an equilibrium configuration, but it is unstable—if the lower ends are moved toward each other, the attraction of the lower ends will be larger than that of the upper ends and the configuration will shift to (c). Configuration (c) is another possible stable configuration.

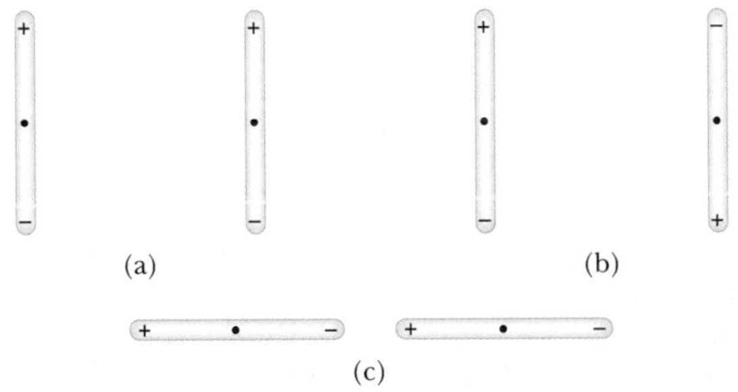


5. Move an object A with a net positive charge so it is near, but not touching, a neutral metallic object B that is insulated from the ground. The presence of A will polarize B, causing an excess negative charge to exist on the side nearest A and an equal magnitude excess positive charge to exist on the side farthest from A. While A is still near B, touch B with your hand. This allows additional electrons to flow from ground through your body and onto B. With A continuing to be near but not touching B, remove your hand from B, thus trapping the excess electrons on B. When A is now removed, B is left with excess electrons causing a net negative charge. This negative charge, by means of mutual repulsion, will now spread uniformly over the entire surface of B.
7. An object's mass decreases very slightly (immeasurably) when it is given a positive charge, because it loses electrons. When the object is given a negative charge, its mass increases slightly because it gains electrons.
9. Electric field lines start on positive charges and end on negative charges. Thus, if the fair weather field is directed into the ground, the ground must have a negative charge.
11. The two charged plates create a region of uniform electric field between them, directed from the positive toward the negative plate. Once the ball is disturbed so as to touch one plate, say the negative one, some negative charge will be transferred to the ball and it will experience an electric force that will accelerate it to the positive plate. Once the charge touches the positive plate, it will release its negative charge, acquire a positive charge, and accelerate back to the negative plate. The ball will continue to move back and forth between the plates until it has transferred all their net charge, thereby making both plates neutral.
13. The electric shielding effect of conductors depends on the fact that there are two kinds of charge—positive and negative. As a result, charges can move within the conductor so that the combination of positive and negative charges establishes an electric field that exactly cancels the external field within the conductor and any cavities inside the conductor. There is only one type of gravitational charge, however—there is no "negative mass." As a result, gravitational shielding is not possible.
15. The electric field patterns of each of these three configurations do not have sufficient symmetry to make the calculations practical. Gauss's law is only useful for calculating the electric field of highly symmetric charge distributions, such as uniformly charged spheres, cylinders, and sheets.
17. No. The balloon induces charge of opposite sign in the wall, causing the balloon and the wall to be attracted to each other. The balloon eventually falls because its charge slowly diminishes after leaking to ground. Some of the balloon's charge could also be lost due to positive ions in the surrounding atmosphere, which would tend to neutralize the negative charges on the balloon.
19. When the comb is nearby, charges separate on the paper, resulting in the paper being attracted. After contact, charges from the comb are transferred to the paper so that it has the same type charge as the comb. It is thus repelled.

### Problems

1. $1.1 \times 10^{-8}$ N (attractive)
3. 91 N (repulsion)
5. **(a)** 36.8 N **(b)** $5.54 \times 10^{27}$ m/s$^2$
7. $5.12 \times 10^5$ N
9. **(a)** $2.2 \times 10^{-5}$ N (attraction) **(b)** $9.0 \times 10^{-7}$ N (repulsion)
11. $1.38 \times 10^{-5}$ N at 77.5° below $-x$ axis
13. 0.872 N at 30.0° below $+x$ axis
15. 7.2 nC
17. $5.5 \times 10^{11}$ N/C (away from the proton)
19. $7.20 \times 10^5$ N/C (downward)

**21.** $1.63 \times 10^5$ N/C
**23.** **(a)** $6.12 \times 10^{10}$ m/s$^2$ **(b)** 19.6 $\mu$s
**(c)** 11.8 m **(d)** $1.20 \times 10^{-15}$ J
**25.** zero
**27.** 1.8 m to the left of the $-2.5$ $\mu$C charge
**33.** **(a)** 0
**(b)** $+5$ $\mu$C inside, $-5$ $\mu$C outside
**(c)** 0 inside, $-5$ $\mu$C outside
**(d)** 0 inside, $-5$ $\mu$C outside
**35.** $1.3 \times 10^{-3}$ C
**37.** **(a)** $4.8 \times 10^{-15}$ N **(b)** $2.9 \times 10^{12}$ m/s$^2$
**39.** **(a)** 858 N$\cdot$m$^2$/C
**(b)** 0 **(c)** 657 N$\cdot$m$^2$/C
**41.** $4.1 \times 10^6$ N/C
**43.** **(a)** 0 **(b)** $k_e q/r^2$ outward
**47.** 57.5 N
**49.** 24 N/C in the $+x$ direction
**51.** **(a)** $E = 2k_e qb(a^2 + b^2)^{-3/2}$ in $+x$ direction
**(b)** $E = 2k_e Qb(a^2 + b^2)^{-3/2}$ in $+x$ direction
**53.** **(a)** 0
**(b)** $7.99 \times 10^7$ N/C (outward)
**(c)** 0
**(d)** $7.34 \times 10^6$ N/C (outward)
**55.** $3.55 \times 10^5$ N$\cdot$m$^2$/C
**57.** $4.4 \times 10^5$ N/C
**59.** **(a)** 10.9 nC **(b)** $5.43 \times 10^{-3}$ N

## CHAPTER 16

### Quick Quizzes

**1.** **(b)**
**2.** Either **(a)** or **(b)** might be true
**3.** **(a)** and **(b)**
**4.** **(c)**
**5.** **(a)** C decreases. **(b)** $Q$ stays the same.
**(c)** $E$ stays the same. **(d)** $\Delta V$ increases.
**(e)** The energy stored increases.
**6.** **(a)** $C$ increases. **(b)** $Q$ increases.
**(c)** $E$ stays the same. **(d)** $\Delta V$ remains the same.
**(e)** The energy stored increases.

### Conceptual Questions

**1.** The statement is false. A uniform field simply means that the field is constant in magnitude and direction. Any electric field does work on a charge as that charge undergoes a displacement in the direction of the field. Thus, a change in potential occurs when one undergoes any displacement along the direction of the field, even if the field is uniform. If we define the potential to be zero at one point, it will be different from zero at neighboring points that are downhill and uphill in the field.

**3.** The work done in pulling the capacitor plates farther apart is transferred into additional electric energy stored in the capacitor. The charge is constant and the capacitance decreases, but the potential difference between the plates increases, which results in an increase in the stored electric energy.

**5.** The power line, if it makes electrical contact with the metal of the car, will raise the potential of the car to 20 kV. It will also raise the potential of your body to 20 kV, because you are in contact with the car. In itself, this is not a problem. If you step out of the car, your body at 20 kV will make contact with the ground, which is at zero volts. As a result, a current will pass through your body, and you will likely be injured. Thus, it is best to stay in the car until help arrives.

**7.** If two points on a conducting object were at different potentials, then free charges in the object would move, and we would not have static conditions, in contradiction to the initial assumption. (Free positive charges would migrate from higher to lower potential locations; free electrons would move rapidly from lower to higher potential locations.) The charges would continue to move until the potential became equal everywhere in the conductor.

**9.** The capacitor often remains charged long after the voltage source is disconnected. This residual charge can be lethal. The capacitor can be safely handled after discharging the plates by short-circuiting the device with a conductor, such as a screwdriver with an insulating handle.

**11.** Field lines represent the direction of the electric force on a positive test charge. If electric field lines were to cross, then at the point of crossing, there would be an ambiguity regarding the direction of the force on the test charge, because there would be two possible forces. Thus, electric field lines cannot cross. It is possible for equipotential surfaces to cross. (However, equipotential surfaces at different potentials cannot intersect.) For example, suppose two identical positive charges are at diagonally opposite corners of a square, and two negative charges of equal magnitude are at the other two corners. Then the planes perpendicular to the sides of the square at their midpoints are equipotential surfaces. These two planes cross each other, at the line perpendicular to the square at its center.

**13.** (a) From $Q = C(\Delta V)$, if you double the potential difference across a capacitor having a fixed capacitance, you will double the charge stored in the capacitor. (b) The energy stored in a capacitor may be expressed as $W = Q^2/2C$. Assuming a parallel-plate capacitor, the capacitance is $C = \kappa\epsilon_0 A/d$, and the stored energy is $W = Q^2 d/2\kappa\epsilon_0 A$. Thus, doubling the plate separation while holding the charge constant will double the energy stored.

**15.** You should use a dielectric-filled capacitor whose dielectric constant is very large. Furthermore, you should make the dielectric as thin as possible, keeping in mind that dielectric breakdown must also be considered.

**17.** Inserting a dielectric into the capacitor increases the capacitance of that device. (a) From $W = C(\Delta V)^2/2$, it is seen that increasing the capacitance while keeping the potential difference constant will increase the energy stored in the capacitor.
(b) From $W = Q^2/2C$, it is seen that increasing the capacitance while holding the charge on the capacitor constant decreases the energy stored in the capacitor.

### Problems

**1.** **(a)** $6.40 \times 10^{-19}$ J **(b)** $-6.40 \times 10^{-19}$ J
**(c)** $-4.00$ V
**3.** $1.4 \times 10^{-20}$ J
**5.** $1.7 \times 10^6$ N/C
**7.** **(a)** $1.13 \times 10^5$ N/C **(b)** $1.80 \times 10^{-14}$ N
**(c)** $4.38 \times 10^{-17}$ J
**9.** **(a)** 0.500 m **(b)** 0.250 m
**11.** **(a)** $1.44 \times 10^{-7}$ V **(b)** $-7.19 \times 10^{-8}$ V
**13.** **(a)** $2.67 \times 10^6$ V **(b)** $2.13 \times 10^6$ V
**15.** **(a)** 103 V
**(b)** $-3.85 \times 10^{-7}$ J; positive work must be done to separate the charges.
**17.** $-11.0$ kV
**19.** $2.74 \times 10^{-14}$ m
**21.** 0.719 m, 1.44 m, 2.88 m. No. The equipotentials are not uniformly spaced. Instead, the radius of an equipotential is inversely proportional to the potential.
**23.** **(a)** $1.1 \times 10^{-8}$ F **(b)** 27 C
**25.** **(a)** 11.1 kV/m toward the negative plate **(b)** 3.74 pF
**(c)** 74.7 pC and $-74.7$ pC
**27.** $2.26 \times 10^{-5}$ m$^2$
**29.** **(a)** 13.3 $\mu$C on each **(b)** 20.0 $\mu$C, 40.0 $\mu$C

31. (a) 2.00 $\mu$F
(b) $Q_3 = 24.0\ \mu$C, $Q_4 = 16.0\ \mu$C, $Q_2 = 8.00\ \mu$C, $(\Delta V)_2 = (\Delta V)_4 = 4.00$ V, $(\Delta V)_3 = 8.00$ V
33. (a) 5.96 $\mu$F
(b) $Q_{20} = 89.5\ \mu$C, $Q_6 = 63.2\ \mu$C, $Q_3 = Q_{15} = 26.3\ \mu$C
35. $Q_1 = 16.0\ \mu$C, $Q_5 = 80.0\ \mu$C, $Q_8 = 64.0\ \mu$C, $Q_4 = 32.0\ \mu$C
37. (a) $Q_{25} = 1.25$ mC, $Q_{40} = 2.00$ mC
(b) $Q'_{25} = 288\ \mu$C, $Q'_{40} = 462\ \mu$C, $\Delta V = 11.5$ V
39. $Q'_1 = 3.33\ \mu$C, $Q'_2 = 6.67\ \mu$C
41. 83.6 $\mu$C
43. $2.55 \times 10^{-11}$ J
45. $3.2 \times 10^{10}$ J
47. $\kappa = 4.0$
49. (a) 8.13 nF (b) 2.40 kV
51. (a) volume $9.09 \times 10^{-16}$ m$^3$, area $4.54 \times 10^{-10}$ m$^2$
(b) $2.01 \times 10^{-13}$ F
(c) $2.01 \times 10^{-14}$ C; $1.26 \times 10^5$ electronic charges
55. $v = 1.67$ m/s upward, $y = 6.24$ cm above bottom plate
57. 6.25 $\mu$F
59. 4.47 kV
61. 0.75 mC on $C_1$, 0.25 mC on $C_2$
65. 50 N

## CHAPTER 17

### Quick Quizzes

1. d, b = c, a
2. c, d
3. b
4. b, d
5. a
6. $I_a = I_b > I_c = I_d > I_e = I_f$
7. B, B

### Conceptual Questions

1. It is made possible by the very high density of charge carriers ($\sim 10^{28}$/m$^3$) in metallic conductors.
3. The gravitational force pulling the electron to the bottom of a piece of metal is much smaller than the electrical repulsion pushing the electrons apart. Thus, free electrons stay distributed throughout the metal. The concept of charges residing on the surface of a metal is true for a metal with an excess charge. The number of free electrons in an electrically neutral piece of metal is the same as the number of positive ions—the metal has zero net charge.
5. A voltage is not something that "surges through" a completed circuit. A voltage is a potential difference that is applied across a device or a circuit. What goes through the circuit is current. Thus, it would be more correct to say, "1 ampere of electricity surged through the victim's body." Although this current would have disastrous results on the human body, a value of 1 (ampere) doesn't sound as exciting for a newspaper article as 10 000 (volts). Another possibility is to write, "10 000 volts of electricity were applied across the victim's body," which still doesn't sound quite as exciting.
7. We would conclude that the conductor is nonohmic.
9. The shape, dimensions, and the resistivity affect the resistance of a conductor. Because temperature and impurities affect the conductor's resistivity, these factors also affect resistance.
11. The radius of wire B is the square root of three times the radius of wire A, to make its cross-sectional area three times larger.
13. Drift velocity might increase steadily as time goes on, because collisions between electrons and atoms in the wire would be essentially nonexistent and the conduction electrons would move with constant acceleration. The current would rise steadily without bound also, because $I$ is proportional to the drift velocity.

### Problems

1. $3.00 \times 10^{20}$ electrons move past in the direction opposite to the current.
3. 3.0 mA
5. 1.05 mA
7. 27 yr
9. (a) $n$ is unaffected (b) $v_d$ is doubled
11. 32 V, is 200 times larger than 0.16 V
13. 0.17 mm
15. (a) 30 Ω (b) $4.7 \times 10^{-4}$ Ω·m
17. silver ($\rho = 1.59 \times 10^{-8}$ Ω·m)
19. 2.50 mA decrease
21. 1.98 A
23. 26 mA
25. (a) $5.89 \times 10^{-2}$ Ω (b) $5.45 \times 10^{-2}$ Ω
27. (a) 3.0 A (b) 2.9 A
29. (a) 1.2 Ω
(b) $8.0 \times 10^{-4}$ (a 0.080% increase)
31. 5.00 A, 24.0 Ω
33. 18 bulbs
35. 11.2 min
37. 34.4 Ω
39. 1.6 cm
41. 295 metric tons/h
43. 26 cents
45. 23 cents
47. \$1.2
49. 1.1 km
51. $1.47 \times 10^{-6}$ Ω·m, differs by 2.0% from value in Table 17.1
53. 0.400 $\mu$A
55. (a) 667 A (b) 50.0 km
57. $3.77 \times 10^{28}$/m$^3$
59. (a) 144 Ω (b) 26 m
(c) To fit the required length into a small space.
(d) 25 m
61. 37 MΩ
63. 0.48 kg/s
65. (a) $2.6 \times 10^{-5}$ Ω (b) 76 kg

## CHAPTER 18

### Quick Quizzes

1. Bulb $R_1$ becomes brighter. Bulb $R_2$ goes out.
2. (b)
3. (a)
4. (a) decrease; (b) decrease; (c) increase; (d) decrease; (e) decrease.

### Conceptual Questions

1. No. When a battery serves as a source and supplies current to a circuit, the conventional current flows through the battery from the negative terminal to the positive one. However, when a source having a larger emf than the battery is used to charge the battery, the conventional current is forced to flow through the battery from the positive terminal to the negative one.
3. The total amount of energy delivered by the battery will be less than $W$. Recall that a battery can be considered to be an ideal, resistanceless battery in series with the internal resistance. When charging, the energy delivered to the battery includes the energy necessary to charge the ideal battery plus the energy that goes into raising the battery's temperature due to $I^2r$ heating in the internal resistance. This latter energy is not available during the discharge of the battery. During discharge, part of the reduced available energy again transforms into internal energy in the internal resistance, further reducing the available energy below $W$.

5. The starter in the automobile draws a relatively large current from the battery. This large current causes a significant voltage drop across the internal resistance of the battery. As a result, the terminal voltage of the battery is reduced, and the headlights dim accordingly.
7. An electrical appliance has a given resistance. Thus, when it is attached to a power source with a known potential difference, a definite current will be drawn. The device can be labeled with both the voltage and the current. Batteries, however, can be applied to a number of devices. Each device will have a different resistance, so the current from the battery will vary with the device. As a result, only the voltage of the battery can be specified.
9. Connecting batteries in parallel does not increase the emf. A high-current device connected to batteries in parallel can draw currents from both batteries. Thus, connecting the batteries in parallel does increase the possible current output and, therefore, the possible power output.
11. The lightbulb will glow for a very short while as the capacitor is being charged. Once the capacitor is almost totally charged, the current in the circuit will be nearly zero and the bulb does not glow.
13. The bird is at rest on a wire whose electric potential is nearly constant along its length. In order to be electrocuted, a large potential difference is required between the bird's feet. The potential difference between the bird's feet is too small to harm the bird.
15. The junction rule is a statement of conservation of charge. It says that the amount of charge that enters the junction in some time interval must equal the charge that leaves the junction in that time interval. The loop rule is a statement of conservation of energy. It says that the potential increases and decreases around a closed loop in a circuit must add to zero.
17. A few of the factors involved are as follows: the conductivity of the string (is it wet or dry?); how well you are insulated from ground (are you wearing thick rubber- or leather-soled shoes?); the magnitude of the potential difference between you and the kite; the type and condition of the soil under your feet.
19. She will not be electrocuted if she holds onto only one high-voltage wire, because she is not completing a circuit. There is no potential difference across her body, as long as she clings to only one wire. However, she should immediately release the wire once it breaks, because she will become part of a closed circuit when she reaches the ground or comes into contact with another object.
21. **(a)** The intensity of both lamps increases because lamp C is short-circuited and there is current (which increases) only in lamps A and B. **(b)** The intensity of lamp C goes to zero because the current in this branch goes to zero. **(c)** The current in the circuit increases because the total resistance decreases from $3R$ (with the switch open) to $2R$ (after the switch is closed). **(d)** The voltage drop across lamps A and B increases, while the voltage drop across lamp C becomes zero. **(e)** The power dissipated increases from $\mathcal{E}^2/3R$ (with the switch open) to $\mathcal{E}^2/2R$ (after the switch is closed).

### Problems

1. 4.92 Ω
3. 73.8 W; your circuit diagram will consist of two 0.800-Ω resistors in series with the 192-Ω resistance of the bulb.
5. **(a)** 17.1 Ω
   **(b)** 1.99 A for 4.00 Ω and 9.00 Ω, 1.17 A for 7.00 Ω, 0.818 A for 10.0 Ω
7. 9.8 Ω
9. **(a)** 0.227 A **(b)** 5.68 V
11. 55 Ω
13. 0.43 A
15. **(a)** Connect two 50-Ω resistors in parallel, and then connect this combination in series with a 20-Ω resistor. **(b)** Connect two 50-Ω resistors in parallel, connect two 20-Ω resistors in parallel, and then connect these two combinations in series with each other.
17. 0.846 A down in 8.00-Ω resistor; 0.462 A down in the middle branch; 1.31 A up in the right-hand branch
19. **(a)** 3.00 mA **(b)** − 19.0 V **(c)** 4.50 V
21. 10.7 V
23. **(a)** 0.385 mA, 3.08 mA, 2.69 mA
    **(b)** 69.2 V with $c$ at the higher potential
25. $I_1 = 3.5$ A, $I_2 = 2.5$ A, $I_3 = 1.0$ A
27. $I_{30} = 0.353$ A, $I_5 = 0.118$ A, $I_{20} = 0.471$ A
29. $\Delta V_2 = 3.05$ V, $\Delta V_3 = 4.57$ V, $\Delta V_4 = 7.38$ V, $\Delta V_5 = 1.62$ V
31. **(a)** 12 s **(b)** $1.2 \times 10^{-4}$ C
33. $1.3 \times 10^{-4}$ C
35. 0.982 s
37. **(a)** heater, 10.8 A; toaster, 8.33 A; grill, 12.5 A
    **(b)** $I_{total} = 31.6$ A, so a 30-A breaker is insufficient.
39. **(a)** 6.25 A **(b)** 750 W
41. **(a)** $1.2 \times 10^{-9}$ C, $7.3 \times 10^{9}$ $K^+$ions. Not large, only $1e/320$ Å$^2$
    **(b)** $1.7 \times 10^{-9}$ C, $1.0 \times 10^{10}$ $Na^+$ ions
    **(c)** 0.83 μA **(d)** $7.5 \times 10^{-12}$ J
43. 9.7 nW
45. 7.5 Ω
47. **(a)** 15 Ω
    **(b)** $I_1 = 1.0$ A, $I_2 = I_3 = 0.50$ A, $I_4 = 0.30$ A, and $I_5 = 0.20$ A
    **(c)** $(\Delta V)_{ac} = 6.0$ V, $(\Delta V)_{ce} = 1.2$ V, $(\Delta V)_{ed} = (\Delta V)_{fd} = 1.8$ V, $(\Delta V)_{cd} = 3.0$ V, $(\Delta V)_{db} = 6.0$ V
    **(d)** $\mathcal{P}_{ac} = 6.0$ W, $\mathcal{P}_{ce} = 0.60$ W, $\mathcal{P}_{ed} = 0.54$ W, $\mathcal{P}_{fd} = 0.36$ W, $\mathcal{P}_{cd} = 1.5$ W, $\mathcal{P}_{db} = 6.0$ W
49. **(a)** 12.4 V **(b)** 9.65 V
51. $I_1 = 0$, $I_2 = I_3 = 0.50$ A
53. 112 V, 0.200 Ω
55. **(a)** $R_x = R_2 - (\frac{1}{4})R_1$
    **(b)** $R_x = 2.8$ Ω (inadequate grounding)
59. $\mathcal{P} = \dfrac{(144\text{ V}^2)R}{(R + 10.0\ \Omega)^2}$

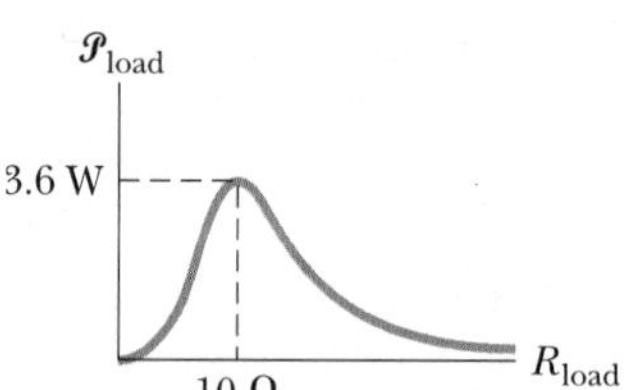


## Chapter 19

### Quick Quizzes

1. $a$ or $c$ might be true
2. c
3. a
4. c
5. b

### Conceptual Questions

1. No. In a uniform magnetic field, the poles of the bar magnetic experience forces that are equal in magnitude but oppositely directed. Thus, the net force exerted on the magnet by the magnetic field is zero.
3. The proton moves in a circular path upward on the page. After completing half the circle, it exits the field and moves in

a straight-line path back in the direction from whence it came. An electron behaves similarly, but the direction of traversal of the circle is downward, and the radius of the circular path is smaller.

**5.** The magnetic force on a moving charged particle is always perpendicular to the direction of motion. There is no magnetic force on the charge when it moves parallel to the direction of the magnetic field. However, the force on a charged particle moving in an electric field is never zero and is always parallel to the direction of the electric field. Therefore, by projecting the charged particle in different directions, it is possible to determine the nature of the field.

**7.** The magnetic field produces a magnetic force on the electrons moving toward the screen that produce the image. This magnetic force deflects the electrons to regions on the screen other than the ones to which they are supposed to go. The result is a distorted image.

**9.** Such levitation would never occur. At the North Pole where Earth's magnetic field is downward, toward the equivalent of a buried south pole, a coffin would be repelled if its south magnetic pole were directed downward. However, equilibrium would be only transitory as any slight disturbance would upset the balance between the magnetic force and the gravitational force.

**11.** If you are moving along with the electrons, you will measure zero current for the electrons, so the electrons would not produce a magnetic field according to your observations. However, the fixed positive charges in the metal are now moving backward relative to you and creating a current equivalent to the forward motion of the electrons when you were stationary. Thus, you will measure the same magnetic field as when you were stationary, but it will be due to the positive charges presumed to be moving from your point of view.

**13.** A compass does not detect currents in wires near light switches for two reasons. Because the cable to the light switch contains two wires, with one carrying current to the switch and the other away from the switch, the net magnetic field is very small and falls off rapidly. The second reason is that the current is alternating at 60 Hz with increasing distance. As a result, the magnetic field is oscillating at 60 Hz, also. This frequency is too fast for the compass to follow, so the effect on the compass reading averages to zero.

**15.** The levitating wire is stable with respect to vertical motion—if it is displaced upward, the repulsive force weakens, and the wire drops back down. Conversely, if it drops lower, the repulsive force increases, and it moves back up. The wire is not stable, however, with respect to lateral movement. If it moves away from the vertical position directly over the lower wire, the repulsive force has a sideways component, which pushes the wire away.

In the case of the attracting wires, the hanging wire is not stable for vertical movement. If it rises, the attractive force increases, and the wire moves even closer to the upper wire. If the hanging wire falls, the attractive force weakens, and the wire falls farther. If the wire moves to the right, it moves farther from the upper wire and the attractive force decreases. Although there is a restoring force component pulling it back to the left, the vertical force component is not strong enough to hold it up and it falls.

**17.** Each coil of the Slinky becomes a magnet, because a coil acts as a current loop. Because the sense of rotation of the current is the same in all coils, each coil becomes a magnet with the same orientation of poles. Thus, all of the coils attract, and the Slinky compresses.

**19.** There is no net force on the wires, but there is a torque. To understand this, imagine a fixed vertical wire, and a free horizontal wire (Fig. ANS Q19.19). The vertical wire carries an upward current, and creates a magnetic field that circles the wire. Each segment of the horizontal wire (of length $\ell$) also carries a current that interacts with the magnetic field according to the equation $F = BI\ell \sin \theta$. Applying the right-hand rule, we see that the horizontal wire experiences an upward force on one side, and an equal downward force on the other. The forces cancel, creating a torque around the point at which the wires cross.

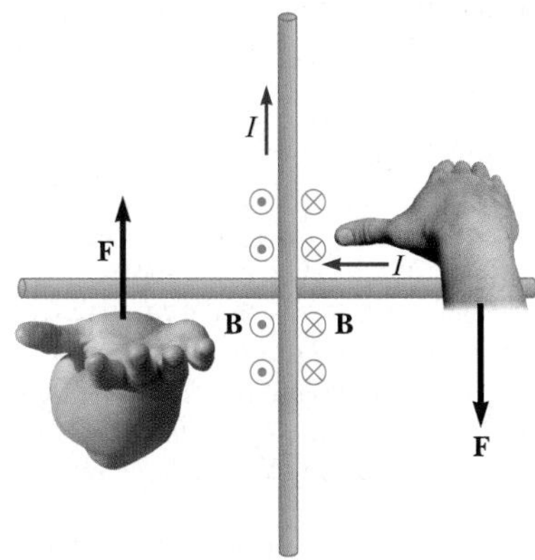


## Problems

**1.** **(a)** horizontal and due east
**(b)** horizontal and 30°N of E
**(c)** horizontal and due east
**(d)** zero force

**3.** **(a)** into the page **(b)** toward the right
**(c)** toward the bottom of the page

**5.** $F_g = 8.93 \times 10^{-30}$ N (downward), $F_e = 1.60 \times 10^{-17}$ N (upward), $F_m = 4.80 \times 10^{-17}$ N (downward)

**7.** $2.83 \times 10^7$ m/s west

**9.** 0.021 T in the $-y$ direction

**11.** $8.0 \times 10^{-3}$ T in the $+z$ direction

**13.** **(a)** to the left **(b)** into the page
**(c)** out of the page **(d)** toward top of page
**(e)** into the page **(f)** out of the page

**15.** 7.50 N

**17.** 0.131 T (downward)

**19.** 0.20 T directed out of the page

**21.** *ab*: 0, *bc*: 0.040 0 N in the $-x$ direction, *cd*: 0.040 0 N in the $-z$ direction, *da*: $0.040\,0\sqrt{2}$ N parallel to the $xz$ plane and at 45° with both $+x$ and $+z$ directions

**23.** $9.05 \times 10^{-4}$ N·m tending to make the left-hand side of the loop move toward you and the right-hand side move away.

**25.** **(a)** 3.97° **(b)** $3.39 \times 10^{-3}$ N·m

**27.** $2.0 \times 10^{-12}$ kg

**31.** 1.77 cm

**33.** **(a)** 5.00 cm **(b)** $8.78 \times 10^6$ m/s

**35.** 20.0 $\mu$T

**37.** 2.0 cm

**39.** 20.0 $\mu$T toward the bottom of the page

**41.** 0.167 $\mu$T out of the page

**43.** **(a)** 4.00 m **(b)** 7.50 nT **(c)** 1.26 m **(d)** zero

**45.** 4.5 mm

**47.** 31.8 mA

**49.** $2.26 \times 10^{-4}$ N away from the center, zero torque

**51.** 1.7 N·m

**53.** **(a)** 0.500 $\mu$T out of the page
**(b)** 3.89 $\mu$T parallel to the $xy$ plane and at 59.0° clockwise from the $+x$ direction

**55.** 2.13 cm

**57.** **(a)** 1.33 m/s
**(b)** the sign of the emf is independent of charge

**59.** $1.41 \times 10^{-6}$ N

**61.** 13.0 $\mu$T toward the bottom of the page
**63.** **(a)** 53 $\mu$T toward the bottom of the page
**(b)** 20 $\mu$T toward the bottom of the page
**(c)** 0
**65.** **(a)** $-8.00 \times 10^{-21}$ kg·m/s **(b)** 8.90°

## CHAPTER 20

### Quick Quizzes

**1.** *b, c, a*
**2.** The left wing tip on the west side of the airplane.
**3.** b
**4.** c
**5.** b

### Conceptual Questions

**1.** According to Faraday's law, an emf is induced in a wire loop if the magnetic flux through the loop changes with time. In this situation, an emf can be induced by either rotating the loop around an arbitrary axis or by changing the shape of the loop.
**3.** As the spacecraft moves through space, it is apparently moving from a region of one magnetic field strength to a region of a different magnetic field strength. The changing magnetic field through the coil induces an emf and corresponding current in the coil.
**5.** If the bar were moving to the left, the magnetic force on the negative charges in the bar would be upward, causing an accumulation of negative charge on the top, and positive charges at the bottom. Hence, the electric field in the bar would be upward.
**7.** If, for any reason the magnetic field should change rapidly, a large emf could be induced in the bracelet. If the bracelet were not a continuous band, this emf would cause high-voltage arcs to occur at any gap in the band. If the bracelet were a continuous band, the induced emf would produce a large induced current and result in resistance heating of the bracelet.
**11.** As the aluminum plate moves into the field, eddy currents are induced in the metal by the changing magnetic field at the plate. The magnetic field of the electromagnet interacts with this current producing a retarding force on the plate, slowing it down. In a similar fashion, as the plate leaves the magnetic field, a current is induced, and once again there is an upward force to slow the plate.
**13.** The energy stored in an inductor carrying a current $I$ is equal to $PE_L = \frac{1}{2}LI^2$. Therefore, doubling the current quadruples the energy stored in the inductor.
**15.** If an external battery is acting to increase the current in the inductor, an emf is induced in a direction to oppose the increase of current. Likewise, if we attempt to reduce the current in the inductor, the emf set up tries to support the current. Thus, the induced emf always acts to oppose the change occurring in the circuit, or it acts in the "back" direction to the change.

### Problems

**1.** $5.9 \times 10^{-2}$ T·m²
**3.** $1.72 \times 10^{-6}$ T·m², 0, $9.15 \times 10^{-7}$ T·m²
**5.** $\Phi_{B,\text{net}} = 0$ **(b)** 0
**7.** **(a)** $3.1 \times 10^{-3}$ T·m² **(b)** $\Phi_{B,\text{net}} = 0$
**9.** 9.82 mV
**11.** 160 A
**13.** 2.7 T/s
**15.** **(a)** $4.0 \times 10^{-6}$ T·m² **(b)** 18 $\mu$V
**17.** 11.6 $\mu$V
**19.** 0.763 V
**21.** **(a)** toward the east **(b)** $4.58 \times 10^{-4}$ V
**23.** **(a)** left to right **(b)** right to left
**25.** left to right
**27.** into the page
**29.** **(a)** right to left **(b)** right to left **(c)** left to right **(d)** left to right
**31.** 28.8 V
**33.** **(a)** 11 Ω **(b)** 76 V
**35.** **(a)** 60 V **(b)** 57 V **(c)** 0.13 s
**37.** 100 V
**39.** **(a)** 2.0 mH **(b)** 38 A/s
**43.** 12 mH
**45.** 1.92 Ω
**47.** 0.140 J
**49.** **(a)** 18 J **(b)** 7.2 J
**51.** negative ($V_a < V_b$)
**53.** **(a)** 20.0 ms **(b)** 37.9 V **(c)** 1.52 mV **(d)** 51.8 mA
**55.** 1.20 $\mu$C
**57.** **(a)** 0.500 A **(b)** 2.00 W **(c)** 2.00 W
**59.** 115 kV
**61.** **(a)** 0.157 mV (end $B$ is positive)
**(b)** 5.89 mV (end $A$ is positive)
**63.** **(a)** 9.00 A **(b)** 10.8 N
**(c)** $b$ is at the higher potential **(d)** No

## CHAPTER 21

### Quick Quizzes

**1.** c
**2.** b
**3.** b
**4.** b, c

### Conceptual Questions

**1.** For best reception, the length of the antenna should be parallel to the orientation of the oscillating electric field. Because of atmospheric variations and reflections of the wave before it arrives at your location, the orientation of this field may be in different directions for different stations.
**3.** The primary coil of the transformer is an inductor. When an AC voltage is applied, the back emf due to the inductance will limit the current flow through the coil. If DC voltage is applied, there is no back emf and the current can rise to a higher value. It is possible that this increased current will deliver so much energy to the resistance in the coil that its temperature rises to the point at which insulation on the wire can burn.
**5.** An antenna that is a conducting line responds to the electric field of the electromagnetic wave—the oscillating electric field causes an electric force on electrons in the wire along its length. The movement of electrons along the wire is detected as a current by the radio and amplified. Thus, a line antenna must have the same orientation as the broadcast antenna. A loop antenna responds to the magnetic field in the radio wave. The varying magnetic field induces a varying current in the loop (Faraday's law), and this signal is amplified. The loop should be in the vertical plane containing the sight line to the broadcast antenna, so the magnetic field lines go through the area of the loop.
**7.** The flashing of the light according to Morse code is a drastic amplitude modulation—the amplitude is changing from a maximum to zero. In this sense, it is similar to the on-and-off binary code used in computers and compact discs. The carrier frequency is that of the light, on the order of $10^{14}$ Hz. The signal frequency depends on the skill of the signal operator, but it is on the order of a single hertz, as the light is flashed on and off. The broadcasting antenna for this modu-

lated signal is the filament of the lightbulb in the signal source. The receiving antenna is the eye.

**9.** The sail should be as reflective as possible so that the maximum momentum is transferred to the sail from the reflection of sunlight.

**11.** Suppose the extraterrestrial looks around your kitchen. Lightbulbs and the toaster glow brightly in the infrared. Somewhat fainter are the back of the refrigerator and the back of the television set, while the television screen is dark. The pipes under the sink show the same weak glow as the walls until you turn on the faucets. Then the pipe on the right gets darker and that on the left develops a gleam that quickly runs up along its length. The food on the plates shines, as does human skin, the same color for all races. Clothing is dark as a rule, but your seat and the chair seat glow alike after you stand up. Your face appears lit from within, like a jack-o-lantern; your nostrils and the openings of your ear canals are bright; brighter still are the pupils of your eyes.

**13.** Radio waves move at the speed of light. They can travel around the curved surface of the Earth, bouncing between the ground and the ionosphere, which has an altitude that is small compared to the radius of the Earth. The distance across the lower 48 states is approximately 5 000 km, requiring a travel time of $(5 \times 10^6 \text{ m})/(3 \times 10^8 \text{ m/s}) \sim 10^{-2}$ s. Likewise, radio waves take only 0.07 s to travel halfway around the Earth. In other words, a speech can be heard on the other side of the world (in the form of radio waves) before it is heard at the back of the room (in the form of sound waves).

**15.** No. The wire will emit electromagnetic waves only if the current varies in time. The radiation is the result of accelerating charges, which can only occur when the current is not constant.

**17.** The resonance frequency is determined by the inductance and the capacitance in the circuit. It is not affected by the resistance in the circuit, so doubling the resistance does not alter the resonance frequency.

### Problems

**1.** **(a)** 141 V **(b)** 20.0 A **(c)** 28.3 A **(d)** 2.00 kW
**3.** 70.7 V, 2.95 A
**5.** 6.76 W
**9.** $4.0 \times 10^2$ Hz
**11.** 17 $\mu$F
**15.** 3.14 A
**17.** $0.450 \text{ T} \cdot \text{m}^2$
**19.** **(a)** 205 mA **(b)** 154 V **(c)** 54.3 V **(d)** 90.0°
**(e)**

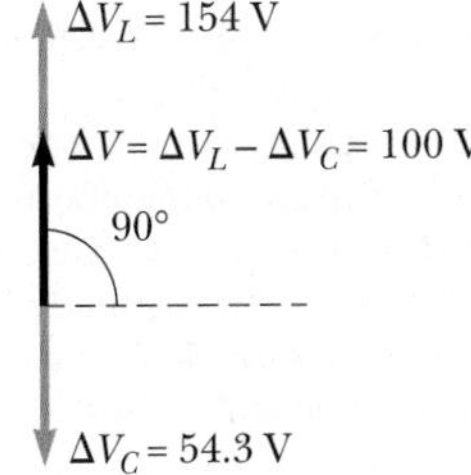


**21.** **(a)** 1.4 k$\Omega$ **(b)** 0.10 A **(c)** 51° **(d)** voltage leads current
**23.** **(a)** 89.6 V **(b)** 109 V
**25.** 1.88 V
**27.** **(a)** 103 V **(b)** 150 V **(c)** 127 V **(d)** 23.6 V
**29.** **(a)** 208 $\Omega$ **(b)** 40.0 $\Omega$ **(c)** 0.541 H
**31.** **(a)** $1.8 \times 10^2$ $\Omega$ **(b)** 0.71 H
**33.** 2.29 $\mu$H
**35.** $C_{min} = 4.9$ nF, $C_{max} = 51$ nF
**37.** 0.242 J
**39.** **(a)** 1 600 **(b)** 30 A
**41.** **(a)** $1.1 \times 10^3$ kW **(b)** $3.1 \times 10^2$ A **(c)** $8.3 \times 10^3$ A
**43.** 1 000 km, there will always be better use for tax money.
**45.** 733 nT
**47.** $2.94 \times 10^8$ m/s
**49.** $E_{max} = 1.01 \times 10^3$ V/m, $B_{max} = 3.35 \times 10^{-6}$ T
**51.** **(a)** 188 m to 556 m **(b)** 2.78 m to 3.4 m
**53.** 60.0 km
**55.** **(a)** 18 turns **(b)** 3.6 W
**57.** 99.6 mH
**59.** 1.7 cents
**61.** **(a)** resistor and inductor **(b)** $R = 10\ \Omega$, $L = 30$ mH
**63.** **(a)** $6.7 \times 10^{-16}$ T **(b)** $5.3 \times 10^{-17}$ W/m$^2$ **(c)** $1.7 \times 10^{-14}$ W
**65.** **(a)** 0.536 N **(b)** $8.93 \times 10^{-5}$ m/s$^2$ **(c)** 33.9 days
**67.** $4.47 \times 10^{-9}$ J

## CHAPTER 22

### Quick Quizzes

**1.** a
**2.** Beams 2 and 4 are reflected; beams 3 and 5 are refracted.
**3.** b
**4.** c

### Conceptual Questions

**1.** Sound radiated upward at an acute angle with the horizontal is bent back toward Earth by refraction. This means that the sound can reach the listener by this path as well as by a direct path. Thus, the sound is louder.

**3.** The color will not change, for two reasons. First, despite the popular statement that color depends on wavelength, it actually depends on the *frequency* of the light, which does not change under water. Second, when the light enters the eye, it travels through the fluid within the eye. Thus, even if color did depend on wavelength, the important wavelength is that of the light in the ocular fluid, which does not depend on the medium through which the light traveled to reach the eye.

**5.** A prism creates a multicolored spectrum when white light passes through it because the white light contains all visible wavelengths. Because the index of refraction varies with wavelength, the different wavelengths emerge from the prism traveling in slightly different directions. This produces the observed rainbow of colors. If only one wavelength of light is allowed to enter the prism, as in the case of the second prism, a single wavelength (and hence color) will emerge from it.

**7.** No, the catalog information is incorrect. The index of refraction is given by $n = c/v$, where $c$ is the speed of light in vacuum and $v$ is the speed of light in the material. Because light travels faster in a vacuum than any other material, it is impossible for the index of refraction of any material to have a value less than 1.

**9.** There is no dependence of the angle of reflection on wavelength, because the light does not enter deeply into the material during reflection—it reflects from the surface.

**11.** A ball covered with mirrors sparkles by reflecting light from its surface. On the other hand, a faceted diamond lets in light at the top, reflects it by total internal reflection in the bottom half, and sends the light out through the top again. Because of its high index of refraction, the critical angle for diamond in air for total internal reflection, $\theta_c = \sin^{-1}(n_{air}/n_{diamond})$, is small. Thus, light rays enter through a large area and exit through a very small area with a much higher intensity. When a diamond is immersed in carbon disulfide, the critical angle is increased to $\theta_c = \sin^{-1}(n_{\text{carbon disulfide}}/n_{\text{diamond}})$. As a result, the light is emitted from the diamond over a larger area and appears less intense.

**13.** The index of refraction of water is 1.333, quite different from that of air, which has an index of refraction of about 1. On the other hand, the index of refraction of liquid helium happens to be much closer to that of air. Consequently, light undergoes less refraction in helium than it does in water.

**15.** The diamond acts like a prism in dispersing the light into its spectral components. Different colors are observed as a consequence of the manner in which the index of refraction varies with the wavelength.

**17.** Light travels through vacuum at a speed of $3 \times 10^8$ m/s. Thus, an image we see from a distant star or galaxy must have been generated some time ago. For example, the star Altair is 16 lightyears away; if we look at an image of Altair today, we know only what Altair looked like 16 years ago. This may not initially seem significant; however, astronomers who look at other galaxies can get an idea of what galaxies looked like when they were much younger. Thus, it does make sense to speak of "looking backward in time."

### Problems

**1.** $3.00 \times 10^8$ m/s
**3.** 114 rad/s for a maximum intensity of returning light
**5.** **(b)** $3.000 \times 10^8$ m/s
**7.** 19.5° above the horizontal
**9.** **(a)** 1.52 **(b)** 417 nm **(c)** $4.74 \times 10^{14}$ Hz **(d)** $1.98 \times 10^8$ m/s
**11.** **(a)** 584 nm **(b)** 1.12
**13.** 111°
**15.** 16.5°
**17.** five times from the right-hand mirror and six times from the left
**19.** 0.388 cm
**21.** $\theta = 30.4°$, $\theta' = 22.3°$
**23.** 6.39 ns
**25.** $\theta = \tan^{-1}(n_g)$
**27.** 3.39 m
**29.** $\theta_{\text{red}} = 48.22°$, $\theta_{\text{blue}} = 47.79°$
**31.** **(a)** $\theta_{1i} = 30°$, $\theta_{1r} = 19°$, $\theta_{2i} = 41°$, $\theta_{2r} = 77°$ **(b)** first surface: $\theta_{\text{reflection}} = 30°$, second surface: $\theta_{\text{reflection}} = 41°$
**33.** **(a)** 31.3° **(b)** 44.2° **(c)** 49.8°
**35.** **(a)** 33.4° **(b)** 53.4°
**37.** 1.38
**39.** 1.000 08
**41.** **(a)** 10.7° **(b)** air **(c)** Sound falling on the wall from most directions is 100% reflected.
**43.** 27.5°
**45.** 22.0°
**47.** **(a)** 53.1° **(b)** $\geq 38.7°$
**49.** **(a)** 38.5° **(b)** $\geq 1.44$
**53.** 24.7°
**55.** 1.93
**59.** $\theta = \sin^{-1}(\sqrt{n^2 - 1}\sin\phi - \cos\phi)$

## CHAPTER 23

### Quick Quizzes

**1.** At *C*
**2.** c
**3.** **(a)** False **(b)** False **(c)** True
**4.** b
**5.** An infinite number
**6.** **(a)** False **(b)** True **(c)** False

### Conceptual Questions

**1.** You will not be able to focus your eyes on both the picture and your image at the same time. To focus on the picture, you must adjust your eyes so that an object several centimeters away (the picture) is in focus. Thus, you are focusing on the mirror surface. But, your image in the mirror is as far behind the mirror as you are in front of it. Thus, you must focus your eyes beyond the mirror, twice as far away as the picture to bring the image into focus.

**3.** A single flat mirror forms a virtual image of an object due to two factors. First, the light rays from the object are necessarily diverging from the object, and second, the lack of curvature of the flat mirror cannot convert diverging rays to converging rays. If another optical element is first used to cause light rays to converge, then the flat mirror can be placed in the region in which the converging rays are present, and it will simply change the direction of the rays so that the real image is formed in a different location. For example, if a real image is formed by a convex lens, and the flat mirror is placed between the lens and the image position, the image formed by the mirror will be real.

**5.** The ultrasonic range finder sends out a sound wave and measures the time for the echo to return. Using this information, the camera calculates the distance to the subject and sets the camera lens. When the camera is facing a mirror, the ultrasonic signal reflects from the mirror surface and the camera adjusts its focus so that the mirror surface is at the correct focusing distance from the camera. But your image in the mirror is twice this distance from the camera, so it is blurry.

**7.** Light rays diverge from the position of a virtual image just as they do from an actual object. Thus, a virtual image can be as easily photographed as any object can. Of course, the camera would have to be placed near the axis of the lens or mirror in order to intercept the light rays.

**9.** We consider the two trees to be two separate objects. The far tree is an object that is farther from the lens than the near tree. Thus, the image of the far tree will be closer to the lens than the image of the near tree. The screen must be moved closer to the lens to put the far tree in focus.

**11.** If a converging lens is placed in a liquid having an index of refraction larger than that of the lens material, the direction of refractions at the lens surfaces will be reversed, and the lens will diverge light. A mirror depends only on reflection, which is independent of the surrounding material, so a converging mirror will be converging in any liquid.

**13.** This is a possible scenario. When light crosses a boundary between air and ice, it will refract in the same manner as it does when crossing a boundary of the same shape between air and glass. Thus, a converging lens may be made from ice as well as glass. However, ice is such a strong absorber of infrared radiation that it is unlikely you will be able to start a fire with a small ice lens.

**15.** The focal length for a mirror is determined by the law of reflection from the mirror surface. The law of reflection is independent of the material of which the mirror is made and of the surrounding medium. Thus, the focal length depends only on the radius of curvature and not on the material. The focal length of a lens depends on the indices of refraction of the lens material and surrounding medium. Thus, the focal length of a lens depends on the lens material.

**17.** If parallel rays strike a flat mirror, they are reflected as parallel rays. Therefore, the focal length of a flat mirror is infinite, $f \rightarrow \infty$. This is consistent with the mirror equation

$$\frac{1}{p} + \frac{1}{q} = \frac{1}{f}$$

Because the image is always as far behind a flat mirror as the object is in front of it, then $q = -p$ and the mirror equation yields $f \rightarrow \infty$.

### Problems

1. on the order of $10^{-9}$ s younger
3. 10.0 ft, 30.0 ft, 40.0 ft
5. 0.268 m behind the mirror; virtual, upright, and diminished; $M = 0.026\,8$
7. **(a)** 13.3 cm in front of mirror, real, inverted, $M = -0.333$ **(b)** 20.0 cm in front of mirror, real, inverted, $M = -1.00$ **(c)** No image is formed. Parallel rays leave the mirror.
9. behind the worshipper, 3.33 m from the deepest point in the niche
11. 5.00 cm
13. 1.0 m
15. 8.05 cm
17. $-20.0$ cm
19. **(a)** concave with focal length $f = 0.83$ m **(b)** Object must be 1.0 m in front of the mirror.
21. 38.2 cm below the upper surface of the ice
23. 3.8 mm
25. **(a)** 120 cm **(b)** $-24.0$ cm **(c)** $-8.00$ cm **(d)** $-3.43$ cm
27. 20.0 cm
29. **(a)** 40.0 cm beyond the lens, real, inverted, $M = -1.00$ **(b)** No image is formed. Parallel rays leave the lens. **(c)** 20.0 cm in front of the lens, virtual, upright, $M = +2.00$
31. **(a)** 13.3 cm in front of the lens, virtual, upright, $M = +\frac{1}{3}$ **(b)** 10.0 cm in front of the lens, virtual, upright, $M = +\frac{1}{2}$ **(c)** 6.67 cm in front of the lens, virtual, upright, $M = +\frac{2}{3}$
33. **(a)** either 9.63 cm or 3.27 cm **(b)** 2.10 cm
35. **(a)** 39.0 mm **(b)** 39.5 mm
37. at distance $2|f|$ in front of lens
39. 40.0 cm
41. 30.0 cm to the left of the second lens, $M = -3.00$
43. 7.47 cm in front of the second lens; 1.07 cm; virtual, upright
45. from 0.224 m to 18.2 m
47. real image, 5.71 cm in front of the mirror
49. $-40.0$ cm
51. 160 cm to the left of the lens, inverted, $M = -0.800$
53. $q = 10.7$ cm
55. 32.0 cm to the right of the second surface (real image)
57. **(a)** 20.0 cm to the right of the second lens; $M = -6.00$ **(b)** inverted **(c)** 6.67 cm to the right of the second lens; $M = -2.00$; inverted
59. **(a)** 1.99 **(b)** 10.0 cm to the left of the lens **(c)** inverted
61. **(a)** 5.45 m to the left of the lens **(b)** 8.24 m to the left of the lens **(c)** 17.1 m to the left of the lens **(d)** by surrounding the lens with a medium having a refractive index greater than that of the lens material

## CHAPTER 24

### Quick Quizzes

1. b
2. The compact disc.

### Conceptual Questions

1. You will not see an interference pattern from the automobile headlights for two reasons. The first is that the headlights are not coherent sources and are therefore incapable of producing sustained interference. Also, the headlights are so far apart in comparison to the wavelengths emitted that, even if they were made into coherent sources, the interference maxima and minima would be too closely spaced to be observable.
3. The result of the double slit is to redistribute the energy arriving at the screen. Although there is no energy at the location of a dark fringe, there is four times as much energy at the location of a bright fringe as there would be with only a single narrow slit. The total amount of energy arriving at the screen is twice as much as with a single slit, as it must be according to the conservation of energy.
5. One of the materials has a higher index of refraction than water, the other lower. The material with the higher index of refraction than water appears black as it approaches zero thickness. There is a 180° phase shift for the light reflected from the upper surface, but no such phase change from the lower surface, because the index of refraction for water on the other side is lower than that of the film. Thus, the two reflections are out of phase and interfere destructively. The material with index of refraction lower than water experiences a phase change for the light reflected from both upper and lower surfaces, so that the reflections from the zero-thickness film will be back in phase, and the film will appear bright.
7. For normal incidence, the extra path length followed by the reflected ray is twice the thickness of the film. For destructive interference, this must be a distance of half a wavelength of the light in the material of the film. For a film in air no 180° phase change occurs in these reflections, so the thickness of the film must be one-quarter wavelength, which is the same as the condition for constructive interference of reflected light. This means that the transmitted light is a minimum when the reflected light is a maximum, and vice-versa.
9. Since the light reflecting at the lower surface of the air film experience a 180° phase change whereas light reflecting from the upper surface of the film does not undergo a phase change, the central spot (where the film has near zero thickness) is dark. If the observed rings are not circular, the curved surface of the lens does not have a true spherical shape.
11. For regional communication at Earth's surface, radio waves are typically broadcast from currents oscillating in tall, vertical towers. These waves have vertical planes of polarization. Light originates from the vibrations of atoms, or electronic transitions within atoms, which represent oscillations in all possible directions. Thus, light is not generally polarized.
13. Yes. In order to do this, first measure the radar-reflectivity of the metal of your airplane. Then, choose a light, durable material that has approximately half the radar-reflectivity of the metal in your plane. Measure its index of refraction, and place onto the metal a coating equal in thickness to one quarter of 3 cm, divided by that index. Sell it quick and then you can sell to the supposed enemy new radars operating at 1.5 cm, which the coated metal will reflect with extra-high efficiency.
15. If you wish to perform an interference experiment, you need monochromatic coherent light. To obtain it, you must first pass light from an ordinary source through a prism or diffraction grating to disperse different colors into different directions. Using a single narrow slit, select a single color and make that light diffract to cover both slits for Young's experiment. The procedure is much simpler with a laser because its output is already monochromatic and coherent.

**17.** Audible sound has wavelengths on the order of meters or centimeters, whereas visible light has wavelengths on the order of half a micrometer. Sound, therefore, diffracts around walls and doorways (roughly meter-sized apertures). Visible light diffracts only through very small angles as it passes ordinary sized objects or apertures, because $\sin\theta = m\lambda/a$ by Equation 24.11, and $\lambda/a$ is extremely small.

**19.** Strictly speaking, the ribs do act as a diffraction grating, but the separation distance of the ribs is so much larger than the wavelength of the x-rays that there are no observable effects.

### Problems

**1.** 1.58 cm
**3.** **(a)** 2.62 mm **(b)** 2.62 mm
**5.** **(a)** 36.2° **(b)** 5.08 cm **(c)** $5.08 \times 10^{14}$ Hz
**7.** **(a)** 55.7 m **(b)** 124 m
**9.** 75.0 m
**11.** 11.3 m
**13.** 148 m
**15.** 91.9 nm
**17.** Any odd-integer multiple of 85.4 nm
**19.** 0.500 cm
**21.** **(a)** 238 nm **(b)** $\lambda$ will increase **(c)** 328 nm
**23.** 4.35 $\mu$m
**25.** 4.75 $\mu$m
**27.** No, the wavelengths intensified are 276 nm, 138 nm, 92.0 nm, . . .
**29.** 4.22 mm
**31.** **(a)** 1.1 m **(b)** 1.7 mm
**33.** 1.20 mm, 1.20 mm
**35.** **(a)** 479 nm, 647 nm, 698 nm **(b)** 20.5°, 28.3°, 30.7°
**37.** 5.49° in first order; 12.3° in second order; and 24.9° in third order
**39.** 44.5 cm
**41.** 10.1 cm
**43.** **(a)** 25.6° **(b)** 19.0°
**45.** **(a)** 1.11 **(b)** 42.0°
**47.** **(a)** 56.7° **(b)** 48.8°
**49.** 31.2°
**53.** 6.89 units
**55.** **(a)** 413.7 nm, 409.7 nm **(b)** 8.6°
**57.** 0.156 mm
**59.** 2.50 mm
**61.** 58.1°
**63.** **(a)** 16.6 m **(b)** 8.28 m
**65.** 1.31
**67.** 0.350 mm
**69.** 115 nm

## Chapter 25

### Answers to Quick Quizzes

**1.** You should choose a blue filter.

### Conceptual Questions

**1.** The observer is not using the lens as a simple magnifier. To use a lens as a simple magnifier, the object distance must be less than the focal length of the lens. Also, a simple magnifier produces a virtual image at the normal near point of the eye, or at a point about $q = -25$ cm. With a large object distance and a relatively short image distance, the magnitude of the magnification by the lens would be considerably less than one. Most likely, the lens is part of a lens combination used as a telescope.

**3.** The image formed on the retina by the lens and cornea is already inverted.

**5.** There will be an effect on the interference pattern—it will be distorted. The high temperature of the flame changes the index of refraction of air for the arm of the interferometer in which the match is held. As the index of refraction varies turbulently, the wavelength of the light in that region will also vary turbulently. As a result, the effective difference in length between the two arms varies, resulting in a wildly varying interference pattern.

**7.** Large lenses are difficult to manufacture and machine with accuracy. Their large weight leads to sagging, which produces a distorted image. In reflecting telescopes, light does not pass through glass; hence, problems associated with chromatic aberrations are eliminated. Large-diameter reflecting telescopes are also technically easier to construct. Some designs use a rotating pool of mercury as the reflecting surface.

**9.** In order to "see" an object, the wavelength of the light in the microscope must be smaller than the size of the object. An atom is much smaller than the wavelength of light in the visible spectrum, so an atom can never be seen using visible light.

### Problems

**1.** 30.0 cm beyond the lens, $M = -1/5$
**3.** Yes, the largest image dimension is only 19 mm.
**5.** $f/1.4$
**7.** $f/8.0$
**9.** 40.0 cm
**11.** 23.2 cm
**13.** **(a)** $-2.00$ diopters **(b)** 17.6 cm
**15.** $+17.0$ diopters
**17.** $m = +3.50$
**19.** **(a)** 4.07 cm **(b)** $m = +7.14$
**21.** **(a)** $|M| = 1.22$ **(b)** $\theta/\theta_0 = 6.08$
**23.** 2.1 cm
**25.** $m = -115$
**27.** $f_o = 90$ cm, $f_e = 2.0$ cm
**29.** **(b)** $-fh/p$ **(c)** $-1.07$ mm
**31.** **(a)** $m = 1.50$ **(b)** $m = 1.90$
**33.** 492 km
**35.** 0.40 $\mu$rad
**37.** 3.09 m
**39.** 9.8 km
**41.** 50.4 $\mu$m
**43.** 449 nm, blue
**45.** 98 fringe shifts
**47.** No. A resolving power of $2.0 \times 10^5$ is needed and that available is only $1.8 \times 10^5$.
**49.** **(a)** $+2.67$ diopters **(b)** 0.16 diopters too low
**51.** **(a)** $+44.6$ diopters **(b)** 3.03 diopters
**53.** **(a)** $1.0 \times 10^3$ lines **(b)** $3.3 \times 10^2$ lines
**55.** $m = 10.7$

## Chapter 26

### Quick Quizzes

**1.** b
**2.** The answer to both questions is no.
**3.** a, e
**4.** **(a)** False **(b)** False **(c)** True **(d)** False
**5.** a

### Conceptual Questions

**1.** One must be careful using such phrases because events that occur at different locations and appear to occur simultaneously to one observer will not appear simultaneous to observers in motion relative to the first. This lack of simultaneity becomes more pronounced at very high speeds.

**3.** This scenario is not possible with light. Light waves are described by the principles of special relativity. As you detect the light wave ahead of you and moving away from you (which would be a pretty good trick—think about it!), its velocity relative to you is $c$. Thus, you will not be able to catch up to the light wave.

5. Present knowledge of the Universe shows that even the "fixed" stars are in motion relative to us and in motion relative to each other. Thus, the stars are not really fixed at all and do not provide us with an absolute frame of reference.
7. The light from the quasar moves at $3.00 \times 10^8$ m/s. The speed of light is independent of the motion of the source or observer.
9. For a wonderful fictional exploration of this question, get a "Mr. Tompkins" book by George Gamow. All of the relativity effects would be obvious in our lives. Time dilation and length contraction would both occur. Driving home in a hurry, you would push on the gas pedal not to increase your speed very much, but to make the blocks shorter. Big Doppler shifts in wave frequencies would make red lights look green as you approached and make car horns and radios useless. High-speed transportation would be both very expensive, requiring huge fuel purchases, as well as dangerous, because a speeding car could knock down a building. When you got home, hungry for lunch, you would find that you had missed dinner; there would be a five-day delay in transit when you watch a live TV program originating in Australia. Finally, we would not be able to see the Milky Way, because the fireball of the Big Bang would surround us at the distance of Rigel, or Deneb.
11. This is not a violation of the concepts of relativity. The theory of relativity states that the result of any measurement of the vacuum speed of light will be independent of the motion of the observer or source. It makes no claim that light must have the same speed in all media.
13. Your assignment: Measure the length of a rod as it slides past you. Mark the position of its front end on the floor and have an assistant mark the position of the back end. Then measure the distance between the two marks. This distance will represent the length of the rod only if the two marks were made simultaneously in your frame of reference.

### Problems

1. **(a)** $t_{OB} = 1.67 \times 10^3$ s, $t_{OA} = 2.04 \times 10^3$ s
   **(b)** $t_{BO} = 2.50 \times 10^3$ s, $t_{AO} = 2.04 \times 10^3$ s
   **(c)** $\Delta t = 90$ s
3. 5.0 s
5. $c(\sqrt{3}/2)$
7. **(a)** $1.3 \times 10^{-7}$ s **(b)** 38 m
   **(c)** 7.6 m
9. **(a)** 2.2 μs **(b)** 0.65 km
11. $0.950c$
13. yes, with 19 m to spare
15. **(a)** 39.2 μs **(b)** accurate to one digit
17. $3.3 \times 10^5$ m/s
19. $0.285c$
21. $0.54c$ to the right
23. $0.357c$
25. $0.998c$ toward the right
27. **(a)** 54 min **(b)** 52 min
29. $c(\sqrt{3}/2)$
31. $0.272c$
33. 18.4 g/cm$^3$
35. 1.98 MeV
37. $2.27 \times 10^{23}$ Hz, 1.32 fm for each photon
39. **(a)** $3.10 \times 10^5$ m/s **(b)** $0.758c$
41. 1.42 MeV/$c$
43. **(a)** $0.80c$ **(b)** $7.5 \times 10^3$ s
   **(c)** $1.4 \times 10^{12}$ m, $0.38c$
45. $0.37c$ in $+x$ direction
47. **(a)** $v/c = 1 - 1.12 \times 10^{-10}$ **(b)** $6.00 \times 10^{27}$ J
   **(c)** \$$2.17 \times 10^{20}$
49. $0.80c$
51. **(a)** $0.946c$ **(b)** 0.160 ly
   **(c)** 0.114 yr **(d)** $7.50 \times 10^{22}$ J
53. **(a)** 7.0 μs **(c)** $1.1 \times 10^4$ muons

## CHAPTER 27

### Quick Quizzes

1. b
2. c
3. c
4. c
5. b

### Conceptual Questions

1. The shape of an object is determined by observing the light reflecting from its surface. In a kiln, the objects will be very hot and will be glowing red. The emitted radiation is far stronger than the reflected radiation, and the thermal radiation emitted is only slightly dependent on the material from which the objects are made. Thus, we have a collection of objects, including the kiln walls, glowing equally with emitted radiation and only weak reflected light compared to the emitted light. This will result in indistinct outlines of the objects when viewed.
3. The "blackness" of a blackbody refers to its ideal property of absorbing all radiation incident on it. If an observed room-temperature object in everyday life absorbs all radiation, we describe it as (visibly) black. The black appearance, however, is due to the fact that our eyes are sensitive only to visible light. If we could detect infrared light with our eyes, we would see the object emitting radiation. If the temperature of the blackbody is raised, Wien's law tells us that the emitted radiation will move into the visible. Thus, the blackbody could appear possibly as black, red, white, or blue, depending on its temperature.
5. All objects do radiate energy, but at room temperature, this energy is primarily in the infrared region of the electromagnetic spectrum, which our eyes cannot detect. (Pit vipers have sensory organs that are sensitive to infrared radiation; thus they can seek out their warm-blooded prey in what we would consider absolute darkness.)
7. Most metals have cutoff frequencies corresponding to photons in or near the visible range of the electromagnetic spectrum. AM radio wave photons will have far too little energy to eject electrons from the metal.
9. We can picture higher-frequency light as a stream of photons of higher energy. In a collision, one photon can give all of its energy to a single electron. The kinetic energy of such an electron is measured by the stopping potential. The reverse voltage (stopping voltage) required to stop the current is proportional to the frequency of the incoming light. More intense light consists of more photons striking a unit area each second, but atoms are so small that one emitted electron never gets a "kick" from more than one photon. Increasing the light intensity will generally increase the size of the current but will not change the energy of the individual ejected electrons. Thus, the stopping potential remains constant.
11. Wave theory predicts that the photoelectric effect should occur at any frequency, provided that the light intensity is high enough. However, as seen in photoelectric experiments, the light must have sufficiently high frequency for the effect to occur.
13. The photons will transfer the same energy to all photoelectrons in the absorption process. However, different electrons may start from different potential energy levels and emerge from the surface with different kinetic energies. The energy

of the incident photons determines only the maximum kinetic energy of the liberated electrons.

**15.** No. Suppose that the incident light frequency at which you first observed the photoelectric effect is above the cutoff frequency of the first metal but less than the cutoff frequency of the second metal. In that case, the photoelectric effect would not be observed at all in the second metal.

**17.** In a collision with an electron or other scattering particle, the photon will impart some energy to that particle. Therefore, the energy of the photon will be decreased and its wavelength increased.

### Problems

**1.** **(a)** $\approx 3000$ K **(b)** $\approx 20\ 000$ K

**3.** **(a)** 999 nm
**(b)** Peak radiation is in the infrared region.

**5.** **(a)** $2.49 \times 10^{-5}$ eV **(b)** 2.49 eV
**(c)** 249 eV

**7.** $2.27 \times 10^{30}$ photons/s

**9.** **(a)** $2.3 \times 10^{31}$ **(b)** $\Delta E/E = 4.3 \times 10^{-32}$

**11.** **(a)** 2.24 eV **(b)** 555 nm
**(c)** $5.41 \times 10^{14}$ Hz

**13.** **(a)** 296 nm **(b)** $1.01 \times 10^{15}$ Hz
**(c)** 2.71 V

**15.** 148 days, incompatible with observation

**17.** $4.8 \times 10^{14}$ Hz, 2.0 eV

**19.** $1.2 \times 10^{2}$ V and $1.2 \times 10^{7}$ V, respectively

**21.** 41.4 kV

**23.** 0.078 nm

**25.** 0.281 nm

**27.** 1.78 eV, $9.47 \times 10^{-28}$ kg·m/s

**29.** 70°

**31.** $1.18 \times 10^{-23}$ kg·m/s, 478 eV

**33.** **(a)** 1.2 eV **(b)** $6.5 \times 10^{5}$ m/s

**35.** **(a)** 1.46 km/s **(b)** $7.28 \times 10^{-11}$ m

**37.** **(a)** $\sim 10^{2}$ MeV
**(b)** No. With kinetic energy much larger than the magnitude of the negative potential energy, the electron would immediately escape.

**39.** $3.58 \times 10^{-13}$ m

**41.** **(a)** 15 keV **(b)** $1.2 \times 10^{2}$ keV

**43.** $2.1 \times 10^{-32}$ m/s

**45.** 116 m/s

**47.** $\approx 5\ 200$ K; clearly, a firefly is not at this temperature, so this cannot be blackbody radiation.

**49.** 18.2°

**51.** 1.36 eV

**53.** 2.00 eV

**55.** **(a)** $0.0220c$ **(b)** $0.9992c$

**57.** **(b)** 3.72 km/s

**59.** **(b)** $2.60 \times 10^{-16}$ m

## Chapter 28

### Quick Quizzes

**1.** d

**2.** **(a)** 9 **(b)** 16

**3.** **(a)** 5 values of $\ell$
**(b)** 9 values of $m_\ell$

**4.** b

### Conceptual Questions

**1.** If the energy of the hydrogen atom were proportional to $n$ (or any power of $n$), then the energy would become infinite as $n$ grew to infinity. But the energy of the atom is inversely proportional to $n^2$. Thus, as $n$ grows to infinity, the energy of the atom approaches a value that is above the ground state by a finite amount, namely the ionization energy 13.6 eV. As the electron falls from one bound state to another, its energy loss is always less than the ionization energy. The energy and frequency of any emitted photon are finite.

**3.** The characteristic x-rays originate from transitions within the atoms of the target, such as an L shell electron making a transition to a vacancy in the K shell. This vacancy in the K shell is caused when an accelerated electron in the x-ray tube supplies energy to the K shell electron to eject it from the atom. If the energy of the bombarding electrons were to be increased, the K shell electron will be ejected from the atom with more remaining kinetic energy. But the energy difference between the K and L shell has not changed, so the emitted x-ray has exactly the same wavelength.

**5.** A continuous spectrum without characteristic x-rays is possible. At a low-accelerating potential difference for the electron, the electron may not have enough energy to eject an electron from a target atom. As a result, there will be no characteristic x-rays. The change in speed of the electron as it enters the target will result in the continuous spectrum.

**7.** The hologram is an interference pattern between light scattered from the object and the reference beam. If anything moves by a distance comparable to the wavelength of the light (or more), the pattern will wash out. The effect is just like making the slits vibrate in Young's experiment, to make the interference fringes vibrate wildly so that a photograph of the screen displays only the average intensity everywhere.

**9.** If the Pauli exclusion principle were not valid, the elements and their chemical behavior would be grossly different because every electron would end up in the lowest energy level of the atom. All matter would therefore be nearly alike in its chemistry and composition, since the shell structures of each element would be identical. Most materials would have a much higher density, and the spectra of atoms and molecules would be very simple, resulting in the existence of less color in the world.

**11.** The three elements have similar electronic configurations, with filled inner shells, plus a single electron in an *s* orbital. Since atoms typically interact through their unfilled outer shells, and the outer shell of each of these atoms is similar, the chemical interactions of the three atoms is also similar.

**13.** Each of the eight electrons must have at least one quantum number different from each of the others. They can differ (in $m_s$) by being spin-up or spin-down. They can differ (in $\ell$) in angular momentum and in the general shape of the wave function. Those electrons with $\ell = 1$ can differ (in $m_\ell$) in orientation of angular momentum.

**15.** The way the total energy of an atom, and in fact other energies, can be negative is through an arbitrary choice of the zero point for potential energies. The electrostatic potential energy between charged particles is arbitrarily chosen to be zero when the particles are separated by an infinite distance. When two particles having opposite signs, such as the electron and proton in a hydrogen atom, are a finite distance apart, the potential energy is less than zero, or negative. If they are sufficiently close, the magnitude of the negative potential energy may exceed the positive kinetic energy, making the total energy negative.

**17.** Stimulated emission is the reason laser light is coherent and tends to travel in a well-defined parallel beam. When a photon passing by an excited atom stimulates that atom to emit a photon, the emitted photon is in phase with the original photon and travels in the same direction. As this process is repeated many times, an intense, parallel beam of coherent light is produced. Without stimulated emission, the excited atoms would return to the ground state by emitting photons at random times and in random directions. The resulting light would not have the useful properties of laser light.

| $n$ | 3 | 3 | 3 | 3 | 3 | 3 | 3 | 3 | 3 | 3 | 3 | 3 | 3 | 3 | 3 |
|---|---|---|---|---|---|---|---|---|---|---|---|---|---|---|---|
| $\ell$ | 2 | 2 | 2 | 2 | 2 | 2 | 2 | 2 | 2 | 2 | 2 | 2 | 2 | 2 | 2 |
| $m_\ell$ | +2 | +2 | +2 | +1 | +1 | +1 | 0 | 0 | 0 | −1 | −1 | −1 | −2 | −2 | −2 |
| $m_s$ | +1 | 0 | −1 | +1 | 0 | −1 | +1 | 0 | −1 | +1 | 0 | −1 | +1 | 0 | −1 |

### Problems

1. 656 nm, 486 nm, and 434 nm
3. (a) $2.3 \times 10^{-8}$ N
   (b) $-14$ eV
5. (a) $1.6 \times 10^{6}$ m/s
   (b) No. $v/c = 5.3 \times 10^{-3} \ll 1$
   (c) 0.46 nm
   (d) Yes. The wavelength is roughly the same size as the atom.
7. (a) 0.212 nm (b) $9.95 \times 10^{-25}$ kg · m/s
   (c) $2.11 \times 10^{-34}$ J · s
   (d) 3.40 eV (e) $-6.80$ eV (f) $-3.40$ eV
11. $8.22 \times 10^{-8}$ N
13. (a) 0.967 eV (b) 0.266 eV
15. (a) 122 nm, 91.1 nm (b) $1.87 \times 10^{3}$ nm, 820 nm
17. 97.2 nm
19. (a) 488 nm (b) 0.814 m/s
21. (d) $n = 2.53 \times 10^{74}$
   (e) No. At such large quantum numbers the allowed energies are essentially continuous.
23. (a) $2.47 \times 10^{15}$ Hz, $f_{\text{orb}} = 8.23 \times 10^{14}$ Hz
   (b) $6.59 \times 10^{3}$ Hz, $f_{\text{orb}} = 6.59 \times 10^{3}$ Hz. For large $n$, classical theory and quantum theory approach each other in their results.
25. $4.42 \times 10^{4}$ m/s
27. (a) $-122$ eV (b) $1.76 \times 10^{-11}$ m
29. (a) 0.026 5 nm
   (b) 0.017 6 nm
   (c) 0.013 2 nm
31. 1.33 nm
33. $n = 3, \ell = 1, m_\ell = +1, m_s = \pm 1/2$; $n = 3, \ell = 1, m_\ell = 0, m_s = \pm 1/2$; $n = 3, \ell = 1, m_\ell = -1, m_s = \pm 1/2$
35. The table at the top of the page summarizes 15 possible states:
37. (a) 30 possible states (b) 36
39. (a) $n = 4$ and $\ell = 2$
   (b) $m_\ell = (0, \pm 1, \pm 2)$, $m_s = \pm \frac{1}{2}$
   (c) $1s^2 2s^2 2p^6 3s^2 3p^6 3d^{10} 4s^2 4p^6 4d^2 5s^2$ = [Kr] $4d^2 5s^2$
41. 0.160 nm
43. L shell: 11.7 keV; M shell: 10.0 keV; N shell: 2.30 keV
45. (a) 10.2 eV (b) $7.88 \times 10^{4}$ K
47. (a) $-8.18$ eV, $-2.04$ eV, $-0.904$ eV, $-0.510$ eV, $-0.325$ eV
   (b) $1.09 \times 10^{3}$ nm and 609 nm
49. (a) The four lowest energies are $-10.39$ eV, $-5.502$ eV, $-3.687$ eV, and $-2.567$ eV.
   (b) The wavelengths of the emission lines are 158.5 nm, 185.0 nm, 253.7 nm, 422.5 nm, 683.2 nm, and 1 107 nm.
   (c) $1.31 \times 10^{6}$ m/s
51. (a) $4.24 \times 10^{15}$ W/m$^2$ (b) $1.20 \times 10^{-12}$ J
55. (a) $E_n = (-1.49 \times 10^{4}\text{ eV})/n^2$
   (b) $n = 4 \rightarrow n = 1$
57. (a) $9.03 \times 10^{22}$ m/s$^2$ (b) $-4.63 \times 10^{-8}$ W
   (c) $\sim 10^{-11}$ s

## Chapter 29

### Answers to Quick Quizzes

1. c
2. b
3. a
4. a and b
5. b

### Conceptual Questions

1. Isotopes of a given element correspond to nuclei with different numbers of neutrons. This results in a variety of different physical properties for the nuclei, including the obvious one of mass. The chemical behavior, however, is governed by the electrons. All isotopes of a given element have the same number of electrons and, therefore, the same chemical behavior.
3. An alpha particle contains two protons and two neutrons. Because a hydrogen nucleus only contains one proton, it cannot emit an alpha particle.
5. In alpha decay, there are only two final particles—the alpha particle and the daughter nucleus. There are also two conservation principles—energy and momentum. As a result, the alpha particle must be ejected with a discrete energy to satisfy both conservation principles. However, beta decay is a three-particle decay—the beta particle, the neutrino (or antineutron), and the daughter nucleus. As a result, the energy and momentum can be shared in a variety of ways among the three particles while still satisfying the two conservation principles. This allows a continuous range of energies for the beta particle.
7. The larger rest energy of the neutron means that a free proton in space will not spontaneously decay into a neutron and a positron. When the proton is in the nucleus, however, the important question is that of the total rest energy of the nucleus. If it is energetically lower for the nucleus to have one less proton and one more neutron, then the decay process will occur to achieve this lower energy.
9. Carbon dating cannot generally be used to estimate the age of a stone, because the stone was not alive to take up carbon from the environment. Only the ages of artifacts that were once alive can be estimated with carbon dating.
11. The protons, although held together by the nuclear force, are repulsed by the electrical force. If enough protons were placed together in a nucleus, the electrical force would overcome the nuclear force, which is based on the number of particles, and cause the nucleus to fission.
   The addition of neutrons prevents such fission. The neutron does not increase the electrical force, being electrically neutral, but does contribute to the nuclear force.
13. The statement is false. Both patterns show monotonic decrease over time, but with very different shapes. For radioactive decay, maximum activity occurs at time zero. Cohorts of people now living will be dying most rapidly perhaps forty years from now. Everyone now living will be dead within less than two centuries, while the mathematical model of radioactive decay tails off exponentially forever. A radioactive nucleus never gets old. It has constant probability of decay however long it has existed.
15. Since the two samples are of the same radioactive nuclide, they have the same half-life; the 2:1 difference in activity is due to a 2:1 difference in the mass of each sample. After five half-lives, each will have decreased in mass by a power of $2^5 = 32$. However, since this simply means that the mass of each is 32 times smaller, the ratio of the masses will still be $(\frac{2}{32}):(\frac{1}{32})$, or 2:1. Therefore, the ratio of their activities will *always* be 2:1.
17. The photon and the neutrino are similar in that both particles have zero mass and zero charge. Both must travel at the

speed of light and are capable of transferring both energy and momentum. They differ in that the photon has spin (intrinsic angular momentum) of $\hbar$ and is involved in electromagnetic interactions, whereas the neutrino has spin $\hbar/2$ and is closely related to beta decays.

### Problems

**1.** $A = 2$, $r = 1.5$ fm; $A = 60$, $r = 4.7$ fm; $A = 197$, $r = 7.0$ fm; $A = 239$, $r = 7.4$ fm
**3.** $1.8 \times 10^2$ m
**5.** **(a)** 27.6 N away from the carbon nucleus **(b)** $4.16 \times 10^{27}$ m/s$^2$ away from the nucleus **(c)** 1.73 MeV
**7.** **(a)** $1.9 \times 10^7$ m/s **(b)** 7.1 MeV
**9.** 8.66 MeV/nucleon for $^{93}_{41}$Nb, 7.92 MeV/nucleon for $^{197}_{79}$Au
**11.** 3.54 MeV
**13.** 0.210 MeV/nucleon greater for $^{23}_{11}$Na, attributable to less proton repulsion
**15.** 0.46 Ci
**17.** **(a)** $9.98 \times 10^{-7}$ s$^{-1}$ **(b)** $1.9 \times 10^{10}$ nuclei
**19.** **(a)** $3.18 \times 10^{-7}$ mol **(b)** $1.92 \times 10^{17}$ nuclei **(c)** $1.08 \times 10^{14}$ Bq **(d)** $8.96 \times 10^6$ Bq
**21.** $4.31 \times 10^3$ yr
**23.** **(a)** $5.58 \times 10^{-2}$ h$^{-1}$, 12.4 h **(b)** $2.39 \times 10^{13}$ nuclei **(c)** 1.9 mCi
**25.** $^{208}_{81}$Tl, $^{95}_{37}$Rb, $^{144}_{60}$Nd
**27.** $^{40}_{20}$Ca, $^{94}_{42}$Mo, $^{4}_{2}$He
**29.** e$^+$ decay, $^{56}_{27}\text{Co} \rightarrow {}^{56}_{26}\text{Fe} + e^+ + \nu$
**31.** 71.4 keV
**33.** 18.6 keV
**35.** $4.22 \times 10^3$ yr
**37.** **(a)** $^{30}_{15}$P **(b)** $-2.64$ MeV
**39.** **(a)** $^{21}_{10}$Ne **(b)** $^{144}_{54}$Xe **(c)** $X = e^+$, $X' = \nu$
**41.** **(a)** $^{13}_{6}$C **(b)** $^{10}_{5}$B
**43.** 1.00 MeV
**45.** **(a)** $^{1}_{0}$n **(b)** Fluorine mass = 18.000 953 $u$
**47.** 18.8 J
**49.** 24 d
**51.** **(a)** $8.97 \times 10^{11}$ electrons **(b)** 0.100 J **(c)** 100 rad
**53.** 46.5 d
**55.** $Q = 3.27$ MeV $> 0$, no threshold energy required
**57.** **(a)** $2.52 \times 10^{24}$ **(b)** $2.29 \times 10^{12}$ Bq **(c)** $1.07 \times 10^6$ yr
**59.** **(a)** $4.0 \times 10^9$ yr **(b)** It could be no older. The rock could be younger if some $^{87}$Sr were initially present.
**61.** 54 $\mu$Ci
**63.** $2.3 \times 10^2$ yr
**65.** $4.4 \times 10^{-8}$ kg/h

## CHAPTER 30

### Answers to Quick Quizzes

**1.** c
**2.** a
**3.** b
**4.** c, e

### Conceptual Questions

**1.** The experiment described is a nice analogy to the Rutherford scattering experiment. In the Rutherford experiment, alpha particles were scattered from atoms, and the scattering was consistent with a small structure in the atom containing the positive charge.

**3.** The largest charge of a quark is $2e/3$, so a combination of only two particles, a quark and an antiquark forming a meson, could not possibly have electric charge up to $+2e$. Only particles containing three quarks, each with a charge of $2e/3$, can combine to produce a total charge of 2e.

**5.** Until about 700 000 years after the Big Bang, the temperature of the Universe was high enough for any atoms that formed to be ionized by ambient radiation. Once the average radiation energy dropped below the hydrogen ionization energy of 13.6 eV, hydrogen atoms could form and remain as neutral atoms for a relatively long period of time.

**7.** In the quark model, all hadrons are composed of smaller units called quarks. Quarks have a fractional electric charge and a baryon number of $\frac{1}{3}$. There are six flavors of quarks; up (u) down (d), strange (s), charmed (c), top (t), and bottom (b). All baryons contain three quarks, and all mesons contain one quark and one antiquark. Section 30.12 has a more detailed discussion of the quark model.

**9.** Baryons and mesons are hadrons, interacting primarily through the strong force. They are not elementary particles, being composed of either three quarks (baryons), or a quark and an antiquark (mesons). Baryons have a nonzero baryon number with a spin of either $\frac{1}{2}$ or $\frac{3}{2}$. Mesons have a baryon number of zero and a spin of either 0 or 1.

**11.** All stable particles other than protons and neutrons have baryon number zero. Since the baryon number must be conserved, and the final states of the kaon decay contain no protons or neutrons, the baryon number of all kaons must be *zero*.

**13.** Yes, but the strong interaction predominates.

**15.** Unless the particles have enough kinetic energy to produce a baryon-antibaryon pair, the answer is *no*. Antibaryons have a baryon number of $-1$; baryons have a baryon number of $+1$; mesons have a baryon number of 0. If such an interaction were to occur, and produce a baryon, the baryon number would not be conserved.

**17.** Baryons and antibaryons contain three quarks, whereas mesons and antimesons contain two quarks. Quarks have a spin of $\frac{1}{2}$; thus three quarks in a baryon can only combine to form a net spin that is half-integral. Likewise, two quarks in a meson can only combine to form a net spin of 0 or 1.

**19.** Nuclear and coal-fired power plants are similar in that both produce high-pressure steam to drive turbines and generate electricity. They differ in the type and amount of fuel consumed, the ways they affect the environment, and in some safety issues. A moderate size nuclear-powered plant consumes a few kilograms of fissionable materials per day, whereas a comparable size coal-powered plant requires a thousand tons or more of coal each day. One byproduct of nuclear plants is thermal pollution in the waters used for cooling. In addition to thermal pollution, coal-fired plants produce emissions that cause acid rain and contribute to the greenhouse effect. Both types of plants have risks of steam explosions. In addition, nuclear-powered plants raise questions on how to dispose of radioactive waste materials.

### Problems

**1.** $^{1}_{0}\text{n} + {}^{235}_{92}\text{U} \rightarrow {}^{141}_{56}\text{Ba} + {}^{92}_{36}\text{Kr} + 3\,{}^{1}_{0}\text{n}$, 3 neutrons are released
**3.** 126 MeV
**5.** **(a)** 16.2 kg **(b)** 117 g
**7.** $2.9 \times 10^3$ km ($\approx$ 1 800 miles)
**9.** 1.01 g
**11.** **(a)** $^{8}_{4}$Be **(b)** $^{12}_{6}$C **(c)** 7.27 MeV
**13.** $3.07 \times 10^{22}$ events/yr
**15.** **(a)** $3.44 \times 10^{30}$ J **(b)** $1.56 \times 10^8$ yr
**17.** **(a)** $4.53 \times 10^{23}$ Hz **(b)** 0.622 fm
**19.** $\sim 10^{-23}$ s
**21.** $\sim 10^{-18}$ m
**23.** **(a)** conservation of electron-lepton number and conservation of muon-lepton number **(b)** conservation of charge **(c)** conservation of baryon number **(d)** conservation of baryon number **(e)** conservation of charge

**25.** $\overline{\nu}_\mu$

**27.** **(a)** $\overline{\nu}_\mu$ **(b)** $\nu_\mu$ **(c)** $\overline{\nu}_e$ **(d)** $\nu_e$
**(e)** $\nu_\mu$ **(f)** $\nu_\mu$ and $\overline{\nu}_e$

**29.** **(a)** not allowed; violates conservation of baryon number
**(b)** strong interaction **(c)** weak interaction
**(d)** weak interaction
**(e)** electromagnetic interaction

**31.** **(a)** not conserved **(b)** conserved
**(c)** conserved **(d)** not conserved
**(e)** not conserved **(f)** not conserved

**33.** **(a)** charge, baryon number, $L_e$, $L_\tau$
**(b)** charge, baryon number, $L_e$, $L_\mu$, $L_\tau$
**(c)** charge, $L_e$, $L_\mu$, $L_\tau$, strangeness number
**(d)** charge, baryon number, $L_e$, $L_\mu$, $L_\tau$, strangeness number
**(e)** charge, baryon number, $L_e$, $L_\mu$, $L_\tau$, strangeness number
**(f)** charge, baryon number, $L_e$, $L_\mu$, $L_\tau$, strangeness number

**35.** $3.34 \times 10^{26}$ electrons, $9.36 \times 10^{26}$ up quarks, $8.70 \times 10^{26}$ down quarks

**37.** **(a)** $\Sigma^+$ **(b)** $\pi^-$ **(c)** $K^0$ **(d)** $\Xi^-$

**39.**

| Reaction | At Quark Level | Net Quarks (Before and After) |
|---|---|---|
| $\pi^- + p \rightarrow K^0 + \Lambda^0$ | $\overline{u}d + uud \rightarrow d\overline{s} + uds$ | 1 up, 2 down, 0 strange |
| $\pi^+ + p \rightarrow K^+ + \Sigma^+$ | $u\overline{d} + uud \rightarrow u\overline{s} + uus$ | 3 up, 0 down, 0 strange |
| $K^- + p \rightarrow K^+ + K^0 + \Omega^-$ | $\overline{u}s + uud \rightarrow u\overline{s} + d\overline{s} + sss$ | 1 up, 1 down, 1 strange |

**(d)** The mystery particle is a $\Lambda^0$ or a $\Sigma^0$.

**41.** a neutron, udd

**43.** 70.45 MeV

**45.** 18.8 MeV

**47.** **(a)** electron-lepton and muon-lepton numbers not conserved
**(b)** electron-lepton number not conserved
**(c)** charge not conserved
**(d)** baryon and electron-lepton numbers not conserved
**(e)** strangeness violated by 2 units

**49.** **(a)** $1.5 \times 10^{24}$ nuclei **(b)** $\approx 0.6$ kg

**51.** **(a)** 1 baryon before and zero baryons after decay. Baryon number is not conserved.
**(b)** 469 MeV, 469 MeV/c
**(c)** 0.999 999 4c

# Index

Page numbers in italics indicate illustrations; page numbers followed by "t" indicate tables.

## A



## B



## C

# D

# Q



# R

## PHYSICAL CONSTANTS

| Quantity | Symbol | Value | SI unit |
|---|---|---|---|
| Speed of light in vacuum | $c$ | $3.00 \times 10^{8}$ | m/s |
| Permittivity of free space | $\epsilon_0$ | $8.85 \times 10^{-12}$ | $C^2/N \cdot m^2$ |
| Coulomb constant, $1/4\pi\epsilon_0$ | $k_e$ | $8.99 \times 10^{9}$ | $N \cdot m^2/C^2$ |
| Permeability of free space | $\mu_0$ | $1.26 \times 10^{-6}$<br>$(4\pi \times 10^{-7}$ exactly) | $T \cdot m/A$ |
| Elementary charge | $e$ | $1.60 \times 10^{-19}$ | C |
| Planck's constant | $h$<br>$\hbar = h/2\pi$ | $6.63 \times 10^{-34}$<br>$1.05 \times 10^{-34}$ | $J \cdot s$<br>$J \cdot s$ |
| Electron mass | $m_e$ | $9.11 \times 10^{-31}$<br>$5.49 \times 10^{-4}$ | kg<br>u |
| Proton mass | $m_p$ | $1.672\ 65 \times 10^{-27}$<br>1.007 276 | kg<br>u |
| Neutron mass | $m_n$ | $1.674\ 95 \times 10^{-27}$<br>1.008 665 | kg<br>u |
| Avogadro's number | $N_A$ | $6.02 \times 10^{23}$ | $mol^{-1}$ |
| Universal gas constant | $R$ | 8.31 | $J/mol \cdot K$ |
| Boltzmann's constant | $k_B$ | $1.38 \times 10^{-23}$ | J/K |
| Stefan-Boltzmann constant | $\sigma$ | $5.67 \times 10^{-8}$ | $W/m^2 \cdot K^4$ |
| Molar volume of ideal gas at STP | $V$ | 22.4<br>$2.24 \times 10^{-2}$ | L/mol<br>$m^3/mol$ |
| Rydberg constant | $R_H$ | $1.10 \times 10^{7}$ | $m^{-1}$ |
| Bohr radius | $a_0$ | $5.29 \times 10^{-11}$ | m |
| Electron Compton wavelength | $h/m_e c$ | $2.43 \times 10^{-12}$ | m |
| Gravitational constant | $G$ | $6.67 \times 10^{-11}$ | $N \cdot m^2/kg^2$ |
| Standard free-fall acceleration | $g$ | 9.80 | $m/s^2$ |
| Radius of Earth (at equator) | $R_E$ | $6.38 \times 10^{6}$ | m |
| Mass of Earth | $M_E$ | $5.98 \times 10^{24}$ | kg |
| Radius of Moon | $R_M$ | $1.74 \times 10^{6}$ | m |
| Mass of Moon | $M_M$ | $7.36 \times 10^{22}$ | kg |

The values presented in this table are those used in computations in the text. Generally, the physical constants are known to much better precision.